# Biotechnology

Second Edition

**Volume 6**

## Products of Primary Metabolism

# Biotechnology

Second Edition

**Fundamentals**

Volume 1
Biological Fundamentals

Volume 2
Genetic Fundamentals and
Genetic Engineering

Volume 3
Bioprocessing

Volume 4
Measuring, Modelling, and Control

**Products**

Volume 5
Recombinant Proteins,
Monoclonal Antibodies, and
Therapeutic Genes

Volume 6
Products of Primary Metabolism

Volume 7
Products of Secondary Metabolism

Volume 8
Biotransformations

**Special Topics**

Volume 9
Enzymes, Biomass, Food and Feed

Volume 10
Special Processes

Volume 11
Environmental Processes

Volume 12
Legal, Economic and
Ethical Dimensions



Distribution:

VCH, P. O. Box 101161, D-69451 Weinheim (Federal Republic of Germany)

Switzerland: VCH, P. O. Box, CH-4020 Basel (Switzerland)

United Kingdom and Ireland: VCH (UK) Ltd., 8 Wellington Court, Cambridge CB1 1HZ (England)

USA and Canada: VCH, 333 7th Avenue, New York, NY 10001 (USA)

Japan: VCH, Eikow Building, 10-9 Hongo 1-chome, Bunkyo-ku, Tokyo 113 (Japan)

ISBN 3-527-28316-1 (VCH, Weinheim)
Set ISBN 3-527-28310-2 (VCH, Weinheim)

A Multi-Volume Comprehensive Treatise

# Biotechnology

Second, Completely Revised Edition

Edited by
H.-J. Rehm and G. Reed
in cooperation with
A. Pühler and P. Stadler

**Volume 6**

# Products of Primary Metabolism

Edited by
M. Roehr

Weinheim · New York
Basel · Cambridge · Tokyo

Series Editors:
Prof. Dr. H.-J. Rehm
Institut für Mikrobiologie
Universität Münster
Corrensstraße 3
D-48149 Münster
FRG

Prof. Dr. A. Pühler
Biologie VI (Genetik)
Universität Bielefeld
P.O. Box 100131
D-33501 Bielefeld
FRG

Dr. G. Reed
1914 N. Prospect Ave. #61
Milwaukee, WI 53202-1401
USA

Prof. Dr. P. J. W. Stadler
Bayer AG
Verfahrensentwicklung Biochemie
Leitung
Friedrich-Ebert-Straße 217
D-42096 Wuppertal
FRG

Volume Editor:
Prof. Dr. M. Roehr
Institut für Biochemische
Technologie und Mikrobiologie
Technische Universität
Getreidemarkt 9/12
A-1060 Wien
Austria



Published jointly by
VCH Verlagsgesellschaft mbH, Weinheim (Federal Republic of Germany)
VCH Publishers Inc., New York, NY (USA)

Executive Editor: Dr. Hans-Joachim Kraus
Editorial Director: Karin Dembowsky
Production Manager: Dipl. Wirt.-Ing. (FH) Hans-Jochen Schmitt

Library of Congress Card No.: applied for

British Library Cataloguing-in-Publication Data:
A catalogue record for this book is available from the British Library

Die Deutsche Bibliothek – CIP-Einheitsaufnahme
**Biotechnology** : a multi volume comprehensive treatise / ed. by H.-J. Rehm and G. Reed. In cooperation with A. Pühler and P. Stadler. – 2., completely rev. ed. – Weinheim ; New York ; Basel ; Cambridge ; Tokyo : VCH.
ISBN 3-527-28310-2 (Weinheim ...)
ISBN 1-56081-602-3 (New York)
NE: Rehm, Hans J. [Hrsg.]

2., completely rev. ed.
Vol. 6. Products of primary metabolism / ed. by M. Roehr. – 1996
ISBN 3-527-28316-1
NE: Roehr, Max [Hrsg.]





Printed on acid-free and chlorine-free paper.


Composition and Printing: Zechnersche Buchdruckerei, D-67330 Speyer.
Bookbinding: J. Schäffer, D-67269 Grünstadt.
Printed in the Federal Republic of Germany

# Preface

In recognition of the enormous advances in biotechnology in recent years, we are pleased to present this Second Edition of "Biotechnology" relatively soon after the introduction of the First Edition of this multi-volume comprehensive treatise. Since this series was extremely well accepted by the scientific community, we have maintained the overall goal of creating a number of volumes, each devoted to a certain topic, which provide scientists in academia, industry, and public institutions with a well-balanced and comprehensive overview of this growing field. We have fully revised the Second Edition and expanded it from ten to twelve volumes in order to take all recent developments into account.

These twelve volumes are organized into three sections. The first four volumes consider the fundamentals of biotechnology from biological, biochemical, molecular biological, and chemical engineering perspectives. The next four volumes are devoted to products of industrial relevance. Special attention is given here to products derived from genetically engineered microorganisms and mammalian cells. The last four volumes are dedicated to the description of special topics.

The new "Biotechnology" is a reference work, a comprehensive description of the state-of-the-art, and a guide to the original literature. It is specifically directed to microbiologists, biochemists, molecular biologists, bioengineers, chemical engineers, and food and pharmaceutical chemists working in industry, at universities or at public institutions.

A carefully selected and distinguished Scientific Advisory Board stands behind the series. Its members come from key institutions representing scientific input from about twenty countries.

The volume editors and the authors of the individual chapters have been chosen for their recognized expertise and their contributions to the various fields of biotechnology. Their willingness to impart this knowledge to their colleagues forms the basis of "Biotechnology" and is gratefully acknowledged. Moreover, this work could not have been brought to fruition without the foresight and the constant and diligent support of the publisher. We are grateful to VCH for publishing "Biotechnology" with their customary excellence. Special thanks are due to Dr. Hans-Joachim Kraus and Karin Dembowsky, without whose constant efforts the series could not be published. Finally, the editors wish to thank the members of the Scientific Advisory Board for their encouragement, their helpful suggestions, and their constructive criticism.

H.-J. Rehm
G. Reed
A. Pühler
P. Stadler

# Scientific Advisory Board

# Contents

# Contributors

Prof. Dr. Hubert Bahl
Abteilung Mikrobiologie
Fachbereich Biologie
Universität Rostock
D-18051 Rostock
FRG
*Chapter 6*

Prof. Dr. Klaus Buchholz
Institut für Technologie der Kohlenhydrate
Technische Universität
Langer Kamp 5
D-38106 Braunschweig
FRG
*Chapter 1a*

Prof. Dr. Peter Dürre
Abteilung Angewandte Mikrobiologie und Mykologie
Universität Ulm
D-89069 Ulm
FRG
*Chapter 6*

Dr. Heinrich Ebner
Piringerhofstraße 13
A-4020 Linz
Austria
*Chapter 12*

Prof. Dr. Nobuyoshi Esaki
Institute for Chemical Research
Kyoto University
Gokasho, Uji
Kyoto 611
Japan
*Chapter 14b*

Dr. Heinrich Follmann
Heinrich Frings GmbH
Jonas-Cahn-Str. 9
D-53115 Bonn
FRG
*Chapter 12*

Dr. Setsuo Furuyoshi
Research Institute of Molecular Genetics
Kochi University
B200 Monobe, Nanoku
Kochi 783
Japan
*Chapter 14b*

Dipl.-Ing. Dr. Ján Kašcák
Pod Brezinou 9
SK-9100 Trenein
Slovakia
*Chapter 8*

Dr. Jiri Komínek
Institut für Biochemische Technologie und Mikrobiologie
Technische Universität
Getreidemarkt 9/12
A-1060 Wien
Austria
*Chapters 8, 9, 10*

Prof. Dr. Naim Kosaric
The University of Western Ontario
Department of Chemical and Biochemical Engineering
London, Ontario, N6A 5B9
Canada
*Chapters 4, 17*

Prof. Dr. Shigeru Nakamori
Department of Bioscience
Fukui Prefectural University
Kenjojima 4-1-1, Matsuoka-Cho
Fukui 910-11
Japan
*Chapter 14b*

Dr. Christian P. Kubicek
Institut für Biochemische Technologie und Mikrobiologie
Technische Universität
Getreidemarkt 9/12
A-1060 Wien
Austria
*Chapters 9, 10, 11*

Prof. Dr. Ing. Hans Joachim Pieper
Universität Hohenheim
Fachgruppe 5
Lebensmitteltechnologie
Garbenstraße 20
D-70599 Stuttgart
FRG
*Chapter 3*

Dr. Akira Kuninaka
Yamasa Corporation
10-1 Aroicho 2-chome,
Chosi, Chiba-ken 288
Japan
*Chapter 15*

Prof. Dr. Hans-Jürgen Rehm
Universität Münster
Institut für Mikrobiologie
Corrensstraße 3
D-48149 Münster
FRG
*Chapter 5*

Dr. Tatsuo Kurihara
Institute for Chemical Research
Kyoto University
Gokasho, Uji
Kyoto 611
Japan
*Chapter 14b*

Prof. Dr. Max Roehr
Institut für Biochemische Technologie und Mikrobiologie
Technische Universität
Getreidemarkt 9/12
A-1060 Wien
Austria
*Chapters 8, 9, 10, 11*

Prof. Dr. Wolfgang Leuchtenberger
Degussa AG
Forschung und Entwicklung
Futtermitteladditive
Kantstraße 2
D-33790 Halle-Künsebeck
FRG
*Chapter 14a*

Prof. Dr. Friedrich Schneider
Institut für Volkswirtschaftslehre
Universität Linz
A-4010 Linz-Auhof
Austria
*Chapter 2*

Prof. Ian S. Maddox
Massey University
Department of Process and Environmental Technology
Palmerston North
New Zealand
*Chapter 7*

Dr. Sylvia Sellmer
Heinrich Frings GmbH
Jonas-Cahn-Str. 9
D-53115 Bonn
FRG
*Chapter 12*

Dr. Thomas Senn
Universität Hohenheim
Fachgruppe 5
Lebensmitteltechnologie
Garbenstraße 20
D-70599 Stuttgart
FRG
*Chapter 3*

Prof. Dr. Kenji Soda
Department of Biotechnology
Faculty of Engineering
Kansai University
3-3-35 Yamate-Cho, Suita
Osaka 564
Japan
*Chapter 14b*

Prof. Dr. Alexander Steinbüchel
Universität Münster
Institut für Mikrobiologie
Corrensstraße 3
D-48149 Münster
FRG
*Chapter 13*

Dipl.-Ing. Dr. Horst Steinmüller
Geymanngang 6
A-4020 Linz
Austria
*Chapter 2*

Dr. Eberhard Stoppok
Institut für Technologie der Kohlenhydrate
Technische Universität
Langer Kamp 5
D-38106 Braunschweig
FRG
*Chapter 1a*

Prof. Ian W. Sutherland
The University of Edinburgh
Institute of Cell and Molecular Biology
Mayfield Road
Edinburgh EH9 3JH
UK
*Chapter 16*

Dr. Jean-Claude de Troostembergh
Cerestar
Research and Development Centre
Havenstraat, 84
B-1800 Vilvoorde
Belgium
*Chapter 1b*

# Introduction

MAX ROEHR

Wien, Austria

*Scientia difficilis sed fructuosa.*

Biotechnology comprises a wide range of disciplines – from classical biotechnology to the so-called new biotechnologies.

*Classical biotechnology* is usually defined as "the utilization of the biochemical potential of cells (in most cases of microbial origin) or enzymes thereof for the industrial production of a great variety of useful substances".

With the *"new" biotechnologies* the situation is somewhat more complex: If defined as the production and industrial utilization of genetically engineered cells, this will be according to the description given above. Frequently, however, the term biotechnology has merely been used to describe the methodology of genetic engineering and similar techniques (see, e.g., US Congress, Office of Technology Assessment, 1984). This rather careless (to be polite) handling of long-established terms, apart from being unfair, has caused considerable confusion harmful not only to biotechnologists but also to geneticists and others; this may be seen, e.g., when ordering one of those many texts on *biotechnology* and being surprised by its contents. A similar example of such misuse of terms has been found by the author in the science column of a prominent newspaper, describing progress in micro-surgery of the human ear as an achievement in microbiology!

The present volume deals with classical products of primary metabolism. Most of these products are so-called high-volume/low-price products. Their economic production depends to a large degree on the properties and costs of the respective raw materials, especially the carbon sources. The volume editor has, therefore, decided to include three chapters on raw materials: The first chapter (1a) deals with sugar-based raw materials, the second (1b) provides data on starch-based raw materials for fermentation applications. It is intended to fill a gap often experienced by workers in the field when designing a certain process and facing the necessity to know properties of raw materials only to be found spread over the literature. In the third chapter basic problems of economics of supply and utilization of the common sugar and starch-based raw materials are discussed.

Most of the authors of the First Edition where kind enough to lend their time and expertise to the new edition, which is gratefully acknowledged. Several chapters of the first edition are treated in other volumes of the series. Other amendments were made to improve the utility of the present volume. For instance, in view of the fact that the former chapter on amino acids mainly treated more basic aspects of amino acid production, a separate chapter (14a) dealing with the industrial production of these important commodities has been provided.

During the preparation of the various chapters, one striking feature became evident: In contrast to the presumption that there had been only little progress in the industry of products of primary metabolism, almost everywhere considerable advances could be identified. This was either due to the necessity of circumventing environmental problems, to engineering progress especially in downstream processing, or simply due to better insights into the physiology, metabolic regulation, and the genetics of the producing cellular systems.

Hopefully, therefore, this volume should also be the basis for establishing more sustainable technologies for the next century. On the other hand, knowledge of the various processes for the production of commodities treated in this volume could likewise be the basis for establishing technical variants in less industrialized countries using indigenous raw materials.

It is thus hoped that the present volume will contribute to further progress in biotechnology and will again find acceptance by many colleagues. These, in turn, are kindly invited to contribute their criticism, to suggest improvements, and to provide additional knowledge and data.

Vienna, June 1996 Max Roehr

# I. Raw Materials and Raw Material Strategies

# 1a Sugar-Based Raw Materials for Fermentation Applications

EBERHARD STOPPOK
KLAUS BUCHHOLZ

Braunschweig, Germany

# 1 Introduction

The principal raw materials for most fermentation processes are carbohydrates providing major sources of energy for the innumerable microorganisms involved in industrial fermentation processes. Pure sugars such as glucose or sucrose are often too expensive to be used industrially so that usually it is necessary to find a cheap carbon source (RIVIÈRE, 1977). The application of pure sugars is restricted to fermentations intended to yield highly purified products, especially where substrate mixtures containing colored carbohydrates would require economically inefficient further processing.

As in any other industrial process economics are of prime importance in choosing a raw material of a fermentation process. The carbon source still causes a significant portion of the overall production costs, accounting for at least 25% of total costs (DALE and LINDEN, 1984). For fermentations producing high-value products, the carbon substrate cost may be considered not that important, but for substrate cost-dominated fermentations, only inexpensive sugar syrups can substantially reduce fermentation product costs.

Natural products fluctuate in price and sometimes in availability according to the market. For this reason it can be misleading to assess the use of potential raw materials only in terms of their price. Apart from cost, the choice of a particular raw material will depend on many other factors, i.e., the type of fermentation, the yield of end product, and the cost of additional nutrient compounds required by the microorganisms.

Carbohydrates usually act as both, sources of energy and building blocks for the synthesis of macromolecules, and also supply with oxygen and hydrogen. More complex raw materials like syrups or molasses contain nitrogen, minerals, vitamins, or further growth substances in addition which can reduce the cost of biomass production (BRONN, 1985, 1987; OLBRICH 1973).

For further information about media for microbial fermentations see CEJKA (1985) and GREASHAM (1993).

Although numerous pure or raw carbohydrates can be obtained from starch, cellulose, or other carbon containing raw materials, this chapter will focus on raw materials based on sugar. Thus, an attempt will be made in the following to describe presently available raw materials derived from the sugar recovery process. The term "sugar" is used here for the disaccharide sucrose and products of the sugar industry essentially composed of sucrose.

# 2 Provenance of Sugar and Raw Materials

During sugar processing from beet or cane a number of sugar containing syrups, highly or less pure crystal sugars, or molasses become available.

The use of pure sucrose or molasses is well known in the fermentation industry but there is a lack of information about a number of further substrates which may be more advantageous for special applications. To understand the provenance of sugar containing raw materials some basic information about the sugar recovery process and the terms used in the sugar industry is essential. This is briefly summarized below and illustrated in Figs. 1 and 2.

During sugar processing sugar is isolated from plant compounds and finally brought to crystallization. The *dry substance* (DS) of juices or syrups consists of *sucrose* (S) and other substances extracted from beet or cane which are simplified to the *nonsucrose* (nonsugar) fraction (NS) in sugar technology. The sucrose content is generally measured by polarimetric methods whereas the solids are determined with a refractometer. (The outdated determination in °Brix found occasionally up to now should be avoided.) In industry, the sucrose content present in percentage terms of dry substance content (also called *solid matter* or *solids content*) is referred to as *purity,* also called *coefficient of purity, exponent of purity, quotient of purity,* or sometimes only *exponent* or *quotient:*

$$p = \frac{\text{Sucrose}}{\text{Dry substance}} \cdot 100\ [\%]$$

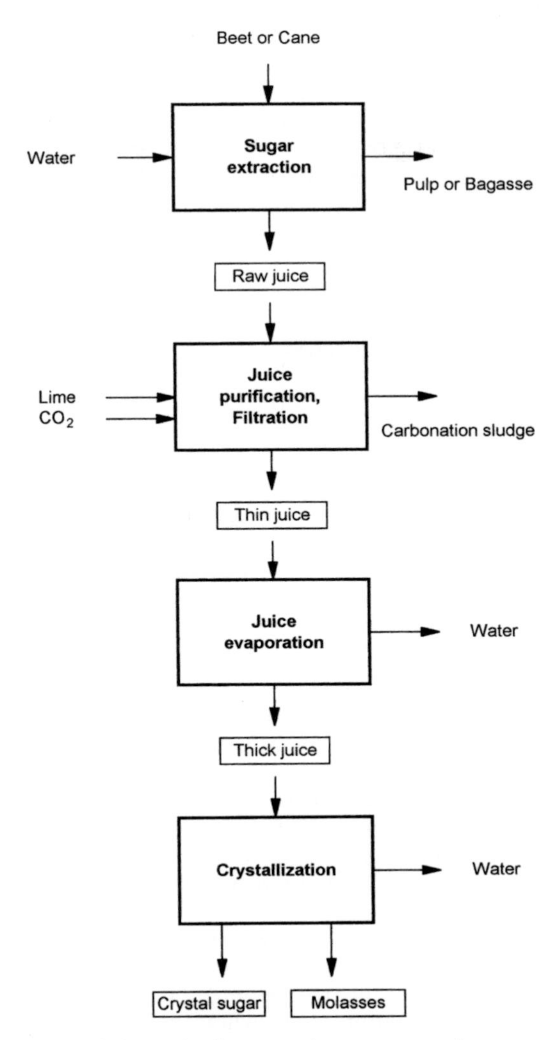


**Fig. 1.** Schematic diagram of sugar processing.

Apart from other properties all intermediate and end products of the sugar recovery process are characterized by their purity which is possibly the most generally employed and the most important term in sugar manufacture.

## 2.1 Products from Extraction, Purification, and Evaporation

In the first step of sugar processing sucrose is extracted by continous extraction systems from sliced beet cossettes or pressed out from sugar cane, resulting in a dark colored *raw juice* with a dry substance concentration of 13–17% and a purity of 89–91% (Fig. 1).

In beet factories the remaining *wet pulp* (exhausted cossettes) is pressed to *pressed pulp* and then conserved by evaporating water to *dried pulp* with a solids content of 90–92%. *Molassed dried pulp* is produced by the addition of 1–3% *molasses* to the pressed pulp prior to drying. In cane factories *bagasse,* the fibrous residue of the cane stalks is obtained after crushing in mills and extraction of the juice.

The raw juice is purified using similar processes in cane or beet sugar factories. An objective of all purification steps is to remove the nonsucrose substances contained in the raw juice and thus to increase the purity of the juice (to increase the ratio of sucrose to total solids) which in turn will allow more complete recovery of sucrose by crystallization. Carbonation of limed juice eliminates 30–40% (wt) of the soluble nonsugar fraction and results in the purified yellowish *thin juice* after filtration. As in the clarification process not all nonsugar compounds are eliminated, a certain amount of nonsugar is carried through the further process.

Evaporation in a series of evaporators concentrates thin juice to the highly viscous brown *thick juice* (solids content 68–74%) with a sucrose concentration of up to 70% and a purity of 93%. For all sugar solutions of higher concentration the term *syrup* is also used.

## 2.2 Products from Crystallization

Thick juice is pumped to the crystallization station – a series of single-effect vacuum pans which exist in many designs and combinations. The sucrose and syrup qualities produced during crystallization by evaporation may vary considerably from factory to factory depending on the crystallization scheme used in the sugar end (sugar house).

During the evaporating crystallization process finally *magma* (*massecuite*) is formed, a mixture of crystals and syrup. The magma is filled batchwise in a screen-basket centrifuge

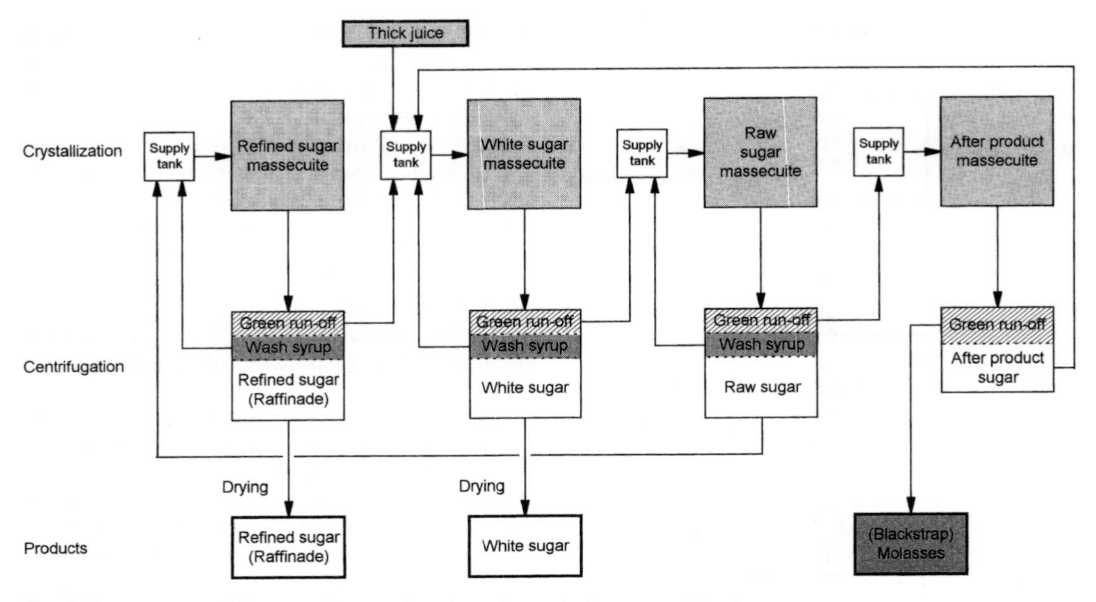


**Fig. 2.** Provenance of intermediate and end products during crystallization.

where the sugar crystals are separated from the *green run-off* (also called *green syrup*), the first mother syrup which is subsequently treated in following steps. Green run-offs may additionally be designed more accurately by the name of their origin, e.g., a *white sugar green run-off* would be derived from the white sugar stage.

For the production of sugar for consumption (white or refined) the crystals remaining in the centrifuge are washed with a fine stream of hot water to remove most of their syrup coating, forming the *wash syrup.* This syrup generally is collected separately and recycled to the crystallizer.

On the one hand very pure sugars are obtained directly by a centrifugation step, on the other hand sugar crystals are dissolved and recycled to the preceding crystallization step for further purification.

As indicated above, the product spectrum depends on the specific crystallization process applied in a sugar factory. Today, most of the beet sugar factories produce white sugar, whereas most cane factories produce raw sugar. This raw sugar is processed further to white sugar in refineries that may be located elsewhere.

A common and frequently used crystallization scheme of a beet factory is a four-product scheme with two end products illustrated, e.g., in Fig. 2.

In the first step *white sugar* is produced, the remaining *green run-off* (*green syrup, A-molasses*) is passed to the following raw sugar stage. (The term A- or B-molasses is used in cane sugar processing for an intermediate run-off and should not be mixed up with final molasses.) The *raw sugar* produced here is dissolved and introduced into the refined sugar crystallization stage where *refined sugar* (*extra white sugar*) is obtained by centrifugation. The *green run-off* of this step is pumped into the white sugar crystallizer together with the *thick juice.*

The *raw sugar green run-off* (*B-molasses*) is taken into the after-product stage. The *after-product sugar* of this step is dissolved and recycled to the white sugar crystallizer in addition to the thick juice. (Alternatively, it could be affined and recycled to the raw sugar crystallizer in addition to the white sugar green run-off).

For economical reasons, the *after-product green run-off* is not crystallized further as its purity is very much reduced. The mother syrup obtained by centrifugation after this last crystallization step is the well known *blackstrap molasses* (also called *final molasses* or *factory molasses*) were principally all nonsugar substances of the thick juice are found. In contrast to other uses of the term "molasses" this type of molasses is defined as a "real molasses", as will be explained below.

If white sugar is processed in refineries the resulting mother syrups are called first syrup, second syrup, etc. and the term *refinery (blackstrap) molasses* is used for the last syrup.

General information on sugar processing including quality considerations is given by PANCOAST and JUNK (1980), CHEN and CHOU (1993), and SCHIWECK and CLARKE (1994). The technology of sugar manufacture is also described comprehensively by SCHNEIDER (1968), MCGINNIS (1982), and HONIG (1953, 1959, 1963).

# 3 Quality Parameters for Sugar

## 3.1 Crystal Sugar

Although sucrose is most diversely used in the form of molasses, pure, commercial-grade sucrose is also used in industrial fermentations. White sugar produced from cane or beet is, because of its crystalline nature, an organic compound of very high purity which is scarcely exceeded by other industrial-scale products. In most cases purity is of the order of at least 99.9%. The main impurity is water and other impurities are inorganic salts and organic compounds such as reducing sugars, oligosaccharides, organic acids, and nitrogenous substances, most of which originate from the raw cane or beet (DUTTON, 1979). Some of the impurities are also formed during sugar production via chemical reactions, e.g., of the reducing sugars with the amino acids in the well-known Maillard reaction, which leads to the formation of pigments ranging in color from yellow to dark brown.

Analytical control techniques play a vital part in international standardization and the sugar industry was probably one of the first to try to achieve harmonization of analytical procedures on a worldwide basis. Although some countries had their own standards for sugar before, the first International Recommended Standard was drawn up under the Food Standards Programme of the FAO/WHO by the Codex Alimentarius Commission in 1969. This standard defines two specifications for white sugar: specification A for sugar described as white sugar and specification B for lower-purity sugars designated as plantation white sugar or mill white sugar.

Methods for the determination of quality parameters were also specified. Most of the analytical methods for sucrose, invert sugars, and related subjects are developed by the International Commission for Uniform Methods in Sugar Analysis (ICUMSA) and are published periodically in its Reports Proceedings.

The ICUMSA methods as published in 1964 have been revised and published as *"Sugar Analysis: ICUMSA Methods"* by the commission (ICUMSA, 1979, 1994). Since white sugar is a commercial product worldwide, quality specifications exist as official, internationally and nationally obligatory regulations or voluntary quality agreements between sugar manufacturers and the sugar industry.

White sugar is purified and crystallized sucrose (saccharose). Its essential composition and quality is specified mainly by polarization, invert sugar content, conductivity ash, loss on drying, color, and content of sulphur dioxide. The term polarization is customarily used in sugar analysis for the optical rotation of a sugar product, measured under defined conditions (ICUMSA) as a percentage of the rotation of pure sucrose, measured under the same conditions. Polarization is carried out in a polarimeter with International Sugar Scale (°Z) complying with ICUMSA definitions.

Specifications for the manufacture of crystal sugar in the European Community (EC), which have been accepted by many other countries, are generally used in international white sugar trading and are given in Tab. 1. In contrast to the two Codex specifications, the EC directive introduced three main classifications for white sugar. They were in close keeping with those of the Codex, but more

**Tab. 1.** Quality Parameters of Crystal Sugar in the EC

| Purified and Crystallized Sucrose of Sound and Fair Marketable Quality | Crystal Sugar Category | | |
|---|---|---|---|
| Specification | 1 | 2 | 3 |
| 1. Polarization [°Z] min. | – | 99.7 | 99.7 |
| 2. Water [%] max. | 0.06 | 0.06 | 0.06 |
| 3. Invert sugar content [%] max. | 0.04 | 0.04 | 0.04 |
| 4. Color type, Brunswick Institute Method, max. units | (2) | (4.5) | (6) |
| Points (0.5 units = 1 point) | 4 | 9 | 12 |
| 5. Conductivity ash [%] max. | (0.018) | (0.0270) | – |
| Points (0.0018% = 1 point) max. | 6 | 15 | – |
| 6. Color in solution, ICUMSA units, max. | (22.5) | (45) | – |
| Points (7.5 units = 1 point) max. | 3 | 6 | – |
| 7. Points total number, max. | 8 | 22 | – |

stringent requirements were introduced for the highest-quality sugar which was designed as "extra white sugar". These EC specifications are restricted to refined sugars that are evaluated by a points system regarding color (Brunswick Institute Method), conductivity ash, and color in solution as indicated in Tab. 1. The content of residual sulphur dioxide is restricted to all crystal sugars.

The main directives for sugar analysis by other organizations are similar to those described above, but they may vary in detail or can be extended by further specifications. In the United States, the quality standard is generally based on the "Bottlers" specifications of the National Soft Drink Association (NSDA, 1975). The National Food Processors Association (1968) has established microbiological standards for dry and liquid sucrose.

DOWLING (1990) described the most frequently used methods of sugar quality testing in the USA and in other countries. Several other countries have their own specifications which may differ in detail from the cited specifications or contain further restrictions.

Raw sugar is one of the main products in international sugar commerce. It is preferably manufactured to white sugar in refineries where the raw material from raw sugar factories is processed instead of thick juice. Beet raw sugar usually is not assigned for human consumption because of its unpleasant taste and salt content, whereas cane raw sugar can be used directly in the production of food or as household sugar.

The economic importance of raw sugar analysis is primarily defined by the polarization based on the International Sugar Scale. Although this method is applicable to all raw sugars it is of limited accuracy for raw sugar of less than 95% purity, where the result is a function of both, purity and nature of impurities present in the sugar. Commercial raw sugar has a polarization of 94–99 °Z. For purchase the importance of other physical and chemical characteristics of raw sugars than polarization has been increasingly recognized. In 1966, the American Sugar Company (now Amstar Corporation) established criteria of raw sugar quality which were revised in 1982 (American Sugar Company, 1968). In 1984, the Amstar Contract provided a new adjustment scale (Amstar Raw Sugar Contract, 1984), while a further new scale was enacted by the Savannah Contract (1984). In addition to polarization testing, the two contracts also include premiums or penalties for moisture, ash, dextran, grain size, and color.

Commonly raw sugar is also characterized by a "safety factor" (SF), i.e., the moisture–nonsugar (non-Pol) relationship of the sugar. This safety factor expresses the stability of raw sugar against microbial deterioration (by providing high osmotic pressure of a potential liquid film). Hence, control of the moisture content of the sugar by the safety factor is essential to provide sufficient storage capacity.

The safety factor for raw sugar must not exceed 1/4 of the nonsucrose, or % moisture/(100 − Pol) = 0.250 or less, whereby exact limits depend on the directives of the company or the consumer.

## 3.2 Liquid Sugar

Refinery products in liquid form have become more and more important. There are two basic types of sucrose-derived liquid sugars: those which are almost all sucrose and those which are partly converted to reducing sugars. They are predominantly used in the food industry. Concentrated liquid sugars are of high quality and offer technical advantages for special applications.

The quality of liquid sugar solutions is characterized by parameters similar to those of crystal sugar. Additionally dry mass and invert sugar content are used for characterization. Only liquid sugar solutions derived from sugar processing are discussed in this chapter, excluding aqueous solutions of saccharides obtained from starch (corn syrups).

Several types of liquid sugars are commercially available, and three main types are distinguished in the EC: (white) sugar solution, (white) invert sugar solution, and (white) invert sugar syrup. The supplementary designation "white" may be added to these liquid sugars if color in solution does not exceed 25 ICUMSA units and conductivity ash is maximal 0.1% on dry matter.

In Tab. 2 the specifications for liquid sugars according to EC Sugar Directives are shown. As discussed for crystal sugars, the EC specification may differ from those of other commissions. In the USA (invert) liquid sugars are distinguished by more graduated specifications.

In spite of these specifications, which are intended to promote standardization, sugar quality varies widely throughout the world. For domestic use, such variations probably have little importance, but a large amount of sugar is used as raw material in food manufacture where specific quality criteria are of importance.

It can finally be stated that crystal and liquid sugars are characterized by internationally accepted methods. Moreover, specific additional methods or directives of a company or a customer can be applied in sugar trade, assuming that every partner agrees.

# 4 Raw Materials as Substrates

A variety of conceivable raw materials such as juices, syrups or green run-offs, and crystal sugars of minor purity become avail-

**Tab. 2.** Quality Parameters of Liquid Sugar in the EC

| Specification | Liquid Sugar Category | | |
|---|---|---|---|
| | Sugar Solution | Invert Sugar Solution | Invert Sugar Syrup |
| 1. Dry matter [%] min. | 62 | 62 | 62 |
| 2. Invert sugar content [% odm[a]] | max. 3 (Ratio fructose–glucose 1 ± 0,2) | min. 3, max. 50 (Ratio fructose–glucose 1 ± 0.1) | min. 50 (Ratio fructose–glucose 1 ± 0.1) |
| 3. Conductivity ash [% odm] max. | 0.1 | 0.4 | 0.4 |
| 4. Color in solution, ICUMSA units, max. | 45 | – | – |
| 5. Sulphur dioxide [mg $kg^{-1}$] max. | 15 | 15 | 15 |

[a] odm: on dry matter

able during sugar processing as indicated in Figs. 1 and 2. With regard to fermentations the term "raw material" is used here for substances still containing a certain amount of nonsucrose, without consideration of the extent of processing in a factory. Consequently, raw juice and molasses are regarded as raw materials, although molasses, from the viewpoint of a factory, is a final product which has been processed much more profoundly than raw juice.

## 4.1 Composition

In biotechnological fermentations the nonsucrose fraction of these raw materials may be advantageous as substances of nutritive value are provided, thus reducing costs of synthetic supplements or growth substances (OLBRICH, 1973). Moreover, nonsucrose includes other fermentable sugars such as raffinose, kestoses, or invert sugar which can be utilized by most microorganisms of industrial importance.

Since most of these intermediate products have no commercial importance, only scarce information on their composition is available from in-process control. However, the composition of all these raw materials basically is very similar. They mainly differ in their content of water and in purity (and hence their nonsucrose content) as well as in some minor chemical alterations during evaporation.

In Fig. 3 the mass flow of the main compounds of thin juice (related to 100 kg beet) to the end products crystal sugar and molasses is illustrated. Fig. 3 clearly shows the reduction of water content during evaporation and crystallization. Furthermore, the important fact is demonstrated that the nonsucrose fraction of thin juice is almost the same in molasses if white sugar is produced. Hence, the composition of the nonsucrose fraction of any syrup or intermediate sugar is very similar to that of molasses (except glutamine) whereas its amount is generally on a distinctly lower level. As the composition of molasses is sufficiently characterized, the composition of nonsucrose of any intermediate product can be roughly evaluated using data obtained with molasses.

The sugar and solids content of raw materials and intermediate products may vary considerably depending on specific process parameters and on the quality of beet or cane manufactured in a factory. In Tab. 3 analytical values of solids and sugar contents of the main raw materials are given. Data are shown for final molasses (from the centrifuges), whereas for commercial purposes it is common practice to dilute the heavy factory molasses to a standard solids content.

In contrast to only slight chemical alterations of most raw materials during evaporation and crystallization, raw juice undergoes a more distinct change in the juice purification process where 30–40% of soluble nonsucrose substances are removed. Raw juice contains almost all colloidally dispersed and molecularly dissolved substances present in the cell juice and, additionally, the products of micro-

**Tab. 3.** Solids and Sugar Content of Raw Material from Beet Sugar Processing

| Source | Solids [%] | Sugar [%] | Purity [%] |
|---|---|---|---|
| Raw juice | 15–18 | 13–16.5 | 89–91 |
| Thin juice | 15–18 | 13.5–17 | 91–93 |
| Thick juice | 68–74 | 61–70 | >91 |
| Green run-off (white and raw) | 78–82 | 63–70 | 77–90 |
| Wash syrup (white and raw) | 73–78 | 65–70 | 83–95 |
| Final molasses | 80–89 | 50–56 | 58–63 |

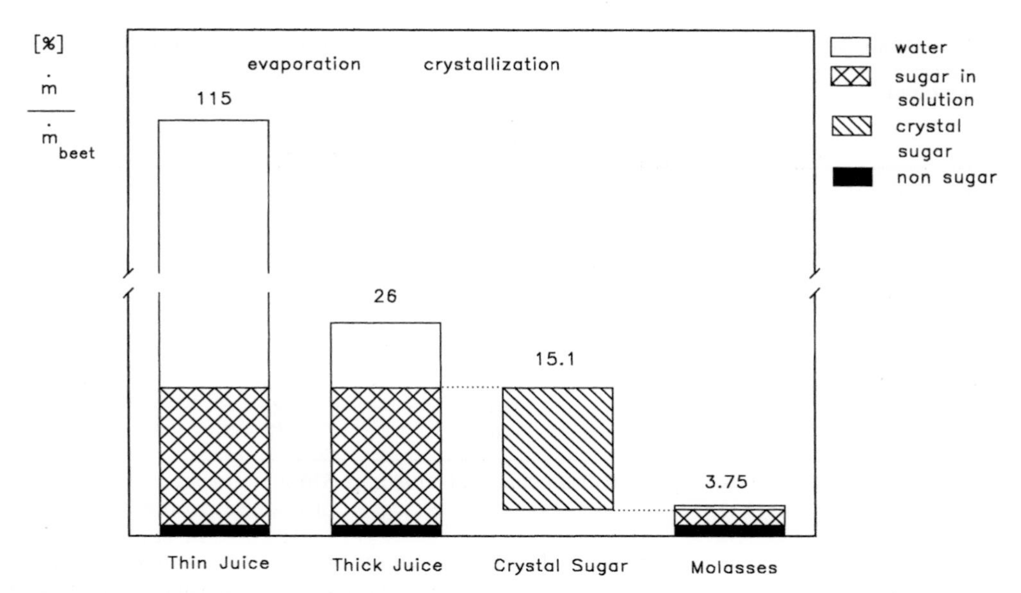


**Fig. 3.** Mass flow of sugar and nonsugar.

bial activity (mainly lactic acid, but also minor amounts of formic, acetic, and propionic acid) during extraction. In Tab. 4 the approximate composition of raw juice is shown. The main sources for nitrogen are proteins, betaine, and amino acids. Besides these main compounds considerable amounts of pantothenic acid (750 μg 100 $g^{-1}$ sucrose) or further growth stimulating compounds are included in raw juice (KLAUSHOFER, 1980).

During raw juice purification some acids are formed and other acids are removed, so that the acids composition differs from that measured in molasses. The amount of acids may vary considerably depending on the beet composition. Tab. 5 gives an example for the acids composition of raw juice, thin juice, thick juice, and molasses in a beet factory (KOWITZ-FREYNHAGEN, 1979; REINEFELD et al., 1980) using classical juice purification. For comparison, data are based on 100 g sugar for the juices. As sugar is removed in the process, the data for molasses are based on a dry matter basis. During juice purification

**Tab. 4.** Analytical Values of Beet Raw Juice

| Quantity | | Quantity | Based on 100 g DS |
|---|---|---|---|
| Solids content [%] | 15–18 | Ash [%] | 2.5 |
| | | Invert sugar [%] | 0.4–0.7 |
| Sugar content [%] | 13–16.5 | Total nitrogen | 0.48 |
| | | Pectin content | 0.37 |
| Purity [%] | 89–91 | Protein content | 0.30 |
| | | Cation content | 35–45 meq |
| pH | 6.2 | Anion content | 40–50 meq |
| | | Amino acid content | 10–20 meq |

SCHNEIDER (1968), p. 249; REINEFELD et al. (1980), p. 563; SCHIWECK and CLARKE (1994), p. 365

**Tab. 5.** Analytical Values of Acid Composition in Raw Materials During Beet Sugar Processing

| Compound | Raw Juice | Thin Juice [mg 100 $g^{-1}$ Sucrose] | Thick Juice | Molasses [mg 100 $g^{-1}$ DS] |
|---|---|---|---|---|
| Lactic acid | 143 | 333 | 345 | 1354 |
| Glycolic acid | 24 | 41 | 65 | 225 |
| Glyceric acid | – | 32 | 56 | 213 |
| Saccharic acids | – | 193 | 279 | 1248 |
| Pyrrolidonecarboxylic acid | – | 618 | 1424 | 6934 |
| Oxalic acid | 611 | 18 | 9 | 24 |
| Phosphoric acid | 420 | traces | traces | traces |
| Malic acid | 122 | 50 | 50 | 213 |
| Citric acid | 737 | 46 | 22 | 51 |

KOWITZ-FREYNHAGEN (1979), p. 41; REINEFELD et al. (1980), p. 570

mainly pyrrolidone carboxylic acid, lactic acid, glycolic acid, glyceric acid, and saccharic acid are formed. Phosphoric acid, oxalic acid, and citric acid are distinctly reduced in concentration by juice purification.

Although nonsucrose substances containing nitrogen constitute the prevailing part of nonsucrose, only scarce innovative literature on the amino acid content of sugar raw materials is available (KUBADINOW and HAMPEL, 1975; REINEFELD et al., 1982a, b; SCHIWECK, 1991; SCHIWECK et al., 1993). Analytical values of amino acids in raw juices are shown in Tab. 6. The data vary considerably according to the different origins of beets, but all raw juices show a predominant amount of glutamine (incl. glutamic acid), followed by aspartic acid (incl. asparagine) and 4-aminobutyric acid.

Without any exception, the amino acid content decreases during juice purification (REINEFELD at al., 1982a; SCHIWECK, 1991). Only glycine and 4-aminobutyric acid are newly formed whereas glutamine decreases by further reactions (formation of pyrrolidone carboxylic acid). Thus 4-aminobutyric acid becomes the predominant amino acid in sugar raw materials.

The general composition of molasses is described sufficiently in current sugar processing handbooks as cited above. More specific molasses compounds are treated by TRESSL et al. (1976), SCHIWECK (1977), FIEDLER et al. (1981), SINDA and PARKKINEN (1980). Analytical values for the composition of molasses in recent literature are frequently based on early investigations published by BINKLEY and WOLFROM (1953), UNDERKOFLER and HICKEY (1954), WHITE (1954), OLBRICH (1956), and HONIG (1963). As up to now molasses is the most important sugar-based raw material with many applications it will be discussed separately in Sect. 5.

Besides the liquid raw materials described above, bagasse and beet pulp present residual solid raw material from sugar extraction. Most of the bagasse production is used in the raw sugar factories to supply fuel for the steam generation. The surplus of bagasse is frequently used to produce fibrous products. The direct use of untreated bagasse in fermentations is rather restricted because of its unsuitable composition. The composition of bagasse and beet pulp (dry matter) is shown in Tab. 7 according to data in literature (ARNTZ et al., 1985; CHEN and CHOU, 1993; KJAERGAARD, 1984; PATURAU, 1989; SCHNEIDER, 1968).

Compared to beet pulp significant differences in the dry matter composition are observed. There are no pectins in bagasse, and the lignin content is much lower in beet pulp than in bagasse. Hence, beet pulp is more suitable to microbial attack increasing its potential use as a substrate for fermentation. The possibilities of establishing biotechnolog-

**Tab. 6.** Analytical Values of Amino Acid Composition in Raw Juices [mg 100 $g^{-1}$ sucrose]

| Amino Acids | KUBADINOW and HAMPEL (1975), p. 397 ∅ Campaign 1974 | REINEFELD et al. (1982b), p. 1116 ∅ Factory A+B | SCHIWECK (1991), p. 799 Factory Offstein 1990 |
|---|---|---|---|
| Glx (Gln+Glu) | 2439 | 1020.4 | 1057 |
| Asx (Asn+Asp) | 472 | 316.1 | 207.3 |
| 4-amino-butyric acid | 139.4 | 140.8 | 165.8 |
| Ser | 188.3 | 100.1 | 130.2 |
| Ile | 139.4 | 72.0 | 59.0 |
| Leu | 149.5 | 61.8 | 44.9 |
| Ala | 94.4 | 67.2 | 80.1 |
| Val | 91.1 | 35.6 | 34.0 |
| Thr | 80.1 | 29.0 | 25.9 |
| Tyr | – | 16.3 | 50.0 |
| Pro | 34.3 | 20.4 | – |
| Met | 13.1 | 23.0 | 13.0 |
| Lys | 23.8 | 11.9 | – |
| Phe | 20.6 | 14.3 | – |
| Gly | 11.1 | 11.6 | 10.9 |

**Tab. 7.** Average Composition of the Dry Matter of Beet Pulp and Bagasse

| Compound | % of Dry Matter | |
|---|---|---|
| | Beet Pulp | Bagasse |
| Cellulose | 18–30 | 27–46 |
| Hemicelluloses | 18–34 | 26–32 |
| Pectins | 15–34 | 0 |
| Lignins | 3– 6 | 18–22 |
| Insoluble ash | 2– 7 | 1– 4 |
| Proteins | 2– 7 | – |
| Sugar | 3– 6 | – |

ical utilization of beet pulp and bagasse are reviewed by KJAERGAARD (1984).

## 4.2 Availability and Applicability in Fermentations

A fermentation substrate must be readily available throughout most of the year. Raw materials produced seasonally are not desired if the harvesting period is short and the material to be used is subject to contamination and spoilage. Thus many industries need a substrate which is relatively stable and can be stored for 6–9 months without decomposition.

Apart from bagasse, pulp, and molasses most of the raw materials produced in a sugar factory are intermediate products of the sugar recovery process, which are subsequently treated to crystal sugar. For many fermentations highly pure carbon sources are not necessary because further medium supplements are often added in technical quality. For cost reasons the use of nonrefined carbohydrate substrates often becomes mandatory in bulk product fermentations. Thus, less pure crystal sugars (after-product sugar, raw sugar) or even sugar containing raw materials seem to be advisable. However, some raw materials from sugar processing are not accessible in abundant amounts and sometimes their applicability might be restricted in case of storage problems. Transport costs have to be considered; they may become significant and prohibitive if too much water is present and will prohibit the use of some waste materials at sites removed from their place of production (RATLEDGE, 1977).

*Raw juice* with a sucrose content of about 15% can serve as an advantageous carbon source for bulk fermentations as it contains some additional fermentative compounds which are partly eliminated in the next processing step. It is produced temporarily during the sugar campaign, varying from 3 months of the year to year-round in several tropical countries. Despite the addition of disinfectants microbial activity cannot be completely avoided and because of its suitable pH, temperature, and water content it cannot be stored without pretreatment at ambient temperature. Hence, industrial application of fresh raw juice for biotechnological use is restricted to the sugar campaign. Moreover, a location close to a (raw) sugar factory would be favorable to avoid long transport times and costs.

Raw juices can be steamed down to the consistence of syrup to make them storable. To overcome the storage problem efforts have been made at a pilot scale to produce a *raw thick juice* (concentrated raw juice) which provided growth comparable to that obtained with thick juice (MANZKE et al., 1992). Parameters influencing the storability of raw thick juice, including $a_w$, pH, $T$ and disinfectants, were investigated by FIEDLER et al. (1993).

The sucrose concentration in *thin juice* is similar to that of raw juice, but nonsucrose is elimated partly as described above. The microbial conditions of fresh thin juice (pH about 9) are excellent since it is practically sterile after strong shifts of pH and heat treatment during the purification process. But, similar to raw juice, problems arise on the storage of unconcentrated thin juice as it will support microbial growth over a wide range of temperature and pH. Both juices are available in any amount and can be purchased by an agreement with a (raw) sugar factory (not refinery).

Most microorganisms found in sugar juices normally will not develop or grow in solutions of sugar with a dry substance content of more than 67% and at a pH of 8.0. Hence the use of *thick juice* (pH about 9) as carbon source for fermentations could distinctly reduce any storage problem occurring with unconcentrated thin juice. The composition of both of these raw materials is nearly identical except for some minor differences in amino acids and organic acids compounds.

Thick juice with a sugar content of 61–70% can be stored for months under appropriate conditions. In spite of its natural resistance to microbial attack, every sanitary precaution must be followed exactly, since an infection could start because of inadvertent dilution of surfaces by condensed moisture or other adverse events (MCGINNIS, 1982). Long-term storage has been investigated by VACCARI et al. (1985, 1988) and LAMBRECHTS et al. (1967). Storage of thick juice for up to 130 d at an industrial scale (>2000 t) was studied by BOHN et al. (1974). As raw juice and thin juice, fresh thick juice is available in every sugar factory (except refineries) only during the campaign. Out of the campaign period thick juice would be disposable only in factories with thick juice storage.

From the crystallization stage some syrups with high purity (wash syrups) or lower purity (green run-offs) become available during the sugar campaign. A reduction in costs may hardly be expected using intermediate products with higher purity than thick juice, as the price generally depends on the amount of crystallizable sucrose. Consequently, only the use of syrups with lower purity such as *white sugar green run-off* ($p$=78–90%) or *raw sugar green run-off* ($p$=76–83%) may be of interest. These syrups are intermediate products that are only produced during the campaign and normally are not stored at a factory. If storage of these syrups is desired, precautions as with thick juice must be taken.

Although a number of raw materials are available from the sugar recovery processs, sucrose is most diversely used in the form of *molasses,* which is the crudest form of sucrose, being the concentrated mother liquor after crystallization from sugar solutions of beet or cane.

Considering the importance of molasses as a biotechnological substrate and the amount of disposable analytical data on it, molasses is characterized in more detail in the following section.

# 5 Molasses

## 5.1 Types of Molasses

Molasses as discussed in the following is a product of the sugar industry based on both cane and sugar beet. It is defined as the end product of sugar manufacture from which no more sugar can be crystallized by conventional methods. Molasses can be defined more precisely as the mother syrup (green run-off) resulting from the final low-grade massecuite of sugar, the purity of which has been reduced to the point where further crystallization of sugar is uneconomical. Molasses, therefore, is a solution of sugar, organic matter, and inorganic matter in water. Even if all known components of molasses were added synthetically, the result would certainly not be molasses. Thus, in accordance with the statement of three experts of sugar analysis nobody really knows what molasses exactly is (WOLFROM et al., 1952).

Both cane and beet molasses vary in composition from year to year and from location to location. Hence, it is common practice of buyers to select material by experiment during the season and then purchase the years's supply for storage in bulk (RHODES and FLETCHER, 1966).

If molasses is sold as raw material (nutrient medium) for the production of ethanol, yeast, citric acid, or other chemical or biochemical processes, general trading conditions or as-agreed specificatons are valid (SCHIWECK and CLARKE, 1994).

Apart from the definition given above, the term "molasses" is also used for other substances which are not in accordance with this definition. OLBRICH (1956, 1970, 1973) distinguishes between real molasses, molasses-like substances, and molasses-similar substances. To avoid any confusion, the different types of molasses are explained as follows.

### 5.1.1 Real Molasses

Real molasses are mother liquors of the sugar recovery process according to the abovementioned definition. They are final effluents (final molasses) and by-products of sugar manufacture. Due to their origin sugar cane molasses and beet molasses are distinguished.

*Blackstrap molasses* (*sugar house blackstrap molasses*) are molasses produced in raw (or white) sugar factories from cane or beet. Often only the term *molasses* is used for molasses from a beet factory whereas *blackstrap* (*molasses*) is preferably used for cane molasses.

*Refinery* (*final*) *molasses* (*refinery blackstrap, barrel syrup*) is derived mainly from cane refineries where (high-grade) white sugar is produced from raw sugar.

Additionally *palm molasses* is obtained from the production of palm sugar in India and Burma in amounts of some importance.

*Desugarization factory molasses* (*discard molasses*) are produced when additional sugar is eliminated from beet molasses by special treatments thus reducing their applicability for biotechnological use.

The resulting molasses of such processes are *Quentin molasses, SCC* (Sugar-Chemical-Company) *molasses* and *Steffen molasses.*

### 5.1.2 Molasses-Like Products

These types of molasses are mother liquors of the recovery process of sugars other than sucrose. *Hydrol* (*corn hydrol*) is the mother liquor of glucose recovery after the conversion of starch into sugar. *Milk sugar* (*factory*) *molasses* results from the production of lactose from whey.

Compared to real molasses these types of molasses are of little importance for biotechnological use.

### 5.1.3 Molasses-Similar Products

In contrast to the final effluents of real molasses these substrates are neither final effluents nor by-products nor mother liquors. They are the immediate object of production and, therefore, main products. Thus these molasses are not molasses in the original sense of the word but notwithstanding have partly found application in biotechnology.

*High-test molasses* (also called *cane high-test molasses*, *invert molasses*, and *cane invert syrup*) is produced by direct evaporation of partly inverted cane juice, while no attempt is made to obtain crystalline sucrose. Thus, strictly speaking, high-test molasses is not molasses but an inverted syrup of controlled composition, purified only moderately, and concentrated to about 80–85% solids. Today, high-test molasses is prepared by using the enzyme invertase to make a partially inverted product.

It is largely used in distilleries and as a commercial sweetener (CHEN and CHOU, 1993). About 95% of the sugars in high-test molasses can be fermented by yeasts. The origin of high-test molasses goes back to the years of depression when a surplus of sugar was produced and the demand for molasses sugar was higher then for sugar produced by blackstrap only (HODGE and HILDEBRANDT, 1954). The composition of high-test molasses can be defined in various ways; a generally accepted composition is shown in Tab. 8.

*Citrus molasses* is produced by concentration of clarified press juice from peels of citrus fruits to a syrup with a solids content of 72%. According to HENDRICKSON and KESTERSON (1967) it consists of 20.5% sucrose, 23.5% invert sugar (total sugar 45%), and 4.7% ash with a pH of 5–7. Citrus molasses is available in locations where citrus fruits are manufactured.

## 5.2 Composition of Beet and Cane Molasses

Both raw and refined sucrose are relatively expensive raw materials for industrial fermentations. In addition to these sources of energy further expensive nutrients for growth must be added when using pure carbon sources. This has led to the widespread use of cane and beet sugar molasses as a major source of carbohydrates for fermentation. Both types of molasses mainly differ in their sucrose content, whereas the amount of total fermentable sugar is similar.

Final molasses arising from the centrifugals accumulates in this separation process to a dry matter content of 80–89% whereas trade molasses may be diluted to a solids content of (minimum) 76.3%. To compare analytical data obtained for molasses from different origins data are often normalized to a specified solids content of 80%.

Molasses is an agricultural product and its composition varies with the variety and maturity of cane or beet as well as with climate and soil conditions. Variations in composition are more distinct with cane than with beet. In addition, processing conditions in the sugar factory may also bring about changes in the composition of molasses. Thus, again only indicative average values can be given for the composition as shown in Tab. 9.

Since the agricultural conditions of beet are different from those of cane there are differences in composition with regard to the sugar

**Tab. 8.** Average Composition of High-Test Molasses

| Quantity | Analytical Values [%] |
|---|---|
| Solids | 80 –85 |
| Sucrose | 15 –35 |
| Reducing sugars | 50 –65 |
| Invert sugars | 40 –60 |
| Total sugars as invert | 75 –80 |
| Organic nonsucrose | 4 – 8 |
| Ash | 2 – 3 |
| $K_2O$ | 0.7 – 1.4 |
| CaO | 0.03– 0.3 |
| MgO | 0.01– 0.03 |
| $P_2O_5$ | 0.2 – 0.6 |
| pH | 5 – 6 |

OLBRICH (1956), p. 51; HODGE and HILDEBRANDT (1954), p. 77; United Molasses Company (1986).

**Tab. 9.** Main Components of Beet and Cane Blackstrap Molasses

| Component | Beet | Cane |
|---|---|---|
| | [% of molasses] | |
| Dry solids | 76–84 | 75–83 |
| | [% of solids] | |
| Sucrose | 58 –64 | 32 –45 |
| Raffinose | 0 – 4.2 | – |
| Glucose | – | 5 –11 |
| Fructose | – | 6 –15 |
| Invert | 0 – 1.2 | – |
| Total sugar (sucrose + reducing sugar) | – | 52 –65 |
| Organic nonsucrose | | |
| nitrogenous | 19 | 5 |
| nitrogen-free | 5 | 10 |
| Ash | 8.5–17.1 | 7 –11 |
| Total nitrogen | 1.7– 2.4 | 0.4– 1.5 |
| pH | 6.2– 8.4 | 4.5– 6.0 |

BINKLEY and WOLFROM (1953); WILEY (1954), p. 314; OLBRICH (1956), p. 12; RÜTER (1975); PATURAU (1989).

and the nonsugar contents. The typical weight ratios of the major components found in both beet and cane molasses differ as follows: In contrast to cane molasses the total sugars of beet molasses consist almost entirely of sucrose with only small quantities of reducing sugars.

As compared with beet molasses, in general cane molasses has a lower content of nonsugar substances and ash. The sugars in cane molasses are principally sucrose, glucose, and fructose. Glucose and fructose are often referred to as reducing sugars and usually reported as such in conventional analyses. Raffinose, in varying amounts, occurs in beet molasses and is not readily utilized by yeasts. A distinction is made between total sugar and fermentable sugar, and the difference may amount to several percent.

Beet and cane molasses vary not only in their contents of utilizable carbohydrates but also with respect to other constituents. The nonsugar components of dry matter are composed of *inorganic* and *organic* substances, and among these *nitrogen-free* organic compounds and *nitrogenous* organic compounds are distinguished.

## 5.2.1 Nitrogen-Free Compounds

In Tab. 10 the composition of inorganic matter (ash) of beet and cane molasses is compared. Cane molasses has higher contents of $SiO_2$, CaO, MgO and $P_2O_5$ than beet molasses. In addition to the ash constituents noted in Tab. 10, molasses also contains many elements in very low concentrations, the so-

**Tab. 10.** Approximate Composition of Beet and Cane Molasses Ash

| Constituent | Beet | Cane |
|---|---|---|
| | [% of ash] | |
| $K_2O$ | 66 –72 | 30 –50 |
| $Na_2O$ | 9.5 –16 | 0.3– 9.0 |
| CaO | 4 – 7 | 7 –15 |
| MgO | 0 – 0.8 | 2 –14 |
| $FeO_3$ | 0.01– 0.45 | – |
| $P_2O_5$ | 0.23– 0.8 | 0.5– 2.5 |
| $SO_3$ | 0.6 – 2.6 | 7 –27 |
| $SiO_2$ | 0 – 1.45 | 1 – 7 |

OLBRICH (1963), p. 540; CHEN and CHOU (1993), p. 408

called trace elements, which must be supplemented when using pure carbon sources (OLBRICH, 1963; BRONN, 1985).

The organic, nitrogen-free compounds of molasses mainly consist of organic acids, gums and other carbohydrates, polysaccharides, wax, sterols, pigments, etc. – with varying amounts in beet and cane molasses.

In Tab. 11 the acids identified in beet and cane molasses are summarized. A number of organic acids are present. The most important in quantity is lactic acid in beet molasses (resulting from microbial processes and invert sugar destruction during juice clarification) and aconitic acid in cane molasses. Under conditions of yeast fermentation all organic acids except formic acid are used as carbon sources. Hence, the yields by yeast from molasses – related to sugar – are higher compared to other, purer substrates.

## 5.2.2 Nitrogenous Compounds

Nitrogen in molasses mainly originates from free amino acids, pyrrolidone carboxylic acid, betaine, nucleotides, and nucleosides and peptides (Tab. 12).

The content of free amino acids as well as the content of pyrrolidone carboxylic acid amounts to 10–11% each (SCHIWECK et al., 1993) or may even increase to 24–50% (OLBRICH, 1956; REINEFELD et al., 1986). Furthermore, the amount of amino acids found after hydrolysis can sum up to an amount comparable to that of free amino acids. From the total nitrogen in molasses "crude protein" is frequently figured as N × 6.3 being the main nitrogenous compound of cane molasses.

In general, beet molasses contain more total nitrogen and organic nitrogenous substances than cane molasses. Betaine (trimethylglycine) is the most abundant nitrogenous compound found only in beet molasses and accounts from 25% to up to 50% of the total nitrogen content. These nitrogen compounds give beet molasses the characteristic odor and taste.

The distribution of amino acids found in beet molasses is shown in Tab. 13. The predominant amino acids analyzed in molasses are 4-aminobutyric acid, glutamine and glu-

**Tab. 11.** Organic Acids in Beet and Cane Molasses

| No. | Component [mg 100 $g^{-1}$ dry matter] | Beet | | | | Cane | |
|---|---|---|---|---|---|---|---|
| | | KOWITZ-FREYNHAGEN (1979) | WALLENSTEIN and BOHN (1963) | DIERSSEN et al. (1956) | SCHIWECK and HABERL (1973) | CHEN and CHOU (1993) | BINKLEY and WOLFROM (1952) |
| 1 | Formic acid | – | 350 | 375 | 488 | – | 120 |
| 2 | Acetic acid | – | – | 925 | 527 | – | 240 |
| 3 | Butyric acid | – | – | 625 | 0 | – | – |
| | Sum of 2+3 | – | 1010 | 1550 | 527 | – | – |
| 4 | Propionic acid | – | – | 250 | – | – | – |
| 5 | Lactic acid | 1576 | 3050 | – | 1595 | – | 600 |
| 6 | Glyceric acid | 176 | – | – | – | – | – |
| 7 | Saccharic acid | 1128 | – | – | – | – | – |
| 8 | Glycolic acid | 245 | – | – | – | – | – |
| 9 | Oxalic acid | 18 | – | – | – | – | – |
| 10 | Malic acid | 204 | 430 | – | 319 | – | – |
| 11 | Citric acid | 32 | 410 | – | 306 | – | – |
| 12 | Aconic acid | – | – | – | – | 1250–6250 | 950–4950 |
| | Sum of 9–12 | – | – | – | – | 1875–7500 | – |

**Tab. 12.** Content of Nitrogen in Beet and Cane Molasses

| Nitrogenous Constituent [% of molasses] | Beet | | | Cane |
|---|---|---|---|---|
| | OLBRICH (1956) | SCHIWECK et al. (1993) | REINEFELD et al. (1986) | BINKLEY and WOLFROM (1953) |
| Total N | 1.24–2.06 | 2.05 | 2.04 | 0.4 –1.5 |
| α-Amino-N | 0.15–0.53 | 0.2 | 0.37 | 0.03–0.05 |
| Betaine N | 0.84–1.12 | 0.53 | 0.85 | 0 |
| Nucleotide/Nucleoside | – | 0.07 | 0.11 | – |
| Peptides | – | 0.2 | 0.04 | – |
| Pyrrolidone carboxylic acid N | 0.14–0.15 | 0.21 | 0.5 | 0 |

**Tab. 13.** Analytical Values of Amino Acid Composition in Beet Molasses

| Amino Acid [% of total amino acids] | REINEFELD et al. (1982a), p. 289 | SCHIWECK (1991), p. 800 Factory Offstein 1990 | STEINMETZER (1991) |
|---|---|---|---|
| Glx (Gln + Glu) | 14 | 15.4 | 15.4 |
| Asx (Asn + Asp) | 22 | 19.9 | 14.3 |
| 4-Aminobutyric acid | 14 | 13.3 | 22.4 |
| Ser | 8.3 | 7.3 | 6.8 |
| Ile | 8 | 5.9 | 7.1 |
| Leu | 7.2 | 5.6 | 6.8 |
| Ala | 6.6 | 8.4 | 7.7 |
| Val | 4.6 | 7.1 | 4.2 |
| Thr | 1.7 | 1.9 | 1.3 |
| Tyr | 0.6 | 0.8 | 9.3 |
| Pro | 3.9 | 0 | – |
| Met | 0.8 | 0.8 | 0.9 |
| Lys | 1.2 | 0 | – |
| Phe | 1.2 | 11.5 | 1.3 |
| Gly | 2.7 | 2.4 | 2.9 |
| Trp | 1.9 | 0 | – |

tamic acid (sum), and asparagine and aspartic acid (sum). Glutamic acid and aspartic acid are also the main amino acids in cane molasses (PATURAU, 1989).

Betaine and some up to now not clearly identified nitrogen containing compounds are not assimilated by most microorganisms. For biotechnological use not the total nitrogen content but content of nitrogen than can be assimilated is of importance. Hence, mainly free and bound amino acids, peptides, ammonia, nitrite, and nitrate are essential to provide nitrogen in molasses.

### 5.2.3 Growth Promoting Products

Molasses are a valuable source of growth substances in sufficient quantities and may be significant in fermentation processes, in particular in the manufacture of yeast. Apart from panthothenic acid, inositol, and biotin molasses also contains further growth substances which may exert a stimulative effect on the growth of microorganisms (Tab. 14).

Beet and cane molasses mainly differ in the amount of biotin which is considerably higher in cane molasses. Should the biotin content of beet molasses be insufficient for optimal growth, it is feasible to add some cane molasses. The composition of high-test molasses, as can also be seen from Tab. 14, differs markedly from that of the traditional molasses.

The contents of minerals and vitamins in raw substrates normally are not sufficient to provide the complete amount needed for the production of baker's yeast. If molasses is replaced by purer carbon sources the costs for the substrate and for additional growth factors increase and must be taken into consideration (BRONN, 1987; SCHIWECK, 1995).

If molasses is completely replaced by thick juice the costs of trace elements and growth factors may rise to 19% of the total substrate costs compared to 1.6% by the sole use of molasses (FRICKO, 1995). Therefore, from an economical point of view only mixtures of thick juice and molasses provide good yeast quality.

In Tab. 15 the additional need for organic growth substances (vitamins) per t of substrate and the cost of these growth substances per t of fresh baker's yeast is shown. A slight increase of additional costs is observed with sucrose containing substrates purer than molasses. The costs for additional growth factors increase drastically with substrates based on glucose because of the requirement of inositol supplementation.

Hence, the increase of costs for additional growth substances (mainly inositol) make the economic efficiency of some substrates very questionable.

## 5.2.4 Undesirable Compounds

Occasionally the occurrence of undesired molasses compounds have also been described. Volatile nitrogenous pyrazines, pyrroles as well as furans and phenols (products of the Maillard reaction) have been detected, all in concentrations in the ppb or ppm range

**Tab. 14.** Growth Substances Contents of Different Molasses

| Molasses Type | Beet | | | Cane | | High-Test |
|---|---|---|---|---|---|---|
| Constituent | OLBRICH (1963) | SCHIWECK (1995) | BRONN (1985) | OLBRICH (1963) | PATURAU (1989) | OLBRICH (1963) |
| [µg/kg]: | | | | | | |
| Biotin | 40–130 | 18–27 | 50 | 1200–3200 | 1000–300 | 300–400 |
| Folic acid | 210 | – | – | 40 | 300–400 | 150 |
| [mg/kg]: | | | | | | |
| Thiamine, Vit. $B_1$ | 1.3–4.0 | – | 2.5 | 1.4–8.3 | 0.6–1.0 | 1.6 |
| Riboflavin, Vit. $B_2$ | 0.41 | – | – | 2.5 | 2.0–3.0 | 0.62 |
| Pyridoxin, Vit. $B_6$ | 5.4 | – | – | 6.5 | 1.0–7.0 | 1.6 |
| Nicotinic acid | 43–51 | – | – | 21 | 17–30 | 2.4 |
| Pantothenic acid | 50–110 | 23–70 | 80 | 21–54 | 20–60 | 2.5–2.7 |
| [g/kg]: | | | | | | |
| Inositol | 5.8–8.0 | 0.7–1.8 | 6.5 | 6.0 | 2.0 | 0.85–1.0 |

**Tab. 15.** Additional Need and Costs of Growth Substances (Vitamins) per t of Baker's Yeast

| Raw Material Used | BRONN (1987) Additional Need of Growth Substances [g $t^{-1}$ substrate] | | | | SCHIWECK (1995) Costs for Additional Growth Substances [\$ $t^{-1}$ of yeast] |
|---|---|---|---|---|---|
| | Biotin | Pantothenic Acid | Inositol | Thiamin | |
| Thick juice | 0.39 | 37 | – | 14.0 | 11.6 |
| Green run-off | 0.34 | 8 | – | 12.0 | 9.2 |
| Molasses | 0.26 | – | – | 9.1 | 7.6 |
| Glucose syrup | 0.40 | 52 | 780 | 14.3 | 65 |
| Sugar from starch | 0.40 | 52 | 780 | 14.3 | 65 |
| Glucose syrup, technical quality | 0.32 | 42 | 630 | 11.5 | 52 |

(TRESSL et al., 1976; FIEDLER et al., 1981). Other unwelcome contaminants were heavy metals (DETAVERNIER, 1982) and biocides (SCHIWECK, 1980). Some of these substances are said to have a suppressant effect on yeast growth. Theoretically, a suppression by volatile acids could take place when present in high concentrations in an undissociated form. As indicated in Tab. 11 the volatile organic acids in molasses occur in very low concentrations, so that normally a suppression is very improbable. This was also confirmed in growth tests carried out by SCHIWECK (1995). Furthermore, it was shown that the values for other compounds found in molasses are much smaller than the inhibition concentrations for these substances, as ascertained by ZAUNER et al. (1979).

The suitability of molasses for industrial fermentations cannot be evaluated according to their origin and chemical composition. The only effective means for such an evaluation is the biological test for a given application.

## 5.3 Characteristics of Molasses

### 5.3.1 Pretreatment

Molasses may contain components which are harmful to a fermentation process or the preparation of yeast. Hence, in processes with molasses as the sole carbon source or when it is used in high amounts as cosubstrate, it is pretreated and inhibitors are partly removed. The choice of pretreatment and the addition of further nutrients depends on the conditions and the requirements of the process. A number of clarification methods can be employed singly or combined for coagulation, flocculation, and removal of undesirable substances from molasses.

For yeast fermentation molasses has to be clarified to remove factors harmful to yeast production. This can be done by a combined physicochemical and chemical treatment. After dilution with water sulphuric acid is added and the molasses is heated to about 90 °C. After addition of 1–2% phosphate the clear solution can be decanted; it is cooled, held in a storage tank and fed to the fermenters as required. Alternatively, precipitations can be carried out under alkaline conditions.

Another commonly used method is centrifugal clarification of diluted molasses after heating. This method is characterized by its simplicity; steam, time, and space are saved and it can be carried out with only very slight losses (OLBRICH, 1963).

Beet and cane molasses normally contain most substances necessary for nutrition. However, nitrogen, phosphorus, and eventually also magnesium may be too low in concentration. In these cases a hydrous solution of $(NH_4)_2SO_4$, which lowers the pH value can be added, or ammonia ($NH_3$) in the form of $NH_4H_2PO_4$, which raises the pH value (RÜTTER, 1975).

For citric acid production molasses must also be pretreated to be a suitable substrate. It is diluted with water, acidified with sulphuric acid, and potassium ferrocyanide is added to precipitate heavy metals. Further nutrients may be added in a way similar to that for yeast fermentation.

### 5.3.2 Physical Properties

Several physical properties of molasses are of more or less importance and should be included in the discussion of molasses composition. The viscosity of the molasses to be pumped is of utmost importance, since pipe friction has been found to increase directly proportional to viscosity.

Molasses show great differences in viscosity, which principally are due to the varying solids content. The viscosity of an individual molasses depends on the temperature and on the water content. A rise in temperature of 10°C, e.g., may reduce viscosity to a half or less, and a reduced dry matter content will also decrease viscosity in a similar range.

Beet molasses is usually of lower viscosity than cane molasses, particularly at ambient temperatures, but in both cases it is affected by the dry matter content.

Molasses exhibits the phenomenon called "critical viscosity", i.e., a sudden and marked increase in viscosity as soon as a specific dry substance content is exceeded. Critical viscosity of cane molasses is usually reached between 81% and 85% of dry matter content.

The specific heat of molasses needs to be known for the calculation of heat transfer in heating and cooling. The specific heat of a substance is defined as the ratio of the heat required to raise the temperature of that substance by 1°C and the heat required to raise the temperature of an identical weight of water, usually from from 14.5°C to 15.5°C. In technical calculations the mean value of the specific heats at the initial and the final temperature, the so-called mean specific heat ($c_m$), is used, which is usually given as 0.5 for molasses. If molasses is to be heated for pumping, preferably hot water coils should be used for steaming to avoid overheating. At any temperature >60°C there is the danger of thermal decomposition. The reaction that might occur is most probably the so-called Maillard reaction between reducing sugars and nitrogen compounds which can result in a spontaneus destruction and explosion. It must be pointed out that heating of undiluted molasses must be performed under strict control of pH and temperature to avoid the exothermal Maillard reaction. For handling molasses a temperature of 40°C is recommended. At this temperature little or no destruction will occur, molasses is relatively stable and may also be stored for a limited time with constant temperature control.

Fore more detailed information about the properties of molasses the reader is referred to general literature, e.g. CHEN and CHOU (1993), MCGINNIS (1982), United Molasses Company (1975), OLBRICH (1963), PANCOAST and JUNK (1980).

## 6 Use of Raw Materials for Fermentations

Raw materials from sugar processing are frequently used as substrates for fermentations with molasses being the major compound. The traditional use of molasses was in the manufacture of alcohol, a process carried out in molasses distilleries, or in the production of yeast. Meanwhile further raw materials from sugar processing are used solely or in combination with other carbon sources or in mixtures with molasses. The purpose of this section is to give current examples for the suitability of raw materials from sugar processing for fermentation processes, including new applications, without claiming to be complete.

Whereas bagasse is preferably used as direct fuel in the cane sugar industry or for pulp and paper production (PATURAU, 1989) beet pulp is more suitable as carbon source for bulk fermentations. Apart from the introduction of beet pulp for the industrial production of alcohol efforts have been made to improve the protein content of beet pulp supplemented with molasses (NIGAM, 1994a, b).

Studies were also performed to directly convert beet pulp to biogas (LESCURE and BOURLET, 1984; STOPPOK and BUCHHOLZ, 1985; STOPPOK et al., 1988).

Some unconventional raw materials such as raw juice concentrate, thick juice, and green run-off were tested for baker's yeast production by KLAUSHOFER (1980). With mixtures of juices and molasses yields almost corresponding to the yields with pure molasses were obtained (KLAUSHOFER 1974, 1980).

The replacement of molasses by thick juice and green run-off for yeast production was studied and evaluated by BRONN (1987) and FRICKO (1995).

An improved method for the production of glycerol was developed using cane molasses as carbon source (KALLE et al., 1985). NAKAGAWA et al. (1990) described the pretreatment of beet thick juice and molasses as carbon sources for itaconic acid. Although sugar solutions of various origins have been used to produce citric acid, molasses mainly is introduced in industrial citric acid production with *Aspergillus niger.* Molasses has to be pretreated with potassium ferrocyanide, an important compound for precipitation, to be suitable as substrate (LEOPOLD et al., 1969).

MANZKE et al. (1992) studied the suitability of raw thick juice for the production of ethanol. The ability of three thermotolerant yeasts to ferment cane juice and molasses to ethanol was investigated by KAR and VISWANATHAN (1985). ABATE et al. (1987) examined ethanol production by a flocculent yeast and found that productivity and ethanol concentration were higher using sugarcane juice and molasses than sucrose medium supplemented with peptone. It was shown by BENETT (1981) that liquors, syrups, and sweetwater of a refinery can be fermented directly to ethanol.

Sugarcane juice and sugarbeet molasses were also used in the production of levan (HAN and WATSON, 1992).

Good lactic acid formation was achieved by supplementation of whey ultrafiltrate with beet molasses (CHIARINI et al., 1992).

SMEKAL et al. (1988) studied the biosynthesis of L-lysine with thick juice and molasses. In experiments comparing sugar beet molasses to other carbon sources HILLIGER et al. (1986) improved the yield of lysine and reduced the costs in fermentations with *Corynebacterium glutamicum.* KHALAF et al. (1980) achieved best lysine production on a medium containing 15% sugar cane molasses. Molasses was hydrolyzed and treated with $Ca^{2+}$ to produce fructose and a good carbon source for glutamic acid and lysine fermentation (Ajinomoto Co., 1976).

The yield of L-amino acids was increased by the addition of methyl glycine to a medium supplemented with cane molasses as carbon source (YOSHIHARA et al., 1991). Sugar beet molasses was used as carbon source and as substrate for bioconversion to fructose diphosphate (COMPAGNO et al., 1992).

HAARD (1988) reported on the formation of the pigment astaxanthin with molasses as a cheap fermentation substrate for the red yeast *Phaffia rhodozyma.* Riboflavin (vitamin $B_2$) was produced in a medium containing beet molasses as the sole carbon source (SABRY et al., 1993).

AFSCHAR et al. (1990) demonstrated that the use of high-test molasses for the production of acetone and butanol may be economic in countries with excess agricultural products, like in Brazil. Fermentations of high-test molasses to 2,3-butanediol with continuous, fed-batch, and batch cultures of *Klebsiella oxytoca* were investigated with regard to product yield and product concentration (AFSCHAR et al., 1991).

A two-stage fermentation process was developed for the simultaneous conversion of sugar or molasses of various types to propionic acid and vitamin $B_{12}$ (QUESADA-CHANTO et al., 1994).

Blackstrap molasses was found to be a better carbon source than glucose for the formation of the antibiotic oxytetracycline. Moreover, it is cheaper than other suitable raw materials (ABOU-ZEID et al., 1993).

Beet molasses had a stimulatory effect on the production of poly-hydroxyalkanoates (PHB) (PAGE, 1989). Stimulation of growth and PHB yield were also observed when mixing 0.5–3.0% beet molasses with 2% sucrose (PAGE, 1990).

Sugarcane molasses was converted to xanthan gum using a strain of *Xanthomonas campestris* (ABD EL-SALAM et al., 1994). Cane molasses was found to be a good carbon source for growth and production of single-cell protein by *Saccharomyces cerevisiae* var. *carlsbergensis* and revealed an increase of 415% SCP content and 1018% amino acids (ABDEL-RAHMAN et al. 1991). Six out of eight fungi studied showed higher mycelial growth and protein contents on molasses as compared to glucose and sucrose (GARCHA et al., 1973).

Numerous further applications for the suitability of sugar-based raw materials, especially molasses, have been reported. The most widely used applications were in the field of production of alcohol, yeast, and organic acids. For more detailed information using raw materials as carbon sources the reader is referred to special publications of sugar manufacture or industrial microbiology (CHEN and CHOU, 1993; OLBRICH, 1963, 1970; PATURAU, 1989; REHM, 1980; RÜTER, 1975).

Acknowledgement

We gratefully acknowledge remarkable support and discussions of BERNHARD EKELHOF in Sect. 2 and worthful help of THORALF SCHULZ in this Section. We also thank HANJO PUKE and KLAUS THIELECKE for their assistance in Sect. 3.

# 7 References

ABATE, C. M., CALLIERI, D. A. S., ACOSTA, S., DE VIE, M. (1987), Production of ethanol by a flocculant *Saccharomyces* sp. in a continuous upflow reactor using sucrose, sugar-cane juice, and molasses as the carbon source, *J. Appl. Microbiol. Biotechnol.* **3**, 401–409.

ABD EL-SALAM, M. H., FADEL, M. A., MURAD, H. A. (1994), Bioconversion or sugarcane molasses into xanthan gum, *J. Biotechnol.* **33**, 103–106.

ABDEL-RAHMAN, T. M. A., TAHANY, M. A., SALAMA, A. M., ALI, M. I. A., ABO-HAMED, M. A. A. (1991), Medium composition influencing single-cell protein and cholesterol production by *Saccharomyces cerevisiae* var. *carlsbergensis, Egypt. J. Physiol. Sci.* **15**, 107–118.

ABOU-ZEID, A. A., KHAN, J. A., ABULNAJA, K. O. (1993), Oxytetracycline formation in blackstrap molasses medium by *Streptomyces rimosus, Zentralbl. Mikrobiol.* **148**, 351–356.

AFSCHAR, A. S., VAZ ROSSELL, C. E., SCHALLER, K. (1990), Bacterial conversion of molasses to acetone and butanol, *Appl. Microbiol. Biotechnol.* **34**, 168–171.

AFSCHAR, A. S., BELLGARDT, K. H., VAZ ROSSELL, C. E., CZOK, A., SCHALLER, K. (1991), The production of 2,3-butanediol by fermentation of high test molasses, *Appl. Microbiol. Biotechnol.* **34**, 582–585.

Ajinomoto Co., Inc. Japan (1976), Carbon source for fermentation, *French Patents,* FR 2349650 771125, FR 76-12808 760429.

American Sugar Company (1968), *DIF Form 2021-56 G: Bulk Raw Sugar Contract.* January 1968, revised in October 1982.

Amstar Raw Sugar Contract (1984), Amstar Corporation, American Sugar Division, 2021-84 G, Effective July 1, 1984.

ARNTZ, H. J., STOPPOK, E., BUCHHOLZ, K. (1985), Anaerobic hydrolysis of beet pulp – discontinuous experiments, *Biotechnol. Lett.* **7**, 113–118.

BENETT, M. C. (1981), Methods of alcohol production available to the cane sugar refiner, *Sugar J.* **44**, 19–21.

BINKLEY, W. W., WOLFROM, M. L. (1953), Composition of cane juice and final molasses, *Adv. Carbohydr. Chem.* **8** (HUDSON, C. S., WOLFROM, M. L., Eds.), 291–314.

BOHN, K., DEPOLT, F., MANZKE, E., MÜLLER, G., SCHICK, R., WEISE, H. (1974), Zuckerfabrikation mit Dicksaftlagerung, *Lebensmittelindustrie* **21**, 35–39.

BRONN, W. K. (1985), Möglichkeiten der Substitution von Melasse durch andere Rohstoffe für die Backhefeherstellung I., *Branntweinwirtschaft* **125**, 228–233.

BRONN, W. K. (1987), Möglichkeiten der Substitution von Melasse durch andere Rohstoffe für die Backhefeherstellung III., *Branntweinwirtschaft* **127**, 190–192.

CEJKA, A. (1985), Preparation of media, in: *Biotechnology,* Vol. 2, 1st Edn. (REHM, H.-J., REED, G., Eds.), pp. 629–698. Weinheim: VCH.

CHEN, J. C., CHOU, C. C. (1993), *Cane Sugar Handbook,* 12th Edn. New York-Chichester-Brisbane-Toronto-Singapore: John Wiley & Sons.

CHIARINI, L., MARA, L., TABACCHIONI, S. (1992), Influence of growth supplements on lactic acid production in whey ultrafiltrate by *Lactobacillus helveticus, Appl. Microbiol. Biotechnol.* **36**, 461–464.

Codex Alimentarius (1969), *Joint FAO/WHO Food Standards Programme. Recommended International Standard for White Sugar. CAC/RS4-1969.* Rome: Food and Agriculture Organization, UN and WHO.

COMPAGNO, C., TURA, A., RANZI, B. M., MARTEGANI, E. (1992), Production of fructose diphosphate by bioconversion of molasses with *Saccharomyces cerevisiae* cells, *Biotechnol. Lett.* **14**, 495–498.

DALE, E., LINDEN, E. (1984), Fermentation substrates and economics, in: *Annual Reports on Fermentation Processes,* Vol. 7 (TSAO, G. T., Ed.), pp. 107–134. Orlando, FL: Academic Press, Inc.

DETAVERNIER, R. (1982), Inorganic non-sugars, in: *ICUMSA, Report Proc. 18th Session,* held in Dublin, 13–18 June, pp. 283–304.

DIERSSEN, G. A., HOLTEGAARD, K., JENSEN, B., ROSEN, K. (1956), Volatile carboxylic acids in molasses and their inhibitory action on fermentation, *Int. Sugar J.* **58**, 35–39.

DOWLING, J. F. (1990), White sugar quality – the overall viewpoint, in: *Raw Quality and White Sugar Quality,* Proc. S.P.R.I. Workshop. (CLARKE, M. A., Ed.), pp. 133–137. Berlin: Verlag Dr. Albert Bartens.

DUTTON, J. V. (1979), Control and standardisation of Sugar, in: *Sugar: Science and Technology* (BIRCH, G. G., PARKER, K. J., Eds.), pp. 231–239. London: Elsevier Applied Science Publ. Ltd.

FIEDLER, A., JAKOB, R., TRESSL, R., BRONN, W. K. (1981), Über das Vorkommen von organischen Säuren und flüchtigen stickstoffhaltigen Substanzen in verschiedenen Rübenmelassen, *Branntweinwirtschaft* **121**, 202–203.

FIEDLER, B., SCHMIDT, P. V., KUNKEL, K. (1993), Microbial studies on the storability of raw thick juice, *Zuckerind.* **118**, 872–876 (in German).

FRICKO, P. (1995), Technologische und ökonomische Bewertung der Backhefeherstellung bei Einsatz von Rübendicksaft, in: *Backhefetechnologie, Hilfsstoffe, Zusätze und zelluläre Schutzfaktoren,* Vortragstexte der VH-Hefetagung 9./10. 5. 1995 in Hamburg (Versuchsanstalt der Hefeindustrie e.V., Ed.), Berlin.

GARCHA, J. S., KAUR, C., SINGH, A., RAHEJA, R. K. (1973), Production of fungal proteins from molasses. Comparative study using three carbon sources, *Indian J. Anim. Res.* **7**, 19–22.

GREASHAM, R. L. (1993), Media for microbial fermentations, in: *Biotechnology,* Vol. 3, 2nd Edn. (REHM, H.-J., REED, G., Eds.), pp. 127–139. Weinheim: VCH.

HAARD, N. F. (1988), Astanxanthin formation by the yeast *Phaffia rhodozyma* on molasses, *Biotechnol. Lett.* **10**, 609–614.

HAN, Y. W., WATSON, M. A. (1992), Production of microbial levan from sucrose, sugarcane juice and beet molasses, *J. Ind. Microbiol.* **9**, 257–260.

HENDRICKSON, R., KESTERSON, J. W. (1967), Zitrusmelasse: Erzeugung und Zusammensetzung, in: *F. O. Lichts International Molasses Report,* Vol. 11, pp. 1–4.

HILLIGER, M., FUCHS, R., KREIBICH, G., MENZ, J., WEISE, H., HARZFELD, G., BÖTTCHER, J. (1986), Lysine fermentation using molasses as carbon source, *German Patent* DD 281415 A5, *Application* 86-294383.

HODGE, H. M., HILDEBRANDT, F. M. (1954), Alcoholic fermentation of molasses, in: *Industrial Fermentations* (UNDERKOFLER, L. A., HICKEY, R. J., Eds.), pp. 73–94. New York: Chemical Publishing Co. Inc.

HONIG, P. (1953, 1959, 1963), Principles of Sugar Technology, Vols. I–III. Amsterdam: Elsevier.

ICUMSA (1979), *Sugar Analysis, ICUMSA Methods.* (SCHNEIDER, F., Ed.), Int. Comm. for Uniform Methods of Sugar Analysis, Peterborough, England.

ICUMSA (1994), *Methods Book.* Publication Department, c/o British Sugar Technical Center, Norwich Research Park, Colney, Norwich NR4 7 UB, England.

KALLE, G. P., NAIK, S. C., LASHKARI, B. Z. (1985), Improved glycerol production from cane molasses by the sulphite process with vacuum or continuous carbon dioxide sparging during fermentation, *J. Ferment. Technol.* **63**, 231–237.

KAR, R., VISWANATHAN, L. (1985), Ethanolic fermentation by thermotolerant yeasts, *J. Chem. Technol. Biotechnol.* **35 B** (4), 235–238.

KHALAF ALLAH, A. M., JANZSO, B., HOLLO, J. (1980), Lysine production with *Brevibacterium* sp. 22 using sugar cane molasses. II. Effect of carbon source, *Acta Aliment.* **9**, 117–128.

KJAERGAARD, L. (1984), Examples of biotechnological utilization of beet pulp and bagasse, in: *Sugar Technology Rewiews,* Vol. 10 (MCGINNIS, R. A., MULLER, E. G., Eds.), pp. 183–237. Amsterdam: Elsevier Science Publishers B.V.

KLAUSHOFER, H. (1974), Die Eignung verschiedener Sirupe der Rübenfabrikation zur Backhefeherstellung, *Z. Zuckerind.* **24**, 173–176.

KLAUSHOFER, H. (1980), Some unconventional raw materials for baker's yeast production, in: *Problems with Molasses in the Yeast Industry,* Proc. Helsinki 31. 8.–1. 9. 1979 (SINDA, E., PARKKINEN, E., Eds.), pp. 67–76. Helsinki: Kauppakirjapaino.

KOWITZ-FREYNHAGEN, P. (1979), Gaschromatographische Untersuchungen über die Veränderungen im Säurespektrum technischer Zuckersäfte während des Zuckergewinnungsprozesses über die Einflüsse der Verfahrensparameter, *Thesis,* Technical University of Braunschweig.

KUBADINOW, N., HAMPEL, W. (1975), Le comportement des acides aminés libres au cours de l'épuration des jus, *Sucr. Belge* **94**, 394–404.

LAMBRECHTS, L., GIBON, R., SIMONART, A. (1967), La conservation du jus dense et son travail ultérieur, *Sucr. Belge* **87**, 139–147.

LEOPOLD, H., CAISOVA, D., BASTL, J. (1969), Effect of potassium ferrocyanide in the preparation of molasses solutions for citric acid fermentation, *Nahrung* **13**, 681–686.

LESCURE, J. P., BOURLET, P. (1984), Bioconversion de pulpes de betteraves et de refus de lavoir, *Ind. Aliment. Agric.* **101**, 601–607.

MANZKE, E., SCHMIDT, P. V., RIECK, R., SENGE, B., STEINER, B. (1992), Raw thick juice: Manufacture, storage and utilization as feedstock in the biotechnological industry, *Zuckerind.* **117**, 984–990 (in German).

MCGINNIS, R. A. (1982), Beet-sugar technology, 3rd Edn. Fort Collins, CO: Beet Sugar Development Foundation.

NAKAGAWA, N., KAMIIE, K., ISHIBASHI, K.-I., HIRONAKA, K. (1990), Pretreatments of beet thick juice and molasses as carbon sources for itaconic acid fermentation, *Res. Bull. Obihiro Univ. Ser. I* **17**, 7–12 (in Japanese).

National Food Processors Association (1968), *Laboratory Manual for Food Canners and Processors*, Vol. 1, 3rd Edn. Westport, CT: AVI Publishing Co.

NIGAM, P. (1994a), Processing of sugar beet pulp in simultaneous saccharification and fermentation for the production of a protein-enriched product, *Process Biochem.* **29**, 331–336.

NIGAM, P. (1994b), Process selection for protein-enrichment; fermentation of the sugar industry by-products molasses and sugar beet pulp, *Process Biochem.* **29**, 337–342.

NSDA (1975), *Quality Specifications and Test Procedures for Bottlers' Granulated and Liquid Sugar.* Washington, D.C.: Natl. Soft Drink Assoc.

OLBRICH, H. (1956), *Die Melasse,* Institut für Gärungsgewerbe, Berlin: Hentschel, Heidrich & Co.

OLBRICH, H. (1963), Molasses, in: *Principles of Sugar Technology,* Vol. III (HONIG, P., Ed.), pp. 511. Amsterdam: Elsevier.

OLBRICH, H. (1970), *Geschichte der Melasse.* Berlin: Verlag Dr. Albert Bartens.

OLBRICH, H. (1973), Melasse als Rohstoffproblem der Backhefeindustrie, *Branntweinwirtschaft* **113**, 53–68.

PAGE, W. J. (1989), Production of poly-$\beta$-hydroxybutyrate by *Azotobacter vinelandii* strain UWD during growth on molasses and other complex carbon sources, *Appl. Microbiol. Biotechnol.* **31**, 329–333.

PAGE, W. J. (1990), Production of poly-$\beta$-hydroxybutyrate by *Azotobacter vinelandii* UWD in beet molasses culture at high aeration, *NATO ASI Ser.,* Ser. E **186** (Novel Biodegrad. Microb. Polym.), 423–424.

PANCOAST, H. M., JUNK, W. R. (1980), *Handbook of Sugars,* 2nd Edn. Westport, CT: AVI Publishing Co.

PATURAU, J. M. (1989), By-products of the cane sugar industry, sugar series 11. Amsterdam: Elsevier Science Publishers B.V.

QUESADA-CHANTO, A., AFSCHAR, A. S., WAGNER, F. (1994), Microbial production of propionic acid and vitamin $B_{12}$ using molasses or sugar, *Appl. Microbiol. Biotechnol.* **41**, 378–383.

RATLEDGE, C. (1977), Fermentations substrates, in: *Annual Reports on Fermentation Processes*, Vol. 1 (PERLMAN, D., Ed.), pp. 49–71. New York-London: Academic Press.

REHM, H. J. (1980), *Industrielle Mirobiologie,* 2nd Edn. Berlin-Heidelberg-New York: Springer-Verlag.

REINEFELD, E., BLIESENER, K. M., KOWITZ-FREYNHAGEN, P. (1980), Studien über die Säurebildung und Säureeliminierung in technischen Zuckersäften, *Zuckerind.* **105**, 563–573.

REINEFELD, E., BLIESENER, K. M., SCHULZE, J. (1982a), Beitrag zur Kenntnis des Verhaltens der Aminosäuren im Verlauf der Zuckergewinnung: Gaschromatographische Untersuchungen an Zuckerfabrikprodukten auf freie und durch Säurehyrolyse freigesetzbare Aminosäuren, *Zuckerind.* **107**, 283–292.

REINEFELD, E., BLIESENER, K. M., SCHULZE, J. (1982b), Beitrag zur Kenntnis des Verhaltens der Aminosäuren im Verlauf der Zuckergewinnung: Untersuchungen über die Auswirkung von Prozeßparametern, *Zuckerind.* **107**, 1111–1119.

REINEFELD, E., BLIESENER, K.-M., SZCZECINSKY, H.-J. (1986), Zum Nachweis und Vorkommen von Nukleotid-Spaltprodukten in technischen Zuckerlösungen, *Zuckerind.* **111**, 1017–1025.

RHODES, A., FLETCHER, D. L. (1966), Nutrition of microorganisms and media for industrial fermentations, in: *Principles of Industrial Microbiology*, pp. 58–76. Oxford-London-Braunschweig: Pergamon Press.

RIVIÈRE, J. (1977), Culture media, in: *Industrial Applications of Microbiology* (MOSS, M. O., SMITH, J. E., Eds.), pp. 27–58. London: Surrey University Press.

RÜTER, P. (1975), Molasses utilization, *FAO Services Bulletin.* Rome: Food and Agriculture Organization of the United Nations.

SABRY, S. A., GHANEM, K. M., GHOZLAN, H. A. (1993), Riboflavin production by *Aspergillus terreus* from beet molasses, *Microbiologia* **9**, 118–124.

Savannah Raw Sugar Contract (1984), Savannah Foods & Industries, Inc., SF & J – 1984.

SCHIWECK, H. (1977), Die Konzentration einiger Melasseinhaltsstoffe in Melassen verschiedener Provenienz, *Branntweinwirtschaft* **117**, 87–93.

SCHIWECK, H. (1980), The influence of the different constituents of molasses on its suitability for yeast production, in: *Problems with Molasses in the Yeast Industry,* Proc. Helsinki 31. 8.–1. 9. 1979 (SINDA, E., PARKINNEN, E., Eds.), pp. 21–37. Helsinki: Kauppakirjapaino.

SCHIWECK, H. (1991), Zuckererzeugung im Spannungsfeld zwischen Rübenqualitität, Energieverbrauch und Produktsicherheit – neuere technologische Entwicklungen, *Zuckerind.* **116**, 793–805.

SCHIWECK, H. (1995), Composition of sugar beet molasses, *Zuckerind.* **119**, 272–282.

SCHIWECK, H., CLARKE, M. (1994), Sugar, in: *Ullmann's Encyclopedia of Industrial Chemistry,* Vol. A25, pp. 345–412. Weinheim: VCH.

SCHIWECK, H., HABERL, L. (1973), Einfluß der Zusammensetzung der Rübenmelasse auf die Hefeausbeute, *Branntweinwirtschaft* **113**, 76–83.

SCHIWECK, H., JEANTEUR-DE BEUKELAER, C., VOGEL, M. (1993), The behaviour of nitrogen containing nonsucrose substances of beet during the sugar recovery process, *Zuckerind.* **118**, 15–23.

SCHNEIDER, F. (1968), *Technologie des Zuckers,* 2nd Edn. Hannover: Verlag M. H. Schaper.

SINDA, E., PARKKINEN, E. (Eds.) (1980), *Problems with Molasses in the Yeast Industry,* Proc. Helsinki 31. 8.–1. 9. 1979. Helsinki: Kauppakirjapaino.

SMEKAL, F., PELECHOVA, J., SIROCKOVA, E., ZDANOVA, N., LEONOVA, T. (1988), Biochemical and production properties of *Corynebacterium glutamicum* strains, *Kvasny Prum.* **34**, 12–14, 19.

STEINMETZER, H. (1991), Utilization of the nonsugar components of sugar beet molasses, *Zuckerind.* **116**, 30–38 (in German).

STOPPOK, E., BUCHHOLZ, K. (1985), Continuous anaerobic conversion of sugar beet pulp to biogas, *Biotechnol. Lett.* **7**, 119–124.

STOPPOK, E., BARZ, U., BUCHHOLZ, K. (1988), Investigations and scale up of anaerobic fermentations of solid beet waste. *5th Int. Symp. Anaerob. Digest.,* Bologna, *Poster Papers* (TILCHE, A., ROZZI, A., Eds.), pp. 799–802. Bologna: Monduzzi Editore.

TRESSL, R., JAKOB, R., KOISSA, T., BRONN, W. K. (1976), Gaschromatographisch-massenspektrometrische Untersuchung flüchtiger Inhaltsstoffe von Melasse, *Branntweinwirtschaft* **116**, 117–119.

UNDERKOFLER, L. A., HICKEV, R. J. (1954), *Industrial Fermentations,* Vol. I. New York: Chemical Publishing Co. Inc.

United Molasses Company (1975), Section D: Properties of molasses, in: *Composition, Properties and Uses of Molasses and Related Products* (United Molasses Trading Co. Ltd., Ed.), pp. 20–30. London.

United Molasses Company (1986), Section A: The composition of molasses, in: *The Analysis of Molasses* (United Molasses Trading Co. Ltd., Ed.), pp. 1–3. London.

VACCARI, G., SGUALDINO, G., MANTOVANI, G., MATTEUZZI, D., SCHEDA, W. (1985), Stabilita chimica e microbiologica di sciroppi zuccherini sottoposti a lunga conservacione, *Ind. Sacc. Ital.* **78**, 152–161.

VACCARI, G., SCHEDA, W., SGUALDINO, G., MANTOVANI, G. (1988), Etudes complémentaires concernant la possibilité de conserver les sirops sucres pendant une longue periode, *Ind. Aliment. Agro-ind.* **105**, 141–147.

WALLENSTEIN, H. D., BOHN, K. (1963), Säuren in Rüben und Säften, *Z. Zuckerind.* **13**, 125–131.

WHITE, J. (1954), *Yeast Technology.* London: Chapman & Hall.

WILEY, A. J. (1954), Food and feed yeast, in: *Industrial Fermentations,* Vol. I. (UNDERKOFLER, L. A., HICKEY, R. J., Eds.), p. 314. New York: Chemical Publishing Co. Inc.

WOLFROM, M. L., BINKLEY, W. W., MARTIN, L. F. (1952), Molasses: important but neglected product of sugar cane, *Sugar* **47**, Vol. 5, 33.

YOSHIHARA, Y., KAWAHARA, Y., YAMADA, Y., IKEDA, S. (1991), Fermentation of amino acids in media supplemented with glycine derivatives, *US Patents* US 91-675961 910327, US 90-472415 090131.

ZAUNER, E., BRONN, W. K., DELLWEG, H., TRESSL, R. (1979), Hemmwirkung von Nebenkomponenten der Rübenmelasse und von Pflanzenschutzmitteln auf die Atmung und Gärung von Hefe, *Branntweinwirtschaft* **119**, 154–163.

# 1b Starch-Based Raw Materials for Fermentation Applications

JEAN-CLAUDE DE TROOSTEMBERGH

Vilvoorde, Belgium

# 1 Introduction

Starch is the most abundant storage carbohydrate in plants. It is present in seeds or tubers in quantities exceeding generally half of the weight on dry basis (DE MENEZES et al., 1978; GRAM et al., 1989; WATSON, 1984). Starch is a source of readily available energy and serves as building blocks for the growing plant. Starch occurs in two forms, a linear molecule called amylose and a branched molecule called amylopectin. Amylose consists of glucose units bound by $\alpha(1,4)$-linkages whereas amylopectins are branched chains bound by $\alpha(1,6)$ linkages to the $\alpha(1,4)$-chain. The proportion of amylopectin to amylose is dependent on the origin of starch (WHISTLER and DADIEL, 1984).

Starch can be extracted relatively easily from different raw materials in high yields and purification grades. In addition, it can be converted by simple processes to give a very broad spectrum of hydrolysis products from maltodextrins to glucose (REEVE, 1992). Starch or starch derivatives can be metabolized almost quantitatively by a very large number of microorganisms. For all these reasons, starch is the fermentation raw material of choice and is widely used by the fermentation industry (CEJKA, 1985).

Starch used in fermentation processes is extracted from various sources depending on the geographic area. For example, corn (maize) is the major source of starch in the United States, wheat, corn, and potato starches are used in Europe, corn and cassava starches in Asia.

Tab. 1 shows the typical composition of plant raw materials used in starch production.

In general starches and starch hydrolysis products originating from different plant raw materials are interchangeable for fermentation applications. Nevertheless, in some cases the presence of trace elements influences the fermentation positively and makes one starch source preferable to another. Most frequently, the choice of the raw material is directed by price considerations which are influenced not only by factors like crop yields, extraction costs, by-product return, waste production, etc. but also by governmental agricultural policies.

Fig. 1 gives a general scheme of the different steps to obtain starch or starch-derived products. Products from glucose oligosaccharides to simple monomers can be obtained from starch. The main technologies involved are enzyme conversions such as liquefaction, saccharification or isomerization, separation processes such as screening, sifting, centrifugation, chromatography or crystallization, and chemical transformations like hydrogenation or epimerization (SCHENK and HEBEDA, 1992).

# 2 Starch

The starch extraction process is dependent on the raw material used. Production of

**Tab. 1.** Typical Composition of Raw Materials for Starch Production

| | Corn[a] | Wheat[b] | Tapioca[c] | Potato[d] |
|---|---|---|---|---|
| Moisture [%, wet basis] | 7–23 | 13–17 | 60–78 | 63–87 |
| Starch [%, dry basis] | 64–78 | 62–72 | 80–89 | 70–80 |
| Protein [%, dry basis] | 8–14 | 11–18 | 2.1–6.2 | 5.3–12.5 |
| Fat [%, dry basis] | 3.1–5.7 | 1.8–2.3 | 0.2–0.7 | 0.15–2.6 |
| Ash [%, dry basis] | 1.1–3.9 | 1.5–2.0 | 0.9–2.4 | 3.35–5.16 |
| Fiber [%, dry basis] | 8.3–11.9 | 2.5–3.2 | 1.7–3.8 | 1.29–9.45 |

[a] WATSON (1984)
[b] GRAM et al. (1989)
[c] DE MENEZES et al. (1978)
[d] MITCH (1983)

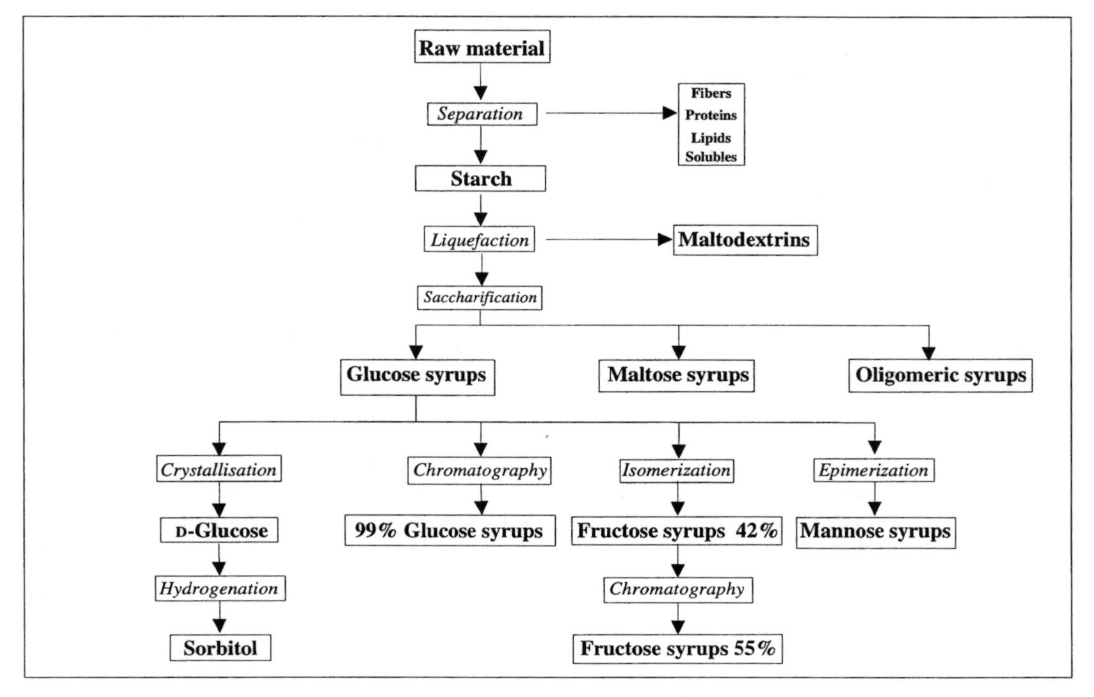


**Fig. 1.** General scheme of production of starch and starch derivatives.

starch from corn (maize), wheat, and tapioca is discussed in more detail below.

## 2.1 Corn Starch Production

A typical starch production process from corn (maize) uses the wet milling process described below (MAY, 1987; WATSON, 1984; ZOBEL, 1992). After cleaning the corn to remove dust, broken corn, etc., the maize is steeped for 30–40 h in sulphited hot water. The purpose of steeping is to soften the kernel, to break the disulfide bridges of the gluten protein, and to extract the solubles from the grain. During steeping, solubles undergo a natural lactic fermentation converting the free sugars to lactic acid. The salts, the free amino acids, and the vitamins are also extracted from the kernel during this process. When steeping is finished, the steep water is separated from the grain and concentrated to 50% solids. This corn steep liquor is used as a nutrient by the fermentation industry mainly for the production of antibiotics and enzymes.

After steeping, the softened corn undergoes a coarse milling which allows the removal of the germ. Because the germ contains about 50% oil, separation is made by flotation due to the lower density of the germ compared to the starch/gluten slurry. Afterwards the slurry is finely ground and the fiber fraction is separated by sieves.

Starch/gluten separation takes place in centrifuges. The separation principle is based on the difference in density between starch and gluten. Starch is washed in a battery of hydrocyclone separators with countercurrent washing and thereafter dried to about 11–13% moisture.

## 2.2 Wheat Starch Production

Many wheat starch extraction processes are described in the literature (KNIGHT and OLSON, 1984; POMERANZ, 1989; ZOBEL, 1992).

Dry milling processes are preferred to wet milling processes because they preserve the gluten vitality. Vital gluten is a very valuable product of the wheat starch industry and is used as dough improver for bread and in pet food.

Wheat is finely milled, then the bran fraction is separated from the wheat flour on screens. A dough is prepared by mixing the wheat flour with water. Sieving and screening is again used to separate the vital gluten from the starch fraction. Many precautions must be taken to preserve the functional properties of gluten for bread making. Heat or contact with oxygen or chemicals are known to induce loss of vitality. The starch is then washed in a hydrocyclone battery to remove the remaining protein and other solubles. Wheat starch contains large starch granules (A-starch) and small granules (B-starch). B-starch has a low density and escapes from the starch washing battery. A-starch is recovered by centrifuges and dried to 12–13% moisture.

## 2.3 Tapioca Starch Production

Tapioca or cassava starch is extracted from the tuberous roots of manioc (CORBISHLEY and MILLER, 1984). This plant is grown mainly in tropical areas. Because the root composition is low in protein and soluble carbohydrates, the production of starch is relatively straightforward. After washing and peeling, the roots are rasped into a pulp thereby liberating 70–90% of the starch granules. After a secondary finer rasping the pulp is screened to remove large particles which are recycled to the raspers. The starch milk is then processed in a similar way to other starches, i.e., by screening, centrifuging, washing, and drying.

## 2.4 Potato Starch Production

After cleaning and washing, the potatoes are milled by rotary saw blade rasps into a coarse potato pulp. Starch is separated from the pulp by a series of rotating screens (two at least), a second finer pulp rasping which is followed by at least two screening stages. The starch milk is concentrated in centrifugal separators or in hydrocyclones during or after wet screening. Washing water is also used to support the separation of starch from the pulp. The starch is then washed and dried (DE WILLIGEN, 1976).

## 2.5 Starch Composition and Use

Tab. 2 shows the composition of corn, wheat, tapioca, and potato starches. In a good quality corn starch, the protein and salt contents should be below 0.4% and 0.35%, re-

**Tab. 2.** Typical Composition of Starch

| | Corn Starch | Wheat Starch | Tapioca Starch | Potato Starch |
|---|---|---|---|---|
| Aspect | White powder | White powder | White powder | White powder |
| Dry substance [%] | 87–89 | 87–89 | 87–89 | 79.6–82.5 |
| pH | 4.0–5.0 | 6.0–7.0 | 4.0–6.0 | 6.8–8.0 |
| Bulk density [g/L] | 500–700 | 500–600 | 450–700 | 600–850 |
| Starch [%, dry basis] | >97 | >97 | >97 | >97 |
| Ash [%, dry basis] | <0.35 | | <0.2 | <0.3 |
| Protein [%, dry basis] | <0.4 | <0.4 | <0.1 | <0.22 |
| Fat [%, dry basis] | <0.12 | | <0.1 | |
| $SO_2$ [ppm] | <50 | <10 | <50 | |
| Sodium [ppm] | 200 | 300 | | |
| Potassium [ppm] | 50 | 200 | | |
| Calcium [ppm] | 50 | 30 | | |
| Magnesium [ppm] | 20 | 200 | | |

spectively, although higher levels are not critical for fermentation applications.

As far as fermentation applications are concerned, starch cannot be used as such because it undergoes gelatinization during sterilization of the fermentation medium thereby developing high viscosity of the broth. Liquefaction in the presence of an α-amylase is necessary before or during sterilization in order to decrease the medium viscosity developing during heating. The heating profile and pH of the starch slurry must be adapted to allow optimal enzyme action; this may also require the presence of calcium ions to enhance enzyme activity.

Liquefied starch can be used in applications where the microorganism involved produces the necessary amylases to hydrolyze the substrate. This is mainly the case with bacilli or fungi.

For example, ATKINSON and MAVITUNA (1991) and VOLOCH et al. (1985) have described the use of starch for the production of 2,3-butanediol by *Bacillus polymyxa* although this process is not applied industrially. Starch can also be used for the fermentative production of acetone and butanol by *Clostridium acetobutylicum* (ATKINSON and MAVITUNA, 1991; CEJKA, 1985; see also Chapter 6, this volume) and for the production of propionic and butyric acid by *Propionibacterium technicum* (BUCHTA, 1983). Yeast strains capable of direct fermentation of manioc starch were developed by MATTOON et al. (1987) by hybridizing strains of *Saccharomyces diastaticus* and *Saccharomyces cerevisiae.* These strains were able to produce 12% ethanol within three days at almost 100% starch conversion. Starch is also being used as brewing adjunct to the wort. In this case hydrolysis is done during wort preparation (STEWART and RUSSELL, 1985; MAIORELLA, 1985).

If liquefied starch is further saccharified with glucoamylase to yield a glucose containing liquor, it can be used for many other applications where hydrolysis does not take place during fermentation. The most important industrial application is the production of ethanol for fuel (REED, 1982; MAIORELLA, 1985).

Zanflo, an exopolysaccharide produced by *Erwinia tahitica*, is produced on a medium based on lactose and hydrolyzed starch as carbon sources; these carbon sources are more suitable than glucose, sucrose, and maltose (ATKINSON and MAVITUNA, 1991; MARGARITIS and PACE, 1985).

Although not a primary metabolite, it is worth mentioning that starch is widely used in industry for the production of enzymes like α-amylases, glucoamylases, pullulanases, and proteases (ATKINSON and MAVITUNA, 1991). Starch can also be used for the production of several antibiotics like tetracyclines by *Streptomyces aureofaciens* (BEHAL, 1987; HOSTALEK and VANEK, 1986), cephamycin by streptomycetes (OMSTEAD et al., 1985), leucomycin by *Streptoverticillium kitasatoensis* (ŌMURA and TANAKA, 1986), lincomycin by *Streptomyces lincolnensis*, novobiocin by *Streptomyces* sp., vancomycin by *Streptomyces orientalis* (BERDY, 1986), and nocardicin A by *Nocardia uniformis* (BUCKLAND et al., 1985).

# 3 Maltodextrins

## 3.1 Maltodextrin Production

Maltodextrins are products resulting from enzyme or acid hydrolysis of starch in a process known as liquefaction (ALEXANDER, 1992). The degree of dextrinization is expressed by the dextrose equivalent (DE) which indicates the reducing power of the molecule as percent of the reducing power of glucose on a dry weight basis. Unhydrolyzed starch has a DE of 0 while glucose has a DE of 100. Although the DE indicates the degree of starch hydrolysis, it does not give any information on the average size of the oligosaccharides. In other words, two maltodextrins may have the same DE but very different molecular-weight distributions of the glucose oligomers.

Starch liquefaction is a rather critical step because starch hydrolysis must take place at the same time as gelatinization to avoid excessive viscosity development. Modern processes use heat-stable α-amylase from *Bacillus licheniformis* or *Bacillus stearothermophi-*

**Tab. 3.** Typical Composition of Glucose Hydrolyzates

| | 14 DE Maltodextrin | 18 DE Maltodextrin |
|---|---|---|
| Aspect | White powder, soluble in water | White powder, soluble in water |
| Dry substance [%] | >95 | >95 |
| pH | 4.0–5.0 | 4.0–5.0 |
| Protein [%] | <0.1 | <0.1 |
| Bulk density [g/L] | 450–600 | 450–600 |
| Dextrose equivalent (DE) | 13–15 | 17–19.9 |
| **Typical carbohydrate composition:** | | |
| Glucose [%] | 1 | 1 |
| Disaccharides [%] | 3 | 5 |
| Trisaccharides [%] | 6 | 10 |
| Polysaccharides [%] | 90 | 84 |
| $SO_2$ [ppm] | <20 | <20 |
| Ash [%] | <0.1 | <0.1 |
| Nitrate [ppm] | <250 | <250 |
| Sodium [ppm] | <300 | <300 |
| Potassium [ppm] | <5 | <5 |
| Magnesium [ppm] | <2 | <2 |
| Calcium [ppm] | <10 | <10 |
| Iron [ppm] | <0.5 | <0.5 |

*lus* which can operate above 100 °C. The enzyme is added to the starch slurry at 30% d.s. and the slurry is heated up rapidly to 105–110 °C for 5–10 min. Afterwards the solution is cooled below 100 °C and incubated for about 2 h to allow further enzyme action.

There are different processes used for starch liquefaction employing different enzyme dosages, temperatures, and incubation times. At the end of liquefaction the DE of the starch hydrolyzate should normally be between 10 and 19.9, depending on the conditions. The molecular-weight distribution of the oligomers is highly influenced by the conditions used. A good liquefaction process should be fast enough to avoid starch retrogradation. Retrograded starch particles cannot be hydrolyzed further by enzymes. During liquefaction proteinaceous and fatty materials coagulate and cause turbidity of the solution. They are removed by filtration on filters precoated with diatomaceous earth or sawdust. After filtration the maltodextrins may be refined by ion exchange demineralization and dried by spray-drying.

## 3.2 Maltodextrins: Composition and Use

Many different types of maltodextrins can be obtained depending on the extent of hydrolysis, and Tab. 3 specifies the composition of two most typical products.

Maltodextrins are easily soluble products which are already liquefied. There are no problems of gelatinization and viscosity development during sterilization and they are available in demineralized or non-demineralized forms.

Maltodextrins are not normally used for the production of primary metabolites but they can be used for the production of antibiotics like cephalosporin C (Atkinson and Mavituna, 1991; Smith, 1985), penicillin (Atkinson and Mavituna, 1991; Swartz, 1985), or streptomycin (Atkinson and Mavituna, 1991; Florent, 1985).

# 4 Liquid Syrups

After starch liquefaction the oligo- and polysaccharides can be further hydrolyzed by different types of enzymes in a process called saccharification. The type of enzyme used will determine the DE and the sugar breakdown of the syrup. Typical saccharification enzymes mainly are $\beta$-amylase, glucoamylase, and debranching enzymes like pullulanase or isoamylase, or combinations of these. In addition, isomerization or epimerization of the syrup can be performed (SCHENK and HEBEDA, 1992).

Concentrated liquid syrups are commercially available with solids contents between 70 and 80%. These syrups must be stored above 50°C, and preferably at 60°C to avoid development of high viscosities or eventual crystallization.

## 4.1 Glucose Syrups

### 4.1.1 Production

Glucose syrups are obtained by the action of glucoamylase on starch liquefacts (TEAGUE and BRUMM, 1992). Glucoamylase hydrolyzes both $\alpha$(1,4)- and $\alpha$(1,6)-linkages of starch, but the latter at a lower rate. Addition of a debranching enzyme promotes saccharification. Commercial glucoamylases are produced by *Aspergillus niger*, pullulanases by *Klebsiella pneumoniae* or *Bacillus pullulanolyticus*.

When a high glucose content in the syrup has to be obtained, parameters like the d.s. of the solution, enzyme dosage, temperature and incubation time must be carefully selected in order to minimize formation of reversion products.

High DE require long incubation times and a careful control of the process. Use of debranching enzymes can boost the glucose level to about 96% on a dry basis.

After saccharification the syrup is normally refined by vacuum filtration to remove insoluble particles and by ion exchange demineralization and decolorization. The syrup is then concentrated to 70% or 75% d.s.

The glucose level of the syrup can be further increased to more than 99% by chromatographic separation techniques (MULVIHILL, 1992).

### 4.1.2 Composition and Use

Tab. 4 shows the composition of the most typical glucose syrups. Syrups with intermediate glucose contents can of course also be produced.

These syrups, commercially also known as dextrose hydrolyzates are the starch-derived syrups most frequently used in fermentation applications. They are easy to handle and contain a high level of fermentables. Syrups containing 85% to 95% glucose are the most commonly used. Higher glucose contents between 95% and 99% are required for special applications like vitamin C production.

Glucose syrups are used for the production of citric acid by *Aspergillus niger* and *Candida* yeast (see Chapter 9, this volume; MILSOM and MEERS, 1985a; KUBICEK and RÖHR, 1986). Modern fermentation processes using *Aspergillus niger* require highly refined syrups. Most critical contaminants are iron and manganese ions which are removed by ion exchange; residual levels should be as low as possible as these ions inhibit the production of the acid in *Aspergillus*.

Glucose syrups are also used for the production of other organic acids like gluconic acid by *Aspergillus niger* and *Gluconobacter* species (MILSOM and MEERS, 1985b; see Chapter 10, this volume), and itaconic acid by *Aspergillus terreus* (MILSOM and MEERS, 1985b; BUCHTA, 1983). Production of lactic acid from glucose syrups is possible (ATKINSON and MAVITUNA, 1991; VICKROY, 1985) but pure glucose (dextrose) is mostly preferred.

In amino acid production mainly glucose syrups are used. Being generally extracted from the fermented liquor by adsorption on ion exchangers, the presence of residual unfermented sugars is less critical to the purification process. Monosodium glutamate (ATKINSON and MAVITUNA, 1991; HIROSE et al., 1985), lysine (ATKINSON and MAVITUNA, 1991; NAKAYAMA, 1985), and threonine

**Tab. 4.** Typical Composition of Glucose Hydrolyzates

| | 80-95 DE Syrup | 97 DE Syrup | 99 DE Syrup |
|---|---|---|---|
| Aspect | Viscous liquid | Viscous liquid | Viscous liquid |
| Dry substance [%] | 70 | 70 | 70–72 |
| pH | 3.5–5.5 | 3.5–5.5 | 3–6 |
| Refractive index (20°C) | 1.46 | 1.46 | 1.46 |
| Density (20°C) | 1.34 | 1.34 | 1.34 |
| Viscosity [mPa s] (20°C) | 500 | 500 | 500 |
| Storage temperature [°C] | 50–60 | 50–60 | 50–60 |
| Dextrose equivalent (DE) | 80–95 | 97 | 99.5 |
| **Typical carbohydrate composition:** | | | |
| Glucose [%] | 75–92 | 95 | 99.5 |
| Disaccharides [%] | 4–12 | 3 | 0.3 |
| Trisaccharides [%] | 1–7 | 1 | 0.1 |
| Polysaccharides [%] | 3–10 | 1 | 0.1 |
| Ash [%, dry basis] | <0.13 | <0.13 | <0.1 |
| $SO_2$ [ppm] | | <20 | <10 |
| Calcium [ppm] | | <10 | <10 |
| Magnesium [ppm] | | <10 | <10 |
| Sodium [ppm] | | <100 | <100 |
| Potassium [ppm] | | <10 | <10 |
| Chloride [ppm] | | | <10 |
| Sulfate [ppm] | | | <30 |
| Iron [ppm] | | | <1 |

(SHIMURA, 1985) mainly are produced from glucose syrups as raw material.

Xanthan polysaccharide production from glucose syrups has been described by ROSEIRO et al. (1992), ATKINSON and MAVITUNA (1991), and EYSSAUTIER (1993). Other polysaccharides like curdlan produced by *Alcaligenes faecalis* or scleroglucan produced by *Sclerotium rolfsii* are made from glucose (MARGARITIS and PACE, 1985).

Glucose syrups are also used for the production of polyhydroxybutyrate by *Alcaligenes eutrophus* (COLLINS, 1987).

Erythritol, a C4 polyol used as non-caloric sweetener in Japan is produced from glucose syrup (SASAKI et al., 1993) and corn steep liquor. Microbial productions of butanol/acetone and 2,3-butanediol from glucose are described in Chapters 6 an 7, this volume, although they are not performed industrially.

The antibiotics producing industry extensively uses glucose syrups for the production of penicillins, cephalosporin C, streptomycin, rifampycin, lincomycin, novobiocin, tylosin, etc. (ATKINSON and MAVITUNA, 1991).

## 4.2 Maltose Syrups

### 4.2.1 Production

Maltose syrups are obtained by the enzymatic action of $\beta$-amylases on starch liquefacts (OKADA and NAKAKUKI, 1992). The most frequently used and least expensive enzyme is malt extract.

$\beta$-amylase releases maltose units from the starch molecule by an exo-attack from the non-reducing end of the starch molecule. It is unable, however, to pass the $\alpha(1,6)$-linkages. Therefore, the maximum maltose content achievable in the syrup is about 50%. Such a syrup is known as "High Maltose Syrup" (HMS). When a debranching enzyme is added to the $\beta$-amylase, higher maltose contents

**Tab. 5.** Typical Composition of Maltose Syrups

| | High Maltose Syrup | Very High Maltose Syrup |
|---|---|---|
| Aspect | Viscous liquid | Viscous liquid |
| Dry substance [%] | 80 | 70–72 |
| pH | 4.0–5.2 | 4.6–5.5 |
| Refractive index (20 °C) | 1.50 | 1.47 |
| Density (20 °C) | 1.40 | 1.35 |
| Viscosity [mPa s] (20 °C) | 8,000–10,000 | 8,000–10,000 |
| Storage temperature (°C) | 50 | 50 |
| Dextrose equivalent (DE) | 42–46 | 56 |
| **Typical carbohydrate composition:** | | |
| Glucose (%) | 2–3 | 2 |
| Disaccharides [%] | 46–52 | 75–79 |
| Trisaccharides [%] | 20–24 | 13–16 |
| Polysaccharides [%] | 21–32 | 6–8 |
| $SO_2$ [ppm] | <20 | <20 |
| Ash [%] | <0.1 | |
| Sodium [ppm] | <200 | <200 |
| Potassium [ppm] | <5 | <5 |
| Calcium [ppm] | <5 | <5 |
| Magnesium [ppm] | <30 | <30 |

between 75% and 85% can be obtained and the syrup is known as "Very High Maltose Syrup" (VHM). The final maltose content is highly dependent on the DE of the liquefact used for saccharification. A very good control of the starch liquefaction step is, therefore, essential to reach the desired maltose level.

### 4.2.2 Composition and Use

The composition of maltose syrups and very high maltose syrups is specified in Tab. 5. Maltose syrups are typically low in glucose content (below 5%) and are therefore suitable in fermentations where a glucose repression effect can be observed as, e.g., in antibiotic fermentations like penicillin, streptomycin, oxytetracyclin, and novobiocin (ATKINSON and MAVITUNA, 1991; HOSTALEK and VANEK, 1986; BERDY, 1986). Maltose syrups are also used in beer manufacturing to increase the sugar content of the wort before fermentation (STEWART and RUSSELL, 1985). Due to their high content in oligomers, maltose syrups containing about 50% maltose have a low fermentability for microorganisms which do not produce amylases while very high maltose syrup exhibits a much higher fermentability.

## 4.3 Special Syrups: Glucose/Maltose/Maltotriose/Maltotetraose

Use of special enzymes or combinations of enzymes can give more specific syrup compositions in terms of glucose, maltose, maltotriose, or maltotetraose contents (OKADA and NAKAKUKI, 1992).

For example, prolonged action of *Bacillus licheniformis* enzyme in saccharification alone will give a 30 DE syrup containing about 10% maltose and 17% maltotriose. Action of *Aspergillus niger* glucoamylase after $\beta$-amylase action will give a 65 DE syrup know as "High Dextrose Maltose Syrup" (HDMS) containing about 30% glucose and 40% maltose.

Other special enzymes can be used to produce specialty syrups. For example, syrups containing more than 50% maltotriose can be obtained by amylopectin incubation with

**Tab. 6.** Typical Composition of Glucose/Maltose/Maltotriose Syrups

| | High Dextrose/ Maltose Syrup (HDMS) | 30 DE Syrup |
|---|---|---|
| Aspect | Viscous liquid | Viscous liquid |
| Dry substance [%] | 78–79 | 80–81 |
| pH | 4–4.5 | 4.8–5.5 |
| Refractive index (20 °C) | 1.49 | 1.50 |
| Density (20 °C) | 1.40 | 1.40 |
| Viscosity [mPa s] (20 °C) | 15,000–18,000 | 30,000–35,000 |
| Storage temperature (°C) | 50 | 50 |
| Dextrose equivalent (DE) | 62–66 | 28–31 |
| **Typical carbohydrate composition:** | | |
| Glucose [%] | 31–35 | 2.5–3.5 |
| Disaccharides [%] | 41–45 | 9.5–12.5 |
| Trisaccharides [%] | 3–6 | 16–18 |
| Polysaccharides [%] | 14–25 | 66–72 |
| $SO_2$ [ppm] | <20 | <20 |
| Ash [%] | <0.1 | <0.1 |
| Sodium [ppm] | <200 | <200 |
| Potassium [ppm] | <5 | <5 |
| Calcium [ppm] | <5 | <5 |
| Magnesium [ppm] | <30 | <30 |

*Streptomyces griseus* or *Bacillus subtilis* amylase (TAKASAKI, 1985). In the same way ROBYT and ACKERMAN (1971) described an amylase from *Pseudomonas stutzeri* which can produce syrups containing up to 50% maltotetraose from starch. It has been reported that amylase from *Bacillus licheniformis* can produce up to 40% maltopentaose while *Klebsiella pneumoniae* amylase produces predominantly maltohexaose.

The composition of 30 DE syrups and HDMS syrups is given in Tab. 6. These specialty syrups can be used in fermentations where a substrate intermediate between starch and simple sugars is used. Optimization of the carbohydrate breakdown can be done to improve metabolite formation.

## 4.4 Fructose Syrups

### 4.4.1 Production

Fructose syrups are produced by enzymatic isomerization of glucose syrups with a microbial glucose isomerase from *Bacillus coagulans, Actinoplanes missouriensis, Flavobacterium arborescens* or *Streptomyces* sp. The principle of the isomerization process has been described by WHITE (1992). After saccharification with glucoamylase the glucose syrups are filtered then refined by demineralization. The pH of the syrup is adjusted to 7.0–8.0 and a magnesium salt is added to the substrate as enzyme activator. The isomerization process is a continuous process using immobilized enzyme technology where the syrup passes continuously through a reactor containing an enzyme bound to a solid carrier. The optimum temperature of isomerization is about 60 °C; the isomerization reaction is controlled by adjustment of the flow rate through the reactor. In practice, reactors are controlled to produce 42% fructose syrups "High Fructose Corn Syrups" (HFCS-42).

Higher fructose levels can be achieved by chromatographic enrichment to 90% fructose and blending with syrups containing 42% fructose to yield 55% fructose (HFCS-55). HFCS-42 syrups are concentrated to 71% d.s. while HFCS-55 is concentrated to 77% d.s.

**Tab. 7.** Typical Composition of Fructose Syrups

| | High Fructose Syrup 42 | High Fructose Syrup 55 |
|---|---|---|
| Aspect | Low viscous liquid | Viscous liquid |
| Dry substance [%] | 71 | 77 |
| pH | 4–5 | 4–5 |
| Refractive index (20°C) | 1.46 | 1.47 |
| Density (20°C) | 1.35 | 1.38 |
| Viscosity [mPa s] (20°C) | 200 | 800 |
| Dextrose equivalent (DE) | 94.5 | 94.5 |
| **Typical carbohydrate composition:** | | |
| Glucose [%] | 51 | 37–40 |
| Fructose [%] | 42 | 55–58 |
| Oligo- and Polysaccharides [%] | 7 | <5 |
| $SO_2$ [ppm] | <20 | <20 |
| Ash [%, dry basis] | <0.05 | <0.05 |
| Sodium [ppm] | <200 | <200 |
| Potassium [ppm] | <5 | <5 |
| Calcium [ppm] | <5 | <5 |
| Magnesium [ppm] | <30 | <30 |

### 4.4.2 Composition and Use

The composition of the two main commercial fructose syrups is listed in Tab. 7. Syrups with lower or higher fructose contents are also produced. The use of fructose syrups competes with that of sucrose as fermentation raw material. It must be pointed out that utilization of fructose syrups in Europe is limited by a quota system.

Fructose is reported to be a suitable substrate for the following fermentations: production of polyhydroxybutyrate by *Alcaligenes faecalis* (COLLINS, 1987), production of ethanol by *Zymomonas mobilis* (ROGERS et al., 1982; see also Chapter 4, this volume), production of acetone/butanol by *Clostridium*, of glutamic acid by *Corynebacterium*, of penicillin and streptomycin (ATKINSON and MAVITUNA, 1991).

## 4.5 Mannose Syrups

Mannose syrups can be produced by catalytic epimerization of glucose syrups or dextrose in aqueous solution, using a hexavalent molybdenum catalyst to give an equilibrium concentration of about 30% mannose (SHIODA, 1992). Mannose can be recovered in high yield and purity by chromatographic separation using ion exchange resins. The combined technology of epimerization and chromatographic separation can yield mannose syrups containing 10–90% mannose, the rest consists of glucose and about 5% oligosaccharides.

DE TROOSTEMBERGH et al., (1993) describe the use of mannose syrup for the production of xanthan gum. Use of mannose syrup as substrate increases the yield and productivity by about 15%. This can be explained by the carbohydrate composition of xanthan gum which contains about 40% mannose.

# 5 Crystalline D-Glucose (Dextrose)

Crystalline D-glucose is often named dextrose to be distinguished from glucose in its liquid form. D-Glucose can be crystallized in the monohydrate form below 50°C or in the anhydrous form above 50°C.

**Tab. 8.** Typical Composition of Crystalline Glucose (Dextrose)

| | Dextrose Monohydrate | Anhydrous Dextrose |
|---|---|---|
| Aspect | White crystalline powder | White crystalline powder |
| Dry substance [%] | >91 | >99 |
| pH | 4–6 | 4–6 |
| Bulk density [g/L] | 700–800 | 900–1000 |
| Dextrose equivalent (DE) | >99.5 | >99.5 |
| Glucose [%, dry basis] | >99.5 | >99.5 |
| Ash [%, dry basis] | <0.05 | <0.05 |
| $SO_2$ [ppm] | <10 | <5 |
| Calcium [ppm] | <5 | <5 |
| Magnesium [ppm] | <5 | <5 |
| Sodium [ppm] | <50 | <50 |
| Potassium [ppm] | <5 | <5 |
| Iron [ppm] | <1 | |

D-Glucose monohydrate crystallization occurs by slow cooling of concentrated saturated glucose syrups from 50 °C to 30 °C over 2-3 days. The crystals are recovered in basket centrifuges, washed with water and dried to less than 9% moisture. The mother liquor may be further evaporated to allow a second crystallization cut.

Anhydrous D-glucose is produced by evaporation crystallization at 65 °C under vacuum in a batch crystallization process. When the desired crystal phase yield is obtained the contents of the crystallizer are centrifuged in basket centrifuges. After washing, the crystals are dried to a very low water content.

Tab. 8 shows the composition of glucose (dextrose) monohydrate and anhydrous glucose. Purities exceed 99.5% glucose on a dry weight basis. Anhydrous D-glucose normally is not used in fermentation applications. Dissolution in water is very critical due to the rapid water uptake of the crystals which results in caking of the powder.

D-Glucose monohydrate is commonly used for laboratory or small-scale fermentations with small quantities required. It can be stored easily, does not require special storage equipment (like syrups), and is stable for months. For large-scale fermentations crystalline dextrose is less attractive than liquid glucose syrups. Exceptions are, e.g., the production of special pharmaceuticals where very pure substrates have to be used, or when the product purification requires a very pure fermentation raw material.

# 6 Sorbitol

Sorbitol is produced by hydrogenation of glucose (dextrose) in the presence of a metal catalyst. Hydrogenation takes place in pressure reactors under hydrogen pressure and a high temperature. The raw material to be hydrogenated must be as pure as possible in order to avoid catalyst poisoning. After reaction the solid catalyst is reused. The reaction mixture is then refined by ion exchange. Tab. 9 gives the composition of sorbitol in liquid and solid forms produced from dextrose.

As far as fermentation applications are concerned, sorbitol is predominantly used for the production of ascorbic acid (vitamin C) by the Reichstein process. In this process sorbitol is oxidized microbiologically into sorbose by *Gluconobacter oxydans* under aerobic conditions (FLORENT, 1986). Sorbose is then purified from the broth by crystallization and used for conversion to vitamin C. During hydrogenation a small amount of mannitol can be formed from glucose by hydrogenation and isomerization. Mannitol is also fermented by *Gluconobacter* sp. and converted to fructose which gives rise to isoascorbic acid. It is

**Tab. 9.** Composition of Sorbitol

| | Sorbitol Syrup | Sorbitol Powder |
|---|---|---|
| Aspect | Low viscous liquid | White powder |
| Dry substance [%] | 70 | >99.5 |
| pH | 5–7 | 5–7 |
| Refractive index (20°C) | 1.45 | – |
| Viscosity [mPa s] (20°C) | 80 | – |
| **Polyol composition:** | | |
| Hexitol [%] | >99 | >99 |
| Mannitol [%] | <2 | <1 |
| $SO_2$ [ppm] | <10 | <10 |
| Calcium [ppm] | <30 | <30 |
| Magnesium [ppm] | <5 | <5 |
| Sodium [ppm] | <20 | <20 |

therefore important to minimize the formation of mannitol during hydrogenation.

# 7 Selection of Starch-Derived Raw Materials for Fermentation

Besides economic considerations the selection of a starch-based raw material must be based on the three main criteria listed below.

## 7.1 Microorganism

The properties of the microbial strain used in a fermentation process is of primary importance for the selection of the raw material. The microorganism must produce amylases in the fermentation broth to be able to ferment starch or maltodextrins. These enzymes must be produced in an amount high enough for complete substrate conversion.

Maltase enzymes are necessary to ferment oligosaccharides such as maltose and maltotriose. Glucose is of course the most suitable substrate for species which do not produce the necessary enzymes to hydrolyze oligo- or polysaccharides.

Another important aspect is the fact that in some cases the substrate must be made available to the microorganism at a certain rate optimizing the production of the desired metabolite. If the substrate concentration in the broth is too high, other by-products might be formed. Alternatively, a repression by the substrate might occur as, e.g., the well-known glucose repression in yeast or in the production of antibiotics. The substrate uptake rate can be not only controlled by selecting the right substrate composition but also by using controlled feeding techniques in order to keep the glucose concentration in the broth at the desired level.

## 7.2 Process Constraints

Laboratory conditions generally allow the use of any type of substrate, while plant conditions may restrict the choice. For example, starch cannot by used as such in fermentation; it must first be liquefied before or during sterilization of the fermentation medium. Starch liquefaction is a relatively critical operation due to the high viscosity developing during liquefaction. Therefore, purpose-made equipment must be available to use starch itself. This is the reason why commonly ready-to-use liquid syrups are preferred in the fermentation industry.

Other types of constraints may also be important in the choice of substrate. For exam-

**Tab. 10.** List of Major European Starch and Starch Derivatives Suppliers to the Fermentation Industry

| Supplier | Address | Fax Number |
|---|---|---|
| Amylum | Burchtstraat 10<br>B-9300 Aalst | (56) 733034 |
| AVEBE | P.O. Box 15<br>NL-9640 AA Veendam | (31) 598764230 |
| Cargill BV | Postbus 34<br>NL-4600 AA Bergen op Zoom | (1640) 54489 |
| Cerestar | Avenue Louise 149, Bte 13<br>B-1050 Bruxelles | (2) 5378554 |
| Roquette Frères | Rue Patou, 4<br>F-59022 Lille Cedex | (20) 132813 |

ple, when liquefied starch is used, oxygen transfer might be limited by the design of the available equipment in terms of low air flow rate or installed power. In such cases a less viscous substrate might be preferable for optimal performance.

## 7.3 Downstream Processing Constraints

A raw material must be selected in order to achieve the highest fermentability of the substrate and the highest conversion into the desired product not only to minimize the cost of the raw material but also to facilitate product recovery from the broth and to minimize the amount of waste produced. An economic comparison is required between the higer cost of a purer raw material and the reduced cost of waste disposal. Also the effects of impurities in the fermentation broth on the recovery yield and on the final product quality must be assessed. Residual carbohydrate impurities may negatively affect the product stability during the recovery process.

## 7.4 Commercial Manufacturers of Starch, Glucose Syrups and Derivatives for the Fermentation Industry

Tab. 10 gives a list of the main European producers of starch and starch-derived carbohydrates for the fermentation industry.

Acknowledgement

The author thanks F. Oudenne for her contribution in collecting the data used in this chapter.

# 8 References

Alexander, R. J. (1992), Maltodextrins: production, properties and applications, in: *Starch Hydrolysis Products: Worldwide Technology, Production, and Applications* (Schenck, F. W., Hebeda, R. E., Eds.), pp. 233–275, Weinheim: VCH.

Atkinson, B., Mavituna, F. (1991), Industrial microbial processes, in: *Biochemical Engineering and Biotechnology Handbook* (Atkinson, B., Mavituna, F., Eds.), 2nd Edn., pp. 1111–1220, New York: Stockton Press.

Behal, V. (1987), The tetracycline fermentation and its regulation, *CRC Crit. Rev. Biotechnol.* **5** (4), 275–318.

Berdy, J. (1986), Further antibiotics with practical application, in: *Biotechnology* (Rehm, H.-J., Reed, G., Eds.), Vol. 4, 1st Edn., pp. 465–507, Weinheim: VCH.

Buchta, K. (1983), Organic acids of minor importance, in: *Biotechnology* (Rehm, H.-J., Reed, G., Eds.), Vol. 3, 1st Edn., pp. 467–478, Weinheim: Verlag Chemie.

Buckland, B. C., Omstead, D. R., Santamaria, V. (1985), Novel $\beta$-lactam antibiotics, in: *Comprehensive Biotechnology* (Moo-Young, M., Ed.), Vol. 3, pp. 49–68, Oxford: Pergamon Press.

Cejka, A. (1985), Preparation of media, in: *Biotechnology* (Rehm, H.-J., Reed, G., Eds.), Vol. 2, 1st Edn., pp. 629–698, Weinheim: VCH.

COLLINS, S. H. (1987), Choice of substrate in polyhydroxybutyrate synthesis, in: *Carbon Substrates in Biotechnology* (STOWELL, J. D., BEARDSMORE, A. J., KEEVIL, C. W., WOODWARD, J. R., Eds.), Vol. 21, pp. 161–168, Oxford: IRL Press.

CORBISHLEY, D. A., MILLER, W. (1984), Tapioca, arrowroot, and sago starches: production, in: *Starch: Chemistry and Technology* (WHISTLER, R. L., BEMILLER, J. N., PASCHALL, E. F., Eds.), 2nd Edn., pp. 469–478, London: Academic Press.

DE MENEZES, T. J. B., ARAKAKI, T., DE LAMO, P. R., SALES, A. M. (1978), Fungal cellulases as an aid for the saccharification of cassava, *Biotechnol. Bioeng.* **20**, 555–565.

DE TROOSTEMBERGH, J. C., BECK, R., DE WANNEMAEKER, B. (1993), *British Patent* GB-001894, EPA 609995A1.

DE WILLIGEN, A. H. A. (1976), The manufacture of potato starch in: *Starch Production Technology* (RADLEY, J. A., Ed.), pp. 135–154, Englewood, NJ: Applied Science Publishers.

EYSSAUTIER, B. (1993), *European Patent* 0 296 965 B1.

FLORENT, J. (1985), Streptomycin and commercially important aminoglycoside antibiotics, in: *Comprehensive Biotechnology* (MOO-YOUNG, M., Ed.), Vol. 3, pp. 137–162, Oxford: Pergamon Press.

FLORENT, J. (1986), Vitamins, in: *Biotechnology* (REHM, H.-J., REED, G., Eds.), Vol. 4, 1st Edn., pp. 115–158, Weinheim: VCH.

GRAM, L. E., MUNCK, L., ANDERSEN, M. P. (1989), A short milling process for wheat, in: *Wheat is Unique* (POMERANZ, Y., Ed.), pp. 445–455, St. Paul: American Association of Cereal Chemists.

HIROSE, Y., ENEI, H., SHIBAI, H. (1985), L-Glutamic acid fermentation, in: *Comprehensive Biotechnology* (MOO-YOUNG, M., Ed.), Vol. 3, pp. 593–600, Oxford: Pergamon Press.

HOSTALEK, Z., VANEK, Z. (1986), Tetracyclines, in: *Biotechnology* (REHM, H.-J., REED, G., Eds.), Vol. 4, 1st Edn., pp. 393–429, Weinheim: VCH.

KNIGHT, J. W., OLSON, R. M. (1984), Wheat starch: production, modification, and uses, in: *Starch: Chemistry and Technology* (WHISTLER, R. L., BEMILLER, J. N., PASCHALL, E. F., Eds.), 2nd Edn., pp. 491–506, London: Academic Press.

KUBICEK, C. P., RÖHR, M. (1986), Citric Acid Fermentation, *CRC Crit. Rev. Biotechnol.* **3**, 331–373.

MAIORELLA, B. L. (1985), Ethanol, in: *Comprehensive Biotechnology* (MOO-YOUNG, M., Ed.), Vol. 3, pp. 861–914, Oxford: Pergamon Press.

MARGARITIS, A., PACE, G. W. (1985), Microbial polysaccharides, in: *Comprehensive Biotechnology* (MOO-YOUNG, M., Ed.), Vol. 3, pp. 1005–1044, Oxford: Pergamon Press.

MATTOON, J. R., KIM, K., LALUCE, C. (1987), The application of genetics to the development of starch-fermenting yeasts, *CRC Crit. Rev. Biotechnol.* **5**, 3, 195–204.

MAY, J. B. (1987), Wet milling: process and products, in: *Corn: Chemistry and Technology* (WATSON, S. A., RAMSTAD, P. E., Eds.), pp. 377–397, St. Paul: American Association of Cereal Chemists.

MILSOM, P. E., MEERS, J. L. (1985a), Citric acid, in: *Comprehensive Biotechnology* (MOO-YOUNG, M., Ed.), Vol. 3, pp. 665–680, Oxford: Pergamon Press.

MILSOM, P. E., MEERS, J. L. (1985b), Gluconic and itaconic acids, in: *Comprehensive Biotechnology* (MOO-YOUNG, M., Ed.), Vol. 3, pp. 681–699, Oxford: Pergamon Press.

MITCH, E. (1983), Potato starch: production and uses, in: *Starch: Chemistry and Technology* (WHISTLER, R. L., BEMILLER, J. N., PASCHALL, E. F., Eds), pp. 479–490, London: Academic Press.

MULVIHILL, P. J. (1992), Crystalline and liquid dextrose products: production, properties, and applications in: *Starch Hydrolysis Products: Worldwide Technology, Production, and Applications* (SCHENCK, F. W., HEBEDA, R. E., Eds.), pp. 121–176, Weinheim: VCH.

NAKAYAMA, K. (1985), Lysine, in: *Comprehensive Biotechnology* (MOO-YOUNG, M., Ed.), Vol. 3, pp. 607–620, Oxford: Pergamon Press.

OKADA, M., NAKAKUKI, T. (1992), Oligosaccharides: production, properties, and applications, in: *Starch Hydrolysis Products: Worldwide Technology, Production, and Applications* (SCHENCK, F. W., HEBEDA, R. E., Eds.), pp. 335–366, Weinheim: VCH.

OMSTEAD, D. R., HUNT, G. R., BUCKLAND, B. C. (1985), Commercial production of cephamycin antibiotics, in: *Comprehensive Biotechnology* (MOO-YOUNG, M., Ed.), Vol. 3, pp. 187–210, Oxford: Pergamon Press.

ŌMURA, S., TANAKA, Y. (1986), Macrolide antibiotics, in: *Biotechnology* (REHM, H.-J., REED, G., Eds.), Vol. 4, 1st Edn., pp. 359–391, Weinheim: VCH.

POMERANZ, Y. (Ed.) (1989), *Wheat is Unique.* St. Paul: American Association of Cereal Chemists.

REED, G. (1982), Production of fermentation alcohol as a fuel source, in: *Prescott & Dunn's Industrial Microbiology* (REED, G., Ed.), 4th Edn., pp. 835–859, Connecticut: The AVI Publishing Co.

REEVE, A. (1992), Starch hydrolysis: processes and equipment, in: *Starch Hydrolysis Products: Worldwide Technology, Production, and Applications* (SCHENCK, F. W., HEBEDA, R. E., Eds.), pp. 79–120, Weinheim: VCH.

ROBYT, J. F., ACKERMAN, R. J. (1971), Isolation, purification and characterisation of a maltotetraose-producing amylase from *Pseudomonas stutzeri, Arch. Biochem. Biophys.* **145**, 105–114.

ROGERS, P. L., LEE, K. J., SKOTNICKI, M. L., TRIBE, D. E. (1982), Ethanol production by *Zymomonas mobilis*, in: *Advances in Biochemical Engineering* (FIETCHER, A., Ed.), Vol. 23, pp. 37–84, Berlin–Heidelberg–New York: Springer-Verlag.

ROSEIRO, J. C., ESGALHADO, M. E., AMARAL COLLACO, M. T., EMERY, A. N. (1992), Medium development for xanthan production, *Process Biochem.* **27**, 167–175.

SASAKI, T., KASUMI, T., KUBO, N., KAINUMA, K., WAKO, K., ISHIZUKA, H., KAWAGUCHI, G., ODA, T. (1993), *European Patent* 0 262 463 B1.

SCHENCK, F. W., HEBEDA, R. E. (1992), *Starch Hydrolysis Products: Worldwide Technology, Production, and Applications,* Weinheim: VCH.

SHIMURA, K. (1985), Threonine, in: *Comprehensive Biotechnology* (MOO-YOUNG, M., Ed.), Vol. 3, pp. 641–651, Oxford: Pergamon Press.

SHIODA, K. (1992), Chromatographic separation processes, in: *Starch Hydrolysis Products: Worldwide Technology, Production, and Applications* (SCHENCK, F. W., HEBEDA, R. E., Eds.), pp. 555–572, Weinheim: VCH.

SMITH, A. (1985), Cephalosporins, in: *Comprehensive Biotechnology* (MOO-YOUNG, M., Ed.), Vol. 3, pp. 163–185, Oxford: Pergamon Press.

STEWART, G. G., RUSSELL, I. (1985), Modern brewing technology, in: *Comprehensive Biotechnology* (MOO-YOUNG, M., Ed.), Vol. 3, pp. 336–381, Oxford: Pergamon Press.

SWARTZ, R. W. (1985), Penicillins, in: *Comprehensive Biotechnology* (MOO-YOUNG, M., Ed.), Vol. 3, pp. 7–47, Oxford: Pergamon Press.

TAKASAKI, Y. (1985), An amylase producing maltotriose from *Bacillus subtilis, Agric. Biol. Chem.* **49**, 4, 1091–1097.

TEAGUE, W. M., BRUMM, P. J. (1992), Commercial enzymes for starch hydrolysis products, in: *Starch Hydrolysis Products: Worldwide Technology, Production, and Applications* (SCHENCK, F. W., HEBEDA, R. E., Eds.), pp. 45–77, Weinheim: VCH.

VICKROY, T. B. (1985), Lactic acid, in: *Comprehensive Biotechnology* (MOO-YOUNG, M., Ed.), Vol. 3, pp. 761–776, Oxford: Pergamon Press.

VOLOCH, M., JANSEN, N. B., LADISCH, M. R., TSAO, G. T., NARAYAN, R., RODWELL, V. W. (1985), 2,3-Butanediol, in: *Comprehensive Biotechnology* (MOO-YOUNG, M., Ed.), Vol. 3, pp. 933–947, Oxford: Pergamon Press.

WATSON, S. A. (1984), Corn and sorghum starches: production, in: *Starch: Chemistry and Technology* (WHISTLER, R. L., BEMILLER, J. N., PASCHALL, E. F., Eds.), 2nd Edn., pp. 417-468, London: Academic Press.

WHISTLER, R. L., DADIEL, J. R. (1984), Molecular structure of starch, in: *Starch: Chemistry and Technology* (WHISTLER, R. L., BEMILLER, J. N., PASCHALL, E. F., Eds.), 2nd Edn., pp. 153–182, London: Academic Press.

WHITE, J. S. (1992), Fructose syrup: production, properties, and applications, in: *Starch Hydrolysis Products: Worlwide Technology, Production, and Applications* (SCHENCK, F. W., HEBEDA, R. E., Eds.), pp. 177–199, Weinheim: VCH.

ZOBEL, H. F. (1992), Starch: sources, production, and properties, in: *Starch Hydrolysis Products: Worlwide Technology, Production, and Applications* (SCHENCK, F. W., HEBEDA, R. E., Eds.), pp. 23–44, Weinheim: VCH.

# 2 Raw Material Strategies – Economical Problems

FRIEDRICH SCHNEIDER
HORST STEINMÜLLER

Linz, Austria

# 1 Introduction

From a biotechnological perspective the market for raw materials is in a state of transition today. On the one hand, we observe more and more severe environmental problems which might seriously affect biotechnological raw material resources and on the other hand, several biotechnological raw materials are not used at all or are only used to a very limited extent, because their market price is much too high for economical use and because they have to compete with artificial raw materials, which might produce adverse environmental effects.

Hence, the aim of this chapter is to provide some information about raw material strategies and the respective economic problems. The market for biotechnological products today and in the future is covered, raw materials, especially carbon sources, for the fermentation industry are described, and the break-even point of sugar prices for selected bioprocesses is discussed. Arguments are presented for raw material prices of sugar feedstocks in Western Europe and the price gap between renewable and non-renewable resources is shown. Finally, two economic instruments (environmental taxes and certificates) are discussed to promote the change from non-renewable to renewable resources. The advantages and disadvantages of these two instruments are discussed in some detail, and it is proven to be necessary to introduce these instruments in order to transform pure market economies into sustainable market economies.

# 2 The Market for Biotechnological Products Today and in the Future

Biotechnology is defined as any technique using living organisms (or parts of organisms) to make or modify products (to improve plants or animals) or to develop microorganisms for specific uses. It is not only a very old industry (beer brewing for more than 3000 years) but also one of the key technologies of the outgoing 20th century (U.S. CONGRESS, 1991).

**Tab. 1.** Production Output of Some Biotechnological Products (assumptions based on informations in *Chemical Business Newsbase,* 1992–1994)

| Product | Annual Production [t] |
|---|---|
| Baker's yeast | 300000 |
| Citric acid | 450000 |
| Lactic acid | 50000 |
| Lysine | 90000 |
| Glutamic acid | 270000 |
| Antibiotics | 30000 |
| Biopolymers | 10000 |

Today the main biotechnological products are ethanol, baker's yeast, organic acids, amino acids and pharmaceuticals like antibiotics, vaccines, and other drugs. Ethanol is the most important product with a worldwide production volume of $30 \cdot 10^9$ L per year. There are three main fields of application: human consumption, use as raw material for the chemical industry and as motor fuel. The largest markets for ethanol are Brazil (annual production ca. $14 \cdot 10^9$ L) and the United States (ca. $5 \cdot 10^9$ L), as in both countries a subsidized alcohol program for combustion engine fuel has been launched. The volume of all other products is 20–30 times smaller (Tab. 1).

The annual production volume is only one figure of interest, the other is the market volume. According to the EC SENIOR ADVISORY GROUP ON BIOTECHNOLOGY (1992) in 1990 the world market was ca. $6 \cdot 10^9$ U.S. $ with a high annual growth rate. Their estimate for the year 2000 is a market volume of ca. $100 \cdot 10^9$ U.S. $.

The principal market segments (Tab. 2) are and will be the pharmaceutical and food industry as well as agriculture. Environmental biotechnology (treatment of waste water, of contaminated soil, etc.) will also increase, but with a much lower growth rate, which reduces its importance in the biotechnological field.

To achieve this, it will be necessary to produce bulk materials like polymers, fuels, etc.

**Tab. 2.** World Market for Biotechnological Products (in million U.S. $) (Senior Advisory Group on Biotechnology, 1992)

| Segment | Market Volume 1991 | Market Volume 2000 |
|---|---|---|
| Agricultural, food, pharma | 2800 | 46700 |
| Environment | 1200 | 2300 |
| Chemicals | 500 | 3300 |
| Others | 100 | 17000 |
| Equipment | 1400 | 27900 |
| Sum | 6000 | 97200 |

besides high-price products (i.e., pharmaceuticals), on which most efforts of new biotechnologies are spent (industrial use of rDNA, cell fusion).

At present, the availability of crude oil is better than expected in the 1970s and the prices of fossil fuels are very low and, therefore, this aim seems to be highly jeopardized.

Nevertheless, there is a glimmer of hope: Firstly, many countries in the world are still deficient in fossil raw materials and, therefore, they are interested in establishing industries exploiting renewable resources for energy production and use as raw materials. Secondly, the problem of overproduction in agriculture in most of the industrialized countries has to be solved.

The present discussion on the impact of $CO_2$ as a causative agent of the greenhouse effect has increased interest in replacing fossil resources by renewable raw materials. It is, therefore, necessary to identify potential renewable feedstocks for basic chemicals and energy. Biomass with a spectrum of products might be an alternative to fossil hydrocarbons such as oil, natural gas, and coal which were the basis of growth in the petrochemical industry during the last few decades.

The role of biotechnology in such a different economical situation is not clearly defined yet, but with the introduction of renewable resources as feedstock and mild process conditions this technology will probably be preferred.

# 3 Raw Materials for the Fermentation Industry

For fermentation processes the raw materials contain carbon and nitrogen, trace minerals, vitamins, and cofactors. In the following, carbon sources produced by agriculture are discussed.

In any fermentation process, a carbohydrate feedstock is required to supply both energy and carbon skeletons. Today carbohydrate feedstocks are best established as fermentation substrates, the most important of which are:

- molasses (a by-product of beet and cane sugar production),
- starch (from starch-producing plants, e.g., maize and wheat),
- starch hydrolyzates, including dextrins and glucose syrups,
- sucrose,
- lactose,
- cereals, beet, potatoes.

In addition, other carbon sources of minor importance are used as fermentation feedstocks:

- oils and fats,
- hydrocarbons (*n*-alkanes),
- sulphite-spent liquor (a by-product of sulphite pulping),
- whey and lactose permeate (by-products of cheese production).

Sugars, the most important raw materials for the fermentation industry, can be obtained from many agricultural sources. The choice of feedstock is based not only on the

**Tab. 3.** Comparison of Conversion Cost (in U.S. $) (calculation based on information by Vogelbusch Inc., Vienna, Austria; personal communication, 1995)

| | 1 $m^3$ EtOH | 1 t Citric Acid | 1 t Lactic Acid | 1 t SCP |
|---|---|---|---|---|
| Annual plant capacity | 30000 $m^2$ | 10000 t | 1500 t | 10000 t |
| Chemicals | 10 | 367 | 158 | 70 |
| Energy | 75 | 60 | 60 | 54 |
| Personnel | 25 | 110 | 42 | 20 |
| Fixed cost | 10 | 73 | 170 | 30 |
| Capital charges | 34 | 258 | 533 | 112 |
| Sum | 154 | 868 | 963 | 286 |

specific requirements of the fermentation but also on local availability. Therefore, in Western Europe beet molasses is used as a common feedstock for many industrial fermentations including ethanol, single-cell protein, citric acid, and amino acids like lysine and glutamic acid. Corn-derived starch and glucose syrups are mainly used by the U.S. fermentation industry. However, in Western Europe the corn starch industry has developed rapidly and their products are now used as fermentation feedstocks as well.

For most of the above mentioned fermentation processes different carbohydrates are possible carbon sources, which is of economical interest for the fermentation companies. At present, in most cases the price is the only factor of decision, but in the future, environmental necessities and additional uses of byproducts will also play a role. This might be of special interest to companies utilizing molasses, as the low raw material price can easily be neutralized by the high cost of waste water treatment.

Besides the use of "classical" carbohydrate sources, there were many attempts to upgrade lignocellulosic materials, mainly agricultural residues like straw, bagasse, etc. A lot of R+D work was done since the late 1970s to develop a new hydrolysis process either using acids or enzymes (SATTLER, 1993). At present, no process runs at a commercial level, but worldwide interest is still alive, as proven by the 8th E.C. Conference on *"Biomass for Energy and Industry"* in October 1994 and the meeting of the Task Group NR. 5 of the IEA in Vancouver in November 1994.

# 4 Break-Even Sugar Prices for Selected Bioprocesses

## 4.1 Conversion Cost of Monomeric Sugar Products

For the evaluation of the different raw material lines it is necessary to know the cost of sugar as feedstock for different conversion processes. In order to determine the break-even sugar price the conversion cost of potential products of monomeric sugars and their achieveable selling prices ex factory have to be compared. The calculation of the conversion cost – within a range of ±20% accuracy – is based on Austrian prices for chemicals, energy and personnel, a depreciation of 12 years and an interest rate of 8% (Tab. 3).

A large number of products can be produced from monomeric sugars and only those were used for the following analysis that already cover a big market volume or promise a tremendous market increase in the future. The respective statements would also be applicable for the production of amino acids for which a precise calculation of production costs is available.

## 4.2 Achieveable Selling Price ex Factory

Although genuine world market prices for some of these products are not known, it is possible to estimate them.

**Ethanol**
Ethanol as a biotechnological product to be used as fuel is in direct competition with petrochemical products. Due to the high fluctuation of crude oil prices in the past and the ongoing discussion about energy vs. $CO_2$ taxes it is difficult to establish a realistic price for ethanol. However, it can be said that it will be competitive as a fuel component with a selling price in the range of other octane enhancers such as methyl-*tert*-butylether (MTBE) or *tert*-butylalcohol (TBA). At present, a price of 0.35 U.S. $ per liter seems to be reasonable in Western Europe.

**Citric acid**
Citric acid is used in the food and the detergent industry to replace phosphorus, especially in liquid detergents. The world market of ca. 700000 t still increases. The market price ex factory for citric acid is relatively stable: ca. 1.5 U.S. $ per kg citric acid monohydrate (CAM).

**Lactic acid**
At present the market for lactic acid is only in the range of 45000–50000 t per year. The market price ranges between 2 and 2.5 U.S. $ per kg pure lactic acid. When expanding on a larger market it will directly compete with citric acid and other organic acids so that it would have to be cheaper than citric acid to enter these new market areas and reach a significant market volume. In this calculation a price reduction of 10% is assumed for lactic acid compared to citric acid. The calculated market price is, therefore, 1.35 U.S. $ per kg.

**Single-Cell Protein (SCP)**
SCP from pentoses is obtained by strains of *Candida* sp. Due to the high RNA content of yeast biomass it can only be used for animal feed and is in direct competition with soy bean meal. However, soy bean meal is a commodity with a fluctuating price. An average price of 0.3 U.S. $ per kg that has been held in Western Europe during the last years was taken as a basis for the calculations. A price of SCP calculated only on the basis of the higher protein content could be 25% higher and, therefore, would be 0.375 U.S. $ per kg.

## 4.3 Break-Even Sugar Price

In the following section the maximal price of sugar is calculated by using the selling price as calculated in Sect. 4.2 and the conversion cost as calculated in Sect. 4.1. This price is called break-even sugar price (BESP) and is calculated following Eq. (1):

$$\text{BESP} = \frac{\text{Product selling price} - \text{Conversion cost}}{\text{Amount of sugar required}} \quad (1)$$

**Ethanol**
According to Gay-Lussac's law the fermentation of ethanol follows the reaction

$$\begin{array}{lll} C_6H_{12}O_6 \rightarrow & 2\,C_2H_5OH & +\,2\,CO_2 \\ \text{Glucose} \rightarrow & \text{Ethanol} & +\,\text{Carbon dioxide} \\ (180\text{ kg}) & (2\times 46\text{ kg}) & (2\times 44\text{ kg}) \end{array}$$

Hence, 92 kg ethanol can be produced from 180 kg glucose equivalent (0.511 kg ethanol per kg glucose). The experimental findings of PASTEUR showed, however, the consumption of part of the sugar by the process itself. This includes heat equivalent, yeast propagation, and formation of residual secondary fermentation products such as glycerol, succinic acid, etc. Thus, at 100% fermentation and distillation efficiencies, the maximum practical result of fermentation is:

$$\frac{0.5111 \cdot (1-0.0528)}{0.7937} = 0.61\text{ L} \quad (2)$$

of absolute alcohol per kg glucose

An optimized full-scale ethanol plant runs with a maximal efficiency of 94%. Thus, for the production of 1 L ethanol 1.75 kg hexoses are needed.

$$BESP_{(Ethanol)} = \frac{350-154}{1750} = 0.11 \text{ U.S. \$ per kg} \quad (3)$$

In a similar way the BESP values can be calculated for citric acid, lactic acid, and single-cell protein.

**Citric acid**
For the production of 1 kg citric acid monohydrate (CAM) 1.20 kg hexoses are needed.

$$BESP_{(CAM)} = \frac{1500-868}{1200} = 0.53 \text{ U.S. \$ per kg} \quad (4)$$

**Lactic acid**
To produce 1 kg lactic acid (LA) 1.13 kg hexoses are required.

$$BESP_{(LA)} = \frac{1350-963}{1130} = 0.34 \text{ U.S. \$ per kg} \quad (5)$$

**Single-cell protein**
For the production of 1 kg marketable product of SCP with 92% dry matter 2 kg of monomeric sugars are needed.

$$BESP_{(SCP)} = \frac{375-286}{2000} = 0.04 \text{ U.S. \$ per kg} \quad (6)$$

These results are summarized in Tab. 4.

**Tab. 4.** Calculation (1994) of Break-Even Sugar Prices (BESP) (in U.S. $ per kg)

| Product | BESP |
|---|---|
| Ethanol | 0.11 |
| Citric acid | 0.53 |
| Lactic acid | 0.34 |
| Single-cell protein | 0.04 |

# 5 Raw Material Prices for Sugar Feedstocks in Western Europe

The fermentation industry always looked for waste and by-products from different sources as feedstock, because of continued interest in reducing raw material costs in the past 20 years. In the following, raw materials will be analyzed which are solely produced for the use as feedstock in the fermentation industry.

As the first step of calculation the production cost for maize, wheat, sugar beet, green fodder (mixture of grass and clover cultivated on arable land) and grass (from permanent grass land) are presented. The basis of these calculations are average Austrian yields and prices for production goods after accession to the EU in 1995 (Tab. 5).

An important factor is the farmer's income including labor income, depreciation, and interest for the required machinery (D+I) as well as profit and risk. In the following calculation a farmer's income without subsidies is used that is equivalent to the average income of an Austrian laborer. Assuming that enough land is owned, working time is the limiting factor.

Maize and wheat are plants containing starch, and glucose as feedstock for the fermentation industry can be obtained by enzymatic hydrolysis of starch. Sucrose is produced from sugar beet or sugar cane in sugar refinery plants. Glucose and fructose are hexoses produced by enzymatic hydrolysis of grass and green fodder. Fodder protein is a by-product obtained from maize (255 kg per ton of maize with 30% raw protein), wheat (323 kg per ton of wheat with 30% raw protein), and grass (500 kg per ton with 31% raw protein for green fodder and 25% for grass from permanent grass land, respectively). The price of fodder protein is calculated based on the price of soybean cake – for sure, a very optimistic approach because of strong competition in the protein market. Pentoses can be obtained as by-products of hemicellulose hydrolysis in grass, but due to the problem of pentose utilization (STEINMÜLLER, 1991) no by-product credit is calculated for pentoses. EIBENSTEINER and STEINMÜLLER (1994)

**Tab. 5.** Calculation (1995) of Variable Production Costs for Different Agricultural Products (in U.S. $ per ha)

| | Maize 30% $H_2O$ | Wheat | Sugar Beet | Green Fodder | Grass |
|---|---|---|---|---|---|
| Yield [t per ha] | 7.5 dried | 5.5 dried | 50 | 16 DS[a] | 9 DS[a] |
| Seed | 150 | 85 | 97 | 40 | — |
| Fertilizers | 246 | 201 | 339 | 455 | 264 |
| Plant protection | 110 | 99 | 323 | — | 17 |
| Insurance | 15 | 19 | 66 | — | — |
| Variable machinery cost | 118 | 119 | 320 | 320 | 220 |
| Threshing (Lohndrusch) | 154 | 125 | 330 | — | — |
| Drying | 367 | 8 | — | — | — |
| Silage | — | — | — | 840 | 472 |
| Sum | 1160 | 657 | 1475 | 1655 | 973 |

[a] Dry substance

**Tab. 6.** Calculation (1995) of Full Costs for Different Agricultural Products (in U.S. $ per ha)

| | Maize 30% $H_2O$ | Wheat | Sugar Beet | Green Fodder | Grass |
|---|---|---|---|---|---|
| Variable cost | 1160 | 657 | 1475 | 1655 | 973 |
| Labor income | 125 | 125 | 417 | 367 | 258 |
| D+I | 125 | 125 | 400 | 350 | 250 |
| Profit and risk | 100 | 100 | 100 | 100 | 100 |
| Total cost | 1510 | 1007 | 2392 | 2472 | 1581 |

**Tab. 7.** Calculation (1995) of Raw Material Prices after Storage

| | Maize | Wheat | Sugar Beet | Green Fodder | Grass |
|---|---|---|---|---|---|
| Total cost per ha [U.S. $] | 1510 | 1007 | 2392 | 2472 | 1581 |
| Hectare yield after storage [t DS[a]] | 6.45 | 4.73 | 50 | 13.6 | 7.65 |
| Cost [U.S. $ per DS[a]] | 234 | 213 | 47.8[b] | 182 | 207 |

[a] Dry substance
[b] Sugar beet is not calculated as dry substance

have reported the possibility to utilize lactic acid produced in silage. A by-product credit of 333 U.S. $ is calculated per ton of pure lactic acid (60 kg lactic acid per ton of grass resp. green fodder can be obtained) dissolved in the press liquor with a concentration of 4%.

Nearly all costs are in the range of 300 U.S. $ per ton of sugar. Only sucrose costs from sugar beet are higher, but there are no problems of by-product utilization and it is possibile to reduce costs by incorporation of green-syrup production in an already existing sugar plant. Production costs of glucose from starch plants are more reliable than those from plants utilizing grass, as in the latter case they are extrapolated from small pilot trials, whereas starch hydrolysis plants run at an industrial scale worldwide.

**Tab. 8.** Calculation (1995) of the Sugar Cost (in U.S. $)

| | Maize | Wheat | Sugar Beet | Green Fodder | Grass |
|---|---|---|---|---|---|
| Raw material cost | 234 | 213 | 47.8 | 182 | 207 |
| Production cost[a] | 40 | 40 | 41.7 | 62 | 62 |
| By-product credit[a] | 58 | 73 | — | 137 | 137 |
| Yield [kg sugar per ton] | 710 | 630 | 170 | 350 | 350 |
| Sugar cost [U.S. $ per kg] | 0.30 | 0.28 | 0.53 | 0.31 | 0.38 |

[a] Calculated per ton of raw material

# 6 Price Gap between Renewable and Non-Renewable Resources

In Fig. 1 the sugar production cost as calculated in Sect. 5 and the achieveable break-even sugar prices of the chosen products presented in Sect. 4.3 are compared.

As can be seen in Fig. 1 sugars for the production of organic acids used in the conventional market segments (food, pharmaceutical, and detergent industry) can only be produced as feedstock for the fermentation industry on a full cost basis, i.e., without subsidies. Should the products have to compete either with fossil energy sources (e.g., ethanol as automotive fuel) or with subsidized agricultural products (e.g., soy bean meal) these sugars are too expensive for a feasible production. The situation is similar with biotechnological products that compete with polymers produced from petrochemical raw materials. The lactic acid polymer (polylactate) could substitute for polyethylene, but the price for polyethylene is about 0.5 U.S. $ per kg, whereas for polylacetate it amounts to nearly 5–6 U.S. $ per kg (Eibensteiner and Steinmüller, 1994) with a lactic acid price of ca. 1.5 U.S. $ per kg. The same is also true for lactic acid substituting for acrylic acid, with a market-price of 1.2 U.S. $ per kg.

Ethanol production from grain or green fodder will only be feasible when the ethanol price is doubled to 0.70 U.S. $ per liter.

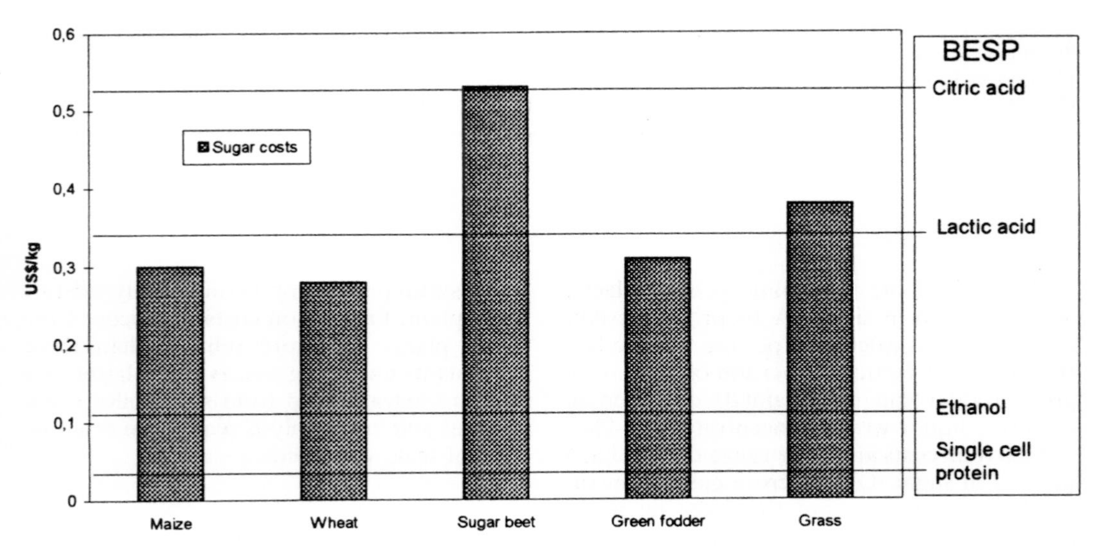


**Fig. 1.** Comparison of sugar production costs and break-even sugar prices (calculation 1995)

# 7 Economic Instruments to Reduce the Price Gap

As outlined in Sect. 6, there is a considerable price gap between renewable and non-renewable resources. Hence, for several years a discussion went on in environmental economics and in politics how to reduce this gap and how to transform market economies into sustainable market economies. From the huge amount of literature in this area only a few of the latest surveys are cited here: OECD (1993a, b, 1994), SCHNEIDER (1994), PEARCE and TURNER (1990), CROPPER and OATES (1992) VON WEIZSÄCKER (1989).

In order to make progress in this matter, two instruments are widely discussed to introduce the ecological objectives into market economies or to strengthen them.

## 7.1 Environmental Taxes

The first widely discussed instrument is the use of environmental taxes. The idea is, e.g., that additional taxes are imposed on non-renewable resources, so that the increase of cost will reduce their utilization. The tax rates should be set in order to start substitution processes and as a final goal, these materials are not used at all or only to a very limited extend. It is very important that the revenues from these taxes be solely used to either reduce other taxes (like taxes on labor) or stimulate investments for the development of biotechnological raw materials or renewable raw materials. Using the environmental tax instrument is dangerous: The considerable amount of revenues these taxes create (at least during their phase of initiation) could be treated as a permanent source of income so that, finally, there would only be a very limited (or even no) incentive to strengthen the substitution effect. As in almost all OECD states high federal and state budget deficits are observed, there is a strong incentive for every government to handle their tax revenues as permanent sources of income. Therefore, the institutional design of such tax arrangements has to be planned very carefully so that such a situation can be avoided. The great advantage of this instument – if introduced in a forseeable and planable way – is that every enterprise and consumer is able to easily calculate how expensive it is to use non-renewable resources and how much money can be saved when switching to renewable resources. Moreover, a strong incentive structure for the innovation of new technologies is created.

## 7.2 Environmental Certificates

The second instrument are environmental certificates. Whereas with environmental taxes the prices are the steering variable, in the case of certificates the quantities are the steering instumental variable. The idea of certificates is the following: For using a certain amount of non-renewable resources an adequate environmental certificate has to be bought. The total quantity of certificates is fixed by a state agency. In the first year, a certain number of certificates can be bought from this governmental institution. In the following years it is reduced to the final quantity of certificates. This decreasing number of certificates has the effect that the price set at the beginning will rise over time. The certificates can be treated like other goods. When switching from non-renewable resources, the certificates can be resold to the governmental agency which saves the customer a lot of money.

Care has to be taken with this instrument in order to avoid a monopoly situation with one firm "collecting" all certificates and preventing other firms from entering the market. The certificates must be resold to the state agencies and they either eliminate them or sell them to other clients. As the number of certificates is reduced every year while the prices increase, all customers will have a strong incentive to use more (renewable) resources. The disadvantages of the certificate instrument are: (1) It might be difficult (or impossible) for new firms to enter the market so that competition is hindered; therefore, it is crucial that the number of certificates sold is restricted to actual production. (2) It can be very difficult to set up standards and quantities for certificates and to solve the control (i.e., monitoring) problem when a great number of customers (e.g., car drivers) use such certificates.

## 7.3 The Implementation of These Instruments

Both instruments are orientated incentives. They fit nicely into market economics by transforming them into sustainable market economics. They also have the great advantage that they encourage producers and/or consumers to replace non-renewable with renewable resources as quickly as possible and to invent new technologies which might support this substitution process.

One may ask how high the rates (in case of taxes) should be and how small the number of certificates for starting the substitution process. This question is difficult to answer in global terms, but it can be argued that the rates (quantities) should be fixed in order to make renewable resources cheaper than non-renewable resources for a period of 5 years.

The use of other instruments like laws and rules, e.g., to prohibit the use of non-renewable resources, is difficult to monitor and might not be very efficient, because the interest of individuals to invent new technologies and to switch to renewable resources is very limited. In general, it should be challenging for a government to use market-oriented environmental instruments to start the transformation process from a pure economy into sustainable market economy. The use of environmental taxes or certificates to switch from non-renewable to renewable resources might be a good starting point for establishing a sustainable market economy.

# 8 References

CROPPER, M. L., OATES, W. E. (1992), Environmental economics: a survey, in: *Journal of Economic Literature,* Vol. XXX/3, pp. 675–740.

EIBENSTEINER, F., STEINMÜLLER, H. (1994), Milchsäuregewinnung aus Silage-Fermentationen, *Scientific Report for the Ministry of Science,* Vienna.

OECD (1993a), *Taxation and Environment: Complementary Policies,* Paris.

OECD (1993b), *Environmental Policies and Industrial Competitiveness,* Paris.

OECD (1994), *Managing the Environment: The Role of Economics Instruments,* Paris.

PEARCE, D. W., TURNER, R. K. (1990), *Economics of Natural Resources and the Environment.* New York: Harvester Wheatsheaf.

SATTLER, J. N. (1993), *Bioconversion of Forest and Agricultural Plant Residues,* pp. 33–73. Wallingford, UK: CAB International.

SCHNEIDER, F. (1994), Ecological objectives in a market economy: three simple questions, but no simple answers? in: *Environmental Standards in the European Union in an Interdisciplinary Framework* (FAURE, M., VERVAELE, J., WEALE, A., Eds.), pp. 93–116. Antwerpen: Maklu-Nomos-Blackstone-Schulthess.

SENIOR ADVISORY GROUP ON BIOTECHNOLOGY (1992), *Report on Biotechnology,* Brussels.

STEINMÜLLER, H. (1991), Enzymatic hydrolysis of wheat straw: a techno-economical study, *Doctorial Thesis,* Technical University, Graz.

U.S. CONGRESS, OFFICE OF TECHNOLOGY ASSESSMENT (1991), *Biotechnology in a Global Economy,* Washington.

WEIZSÄCKER, E. U., VON (1989), *Erdpolitik: Ökologische Realpolitik an der Schwelle zum Jahrhundert der Umwelt.* Darmstadt: Wissenschaftliche Buchgesellschaft.

# II. Products of Primary Metabolism

# 3 Ethanol – Classical Methods

THOMAS SENN
HANS JOACHIM PIEPER
Stuttgart, Germany

# List of Abbreviations

AAQ: Autoamylolytical quotient (Sect. 13.2.3)
BAA: Bacterial $\alpha$-amylase from *Bacillus subtilis*
BAB: Bacterial $\alpha$-amylase expressed by *Bacillus licheniformis*
DBM: Dried distillers' barley malt
DDGS: Dried distillers' grains with solubles
DDS: Dried distillers' solubles
DMP: Dispersing mash process
DP: Degree of polymerization (oligosaccharides)
DS: Dry substance
FAA: Fungal $\alpha$-amylase from *Aspergillus oryzae*
FS: Fermentable substance (Sect. 13.2.2.1)
GAA: Glucoamylase from *Aspergillus niger*
GAR: Glucoamylase from *Rhizopus* sp.
GLS-Process: Mashing process according to Große-Lohmann-Spradau
hlA: Hectoliters of ethanol (100%)
HPCP: High pressure cooking process
KMV: Cold mash process, milling and mashing process at saccharification temperature
lA: Liters of ethanol (100%)
MMP: Milling and mashing process at higher temperatures
NNE: Non-nitrogenous extract
OS: Organic substance
SRP: Stillage recycling process
TBA: Thermostable bacterial $\alpha$-amylase from *Bacillus licheniformis*
VP: Verlinden Process, free from bacteria

# 1 Starch Containing Raw Materials

## 1.1 Potatoes

Potatoes represent the most widely used starchy raw material in ethanol production in Germany and Eastern Europe. On average potatoes consist of ca. 75% water and 25% dry substance. An average analysis is given in Tab. 1. Regarding the suitability for ethanol production, the starch content of potatoes is the most important criterion. In addition to starch, potatoes contain low amounts of sugars, mainly sucrose, glucose, and fructose. Sugar and starch contents depend on the variety and level of ripeness of the potatoes. The starch content also depends on climatic, growth, as well as on storage conditions. The loss of starch during storage, e.g., amounts to about 8% after 6 months and 16.5% after 8 months of storage in a regular operating storage cellar.

As shown in Tab. 1, potatoes contain pectin. This pectin content is responsible for the methanol content of spirits produced from potatoes. Milling of potatoes leads to the release of pectin esterases which immediately start cleavage of the methyl ester bonds of pectin. Using a pressure cooking process in mashing potatoes, pectin esterases are inactivated by heat, but a virtually complete thermal de-esterification takes place. This is why the methanol content in raw spirits obtained from pressure cooking processes is higher compared to raw spirits produced by pressureless processes (BOETTGER et al., 1995).

**Tab. 1.** Average Analysis of Potatoes

| Component | Percentage [%] |
|---|---|
| Water | 72.0 – 80.0 |
| Starch | 12.0 – 21.0 |
| Sugar (reducing) | 0.07 – 1.5 |
| Dextrin and pectin | 0.2 – 1.6 |
| Pentosans | 0.75 – 1.0 |
| Nitrogenous components | 1.2 – 3.2 |
| Fat | 0.1 – 0.3 |
| Crude fiber | 0.5 – 1.5 |
| Ash | 0.5 – 1.5 |

## 1.2 Wheat

Wheat is often used in German grain distilleries, because it yields an especially mild and smooth distillate. The starch content of wheat is usually about 60%, leading to ethanol yields of about 38 lA per 100 kg wheat (Tabs. 2 and 3). If wheat containing more than 13% raw protein is used for ethanol production, fermentation problems may occur. If wheat with a high protein content is processed without pressure, the mashes tend to foam during fermentation. Often these mashes can only be fermented if an antifoam agent (e.g., silicone antifoam) is used in fermentation. Tabs. 2 and 4 show the composition of wheat grains.

The possibility of using a certain pressureless processing of wheat depends on the activity of the autoamylolytic enzyme system, which can be measured by determination of the autoamylolytical quotient (AAQ) (see Sect. 13.2.3). Fig. 1 shows the AAQ of several important varieties of wheat. The pressureless processing of wheat is possible without any problems if the lot of wheat used has an AAQ of 95% or higher. In general, the processing of waxy wheat is problematic and re-

**Tab. 2.** Components of Wheat Grains

| Component | Percentage [%] |
|---|---|
| Seed coat | 15.0 |
| Endosperm | 83.0 |
| Germ | 2.0 |

**Tab. 3.** Average Analysis of Wheat

| Component | Percentage [%] |
|---|---|
| Water | 13.2 |
| Crude protein | 11.7 |
| Crude fat | 2.0 |
| NNE | 69.3 |
| Crude fiber | 2.0 |
| Ash | 1.8 |

**Tab. 4.** Composition of Wheat Components [% of DS]

| Component | Protein | Ash | Fat | Carbo-hydrates |
|---|---|---|---|---|
| Seed coat | 7 – 12 | 5 – 8 | 1 | 80 – 88 |
| Aleuron layer | 24 – 26 | 10 – 12 | 8 – 10 | 52 – 58 |
| Endosperm | 10 – 14 | 0.4 – 0.6 | 1.8 – 1.2 | 83 – 87 |
| Germ | 24 – 28 | 4 – 5 | 8 – 12 | 55 – 64 |

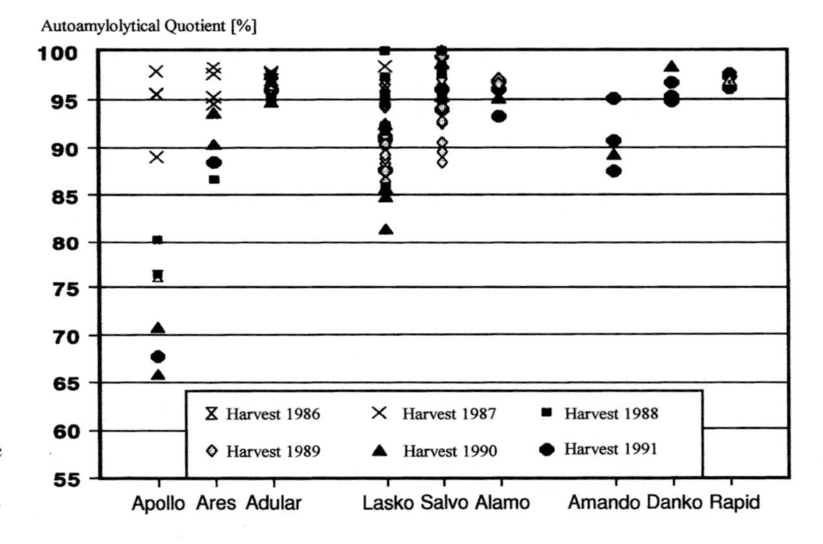


**Fig. 1.** Autoamylolytical quotient of wheat, triticale and rye varieties, depending on the year of harvest.

quires a very effective decomposition of the raw material (Sect. 8.1).

## 1.3 Rye

The starch content of rye is about 2–4% lower than that of wheat. A batch of rye with a high starch content yields about 37 lA per 100 kg of rye. Rye is also an important raw material in German grain distilleries. An average analysis is given in Tab. 5.

**Tab. 5.** Average Analysis of Rye

| Component | Percentage [%] |
|---|---|
| Water | 13.7 |
| Crude protein | 11.6 |
| Crude fat | 1.7 |
| NNE | 69.0 |
| Crude fiber | 2.1 |
| Ash | 1.9 |

Virtually all varieties of rye contain a highly active autoamylolytical enzyme system (Fig. 1) and they are, therefore, suitable for pressureless processes. However, as shown in Tab. 5, rye may contain pentosans, depending on climatic and growth conditions as well as on variety. These pentosans often lead to high viscosities in rye mashes, resulting in problems during the mashing and the fermentation (Sect. 8.3).

## 1.4 Triticale

Triticale, a hybrid of wheat and rye, which has been used in ethanol production since just

**Tab. 6.** Composition of Triticale (KLING and WÖHLBIER 1983)

| Component of DS [%] | |
|---|---|
| OS | 97.2 |
| Crude protein | 17.3 |
| Crude fat | 1.8 |
| Crude fiber | 3.1 |
| NNE | 73.8 |
| Ash | 2.3 |

a few years, is a very important raw material. The starch content is about 60% of original substance leading to ethanol yields of 38 lA per 100 kg of triticale. Triticale does not contain considerable amounts of pentosans, and so there are no problems regarding mash viscosity.

Some varieties of triticale exhibit high autoamylolytical enzyme activity and, therefore, it is possible to process triticale without using any additional saccharifying enzyme. To examine single lots of triticale, it is necessary to determine the AAQ as described in Sect. 13.2.3. Not only is it possible to saccharify the starch in triticale, but the same amount of starch from other grains or from potatoes can additionally be saccharified. Triticale is a potentially rich source of the saccharifying enzymes needed in distillery. The composition of triticale has rarely been examined; the data shown in Tab. 6 are approximations.

## 1.5 Corn (Maize)

### 1.5.1 Dried Storable Corn Grain

Corn is a very important raw material for ethanol production in the USA and South America. In Europe great amounts of corn are also used for ethanol production. An averarge analysis is shown in Tab. 7. The suitability of corn for ethanol production depends on the contents of starch and horny endosperm. A high content of horny endosperm leads to problems in ethanol production using milling processes (Sect. 5.4.1). European corn, with a starch content of about 62–65%, yields at least 40 lA per 100 kg of corn. Tab. 8 shows a typical analysis of corn grain from a Southern German distillery. When purchasing corn, its water content and cleanness must be considered. The fat contained in corn prevents foam formation during the fermentation.

### 1.5.2 Corn Grain Silage

In many parts of Europe, due to climatic conditions, corn does not ripen sufficiently to be harvested as natural dried corn grain. Very often corn grain has to be dried artificially which leads to high costs for drying. These costs can be reduced if corn is used in distilleries as corn grain silage (PIEPER and PÖNITZ, 1973). Corn grain silage can effectively

**Tab. 7.** Average Analysis of Corn Grain [% of DS] (KLING and WÖHLBIER, 1983)

| Component | Naturally Dried Mean Value (n = 496) | Artificially Dried Mean Value (n = 12) |
|---|---|---|
| | Mean s | Mean s |
| OS | 98.3 ± 0.6 | 98.2 ± 0.6 |
| Crude protein | 10.8 ± 1.1 | 10.7 ± 1.0 |
| Crude fat | 4.7 ± 0.8 | 4.7 ± 0.6 |
| Crude fiber | 2.6 ± 0.8 | 2.6 ± 0.3 |
| NNE | 80.2 ± 2.2 | 80.2 ± 1.2 |
| Ash | 1.7 ± 0.6 | 1.8 ± 0.6 |
| DS | ca. 88 | 88.9 ± 2.9 |

**Tab. 8.** Average Analysis of Dried Storable Corn Grain from a Distillery in Southern Germany (PIEPER and PÖNITZ, 1973)

| Component | Minimum [%] of OS | Maximum [%] of OS | Mean Value [%] of OS n=3 |
|---|---|---|---|
| Water | 14.8 | 15.2 | 15.1 |
| Ash | 1.4 | 1.5 | 1.5 |
| Crude protein | 8.3 | 8.5 | 8.4 |
| Crude fiber | 1.9 | 2.1 | 2.0 |
| Fat | 3.6 | 3.9 | 3.7 |
| Starch | 62.2 | 63.1 | 62.6 |
| Other NNE | 6.3 | 7.4 | 6.7 |

**Tab. 9.** Average Analysis of Corn Grain Silage form a Distillery in Southern Germany (PIEPER and PÖNITZ, 1973)

| Component | Minimum [%] of OS | Maximum [%] of OS | Mean Value [%] of OS n=6 |
|---|---|---|---|
| Water | 41.6 | 41.6 | 41.6 |
| Ash | 1.0 | 1.1 | 1.1 |
| Crude protein | 5.7 | 5.8 | 5.8 |
| Crude fiber | 1.4 | 1.7 | 1.6 |
| Fat | 2.4 | 2.6 | 2.5 |
| Starch | 42.3 | 42.5 | 42.4 |
| Other NNE | 5.0 | 5.5 | 5.0 |

be processed using HPCP or better using DMP with stillage recycling (Sect. 5.4.2). A typical analysis of corn grain silage from a distillery in Southern Germany is shown in Tab. 9.

If corn grain silage is used in ethanol production, one has to take care that a pure lactic acid fermentation takes place. Minimal concentrations of butyric acid in the stillage, due to a contamination of silage with butyric acid bacteria, lead to a total breakdown of the fermentation since butyric acid is strongly toxic for yeasts.

## 1.6 Barley

In ethanol production, barley is mostly used as malting grain. Since it grows very well in Eastern Europe, it is also an interesting raw material in ethanol production. There are two notable disadvantages of barley as a raw material in distilleries: the husks surrounding the kernels and the content of glucans which leads to high viscosities in mashes. Therefore, special processing is necessary in preparing mashes from barley (Sect. 5.5). Tab. 10 shows an average analysis of barley. Consisting of about 55% starch, barley yields about 35 lA per 100 kg FS. Compared to distillates from wheat, potable distillates produced from barley are smooth, but have a more powerful grain taste.

## 1.7 Sweet Sorghum

Sweet sorghum is rarely used in Europe for ethanol production. About its composition there are no reliable data available in the literature regarding ethanol production. But within the last 20 years the growth of sweet

**Tab. 10.** Average Analysis of Barley [% of DS] (KLING and WÖHLBIER, 1983, modified)

| Component | Mean Value (n = 1249) Mean s |
|---|---|
| OS | 97.2 ± 0.7 |
| Crude protein | 11.8 ± 1.7 |
| Crude fat | 2.2 ± 0.5 |
| Crude fiber | 5.3 ± 1.5 |
| NNE | 77.9 ± 2.6 |
| Starch as a part of it | 63.2 |
| Ash | 2.8 ± 0.7 |
| DS | ca. 87% |

sorghum for ethanol production in Austria and Germany has been investigated (SALZBRUNN, 1982; DIEDRICH et al., 1993). Sweet sorghum is a native plant of subtropical and tropical regions, but also grows in certain parts of Austria and Germany reaching a height of 3–3.5 m. Trial plots yielded 5–8.8 t of fermentable sugars per ha, depending on the cultivation site. To obtain the sugar-containing juice from the sweet sorghum plant it can either be extracted with water or it can be pressed out using roller mills. For conservation it can be concentrated up to 80° Bx using a downflow evaporator (SALZBRUNN, 1982).

## 1.8 Sorghum Grain

The worldwide production of sorghum grain takes the fourth place of all varieties of grains. Sorghum grain exists in yellow and brown colored types that show no significant differences in composition. An average analysis of sorghum is shown in Tab. 11. Sorghum kernels are round with a diameter of 5–7 mm. Sorghum contains about 62–65% starch and yields about 40 lA per 100 kg of sorghum. These data are comparable to corn.

When purchasing sorghum grain, one should check its cleanness, especially it should be free from sand and corn weevils. Due to the fact that sorghum starch is waxy, it is not easily decomposed. Therefore, sorghum should either be processed with HPCP, or preferably by DMP and stillage recycling.

**Tab. 11.** Average Analysis of Sorghum Grain

| Component | Percentage [%] |
|---|---|
| Water | 11 – 12 |
| Crude protein | 9 – 12 |
| Fat | 3 – 4.5 |
| NNE | up to 71 |
| Crude fiber | 3 |
| Ash | 1.5 – 3 |
| Starch and sugar | 58 – 63 |

## 1.9 Manioc

Manioc is a tropical plant, forming starch-containing roots. Since manioc roots do not keep well, they should be processed immediately or, alternatively manioc starch can be produced from dried roots. It is also possible to mill the roots and to dry the obtained manioc flour. Average analyses of different manioc products are given in Tab. 12.

Manioc contains toxic levels of a cyanogenic glucoside (up to 2.8 g per kg DS), and, therefore, it is recommended in ethanol production to process manioc using HPCP. In this way manioc products are detoxified by deaeration. It is also possible to detoxify manioc products by adding sodium thiosulphate, and hence it should be possible to process manioc using other mashing processes. There are increasingly more varieties of manioc being cultivated which are free from cyanogenic glucosides.

Sometimes manioc flour or manioc starch contain up to 20% sand. If this is not detected before the material is processed, the sand will settle down in the fermentation tank and take the yeast with it, leading to drastical disruptions during fermentation (KREIPE, 1982).

# 2 Technical Amylolysis

To reach an almost total degradation of starch to fermentable sugars in technical processes, two main groups of amylolytical enzymes are required: one group comprises

**Tab. 12.** Average Analysis of Manioc Products

| Component | Manioc Root | Manioc Starch | Manioc Flour |
|---|---|---|---|
| Water | 70.3 | 12.6 | 14.0 |
| Protein | 1.1 | 0.6 | 1.2 |
| Fat | 0.4 | 0.2 | 0.4 |
| Fiber | 1.1 | 0.2 | 2.0 |
| Starch | 21.5 | – | 74.3 |
| Ash | 0.5 | 0.3 | 1.4 |
| Non-nitrogenous components (including starch) | | 86.1 | 81.0 |

liquefying $\alpha$-amylases, the other group saccharifying glucoamylases, $\beta$-amylases, and $\alpha$-amylases.

## 2.1 Enzymatic Starch Liquefaction

Technical liquefying enzymes are virtually all $\alpha$-amylases ($\alpha$-1,4-glucane4-glucanohydrolase, E.C. 3.2.1.1) that split $\alpha$-1,4 bonds in amylose and amylopectin. $\alpha$-Amylase is an endo-acting enzyme and its action is often considered to be random, i.e., the enzyme has equal preference for all $\alpha$-1,4 linkages except those adjacent to the ends of the substrate chain and those in the vicinity of branch points. The $\alpha$-1,6 glycosidic bonds are not hydrolyzed. The properties as well as the action of $\alpha$-amylase depend on the microorganisms or plants from which it is derived. However, all $\alpha$-amylases rapidly decrease the viscosity of starch solutions.

### 2.1.1 Thermostable Bacterial $\alpha$-Amylase of *Bacillus licheniformis* (TBA)

TBA was isolated, purified, and characterized by CHIANG et al. (1979). The characteristics of TBA in this work were determined with soluble lintner starch as substrate. The optimum pH for the purified enzyme is between 6 and 7 as shown by CHIANG et al. (1979), the optimum temperature is 85 °C. It was further shown, that upon hydrolyzing corn starch with TBA, mainly maltotriose, maltopentaose, and maltohexaose were formed. Without substrate and without added calcium ions stability of TBA decreases rapidly at temperatures above 65 °C. Consequently TBA may be added during mashing processes only when the substrate is present. Using a technical enzyme preparation of $\alpha$-amylase from *B. licheniformis,* ROSENDAL et al. (1979) also showed that the optimum pH for the hydrolysis of soluble starch by TBA lies between 6 and 7. The enzyme used in this work was absolutely stable at 90 °C. An investigation of the action of technical amylolytic enzymes using corn mash as substrate was described by SENN (1988). An optimum pH range from 6.2 to 7.5 was found, with pH values below 5.6 leading to a rapid decrease in enzyme activity. The optimum temperature for TBA in this work was 80–85 °C. Furthermore, it could be shown that enzyme activity also depends on the proportion of horny to floury endosperm of the processed corn. The higher the proportion of horny endosperm, the lower the enzyme activity determined in such mashes. This shows that it is more difficult to digest starch from horny than from floury endosperm.

Liquefaction of corn mashes using TBA yields mainly starch fragments with a degree of polymerization of more than 10 glucose units as well as maltose and glucose. But the content of glucose and maltose does not rise to more than 5 g $L^{-1}$ mash for each component after 30 min of liquefaction. During 4 h of liquefaction there is no further progress in degradation. If fermentable sugars are metabolized by fermentation, the starch fractions DP 4 to DP 7 rise, but fermentable sugars can not be determined (SENN, 1992).

### 2.1.2 Bacterial α-Amylase of *Bacillus subtilis* (BAA)

Determined in soluble starch as substrate BAA shows an optimum pH value between 5.3 and 6.4, and an optimum temperature of 50°C (ROBYT, 1984).

FOGARTY and KELLY (1979) reported that with starch as substrate BAA produces doubly branched limit dextrins. Furthermore, two highly branched dextrins containing 9 and 10 glucose units were isolated, and both were shown to be mixtures of 4 triply branched dextrins. These low molecular branched limit dextrins are very difficult to hydrolyze with glucoamylase from *Aspergillus niger.* That is why starch degradation often remains incomplete if BAA is used for liquefaction.

Using corn mash as substrate the optimum conditions for BAA are a pH between 5.8 and 6.8 and a temperature of 55–60°C (SENN, 1988; SENN and PIEPER, 1991). Under these conditions BAA is stable up to 65°C if the pH is adjusted to between 6.2 and 6.4.

### 2.1.3 Bacterial α-Amylase Expressed by *Bacillus licheniformis* (BAB)

BAB, a new technical enzyme produced with a genetically engineered strain of *B. licheniformis* (Liquozym, NOVO Nordisk, Denmark) has been available for the past years (KLISCH, 1991). BAB is characterized by its tolerance to low pH values down to 4.8–4.5. But it is only possible to liquefy cereal mashes using BAB. Liquefaction of cereal mashes is very effective; in mashes from potatoes it works insufficiently. This enzyme is thermostable up to 90°C. Due to its pH tolerance, BAB is the optimum liquefaction enzyme in processes with included recycling of stillage.

### 2.1.4 Fungal α-Amylase of *Aspergillus oryzae* (FAA)

As reported by FOGARTY and KELLY (1979), FAA contains only a few amino acid residues. Therefore, FAA is relatively stable in the acid pH range. The optimum conditions for this enzyme have been reported to be a pH value between 5.5 and 5.9 and a temperature of 40°C. Depending on the stability of FAA the pH value can range from 5.5 to 8.5 (FOGARTY and KELLY, 1979). Furthermore, as reported by TAKAYA et al. (1978), FAA is able to attack native starch granules. At a pH of 7.2 and a temperature of 37°C after 60 h more than 40% of starch weighed in was dextrinized.

Using corn mash for determination of enzyme properties and under technical conditions, the optimum pH ranges from 5.0 to 6.0. At a pH of 4.5, FAA displays 50% of its activity measured under optimum conditions (SENN, 1988). The optimum temperature is reported to lie within the range of 50–57°C.

The use of FAA promotes a quite effective further decrease in viscosity at the saccharification temperature combined with a more effective dextrinization of starch. This supports a total degradation of starch. The pH tolerance of FAA guarantees that the enzyme, for a certain time, is active during the fermentation until the pH falls below 4.5.

## 2.2 Enzymatic Starch Liquefaction and Saccharification

Malt is the classical source of amylolytical enzymes used in alcohol production technology. It contains both liquefying and saccharifying enzymes. The amylolytical components of malt are

- α-amylase (Sect. 2.1),
- β-amylase (α-1,4-glucan maltohydrolase, EC 3.2.1.2),
- limit dextrinase,
- *R*-enzyme.

As reported by SARGEANT and WALKER (1977), α-amylase from malt can hydrolyze native starch granules.

$\beta$-Amylase is an exo-acting enzyme and hydrolyzes starch yielding maltose. Starch molecules are attacked from the non-reducing end of the glucose chains. $\beta$-Amylase hydrolyzes only $\alpha$-1,4 linkages and is unable to bypass $\alpha$-1,6 glycosidic linkages in amylopectin. Degradation of branched amylopectin, therefore, is incomplete. Action of $\beta$-amylase on amylopectin results in a 50–60% conversion to maltose and the formation of $\beta$-limit dextrin containing all $\alpha$-1,6 linkages. The optimum conditions for $\beta$-amylase are a temperature of 50°C and a pH of 5.0. $\beta$-Amylase is stable within a pH range of 4.0–6.0.

Limit dextrinase from malt has an optimum pH of 5.1 and an optimum temperature of 40°C. This enzyme is unable to cleave $\alpha$-1,6 linkages in substrates that do not contain a sufficient number of $\alpha$-1,4 linkages. Limit dextrinase is also unable to dextrinize amylopectin or $\beta$-limit dextrin; it mainly debranches and dextrinizes $\alpha$-limit dextrins. It was shown by HARRIS (1962) that there is a higly significant correlation between ethanol yield using pressureless processes and the limit dextrinase content of the malt used.

Another debranching enzyme from malt is the *R*-enzyme. This enzyme cleaves $\alpha$-1,6 linkages in amylopectin and $\beta$-limit dextrin; $\alpha$-limit dextrins are not attacked by *R*-enzyme. Debranching of amylopectin and $\beta$-limit dextrin is, however, incomplete, since the *R*-enzyme needs 5 or more glucose units between two $\alpha$-1,6 linkages in order to cleave them. The optimum conditions for *R*-enzyme are 40°C and a pH of 5.3 (HARRIS, 1962).

All of these amylolytical enzymes from malt work together and act very fast. After only 15 min of action on the substrate a maltose–dextrin equilibrium is reached with about 66% maltose, 4% glucose, 10% maltotriose, and 20% limit dextrins. Saccharification is not completed during the mashing process using exclusively malt. Saccharification is only completed if maltose is metabolized by yeast fermentation. Hence, there are two steps in saccharification: a first step of the main saccharification reaching an equilibrium and a second step of residual saccharification during the fermentation ("secondary fermentation").

### 2.2.1 Green Malt

The use of green malt, manufactured in distilleries, for liquefaction and saccharification in classical distillation technology has a long tradition. It was often used in potato distilleries up to 1970. Nowadays, since technical enzyme preparations are available commercially, the use of green malt is too expensive. However, it is of great interest for Eastern European countries with limited foreign currency reserves where technical enzymes are not available, and barley is grown without any problems.

The malting process in distilleries comprises two main stages: the steeping of barley and the germination process. Both of these production steps can be carried out in one apparatus by using a Galland malting drum (NARZISS, 1976; SCHUSTER, 1962; LEWIS and YOUNG, 1995). This pneumatic malting system, which can take a batch size of up to 15 t of barley, is the optimum related to the needs of distilleries.

In order to use green malt in distilleries, it is necessary to grind it thoroughly. A special apparatus (KREIPE, 1981) is in use to produce the malt slurry with a green malt to water ratio of 1:3. This apparatus consists of a vessel which is filled up with the required amount of water. A centrifugal pump, fitted as a centrifugal mill, recirculates the water and the malt slurry. Green malt is then added to the water, avoiding clots. Methanal (formaldehyde) may also be added for disinfection. The malt slurry is thoroughly ground after 30–40 min.

### 2.2.2 Kiln-Dried Malt

To produce a distillers' kilned malt, it is important to use low temperatures in kilning to save enzyme activity. This reduces the moisture content of green malt initially to 10–12% by passing a large excess of air for 12 h at a temperature of 40–50°C through the grain. Subsequently the moisture content is further reduced to 4–5% by raising the air temperature to 55–60°C.

Kilned malt must be thoroughly ground before use in a mashing process. The milled malt is then mixed with water in a ratio of 1:3

**Tab. 13.** Evaluation of Dried Distillers' Malt (DREWS and PIEPER, 1965)

| Activity of $\alpha$-Amylase SKB-Units per g Malt DS | Evaluation |
|---|---|
| >64 | excellent |
| 53 – 64 | good |
| 41 – 52 | sufficient |
| <41 | insufficient |

and at 50°C to bring enzymes into solution. Methanal (formaldehyde) may be added for disinfection.

The $\alpha$-amylase activity of malt is determined by the SKB method, according to SANDSTED, KNEEN, and BLISH (PIEPER, 1970). DREWS and PIEPER (1965) recommend the evaluation presented in Tab. 13.

## 2.2.2.1 Barley as a Malting Grain

Barley is the most widely used grain in malting. Barley used for making distillers' malt is of smaller size and higher nitrogen content than barley used for brewer's malt. Before malting, barley must be cleaned and free from weed seeds and broken grains. For steeping, barley usually is treated in water twice for 2–4 h, and exposed to air for 20–24 h after each steeping. The temperature is adjusted to 10–12°C, and must not exceed 15°C. The germination period lasts for 6–8 d and the grain is turned twice daily (SCHUSTER, 1962). By adding gibberellic acid to the barley when the germination period is started, the germination time can be reduced to 4–6 d while reaching the same or higher enzyme activities (PIEPER, 1968).

## 2.2.2.2 Other Grains in Malting

Wheat can also be very effectively used as a malting grain (PIEPER, 1984). The batches of kiln dried wheat malt examined in this work exhibited $\alpha$-amylase activities of between 117 and 165 SKB-units per g of malt DS, which is three times the activity of a good kilned barley malt. The ethanol yield of wheat malt, which is more than 38 lA per 100 kg malt, is impossible to reach with barley malt. PIEPER (1984) further found that the use of wheat dried malt for saccharification of wheat mashes yields 67 lA per 100 kg of starch. The wheat-to-malt ratio was 10:1. The excellent ethanol yield was reached by keeping a saccharification rest of 30 min at 55°C and a pH of 5.5. There are, however, some difficulties in malting wheat, especially during the germination stage. Germination is manifested by the growth of roots and the shoot. The growing roots have a tendency to break easily, leading to losses in enzyme activity and increasing the risk of infections during germination. This is probably the reason why wheat is generally not used in manufacturing distillers' malt.

Triticale is another grain which can very effectively be used in malting for distillers use (THOMAS, 1991). To reach optimum enzyme activities, triticale was steeped at 15°C for 48 h. Therefore, a pneumatic malting system was used, and triticale was steeped in water twice for 4 h. For the rest of the time triticale was aerated and sprinkled with water at 15°C. At the end of this steeping procedure, the tip of the root was just breaking through the husk, and the water content was 38–42%. Triticale was then transferred to the germination chamber and allowed to germinate at 15°C. During germination triticale was aerated with humidified air. Kilning was carried out at temperatures between 40°C and 50°C.

After malting under these conditions, $\alpha$-amylase activities in triticale malts reached 170 SKB-units per g of malt DS after only 4 d of germination. The optimum conditions for the use of triticale malt in saccharification are 55°C and a pH of 5.2–5.5. By saccharifying corn mashes using kilned triticale malt (with a corn–malt ratio of 10:1), ethanol yields from about 67 lA per 100 kg starch could be reached. These yields of ethanol from corn cannot be obtained with using barley malt. Hence, triticale is a very important grain in malting for distillers' use.

## 2.3 Enzymatic Starch Saccharification

Glucoamylase (EC 3.2.1.3) is an exo-acting enzyme, hydrolyzing α-1,4 α-1,6, and α-1,3 glycosidic linkages in amylose and amylopectin. The rates of hydrolysis depend on the molecular size and structure of the substrates (FOGARTY and KELLY, 1979). Thus glucoamylase from *Aspergillus niger*, e.g., hydrolyzes isomaltose at a lesser rate than maltose: These authors show, that branched substrates are hardly degraded by glucoamylases derived from several fungi. This may be a problem in alcohol production technology, if an α-amylase is used for liquefaction that yields double or triple branched α-limit dextrins.

### 2.3.1 Glucoamylase of *Aspergillus niger* (GAA)

Two structurally different glucoamylases from *Aspergillus niger*, glucoamylase 1 and glucoamylase 11, have been characterized (FOGARTY and KELLY, 1979). The enzymes differ mainly in amino acid composition. Both enzymes, examined with soluble starch as substrate, were found to have a pH optimum of 4.5–5.0 and an optimum temperature of 60°C; the isoelectric point is given for GAA1 as 3.4, and for GAA 11 as 4.0.

Using corn mash as substrate, the optimum range of pH value reaches from 5.0 down to 3.4 (SENN and PIEPER, 1991). Thus, GAA is stable during fermentation. With respect to temperature, GAA in this work was found to be stable up to 70°C with an optimum at 65°C.

### 2.3.2 Glucoamylase of *Rhizopus* sp. (GAR)

The optimum conditions for GAR are a temperature of 40°C and a pH value ranging from 4.5 to 6.3 (FOGARTY and KELLY, 1979). The manufacturers of technical GAR products report the optimum conditions as 40–60°C and a pH range from 4.0 to 5.5.

Two kinds of glucoamylases were isolated from a *Rhizopus* sp. Glucoamylase 1 shows strong debranching activity and is able to degrade raw starch, while glucoamylase 11 generally shows low activities in both cases. This special debranching activity of glucoamylase 1 from a *Rhizopus* sp. is very useful in achieving an almost total conversion of starch to fermentable sugars in pressureless processes.

Using corn mash as substrate, the optimum conditions for GAR were found to be 55–60°C and a pH of 4.4–5.4 (SENN and PIEPER, 1991); the enzyme was stable at a pH as low as 3.8. To save supplementary hemicellulolytic and proteolytic activities in technical GAR preparations, the temperature of mashes should not be higher than 52°C when the enzyme is added.

### 2.3.3 Enzyme Combinations

In practice, single enzymes are rarely used for saccharification of mashes. Due to the different characteristics of the various enzymes, it is important to know which enzymes may be combined successfully in mashing processes and fermentation. As reported by SENN (1992), different combinations of technical enzymes may exhibit either complementary or inhibitory effects. To examine these effects, starch degradation in corn mashes was followed using several technical enzyme combinations. During mashing and fermentation processes the content of saccharides and oligosaccharides up to DP10 (degree of polymerization) in mashes was measured by HPLC. The mashing processes in these examinations where carried out with saccharification rests of about 30 min, 360 min, and without any saccharification rest.

With a saccharification rest of 6 h, the combination of GAA and FAA, often used in practice, leads to a rapid degradation of the fraction with high molecular weight, with a rapid increase in glucose and maltose concentrations. However, further degradation of the fraction DP > 10 is quite slow and remains incomplete. After 24 h of fermentation, the amounts of fermentable sugars are very low, resulting in a slow and incomplete saccharification and fermentation (Fig. 2).

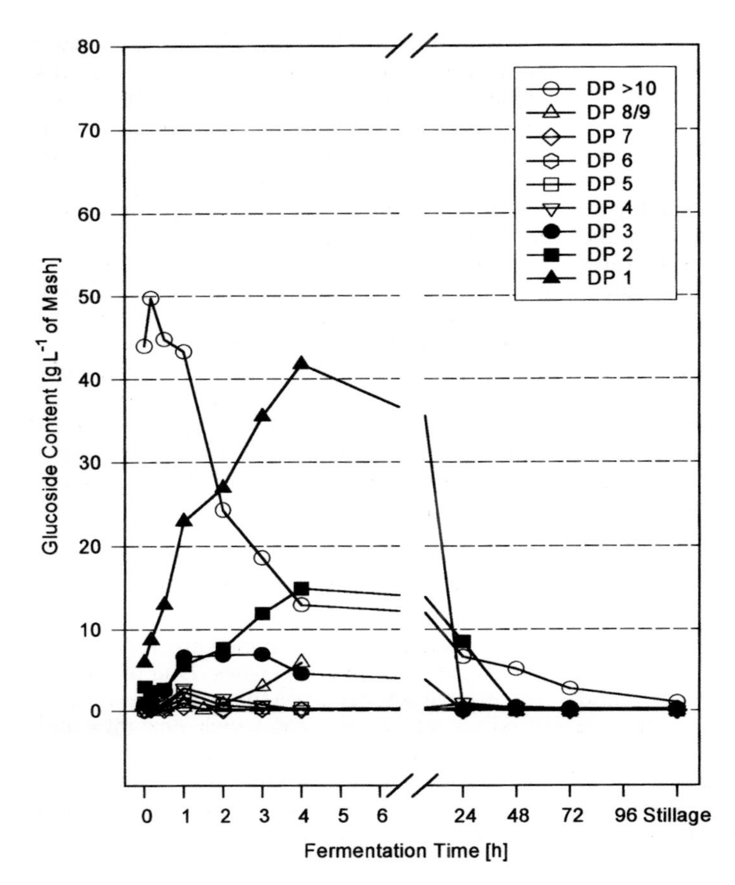


**Fig. 2.** Saccharification of corn mash using a combination of GAA and FAA. Yeast added after 6 h of saccharification; distillation after 72 h of fermentation.

If the saccharification rest lasts for only 30 min, the maltose and DP>10 fractions increase again during the second day of fermentation, leading to a sluggish fermentation, too.

The combined saccharification with GAR and FAA gives a significantly better degradation of the DP>10 fraction than GAR alone (Fig. 3). The additional use of GAR together with GAA and FAA shows an inhibition, because degradation of starch is significantly slower than with the supply of GAA and FAA (Fig. 4.5).

"OPTIMALT®" (Solvay Enzymes, Nienburg) is an enzyme combination containing GAR, kiln dried distillers barley malt (DBM), GAA, and FAA, developed at the Versuchs- und Lehrbrennerei, Hohenheim University. When it is used in saccharification, the concentration of fermentable sugars in mashes rises rapidly; this is never achieved with other enzyme combinations (Fig. 6).

Even without a saccharification rest, the mashes contain sufficient amounts of fermentable sugars during the whole fermentation. Although the amount of DBM in this combination is only 3 kg $t^{-1}$ starch, it has the same effect as when DBM is used alone; after a saccharification period of only 4 h the DP 1–DP 3 fractions are present in significant amounts. Hence, OPTIMALT® ensures a fast and almost complete degradation of starch during saccharification and fermentation, even without any saccharification rest (Fig. 7).

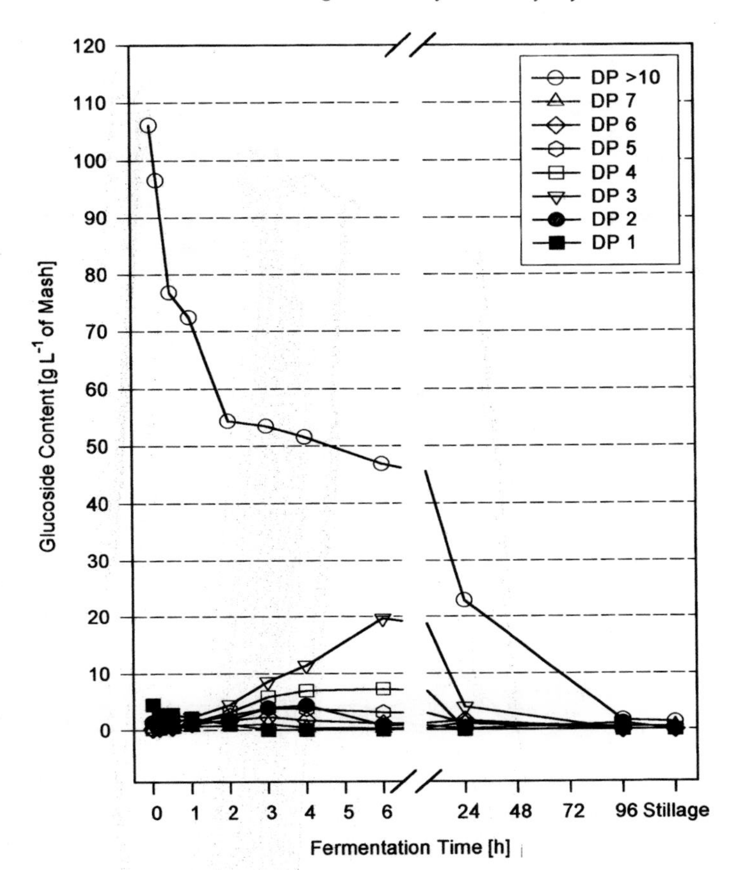


**Fig. 3.** Saccharification of corn mash using a combination of GAR and FAA. Yeast added when saccharification is started; distillation after 96 h of fermentation.

# 3 Starch Degradation by Autoamylolysis

It has long been shown that some native cereal grains (wheat, rye) contain autoamylolytic activities. These enzyme activities were often used in the traditional pressureless process called "cold mash process (Kaltmaischverfahren)" prior to 1940 (Sect. 4.3.1.1). Nevertheless, it was impossible to develop a reliable technical process using these autoamyloytical activities due to the lack of reliable quantitative methods for the determination and examination of autoamylolytic acitivities in different charges of raw materials. Such methods (Sect. 13.2.3) were developed in 1989 at Hohenheim University, reliable technical processes using autoamylolytical activities have also been developed (Sect. 8).

To examine the autoamylolytical activity of raw materials, RAU et al. (1993) defined the so-called Autoamylolytical Quotient (AAQ). This AAQ is determined by carrying out two separate fermentation tests with the same raw material. The first fermentation test runs using technical enzymes to determine the maximum ethanol yields obtainable with the raw material used. The second fermentation test is carried out without the addition of technical enzymes or malt to determine the ethanol yield obtained under autoamylolytic conditions. AAQ then is, related to the raw material used, defined following Eq. (1):

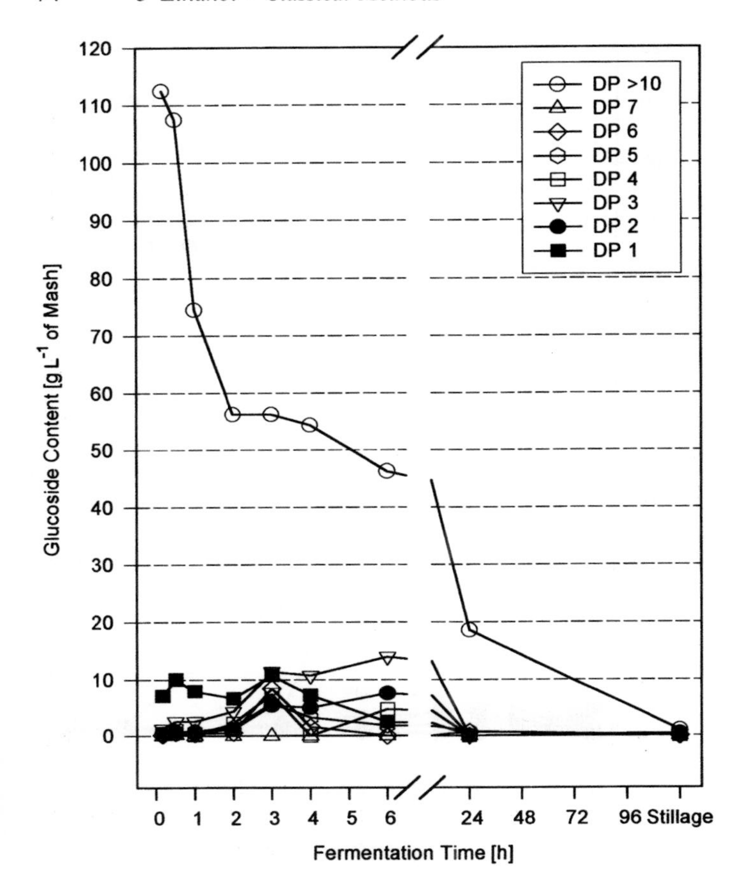


**Fig. 4.** Saccharification of corn mash using a combination of GAR, GAA and FAA. Yeast added after 30 min of saccharification; distillation after 72 h of fermentation.

$$\text{AAQ} = \frac{\text{Ethanol yield [lA/100 kg FS] without technical enzymes} \cdot 100}{\text{Ethanol yield [lA/100 kg FS] using technical enzymes}} \tag{1}$$

The mashes used in the fermentation tests with additional technical amylases were liquefied at 65 °C using thermostable $\alpha$-amylase from *Bacillus licheniformis*. This is not feasible when mashing under autoamylolytic conditions, since the autoamylolytical enzyme system does not persist at 65 °C (RAU, 1989). Total gelatinization of starch in these mashes is required to reach a complete degradation of starch to fermentable sugars, and gelatinization requires a temperature of about 65 °C, e.g., in wheat mashes. This problem can be solved by using a certain time and temperature program in mashing processes, which depends on the raw material. The enzymes in native wheat kernels which are dried for storage are $\alpha$-amylases, $\beta$-amylase, and limit dextrinase (MARCHYLO et al., 1984; LABERGE and MARCHYLO, 1983; MANNERS and SPERRA, 1966).

## 3.1 Wheat

The optimum conditions for the wheat autoamylolytic enzyme system are 55 °C and a pH of 5.3–5.5 (RAU, 1989). Due to the gelatinization temperature of wheat starch, wheat mashes must be heated to 64 °C. To protect

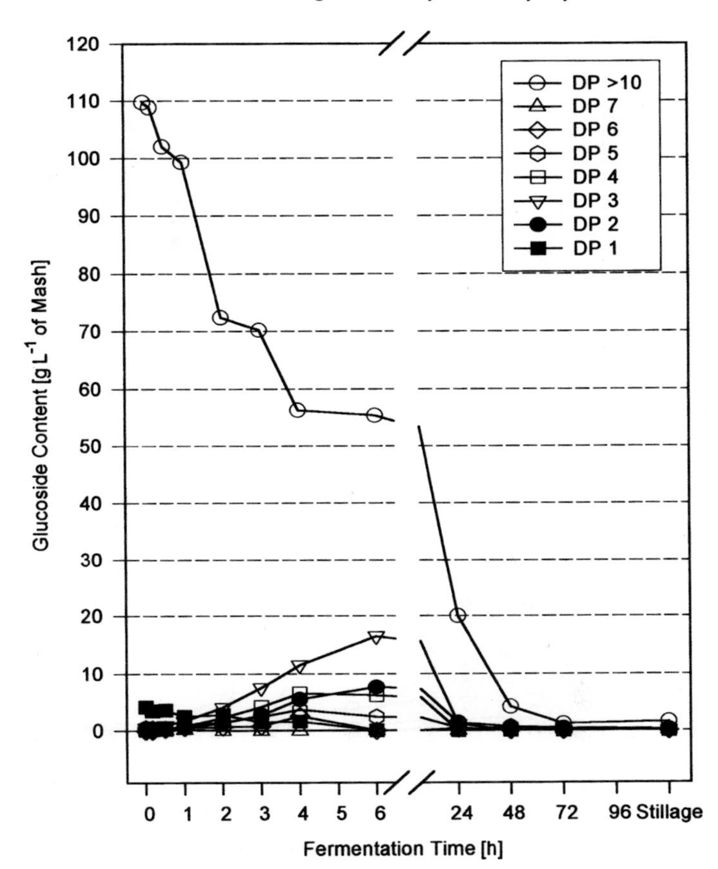


**Fig. 5.** Saccharification of corn mash using a combination of GAR, GAA and FAA. Yeast added when saccharification is started; distillation after 72 h of fermentation.

the enzyme system during the mashing process this temperature may be kept for only 10 min. Then the mash must be cooled down to 55 °C for a saccharification rest of 30 min at a pH of 5.3–5.4. The best variety of wheat for use in autoamylolytic processes is "Alamo", with an AAQ > 95, independent of climate and growth conditions in different years of harvest.

## 3.2 Rye

Almost all varieties of rye show an AAQ > 95 (AUFHAMMER et al., 1993). The optimum conditions of the rye autoamylolytic enzyme system are the same as for wheat. But due to the content of pentosans, rye mashes often become very viscous. To avoid problems in mash treatment and fermentation, the viscosity of rye mashes should be reduced. This can be done either by using pentosanases, which are costly, or by using a certain time and temperature program (Sect. 6.2).

## 3.3 Triticale

The autoamylolytic enzyme system of triticale is maximally active at 55–60 °C and at a pH of 5.0–5.8 (THOMAS et al., 1991; SENN et al., 1993). To reach a sufficient gelatinization of starch from triticale the triticale mashes must be heated to only 60 °C for 60 min after the starch has been completely released. The autoamylolytic system of triticale is stable at this temperature. A maximum ethanol yield is

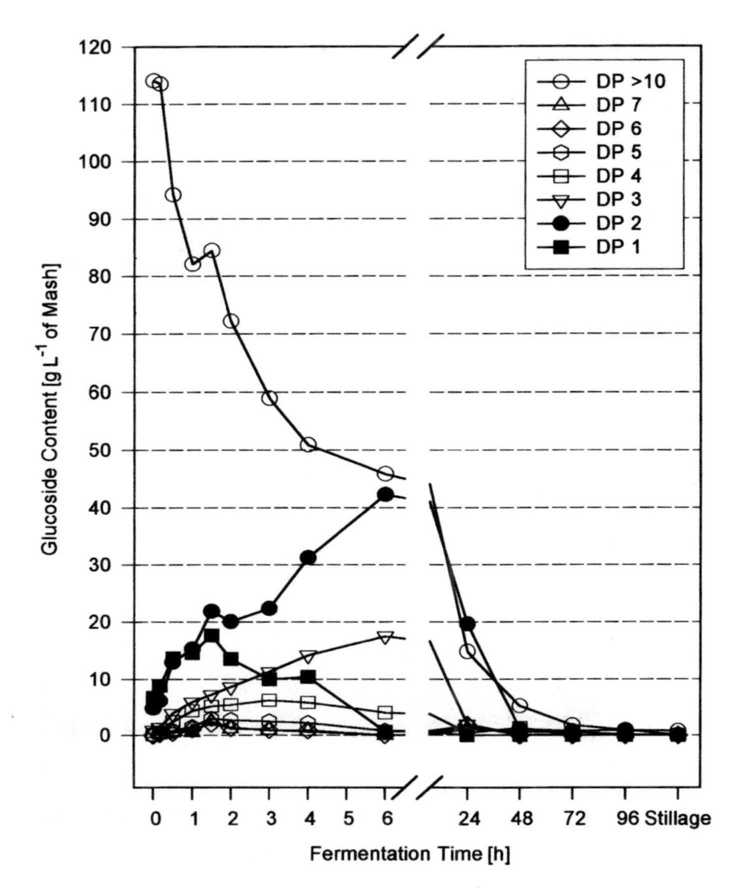


**Fig. 6.** Saccharification of corn mash using the enzyme combination OPTIMALT®. Yeast added after 30 min of saccharification; distillation after 96 h of fermentation.

achieved at a mash temperature of 60 °C by adjusting the pH to 5.8 for 30 min and then lowering it to 5.2 for 30 min.

Starch degradation by autoamylolysis is very different from starch degradation by the action of technical enzyme preparations (SENN, 1995). In this work, degradation of starch in corn mashes was compared with autoamylolysis of starch in mashes from triticale (Fig. 8). After liquefaction, 1 L of corn mash contained about 110 g oligosaccharides in the DP > 10 fraction and about 5 g of directly fermentable sugars (maltose and glucose). In contrast, triticale mashes (autoamylolytically processed) had a DP > 10 content of 30 g $L^{-1}$ mash and a directly fermentable sugar content of about 80 g $L^{-1}$ mash. This fraction of directly fermentable sugars in corn mashes reached a maximum of about 40 g $L^{-1}$ mash when saccharification was carried out with the enzyme combination OPTIMALT®.

The additional use of technical saccharification enzymes does not affect the process of starch degradation if mashes are processed under autoamylolytic pH and temperature conditions (Fig. 9). Autoamylolytic starch degradation leads only to small increases of the DP3 to DP7 fractions at the end of fermentation. But the addition of the $\alpha$-amylase from *B. licheniformis* to the autoamylolytic process changes the situation and starch degradation is complete by the end of fermentation. This clearly shows the effectiveness of the autoamylolytic enzyme system. Further studies on the autoamylolytic properties of triticale which depend on growth conditions have been reported by AUFHAMMER et al. (1993, 1994).

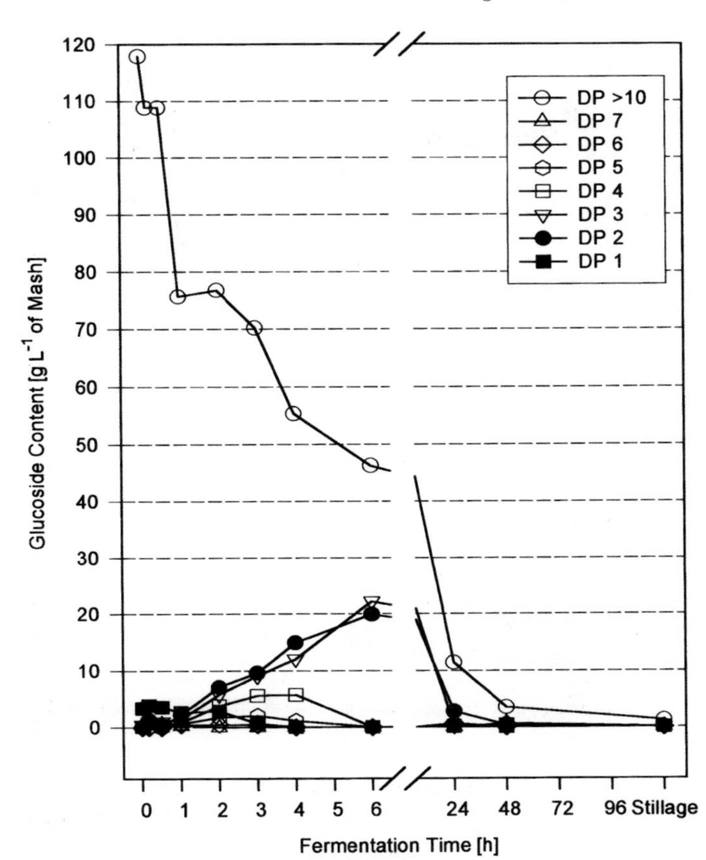


**Fig. 7.** Saccharification of corn mash using the enzyme combination OPTIMALT®. Yeast added when saccharification is started; distillation after 72 h of fermentation.

# 4 Mashing Processes

## 4.1 Mashing Equipment

### 4.1.1 Wet Cleaning of Potatoes

Before processing potatoes, they must be cleaned and free from sand, stones, soil, and potato foliage. For this purpose washing rolls with a minimum length of about 3 m and a minimum diameter of 1.5 m are used. The potatoes pass through the turning wash roll with countercurrent flow of warm water. After this, the potatoes pass through a stone catcher and are then delivered to an elevator which also has a countercurrent water flow. The elevator delivers the potatoes to the storage tank where they are weighed. Cleaning of the potatoes starts by washing them out from the potato storage cellar. From this washing channel, potatoes are pumped with a special centrifugal potato pump to the washing roll, which often requires the use of elevators. Water consumption in this washing process reaches 3–5 times the volume of the potatoes. Normally the water used in cooling down the mashes is used again for washing the potatoes, thereby minimizing the consumption of fresh water.

### 4.1.2 Grinding Raw Materials

One of the fundamentals of pressureless mashing processes is the thorough grinding of raw materials, which is usually done with

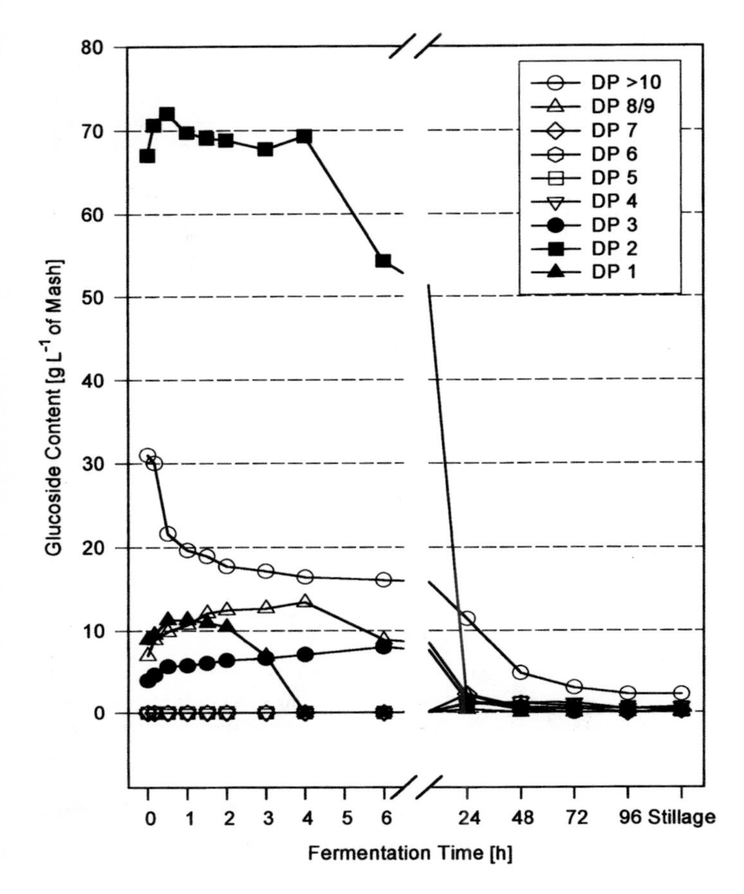


**Fig. 8.** Saccharification of triticale mash under autoamylolytical conditions. Yeast added after 30 min of saccharification; distillation after 96 h of fermentation.

hammer mills or dispersing machines and leads to a better digestion of starch.

### 4.1.2.1 Mills

For milling the raw materials usually only hammer mills are used in practice. These mills can be used under dry or wet conditions. When milling cereals under dry conditions, it is necessary to fit the mills wih a dust collector to avoid the settling of dust throughout the distillery. An advantage of dry processing is that milling can be done overnight, storing the meal in a hopper. To reach a sufficient degree of disintegration in the hammer mill a 1–1.5 mm screen is needed. Wet milling increases the throughput of raw materials but decreases the degree of disintegration. Therefore, water is added together with the material to be ground into the milling chamber.

An alternative to these two possibilities is to mill under dry conditions, simultaneously adding water to rinse out the meal, and then to pump it directly into the mash tub, using an eccentric screw pump situated below the mill. The rinsing water is delivered only to the meal chamber of the mill. Only the raw material is delivered to the milling chamber and, therefore, milling is carried out under dry conditions, while the formation of dust is completely avoided.

In practice, hammer mills are often fitted with 1.5 mm slot screens. These slot screens are more abrasion-resistant than perforated screens but are less efficient in milling. However, these slot screens are a good alternative when milling potatoes, since milling potatoes

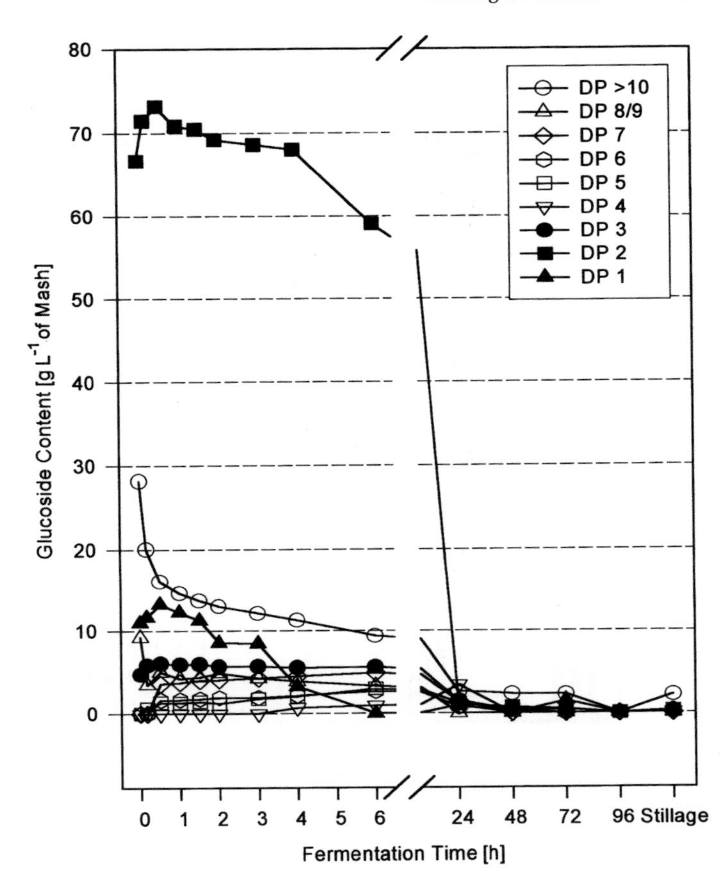


**Fig. 9.** Saccharification of triticale mash using the enzyme combination OPTIMALT®. Yeast added after 30 min of saccharification; distillation after 96 h of fermentation.

results in more abrasive wear than milling cereals due to the load of soil.

### 4.1.2.2 Dispersing Machines

The main purpose of disintegrating raw materials for alcohol production is to release starch from cell material. Therefore, ideally each single cell should be broken up, which is impossible to achieve using a mill. In ethanol production technology two kinds of rotor-stator dispersing machines are in use: in-line machines working in the continuously running Supramyl process (MISSELHORN, 1980a) and batch machines from the ULTRA-TURRAX type used in the dispersing mash process (Fig. 10). The use of in-line machines ensures that all of the mash passes the dispersing head, but these machines are damaged by stones and sand delivered together with mash. If, in contrast, a mash tub is fitted with an ULTRA-TURRAX for batch processing, this makes the passing of mash through the dispersion head a statistical problem. Both systems lead to the same and very effective disintegration of raw materials; however, to drive one in-line machine, an electric motor power of 55 kW is needed in the Supramyl process, while working batchwise an electric motor power of only 17.5 kW is required for one of these machines. Thus, the investment needed for a batch machine is a third of that needed for an in-line machine.

This good effect of disintegration of raw material is possible, because the action of dispersing machines is not comparable with milling. There are four effects of disintegration

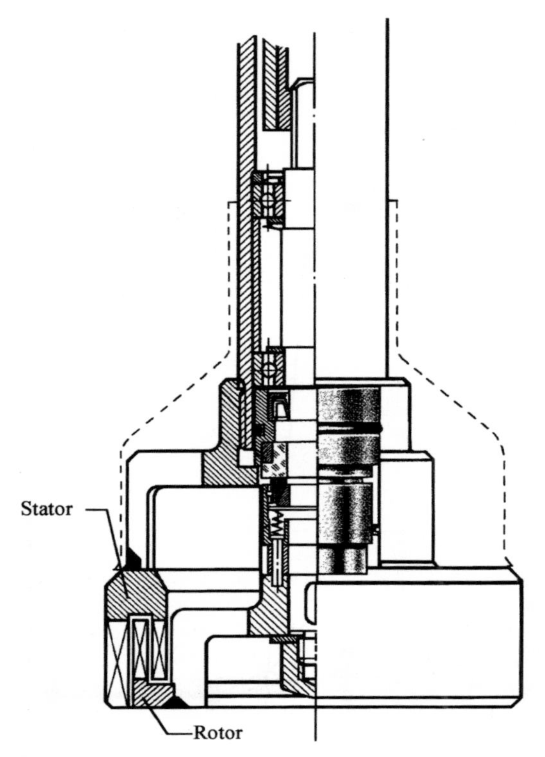


**Fig. 10.** Rotor-Stator dispersing machine (ULTRA-TURRAX type, IKA Maschinenbau, D-79219 Staufen).

that take place simultaneously and are overlapping:

- the mechanical shearing effect between the toothed rotors and stators,
- compression and decompression power, which leads to an intensive loosening of the structure of cell material,
- dispersion effect by splitting the mash stream into many single streams while passing the dispersion head,
- dispersion effect by microcavitation in very small regions with a high energy density.

By using an ULTRA-TURRAX in batch processing it is possible to process cereals and potatoes without previous grinding or steeping. And by generating high-frequency oscillations in the mash tub, a virtually complete release of starch is achieved and encrustations do not form in the mash tub.

### 4.1.3 Mash Tubs

A modern mash tub should comply with the following criteria:

- made from stainless steel,
- fitted with an effective and energy saving agitator,
- fitted with a sufficient cooling surface,
- easy and reliable cleaning and disinfection.

A mash tub should be cylindrical with a slightly domed bottom and top in order to achieve a good effect of agitation and to permit total discharging.

Most of the common mash tubs (Fig. 11) are fitted with impeller or propeller agitators. PIEPER et al. (1990) showed that it is better to use a pitched-blade turbine for this purpose. The use of pitched-blade turbines saves about 50% of energy input for agitation of mashes compared to an impeller agitator. Furthermore these turbines guarantee an even distribution of whole cereal grains in the mash, starting a dispersing mash process, that allows, e.g., processing of cereals without previous grinding and steeping.

Agitators are commonly equipped with a drive from either the top or from the bottom. Selecting a drive from the top requires a drive shaft with a length corresponding to the height of the mash tub, and selecting a drive from the bottom in practice leads to problems with the seal.

For cooling mashes, the most common mash tubs are fitted with a cooling coil that is usually made of copper or stainless steel. To have a sufficient cooling effect with cooling water at 10 °C, it is necessary to have a cooling coil of about 4 $m^2\ m^{-3}$ mash.

Cooling is not only affected by the cooling surface and water temperature, but also by the design of the cooling coil, which must permit a good circulation of mash around the whole coil in the mash tub. Therefore, a sufficient distance between the wall of mash tub and the cooling coil and between the single

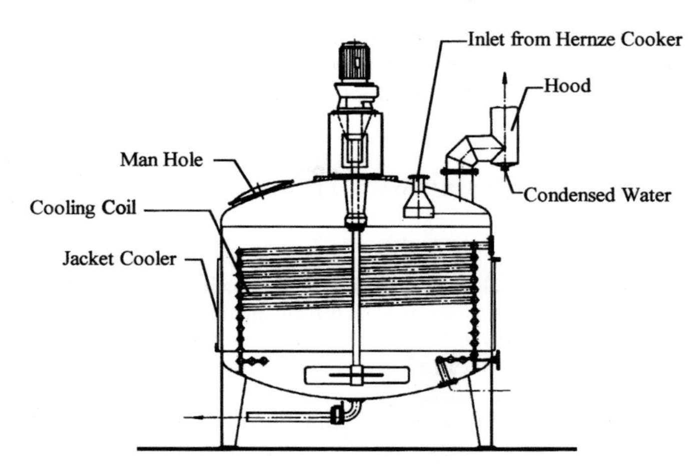


**Fig. 11.** Mash tub (KREIPE, 1981).

turns of the coil must be maintained. When designing a cooling coil, it must be ensured that there are no dirt-holding spaces around the spacers and fittings.

### 4.1.4 Heat Exchangers

In order to save energy, time, and cooling water some distilleries have started to use heat exchangers separated from the mash tub for heating and cooling mashes. Four types of heat exchangers are in use: spiral-plate heat exchangers, tubular heat exchangers, plate-type heat exchangers, and spiral-tubular heat exchangers.

In everyday practice spiral-plate heat exchangers have a strong tendency to form encrustations, both during the heating up of mashes before liquefaction and during the cooling down of mashes. This seems to be due to the low flow rates of mash between the spiral plates, which leads to sedimentation of mash solids and further lumping. Thus, the cross-sectional area gets smaller, and it is necessary to clean these spiral-plate heat exchangers regularly with water, using high flow rates to rinse out mash solid agglomerations, and afterwards with a hot sodium hydroxide solution (2% NaOH in water) to lower the risk of contamination which is ever present in distilleries.

The use of tubular heat exchangers is more successful. The tendency to form encrustations is quite low and they are quite easy to clean since cleaning balls can be pumped through the pipes. But in heating up cereal mashes there are problems too, due to the many 180° elbow fittings, which are design features of tubular heat exchangers. When cereal mashes, especially corn mashes, are heated up, gritty particles of the mash often show a tendency to sediment in the regions of the elbow fittings. Due to the degree of disintegration of the raw material pumped through the pipes, this sometimes results in clogging up the heat exchanger.

In the last few years, spiral-tubular heat exchangers have been installed in distilleries. Due to the design of these heat exchangers, in which spiral bent coils are welded to one another without elbow fittings, have the same advantages as tubular heat exchangers, but there is no problem treating cereal mashes. They are effective heat exchangers and they are easy to clean.

It is also possible to use plate-type heat exchangers in distilleries. But for heating and cooling cereal mashes these must be non-clogging plate-type heat exchangers. These are fitted with corrugated plates generating corrugated rectangular ducts. The single plates are kept apart only by the seals. Over the entire length of the plates there is the same slit width between the plates. This slit width sould be 10–12 mm to permit the use of these non-clogging plate-type heat exchangers in distilleries. It is very important that the mash flow rate is higher than 0.3 m $s^{-1}$ to avoid sedimentation and agglomeration of mash solids inside the heat exchanger.

#### 4.1.4.1 Processing with Heat Exchangers

Using heat exchangers in distilleries to heat up mashes, a rise in mash temperature to more than 55–60 °C inside the heat exchangers should be avoided. If gelatinization of starch takes place inside heat exchangers, the formation of encrustations can hardly be avoided, even if liquefaction enzymes are added.

If heat exchangers are installed in distilleries, they can be used in two different ways. After reaching a sufficient disintegration and liquefaction the mash can be pumped directly into the fermentation tanks, passing the heat exchanger, where it is cooled down to set temperature. In this case, saccharifying enzymes and yeast mash are added to the mash in the fermentation tanks.

Using the second method of processing, mash is first pumped through the heat exchanger and then back to the mash tub. During this circulation the mash is cooled down until the saccharification temperature is reached in the mash tub. If at this temperature saccharification enzymes are added, the mash is pumped to the heat exchanger and fermentation tank, where it is cooled down to set temperature. Yeast mash is also added after the mash has cooled.

### 4.1.5 Henze Cooker

Up to 30% of the distilleries in Germany are still working with a Henze cooker. The design of these pressure vessels is described in detail by KREIPE (1981) (Fig. 12).

The Henze cooker is the most popular pressure vessel used in European distilleries. It is manufactured from a cylindrical and a conical part with certain dimensions to achieve a virtually even distribution of vapor. The optimal dimensions are:

cylindrical height : conical height : cylindrical diameter = 1 : 2 : 1.3

If the dimensions of Henze cookers are of these proportions, it is not necessary to install a stirrer if no milled raw materials are processed. Today, these pressure vessels should be made from stainless steel to avoid corrosion.

The Henze cooker has to be fitted with a steam valve to inject live steam, and a blowdown outlet valve at the bottom of the cooker. About two steam inlet valves in the conical region and an additional deaeration valve at the top of the cooker are required. In addition, the cooker should be fitted with a manometer and a safety valve. The blowdown tube, leading mash from the cooker to the mash tub, should be sized to guarantee that the cooker is blown down within 20 min; otherwise intense caramelization takes place.

## 4.2 Pressure Boiling Processes

The term "Pressure Boiling Process" is applied to all processes in which the release of starch from raw materials and its gelatinization with water take place at temperatures above 100 °C. This process was developed before 1900 and is still in use today. It is commonly used for the processing of potatoes, corn, cassava, millet, and other starch-containing raw materials. Until 1975 it was virtually the only method used in Europe.

### 4.2.1 High Pressure Cooking Process (HPCP)

This method is widely used because of its applicability for almost all starchy raw materials. In general, there is no need for a size reduction of the raw materials. We find an almost complete release of starch from the raw materials which is gelatinized to a great extent. In addition, the use of high temperature and pressure is relatively safe in everyday operations.

The almost complete decomposition of raw materials in HPCP is achieved by the use of

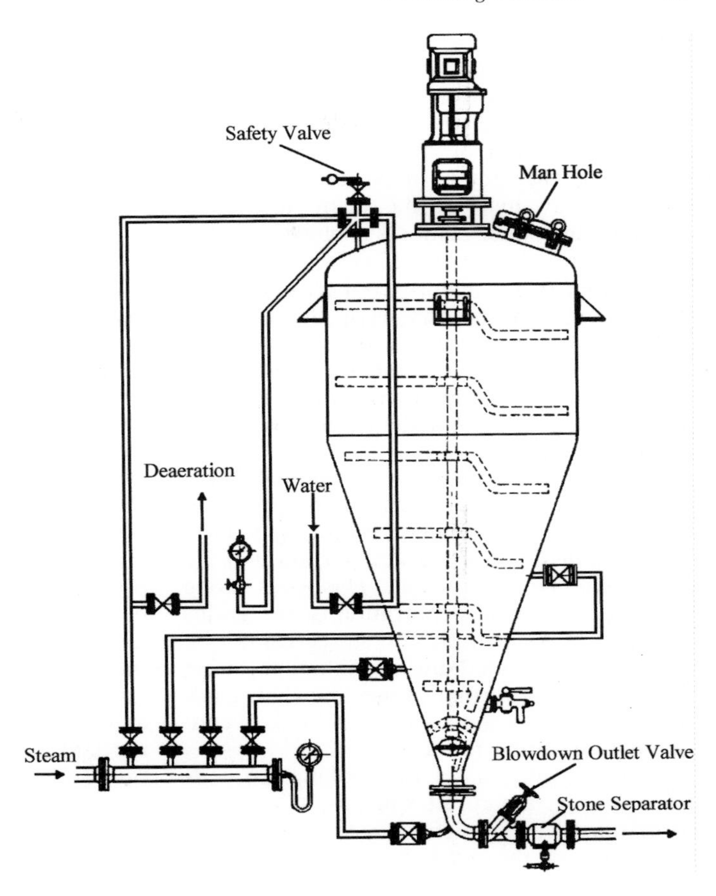


**Fig. 12.** Henze cooker (KREIPE, 1981).

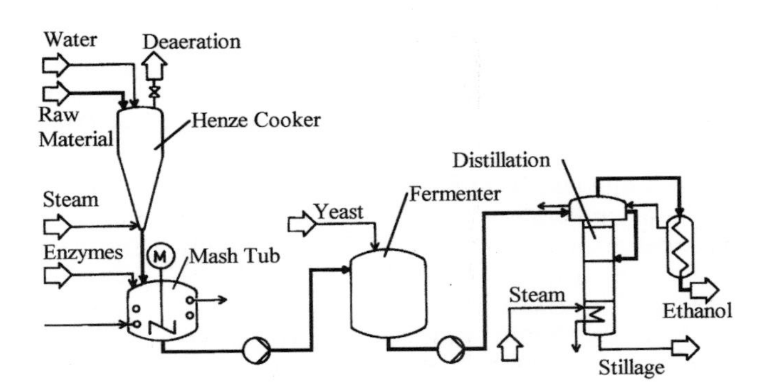


**Fig. 13.** High pressure cooking process (PIEPER and BOHNER, 1985).

high temperature and pressure in the presence of water in a Henze cooker (Figs. 12 and 13). In this process step, whole grains or whole potatoes are placed in the Henze cooker together with the required amount of water. Steam is introduced from the bottom of the Henze cooker to the material to be cooked, thereby raising the pressure slowly to an overpressure of 4.5–6 bar (140–160 °C), depending on the raw material used. The contents of the cooker are kept at this pressure for 40–60 min.

With this processing mode, pressure and temperature cause an extensive dissolution and gelatinization of starch in water, while preserving the outer shape of the grains. Once the raw material has been sufficiently cooked, the cooker contents is blown into the masher through a blow-off valve situated at the bottom of the Henze cooker. Only then does an extensive breakdown of the cell association and consequently the release of starch take place, due to the sudden pressure drop in the valve. Any remaining starch granules are exposed and immediatly gelatinized at this point.

The determination of the optimal time-point for blowing out the cooker contents is difficult. If the cooking process is stopped too early, the dissolution and gelatinization of the starch will be insufficient. This goes along with an inadequate penetration of the grains with water. As a result, the starch cannot be completely degraded during cooking, blowing out, and during further processing. This leads to ethanol losses and an increased risk of contamination during fermentation.

If, on the other hand, the cooking proceeds for too long, the result will be an increased formation of melanoidines and a severe caramelization of the mash due to the Maillard reaction. Consequently, losses of starch and sugars will occur. Furthermore, melanoidines inhibit the fermentation.

The mash is then cooled down in the mash tub and liquified by putting thermostable $\alpha$-amylase from *B. licheniformis* into the masher before blowing out is started. Otherwise, bacterial $\alpha$-amylase from *B. subtilis* can be added when a temperature of 75 °C is reached. For the reasons mentioned in Sect. 2.1, preferably $\alpha$-amylase from *B. licheniformis* should be used in this processing mode.

Once the temperature of the mash has dropped to about 60–55 °C, glucoamylase from *A. niger*, which may be accompanied by fungal $\alpha$-amylase from *A. oryzae*, is added for the saccharification of liquified starch. The mash is further cooled down to set temperature for fermentation. Yeast mash is usually used for the addition of yeast at 35 °C. The mash is pumped into a fermentation vessel and after 3 d of fermentation at 32–36 °C, the mash is distilled. The resulting stillage is used as feedstuff or as fertilizer.

The high energy consumption for HPCP, which amounts to about 7 MJ per lA for the mashing process only, led to the design of a process that could be pressureless and safe.

### 4.2.2 Bacteria-Free Fermentation Process of Verlinden (Verlinden Process, VP)

The Verlinden Process (VP) developed in 1901 in Belgium also uses a Henze cooker. In general, this process is only used for cereals. The pressure cooking conditions are the same as given for the HPCP (Sect. 4.2.1). But in this process mash tubs are not used. After the usual pressure treatment of mashes, they are blown out directly into special fermentation tanks fitted out with agitators. The vapor of this operation sterilizes the fermentation tanks. The mash which is agitated in the fermentation tank is cooled down by sprinkling the tanks. If microbial enzymes for liquefaction and saccharification are used, they are added to the mash at the usual temperatures. Dried distillers malt ore, a green malt slurry, must be disinfected with methanal (formaldehyde) before adding to the mashes.

Although this VP gives a certain guarantee that the fermentation runs free of contamination, the disadvantages of this process are obvious. The cooling rate is very low since the cooling surface is small, and sprinkling leads to the formation of hard scale. Furthermore, water consumption is very high and the process takes a long time. To start fermentation for the VP, pure culture yeast or fresh baker's yeast should be used in order to minimize the risk of contamination; however, this yeast is not ideally suited to the raw material that is processed. In addition, the fermentation equipment is expensive and the consumption of energy is high compared with other processes. Practical experience showed, that the VP does not yield more ethanol than the HPCP and is liable to the same risks of contamination.

## 4.3 Pressureless Breakdown of Starch

### 4.3.1 Infusion Processes

#### 4.3.1.1 Milling and Mashing Process at Saccharification Temperature

Prior to 1950 this process was known as "Kaltmaisch-Verfahren" (KMV, cold mash process). It can be used to process wheat, rye, and triticale. This process is based on the enzymatic activities of used raw materials. These activities lead to dissolution and partial saccharification of starch without previous liquefaction (KREIPE, 1981). A precondition for the successful application of this process is that the raw materials are very finely ground (KREIPE, 1981). The milling of grains is usually carried out under dry conditions. The mill should be fitted with a dust collector, and the ground material is loaded into a storage hopper situated over the mash tub.

The thoroughly ground cereals (wheat, rye, triticale) are stirred into the mash tub which contains the necessary amount of cold water (Fig. 14). The water temperature must be below 15 °C, and agglomeration must be prevented. The mash is kept cold overnight and left to soak. The pH is adjusted to between 5.6 and 5.8. If microbial enzymes for starch liquefaction are used in the process, they can be added while stirring in the ground grain.

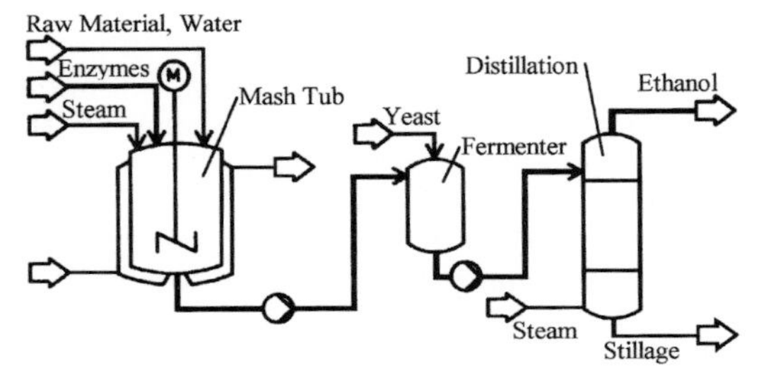


**Fig. 14.** Milling and mashing process, "cold mash process (Kaltmaischverfahren)" (PIEPER and BOHNER, 1985).

The next morning the soaked mash is heated to 50 °C by direct steam and kept at this temperature for about 30 min for a protein catabolic rest. Then the mash is heated to 58–60 °C, and this temperature is maintained for 60 min to provide a digestion, liquefaction, and saccharification rest. Afterwards the mash is cooled down to set temperature.

For liquefaction and saccharification distillers barley malt is normally added to the mash when heating up is started. If microbial enzymes are used for saccharification, they can also be added at this time, provided they are stable at 60 °C. At 35 °C, fermented yeast mash is added to the sweet mash. After 3 d of fermentation the mash is distilled and the stillage is used as feedstuff or fertilizer.

Some distilleries have started to use soaking water with higher temperatures in order to save steam consumption in the mashing process. The temperature may not exceed 50 °C, and soaking overnight is impossible because of the risk of infection. It is, however, recommended to allow a soaking rest of 1–2 h, otherwise the risk of contamination increases rapidly.

A disadvantage of this KMV process is that unfortunately the autoamylolytic enzyme system in cereal grains is not stable at 60 °C for more than 10–15 min. In addition, the enzyme activities vary greatly between different charges of raw material (Fig. 1), and with different charges of raw materials different particle-size distributions result in spite of the use of the same mill. Both of these factors lead to an incomplete release and degradation of starch during the mashing processes. This poor breakdown of starch and the low maximum temperature used in this process give rise to an increased risk of contamination during fermentation. Therefore, this process is rarely used now.

#### 4.3.1.2 Große-Lohmann-Spradau (GLS) Process

This process was developed to reduce energy consumption in distilleries using pressure boiling processes. In these distilleries, the old Henze cooker is often used as the first of two mash tubs. This GLS process is mostly used

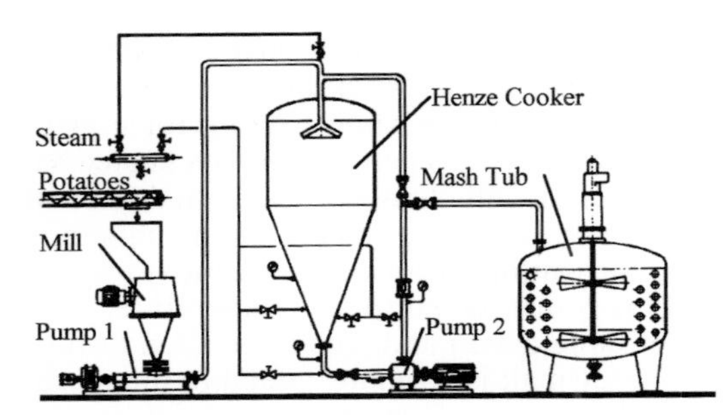


**Fig. 15.** Mashing process according to Große-Lohmann-Spradau (KREIPE, 1982).

to process potatoes, or sometimes corn (Fig. 15).

To run this process a very efficient pump, usually a centrifugal pump, is installed between mash tub 1 and mash tub 2. This pump permits the pumping of high amounts of mash from the bottom of mash tub 1 (normally an old Henze cooker) back to the top of mash tub 1. A second pipe allows pumping of the mash into mash tub 2.

To start the GLS process, raw materials are milled by a hammer mill. In these mills, slot sieves with a slot width of about 1–1.5 mm are often used. The use of perforated metal screens with a clear space of 1.5–2 mm in these hammer mills leads to better results in fineness of grinding. Milling can be done under dry or wet conditions. For grinding under wet conditions, water is poured into the grinding chamber together with the raw materials. This kind of milling leads to a faster but less efficient grinding compared to dry conditions. Using dry conditions, water is used only to rinse out the ground material from the mill.

To start the GLS process, water with a maximum temperature of 50 °C is filled into mash tub 1 and liquefying enzymes are added. Then the circulating pump is started, and the water is pumped from the bottom of mash tub 1 back to the top of it. After grinding, the mash is also pumped into mash tub 1. The mash flows are brought together directly over mash tub 1 and pumped over a mixing plate installed in the head area of this vessel. The mash is heated with live steam to 90 °C while both pumps are running. When mash tub 1 is filled up to its maximum level, the milling and pumps are stopped and the temperature is maintained at 90 °C for 1 h. After this liquefaction rest, the mash is pumped into mash tub 2 and cooled down. When a temperature of about 55 °C is reached, saccharifying enzymes are added and a saccharifying rest of 15–30 min may be carried out. Afterwards, cooling is continued and at 35°C the yeast mash is added.

This GLS process permits good processing of potatoes. The result of the process, however, depends strongly on the milling efficiency, especially for processing corn. It is impossible to reach a total release of starch from plant cells with a reasonable expenditure of energy by using mills. Undigested starch leads to problems mainly during fermentation because of higher viscosities and risks of contamination.

## 4.3.1.3 Milling and Mashing Process at Higher Temperatures (MMP)

The milling and mashing process at higher temperatures (MMP) is often used in distilleries that have invested in new mashing equipment since 1980. It permits the processing of all starchy raw materials. The flowsheet of this process is shown in Fig. 14.

At the start of MMP, the starchy raw materials are usually ground in a hammer mill normally fitted with a 1.5 mm screen and a dosing pump for liquefaction enzymes. Cereals are mostly milled under dry conditions, using water to rinse out ground material. Liquefaction enzymes are added together with rinsing water which may have a temperature of up to 55 °C to reduce energy consumption. For milling potatoes, the amount of rinsing water used in the mill is very low, while the ratio of cereals to be ground to water has to be at least 1:1. Mash is delivered from the bottom outlet of the mill to the mash tub with an eccentric screw pump.

In the mash tub, after pH adjustment to 6.0–6.2, the mash is heated to liquefaction temperature with live steam. Heating can also be done in the delivery line using a temperature-controlled steam injector. The liquefaction temperature depends on the raw material (potatoes: 90–95 °C; maize: 80–90 °C; wheat, rye, triticale: 65–70 °C). After a liquefaction rest of about 30 min, the mash is cooled down to saccharification temperature, and saccharification enzymes are added. The pH should be adjusted to 5.3–5.5 using concentrated sulfuric acid to optimize the activity of the enzymes added to the mash. Afterwards the mash is cooled down to set temperature, and usually yeast mash is pumped into the mash.

Sometimes, to save energy, the hot stillage coming out of the distillation column is used to heat up the mash by directing it through the cooling coil located in the mash tub. It is also possible to heat up potato mashes in the delivery line with hot stillage between the hammer mill and the mash tub using tubular heat exchangers or spiral-plate heat exchangers. But the temperature of the mash may not exceed 55 °C to prevent gelatinization of starch. Although liquefying enzymes are added to the mash, encrustations can not be avoided if the gelatinization temperature is exceeded in the heat exchanger. In practice, spiral-plate heat exchangers in particular have a tendency to build up encrustations. Since the temperature may not exceed 55 °C in a heat exchanger, it should not be employed while processing cereals; this temperature can easily be reached using the warmed up cooling water for rinsing out the ground material from the mill.

The MMP depends entirely on the effect of milling and the efficiency of the enzymes used, especially for saccharification. Incomplete release and degradation of starch in the process followed by increasing risks of contamination and losses in ethanol yield may result if milling and enzyme action are not optimal.

## 4.3.2 Recycling Processes

### 4.3.2.1 Stillage Recycling Process (SRP)

In the endeavor to reduce energy consumption in pressureless processes, PIEPER and JUNG (1982) developed the stillage recycling process (SRP). This process allows optimum utilization of thermal energy contained in the stillage, but this process may only be used with wheat, rye or triticale. The flow sheet of the SRP is shown in Fig. 16.

The recycling procedure of the stillage is important for the SRP. The ethanol-free stillage coming out from the continuous mash still, with a temperature of about 102 °C, is delivered into an intermediate vessel. Driven by the hydrostatic head, the stillage is led to a centrifugal decanter for separation of the solids (seed shells, germs, endosperm residues) from the liquid fraction. The stillage is sepa-

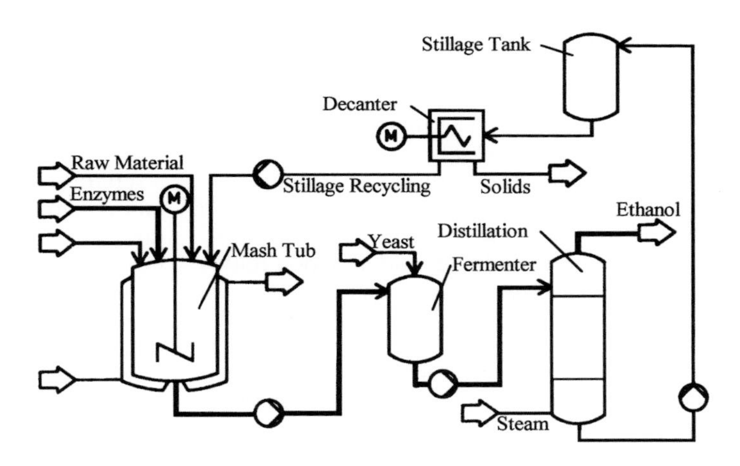


**Fig. 16.** Stillage recycling process (PIEPER and BOHNER, 1985).

rated into two phases: a solid phase, containing about 30% DS, and a liquid phase with a low solids concentration. The solid phase is used as feedstuff, and the liquid phase is directly recycled in the mashing process to substitute for needed process water. The residual 20–30% of the liquid phase which is not needed in the process is also used as feedstuff. The liquid phase of stillage is now used as process liquid in the following mashing process. The mash is fermented, and with distilling of this mash the stillage cycle is closed.

To run the SRP, the liquid phase of stillage obtained from the centrifugal decanter is pumped into the mash tub. While agitating, the pH is adjusted to 6.3–6.4 using technical calcium hydroxide. The temperature of stillage in the mash tub reaches 80–85 °C if the process is started in this manner.

One half of the thoroughly fine milled raw material is loaded into the mash tub while the agitator is running. Because of the high temperature of the stillage, charging must be done slowly and evenly to avoid agglomeration. This is necessary since the high temperature results in a rapid gelatinization of starch charged to the mash. Just after beginning to load raw material into the mash tub, thermostable $\alpha$-amylase from *B. licheniformis* is added to the mash. This prevents an unwanted increase in viscosity. Upon adding the raw material to the mash, the temperature slowly decreases and the efficiency of the thermostable $\alpha$-amylase slowly diminishes. Therefore, when 50% of the raw material has been added, charging is stopped and, if necessary, the mash is cooled down to 74 °C. At this temperature bacterial $\alpha$-amylase from *B. subtilis* is added, and the second half of the raw material is charged. The mash normally has a temperature of 70 °C and a pH of 6.0 if loading of raw material is done properly.

The mash is then cooled down to 57 °C and the pH is adjusted to 5.5. After that distillers' dried malt immediately is added. Upon reaching 55 °C, glucoamylase from *A. niger* is added to the mash. After further lowering the pH to 5.0, fungal $\alpha$-amylase from *A. oryzae* is added. A temperature of 55 °C is maintained for 1 h as a saccharification rest. Following the saccharification rest, the mash is cooled down to set temperature and yeast mash is added. The mashes produced with the SRP can be completely fermented within 44 h.

### 4.3.2.2 Dispersing Mash Process Developed at Hohenheim University (DMP)

The dispersing mash process (DMP) with stillage recycling as one of its characteristic components, was developed by PIEPER and SENN (1987). The flowsheet depicting the process is shown in Fig. 17.

The DMP allows processing of all starchy raw materials, as well as Jerusalem artichokes and sugar beet. It is characterized by three important features:

(1) decomposition of raw materials employing a rotor-stator dispersion machine;
(2) use of stillage recycling to reduce energy consumption and to optimize fermentation efficiency;
(3) use of OPTIMALT®, an optimum enzyme combination for saccharification of mashes from pressureless processes; this also leads to an optimum decantation of stillage.

The DMP permits processing of cereals and potatoes without milling or after just a coarse milling. If there are more than one or two mashes processed per day in one mash tub, it is recommended to use a hammer mill fitted with a 4 mm screen to grind the material. This reduces the time used for releasing starch from the raw material.

The stillage used for stillage recycling must be separated from the solids contained in the stillage. This can easily be done by employing a separation tank. The stillage from the distillation column is delivered into this separation tank for storage. After about 3–5 h, the stillage is sedimented and clearly consists of two phases. The solid-free phase is used in the process again. It amounts to 50% (wheat, rye, triticale) and to 70% (corn) of the total stillage. The solids containing phase of stillage, which is barely a fluid, is used as feedstuff or as fertilizer.

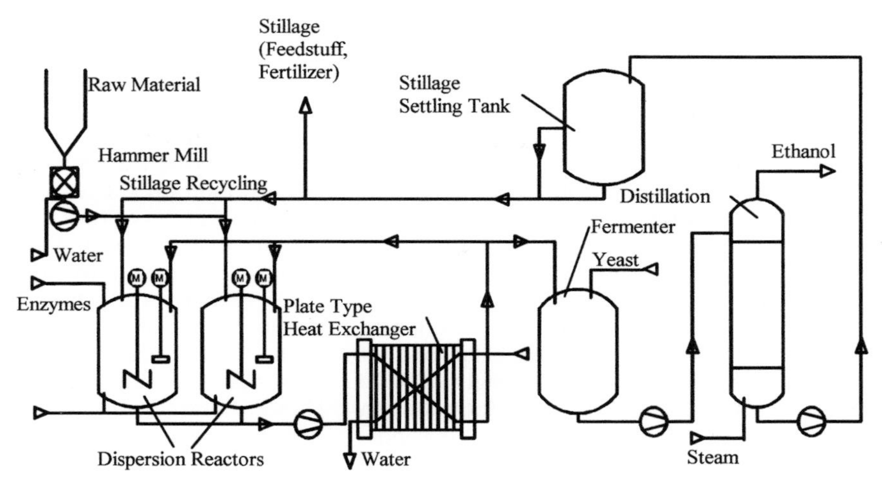


**Fig. 17.** Dispersing mash process.

When processing potatoes, the unseparated original stillage is used for recycling in an amount of about 15% of total stillage.

At the start of the DMP, the necessary amount of solid-free stillage is pumped into the dispersing reactor, which is used as mash tub. The pH is adjusted using hydrated lime (calcium hydroxide) depending on the properties of the liquefying enzymes used.

The coarse-milled raw material is pumped into the dispersion reactor using as little water as possible for pumping, which should have a maximum temperature of 60 °C. At the start of the process, the temperature of the mash is maintained at liquefaction temperature by injecting live steam. The rotor-stator dispersion machine is turned on when charging is completed and the mash is treated until a sufficient release of starch is guaranteed. The degree of disintegration is determined using a special hydrosizer (Sect. 13.3.1). When the hydrosizer shows that sufficient degree of disintegration, dispersing is stopped and the mash is cooled to saccharification temperature by using either a special plate-type heat exchanger outside or a cooling coil inside the dispersion reactor. For saccharification at 53 °C OPTIMALT® is added to the mash. After a short saccharification rest, which is not necessarily required, the mash is directly pumped into the fermentation tank, passing the plate-type heat exchanger where the mash is cooled down to set temperature. Yeast mash is admixed when the mash stream has passed the cooler. If a cooled coil is in use, the mash is cooled down to set temperature inside the dispersion reactor and yeast mash is added.

Several dispersion reactors can be loaded in succession with mash, dispersed and cooled. This multi-batch processing leads to a quasi-continuous processing which, however, allows the individual determination of the degree of disintegration for every single mash. The dispersion reactor may have a volume of up to 15 $m^3$ to process up to 3 t of cereals for instance in one mashing.

The DMP leads to a virtually total release of starch from plant cells. In addition, the use of OPTIMALT® for saccharification results in a complete degradation of starch, high ethanol yields, and good settling properties of the stillage. The use of stillage recycling guarantees a rapid start of the fermentation, a reduced length of the fermentation, and optimum conditions for yeast propagation. Hence, DMP is an environmentally acceptable and energy saving process that yields up to 66 lA per 100 kg starch in industrial plants.

# 5 Processing Potatoes

Potatoes used in alcohol production should be in good condition and free from plant diseases. This is a problem, since industrial potatoes are harvested only from September until the end of October, and it is impossible to process all the harvested potatoes in that time. Therefore, it is necessary to store potatoes before processing, for which special store houses are in use. They must be fitted with aeration and washing channels. In one of these store houses with, e.g., a length of 31 m and a width of 15 m, $2 \cdot 10^3$ t of potatoes can be stored with a bed depth of 6.5 m. It is possible to store potatoes up to a bed depth of about 12 m. To maintain potatoes in good condition during storage, it is necessary to keep them cool in the storage houses. Therefore, air is blown into the aeration channels below the bed of potatoes. Cold air is passed through the bed of potatoes from the bottom to the top of the storage house and leaves through air registers on top of the roof. In autumn this aeration is done at night to use as cool air as possible. Potatoes should be kept at temperatures between +4°C and +7°C. During the loading period aeration of the storing houses is also necessary to avoid formation of condensed water on the surface of cold potatoes (KREIPE, 1981).

Before processing potatoes they must be washed thoroughly, as described in Sect. 4.1.1, to minimize the risk of contamination with, e.g., spores of a *Clostridium* sp. contained in soil adhering to potatoes. Growth of these microbes in mashes may result in the formation of propenal (acrolein) during fermentation.

If the HPCP is used to process potatoes, washed potatoes are loaded into the Henze cooker. They are first steamed from the top of the Henze cooker using direct steam, while condensate draining off at the bottom. After about 15 min, when the cone of the Henze cooker is warmed up, steaming from the top is stopped and steaming from the bottom is started. This is done by closing the condensate outlet valve and opening the air outlet valve a little at the top of the Henze cooker. In this way the pressure is increased to 5 bar in 30–40 min. The potatoes are kept at this pressure for about 20–25 min, and then the content of the Henze cooker is blown out into the mash tub. It is possible to adapt this steaming process to different kinds of potatoes by varying the time the potatoes are kept at maximum pressure.

In pressureless processes, washed potatoes first have to be ground thoroughly. For this purpose hammer mills can be used; however, employing a rotor-stator dispersion mashine results in a better degree of disintegration. Using DMP it is also possible to process potatoes in a mash tub without previous grinding. Hence, potatoes can be loaded into the mash tub at a rate of about 10 t $h^{-1}$, the same rate that can be reached by milling potatoes. A sufficient degree of disintegration is reached if by mash hydrosizing (Sect. 13.3.1) the coarse fraction is $<1$ mL and the sum of coarse and suspended fractions is $<1.5$ mL. Since BAB (see Sect. 2.1.3) does not work in potato mashes, the pH has to be adjusted to 6.0–6.2 for liquefaction. To complete the gelatinization of potato starch, the mash temperature must reach at least 80–85 °C. But heating up potato mashes to 92–94 °C lowers the viscosity of mashes and cooling is accelerated. Low viscosities also accelerate the fermentation and reduce the risk of contamination. A further acceleration of fermentation can be achieved by using recycled stillage in the process. In mashing 7.5 t of potatoes, 2 $m^3$ of original stillage, draining from the still, can be used.

# 6 Processing Grain

If HPCP is used for processing grain it is important to load enough water into the Henze cooker before starting the steaming process. Normally, therefore, about 300 L water are used per 100 kg of grain of any variety. The required amount of water is pumped into the Henze cooker, and the steaming valve at the bottom is opened a litte to suspend the grain. Then the grain is loaded into the Henze cooker and after closing, the bottom steaming valve is opened. The grain has to be evenly distributed throughout the whole volume as it is very important that all the single starch granules come into contact with a sufficient amount of hot water to reach a complete gelatinization. Therefore, the side steam valves should also be opened.

Steaming should take about 40–45 min to raise the pressure up to 5 bar. The content of the cooker should be kept for about 40–45 min at this pressure. The time needed to hold different grains at maximum pressure can vary from 25 to 60 min. Therefore, the progress in steaming should be examined by taking a mash sample from the sample outlet at the bottom of the Henze cooker. This sample is poured into a sieve and is visually examined. The seed coats should be completely separated from endosperm fragments before the cooker is blown out during about 20 min.

The Henze cooker should be fitted with an agitator for processing milled grain products with HPCP. The agitator is necessary to obtain a good suspension of the raw material in water. The water temperature may not exceed 50 °C. Otherwise, the milled product becomes lumpy and gelatinization and solubilization of starch remain incomplete. The time needed to keep ground material at maximum pressure can be reduced to 10–20 min.

It is necessary to use different conditions in processing different varieties of grain in pressureless processes. Hammer mills or dispersion machines are used when milling grain for alcohol production.

Milling of grain for alcohol production is normally carried out with hammer mills fitted with a 1.5 mm screen. The degree of disintegration reached by this method is sufficient for ethanol yields of up to 62 lA per 100 kg FS if processed grain shows high autoamylolytical activity. If a sufficient autoamylolytical activity is not present in the grain to be processed – and in practice this is the normal case – the losses in ethanol yield increase. The mill has to be fitted with a 1.0 mm or 0.5 mm screen to reach an almost total digestion of starch and ethanol yields of about 64–65 lA per 100 kg FS during mashing, especially if the grain or corn is waxy or dried after harvesting at high temperatures, inactivating the autoamylolytical enzyme system. But this leads to a high energy consumption and a low throughput in milling, and is, therefore, not economical. Furthermore, it is impossible to grind corn with 0.5 mm screen, due to the fat content of corn which leads to the clogging of screens.

If dispersion machines are used for the digestion of grain, the grain is first coarse milled using a hammer mill fitted with a 4 mm screen. The coarse ground material is pumped into the mash tub and dispersed there. A sufficient degree of disintegration is reached if by hydrosizing, the ratio between the coarse and the sum of the coarse and suspended fractions is $<1$ mL : 3 mL. This guarantees ethanol yields of 64–65 lA per 100 kg FS, if no further mistakes are made during mashing and the fermentation process. Using DMP the mash is dispersed after the mash tub is filled up. Then the mash is heated up, and while dispersing it is kept at liquefaction temperature until disintegration is completed. If only one or two mashes are processed per day, grain can be loaded directly into the mash tub without previous grinding. But dispersing the mashes takes twice the time of that required with coarse ground material; otherwise, it is not necessary to invest in a mill. This is quite important too, since it needs no daily cleaning of the hammer mill.

The use of BAB for liquefaction of mashes is recommended using stillage recycling in grain processing. Liquefaction of grain mashes can thereby be carried out at a pH of 5.0, making it possible to run the whole mashing process at this single pH value.

## 6.1 Wheat

To complete gelatinization of wheat starch, a temperature of about 65 °C is required. If, due to contamination of the raw material, there is a great risk of contamination in mashes, the liquefaction temperature may be raised to 75 °C. Higher mash temperatures lead to a better pasteurization effect, but results in losses of ethanol of about 2–3 lA per 100 kg FS. A protein catabolic rest during the increase of the mash temperature is not required. A maximum temperature of 65 °C has to be kept for at least 30 min to reach a sufficient liquefaction. During cooling of wheat mashes, a saccharification rest is not required, but it leads to a quick start of fermentation. A saccharification rest may be kept at 52–55 °C for about 15 min.

## 6.2 Rye

The processing of rye is affected by the content of pentosans. A high content of pentosans leads to high viscosities in mashes which become evident at the beginning of the process, when ground rye is solubilized in water. These high viscosities persist during the mashing process and the fermentation. This problem can be solved by using pentosanases, which are rather expensive, or by running a certain mashing program (QUADT, 1994). If pentosanases are employed in the process, they are added to the mash together with liquefying enzymes; the liquefaction temperature may not exceed 60 °C, since pentosanases are not stable at temperatures higher than 60 °C.

The optimum temperature for liquefaction of rye mashes is 60 °C (QUADT, 1994). Temperatures of more than 70 °C result in losses of ethanol yields of up to 3 lA per 100 kg FS. To process rye without using pentosanases, a pentosan catabolic rest at the beginning of process is required. This catabolic rest has to be carried out at 50 °C and a pH of 5.0. Therefore, a needed amount of stillage and/or water is filled into the mash tub to reach a temperature of about 50 °C. If only stillage is used, it must be cooled down. Then the pH is adjusted to 4.6–4.8, the milling of rye is started using a 4 mm screen, and the ground material is pumped into the mash tub using warm water. If some ground rye is pumped into the mash tub, BAB is added to the mash as a liquefying enzyme. The temperature of 50 °C and a pH of 5.0 is maintained while filling up the mash tub and during further dispersion of the mash. This pentosan catabolic rest should be kept for 30 min to reach a sufficient degradation of pentosans by the action of enzymes contained in rye. The mash is then heated to 60 °C for liquefaction while disintegration of rye is completed by dispersing. Liquefaction at 60 °C should be carried out for at least 30 min or until disintegration is completed, if the latter lasts longer. After that the mash is treated as usual; with this mashing program a further addition of pentosanases only shows marginal effects.

This mashing program described above can also be used if only milling is used for disintegration. However, milling rye with a 1.5 mm screen results in lower degrees of disintegration followed by losses in ethanol yields of about 3 lA per 100 kg FS.

## 6.3 Triticale

The way to process triticale is the same as described for wheat (Sect. 6.1), because there is no problem with pentosans in mashing triticale. Triticale is a very important raw material for distilleries, due to the high autoamylolytical activities in some varieties. How to make use of these effects is shown in Sect. 8.

## 6.4 Corn

### 6.4.1 Dried Storable Corn Grain

Processing corn for ethanol production is influenced by the high temperature needed for gelatinization of corn starch and the presence of varying amounts of horny endosperm in the corn. The horny endosperm of corn is elastic. This leads to problems in milling since milled corn remains gritty, even if a 1.5 mm screen is used. The use of smaller screens is not recommended due to the fat content of corn, which leads to very small throughputs in milling or to clogging of screens. Gritty particles obtained from milling will not completely dissolve during mashing and fermentation. This leads to slow fermentations and losses in ethanol yields. A virtually complete release of starch is only guaranteed using HPCP or DMP. However, it is impossible to avoid ethanol losses using HPCP, due to the Maillard reaction and caramelization during the steaming process.

To liquefy corn mashes, it is necessary to heat them to 80°–85 °C for at least about 30 min. Up to 90% of the water needed can be substituted by stillage if the DMP is used. If stillage recycling is carried out in the process, it is recommended to add BAB to liquefy the mashes at a pH of 5.0. After loading the stillage and/or hot water obtained from cooling facilities into the dispersing mash tub, coarsely ground corn is pumped into the mash tub. If only one or two mashes are processed a day, it is possible to load corn into the mash tub without previous grinding. In either case, the mash is immediately heated up to 80 °C and dispersed at this temperature until the mash hydrosizer shows sufficient results (Sect. 13.3.1). Then the mash is treated as usual. The fat content of corn, disadvantageous in milling, is an advantage in fermentation. There is no foaming, and the fermentation tanks can be filled up almost completely.

### 6.4.2 Corn Grain Silage

Corn grain silage may be processed with the HPCP. A maximum pressure of 5 bar in the steaming process has to be maintained for about 20 min under the conditions described above. In either case, using HPCP or pressureless processes, the amount of water in the mashing process requires correction for the water content in corn grain silage. The ratio of starch to water in the process should be about 1:6.

When processing corn grain silage with pressureless processes, the raw material must be ground either by milling or dispersing. If hammer mills are used, the problems which arise are the same as those mentioned in Sect. 6.4.1. The way to process corn grain silage using DMP is also the same as for storable dried corn grain. When working with a pressureless process, it only has to be taken into consideration, that the pH value has to be observed when loading the corn silage into the mash tubs. Lactic acid, contained in the silage, leads to an acceleration of the fermentation. By using stillage recycling in the process, it is possible to reduce the fermentation period to only 2 d. This is impossible when processing dried corn grain.

Ethanol yields with corn grain silage amount to about 61 lA per 100 kg starch using HPCP (PIEPER and PÖNITZ, 1973) and about 63 lA per 100 kg starch when working with pressureless infusion processes (TREU, 1991). Using DMP for processing corn grain silage, ethanol yields amount to more than 64 lA per 100 kg FS. However, ethanol yields reported in the literature which are not related to FS must be regarded with caution (Sect. 13.2). That is especially true for corn grain silage, which contains mono- and disaccharides, and lactic acid – substances leading to errors in other analytical methods.

## 6.5 Barley

The husks of barley cause some problems in processing barley with pressureless processes. Using screens smaller than 4 mm in milling, the throughput is greatly reduced. A further problem is the formation of pearl barley in milling, even if 1.5 mm screens are used. Pearl barley swells strongly during the mashing process, and the starch contained in it can not be released. Therefore, a dispersion step is essential in processing barley. A third problem in mashing is the $\beta$-glucan content in barley. $\beta$-Glucan leads to high viscosities in barley mashes from the beginning of the mashing process. If the mashes are heated up to starch gelatinizing temperature, viscosity increases and cannot be controlled. The breakdown of $\beta$-glucan can be carried out by using $\beta$-glucanases, which are quite expensive, or by using a certain mashing and temperature program thereby utilizing $\beta$-glucanases contained in the barley.

It is impossible to process barley, even using DMP, without previous grinding (HEIL et al., 1994). This is done with a 4 mm screen in a hammer mill. The necessary amount of water and/or stillage is loaded into the mash tub and adjusted to a pH of 5.2 and a temperature of 40–50 °C. For liquefaction, the addition of BAB to barley mashes is recommended. Then ground barley from the hammer mill is pumped into the mash tub and dispersion is started. To break down $\beta$-glucan, the mash has to be maintained at the aforementioned pH and temperature conditions for about 30 min. After that the mash is heated to a liquefaction temperature of 80 °C and kept at this temperature for at least 30 min. When the result of mash hydrosizing is satisfactory (Sect. 13.3.1) the dispersion machine is stopped and the barley mashes are treated as usual.

# 7 Processing Tropical Raw Materials

## 7.1 Sweet Sorghum

Sweet sorghum is a raw material containing fermentable sugars. Since it does not keep well after harvest, it cannot be imported from tropical countries. There have been some attempts to grow sweet sorghum in Austria and Germany (DIEDRICH et al., 1993; SALZBRUNN, 1982). To obtain ethanol from sweet sorghum it is first necessary to produce a sugar juice from the plant stems. This can be done in two ways: The first possibility is water extraction with a countercurrent extraction plant; the second way is to press the sugar juice from the plant stems using a roller mill. In both cases it was possible to recover about 91–95% of total sugar contained in the plants.

Fermentation caused no problem, but it was necessary to pasteurize the sugar juice to avoid contaminations. Batch fermentation was finished within 40 h. An ethanol yield of about 58 lA per 100 kg of sugar can be expected (SALZBRUNN, 1982).

The sugar juice obtained by extraction or pressing does not keep well. Therefore, it is necessary to concentrate the juice up to 80°Bx or 74% DS using a downflow evaporator. After chilling the sirup is extremely viscous, or even solidifies and is storable.

## 7.2 Sorghum Grain

Sorghum grain should only be processed using HPCP because of the waxy structure of the starchy endosperm (KREIPE, 1981). However, the experience with corn using pressureless processes, especially DMP, demonstrates that it is possible to process waxy raw materials with pressureless processes very effectively. There is no doubt that it should be possible to process sorghum grain the same way as corn. Using DMP and stillage recycling, ethanol production from sorghum grain saves much energy and is environmentally friendly.

## 7.3 Manioc

Manioc products should also be processed by the HPCP (KREIPE, 1981). About 350 L of water are needed for 100 kg of manioc in the cooking process. The pressure cooking conditions have been given as 3.5–4 bar for about 30 min. Saccharification has to be carried out using bacterial amylases, since saccharification using malt remains incomplete. During pressure cooking it is possible to blow out hydrocyanic acid using the deaeration valve.

Worldwide more and more varieties of manioc are cultivated which do not contain cyanogenic glucosides. Therefore, it should be possible to use pressureless processes. Furthermore, it should also be possible to blow out hydrocyanic acid by keeping manioc mashes at boiling temperature for about 10–20 min; the cyanogenic glucosides are not stable at this temperature. This could be a way to process manioc with more energy savings.

# 8 Mashing Processes Using Autoamylolytical Activities in Raw Materials

As shown in Sect. 6, it is necessary to heat cereal mashes up to temperatures, some degrees higher than gelatinization temperature in order to reach a complete gelatinization of cereal starch. However, the autoamylolytical enzyme system is not stable at these temperatures. Therefore, if the autoamylolytical enzyme system is to be used in mashing processes, a way must be found to complete gelatinization of starch without inactivating the enzyme system. This can be achieved by using a certain time, temperature, and pH combination during mashing processes (RAU et al., 1993; SENN et al., 1991). But one has to take into consideration, that this mashing program differs with different raw materials and combinations of raw materials.

Processing under autoamylolytical conditions is more labor intensive than other processes. The conditions (temperature, pH, and time) for an autoamylolytical process, wich are described in the following sections, have to be followed exactly, and the cleaning procedures in the distillery are very important. This is due to low maximum temperatures that are maintained only for minutes. But using stillage recycling and taking care of the yeast cultivation guarantees a rapid start of the fermentation and the best chance to avoid contamination.

Besides all this, a only 20% of the normal amount of saccharifying enzymes used assures processing under disadvantageous conditions.

## 8.1 Processing Wheat

It is possible to process wheat with an AAQ of 95% or higher without the addition of enzymes. It is necessary to grind the material thoroughly or to use a hammer mill fitted with a 4 mm screen, combined with further use of the dispersing mash process. In either case, water and/or stillage is pumped into the mash tub, and after adjusting the pH to between 5.2 and 5.5, wheat is ground in the hammer mill and pumped into the mash tub; BAB is then added as liquefying enzyme. The mash temperature in the mash tub may not exceed 50 °C in this stage of the process. The mash is maintained under these conditions for a minimum of 30 min while it is being dispersed.

If mash hydrosizing shows that a sufficient degree of disintegration has occurred the dispersion machine is stopped and heating of the mash is started. The mash is heated slowly by about 1 °C per min. Heating is stopped at exactly 64 °C, at which temperature the mash is kept for 5–10 min and then rapidly cooled down to a saccharification temperature of 53–55 °C and the pH is adjusted to 5.3. The mashes are maintained under these conditions for 20–30 min for saccharification. When the saccharification rest is completed, the mash is cooled down to set temperature and fermented as usual.

## 8.2 Processing Triticale

With an autoamylolytical process it is also necessary to grind the raw material thoroughly or to use a dispersing mash process. An AAQ of at least 95% is required.

The autoamylolytical enzyme system of triticale, in contrast to that of wheat, is rapidly inactivated at 64 °C. However, gelatinization of triticale starch is completed at 60–62 °C, permitting an autoamylolytical processing of triticale.

The required amount of water and/or stillage is pumped into the mash tub and the pH is adjusted to between 5.2 and 5.5. The mash temperature may not exceed 50 °C at this stage of the process. Then the triticale is ground in a hammer mill and also pumped into the mash tub, with BAB added for liquefaction. When milling of the raw material is completed, the mash is maintained at these conditions for about 30 min. During this time disintegration of triticale is completed by use of the dispersing machine, until mash hydrosizing shows a sufficient degree of disintegration. Afterwards the mash is slowly heated (1 °C per min) to 62 °C as a maximum. The mash is maintained at this temperature for 5–10 min and then rapidly cooled down to the saccharification temperature of 53–55 °C and the pH is adjusted to between 5.3 and 5.5. After a saccharification rest of 20–30 min the mash is corled down to set temperature and fermented as usual.

The gelatinization temperature should be as low as possible to save the autoamylolytical enzyme system of triticale during mashing. Therefore, it is helpful to use microexamination to check starch gelatinization. This examination should be carried out for each new portion of triticale that is worked up. If possible, the gelatinization temperature should be lowered to 60 °C, or, if really necessary, it may be raised to 64 °C for 5 min only.

## 8.3 Processing Rye

Using the autoamylolytical process for rye, an additional problem is encountered due to the pentosan content of rye (see Sect. 6.3); on the other hand, virtually all varieties of rye have an AAQ higher than 95%.

The pH value of the needed amount of water and/or stillage is adjusted to 5.0 before and after the ground rye is pumped into the mash tub in order to avoid adding pentosanases during the autoamylolytical mashing process. The mash is maintained at 40–45 °C for not more than 20 min at this pH value. If DMP is used, the mash is dispersed under these conditions until a sufficient degree of disintegration is reached. Then it is heated to 60 °C to complete gelatinization of starch for a maximum of 10 min. The gelatinization temperature can be raised to 62 °C if needed for starch gelatinization; however, the mash should not be kept longer than 5 min at this temperature.

Microexamination of the mash should be used to determine whether temperatures higher than 60 °C are needed. To complete saccharification, the mash is cooled to 52–55 °C and the pH is adjusted to 5.3. After a saccharification rest of 20–30 min, the mash is cooled down to set temperature and fermented as usual.

## 8.4 Saccharification of Raw Materials with Weak Autoamylolytical Activities (Wheat, Corn, Potatoes)

It is not only possible to use the described autoamylolytical activities in triticale to saccharify starch from triticale, it is also possible to use portions of triticale with an AAQ higher than 95% for saccharification of starch from raw materials such as potatoes, wheat, and corn, with weak or no autoamylolytical activities (THOMAS, 1991; SENN et al., 1991). For this purpose it is necessary to prepare two mashes separately in two mash tubs, with one of these mash tubs needing only half the capacity of the other one.

Generally, in one of these mash tubs the gelatinization and liquefaction of the raw material with weak autoamylolytical activities is carried out under the required conditions (Sect. 4.5); the second mash tub is used to prepare the gelatinized and liquefied mash from triticale as described in Sect. 7.2. These mash preparations can be carried out simultaneously, and when both mashes are cooled down to saccharification temperature (53 °C), they are mixed in the bigger mash tub for saccharification.

This process is quite complicated and difficult. But for production of fuel from renewable resources, it is of great importance. Potatoes, e.g., yield twice the amount of starch per ha compared with cereal crops. Also, the cultivation of potatoes and triticale results in a complete rotation of crops and permits agricultural starch production on all the area of arable land. Furthermore, processing in this way avoids the costs for saccharification enzymes. This reduction in costs for ethanol production is important, even if only 20% of the normal concentrations of saccharification enzymes are used (Sect. 8).

Ethanol yields of these mixed mashes depend strongly on the ratio of starch from weak autoamylolytical sources to starch from triticale as shown in Figs. 18 and 19. As shown for potatoes and corn, it is necessary to have 50% of the starch from triticale in these mixed mashes to reach optimum ethanol yields of more than 64 lA per 100 kg FS.

# 9 Yeast Mash Treatment

The yeast needed for fermentation is normally cultivated in the distilleries as a separately fermented yeast mash, usually using the sulfuric acid method. The volume of yeast mash needed daily in distilleries ranges from 5–10% of the mash volume. To accelerate the

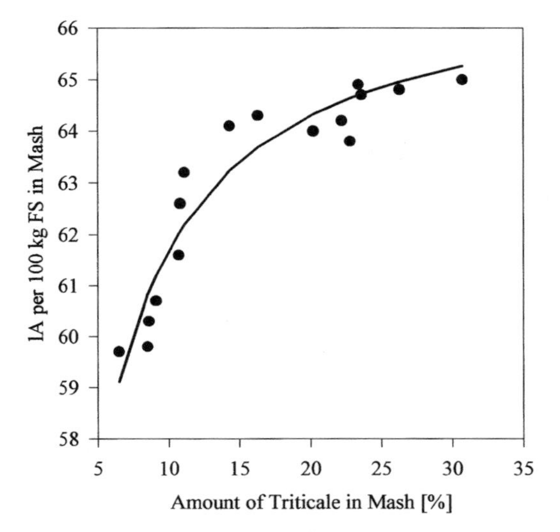


**Fig. 18.** The relationship between ethanol yields from potatoes using triticale for saccharification and the amounts of triticale in the mash.

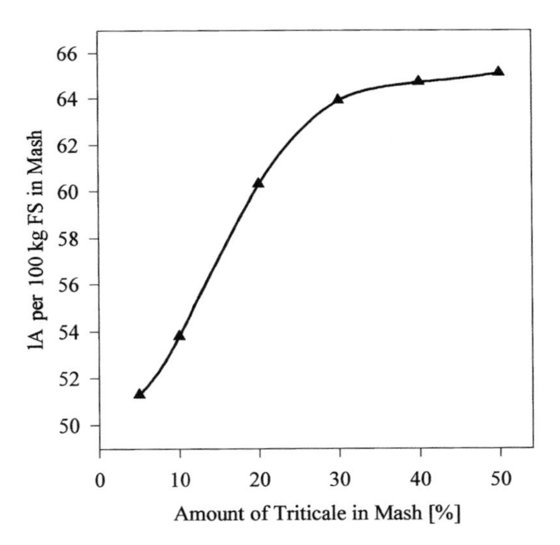


**Fig. 19.** The relationship between ethanol yields from corn, using triticale for saccharification, and the amounts of triticale in the mash.

fermentation and to suppress contamination in fermentations, 10% of yeast mash should be used.

A dried distillers' pure-culture yeast is usually used. This distillers pure-culture yeast is added to sweet mash which was pumped into the yeast mash tub. Then the pH is adjusted to between 3.0 and 3.5 and the yeast mash is cooled to 20–22 °C. During cultivation for about 18–24 h the temperature of the mash should not exceed 27 °C. This can be regulated by changing the set temperature of the yeast mash or by using the cooling facilities of the yeast mash tub.

Generally, the extract content of yeast mashes during yeast mash fermentation should decrease to about 50% of the extract content of the sweet mash. Thus the fermented (ripened) yeast mash should have an extract content of 7–10% mas when it is added to sweet mashes. This guarantees a rapid start of the fermentation and ensures an optimum protection against contaminations.

The use of stillage recycling in ethanol production also leads to an accelerated fermentation in yeast mashes, and, therefore, it may be necessary to cool yeast mashes more intensively. However, stillage recycling allows the lowering of pH values in yeast mashes down to 2.5. This pH does not effect on yeast growth, but leads to an effective protection against bacterial contamination in yeast mashes. But the pH value is not lowered below 3.0 at the start. To reach a pH of 2.5, the pH-value of yeast mash should be successively lowered during the first three days of yeast cultivation.

The pH adjustment of yeast mashes is done by adding concentrated sulfuric acid to the mash. Therefore, this method is called the sulfuric acid method.

# 10 Fermentation

## 10.1 Batch Fermentation

The fermentation of mashes is normally carried out in cylindrical fermentation vessels. These fermenters should be

- made from stainless steel,
- standing cylindrical fermenters, designed higher than the diameter of the vessel,

- fitted with manholes at the bottom of the cylindrical part of fermenter and at the top,
- fitted with an inspection glass above the bottom manhole, a thermometer and a carbon dioxide collection tube at the top.

If the volume of the fermenter is greater than 40 $m^3$ it is useful to fit it with a cooling coil, which must be designed to enable a rapid and effective cleaning. Such a cooling facility inside the fermenters is quite effective and permits a better temperature control during the fermentation. While the fermentation temperature in smaller fermenters or in fermenters without any cooling facility is regulated by the selection of set temperature, fermentation can be started at a higher set temperature if a cooling coil is installed in the fermenter. Using, e.g., a fermenter with a volume of about 30 $m^3$, the set temperature has to be reduced to 20–22 °C. If a cooling coil is installed into the fermenter, set temperature of 28–30 °C can be maintained. These higher set temperatures lead to an accelerated start of the fermentation and, as a consequence, the duration of the fermentation is reduced as well as the risk of contamination. If, however, the yeast mash used in the fermentation is not absolutely free from contamination, higher set temperatures result in an increased risk of contamination. A cooling coil in fermenters reduces the consumption of cooling water during the mashing process, while water consumption during the fermentation increases.

A perforated water tubing is often installed on the top so that water can be sprinkled on the surface of the fermenters. But this kind of spray cooling is quite ineffective and leads to encrustations on the surface of fermenters.

Fermentation can be started with yeast mash in two ways. The first possibility is the addition of yeast mash to sweet mash in the mash tub at about 33 °C. This inoculated mash is then cooled down to set temperature and pumped to the fermenter and the yeast mash tub again. The second way to start fermentation, usually used if the distillery is equipped with a heat exchanger, is to pump about 90–95% of fermented (ripened) yeast mash into the fermenter. After this, or while the yeast mash is being pumped into the fermenter, the sweet mash from the mash tub is pumped into the fermenter by passage through the heat exchanger to reach set temperature. And the yeast mash tub is refilled too, with sweet mash cooled down to set temperature.

Generally the set temperature is selected to reach a fermentation temperature of 34 °C after about 24 h; the fermentation temperature should not exceed 36 °C during the entire fermentation. A cooling facility may be used in fermenters to maintain the fermentation temperature at 34 °C after this temperature is reached.

In practice it is important to follow the course of the fermentation analytically. Thus extract content, pH and temperature should be measured and documented at least once a day. Microscopic examinations should also be carried out daily.

Normally, after 24 h of fermentation about 50% of fermentable extract contained in the mashes is metabolized and the ethanol content has reached about 4% vol. At the start of fermentation, the pH is 5.2, which decreases over about 24 h to 4.6–4.8. During further fermentation the pH decreases to about 4.2–4.5. At the end of fermentation the pH increases by about 0.2 pH units if the mash is free from contamination. If the pH falls below 4.0, this is normally due to contamination.

It is impossible to give generally valid data concerning extract contents (degree of fermentation) as a criterion for the completion of the fermentation. These data depend on the mashing process used (pressure cooking or pressureless), the raw material used, the ratio of stillage recycling used in the process, and the extract content of sweet mashes. Therefore, it is essential to follow fermentations using analytical methods in each distillery. Reliable data for a certain distillery can be obtained, if these determinations provide consistent data, and if the same mashes yield more than 64 lA per 100 kg FS. In practice, the degree of fermentation varies between −1.5 and +4 % mas. Tab. 14 shows the data for the classical HPCP (KREIPE, 1981).

Stillage recycling, used in the mashing process, leads to an additional important effect in the fermentation (SENN, 1988). The start of the fermentation is clearly accelerated if stillage recycling is used; corn mash pre-

**Tab. 14.** Degree of Fermentation (HPCP) (KREIPE, 1981)

| Raw Material | %mas |
|---|---|
| Potatoes | 0.7–1.4 |
| Corn, sorghum | <0.0 |
| Rye | 0.5–1.0 |
| Wheat | 0.0–0.5 |
| Manioc | <0.0 |

pared without stillage has a long lag phase of more than 4 h. If stillage amounts to 50% of the process liquid, the lag phase is reduced to 1.5–2 h. Using exclusively stillage as process liquid, the fermentation starts with a small or no lag phase. This rapid start of fermentation using stillage recycling clearly reduces the risk of contamination. Not only is the start of the fermentation accelerated, but also the duration of the fermentation is reduced. Thus, processing wheat, rye, or triticale, it is possible to conduct a fermentation of only 40 h without problems. With a set temperature of, e.g., 28 °C, it is possible to finish this fermentation within 30 h. Fermentation of corn mashes takes about 60 h, while fermentation of corn grain silage takes 40 h.

## 10.2 Suppression of Contaminants

In recent years more and more distilleries have been affected by the formation of propenal (acrolein) in mashes during fermentation. Propenal is enriched in raw distillates and results in a drastic reduction in product quality. Therefore, it was necessary to develop methods to avoid formation of propenal in mashes. KRELL and PIEPER (1995) tried to avoid propenal (acrolein) formation right from the beginning of the fermentation while BUTZKE and MISSELHORN (1992) tried to reduce the propenal content in raw distillates during the distillation process by chemical oxidation.

Since the formation of propenal (acrolein) in mashes is due to the growth and metabolism of sporulating bacteria (KRELL and PIEPER, 1995), it is necessary to create conditions in mashes, that lead to pasteurization of the mashes during the mashing process and that prevent germination of bacterial spores during the fermentation. Therefore, mashes are kept at more than 65 °C for a minimum of 30 min to reduce the microbial count as far as possible. After liquefaction and pasteurization, the mashes are cooled to saccharification temperature. But, before 60°C are reached, the pH of the mashes is adjusted to 4.0 to avoid germination of spores during further processing, using concentrated sulfuric acid. This kind of specific acidification of mashes was found to stop propenal (acrolein) formation effectively in a German distillery processing 15 t of potatoes per fermenter, and at the same time, ethanol yields increased from 61.9 lA per 100 kg FS (n = 89) to 64.2 lA per 100 kg FS (n = 104).

Specific acidification of mashes is very effective in avoiding contamination of mashes due to the germination of spores. But there is only a weak effect in avoiding contamination that results from contaminated mash residues in pipes, heat exchangers and other regions that are difficult to clean. But, if the mashes have a pH value of 4.0 it is possible to use sulfur dioxide ($SO_2$) to reduce the risk of contamination drastically, as done in wine making technology (KRELL and PIEPER, 1995). The use of $SO_2$ permits the processing of all starchy raw materials virtually free from contamination, if the following steps are carried out:

- acidification of the mash to a pH of 4.0 above 60 °C using concentrated sulfuric acid (about 1 L $m^{-3}$ mash is needed),
- acidification of yeast mash to a pH of 2.5.

If these measures are not sufficient to suppress contamination in mashes, an additional step is used:

- addition of $SO_2$ (50 g $m^{-3}$) or $K_2S_2O_5$ (100 g $m^{-3}$) to the mashes.

# 11 Distillation

## 11.1 Distillation of Raw Spirit from Mashes

Under the conditions of the German state monopoly on ethanol, it is necessary to distill a raw spirit from fermented mashes with an ethanol concentration of between 82% and 87% by volume, if it is delivered to the state monopoly. The stills used in distilleries to produce raw spirit for the state monopoly nowadays are completely manufactured from stainless steel. To reach the required ethanol concentration in raw spirit, a distillation apparatus is needed, that consists of four sections:

- the mash column, to distill off ethanol from the mash,
- the enriching section, to increase ethanol concentration in the raw distillate,
- the dephlegmator, to enrich ethanol in the vapor phase by partial condensation,
- the cooling section, to condense and cool down the ethanol vapor.

The mash column (Fig. 20) can either be heated by direct steam injection or indirectly, using a heater assembly which consists of heating pipes or heating plates. Mash columns are usually constructed with bubble plates because of the solids contained in mashes. Depending on the efficiency of distillation, bubble plates should be at least 300 mm apart. It is sufficient to use a single bubble cap on the distillation plates if the diameter of the column is not larger than 800 mm. When using columns with a diameter of more than 1 m, it is necessary to fit the plates with more bubble caps to maintain distillation efficiency. The diameters of these plates should not be too small, to avoid clogging if several bubble caps are used on a distillation plate. At the front stills should be fitted with sight glasses, and, if the column diameters exceed 700 mm, stills should additionally be fitted on the backside with openings for cleaning between each plate

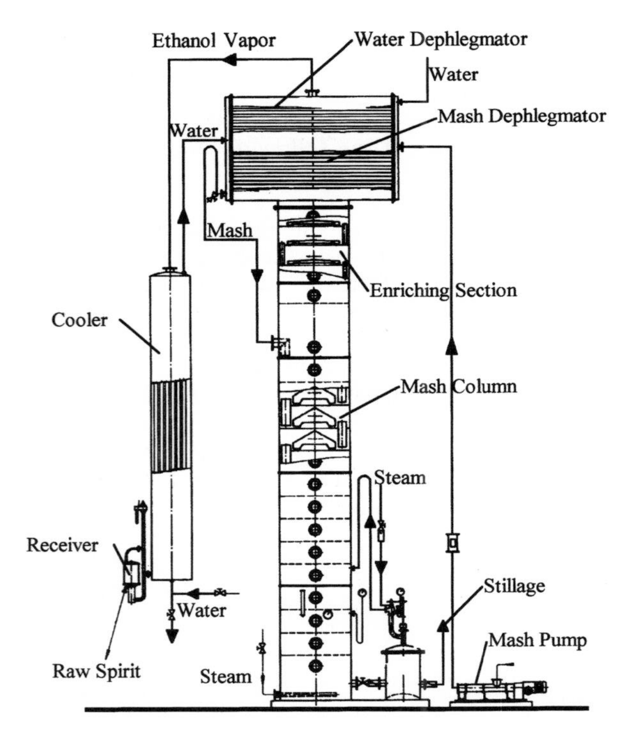


**Fig. 20.** Mash distillation column; mash and water dephlegmator (KREIPE, 1981).

(KREIPE, 1981). To ensure that the stillage leaving the mash column is free from ethanol, about 13–15 bubble plates are needed.

The enriching section of raw spirit stills is normally fitted with 4–6 bubble plates, with a distance of about 200 m between them. It is important to have a sufficient distance between the very top bubble plate of the mash column and the bottom plate of the enriching section to avoid that foam or mash is carried over to the enriching section by the vapor stream. The enriching section should also be fitted with sight glasses and cleaning openings, as described above.

Lying rectangular dephlegmators are in use to obtain an effective enrichment of ethanol in the vapor phase. The inside of these dephlegmators consist of a tube system, which is easy to clean. The mash that is distilled passes two thirds of these pipes. One third is used as a dephlegmator system using cooling water. In this manner, passing through the tubing system of the dephlegmator, the mash is heated to about 80 °C, while it is also used as a cooling agent for dephlegmation. The topmost third of the tubing system in these types of dephlegmators is used as a water dephlegmation system, to complete the dephlegmation at an enrichment ratio needed to reach a distillate with a sufficient ethanol concentration. A sufficient ethanol concentration of raw spirit is reached, if the ethanol vapor phase leaving the dephlegmator has a temperature of 78 °C or lower.

In 1992, a new generation of dephlegmators was developed, designed as pure mash dephlegmators (Fig. 21). These dephlegmators are cylindrical with a diameter that is somewhat greater than the diameter of the mash column and the enriching section. Internally there is a doubled heating coil through which the mash passes, and a water container situated in the middle of the heating coil. The length of the coil is sufficient to lead to the necessary enrichment ratio, if only fermented mash is used as a cooling agent in dephlegmation. The water container in this new kind of dephlegmator is used only as a reserve facility for extreme situations such as unusually low ethanol concentrations in mashes.

After leaving the top of the dephlegmator, the ethanol vapor is condensed, and the raw spirit obtained is cooled to about 20 °C, usually by using tubular coolers. Often the cooling

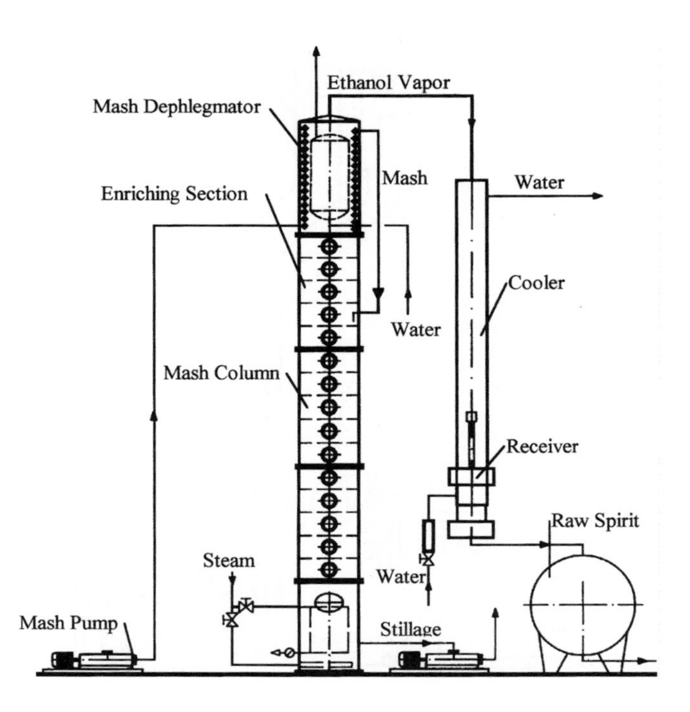


**Fig. 21.** Mash distillation column; pure mash dephlegmator (J. Carl, D-73002 Göppingen).

water needed for distillation is first used as a cooling agent in the tubular cooler and then, after it has warmed up there, it is used in the dephlegmator. This saves some cooling water, but regulation of the still is easier and more effective if cooling water used in the cooler and the dephlegmator is separately controlled. It is pointless to couple cooling and dephelegmation, since the cooling of the product is without any influence on the dephlegmation process. In modern stills, both processes consuming cooling water are separately controlled. The energy consumption of the mash still described above amounts to about 2 kg of vapor per lA (KREIPE, 1981).

## 11.2 Rectification of Product Spirit from Raw Spirit

Rectification of raw spirit, when distilleries produce their own potable spirit, is normally carried out using a batch rectifying apparatus. These rectifying stills are manufactured completely from copper. Only the pipes collecting the top fraction of the heads are from stainless steel to avoid corrosion of copper. These rectification stills (Fig. 22) in principle consist of:

- a reboiler (still pot),
- a rectifying column,
- a main condenser,
- an aldehyde condenser,
- a product cooler,
- a receiver for the top of the heads,
- a receiver for the heads, the product fraction and the tailings,
- a fusel oil decanter.

Before raw spirit is rectified, it is pumped into the reboiler and diluted there to about 40% ethanol by volume. The reboiler should be only filled up to about 80% of its volume. To heat up diluted raw spirit, the reboiler is fit-

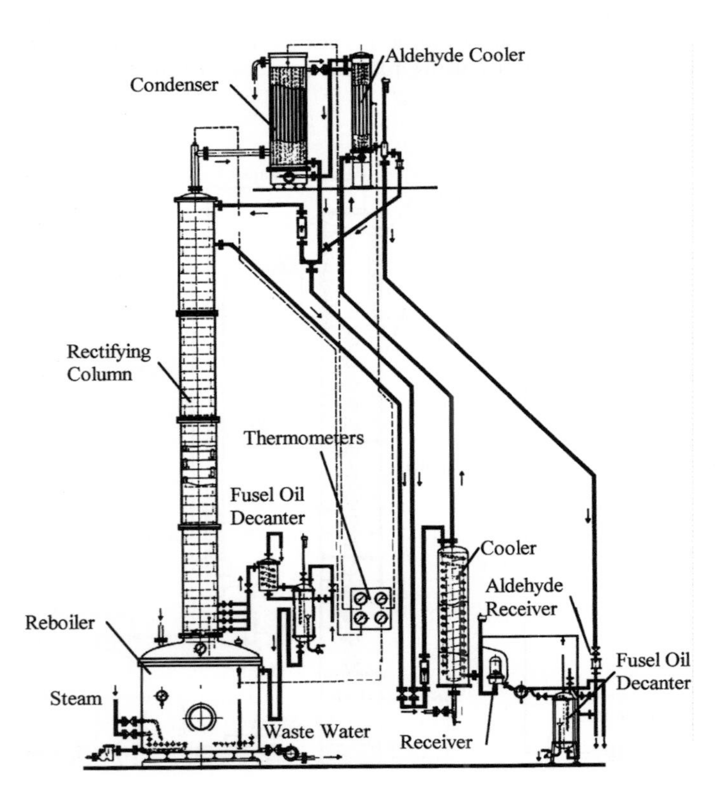


**Fig. 22.** Batch rectifying apparatus (KREIPE, 1981).

ted with a heating coil for indirect heating. In addition, the reboiler is fitted with pressure and temperature indicators, a sight glass, a manhole and sometimes a still dome.

The rectifying column consists of at least 45 sieve plates in order to enrich ethanol to at least 94% by volume or more. The ethanol vapor leaves the rectifying column at the top and is led to the main condenser. The main condenser and the aldehyde condenser are manufactured mostly as tubular coolers. The less volatile components condense in the main condenser and flow back to the top of the rectifying column. The highly volatile components are led to the aldehyde condenser and condense there to flow back to the top of the column. Alternatively both condensation products can be drawn off from the condensers and collected in separate receivers. At the beginning of rectification, the top of the heads are drawn off from the aldehyde condenser and are collected separately using the top of the heads receiver. After separation of these tops, the product is drawn off as liquid from the rectifying column. A plate, some plates below the top plate, is used for this purpose. During further rectification, this product is fractionated into heads, product, and tailings. The tailings are fed to the fusel oil decanter where they are diluted with cold water to separate the fusel oils.

Vapor and water pressure must be precisely controlled to obtain a sufficient degree of rectification. Furthermore, it is necessary to install flow meters for steam used in heating and for cooling water. Flow meters are also needed in product and reflux tubes (downpipes) to adjust the reflux ratio during the rectification process, since rectification efficiency depends greatly on the reflux ratio.

In practice, the rectifying stills in use are of various designs. It is possible to construct a rectifying column directly above or beside the reboiler. Or, e.g., it is feasible to install a second product cooler, to permit cooling of the product fraction separate from the cooling of heads and tailings. Another possibility is the installation of draining tubes at the bottom of the rectifying column to draw off fusel oil-containing fractions into a second fusel oil decanter. Further information on the design of rectifying plants and rectification theory have been reported by KREIPE (1981) and KIRSCHBAUM (1969).

This kind of batch fractionated rectification leads to an energy consumption of about 2.5 kg of vapor per lA. Then energy consumption would increase to 5–6 kg vapor per lA, if distillation and rectification are carried out separately.

## 11.3 Distillation and Rectification of the Alcohol Product from Mashes

Because of the energy consumption in distillation and rectification, it is necessary to use a continuous and combined distillation and rectification still in large-scale plants for ethanol production. KREIPE (1981) and KIRSCHBAUM (1969) have reported details for such plants. Investment in such plants is too expensive, and their capacities are too high for distilleries working in the classical tradition. In 1993, a distillation and rectification still was developed that may be of interest to distilleries producing their own potable neutral product distillates. This still can also be used at a large scale.

As shown in Fig. 23, fermented mash is pumped through the cooling coil of the dephlegmator on top of the rectification column and is heated up there. The heated mash is fed to the degassing section of the mash column, and while flowing downwards, ethanol is stripped from the mash. The stillage leaves the mash colunm through the bottom. The degassing section is used to strip high-volatile components which are led to a dephlegmator. The phlegma is fed back to the mash column and the volatile components are condensed and cooled in a tubular cooler, from where the headings are drawn off into a separate receiver.

The ethanol vapor is fed to the rectification column, where rectification takes place as described above, but continuously. From a plate near the top of the rectification column, product spirit is fed to an end column as a liquid draw. This end column is used to separate residual aroma from the product spirit as far as

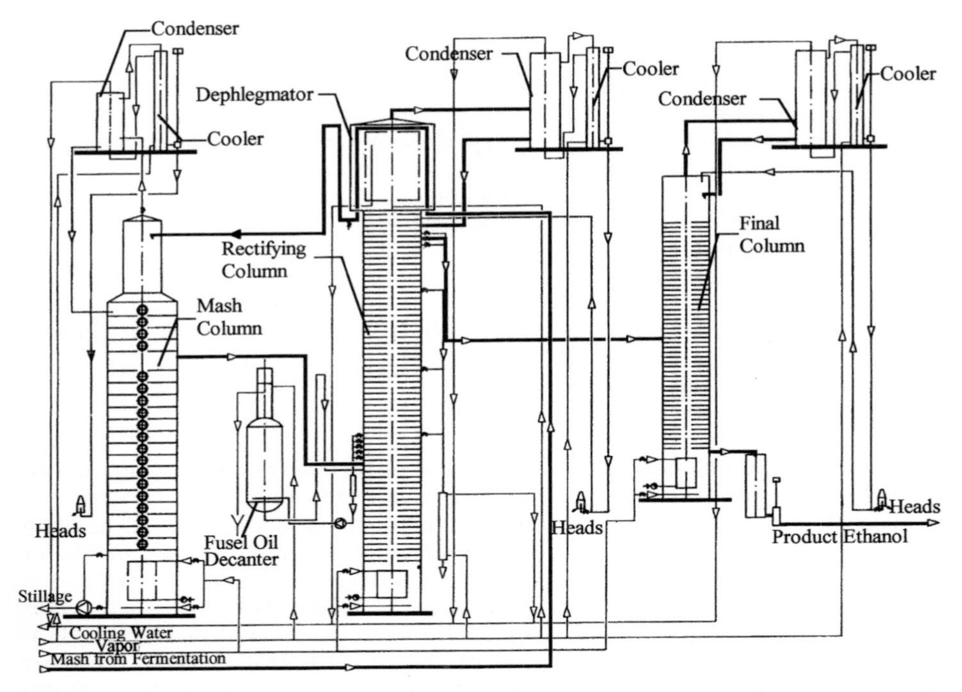


**Fig. 23.** Continuous distillation of product ethanol from mash (J. Carl, D-73002 Göppingen).

possible to yield virtually neutral spirit with about 96% by volume.

Energy consumption for this type of continuous distillation and rectification also amounts to 5 kg of vapor per lA, but the capacity is much higher compared to using a batch rectifying still.

# 12 Stillage

## 12.1 Stillage as a Feedstuff

About 90–140 L of stillage are obtained from each hectoliter of fermented mash depending on the distillation equipment. For a continuous mash distillation stillage amounts to 1,100–1,400 L per 100 lA.

While in the USA more than 85% of grain stillage is dried, in Europe stillage is fed as a liquid as it is obtained from distillation. This is due to the smaller scale of the distilleries which does not warrant the costs of investment for stillage drying plants. In addition to the costs, drying of stillage consumes the same amount of energy as is needed for ethanol production itself (PIEPER, 1983). It should also be considered that it is easier to dry corn stillage than it is to dry stillages from rye, potato, or wheat mashes.

Before stillage is dried, it has to be separated. Concentration of stillage is more successful, the lower the solids content. So before drying, stillage is first separated into thin stillage and solids, called grains, using sieves or decanters. When sieves are used, the grains obtained are about 15–20% DS and can be further dried to about 30–35% DS with a certain press. The grains obtained from a decanter reach a DS of about 25–30%. Thin stillage is further clarified using centrifuges or settling basins. The centrifuge cake obtained from centrifuges has about 20% of DS, of which

50% consists of protein. The clarified thin stillage is then concentrated up to 35–40% DS, and in some cases to 60% DS. Using spray dryers or drum dryers this concentrated sirup is dried to about 90–95% DS. If only clarified thin stillage is dried, the product is called "Dried Distillers' Solubles" (PIEPER, 1983).

To yield a product containing the total DS from the original stillage, the concentrated sirup of thin stillage has to be homogeneously mixed with the solid fraction. This is a difficult process, that is only successful, if previously dried and hot stillage is also added to this mixture. "Distillers' Dried Grains with Solubles" are obtained when this mixture is dried in special drum dryers to 92–94% DS.

Stillage and stillage products used as feedstuff can be classified as follows (PIEPER, 1983):

- Whole stillage: stillage obtained from distillation.
- Thin stillage: liquid phase obtained by sieving, centrifugation, or settling.
- Grains: solid phase obtained by sieving, centrifugation, or settling.
- Centrifuge cake; obtained by clarification of thin stillage.
- Semi-solid distillers' solubles: concentrated thin stillage produced by evaporation.
- Dried distillers' solubles: dried concentrated thin stillage.
- Distillers' dried grains, light grains: dried grains, obtained by sieving, centrifugation or settling.
- Distillers' dried grains with solubles, dark grains: mixture of semi-solid distillers' solubles and grains, dried with a drum dryer or mixture of light grains and dried distillers' solubles.

As stillage is a very important feedstuff, the nutrient content of stillages of different origins, which varies widely, is shown in Tabs. 15–21. Data for digestibility coefficients are given for information only.

Some additional data of the composition of stillage from wheat and corn, obtained using DMP with stillage recycling, are shown in Tabs. 22 and 23. This process yields two fractions of stillage, which are stored in a sedimentation tank and settled there. The thin stillage fraction is clarified and used in the process again, and amounts to about 50% of total stillage. The other fraction is thick stillage, which contains virtually all the solids. It is used as feedstuff or fertilizer.

**Tab. 15.** Nutrient Content [% of DS] and Digestibility [%] of Stillage from Potatoes (KLING and WÖHLBIER, 1983)

| Component | Whole Stillage n=27 | | DDGS n=7 | | DDS n=1 |
|---|---|---|---|---|---|
| OS | 87.4±3.1 | | 85.6±1.7 | | 91.8 |
| Raw protein | 27.0±3.4 | | 27.3±2.1 | | 11.7 |
| Raw fat | 2.7±3.4 | | 2.1±1.3 | | 2.1 |
| Crude fiber | 8.1±3.8 | | 9.8±3.4 | | 9.6 |
| NNE | 49.9±6.8 | | 46.5±3.6 | | 68.5 |
| ASh | 12.6±3.1 | | 14.4±1.7 | | 8.2 |
| DS | ca. 6% | | ca. 90% | | ca. 93% |
| Digestibility | Cow n=2 | Pig n=2 | Cow n=5 | Pig n=7 | |
| OS | 63 | 70 | 72 | 68 | |
| Raw protein | 56 | 58 | 58 | 47 | |
| Raw fat | 50 | 72 | 40 | 56 | |
| Crude fiber | 37 | 77 | 74 | 66 | |
| NNE | 71 | 79 | 80 | 84 | |

**Tab. 16.** Nutrient Content [% of DS] and Digestibility [%] of Stillage from Wheat (KLING and WÖHLBIER, 1983)

| Component | Whole Stillage n=1 | DDGS n=2 | DDS n=6 | Light Grains n=6 |
|---|---|---|---|---|
| Os | 91.4 | 92.5 | 91.8±2.5 | 97.6±0.2 |
| Raw protein | 34.8 | 32.0 | 41.2±4.6 | 36.4±7.5 |
| Raw fat | 2.2 | 6.4 | 0.9±0.3 | 4.4±0.8 |
| Crude fiber | 3.4 | 6.5 | 2.4±0.3 | 13.4±1.7 |
| NNE | 51.0 | 44.6 | 47.2±3.1 | 43.9±4.7 |
| Ash | 8.6 | 7.5 | 8.2±2.5 | 2.4±0.2 |
| DS | ca. 4.2% | ca. 91% | ca. 89% | ca. 90% |
| Digestibility | Cow n=1 | Cow n=1 | Pig n=4 | |
| OS | 58 | 57 | 67 | |
| Raw protein | 54 | 55 | 68 | |
| Raw fat | 50 | 47 | 100 | |
| Crude fiber | 43 | 44 | 20 | |
| NNE | 62 | 61 | 72 | |

**Tab. 17.** Nutrient Content [% of DS] and Digestibility [%] of Stillage from Rye (KLING and WÖHLBIER, 1983)

| Component | Whole Stillage n=1 | DDGS n=3 | DDS n=3 | Light Grains n=4 |
|---|---|---|---|---|
| OS | 96.5 | 97.0 | 91.5 | 92.4±10.8 |
| Raw protein | 42.4 | 22.0 | 38.6 | 25.9± 2.2 |
| Raw fat | 3.5 | 5.1 | 0.8 | 6.5± 0.4 |
| Crude fiber | 5.9 | 15.1 | 2.5 | 13.9± 2.0 |
| NNE | 44.7 | 54.8 | 49.6 | 46.1± 9.8 |
| Ash | 3.5 | 3.0 | 8.5 | 7.7±10.8 |
| DS | ca. 8% | ca. 89% | ca. 95% | ca. 91% |
| Digestibility | Sheep n=2 | Sheep n=4 | | |
| OS | 65 | 51± 8 | | |
| Raw protein | 60 | 59± 7 | | |
| Raw fat | 58 | 62± 3 | | |
| Crude fiber | 50 | 50±18 | | |
| NNE | 71 | 49± 6 | | |

## 12.2 Stillage as a Fertilizer

If stillage is used as a fertilizer it should be spread as fresh stillage. Placing of fresh stillage is virtually odorless. If stillage cannot be spread fresh, and it is necessary to store it, preserving agents have to be added. Stillage can create a very unpleasant smell if it is stored without addition of preserving agents.

With respect to use of stillage from potatoes, MATTHES (1995) calculated the following nutrients content of fertilizer:

**Tab. 18.** Nutrient Content [% of DS] and Digestibility [%] of Stillage from Barley (KLING and WÖHLBIER, 1983)

| Component | Whole Stillage n=8 | DDGS n=2 | DDS n=8 | Light Grains n=1 |
|---|---|---|---|---|
| OS | 97.9±0.2 | 94.5 | 94.5±0.6 | 96.9 |
| Raw protein | 31.3±0.6 | 22.7 | 26.8±4.6 | 19.8 |
| Raw fat | 10.2±1.2 | 4.0 | 5.2±2.2 | 8.4 |
| Crude fiber | 13.7±0.8 | 11.0 | 10.4±5.0 | 18.6 |
| NNE | 42.7±1.4 | 56.8 | 52.1±8.7 | 50.0 |
| Ash | 2.1±0.2 | 5.5 | 5.5±0.6 | 3.1 |
| DS | ca. 26% | ca. 93% | ca. 91% | ca. 93% |
| **Digestibility** | **Sheep n=4** | | | |
| OS | 66±2 | | | |
| Raw protein | 81±1 | | | |
| Raw fat | 88±1 | | | |
| Crude fiber | 34±8 | | | |
| NNE | 63±2 | | | |

**Tab. 19.** Nutrient Content [% of DS] and Digestibility [%] of Stillage from Corn (KLING and WÖHLBIER, 1983)

| Component | Whole Stillage n=12 | DDGS n=24 | | DDS n=24 | | Light Grains n=14 |
|---|---|---|---|---|---|---|
| OS | 95.3± 1.3 | 95.2±2.2 | | 89.8±7.7 | | 96.7±1.5 |
| Raw protein | 25.5± 7.0 | 28.5±5.3 | | 28.2±5.5 | | 26.6±5.8 |
| Raw fat | 11.7± 3.6 | 7.7±3.4 | | 5.4±3.0 | | 7.9±7.8 |
| Crude fiber | 10.6± 4.7 | 11.9±3.2 | | 5.3±3.4 | | 12.3±2.9 |
| NNE | 47.6±12.5 | 47.2±6.5 | | 51.2±5.5 | | 49.9±8.8 |
| Ash | 4.7± 1.3 | 4.8±2.2 | | 10.1±7.7 | | 3.3±1.5 |
| DS | ca. 8.5% | ca. 90% | | ca. 92% | | ca. 92% |
| **Digestibility** | **Sheep n=2** | **Cow n=1** | **Pig n=4** | **Cow n=2** | **Chicken n=1** | **Chicken n=1** |
| OS | 67 | 72 | 80 | 75 | 69 | 58 |
| Raw protein | 64 | 67 | 78 | 73 | 80 | 78 |
| Raw fat | 89 | 91 | 92 | 81 | 90 | 56 |
| Crude fiber | 59 | 69 | 81 | 32 | 0.0 | 36 |
| NNE | 71 | 75 | 91 | 89 | 60 | 51 |

- 2.7 kg m$^{-3}$ N
- 1.1 kg m$^{-3}$ $P_2O_5$
- 4.2 kg m$^{-3}$ $K_2O$

Further the use of fresh potato stillage results in a cost saving of 0.93 DM per hlA produced in a distillery, compared with the use of mineral fertilizers. But if stillage has to be stored there are additional costs that amount up to 5.45 DM per hlA (MATTHES, 1995).

For the use of fresh grain stillage from a distillery processing with stillage recycling

**Tab. 20.** Nutrient Content [% of DS] of Stillage from Sorghum Grain (KLING and WÖHLBIER, 1983)

| Component | Whole Stillage n=8 | DDGS n=2 | DDS n=2 |
|---|---|---|---|
| OS | 94.7±1.3 | 93.7 | 86.2 |
| Raw protein | 29.8±3.4 | 26.6 | 21.2 |
| Raw fat | 8.6±2.7 | 7.3 | 6.6 |
| Crude fiber | 8.8±1.7 | 9.1 | 3.8 |
| NNE | 47.5±6.4 | 49.9 | 54.7 |
| Ash | 5.3±1.3 | 6.3 | 13.8 |
| DS | ca. 17% | ca. 93% | ca. 92% |

**Tab. 21.** Nutrient Content [% of DS] of Stillage from Manioc (KLING and WÖHLBIER, 1983)

| Component | Whole Stillage n=1 | DDGS n=1 | DDS n=1 |
|---|---|---|---|
| OS | 93.2 | 90.7 | 91.1 |
| Raw protein | 14.4 | 12.3 | 28.4 |
| Raw fat | 1.8 | 1.6 | 5.3 |
| Crude fiber | 2.8 | 5.6 | 2.1 |
| NNE | 74.2 | 71.2 | 55.3 |
| Ash | 6.7 | 9.3 | 8.9 |
| DS | ca. 3.3% | ca. 92% | ca. 95% |

**Tab. 22.** Composition of Stillage [% of OS] from Wheat Obtained from DMP with Stillage Recycling

| Component | Whole Stillage n=2 | Thin Stillage n=2 | Thick Stillage n=2 |
|---|---|---|---|
| Raw protein | 2.40 | 1.34 | 3.38 |
| Raw fat | 0.35 | 0.00 | 0.82 |
| Crude fiber | 0.43 | 0.004 | 1.14 |
| NNE | 2.26 | 1.68 | 4.15 |
| Ash | 0.40 | 0.38 | 0.41 |
| DS | 6.2 | 3.4 | 9.9 |

**Tab. 23.** Composition of Stillage [% of DS] from Corn Obtained from DMP with Stillage Recycling

| Component | Whole Stillage n=2 | Thin Stillage n=2 | Thick Stillage n=2 |
|---|---|---|---|
| Raw protein | 31.5 | 21.2 | 41.4 |
| Raw fat | 15.1 | – | 14.3 |
| Crude fiber | 12.6 | – | 9.0 |
| NNE | 34.3 | – | 30.3 |
| Ash | 6.5 | – | 5.0 |
| DS | 3.7 | 2.1 | 5.8 |

(settling of stillage in a sedimentation tank), the savings are greater because the amount of stillage is halved, and the protein content of thick stillage is about 50% higher compared with whole stillage obtained from the distillation. This reduces fertilizer costs compared with the fertilizer from stillage of potatoes. In addition, if stillage has to be stored, one has to store half the amount only, thereby reducing the storage costs.

There are some advantages to the use of stillage as fertilizer compared with mineral fertilizers. Stillage contains only a part of the total nitrogenous nutrients in the form of immediately available nitrogen. Most of the nitrogen is fixed in proteins, and is slowly mineralized during the growth period. Thus, there is only a low risk of rinsing out nitrogen to the ground water. But further tests of the use of stillage as a fertilizer are still necessary.

# 13 Analytical Methods

## 13.1 Introduction

It is absolutely necessary to follow conversion and distillation processes using analytical methods. There is no other way to ensure an effective and careful process and to avoid contamination and ethanol losses due to incomplete conversion. The most important analytical methods used in distilleries are described in this section.

## 13.2 Analysis of Raw Materials

### 13.2.1 Starch Content of Potatoes

In practice it is very difficult to determine the exact starch content of potatoes, since it is virtually impossible to prepare an average sample under practical conditions. Furthermore, the starch content varies among individual potatoes.

Special balances are used to determine the density of potatoes. In this procedure exactly 5,050 g of wet potatoes are weighed out, and then these potatoes are weighed again under water. The density of potatoes can be calculated using Eq. (2):

$$\begin{aligned}\text{Density} &= \frac{\text{Absolute weight}}{\text{Volume}}\\ &= \frac{\text{Absolute weight}}{\text{Loss of weight}}\\ &= \frac{5{,}050\ \text{g}}{5{,}050\ \text{g} - \text{Loss of weight}}\end{aligned} \tag{2}$$

The relation between the density of potatoes and their dry substance or starch content can be regarded as constant. Therefore, it is possible to calculate the starch content of potatoes using the tables first calculated by BEHREND (KREIPE, 1981; ADAM et al., 1995). The most widely used balances for this purpose are those of PAROW or ECKERT. Both balances are scaled to weigh out 5,050 g potatoes, and a second scale directly shows the starch content when the potatoes are weighed under water.

This method is not very exact, but good enough for ethanol production purposes; these balances are also often used in the starch and food industry. If large potatoes are weighed with these balances, it is necessary to cut through them, since large potatoes often have internal cavities.

### 13.2.2 Starch Content of Grain

Normally the starch determination is carried out using polarimetric methods. These methods consist of the following steps:

- release of starch from cell material and solubilization of starch;
- separation of optically active substances, that may interfere with the measurement of starch;
- measurement of the angle of polarization due to the concentration of solubilized starch.

These methods have been described by EWERS (1909), EARL and MILLNER (1944), and CLENDENNING (1945). It was shown by SENN and PIEPER (1987) that a variation in clarifying agents used for the separation of optically active substances interferes with the measurement of starch. Therefore, it was suggested by SENN and PIEPER (1987) to determine instead the content of fermentable substances in grain for ethanol production purposes.

#### 13.2.2.1 Determination of Fermentable Substance in Grain (FS)

The fermentable substance (FS) is defined as the sum of the glucose and maltose contents of the raw material, calculated as starch, that can be determined using HPLC after the raw material is completely digested and dispersed as well as liquified and saccharified by addition of technical enzymes.

The determination of FS is carried out using the following procedure:

- 10 g of thoroughly milled grain are weighed in a mash beaker of a laboratory masher with a precision of 1 mg.
- 300 mL of $H_2O$ are added.
- The pH is adjusted to 6.0–6.5.
- Each charge is dispersed for 1.5 min, using a laboratory dispersing machine (e.g., ULTRA TURRAX™).
- Then the mash beaker must be placed in the laboratory masher and is stirred there.
- 0.2 mL of a thermostable $\alpha$-amylase from *B. licheniformis* is added (e.g., Termamyl 60 L, NOVO-Industri, Kopenhagen).
- The laboratory masher is heated up to 95 °C.
- While stirring the charges are maintained at 95 °C for 60 min.
- During this rest, after about 30 min, the charges are dispersed again for about 3 min.
- After this rest the masher is cooled to 52–53 °C (temperature inside the beakers).
- The pH is adjusted to 5.0.
- For saccharification the following enzymes are added:
  - 0.2 mL fungal $\alpha$-amylase from *A. oryzae* (e.g., Fungamyl L 800L, NOVO-Industri, Kopenhagen),
  - 2.0 mL glucoamylase from *A. niger* (e.g., Optisprit-L, Solvay Enzymes, Hannover),
  - 0.1 g glucoamylase from *Rhizopus* sp. (e.g., Optilase G 150, Solvay Enzymes, Hannover).
- A saccharification rest is done over night to complete saccharification, for which the beakers have to be covered.
- Then the charges are quantitatively rinsed into a 1,000 mL measuring flask.
- The measuring flask is kept at a constant temperature and exactly filled up.
- The sample is then filtered using folded filters (MN 615 ¼).
- The sample is filtered again using a 0.45 μm membrane filter.
- 20 μL of this sample are then injected into the HPLC system.

For the determination of glucose and maltose the following chromatographic equipment and conditions are used:

- Pump: Bischoff, Model 2200
- Chromatographic column: Biorad, HPX 87 H, 30 cm
- Temperature: 50 °C
- Eluent: purified water
- Flow rate: 0.7 mL $min^{-1}$
- Detection: RI-Detector, ERC-7510
- Sample amount: 10 μL

Using the results obtained from the HPLC determination, the FS content of the raw materials is calculated following Eq. (3):

$$FS = \frac{\text{Glucose } [\text{g L}^{-1}] \cdot 0.899 + \text{Maltose } [\text{g L}^{-1}] \cdot 0.947}{\text{Weighed portion of raw material } [\text{g}]} \cdot 100$$

This method for the determination of FS can be used for all starch containing raw materials, weighing about 1–7 g of starch into the mash beakers. This method can also be used to examine stillage. Organic acids and alcohols can also be determined with a Biorad HPX-87-H column. Therefore, the analysis of stillage can be used to examine the quality of the ethanol production processes, showing, e.g., if there is residual FS in the stillage, contamination of mashes (organic acids), and ethanol loss in distillation.

### 13.2.3 Autoamylolytical Quotient (AAQ)

For processing wheat, rye, or triticale, it is important to know the AAQ, which gives information on the activity of the autoamylolytical enzyme system. The AAQ is defined as the percentage yield of ethanol obtained without the addition of saccharifying enzymes, compared with the ethanol yield with addition of an optimum combination of technical enzymes (see Eq. (1)).

In order to calculate the AAQ, it is necessary to carry out fermentation experiments in the laboratory, using the following procedures to examine wheat, rye, and triticale.

Fermentation test using technical enzymes:

- 80.0 g of thoroughly milled grain are weighed in a mash beaker of a laboratory masher.
- 300 mL of $H_2O$ are added.
- The pH is adjusted to 6.0–6.5.
- Each charge is dispersed for 1.5 min, using a laboratory dispersing machine (e.g., ULTRA TURRAX™).
- Then the mash beaker is placed in a laboratory masher and stirred.
- 0.65 mL of a thermostable $\alpha$-amylase from *B. licheniformis* is added (e.g., Termamyl 60L, NOVO-Industri, Kopenhagen).
- The laboratory masher is heated to the liquefaction temperature of 65 °C.
- While stirring, the charges are maintained at liquefaction temperature for 30 min.
- After this rest the masher is cooled to 52–53 °C (temperature measured inside the beakers).
- The pH is adjusted to 5.0–5.2.
- For saccharification the following enzymes are added:
  - fungal $\alpha$-amylase from *A. oryzae* (0.1 mL $kg^{-1}$ FS, e.g., Fungamyl L 800L, NOVO-Industri, Kopenhagen),
  - glucoamylase from *A. niger* (0.16 mL $kg^{-1}$ FS, e.g., Optisprit-L, Solvay Enzymes, Hannover),
  - glucoamylase from *Rhizopus* sp. (0.4 g $kg^{-1}$ FS, e.g., Optilase G 150, Solvay Enzymes, Hannover).
- A saccharification rest of 30 min is maintained at 52–53 °C.
- The mash is then cooled to 30 °C, and rinsed quantitatively into an Erlenmeyer flask.
- 1.0 g of a dried distillers', pure-culture yeast, which has been rehydrated before the test is added.
- The Erlenmeyer flask is then closed with a fermentation tube and placed into a water bath at 30 °C. The mash is fermented there for 3 d.

Autoamylolytical fermentation test without using any technical enzymes:

- 80.0 g of thoroughly milled grain are weighed in a mash beaker of a laboratory masher.
- 300 mL of $H_2O$ are added.
- The pH is adjusted to 5.5.
- Each charge is dispersed for 1.5 min, using a laboratory dispersing machine (e.g., ULTRA TURRAX™).
- The mash beaker is placed in a laboratory masher and stirred.
- The laboratory masher is heated to the liquefaction temperature of 60 °C.
- While stirring, the charges are maintained at liquefaction temperature for 60 min.
- After the rest, the pH is adjusted to 5.0–5.2.
- Cooling to 30 °C and further processing is done as in the fermentation test using technical enzymes, except that no technical enzymes are used.

When the fermentation is completed, the mashes are distilled using a laboratory still, ethanol is measured by araeometry or densimetry, and the AAQ is calculated.

If the AAQ is higher than 95%, it is possible to process the raw material using one of the autoamylolytical processes described in Sect. 7. Fig. 24 shows the relation between ethanol yields from grains and the AAQ measured in these grains.

## 13.3 Analysis of Mashes

### 13.3.1 Mash Hydrosizing

Mash hydrosizing which is an easy and effective method to examine the degree of de-

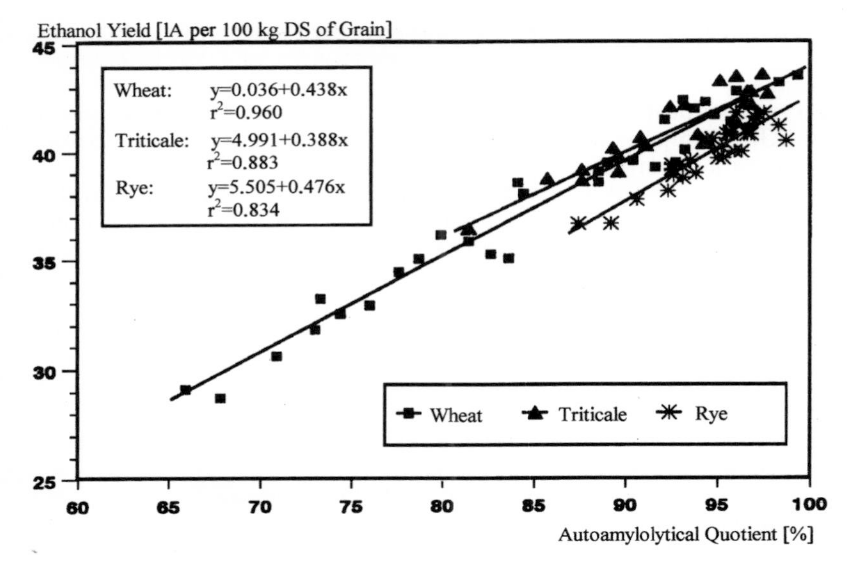


**Fig. 24.** Ethanol yields from wheat, triticale and rye: relation to autoamylolytical quotient.

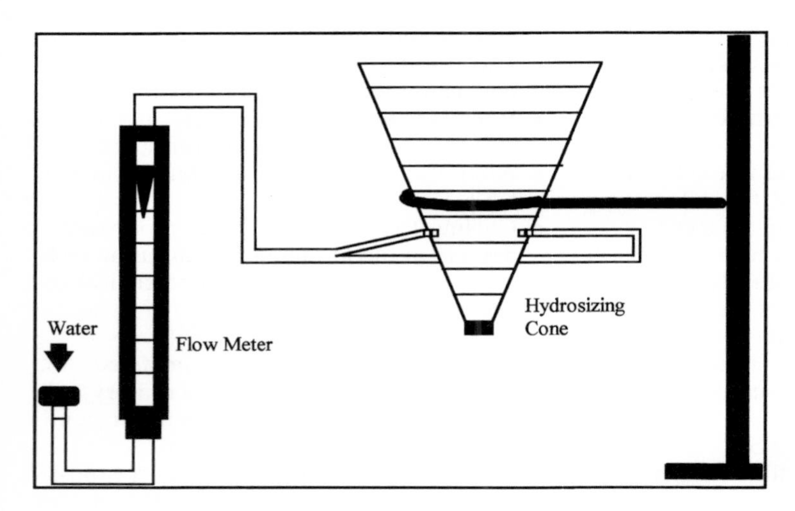


**Fig. 25.** Mash hydrosizer.

composition of raw materials during the mashing process was developed by PIEPER and HOTZ (1988). The mash hydrosizer consists of a sediment cone fitted with two water nozzles to drain water into the cone. The outlets of the nozzles are aligned horizontally to create a rotating upward flow (Fig. 25). The flow rate of water drained into the hydrosizer depends on the raw material for the mashes (Tab. 24).

**Tab. 24.** Flow Rates of Water Used in Hydrosizing of Mashes from Different Raw Materials

| Raw Material | Flow Rate [L $h^{-1}$] |
|---|---|
| Potatoes | 180 |
| Wheat, rye, triticale | 260 |
| Barley | 280 |
| Corn | 300 |

To examine a mash, water flow is started and 0.5 L of mash are poured into the sediment cone from the top. Then the water flow is adjusted exactly, and the mash is hydrosized for 10 min, during which time the solids are separated into three fractions:

- the light fraction, which is rinsed out of the sediment cone on the top, containing small and light particles, already digested;
- the suspended fraction, remaining in the hydrosizer and containing only small amounts of starch that is not released from the cell material;
- the coarse fraction, which settles to the bottom of the sediment cone, even though water is still running; this fraction contains the heavy particles, that still contain large amounts of starch included in the cell material.

After 10 min the flow of water is stopped and the suspended fraction settles to the bottom of the hydrosizer. The sediment cone is scaled in mL, and so the amount of the coarse fraction and the sum of the suspended and coarse fractions are read off as results. To guarantee optimum ethanol yields from all starchy raw materials, hydrosizing should give the following maximum values:

Coarse fraction: $<1.5$ mL
Coarse fraction + suspended fraction:
$<5.0$ mL
If barley is processed:
Coarse fraction + suspended fraction:
$<15.0$ mL

### 13.3.2 Extract of Mashes

Only saccharometers according to Plato should be used to measure the extract content of mashes. These araeometers are scaled in %mas at 20°C and are fitted with a thermometer and a temperature correction scale. It is necessary to filter the mashes using folded filters before the extract content of mashes can be measured. The filtered mash is filled into a measuring cylinder, standing in a perpendicular position. The measured extract content represents the total content of solubilized substances, whether or not they are fermentable, and is not equivalent to the content of fermentable sugars. Therefore, the extract content depends not only on the starch content of the raw materials, but also on the mashing process.

The extract content in sweet mashes normally amounts to about 17%–18%mas using the HPCP. With pressureless mashing processes, the extract content amounts to 14%–16%mas if no stillage recycling is used. If stillage recycling is used, the extract content depends on the proportion of stillage recycling and amounts to 16%–20%mas in sweet mashes. The extract contents, measured when fementation is completed and when no stillage recycling is used in ethanol production, are shown in Tab. 14.

Using stillage recycling, the degree of fermentation depends on the proportion of stillage recycling used. However, using the same conditions and the same raw material in ethanol production results in the same degree of fermentation. So it is also possible to use measurements of the extract content and the degree of fermentation for the evaluation of fermentation efficiency.

### 13.3.3 pH of Mashes

The efficiency of the enzymatic treatment and the yeast fermentation depends greatly on the pH of mashes. Therefore, it is necessary to measure and to adjust the pH during the mashing processes and the fermentation depending on the enzymes used. The optimum pH values required for different processes and raw materials are described in Sects. 2 and 5.8. For the measurement of pH values only pH meters with a pH electrode should be used. The often used pH indicator paper is not precise enough, and the resulting color change depends not only on the pH value, but also on the color of the mashes.

### 13.3.4 Content of Ethanol in Mashes and Distillates

The ethanol content of mashes is determined by a laboratory distillation. The obtained distillate is measured using alcoholometers. In alcoholometry only alcoholometers authorized by the EU according to DIN 12803 should be used. These araeometers are calibrated to show the volume concentration of ethanol at 20°C (%vol) in a mixture of ethanol and water. Additionally these alcoholometers are fitted with a thermometer. It is possible to measure ethanol concentrations in mixtures of ethanol and water between 0°C and 30°C by use of the official EU ethanol tables.

At the end of fermentation, mash solids start settling. A sample taken from a sampling cock is not an average sample. The best way to get reliable data is to filter the mashes, using folded filter papers, before distillation. Afterwards, 200 mL of mash filtrate are filled into the boiling flask of a laboratory still and diluted with 200 mL of $H_2O$; the pH is adjusted to 7.0. Then the diluted mash filtrate is distilled into a 200 mL one-mark volumetric flask, used as a receiving flask. The volumetric flask is removed from the lab still and the temperature is adjusted before it is filled up to the mark with distilled water. Then the ethanol concentration is measured using an EU-approved alcoholometer. The ethanol concentration measured in the laboratory distillate is equal to the ethanol content in mash filtrates. Therefore, the ethanol content in mash is calculated from the ethanol content determined in the mash filtrate times the spent grains factor. The grains factor is 0.97 for mashes from wheat, rye, triticale, corn, and potatoes, and 0.95 for barley mashes (KREIPE, 1981).

## 13.3.5 Microexamination

Microscopic examinations of mashes and yeast mashes should be carried out daily in a distillery. For this purpose, a microscope should be used that magnifies 1,000 times, preferably with phase-contrast. Microexamination is the only reliable method to determine whether the mashes are contaminated, and in addition shows if the yeast is in suitable condition.

It is not necessary to precisely count the microorganisms seen in the microscope; it is sufficient to classify the mashes as shown in Tab. 25. For this, a magnification of 1:1,000 is needed, and it is necessary to examine five or more different fields of view.

## 13.4 Analysis of Yeast Mashes

Yeast mash treatment is described in Sect. 9. It is important to carry out microexaminations and extract determinations daily and to watch the temperature and the pH value.

In a yeast mash there may be no bacteria visible in microexaminations. The pH of the yeast mashes should be adjusted to 2.0 for about 2 h before the yeast mash is used for inoculation of sweet mash, if bacteria are present in yeast mashes. Otherwise, the pH of yeast mashes is adjusted to 2.5 (Sect. 9).

During fermentation of the yeast mash, the extract content should be reduced to about 50% of the extract content of the sweet mash, that was used in its preparation. The temperature of the yeast mash has to be kept below 27°C until it is used for inoculation.

**Tab. 25.** Mash Classification Using Microexamination

| Visual Examination | Contamination |
|---|---|
| No bacteria | O.K. |
| Single bacteria are visible | Technically free from contaminations |
| Number of bacteria up to 20% of yeast cells | Lightly contaminated |
| Up to number of bacteria = number of yest cells | Strongly contaminated |
| Number of bacteria > number of yeast cells | Cultivation of bacteria |

## 13.5 Analysis of Stillage

### 13.5.1 Content of Ethanol in Stillage

The stillage should be free of ethanol. If ethanol is present in stillage, this is due to a loss of ethanol and it shows that the distillation was not carried out with enough care. If ethanol is found in the stillage, usually the mash flow in the still is too high.

The determination of ethanol in stillage is carried out as described in Sect. 13.3.5. The stillage is not filtered before the sample is filled into the boiling flask of the laboratory still. Instead the samples of stillage have to be withdrawn directly from the bottom of the mash column. These samples must be filled immediately into a sample container that can be closed tightly. When the sample of stillage is cooled down in this tight container, it can be used for ethanol determination.

The ethanol contents measured in stillage are normally below 0.5% vol. It is not possible to get reliable data using a laboratory still, since stillage can not be compared with a mixture of ethanol and water; during the laboratory distillation of stillage, a lot of impurities (fatty acids, esters, aldehydes, etc.) are also distilled. This also occurs in the distillation of mashes, but the low ethanol content increases the error of measurement using araeometry or densimetry. Better results in the ethanol determined in stillage can be achieved if HPLC is used, as described in Sect. 13.5.2.

### 13.5.2 Content of Starch and Fermentable Sugars in Stillage

Normally the determination of residual starch and fermentable sugars in stillage is carried out using a fermentation test. The stillage is treated with thermostable $\alpha$-amylases at 90 °C and with saccharifying enzymes at 55 °C as described in Sect. 13.2.3. After that, the stillage is fermented and distilled as described above, and the ethanol content is measured using alcoholometers. This method shows the same disadvantages as described in Sect. 13.5.1. Since, as a rule, the content of FS in stillage amounts to not more than 2% mas, it is not sure if the FS is metabolized completetely or if the fermentation starts at all after the yeast is added.

Reliable results of tests for residual FS contents in stillage can be achieved if the HPLC method for the determination of FS in grain is used. For this, 300 g of stillage are weighed into the mash beaker, and there is no further dilution of stillage with this method. This is the only change in the method; further treatment is carried out as described in Sect. 13.2.2.1.

It is also possible to inject a sample of stillage already filtered, without further digestion, into the HPLC system, for the detection of ethanol and free fermentable sugars in the stillage. Using a Biorad column HPX-87-H, as described, some acids can also be detected. The HPLC method shows reliable results for residual contents of ethanol, free fermentable sugars and FS, and contamination during the fermentation. The FS determined in this way is a reliable indicator of the effectivity of the mashing process as well as of the fermentation and distillation, if stillage is examined once before and once after further digestion.

# 14 Energy Consumption and Energy Balance in Classical Processes

The energy consumption in ethanol production depends greatly on the process used. The energy used in some classical processes for ethanol production from potatoes, corn, and wheat is shown in Tabs. 26 and 27. The energy consumption also depends on the ratio of stillage recycling. Due to the fact that potatoes contain about 80–100% of the process liquid needed, processing of potatoes leads to the highest energy consumption in this comparison. It was found that DMP saves again 30% of energy consumption in the mashing process, compared with the process described by GOSLICH (1990). So it is possi-

**Tab. 26.** Energy Consumption in Ethanol Production from Potatoes and Corn [MJ per hlA]

| Raw Material | Potatoes | Corn | |
|---|---|---|---|
| Ratio of Stillage Recycling | 15% | 30% | 50% |
| Mashing process | | | |
| Electrical energy | 30 | 40 | 40 |
| Thermal energy | 380 | 170 | 150 |
| Σ Mashing process | 410 | 210 | 190 |
| Distillation[a] | 700 | 700 | 700 |
| Total | 1110 | 910 | 890 |

[a] 250 kg of steam per hlA = 700 MJ per hlA for distillation of raw spirit (85 %vol.). These data depend strongly on the distillation equipment used.

**Tab. 27.** Energy Consumption in Ethanol Production from Wheat [MJ per hlA]

| Raw Material | Wheat, DMP | | HPCP[a] | Wheat[b] |
|---|---|---|---|---|
| Ratio of Stillage Recycling | 30% | 50% | – | – |
| Mashing process | | | | |
| Electrical energy | 40 | 40 | 20 | 60 |
| Thermal energy | 80 | 0 | 700 | 0 |
| Σ Mashing process | 120 | 40 | 720 | 60 |
| Distillation[c] | 700 | 700 | 700 | 700 |
| Total | 820 | 740 | 1420 | 760 |

[a] PIEPER and BOHNER (1985).
[b] Infusion process, using a spiral heat exchanger for heat recovery from stillage to mash (GOSLICH, 1990).
[c] 250 kg of steam per hlA = 700 MJ per hlA for a conventional distillation of raw spirit (85% vol.). These data depend strongly on the distillation equipment used.

ble to run a mashing process for ethanol production from wheat, rye, or triticale, consuming only 40 MJ per hlA, if DMP is used. This data were not just estimated but rather measured in practice, and it is reproducible in all plants using DMP.

If these data for energy consumption are compared with data from large-scale ethanol production (MISSELHORN, 1980b), it can be seen that it is possible, using DMP and other classical methods, to produce ethanol from starchy raw materials with energy inputs that are lower than those calculated for large-scale plants (Tab. 28). Additionally one has to take into account than at a large-scale it is possible to produce dehydrated ethanol with the same energy that is needed at a small scale to produce raw spirit (85% vol).

Virtually all the energy balances for ethanol production from agricultural starchy raw materials are calculated with data from large-scale plants. Thus the data given above is of great significance for energy balances. As shown in Tab. 29, energy data for DMP is of

**Tab. 28.** Energy Consumption of Large-Scale Ethanol Production (MISSELHORN, 1980b)

| | Thermal Energy [MJ per hlA] | Electrical Energy [MJ per hlA] |
|---|---|---|
| Cleaning | 2.2 | 7.2 |
| Mashing process, batch | 770 | 14.4 |
| Mashing process, continuous | 242 | 7.2 |
| Fermentation | 2.2 | 3.6 |
| Distillation, conventional dehydrated ethanol | 935 | 10.8 |
| Distillation, multi-pressure system dehydrated ethanol | 550 | 14.4 |
| Total | | |
| Batch process conventional distillation | 1745 | |
| Continuous process multi-pressure distillation | 828 | |

**Tab. 29.** Energy Balance for Ethanol Production from Wheat or Triticale [GJ ha$^{-1}$] (Deutscher Bundestag, 1986 (FAL); SCHÄFER, 1995)

| | FAL[a] | Triticale, DMP | Wheat, DMP |
|---|---|---|---|
| | Input | | |
| Soil treatment | | 3.40 | 3.35 |
| Fertilizers + biocides | | 11.00 | 11.00 |
| Grain harvest | | 1.06 | 0.99 |
| Straw pressing | | 1.70 | 1.54 |
| Σ Agriculture | 26.2 | 17.16 | 16.88 |
| Mashing process | | 1.11 | 1.03 |
| Fermentation + enzymes | | 0.56 | 0.51 |
| Multipressure distillation | | 16.06 | 14.98 |
| Σ Conversion | 29.9 | 17.73 | 16.52 |
| Σ Input | 56.1 | | |
| With straw recovery | | 34.89 | 33.40 |
| Without straw recovery | | 33.19 | 31.86 |
| | Output | | |
| Ethanol | 44.5 | 58.69 | 54.76 |
| Stillage | 10.1 | 7.27 | 6.44 |
| Straw | 88.1 | 117.76 | 106.59 |
| Biogas | 9.6 | | |
| Σ Output | 152.3 | 183.72 | 167.79 |
| | Output : Input | | |
| Bioethanol | 0.96 | 1.77 | 1.72 |
| Bioethanol + stillage | 1.14 | 1.99 | 1.92 |
| Bioethanol, stillage + straw | 2.71 | 5.27 | 5.02 |

[a] Bundesforschungsanstalt für Landwirtschaft.

considerable importance in the energy balance of ethanol production. This energy balance (SCHÄFER, 1995) was calculated from empirically determined data for the growth of wheat and triticale in agriculture and from the conversion data derived from DMP. Compared with the data from large-scale plants (calculated), the ratio of energy output to input increases from 0.96 (FAL, see Tab. 29) to 1.77 (triticale, DMP). If the use of stillage as a fertilizer, and of straw (fuel value), is included in this calculation, the energy output to input ratio increases to 5.27.

Given these data on energy consumption and energy balance, it should be possible to produce bioethanol as a fuel component under much better conditions than currently thought. One also has to consider that if the stillage is used to produce biogas before it is used as a fertilizer, ethanol production is running energetically autarkical. In this manner it is possible to produce a fuel component within the $CO_2$ circle and surplus energy.

# 15 References

ADAM, W., BARTELS, W., CHRISTOPH, N., STEMPFL, W. (1995), *Brennereianalytik,* Vol. 1–2. Hamburg: Behr's Verlag.

AUFHAMMER, W., PIEPER, H. J., STÜTZEL, H., SCHÄFER, V. (1993), Eignung von Korngut verschiedener Getreidearten zur Bioethanolproduktion in Abhängigkeit von der Sorte und den Aufwuchsbedingungen, *Bodenkultur* **44**, 183–194.

AUFHAMMER, W., PIEPER, H. J., KÜBLER, E., SCHÄFER, V. (1994), Eignung des Korngutes von Weizen und Triticale für die Bioethanolproduktion in Abhängigkeit vom Verarbeitungszeitraum nach der Reife, *Bodenkultur* **45**, 177–187.

BOETTGER, A., ROESNER, R., PIEPER, H. J. (1995), Untersuchungen zur Inhibierung von Pektinesterase durch Tenside und Gerbstoffe bei der industriellen Alkoholgewinnung aus Kartoffeln, in: *Proc. DECHEMA-Jahrestagungen* 1995, Vol. 1, pp. 324–325.

BUTZKE, C., MISSELHORN, K. (1992), Zur Acroleinminimierung in Rohsprit, *Branntweinwirtschaft* **132**, 27–30.

CHIANG, J. P., ALTER, J. E., STERNBERG, M. (1979), Purification and characterization of a thermostable $\alpha$-amylase from *Bacillus licheniformis, Starch/Stärke* **31**, 86–92.

CLENDENNING, K. A. (1945), *Can. J. Res., Sect. B* **23**, 113.

Deutscher Bundestag (1986), Antwort der Bundesregierung auf die große Anfrage der Abgeordneten Carstensen (EIGEN et al.) Nachwachsende Rohstoffe, *Drucksache 10/5558.*

DIEDRICH, J., KAHNT, G., GRONBACH, G. (1993), Bioalkohol und Zellulose aus Topinambur und Zuckerhirse, *Landwirtsch. Wochenbl. Baden-Württemberg* **160**, Heft 4, 16–18.

DREWS, B., PIEPER, H. J. (1965), Weitere Untersuchungen über den Einfluß von Gibberellinsäure bei der Herstellung von Brennmalzen, *Branntweinwirtschaft* **105**, 29–31.

EARL, F. E., MILNER, R. T. (1944), *Cereal Chem.* **21**, 567.

EWERS, E. (1909), *Z. Unters. Nahr. Genussm.* **18**, 244.

FOGARTY, W. M., KELLY, C. T. (1979), Starch degrading enzymes of microbial origin, in: *Progress in Industrial Microbiology* Vol. 15 (BULL, M. J., Ed.). Amsterdam: Elsevier.

GOSLICH, V. (1990), Eine moderne Brennereianlage mit einem kontinuierlichen Maischverfahren und Schlemperecycling, *Branntweinwirtschaft* **130**, 254–260.

HARRIS, G. (1962), The enzyme content and enzymatic conversion of malt, in: *Barley and Malt, Biology Biochemistry, Technology* (COOK, A. H., Ed.). New York: Academic Press.

HEIL, M., SENN, T., PIEPER, H. J. (1994), Alkoholproduktion aus Gerste nach dem Hohenheimer Dispergier-Maischverfahren mit Schlemperecycling, *Handb. Brennerei Alkoholwirtsch.* **41**, 333–353.

KIRSCHBAUM, E. (1969), *Destillier- und Rektifiziertechnik,* 4. Edn. Berlin-Heidelberg: Springer-Verlag.

KLING, M., WÖHLBIER, W. (1983), *Handels-Futtermittel* Vol. 2a and Vol. 2b. Stuttgart: Eugen Ulmer.

KLISCH, W. (1991), Vergleichende Untersuchungen der Verflüssigungswirkung einer neuen Alpha-Amylase von NOVO-Nordisk mit klassischen NOVO-Enzymen, *Branntweinwirtschaft* **131**, 342–344.

KREIPE, H. (1981), Getreide und Kartoffelbrennerei, in: *Handbuch der Getränketechnologie.* Stuttgart: Eugen Ulmer.

KREIPE, H. (1982), Der drucklose Stärkeaufschluß in Theorie und Praxis, *Handb. Brennerei Alkoholwirtsch.* **29**, 248–272.

KRELL, U., PIEPER, H. J. (1995), Entwicklung eines Betriebsverfahrens zur acroleinfreien Alkoholproduktion aus stärkehaltigen Rohstoffen, *Handb. Brennerei Alkoholwirtsch.* **42**, 371–391.

LABERGE, D. E., MARCHYLO, B. A. (1983), Heterogeneity of the $\beta$-amylase enzymes of barley, *J. Am. Soc. Brew. Chem.* **41**, 120–122.

LEWIS, M. J., YOUNG, T. W. (1995), *Brewing.* London: Chapman and Hall.

MANNERS, D. J., SPERRA, K. L. (1966), Studies on carbohydrate-metabolizing enzymes. Part XIV: The specifity of *R*-enzyme from malted barley, *J. Inst. Brew.* **72**, 360–365.

MARCHYLO, B. A., KRUGER, J. E., MACGREGOR, A. W. (1984), Production of multiple forms of $\alpha$-amylase in germinated, incubated, whole, de-embryonated wheat kernels, *Cereal Chem.* **61**, 305–310.

MATTHES, F. (1995), Bewertung von Schlempe als Bodendünger, *Handb. Brennerei Alkoholind.* **42**, 393–402.

MISSELHORN, K. (1980a), Supramyl – ein Verfahren zur Herstellung von Energiealkohol, *Chem. Rundsch.* Nr. **38**.

MISSELHORN, K. (1980b), Äthanol als Energiequelle und chemischer Rohstoff, *Branntweinwirtschaft* **120**, 2–10.

NARZISS, L. (1976), *Die Bierbrauerei,* Vol. 2, Technologie der Malzbereitung. Stuttgart: Ferdinand Enke Verlag.

PIEPER, H. J. (1968), Technologische Untersuchungen über Herstellung, Eigenschaften und Wirksamkeit von Gibberelinsäure-Grünmalzen, *Branntweinwirtschaft* **108**, 319–322, 345–352, 377–380.

PIEPER, H. J. (1970), *Mikrobielle Amylasen bei der Alkoholgewinnung.* Stuttgart: Eugen Ulmer.

PIEPER, H. J. (1983), Gärungstechnologische Alkoholproduktion, in: *Handels-Futtermittel,* Vol. 2A (KLING, M., WÖHLBIER, W., Eds.), pp. 91–106. Stuttgart: Eugen Ulmer.

PIEPER, H. J. (1984), Brennereitechnologische Erfahrungen mit hochaktiven Weizen-Darrmalzen als Verzuckerungsmittel zur Gewinnung von Kornbranntwein nach einem vereinfachten Kaltmaischverfahren, *Handb. Brennerei Alkoholwirtsch.* **31**, 261–279.

PIEPER, H. J., BOHNER, K. (1985), Energiebedarf, Energiekosten und Wirtschaftlichkeit verschiedener Alkoholproduktionsverfahren für Kornbranntwein unter besonderer Berücksichtigung des Schlempe-Recyclingverfahrens (SRV), *Branntweinwirtschaft* **125**, 286–293.

PIEPER, H. J., HOTZ, U. (1988), Eine einfache Betriebskontrollmethode zur Feststellung des Zerkleinerungsgrades von Getreide in Maischen unter besonderer Berücksichtigung der Erfassung des Anteils an nicht aufgeschlossener Stärke, *Handb. Brennerei Alkoholind.* **35**, 277–290.

PIEPER, H. J., JUNG, O. (1982), Energiesparende Alkoholgewinnung durch Schlemperecycling (SRC-Verfahren), in: *Proc. 5. Symp. Techn. Mikrobiol.* Berlin: Verlag Versuchs- und Lehranstalt für Spiritusfabrikation und Fermentationstechnologie im Institut für Gärungsgewerbe und Biotechnologie.

PIEPER, H. J., PÖNITZ, H. (1973), Zur Gewinnung von Gärungsalkohol aus siliertem Körnermais, *Chem. Mikrobiol. Technol. Lebensm.* **2**, 174–179.

PIEPER, H. J., SENN, T. (1987), Das Ganzkorn-Maischverfahren – Ein neues druckloses Verfahren zur Gewinnung von Gärungsalkohol aus stärkehaltigen Rohstoffen, *Handb. Brennerei Alkoholwirtsch.* **34**, 275–291.

PIEPER, H. J., BOHNER, K., KNOLL, A. (1990), Untersuchung der Fluidumwälzung und Wärmeübertragung in einem Maischapparat, *Thesis,* University of Hohenheim, Stuttgart.

QUADT, A. (1994), Alkoholproduktion aus Roggen nach dem Hohenheimer Dispergier-Maischverfahren, *Thesis,* University of Hohenheim, Stuttgart.

RAU, T. (1989), Das autoamylolytische Enzymsystem des Weizens, seine quantitative Erfassung und technologische Nutzung bei fremdenzymreduzierter Amylolyse unter besonderer Berücksichtigung der Ethanolproduktion, *PhD Thesis,* University of Hohenheim, Stuttgart.

RAU, T., THOMAS, L., SENN, T., PIEPER, H. J. (1993), Technologische Kriterien zur Beurteilung der Industrietauglichkeit von Weizensorten unter besonderer Berücksichtigung der Alkoholproduktion, *Dtsch. Lebensm. Rundsch.* **89**, 208–210.

ROBYT, J. F. (1984), Enzymes in the hydrolysis and synthesis of starch, in: *Starch, Chemistry and Technology,* 2nd Edn. Orlando, FL: Academic Press, Inc.

ROSENDAL, P., NIELSEN, B. H., LANGE, N. K. (1979), Stability of bacterial $\alpha$-amylase in the starch liquefaction process, *Starch/Stärke* **31**, 368–372.

SALZBRUNN, W. (1982), Äthanol aus Zuckerhirse in Österreich, in: *Proc. 5. Symp. Techn. Mikrobiol.,* pp. 298–304. Berlin: Verlag Versuchs- und Lehranstalt für Spiritusfabrikation und Fermentationstechnologie im Institut für Gärungsgewerbe und Biotechnologie.

SARGEANT, J. G., WALKER, T. S. (1977), Adsorption of wheat $\alpha$-amylase isoenzymes to wheat starch, *Starch/Stärke* **29**, 160–163.

SCHÄFER, V. (1995), Effekte von Aufwuchsbedingungen und Anbauverfahren auf die Eignung von Korngut verschiedener Getreidebestände als Rohstoff für die Bioethanolproduktion, *PhD Thesis,* University of Hohenheim, Stuttgart.

SCHUSTER, K. (1962), Malting technology, in: *Barley and Malt, Biology, Biochemistry, Technology* (COOK, A. H., Ed.). New York: Academic Press.

SENN, T. (1988), Zur biotechnischen Amylolyse schwer aufschließbarer Getreidearten bei der industriellen Bioethanolproduktion, *PhD Thesis,* University of Hohenheim, Stuttgart.

SENN, T. (1992), Examinations in starch degradation using technical enzyme preparations in bioethanol production, in: *Proc. DECHEMA Biotechnol. Conf.* Vol. 5, Part A, pp. 155–160. Weinheim: VCH.

SENN, T. (1995), Autoamylolytischer Stärkeabbau bei der Bioethanolproduktion aus Triticale, in: *Proc. DECHEMA Jahrestagungen* 1995, Vol. 1, pp. 328–329.

SENN, T., PIEPER, H. J. (1987), Das Ganzkornmaischverfahren: Ein neues druckloses Verfahren zur Bioethanolgewinnung aus schwer aufschließbaren stärkehaltigen Rohstoffen, in: *Proc. 8. Filderstädter Colloquium* 1986 – Alkoholtechnologie, Verband der Lebensmitteltechnologen e.V., Filderstadt, pp. 189–206.

SENN, T., PIEPER, H. J. (1991), Untersuchungen zur Ermittlung der Wirkungsoptima technischer Enzympräparate in Maismaische-Substrat bei der Alkoholproduktion, *Branntweinwirtschaft* **131**, 214–222.

SENN, T., THOMAS, L., PIEPER, H. J. (1991), Bioethanolproduktion aus Triticale unter ausschließlicher Nutzung des korneigenen Amylase-Systems, *Wiss. Z. TH Köthen* **2**, 53–60.

SENN, T., THOMAS, L., PIEPER, H. J. (1993), Zur Bedeutung autoamylolytischer Eigenschaften von Triticale-Sorten für die Gärungsalkoholgewinnung, *Handb. Brennerei Alkoholwirtsch.* **40**, 383–401.

TAKAYA, T., SUGIMOTO, Y., IMO, E., TOMINAGA, Y., NAKATINI, N., FUWA, H. (1978), Degradation of starch granules by $\alpha$-amylases of fungi, *Starch/Stärke* **30**, 289–293.

THOMAS, L. (1991), Enzymtechnische Untersuchungen mit Triticale zur technischen Amylolyse unter besonderer Berücksichtigung der fremdenzymfreien Bioethanolproduktion, *PhD Thesis,* University of Hohenheim, Stuttgart.

THOMAS, L., SENN, T., PIEPER, H. J. (1991), Bioethanol aus Triticale, GIT *Fachz. Lab.* **35**, 1087–1089.

TREU, H. (1991), Verarbeitung von Körnermaissilage in Österreich, *Branntweinwirtsch.* **131**, 238–239.

# 4 Ethanol – Potential Source of Energy and Chemical Products

NAIM KOSARIC

London, Ontario, Canada

# 1 Yeast Fermentation

Yeasts are capable to utilize a variety of substrates (Tab. 1). In general, they are able to grow and efficiently ferment ethanol at pH values of 3.5–6.0 and temperatures of 28–35 °C. Though the initial rate of ethanol production is higher at increased temperatures (~40 °C) the overall productivity of the fermentation is decreased due to ethanol product inhibition (JONES et al., 1981).

Yeasts, under anaerobic conditions, metabolize glucose to ethanol primarily by the Embden-Meyerhof pathway. The overall net reaction involves the production of 2 mol each of ethanol, $CO_2$, and ATP per mol of glucose fermented. Therefore, on a weight basis, each gram of glucose can theoretically give rise to 0.51 g alcohol. The yield attained in practical fermentations, however, does not usually exceed 90–95% of the theoretical value. This is due to the requirement for some nutrients to be utilized in the synthesis of new biomass and other cell maintenance-related reactions. Side reactions also occur in the fermentation (usually to glycerol and succinate) which may consume up to 4–5% of the total substrate. If these reactions could be eliminated, an additional 2.7% yield of ethanol from carbohydrate would result (OURA, 1977).

Fig. 1 represents a simplified scheme for the anaerobic and aerobic catabolism of *Saccharomyces cerevisiae.* The Embden-Meyerhof-Parnas pathway for anaerobic metabolism of glucose to ethanol is shown in Fig. 2. The individual reactions and thermodynamics

**Tab. 1.** The Ability of *Saccharomyces* and *Kluyveromyces* Species to Ferment Sugars (JONES et al., 1981)[a]

| Carbon Number of Basic Subunit | Type of Basic Subunit | Sugar | Basic Unit | Yeast | | |
|---|---|---|---|---|---|---|
| | | | | *S. cerevisiae* | *S. uvarum (carlsbergensis)* | *Kluyveromyces fragilis* |
| 6 | aldoses | glucose | glucose | + | + | + |
| | | maltose | glucose | + | + | − |
| | | maltotriose | glucose | + | + | − |
| | | cellobiose | glucose | − | − | − |
| | | trehalose | glucose | +/− | +/− | − |
| | | galactose | galactose | + | + | + |
| | | mannose | mannose | + | + | + |
| | | lactose | glucose, galactose | − | − | + |
| | | melibiose | glucose, galactose | − | + | |
| | ketoses | fructose | fructose | + | + | + |
| | | sorbose | sorbose | − | − | − |
| | aldoses and ketoses | sucrose | glucose, fructose | + | + | + |
| | | raffinose | glucose, fructose galactose | +/− | + | +/− |
| | deoxy-sugars | rhamnose | 6-deoxymannose | − | − | − |
| | | deoxyribose | 2-deoxyribose | +/− | +/− | +/− |
| 5 | aldoses | arabinose | arabinose | − | − | − |
| | | xylose | xylose | − | − | − |

[a] New taxonomy: *S. uvarum* included in *S. cerevisiae, K. fragilis* changed to *K. marxianus.*

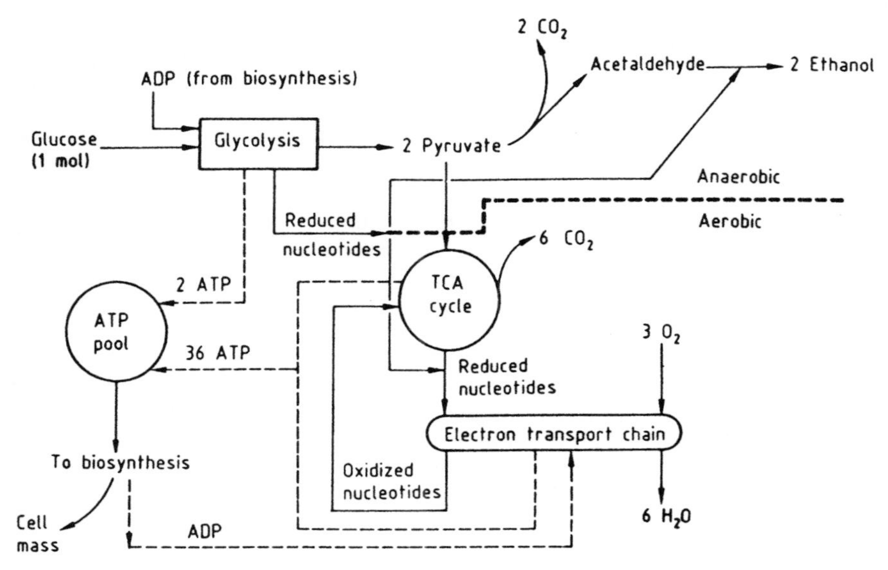


**Fig. 1.** Simplified chart of anaerobic and aerobic catabolism of *Saccharomyces cerevisiae* ADP: adenosine diphosphate; ATP: adenosine triphosphate; TCA: tricarboxylic acid (citric acid).

of glycolysis and alcoholic fermentation are shown on Tab. 2.

A small concentration of oxygen must be provided to the fermenting yeast as it is a necessary component in the biosynthesis of polyunsaturated fats and lipids. Typical amounts of $O_2$ to be maintained in the broth are 0.05–0.10 mm Hg oxygen tension. Any values higher than this will promote cell growth at the expense of ethanol productivity (i.e., the Pasteur effect).

The relative requirements for nutrients not utilized in ethanol synthesis are in proportion to the major components of the yeast cell. These include carbon, oxygen, nitrogen, and hydrogen. Small quantities of phosphorus, sulfur, potassium, and magnesium must also be provided. Minerals (i.e., Mn, Co, Cu, Zn) and organic growth factors (amino acids, nucleic acids, and vitamins) are required in trace amounts.

Many feedstocks under consideration for large-scale ethanol production supply all nutrients necessary for yeast growth in addition to carbohydrate for bioconversion. Additional supplementation with nutrients may be required in some cases. These nutrients may be provided as individual components such as ammonium salts and potassium phosphate or from a low-cost source such as corn steep liquor.

Yeasts are very susceptible to ethanol inhibition. Concentrations of 1–2% (w/v) are sufficient to retard microbial growth and at 10% (w/v) alcohol, the growth rate of the organisms is nearly halted (Brown et al., 1981).

Over long fermentation times, ethyl alcohol exhibits traditional non-competitive Michaelis-Menten inhibition on microbial growth (Aiba et al., 1968), however, studies by Brown et al. (1981) indicate that the immediate effects of this inhibition are more complex. Addition of ethanol to log phase yeast cultures results in a rapid reduction of growth rate (possibly due to effects on protein synthesis), a decrease in cell viability (through irreversible denaturation of enzymes), and to a much lesser extent ethanol lowers the rate of its own synthesis. Observations that the extent of ethanol tolerance for certain yeast strains is dependent upon the fatty acyl composition of their plasma membranes (Thomas and Rose, 1979) would indicate that the fatty acyl composition favors or inhibits excretion of ethanol from the plasma.

In Tab. 3 and Fig. 3 some of the kinetic

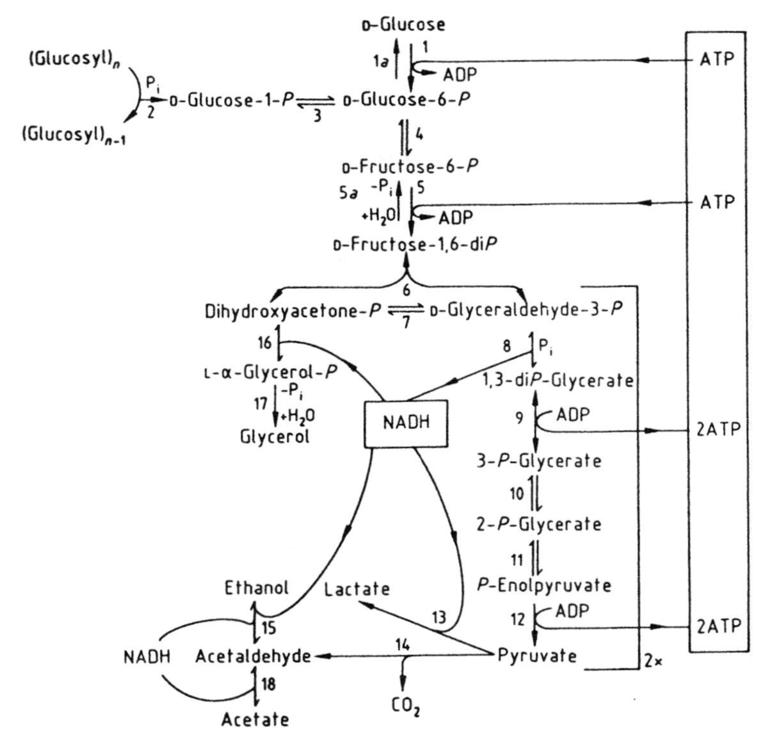


**Fig. 2.** Embden-Meyerhof-Parnas scheme of glycolysis.
Glycolysis:
Glucose + 2 $P_i$ + 2 ADP → 2 Lactate + 2 ATP (classical mammalian muscle or brain)
Glucose + $P_i$ → α-Glycerol phosphate + pyruvate (insect flight muscle, striated muscle)
Fermentation:
Glucose + 2 $P_i$ + 2 ADP → 2 Ethanol + 2 $CO_2$ + 2 ATP + 2 $H_2O$ (1st form)
Glucose + $HSO_3^-$ → Glycerol + Acetaldehyde · $HSO_3$ + $CO_2$ (2nd form, no net ATP)
Glucose + ($P_i$) → α-Glycerol phosphate + Acetaldehyde + $CO_2$
↓
Glycerol + $P_i$ (3rd form, no net ATP)

models and inhibition constants are compared which have been proposed to quantify the effects of ethanol inhibition. As can be seen, there is a great variability between these models. It has been demonstrated (NOVAK et al., 1981; HOPPE and HANSFORD, 1982) that ethanol which is produced during the fermentation (autogenous ethanol) is more inhibitory to cell growth than that added artificially from an exogenous source. Inhibition studies on fermentation data for autogenously produced ethanol are most likely better representations of the true phenomenon.

# 2 Ethanol Fermentation with Bacteria

A great number of bacteria are capable of ethanol formation (BUCHANAN and GIBBONS, 1974). Many of these microorganisms, however, generate multiple end products in addition to ethyl alcohol. These include other alcohols (butanol, isopropyl alcohol, 2,3-butanediol), organic acids (acetic, butyric, formic, and lactic acid), polyols (arabitol, glycer-

**Tab. 2.** Reactions and Thermodynamics of Glycolysis and Alcoholic Fermentation

| Reaction Number | Equation[a] | Name of Enzyme | Characteristic Inhibitor | $\Delta G^{0'\,b}$, [kJ mol$^{-1}$] |
|---|---|---|---|---|
| 1 | Glucose + ATP $\xrightarrow{Mg^{2+}}$ Glucose 6-*P* + ADP | hexokinase<br>glucokinase | | −14.32 |
| 1a | Glucose 6-*P* + $H_2O$ $\xrightarrow{Mg^{2+}}$ Glucose + $P_i$ | glucose 6-phosphatase<br>nonspecific phosphatases | glucose, orinase | −16.83 |
| 2 | Glycogen + *n* $P_i$ ⇌ *n* Glucose 1-*P* | α-1,4-glucan phosphorylase | | 3.06 |
| 3 | Glucose 1-*P* $\xrightleftharpoons{\text{Glucose 1,6-di}P}$ Glucose 6-*P* | phosphoglucomutase | F, organophosphorus inhibitors | −7.29 |
| 4 | Glucose 6-*P* ⇌ Fructose 6-*P* | phosphoglucose (glucose phosphate) isomerase | 2-deoxyglucose 6-phosphate | 2.09 |
| 5 | Fructose 6-*P* + ATP $\xrightarrow[\text{(ADP, AMP)}K^+]{Mg^{2+}}$ Fructose 1,6-di*P* + ADP + $H^+$ | phosphofructokinase | ATP, citrate | −14.24 |
| 5a | Fructose 1,6-di*P* + $H_2O$ $\xrightarrow{Mg^{2+}}$ Fructose 6-*P* + $P_i$ | fructose diphosphatase<br>nonspecific phosphatases | AMP, fructose 1,6-diphosphate, $Zn^{2+}$, $Fe^{2+}$ | −16.75 |
| 6 | Fructose 1,6-di*P* ⇌ Dihydroxyacetone *P* + Glyceraldehyde 3-P | (fructose phosphate) aldolase | chelating agents (microbial enzymes only) | 23.99 |
| 7 | Dihydroxyacetone P ⇌ Glyceraldehyde 3-*P* | triose phosphate isomerase | | 7.66 |
| 8 | 2 × (Glyceraldehyde 3-*P* + $P_i$ + $NAD^+$ ⇌ 1,3-Diphosphoglycerate + NADH + $H^+$) | glyceraldehyde phosphate dehydrogenase; triose phosphate dehydrogenase | $ICH_2COR$<br>D-threose 2,4-diphosphate | 2 × (6.28) |
| 9 | 2 × (1,3-Diphosphoglycerate + ADP + $H^+$ $\xrightarrow{Mg^{2+}}$ Phosphoglycerate + ATP) | phosphoglycerate kinase | | 2 × (−28.39) |

| Reaction Number | Equation[a] | Name of Enzyme | Characteristic Inhibitor | $\Delta G^{0'}$ [b] [kJ mol$^{-1}$] |
|---|---|---|---|---|
| 10 | 2 × (3-Phosphoglycerate $\xrightleftharpoons{\text{Glycerate 2,3-di}P}$ 2-Phosphoglycerate) | phosphoglyceromutase | | 2 × (4.43) |
| 11 | 2 × (2-Phosphoglycerate $\xrightleftharpoons{Mg^{2+} \text{ or } Mn^{2+}}$ Phosphoenolpyruvate) | enolase (phosphopyruvate hydratase) | $Ca^{2+}$<br>$F^-$ plus $P_i$ | 2 × (1.84) |
| 12 | 2 × (Phosphoenolpyruvate + ADP + $H^+$ $\xrightarrow[K^+(Rb^+, Cs^+)]{Mg^{2+}}$ Pyruvate + ATP) | pyruvate kinase | $Ca^{2+}$, vs. $Mg^{2+}$<br>$Na^+$ vs. $K^+$ | 2 × (−23.95) |
| 13 | 2 × (Pyruvate + NADH + $H^+$ ⇌ Lactate + $NAD^+$) | lactate dehydrogenase | oxamate | 2 × (−25.12) |
| 14 | 2 × (Pyruvate + $H^+$ → Acetaldehyde + $CO_2$) | pyruvate (de)carboxylase | | 2 × (−19.76) |
| 15 | 2 × (Acetaldehyde + NADH + $H^+$ ⇌ Ethanol + $NAD^+$) | alcohol dehydrogenase | ($HSO_3^-$) | 2 ± (−21.56) |
| Sums: | $(Glucose)_n + H_2O \rightarrow 2$ Lactate + 2 $H^+$ + $(Glucose)_{n+1}$ | | | −219.40 |
| | $(Glucose)_n + 3\ P_i + 3$ ADP → 2 Lactate + 3 ATP + $(Glucose)_{n-1}$ | | glycolysis (muscle) | −114.38[c] |
| | Glucose → 2 Lactate + 2 $H^+$ | | | −198.45 |
| | Glucose + 2 $P_i$ + 2 ADP → 2 Lactate + 2 ATP | | glycolysis or lactate fermentation | −124.52[c] |
| | Glucose → 2 Ethanol + $CO_2$ | | | −234.88 |
| | Glucose + 2 $P_i$ + 2 ADP → 2 Ethanol + 2 $CO_2$ + 2 ATP | | alcoholic fermentation | −156.92[c] |

[a] Cosubstrates or coenzymes shown above, activators below arrow; P: phosphate.
[b] $\Delta G^{0'}$ values refer to pH 7.0 with all other reactants including $H_2O$ at unit activity; the free energy of formation of glucose in aqueous solution equals 910.88 kJ mol$^{-1}$, its $\Delta G^0$ of combustion to $CO_2 + H_2O$ is 2872 kJ mol$^{-1}$, and $\Delta G^{0'}$ for $(glucose)_n + H_2O \rightarrow (glucose)_{n-1}$ equals −21.06 kJ mol$^{-1}$.
[c] From this table.

**Tab. 3.** Inhibition Kinetic Constants Associated with Data Procuded by Different Authors (HOPPE and HANSFORD, 1982)

| Parameter | Added Ethanol | | | Autogenous Ethanol | | |
|---|---|---|---|---|---|---|
| | EGAMBERDIEV and IERUSALIMSKII (1968) | AIBA and SHODA (1969) | BAZUA and WILKE (1977) | PIRONTI (1971) | CYSEWSKI (1976) | HOPPE and HANSFORD (1982) |
| °C | 28 | 30 | 35 | 30 | 35 | 30 |
| $\mu_{max}$ $[h^{-1}]^a$ | 0.31 | 0.43 | 0.64 | 0.26 | 0.58 | 0.64 |
| $K_s$ $[g\ L^{-1}]^b$ | – | – | 0.24 | 15.5 | 4.9 | 3.3 |
| $K_p$ $[g\ L^{-1}]^c$ | 20.6 | 55 | 40 | 13.7 | 5.0 | 5.2 |
| $Y_{p/s}{}^d$ | 0.39 | 0.35 | 0.52 | 0.47 | 0.44 | 0.43 |

[a] Maximum specific growth rate
[b] Substrate saturation constant
[c] Product inhibition constant
[d] Product yield coefficient

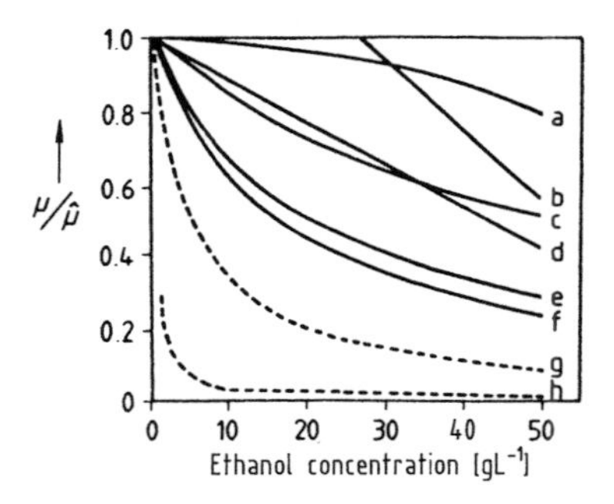


**Fig. 3.** Comparison of the effect of various ethanol inhibition functions (HOPPE and HANSFORD, 1982).
(a) BAZUA and WILKE (1977), continuous; (b) HOLZBERG et al. (1967), batch; (c) AIBA and SHODA (1969), continuous; (d) GHOSE and TYAGI (1979), continuous; (e) EGAMBERDIEV and IERUSALIMSKII (1968), batch; (f) AIBA and SHODA (1969), batch; (g) HOPPE (1981), continuous; (h) STREHAIANO et al. (1978), batch.
$\mu$: specific growth rate; $\hat{\mu}$ maximum specific growth rate; source of ethanol: —— added, ---- autogenous.

ol, and xylitol), ketones (acetone) or various gases (methane, carbon dioxide, hydrogen). The microbes which are capable of producing ethanol as the major product (i.e., a minimum of 1 mol ethanol produced per mol of glucose utilized) are shown in Tab. 4.

Many bacteria (i.e., Enterobacteriaceae, Spirochaeta, *Bacteroides,* etc.), as yeasts, metabolize glucose by the Embden-Meyerhof pathway. The Entner-Doudoroff pathway is an additional means of glucose consumption in many bacteria. Glucose is phosphorylated and then oxidized to 6-phosphogluconate. At this point, dehydration occurs to form 2-keto-3-deoxy-6-phosphogluconate (KDPG) which is then cleaved by KDPG-aldolase. The net yield is 2 mol pyruvate formed from 1 mol glucose and the generation of 1 mol ATP.

Multiple end products may be produced by organisms which conduct mixed acid-type fermentations such as the "enteric" group of facultative anaerobic bacteria. The possible routes of these complex pathways are illustrated in Fig 4. Phosphoenolpyruvate produced in the Embden-Meyerhof pathway may be further broken down to such diverse products as ethanol, formate, acetate, succinate, lactate, and 2,3-butanediol.

As can be seen from Tab. 4, a number of bacteria are able to produce relatively high yields of ethanol. Although some mesophilic *Clostridia* strains are capable of yielding higher concentrations, only *Zymomonas mobilis* can be regarded as a strict ethanol producer.

LEE et al. (1981) studied the fermentation kinetics of *Z. mobilis* ZM4 on artificial media containing either glucose, fructose, or sucrose as carbon source. Kinetic data for growth of

**Tab. 4.** Bacterial Species Producing Ethanol as the Main Fermentation Product (WIEGEL, 1980)

| Mesophilic Organisms | | mmol Ethanol Produced per mmol Glucose Metabolized |
|---|---|---|
| *Clostridum sporogenes* | | up to 4.15[a] |
| *Clostridium indolis* (pathogenic) | | 1.96[a] |
| *Clostridium sphenoides* | | 1.8[a] (1.8)[b] |
| *Clostridium sordelli* (pathogenic) | | 1.7 |
| *Zymomonas mobilis* (syn. *anaerobica*) | | 1.9 (anaerobe) |
| *Zymomonas mobilis* ssp. *pomaceae* | | 1.7 |
| *Spirochaeta aurantia* | | 1.5 (0.8) |
| *Spirochaeta stenostrepta* | | 0.84 (1.46) |
| *Spirochaeta litoralis* | | 1.1 (1.4) |
| *Erwinia amylovora* | | 1.2 |
| *Leuconostoc mesenteroides* | | 1.1 |
| *Streptococcus lactis* | | 1.0 |
| *Sarcina ventriculi* (syn. *Zymosarcina*) | | 1.0 |
| **Thermophilic Organisms** | **$T_{max}$ [°C]** | **mmol Ethanol Produced per mmol Glucose Utilized** |
| *Thermoanaerobacter ethanolicus* (gen. nov.) | 78 | 1.9 |
| *Clostridum thermohydrosulfuricum* | 78 | 1.6 |
| *Bacillus stearothermophilus* | 78 | 1.0 (anaerobic above 55°C) |
| *Thermoanaerobium brockit* | 78 | 0.95 |
| *Clostridium thermosaccharolyticum* (syn. *tartarivorum*) | 68 | 1.1 |
| *Clostridium thermocellum (thermocellulaseum)* | 68 | 1.0 |

[a] In the presence of high amounts of yeast extract.
[b] Values in brackets were obtained with resting cells.

this bacterium are presented in Tab. 5. It is apparent that the specific rates of sugar uptake and ethanol production are at a maximum when utilizing the glucose medium. Major differences are observed in cell yield with decreased values for growth in fructose and sucrose medium. Substrate inhibition of this bacterial species is not severe as it has been shown that growth will occur up to a glucose concentration of 40% (w/v) (SWINGS and DELEY, 1977).

Continuous cultivation of *Z. mobilis* on glucose media has been investigated by ROGERS et al. (1980). Their results are presented in Fig. 5. With a glucose feed concentration of 100 g $L^{-1}$, stable growth was achieved with ethanol concentrations up to 49 g $L^{-1}$. Complete utilization of a 100 g $L^{-1}$ glucose solu-

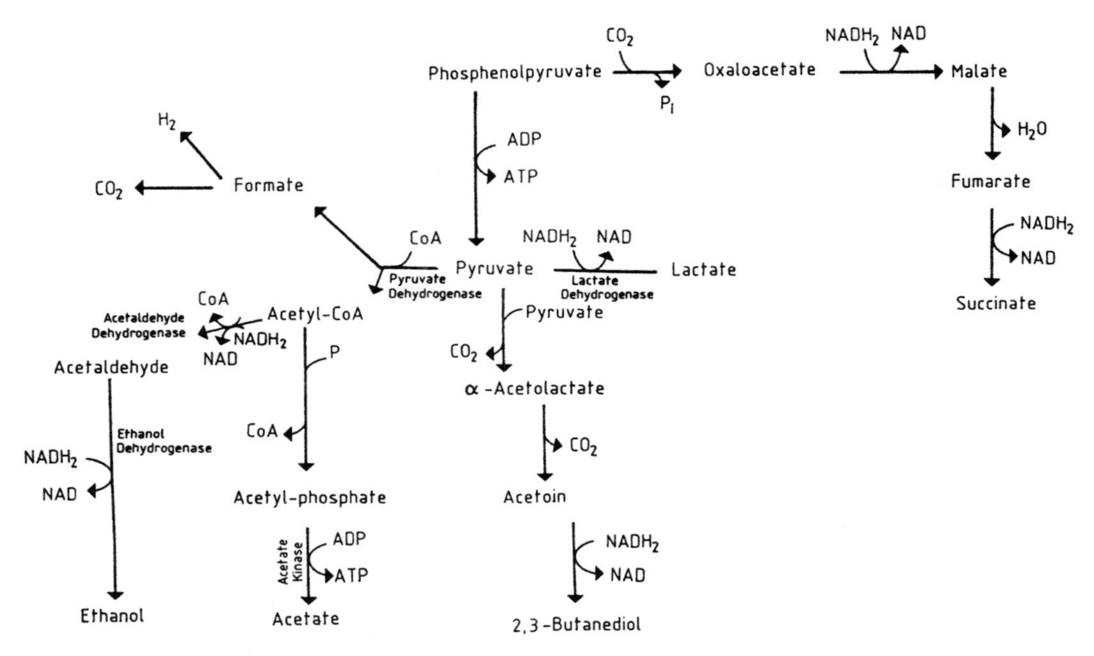


**Fig. 4.** Pathways for the formation of products of the mixed acid and 2,3-butanediol fermentations (ROSENBERG, 1980).

**Tab. 5.** Kinetic Parameters for Growth of *Zymomonas mobilis* Strain ZM4 in Batch Culture with Different Carbon Substrates (Initial Concentration 250 g $L^{-1}$) (LEE et al., 1981)

| Kinetic Parameter | Substrate | | |
|---|---|---|---|
| | Glucose | Fructose | Sucrose |
| Specific growth rate, $\mu$ [$h^{-1}$] | 0.18 | 0.10 | 0.14 |
| Specific substrate consumption rate, $q_s$ [g $g^{-1}$ $h^{-1}$] | 11.3 | 10.4 | 10.0[a] |
| Specific ethanol production rate, $q_P$ [g $g^{-1}$ $h^{-1}$] | 5.4 | 5.1 | 4.6 |
| Cell yield, $Y$ [g $g^{-1}$] | 0.015 | 0.009 | 0.014 |
| Ethanol yield, $Y_{P/S}$ [g $g^{-1}$] | 0.48 | 0.48 | 0.46 |
| Ethanol yield [% of theoretical] | 94.1 | 94.1 | 90.2[b] |
| Maximum ethanol concentration [g $L^{-1}$] | 117 | 119 | 89 |
| Time period of calculation of maximum rates [h] | 0–19 | 0–28 | 0–15 |

[a] Based on changes in total reducing sugar after inversion.
[b] Not corrected for levan formation.

tion was achieved at a dilution rate of 2.0 $h^{-1}$ in a *Z. mobilis* bioreactor employing cell recycle by means of membrane filtration (ROGERS et al., 1980). Volumetric ethanol productivity was reported to be 120 g $L^{-1}$ $h^{-1}$ with a steady state ethanol concentration of 48 g $L^{-1}$.

In the United States, at the University of Florida, INGRAM (1993) reported a development of series of genetically engineered bacteria which efficiently ferment all of the sugars present in lignocellulose. This was done by inserting a portable, artificial operon containing the *Z. mobilis* genes for alcohol dehy-

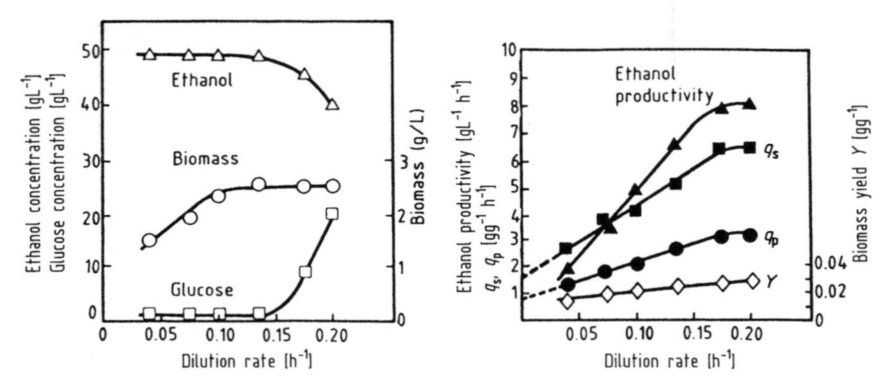


**Fig. 5.** Steady state data and kinetic parameters for growth of *Zymomonas mobilis* in continuous culture on 10% glucose medium (ROGERS et al., 1980).

drogenase and pyruvate decarboxylase into other bacteria with a native ability to metabolize different sugars. Organisms have been developed which can ferment cellobiose, cellotriose, xylobiose, xylotriose, maltose, maltotriose, and other oligomeric sugars. The depolymerization of monomeric sugars prior to fermentation was not required. Efficiencies exceeded 90% of theoretical yields.

## 2.1 Thermophilic Organisms

The advantages of using thermophilic microbes in the production of fuel alcohol are numerous. Since ethanol producing yeasts growing at an optimal fermentation temperature of 50 °C have to be discovered yet, it is apparent that thermophilic bacteria (Tab. 4) are necessary for these processes.

According to WIEGEL (1980) some benefits of high-temperature fermentation are as follows:

- Thermophiles exhibit high catabolic activity at the temperatures optimal for their growth. This results in shorter fermentation times, higher productivity, and an overall increase in the efficiency of fermentation.
- The solubility of oxygen and other gases in the fermentation broth decreases with increasing temperature. This phenomenon supports the establishment and long-term maintenance of anaerobic conditions. The optimum temperature of extreme thermophiles, e.g., is 66–69 °C and at this temperature, the solubility of $O_2$ in the media is 80% lower than at 30 °C.
- Substrates with low solubilities at ambient temperatures would exhibit greater solubility at optimal fermentation temperatures. As such, it is possible that substrate availability would no longer be the rate-limiting step for the process.
- The viscosity of the fermentation broth decreases with increasing temperature. Therefore, the energy required to maintain proper agitation of the growth media is lowered.
- The recovery of ethanol is enhanced at high temperatures. This fact may be utilized in combination with such continuous ethanol removal processes as Vacuferm. The increase of alcohol in the gaseous phase decreases the required degree of vacuum necessary for efficient product recovery.
- The metabolic activity of microbes and frictional effects of agitation serve to generate large amounts of heat. Thus, additional energy to maintain the vessels at the desired temperature as well as the cooling requirements after sterilization are minimized.
- Compared to mesophilic bacteria, sterile conditions are not as essential in a thermophilic process. No obligate thermophilic pathogens are known at present. However, contamination may occur by thermophilic fungi and other bacteria.

Interest has also been centered upon the extreme thermophilic chemoorganotrophic anaerobe, *Thermoanaerobacter ethanolicus* (WIEGEL and LJUNGDAHL, 1981). In addition to its extreme thermophilic nature (temperature optimum of 69°C), *T. ethanolicus* has two basic advantages over other organisms. (1) It exhibits a very broad pH optimum of 5.5–8.5 (growth will occur at pH 4.5–9.5), and (2) it is able to ferment carbohydrates to an almost quantitative yield. *T. ethanolicus* is capable of utilizing a wide range of substrates to produce ethanol. These include starch, cellobiose, lactose, and various pentose sugars. Relative merits of mesophilic and thermophilic fermentation for alcohol production were also discussed by MISTRY (1991).

## 2.2 Bacteria vs. Yeast

Comparison of fermentations conducted by yeast and bacteria can be made on the basis of the kinetics of both systems. The differences observed reflect the metabolic dissimilarities in these groups of organisms.

From Tab. 6 it can be seen that for batch growth at the conditions indicated, *Zymomonas mobilis* conduct a more efficient fermentation than *Saccharomyces uvarum.* Key kinetic parameters indicating the superiority of this bacterial species include specific growth rate, $\mu$ (2.4 times higher than yeast), specific ethanol production rate, $q_p$ (2.9-fold increase), and specific glucose uptake rate $q_s$ (2.6 times higher than for *S. uvarum*).

For maximum ethanol productivity in continuous systems, bacteria are superior to yeasts. The greatest productivity achieved by a yeast system (cell recycle with vacuum) was only 68% of that reported for a *Z. mobilis* process employing cell recycle (Tab. 7; ROGERS et al., 1980). The data presented in Tab. 7 indicate that for cell recycle systems operating at the same glucose input concentration (100 g $L^{-1}$), bacterial ethanol productivity is 4.1 times that of yeast.

In yeast fermentations oxygen is required for cell wall synthesis, stabilization of lipid structures, and general maintenance of cellular processes. However, aerobic conditions also lead to a decrease in ethanol yield and a subsequent increase in biomass concentration due to the Pasteur effect. Since many bacteria are strict anaerobes, with them higher ethanol productivities and lower biomass production are possible. Lower bacterial cell concentration is also a consequence of the reduced energy available for growth under anaerobic conditions (1 mol ATP per mol glucose consumed via the Entner-Doudoroff pathway vs. 2 mol ATP via the Embden-Meyerhof pathway using yeast).

Data have been reported (ROGERS et al., 1980), which show that the ethanol tolerance of *Z. mobilis* is equivalent or higher than that

**Tab. 6.** Kinetic Parameters for *Zymomonas mobilis* and *Saccharomyces uvarum* on 250 g $L^{-1}$ Glucose Media in Non-Aerated Batch Culture (30°C, pH 5.0) (ROGERS et al., 1980)

| Kinetic Parameter | *Z. mobilis* | *S. uvarum* |
|---|---|---|
| Specific growth rate, $\mu$ [L $h^{-1}$] | 0.13 | 0.055[c] |
| Specific glucose uptake rate, $q_s$ [g $g^{-1}$ $h^{-1}$] | 5.5 | 2.1[c] |
| Specific ethanol production rate, $q_p$ [g $g^{-1}$ $h^{-1}$] | 2.5 | 0.87[c] |
| Cell yield, $Y$ [g $g^{-1}$][a] | 0.019 | 0.033 |
| Ethanol yield, $Y_{p/s}$ [g $g^{-1}$][a] | 0.47 | 0.44 |
| Relative ethanol yield [%][a,b] | 92.5 | 86 |
| Maximal ethanol concentration [g $L^{-1}$] | 102 | 108 |

[a] Based on the difference between initial and residual glucose concentrations.

[b] A molar reaction stoichiometry of 1 Glucose → 2 Ethanol + 2 $CO_2$ has been assumed for a theoretical yield.

[c] Kinetic parameters calculated for a fermentation run between 16 and 22 h, the culture growing fully anaerobically.

**Tab. 7.** Comparative Ethanol Productivities of Continuous Systems (ROGERS et al., 1980)

| Organism | System | Optimal Growth Conditions | | | | Maximal Productivity |
|---|---|---|---|---|---|---|
| | | Input Glucose [g L$^{-1}$] | Dilution Rate [h$^{-1}$] | Cell Concentration [g L$^{-1}$] | Ethanol Concentration [g L$^{-1}$] | [g L$^{-1}$ h$^{-1}$] |
| *Saccharomyces cerevisiae* ATCC 4126 | no recycle | 100 | 0.17 | 12 | 41 | 7.0 |
| *S. cerevisiae* ATCC 4126 | cell recycle | 100 | 0.08 | 50 | 43 | 29 |
| *S. cerevisiae* NRRL Y-132 | cell recycle | 150 | 0.53 | 48 | 60.5 | 32 |
| *S. uvarum* ATCC 26602 | cell recycle | 200 | 0.60 | 50 | 60 | 36 |
| *S. cerevisiae* ATCC 4126 | vacuum (6.7 kPa) | 334 | | 50 | 100–160[b] | 40 |
| *S. cerevisiae* ATCC 4126 | cell recycle Vacuum (6.7 kPa) | 334 | 0.23[a] | 124 | 110–160[b] | 82 |
| *Zymomonas mobilis* ATCC 10988 | cell recycle | 100 | 2.7 | 38 | 44.5 | 120 |

[a] Based on bleed rate from clarifier.
[b] Concentration in vapor stream from vacuum fermenter.

of certain strains of *S. cerevisiae.* It is necessary to specify the carbon source which the organism utilizes before comparisons of this type can be made. Future progress in ethanol tolerance and the range of possible substrates used in these fermentation systems may be enhanced through genetic recombination techniques. Due to their prokaryotic nature, manipulation of the genetic material in bacterial species is carried out with greater ease than in yeasts.

# 3 Immobilized Cell Systems

Mass transfer limitations resulting from cell immobilization greatly affect the kinetics of ethanol production in these systems. Where entrapment in alginate beads is involved, diffusion of substrate, ethanol, and $CO_2$ can be enhanced by employing structurally stable beads of small diameter. Small beads minimize the interfacial areas between solid and liquid phases, and thus permit the maintenance of high concentrations of viable cells. Disruption of bead structure and subsequent cell losses result from the accumulation of $CO_2$ within beads due to inadequate diffusion.

GROTE et al. (1980) have clearly demonstrated the dependence of ethanol productivity and the specific rates of glucose uptake ($q_s$) and ethanol production ($q_p$) upon the dilution rate of immobilized bioreactors (Figs. 6 and 7). Dilution rates of approximately 0.8 h$^{-1}$ were found to be optimal for the bacterial fermentations investigated. The high volumetric ethanol productivities characteristic for immobilized systems are shown in Tab. 8. From these data it is evident that in bacterial fermentations the yield of ethanol is higher than

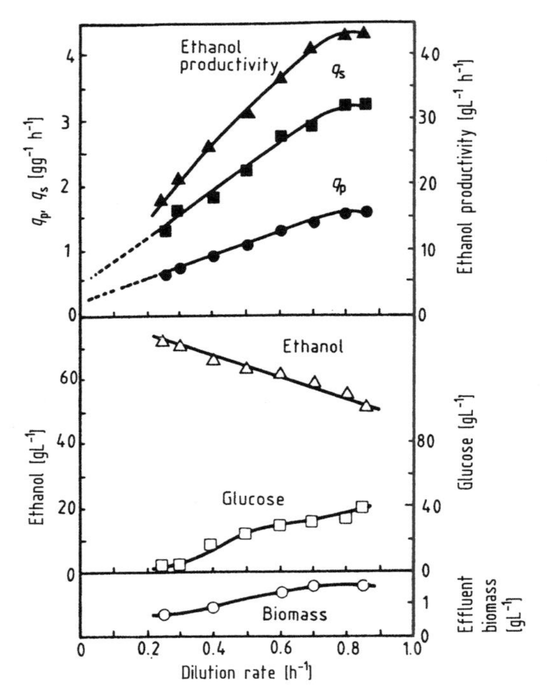


**Fig. 6.** Effect of dilution rate on fermentation with Ca-alginate immobilized cells (GROTE et al., 1980).

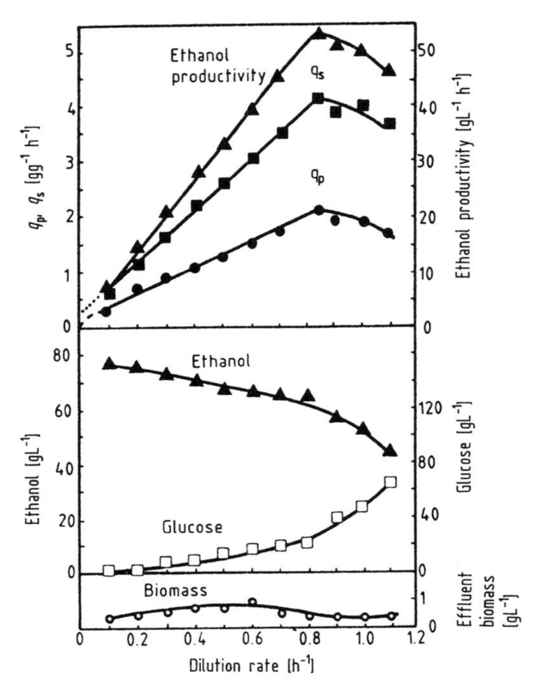


**Fig. 7.** Effect of dilution rate on fermentation with $\kappa$-carrageenan immobilized cells (GROTE et al., 1980).

in yeast fermentations. The most significant advantage of immobilized cell systems probably is the ability to operate with high productivity at dilution rates exceeding the maximum specific growth rate ($\mu_{max}$) of the microbe.

Several theories have been proposed to explain the enhanced fermentation capacity of microorganisms as a result of immobilization. (1) A reduction in the ethanol concentration in the immediate microenvironment of the organism due to the formation of a protective layer or specific adsorption of ethanol by the support may act to minimize end product inhibition. (2) Substrate inhibition may be diminished by a gel matrix, if the rate of fermentation meets or exceeds the rate of glucose diffusion to the cell. (3) Alteration of the cell membrane during the process of immobilization provides improved transfer of substrate into and of product out of the microbe.

The effect of temperature on the rate of ethanol production is markedly different for free and for immobilized systems (Fig. 8). A constant increase in rate is observed with free *Saccharomyces cerevisiae* as temperature is increased from 25 °C to 42 °C. With cells immobilized in sodium alginate a maximum occurs at 30 °C. The lower temperature optimum for immobilized systems may result from the diffusional limitations of ethanol within the support matrix. At higher temperatures, ethanol production exceeds its rate of diffusion so that accumulation occurs within the beads. The inhibitory levels then cause the declines observed in ethanol production rate.

Significant differences are also apparent with regard to the effect of pH on the fermentation rate (Fig. 9). The narrow pH optimum characteristic for free cell systems is replaced by an extremely broad range upon immobili-

**Tab. 8.** Ethanol Productivity in Immobilized Systems (*Saccharomyces cerevisiae* and *Zymomonas mobilis*)

| System | Feed Sugar Concentration [g L$^{-1}$] | Feed Sugar Utilization [%] | Dilution Rate [h$^{-1}$] | Maximal Ethanol Productivity [g L$^{-1}$ h$^{-1}$] | Reference |
|---|---|---|---|---|---|
| *S. cerevisiae* carrageenan | glucose 100 | 86 | 1.0 | 43 | WADA et al. (1979) |
| *S. cerevisiae* Ca-alginate | glucose 127 | 63 | 4.6 | 53.8 | WILLIAMS and MUNNECKE (1981) |
| *S. cerevisiae* Ca-alginate | molasses 175 | 83 | 0.3 | 21.3 | LINKO and LINKO (1981b) |
| *S. cerevisiae* carrier A | molasses 197 | 74 | 0.35 | 25 | GHOSE and BANDYOPADHYAY (1980) |
| *Z. mobilis* Ca-alginate | glucose 150 | 75 | 0.85 | 44 | GROTE et al. (1980) |
| *Z. mobilis* Ca-alginate | glucose 100 | 87 | 2.4 | 102 | MARGARITIS et al. (1981) |
| *Z. mobilis* carrageenan | glucose 150 | 85 | 0.8 | 53 | GROTE et al. (1980) |
| *Z. mobilis* flocculation | glucose 100 | | | 120 | ACCURI et al. (1980) |
| *Z. mobilis* borosilicate | glucose 50 | | | 85 | ARCURI et al. (1980) |
| *Z. mobilis* carrageenan-locust bean gum | whey-lactose 50 | 89 | | 178 | LINKO and LINKO (1981a) |

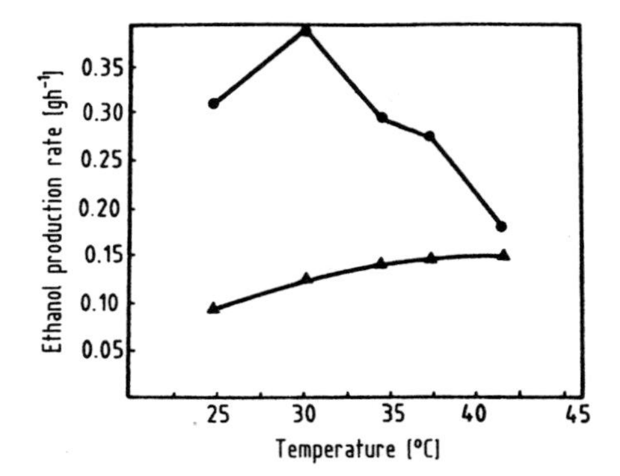


**Fig. 8.** Relationship between temperature and ethanol production rate for free yeast cell suspensions (—▲—▲—) and immobilized yeast cells (—●—●—) (WILLIAMS and MUNNECKE, 1981).

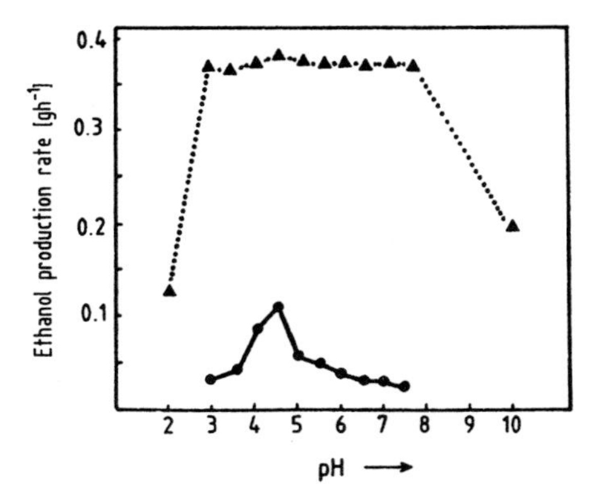


**Fig. 9.** Relationship between pH and ethanol production rate for free yeast cell suspensions (—●—●—) and immobilized yeast cells (—▲—▲—) (WILLIAMS and MUNNECKE, 1981).

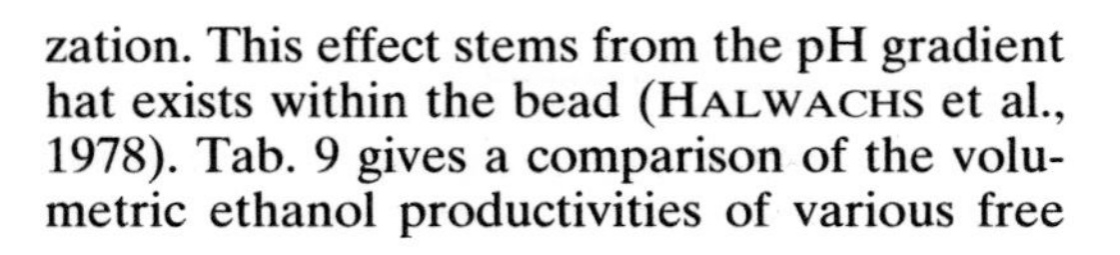
zation. This effect stems from the pH gradient hat exists within the bead (HALWACHS et al., 1978). Tab. 9 gives a comparison of the volumetric ethanol productivities of various free and immobilized systems. In terms of ethanol productivity, immobilized cell systems are superior to free cell methods.

**Tab. 9.** Ethanol Productivity in Free and Immobilized Systems (with *Saccharomyces cerevisiae*) (TYAGI and GHOSE, 1982)

| Process | Substrate | Volumetric Ethanol Productivity [g $L^{-1}$ $h^{-1}$] |
|---|---|---|
| Batch | molasses | 2.0 (excluding downtime of fermenter) |
| Continuous (free cell) | molasses | 3.35 |
| Continuous (immobilized) | molasses | 28.6 |

# 4 Substrates for Industrial Alcohol Production

## 4.1 Sugar Crops

### 4.1.1 Sugarcane

Although sugarcane is grown primarily for sucrose and molasses production, it is also used as a raw material in ethanolic fermentations. It has a high biomass yield of desirable composition as shown in Fig. 10. The crop may be harvested over an extended period of time due to its long growing season.

The fermentable carbohydrates from sugarcane may be directly utilized in the form of cane juice (central or autonomous distilleries)

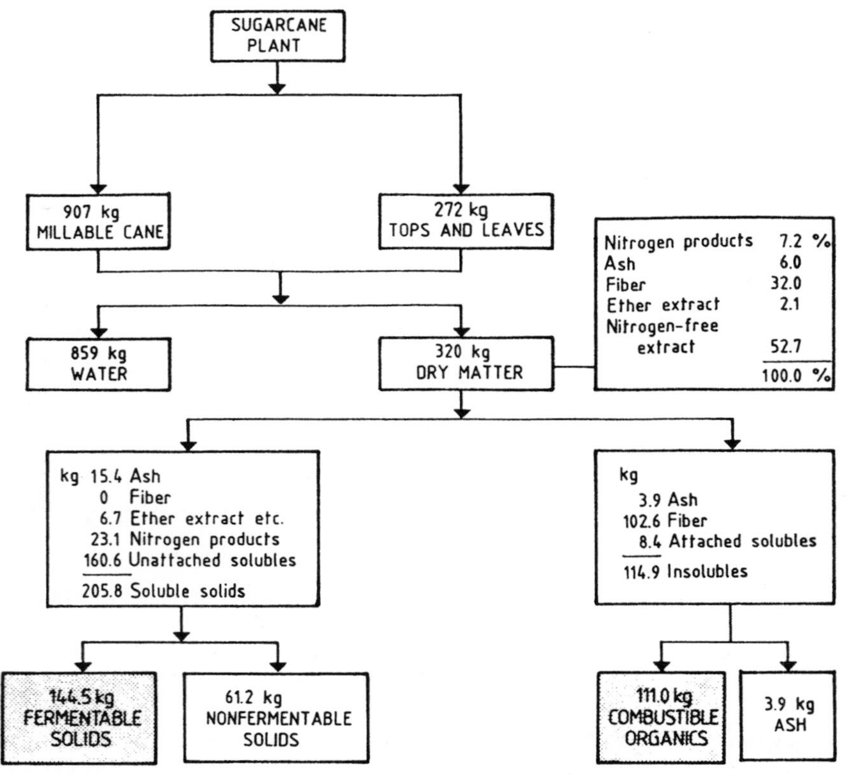


**Fig. 10.** Typical composition of sugarcane. Fermentable solids include sucrose, starches, and various other carbohydrates; products in shaded boxes are usable for fuel (NATHAN, 1978).

or in conjunction with a sugar factory from black strap molasses (annexed distilleries) (KOSARIC et al., 1980).

Cane juice is prepared by crushing the raw cane and after extraction, clarifying with milk of lime and $H_2SO_4$ to precipitate the inorganic fraction (PROUTY, 1980). The resulting extract is a green, sticky fluid, slightly more viscous than water, with an average sucrose content of 12–13%. It may then be evaporated to the desired concentration and used directly in the fermentation. A major disadvantage in the utilization of sugarcane juice is its lack of stability over extended periods of storage.

Black strap molasses is the non-crystallizable residue remaining after the sucrose has been crystallized from cane juice. This heavy viscous material is composed of sucrose, glucose, and fructose (invert sugars) at a total carbohydrate concentration of 50–60% (w/v). Molasses may be easily stored for long periods of time and diluted to the required concentration prior to use.

### 4.1.2 Sugar and Fodder Beets

Sugar beet is a more versatile crop than sugarcane since it can tolerate a wide range of soil and climatic conditions. This allows for its successful growth throughout nearly one-half of the United States, Europe, Africa, Australia, and New Zealand. Although the fermentable carbohydrate content is lower than that of sugarcane (Fig. 11), production of this crop per unit area may exceed sugarcane capabilities depending upon the conditions of cultivation.

As for sugarcane, beet molasses is generated in large volumes from the sucrose recovery operation. In Tab. 10 the composition of cane and beet molasses are compared. These raw materials contain sufficient nitrogen and other organic and inorganic nutrients such that little, if any, fortification is required prior to fermentation. Additional benefits from sugar beet come from a high yield of crop coproducts such as beet tops and extracted pulp. The pulp is bulky and palatable and has a

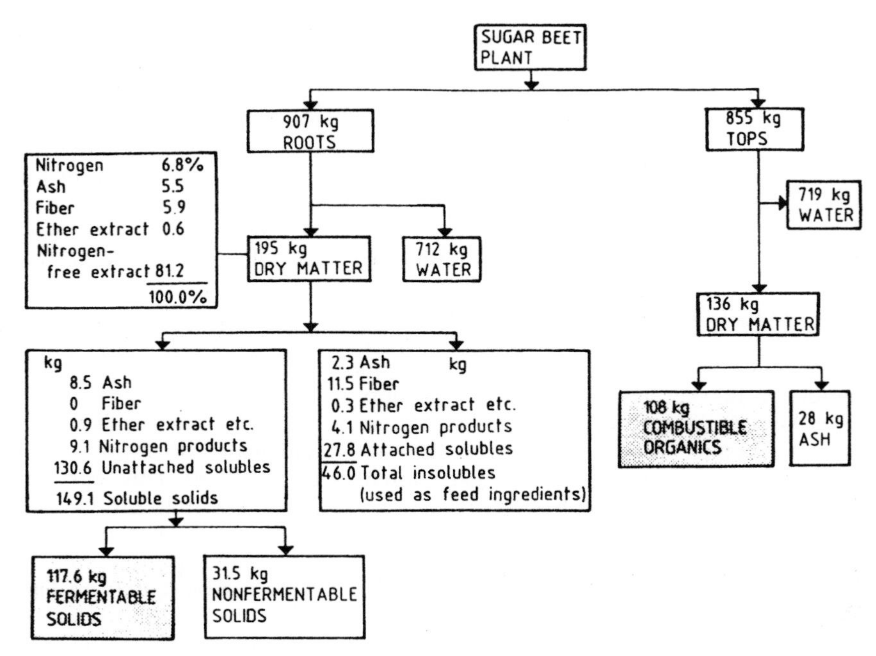


**Fig. 11.** Typical composition of sugar beets. Fermentable solids include sucrose starches, and various other carbohydrates; products in shaded boxes are usable for fuel (NATHAN, 1978).

**Tab. 10.** Cane and Beet Molasses – Principal Values at 75% Dry Matter (BAKER, 1979)

| Component | | Cane | Beet |
|---|---|---|---|
| Total sugars | [%] | 48/56 | 48/52 |
| Non-sugar organic matter | [%] | 9/12 | 12/17 |
| Sulfated ash | [%] | 10/15 | 10/12 |
| Total organic matter[a] | [%] | 60/65 | 63/65 |
| Protein, i.e. N×6.25 | [%] | 2/4 | 6/10 |
| Sodium | [%] | 0.1/0.4 | 0.3/0.7 |
| Potassium | [%] | 1.5/5.0 | 2.0/7.0 |
| Calcium | [%] | 0.4/0.8 | 0.1/0.5 |
| Chlorine | [%] | 0.7/3.0 | 0.5/1.5 |
| Phosphorus | [%] | 0.6/2.0 | 0.02/0.07 |
| Biotin | [mg kg$^{-1}$] | 1.2/3.2 | 0.04/0.13 |
| Folic acid | [mg kg$^{-1}$] | ca. 0.04 | ca. 0.2 |
| Inositol | [mg kg$^{-1}$] | ca. 6,000 | 5,800/8,000 |
| Ca-pantothenate | [mg kg$^{-1}$] | 54/6.5 | 50/100 |
| Pyridoxine | [mg kg$^{-1}$] | 2/6.5 | ca. 5.4 |
| Riboflavin | [mg kg$^{-1}$] | ca. 2.5 | ca. 0.4 |
| Thiamine | [mg kg$^{-1}$] | ca.1.8 | ca. 1.3 |
| Nicotinic acid | [mg kg$^{-1}$] | 20/800 | 20/45 |
| Choline | [mg kg$^{-1}$] | 600/800 | 400/600 |

[a] Total organic matter is total solids less sulfated ash.

high feed value in wet or dry form. The tops may be returned to the soil for erosion control and nutrient replacement.

### 4.1.3 Fruit Crops

Yeast can ferment fruit sugar (usually 6–12% fructose) without pretreatment. Fruits with a high fructose and sucrose content are grapes, peaches, apricots, pears, and pineapples. Though the potential of this raw material is relatively high, its use as a feedstock for fuel alcohol production is not likely due to its significant market value for human consumption. Also, fruits are very perishable and sensitive to damage which may lead to substantial losses of their fermentable sugar content dependent upon the extent of spoilage.

## 4.2 Industrial and Food Processing Wastes

Industrial and food operation waste streams require some form of treatment to reduce the environmental impact of their disposal. Anaerobic fermentation would not only generate fuel ethanol but also reduce the biological oxygen demand (*BOD*) of the raw material to an acceptable level. Disadvantages in the utilization of waste materials include the relatively small-scale and widely scattered locations of their production as well as their low carbohydrate concentrations.

### 4.2.1 Waste Sulfite Liquors (WSL)

Approximately $100 \cdot 10^6$ t a$^{-1}$ WSL are produced as a by-product of the worldwide pulp and paper operations (FORAGE and RIGHELATO, 1979). This represents ca. 9,180 L WSL per ton of pulp produced. The material arises from the treatment of wood for pulp and from paper production.

As can be seen from Tab. 11, the chief components of WSL are lignosulfonates and pentose sugars. The $BOD_5$ value for this material is extremely high (25,000–50,000 ppm). Alcoholic fermentation decreases *BOD* by 90%,

**Tab. 11.** Chemical Composition of Organic Dry Substance in a Spent Spruce Sulfite Liquor (DETROY and HESSELTINE, 1978)

| Component | [%] |
|---|---|
| Lignonsulfonic acids | 43 |
| Hemilignin compounds | 12 |
| Incompletely hydrolyzed hemicellulose compounds and uronic acids | 7 |
| Monosaccharides | |
| D-Glucose | 2.6 |
| D-Xylose | 4.6 |
| D-Mannose | 11.0 |
| D-Galactose | 2.6 |
| L-Arabinose | 0.9 |
| Acetic acid | 6 |
| Aldonic acids and substances not investigated | 10 |

however, the reduction of total organic carbon is low due to the recalcitrant nature of compounds such as lignins, hemilignins, and unhydrolyzed hemicelluloses.

Pretreatment of this waste before fermentation is minimal. Steam or aeration stripping at pH 1.5–3.0 is required to remove $SO_2$ which would otherwise inhibit microbial growth. The pH must then be adjusted to optimum and the media supplemented with nitrogen and phosphate nutrients.

### 4.2.2 Whey

Whey is the liquid effluent generated by the cheese and casein manufacturing industries. The composition is shown in Tab. 12. Sweet whey is a by-product from the manufacture of various hard and soft cheeses whereas acid whey is generated from the production of cottage cheese.

### 4.2.3 Food Industry Wastes

The characteristics of the food processing industry make large-scale utilization of waste streams improbable. The factors involved in the availability of such residues are as follows:

- Highly disperse points of origin;
- seasonal aspects of production (average operation of fruit and vegetable processing

**Tab. 12.** Comparison of Sweet and Acid Whey Composition (DREWS, 1975)

| Component | Composition in | |
|---|---|---|
| | Sweet Whey [%] | Acid Whey [%] |
| Dry matter | 6–7 | 5–6 |
| Ash | 0.5–0.7 | 0.7–0.8 |
| Crude protein | 0.8–1.0 | 0.8–1.0 |
| Nitrogenous compounds as % of total nitrogen | | |
| genuine protein | 52.5 | 43.9 |
| peptides | 31.3 | 33.1 |
| amino acid | 2.5 | 6.1 |
| creatin | 2.6 | 2.5 |
| ammonia | 1.0 | 2.3 |
| urea | 9.1 | 10.3 |
| purines | 1.0 | 1.8 |
| Lactose | 4.5–5.0 | 3.8–4.2 |
| Lactic acid | traces | up to 0.8 |
| Citric acid | 0.1 | 0.1 |
| pH | 4.5–6.7 | 3.9–4.5 |

plants is about 65% of each year and 75% of total processing is completed in slightly more than 4 months);
- high variability in both the composition and characteristics of waste streams (solid and liquid);
- predominant need for the addition of nitrogen and added nutrients at added cost;
- generally low sugar concentration ($\sim$4%) which is not sufficient for economical ethanol recovery;
- extensive competition with established feed and by-product markets which provide greater cash return for producers of the waste.

## 4.3 Starches

A variety of starches can be used for ethanol production by fermentation, e.g., grains, cassava (manioc, tapioca), sweet potato, sweet sorghum, and Jerusalem artichoke. Selection of an appropriate substrate depends on a number of factors, not the least of which is the geographical climate of the intended production site. Thus, while corn, wheat, rice, potatoes, and sugar beets are the most common feedstocks in Europe and North America, sugarcane, molasses, cassava, babassu nuts, and sweet potatoes appear to provide the most promising supply of ethanol for tropical countries such as Brazil.

### 4.3.1 Corn

According to MIRANOWSKI (1981), corn is the most viable feedstock for manufacture of ethanol in the United States. The advantages of corn include:

- a relatively high yield;
- a broad geographical cultivation range;
- corn, like sugarcane, has a C4 photosynthetic mechanism that is inherently quite efficient;
- the energy output–input ratio for corn is higher than for other major crops, with the exception of sugarcane;
- annual production of corn biomass for all purposes probably exceeds $300 \cdot 10^6$ t (dry basis) in the United States, about 40% of which are residues that presently have little commercial value.

The immediate availability of corn is an important consideration. Extremely efficient systems are already in place for commercially handling corn from seed through culture, harvest, collection, storage, grading, transport, marketing, and shipping, all at very low cost (KELM, 1980).

As an energy feedstock, three main alternatives exist for the utilization of corn:

(1) The entire corn plant can be harvested for silage and used for energy production.
(2) Utilize corn grain for ethanol production and return the residue to the soil or use it for livestock feed.
(3) Use semi-dried corn plant residues after harvesting of the grain to produce furfural, SNG (substitute natural gas), ammonia, or simple sugars.

In evaluating the potential of corn (and any other food crop) for the production of energy, the moral issue of food vs. fuel must be considered. Approximately 66% of the grain produced in the United States is used as food or feed. The proportion of low-quality ("distressed") grain unsuitable for utilization as food may be as high as 5% of the annual grain production. This material may be suitable for fuel alcohol production.

### 4.3.2 Cassava

Cassava (*Manihot esculenta*), also called manioc or tapioca, is cultivated in many tropical countries. Brazil, Indonesia, and Zaire are the most important producers. Cassava roots generally contain 20–35 wt.% starch and 1–2 wt.% protein, although strains with up to 38% starch were developed (CHAN, 1969). The composition of cassava is shown in Tab. 13.

At a productivity level of 30 t $ha^{-1}$ $a^{-1}$ with 25 wt.% starch and a conversion efficiency of 70%, ethanol yields from cassava are 3,440 L $ha^{-1}$. Ethanol yields as high as

**Tab. 13.** Composition of Bitter Cultivars of Cassava (DE MENEZES et al., 1978)

| Component | g per 100 g Dry Matter |
|---|---|
| Starch | 80–89 |
| Total sugars | 3.6–6.2 |
| Reducing sugar | 0.1–2.8 |
| Pentosans | 0.1–1.1 |
| Fiber | 1.7–3.8 |
| Protein | 2.1–6.2 |
| Fat | 0.2–0.7 |
| Ash | 0.9–2.4 |

7,600 L $ha^{-1}$ $a^{-1}$ have been reported (ANONYMOUS, 1980b).

The advantages of cassava as an energy crop for fuel alcohol production include:

- Potential for high alcohol yields per ha of land.
- Lower soil quality required than for sugarcane, therefore, vast areas of little used land available for cultivation.
- High tolerance to draught and disease.
- Available for 48 h before serious deterioration between crop lifting and processing.
- Cassava chips can be easily dried by boiler gases to a moisture content <20% for stable storage of up to one year.
- Using amylolytic enzymes, good conversion yields are attainable at reasonable cost.
- In mash fermentations, no acids or nutrients are required so that the process can be operated at pH values near neutral; corrosion is thus minimized.
- The high crude protein content of cassava leaves and stems (17%) makes production of a valuable animal feed by-product feasible.

## 4.3.3 Sweet Potato

With a higher starch yield per unit land cultivated than corn, sweet potato (*Ipomoea batatas*) represents a fuel crop of significant potential. Sweet potato powder (SPP), generated by freeze drying, then grinding and screening the tubers, has a starch content of 64.4% on a dry weight basis (AZHAR, 1981).

## 4.3.4 Sweet Sorghum

A member of the grass family, sweet sorghum (especially *Sorghum saccharatum*) is a valuable energy crop containing both starches and sugars. Its composition is given in Fig. 12.

With currently available cultivars, ethanol yields of 3,500–4,000 L $ha^{-1}$ can be obtained from the fermentable sugars alone. An additional 1,600–1,900 L can be produced from stalk fibers. More than 17,000 lines of sorghum are known to exist. With hybrid strains, it is anticipated that yields may be increased 30% above present levels (MCCLURE et al., 1980).

Among the attributes of this plant that make it a viable candidate for energy production are:

- its adaptability to the majority of the world's agricultural regions;
- its resistance to draught;
- its efficient utilization of nutrients.

## 4.3.5 Jerusalem Artichoke

The Jerusalem artichoke (*Helianthus tuberosus,* in Europe often called "Topinambur") is related to the sunflower and contains a widespread root system that produces tubers. The plant is native to North America and was introduced to Europe in the 17th century. It grows better in poor soils than most crops. The plant is resistant to pests and common plant diseases as well as to cold temperatures, which makes it a hardy perennial for cold climates.

The interest in this crop stems mainly from its high productivity and relatively high carbohydrate content. The average yield under optimum conditions is 9 t $ha^{-1}$ (d.m.) of forage and 45 t $ha^{-1}$ (80% moisture) of tubers with approximately 18% (wet basis) total reducing sugar after inulin hydrolysis (CHUBEY, B. B., Agriculture Canada, Morden, Manitoba, personal communication, 1981).

The carbohydrates in Jerusalem artichokes are mainly fructofuranose units in the form of inulin which consists of linear chains of approximately 35 D-fructose molecules united

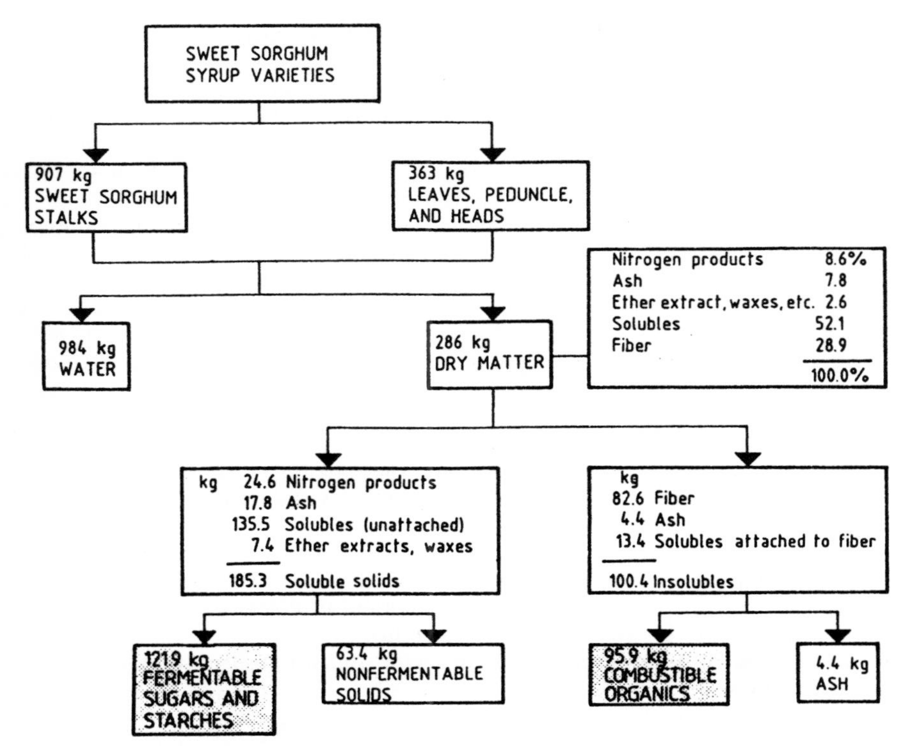


**Fig. 12.** Estimated composition of sweet sorghum, syrup varieties (NATHAN, 1978). Average yields per ha: stalks 47.4 t; leaves, peduncles, and heads 19.3 t; total 66.7 t.

by $\beta$-(2,1) linkages and terminated by a D-glucose molecule which is linked to fructose by an $\alpha$-(1,2) bond as in sucrose (BACON and EDELMAN, 1951; EIHE, 1976). Breakdown products of inulin and other similar carbohydrates are also present but studies have shown that the fructose–glucose ratios seldom fall below 80%:20% (KIERSTAN, 1980).

Because of the high carbohydrate content in the tubers, Jerusalem artichoke can serve as a good source of fermentable sugars for production of ethanol. At 80–90% conversion efficiency, it has been suggested that ethanol yields of 3,900–4,500 L $h^{-1}$ are attainable (HAYES, 1981). The amount of alcohol obtainable per ha of sugar beet, corn and wheat, and Jerusalem artichoke would yield 1.7, 2.0 and 3.7 times more alcohol, respectively (STAUFFER et al., 1975). Pulp obtained after extraction of juice for alcohol fermentation can serve as a nutritional animal feed.

## 4.3.6 Starch Saccharification

### 4.3.6.1 Enzymatic Hydrolysis of Starch

Two main groups of enzymes are involved in the enzymatic degradation of starch. The first consists of enzymes which split the $\alpha$-(1,4) bonds between glucose residues. This group can be further classified into

- endo-enzymes which produce random or internal breaks, and
- exo-enzymes which act from chain ends.

The exo-enzymes catalyze the specific hydrolysis of the $\alpha$-(1,6) interchain linkages of amylopectin.

*α-Amylase* (α-1,4-glucan 4-glucanohydrolase, E.C. 3.2.1.1)

This widely distributed enzyme found in plants, mammalian tissues and microorganisms, causes random hydrolysis of α-(1,4) linkages in starch. The initial conversion of starch to molecules of medium size is then succeeded by a slower reaction generating maltose, maltotriose, glucose, and some oligomeric "dextrins". These latter are formed because α-(1,6) bonds are not only resistant to the action of α-amylase but confer stability to adjacent α-(1,4) bonds. The molecular weight of these enzymes is approximately 50,000.

*β-Amylase* (α-1,4-glucan maltohydrolase, E.C. 3.2.1.2)

Found in most higher plants, this enzyme is also produced by microbes of the genus *Bacillus*. β-Amylase attacks from the non-reducing ends of starch molecules, hydrolyzing alternate α-(1,4) glycosidic linkages to generate maltose.

Because this enzyme is unable to hydrolyze the α-(1,6) branch points in starch, the end products of its action on this substrate are maltose and β-limit dextrins.

*Glucoamylase* (amyloglucosidase, α-1,4-glucan glucohydrolase, E.C. 3.2.1.3)

This non-specific, exo-acting carbohydrase cleaves α-1,3, α-1,4 and α-1,6 glycosidic bonds, generating glucose monomers from the non-reducing ends of starch chains. Glucoamylase is unable, however, to completely hydrolyze starch to glucose (MARSHALL and WHELAN, 1970). The total breakdown of starch requires the participation of an endo-acting enzyme.

Glucoamylase can be separated into two fractions, glucoamylase I and II. Glucoamylase I possesses high debranching activity, absorbs directly onto starch, digesting it easily. Its action is accelerated by the presence of α-amylase. In contrast, glucoamylase II cannot be absorbed onto starch and exhibits only weak debranching activity. The slow hydrolysis produced by this fraction is not accelerated by α-amylase. Glucose, and especially maltose, were observed to inhibit both the digestion of starch by glucoamylase and the adsorption of the enzyme onto it.

Saccharification by enzymatic means is generally easier with cereal starches than with root starches. Raw cassava starch, however, is an exception in that it is as easily digested by amylases as corn starch.

### 4.3.6.2 Acid Hydrolysis of Starch

In acid hydrolysis, the breakdown of starch to glucose is accompanied by further degradation of the sugar to 5-hydroxymethyl furfural, levulinic acid and formic acid (KERR, 1944). Acid concentration and type, temperature and starch concentration have been shown to be key factors in the relative yields of glucose by-products (AZHAR, 1981). Although acid hydrolysis of cassava starch has been shown (DE MENEZES, 1978) to provide more than 98.8% of reducing sugars, the procedure is not recommended because of low alcohol yields (approximately 75% of the theoretical value) due to the presence of non-fermentable and/or inhibitory by-products.

## 4.4 Lignocellulose

The potential of utilizing global reserves of lignocellulosic materials for conversion to useful fermentation products such as fuel ethanol has generated extensive interest in the past few decades. By far, world production and present stocks of cellulose-based biomass is larger in volume than any other carbohydrate source. Given an average annual value for incident solar energy reaching the earth's surface estimated to be $3.67 \cdot 10^{21}$ kJ $a^{-1}$ (HOLDREN and EHRLICH, 1974), global photosynthesis (with an efficiency of 0.07%) could convert $2.57 \cdot 10^{18}$ kJ $a^{-1}$ to cellulose containing biomass. This efficiency would yield an annual net production of $1.8 \cdot 10^{11}$ t of biodegradable material, 40% of which is cellulose (WHITTAKER, 1970). A production of $1{-}1.25 \cdot 10^{11}$ t $a^{-1}$ of terrestrial dry mass together with 0.44–0.55 · $10^{11}$ t $a^{-1}$ in the oceans is estimated (SLESSER and LEWIS, 1979).

## 4.4.1 Characteristics of Lignocellulosic Material

Cellulose is a linear homopolymer of anhydroglucose units linked by $\beta$-(1,4) glucosidic bonds. The length of the macromolecule varies greatly as to the source and degree of processing it has undergone. Newsprint, e.g., exhibits an average DP (degree of polymerization) of about 1,000 while cotton is found to have a DP of approximately 10,000 (CALLIHAN and CLEMMER, 1979).

It is not this primary structure which makes cellulose such a hydrolysis-resistant molecule. It rather seems to be the effect of secondary and tertiary configurations of the cellulose chain as well as its close association with other protective polymeric structures within the plant cell wall such as lignin, starch, pectin, hemicelluloses, proteins, and mineral elements.

The hemicellulose and lignin components of the woody fiber are located between the microfibrils or inter-laminar spaces. Hemicelluloses are heteropolymers of galactose, mannose, xylose, arabinose, and various other sugars as well as their uronic acids. Next to cellulose, they are the most abundant organic material on earth. For the economic viability of any process involving the utilization of lignocellulosics, a means of hemicellulose recovery and/or assimilation must be included.

The lignin component of cellulosic-based biomass is responsible to a great extent for the difficulties inherent in cellulose hydrolysis. This macromolecule of phenolic character is the dehydration product of 3 monomeric alcohols, *trans-p*-coumaryl alcohol, *trans*-coniferyl alcohol, and *trans*-sinapyl alcohol. The relative amounts of each varies with the source. The lignin matrix forms a protective sheath around the cellulose microfibrils. While lignin accounts for some degree of protection to the microfibrils, the conformation of native cellulose also influences hydrolysis.

When cotton cellulose (in its native form) is treated with dilute acid, partial hydrolysis occurs rapidly with about 15% of the cellulose chain being degraded to glucose. The remaining 85% which is resistant to hydrolysis is found to exist as rod-shaped particles about 400 Å long and 100 Å wide (CALLIHAN et al., 1979). These particles correspond to the levelling off degree of polymerization (LoDP) for cellulose and are postulated to exist as regions of highly ordered crystallinity whereas the easily degraded regions are amorphous in nature and quickly hydrolyzed.

Two models have been introduced to explain the above and other observed experimental evidence (Fig. 13). The fringed fibrillar model (SCALLAN, 1971) postulates that cellulose molecules in the elementary fibril are fully extended with the molecular direction completely in line with the fibril axis. Crystalline regions are 500 Å in length and are intermittent with each amorphous region. The folding chain model as developed by CHANG (1971) is visualized as the cellulose molecule folding back and forth along the fibrillar axis with a fold length of LoDP. One repeating unit (termed platellite) corresponds to about 1,000 DP and is designated as the chain length between points a and b in Fig. 13. Platellite units are joined in sequence by single-stranded glucose chains which are easily hydrolyzed. A conventional crystallite is comprised of several platellites packed in registry with amorphous regions at the ends and crystalline regions towards the center. The

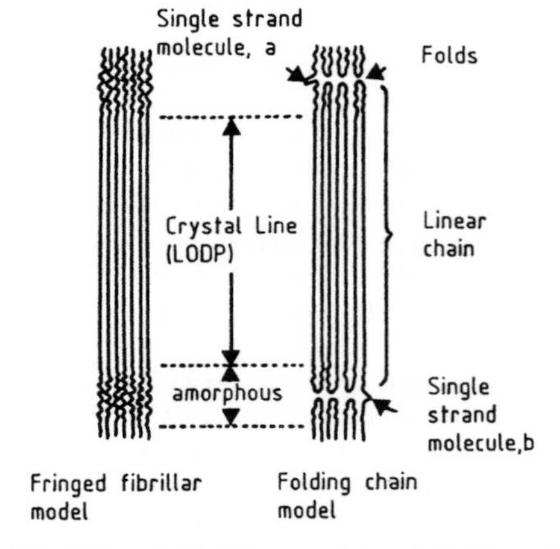


**Fig. 13.** Fringed fibrillar model and folding chain model of cellulose (CHANG et al., 1981).

amorphous regions are said to be comprised of the areas rich in deflected $\beta$-glucosidic linkages found at each fold in the cellulose chain with 3 monomers per 180° turn. These areas would be more susceptible to hydrolytic attack than the linear crystalline areas due to the thermodynamic properties of the deflected covalent bonds and lack of H-bonding.

Elementary fibrils formed by the close packing of the crystallite units are further associated into larger groups known as microfibrils. The microfibril is about 25 nm across and infinitely long. It is these tightly packed structures that are further surrounded by the lignin matrix.

## 4.4.2 Pretreatment

A number of physical and/or chemical methods can be used to separate cellulose from its protective sheath of lignin and increase the surface area of the cellulose crystallite by size reduction and swelling.

### 4.4.2.1 Milling

It has been found that while hammer milling produces desirable size reduction and increases the bulk density of the cellulosic material, it does not increase the susceptibility of the cellulose to hydrolysis (MANDELS et al., 1974). Compression (two roll) milling is a rapid method (5 min and less) to disrupt crystallinity, lower the DP of the cellulose, and increase the bulk density of the material. Increase in bulk density allows cellulose slurry concentrations of 20–30% to be prepared without detrimental agitation or mass transfer effects (SPANO et al., 1979).

Ball milling operations have also been utilized to overcome the lignin barrier. Cellulosic materials have been reduced to sizes of 400 mesh and less by this method (WILKE and MITRA, 1975) and no other pretreatment was found necessary for efficient attack by cellulases (COWLING and KIRK, 1976).

Gentle wet milling has been employed with simultaneous enzymatic degradation of cellulose feedstocks (KELSEY and SHAFIZADEN, 1980). This co-current attrition and hydrolysis has been found by NEILSON et al. (1982) to increase the extent of glucose liberation by 80% over that of conventionally ball milled CF-11 cellulose.

### 4.4.2.2 Steam Explosion

Steam explosion involves the steam heating of green wood chips to approximately 180–200 °C for 5–30 min in a continuous operation, or at higher temperatures (245 °C) for a shorter time (0.5–2 min) in the batch mode. These methods are known as the Stake and Lotech processes, respectively (WAYMAN, 1980). During steam treatment, acids are formed by the decomposition of hemicellulose under high temperature and pressure which then serve to catalyze the depolymerization of intact hemicelluloses and lignin (termed "autohydrolysis"). Upon completion of the heating cycle, at which point the lignin is sufficiently softened, the reaction vessel is abruptly discharged to atmospheric pressure which causes an explosion of the woody cells. This removes lignin from its close association with cellulose and increases the surface area available for catalytic hydrolysis. Residual lignin and hemicelluloses are then easily extracted from the treated product.

Tab. 14 illustrates the enhanced enzymatic hydrolysis parameters for steam exploded hardwoods and crop residues. Increased rates of hydrolysis were not observed for soft woods or municipal wastes. Further advantages of steam explosion include the limitation of undesirable by-product formation and the lack of a need for corrosion proof equipment (BUCHHOLZ et al., 1981).

### 4.4.2.3 Use of Solvents

With the use of appropriate solvents it is possible to selectively remove either lignin or cellulose from the native matrix. This not only serves to disassociate cellulose from its protective lignin covering but also destroys the crystalline structure of native cellulose by successive dissolution and regeneration to a highly active form.

**Tab. 14.** Effect of Steam Pretreatment on the Enzymatic Hydrolysis[a] of Various Cellulosic Substrates (SPANO et al., 1979)

| Substrate | Pre-treatment | Total Reducing Sugars [mg mL$^{-1}$] | |
|---|---|---|---|
| | | 4 h | 24 h |
| Hardwoods | | | |
| Poplar | None | 1.4 | 2.4 |
| | Steam | 15.3 | 25.8 |
| Aspen | None | 1.8 | 3.0 |
| | Steam | 12.8 | 24.8 |
| Agriculture residues | | | |
| Corn stover | None | 4.9 | 7.8 |
| | Steam | 15.7 | 22.5 |
| Sugarcane bagasse | None | 1.7 | 2.5 |
| | Steam | 9.5 | 16.1 |
| Urban waste | None | 10.5 | 18.0 |
| | Steam | 6.2 | 10.8 |
| Softwoods | | | |
| Eastern spruce | None | 2.0 | 3.8 |
| | Steam | 3.5 | 6.4 |
| Douglas fir | None | 1.6 | 3.2 |
| | Steam | 2.8 | 4.3 |

[a] *Trichoderma reesei* cellulase (QM9414), 19 IU per g substrate, 5% substrate slurries, pH 4.8, 50°C, steamed substrates washed prior to enzymatic hydrolysis.

The cost of solvent pretreatment steps has been estimated as the second most important expense in a cellulose saccharification process. Thus, if solvent can be recovered and recycled, this would improve the economic characteristics of the entire operation.

With this recycling aspect in mind, studies have been performed utilizing the solvent Cadoxen, an aqueous alkaline solution of ethylene diamine and cadmium oxide (LADISCH et al., 1978; TSAO, 1978). After cellulose solubilization, the addition of excess water to the solution causes the reprecipitation of the cellulose as a soft floc which is then easily removed and hydrolyzed. The Cadoxen solvent may be recycled. Other solvents include $H_3PO_4$ and alkali.

#### 4.4.2.4 Swelling Agents

Pretreatment agents have been studied for their ability to swell the cellulose matrix and thus open the interior of the fibril to easy attack by the enzymes.

Two types of swelling may take place, intercrystalline and intracrystalline. Intercrystalline swelling is induced by the presence of water and is necessary for any microbial activity to occur. Intracrystalline swelling requires a chemical reagent capable of breaking down the H-bonding of adjacent glucose molecules in the cellulose matrix. These swelling agents include concentrated NaOH, organic bases (i.e., amines), and certain metal salts such as $SnCl_4$ (CHANG et al., 1981). Many of the solvents previously discussed also exhibit swelling properties, however, in general, swelling agents improve hydrolysis to a much lesser extent than solvents (MILLETT et al., 1954) and as well require higher quantities of chemical reagents. The application of heat in conjunction with swelling pretreatment may enhance later hydrolysis to a much greater extent than pretreatment at room temperature (MANDELS et al., 1974).

The greatest drawback of these chemical pretreatments (as opposed to physical methods) is that the treated product is at a very low bulk density and, as such, suspensions of 4–5% are too thick to agitate or transport. The processing of cellulose concentrations at this level is not economical for ethanol recovery.

#### 4.4.2.5 Lignin-Consuming Microbes

There are a number of microorganisms which produce enzymes required for lignin degradation (CRAWFORD and CRAWFORD, 1980) (Tab. 15). The use of these organisms may provide important by-product credits in ethanol production (e.g., SCP) as well as remove the protective lignin coat. Much less harsh conditions are required for this type of biological pretreatment than for other chemical methods. As such, side-reactions are reduced and production of inhibitory agents is limited.

**Tab. 15.** Examples of Microorganisms Capable of Degrading Lignin (BISARIA and GHOSE, 1981)

| Fungi | Bacteria |
|---|---|
| *Paecilomyces* sp. | *Nocardia* sp. |
| *Allescheria* sp. | *Streptomyces* sp. |
| *Preussia* sp. | *Pseudomonas* sp. |
| *Chaetomium* sp. | *Flavobacterium* sp. |
| *Poria* sp. | |

*Poria monticola* and other brown-rot fungi have exhibited a non-enzymatic system involved in lignin degradation and cellulose swelling. This system, based on some association between $Fe^{2+}$ and $H_2O_2$, appears to have potential as a pretreatment method (KOENIGS, 1975).

## 4.4.3 Acid Hydrolysis

Commercial operations which have produced ethanol from cellulosic material have classically utilized acid hydrolysis for the release of fermentable sugars from the native feedstock.

Acid hydrolysis may be categorized under two general approaches, that of high acid concentration at a low temperature or that of low concentration at a high temperature. The relative advantages/disadvantages of each stem from a trade-off between an increased rate and overall yield of hydrolysis vs. the degradation of glucose to undesirable by-products.

### 4.4.3.1 Concentrated Acid

Crystalline cellulose is completely soluble in 72% $H_2SO_4$ or 42% HCl solutions at relatively low temperatures (10–45 °C) (OSHIMA, 1965). The polymer is depolymerized to yield oligosaccharides, the bulk of which is cellulotetraose. Little, if any, glucose monomers are released at this stage. After dissolution in the concentrated acid, the oligomer mixture is then diluted to a lower concentration and heated to about 100–200 °C for 1–3 h. This step converts oligomeric glucose chains to their monomeric constituents.

Acid hydrolysis kinetics at high acid concentrations do not depend upon the structural details or crystallinity of the cellulose substrate. As such, yields >90% of potential glucose may be obtained (GRETHLEIN, 1978).

The main drawback in the use of concentrated acids is that the acid must be recovered and recycled to be economically effective. Such recovery operations are generally cost intensive. The use of HCl has a technical advantage over $H_2SO_4$ in this regard since it is a volatile acid and may be recovered by vacuum stripping methods. Though vacuum recovery of $H_2SO_4$ is not feasible, its use may be integrated with another system with profitable results. An example of this would be the use of the neutralization product $CaSO_4$ in the manufacture of gypsum. Additional expenses arise in concentrated acid processes due to the requirement for corrosive-resistant vessels and large reactor volumes per unit of production due to long reaction times.

### 4.4.3.2 Dilute Acid

While concentrated acid hydrolyzes cellulose rapidly with little or no requirement for pretreatment, the yield of fermentable sugars is usually quite low due to degradation of the glucose as it is released from the polymer. Dilute acid processes yield less degradation products, however, the rate of hydrolysis is lower due to the effect of the resistant crystalline regions within cellulose.

SAEMAN (1945) of the U.S. Forest Products Laboratory predicted the time course of cellulose hydrolysis catalyzed by 0.4–1.6% $H_2SO_4$ (aq.). The kinetics were found to be 1st order and involved 2 reaction steps:

$$\text{Cellulose} \xrightarrow{K_1} \text{Sugar} \tag{1}$$

$$\text{Sugar} \xrightarrow{K_2} \text{Decomposition products} \tag{2}$$

The rate constants ($K_1$, $K_2$) are related to temperature by traditional Arrhenius reaction kinetics. Thus, the fraction of net reducing sugars liberated over that of the potential from the cellulose feedstock may be described as a function of time for any temperature profile and acid concentration by Eq. (3).

$$\frac{C_B}{a} = \frac{K_1}{K_2 - K_1} (e^{K_1 t} - e^{K_2 t}) \qquad (3)$$

where $C_B$ is the net reducing sugar formed, $a$ is the initial cellulose concentration (as glucose equivalents per 100 g starting material), $t$ time, and $K_1$, $K_2$ reaction rate constants.

The activation energies for the decomposition of cellulose and glucose are 189,000 and 137,000 kJ mol$^{-1}$, respectively (GHOSE and GHOSH, 1978) and thus, the values of $K_1$ and $K_2$ are nearly equivalent. Although $C_B\,a^{-1}$ is found to increase with increasing temperature and acid concentration, substantial degradation products are formed under these conditions. Fig. 14 shows acid hydrolysis curves at 180°C using a 0.6% $H_2SO_4$ solution. The ordinate indicates the release of reducing sugars as a percentage of the potential glucose in the cellulose chain. If no degradation products are formed, the reaction would proceed according to curve B (ideal kinetics). Curve A shows the time course for material obeying Saeman kinetics (see Eq. (3)). The real behavior of wood saccharification processes lies in the area between these two limits as illustrated by the broken line. Technical processes which utilize dilute $H_2SO_4$ yield approximately 65–80% of the total reducing sugar.

Kinetic data have shown that the rate of sugar release using dilute acid can be substantially improved by operating at very high temperatures (~500°C) and for short times (YU and MILLER, 1980). The detrimental effects of hemicellulose side reactions which form compounds inhibitory to the subsequent fermentation is a major drawback in these high-temperature processes.

To approach the ideal hydrolysis curve B in Fig. 14, sugar should be removed as soon as it is liberated from the cellulosic chain. To achieve this, an infinite amount of extraction liquid must be provided to the hydrolysis medium. This would amount to a sugar concentration of near zero in the processed stream. WAYMAN (1980) was able to minimize this effect by carrying out a number of hydrolysis stages on steam exploded aspen using 2% $H_2SO_4$ at 190°C for a period of 20 min at each stage. Each limited hydrolysis step liberated about 20% of monomers from the cellulose chain. If free sugar residues are removed at the termination of each partial effect, the overall yield of reducing sugars can reach values in excess of 90% after 5 cycles (Fig. 15).

Overall costs for processes which utilize dilute acids are much less than that for concentrated acids. Therefore, acid recycle is not necessary for the economic viability of the process.

Generally, disadvantages with dilute acid include low sugar yields, high energy consumption due to hydrolysis at elevated tem-

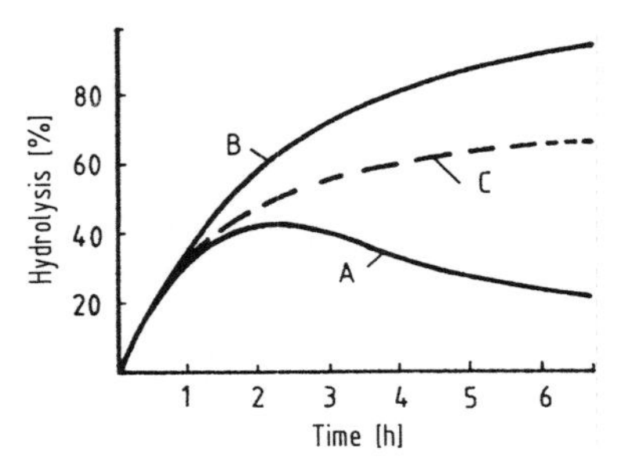


**Fig. 14.** Formation of glucose by the dilute acid hydrolysis of cellulose (WETTSTEIN and DEVOS, 1980).
A: hydrolysis kinetics according to SAEMAN (1945); B: ideal hydrolysis kinetics (no degradation); C: real hydrolysis kinetics.

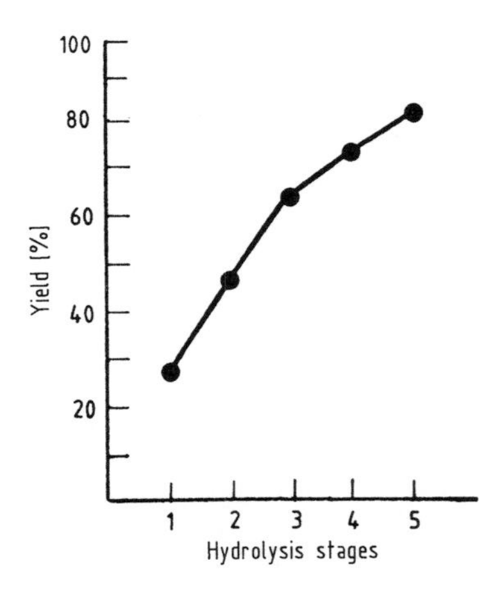


**Fig. 15.** Multistage hydrolysis (yield in % of theoretical) of autohydrolyzed, extracted aspen, employing 2% $H_2SO_4$ on fiber at 190°C for 20 min for each stage (WAYMAN, 1980).

peratures and pressures, and (though less so than concentrated acids) the need for corrosion-resistant materials.

### 4.4.4 Enzymatic Hydrolysis

Cellulases, microbial enzymes capable of cellulose hydrolysis, are in reality a number of several different synergistic components. The cellulase complex is found to consist of 3 basic components which may be present in multiple forms, often as isoenzymes (BISARIA and GHOSE, 1981). These groups of enzymes are classified as

(1) *endo*-β-(1,4)-glucanases (CMC$^{ase}$) consist of several components varying in their degree of randomness, they are postulated to cleave β-(1,4) glucosidic linkages in native cellulose;
(2) *exo*-β-glucanases (Avicel$^{ase}$) consist of specifically two groups:
  - β-(1,4) glucanase glucohydrolase which removes glucose units from the non-reducing end of the cellulose chain,
  - β-(1,4) glucan cellobiohydrolase which removes cellobiose units from the non-reducing end of the chain;
(3) β-(1,4) glucosidase hydrolyzes cellobiose and short-chain oligosaccharides to glucose.

The purification and characterization of these proteins has been carried out with a number of organisms from different sources. It is difficult to generalize on the physical properties of the enzymes since they are subject to a high degree of variation. Their molecular weights commonly range from 12,000–80,000 with β-glucosidase generally being the larger macromolecule.

Cellulases are induced enzymes and are produced only when the organism is grown in the presence of cellulose, cellobiose, lactose, sophorose, or other glucans which contain β-(1,4) linkages (GRATZALI and BROWN, 1979). These enzymes are also highly regulated by end-product inhibition. This phenomenon has led to interest in the continuous removal of hydrolysis products during their formation. Denaturation by shearing is a common drawback of cellulase enzymes (REESE and ROBBINS, 1981), especially at air–liquid interfaces (KIM et al., 1982). Stabilizing agents such as surfactants are effective in reducing the extent of this deactivation.

#### 4.4.4.1 Mechanism of Enzymatic Hydrolysis

A number of comprehensive reviews have been published regarding the mode of attack by cellulases on crystalline cellulose (GHOSE and GHOSH, 1978; BISARIA and GHOSE, 1981; CHANG et al., 1981). The specific theories which have arisen since REESE's original concept (REESE et al., 1950) are quite diverse (ERIKSSON, 1969; REESE, 1977; WOOD, 1980) but the underlying basis seems to be the same. The mechanism involves the cleavage of internal glycosidic bonds by *endo*-glucanase followed by the synergistic attack of *endo*- and *exo*-glucanases. Final hydrolysis of the product oligosaccharides is catalyzed by β-glucosidase. This sequential association between *endo*- and *exo*-glucanase has been demonstrated with the use of electron microscopy by WHITE and BROWN (1981).

REESE (1977) has speculated on the existence of a special member ($C_1$) of the endoglucanase component ($C_x$). This enzyme is not only capable of splitting covalent glycosidic linkages but also of swelling crystalline cellulose by the disruption of hydrogen bonds. Fig. 16 schematically illustrates the synergistic nature of these components.

#### 4.4.4.2 Comparison of Enzymatic and Acid Hydrolysis

Chemical or biochemical degradation of cellulosic substrates may be compared in terms of a number of properties inherent in each process as summarized in Tab. 16.

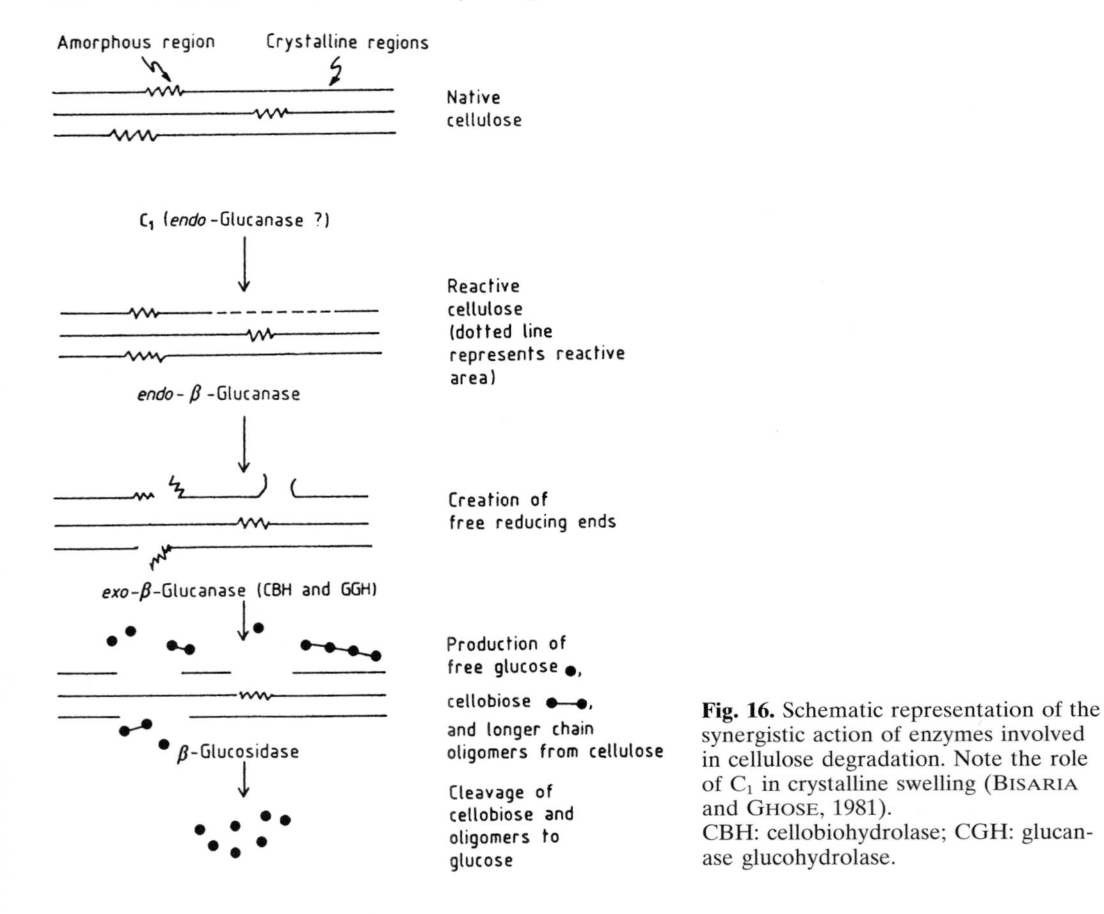


**Fig. 16.** Schematic representation of the synergistic action of enzymes involved in cellulose degradation. Note the role of $C_1$ in crystalline swelling (BISARIA and GHOSE, 1981).
CBH: cellobiohydrolase; CGH: glucanase glucohydrolase.

**Tab. 16.** Comparison of Enzymaticand Acid Hydrolysis Processes for Cellulosic Materials

| Acid | Enzyme |
|---|---|
| 1. Non-specific catalyst, therefore will delignify material as well as hydrolyze cellulose | Specific macromolecular catalyst, therefore extensive physical and chemical pretreatment of material necessary to make cellulose available for degradation |
| 2. Decomposition of hemicellulose to inhibitory compounds (i.e., furfural) | Production of clear sugar syrup ready for subsequent anaerobic fermentation |
| 3. Harsh reaction conditions necessary and, therefore, increased costs for heat- and corrosion-resistant equipment | Run under mild conditions (50 °C, atmospheric pressure, pH 4.8) |
| 4. High chemical costs require catalyst recovery and reuse | Cost to produce cellulases is the most expensive step in the process, therefore, recycle is necessary |
| 5. Rate of hydrolysis is high | Lower rate of hydrolysis |
| 6. Overall yield of glucose is low due to degradation | High glucose yield depending upon system and pretreatment |

# 5 Fermentation Modes of Industrial Interest

## 5.1 Batch Process

Today, most ethanol is produced by the same processes developed in the beverage industry more than a hundred years ago. These methods are based on the simple batch fermentation of carbohydrate feedstocks.

The general characteristics of batch systems are well known. Usually the time required to completely utilize the substrate is 36–48 h. The temperature is held at 10–30 °C and initial pH is adjusted to 4.5. Depending upon the nature of the carbohydrate material, conversion efficiency lies in the range of 90–95% of the theoretical value with a final ethanol concentration of 10–16% (w/v).

Batch technology has been preferred in the past due to the ease of operation, low requirements for complete sterilization, use of unskilled labor, low risk of financial loss, and easy management of feedstocks. However, inherent disadvantages of this system such as decreased fermenter productivity due to long turn-around times and initial growth lag have led to interest in variations of the process.

In an effort to increase fermenter productivity yet retain the simplicity of a batch process, cell recycle has been employed in many cases. This technique does not increase the efficiency of sugar-to-ethanol conversion, however, the time required for the fermentation to run to completion is reduced by as much as 60–70% over traditional batch methods.

The sugar solution supplemented with yeast nutrients is added to the fermenter and the fermenter is inoculated with a rapidly growing culture of yeast from the seed tank. A maximum in ethanol productivity is reached after 14–20 h. Ethanol production then continues at a decreasing rate until about 95% of the sugar is utilized.

Usually several fermenters are operated at staggered intervals to provide a continuous feed to the distillation system. The overall productivity for this process is about 1.8–2.5 kg ethanol produced per $m^3$ fermenter volume per hour (ROSE, 1976). To increase the efficiency of the fermentation step, the "Melle Boinot process" is utilized in most Brazilian distilleries. This process involves centrifugal recuperation of the live yeast from the fermented beer (normally 10–15% by volume of the total) and reinoculating to other fermenters. Tab. 17 presents data for alcohol produced from annexed and autonomous distilleries. Fig. 17 illustrates the steps required to produce ethanol from cane juice-molasses in Brazil.

## 5.2 Semicontinuous Processes

The semicontinuous processes (YAROVENKO, 1978) comprise the so-called outflow–

**Tab. 17.** Productivity Factors for Ethanol Production from Sugarcane (LINDEMAN and ROCCHICCIOLI, 1979)

| Productivity Factor | Alcohol Indirectly from Final Molasses | Alcohol Directly from Sugarcane Juice |
|---|---|---|
| Sugarcane yield in 1.5–2 year cycle (south-central region) | 63 t $ha^{-1}$ | 63 t $ha^{-1}$ |
| Average sucrose yield (13.2 wt.%) | 8.32 t $ha^{-1}$ | 8.32 t $ha^{-1}$ |
| Crystal sugar production | 7.0 t $ha^{-1}$ | — |
| Final molasses or cane juice production | 2.21 t $ha^{-1}$ | 66.2 t $ha^{-1}$ |
| Fermentable sugars, molasses, or juices | 1.32 t $ha^{-1}$ | 8.73 t $ha^{-1}$ |
| Alcohol yield at 100% global efficiency | 675 kg $ha^{-1}$ | 4,460 kg $ha^{-1}$ |
| Alcohol yield with reasonable 85% global efficiency | 11.5 L $t^{-1}$ cane or 730 L $ha^{-1}$ | 75 L $t^{-1}$ cane or 4,800 L $ha^{-1}$ |

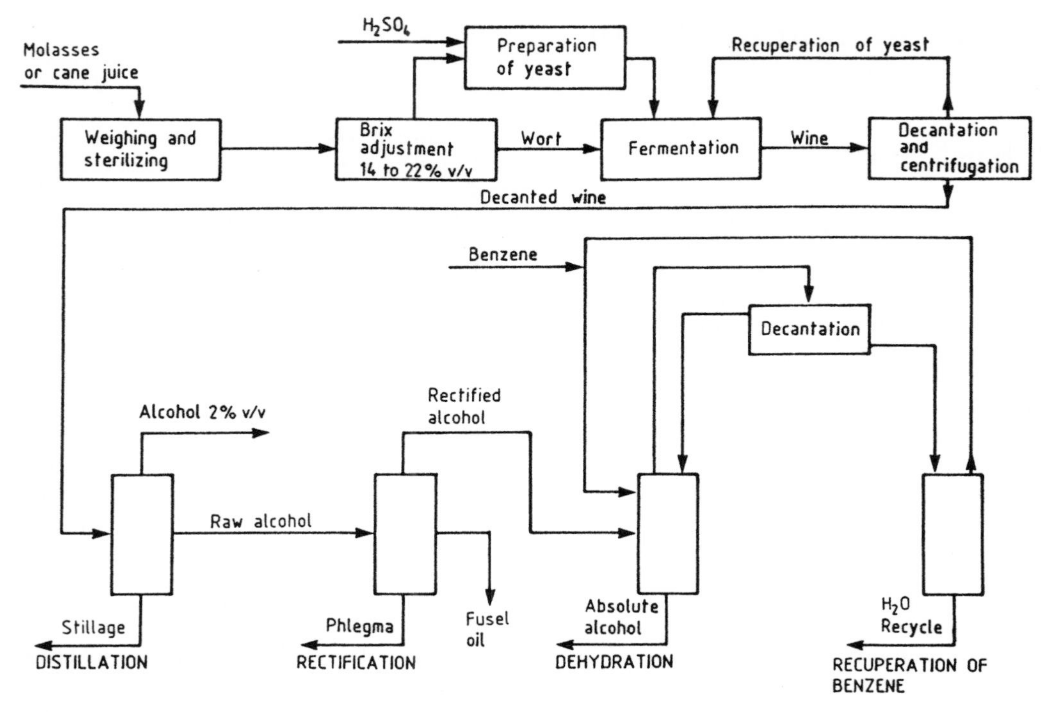


**Fig. 17.** Block diagram for ethanol distillery (LINDEMAN and ROCCHICCIOLI, 1979).

inflow and overflow processes and also a wide range of battery and cyclic fermentation variants. Characteristic of these processes is the continuous feed liquid flow after addition of a nutritive wort and of a seeding culture, such as *Saccharomyces cerevisiae.* The treated medium flows by gravity or is pumped as a seeding culture from the first vessel into the second, while fermentation is continued in the first. Then, in succession, the second fermenter is filled up with the nutritive wort and left for fermentation. Thereafter, the third vessel is filled with the seeding culture and a prolonged nutritive inflow charge is secured. This process continues until all the vessels are charged.

Using this system, the fermentation time is shortened because of higher yeast activity and concentration.

## 5.3 Continuous Processes

Continuous ethanol production eliminates much of the unproductive down-time associated with batch culture. This includes stripping and cleaning the apparatus, recharging with media, and time required for the lag phase. In addition, since the microbe may be essentially "locked" in the exponential phase of its growth cycle, then the overall time-dependent productivity of ethanol formation is increased. This allows for a higher output per unit volume of equipment and as such, cost savings may be made in the construction of smaller fermenters (MAIORELLA et al., 1981).

Two main disadvantages in continuous systems include concern over contamination (either internal through mutation or external by an invading microbe) and problems in maintaining a high fermentation rate. Low rates of fermentation have been shown to be connected with cell death caused by a lack of oxygen. This oxygen requirement may be eliminated by the addition of certain substances (i.e., Tween 80, ergosterol, or linolenic acid) to the fermentation medium or by propagation of the yeast under aeration prior to anaerobiosis (BERAN, 1966).

Specific ethanol productivity in a simple continuous fermenter is ordinarily limited by ethanol inhibition and a low cell concentration. As sugar content of the feed stream is increased, ethanol productivity decreases due to product inhibition effects. At lower concentrations of carbohydrate, inhibition is seen to decrease, however, cell mass concentration also falls off. Thus, an optimum fermenter productivity is reached at about a 10% glucose feed as seen in Fig. 18.

Cell recycle has been utilized in continuous systems to overcome low cell density limitations. With much of the microbial biomass returned to the fermenter an extremely high cell concentration may be maintained. To retain cell viability, a fraction of biomass is removed on a continuous basis. Densities as high as 83 g $L^{-1}$ may be maintained in the fermenter for such a system (DEL ROSARIO et al., 1979) and productivities of up to 30–50 g $L^{-1}$ $h^{-1}$ ethanol are possible.

Complexities arise due to the requirement for some separation device in this process.

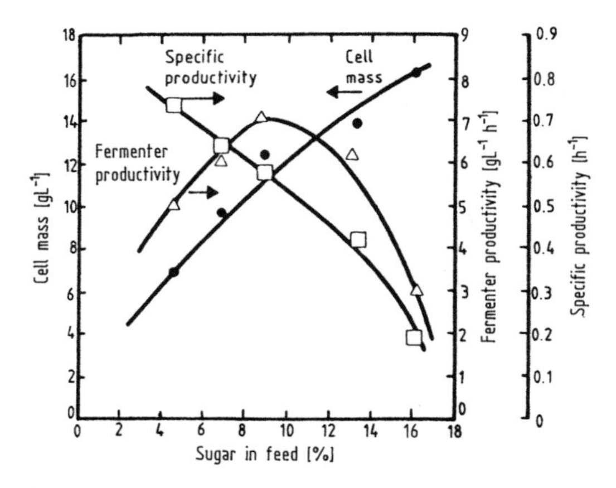


**Fig. 18.** Effect of glucose concentration on continuous fermentation; conditions at "complete" substrate utilization (CYSEWSKI and WILKE, 1978).

Mechanical centrifuges increase capital costs and require substantial maintenance. Other methods have been developed to simplify the cell concentration step such as settling tanks (WASH and BUNGAY, 1979) with the possible addition of agents which will enhance flocculation (WEEKS et al., 1982).

A multistage continuous ethanol fermentation from molasses with recycle of the yeast is used by the Danish Distilleries Ltd. of Grena (ROSEN, 1978). Fig. 19 shows the process. The received molasses is stored in 2–3 tanks, each approximately 1,500 $m^3$, and is then pumped to intermediate containers. From there, the raw material is taken to a set of metering pumps, for molasses, water, $H_2SO_4$, and $(NH_4)_2HPO_4$. The pH is adjusted to 5. This mixture is sterilized in a plate heat exchanger at 100°C. Two or three fermenters are used, with a total volume of about 170 $m^3$. The fermented wort is centrifuged prior to distillation and the yeast returned to fermenter 1.

When the fermentation is started, molasses, water, and ordinary baker's yeast are added to fermenter 1, which is aerated until the yeast has propagated sufficiently to reach the total quantity desired. Approximately 0.02–0.03 L of air per L of liquid per min has to be infused.

Characteristic fermentation data are as shown in Tab. 18. The yield calculated on molasses amounts to approximately 28.29 L alcohol per 100 kg molasses, or a maximum of approximately 65 L per 100 kg fermentable sugar.

Alcon Biotechnology Ltd. (Alcon, 1980) have developed a rapid continuous fermentation process for ethanol production. This process operates with any clean fermentable sugar feedstock including sucrose from sugarcane, sugar beet, and molasses and any hydrolyzed starch from such sources as corn, sorghum and cassava. A simplified flow sheet is shown in Fig. 20. The process uses a single stirred tank fermenter in which yeast is maintained at a high concentration (up to 45 g $L^{-1}$) by recycling.

The fermenter is stirred either by liquid recirculation or by a conventional agitator. The temperature and pH are controlled automatically. The fermenter is equipped with dosing

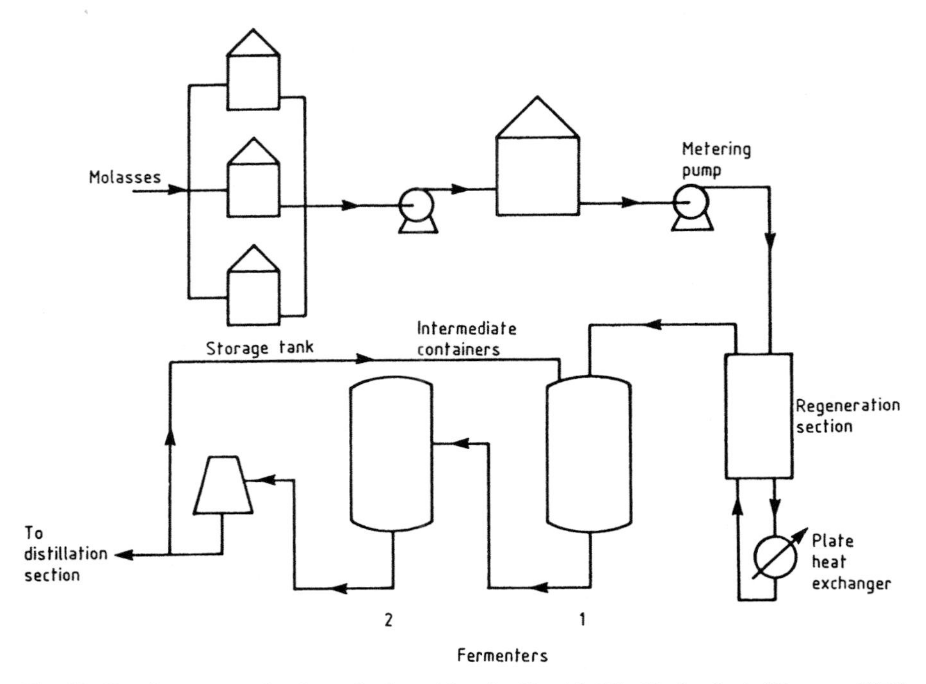


**Fig. 19.** Continuous production of ethanol by the Danish Distilleries Ltd. (ROSEN, 1978).

**Tab. 18.** Performance Data for the Danish Distilleries Process (ROSEN, 1978)

| Component | Fermenter 1[a] | Fermenter 2 |
|---|---|---|
| Yeast, dry matter [g $L^{-1}$] | 10 | 10 |
| pH | 4.7 | 4.8 |
| Alcohol [vol.%] | 6.1 | 8.4 |
| Residual sugar [%] | 1.0 | 0.1 |
| Temperature [°C] | 35 | 35 |

[a] Residence time in each fermenter: 10.5 h; afflux: 600 kg molasses diluted in $22 \cdot 10^3$ L $h^{-1}$.

systems for the addition of antifoam, alkali, and nutrients. The oxygen level is controlled. Yeast is separated by gravity settling. For clarified beet juice at sugar concentrations of 14% (w/v) the ethanol yield is about 93% of the theoretical value at dilution rates of 0.14–0.18 $h^{-1}$. The ethanol concentration in the outlet was 8.5–8.9% (w/v). The capital cost of this system should be low as only one fermenter and gravity settling are used.

A plant using continuous tower fermenters has been under consideration in Australia (ANONYMOUS, 1980a). Cereal, molasses, cane juice, sugar beet, cassava, or coarse grains are being considered as substrates.

The fermenter consists of a vertical cylindrical tower with a conical bottom. The tower is topped by a large-diameter yeast settling zone fitted with baffles. The sugar solution is dumped into the base of the tower which contains a plug of flocculent yeast. Reaction proceeds progressively as the beer rises, but yeast tends to settle back and be retained. High cell densities of 50–80 g $L^{-1}$ are achieved without

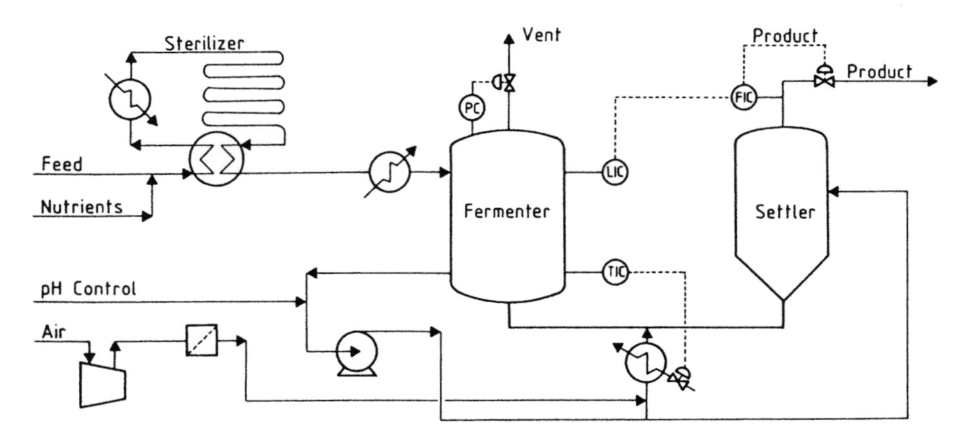


**Fig. 20.** Flowsheet of the ALCON continuous fermentation process.
FIC: flow indicating controller; LIC: level indicating controller; TIC: temperature indicating controller; PC: pressure controller.

an auxiliary mechanical separator. Residence times of below 4 h have been possible with sugar concentrations of up to 12% (w/v) sucrose giving 90% sugar utilization and 90% conversion to ethanol, i.e., up to 5% (w/v) alcohol.

Productivities of up to 80 times higher than for a simple batch process have been achieved in the tower fermenter. Some apparent difficulty is associated with providing the desired oxygen concentration in the fermenter. A major drawback of this APV Co. Ltd. system is the long time required for initial start up. Two or three weeks are required to build up the desired high cell density and achieve stable operation. This is compensated by the very long (more than 12 months) operating times between shutdowns. Capital and operating costs are predicted to be far lower than for conventional batch processes.

## 5.4 Solid Phase Fermentation (Ex-Ferm Process)

A somewhat innovative process for ethanol production was developed in Guatemala (ROLZ, 1978, 1979, 1981) with the objective of reducing costs. It employs fresh or dried sugarcane pieces as a raw substrate. It is a mixed solid–liquid phase system where sucrose extraction and fermentation are conducted simultaneously. Liquid from the first fermenter (packed bed fermenter) is usually used in a second cycle of extraction-fermentation. This could be repeated until ethanol tolerance is reached or until microbial contamination is too high. A more complete juice extraction is claimed (2.2 L per $10^3$ kg) and no separate extraction equipment but modifications of the fermenter are required.

## 5.5 Vacuum Fermentation

As alcohol is toxic to cell growth, processes by which it may be removed upon formation would greatly enhance the productivity of the system. CYSEWSKI and WILKE (1977) have taken advantage of the high volatility of ethanol by running the fermentation under sufficient vacuum to boil off the product at temperatures conductive to yeast growth. A fermenter productivity of 82 g $L^{-1}$ $h^{-1}$ was achieved from a 33.4% (w/v) glucose feed (12-fold that of a conventional continuous process).

## 5.6 Simultaneous Saccharification and Fermentation (SSF) Process

The SSF process for ethanol production was conceptualized in the late 1970s by

GAUSS et al. (1976), TAKAGI et al. (1977), and BLOTKAMP et al. (1978). Employing fermentative microorganisms in combination with cellulose enzymes, sugar accumulation in the fermenter is minimized. In this way, feedback inhibition by product sugars is reduced and higher hydrolysis rates and yields are possible than for saccharification without fermentation at high substrate loadings.

SPINDLER et al. (1991) evaluated 4 woody crops, 3 herbaceous crops, corn cobs, and corn stover in the SSF process as substrates pretreated with dilute acid for ethanol production by *S. cerevisiae* and *Bretanomyces clausenii.* In all cases, SSFs demonstrated faster rates and higher conversion yields than the saccharification without fermentation (SAC-s) processes.

# 6 Industrial Processes

## 6.1 Ethanol from Corn

A flow diagram for a conventional fermentation plant producing $76.0 \cdot 10^3$ $m^3$ $a^{-1}$ anhydrous ethanol from $816.5 \cdot 10^3$ kg $d^{-1}$ corn is shown in Fig. 21.

Corn from storage is fed to a grinder where the kernel size is reduced to expose the interior portion of the grain. Water is added, pH adjusted, and the ground grain is then cooked to solubilize and gelatinize the starch. After cooking and partial solubilization, fungal amylase is added. Yeast, which has also been propagated in the plant is added and the fermentation is allowed to continue for approximately 48 h at a temperature of 32 °C. During this time about 90% of the original starch in the grain is converted to alcohol. The processing proceeds on a cyclic or batch basis, with some vessels containing fermenting mash while other vessels are being filled, emptied, or sterilized.

Once fermentation is completed, the mixture is fed to the beer still where essentially all of the alcohol is distilled overhead to about 50 vol.%. The diluted alcohol is purified by further distillation, which removes fermentation by-products (aldehydes, ketones, fusel oils), yielding 95 vol.% alcohol. If it is desired to produce anhydrous ethanol, the 95 vol.% ethanol is fed to a dehydration section consisting of an extractive distillation with benzene.

In the beer still, the remaining water with dissolved and undissolved solids is drawn from the bottom of the still and fed to a centrifuge. The liquid phase from the centrifuge is concentrated to 50% dissolved solids in a multiple-effect evaporator and mixed with the solids from the centrifuge. This mixture is then dried in a fluidized transport-type dryer to 10% moisture and is used as cattle feed. The cattle feed contains all the proteins that were originally present in the grain, plus the additional proteins from the yeast, resulting in a product containing 28–36% protein by weight.

In addition to alcohol and cattle feed, the original $816.5 \cdot 10^3$ kg of corn yield $175.0 \cdot 10^3$ kg of carbon dioxide and 95 kg of by-product aldehydes, ketones, and fusel oils.

Ethanol production can be incorporated into a wet milling corn processing plant. In this case, the substrate is the isolated corn starch and the by-product is gluten feed.

## 6.2 Ethanol from Cassava Root

The first commercial plant in the world producing ethanol from cassava root commenced production in 1977, with an ethanol capacity of 60 $m^3$ $d^{-1}$ (ANONYMOUS, 1978). It is expected that additional plants will be built in the future using similar processing.

The processing steps used to obtain ethanol from cassava roots are shown in Fig. 22. As the fresh roots are received, they are weighed, washed, peeled, and ground into a mash. Part of this mash, the quantity determined by the plant operating plans, is then side-streamed and dried, producing a meal. Unlike sugarcane, which starts to decay and ferment naturally soon after cutting, dried cassava can be stored for as long as one year without a significant loss of starch, as well as bitter cassava. Bitter cassava varieties contain cyanogenic glucosides which do not cause any trouble as

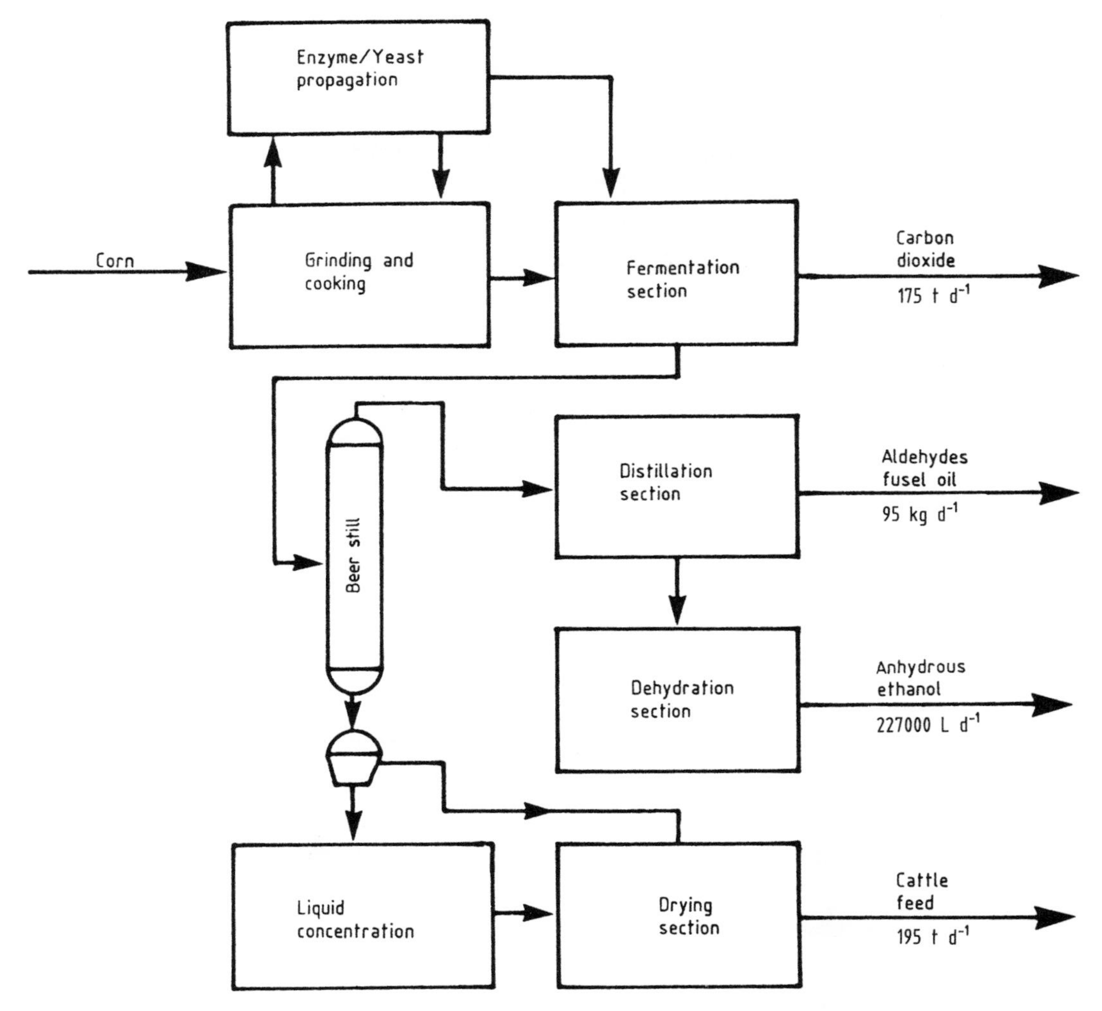


**Fig. 21.** Proposed flow diagram for a conventional fermentation plant producing 76 million liters of anhydrous ethanol per year from corn (SCHELLER, 1976).

they are flashed off with the steam during the mash cooking process (JACKSON, 1976).

The starch molecules are broken down with the aid of $\alpha$-amylase which is added in two steps. The preliminary addition is required to decrease the viscosity of the mash and so facilitate the cooking operation. Glucoamylase is then employed to achieve the final conversion of the liquefied starch material into glucose. This is the time limiting step in the overall process. The final steps in the process of recovering the alcohol from the fermented material are the same as those for sugarcane juice.

Alcohol yields from cassava are in the range of 165–180 L $t^{-1}$. However, as sugarcane harvests can be up to 90 t $ha^{-1}$, the alcohol yield per unit cultivation area is higher for sugarcane under present agricultural methods. Also, the greatest advantage from sugarcane processing is in the use of its fiber (bagasse) as a fuel. A cane stalk contains roughly

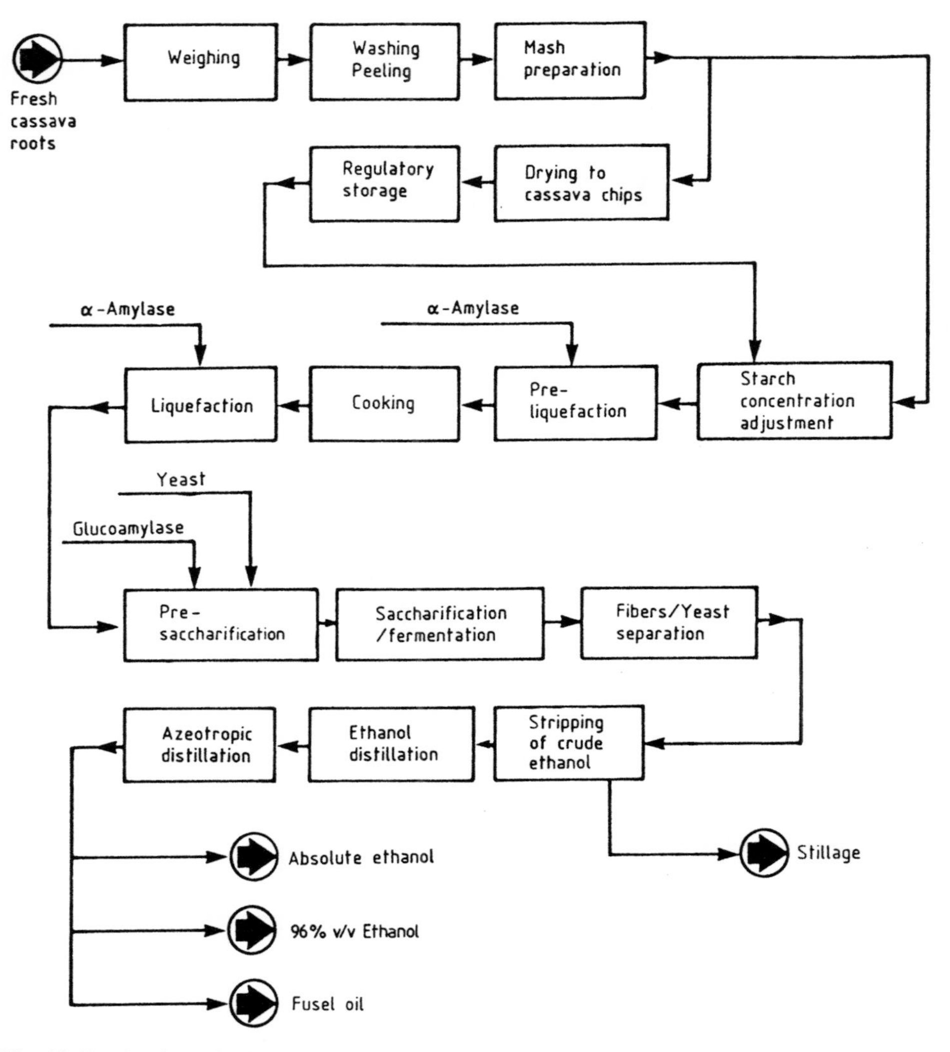


**Fig. 22.** Production of ethanol from cassava root (LINDEMAN and ROCCHICCIOLI, 1979).

the same weight of dry fiber as sugars and, therefore, is more than adequate in providing the energy for alcohol processing. This is not the case with cassava, where the dry fiber content is about 3.5%. Also, because of the necessary conversion of starch to fermentable sugars, a greater energy input is necessary in its processing prior to fermentation.

For batch alcohol manufacture from cassava, the steam requirement is about 1.8 kg steam per kg cassava, which must come from an external source. By continuous processing in which efficient heat recovery systems are used, a considerable fuel reduction can be achieved. Therefore, the biggest problem in using cassava is the high energy requirement.

## 6.3 Ethanol from Potatoes

The Danish Distilleries Ltd. of Aalborg has developed a semicontinuous process for the

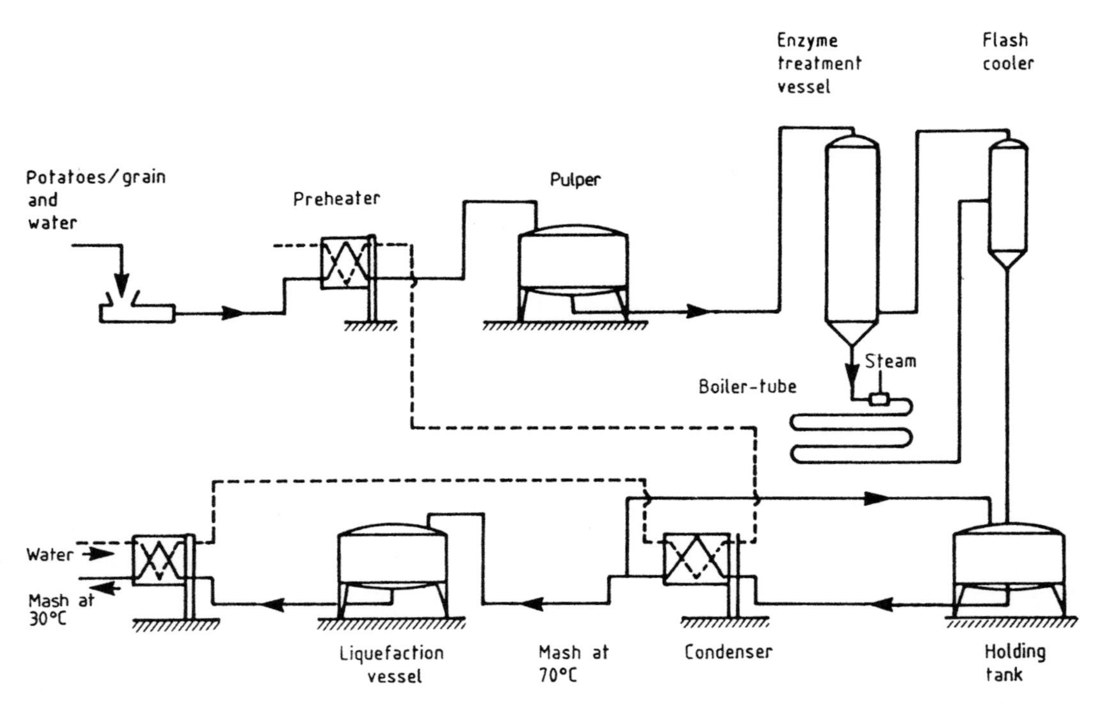


**Fig. 23.** Danish Distilleries Ltd.'s semicontinuous production of alcohol from potatoes/grains (ROSEN, 1978).

production of ethanol from potatoes and grain (ROSEN, 1978). In this process (Fig. 23), the potatoes (or grain) are transported on a conveyor belt to a belt weigher.

Comminution of the potatoes or grain and the blending of grain and water take place in a pulper. The enzyme used is Termamyl 60 prepared by Novo Industri. After enzymatic treatment, the potatoes or grain are heated to 90–95 °C by flash steam in a condenser. The raw material is then transported from the bottom of this apparatus to a boiler tube which is approximately 50 m in length. Heating to a temperature of approximately 150 °C is achieved by 10.1 MPa steam. Retention time at 150 °C is approximately 3 min and the rate of speed in the boiler tube is around 0.3 m $s^{-1}$. The mixture is flashed to atmospheric pressure in a single step, and the steam released is used to preheat the starchy compound. The mash is cooled to 70 °C for liquefaction with commercial amylase preparations of bacterial origin. If pH regulation is required, it is accomplished with slaked lime. Intermission at 70 °C is approximately 25 min. The mash is cooled to 30 °C and is then pumped to the yeast vessels, where batch fermentation to ethanol is carried out in a conventional manner. All temperature and level regulations are automatic, so that the process can be maintained by one operator.

## 6.4 Ethanol from Jerusalem Artichoke Tubers (Topinambur)

A process to produce $360 \cdot 10^3$ kg $a^{-1}$ ethanol from Jerusalem artichoke tubers was designed by KOSARIC et al. (1982). A simplified flowsheet of the process is shown in Fig. 24.

The juice from the ground tubers is obtained by two-step expression with macerating. The final juice contains approximately 12–15% of carbohydrates. The juice is hydrolyzed enzymatically by maintaining it at 56 °C for 2 h at pH 3.8. There is enough enzyme

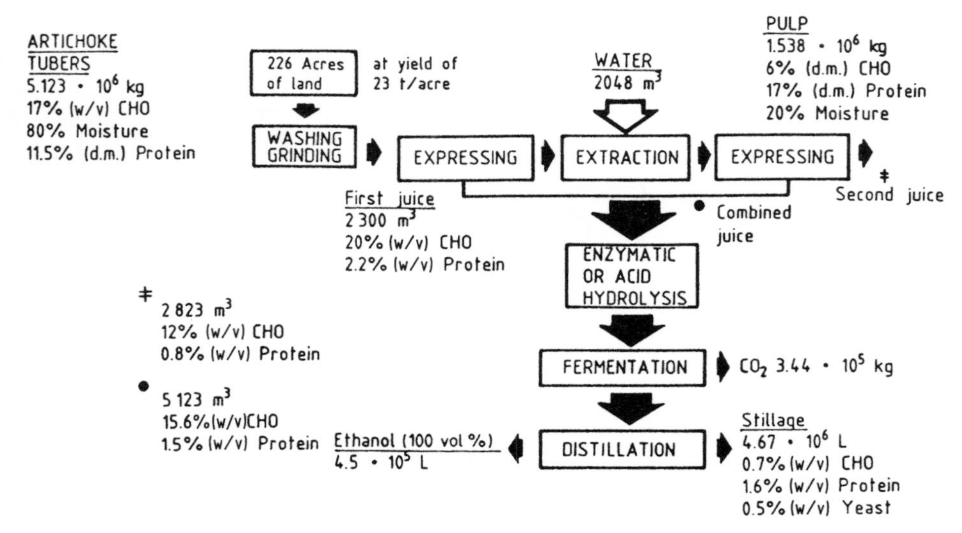


**Fig. 24.** Flowsheet for production of $4.5 \cdot 10^5$ L ethanol (100 vol.%) per year from Jerusalem artichoke tubers (KOSARIC et al., 1982).

present in the plant, particularly if harvested in spring. After cooling, the juice is fermented in the batch mode with recycle of the yeast. The ethanol yield of about 90% is achieved after 28 h. Complete hydrolysis is not required as a special strain of yeast with inulase activity is used.

## 6.5 Ethanol from Cellulose

### 6.5.1 Dilute Sulfuric Acid Process

One of the earliest commercial hydrolysis processes was a dilute sulfuric acid process developed by EWEN and TOMLINSON during World War I. Two plants in the United States producing sugars from wood were operative. For economic reasons, both plants were closed at the end of World War I.

Dilute sulfuric acid hydrolysis of wood was re-examined by the Forest Products Laboratory, U.S. Department of Agriculture, at the request of the War Production Board in 1943. A full-scale wood hydrolysis plant was built in Springfield, Oregon, but it did not commence operation until after the end of World War II (HOKANSON and KATZEN, 1978). A schematic diagram of the plant is shown in Fig. 25. The process is basically a semicontinuous process in which the hydrolyzate percolates through the "chip" bed, which avoids prolonged exposure of sugars to acid at high temperatures and so reduces degradation. Optimum conditions are listed in Tab. 19.

Dilute hydrolysis solution from a previous batch is pumped into the top of the hydrolyzer containing the wood wastes. After the dilute hydrolyzate is charged, hot water and sulfuric acid are added and the temperature is raised to 196 °C (1,520 kPa vessel pressure). After 70 min of pumping, a strong hydrolysis solution starts to flow out of the bottom of the hydrolyzer. Two letdown flash stages are used, the first stage operating at 456.0 kPa and the second stage at atmospheric pressure. The condensates from the flash vapor from both stages contain furfural and methanol, which are recovered. The underflow from the two flash stages is the sugar-containing solution. At the end of the percolation cycle, the lignin-rich residue is discharged, recovered, and used as a fuel.

The flash condensate passes to a distillation tower for recovery of methanol. From the

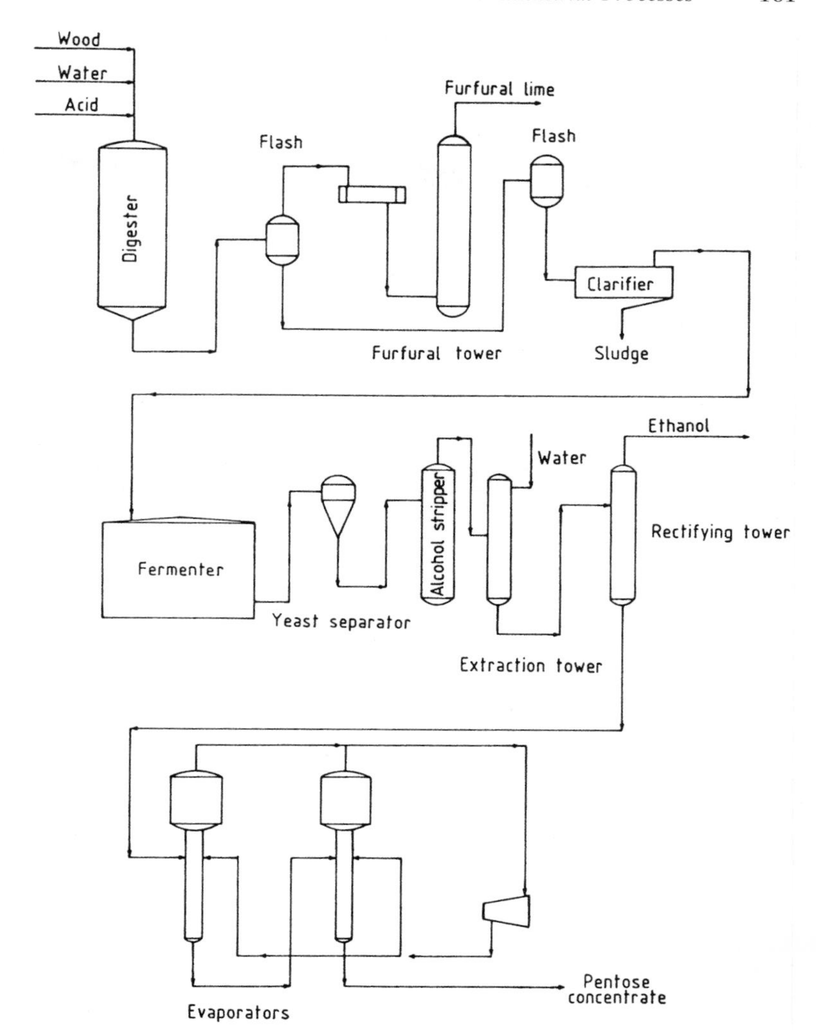


**Fig. 25.** Ethanol production from wood (HOKANSON and KATZEN, 1978).

**Tab. 19.** Optimum Conditions for Ethanol Production from Wood Hydrolyzate (HOKANSON and KATZEN, 1978)

| Parameter | Optimum Conditions |
|---|---|
| Acid concentration in total water | 0.53% |
| Maximum temperature of percolation | 196 °C |
| Rate of temperature rise | 4 °C $min^{-1}$ |
| Percolation time | 145–190 min |
| Ratio of total water–oven-dried wood | 10 |
| Percolation rate | 8.69–14.44 L $min^{-1}$ $m^{-3}$ |

base of this tower the bottoms pass to a second distillation tower for recovery of the furfural–water azeotrope. The hot acid hydrolyzate solution is neutralized with a lime slurry and the precipitated calcium sulfate is separated in a clarifier. Calcium sulfate sludge is concentrated to about 50% solids and is trucked to a disposal area. Neutralized liquor is blended with recovered yeast from a previous fermentation and passed to fermentation tanks. From the fermenters, the fermented liquor passes to yeast separators for recovery of the yeast for recycling (*Saccharomyces cerevisiae*).

Ethanol is recovered in a series of distillation towers and is finally rectified to approximately 95 vol.%. Bottoms from the beer stripping tower contain pentoses. Instead of disposing of this stream, it is economically feasible to concentrate the sugars to a 65% solution for sale as a feed supplement or for conversion into furfural. The plant was closed for economic reasons after World War II.

Inventa AG (Switzerland) has developed a dilute acid hydrolysis process (MENDELSOHN and WETTSTEIN, 1981). The technology is based on the wood saccharification process operated until 1956 in Domat/Ems, which produced $10 \cdot 10^6$ kg $a^{-1}$ of fuel grade ethanol using locally available softwoods. Fig. 26 shows the flowsheet of this process (MENDELSOHN and WETTSTEIN, 1981).

The reactors operate on wood chips at about 1,000 kPa and at 140–180 °C. Hemicellulose and cellulose from wood are hydrolyzed with dilute (0.6 wt.%) acid. The solution of acid and sugars, the wort, is removed from the reactor and collected in a flash tank. The lignocellulose cake leaving the reactor is used as a fuel. The vaporized furfural is collected in a recovery section. The cooled wort is then neutralized with limestone ($CaCO_3$) and the gypsum is separated. The wort is cooled to the temperature required for fermentation and the nutrients are added. The fermentation takes place in a series of fermenters split into two parallel lines for operating flexibility.

A yeast adapted to wood hydrolyzates is used. The fermenters are stirred by gas ($CO_2$) recycled to the process. The mash with an ethanol concentration of about 2% is separated and a concentrated yeast suspension is recycled to the fermenter. Fuel-grade ethyl alcohol produced from the fermented wort is concentrated by distillation. Yield of alcohol is 240 L $t^{-1}$ of wood dry matter. Steam consumption is $14–16 \cdot 10^3$ kg $t^{-1}$ ethanol. The Inventa process can be energy-autonomous, with wood the only resource required.

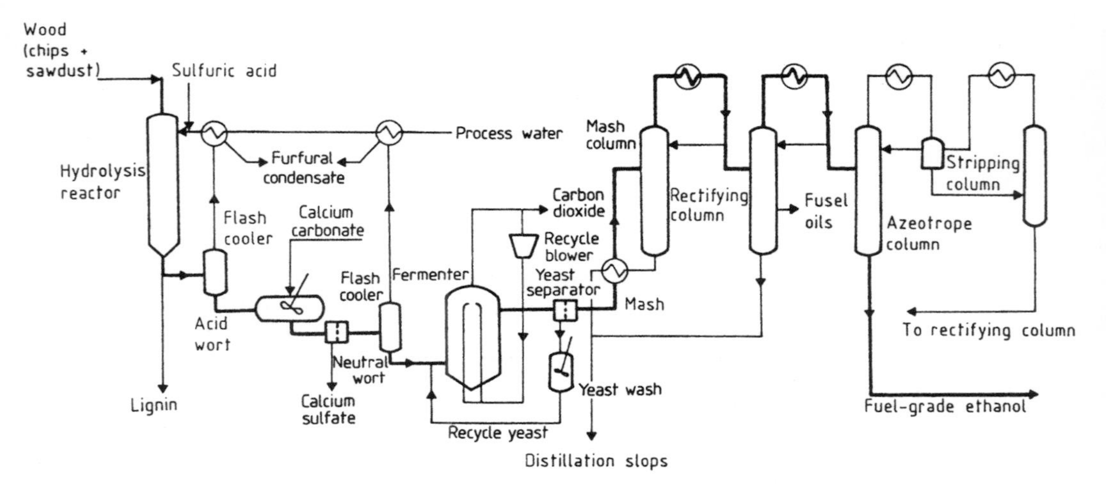


**Fig. 26.** Flowsheet for ethanol production by continuous fermentation (MENDELSOHN and WETTSTEIN, 1981).

## 6.5.2 Strong Acid Hydrolysis Process

For strong acid hydrolysis the proposal based on the Hokkaido Strong Acid Hydrolysis Pretreatment Process is worth mentioning.

The flowsheet in Fig. 27 shows the process. Undried wood is first introduced into a prehydrolysis digester column in which dilute sulfuric acid is used to remove the hemicellulose. Lignin cellulose particles then enter a pressurized feeder and are transported by recycling strong acid hydrolysis solution to the top of the second digester. In the second countercurrent column, the cellulose is hydrolyzed at room temperature by 70–80% sulfuric acid. The glucose–sulfuric acid solution leaving the top of the column is separated by electrodialysis membranes. The glucose retained by the membrane is neutralized and deionized before fermentation. The sulfuric acid permeated from the electrodialysis is evaporated and reconcentrated for recycle. Lignin is separated from the strong acid exiting the bottom of the second digester by filtration and washing.

## 6.5.3 Ethanol Production from Agricultural Residues via Acid Hydrolysis

Two processes based on cornstalks have been employed (SITTON et al., 1979; FOUTCH et al., 1980): dilute acid treatment with sugar yields of up to 50% and concentrated acid treatment with even higher yields. The two-stage process involves a dilute sulfuric acid treatment for pentosan conversion followed by treatment with concentrated sulfuric acid for hexosan conversion. A high yield, low acid utilization and no degradation of pentoses when contacted with concentrated acid occur.

Fig. 28 shows the flow diagram for the hydrolysis process and the mass balance for a plant with a capacity of $17 \cdot 10^3$ $m^3$ $a^{-1}$ ethanol. Ground corn stover (20 mesh) is treated with 4.4% sulfuric acid at 100 °C for 50 min. The mixture is then filtered and the xylose-rich liquid is processed by electrodialysis for acid recovery. The solids are dried and impregnated with 85% $H_2SO_4$. Water is then added to the powder to form an 8% $H_2SO_4$ concentration, and hydrolysis is carried out at

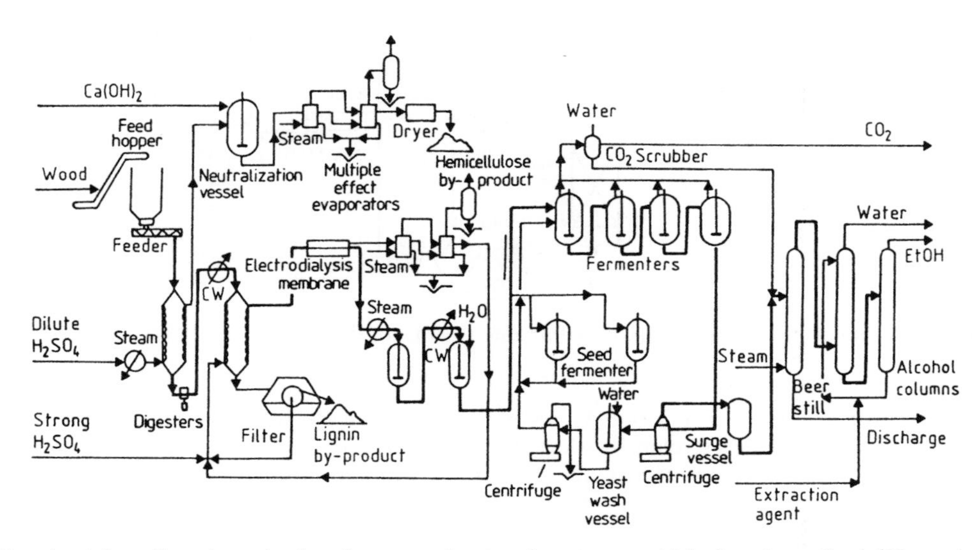


**Fig. 27.** Flowsheet for ethanol production from wood using the strong acid hydroysis method (YU and MILLER, 1980).

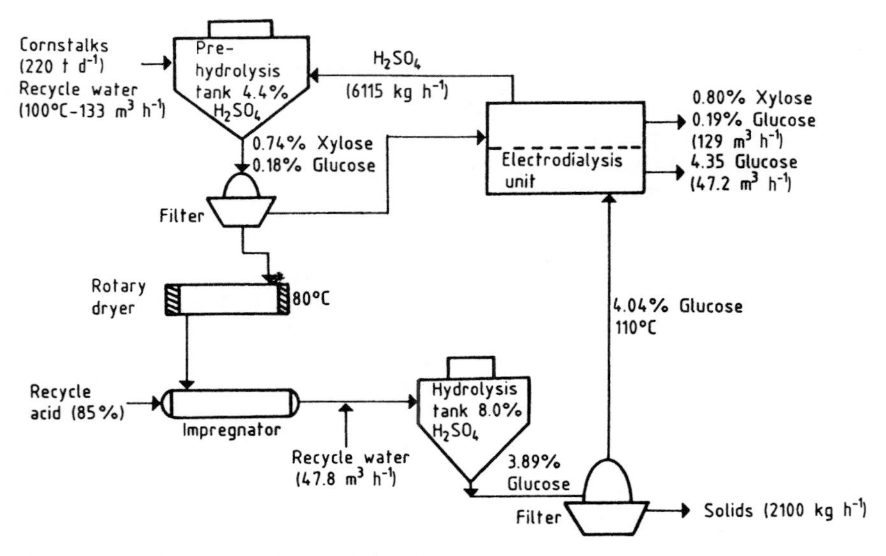


**Fig. 28.** Flowsheet for acid hydrolysis of cornstalks (FOUTCH et al., 1980).

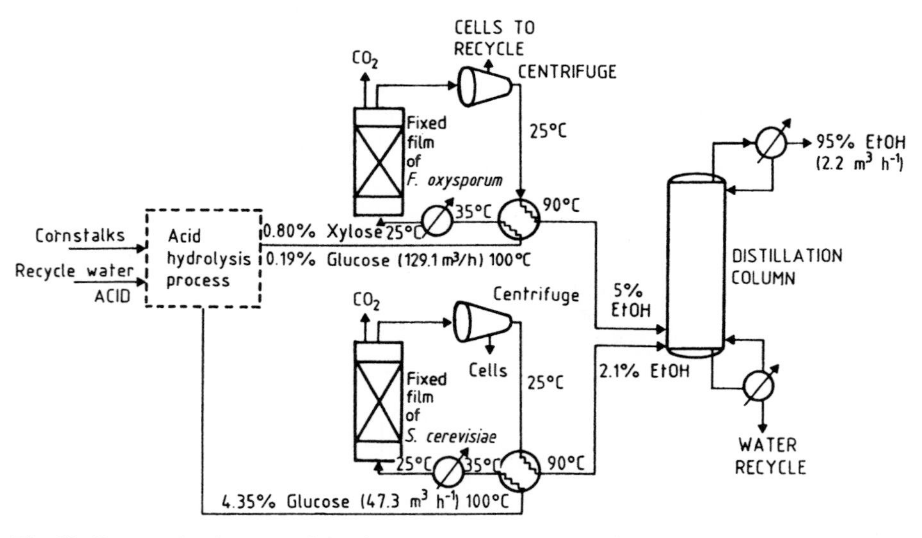


**Fig. 29.** Process for fermentation of hydrolyzate to ethanol (FOUTCH et al., 1980).

110°C for about 10 min. Acid is recovered from the glucose-rich stream by electrodialysis. The remaining solids can be dried and returned to the soil or used as a source of fuel. After hydrolysis xylose and glucose are obtained. The yield of xylose is 95% and the yield of glucose is 89%.

Glucose is converted to ethanol by *Saccharomyces cerevisiae* in the immobilized cell reactor. *Fusarium oxysporum* converts xylose to ethanol in another immobilized cell reactor. Cell overgrowth is removed every two weeks by sparging the reactors with compressed carbon dioxide. The ethanol in the reactor effluent is separated by distillation. Fig. 29 shows the fermentation process for a plant with a capacity of $17 \cdot 10^3$ $m^3$ $a^{-1}$.

## 6.5.4 Ethanol from Newspaper via Enzymatic Hydrolysis

Research has been carried out at the Natick Development Center (NDC) and at the University of California (Berkeley) to design and evaluate a process for the enzymatic hydrolysis of newsprint. Newspaper was chosen as the substrate because it has a high cellulose content, its composition is consistent, and it is readily available.

The process is shown in Fig. 30. A major part of the overall process cost is the production of enzymes. Another major part of the overall process cost is substrate pretreatment. Ball milling has been found to be the most effective physical method of making the maximum amount of cellulose in the substrate available for enzymatic hydrolysis, but it is expensive.

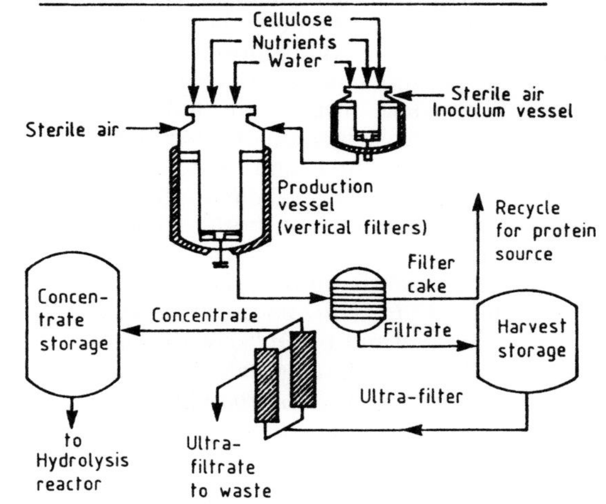


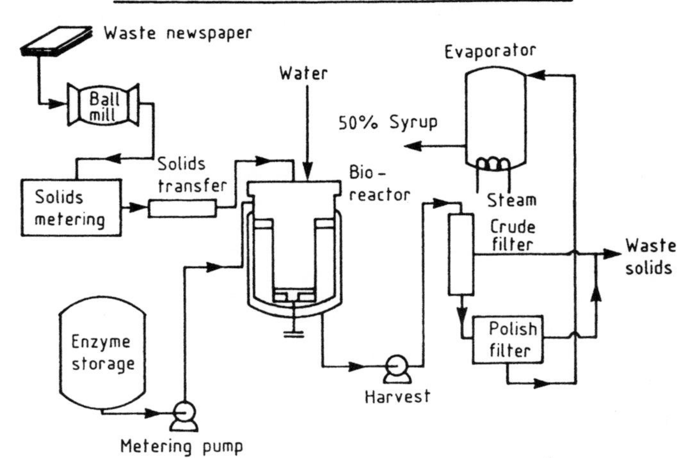


**Fig. 30.** Enzymatic hydrolysis of newsprint by Natick Development Center (NDC) (SPANO, 1976).

## 6.5.5 Ethanol from Municipal Solid Waste via Acid Hydrolysis

Municipal solid waste could be also used for ethanol production. A composition of this waste is given in Tab. 20 (ACKERSON, 1991). The cellulose containing components in this waste are essentially the raw materials for conversion to sugars by acid hydrolysis. Single and two-step hydrolysis, with nearly 100% yields have been reported (ACKERSON, 1991). The resulting sugar solution has been successfully fermented to ethanol. Best conditions for hydrolysis were obtained at high acid concentrations (80% $H_2SO_4$ or 41% HCl) and a temperature of 40°C.

The hydrolyzates from lignocellulose contain not only readily fermentable hexoses but also a high proportion of pentoses. By utilizing also the pentose component (containing predominantly xylose from hemicellulose) a potential 30% increase in yield of ethanol can be achieved (LAWFORD and ROUSSEAU, 1991). Fig. 31 represents a schematic for the production of ethanol from both cellulose and hemicellulose.

In order to efficiently utilize xylose, genetically engineered *E. coli* bacteria (ATCC 11303) carrying plasmid pLOI297 with genes for pyruvate decarboxylase and alcohol dehydrogenase cloned from *Zymomonas,* were used (LAWFORD and ROUSSEAU, 1991). Further work on the production of ethanol from xylose by bacteria carrying genes from *Zymomonas* is summarized in Tab. 21.

The results of the research conducted by LAWFORD and ROUSSEAU are presented in

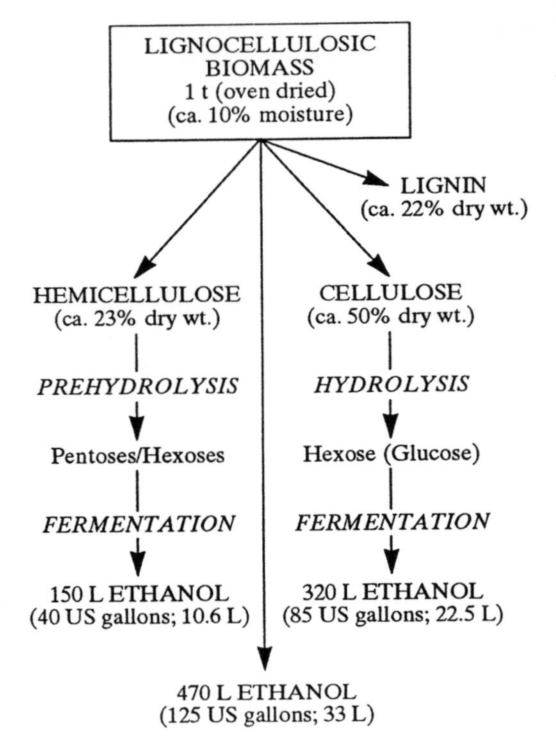


**Fig. 31.** Processing lignocellulosic feedstocks for the production of fuel alcohol and the theoretical maximum yield of ethanol from biomass.

Tabs. 22 and 23. Conditions were examined in pH-controlled stirred tank reactors for their effect on growth and metabolism, such as pH, xylose concentration, nutrient-rich and chemically-lean media, utilization of $C_6/C_5$ mix-

**Tab. 20.** The Composition of Selected Biomass Materials (ACKERSON, 1991)

| Material | Dry Weight of Material [%] | | |
|---|---|---|---|
| | Hemicellulose | Cellulose | Lignin |
| Tanbark oak | 19.6 | 44.8 | 24.8 |
| Corn stover | 28.1 | 36.5 | 10.4 |
| Red clover hay | 20.6 | 36.7 | 15.1 |
| Bagasse | 20.4 | 41.3 | 14.9 |
| Oat hulls | 20.5 | 33.7 | 13.5 |
| Newspaper | 16.0 | 61.0 | 21.0 |
| Processed MSW | 25.0 | 47.0 | 12.0 |

**Tab. 21.** Prodcution of Ethanol from Xylose by Bacteria Carrying Genes from *Zymomonas*[a] (LAWFORD and ROUSSEAU, 1991)

| Host Organism | Genes | Substrate | Maximal EtOH Efficiency [% by wt.] | [%] | Investigators |
|---|---|---|---|---|---|
| *E. coli* JM 101 (pZAN4) | *pdc, adh*B | xylose | 0.37 | 71 | NEALE et al. (1988) |
| *E. coli* B | "PET" plasmid | xylose | 4.2 | 102 | INGRAM et al. (1989) |
| (pLOI 297) | *pdc, adh*B | xylose | 3.9 | 96 | OHTA et al. (1990) |
| *E. coli* S17-1 | "PET" plasmid | xylose | 1.0 | 39 | BECK et al. (1990) |
| (pLOI 308-10) | *pdc, adh*B | glu:xyl 1:1 | 1.6 | 63 | |
| *Erwinia chrysanthemi* | *pdc* | xylose | 0.5 | 87 | TOLAN and FINN (1987a) |
| *Klebsiella* | | arabinose, | | 71 | |
| *planticola* | *pdc* | xylose | 2.5 | 78 | TOLAN and FINN (1987b) |
| | | $C_5$-sugar mix | 3.2 | | |

[a] The majority of these fermentations were conducted anaerobically in batch mode using buffered, nutrient-rich, complex media; *pdc* is the gene for pyruvate decarboxylase; *adh*B is the gene for alcohol dehydrogenase II – both from *Zymomonas mobilis* CP4.
These genes constitute the "PET" operon ("**P**ortable **ET**hanol **P**athway; INGRAM, 1990).

**Tab. 22.** Ethanol Production from Hardwood Hemicellulose Hydrolyzate by Recombinant *E. coli* ATCC 11303 (pLOI 297) (LAWFORD and ROUSSEAU, 1991)

| Conditions[a] Medium Comp. Expt. | Xylose | Products | | Productivity Yield | | | | |
|---|---|---|---|---|---|---|---|---|
| | | Biomass[b] | EtOH | $O_p$ | $O_p^{max}$ | $q_b$ [gP | $Y_{p/s}$ | Conversion Efficiency |
| | [g L$^{-1}$] | [gdw L$^{-1}$] | [g L$^{-1}$] | [gP L$^{-1}$ | h$^{-1}$] | g cell$^{-1}$ h$^{-1}$] | [g g$^{-1}$] | [%] |
| Aspen prehydrolyzate +complex nutrients (LB) 13 mM Acetate (no $Ca(OH)_2$) APH3 | 36.1 | 0.5 | 15.5 | 0.29 | 0.34 | ND | 0.45 | 92 |
| Aspen prehydrolyzate +complex nutrients (LB) 103 mM Acetate (+$Ca(OH)_2$[c] APH4 | 35.9 | 0.5 | 16.9 | 0.60 | 0.78 | ND | 0.47 | 92 |
| Aspen prehydrolyzate +mineral salts (+$Ca(OH)_2$)[c] APH5 | 31.0 | 0.5 | 14.9 | 0.26 | 0.65 | ND | 0.48 | 94 |

[a] The pH was controlled at 7.0 with KOH; the acetate concentration was determined by HPLC analysis and did not change during the fermentation; the monomer sugars in the "prehydrolyzate" were predominantly xylose; the process yield ($Y_{p/s}$) was based on the amount of fermentable sugar in the medium; P: ethanol; ND: not determined.
[b] Approximate initial cell density (inoculum; g dry substance of cells per liter).
[c] Powdered $Ca(OH)_2$ was added to the aspen "prehydrolyzate" to pH 10 and neutralized to pH 7 with 1N $H_2SO_4$ followed by centrifugation to remove insolubles.

**Tab. 23.** Continuous Fermentations in Ethanol Production by Recombinant *E. coli* ATCC 11303 (pLOI297) with Chemically Lean Mineral Salts Medium (LAWFORD and ROUSSEAU, 1991)

| Conditions | | Substrates | | | Product | Productivity | | Yield | |
|---|---|---|---|---|---|---|---|---|---|
| pH | Dilution Rate $[h^{-1}]$ | Concentration $[g\,L^{-1}]$ | | Utilization [%] | EtOH $[g\,L^{-1}]$ | $q_0$ $[gP\,g\,cell^{-1}\,h^{-1}]$ | $O_p$ $[gPL^{-1}h^{-1}]$ | $Y_{p/s}$ $[g\,g^{-1}]$ | Conversion Efficiency [%] |
| Mineral Salts Medium | | | | | | | | | |
| Glucose | | | | | | | | | |
| 6.8 | 0.100 | 31 | | 100 | 9.7 | 1.26 | 0.97 | 0.31 | 61 |
| 6.3 | 0.081 | 36 | | 94 | 12.3 | 0.72 | 1.00 | 0.36 | 71 |
| 6.0 | 0.137 | 31 | | 100 | 14.1 | 0.71 | 1.93 | 0.45 | 88 |
| Xylose | | | | | | | | | |
| 6.0 | 0.97 | 33 | | 54 | 7.8 | 0.54 | 0.76 | 0.44 | 86 |
| Mixture | | Glu | Xyl | | | | | | |
| 6.3 | 0.100 | 21.4 | 12.0 | 100 | 11.2 | 0.65 | 1.12 | 0.34 | 67 |
| 6.3 | 0.214 | 21.4 | 12.0 | 76 | 11.4 | 1.26 | 2.44 | 0.45 | 88 |
| 6.0 | 0.106 | 22.9 | 12.2 | 96 | 11.4 | 0.86 | 1.21 | 0.34 | 67 |
| 6.0 | 0.149 | 22.9 | 12.2 | 81 | 8.8 | 0.62 | 1.31 | 0.31 | 61 |

tures, sensitivity to oxygen and tolerance of acetic acid. The recombinant converted a hardwood (aspen) hemicellulose hydrolyzate (3.5% xylose) to ethanol at an efficiency of 94%. A maximum volumetric productivity of 0.76 $g\,L^{-1}\,h^{-1}$ was observed in $Ca(OH)_2$-treated prehydrolyzate, which was prepared by the Bio-Hol process using a Wenger extruder with $SO_2$ as a catalyst.

## 6.6 Ethanol from Waste Sulfite Liquor (WSL)

The process of fermenting the sugars present in WSL was first investigated at a large scale in Sweden in 1907 by WALLIN and EKSTROM. In 1940, the world production was over $76.0 \cdot 10^6$ L with Sweden producing approximately half of the total. During the period of 1948–1950, 32 Swedish mills produced ca. $114 \cdot 10^6\ L\,a^{-1}$ of 95% alcohol. Today, only a few plants are left in operation in Sweden. This reduction reflects the shift from sulfite pulping to Kraft processes (SHREVE and BRINK, 1977), and the fierce competition with the petroleum industry for the alcohol market. During World War II, two plants were built in North America, one at Ontario Paper Co. in Thorold, Ontario, with a capacity of $7.6 \cdot 10^3\ L\,d^{-1}$ of 95% ethanol, and the other at Puget Sound Pulp and Timber Co. in Bellingham, Washington, with a capacity of $30.0 - 10^3\ L\,d^{-1}$ of 95% ethanol. Since then, Commercial Alcohols Ltd. has constructed a sulfite alcohol plant at Gatineau, Quebec, which is similar in size to the Bellingham plant.

In the plant at Bellingham, the WSL from the flow tank is drained, steam stripped, and stored at about 90°C. Fermentation is carried out continuously in 7 interconnected fermenters of 300 $m^3$ each. Yeast and ammonia or urea are added to the liquor entering the first fermenter. The liquor overflows from the first into the second and so on to the last fermenter. Approximately 95% of the fermentable hexose sugars are utilized in a cycle of 15–20 h. About 80% of sugars are fermented in the first two tanks and another 15% in the remaining tanks by *Saccharomyces cerevisiae.* The fermented liquor contains about 1% yeast by volume and is separated by centrifugation into two streams, one containing 15% by volume which is returned to the first fermenter and one clear liquor stream, which flows into beer storage prior to distillation.

Part of the fermented liquor by-passes around the centrifuge to the beer storage. This process provides a continuous purge of dead yeast and insoluble solids that would otherwise overload the centrifuges. The clarified liquor (beer) is passed by gravity to a storage tank and distilled. Finally, the alcohol is rectified to 95 vol.%. About 95 L of fusel oil and 950 L of methanol are separated daily.

### 6.7 Ethanol from Whey

At present, the commercial fermentation of whey is done by batch as well as continuously with the recycle mode. Carbery Milk Products in Ireland produces alcohol from whey by a batchwise mode (HANSEN, 1980). The whey from the cheddar cheese factory passes through a sieve after which it is separated. The whey continues on to a complex ultrafiltration section with a total capacity of $600 \cdot 10^3$ L $d^{-1}$ of whey. The protein concentrate is spray-dried. The liquid permeate then continues to 6 fermentation tanks, each with a capacity of about 200 $m^3$. The lactose is converted to alcohol at 86% of the theoretical yield. The plant produces $14 \cdot 10^3$ L $d^{-1}$ of 96.5% alcohol.

Economic analysis of alcohol production from whey has shown that it can compete with the production of synthetic alcohol from ethylene. A plant for the production of potable and industrial alcohol has operated in the United States since 1977 (BERNSTEIN et al., 1977).

Because glucose and galactose are more universally fermentable sugars than lactose, it is suggested that $\beta$-galactosidase-treated whey would make a better substrate for industrial fermentation. This aspect has been investigated by O'LEARY et al. (1977). In a subsequent study, REESE (1975) found that an ethanol yield of 6.5% could be obtained using *Saccharomyces cerevisiae* with a lactase-hydrolyzed acid whey permeate containing 30–35% total solids.

Cheese whey can also be used to replace water in the extraction of sugars from different plants, e.g., Jerusalem artichoke, for subsequent ethanol fermentation (KOSARIC and WIECZOREK, 1982).

## 7 By-Products of Ethanol Fermentation

### 7.1 Waste Biomass

Due to the anaerobic nature of the ethanolic fermentation, the overall synthesis of biomass is limited. In general, a 10% substrate feed with 95% conversion to alcohol will yield 5.0 g $L^{-1}$ of dried cell mass. Therefore, separation of the fermenting organisms for cell recovery may not prove to be economically feasible.

For recycle processes, a return of 35–40% of the total biomass in the broth is all that is required to meet fermentation demands. Since the concentration step has been carried out, the remainder may be utilized in by-product markets.

After concentration, microbial biomass may be dried and utilized as a high protein food or feed supplement. Other novel uses for spent brewer's yeast have been studied by KOIVURINTA et al. (1980). These include their application in various food systems for the stabilization of water–oil emulsions and as a substitute for egg-white because of its foaming properties.

### 7.2 Stillage

Stillage consists of the non-volatile fraction of material remaining after alcohol distillation. The composition of this effluent varies as a function of location and the type of feedstock used to produce it. The characteristics of stillage arising from representative crops are shown in Tab. 24. In general, these slops consist of less than 10% solids of which 90% is protein from the feedstock and spent microbial cells. The remainder is made up of residual sugars, residual ethanol, waxes, fats, fibers, and mineral salts.

With conventional techniques, ethanol fermentation is favored by a low concentration of reactants (i.e., 12–20% of aqueous feedstock solutions). Thus, the generation of stillage may amount to 10 times the production volume of alcohol. This fact, coupled with the

**Tab. 24.** Mean Composition of In-Nature Stillages Produced in Brazilian Ethyl Alcohol Distilleries (COSTA-RIBEIRO and CASTELLO-BRANCO, 1981)

| Parameter | Type of Stillage | | | |
|---|---|---|---|---|
| | Molasses [g $L^{-1}$] | Cane-Juice [g $L^{-1}$] | Mixed [g $L^{-1}$] | Mandioca [g $L^{-1}$] |
| Total solids | 81.5 | 23.7 | 52.7 | 22.5 |
| Volatile solids | 60.0 | 20.0 | 40.0 | 20.0 |
| Fixed solids | 21.5 | 3.7 | 12.7 | 2.5 |
| Carbon (as C)[a] | 18.2 | 6.1 | 12.1 | 6.1 |
| Reducing substances | 9.5 | 7.9 | 8.3 | 6.8 |
| Crude protein[b] | 7.5 | 1.9 | 4.4 | 2.5 |
| Potassium (as $K_2O$) | 7.8 | 1.2 | 4.6 | 1.1 |
| Sulfur (as $SO_4$) | 6.4 | 0.6 | 3.7 | 0.1 |
| Calcium (as CaO) | 3.6 | 0.7 | 1.7 | 0.1 |
| Chlorine (as NaCl) | 3.0 | 1.0 | 2.0 | 0.1 |
| Nitrogen (as N) | 1.2 | 0.3 | 0.7 | 0.4 |
| Magnesium (as MgO) | 1.0 | 0.2 | 0.7 | 0.1 |
| Phosphorus (as $P_2O_5$) | 0.2 | 0.01 | 0.1 | 0.2 |
| BOD | 25.0 | 16.4 | 19.8 | 18.9 |
| COD | 65.0 | 33.0 | 45.0 | 23.4 |
| Acidity[c] | 4.5 | 4.5 | 4.5 | 4.5 |

[a] Carbon content = organic solids content: 3.3
[b] Crude protein content = nitrogen content × 6.25
[c] Expressed in pH units

high *BOD* value of the effluent (~20 g $L^{-1}$) has led to concern over its pollution potential. JACKMAN (1977) estimated that in a traditional cane molasses distillery (100,000 L $d^{-1}$ anhydrous ethanol) the resulting pollution load would be equivalent to a city population of $1.7 \cdot 10^6$ persons. Therefore, any processes which would utilize this effluent as by-product would also generate credit as a means of pollution control.

Recovery processes for stillage are numerous. Fig. 32 illustrates some of the main process routes and the possible applications of resulting products. In Brazil, some stillage has been returned to the canefields as fertilizer and for irrigation (CASTELLO-BRANCO et al., 1980). This may be a desirable alternative to disposal since the high potash and organic matter of the effluent can be beneficial to a certain level in crop production. Problems encountered by this application involve increases in solid acidity and salt concentration as well as the extremely pungent odor of the putrefying liquid.

Wet stillage has a high protein content and as such can be fed directly to local feedstock in the form of a warm slurry. Since the overall nutrient composition of this material is limited, it would require fortification with other feed sources to ensure a balanced diet (WYVILL and BATTAGLIA, 1981). Evaporation of the effluent to a thick syrup or powder would extend its storage capacity. Care must be taken to ensure that salt concentrations do not exceed a level detrimental to livestock health.

If evaporation is used to concentrate the stillage, the resulting water or condensate may be recycled to the process stream. This would result in water savings and a decrease in effluent volumes. The use of raw stillage for re-extraction of the feedstock is feasible. However, the build-up of inhibition products and the increase in osmotic pressure due to non-fermentable products would need to be carefully controlled.

Operations in Taiwan have utilized new stillage as feedstock in the production of *To-*

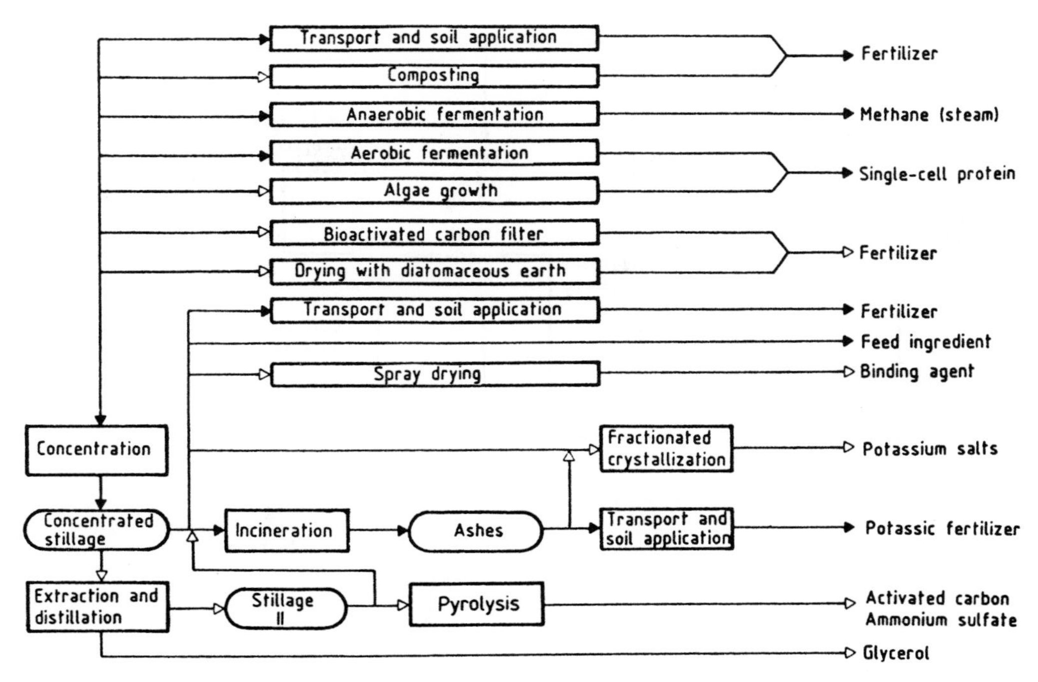


**Fig. 32.** Stillage recovery processes (COSTA-RIBEIRO and CASTELLO-BRANCO, 1981). → main products; —▷ other products.

*rula* yeast (*Candida utilits*) for almost a decade (CHANG and YANG, 1973). Interest has been shown in the use of other fungi for this type of process (GONZALES and MURPHY, 1980).

Fig. 33 illustrates an integrated system developed by SANG et al. (1980) which involves continuous yeast growth on raw alcohol slops and subsequent culture of mold biomass on yeast wastes. These researchers were able to reduce the overall *BOD* of stillage by 90% with a simultaneous yeast productivity of 4.2 g $L^{-1}$ $h^{-1}$.

Factors which will greatly affect the potential of microbial protein production from stillage include the need for exogenous nutrient fortification (i.e., N and P) at additional costs, high initial capital required for plant construction, and a high energy demand for unit processes (especially fermenter cooling in tropical countries).

Waste effluents from distillery operations are also amenable to anaerobic digestion. Operating at mesophilic temperatures (32 °C) and adequate hold-up times, 95% of the original *BOD* value can be eliminated. Sludge gas with a 65% (v/v) methane content may be produced at a yield of 580–720 L $kg^{-1}$ *BOD* removed. The net heat value of this product is ca. 25 MJ $m^{-3}$ and as such may provide boiler steam to drive the unit processes of the distillery (JACKMAN, 1977).

An additional credit is that nitrogen remains in the effluent of the digester as ammonium salts and as such is a good source of organic nitrogen for crop fertilization. The increased pH of digested stillage (final pH 7.4) also makes it more suitable for direct land application (SANCHEZ-RIERA et al., 1982).

## 7.3 Carbon Dioxide

For every $m^3$ of ethanol formed, about 760 kg of $CO_2$ are liberated from the fermentation broth. Of this total, 70–80% can be re-

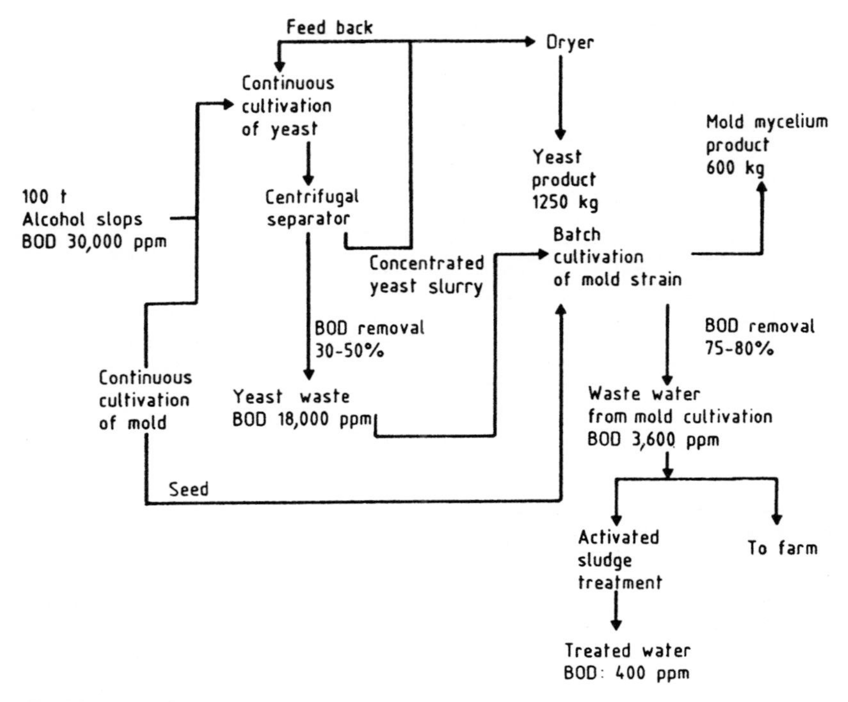


**Fig. 33.** Flow diagram of a proposed process for complete treatment of alcohol slops (SANG et al., 1980).

covered in a closed system. After purification to remove aldehydes and alcohols, the gas may be stored in cylinders or further compressed to solid or liquid form.

The market value of $CO_2$ is relatively variable. It is not economical to transport it a great distance from the plant. In the gaseous state, it may be used to carbonate soda beverages or to enhance the agricultural productivity of greenhouse plants. Liquid $CO_2$ is frequently used in fire extinguishers, refrigeration processes, and as a feedstock in the chemical industry. The primary use of solid $CO_2$ is as a refrigerant.

## 7.4 Fusel Oils

Fusel oils are formed from $\alpha$-keto acids, derived from or leading to amino acids. The overall composition is found to be an isomeric mixture of primary methyl butanols and methyl propanols, the majority of which is isoamyl alcohol (MAIORELLA et al., 1981). Tab. 25 compares the composition of fusel oils from different sources. Yields of up to 20 L may be attained per $m^3$ of ethanol generated using standard sugar sources, however, this value depends upon the pH in the fermenter.

Fusel oils are found to come of the distillation tower at relatively high temperatures and must be completely removed, or plugging of the column may result (COMBES, 1981). Due to its similarity to the major components in gasoline, the use of this by-product would be as a further fuel extender or as an industrial solvent.

**Tab. 25.** Average Composition of Various Fusel Oils [vol.%] (PFENNINGER, 1963)

| Component | Molasses Fusel Oil | Cereal Fusel Oil | Potato Fusel Oil | Fruit Fusel Oil | Baker's Yeast Fusel Oil | Sulfite Liquor Fusel Oil |
|---|---|---|---|---|---|---|
| *n*-Propylalcohol | 8.6 | 9.1 | 16.4 | 10.2 | 11.0 | 6.8 |
| *sec*-Butylalcohol | – | – | – | 2.4 | 6.2 | – |
| Isobutylalcohol | 20.6 | 19.2 | 15.9 | 21.4 | 23.8 | 22.3 |
| *n*-Butylalcohol | 0.5 | 0.2 | 1.2 | 2.2 | 2.3 | 0.7 |
| Opt. act. amylalcohol | 31.3 | 19.0 | 13.6 | 13.6 | 19.0 | 13.1 |
| Isoamylalcohol | 39.1 | 52.4 | 52.9 | 56.4 | 37.7 | 56.1 |
| *n*-Amylalcohol | – | – | – | – | – | 1.0 |

# 8 Economic and Energy Aspects of Ethanol Fermentation

A number of economic analyses related to large-scale production of ethanol are available (SCHELLER and MOHR, 1976; SCHELLER, 1976; DE CARVALHO et al., 1977; LIPINSKY et al., 1977; Intergroup Consulting Economists, Ltd., 1978; FAYED et al., 1981; MISSELHORN, 1980; Novo Industri, 1979; SLESSER and LEWIS, 1979; ACKERSON, 1991; MISTRY, 1991; BRIDGEWATER and DOUBLE, 1991, 1994; MARSH and CUNDIFF, 1991; DUNNE, 1994). The estimated figures as presented vary and rapidly become outdated as inflation increases energy costs, capital costs, operating expenses, and crop prices.

A distribution of costs for ethanol production from various raw materials is shown in Fig. 34. According to MISTRY (1991) feed and processing costs vary depending on the raw material (Tab. 26). For different processes, the costs are broken down as shown in Fig. 35. The characteristics of the above processes are presented in Tab. 27. Capital and operating costs for plants operating at different processes are presented Tabs. 28 and 29. The production costs and cost contributions based on data evaluated for the 5 different processes presented above, are shown in Tabs. 30 and 31. The cost contributions, except that of feed, are shown in Fig. 36.

**Tab. 26.** Distribution of Ethanol Production Cost (MISTRY, 1991)

| Feed | Feed Cost Contribution [%] | Processing Cost Contribution [%] |
|---|---|---|
| Sugarcane | 50–83 | 17–50 |
| Molasses | 78–83 | 17–22 |
| Sugar beet | 50–68 | 32–50 |
| Cassava | 60–75 | 25–40 |
| Corn | 53–87 | 13–47 |
| Wheat | 40–75 | 25–60 |
| Maize | 60–70 | 30–40 |
| Wood wastes | 9–42 | 58–91 |
| Algae | 62 | 38 |
| Wheat straw | 9–23 | 77–91 |

A cost analysis for conversion of sweet sorghum to ethanol was presented by MARSH and CUNDIFF (1991) for a U.S. scenario (Piedmont and Louisiana). Cost to provide year-round, readily fermentable, sweet sorghum feedstock to a central plant in the Piedmont was determined to vary from $0.52–0.79 per liter ($1.96–2.98 per gallon) ethanol potential. Cost to provide feedstock employing the established sugarcane-to-sugar industry in Louisiana is shown to be $0.66 per liter ($2.50 per gallon). Assuming cellulose feedstock is valued at $42.00 $t^{-1}$ dry matter delivered to the ethanol production facility, sweet sorghum by-products are reported to have more value as cattle feed (MARSH and CUNDIFF, 1991).

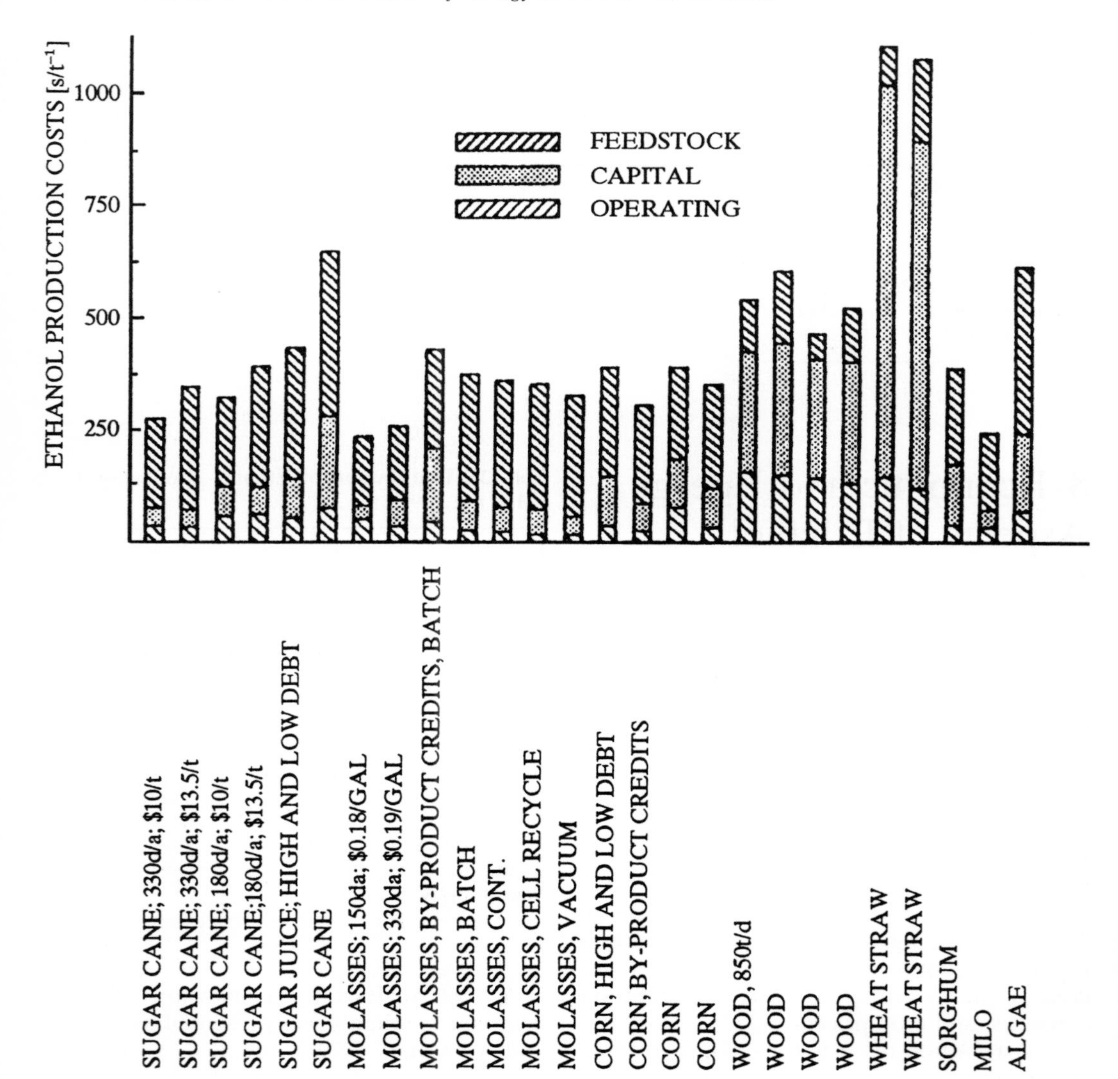


**Fig. 34.** Distribution of ethanol production costs.
Capital structure:
60% debt, 40% common equity
9% interest on debt
return in equity 15%
economic life of facility 30 years
financial tax life 20 years
taxes 53%, facility construction time 2–3 years.

According to DUNNE (1994), the energy inputs and economics from conventional agricultural crops are shown in Tab. 32. The estimated ethanol conversion costs for selected plant sizes are shown in Tab. 33. The estimated feedstock and conversion costs for selected crops are shown in Tab. 34.

An economic analysis for production of

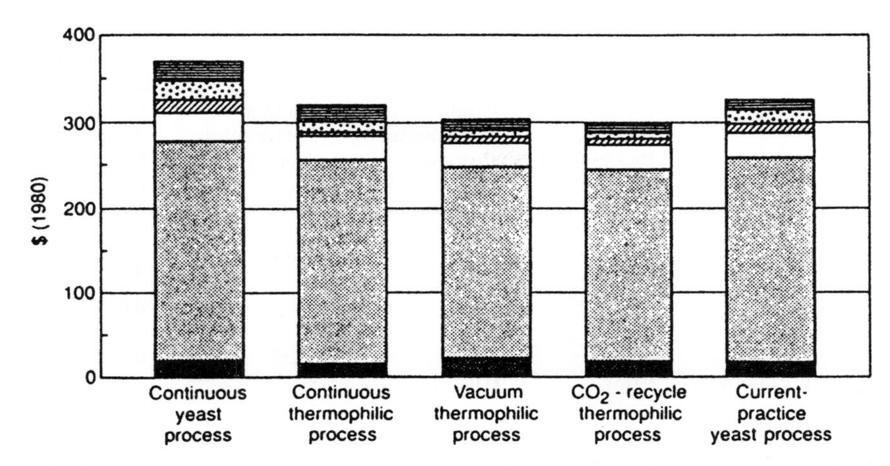


**Fig. 35.** Histograms showing cost contributions in producing 1 $m^3$ ethanol; ■ fixed charges, ▦ feed, □ steam, ▨ water, ⊡ chemicals, ▤ stillage treatment.

**Tab. 27.** Technical Parameters Used in Different Processes for Ethanol Production (MISTRY, 1991)

| Lowsheet | 1 Continuous Yeast Process | 2 Continuous Thermophilic Process | 3 Vacuum Thermophilic Process | 4 $CO_2$ Recycle Thermophilic Process | 5 Current Practice Yeast Process |
|---|---|---|---|---|---|
| Sucrose concentration to sterilizer [wt.%] | 10.0 | 10.0 | 16.6 | 16.6 | 16.8 |
| Sucrose concentration to anaerobic stage [wt.%] | 7.7 | 8.7 | 15.0 | 15.0 | (16.8)[a] |
| Temperature [°C] | 35 | 70 | 70 | 70 | 35 |
| Biomass yield [kg kg $S^{-1}$] | 0.128 | 0.59 | 0.039 | 0.039 | 0.108 |
| Ethanol yield [kg kg $S^{-1}$] | 0.409 | 0.435 | 0.454 | 0.454 | 0.433 |
| Specific productivity, $q_P$ [kg kg biomass $h^{-1}$] | 0.63 | 1.3 | 1.3 | 1.3 | 0.40 |
| Volumetric productivity, $r_P$ [kg $m^{-3}$ $h^{-1}$] | 31.5 | 31.5 | 31.5 | 31.5 | 31.5 |
| Biomass concentration [kg dry mass $m^{-3}$] | 50 | 24.2 | 24.2 | 24.2 | 78.2 |
| Biomass recycle [%] | 75 | 75 | 75 | 75 | 75 |
| Ethanol conc. in beer, $P_w$ [wt.%] | 4.0 | 4.0 | 4.0 | 4.0 | 7.8 |
| Sucrose conversion [%] | 98 | 98 | 98 | 98 | 98 |

[a] Aerobic and anaerobic stages are not separate for process 5.

**Tab. 28.** Components of Capital Costs of Ethanol Production in 1980 (MISTRY, 1991)

| Flowsheet | 1 Continuous Yeast Process [$] | 2 Continuous Thermophilic Process [$] | 3 Vacuum Thermophilic Process [$] | 4 $CO_2$ Recycle Thermophilic Process [$] | 5 Current Practice Yeast Processes [$] |
|---|---|---|---|---|---|
| Evaporator | – | – | 0.270 | 0.270 | 0.300 |
| Sterilizer | 0.130 | 0.092 | 0.058 | 0.058 | 0.083 |
| Aerobic fermenter | 0.101 | 0.065 | 0.035 | 0.035 | 0.133 |
| Anaerobic fermenter | 0.134 | 0.134 | 0.133 | 0.133 | |
| Fermenter cooler | 0.051 | 0.020 | – | – | 0.058 |
| Fermenter reboiler | – | – | 0.015 | 0.014 | – |
| Beer storage vessel | 0.054 | 0.054 | 0.053 | 0.053 | 0.053 |
| Centrifuge(s) | 0.564 | 0.346 | 0.184 | 0.186 | 0.279 |
| Compressor[a] | – | – | 0.350 | 0.180 | – |
| Cooler/condenser | – | 0.032 | 0.069 | 0.072 | – |
| Absorption column | 0.018 | 0.023 | 0.024 | 0.035 | 0.018 |
| Distillation column | 0.123 | 0.217 | 0.100 | 0.105 | 0.069 |
| Distillation reboiler | 0.061 | 0.051 | 0.048 | 0.049 | 0.051 |
| Distillation conditioner | 0.117 | 0.117 | 0.117 | 0.117 | 0.112 |
| Total equipment cost | 1.352 | 1.060 | 1.455 | 1.306 | 1.156 |
| Total plant cost[b] | 5.408 | 4.240 | 5.820 | 5.224 | 4.624 |

[a] For vacuum thermophilic process: cost = cost of vapor compressor + cost of vacuum pump.
[b] Total plant cost = 4 × total equipment cost.

ethanol from municipal solid waste was presented by ACKERSON (1991). A facility has been designed to produce $75.7 \cdot 10^6$ L ($20 \cdot 10^6$ gallons) of ethanol per year, utilizing a strong acid for hydrolysis in a process prior to fermentation by *S. cerevisiae.*

In this process, municipal solid waste (MSW, with a composition as shown in Tab. 35) was collected and delivered to the plant size as needed. The feedstock was prepared by removing plastic, metal and glass, followed by shredding and grinding. The cost for glass and metal removal is not included. Including all the other costs (acid hydrolysis, ethanol fermentation and distillation) the total capital cost is reported to be $35 million including all utilities, storage and off-sites.

The annual operating costs are shown in Tab. 36. A lignin boiler is used to reduce the energy requirements, so that energy costs are reduced to only $0.02 per liter ($0.08 per gallon). Fixed charges are computed as a percentage of the capital investment and in total $5.6 million per year. At an ethanol price of ca. $0.40 per liter ($1.50 per gallon), revenues are generated in the amount of $30 million, yielding a pre-tax profit of $18.5 million per year (ca. $0.25 per liter, resp. $0.93 per gallon) or 53% per year. However, utilization of pentoses is not considered in this analysis.

**Tab. 29.** Production Costs in 1980 (MISTRY, 1991)

| Flowsheet | 1 Continuous Yeast Process | 2 Continuous Thermophilic Process | 3 Vacuum Thermophilic Process | 4 $CO_2$ Recycle Thermophilic Process | 5 Current Practice Yeast Processes |
|---|---|---|---|---|---|
| Plant cost [MS] | 5.409 | 4.420 | 5.810 | 5.224 | 4.624 |
| Annual running cost [M$] | 14.637 | 12.628 | 11.658 | 11.745 | 12.822 |
| Production cost[a,b] [M$ $m^{-3}$] | 368 | 316 | 298 | 298 | 322 |

[a] Production cost is evaluated by taking 15% of total plant cost as the fixed capital *per annum,* and for a 350 day year.
[b] Cost per $m^3$ anhydrous ethanol ≙ cost per 1.05 $m^3$ of 93.5 wt.% ethanol.

**Tab. 30.** Annual Running Costs (1980) for Flowsheets 1–5, Based on Production of 120 $m^3$ $d^{-1}$ Anhydrous Ethanol (MISTRY, 1991)

| Flowsheet | 1 Continuous Yeast Process [$] | 2 Continuous Thermophilic Process [$] | 3 Vacuum Thermophilic Process [$] | 4 $CO_2$ Recycle Thermophilic Process [$] | 5 Current Practice Yeast Processes [$] |
|---|---|---|---|---|---|
| Cane[a] | 10.820 | 10.100 | 9.450 | 9.494 | 10.085 |
| Chemicals[b] | 0.902 | 0.450 | 0.290 | 0.292 | 0.727 |
| Steam[c] | 1.423 | 1.158 | 1.191 | 1.207 | 1.197 |
| Water[d] | 0.639 | 0.148 | 0.310 | 0.277 | 0.406 |
| Stillage treatment[e] | 0.853 | 0.762 | 0.417 | 0.475 | 12.822 |
| Total | 14.637 | 12.628 | 11.658 | 11.745 | 12.822 |

[a] 1,200 kg of 20 wt.% sucrose solution costs 15 $ (1980).
[b] Chemical cost = 0.075 $kg^{-1}$ biomass + 3 S $m^{-3}$ anhydrous production capacity.
[c] Steam cost = 0.0035 $ $kg^{-1}$ HP steam, based on the use sugarcane bagasse.
[d] Water cost = 0.04 $ $m^{-3}$.
[e] Stillage treatment cost = 1 $ $m^{-3}$.

Acid recovery is included, but fermentation of xylose is not provided. According to other data (LAWFORD and ROUSSEAU, 1991), a 30% increase in alcohol yield would be expected if xylose was also fermented to ethanol. One could estimate, therefore, that considerably better economics would be achieved by utilization of the pentoses (available in the acid hydrolyzate), for ethanol production.

KOSARIC et al. (1982) have made an economic evaluation for the small-scale production of ethanol (farm size) for Jerusalem artichoke as raw material. Tab. 37 shows the costs of ethanol production for different plant sizes. The cost of ethanol produced in the plant with a capacity of $4 \cdot 10^6$ kg $a^{-1}$ was estimated to be about 40% of the present price of gasoline in Canada (1995).

**Tab. 31.** Cost Contributions (1980) in Producing 1 $m^3$ Ethanol (MISTRY, 1991)

| Flowsheet | 1 Continuous Yeast Process [$] | 2 Continuous Thermophilic Process [$] | 3 Vacuum Thermophilic Process [$] | 4 $CO_2$ Recycle Thermophilic Process [$] | 5 Current Practice Yeast Processes [$] |
|---|---|---|---|---|---|
| Fixed charges | 19 | 15 | 21 | 19 | 17 |
| Feed | 258 | 241 | 225 | 226 | 240 |
| Steam | 35 | 20 | 28 | 29 | 29 |
| Water | 15 | 4 | 7 | 7 | 10 |
| Chemicals | 22 | 11 | 7 | 7 | 17 |
| Stillage treatment | 20 | 18 | 10 | 11 | 10 |
| Total | 368 | 316 | 298 | 298 | 322 |

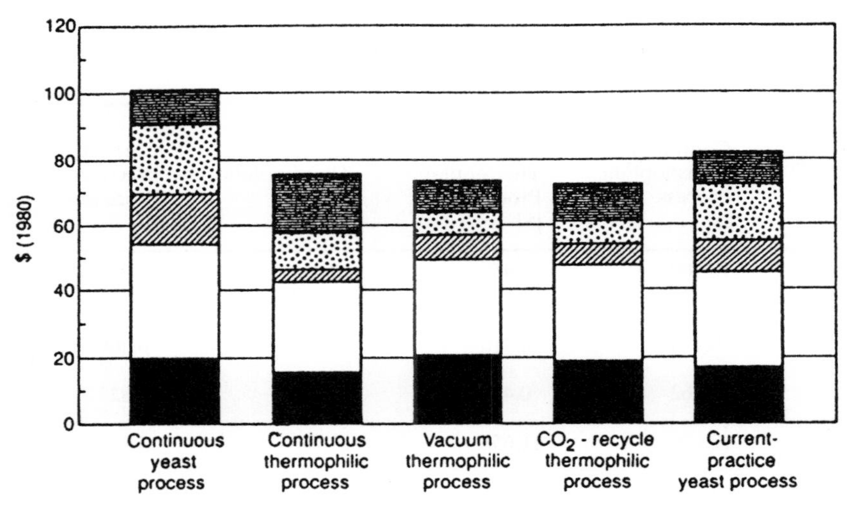


**Fig. 36.** Histograms showing cost contributions, except that of feed, in producing 1 $m^3$ ethanol; ■ fixed charges, □ steam, ▨ water, ⚄ chemicals, ▤ stillage treatment.

## 8.1 Ethanol from Jerusalem Artichokes (A Case Study)

The cost for producing Jerusalem artichokes at the farm level and the cost of processing the Jerusalem artichokes (tops and tubers) to ethanol have been investigated by BAKER et al. (1990, 1991). In this study, various ethanol yields from Jerusalem artichoke feedstock and conversion efficiencies have been considered for implications of different processing plant sizes. Scenarios for Quebec and Western Canada have been presented.

The cost of production varies by system of production (i.e., tubers or tops), by land price, and by region (Tabs. 38 and 39). The total cost per ha for tuber production varied from a high of $4,432 in Eastern Canada to a low of $3,502 in Quebec. As would be ex-

**Tab. 32.** Energy and Financial Inputs Required to Produce Liquid Fuels[a] from Conventional Agriculture Crops (DUNNE, 1994)

| Crop | Energy Input [MJ $L^{-1}$] | Feedstock Cost | By-Product Value | Feedstock Cost less By-Product Value |
|---|---|---|---|---|
| Winter wheat | 8.75 | 22.8 | 11.0 | 11.8 |
| Spring wheat | 8.39 | 24.1 | 11.0 | 13.1 |
| Winter barley | 10.09 | 28.1 | 14.9 | 13.2 |
| Spring barley | 8.21 | 23.0 | 14.9 | 8.1 |
| Winter oats | 12.53 | 32.0 | 24.7 | 7.3 |
| Spring oats | 12.49 | 38.9 | 24.7 | 14.2 |
| Sugar beet[b] | 4.64 | 19.0 | 7.5 | 11.5 |
| Sugar beet[c] | 3.93 | 16.1 | 9.3 | 6.7 |
| Fodder beet[b] | 4.96 | 22.1 | 9.4 | 12.7 |
| Fodder beet[c] | 3.97 | 17.6 | 11.3 | 6.3 |
| Potatoes | 16.99 | 85.1 | 11.1 | 74.0 |
| Swedes | 13.18 | 51.5 | 17.5 | 34.0 |
| Grass 1 cut | 17.36 | 37.8 | 80.0 | −42.2 |
| Grass 2 cuts | 25.80 | 51.1 | 80.00 | −28.9 |
| Grass 3 cuts | 30.87 | 59.5 | 80.00 | −20.5 |
| Grass 4 cuts | 39.10 | 71.3 | 80.00 | − 8.7 |
| Oilseed rape | 19.03 | 41.7 | 10.6 | 31.1 |

[a] Liquid fuel is ethanol for all crops except oilseed rape. For oilseed rape the liquid fuel is vegetable oil.
[b] Roots only.
[c] Roots and tops.

**Tab. 33.** Estimated Ethanol Conversion Cost for Selected Plant Sizes (DUNNE, 1994)

| Plant Size (1/day) | Capital Costs (IR pence/1) | Other Costs (IR pence/1) | Total Costs (IR pence/1) |
|---|---|---|---|
| 25,000 | 29.1 | 18.2 | 47.3 |
| 50,000 | 21.8 | 15.9 | 37.7 |
| 100,000 | 16.5 | 15.3 | 31.8 |
| 200,000 | 12.6 | 14.1 | 26.7 |

pected, the cost of production decreases as the land value decreases. The lowest total cost per ton of tubers on a fresh matter basis was $71.81 in Western Canada. The advantage that Western Canada has over Quebec in tuber production is in terms of the yield (43 t $ha^{-1}$ vs. 41 t $ha^{-1}$).

The total cost of production for tops in large bales per ha varied from a high of $1,493 to a low of $1,195 in Quebec on the least expensive land. Translating these costs per ha into costs per ton on a fresh matter basis, the lowest cost per ton can be found in Western Canada ($14.82). The yield advantage for tops in Western Canada, 100 t $ha^{-1}$, as compared to Quebec, 41 t $ha^{-1}$, plays an important part in providing such a low cost per ton.

The supply price which has been estimated includes a return to the operator. The return to the operator is based on a return of 3% of the value of land and buildings used to pro-

**Tab. 34.** Estimated Feedstock and Conversion Costs for Selected Crops (DUNNE, 1994)

| Feedstock | Feedstock Cost (IR pence/1)[a] | Feedstock Cost less By-Product Credit (IR pence/1)[a] |
|---|---|---|
| Winter wheat | 22.8 | 11.8 |
| Spring barley | 23.0 | 8.1 |
| Sugar beet (roots only) | 19.0 | 11.5 |
| Sugar beet (roots and tops) | 16.1 | 6.7 |
| | Cost of Feedstock Plus Conversion[b] | |
| Winter wheat | 49.5–70.1 | 38.5–59.1 |
| Spring barley | 49.7–70.3 | 34.8–55.4 |
| Sugar beet (roots only) | 45.7–66.3 | 38.2–58.8 |
| Sugar beet (roots plus tops) | 42.8–63.4 | 33.4–54.0 |

[a] Feedstock cost only as per Tab. 32.
[b] Feedstock cost as per Tab. 32 plus the processing costs for both the largest and smallest plants as per Tab. 33.

**Tab. 35.** Municipal Solid Waste Composition (wt.% as discarded) (ACKERSON, 1991)

| Category | Summer | Fall | Winter | Spring | Average |
|---|---|---|---|---|---|
| Paper | 31.0 | 38.9 | 42.2 | 36.5 | 37.4 |
| Yard waste | 27.1 | 6.2 | 0.4 | 14.4 | 13.9 |
| Glass | 17.7 | 22.7 | 24.1 | 20.8 | 20.0 |
| Metal | 7.5 | 9.6 | 10.2 | 8.8 | 9.8 |
| Wood | 7.0 | 9.1 | 9.7 | 8.2 | 8.4 |
| Textiles | 2.6 | 3.4 | 3.6 | 3.1 | 3.1 |
| Leather and rubber | 1.8 | 2.5 | 2.7 | 2.2 | 2.2 |
| Plastics | 1.1 | 1.4 | 1.5 | 1.2 | 1.2 |
| Miscellaneous | 3.1 | 4.0 | 4.2 | 3.7 | 3.4 |

duce either the tubers or the tops from Jerusalem artichoke. The returns for tubers are higher than for tops because of the greater amount of machinery required for this type of production. The returns varying by region and type of production are given in Tab. 40.

The farm-level costs of production can be presented in terms of $ per liter of ethanol by taking into account a range of possible conversion factors, and assuming possible yields. These costs for Quebec and Western Canada for tubers are given in Tabs. 41 and 42, respectively.

Assuming a value of land in Quebec of $2,500 per ha, the feedstock cost per liter ethanol from tubers ranges from a high of $1.58 to a low of $0.33 and with a land value of $675 per ha the range is $1.45 to $0.31. In Western Canada the cost per liter ethanol from tubers ranged from $1.62 on the highest valued land to $0.34 and from $1.59 to $0.33 on the cheaper land. In the case of ethanol production from tops, the range of costs per liter in Quebec are $0.55 to $0.14 on the most expensive land and $0.44 to $0.11 on the lowest valued land. In Western Canada, producing tops on the best land leads to a cost range for ethanol production of $0.29 to $0.11 per liter, and on the lower quality land the range is $0.27 to $0.10.

**Tab. 36.** Economics of a $75.6 \cdot 10^6$ L $a^{-1}$ Ethanol Facility[a] (ACKERSON, 1991)

| *Capital cost:* | | Million $ |
|---|---|---|
| Feedstock preparation | | 3.0 |
| Hydrolysis | | 5.0 |
| Acid recovery | | 8.5 |
| Fermentation and purification | | 8.0 |
| Utilities/offsites | | 6.5 |
| Engineering | | 4.0 |
| | | 35.0 |
| *Operating cost:* | Million $ per year | $ per liter (per gallon) |
| MSW | – | – – |
| Utilities | 1.5 | 0.021 (0.08) |
| Chemicals | 1.9 | 0.034 (0.09) |
| Labor | 2.5 | 0.034 (0.13) |
| Fixed Charges | | |
| Maintenance (4%) | 1.4 | 0.018 (0.07) |
| Depreciation (10%) | 3.5 | 0.048 (0.18) |
| Taxes and insurance (2%) | 0.7 | 0.005 (0.02) |
| Pre-tax profit (53%) | 18.5 | 0.246 (0.93) |
| | 30.0 | 0.396 (1.50) |

[a] $20 \cdot 10^6$ gallons per year.

**Tab. 37.** Cost Comparison for Production of Ethanol (100 vol.%) for Different Plant Sizes (KOSARIC et al., 1982b)

| Costs | Plant Size | | |
|---|---|---|---|
| | $3 \cdot 10^5$ kg | $3.6 \cdot 10^5$ kg | $4 \cdot 10^6$ kg |
| Fixed operating costs (depreciation[a], maintenance and repairs, labor, taxes, interest) | 0.75 | 0.63 | 0.30 |
| Direct operating costs (supplies, steam, power, water) | 0.20 | 0.20 | 0.29 |
| Raw material cost[b] (Jerusalem artichokes) | 0.17 | 0.17 | 0.17 |
| Cost of 1 kg ethanol | 1.12 | 1.00 | 0.76 |
| By-product credits | | | |
| Pulp (17% protein d.m.) | 0.39 | 0.39 | 0.39 |
| Stillage (protein 18.5 kg 1,000 $L^{-1}$) | 0.16 | 0.16 | 0.16 |
| Net cost of ethanol $ $L^{-1}$ | 0.55 | 0.43 | 0.21 |

[a] Buildings at 20 years amortization, equipment at 100 years amortization.
[b] Taken at U.S. $ 10 per ton.

**Tab. 38.** Annual Cost of Production of Tubers (Quebec)

| Land cost per ha | Costs [$] | | | |
|---|---|---|---|---|
| | 2,500 | 1,750 | 1,100 | 675 |
| Total variable cost | 1,719 | 1,719 | 1,719 | 1,719 |
| Total fixed cost | 1,029 | 906 | 800 | 731 |
| Storage and transportation | 1,052 | 1,052 | 1,052 | 1,052 |
| Total cost per ha | 3,800 | 3,677 | 3,571 | 3,502 |
| Cost per ton (fresh matter) | 92.68 | 89.68 | 87.10 | 85.41 |
| Tons per ha (fresh matter) | 41 | 41 | 41 | 41 |
| Land cost per ha (Western Canada) | | | 1,100 | 500 |
| Cost per ton (fresh matter) | | | 73.19 | 71.81 |
| Tons per ha (fresh matter) | | | 53 | 53 |

**Tab. 39.** Annual Cost of Production of Tops in Large Bales (Quebec)

| Land cost per ha | Costs [$] | | | |
|---|---|---|---|---|
| | 2,500 | 1,750 | 1,100 | 675 |
| Total variable cost | 503 | 503 | 503 | 503 |
| Total fixed cost | 792 | 669 | 563 | 494 |
| Storage and transportation | 198 | 198 | 198 | 198 |
| Total cost per ha | 1,493 | 1,370 | 1,264 | 1,195 |
| Cost per ton (fresh matter) | 46.41 | 33.41 | 30.83 | 29.15 |
| Tons per ha (fresh matter) | 41 | 41 | 41 | 41 |
| Land cost per ha (Western Canada) | | | 1,100 | 500 |
| Cost per ton (fresh matter) | | | 15.78 | 14.82 |
| Tons per ha (fresh matter) | | | 100 | 100 |

**Tab. 40.** The Operator's Return for Jerusalem Artichoke

| Region | Return [$ ha$^{-1}$] | | | |
|---|---|---|---|---|
| *Quebec* | | | | |
| Land value | 2,500 | 1,750 | 1,100 | 675 |
| Tubers | 149.05 | 126.55 | 107.05 | 94.30 |
| Tops (large) | 120.68 | 98.18 | 78.68 | 65.93 |
| Tops (small) | 123.25 | 100.76 | 81.26 | 68.51 |
| *Western Canada* | | | | |
| Land values | | | 1,100 | 500 |
| Tubers | | | 160.38 | 142.38 |
| Tops (large) | | | 110.18 | 92.18 |
| *Eastern Canada* | | | | |
| Land values | | 1,900 | 1,100 | |
| Tubers | | 392.04 | 368.04 | |
| Tops (large) | | 282.78 | 258.78 | |
| Tops (small) | | 294.08 | 270.08 | |

**Tab. 41.** Ethanol Feedstock Cost of Production Using Jerusalem Artichoke Tubers

| Land Price [$ ha$^{-1}$] | 2,500 | | | | | 675 | | | | |
|---|---|---|---|---|---|---|---|---|---|---|
| Yield [ha$^{-1}$) | 30.00 | 41.00 | 50.00 | 60.00 | 76.00 | 30.00 | 41.00 | 50.00 | 60.00 | 76.00 |
| Conversion Factor [L t$^{-1}$] | Quebec [$ L$^{-1}$] | | | | | Quebec [$ L$^{-1}$] | | | | |
| 80.00 | 1.58 | 1.16 | 0.95 | 0.79 | 0.62 | 1.46 | 1.07 | 0.88 | 0.73 | 0.58 |
| 100.00 | 1.27 | 0.93 | 0.76 | 0.63 | 0.50 | 1.17 | 0.85 | 0.70 | 0.58 | 0.46 |
| 120.00 | 1.06 | 0.77 | 0.63 | 0.53 | 0.42 | 0.97 | 0.71 | 0.58 | 0.49 | 0.38 |
| 140.00 | 0.90 | 0.66 | 0.54 | 0.45 | 0.36 | 0.83 | 0.61 | 0.50 | 0.42 | 0.33 |
| 150.00 | 0.84 | 0.62 | 0.51 | 0.42 | 0.33 | 0.78 | 0.57 | 0.47 | 0.39 | 0.31 |
| Land Price [$ ha$^{-1}$] | 1,100 | | | | | 500 | | | | |
| Yield [ha$^{-1}$] | 30.00 | 41.00 | 53.00 | 60.00 | 76.00 | 30.00 | 41.00 | 53.00 | 60.00 | 76.00 |
| Conversion Factor [L$^{-1}$] | Western Canada [$ L$^{-1}$] | | | | | Western Canada [$ L$^{-1}$] | | | | |
| 80.00 | 1.62 | 1.18 | 0.91 | 0.81 | 0.64 | 1.59 | 1.16 | 0.90 | 0.79 | 0.63 |
| 100.00 | 1.29 | 0.95 | 0.73 | 0.65 | 0.51 | 1.27 | 0.93 | 0.72 | 0.63 | 0.50 |
| 120.00 | 1.08 | 0.79 | 0.61 | 0.54 | 0.43 | 1.06 | 0.77 | 0.60 | 0.53 | 0.42 |
| 140.00 | 0.92 | 0.68 | 0.52 | 0.46 | 0.36 | 0.91 | 0.66 | 0.51 | 0.45 | 0.36 |
| 150.00 | 0.86 | 0.63 | 0.49 | 0.43 | 0.34 | 0.85 | 0.62 | 0.48 | 0.42 | 0.33 |

**Tab. 42.** Ethanol Feedstock Cost of Production Using Jerusalem Artichoke Tops

| Land Price [$ ha$^{-1}$] | 2,500 | | | | | 675 | | | | |
|---|---|---|---|---|---|---|---|---|---|---|
| Yield [t ha$^{-1}$] | 30.00 | 41.00 | 55.00 | 80.00 | 100.00 | 30.00 | 41.00 | 55.00 | 80.00 | 100.00 |
| Conversion Factor [L t$^{-1}$] | Quebec [$ L$^{-1}$] | | | | | Quebec [$ L$^{-1}$] | | | | |
| 90.00 | 0.55 | 0.40 | 0.30 | 0.21 | 0.17 | 0.44 | 0.32 | 0.24 | 0.17 | 0.13 |
| 100.00 | 0.50 | 0.36 | 0.27 | 0.19 | 0.15 | 0.40 | 0.29 | 0.22 | 0.15 | 0.12 |
| 110.00 | 0.45 | 0.33 | 0.25 | 0.17 | 0.14 | 0.36 | 0.26 | 0.20 | 0.14 | 0.11 |
| Land Price [$ ha$^{-1}$] | 1,100 | | | | | 500 | | | | |
| Yield [t ha$^{-1}$] | 60.00 | 80.00 | 100.00 | 120.00 | 130.00 | 30.00 | 41.00 | 53.00 | 60.00 | 76.00 |
| Conversion Factor [L t$^{-1}$] | Western Canada [$ L$^{-1}$] | | | | | Western Canada [$ L$^{-1}$] | | | | |
| 90.00 | 0.29 | 0.22 | 0.18 | 0.15 | 0.13 | 0.27 | 0.21 | 0.16 | 0.14 | 0.13 |
| 100.00 | 0.26 | 0.20 | 0.16 | 0.13 | 0.12 | 0.25 | 0.19 | 0.15 | 0.12 | 0.11 |
| 110.00 | 0.24 | 0.18 | 0.14 | 0.12 | 0.11 | 0.27 | 0.17 | 0.13 | 0.11 | 0.10 |

These feedstock costs could be compared with other sources presented in Tab. 43. The feedstock requirements and output for a plant with a capacity of $100 \cdot 10^6$ L are presented in Tab. 44. A sensitivity analysis of the results was also done showing that the final cost for ethanol production changes as each of the number of variables is changed independently.

The ethanol cost is quite sensitive to both the feedstock cost and the value of soymeal upon which the by-product credit is based. Each change in input cost of $4 per ton results in a change of about 5 cents per liter in cost of production, each change of $20 per ton in the soymeal price translates into a change of 1.4 cents per liter in cost of production. Costs are not as sensitive to other finan-

**Tab. 43.** Feedstock Costs for an Ethanol Industry Using Alternative Biomass Sources

| Source | 1989 Costs[a] [$ $t^{-1}$] | Conversion Factor [L $t^{-1}$] | Feedstock Costs [$ $L^{-1}$] |
|---|---|---|---|
| Corn | 152.65 | 386.9 | 0.39 |
| Barley | 131.45 | 347.2 | 0.38 |
| Wheat (feed) | 130.45 | 347.2 | 0.38 |
| Wheat (CWRS1) | 247.49 | 347.2 | 0.71 |

[a] Feedstock prices were obtained from Agriculture Canada, Policy Branch, Ottawa (1990).

**Tab. 44.** Feedstock Requirements and Output for a $100 \cdot 10^6$ L Plant

| Inputs/Outputs | Mass [t] | Value [$10^6$ $] |
|---|---|---|
| **Inputs:** | 293,571 | |
| *Jerusalem Artichoke Tops* (DMB) | | |
| Fresh weight basis, moisture 70% | 978,571 | 14.97 |
| Cost (FWB) @ 15.30 [$ $t^{-1}$] | | |
| **Outputs:** | | |
| Ethanol (anhydrous) | 79,000 | |
| Feed by-products (DMB) | 139,422 | |
| 48% Soy meal price @ 240.00 [$ $t^{-1}$] | | |
| 14% Moisture; protein 20.64% | 162,119 | |
| By-product value @ 103.22 [$ $t^{-1}$] | | 16.73 |
| $CO_2$ @ 150.00 [$ $t^{-1}$] | 68,256 | 10.24 |
| **Inputs:** | | |
| *Jerusalem Artichoke Tubers* (DMB) | 211,513 | |
| Fresh weight basis, moisture 80% | 1,057,564 | |
| Cost (FWB) @ 72.50 [$ $t^{-1}$] | | 76.67 |
| **Outputs:** | | |
| Ethanol (anhydrous) | 79,000 | |
| Feed by-products (DMB) | 58,581 | |
| 48% Soy meal price @ 240.00 [$ $t^{-1}$] | | |
| 14% Moisture; protein 31.21% | 68,118 | |
| By-product value @ 156.03 [$ $t^{-1}$] | | 10.63 |
| $CO_2$ @ 150.00 [$ $t^{-1}$] | 68,256 | 10.24 |

cial parameters. Changes to the opportunity cost charged for capital or the amortization period do not add dramatically to cost. Adding another $10 million to the investment would add about 2.5 cents per liter to the cost of production.

The impact of the plant size on the final production costs per liter was also investigated. The total cost of ethanol before by-product credits is $0.47 per liter of which about 18 cents are direct inputs for which there would be little size effect. If the remainder ($0.29 per liter) is scaled by size of plant, the changes in the production cost per liter according to plants of different sizes would be as in Tab. 44.

## 8.2 Energetics

### 8.2.1 Ethanol from Corn

The consumed energy for corn and ethanol production and that obtained through products is shown in Tab. 45. A positive energy balance was achieved by taking into account the energy that could be obtained and used in the process through burning of the whole quantity of stalks, cobs, and husks. SCHELLER and MOHR (1976) proposed for practical reasons to use about 75% cobs, stalks, and husks and to leave the rest in fields for soil conditioning. In that case, the net energy production is still about 12.58 MJ $L^{-1}$ ethanol if the energy deficit for by-product production is not included, or more than 7.53 MJ $L^{-1}$ ethanol if that deficit is included. This analysis is one of the most optimistic. Removing the stalks, cobs, and husks creates a problem because it means mining of soil, and in a short period of time it would cause environmental problems.

Taking into account that the energy crisis is caused by the shortage of crude oil it is interesting to compare how much petroleum energy is used for growing corn and for ethanol production. These data are shown in Tab. 46. Three different processes are compared. For each process, the petroleum-type energy input is higher than output through ethanol and for each liter of ethanol produced, more than 1 L of petroleum energy equivalent would be used.

### 8.2.2 Ethanol from Sugarcane and Cassava

In Tab. 47, energy input and output are shown for production of anhydrous ethanol from sugarcane and cassava in plants with a capacity of 150 $m^3$ $d^{-1}$ ethanol. Taking into

**Tab. 45.** Overall Energy Balance for Grain Alcohol Production from Corn (SCHELLER and MOHR, 1976)

| Energy Production | kJ per 35.42 L of Corn | kJ per L of Ethanol |
|---|---|---|
| Ethanol | 206,698 | 21,052 |
| Aldehydes, fusel oil | 2,953 | 300 |
| Stalks, cobs, husks | 453,354 | 46,525 |
| Total | 663,005 | 67,525 |
| Energy consumption | | |
| Farming operation | 125,731 | 12,805 |
| Transportation of stalks, etc. | 4,265 | 434 |
| Alcohol plant | 295,078 | 30,052 |
| Total | 425,074 | 43,291 |
| Net energy production | 237,932 | 34,234 |
| Net energy loss | | |
| By-product grain production | 49,664 | 5,060 |

**Tab. 46.** Use of Petroleum Type Energy for Growing Corn and for Alcohol Production (SCHRUBEN, 1980)

| Input | Today [kJ L$^{-1}$] | In One Year[a] [kJ L$^{-1}$] | Under study[b] [kJ L$^{-1}$] |
|---|---|---|---|
| To grown corn | 9,688 | 9,688 | 9,688 |
| To cook and convert | 3,990 | 3,990 | 3,990 |
| Germ recovery | 624 | 624 | 624 |
| Distlling | 7,906 | 7,906 | 5,534 |
| Gluten recovery | 731 | 731 | 731 |
| Feed recovery[c] | 7,303 | 4,416 | 4,416 |
| Total electrial energy | 1,061 | 1,061 | 1,061 |
| Total oper liter | 31,303 | 28,416 | 26,044 |
| Subtract electrical energy[d] | 1,061 | 1,061 | 1,061 |
| Petroleum type energy input | 30,242 | 27,355 | 24,983 |
| Output[e] | | | |
| Alcohol (if LHV) | 21,052 | 21,052 | 21,052 |
| (if HHV) | 23,391 | 23,391 | 23,391 |
| Net balance for this plant | | | |
| (if LHV) | −9,190 | −6,303 | −3,931 |
| (if HHV) | −6,851 | −3,964 | −1,592 |
| Liters petroleum energy equivalent used for each liter of alcohol produced | | | |
| (at LHV) | 1,436 | 1,299 | 1,187 |
| (at HHV) | 1,293 | 1,169 | 1,068 |

[a] By using a recompression evaporator with an electric motor.
[b] By using ether for dehydrating.
[c] If stillage is fed to livestock direct from the distillery, then some of the energy would not be used for feed recovery. However, this alcohol plant does use the energy and since it is not available for other users, it must be included for a proper comparison.
[d] Although recompression equipment is to be powered by electricity, no allowance was made by the distiller for increased use. However, electricity was assumed to be generated using nonpetroleum-type fuels (coal or nuclear) although a portion or all may be generated using oil or natural gas.
[e] High heating values (HHV) were used throughout the application which tends to maximize the kJ value of the alcohol output. The low heating value (LHV) is also presented here to provide another comparison.

consideration the energy balance for ethanol production from sugarcane, it is supposed that total electric energy is generated on site and sugarcane bagasse is used as a fuel for steam generation. In the process with cassava, external electric power is used and wood is the fuel source for process steam generation.

## 8.2.3 Ethanol from Wood

If we consider the experimental production of ethanol from aspen wood chips (WAYMAN, 1979) following autohydrolysis and caustic extraction, the ratio of output/input energy is 3.2 and 3.7 for a process assuming acid hydrolysis and enzymatic hydrolysis, respectively (Tab. 48). It can also be seen from the energy balance that all energy required to operate the process can be supplied by the lignin that is recovered. For this calculation the author assumed a 95% yield of ethanol from wood sugars obtained after hydrolysis.

**Tab. 47.** Energetics of Ethanol Production from Sugarcane and Cassava[a] (DE CARVALHO et al., 1977)

| Raw Material | Case | Energy [$4.19 \cdot 10^6$ kJ] | | | | | Net Energy Ratio (Output/Input) |
|---|---|---|---|---|---|---|---|
| | | Input Output | Agri-culture | Distillery | Trans-portation | Total | |
| Sugarcane | Total on-site generation of electric power (sugarcane bagasse as fuel) | 5.59 | 0.42 | 0.017 | 0.26 | 0.70 | 8.0 |
| Cassava | External supply of electric power (wood as fuel) | 5.57 | 0.30 | 0.43 | 0.21 | 0.94 | 5.9 |
| Cassava | Total on-site generation of electric power (wood as fuel) | 5.57 | 0.30 | 0.045 | 0.27 | 0.62 | 9.0 |

[a] Basis: 1 $m^3$ anhydrous ethanol.

**Tab. 48.** Energy Balance for Conversion of Aspen Wood Chips to Ethanol[a] (WAYMAN, 1979)

| | Gross Energy Recovery by | |
|---|---|---|
| | Acid Hydrolysis | Enzyme Hydrolysis [$10^6$ kJ] |
| Ethanol | 6.84 | 8.06 |
| Lignin | 4.41 | 4.41 |
| Volatiles | + | + |
| Total | 11.25 | 12.47 |
| Energy recovery | 52.35% | 58.03% |
| Energy required | | |
| Autohydrolysis | 0.70 | 0.70 |
| Caustic extraction | 0.12 | 0.12 |
| Hydrolysis | 0.70 | 0.23 |
| Distillation | 1.97 | 2.32 |
| Total | 3.49 | 3.37 |
| Net energy recovery | 36.1% | 42.3% |

[a] Basis: $10^3$ kg of aspen wood chips (dry); heat of combustion of aspen chips: $21.48 \cdot 10^6$ kJ.

## 8.2.4 Ethanol from Cornstalks

The requirement and availability of energy in the process for the production of ethanol (96%) through acidic conversion of cornstalks to sugars following fermentation and distillation are presented in Tab. 49. A production of 1,708 kg $h^{-1}$ ethanol is taken as a basis for this energy balance. A positive energy balance was obtained assuming that almost all necessary energy can be supplied from fermenters (cooling) and from remaining solids.

**Tab. 49.** Energy Balance for Conversion of Cornstalks to Ethanol (SITTON et al., 1989)

| Item | Heat Available [$10^6$ kJ h$^{-1}$] | Temperature [°C] | Heat Required [$10^6$ kJ h$^{-1}$] |
|---|---|---|---|
| Acid mixing in first hydrolysis tank | 5.80 | 100 | – |
| Acid reaction in impregnator | 1.84 | 80 | – |
| Acid mixing in second hydrolysis tank | 1.42 | 110 | – |
| Feed to second hydrolysis | – | 80–110 | 5.80 |
| Cooling feeds to fermenters | 56.39 | 110–25 | – |
| Heating feeds from fermenters | – | 25–90 | 48.90 |
| Heat supplied to reboiler | – | 100 | 35.20 |
| Heat recovered from condenser | 7.38 | 84 | – |
| Energy from remaining solids | 39.00 | – | 0.84 |
| Energy required for pumps and centrifuges | – | – | – |
| Total | 111.83 | | 90.74 |

A net energy yield of this process is the ethanol produced (2,165 L h$^{-1}$ or $1.26 \cdot 10^9$ kJ d$^{-1}$).

# 9 Ethanol as a Liquid Fuel

A comprehensive study of ethanol as a liquid fuel was conducted by KOSARIC (1984). Foreign crude oil imports currently provide the raw material for the production of half of the liquid fuels consumed in North America (U.S. Department of Energy, 1980). Recent events have dramatically illustrated the considerable economic cost, instability, and economic vulnerability of such imports. Even though oil prices have recently fallen, there is no security in believing that they will stay at these levels. Ethanol is a liquid fuel that can substitute for some petroleum products now, and increasingly so in the future years.

A concerted effort on behalf of government and the private sector is necessary to promote an ethanol-based fuel industry. Legislation, specifically dealing with programs to set aside land, to allow for the cultivation of biomass on marginal lands, is an essential element to developing this industry.

In Brazil, an effort of government and the private sector was launched in 1975 to develop an alcohol fuel industry. OPEC's 1973 price hike increased Brazil's debt burden significantly. Petroleum products, of which more than 80% were imported, accounted for about 42% of the nation's energy consumption and more than 50% of foreign exchange outlays (Renewable Energy News, 1983).

The Brazilian government launched a program called PRO-ALCOOL, and encouraged the private sector to play a key role in the development of alcohol production technology. Consequently, from 1975–1980, 300 private distilling projects received 90% start-up financing (PISCHINGER and PINTO, 1979).

During the same time period, 10 regional technology research centers collaborated with domestic automobile manufacturers to perfect an efficient alcohol burning engine. Engine conversion technologies were transferred to 500 private enterprises throughout Brazil and mass production of alcohol vehicles was started in 1979. The Brazilian effort showed promise, as 250,000 alcohol burning vehicles and 80,000 automobiles converted to run on a 20% ethanol–gasoline mixture, became functional in the transportation sector.

Even with such advances, certain problems became clear and evident: The alcohol burning engines did not meet the requirements of long-term operation, and there was serious doubt whether ethanol production would meet the demands of the transportation sector. The government provided an aide package to producers for $1 billion, and a research effort with the major car manufacturers to develop a second generation alcohol burning engine. These measures caused an increase in alcohol vehicle sales by 30%, and increased ethanol production by 62% (CONCEIRO, N. P., Head Powertrain Engineering, General Motors do Brasil S.A., personal communication, 1983).

In Canada, Manitoba is the only province where alcohol fuels are commercially available: Since 1981 a 10% ethanol–gasoline blend is available at approximately the same price as regular gasoline (Renewable Energy News, 1983).

In Ontario, there is presently a revived interest to develop an industry to supply fuel ethanol. Ethanol production and use is likely to be initiated by the farming sector in Ontario. Farms can produce ethanol because they have the raw material for its synthesis. The ethanol can be used in farm diesels like trucks, tractors, and combines. On-farm production of fuel gives the farmer:

- insurance of fuel supply,
- ability to use wet, spent grains, and
- investment and fuel tax brakes.

Ethanol production technology is advanced to the point that small-scale ethanol plants have been designed with "step-by-step" instructions for the prospective producer (BERGLUND and RICHARDSON, 1982; CHAMBERS, et al., 1979; DPRA, 1980; GIRD, 1980; LEEPER et al., 1982; LINCOLN, 1980; MCATEE et al., 1982).

Canada, with its vast land area and agricultural potential, is in a position to provide the necessary raw material(s) for ethanol production. The particular crop(s) that might be chosen will depend on many economic variables. However, it is evident that the agricultural sector has a role whose potential is just beginning to be realized. A farm-based ethanol fermentation industry can provide a starting point for future large-scale ethanol production.

## 9.1 Characteristics of Ethanol and Gasoline–Ethanol Blends as Motor Fuel

It is interesting that gasoline and alcohol, which in many respects are quite different, have a nearly equal energy of combustion per unit volume of stoichiometric mixture. It may, therefore, be concluded that these fuels used in an engine, under the same conditions, with the same fraction of the stoichiometric mixture for both, and with fully vaporized fuel, will provide nearly the same power. Hence, the power of an engine cannot be significantly altered by changing these fuels, for similar charge conditions. Equal amounts of energy and power require the use of an approximately 60% greater weight of ethyl alcohol than gasoline (BOLT, 1980).

The vapor pressure of alcohol is greater than that of gasoline, and the latent heat of vaporization is higher, which is primarily responsible for the increased power outputs using alcohol. As a result of the lower vapor pressure and high latent heat of vaporization, it is harder to start a cold engine on ethanol than on gasoline. To correct this deficiency, it has been common to add ether, benzene, or gasoline to the alcohol, each of which serves to increase the vapor pressure, or simply start the engine with gasoline and then switch to alcohol.

Sir HENRY RICARDO tested a variety of fuels in a very comprehensive program, using a variable compression ratio, single cylinder engine. Fig. 37 is a replot of the test results with RICARDO's engine for ethyl alcohol of 198 proof, and gasoline. The increased mean effective pressure obtained using ethanol at all mixture ratios is the most dramatic difference between the two fuels (MENRAD and LOECK, 1979). The increase in pressure is principally due to the greater volumetric efficiency, which results from the high latent heat of vaporization of alcohol and the greater mass of fuel per unit mass of air. Specifically, the higher latent heat of vaporization of ethanol as compared to gasoline produces a greater lowering of temperature upon evaporation, in the intake manifold, resulting in a cooler and denser fuel–air mixture. Hence, a

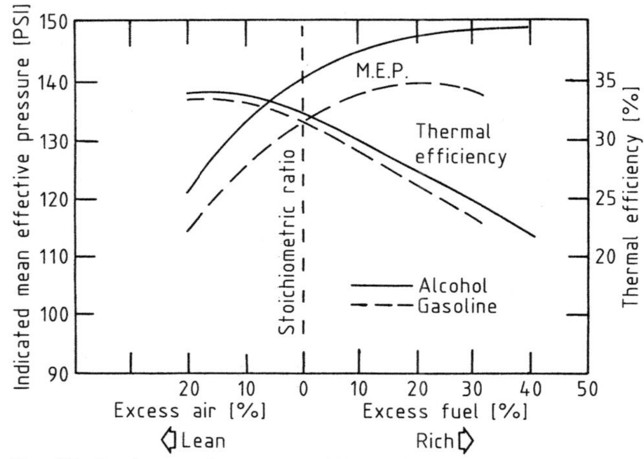


**Fig. 37.** Engine performance with ethyl alcohol and gasoline.

greater charge of air and fuel is taken into the cylinder during the suction stroke and the volumetric efficiency, as mentioned above, increases.

### 9.1.1 Exhaust and Evaporative Emissions

Nitrogen oxides are produced during the combustion process and increase with higher temperatures. The lower flame temperature of ethanol produces substantially less $NO_x$ than hydrocarbon fuels, at any equivalence ratio. When alcohol is used in gasoline as a blend, a reduction in $NO_x$ emissions roughly proportional to the alcohol concentration is observed (BECHTOLD and PULLMAN, 1979; MCCALLUM et al., 1982). When no adjustment of the fuel metering system is made, alcohol–gasoline blends may either increase or reduce $NO_x$ emissions, depending on whether the blend leaning effect moves the fuel–air equivalence ratio closer or further from the approximate value of 0.9, at which $NO_x$ typically peaks. The change is usually slight for moderate blend levels.

Fuel evaporative emissions from alcohol blends are typically much higher than those from straight gasoline (General Motors Research Laboratories, 1980). A 10% ethanol–gasoline blend may increase evaporative emissions of 49% to 62% relative to straight gasoline (MCCALLUM et al., 1982).

One major exhaust problem from the use of ethanol is that of aldehyde emissions (especially acetaldehyde), which are not currently regulated. Aldehydes are photochemically reactive and highly irritating to mucous membranes. Aldehyde emissions from neat ethanol may be from 2–4 times of those from gasoline (WAGNER, 1980).

Carbon monoxide emissions are usually reduced when using ethanol–gasoline blends. This is not directly the result of the ethanol fuel, but rather the fact that CO emissions – and unburned fuel emissions – decrease when combustion occurs at lean equivalence ratios. Therefore, the lower CO emissions associated with ethanol–gasoline blends are mainly due to the blend leaning effect. It may be generally stated that with alcohol fuels, CO emissions are not reduced to the same extent as $NO_x$ emissions, and are similar to those from gasoline powered engines (MENRAD, 1979).

### 9.1.2 Ignition, Cold Start-Up, and Driveability

Partial-load operation shows that alcohol-operated engines require less ignition advance than gasoline engines (MENRAD, 1979). This is mainly due to the higher compression ratio that can be attained with alcohol fuels, and the higher rate of flame propagation.

The most frequently encountered difficulty with ethanol fuel is the poor cold start-up of the engine, especially evident in northern climates during winter. The failure of engines operating on ethanol to start under cold ambient conditions is a direct consequence of the inability of the initial spark ignition energy to produce a self-propagating flame with the required mass burning velocity (RAJAN, 1979).

To alleviate the problem of cold start-up, a fuel vaporizer has been developed to facilitate cold starting of ethanol-fuelled engines. The vaporizer ultimately found to give the best results was designed with a double concentrically wound cell. This design enables vehicle cold starts at temperatures as low as −10°C.

Generally, the poor driveability characteristics of some ethanol–gasoline mixtures is related to an intolerable leaning of the fuel–air ratio. This might occur by the simple mechanism of the blend leaning effect. Newer vehicles equipped with feedback-fuel metering systems maintain a constant fuel–air stoichiometry, and the results of the leaning effect are minimized.

## 9.1.3 Water-Tolerance of Ethanol–Gasoline Blends

Most alcohol is not totally pure and contains 5% water. This is due to the fact that ethanol forms a low-boiling azeotrope during distillation and the 5% water content cannot be separated out. Since distillation cannot increase the concentration of ethanol directly, the addition of benzene forms a ternary mixture of benzene, water and ethanol with a lower boiling point of 64.85 °C, from which the water can be removed by distillation. Benzene and alcohol have a low boiling binary (67.8 °C), hence benzene can be separated by distillation. Anhydrous ethanol is obtained commercially in this manner as 200 proof ethanol.

Gasoline and anhydrous ethanol are miscible in all proportions and over a wide range of temperatures. However, even small

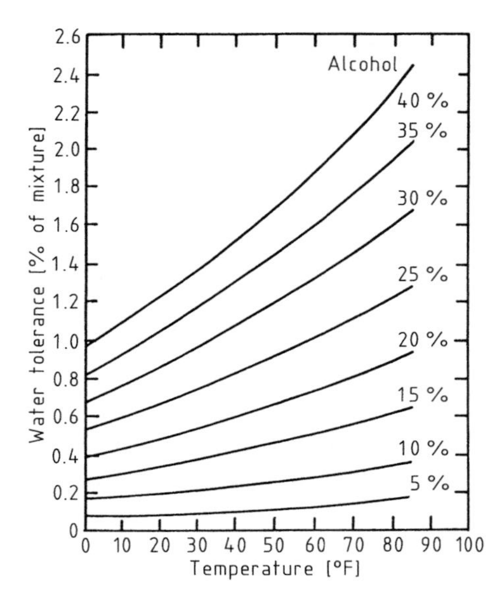


**Fig. 38.** Water tolerance of alcohol–gasoline blends.

amounts of water in a gasoline–ethanol blend will cause separation of the ethanol and gasoline. Fig. 38 shows the water tolerance of alcohol–gasoline blends over a wide temperature range. A 20% alcohol–gasoline blend at room temperature can, at most, tolerate 0.3% water before phase separation. Certain blending agents can be added to the mixture which will increase the water tolerance. Some of these include benzol, benzene, acetone, and butyl alcohol (BOLT, 1980; TERZONI, 1981).

## 9.1.4 Lubrication

A problem which may result in any alcohol-fuelled engine is dilution of the lubricating oil with alcohol condensed from blow-by gases or that which finds its way past the piston rings. This can have adverse effects on the wear of bearings, camshafts and other parts which rely greatly on lubrication to function properly.

It is known that both gasoline and diesel fuel have some lubricating quality, while ethanol has none. Alcohol presents a particular problem since it is known that it has a tendency to wash cylinder walls. This would not be a problem if it could be ensured that ethanol stays in its vapor state during operation of the engine.

Oil of the castor bean is one of the few lubricants that blends with alcohol. 5 vol.% castor oil in ethanol provides the same lubrication as diesel fuel (KIRIK, 1981). However, castor oil and ethanol do not mix easily and vigourous mixing is required. Furthermore, castor oil solidifies between 0°C and 30°C, and hence might present problems in cold climates.

### 9.1.5 Corrosion and Materials Compatibility for Alcohol-Fuelled Vehicles

Little is known about the wear of ethanol-fuelled engines. At present, the most likely explanation seems to be that a combination of corrosion and erosion is causing wear, though other mechanisms may be present (Mueller Associates Inc., 1981). When ethanol burns with insufficient air, acetic acid is formed, which promotes corrosion in the engine (KIRIK, 1981); the smell of vinegar in the exhaust can be used as a signal to adjust the fuel–air ratio.

Ethanol has adverse effects on some gasket materials, certain rubber materials, and some metals. It acts as a good cleaning agent for the deposits left behind by gasoline and for rust. If ethanol is stored in a tank which previously held gasoline, it can be expected that any deposited residue will be removed from the tank walls. This means that fuel filters in dispensing equipment and vehicles may become restricted or plugged soon after the change of fuels. The initial cleansing activity loosens the majority of the material which will be removed and, thus, residue loosening is less a problem as time goes on.

Steel is not greatly affected by ethanol. However, if the water content in an alcohol–gasoline blend is high enough to cause phase separation, the steel in contact with the water-rich phase will show corrosion.

Anhydrous ethanol and ethanol–gasoline blends exert adverse effects on most elastomeric parts such as fuel-pump diaphragms and fuel hoses. These parts deteriorate gradually making more frequent replacement than with gasoline necessary. Fiberglass shows softening and blistering when coming in contact with ethanol, indicating destruction of the laminate. Polyurethane is softened and cracked by contact with ethanol. Nylon, high-density polyethylene and polypropylene are not affected significantly by ethanol.

Elastomers swell and shrink in any organic liquid depending on the particular elastomer and the organic liquid. Therefore, it can be expected that many parts made of elastomers will behave differently if ethanol is used in a fuel system designed for gasoline or diesel fuel.

### 9.1.6 Safety of Alcohol

In general, ethanol is considered to be less dangerous than gasoline or diesel fuel (Mueller Associates Inc., 1981). Since alcohol does not contain light fractions, alcohol fires do not start as readily as gasoline fires. However, alcohol burns with a nearly invisible flame, especially in sunlight, and without the large amount of smoke generated by gasolines. Therefore, detection of ethanol fires might be more difficult.

Due to the high vapor pressure of ethanol, flammable mixtures of air and anhydrous ethanol fuel vapors will be present in fuel tanks under most ambient conditions. Alcohol and air mixtures of 4–13.6% alcohol are explosive; the range for gasoline being only 2–5% (KIRIK, 1981). To circumvent this problem, collapsible fuel tanks can be made, as is already done for safety in racing cars.

## 9.2 Modifications and Conversions of Existing Internal Combustion Engines to Utilize Ethanol and Ethanol–Gasoline Blends

### 9.2.1 Research

The utility of alcohols as substitute fuel for the internal combustion engine has been under investigation for some time. As early as

1897, NIKOLAS A. OTTO used pure ethanol in his first engine (ROTHMAN, 1983). In the 1920s and 1930s in Germany and in Switzerland, fuel ethanol was used extensively. Both, beneficial and detrimental characteristics of ethanol as a motor fuel have long been noted and widely recognized.

The General Motors Research Laboratories in Warren, Michigan, have conducted research using ethanol fuel in a single-cylinder engine, similar to the Ricardo E.6/T engine. The researchers found that engine thermal efficiency increased by 3% with ethanol, compared to gasoline at the same compression ratio (BRINKMAN, 1981). Furthermore, increasing the compression ratio from 7.5:1 to 18:1 with ethanol increased efficiency by 18% over gasoline.

The U.S. Department of Energy conducted research on the influence of a 20% ethanol–indoline blend on the steady-state performance and emission characteristics of a carbureted spark-ignited engine. Data obtained revealed that a 2% increase in engine torque was observed with the blend and a 1.7% rise in brake thermal efficiency, compared to the base fuel. CO emissions remained the same while $NO_x$ and HC emissions decreased with the blend, compared to the base fuel (indoline). From the results obtained, it was concluded that for the same engine conditions, the substitution of practical level alcohol blends (10–15% by volume) will have little effect on steady-state performance and currently regulated emission characteristics (CO, HC and $NO_x$) (ADT and RHEE, 1978).

The Coordinating Research Council Inc. (C.R.C.I.) conducted tests to determine the effects of 10% ethanol–gasoline blends on emissions, fuel economy, and driveability of 14 1980-model year cars. They found that a 10% ethanol addition to gasoline resulted in statistically significant changes in emissions, driveability, and efficiency (BERNSTEIN et al., 1982).

Exhaust and evaporative emissions from a 1974 Brazilian Chevrolet Opala were measured (FUREY and JACKSON, 1977) using gasoline and various ethanol–gasoline mixtures. A 20% ethanol–gasoline mixture reduced exhaust hydrocarbon and CO emissions, but increased aldehyde and $NO_x$ emissions. It appears that leaning of the air–fuel mixture, due to ethanol addition, was the primary cause of the exhaust emission changes.

The U.S. Environmental Protection Agency conducted evaporative, exhaust, and performance characteristics on 11 test vehicles and found a 2% decrease in fuel economy (FUREY and KING, 1980; LAWRENCE, 1979). The Japan Automobile Research Institute has demonstrated that $NO_x$ emissions decrease in I.C. (international combustion) engines as the ratio of alcohol in gasoline increases (MATSUNO et al., 1979).

Nebraska, which was one of the first states to implement GASOHOL (10% anhydrous ethanol with regular-grade unleaded gasoline), used 45 vehicles in a 3.2 million km road test to compare GASOHOL with regular grade unleaded gasoline, and GASOHOL cars obtained up to 5.3% more miles per gallon, and 8.07 more miles per Btu than the cars using unleaded fuel (SCHELLER, 1977). GASOHOL performance was satisfactory under all conditions of weather and driving.

KOSARIC (1984) studied combustion of varying gasoline–ethanol and ethanol–water fuel blends using a Ricardo E.6/Y variable compression, spark-ignited engine at compression ratios ranging from 8.5–15.8. The investigation revealed that the engine could be operated at considerably higher compression ratios when ethanol fuels were used instead of gasoline fuels. Neat ethanol compared to straight gasoline had a 25% higher power output (@ 1,600 rpm and CR 8.5:1); the power output obtained using neat ethanol at CR12.5 and 1,600 rpm represents a 57% increase relative to straight gasoline at CR 8.5:1, and a 26% increase relative to neat ethanol at CR 8.4:1.

The best fuel economy was obtained with gasoline–ethanol blends at high compression; the lowest fuel consumption of all blends tested was recorded using a 40% ethanol–gasoline blend at CR 11.9:1. Ethanol–water fuel mixtures generally showed higher thermal efficiencies than gasoline–ethanol fuel blends; a 36% increase in thermal efficiency was obtained using neat ethanol relative to straight gasoline, at CR 8.5:1. Due to the performance characteristics obtained, ethanol with a 20% water content was recommended as the fuel blend of choice for operation in gasoline-powered engines at low compression (8.5:1).

### 9.2.2 Applications

Gasoline powered engines can be modified to run on ethanol or ethanol–gasoline blends. Furthermore, some car manufacturers such as Volkswagen, General Motors, Volvo, and Daimler Benz have designed and developed alcohol engines. As might be expected, such conversions have varying levels of sophistication. Chevrolet Oldsmobile vehicles of 1979, e.g., can run on GASOHOL, with no engine modification; GM actually encouraged the use of GASOHOL at that time.

The main principle of the simple conversion is to aid in the vaporization of ethanol by starting the engine with gasoline and switching to ethanol after about 5 min running time with gasoline. A second fuel tank for ethanol can be used and an appropriate mechanism to switch from the gasoline to the ethanol tank incorporated into the fuel delivery system of the engine. Some conversion systems are offered commercially. The diameter of the main metering jet of the carburetor needs to be increased by 25% due to the higher volumetric flow of ethanol (1.5 ×) relative to gasoline and, therefore, the fuel–air ratio needs to be appropriately adjusted. The resulting engine power is approximately the same as that for gasoline.

An improved conversion uses high compression as well as some of the above mentioned modifications. Power output is increased by about 10% over a similar gasoline engine, and ethanol consumption is only 10–20% higher by volume (KIRIK, 1981). Some of the changes include fitting new piston heads to increase the compression ratio to 12.5:1, and fitting larger carburetor jets to accommodate for the increased alcohol volume.

Starting from cold is similar to that of the simple conversion, but gasoline as a starting fuel is not recommended; an alcohol–ether mixture from 15–50% ether by volume, might serve as a starting fluid. Propane can be injected for cold starting, or one of the external heat sources can be applied to vaporize alcohol.

Daimler Benz have used both methanol and ethanol as a fuel in automotive engines. Particular importance was attached to the cold start characteristics. A separate preheating system, with an additional cold-start enrichment device was combined with mechanical fuel injection to provide optimum starting from cold at temperatures down to −25 °C (FRICKER, Daimler Benz AG, Stuttgart, personal communication, 1983). Keeping the distance to the knock limit comparable to the gasoline engine with a CR of 9:1, the CR for ethanol can be increased to 11. This optimizes the ignition timing for ethanol operation and employs spark plugs with a suitable heat range. These additional changes result in a considerably higher power output and better torque. The specific fuel consumption is higher with alcohol compared to gasoline, but the thermal efficiency with alcohol fuel is better.

BLASER (1983) has developed a device to permit diesel engines to run on wet ethanol, and also to improve the efficiency of spark-ignition engines running on ethanol less than 100% pure. It is claimed that engines fitted with this device would have the same efficiency as those running on gasoline or diesel oil, and that problems with cold starting, phase separation, and lubricant dilution are virtually eliminated. The device, which is 2″ high and 1″ long, is a container for ethanol. It is attached to an adapter which also holds a spark plug. The adapter is mounted on the head of each engine cylinder, with the tip of the device – containing ethanol – inside the combustion chamber. The engine is started using a glowplug to vaporize the ethanol; on the intake stroke, air is drawn from the manifold and alcohol is drawn from the device. The ethanol vaporizes and mixes with air, and on the compression stroke is compressed for ignition by the spark plug.

A fuel injection system has also been used in a high compression (18:1), spark-ignition engine. The heat of compression is used to evaporate the alcohol and since inadequate heat is left over to ignite the alcohol, a spark-ignition system is added. The basic principles outlined above are utilized in the Brandt System of engine design (KIRIK, 1981). The main objective of injection is to create a finely atomized droplet size which has intimate contact with air in the engine. Since the heat of compression is available immediately as the engine turns over, the Brandt type engine has no cold start-up problems.

Using the same principle as Brandt, the Moneurgaro division of Chenesseau of France has developed the cool compression injection system (C.C.I.S.) (AGACHE, 1983). This system employs a novel combustion chamber design, an internal carburetor, and introduces ethanol to the combustion chamber under high pressure and in a finely fractionized form. Although specifics of the C.C.I.S. system are proprietary, Moneurgaro has made some of the main characteristics of the system available.

Volkswagenwerk AG (1981) of Germany has developed a lambda sensor-controlled continuous injection system (CIS) for both ethanol and methanol operation. The CIS was used in the VW Jetta; 40 of the vehicles were tested. An efficient starting of the engine is guaranteed up to $-5$°C. Volkswagen of America Inc. (1981) has build and optimized a Rabbit for operation on neat ethanol. This vehicle is optimized with respect to fuel economy, driveability, and performance.

The Rabbit and Jetta use the same continuous flow fuel injection system, with lambda feedback control via an oxygen sensor in the exhaust manifold. The continuous flow fuel injection system is mechanically operated but is not driven by the vehicle engine. The heart of the system is the mixture control unit. The volume of intake air is measured by an air flow sensor in the mixture control unit upstream of the throttle plate. Corresponding to the volume of air measured, a fuel distributor apportions an amount of fuel to the individual engine cylinders. The metered fuel is fed to the injection valves which continuously spray the fuel in atomized form in front of the intake valves. In addition, fuel is regulated by the frequency valve at the fuel distributor by use of a closed loop system. This system employs an electronic control unit and a lambda sensor which produces an optimum fuel–air mixture with regard to engine power, fuel consumption, and exhaust gas composition. Fuel is supplied to the mixture control unit by a high pressure pump at a constant flow rate; the unused fuel is circulated back to the tank. Cold starting is enhanced with an auxiliary gasoline fuel supply system. An underhood reservoir supplies gasoline to a fuel pump through a filter. The gasoline, regulated at 30 psi, is fed to the cold start valve when the key is turned on. The cold start valve function is initiated when the ignition switch is turned to "start". The fuel pump runs to assure enough fuel for starting even under long and repetitive cranking conditions, and for the additional cold start at low temperatures.

General Motors of Brazil has developed an ethanol engine which is being marketed at a large scale in Brazil. The engine uses gasoline injection for cold starting. At engine temperatures below 25 °C, a gasoline injection pump and relay timer are activated via the ignition switch during the cranking period, and lasts for 2 s. The volume of gasoline injected over the 2 s period is approximately 20 $cm^3$. The gasoline system is automatically shut off at engine temperatures above 25 °C.

A similar system to the one mentioned above was developed by the Brazilian firm M.W.M., for the PID-alcohol engine (CUCEIRO, General Motors do Brasil, personal communication, 1983). The PID-alcohol engine was developed from a conventional diesel engine which has its main injection system adapted to work with ethanol and received an additional injection system to work with diesel as the combustion igniter fuel. It incorporates, therefore, two injection pumps and two injectors per cylinder, located on the combustion chamber. The additional injection system is required because ethanol does not auto-ignite and diesel fuel is used to act as the combustion initiator. This injection system operates during cold starting.

The Toyota Motor Corporation has modified the mass-produced Starlet model to utilize anhydrous ethanol, at the request of the New Guinea government. Some of the major alterations include an increase in the CR from 9:1 to 11:1, the addition of a sub-fuel 2.4 L tank with a cold start injector installed in the intake manifold. Furthermore, to ensure good driveability during engine warm-up, a cold mixture heater (CMH) and hot air intake (HAI) have been made available, which are automatically operated (Toyota, 1983).

## 9.3 Comparison of Ethanol with Other Motor Fuels

A comparative study of methanol, ethanol, isopropanol, and butanol as pure gasoline blended motor fuels was presented by KELKAR et al. (1989). An International Harvester Silver Diamond engine (Tab. 50) was used for testing different fuels. The cooling arrangement of the engine was modified by removing the radiator and water pump and the carburetor was fitted with a load needle rather than a fixed jet. To ensure enough alcohol supply to the engine, the main jet tube was redrilled to give 35% increase in area.

The experimental results showed that a 10% alcohol–90% gasoline mixture may be used in engines designed for gasoline without modifications. Gasoline blends up to 50/50 with any of the alcohols tested may be used if carburetor adjustment and ignition timing modifications are made to the engines designed for pure gasoline. However, a spark-ignition engine designed to use gasoline must be modified to use pure alcohol. Changes required for the use of methanol, ethanol, isopropanol, and butanol use are technically similar, but the fuels could not be used interchangeably without carburetor adjustment. The basic engine changes included an increase in the compression ratio, enlarging the carburetor jets, and adjusting the ignition timing.

A comparison of production costs of liquid fuels from biomass has been presented by BRIDGEWATER and DOUBLE (1991, 1994). A computer-aided economic analysis is presented for the following processes:

- Mobil's Methanol to Gasoline (MTG) (Lurger gasifier),
- gasoline from wood by the Mobil's Methanol to Olefrus Gasoline and Diesel (MOGD),
- diesel from wood by SMDS (Shell Middle Distillate Synthesis),
- methanol from wood by gasification,
- fuel alcohol from wood (blend of methanol with higher alcohols produced by a modified methanol synthesis process),
- direct thermochemical conversion of wood to gasoline blending stock,
- ethanol from softwood (fermentation of hydrolyzate),
- methanol from RDF (Reception of Refuse-Derived Fuel),
- methanol from straw,
- ethanol from straw (fermentation of hydrolyzate),
- ethanol from beet (fermentation),
- ethanol from wheat (fermentation).

The absolute cost of fuels at 1,000 t $d^{-1}$ feed rate are summarized and ranked in Tab. 51 with feed costs at typical values.

In terms of absolute fuel costs, thermochemical conversion offers the lowest cost products, with the least complex processes being advantageous. Biochemical routes, according to this analysis, are the most costly in absolute terms.

The most attractive processes, by comparing production costs to product values, are alcohol fuels. Direct production of highly aromatic gasoline by direct liquefaction of wood through pyrolysis and zeolites is better, but the process is so far the least developed. For hydrocarbons, the direct liquefaction processes appear to offer considerable advantages, although the state of development is less advanced.

Of the indirect thermochemical routes, the MTG appears to be the most attractive for gasoline and the Shell SMDS for diesel. Both of these processes also have the advantage of producing high-quality products: MTG gasoline is highly aromatic (high octane number),

**Tab. 50.** Specifications for the International Harvester, Silver Diamond Testing Engine (KELKAR et al., 1989)

| Engine Type | Reciprocating, Spark Ignition Engine |
|---|---|
| Number of cylinders | 6 |
| Cylinder arrangement | in-line, vertical |
| Bore × stroke | 3 9/16″ × 4″ |
| Displacement | 240.00 cubic inches compression |
| Ratio | 1. original head: 6.77:1<br>2. modified head 7.76:1 |
| Type of cooling | water-cooled |

**Tab. 51.** Estimated Fuel Cost[a] (BRIDGEWATER and DOUBLE, 1994)

| Rank | Product | Feed | Route | Uncertainty | Cost [£ GK$^{-1}$] |
|---|---|---|---|---|---|
| | Natural gas | | | | 3.0 |
| | Diesel | | | | 4.1 |
| | Methanol | | | | 5.6 |
| | Gasoline | | | | 5.8 |
| 1 | Methanol | straw | gasification | low-moderate | 7.0 |
| 2 | Gasoline | wood | pyrolysis + zelolites | high | 8.6 |
| 3 | Diesel | wood | gasification + SMDS | low-moderate | 8.8 |
| 4 | Gasoline | wood | pyrolysis + hydrotreating | high | 8.8 |
| 5 | Methanol | wood | gasification | low-moderate | 9.1 |
| 6 | Gasoline | wood | liquefaction + hydrotreating | high | 9.4 |
| 7 | Methanol | RDF | gasification | moderate | 9.9 |
| 8 | Fuel alcohol | wood | gasification | moderate | 9.8 |
| | Ethanol | | | | 12.0 |
| 9 | Gasoline | wood | gasification + MTG | low | 12.4 |
| 10 | Ethanol | wood | $H_2SO_4$ hydrolysis + fermentation | low | 14.4 |
| 11 | Gasoline | wood | gasification + MOGD | moderate | 14.6 |
| 12 | Diesel | wood | gasification + MOGD | moderate | 14.6 |
| 13 | Ethanol | wheat | fermentation | low | 14.9 |
| 14 | Ethanol | wood | enzyme hydrolysis + fermentation | moderate | 19.6 |
| 15 | Ethanol | beet | fermentation | low | 25.2 |

[a] Basis: 1,000 t d$^{-1}$ daf feed, feedstock costed at typical cost.

while SMDS diesel has a cetane number far in excess of current specifications, allowing engines on this fuel to be much more efficient than current diesels. The SMDS process has the advantage of producing jet fuel of high quality.

# 10 Present and Potential Markets for Ethanol

In the United States alone, fuel alcohol has grown from its infancy in 1979 to approximately 11.10$^9$ L (2.9 · 10$^9$ gallons) of production capacity in 1991 (HAIGWOOD 1991). Its price was based on the price of wholesale gasoline plus available federal state tax incentives. These incentives allowed ethanol, with production costs of $0.26–0.40 per liter ($1.00–1.25 per gallon) to compete with gasoline prices of $0.11–0.17 per liter ($0.40–0.65 per gallon).

Even though the cost of ethanol production by fermentation is still too high to openly compete (without subsidies) with gasoline, the environmental factors are of primary importance for development of this industry. The ecological impacts of ethanol fuels for the internal combustion engines can be summarized as follows:

- $CO_2$: Bioethanol is a renewable energy which does not contribute to the greenhouse effect. In this case, the carbon cycle is very short because the $CO_2$ given off by fermentation and combustion is reabsorbed by the plants through photosynthesis to vegetation (reduction may be 2- to 3-fold).
- CO and HC: It was shown (POITRAT, 1994) that mixtures with 5–7% ethanol can reduce CO emissions in a proportion of 15–40%. HC emissions may be reduced in a proportion of 2–7%.

- $SO_x$: Biofuels do not contain sulfur and do not contribute to acid rain.
- $NO_x$: Biofuels combustion may increase nitrogen oxides in the exhaust by up to 10%. However, lower combustion temperatures and optimum engine tuning can lead to a drop in $NO_x$ emissions.
- Aromatics: Ethanol does not contain aromatics (unlike some gasolines which contain up to 45% aromatics).
- Aldehydes: Emissions of certain aldehydes/ketones are slightly increased, among which is acetaldehyde. Vehicles equipped with a catalytic converter have shown reductions in aldehyde emissions.
- Nitrates: Maintaining a cover of vegetation during the winter months limits the risk of leaching out fertilizing elements contained in the soil, particularly nitrates.

Maintaining or creating employment by the development of a biofuels industry is also an incentive for development of this industry. Depending on the feedstock used 10–17 direct or indirect employees could be created or maintained per 1,000 t of ethanol production capacity. The agricultural area per employee would be 19–20 ha for sugar beet as feedstock and 30–35 ha for wheat. The ethanol produced per employment would be 100 t with sugar beet and 60–67 t with wheat.

In addition to the use of ethanol as a pure or gasoline-blended fuel for internal combustion engines, ethanol can also be used in many other very important industrial applications (HAIGWOOD, 1991), such as for:

**Coal desulfurization.** In this process, ethanol is added to finely ground coal. The coal is heated and treated with ethanol in the presence of a reaction accelerator. This process removes both the organic and mineral sulfur in a single step, at up to 90% efficiency. The market for this process alone can generate a demand in the U.S.A. of over $3.78 \cdot 10^9$ L ($1 \cdot 10^9$ gallons) of ethanol.

**Biodegradable plastics.** The use of ethanol in the production of biodegradable plastics could create a substantial market for ethanol and, at the same time, reduce the cost of manufacturing biodegradable plastics. Ethanol is used here as a substrate for microorganisms which produce either polyhydroxy butyrate (PHB) or polyhydroxy valerate (PHV). When combined, the two create PHBV, which is a biodegradable plastic. Ethanol as a carbon source in the process could reduce the cost from $33.04 to 2.64 per kg if fuel ethanol excise tax subsidies are applied.

**Cattle feeding.** There is indication that dietary ethanol improves the tenderness of beef without adding to the fat content. This could help the cattle livestock industry to produce tender "low-fat" beef. It has been estimated that an ethanol market of $1{,}900 \cdot 10^6$ L ($500 \cdot 10^6$ gallons) per year could be created in this way.

**Diesel engine.** Research in heat (100% ethanol) fuel alcohol engines is progressing, whereby ethanol is injected with diesel fuel into the engine in presence of peroxides, which allow the blending of ethanol and diesel. Federal emission standards (for buses and trucks) are easier met with these blends.

# 11 References

ACKERSON, M. D. (1991), *16th IGT Conf. Energy from Biomass and Wastes,* Washington, D.C., 725.

ADT, R. R., RHEE, K. T. (1978), Effects of blending alcohol with gasoline on automotive engine steady-state performance and regulated emissions characteristics, U. S. Department of Energy, *Report No. eY-76-5-05-5216,* 8 pp.

AGACHE, G. (1983), *The Cool Compression Injection System*, Monteurgaro, Chenesseau, Orleans Cedex, France, 22 pp.

AIBA, S., SHODA, M. (1969), *J. Ferment. Technol.* **47**, 790.

AIBA, S., SCHODA, M., NATATANI, M. (1968), *Biotechnol. Bioeng.* **10**, 845.

Alcon Biotechnology Ltd. (1980), 20 Eastbourne Terrace, London.

ANONYMOUS (1978), *Chem. Eng. News* **56** (32), 22.

ANONYMOUS (1980a), Continuous alcohol fermentations in Australia, *Int. Sugar J.* **82**, 44.

ANONYMOUS (1980b), *Chem. Eng.* **25**, Aug. 11.

ARCURI, E. J., WORDEN, R. M., SHUMATE, S. E. II. (1980), *Biotechnol. Lett.* **1** (11), 499.

AZHAR, A. (1981), *Biotechnol. Bioeng.* **23**, 879.

BACON, J. S. D., EDELMAN, J. (1951), *Biochem. J.* **48**, 114.

BAKER, B. P. (1979), The availability, composition and properties of cane molasses, in: *Problems with Molasses in the Yeast Industry* (SINDA, E., PARKKINEN, E., Eds.), p. 126, Symposium, Aug. 31–Sept. 1, Helsinki, Finland.

BAKER, L., THOMASSIN, P. J., HENNING, J. C. (1990), The economic competitiveness of Jerusalem artichoke (*Helianthus tuberosus*) as an agricultural feedstock for ethanol production for transportation fuels, *Can. J. Agric. Econ.* **38**, 981–990.

BAKER, L., HENNING, J. C., THOMASSIN, P. J. (1991), Jerusalem artichoke costs of production in Canada: Implications for the production of fuel ethanol. *Pro. Int. Congress Food and Non-Food Applications of Inulin and Inulin-Containing Crops,* Feb. 18–21, Wagewugla, The Netherlands.

BAZUA, C. D., WILKE, C. R. (1977), *Biotechnol. Bioeng. Symp.* **7**, 105.

BECHTHOLD, R., PULLMAN, B. (1979), Driving comparisons of energy economies and emissions from an alcohol and gasoline fuelled vehicle. *Proc. 3rd Int. Symp. Alcohol Fuels Technol.,* May 28–31, Asilomar, CA, 13 pp.

BECK, M. J., JOHNSON, R. D., BAKER, C. S. (1990), *Appl. Biochem. Biotechnol.* **24/25**, 415.

BERAN, K. (1966), *Theoretical and Methodological Basis of Continuous Culture of Microorganisms* (MALEK, I., FENCL, S., Eds.), p. 369. New York: Academic Press.

BERGLUND, G. R., RICHARDSON, J. G. (1982), Design for a small-scale fuel alcohol plant, *Chem. Eng. Progr.* **8** (78), 60–67.

BERNSTEIN, S., TZENG, C. H., SISSON, D. (1977), *Biotech. Bioeng. Symp.* **7**, 1.

BERNSTEIN, L. S., BRINKMAN, N. D., CARLSON, R. R. (1982), Performance evaluation of 10% ethanol–gasoline blends in 1980 model year U.S. cars. *Society of Automotive Engineers Fuels and Lubricants Meeting,* Oct. 18–21, Toronto, Ontario, 17 pp.

BISARIA, V. S., GHOSE, T. K. (1981), *Enzyme Microbiol. Technol.* **3**, 90.

BLASER, R. (1983), *Alcohol Engine Conversion Project,* 14 pp. Vienna, IL: Southern Illinois College.

BLOTKAMP, P. J., TAKAGI, M., PEMBERTON, M. S., EMERT, G. H. (1978), *Proc. 84th Natl. Mtg. AIChE*, Atlanta Georgia, February.

BOLT, J. A. (1980), A survey of alcohol as a motor fuel. Society of Automotive Engineers Inc., *Progress in Technology Series,* No. **19**, 32 pp.

BRIDGEWATER, A. V., DOUBLE, J. M. (1991), *Fuel* **70**, 1209.

BRIDGEWATER, A. V., DOUBLE, J. M. (1994), *Int. J. Energy Res.* **18**, 79.

BRINKMAN, N. D. (1981), Ethanol fuel – a single cylinder engine study of efficiency and exhaust emissions. General Motors Research Laboratories, Warren, MI, *Research Publication No. GMR-3483R-F&L-719,* 16 pp.

BROWN, S. W., OLIVER, S. G., HARRISON, D. E. F., RIGHELATO, R. C. (1981), *Appl. Microbiol. Biotechnol.* **11**, 151.

BUCHANAN, R. E., GIBBONS, N. G. (1974), in: *Bergey's Manual of Determinative Bacteriology,* 8th Edn., Baltimore, ML: Williams and Wilkins Co.

BUCHHOLZ, K., PULS, J., GODELMANN, B., DIETRICHS, H. H. (1981), *Process Biochem.* **16** (1), 37.

CALLIHAN, C. D., CLEMMER, J. E. (1979), in: *Microbial Biomass* (ROSE, A. H., Ed.), p. 271. New York: Academic Press.

CASTELLO-BRANCO, J. R. C., LACAZ, P. A. A., COSTA-RIBEIRO, C. A. L. (1980), *17th Int. Soc. Sugar Cane Technologists Congress*, Manila, February.

CHAMBERS, R. S., HEREEDEN, R. A., JOYCE, J. J., PENNER, P. S. (1979), *Science* **206**, 789.

CHAN, S. K. (1969), *Investigation of the Federal Experiment Station,* Serdang, Malaysia.

CHANG, M. (1971), *J. Polym. Sci.* **C-36**, 343.

CHANG, C. T., YANG, W. L. (1973), *Taiwan Sugar*, September.

CHANG, M. M., CHOU, T. Y. C., TSAO, G. T. (1981), *Adv. Biochem. Eng.* **20**, 15.

COMBES, R. S., (1981), Proceedings of moonshine to motor fuel; *A Workshop on the Regulatory Compliance for Fuel Alcohol Production,* Atlanta, GA, March, p. 184.

COSTA-RIBEIRO, C., CASTELLO-BRANCO, J. R. (1981), *Process Biochem.* **16** (4), 8.

COWLING, E. G., KIRK, T. K. (1976), *Biotechnol. Bioeng. Symp.* **6**, 95.

CRAWFORD, D. K., CRAWFORD, R. K. (1980), *Enzyme Microbiol. Technol.* **2**, 11.

CYSEWSKI, G. R. (1976), Fermentation kinetics and process economics for the production of ethanol. *Ph. D. Thesis*, University of California, Berkeley.

CYSEWSKI, G. R., WILKE, C. R. (1977), *Biotechnol. Bioeng.* **19**, 1125.

CYSEWSKI, G. R., WILKE, C. R. (1978), *Biotechnol. Bioeng.* **20**, 1421.

DE CARVALHO, A. V., JR., MILFONT, W. N., JR., YANG, V., TRINIDADE, S. E. (1977), in: *Int.*

*Symp. Alcohol Fuel Technol.-Methanol and Ethanol,* Nov. 21–23, Wolfsburg, Germany.

De Menezes, T. J. B. (1978), *Process Biochem.* **13** (9), 24, 26.

Del Rosario, E. J., Hoon-Lee, K., Rogers, P. L. (1979), *Biotechnol. Bioeng.* **21**, 1477.

Detroy, R. W., Hesseltine, C. W. (1978), *Process Biochem.* **13** (9), 2.

Development Planning and Research Associates Inc. (1980), *Small Scale Ethanol Production for Fuels,* P. O. Box 727, Manhattan, KS, 66502.

Drews, S. M. (1975), Sonderheft *Ber. Landwirtsch.* **192**, 599.

Dunne, W. (1994), *Int. J. Energy Res.* **18**, 71.

Egamberdiev, N. B., Ierusalimskii, N. D. (1967), *Microbiologiya* **37**, 689.

Eihe, E. P. (1976), *Izv. Akad. Nauk Latv. SSR* **3** (344), 77.

Eriksson, K. E. (1969), *Adv. Chem. Ser.* **95**, 58.

Fayed, M. E., Chao, C. C., Anderson, N. E. (1981), *Proc. 2nd World Congr. Chem. Eng.,* Oct., Montreal, Canada, pp. 4–9.

Forage, A. J., Righelato, R. C. (1979), in: *Microbial Biomass* (Rose, A. H., Ed.), p. 289. New York: Academic Press.

Fouth, G. L., Magruder, G. L., Gaddy, J. L. (1980), *Agricultural Energy,* Vol. 1 *Biomass Energy Crop Production,* p. 299. St. Joseph, MI: American Society of Agricultural Engineers.

Furey, R. L., Jackson, M. W. (1977), Exhaust and evaporative emisions from a Brazilian chevrolet fuelled with ethanol–gasoline blends. *12th Intersociety Energy Converison Engineering Conf.,* Aug. 28–Sept. 2, Washington, DC, 8 pp.

Furey, R. L., King, J. B. (1980), Evaporative and exhaust emissions from cars fuelled with gasoline containing ethanol or methyl tert-butyl ether, *Congress and Exposition Cobo Hall,* Feb. 25–9, Detroit, MI, 17 pp.

Gauss, W. F., Suzuki, S., Takagi, M. (1976), *U.S. Patent* 3990944.

General Motors Research Laboratories (1980), GASOHOL – Its production, economics and potential, *Search* **15**, No. 1, 6 pp.

Ghose, T. K., Bandyopadhyay, K. K. (1980), *Biotechnol. Bioeng.* **22**, 1489.

Ghose, T. K., Ghosh, P. (1978), *J. Appl. Chem. Biotechnol.* **28**, 309.

Ghose, T. K., Tyagi, R. D. (1979), *Biotechnol. Bioeng.* **21**, 1387.

Gird, J. W. (1980), *On-Farm Production and Utilization of Ethanol Fuel.* University of Maryland, Agricultural Engineering Dept., publ. FACTS #126, 39 pp.

Gonzalez, I. M., Murphy, N. F. (1980), *J. Agric. Univ. P. R.* **64**, 138.

Gratzali, M., Brown, R. D. (1979), *Adv. Chem. Ser.* **181**, 237.

Grethlein, H. E. (1978), *J. Appl. Chem. Biotechnol.* **28**, 296.

Grote, W., Lee, K. J., Rogers, P. L. (1980), *Biotechnol. Lett.* **2**, 481.

Haigwood, B. (1991), *16th IGT Conf. Energy from Biomass and Wastes,* Washington, DC, 767.

Halwachs, W., Wandrey, C., Schügerl, K. (1978), *Biotechnol. Bioeng.* **20**, 541.

Hansen, R. (1980), *Nordauropeish Mejeri-Tiolsskrift* Nr. 1–2, 10.

Hayes, R. D. (1981), *Proc. 3rd Bioenergy R and D Seminar,* March 24–25, Ottawa, Canada.

Hokanson, A. E., Katzen, R. (1978), *Chem. Eng. Progr.* **74** (1), 67.

Holdren, J. P., Ehrlich, P. R. (1974), *Am. Sci.* **62**, 282.

Holzberg, I., Finn, R. K., Steinkraus, K. H. (1967), *Biotechnol. Bioeng.* **9**, 413.

Hoppe, G. K. (1981), as cited in Hoppe and Hansford (1982).

Hoppe, G. K., Hansford, G. S. (1982), *Biotechnol. Lett.* **4**, 39.

Ingram, L. O. (1990), Genetic modification of *Escherichia coli* for ethanol production, *Energy from Biomass & Wastes XIV.* Chicago, IL.: Institute of Gas Technology.

Ingram, L. O. (1993), *Symp. Bioremediation and Bioprocessing,* ACS, Denver, CO, 291.

Ingram, L. O., Alterthum, F., Ohta, K., Beall, D. S. (1989), in: *Developments in Industrial Microbiology,* Vol. **32,** New York: Elsevier Science Publ.

Inter Group Consulting Economists Ltd. (1978), A report prepared for the Government of Canada, Interdepartmental Steering Committee on Canadian Renewable Liquid Fuels, Ottawa, Canada.

Jackman, E. A. (1977), *Chem. Eng.* April, 230.

Jackson, E. A. (1976), *Process Biochem.* **11** (5), 29.

Jones, R. P., Pamment, N., Greenfield, P. F. (1981), *Process Biochem.* **6** (4), 41.

Kelkar, A. D., Hooks, L. E., Knofzynski, C. (1989), *Proc. 23rd Intersc. Conversion Engin. Conf.,* Denver, CO, 381.

Kelm, C. R. (1980), *Ind. Eng. Chem. Prod. Res. Dev.* **19**, 483.

Kelsey, R. G., Shafizaden, F. (1980), *Biotechnol. Bioeng.* **22**, 1025.

Kerr, R. W. (1944), *Chemistry and Industry of Starch: Starch Sugars and Related Compounds.* New York: Academic Press.

Kierstan, M. (1980), *Process Biochem.* **15** (4), 2, 4, 32.

KIM, M. H., LEE, S. B., RYU, D. D. Y., REESE, E. T. (1982), *Enzyme Microbiol. Technol.* **4**, 99.

KIRIK, M. (1981), Utilization of ethanol as a motor fuel. *Int. Conf. Agric. Engin. and Agro-Industries in Asia,* Nov. 10–13, 15 pp.

KOENIGS, J. W. (1975), *Biotechnol. Bioeng. Symp.* **5**, 151.

KOIVURINTA, J., JUNNILA, M., KOIVISTONEN, P. (1980), *Lebensm. Wiss. Technol.* **13**, 118.

KOSARIC, S. (1984), An experimental study to evaluate engine performance using various gasoline–ethanol and wet ethanol fuel blends. *Master of Engineering Thesis,* University of Western Ontario, London, Canada.

KOSARIC, N., WIECZOREK, A. (1982), *Paper* presented at the Spring National Meeting of the AIChE, Orlando, FL, February 28–March 3.

KOSARIC, N., NG, D. C., RUSSELL, I., STEWART, G. S. (1980), *Adv. Appl. Microbiol.* **26**, 147.

KOSARIC, N., WIECZOREK, A., DUVNJAK, L., KLIZA, S. (1982), *Bioenergy R and D Seminar,* March 29–31, Winnipeg, Canada.

LADISCH, M. R., LADISCH, C. M., TSAO, G. T. (1978), *Science* **201**, 743.

LAWFORD, H. G., ROUSSEAU, J. D. (1991), *16th IGT Conf. Energy from Biomass and Wastes,* Washington, DC, 583.

LAWRENCE, R. (1979), Emissions from GASOHOL fueled vehicles. *Proc. 3rd Int. Symp. Alcohol Fuels Technol.,* 10 pp., May 29–31, Asilomar, CA.

LEE, K. J., SKOTNICKI, M. L., TRIBE, D. E., ROGERS, P. L. (1981), *Biotechnol. Lett.* **3** (5), 207.

LEEPER, S. A., DAWLEY, L. J., WOLFRAM, J. H., BERGLUND, G. R., RICHARDSON, J. G., MCATEE, R. E. (1982), *Design, Construction, Operation and Costs of a Modern Small-Scale Fuel Alcohol Plant,* 16 pp., Idaho Falls, ID: EG & G Idaho Inc.

LINCOLN, J. W. (1980), *Driving Without Gas,* 150 pp. Charlotte, VT: Garden Way Publishing.

LINDEMAN, L. R., ROCCHICCIOLI, C. (1979), *Biotechnol. Bioeng.* **21**, 1107.

LINKO, P., LINKO, Y. Y. (1981a), *74th Ann. AIChE Meeting,* Nov. 8–12, New Orleans, LA.

LINKO, Y., LINKO, P. (1981b), *Biotechnol Lett.* **3** (1), 21.

LIPINSKY, E. S., SHEPPARD, W. J., OTIS, J. L., HELPER, E. W., MCCLURE, T. A., SCANTLAND, D. A. (1977), *System Study of Fuels from Sugarcane, Sweet Sorghum, Sugar Beets and Corn,* Vol. 5. A final report prepared for the ERDA, Battelle, Columbus, OH.

MAIORELLA, B., WILKE, C. R., BLANCH, H. W. (1981), *Adv. Biochem. Eng.* **20**, 43.

MANDELS, M., HONTZ, L., HYSTROM, J. (1974), *Biotechnol. Bioeng.* **16**, 1471.

MARGARITIS, A., BAJPAI, P., WALLACE, B. (1981), *Biotechnol. Lett.* **3** (11), 613.

MARSH, L. S., CUNDIFF, J. S. (1991), *16th IGT Conf. Energy from Biomass and Wastes,* Washington, DC, 643.

MARSHALL, J. J., WHELAN, W. J. (1970), *FEBS Lett.* **9**, 85.

MATSUNO, M., NAKANO, Y., KCHI, K. (1979), Alcohol engine emissions – emphasis on unregulated compounds. *Proc. 3rd Int. Symp. Alcohol Fuels Technol.,* May 29–31, 12 pp., Asilomar, CA.

MCATEE, R. E., WOLFRAM, J. H., DAWLEY, L., SCHMITT, R. C. (1982), *Operation and Testing Experience with the Doe Small-Scale Fuel Alcohol Plant.* U.S. Department of Energy, Idaho Operations Office, DOE Contract No. DE-Ac07-76ID01570, 43 pp.

MCCALLUM, P. W., TIMBARIO, T. J., BECHTOLD, R. L., EKLUND, E. E. (1982), Methanol–ethanol: alcohol fuels for highway vehicles, *Chem. Engin. Prog.* **78**, No. 8, 8 pp.

MCCLURE, T. A., ARTHUR, M. F., KRESOVICH, S., SCANTLAND, D. A. (1980), *Proc. IV. Int. Symp. Alcohol Fuels Technol.,* Oct., Sao Paulo, Brazil, p. 123.

MENDELSOHN, H. R., WETTSTEIN, P. (1981), *Chem. Eng.* **88** (12), 62.

MENRAD, H. K. (1979), A motor vehicle power plant for ethanol and methanol operation. *Proc. 3rd Int. Symp. Alcohol Fuels Technol.,* May 29–31, Asilomar, CA, 13 pp.

MENRAD, H. K., LOECK, H. (1979), Results from basic research on alcohol powdered vehicles, *Report No. B-49.* Volkswagenwerk AG, Wolfsburg, Germany, 6 pp.

MILLETT, M. A., MOORE, W. E., SEAMAN, J. F. (1954), *Ind. Eng. Chem.* **46**, 1493.

MIRANOWSKI, J. A. (1981), in: Proceedings of moonshine to motor fuel: *A Workshop on Regulatory Compliance for Fuel Alcohol Production.* March 3–4, Atlanta, GA.

MISSELHORN, K. (1980), According to ECC Report "Working Group Alcohol", *Ethanol as a Fuel, 3rd draft,* February 1981.

MISTRY, P. B. (1991), *16th IGT Conf. Energy from Biomass and Wastes,* Washington, DC, 669.

Mueller Associates Inc. (1981), *Vehicle Modification for Alcohol Use.* Completed under contract to U.S. Department of Energy, No. De-AC05-79CS56051, Baltimore, MD, 33 pp.

NATHAN, R. A. (1978), *Fuels from Sugar Crops.* USDOE Technical Information Centre, Oak Ridge, TN.

NEALE, A. D., SCOPES, R. K., KELLY, J. M. (1988), *Appl. Microbiol. Biotechnol.* **29**, 162.

NEILSON, M. J., KELSEY, R. G., SHAFTZADEH, F. (1982), *Biotechnol. Bioeng.* **24**, 193.

NOVAK, M., STREHAIANO, P., MORENO, M., GOMA, G. (1981), *Biotechnol. Bioeng.* **23**, 201.

Novo Industri (1979), Novo Report by BENT VABO (9. 11. 79).

OHTA, K., ALTERTHUM, F., INGRAM, L. O. (1990), *Appl. Environ. Microbiol.* **56**, 463.

O'LEARY, V. S., GREEN, R., SULLIVAN, B. C., HOLSINGER, V. H. (1977), *Biotechnol. Bioeng.* **19**, 1019.

OSHIMA, M. (1965), *Wood Chemistry Process Engineering Aspects.* Noyes Dev. Corp., New York.

OURA, E. (1977), *Process Biochem.* **12** (3), 19.

PFENNINGER, H. (1963), *Z. Lebensm. Unters. Forsch.* **120**, 100.

PIRONTI, F. F. (1971), Kinetics of alcohol fermentation. *Ph.D. Thesis*, Cornell University, Ithaca, New York.

PISCHINGER, F. F., PINTO, N. L. M. (1979), Experience with the utilization of ethanol/gasoline and pure ethanol in Brazilian passenger cars. *Proc. 3rd Int. Symp. Alcohol Fuels Technol.,* May 29–31, Asilomar, CA, 7 pp.

POIRAT, E. (1994), *Int. J. Solar Energy* **15**, 163.

PROUTY, J. L. (1980), *Proc. IV. Ing. Symp. Alcohol Fuels Technol.,* Sao Paulo, Brazil, Oct. 57.

RAJAN, S. (1979), Factors influencing cold starting of engines operating on alcohol fuel. *Proc. 3rd Int. Symp. Alcohol Fuels Technol.,* May 29–32, Asilomar, CA, 12 pp.

REESE, E. T. (1975), *Biotechnol. Bioeng. Symp.* **5**, 77.

REESE, E. T. (1977), *Recent Adv. Phytochem.* **11**, 311.

REESE, E. T., ROBBINS, F. J. (1981), *J. Colloid Interface Sci.* **83**, 393.

REESE, E. T., SIN, R. G. H., LEVINSON, H. S. (1950), *J. Bacteriol.* **59**, 485.

Renewable Energy News (1983), *Brazilian Ethanol. A Success.* September, Vol. 6, No. 6, Sect. 2, 20 pp.

ROGERS, P. L., LEE, K. J., TRIBE, D. E. (1980), *Process Biochem.* **15** (6), 7.

ROLZ, C. (1978), *Desarrolo de technologia para producir etanol para carburante a partir directamene de la cana de azucar. II. El proceso Exferm,* Informe Tecnico ICAITI, Guatemala, p. 78.

ROLZ, C. (1979), *U.S. Patent Appl.* 021244.

ROLZ, C. (1981), *Enzyme Microbiol. Technol.* **3**, 19.

ROSE, D. (1976), *Process Biochem.* **11** (3), 10, 36.

ROSEN, K. (1978), *Process Biochem.* **13** (5), 25.

ROSENBERG, S. L. (1980), *Enzyme Microbiol. Technol.* **2**, 185.

ROTHMAN, H. (1983), *Energy from Alcohol, The Brazilian Experience,* 8 pp. Lexington, KY: The University of Kentucky Press.

SEAMAN, J. F. (1945), *Ind. Eng. Chem. Anal. Ed.* **37**, 43.

SANCHEZ-RIERA, F., VALZ-GIANINENT, S., CALLIERI, D., SINERIZ, F. (1982), *Biotechnol. Lett.* **4**, 127.

SANG, S. L., WANG, L. H., HWANG, P. T., CHUANG, Y. T., KUO, Y. C., CHANG, C. Y. (1980), *Proc. IV. Int. Symp. Alcohol Fuels Technol.,* Oct., Sao Paulo, Brazil.

SCALLAN, A. (1971), *Text. Res. J.* **41**, 647.

SCHELLER, W. A. (1976), Fermentation in cereal processing, *61st Natl. Meet. Am. Ass. Cereal Chem.,* Oct., New Orleans, LA.

SCHELLER, W. A. (1977), *Int. Symp. Alcohol Fuel Technol. – Methanol/Ethanol,* Nov. 21–23, Wolfsburg, Germany.

SCHELLER, W. A., MOHR, B. J. (1976), *Am. Chem. Soc. Div. Fuel Chem. Prep.* **21** (2), 28.

SCHRUBEN, L. W. (1980), Presented at the *Kansas Agriculture Experiment Station Conference,* Manhattan, KS, Jan. 8.

SHREVE, R. N., BRINK, J. (Eds.) (1977), *Chemical Process Industries,* 4th Edn., p. 558. New York: McGraw-Hill.

SITTON, C., FOUTCH, G. C., BOOK, N., GADDY, J. L. (1979), *Process Biochem.* **14** (9), 7.

SLESSER, M., LEWIS, C. (1979), *Biological Energy Resources.* E & FN Spon Ltd., London.

SPANO, L. A. (1976), *Symp. Clean Fuels Biomass, Sewage, Urban Refuse, Agric. Wastes,* January, Orlando, FL, p. 325.

SPANO, L., TASSINAARI, T., RYU, D. D. Y., ALLEN, A., MANDELS, M. (1979), *29th Can. Chem. Eng. Conf.,* Oct., Sarnia, Canada.

SPNDLER, D. D., WYMAN, C., GROHMANN, K. (1991), *16th IGT Conf. Energy from Biomass and Wastes,* Washington, DC, 623.

STAUFFER, M. D., CHUBEY, B. B., DORRELL, D. G. (1975), Canadex field crops. *Agriculture Canada Report,* p. 164.

STREHAIANO, P., MORENO, M., GOMA, G. (1978), *C. R. Acad. Sci. Ser.* D **286**, 225.

SWINGS, J., DELEY, J. (1977), *Bacteriol. Rev.* **41** (2), 1.

TAKAGI, M., ABE, S., SUZUKI, S., EMERT, G. H., GROHMAN, K. (1977), *Proc. Bioconversion Symp.,* IIT, Delhi, 551.

TERZONI, G. (1981), *Water Tolerability of Ethanol-Gasoline Blends,* Ass. Sci. Res. ENI Gp. co., S. Donato Milanese, Italy, Publ. No. B-17, 5 pp.

THOMAS, D. S., ROSE, A. H. (1979), *Arch. Microbiol.* **122**, 49.

TOLAN, J. S., FINN, R. K. (1987a), *Appl. Environ. Microbiol.* **53**, 2033.

TOLAN, J. S., FINN, R. K. (1987b), *Appl. Environ. Microbiol.* **53**, 2039.

Toyota Motor Corporation (1983), *Instruction Manual for the E100 Starlet.* Tokyo, Japan, 7 pp.

TSAO, G. T. (1978a), *Process Biochem.* **13** (10), 12.

TYAGI, R. D., GHOSE, T. K. (1982), *Biotechnol. Bioeng.* **24**, 781.

U.S. Department of Energy (1980), Fuel from Farms, No. SERI/SP-451-519.

Volkswagen of America Inc. (1981), *Report on the Rabbit Concept Vehicle for Neat Ethanol Operation.* Report completed by: Alcohol Energy Systems Inc., 1050-H East Duane Ave., Sunnyvile, CA, 39 pp.

Volkswagenwerk AG (1981), Development of a fleet concept for methanol and ethanol fuel with lambda-sensor-controlled continuous injection system, *Report No. FET 8175 N/4,* Stuttgart, 49 pp.

WADA, M., KATO, J., CHIBATA, I. (1979), *Eur. J. Appl. Microbiol. Biotechnol.* **8**, 241.

WAGNER, T. D. (1980), Practically of alcohols as motor fuel, Society of Automotive Engineers, *Progress in Technology Series* **19**, 17 pp.

WASH, T., BUNGAY, H. (1979), *Biotechnol. Bioeng.* **21**, 1081.

WAYMAN, M. (1980), *Proc. IV. Int. Symp. Alcohol Fuels Technol.,* Oct., Sao Paulo, Brazil.

WAYMAN, M., LORA, J. H., GULBINAS, E. (1979), *J. Am. Chem. Soc.* **90**, 183.

WEEKS, M. G., MUNRO, P. A., SPEDDING, P. L. (1982), *Biotechnol. Lett.* **4**, 85.

WETTSTEIN, P., DEVOS, J. (1980), *Proc. IV. Int. Symp. Alcohol Fuels Technol.,* Oct. 31, Sao Paulo, Brazil.

WHITE, A. R., BROWN, R. M., JR. (1981), *Proc. Natl. Acad. Sci. U.S.A.* **78**, 1047.

WHITTAKER, R. H. (1970), Communities and Ecosystems. New York: Macmillan.

WIEGEL, J. (1980), *Experientia* **36**, 1434.

WIEGEL, J., LJUNGDAHL, L. G. (1981), *Arch. Mikrobiol.* **128**, 343.

WILEK, C. R., MITRA, G. (1975), *Biotechnol. Bioeng. Symp.* **5**, 253.

WILLIAMS, D., MUNNECKE, D. (1981), *Biotechnol. Bioeng.* **23**, 1813.

WOOD, T. M. (1980), *6th Int. Ferment. Symp.,* London, Canada, p. 87.

WYVILL, J. C., BATTAGLIA, G. M. (1981), Proceedings of moonshine to motor fuel; *A Workshop on Regulatory Compliance of Fuel Alcohol Production,* Atlanta, GA, March, p. 192.

YAROVENKO, V. L. (1978), *Adv. Biochem. Eng.* **9**, 1.

YU, J., MILLER, S. F. (1980), *Ind. Eng. Chem. Prod. Res. Dev.* **19**, 137.

# 5 Microbial Production of Glycerol and Other Polyols

HANS-JÜRGEN REHM

Münster, Federal Republic of Germany

# 1 Glycerol

## 1.1 Introduction and History

Glycerol (1,2,3-propane triol) is the simplest alcohol with two primary and one secondary alcohol groups. The two primary alcohol groups are more reactive than the secondary alcohol group.

Glycerol was discovered in 1779 by SCHEELE when saponifying olive oil. In 1813, CHEVREUL demonstrated that fats are glycerol esters of fatty acids. He called the compound glycerol, derived from Greek "*glykys*" (sweet).

PASTEUR already observed that glycerol always is a by-product of the alcoholic fermentation, in concentrations of 2.5–3.6% of the ethanol content (PASTEUR, 1858, 1860).

In 1872, FRIEDEL was the first to find a chemical synthesis of glycerol. Most chemical processes for glycerol production use propylene from which the intermediates allylchloride, acrolein, or propylene oxide are obtained, and finally glycerol is synthesized. Fig. 1 shows the synthetic pathway.

During synthesis, large amounts of chlorinated by-products are produced, that cause some environmental problems. For this reason, the biotechnological production of glycerol has been discussed. Recently, a direct synthesis of glycerol from $CO_2$ and $H_2$ has become available, but this process is not of practical importance.

Traditionally, glycerol was a by-product of fat hydrolysis in the soap industry. Today this process is of minor importance in industrial nations (RÖMPP, 1981, and literature cited therein). There are some biotechnological processes for the production of glycerol in addition to chemical synthetic processes.

Before the discovery of the biochemistry of glycerol formation, the first processes for biotechnological production of glycerol (especially sulfite processes) had been developed and used in Germany (CONNSTEIN and LÜDECKE, 1915/16, 1919). Some time later, COCKING and LILLY (1922) in England and EOFF et al. (1919) in the United States obtained patents for glycerol production with yeasts. They recommended production under alkaline conditions.

A process including yeast recovery from the fermented mash by centrifugation during fermentation as well as the recovery of sulfur dioxide was described by HARRIS and HAJNY (1960).

NICKERSON and CARROL (1945) found that a sugar tolerant yeast, *Zygosaccharomyces,* was able to convert 22% of the metabolized sugar into glycerol in the absence of

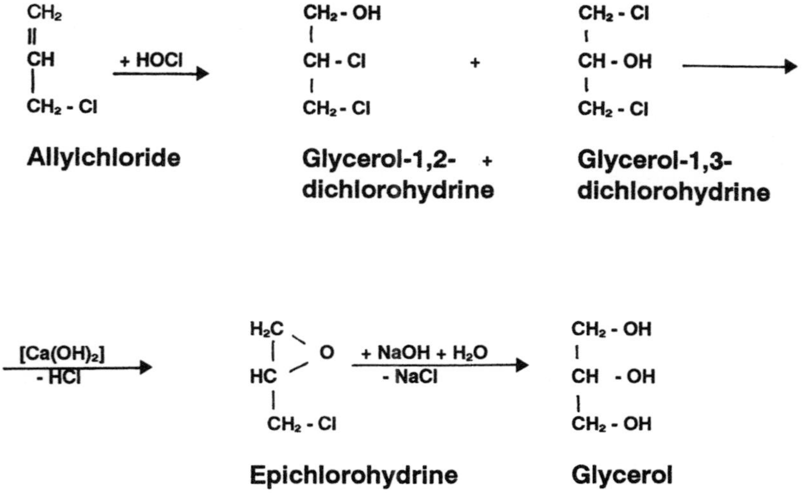

$CH_2{=}CH{-}CH_2{-}Cl$ (Allylchloride) $\xrightarrow{+\,HOCl}$ $CH_2OH{-}CHCl{-}CH_2Cl$ (Glycerol-1,2-dichlorohydrine) + $CH_2Cl{-}CHOH{-}CH_2Cl$ (Glycerol-1,3-dichlorohydrine) $\xrightarrow[-\,HCl]{[Ca(OH)_2]}$ $H_2C{-}O{-}CH{-}CH_2{-}Cl$ (Epichlorohydrine) $\xrightarrow[-\,NaCl]{+\,NaOH\,+\,H_2O}$ $CH_2OH{-}CHOH{-}CH_2OH$ (Glycerol)

**Fig. 1.** Synthesis of glycerol from propyleneallylchloride.

bisulfite, alkali, and other steering agents. This became the basis not only for the production of glycerol but also for the production of other polyhydroxy alcohols (SPENCER and SALLANS, 1956; SPENCER et al. 1957; BISPING et al., 1990a; HÖÖTMANN et al. 1991).

Recently, glycerol production with immobilized microorganisms has been described (BISPING and REHM, 1982, 1986; BISPING et al., 1989; HECKER et al., 1990), and this may have a chance to lead to a commercially viable process. Glycerol production with the green alga *Dunaliella* may also be of interest for new processes (see FRANK and WEGMANN, 1974, and especially BEN-AMOTZ and AVRON, 1981).

Application of a light vacuum during the fermentation process is also a new idea for getting higher glycerol concentrations in the mash (VIRKAR and PANESAR, 1987).

For reviews on the production of glycerol see REHM (1967, 1980), VIJAIKISHORE and KARANTH (1988a), REHM (1988), DELLWEG et al. (1992); on polyol metabolism see SPENCER and SPENCER (1978) and JENNINGS (1984); on glycerol metabolism see LIN (1976, 1977).

## 1.2 Microorganisms

Most strains of *Saccharomyces cerevisiae* produce a few percent of glycerol during alcoholic fermentation. For all processes using sulfite strains of *S. cerevisiae* have been used.

For processes with osmotolerant yeasts *Zygosaccharomyces acidifaciens* (NICKERSON and CARROL, 1945), *Saccharomyces rouxii, S. mellis, Pichia etchelsii, P. farinosa,* and other *Pichia* species, species of *Hansenula, Debaryomyces, Candida, Torula,* and others are suitable organisms (SPENCER and SPENCER, 1978; REHM, 1980, 1988). For the physiology of osmotolerant yeasts and fungi, see REED and NAGODAWITHANA (1991) and BLOMBERG and ADLER (1992).

Many of the osmotolerant species produce other polyhydroxy alcohols in addition to glycerol, e.g., arabitol and erythritol (ONISHI, 1960a). Polyols are also produced by many other fungi (JENNINGS, 1984; OMAR et al., 1992).

Under suitable conditions some bacteria also produce glycerol. In Chapter 7 of this volume the fermentation of glycerol during 2,3-butanediol fermentation is described.

*Bacillus subtilis* (Ford strain) has been considered for practical production of glycerol (NEISH et al., 1945). *Lactobacillus lycopersici* has also been suggested for glycerol production (NELSON and WERKMAN, 1935).

Glycerol production with a halotolerant marine alga, *Dunaliella tertiolecta,* has been described by CRAIGIE and MCLACHLAN (1964) and by BEN-AMOTZ et al. (1982). The genus *Dunaliella* of the order of Volvocales includes a variety of well-defined species of unicellular green algae. The cells are ovoid in shape, 4–10 μm wide and 6–15 μm long. The cells are motile due to the presence of two equally long flagella at each cell. The chloroplast occupies about half of the cell volume.

*Dunaliella* demonstrates a remarkable degree of environmental adaptation to NaCl, from seawater concentrations (0.1 mol/L NaCl) to saturated salt solutions (>5 mol/L NaCl). Under optimal growth conditions the alga has a doubling time of about 5 h that is prolonged to approximately 3 days if the concentration of NaCl is increased to saturation (BEN-AMOTZ and AVRON, 1981).

## 1.3 Biochemistry and Regulation

Glycerol is formed via the fructose bisphosphate pathway. NEUBERG and REINFURTH (1918a, b) added sulfite to fermenting yeasts and found that acetaldehyde binds to sulfite forming a complex that cannot react with $NADH_2$. This $NADH_2$ reacts with dihydroxyacetone phosphate yielding glycerol. The regulation is incomplete, as ethanol is still formed with a lower yield. This fermentation pathway is called "2. Neubergsche Gärung" (Neuberg's second type of fermentation) (Fig. 2). RADLER and SCHÜTZ (1982) observed great differences in the activity of glycerol-3-phosphate dehydrogenase and only small variations in the activity of the alcohol dehydrogenase in *Saccharomyces cerevisiae* strains,

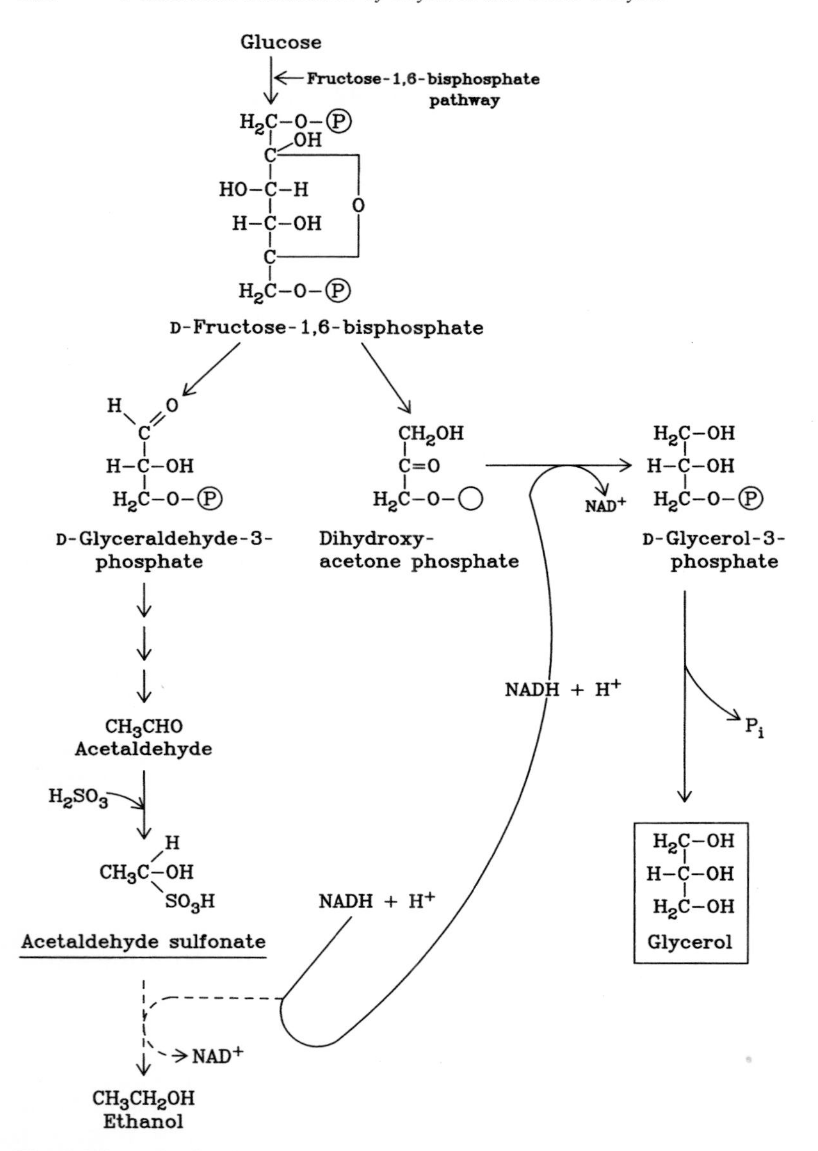


**Fig. 2.** Biosynthesis of glycerol in the presence of sulfurous acid (Neuberg's second type of fermentation) (from REHM, 1980, with permission).

which formed high and low amounts of glycerol.

NEUBERG and HIRSCH (1919) also observed another glycerol forming pathway, "3. Neubergsche Gärung " (Neuberg's third type of fermentation) (Fig. 3). Under the influence of alkaline salts, e.g., carbonates, bicarbonates, and phosphates, 2 mol glycerol, 1 mol acetic acid, 1 mol ethanol, and 2 mol $CO_2$ are produced from 2 mol glucose. The authors proposed the occurrence of a Cannizzaro reaction in which one molecule of acetaldehyde is oxidized to acetate, while a second molecule simultaneously is reduced to etha-

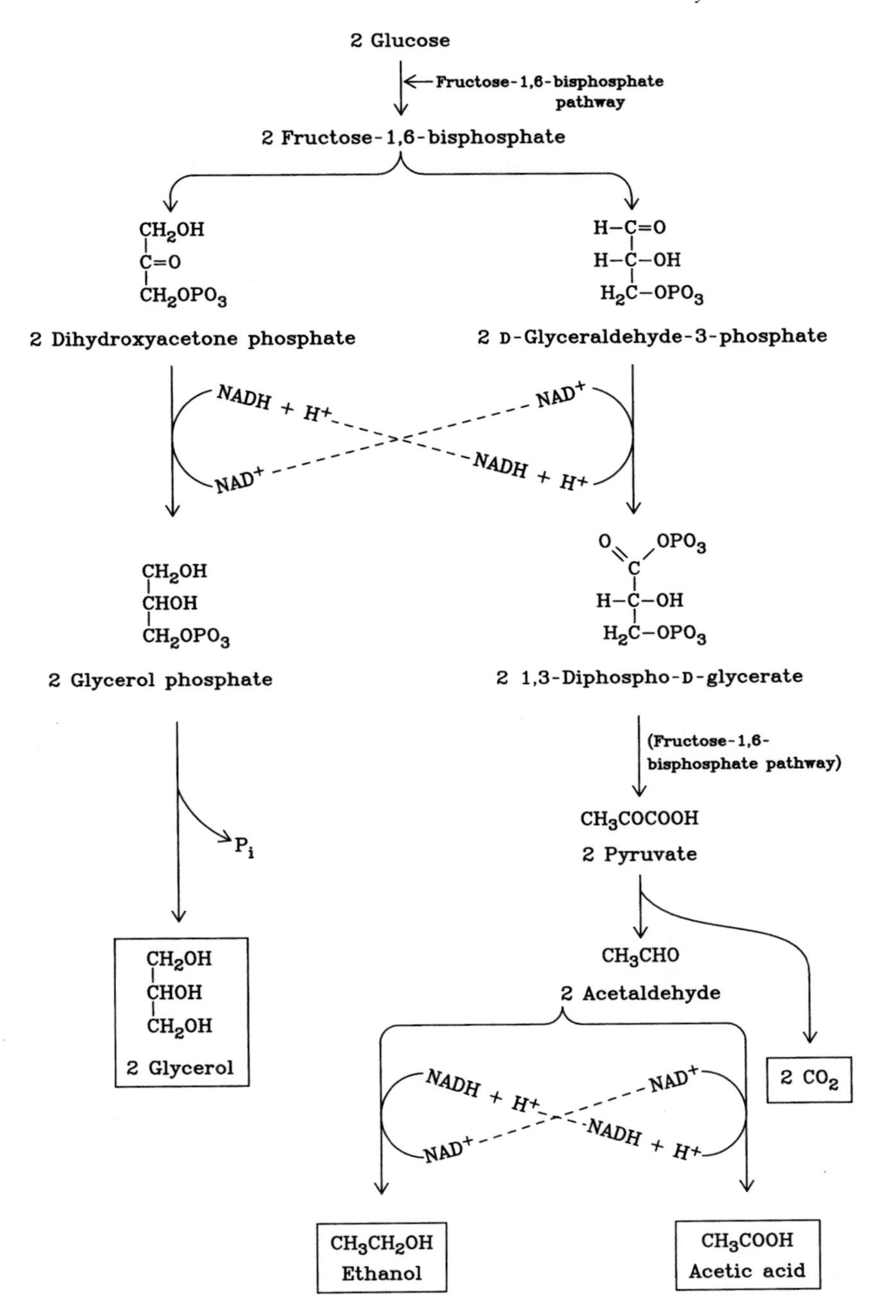


**Fig. 3.** Formation of glycerol under alkaline conditions (Neuberg's third type of fermentation) (from REHM, 1980, with permission).

nol. This reaction does not occur in the absence of living yeast cells. It represents an alternative to the normal metabolism induced by alkalis (FREEMAN and DONALD, 1957).

BLACK (1951) isolated an NAD-dependent acetaldehyde dehydrogenase with a high affinity to acetaldehyde and a pH-optimum of 8.75. This enzyme causes increased production of acetic acid with $NAD^+$ as hydrogen acceptor originating from the production of ethanol. Now there is not enough $NAD^+$ available for the oxidation of D-glyceraldehyde-3-phosphate by glyceraldehyde-3-phosphate dehydrogenase. The $NAD^+$ needed is produced by hydrogenation of dihydroxyacetone phosphate to D-glycerol-3-phosphate.

SPENCER et al. (1956) observed that the metabolic pathway to glycerol in osmotolerant yeasts producing other polyhydroxy alcohols is the same as in *Saccharomyces cerevisiae.* Glycerol formed from 1-$^{14}$C-glucose is labeled in the $-CH_2OH$ carbon atoms and from 2-$^{14}$C-glucose in the $-CHOH$ carbons (see Fig. 11).

*Dunaliella tertiolecta* has to overcome extreme osmotic differences in its environment. The excretion of glycerol seems to play an important role in the protection of the cells against osmotic changes. The pathway of glycerol formation and its regulation were investigated by FRANK and WEGMANN (1974) and by MUTSCHLER and WEGMANN (1974).

During photosynthesis in the presence of $NaHCO_3$ and in a medium of high salt concentrations *Dunaliella* cells produce large amounts of glycerol. At low NaCl concentrations lactate is produced in larger amounts besides a number of photosynthetic intermediates. A these low NaCl concentrations intermediates of the Calvin cycle, such as glyceraldehyde-3-phosphate or dihydroxyacetone phosphate, can easily leave the chloroplasts and can be metabolized in the cytoplasm to lactate, several amino acids, carbohydrates, and other compounds. With increasing salt concentrations in the cytoplasm the enzymes are more and more inhibited so that only a few reactions take place in the cytoplasm; now glycerol formation in the chloroplasts may be increased.

This theory and the proposed pathway of glycerol formation are shown in Fig. 4 (according to MUTSCHLER and WEGMANN, 1974).

## 1.4 Processes for Glycerol Production

### 1.4.1 Glycerol Production with Addition of Sulfite

In all processes with yeasts often resulting in high yields of glycerol, it is very difficult to recover glycerol from the culture liquids, especially when the liquids contain cheap and unrefined raw materials. These difficulties represent the major problem for all microbial glycerol processes.

#### 1.4.1.1 Glycerol Production with Free Cells of Yeasts

The "Protol process" is the most important microbial process for glycerol production with addition of sulfite. During World War I and II, large amounts of glycerol were produced by this process in Germany. It was developed by CONNSTEIN and LÜDECKE (1915/16) (see also CONNSTEIN and LÜDECKE, 1919).

*Saccharomyces cerevisiae* is grown in a culture liquid containing 10% beet sugar (as molasses), 3% $Na_2SO_3$, and the usual salts. After sterilization the solution is inoculated with 1% baker's compressed yeast. Anaerobic fermentation in vessels of about 1000 $m^3$ is finished after 2–3 days at 30–35 °C. At the end of fermentation the mash contains 3% glycerol, 2% ethanol, and about 1% acetaldehyde. The maximum yield of glycerol in this process is up to about 20% of the metabolized sugar. A large amount of cell mass of *S. cerevisiae* is produced during this process. These yeasts can be reused for inoculation after acidification of the mash. The remaining yeast cells are centrifuged or otherwise dewatered and used as fodder yeast. This process was used in Germany during World War I. 6000 t of beet molasses per month were fermented to yield 2000 t of crude glycerol and 1000 t of "dynamite" glycerol (see BERNHAUER, 1957).

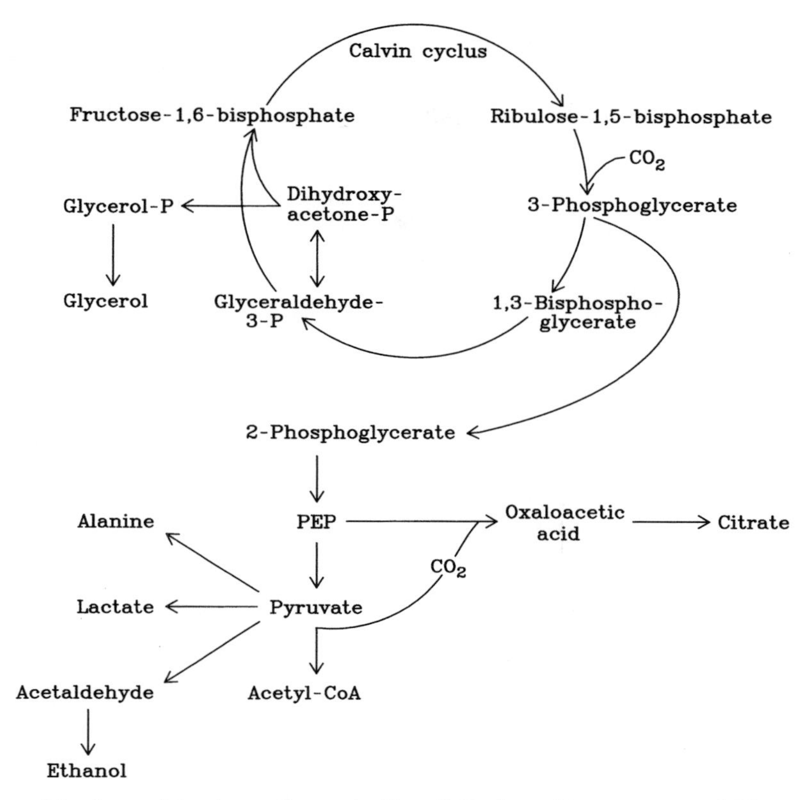


**Fig. 4.** Scheme of the proposed *in vivo* and *in vitro* pathways in *Dunaliella* (from MUTSCHLER and WEGMANN, 1974, with permission).

A similar process using acid-hydrolyzed corn starch with addition of $Na_2SO_3$ yielded 30% glycerol from the metabolized sugar (LEES, 1944). In this process aeration is used to reactivate yeasts injured by the alkaline substrate and the presence of sulfite. The continuous aeration shortens the fermentation time to about 22 h. It is necessary to inoculate large amounts of yeast so that the process can proceed quickly.

In England, COCKING and LILLY (1919, 1922) developed processes at slightly acid pH values by adding a mixture of $Na_2SO_3$ and $NaHSO_3$. This mixture is not as toxic for yeasts as the sole application of $Na_2SO_3$. $NaHSO_3$ is added slowly to the culture liquid during fermentation; 2 mol $NaHSO_3$ and 1 mol $NaHCO_3$ are formed according to the following equation:

$$Na_2SO_3 + NaHSO_3 + CO_2 + H_2O \rightarrow 2\ NaHSO_3 + NaHCO_3$$

Only the 2 mol of $NaHSO_3$ form an addition compound with 2 mol acetaldehyde. The process can be carried out at mildly alkaline conditions by the addition of $Na_2CO_3$. This prolongs fermentation, but increases the yield of glycerol (GRUNTE et al., 1959).

Other patents recommend the addition of sulfites at pH values at which the sulfites are more easily dissolved; e.g., $CaSO_4$ at pH 5.0, $MgSO_4$ at pH 4.7, and $(NH_4)_2SO_4$ at pH 6.5. These processes have not been used in practice yet.

Glycerol may also be produced in a continuous fermentation by *S. cerevisiae* v. *ellipsoideus*. The medium containing 5% sugar is adjusted to pH 6.5–8.0 by the addition of

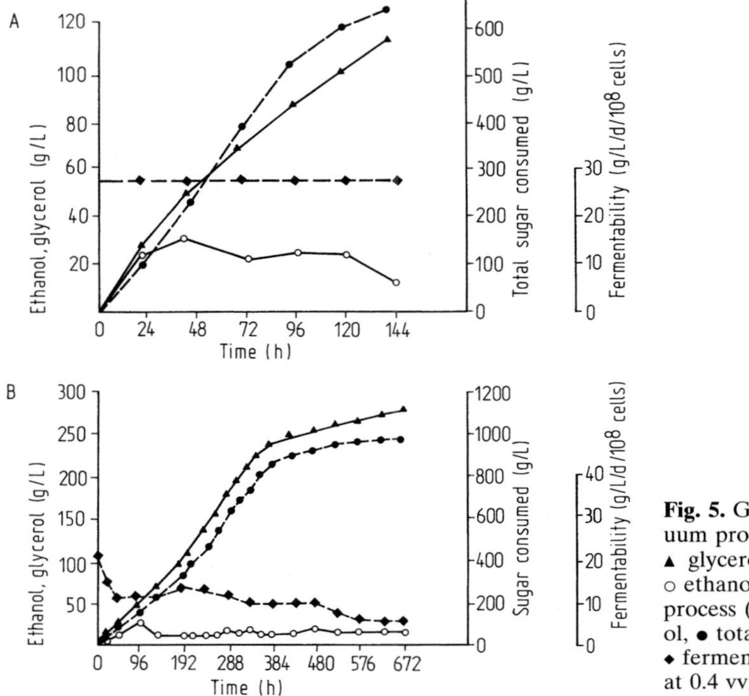


**Fig. 5.** Glycerol production. (A) Vacuum processes (KALLE et al., 1985); ▲ glycerol, ● total sugar consumed, ○ ethanol, ◆ fermentability. (B) $CO_2$-process (KALLE et al., 1985); ▲ glycerol, ● total sugar consumed, ○ ethanol, ◆ fermentability. The $CO_2$ was sparged at 0.4 vvm.

$Na_2CO_3$. The yield of glycerol was 51%, based on the percentage of fermented sugar (SALA and DE AYALA, 1959).

HARRIS and HAJNY (1960) developed a variant of the bisulfite process. The concentration of bisulfite was controlled continuously, and the yeast was recovered by centrifugation and returned to the culture vessel. $SO_2$ was also recovered and reused. In this continuous process 25% glycerol, 17% ethanol, and 11% acetaldehyde were yielded, calculated as a percentage of the metabolized sugar.

The yields of glycerol can be increased when the yeasts are adapted to sulfite (POLGAR et al., 1964). Mutants of *S. cerevisiae*, obtained by UV-irradiation also produced increasing yields of glycerol (up to 250%) (WRIGHT et al., 1957).

In a short note it was described that an aeration of 1.40 vvm yields 92–96 g/L glycerol and 10.7–12.5 g/L ethanol in a sulfite process with cane molasses at a pH of 6.8–7.0 (KALLE and NAIK, 1987).

Improved glycerol production from cane molasses by the sulfite process with vacuum or continuous $CO_2$ sparging during fermentation has been described by KALLE et al. (1985). In this process a glycerol concentration of up to 230 g/L and a productivity of 15 g/L per day were achieved by fermentation under vacuum (80 mm) or with continuous sparging of $CO_2$ (0.4 vvm). Under these conditions the sulfite concentration could be reduced (20 g/L) and the ethanol concentration in the mash was kept below 30 g/L. The process was run for up to 20 days. It was carried out with *S. cerevisiae* and an inoculation of 10% (v/v) in a glass jar fermenter of 2 L working volume. The amount of yeast cells in the mash was about $2.7 \cdot 10^8$ cells/mL. Fig. 5 shows the glycerol production with the vacuum process (A) and the $CO_2$ process (B).

## 1.4.1.2 Glycerol Production with Immobilized Cells of *Saccharomyces cerevisiae*

Experiments with *S. cerevisiae,* immobilized in κ-carrageenan showed that the production of glycerol could be realized by this method. In a fixed-bed reactor a yield of 27.4 g/L glycerol was obtained at 30 °C after a fermentation time of 118 h. The final $Na_2SO_3$ concentration was 55 g/L, the titer of the fully grown cells was only $5 \cdot 10^4$ cells/mL (BISPING and REHM, 1982). At 21 °C a yield of 25.9 g/L was obtained.

κ-Carrageenan is sufficiently stable in the presence of sulfite, but it is too expensive for practical application. When cells of *S. cerevisiae* were entrapped in calcium alginate beads, the stability of alginate against $Na_2SO_3$. $Na_2HSO_3$, a mixture of $Na_2HSO_3$ and MgO, or $(NH_4)_2SO_3$ was not sufficient (BISPING and REHM, 1984a, b). The addition of $MgSO_3$ led to a glycerol yield of 16.3 g/L. Entrapment in polyacrylamide hydrazide showed with *S. cerevisiae* a good stability of the particles in a fixed-bed column reactor. With these particles a glycerol yield of 32 g/L was achieved (BISPING and REHM, 1984b).

Adsorption of *S. cerevisiae* on sintered glass Raschig rings may become a commercially viable process (BISPING and REHM, 1986). Sintered glass with a porosity of 60% and pore dimensions of 60–100 μm has a good adsorption capacity for *S. cerevisiae* (see REHM, 1988, Fig. 5). The sintered glass with immobilized yeasts was used in a fixed-bed loop reactor with a working volume of 8 L (total volume 20 L) (Fig. 6). This reactor type was first described by HEINRICH and REHM (1981). A semi-continuous fermentation led to glycerol yields of 26.2–29.5 g/L (each fermentation lasted about 38 h and could be carried out with non-sterilized substrates) (see REHM, 1988, Fig. 7).

The titers of the full-grown yeasts decreased during the different fermentations to $5 \cdot 10^5 - 6 \cdot 10^5$ cells/mL. This fact is important for recovery of glycerol from the mash.

Further investigations using a vacuum distillation technique and the addition of $CO_2$ gas led, at higher temperatures, to an increased concentration of glycerol in the mash of up to 85 g/L with a yield of 67%, as a percentage of metabolized sugar (HECKER et al., 1990). This highest glycerol concentration was achieved at 30 °C and an average $CO_2$ gassing rate of 0.4 vvm. Fig. 7 shows some results (BISPING et al., 1989).

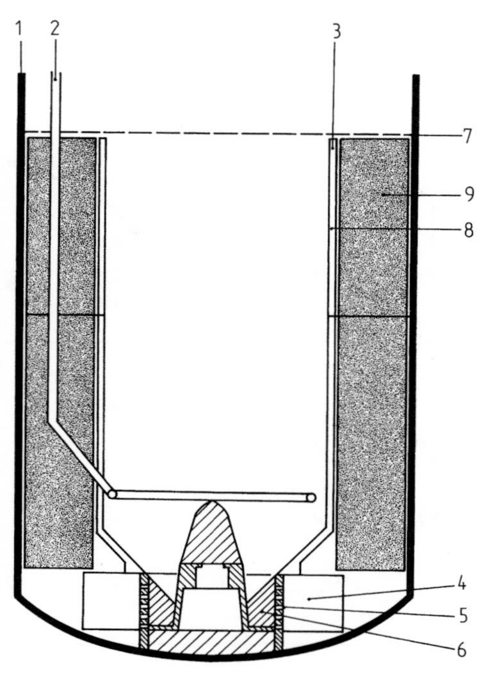

**Fig. 6.** Fixed-bed loop reactor (13 L). (1) Reactor wall, (2) aeration tube, (3) draft tube, (4) guide, (5) emulsion unit, (6) turbine impeller, (7) liquid level at 13 L working volume, (8) glass carrier unit, (9) draft tube (from BISPING and REHM, 1986, with permission).

Fed-batch fermentations with free cells indicated an inhibition mechanism of the glycerol produced, affecting the fermentation capacity of the yeast strain used (BISPING et al., 1989).

Recent investigations have demonstrated that glycerol can be produced in continuous culture with *S. cerevisiae* adsorbed on Raschig glass sinter tubes. At a flow rate of $D = 0.060$ $h^{-1}$, a concentration of 24–25 g/L in the effluent was measured with a volume productivity of 35–37 g/L per day (see Fig. 8) (HECKER et al., 1990). A constant production rate was maintained for nearly 9 months.

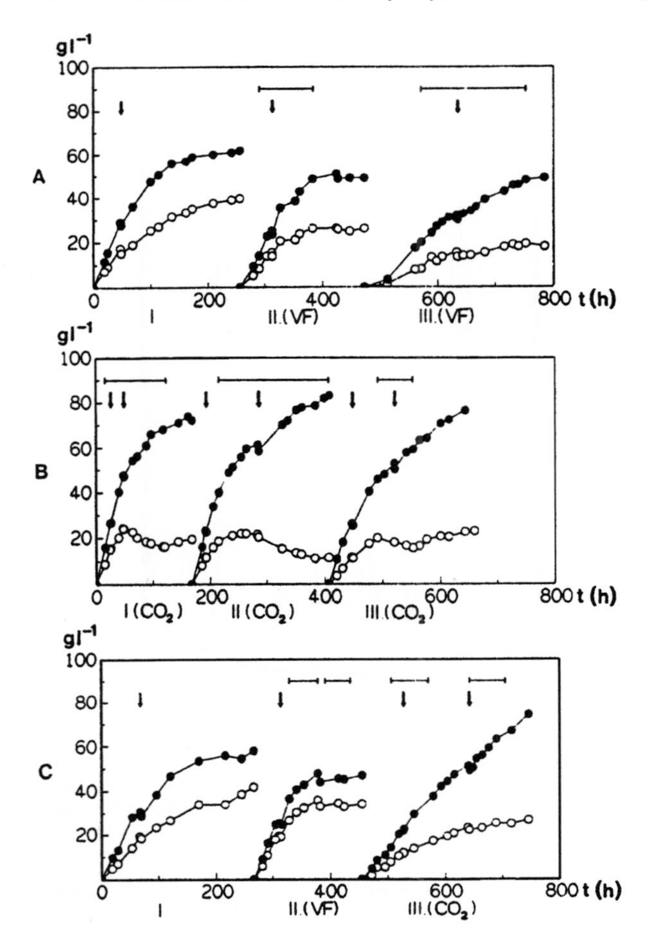


**Fig. 7A–C.** Semicontinuous fed-batch glycerol fermentations with immobilized cells. (A) Volume = 0.8 L, temperature (T) = 24 °C (I, II), 30 °C (III); (B) Volume = 0.8 L, T = 30 °C; (C) Volume = 8 L, T = 24 °C (I, II), 30 °C. • glycerol, ○ ethanol, VF = vacuum fermentation, periods as indicated (⊢⊣); $CO_2$ = $CO_2$ gassing fermentation, periods as indicated (⊢⊣), ↓ = feeding time (100 g/L sucrose and 40 g/L $Na_2SO_3$ (from BISPING et al., 1989, with permission).

Alterations of enzyme activities in immobilized *S. cerevisiae* are described in HILGE-ROTMANN and REHM (1990) and REHM and OMAR (1993).

An interesting method for on-line monitoring of glycerol during industrial fermentation with *S. cerevisiae* has been described by RANK et al. (1995). Using a biosensor with glycerokinase (500 units) immobilized on 80–120 mesh controlled pore glass with a pore size of 50 nm (Serva), amounts of 1 g/L glycerol were monitored.

## 1.4.2 Glycerol Production in the Presence of Alkalis

Besides the various fermentations of hexoses with *S. cerevisiae* in the presence of sulfites and bisulfites, characterized by Neuberg's second type of fermentation, glycerol formation is also stimulated by alkalis and alkaline salts. In this case, fermentation is in accordance with Neuberg's third type of fermentation (see Fig. 3). These processes have not been commercialized at present.

The alkaline fermentation route to glycerol was studied by NEUBERG and FÄRBER (1917), NEUBERG and HIRSCH (1919), and NEUBERG et al. (1920). A very interesting al-

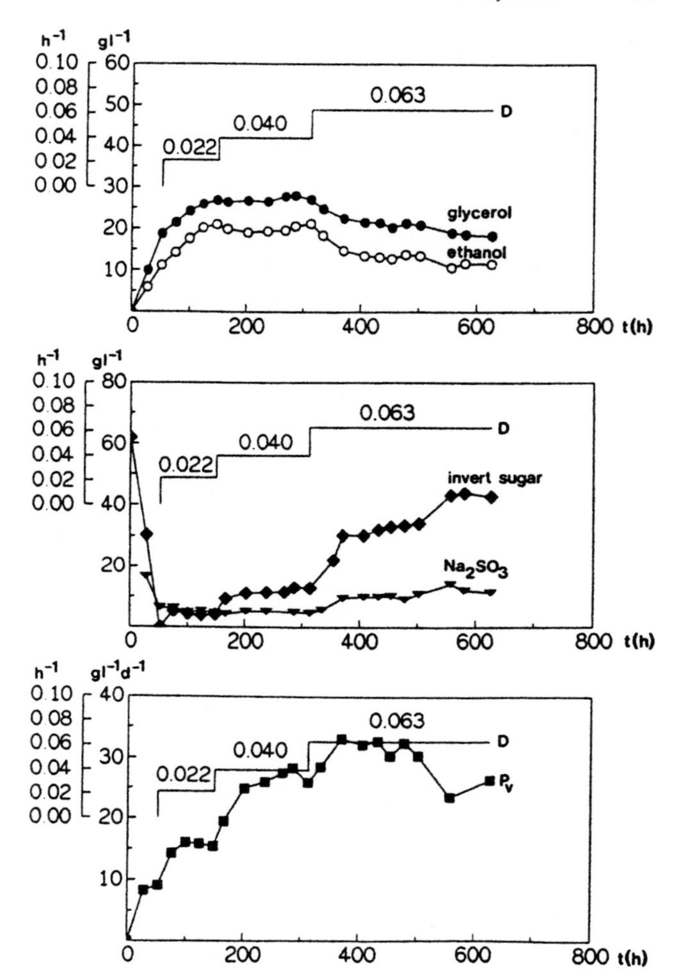


**Fig. 8.** Continuous glycerol production with immobilized cells at various dilution rates (D); the continuous feeding of fresh production medium (100 g/L sucrose and 40 g/L $Na_2SO_3$ was carried out after an initial batch phase; T=24°C; ● glycerol; ○ ethanol; ◆ invert sugar; ▼ $Na_2SO_3$; ■ volumetric productivity of glycerol (Pv) (from HECKER et al., 1990, with permission).

kaline fermentation process is associated with the work of EOFF et al. (1919) in which media based on blackstrap molasses and other sources of sugars were fermented in the presence of sodium carbonate. Blackstrap molasses is fermented by *Saccharomyces ellipsoideus*, and sodium carbonate is added stepwise (up to 5% of the mash) so that the pH of the substrate is always alkaline. The molasses mash can contain up to 20% of sugar. A yield of 20–25% glycerol could be achieved.

For better recovery of glycerol it is possible to distill the alcohol from the mash after fermentation and to ferment the mash once more after adding new molasses. From this second fermentation glycerol can be recovered (US Patent, 1943).

In another process (Canadian Patent 1942; Danish Patent, 1942) yeast is propagated at a slightly acid pH in a molasses medium and with aeration. Glycerol fermentation starts after a good crop of yeast has been grown, and new molasses is added together with $Na_2CO_3$. The process requires large amounts of yeast. 10–50% of yeast (dry wt.) are needed – calculated as a percentage of the sugar concentration of about 20%.

These alkaline fermentations are improved by evaporating the volatile metabolic products during fermentation. This can be done by aerating the mash. The volatiles are recovered in scrubbing towers and cleaned. Glycerol remains in the mash and can be recovered more easily in the absence of volatile components (US Patent, 1943).

FREEMAN and DONALD (1957) screened 41 strains of yeast and determined the glycerol yields. They varied from 10.5 to 24.2% with corresponding ethanol yields of 24.2 to 32%, both based on the percentage of fermentable hexoses. 1,2-Propanediol and 1,3-propanediol are by-products of the alkaline fermentation. They are probably produced by reduction of glycerol. Both metabolites are formed subsequently to the main glycerol fermentation.

### 1.4.3 Glycerol Production with Osmotolerant Yeasts

Many attempts have been made to produce glycerol with osmotolerant yeasts. With these yeasts high yields of glycerol and other polyols can be achieved in the absence of sulfites or alkaline agents (NICKERSON and CARROL, 1945; BLOMBERG and ADLER, 1992).

NICKERSON and CARROL worked with *Zygosaccharomyces acidifaciens (Saccharomyces acidifaciens)* that produced 20% glycerol, based on the weight of fermented sugar.

The natural sources of osmophilic yeasts are fermented honey, brood comb pollen, flower parts, and other sources. Most of the osmophilic yeasts belong to the species *Saccharomyces rouxii* and *S. mellis* (SPENCER and SPENCER, 1978). Osmophilic yeasts tolerate a high sodium chloride concentration. This minimizes contamination and permits fermentation under non-sterile conditions. The yeasts are able to convert as much as 60% of the sugars into polyols (SPENCER and SALLANS, 1956). The conversion rates with osmophilic yeasts are extremely slow. For an economically feasible process the yeasts should selectively produce glycerol. Mutants with a very low activity of alcohol dehydrogenase and with a very low respiratory activity should be developed for effective glycerol production.

The fermentation with osmophilic yeasts has been carried out at sugar concentrations of 20–40%. Yeast extract, urea, $(NH_4)_2SO_4$, ammonium tartrate, or corn steep liquor should be added as nitrogen source. Fast growing strains produce a lot of arabitol and only small amounts of glycerol. Strains with slow but continuous growth produce glycerol and erythritol, and very slowly developing strains produce larger amounts of erythritol (SPENCER et al., 1957). *Pichia miso* and *Debaryomyces mogii* also produce glycerol, besides D-arabitol and erythritol (ONISHI, 1961).

ONISHI (1961) added 18% NaCl to the culture liquid to produce a substrate with a high osmotic pressure. Yeasts of the genus *S. rouxii* could be divided into salt-tolerant and sugar-tolerant yeasts (ONISHI, 1957). VIJAIKISHORE and KARANTH (1984) achieved 17–18% yields of glycerol with osmophilic yeasts. The glucose concentration was 10%, and in six batch fermentations the osmophilic yeasts were harvested by centrifugation and then reused.

A strain of *Torulopsis magnoliae* converted glucose to glycerol with a 30–40% yield in a glucose–yeast extract medium (HAJNY et al., 1960). In a slow fed-batch fermentation the conversion efficiency was approximately 1 mol glycerol produced per mol glucose utilized following the growth phase. The glycerol production phase was extended severalfold by periodic glucose additions. A small amount of phosphate was supplied during the extended fermentation to maintain an active culture (BUTTON et al., 1966). The authors postulated the following conversion of glucose to glycerol:

$$2\ \text{Glucose} + 5\ O_2 \rightarrow 2\ \text{Glycerol} + 6\ CO_2 + 4\ H_2O$$

The final product should be fairly pure and of a reasonably high concentration.

An osmophilic yeast, *Pichia farinosa,* immobilized in calcium alginate or κ-carrageenan, showed an average glycerol production rate of 0.07 g/L per hour in shake flasks. A continuous fermentation in a fluidized bed reactor under steady state operation led to a glycerol concentration of 13.5 g/L in the product stream (VIJAIKISHORE and KARANTH, 1986a, b).

RÖHR et al. (1987) have shown an intermediate accumulation of glycerol, erythritol, arabitol, and mannitol in a total concentration of up to 9 g/L during a citric acid fermentation with *Aspergillus niger* (150 g/L sucrose). In later fermentation stages a partial reconsumption by the mold occurred.

HONECKER et al. (1989) observed polyol formation by *A. niger* depending on the concentration of sucrose. A nitrogen-limited submerged culture of *A. niger* produced 4.9 g/L glycerol, 1.6 g/L erythritol, 0.3 g/L arabitol, but not mannitol in a 240 g/L sucrose medium. The polyol production also depended on the concentration of oxygen. A phosphate-limited culture produced the highest concentrations of polyols.

In further experiments it could be shown, that *A. niger* especially produced glycerol at 40–60% sucrose concentration (free cells) and 30–50% sucrose (Ca-alginate immobilized cells). The yields of glycerol amounted from 1.0 to 1.6 g/L per day in addition to much smaller amounts of arabitol and erythritol (BISPING et al., 1990b).

OMAR et al. (1992) described the production of 1.0 up to 10 g/L polyols after 7 days of batch cultivation of three different strains of *A. niger* with 16 g/L sucrose as substrate. The values of 10 g/L were achieved with Ca-alginate immobilized cells.

*Pichia farinosa,* immobilized in sintered glass Raschig rings, produced 8.1 g/L glycerol per day. The highest concentration achieved in batch culture was 86 g/L; in a fed-batch fermentation with free cells the maximum glycerol concentration of 118 g/L was obtained after 14 days (BISPING et al., 1990a). In semicontinuous fermentations with immobilized cells at an aeration rate of 3 vvm (20% glucose) at the first and second stage and 1.5 vvm (30% glucose) at the third stage, *P. farinosa* produced 7.5, 8.6, and 8.5 g/L glycerol per day.

### 1.4.4 Glycerol Production with Bacteria

In a glucose–peptone–yeast extract medium *Lactobacillus lycopersici* produces approximately 20 kg glycerol, 18 kg ethanol, and 40 kg lactic acid from 100 kg glucose. This process requires three weeks and, therefore, it is not of practical importance (NELSON and WERKAMN, 1935).

Some special strains of *Bacillus subtilis* (e.g., the "Ford strain", *Bacillus licheniformis*) produce 29.5% glycerol, 28.1% 2,3-butylene glycol, 11.6% lactic acid, 2.2% ethanol, formic acid, and $CO_2$ after 8 days in a glucose–yeast extract medium containing carbonate. In this case, the fermentation time also is too long for industrial application (NEISH et al., 1945).

### 1.4.5 Glycerol Production with Algae

The green algae species *Dunaliella tertiolecta* and *D. bardawil* produce glycerol in a medium with high concentrations of NaCl. The intracellular glycerol concentration is linearly proportional to the salt concentration in which *Dunaliella* is grown; at 5 mol/L NaCl, the intracellular glycerol concentration is about 7 mol/L, equivalent to a 56% glycerol solution (BEN-AMOTZ and AVRON, 1981).

Many reports have been published on glycerol production with this organism as a result of interest in the production of glycerol from renewable resources, in this case from $CO_2$ and sunlight (BEN-AMOTZ et al., 1982; BEN-AMOTZ, 1983). As shown in detail by CHEN and CHI (1981) the special features of this bioconversion route are:

- utilization of $CO_2$ as a cheap and renewable resource,
- utilization of solar energy as the major energy source,
- possible recovery of valuable by-products such as protein (70%) and $\beta$-carotene.

Fig. 9 shows a flow sheet diagram for glycerol production with *Dunaliella* (CHEN and CHI, 1981).

Tab. 1 shows the glycerol, $\beta$-carotene, and algal meal production of *D. bardawil* (BEN-AMOTZ and AVRON, 1981).

Some results reported in the literature (HELLEBUST, 1965) suggest significant re-

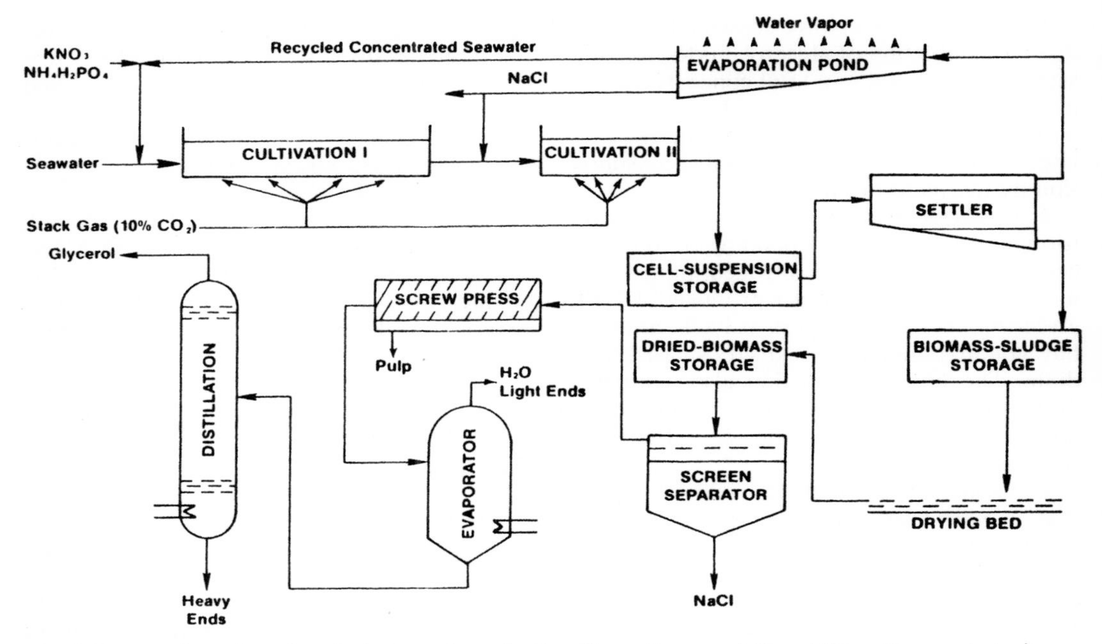


**Fig. 9.** Process flow sheet for algal glycerol production (from CHEN and CHI, 1981, with permission).

**Tab. 1.** Glycerol, β-Carotene, and Algal Meal Production by *Dunaliella bardawil*

| Basis for Caclulation | Productivity [g/m²/d] | | |
|---|---|---|---|
| | Glycerol | β-Carotene | Feed |
| 1. Theoretical | | | |
| a) Growth potential limited | | | |
| 10 cells/L/day | | | |
| $200 \cdot 10^{-2}$ g glycerol/cell | | | |
| 100 L/m² [a] | 20 | 2.5[b] | 27[b] |
| b) Solar radiation limited | | | |
| $1.56 \cdot 10^3$ kJ/m²/day[c] | | | |
| 0.05 stored energy/available energy[d] | | | |
| 0.06 g algae/kJ[c] | | | |
| 0.4 g glycerol/g algae | 15 | 1.9[b] | 21[b] |
| 2. Observed | | | |
| 0.2 g glycerol/L | | | |
| 100 L/m² | | | |
| ⅓ harvested/day | 6.6 | 0.5 | 9.1 |

[a] Assuming a pond 10 cm in depth
[b] Assuming the algae contain 40% glycerol, 5% β-carotene and 55% algal meal on a dry weight basis
[c] Average yearly solar energy at 30° latitude
[d] Photosynthetic conversion efficiency
[e] Heat of combustion of *Dunaliella* is about 16.75 kJ/g algae

lease of glycerol by *Dunaliella* cells depending on parameters such as temperature, light intensity, light quality, salinity, and growth rates.

Until now, all processes proposed for industrial glycerol production with *Dunaliella* cultures have assumed that there is no leakage of the polyol by this alga (BROWN et al., 1982).

GRIZEAU and NAVARRO (1986) immobilized *D. tertiolecta* in Ca-alginate and found that Ca-alginate immobilized cells were reusable for glycerol production. In a hypersaline medium with up to 4 mol/L NaCl they also found an extracellular glycerol production to a maximum of about 5 g/L.

The duration of each batch culture was 4 days and the algae had been immobilized for 2 months. In this time the glycerol production of each batch culture was constant.

## 1.5 Recovery of Glycerol

Glycerol recovery from the crude, fermented mash is very difficult, even though glycerol is completely soluble in water. The high boiling point of glycerol is a further hindrance to a simple recovery. These difficulties affect all biotechnological glycerol processes.

Many recovery methods for microbial glycerol mashes have been described. The most important of these are the following methods (BERNHAUER, 1957; WEIXL-HOFMANN and VON LACROIX, 1962):

(1) The yeast is separated by centrifuging from the mash, obtained by the so called "Protol process". The ethanol is distilled, and sulfites, sulfates, and phosphates are precipitated with calcium salts. Now the mash is filtered, neutralized, and vacuum distilled. A second vacuum distillation leads to pure glycerol that can be used for dynamite production. After the second vacuum distillation, about 40% of the glycerol are lost.
A vacuum distillation is not always necessary. The glycerol can be recovered directly from the mash after separation of inorganic salts by ion exchanging, decolorizing with activated carbon, concentrating, and other cleaning processes.

(2) Direct recovery of glycerol by steam distillation. This process destroys about 50% of the glycerol.

(3) In a complicated process the concentrated mash is sprayed at high temperatures. The glycerol can be recovered from the exhaust gases.

(4) Extraction of the mash with different solvents, e.g., ether, dioxane, butanol, *iso*-amyl alcohol, aniline, has also been recommended for recovery of glycerol. The mash must be cleaned prior to the extraction.

## 1.6 Applications of Glycerol

Crude products containing 78–85% of glycerol are used in industry as intermediates for the production of colors, as a basis for cosmetics, and for other purposes. Distilled glycerol is used for glycerol trinitrate (dynamite) production in the chemical industry, for ointments in the pharmaceutical industry, to increase the moisture of tobaccos, as an antifreeze solution etc. The esters of glycerol are important for the fat and oil industry.

Glycerol can be transformed to dihydroxy acetone with *Acetobacter suboxidans* (US Patent, 1978). In a submerged fermentation the bacteria produce dihydroxy acetone in yields of 75%–90% from a 5%–15% solution of glycerol.

The fermentation was conducted at 30 °C, at pH values from 3.3 to 4.3, for 24–48 h, and the medium was fortified with yeast extract and hydrolyzed fish meal.

## 1.7 Further Aspects of Glycerol Production

The problem of biotechnological glycerol production hinges on the problem of glycerol recovery.

Accordingly, only processes that permit easy recovery of glycerol have a chance of commercialization. Processes should lead to a fermentation solution with

- high concentrations of glycerol,
- absence of by-products, e.g., ethanol, other polyols,
- low amounts of yeast cells,
- absence of residual substances of the original culture liquid, e.g., sulfite, salts, molasses, etc.

A process will have a good chance for industrial use if most of these conditions are met.

High concentrations can be obtained by processes with continuous vacuum distillation. In these processes the ethanol also can be distilled. Processes with low amounts of yeast cells and low concentrations of other substances can be developed with immobilized microorganisms.

New strains of microorganisms that are very tolerant to high concentrations of glycerol, do not produce other polyols and are blocked in the ethanol production pathway, should be developed or constructed by recombinant DNA techniques.

# 2 Other Polyols

The ability to produce other polyols is widely distributed in different microorganisms. The production of 2,3-butanediol is described in Chapter 7 of this volume. Arabitol, sorbitol, erythritol, and also mannitol and xylitol are of some industrial interest. Especially erythritol and arabitol are formed by osmophilic yeasts. The formation of polyhydroxy alcohols is an integral part of the normal metabolism of various yeasts, but it can be influenced by the growth conditions. Fig. 10 shows the formation of polyhydroxy alcohols in the metabolism of *Saccharomyces rouxii*

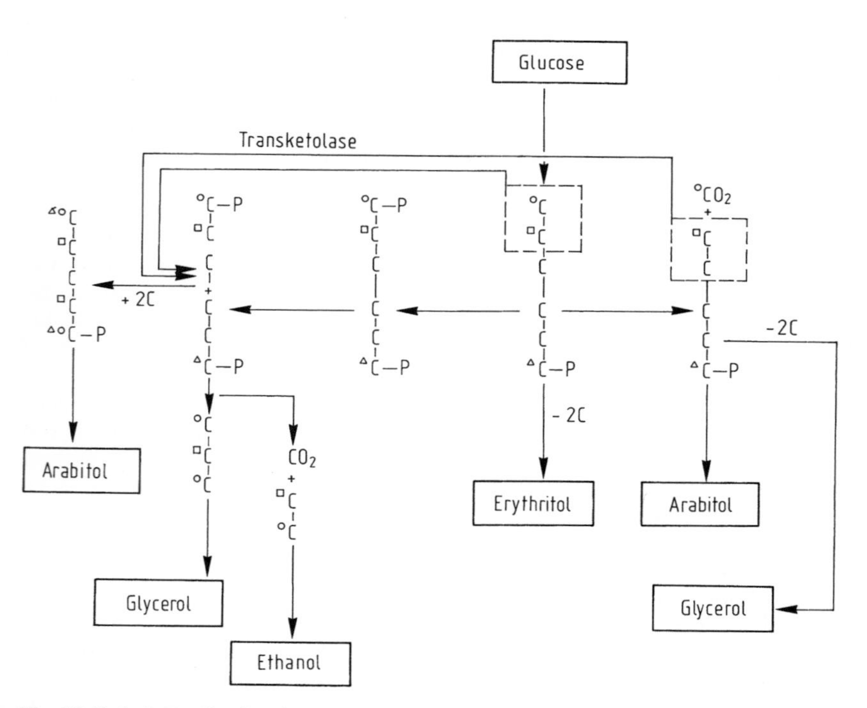


**Fig. 10.** Label distribution in potential carbon skeletons of arabitol, erythritol, glycerol, and ethanol formed from 1-[$^{14}$C], 2-[$^{14}$C], and 6-[$^{14}$C]-glucose by *Saccharomyces rouxii* and *Torulopsis magnoliae.* P indicates a phosphate residue. The carbon atoms are tagged with different symbols (○: 1-[$^{14}$C], □: 2-[$^{14}$C], △: 6-[$^{14}$C]) to help the reader follow the fate of the carbon atoms (from SPENCER and SPENCER, 1978, with permission).

and *Torulopsis magnoliae,* as summarized by SPENCER and SPENCER (1978). A transketolase is necessary for the production of erythritol and arabitol. It would be desirable to have further experiments performed, particularly on the enzymology of the pathways.

JENNINGS (1984) has reviewed polyol metabolism in fungi. In this review all publications concerning the enzyme activities in polyol metabolism have been discussed. From numerous studies, it is clear now that a wide variety of fungi produce increased amounts of glycerol and other polyols in response to a decrease of a high osmotic pressure of the medium. A lower osmotic potential of the medium results in production of lower glycerol concentrations and higher concentrations of polyhydroxic alcohols, e.g., arabitol. When *S. rouxii* was grown in the presence of 3.0 mol/L NaCl in medium containing 0.5 mol/L glucose, 70% of the total polyols formed was glycerol. In the presence of 1.7 mol/L NaCl and 1.6 mol/L glucose the amount of glycerol was only about 50%, with a high proportion of arabitol (ONISHI, 1960b; JENNINGS, 1984). The yeasts ferment 20–40% solutions of sugar in 4–10 days. The polyols can be recovered directly from the culture liquid after separation of the yeasts.

*Pichia miso, Torulopsis famata, Candida polymorpha, Hansenula subpelliculosa, Trigonopsis variabilis,* and *Endomyces chodatii* (REHM, 1980, 1988), as well as a great number of molds (JENNINGS, 1984) also produce different amounts of polyols.

The facts described here, especially related to the regulation of the cell osmotic pressure often lead to the production of several polyols.

In contrast, the cells of *Debaryomyces hansenii* grown in the presence of 2.7 mol/L NaCl may produce arabitol as the major polyol after the cells have entered the stationary phase (ADLER and GUSTAFSSON, 1980).

SPENCER and SHU (1957) demonstrated the influence of oxygen on the production of arabitol by *S. rouxii* (see REHM, 1988, Fig. 12). As the oxygen concentration was increased starting from a low value, the polyol–ethanol ratio also increased.

UEDA (1974) produced arabitol from glucose with *Pichia ohmeri.* The substrate contains 2% glucose, 0.1% yeast extract, 0.2% urea, 0.1% $KH_2PO_4$, 0.1% $MgSO_4$, and 0.5% $CaCO_3$. This solution is inoculated with 5% yeast cells and fermented for 120 h (30 °C, 1.0 vvm aeration, 200 rpm). The sugar concentration must be kept constant at 4–5% glucose by addition of a 70% glucose concentration. The yield is 8.6 g D-arabitol/100 mL, i.e., 50% transformation yield.

BISPING and BAUMANN (1995) immobilized cells of the osmophilic yeast *P. farinosa* in sintered glass Raschig rings for the production of polyols. In fermentations using 20% glucose, immobilization led to a decrease of maximal glycerol and arabitol concentrations from 64.8 to 48 g/L and 42 to 26 g/L, respectively. At the same time, the glycerol–arabitol ratio was increased from 1.55 to 1.84. Concerning the formation of by-products, immobilization caused an increase of ethanol formation and a decrease of citric acid formation. Investigation of the activities of glucose-6-phosphate dehydrogenase (GPDH), phosphofructokinase (PFK), and triosephosphate isomerase (TPI) showed that GPDH activity was increased by immobilization. PFK and TPI activities of immobilized cells were distinctly higher in comparison to free cells. In fermentations using 30% glucose media similar effects were observed. The maxima of glycerol and arabitol concentrations also decreased. At the same time the glycerol–arabitol ratio was increased from 1.5 to 1.8. Concerning the production of by-products, immobilization caused a constant production of ethanol during the whole time of fermentation. The activities of GPDH, PFK, and TPI were also increased by immobilization. The PFK–GDPH ratio was increased by immobilization from 0.23 to 1.55 (20% glucose medium) and from 0.23 to 1.65 (30% glucose medium), respectively. The results indicate that in the physiology of *P. farinosa* cells there is a shift to increased anaerobic metabolism by immobilization in sintered glass Raschig rings. This might be caused by oxygen limitation in the carrier material.

In 11 days *Moniliella tomentosa pollinis* produced a yield of 32% polyols from sucrose. These polyols contained 87% erythritol, 12% glycerol, and 1% ribitol (see BUCHTA, 1994).

From a 40% glucose substrate a mutant of *Aureobasidium* sp. produced 2 g/L erythritol in a continuous culture.

*Candida lipolytica* produces polyols from *n*-alkanes, i.e., mannitol (HATTORI and SUZUKI, 1974a, b). The production of erythritol by a glycerol-requiring auxotrophic mutant of *C. zeylanoides* is also interesting. This mutant produces citric acid at high pH values. At lower pH values citric acid production was repressed, and erythritol formation increased. In an aerated fermentation 180 mg erythritol were excreted after 160 h. Mannitol instead of erythritol was obtained at increased phosphate concentrations (0.1–0.2% $KH_2PO_4$) (HATTORI and SUZUKI, 1974a, b). No information about the metabolism of polyol formation is available for this microorganism.

With 5% alkanes ($C_{12}$–$C_{18}$) at a pH of 3.5, 24–25 °C, 1.0 vvm aeration and 1750 rpm *C. lipolytica* produces 12 g/L mannitol in 8–9 days, which can be cristallized directly from the fermentation broth (DE ZEEUW and TYMAN, 1972).

With heterofermentative *Leuconostoc mesenteroides,* immobilized on reticulated polyurethane foam, mannitol was produced continuously with a stability for 6 days (SOETAERT et al., 1990).

For the biosynthesis of mannitol there is evidence for a cycle from fructose-6-P to mannitol-1-P, to mannitol, and finally to fructose (HULT and GATENBECK, 1978).

Recently, the microbial production of sorbitol has been discussed. This sugar is resorbed by man much slower than glucose and it is metabolized via fructose independently of insulin. Therefore, this sugar is used in large amounts for diabetics. Sorbitol is also used for the microbial oxidation to sorbose in the synthesis of L-ascorbic acid (Reichstein synthesis). At present, it is produced by katalytic hydration of D-glucose. Sorbitol production amounts to 400 000 t/a.

A strain of *Candida boidinii* produced maximum amounts of D-sorbitol, 8.8 g/L and 19.1 g/L, that were obtained from 10 and 20 g/L D-fructose, respectively.

*Zymomonas mobilis* produces sorbitol besides gluconic acid with a glucose–fructose oxidoreductase. This enzyme oxidizes glucose to gluconic acid and reduces fructose to sorbitol. The electrons are transferred by NADP, which is bound to the enzyme. The gluconolactone is hydrolyzed by a gluconolactase to gluconic acid, that can be metabolized by *Zymomonas;* sorbitol cannot be used by this bacterium. In a patent (DE SILVEIRA et al., 1994) the biotransformation of glucose and fructose to gluconic acid and sorbitol by *Zymomonas mobilis* to nearly 100% has been described.

IKEMI et al. (1990) have constructed a membrane bioreactor with a coenzyme regeneration system. The coenzyme, oxidized in the production of sorbitol from glucose by *Candida tropicalis,* was enzymatically regenerated by conjugation with glucose dehydrogenase, associated with the coproduction of gluconic acid from glucose. Substrate conversion was 85%, 100 g/L sorbitol were produced; an equimolar gluconic acid was coproduced for more than 800 h.

In an enzyme membrane reactor with an NAD-dependent glucose dehydrogenase and an NAD-dependent sorbitol dehydrogenase permeabilized cells of *Z. mobilis* transformed in 5–10 h a glucose–fructose medium to more than 95% into sorbitol and gluconic acid (see BRINGER-MEYER and SAHM, 1988; CHUN and ROGERS, 1988).

In a hollow fiber membrane reactor yields of 10–20 g/L per h (0.25–0.59/g cell dry wt. per h) sorbitol and gluconic acid were obtained (PATERSON et al., 1988).

By complete conversion of L-sorbose a maximum amount of 20 g/L L-iditol was obtained (TANI and VONGSUVANLERT, 1987). A *Corynebacterium* sp. reduces L-arabinose to L-arabitol and D-ribose to ribitol with an aldopentose reductase (YOSHITAKE et al., 1976).

Microbial polyol dehydrogenases are used in conventional assays and in biosensors for analytical purposes. D-Mannitol, D-arabitol and D-sorbitol are oxidized by these enzymes to the corresponding keto sugars (SCHNEIDER and GIFFHORN, 1994, see literature therein).

A good overview of the relationships between L-arabinose, L-arabitol, L-xylulose, xylitol, D-xylulose, and other related substances by is given by VAN DER VEEN et al. (1994).

Xylitol is a sweetener reducing caries that can be digested without insulin. It is produced from D-xylose by *Candida pelliculosa* containing a xylose reductase and by permeabilized cells of *Methanobacterium* sp. that are coimmobilized in a photocross linkable resin (NISHIO et al., 1989).

Xylitol has been reported to be produced by other yeasts, especially by *C. boidinii, C. guilliermondii, C. tropicalis, Pachysolen tannophilus,* by fungi such as *Petromyces albertensis,* by bacteria such as *Enterobacter liquefaciens, Corynebacterium* sp., and *Mycobacterium smegmatis.* In all reports it was shown that the production yields were high (more than 90%), and the production rates were low (0.32–1.67 g/L per h) (see ONISHI and SUZUKI, 1969, 1973; HORITSU et al., 1992, and literature cited therein).

HORITSU et al. (1992) found that a maximum production rate of 2.67 g/L per h was obtained when the initial concentrations of D-xylose and yeast extract were 172.0 and 21.0 g/L, respectively, and $K_La$ was 451.50 $\lambda^{-1}$ by 90% oxygen gas.

Xylitol formation by a recombinant *S. cerevisiae* strain containing the *XYL1* gene from *Pichia stipitis* was close to 1 g xylitol per gram of consumed xylose with glucose or ethanol as cosubstrates. A decreased aeration rate increased the xylitol yield based on consumed cosubstrate while the rate of xylitol production decreased. The transformant utilized the cosubstrate more efficiently than a reference strain, in terms of utilization rate and growth rate. This implied that the regeneration of $NAD(P)^+$ during xylitol formation by the transformant balanced the intracellular redox potential (HALLBORN et al., 1994) (see Fig. 11).

The conversion of high amounts of xylulose into xylitol required the addition of ethanol to the food solution. Under conditions of oxygen limitation, acetic acid accumulated in the fermentation broth, causing poisoning of the yeasts at low extracellular pH. Acetic acid toxicity could be avoided by either increasing the pH from 4.5 to 6.5 or by more effective aeration, leading to the further metabolism of acetic acid into cell mass. The best xylitol–ethanol yield of 2.4 g/g was achieved with oxygen limitation.

Under anaerobic conditions ethanol could not be used as a cosubstrate, because for maintenance requirements from ethanol the cells cannot produce ATP anaerobically (MEIANDER et al., 1994).

A 3-stage process has been described (ONISHI and SUZUKI, 1973). In the first step, arabitol is fermented from glucose with *Debaryomyces sake.* After sterilization the unfiltered fermentation broth is inoculated with *Acetobacter suboxydans* and aerobically fermented at 30 °C. Xylose can be produced in high yields at this second stage. Now the broth is sterilized and inoculated with *Candida guilliermondii* var. *soja.* That yeast forms 2 g xylitol/100 mL. The whole yield of xylitol, calculated on the used glucose, is 13.3%.

Recently, a *S. cerevisiae* strain with the xylose reductase of *Pichia stipitis* has been constructed. In a fed-batch fermentation, the best xylitol–ethanol yield of 2.4 g/L was achieved under oxygen-limited conditions (MEINANDER et al., 1994).

# 3 Conclusion and Future Aspects

As demonstrated by this review glycerol can be produced with economical yields with *S. cerevisiae* and such processes can be reactivated quickly, whenever required. The problems of glycerol recovery from the mashes have not been solved yet.

During the past ten years new processes with new methods and new ideas have been under development. Semicontinuous and continuous processes, especially with immobilized microorganisms, are possible and can partly solve the problems of recovery of glycerol from the mashes in two ways: (1) in producing a cleaner mash free from microorganisms and many underived salts, and (2) by obtaining a higher concentration of glycerol in the mash.

In the future, such processes may perhaps have a chance of practical application besides or instead of chemical synthetic processes for glycerol.

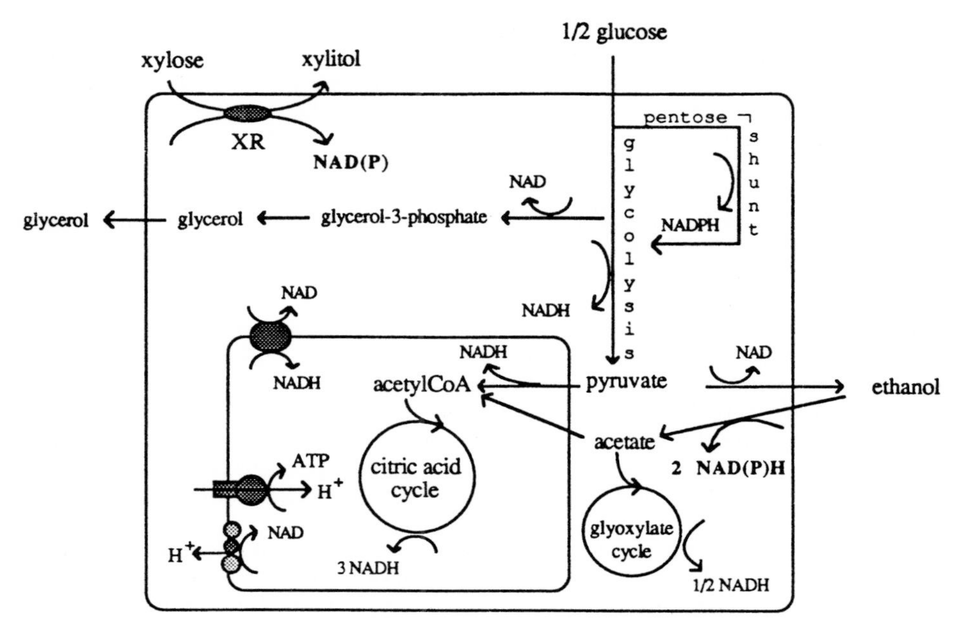


**Fig. 11.** Model of redox reactions possibly involved in xylitol fermentation with the XYL 1 transformant (from HALLBORN et al., 1994, with permission).

In spite of the fact, that most of the polyols other than glycerol described here have a large field of application, cheap biotechnological processes must be worked out and be introduced into practice.

# 4 References


ADLER, L., GUSTAFSSON, L. (1980), Polyhydric alcohol production and intracellular amino acid pool in relation to halotolerance of the yeast *Debaryomyces hansenii. Arch. Microbiol.* **124**, 123–130.

BEN-AMOTZ, A. (1983), Accumulation of metabolites by halotolerant algae and its industrial potential. *Annu. Rev. Microbiol.* **37**, 95–119.

BEN-AMOTZ, A., AVRON, M. (1981), Glycerol and beta-carotene metabolism in the halotolerant algae *Dunaliella* – a model system for biosolar energy conversion. *Trends Biochem. Sci.* **6**, 297–299.

BEN-AMOTZ, A., SUSSMAN, J., AVRON, M. (1982), Glycerol production by *Dunaliella. Experientia* **38**, 49–52.

BERNHAUSER, K. (1957), in: *Ullmans Encyclopädie der technischen Chemie,* Vol. 8, p. 200, München: Urban & Schwarzenberg.

BISPING, B., BAUMANN, U. (1995), Effects of immobilization on polyol production by *Pichia farinosa. Abstract ECB 7,* 19–23 February, Nice.

BISPING, B., REHM, H. J. (1982), Glycerinbildung durch in κ-Carrageenan immobilisierte Zellen von *Saccharomyces cerevisiae,* in: *5. Symposium technischer Mikrobiologie,* pp. 354–360, Berlin: Verlag Versuchs- und Lehranstalt für Spiritusfabrikation und Fermentationstechnologie im Institut für Gärungsgewerbe und Biotechnologie.

BISPING, B., REHM, H. J. (1984a), Glycerinbildung durch immobilisierte Zellen, in: *DECHEMA-Monographien,* Vol. 95, pp. 63–76, Weinheim–Deerfield Beach–Basel: Verlag Chemie.

BISPING, B., REHM, H. J. (1984b), Production of glycerol by immobilized yeast cells, in: *3rd Europ. Congr. Biotechnol.* Sept. 10–14. München, Vol. II, pp. 125–131, Weinheim–Deerfield Beach–Basel: Verlag Chemie.

BISPING, B., REHM, H. J. (1986), Glycerol production by cells of *Saccharomyces cerevisiae* immobilized in sintered glass. *Appl. Microbiol. Biotechnol.* **23**, 174–179.

BISPING, B., HECKER, D., REHM, H. J. (1989), Glycerol production by semicontinuous fed-batch fermentation with immobilized cells of *Saccharomyces cerevisiae*. *Appl. Microbiol. Biotechnol.* **32**, 119–123.

BISPING, B., BAUMANN, U., REHM, H. J. (1990a), Production of glycerol by immobilized *Pichia farinosa*. *Appl. Microbiol. Biotechnol.* **32**, 380–381.

BISPING, B., HELLFORS, H., HONECKER, S., REHM, H. J. (1990b), Formation of citric acid and polyols by immobilized cells of *Aspergillus niger*. *Food Biotechnol.* **4** (1), 17–23.

BLACK, S. (1951), Yeast aldehyde dehydrogenase. *Arch. Biochem. Biophys.* **34**, 86–97.

BLOMBERG, A., ADLER, L. (1992), Physiology of osmotolerance in fungi. *Adv. Microbiol. Physiol.* **33**, 145–212.

BRINGER-MEYER, S., SAHM, H. (1988), Metabolic shifts in *Zymomonas mobiles* in response to growth conditions. *FEMS Microbiol. Rev.* **54**, 131–142.

BROWN, F. F., SUSSMAN. J., AVRON, M., DEGANI, H. (1982), *Biochim. Biophys. Acta* **690**, 165–173.

BUCHTA, K. (1994), Primärmetabolite, in: *Handbuch der Biotechnologie* (PRÄVE, P., FAUST, U., SITTIG, W., SUKATSCH, D. A., Eds.), pp. 593–623, 4. Edn., München–Wien: R. Oldenbourg.

BUTTON, D. K., GARVER, J. C., HAJNY, G. J. (1966), Pilot plant glycerol production with a slow-feed osmophilic yeast fermentation. *Appl. Microbiol.* **14**, 292–294.

*Canadian Patent* (1942), No. 408 881.

CHEN, B. J., CHI, C. H. (1981), Process development and evaluation for algal glycerol production. *Biotechnol. Bioeng.* **23**, 1267–1287.

CHUN, U. H., ROGERS, P. L. (1988), The simultaneous production of sorbitol from fructose and gluconic acid from glucose using an oxidoreductase from *Zymomonas*. *Appl. Microbiol. Biotechnol.* **29**, 19–24.

COCKING, A. T., LILLY, C. H. (1919), Improvements in the production of glycerine by fermentation. *British Patent,* No. 164 034.

COCKING, A. T., LILLY, C. H. (1922), *U. S. Patent,* No. 1 425 838.

CONNSTEIN, W., LÜDECKE, K. (1915/16), *German Patent* (DR PAT.), No. 298 593-96.

CONNSTEIN, W., LÜDECKE, K. (1919), Über Glyceringewinnung durch Gärung. *Ber. Dtsch. Chem. Ges.* **52**, 1385–1391.

CRAIGIE, J. S., MCLACHLAN, J. (1964), Glycerol as a photosynthetic product in *Dunaliella tertiolecta* Butcher. *Can. J. Bot.* **42**, 777–778.

*Danish Patent* (1942), No. 62 582.

DE SILVEIRA, M. M., LOPESDA COSTA, J. P. C., JONAS, R. (1994), Processo de Produçao e Recuperaçao de Sorbitol e Acido Gluconico ou Gluconato. PI No: 9403981-0 (Brasil).

DE ZEEUW, J. R., TYNAN III, E. J. (1972), Fermentative Verfahren zur Herstellung von D-Mannit. DOS 2 203 467.

DELLWEG, H., SCHMID, R. D., TROMMER, W. E. (Eds.) (1992), Römpp Lexikon Biotechnologie, pp. 337–338. Stuttgart, New York: Thieme Verlag.

EOFF, J. R., LINDER, W. V., BEYER, G. F. (1919), Production of glycerine from sugar by fermentation. *Ind. Eng. Chem.* **11**, 842–845.

FRANK, G., WEGMANN, K. (1974), Physiology and biochemistry of glycerol synthesis in *Dunaliella*. *Biol. Zentralbl.* **93**, 707–723.

FREEMAN, G. G., DONALD, G. M. S. (1957), Fermentation processes leading to glycerol. III. Studies on glycerol formation in the presence of alkalis. *Appl. Microbiol.* **5**, 197, 211, 216–220.

GRIZEAU, D., NAVARRO, J. M. (1986), Glycerol production by *Dunaliella tertiolecta* immobilized within Ca-alginate beads. *Biotechnol. Lett.* **8**, 261–264.

GRUNTE, V., OSIPOV, L., KREITSBERG, I. (1959), *Uch. Zap. Rizh. Politekh. Inst.* **2**, 85.

HAJNY, G. J., HENDERSHOT, W. F., PETERSON, W. H. (1960), Factors affecting glycerol production by a newly isolated osmophilic yeast. *Appl. Microbiol.* **8**, 5–11.

HALLBORN, J., GORWA, M. F., MEINANDER, N., PENTTILÄ, M., KERÄNEN, S., HAHN-HÄGERDAL, B. (1994), The influence of cosubstrate and aeration on xylitol formation by recombinant *Saccharomyces cerevisiae* expressing the *XYL1* gene. *Appl. Microbiol. Biotechnol.* **42**, 326–333.

HARRIS, J. F., HAJNY, G. J. (1960), Glycerol production: a pilot-plant investigation for continuous fermentation and recovery. *J. Biochem. Microbiol. Technol. Eng.* **2**, 9–24.

HATTORI, K., SUZUKI, I. (1974a), *Agric. Biol. Chem.* **38**, 581.

HATTORI, K., SUZUKI, I. (1974b), *Agric. Biol. Chem.* **38**, 1203.

HECKER, D., BISPING, B., REHM, H. J. (1990), Continuous glycerol production by the sulfite process with immobilized cells of *Saccharomyces cerevisiae*. *Appl. Microbiol. Biotechnol.* **32**, 627–632.

HEINRICH, M., REHM, H. J. (1981), Growth of *Fusarium moniliforme* on *n*-alkanes: Comparison of an immobilization method with conventional processes. *Eur. J. Appl. Microbiol. Biotechnol.* **11**, 139–145.

HELLEBUST, J. A. (1965), *Limnol. Ozeanogr.* **10**, 192–206.

HILGE-ROTMANN, B., REHM, H. J. (1990), Comparison of fermentation properties and specific enzyme activities of free and calcium-alginate entrapped *Saccharomyces cerevisiae. Appl. Microbiol. Biotechnol.* **33**, 54–58.

HONECKER, S., BISPING, B., ZHU, Y., REHM, H. J. (1989), Influence of sucrose concentration and phosphate limitation on citric acid production by immobilized cells of *Aspergillus niger. Appl. Microbiol. Biotechnol.* **31**, 17–24.

HÖÖTMANN, U., BISPING, B., REHM, H. J. (1991), Physiology of polyol formation by free and immobilized cells of the osmotolerant yeast *Pichia farinosa. Appl. Microbiol. Biotechnol.* **35**, 258–263.

HORITSU, H., YAHASHI, Y., TAKAMIZAWA, K., KAWAI, K., SUZUKI, T., WATANABE, N. (1992), Production of xylitol from D-xylose by *Candida tropicalis:* Optimization of production rate. *Biotechnol. Bioeng.* **40**, 1085–1091.

HULT, K., GATENBECK, S. (1978), *Eur. J. Biochem.* **88**, 607.

IKEMI, M., KOIZUMI, N., ISHIMATSU, Y. (1990), Sorbitol production in charged membrane bioreactor with coenzyme regeneration system: I. Selective retainment of NADP(H) in a continuous reaction. *Biotechnol. Bioeng.* **36**, 149–154.

JENNINGS, D. H. (1984), Polyol metabolism in fungi. *Adv. Microb. Physiol.* **25**, 149–193.

KALLE, G. P., NAIK, S. C. (1987), Effect of controlled aeration on glycerol production in a sulphite process by *Saccharomyces cerevisiae. Biotechnol. Bioeng.* **29**, 1173–1175.

KALLE, G. P., NAIK, S. C., LASHKARI, B. Z. (1985), Improved glycerol production from cane molasses by the sulfite process with vacuum or continuous carbon dioxide sparging during fermentation. *J. Ferment. Technol.* **63**, 231–237.

LEES, T. M. (1944), *Iowa State Coll. J. Sci.* **19**, 38.

LIN, E. C. C. (1976), Glycerol dissimilation and its regulation in bacteria. *Annu. Rev. Microbiol.* **30**, 535–578.

LIN, E. C. C. (1977), Glycerol utilization and its regulation in mammals. *Annu. Rev. Biochem.* **46**, 765–795.

MEINANDER, N., HAHN-HÄGERDAHL, B., LINKO, M., LINKO, P., OJAMO, M. (1994), Fed-batch xylitol production with recombinant XYL-1-expressing *Saccharomyces cerevisiae* using ethanol as a co-substrate. *Appl. Microbiol. Biotechnol.* **42**, 334–339.

MUTSCHLER, D., WEGMANN, K. (1974), Glycerol biosynthesis in a cell free *Dunaliella* preparation. *Biol. Zentralbl.* **93**, 725–729.

NEISH, A. C., BLACKWOOD, A. C., LEDINGHAM, G. A. (1945), Dissimilation of glucose by *Bacillus subtilis* (Ford-strain). *Can. J. Res. Sect. B.* **23**, 290–296.

NELSON, M. E., WERKMAN, C. H. (1935), Dissimilation of glucose by heterofermentative lactic acid bacteria. *J. Bacteriol.* **30**, 574–575.

NEUBERG, C., FÄRBER, E. (1917), Über den Verlauf der alkoholischen Gärung bei alkalischer Reaktion. *Biochem. Z.* **78**, 238–263.

NEUBERG, C., HIRSCH, J. (1919), Die dritte Vergärungsform des Zuckers. *Biochem. Z.* **100**, 304–322.

NEUBERG, C., REINFURTH, E. (1918a), Die Festlegung der Aldehydstufe bei der alkoholischen Gärung. *Biochem. Z.* **89**, 365–414.

NEUBERG, C., REINFURTH, E. (1918b), Natürliche und erzwungene Glycerinbildung bei der alkoholischen Gärung. *Biochem. Z.* **92**, 234–266.

NEUBERG, C., HIRSCH, J., REINFURTH, E. (1920), Die drei Vergärungsformen des Zuckers, ihre Zusammenhänge und Bilanz. *Biochem. Z.* **105**, 307–336.

NICKERSON, W. J., CARROL, W. R. (1945), On the metabolism of *Zygosaccharomyces. Arch. Biochem.* **7**, 257–271.

NISHIO, N., SUGAWA, K., HAYASE, N., NAGAI, S. (1989), Conversion of D-xylose into xylitol by immobilized cells of *Candida pelliculosa* and *Methanobacterium* sp. HU. *J. Ferment. Bioeng.* **67**, 356.

OMAR, S. H., HONECKER, S., REHM, H. J. (1992), A comparative study on the formation of citric acid and polyols and the morphological changes of three strains of free and immobilized *Aspergillus niger. Appl. Microbiol. Biotechnol.* **36**, 518–524.

ONISHI, H. (1957), *Bull. Agric. Chem. Soc. Jpn.* **21**, 151.

ONISHI, H. (1960a), *Rep. Noda. Inst. Sci. Res.* **4**, 1.

ONISHI, H. (1960b), *Bull. Agric. Chem. Soc. Jpn.* **24**, 131, 226.

ONISHI, H. (1961), *Agric. Biol. Chem.* **25**, 341.

ONISHI, H., SUZUKI, T. (1969), Microbial production of xylitol from glucose. *Appl. Microbiol.* **18**, 1031–1035.

PASTEUR, L. (1858), Production constante de glycérine dans la fermentation alcoholique. *C. R. Hebd. Séances Acad. Sci.* **46**, 857.

PASTEUR, L. (1860), Memoire sur la fermentation alcoholique. *Ann. Chim. Phys.* (3e Ser) **58**, 323–426.

PATERSON, S. L., FAME, A. G., FELL, C. J. D., CHUN, U. H., ROGERS, P. L. (1988), Sorbitol and gluconate production in a hollow-fiber membrane reactor by immobilized *Zymomonas mobilis. Biocatalysis* **1**, 217–229.

POLGAR, T. T., BROWER, N. H., SEIDMANN, I. (1964), *U. S. Patent* No. 3 158 550.

RADLER, F., SCHÜTZ, H. (1982), Glycerol production of various strains of *Saccharomyces. Am. J. Enol. Vitic.* **33**, 36–40.

RANK, M., GRAM, J., STERN NIELSEN, K., DANIELSSON, B. (1995), On-line monitoring of ethanol, acetaldehyde and glycerol during industrial fermentations with *Saccharomyces cerevisiae. Appl. Microbiol. Biotechnol.* **42**, 813–817.

REED, G., NAGODAWITHANA, T. (1991), *Yeast Technology,* 2nd Edn., New York: Van Nostrand Reinhold.

REHM, H. J. (1967), Industrielle Mikrobiologie. Berlin–Heidelberg–New York: Springer-Verlag.

REHM, H. J. (1980), *Industrielle Mikrobiologie,* 2nd Edn., Berlin–Heidelberg–New York: Springer-Verlag.

REHM, H. J. (1988), Microbial production of glycerol and other polyols, in: *Biotechnology,* Vol. 6b, 1st Edn., (REHM, H. J., REED, G., eds.), pp. 51–69, Weinheim: VCH.

REHM, H. J., OMAR, S. H. (1993), Special morphological and metabolic behavior of immobilized microorganisms, in: *Biotechnology,* Vol. 1, 2nd Edn., (REHM, H. J., REED, G., Eds.). Weinheim: VCH.

RÖHR, M., KUBICEK, C. P., ZEHENTGRUBER, O., ORTHOFER, R. (1987), Accumulation and partial re-consumption of polyols during citric acid fermentation by *Aspergillus niger. Appl. Microbiol. Biotechnol.* **27**, 235–239.

RÖMPP (1981), Chemie Lexikon, 8th Edn., Vol. 2, (NEUMÜLLER, O. A., Ed.) pp. 1512–1513, Stuttgart: Franckh'sche Verlagsbuchhandlung.

SALA, J. P., DE AYALA, E. B. (1959), *French Patent,* No. 1 178 479.

SCHNEIDER, K. H., GIFFHORN, F. (1994), Overproduction of mannitol dehydrogenase in *Rhodobacter sphaeroides. Appl. Microbiol. Biotechnol.* **41**, 578–583.

SOETAERT, W., DOMER, J., VANDAMME, E. J. (1990), Production of mannitol by *Leuconostoc mesenteroides* immobilized on reticulated polyurethane foam, in: *Physiology of Immobilized Cells,* p. 307, Amsterdam: Elsevier Science Publishers.

SPENCER, J. F. T., SALLANS, H. R. (1956), Production of polyhydric alcohols by osmophilic yeasts. *Can. J. Microbiol.* **2**, 72–79.

SPENCER, J. F. T., SHU, P. (1957), *Can. J. Microbiol.* **3**, 559.

SPENCER, J. F. T., SPENCER, D. M. (1978), Production of polyhydroxy alcohols by osmotolerant yeasts, in: *Economic Microbiology,* Vol. 2, Primary Products of Metabolism (ROSE, A. H., Ed.), pp. 393–425, London–New York–San Francisco: Academic Press.

SPENCER, J. F. T., ROXBURGH, J. M., SALLANS, H. R. (1956), *Can. J. Microbiol.* **2**, 72.

SPENCER, J. F. T., ROXBURGH, J. M., SALLANS, H. R. (1957), Factors influencing the production of polyhydric alcohols by osmophilic yeasts. *J. Agric. Food Chem.* **5**, 64–67.

TANI, Y., VONGSUVANLERT, V. (1987), Sorbitol production by a methanol yeast, *Candida boidinii* (*Kloeckera* sp.) No. 2201. *J. Ferment. Technol.* **65**, 405–411.

UEDA, K. (1974), Mikrobiologisches Verfahren zur Herstellung von D-arabitol. DAS 1 926 178.

*U. S. Patent* (1943), No. 2 414 838.

*U. S. Patent* (1978), No. 4 076 589.

VAN DER VEEN, P., ARST Jr., H. N., MICHEL, J., FLIPPHI, A., VISSER, J. (1994), Extracellular arabinases in *Aspergillus nidulans:* the effect of different *cre* mutations on enzyme levels. *Arch. Microbiol.* **162**, 433–440.

VIJAIKISHORE, P., KARANTH, N. G. (1984), Glycerol production by fermentation. *Appl. Biochem. Biotechnol.* **9**, 243-254.

VIJAIKISHORE, P., KARANTH, N. G. (1986a), *Process Biochem.,* 54–57.

VIJAIKISHORE, P., KARANTH, N. G. (1986b), Glycerol production by immobilized cells of *Pichia farinosa. Biotechnol. Lett.* **8**, 257–260.

VIRKAR, P. D., PANESAR, M. S. (1987), Glycerol production by anaerobic vacuum fermentation of molasses on pilot scale. *Biotechnol. Bioeng.* **29**, 773–774.

WEIXL-HOFMANN, H., VON LACROIX, J. E. (1962), Glyceringärung, in: *Die Hefen,* Vol. 2, pp. 674–691. Nürnberg: Hans Carl Verlag.

WRIGHT, R. E., HENDERSON, W. F., PETERSON, W. H. (1957), *Appl. Microbiol.* **5**, 272.

YOSHITAKE, J., SHIMAMURA, M., IMAI, T. (1976), *Agric. Biol. Chem.* **40**, 1485–1491.

# 6 Microbial Production of Acetone/Butanol/Isopropanol

PETER DÜRRE

Ulm, Federal Republic of Germany

HUBERT BAHL

Rostock, Federal Republic of Germany

# 1 Introduction

The solvents acetone, 1-butanol, and 2-propanol are natural products of a few anaerobic bacteria. During the first half of this century, their synthesis was mainly achieved by fermentation. Acetone–butanol fermentation became in volume the second largest fermentation process in the world, only exceeded by ethanol fermentation of yeast. After 1950, the importance of the process declined rapidly, because the production of acetone and butanol from oil became economically more favorable. However, the oil crisis in 1973 revived the interest in this fermentation and in the bacteria performing it. This chapter tries to review the enormous progress made in the last few years with respect to biochemistry, genetics, and physiology of the respective microorganisms and new developments in process technology.

A number of reviews on various aspects of this topic have been published in the last decade (AWANG et al., 1988; BAHL and GOTTSCHALK, 1988; DÜRRE et al., 1992; ENNIS et al., 1986a; JONES and WOODS, 1986; LENZ and MOREIRA, 1980; MCNEIL and KRISTIANSEN, 1986; MOREIRA, 1983; ROGERS, 1986); they should be consulted for additional information.

Some of the physical properties of acetone, 1-butanol, and 2-propanol (isopropanol) are summarized in Tab. 1. Acetone is an important intermediate in the manufacture of methacrylates and methyl isobutyl ketone and a solvent for resins, paints, varnishes, laquers, and cellulose acetate; it is miscible in all proportions with water. 1-Butanol is a precursor of butyl acetate and dibutyl phthalate and like acetone a good solvent. Its solubility in water is 8% (w/w). 2-Propanol or isopropanol is used in antifreeze composition, as a solvent, e.g., in quick-drying oils and ink, and in cosmetics such as hand lotions and after-shave lotions. Like acetone, it is completely miscible with water.

# 2 History

Butanol formation by fermentation was first noticed by PASTEUR in connection with his discovery of the butyrate fermentation (PASTEUR, 1861a, 1862). The causative organism was named *vibrion butyrique,* and the term "anaerobic" was coined (PASTEUR, 1861b, 1863), when it was realized that the bacteria responsible for this fermentation were killed in air.

A more detailed study on butanol producing bacteria was then conducted by FITZ. He described an organism fermenting glycerol to butanol, butyric acid, $CO_2$, $H_2$, and small amounts of acetic acid, ethanol, lactate, and propanediol (FITZ, 1876, 1878, 1882). It was an anaerobic, sporulating, rod-shaped bacterium, and a drawing showed typical clostridial forms (FITZ, 1878). Lactose and starch did not lead to active fermentation, but mannitol and sucrose could serve as additional substrates. FITZ already observed that from the products formed, butanol was the most toxic compound, and he finally described the isola-

**Tab. 1.** Physical Properties of Acetone, Butanol, and Isopropanol

| Property | Acetone | 1-Butanol | 2-Propanol (Isopropanol) |
|---|---|---|---|
| Molecular weight | 58.08 | 74.12 | 60.1 |
| Melting point at 101.3 kPa | −94.6°C | −90.2°C | −88.5°C |
| Boiling point at 101.3 kPa | 56.1°C | 117.7°C | 82.3°C |
| Specific gravity at 20°C | 0.807 | 0.813 | 0.786 |
| Heat of vaporization | 29.1 kJ $mol^{-1}$ | 43.8 kJ $mol^{-1}$ | 39.8 kJ $mol^{-1}$ |
| Heat of combustion | 1787 kJ $mol^{-1}$ | 198.2 kJ $mol^{-1}$ | 2005.8 kJ $mol^{-1}$ |
| Vapor pressure at 20°C | 24.7 kPa | 0.63 kPa | 4.4 kPa |

Data taken from NELSON and WEBB, 1978; SHERMAN, 1978; PAPA, 1982

**Tab. 2.** History of Description and Isolation of Solvent-Forming Bacteria

| Designation of Microorganism | Solvents Formed | Year of Publication | Author(s) |
|---|---|---|---|
| *Vibrion butyrique* | Butanol | 1861a, b; 1862 | PASTEUR |
| *Bacillus butylicus* | Butanol | 1878; 1882 | FITZ |
| *Clostridium butyricum* I, II, III | Butanol | 1887 | GRUBER |
| *Bacille amylozyme* | Butanol[a] | 1891 | PERDRIX |
| *Bacillus butyricus* | Butanol | 1892 | BOTKIN |
| *Bacillus orthobutylicus* | Butanol | 1893 | GRIMBERT |
| *Granulobacter butylicum* | Butanol | 1893 | BEIJERINCK |
| *Granulobacter saccharobutyricum* | Butanol | 1893 | BEIJERINCK |
| *Amylobacter butylicus* | Butanol | 1895 | DUCLAUX |
| *Bacillus butylicus* (Fitz) | Butanol | 1897 | EMMERLING |
| *Granulobacillus saccharobutyricus immobilis liquefaciens* | Butanol | 1899; 1900 | SCHATTENFROH and GRASSBERGER |
| *Clostridium pastorianum* | Butanol, isobutanol | 1902 | WINOGRADSKY |
| *Bacillus macerans* | Acetone | 1905 | SCHARDINGER |
| *Clostridium americanum* | Butanol, isopropanol | 1906a, b | PRINGSHEIM |
| *Clostridium* sp. | Butanol | 1907 | SCHARDINGER |
| *Bacillus butylicus* (Fitz) | Butanol | 1908 | BUCHNER and MEISENHEIMER |
| *Bacillus amylobacter* A.M. et Bredemann | Butanol, propanol | 1909 | BREDEMANN |
| *Clostridium acetobutylicum* (Weizmann) | Acetone, butanol | 1926 | MCCOY et al. |

[a] The original description reported amylalcohol as a product, but due to the determination procedure butanol is much more likely (PRINGSHEIM, 1906a).

tion of pure cultures of *Bacillus butylicus* from cow feces and hay (FITZ, 1877, 1878, 1882). This pioneering work stimulated many other scientists to investigations on butanol producing anaerobic bacteria. Among them were a number of famous microbiologists (Tab. 2).

BEIJERINCK (1893) described two different strains yielding appreciable quantities of butanol. One of them, producing butyrate additionally, was designated *Granulobacter saccharobutyricum* and was probably identical to *Bacillus butylicus.* The other species, *Granulobacter butylicum,* produced isopropanol along with butanol which, however, had not been discovered before 1920 (FOLPMERS, cited in OSBURN and WERKMAN, 1935). This bacterium was the later *Clostridium butylicum* which was shown to belong to the species *C. beiierinckii* (GEORGE et al., 1983). Pure cultures of butyrate producers additionally forming small amounts of butanol were obtained by WINOGRADSKY in 1902.

The biological formation of isopropanol (together with butanol) was first reported in 1906 for *C. americanum* (PRINGSHEIM, 1906a, b). Acetone was discovered as a product of microbial activity by SCHARDINGER (1905). The responsible organism was named *Bacillus macerans,* and ethanol, acetate, and formate were produced in addition. Thus, the two products butanol and acetone were found in rather different microorganisms. The discovery of their common formation by a single species was connected with the efforts of the chemical industry to produce synthetic rubber. Butanol was considered a precursor of butadiene, the starting material for synthetic rubber production. The British company Strange and Graham Ltd. got interested in such a project and contracted PERKINS and WEIZMANN (University of Manchester) and FERNBACH and SCHOEN (Institut Pasteur) in 1910 to study the formation of butanol by microbial fermentation (see GABRIEL, 1928). FERNBACH isolated an acetone–butanol producer in 1911; WEIZMANN terminated his cooperation with the company in 1912, but continued his work at the University of Manchester. He succeeded in isolating an organism, later named *Clostridium acetobutylicum,* that produced acetone and butanol from starchy materials in better yields than the organism of FERNBACH. Patent applications were filed for the Fernbach process in 1911 and 1912 (FERNBACH and STRANGE, 1911, 1912) and for the Weizmann process in 1915 (WEIZMANN, 1915). Production of acetone and butanol by Strange and Graham Ltd. based on the Fernbach process began in 1913. Potatoes were used as raw material. Following the outbreak of World War I the interest turned from butanol to acetone which was required in large amounts for the production of smokeless powder. Strange and Graham Ltd. supplied the British government with acetone produced in their plant at King's Lynn. WEIZMANN continued research work on his process; pilot-scale studies were performed and the results convinced the government of the superiority of this process. Finally, a production plant was built at the Royal Naval Cordite Factory at Poole. There and in some other distilleries acetone was produced. In addition, the Weizmann process was also introduced into the Strange and Graham Ltd. plant.

Because of the German blockade grain and maize could not be imported to Great Britain in the quantities required. Starchy materials could not be supplied to the fermentation industry anymore, and acetone–butanol production had to be abandoned in Great Britain. However, the know-how was transferred to Canada, and a fermentation plant went into operation in Toronto in August 1916. There and in a plant at Terre Haute (Indiana, USA) acetone was produced until the end of the war when all plants were closed. There was little use for butanol produced at that time, so it was stored in large vats. However, as a result of the rapid growth of the automobile industry, increasing amounts of lacquers were needed for which butanol and its ester, butyl acetate, are excellent solvents. In addition, Prohibition endangered the supply of amyl alcohol, which had been used as a solvent for lacquers and which had been obtained as a by-product of the manufacture of spirits (WALTON, 1945). Soon the situation was reversed as compared to war times: 1-butanol was the product wanted, and acetone became less useful. Commercial Solvents Corp. of Maryland (USA) was founded, they bought the Terre Haute plant in 1919 and started butanol production in 1920. Difficulties with the process emerged in 1923, when the solvent yield went down considerably – due to bacteriophage infection (OGATA and HONGO, 1979). It took almost one year to overcome these problems. A second plant was opened by Commercial Solvents Corp. at Peoria in 1923 with 32 fermenters with a volume of about 190000 L each. It was extended to 96 fermenters in the following years. Production was in the range of 100 tons of solvents per day (JONES and WOODS, 1986).

Strange and Graham Ltd. started production of solvents at King's Lynn again in 1923. The factory, however, was soon destroyed by an explosion, and Strange and Graham Ltd. went into liquidation. The story of the acetone–butanol process in Great Britain continued in 1935 when Commercial Solvent Corp. and the British Distillor's Co. built a plant at Brombrough. The plant was operated with molasses, the substrate that also had replaced grain as raw material in the American plants in the early 1930s. By this replacement the acetone–butanol process was saved at that time, because molasses was much cheaper than grain. In addition, strains of *Clostridium acetobutylicum* had been developed fermenting up to 6.5% sugar and producing 2% solvents. This resulted in a considerable reduction of the costs of product recovery. Until 1936, the patents of WEIZMANN (1915, 1919) prevented a further extension of the acetone–butanol process. These patents expired in 1936, and new fermentation plants were built in the United States, in the former USSR, in Japan, India, Australia, and South Africa. These plants operated through World War II. In 1945, 66% of the 1-butanol and 10% of the acetone produced in the United States came from the fermentation industry. Starting around 1950, the fermentation process went into sharp decline because of the severe competition of the growing petrochemical industry and of steeply rising prices of molasses and grain. Soon all plants in the Western countries were closed; only the plant in South Africa operated until 1982 (JONES and WOODS, 1986); a few plants may still be operating in Russia and in the People's Republic of China.

A critical parameter in fermentations using *C. acetobutylicum* is the tendency of the organism to degenerate, i. e., to lose the ability to produce solvents and to sporulate (KUTZENOK and ASCHNER, 1952; FINN and NOWREY, 1959). Therefore, sporulation is used as a convenient method to store the organism and to maintain its solvent-forming ability. Different methods of spore preservation have been reported. Spores can be kept in sterile dry sand or soil (SPIVEY, 1978; DAVIES and STEPHENSON, 1941; BEESCH, 1952), lyophilized (LAPAGE et al., 1970), or stored in milk medium (BAHL et al., 1982a) or fermentation broth (MONOT et al., 1982; LIN and BLASCHEK, 1983). A heat shock for 90 s at 70°C of a freshly inoculated culture helps to induce sporulation. A similar procedure (1 min at 90°C) will ensure germination of a spore preparation. The mechanisms responsible for clostridial degeneration are not yet understood. However, a recent publication has described the isolation of transposon-induced degeneration-resistant mutants (KASHKET and CAO, 1993) which will allow an analysis of this phenomenon at the molecular level.

Batch culture fermentation with *C. acetobutylicum* became economically unfavorable because of high substrate costs, low solvent yields, large volumes of waste, high amounts of energy required for product recovery (distillation), and the low butanol tolerance of the organism. Starting at concentrations of 200 mM growth is completely inhibited by butanol, whereas acetone and ethanol in similar concentrations show no effect (MOREIRA et al., 1981). The sudden increase of crude oil prices in 1973 triggered a revival of the interest in the biotechnological production of 1-butanol and acetone. So far, it has not led to a process able to compete with the oil-based production of solvents. Acetone is currently produced by the cumene hydroperoxide process or by catalytic dehydrogenation of isopropanol (NELSON and WEBB, 1978). 1-Butanol is synthesized from propylene by the oxo-process or from acetaldehyde by the aldol process (SHERMAN, 1978). 2-Propanol or isopropanol is generally manufactured from propene, either by the indirect hydration process or by catalytic hydration (PAPA, 1982).

# 3 Biochemistry and Genetics of Solvent-Producing Clostridia

## 3.1 Microorganisms

Solvents such as acetone, butanol, and isopropanol are produced in more than trace amounts only by few bacterial species. Most of them are members of the genus *Clostridium* (BAHL and DÜRRE, 1993). However, butanol as a major fermentation product has also been detected with *Butyribacterium methylotrophicum* (GRETHLEIN et al., 1991) and *Hyperthermus butylicus* (ZILLIG et al., 1991). The most prominent species of clostridia, of course, is *C. acetobutylicum*. In times when dozens of patents were obtained for certain aspects of the acetone–butanol fermentation process, a variety of partly exotic names was given to the many production strains included in these patents. Some of the names are depicted in Tab. 3; they are now invalid. Clostridia recognized of producing solvents are summarized in Tab. 4. In addition to *C. acetobutylicum,* the species *C. aurantibutyricum, C. beijerinckii, C. butyricum, C. cadaveris, C. chauvoei, C. felsineum, C. pasteurianum, C. puniceum, C. roseum, C. sporogenes, C. tetani, C. tetanomorphum,* and *C. thermosaccharolyticum* are shown. The products formed by these species vary slightly (Tab. 5). *C. aurantibutyricum* and strains of *C. beijerinckii* (GEORGE et al., 1983; CHEN and HIU, 1986) as well as an aggregate-forming variant of *C. butyricum* (ZOUTBERG et al., 1989b) also form isopropanol which is not produced by the other species. With glucose as substrate, butanol and acetone are recovered in a ratio of 2:1 from the fermentation broth of *C. acetobutylicum* and in a ratio in the order of 10:1

**Tab. 3.** A Small Collection of Names of Solvent-Producing Bacteria Mentioned in Patents

| Name of Organism | Author(s) | Year | U. S. Patent No. |
|---|---|---|---|
| *Bacillus butylaceticum* | FREIBERG | 1925 | 1 537 597 |
| *Bacillus acetobutylicum* | FUNK | 1925 | 1 538 516 |
| *Clostridium saccharobutylicum* gamma | IZSAK and FUNK | 1933 | 1 908 361 |
| *Clostridium saccharobutylacetonicum* | LOUGHLIN | 1935 | 1 992 921 |
| *Clostridium saccharoacetobutylicum* beta, gamma | ARZBERGER | 1936 | 2 050 219 |
| *Clostridium saccharoacetobutylicum* α | WOODRUFF et al. | 1937 | 2 089 522 |
| *Clostridium invertoacetobutylicum* | LEGG and STILES | 1937 | 2 089 562 |
| *Bacillus tetryl* | ARROYO | 1938 | 2 113 471 |
| *P-bacillus* | MÜLLER | 1938b | 2 123 078 |
| *Clostridium propylbutylicum* alpha | MÜLLER | 1938a | 2 132 039 |
| *Clostridium saccharobutylacetonicum liquefaciens* | ARZBERGER | 1938 | 2 139 108 |
| *Clostridium saccharobutylacetonicum liquefaciens* gamma, delta | CARNARIUS and McCUTCHAN | 1938 | 2 139 111 |
| *Clostridium madisonii* | McCOY | 1946 | 2 398 837 |
| *Clostridium saccharoperbutylicum* | BEESCH | 1948 | 2 439 791 |
| *Clostridium saccharoperbutylacetonicum* | HONGO | 1960 | 2 945 786 |

in the broth of *C. puniceum* and *C. beijerinckii* (HOLT et al., 1988; CHEN and HIU, 1986). Butanol and equal amounts of ethanol, but no acetone or isopropanol, are produced by *C. tetanomorphum* (GOTTWALD et al., 1984). Cultures of *C. acetobutylicum* continue to produce solvents after growth has ceased, while with *C. puniceum* and *C. tetanomorphum* solvent production is only observed during growth. When *C. pasteurianum* is grown in media of high sugar content or with glycerol as a substrate under phosphate limitation, high butanol concentrations are yielded (130 and 45 mmol $L^{-1}$, respectively) (HARRIS et al., 1986; DABROCK et al., 1992). Small amounts of butanol are among the products of a number of other *Clostridium* species (CATO et al., 1986; HIPPE et al., 1992). Enrichment and isolation of solvent producing clostridia is relatively easy. Soil samples from potatoes, other root crops, or roots of beans are good sources (CALAM, 1979). Enrichment can be done in stabbed potato tubers (VELDKAMP, 1965) or maize mash medium (WEIZMANN, 1919). Isolation of pure cultures can be achieved by applying techniques commonly used with strictly anaerobic bacteria (BREZNAK and COSTILOW, 1994).

Since *C. acetobutylicum* produces and tolerates the highest solvent concentrations and since it has been employed in the industrial process, the further discussions will primarily deal with this microorganism. *C. beiierinckii* will be included as a model organism for isopropanol formation. *C. acetobutylicum* is a gram-positive, straight rod measuring 0.6–0.9 by 2.4–4.7 μm. The vegetative cells are motile with peritrichous flagella. Subterminal ovoid spores are formed (Fig. 1). The optimum growth temperature is 37°C, and biotin and *p*-aminobenzoate are required as growth factors (RUBBO et al., 1941). Further physiological properties have been summarized in other review articles (CATO et al., 1986; HIPPE et al., 1992).

A note of caution must be added as to comparison of experimental data obtained with different strains of *C. acetobutylicum* and *C. beijerinckii.* About 50 of such strains are currently in use or available from the various culture collections. Although some of them were supposed to be identical, this has been disproven by studying their physiological characteristics and by hybridization analyses of cloned genes (SAUER and DÜRRE, 1993; WOOLLEY and MORRIS, 1990; YOUNG et al., 1989). A recent investigation has provided unequivocal evidence that of the mostly used strains only ATCC 824, DSM 792, and DSM 1731 are true *C. acetobutylicum.* Strain

**Tab. 4.** Some Properties of Butanol-Producing Clostridia

| | *C. aceto-butylicum* | *C. auranti-butyricum* | *C. beijer-inckii* | *C. buty-ricum* | *C. cada-veris* | *C. chau-voei* | *C. fel-sineum* |
|---|---|---|---|---|---|---|---|
| Gelatin hydrolyzed | – | + | – | – | + | + | + |
| Motility | + | + | + | ± | ± | + | + |
| Lipase produced | – | + | – | – | – | – | – |
| Starch hydrolyzed | + | + | ± | + | – | – | ± |
| Sugars utilized | | | | | | | |
| Glucose | + | + | + | + | + | + | + |
| Fructose | + | + | + | + | ± | – | + |
| Lactose | + | + | + | + | – | + | + |
| Maltose | + | + | + | + | – | + | ± |
| Sucrose | + | + | + | + | – | + | + |

**Tab. 4.** (continued)

| | *C. pasteur-ianum* | *C. puni-ceum* | *C. roseum* | *C. sporo-genes* | *C. tetani* | *C. tetano-morphum* | *C. thermo-saccharo-lyticum* |
|---|---|---|---|---|---|---|---|
| Gelatin hydrolyzed | – | + | + | + | + | – | – |
| Motility | ± | + | ± | ± | ± | + | ± |
| Lipase produced | – | – | – | + | – | + | |
| Starch hydrolyzed | – | + | – | – | – | – | – |
| Sugars utilized | | | | | | | |
| Glucose | + | + | + | + | + | + | + |
| Fructose | + | ± | + | – | – | + | + |
| Lactose | – | ± | + | – | – | – | + |
| Maltose | + | + | – | – | – | + | + |
| Sucrose | + | + | + | – | – | | + |

Data taken from CATO et al., 1986; CHEN and HIU, 1986; FREIER-SCHRÖDER et al., 1989; GEORGE et al., 1983; GOTTWALD et al., 1984; HARRIS et al., 1986; HOLT et al., 1988; MCCLUNG and MCCOY, 1935; MCCOY and MCCLUNG, 1935; NAKAMURA et al., 1979; PETERSEN, 1991; WILDE et al., 1989.

NCIMB 8052 is a member of the *C. beiier-inckii* group, whereas strain P 262 belongs to a different species (KEIS et al., 1994; JOHNSON and JONES, personal communication).

## 3.2 Utilization of Substrates

A great variety of mono- and disaccharides and related compounds are used by *C. acetobutylicum* for growth. In addition to the substrates listed in Tab. 3, these are: cellobiose, D-mannose, D-galactose, D-gluconate, D-galacturonate, D-glucosamine, D-ribose, D-xylose, L-arabinose, L-rhamnose, and glycerol. Furthermore, polysaccharides as starch and xylan can be fermented. The process of substrate acquisition can be considered in two or three stages: (*1*) degradation of polymeric substrates by secreted enzymes, (*2*) uptake of the degradation products or other low-molecular weight substrates, and (*3*) intracellular metabolism of the carbohydrates.

### 3.2.1 Degradation of Polymers

Despite the fact that starch was the first industrial substrate for the production of acetone–butanol by fermentation, only little information is available on the extracellular enzymes involved in the degradation of this polymer. During growth of *C. acetobutylicum* on starch an $\alpha$-amylase and a glucoamylase, ori-

**Tab. 5.** Solvent Production by Some Clostridial Species Grown in Complex Media Contaning 2% (w/v) Glucose[a]

| Organism | Products Formed [mmol L$^{-1}$] | | | |
|---|---|---|---|---|
| | Acetone | Butanol | Ethanol | Isopropanol |
| *C. acetobutylicum* | 14.0 | 30.2 | 5.0 | – |
| *C. aurantibutyricum* | 20.5 | 45.4 | – | 4.5 |
| *C. beijerinckii* | 6.0 | 67.9 | – | – |
| *C. beijerinckii* | – | 44.8 | – | 9.8 |
| *C. butyricum*[a] | – | 17 | – | 7 |
| *C. cadaveris* | | 11.2 | – | – |
| *C. pasteurianum* | – | 22 | 8 | – |
| *C. puniceum* | 16.8 | 75.6 | – | – |
| *C. sporogenes* | – | 11.2 | – | – |
| *C. tetanomorphum* | – | 47.1 | 42.7 | – |
| *C. thermosaccharolyticum* | – | 40.0 | 85.0 | – |

Data taken from DABROCK et al., 1992; FREIER-SCHRÖDER et al., 1989; GEORGE et al., 1983; GOTTWALD et al., 1984; HOLT et al., 1988; ZOUTBERG et al., 1989a.
[a] An aggregate-forming variant with 2.7% glucose has been used.

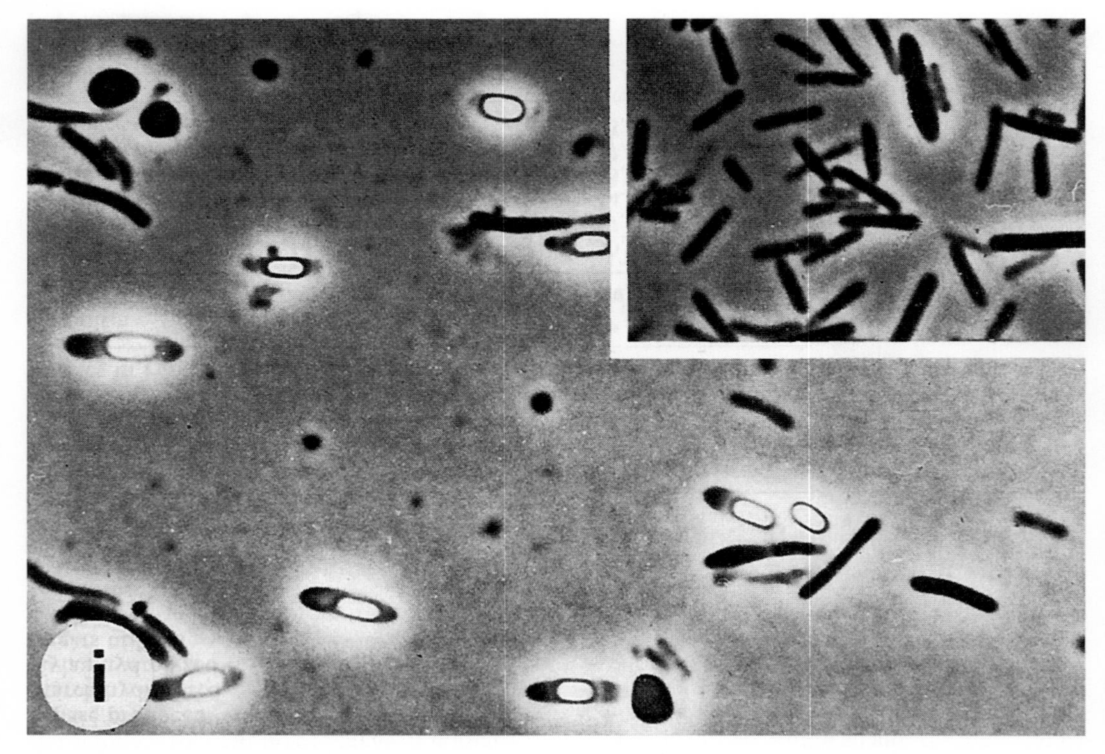

**Fig. 1.** Phase-contrast photomicrograph of *Clostridium acetobutylicum* (courtesy by H. HIPPE).

ginally referred to as a maltase, are produced (HOCKENHULL and HERBERT, 1945; FRENCH and KNAPP, 1950; SCOTT and HEDRICK, 1952; ENSLEY et al., 1975). The synthesis of these enzymes is generally subjected to catabolite repression by glucose and induced by starch or its degradation products (HOCKENHULL and HERBERT, 1945; CHOJECKI and BLASCHEK, 1986). Recently, the $\alpha$-amylase from *C. acetobutylicum* ATCC 824 has been purified and characterized (PAQUET et al., 1991). This enzyme has a molecular weight of 84 kDa, an isoelectric point of 4.7, an optimal pH of 5.6, and is very sensitive to thermal inactivation. Higher activities were found with high-molecular weight substrates as compared to low-molecular weight maltooligosaccharides. Glycogen and pullulan were slowly hydrolyzed, whereas dextrans and cyclodextrins were not attacked. Interestingly, the product of an amylase gene of *C. acetobutylicum* cloned by VERHASSELT et al. (1989) has a molecular mass of only 53.9 kDa. This gene is not identical to a truncated open reading frame with significant similarity to the $\alpha$-amylase gene of *Bacillus subtilis* identified by GERISCHER and DÜRRE (1990). Therefore, it is likely that *C. acetobutylicum* produces different amylases for the utilization of starch.

Xylan can serve as sole carbon source for *C. acetobutylicum* (LEE et al., 1985a; LEMMEL et al., 1986). However, growth is slow, and larch wood xylan is only partially hydrolyzed. Two endoxylanases and a $\beta$-D-xylosidase of *C. acetobutylicum* were purified and characterized (LEE et al., 1985a, 1987). The smallest oligosaccharides degraded by xylanase A and B are xylohexaose and xylotetraose, respectively. Xylanase A also exhibits carboxymethyl cellulase activity.

*C. acetobutylicum* does not grow on cellulose; nevertheless, two activities of the cellulase complex can be detected in certain strains of this organism (ALLCOCK and WOODS, 1981). LEE et al. (1985b) showed that two strains out of 21 tested produced extracellular endo- 1,4-$\beta$-glucanase and cellobiase ($\beta$-D-glucosidase) activities during growth on cellobiose.

### 3.2.2 Uptake of Mono- and Disaccharides

Very little is known on substrate transport in *C. acetobutylicum.* Phosphotransferase systems (PTS) are apparently responsible for uptake of glucose and fructose (VON HUGO and GOTTSCHALK, 1974; HUTKINS and KASHKET, 1986):

$$\text{Glucose} + \text{P-HPr} \xrightarrow{\text{Enzyme II}} \text{Glucose-6-phosphate} + \text{HPr}$$

$$\text{Fructose} + \text{P-HPr} \xrightarrow{\text{Enzyme II}} \text{Fructose-1-phosphate} + \text{HPr}$$

P-HPr is a phosphorylated protein generated from HPr with phosphoenolpyruvate (PEP) as the source of energy-rich phosphate:

$$\text{PEP} + \text{HPr} \xrightarrow{\text{Enzyme I}} \text{P-HPr} + \text{Pyruvate}$$

A detailed biochemical study of the glucose PTS of *C. acetobutylicum* has been provided recently (MITCHELL et al., 1991). The presence of the four components of the system, enzyme I, HPr, enzyme $\text{IIA}^{\text{glu}}$, and enzyme $\text{II}^{\text{glu}}$, was demonstrated. The *C. acetobutylicum* PTS therefore displays the same architecture as in other bacteria. Other substrates of *C. acetobutylicum* might be taken up either also by PTS as in the case of glucitol (MITCHELL, 1992), by ABC transport sytems (HIGGINS, 1992), or by symport mechanisms driven by the transmembrane proton gradient. Disaccharides such as sucrose or maltose might then be cleaved by appropriate phosphorylases, e.g.,

$$\text{Maltose} + \text{Phosphate} \xrightarrow{\text{Maltose Phosphorylase}} \text{Glucose} + \text{Glucose-1-phosphate}$$

Free glucose can be converted to glucose-6-phosphate by hexokinase.

### 3.2.3 Intracellular Sugar Metabolism to Pyruvate

*Clostridium* species generally employ the Embden-Meyerhof-Parnas (EMP) pathway

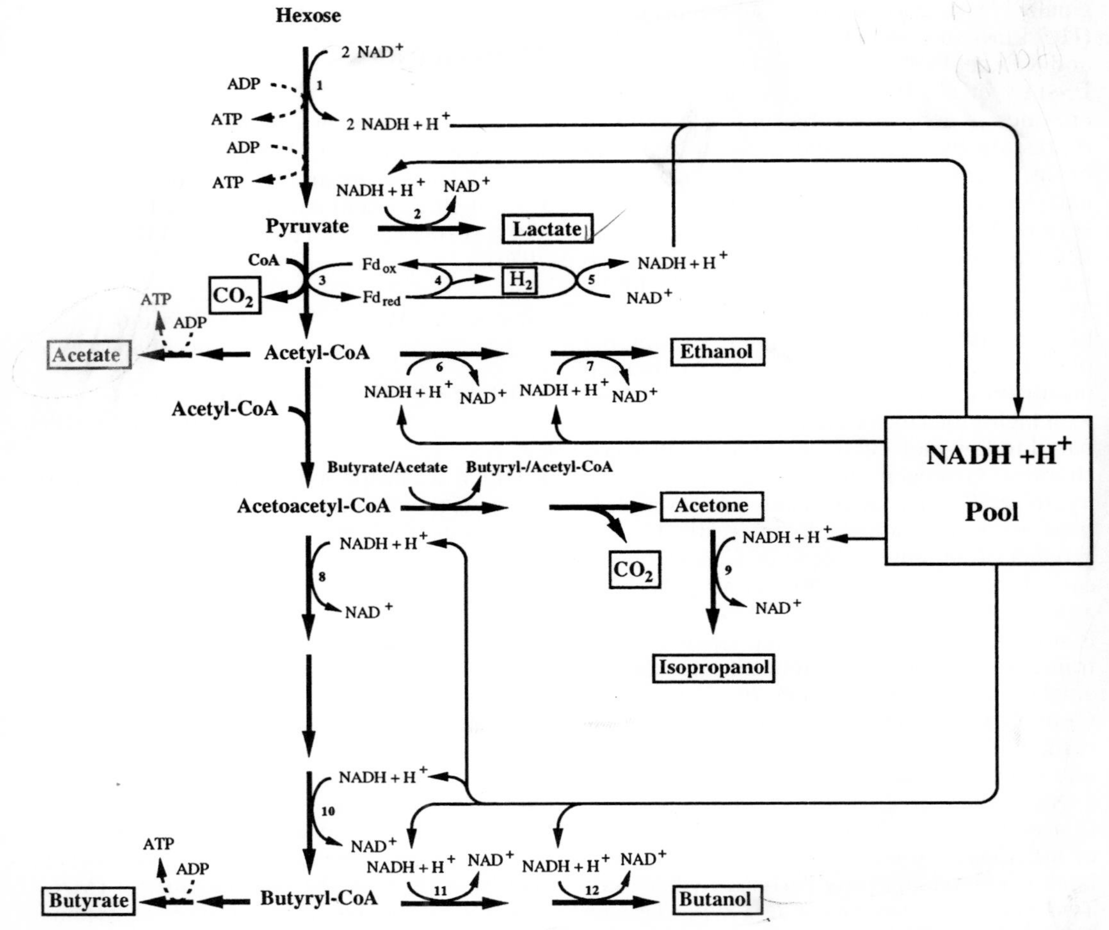


**Fig. 2.** Carbon (thick arrows) and electron (thin arrows) flow during hexose fermentation in *Clostridium acetobutylicum* or *C. beijerinckii.* Enzymes involved in oxidation–reduction reactions are numbered as follows: (1) glyceraldehyde-3-phosphate dehydrogenase; (2) lactate dehydrogenase; (3) pyruvate:ferredoxin oxidoreductase; (4) hydrogenase; (5) NADH:ferredoxin oxidoreductase; (6) acetaldehyde dehydrogenase; (7) ethanol dehydrogenase; (8) $\beta$-hydroxybutyryl-CoA dehydrogenase; (9) isopropanol dehydrogenase; (10) butyryl-CoA dehydrogenase; (11) butyraldehyde dehydrogenase; (12) butanol dehydrogenase. In certain alcohol dehydrogenease reactions some strains use $NADP^+/NADPH + H^+$ as electron acceptor/donor instead of $NAD^+/NADH + H^+$, which are generally shown.

for the degradation of hexose phosphates (THAUER et al., 1977; GOTTSCHALK, 1986; ROGERS and GOTTSCHALK, 1993). This is apparently also true for *C. acetobutylicum.* A corresponding scheme is shown in Fig. 2. Sugar acids such as gluconate are degraded via a modified Entner-Doudoroff pathway (ANDREESEN and GOTTSCHALK, 1969). D-Gluconate is dehydrated to yield 2-keto-3-deoxygluconate which subsequently is phosphorylated and cleaved to yield pyruvate and 3-phosphoglyceraldehyde (Fig. 3).

Pentoses are converted into ribose-5-phosphate and xylulose-5-phosphate which are fed into the pentose phosphate cycle; the resulting hexose phosphates are subsequently catabolized by the EMP pathway. Enzymes such as transketolase and transaldolase have been

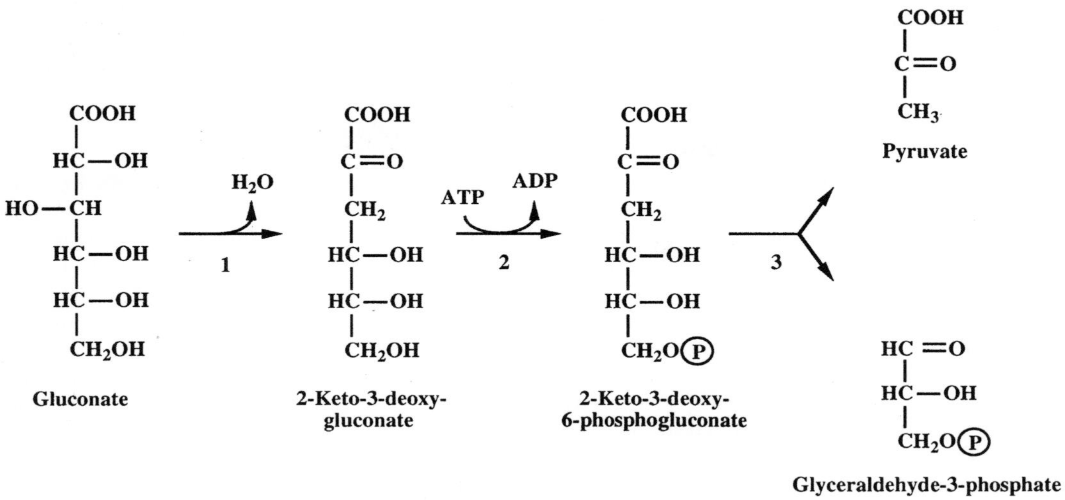

Gluconate → 2-Keto-3-deoxy-gluconate → 2-Keto-3-deoxy-6-phosphogluconate → Pyruvate + Glyceraldehyde-3-phosphate

**Fig. 3.** Degradation of gluconate by *Clostridium* species via the modified Entner–Doudoroff pathway. (1) gluconate dehydratase; (2) 2-keto-3-deoxygluconate kinase; (3) 2-keto-3-deoxy-6-phosphogluconate aldolase.

detected in species closely related to *C. acetobutylicum* ("*C. butylicum*"). The results of tracer studies were in agreement with the operation of the pentose phosphate cycle and the EMP pathway (CYNKIN and DELWICHE, 1958; CYNKIN and GIBBS, 1958; VOLESKY and SZCZESNY, 1983).

## 3.3 Formation of Products

### 3.3.1 Formation of Acids

Lactate becomes a major fermentation product of *C. acetobutylicum* under conditions of growth limiting iron concentrations (10 μM) and pH values higher than 5 (BAHL et al., 1986). Inhibition of the hydrogenase by carbon monoxide or potassium cyanide causes a similar effect (HANSON and RODGERS, 1946; KATAGIRI et al., 1960; KEMPNER and KUBOWITZ, 1933; KUBOWITZ, 1934). The lactate dehydrogenase (Ldh) responsible for reduction of pyruvate to lactate has been purified. It consists of 4 identical subunits with a molecular mass of 36 kDa, has a pH optimum of 5.8, uses NADH as a coenzyme, and is specifically activated by fructose-1,6-diphosphate (FREIER and GOTTSCHALK, 1987). Lactate might be taken up again and subsequently consumed, if the pH decreases to values below 4.5 and if the iron concentration is no longer growth-limiting. For consumption, however, acetate is needed as a co-substrate. Conversion of lactate into pyruvate seems to be mediated by a different lactate dehydrogenase (DIEZ-GONZALEZ et al., 1994). A lactate dehydrogenase gene has been cloned from *C. acetobutylicum* strain B643. The recombinant plasmid encoded a 38 kDa enzyme that was activated by addition of fructose-1,6-diphosphate (CONTAG et al., 1990).

The central enzyme for the breakdown of pyruvate is pyruvate:ferredoxin oxidoreductase (Pfo) (Fig. 2). The enzyme from *Clostridium acetobutylicum* was purified and characterized (MEINECKE et al., 1989). The molecular mass was found to be 123 kDa per monomer in SDS polyacrylamide gel electrophoresis. The subunit composition of the native enzyme remained undetermined due to the high oxygen sensitivity (50% inactivation in 1 h). The protein contained 2.9 mol sulfur, 4.1 mol iron, and 0.4 mol thiamine pyrophosphate per mol monomer, suggesting the presence of one 4Fe-4S cluster and one thiamine pyrophosphate in this subunit. The apparent $K_m$ values for pyruvate and coenzyme were determined to be 322 μM and 3.7 μM, respectively. Products of the reaction are acetyl-CoA, $CO_2$, and reduced ferredoxin.

Acetyl-CoA is partly converted to acetate and partly to butyrate. The ratio in which these products are formed varies to some extent. Acetate formation involves the enzymes phosphotransacetylase and acetate kinase, whereas butyrate production is catalyzed by subsequent action of thiolase, L(+)-3-hydroxybutyryl-CoA dehydrogenase, crotonase, butyryl-CoA dehydrogenase, phosphotransbutyrylase, and butyrate kinase (GAVARD et al., 1957; TWAROG and WOLFE, 1962; VALENTINE and WOLFE, 1960; ANDERSCH et al., 1983; HARTMANIS and GATENBECK, 1984). Production of both acids is associated with ATP synthesis in the kinase reactions. In Tab. 6, some characteristics of enzymes involved in pyruvate breakdown by *C. acetobutylicum* are summarized.

Both, phosphotransacetylase (Pta) and acetate kinase (Ack), have not been purified from *C. acetobutylicum.* Their activity has been measured in crude extracts. Strong expression could only be found during the acidogenic growth phase (ANDERSCH et al., 1983; BALLONGUE et al., 1986; HARTMANIS and GATENBECK, 1984). However, in another report such a strict activity pattern is not denoted (HÜSEMANN and PAPOUTSAKIS, 1989). Phosphotransacetylase from *C. beijerinckii* has been partially purified. Its molecular mass is 56–57 kDa (CHEN, 1993).

Formation of $C_4$-compounds starts by condensation of two acetyl-CoA molecules to yield acetoacetyl-CoA. This reaction is catalyzed by thiolase (Thl). The respective enzyme has been purified from *C. acetobutylicum* and consists of 4 identical subunits with a molecular mass of 44 kDa (WIESENBORN et al., 1988). The apparent $K_m$ values for acetyl-CoA, sulfhydryl-CoA, and acetoacetyl-CoA are 270 μM, 4.8 μM, and 32 μM, respectively. The structural gene of this enzyme has been cloned (PETERSEN and BENNETT, 1991). Recently, a different gene for a second thiolase from *C. acetobutylicum* as well as *C. beijerinckii* has been detected and sequenced (WINZER, MINTON, and DÜRRE, unpublished observations). It is tempting to speculate that these different thiolases might be alternatively active during acidogenic resp. solventogenic growth phases. The presence of two different thiolases has also been described for another solvent forming *Clostridium, C. pasteurianum* (BERNDT and SCHLEGEL, 1975). A report by a Japanese group might indicate that one of the thiolases from *C. acetobutylicum* ATCC 824 exhibits coenzyme A transferase activity in addition (NAKAMURA et al., 1990). A thiolase with a native molecular mass of 100–120 kDa has also been purified from two *C. beijerinckii* strains (CHEN, 1993).

Reduction of acetoacetyl-CoA to butyryl-CoA is catalyzed by the subsequent action of 3-hydroxybutyryl-CoA dehydrogenase (Hbd), crotonase (Cch), and butyryl-CoA dehydrogenase (Bcd). Hbd has not been purified from *C. acetobutylicum.* Determination of its activity in crude extracts showed a rapid decrease several hours after the onset of solventogenesis (HARTMANIS and GATENBECK, 1984). The enzyme has been purified from *C. beijerinckii* NRRL B593. The subunit and native molecular masses were reported to be 31 and 213 kDa, respectively (COLBY and CHEN, 1992). The amino terminus of the protein proved to be almost identical to the deduced amino acid sequence of a *hbd* structural gene that had been cloned and sequenced from strain P262 (COLBY and CHEN, 1992; YOUNGLESON et al., 1989b). Crotonase (crotonyl-CoA hydratase) has been obtained in homogenous form from an unspecified *C. acetobutylicum* strain (WATERSON et al., 1972), possibly strain NRRL B528 (CHEN, 1993). The enzyme is a tetramer, consisting of 4 identical subunits with a molecular mass of 40 kDa. It has a limited substrate specificity, being active only with $C_4$- and $C_6$-enoyl CoA, and shows a remarkable sensitivity towards high concentrations of crotonyl-CoA (WATERSON et al., 1972). Its activity pattern during growth is almost identical to that of Hbd (HARTMANIS and GATENBECK, 1984). Little information is available on the last enzyme of the reaction sequence, butyryl-CoA dehydrogenase. Enzymatic determinations in crude extracts of *C. acetobutylicum* were of poor reproducibility, which might indicate a high oxygen sensitivity (HARTMANIS and GATENBECK, 1984).

**Tab. 6.** Enzymes of *C. acetobutylicum* and *C. beijerinckii* Involved in Pyruvate Degradation

| Enzyme | Host | Coenzyme | Purified | Native Molecular Mass [kDa] | Subunit Size [kDa] | Composition | Cloned | Sequenced |
|---|---|---|---|---|---|---|---|---|
| Lactate DH[a] | *C. acetobutylicum* | NADH | + | NR[b] | 36 | $\alpha_4$ | + | – |
| Pyruvate:ferredoxin OR[c] | *C. acetobutylicum* | TPP[d], CoASH | + | NR | 123 | NR | – | – |
| Phosphotransacetylase | *C. beijerinckii* | CoASH | ±[e] | NR | 56–57 | NR | – | – |
| | *C. acetobutylicum* | | – | | | | – | – |
| Acetate kinase | *C. acetobutylicum* | ADP | – | | | | – | – |
| Acetaldehyde DH | *C. acetobutylicum* | NADH | – | | | | – | – |
| Alcohol DH | *C. acetobutylicum* | NADPH | + | NR | 44 | NR | + | + |
| Thiolase | *C. acetobutylicum* | CoASH | + | NR | 44 | $\alpha_4$ | + | – |
| | *C. beijerinckii* | CoASH | + | 100–120 | NR | NR | – | – |
| β-Hydroxybutyryl-CoA DH | *C. beijerinckii* | NADH | + | 213 | 31 | NR | – | – |
| | *C. acetobutylicum* | NADH | – | | | | + | + |
| Crotonase | *C. acetobutylicum* | | + | 158 | 40 | $\alpha_4$ | – | – |
| Butyryl-CoA DH | *C. acetobutylicum* | NR | – | | | | – | – |
| Phosphotransbutyrylase | *C. acetobutylicum* | CoASH | + | 264 | 31 | NR | + | + |
| | *C. beijerinckii* | CoASH | + | 205 | 33 | NR | – | – |
| Butyrate kinase | *C. acetobutylicum* | ADP | + | 85 | 39 | $\alpha_2$ | + | + |
| Hydrogenase | *C. acetobutylicum* | | – | | 63 | | + | + |
| NADH:ferredoxin OR | *C. acetobutylicum* | NADH | – | | | | – | – |
| NADPH:ferredoxin OR | *C. acetobutylicum* | NADPH | – | | | | – | – |
| NADH:rubredoxin OR | *C. acetobutylicum* | NADPH, FAD | + | NR | 41 | NR | – | – |
| CoA transferase | *C. acetobutylicum* | | + | 93 | 23, 24 | $\alpha_2\beta_2$ | + | + |
| | *C. beijerinckii* | | + | 85 | 23, 28 | NR | – | – |
| Acetoacetate decarboxylase | *C. acetobutylicum* | | + | 330 | 28 | $\alpha_{12}$ | + | + |
| | *C. beijerinckii* | | + | 200–230 | NR | NR | – | – |
| Aldehyde/alcohol DH (E) | *C. acetobutylicum* | NR | – | | 96 | | + | + |
| Butyraldehyde DH | *C. acetobutylicum* | NADH | + | 115 | 56 | $\alpha_2$ | – | – |
| | *C. beijerinckii* | NADH | + | 100 | 55 | $\alpha_2$ | – | – |
| Butanol DH (I) | *C. acetobutylicum* | NADH | + | NR | 43 | NR | + | + |
| Butanol DH (II) | *C. acetobutylicum* | NADH | + | 82 | 43 | $\alpha_2$ | + | + |
| Primary/secondary alcohol DH | *C. beijerinckii* | NADPH | + | 100 | 38 | NR | + | + |
| Alcohol DH | *C. beijerinckii* | NAD(P)H | + | 80 | 40, 43.5 | $\alpha_2$, $\alpha\beta$, $\beta_2$ | – | – |
| Alcohol DH | *C. beijerinckii* | NADPH | – | | | | | |

[a] Dehydrogenase
[b] Not reported
[c] Oxidoreductase
[d] Thiamine pyrophosphate
[e] Partially purified

The final steps of butyrate formation are catalyzed by phosphotransbutyrylase (Ptb) and butyrate kinase (Buk). Ptb activity is mainly found during the acidogenic fermentation phase (ANDERSCH et al., 1983; HARTMANIS and GATENBECK, 1984). The enzyme has been purified from *C. acetobutylicum* strain ATCC 824 and from *C. beijerinckii* (WIESENBORN et al., 1989a; THOMPSON and CHEN, 1990). Both proteins consist of identical subunits with a molecular mass of 31 and 33 kDa, respectively. However, sizes of the native enzymes differ considerably (264 vs. 205 kDa). This might be due to variations in the determination procedures. Ptb is very sensitive to pH changes and almost completely inactive in the butyryl phosphate-forming direction at a pH of about 6. The enzyme from *C. beijerinckii* also reacts with acetoacetyl-CoA in the presence of phosphate, and acetoacetyl phosphate might be a product (THOMPSON and CHEN, 1990). Whether this reaction is of physiological relevance is not yet known. Butyrate kinase has been purified from *C. acetobutylicum* ATCC 824. The native enzyme is a dimer of two apparently identical subunits that have a molecular mass of 39 kDa (HARTMANIS, 1987). Its relative activity with acetate is only 6% of that with butyrate. The genes for both enzymes have been cloned and sequenced from *C. acetobutylicum* ATCC 824 (CARY et al., 1988; WALTER et al., 1993) and NCIMB 8052 (now grouped with *C. beijerinckii,* as mentioned above) (OULTRAM et al., 1993). The two genes are contiguous on the chromosome and most likely form an operon (WALTER et al., 1993).

During acid production, reduced ferredoxin is oxidized by hydrogenase and $H_2$ is produced. Hydrogenase activity has been demonstrated in cell extracts of *C. acetobutylicum* (ANDERSCH et al., 1983; KIM and ZEIKUS, 1985). The respective gene has recently been cloned and sequenced from strain P 262 (SANTANGELO et al., 1994). The NADH generated in the 3-phosphoglyceraldehyde dehydrogenase reaction is oxidized in the $\beta$-hydroxybutyryl-CoA and butyryl-CoA dehydrogenase reactions. This results in the following redox balance:

$$\text{Glucose} + 2\ \text{NAD}^+ \rightarrow 2\ \text{Pyruvate} + 2\ \text{NADH} + 2\ \text{H}^+$$
$$2\ \text{Pyruvate} \rightarrow 2\ \text{Acetyl-CoA} + 2\ \text{CO}_2 + 2\ \text{H}$$
$$2\ \text{Acetyl-CoA} + 2\ \text{NADH} + 2\ \text{H}^+ \rightarrow \text{Butyrate} + 2\ \text{NAD}^+$$

---

$$\text{Glucose} \rightarrow \text{Butyrate} + 2\ \text{CO}_2 + 2\ \text{H}_2$$

However, not only 2 $H_2$/glucose but approximately 2.3 $H_2$/glucose are produced during acid formation. The additional 0.3 $H_2$ originate from NADH; it is oxidized by NADH:ferredoxin oxidoreductase which has been studied in *C. acetobutylicum* to some extent (PETITDEMANGE et al., 1976, 1977). The enzyme is activated by acetyl-CoA and inhibited by NADH. Hydrogenase in turn utilizes the electrons carried by reduced ferredoxin and, together with protons, forms molecular hydrogen. Since part of the NADH of the above equation is oxidized by $H_2$ evolution, not all acetyl-CoA is converted to butyrate. This is the reason why acetate is a fermentation product in addition to butyrate. There are thermodynamic limitations as to the extent of $H_2$ evolution from NADH, because the redox potential of the couple NADH/ $NADH^+ + H^+$ is more positive than that of $H_2/2\ H^+$ (THAUER et al., 1977). Another enzyme, NADPH:ferredoxin oxidoreductase, can also utilize reduced ferredoxin in the controlled production of NADPH. This may be the only route for the production of NADPH required for biosynthetic reactions, since most clostridia appear to lack the enzymes necessary for the NADPH-yielding oxidation of glucose-6-phosphate (JUNGERMANN et al., 1973). A NADH:rubredoxin oxidoreductase with a molecular mass of 41 kDa and FAD as a prosthetic group has also been purified from *C. acetobutylicum* ATCC 824 (PETITDEMANGE et al., 1979). The physiological function of this enzyme is still unknown. It is induced by acetate (and, to a lower extent, by butyrate) only at low pH values. Maximal activity has been observed at pH 4.8 (BALLONGUE et al., 1986). After the onset of solventogenesis the enzymatic activity rapidly decreases (MARCZAK et al., 1983, 1984). The level of rubredoxin in the cells fluctuates accordingly, while the concentration of ferredoxin remains almost constant throughout the fermentation (MARCZAK et al., 1985). Under iron limitation ferredoxin can no longer be synthesized. Instead, flavodoxin is formed, the gene of which has been cloned from *C. acetobutylicum* P 262 (SANTANGELO et al., 1991).

### 3.3.2 Formation of Solvents

Several conditions must be met to ensure a reproducible metabolic shift from acidogenesis to solventogenesis in *C. acetobutylicum.* These are a low pH, excess of substrate, threshold concentrations of butyrate and/or acetate, and a suitable growth-limiting compound such as phosphate or sulfate (BAHL and GOTTSCHALK, 1984). In strain NCIMB 8052 (now considered to be a member of the *C. beijerinckii* group) formation of acetone and butanol could be initiated at neutral pH by adding high concentrations (100 mmol $L^{-1}$ each) of acetate and butyrate (HOLT et al., 1984). Before the onset of solventogenesis a new set of enzymes is synthesized. Acetone production is catalyzed by acetoacetyl-CoA:acetate/butyrate coenzyme A transferase (CoA transferase or Ctf) and acetoacetate decarboxylase. In *C. beijerinckii* strains able to form isopropanol, a primary/secondary alcohol dehydrogenase subsequently converts acetone into isopropanol. Butanol is made from butyryl-CoA by action of various butyraldehyde and butanol dehydrogenase activities. Ethanol on the other hand is constitutively produced by *C. acetobutylicum* throughout the fermentation (GERISCHER and DÜRRE, 1992). Its synthesis is brought about by a NADPH-dependent alcohol dehydrogenase and a specific acetaldehyde dehydrogenase that does not react with butyryl-CoA (BERTRAM et al., 1990).

CoA transferase of *C. acetobutylicum* ATCC 824 has been purified to homogeneity (WIESENBORN et al., 1989b). The native enzyme was a heterotetramer of two different subunits with a molecular mass of 23 kDa and 28 kDa, respectively. However, cloning and sequencing of the respective genes (*ctfA* and *ctfB*) led to deduced values of 22.7 kDa and 23.7 kDa for strain ATCC 824 (CARY et al., 1990; PETERSEN et al., 1993) and 23.6 kDa for both subunits in strain DSM 792 (GERISCHER and DÜRRE, 1990; FISCHER et al., 1993). Similar data have been determined with purified CoA transferase from *C. beijerinckii* NRRL B593 (CHEN, 1993). CoA transferase plays a major role in the uptake of acids during solventogenesis (preferentially butyrate) that are subsequently converted to acetone and butanol (ANDERSCH et al., 1983; HARTMANIS et al., 1984). To some extent, butyrate can be taken up in batch cultures via the reversed reactions of butyrate kinase and phosphotransbutyrylase (HÜSEMANN and PAPOUTSAKIS, 1989). A surprising feature of CoA transferase are the high $K_m$ values for acetate and butyrate (1.2 mol $L^{-1}$ and 0.6 mol $L^{-1}$, respectively). They have been suggested to reflect a gradational response to the progressive toxic effects of increasing levels of these acids (WIESENBORN et al., 1989b). $K_m$ values for the corresponding enzyme from *C. beijerinckii* are significantly lower (CHEN, 1993).

Acetoacetate decarboxylase (Adc) of *C. acetobutylicum* has been the subject of extensive investigations in the 1960s aiming to elucidate its reaction mechanism (FRIDOVICH, 1972; WESTHEIMER, 1969). Purification of the native enzyme revealed that it had a molecular mass of 340 kDa and consisted of 12 identical subunits with a molecular mass of 29 kDa (TAGAKI and WESTHEIMER, 1968). Later investigations showed the respective data to be rather 330 kDa and 28 kDa (GERISCHER and DÜRRE, 1990; PETERSEN and BENNETT, 1990), confirmed by the deduced amino acid sequence after cloning and sequencing the structural gene, which forms a monocistronic operon (GERISCHER and DÜRRE, 1990; 1992; PETERSEN and BENNETT, 1990; PETERSEN et al., 1993). This operon is adjacent, but convergently arranged to the *ctf* genes that form a common transcription unit together with the gene for an aldehyde/alcohol dehydrogenase (FISCHER et al., 1993; NAIR et al., 1994). The reaction mechanism of Adc involves the intermediate formation of a Schiff base between the substrate acetoacetate and a lysine residue at the active site (WARREN et al., 1966). Native acetoacetate decarboxylase from *C. beijerinckii* NRRL B592 and NRRL B593 has a considerably smaller molecular mass of 200–230 kDa

(CHEN, 1993). The latter strain belongs to the few clostridia able to further reduce acetone to isopropanol (GEORGE et al., 1983). The enzyme catalyzing this reaction is a primary/secondary alcohol dehydrogenase with a native molecular mass of 100 kDa, consisting of identical subunits with a molecular mass of 38 kDa (CHEN, 1993; HIU et al., 1987). It is NADPH dependent and, although it converts acetaldehyde to ethanol *in vitro,* reduction of acetone to isopropanol is the preferred reaction. The structural gene has been cloned and sequenced. The deduced amino acid sequence shows 75% identity to the thermostable alcohol dehydrogenase of *Thermoanaerobium brockii,* an anaerobe producing ethanol (CHEN, 1993).

Butanol formation is catalyzed by butyraldehyde (Bad) and butanol dehydrogenase (Bdh) activities. A butyraldehyde dehydrogenase has been purified from *C. acetobutylicum* NRRL B643 and from C. beijerinckii NRRL B592 (PALOSAARI and ROGERS, 1988; YAN and CHEN, 1990). Molecular size and subunit composition were similar for the two enzymes. Bad from *C. acetobutylicum* was a homodimer of two 56 kDa subunits, yielding a native enzyme with a molecular mass of 115 kDa. The respective data for *C. beijerinckii* are 55 kDa and 100 kDa. Both enzymes prefer NADH as a coenzyme over NADPH, and butyryl-CoA is a better substrate than acetyl-CoA. Recently, in *C. acetobutylicum* a gene has been cloned and sequenced that has high homology with the *adhE* gene of *Escherichia coli* known to encode a protein with aldehyde and alcohol dehydrogenase domains. The respective clostridial enzyme is believed to play a role in butanol synthesis at the onset of solvent formation (FISCHER et al., 1993; NAIR et al., 1994).

For a long time, conflicting data have been reported on alcohol dehydrogenase in *C. acetobutylicum.* The situation was clarified in 1987 by a report providing evidence for at least two different enzyme activities in this organism (DÜRRE et al., 1987). An NADPH-dependent enzyme was purified from strain DSM 792. It consists of identical subunits with a molecular mass of 44 kDa, it is very unstable and has a pH optimum between 7.8 and 8.5 (MICHELS, 1990). The physiological role of this alcohol dehydrogenase is probably in ethanol formation in cooperation with a specific acetaldehyde dehydrogenase as suggested from studies with transposon-induced mutants (BERTRAM et al., 1990). Activity of the NADPH-dependent alcohol dehydrogenase is found throughout the fermentation (DÜRRE et al., 1987) confirming the constitutive nature of the enzyme which is in agreement with the constitutive formation of low ethanol levels (GERISCHER and DÜRRE, 1992). This alcohol dehydrogenase probably regulates the pool of reduced nicotinamide adenine dinucleotide phosphate which is supplied by NADPH:ferredoxin oxidoreductase. From *C. acetobutylicum* P 262 a gene encoding a NADPH-dependent alcohol dehydrogenase has been cloned and sequenced (YOUNGLESON et al., 1988, 1989a). However, due to the different taxonomic grouping of P 262 it is questionable whether the respective enzyme has the same physiological function as in strain DSM 792.

Two primarily NADH-dependent butanol dehydrogenase isozymes have been purified from *C. acetobutylicum* ATCC 824 (PETERSEN et al., 1991; WELCH et al., 1989). Both native enzymes are homodimers with subunits of molecular masses of about 42 kDa. Bdh I had only a 2-fold higher activity with butyraldehyde than with acetaldehyde, whereas Bdh II was reported to have a 46-fold higher activity with butyraldehyde than with acetaldehyde. Bdh II required $Zn^{2+}$ for activity and had a pH optimum of 5.5, a value around which the cells switch from acidogenesis to solventogenesis. Thus, Bdh II is probably the major alcohol dehydrogenase involved in butanol production. The genes of both isozymes (*bdhA* and *bdhB*) have been cloned and sequenced (PETERSEN et al., 1991; WALTER et al., 1992). They are arranged in monocistronic operons each, that are contiguous on the chromosome and are controlled by different promoters. Recently, another gene of *C. acetobutylicum* with high homology to the *E. coli adhE* gene has been cloned and sequenced (FISCHER et al., 1993; NAIR et al., 1994). The respective *E. coli* gene product has aldehyde and alcohol dehydrogenase domains (GOODLOVE et al., 1989) and its clostridial analog is believed to be involved in butanol synthesis (FISCHER et al., 1993; NAIR et al., 1994). The clostridial *adhE* (or *aad*) gene is part of an operon additionally containing the *ctfA* and *ctfB* genes. This transcription unit (*sol* operon) is controlled by two different promoters. Thus, at least three different butanol dehydrogenases might be involved in butanol production in *C. acetobutylicum.*

A similar situation has been found with *C. beijerinckii.* Strain NRRL B592 producing acetone, butanol, and ethanol, but not isopropanol, expresses two types of alcohol dehydrogenases with distinct coenzyme specificity. One of them is NADPH-dependent whereas the other can use either NADH or NADPH (CHEN, 1993). This latter enzyme has a native molecular mass of 80 kDa, consists of two subunits with molecular masses of 40 kDa and 43.5 kDa, and yields three distinct activity bands after electrophoresis of purified protein under non-denaturing conditions. These three species might be dimers of the composition $\alpha_2$, $\alpha\beta$, and $\beta_2$ CHEN, 1992).

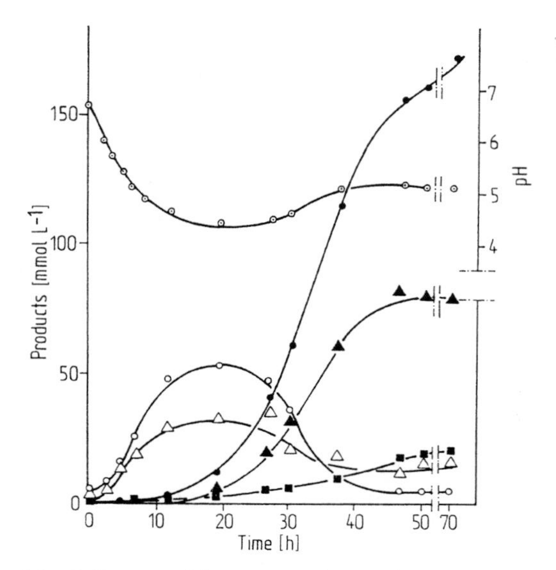


**Fig. 4.** Course of the acetone–butanol fermentation in batch culture. Butanol, ●; acetone, ▲; ethanol, ■; butyrate, ○; acetate, △; pH, ⊙ (BAHL et al., 1982b).

## 3.3.3 Regulation of Product Formation

A batch culture of *C. acetobutylicum* classically shifts from acid to solvent formation towards the end of growth at pH values below 5 (Fig. 4). The shift can reproducibly be achieved in complex media (OXFORD et al., 1940) and is often associated with sporulation (LONG et al., 1984a, b). The ability to initiate and complete the process of sporulation is not a prerequisite of solvent production. In agreement with this an asporogenous mutant is selected in solvent producing continuous cultures (MEINECKE et al., 1984). Although the signal for initiating sporulation is not nutrient limitation as in *Bacillus* species (LONG et al., 1984b) the process of spore formation seems to be identical since analogs of the sporulation-specific *Bacillus sigma* factor genes have been cloned and sequenced from *C. acetobutylicum* (SAUER et al., 1994).

The shift to solventogenesis in *C. acetobutylicum* and *C. beijerinckii* is characterized by a decrease of the activity of some acidogenic enzymes (ANDERSCH et al., 1983; HARTMANIS and GATENBECK, 1984). At about the same time the enzymes needed for solventogenesis are induced or derepressed (ANDERSCH et al., 1983; DÜRRE et al., 1987; HARTMANIS and GATENBECK, 1984; YAN et al., 1988). Induction requires synthesis of new mRNA and protein as shown by experiments with *C. acetobutylicum* using the transcription or translation inhibitors rifampicin and chloramphenicol (BALLONGUE et al., 1985; PALOSAARI and ROGERS, 1988; WELCH et al., 1992). mRNA analyses provided evidence that induction or derepression started several hours before solvents could be detected in the medium. Induction of the *adc, bdhA, bdhB,* and *sol* operons at the mRNA level showed a rapid increase up to a maximum shortly before the production of acetone and butanol followed by a massive decrease of transcripts (FISCHER et al., 1993; GERISCHER and DÜRRE, 1992; WALTER et al., 1992). During the shift from acidogenesis to solventogenesis also the synthesis of five heat shock proteins including GroEL and DnaK is induced (PICH et al., 1990; BAHL, 1993). This stress response is linked to onset of solventogenesis and sporulation in a yet unknown way. Genes encoding several of these proteins have been cloned and sequenced (BEHRENS et al., 1993; NARBERHAUS and BAHL, 1992; NARBERHAUS et al., 1992; SAUER and DÜRRE, 1993).

Despite the enormous progress made in elucidation of gene structure and mRNA regulation, little is known on the signal(s) that initiate solventogenesis (and possibly sporulation and stress response). Involvement of alternate *sigma* factors in transcription of genes encoding solventogenic enzymes is unlikely, since the RNA polymerase of *C. acetobutylicum* has been purified (PICH and BAHL, 1991) and, after isolation from acidogenic and solventogenic cells, showed the same composition including only the vegetative *sigma* factor (BAHL, 1993). Recently, a model has been proposed that postulates the ATP pool to be a signal for acetone formation and the NAD(P)H pool to be a signal for butanol production (GRUPE and GOTTSCHALK, 1992). However, since CoA transferase (required for acetone production) and AdhE (needed for butanol formation) are encoded in a common transcription unit (FISCHER et al., 1993), this model must at least be modified, although the pools of ATP and reducing equivalents certainly play a major role in regulation. On the other hand it might be possible that DNA topology directly responds to changes in the environment (pH, salt concentration, osmotic pressure, temperature) and thus provokes transcription of certain genes. The degree of supercoiling of DNA isolated from acidogenic and solventogenic cells was different (WONG and BENNETT, 1994). Supercoiling is controlled by topoisomerase I and DNA gyrase, the genes of which have been cloned and sequenced from *C. acetobutylicum* DSM 792 (ULLMANN and DÜRRE, 1994). Transcription analysis and targeted inactivation of the *gyr* genes should provide a conclusive answer to this problem.

An additional level of regulation might be present as indicated by studies with a transposon-induced, acetone- and butanol-negative mutant of *C. acetobutylicum* (BERTRAM et al., 1990). Genetic analysis revealed that the transposon Tn*916* had inserted in front of a tRNA gene thus causing massive reduction of expression (SAUER and DÜRRE, 1992). This tRNA recognizes the rarely used ACG codon which, however, is present in all genes required for acetone and butanol synthesis. In addition, it has been found in inducible genes for sporulation, autolysis, and uptake or metabolism of minor C or N substrates (SAUER and DÜRRE, 1992). This codon distribution could either reflect an evolutionary process with low mutational pressure on weakly expressed genes or represent a novel translational control mechanism. Such a system has been found in the genus *Streptomyces* where a rare tRNA controls production of aerial mycelium and antibiotics (LESKIW et al., 1991).

## 3.4 Strain Improvement

Various research activities are devoted to the solution of cultivation and product recovery problems from which the acetone–butanol fermentation suffers most. In addition, knowledge on the physiology and genetics of *C. acetobutylicum* and *C. beijerinckii* is rapidly increasing. Besides the work concerned with correlation of culture conditions to the physiology of the organism and with the regulatory mechanisms controlling the key enzymes of the different metabolic pathways, there is great interest in removing some limiting features of *C. acetobutylicum.* An improvement of the microorganisms could include: increased resistance to butanol, fewer end products (e.g., no acetone), and broader substrate range (e.g., cellulose).

Butanol toxicity is the limiting factor with respect to the maximum amount of solvents (acetone, butanol, ethanol) to be achieved in the fermentation broth (about 20 g $L^{-1}$). Improvement in this area is critical for the economics of the process (reduction of product recovery costs). It is generally accepted that the cell membrane is one target of alcohol attack (INGRAM, 1986). Owing to their amphiphatic character, alcohols dissolve in the membrane lipids, increase their permeability (DOMBEK and INGRAM, 1984), and affect membrane fluidity (VOLLHERBST-SCHNECK et al., 1984). Thus, high concentrations of butanol lead to a complete abolition of the pH gradient, lower the intracellular level of ATP, cause the release of intracellular metabolites, and inhibit sugar uptake (MOREIRA et al., 1981; GOTTWALD and GOTTSCHALK, 1985; BOWLES and ELLEFSON, 1985; HUTKINS and KASHKET, 1986). These effects clearly demonstrate the importance of the cell membrane with respect to butanol tolerance. Other sites of interference, e.g., inhibition of glycolytic enzymes (HERRERO, 1983), cannot be excluded. In view of the complex effects of butanol, it might be very difficult to isolate strains of *C. acetobutylicum* with a substantial higher butanol tolerance. Chemical mutagenesis has been used to obtain butanol-tolerant strains that, however, produced only little more butanol (if at all) than the wild type (HERMANN et al., 1985; LEMMEL, 1985; LIN and BLASCHEK, 1983).

Non-production of the less desirable solvent acetone can be achieved by selecting for mutants in the presence of 2-bromobutyrate (JANATI-IDRISSI et al., 1987). Low acetone yields also result from growth on whey or under iron limitation (BAHL et al., 1986). Improvements in substrate utilization have been attempted by isolating mutants with enhanced amylolytic activity (ANNOUS and BLASCHEK, 1991) or by transformation of endoglucanase genes from *C. cellulovorans* or *C. thermocellum* into *C. acetobutylicum* (KIM et al., 1994; OULTRAM et al., 1990). A class of spontaneous mutants of *C. acetobutylicum* NRRL B643 resistant to allyl alcohol produced considerable quantities of butyraldehyde in addition to acetone and butanol which could be of interest for chemical syntheses (ROGERS and PALOSAARI, 1987).

Meanwhile, the repertoire of genetic and recombinant DNA techniques to be used with *C. acetobutylicum* has increased impressively. A variety of shuttle vectors (with either *E. coli* or *B. subtilis* as alternative host) are available (AZEDDOUG et al., 1992; LEE et al., 1992; MINTON et al., 1993; STRÄTZ et al., 1994; TRUFFAUT and SEBALD, 1988; TRUFFAUT et al., 1989; YOON et al., 1991; YOSHINO et al., 1990), transformation and electroporation protocols have been worked out (LIN and BLASCHEK, 1984; MERMELSTEIN et al., 1992; OULTRAM et al., 1988a; REID et al., 1983; REYSSET et al., 1988; REYSSET and SEBALD, 1993), DNA transfer by conjugation is easily performable (OULTRAM and YOUNG, 1985; OULTRAM et al., 1987, 1988b; REYSSET and SEBALD, 1985; WILKINSON and YOUNG, 1994; WILLIAMS et al., 1990; YOUNG, 1993; YU and PEARCE, 1986), the isolation of a filamentous phage led to the construction of a phagemid (KIM and BLASCHEK, 1991, 1993), and transposon mutagenesis is well established (BERTRAM and DÜRRE, 1989; BERTRAM et al., 1990; DÜRRE, 1993; WOOLLEY et al., 1989; YOUNG, 1993). Very important was the finding that DNA to be transformed into *C. acetobutylicum* must be methylated to avoid restriction (MERMELSTEIN and PAPOUTSAKIS, 1993). Protease-deficient mutants will be helpful for the study of protease-labile proteins (SASS et al., 1993) and DNase-negative mutants might be helpful in genetic manipulations (BURCHHARDT and DÜRRE, 1990).

Recently, an artificial operon (*ace* operon) has been constructed containing the *adc, ctfA,* and *ctfB* genes, all transcribed from the *adc* promoter. After transformation into a acetone–butanol-negative mutant of *C. acetobutylicum,* the recombinant strain regained the ability to form acetone (MERMELSTEIN et al., 1993). When introduced into the wild type the *ace*-containing plasmid also led to an increase in solvent formation, although plasmid without insert caused a significant (but somewhat lower) stimulation of acetone, butanol, and ethanol production. Thus, genetic manipulations to alter the product formation capabilities are now available and might lead to a reintroduction of this fermentation at an industrial scale.

# 4 New Developments of the Fermentation Process

The industrial acetone–butanol fermentation process, which was in use for over forty years, is documented well and in detail in the literature (BEESCH, 1952; PRESCOTT and DUNN, 1959; ROSS, 1961; HASTINGS, 1978; SPIVEY, 1978; JONES and WOODS, 1986;

BAHL and GOTTSCHALK, 1988). This traditional commercial batch process of acetone–butanol fermentation is no longer in use. The main factors influencing the unfavorable economics were high raw material and transportation costs and intrinsic limitations (low solvent yields and low final concentrations, undesirable solvent ratios) resulting in high processing costs and waste disposal problems.

The substrate caused about 60% of the overall production costs (ROSS, 1961). The price for maize starch and molasses increased in the years after World War II to a level at which the fermentation could no longer compete with the synthetic route that used cheap oil as a feedstock. Although in some cases alternative substrates were available, transportation costs of the bulky material prevented its use. In addition, a theoretical calculation taking the biochemical pathways of *C. acetobutylicum* into account showed that the maximum possible yield of solvents is 0.38 g per g of glucose converted (LEUNG and WANG, 1981). In practice, this yield was not always reached, particularly with molasses as substrate. Also, the productivity of a typical batch fermentation process was low (0.2–0.6 kg $m^{-3}$ $h^{-1}$ solvents). The solvents represented a mixture of acetone, butanol, and ethanol, whereas often only one product, e.g., butanol as desired.

Another major drawback is the low butanol tolerance of *C. acetobutylicum.* A maximum total solvent concentration of 20 kg $m^{-3}$ can be achieved in the fermentation broth. The recovery of solvents from this dilute solution (normally by distillation) resulted in substantial costs. The fuel expenses for steam generation amounted to about 15–20% of the production costs; 65% of the steam was used for distillation. Due to the low product concentrations the acetone–butanol fermentation generates especially large volumes of waste. The disposal of such waste often was a problem (SPIVEY, 1978).

After the era of cheap oil has gone – made obvious through the oil crisis of 1973/74 – there is a renewed interest in fermentation processes for the production of fuels and chemicals from biomass. This holds also for the acetone–butanol fermentation. The following section will emphasize recent research related to the improvement of the fermentation process. Included are topics such as alternative substrates, continuous culture, immobilized cells, cell recycling, and product recovery. Progress in several of these areas will be crucial for renewed economic viability of acetone–butanol fermentation. Three parameters are important for the evaluation of new processes in comparison to the traditional batch process: (1) the final solvent concentration obtained [g solvents $L^{-1}$], (2) the yield [kg solvents $kg^{-1}$ sugar], and (3) the productivity [g solvents $L^{-1}$ $h^{-1}$].

## 4.1 Continuous Culture

A fermentation process that can be operated continuously has some advantages compared to a batch process. Only one series of inoculum cultures is needed for a long production period. A "dead season" necessary for filling, sterilization, cooling, and cleaning of the equipment is largely decreased, and the volume of the fermenter vessel can be reduced without a loss of production capacity (higher productivities). The use of a continuous process for acetone–butanol fermentation might provide an additional advantage. The fermentation time can be shortened by eliminating the first acidogenic phase when the parameters are known to keep the cells in the solventogenic phase constantly for a long time. Despite all potential advantages over a batch process there are only few reports about commercial continuous production of solvents by *C. acetobutylicum.* DYR et al. (1958) reported a continuous acetone–butanol production in a three to five vessel flow system. Although the productivity was three times higher compared to the batch process the substrate concentration and the solvent yields were low, and the acids formed in the first stage were not converted into solvents in the following stages. YAROVENKO (1964) described a similar fermentation carried out in a pilot plant consisting of a chain of eleven fermenters. Detailed information of the conditions of these continuous fermentations was not given.

To be economically attractive a continuous process must at least achieve the final concentration and yield of solvents that can be obtained in batch culture. In addition, significantly higher productivities are necessary.

In recent years, the continuous culture of *C. acetobutylicum* has been mainly used as a research tool to define parameters responsible for changes in the physiology and the activity of this microorganism. Under steady-state conditions a constant environment is provided and the influence of a single parameter and its interactions with other factors can be determined. The following fundamental areas of acetone–butanol fermentation were examined: the effect of medium components and acidic fermentation products on solvent production, the influence of temperature, culture pH, dilution rate ($D$), maximum attainable solvent concentration and yield, and the stability of a continuous culture with respect to the ability of solvent production.

Tab. 7 gives a summary of results obtained with chemostat cultures of *C. acetobutylicum.* A direct comparison of the results is difficult because of differences in the strains used, in medium composition, and in fermentation conditions.

Generally, it can be concluded that no single growth-limiting factor specifically induces solvent production in a chemostat. However, some nutrients have been shown to be more suitable for growth limitation and production of solvents in high yields than others. In glucose-, nitrogen-, or magnesium-limited chemostats steady-state solvent production was low or difficult to maintain, and an application of these kinds of limitations to an industrial process seems unlikely (GOTTSCHAL and MORRIS, 1981b; BAHL et al., 1982a; ANDERSCH et al., 1982; MONOT and ENGGASSER, 1983; JÖBSES and ROELS, 1983; STEPHENS et al., 1985; ROOS et al., 1984; MONOT et al., 1983; BAHL and GOTTSCHALK, 1984). Phosphate and sulfate belong to the group of suitable growth-limiting factors (BAHL et al., 1982b; BAHL and GOTTSCHALK, 1984).

The pH value had an important influence on the product pattern of *C. acetobutylicum* in continuous culture. Significant solvent production was only observed at pH values of 5 and below. The optimal pH may vary with respect to the strain used, e.g., *C. acetobutylicum* DSM 1731: pH 4.3 (BAHL et al., 1982a, b); ATCC 824: pH 4.8–5.0 (MONOT and ENGASSER, 1983). The temperature applied was in the same range as in the batch process (33–37 °C) and had minor influence. Increased levels of butyrate and/or acetate are able to induce solvent production in batch culture (GOTTSCHAL and MORRIS, 1981a). In agreement with these results, acids have to be present in continuous culture at threshold concentrations (about 10 mmol $L^{-1}$) to make solvent production possible (BAHL and GOTTSCHALK, 1984).

All experiments in continuous culture to test the influence of the dilution rate showed that high solvent concentrations could be obtained at a low dilution rate ($D \leq 0.05\ h^{-1}$) (LEUNG and WANG, 1981; BAHL et al., 1982b). Although higher dilution rates resulted in improved productivity, this was at the expense of reduced substrate turnover and solvent concentration.

Using a two-stage phosphate-limited chemostat BAHL et al. (1982b) reported solvent concentrations of 18.2 g $L^{-1}$ with a yield of 0.34 g solvents per g glucose and a productivity of 0.44 g $L^{-1}$ $h^{-1}$ (Tab. 8). The first stage was run at $D = 0.125\ h^{-1}$ (pH 4.3, 37 °C), and the second stage was operated at $D = 0.03\ h^{-1}$ (pH 4.3, 33 °C). Compared to a batch culture using the same strain (*C. acetobutylicum* DSM 1731) and medium, the productivity doubled without a loss of substrate utilization, solvent concentration, or yield (BAHL et al., 1982b).

Although the economics remain to be established this two-stage process seems to be promising. Performance of the fermentation in this way has the advantage that the conditions at the stages (e.g., temperature, dilution rate) can be optimized with respect to either growth or solvent formation. At the first stage the cells are growing under conditions under which solvent formation is induced. The second stage is primarily devoted to the conversion of the residual sugars to solvents. Since the limiting growth factor is exhausted, growth is not possible here. This principle seems to have an additional advantage with respect to culture stability. Culture degeneration and difficulties in maintaining steady-

**Tab. 7.** Production of Solvents by *Clostridium acetobutylicum* in Chemostats

| Fermentation Process Nutrient Limitation | Strain | Substrate Concentration [g L$^{-1}$] | Temperature [°C] | pH | $D$ [h$^{-1}$] | Results Solvents [g L$^{-1}$] | Productivity [g L$^{-1}$ h$^{-1}$] | Yield [g/g] | Comments | References |
|---|---|---|---|---|---|---|---|---|---|---|
| **Glucose** | | | | | | | | | | |
| 2.7 g L$^{-1}$ | NCIB 8052 | 2.7 | 35 | 5.7 | 0.08 | – | – | – | No solvent production | Gottschal and Morris (1981b) |
| 3.4 g L$^{-1}$ | DSM 1731 | 3.4 | 37 | 4.3 | 0.13 | 0.07 | 0.01 | 0.02 | Induction of solvent formation by lowering the pH below 5.0; optimum: pH 4.3 | Bahl et al. (1982a) |
| **Nitrogen** | | | | | | | | | | |
| 0.4 g L$^{-1}$ Ammonium chloride | NCIB 8052 | 2.7 | 35 | 5.7 | 0.08 | – | – | – | No solvent production at relatively high pH values and low sugar concentrations | Gottschal and Morris (1981b) |
| 2 g L$^{-1}$ Ammonium sulfate | DSM 1731 | 54 | 37 | 5.2 | 0.22 | 2.50 | 0.95 | 0.08 | Solvent production at pH values from 5.2 to 4.3 and at high glucose concentrations | Andersch et al. (1982) |
| 0.2 g L$^{-1}$ Ammonium acetate | ATCC 824 | 45.5 | 35 | 5.0 | 0.04 | 8.00 | 0.30 | 0.29 | Solvent production at low dilution rate, low pH, and high glucose concentrations | Monot and Engasser (1983) |
| **Sulfate** | | | | | | | | | | |
| 0.05 g L$^{-1}$ Magnesium sulfate 0.01 g L$^{-1}$ Manganese sulfate | DSM 1731 | 54 | 37 | 4.3 | 0.10 | 4.3 | 0.43 | 0.32 | Yield comparable to batch process; higher solvent concentration possible at lower dilution rates | Bahl and Gottschalk (1984) |

| | | | | | | | | | | |
|---|---|---|---|---|---|---|---|---|---|---|
| **Phosphate** | | | | | | | | | | |
| Two-stage chemostat | | | stage I | | | | | | | |
| 0.1 g $L^{-1}$ Monopotassium phosphate | DSM 1731 | 54 | 37 | 4.3 | 0.125 | 18.2 | 0.44 | 0.34 | Successful laboratory-scale two-stage system for continuous solvent production, no culture degeneration over a period of one year | BAHL et al. (1982b) |
| | | | stage II 33 | 4.3 | 0.03 | | | | | |
| **Magnesium** | | | | | | | | | | |
| 0.02 g $L^{-1}$ Magnesium sulfate | DSM 1731 | 54 | 37 | 4.3 | 0.06 | 0.8 | 0.05 | 0.07 | Poor solvent production at low pH, low dilution rate, and excess of substrate | BAHL and GOTTSCHALK (1984) |

**Tab. 8.** Concentration of Products and Consumption of Substrate in a Two-Stage Continuous Culture of *Clostridium acetobutylicum* under Phosphate Limitation (BAHL et al., 1982b)

| | Stage 1 | Stage 2 |
|---|---|---|
| Process Parameter | | |
| pH | 4.3 | 4.3 |
| Temperature [°C] | 37 | 33 |
| Dilution rate [$h^{-1}$] | 0.125 | 0.03 |
| Products [g $L^{-1}$] | | |
| Butanol | 3.6 | 12.6 |
| Acetone | 1.6 | 4.8 |
| Ethanol | 0.3 | 0.8 |
| Butyrate | 1.3 | 0.8 |
| Acetate | 0.4 | 0.5 |
| Consumption of substrate [%] | 40 | 99.7 |
| Yield [g solvent/g glucose] | 0.25 | 0.34 |
| Productivity [g solvent $L^{-1}$ $h^{-1}$] | 0.08 | 0.44[a] |

[a] Productivity of the overall process

state solvent production over a longer period (i. e., >200 h) have often been encountered (STEPHENS et al., 1985; GOTTSCHAL and MORRIS, 1981a). High solvent concentrations seem to be responsible for this effect (FICK et al., 1985). On the other hand, long-term solvent production with *C. acetobutylicum* in the two-stage phosphate-limited chemostat has been reported. Running the first stage under conditions of low solvent concentrations (high dilution rate) but solvent formation is absolutely required because of the effect of acids at a low pH (4.3), the selection of such non-producing strains is apparently prevented. It is interesting that under these conditions an asporogenous strain of *C. acetobutylicum* was selected with an unchanged capability of producing solvents (MEINECKE et al., 1984). Other continuous flow fermentation processes (excess substrates, LEUNG and WANG, 1981; turbidostat, GOTTSCHAL and MORRIS, 1982; pH-auxostat, STEPHENS et al., 1985) were also used, and the results support the view that a continuous acetone–butanol fermentation is feasible.

## 4.2 Cell Immobilization and Cell Recycling

The low productivity and the low final concentration of solvents in the fermentation broth are the major drawbacks. Improvement in this field is necessary before a reintroduction of the acetone–butanol fermentation process can be envisaged.

Cell immobilization is a technique to confine the biocatalysts within the fermenter system. Thereby higher cell densities are possible resulting in higher productivities (smaller reactor volumes possible). Other advantages include the use of simpler non-growth media and easier separation of the cells from the products. On the other hand, the activity of the cells may be affected due to immobilization conditions.

Immobilization of vegetative cells and spores of *C. acetobutylicum* has been described. Calcium alginate (spherical beads or spiral wound flat sheets) was used to entrap the organism for a continuous solvent production in a glucose medium (HÄGGSTRÖM and ENFORS, 1982; FÖRERG et al., 1983). Packed-bed reactors, continuous stirred tank reactors, and fluidized column reactors served as reaction vessels.

Generally, higher solvent productivities (2.4–2.8 g $L^{-1}$ $h^{-1}$) have been obtained. However, substrate utilization and solvent concentrations were low (1.5–4.5 g $L^{-1}$). Another problem is the rapid loss of activity of *C. acetobutylicum* in non-growth media. The pulsewise addition of nutrients to the fermentation medium resulted in some improvement (FÖRBERG et al., 1983).

In addition to the entrapment technique, adsorption of cells to a solid surface can be used for immobilization. The successful adsorption of *C. acetobutylicum* to beech wood shavings (FÖRBERG and HÄGGSTRÖM, 1985), arranged as parallel sheets, is promising since it is a simple, cheap, and non-toxic immobilization method. This process using the intermittent nutrient dosage method was run for over one month with low cell leakage, but again the solvent concentrations were low (5–6 g $L^{-1}$). To overcome this problem the use of an immobilized sporulation-deficient strain of *C. acetobutylicum* P 262 has been reported (LARGIER et al., 1985). At a dilution rate of 0.42 $h^{-1}$ the solvent concentration reached 15 g $L^{-1}$ with a productivity of 3.02 g $L^{-1}$ $h^{-1}$ and a yield of 0.44 kg solvents per kg sucrose. The 5-fold increase in solvent concentration at a similar productivity represents major progress as compared to immobilized wild type cells.

These results clearly show that solvent production with immobilized cells is principally possible. Advances in other fields, e.g., strain improvement are necessary for the application of such a process.

Cell recycling is an alternative to increase productivity by the reuse of productive cells in a continuous culture system. Using this method the solvent productivity of *C. acetobutylicum* could be increased considerably (up to 3 g $L^{-1}$ $h^{-1}$) without a decrease in the concentration of solvents (up to 22 g $L^{-1}$) (AFSCHAR et al., 1985; SCHLOTE and GOTTSCHALK, 1986). Instability of the cells with respect to solvent formation was again observed after a long operation period at high solvent concentrations. AFSCHAR et al. (1985) could maintain a long-term cultivation under stable conditions by combining cell recycling with a two-stage fermentation. At the first stage the growing cells were not exposed to high solvent concentrations which otherwise favor the selection of non-producing strains. SCHLOTE and GOTTSCHALK (1986), on the other hand, observed no change in the activity of the cells over a period of three months in a phosphate-limited chemostat with cell recycling. The special conditions of phosphate limitation and the low pH of 4.3 might have been responsible for this stability of the culture.

Industrial application of this technique will depend on the availability of filtration modules that are easy to handle and not affecting the activity of the *C. acetobutylicum* cells. A wide variety of new membrane types in different configurations (hollow fiber, tubular, flat sheet) are available now which might lead to a more practicable process, in conjunction with a better knowledge of the physiology of *C. acetobutylicum* in high density suspensions.

In Tab. 9 the solvent concentrations, yields, and productivities obtained in different continuous acetone–butanol fermentations with cell recycle or immobilized cells are summarized.

## 4.3 Product Recovery

The traditional recovery by distillation of the accumulated solvents present in the fermentation broth in relatively low concentrations results in high costs and is one of the major drawbacks to the acetone–butanol fermentation. As long as strains of *C. acetobutylicum* tolerating higher concentrations of butanol are not available improvement of this fermentation might come from a more economic recovery process. Furthermore, as described above, substantial progress has been made in the development of continuous processes. However, the high productivities of such processes were often accompanied by reduced substrate utilization and low product concentration. These problems might be solved by recycling the fermenter effluent to allow further substrate turnover, and then to avoid product inhibition by an effective product removal technique. Such techniques could include the use of membranes (reverse osmosis, pervaporation), adsorbents, liquid–liquid extraction, and chemical recovery methods. A

**Tab. 9.** Solvent Concentration, Yield, and Productivity in Continuous Acetone–Butanol Fermentations by *Clostridium acetobutylicum* with Cell Recycle or Immobilization

| Operation Mode | Concentration [g L$^{-1}$] | Yield [g g$^{-1}$] | Productivity [g L$^{-1}$ h$^{-1}$] | Reference |
|---|---|---|---|---|
| Immobilized cells Single-stage, complex medium, intermittent feeding | 1.00 | 0.20 | 0.70 | FÖRBERG et al., 1983 |
| Immobilized cells Single-stage, complex medium, steady state | 15.42 | 0.34 | 3.02 | LARGIER et al., 1985 |
| Immobilized cells Two-stage, complex medium, steady state | 3.94 | 0.21 | 4.02 | FRICK and SCHÜGERL, 1986 |
| Immobilized cells Single-stage, complex medium, steady state | 4.10 | 0.23 | 4.10 | QURESHI and MADDOX, 1987 |
| Cell recycle Two-stage, synthetic medium, steady state | 12.00 | 0.30 | 3.00 | AFSCHAR et al., 1985 |
| Cell recycle Single-stage, synthetic medium, steady state | 13.00 | 0.29 | 6.50 | PIERROT et al., 1986 |
| Cell recycle Single-stage, synthetic medium, phosphate-limited, steady state | 21.71 | 0.32 | 2.17 | SCHLOTE and GOTTSCHALK, 1986 |

detailed overview of these technologies and their future prospects is given by ENNIS et al. (1986a). Although most of the work aimed at the possible use of alternative solvent recovery processes and is related to ethanol fermentation, there is great interest to apply the results to the acetone–butanol fermentation as well (ENNIS et al., 1986b; DADGAR and FOUTCH, 1985; TAYA et al., 1985; GARCIA III et al., 1984, 1986; GROOT et al., 1992). Solvent extraction can be performed either *in situ* in the fermentation vessel or in a bypass outside the fermenter in a recycle stream of the fermentation broth (LINDEN et al., 1986).

Several *in situ* adsorption systems have been tried in acetone–butanol fermentation. One prerequisite of the extractants used is that they do not affect cell growth and activity. Corn oil, paraffin oil, kerosene, dibutyl phthalate (WANG et al., 1979), and oleyl alcohol (*cis*-9-octadecene-1-ol) (TAYA et al., 1985) proved to be good extractants for butanol. By automatic withdrawing and feeding operations of oleyl alcohol regulated by the volume of gas evolved, the glucose concentration used in a fed-batch extractive fermentation system of *C. acetobutylicum* was 120 g L$^{-1}$ and the solvent concentration was held below 2 g L$^{-1}$. The total amount of butanol produced was 20.4 g L$^{-1}$. Other adsorbents used were activated carbon or silicalite (MADDOX, 1983). 85 mg of butanol could be adsorbed per g silicalite and later released by thermal desorption. The use of a pervaporation system in batch culture resulted in an increase in glucose utilization and in continuous culture with immobilized cells in higher glucose conversion and higher productivity (GROOT et al., 1984a, b).

GARCIA III et al. (1986) examined the separation of butanol from the fermentation broth by reverse osmosis. They concluded that the integration of membrane technology could overcome the problems of low productivity and dilute product concentration associated with the acetone–butanol fermentation. Development in membrane technology is one factor influencing the economic feasibility of alternative product recovery processes for large-scale solvent production.

Which of the different solvent recovery methods will finally be the most suitable for acetone–butanol fermentation is still an open question. Recently, five technologies for *in situ* product recovery (stripping, adsorption, liquid–liquid extraction, pervaporation, membrane solvent extraction) have been directly compared using otherwise identical fermentation conditions (GROOT et al., 1992). From these, pervaporation and liquid–liquid extraction were found to have the greatest potentials. MADDOX et al. (1994) investigated four product removal techniques and showed that gas stripping allowed the highest solvent production and productivity.

## 4.4 Alternative Fermentation Substrates

Starch and molasses were the traditional raw materials for the biological production of acetone and butanol. However, a wide variety of other fermentable sugars from other sources, including wastes, are potential substrates for this fermentation process. Examples of alternative resources are apple pomace containing fructose, glucose, sucrose as fermentable sugars (VOGET et al., 1985), whey (lactose) (MADDOX et al., 1994), Jerusalem artichokes (fructan) (MARCHAL et al., 1985), and lignocellulose (xylan and cellulose) (MADDOX and MURRAY, 1983; YU and SADDLER, 1983; YU et al., 1985; FOND et al., 1983). Although lignocellulose as raw material is potentially cheap and abundant more fundamental physiological, biochemical, and genetic research has to be done before it can be used in an industrial acetone–butanol fermentation process (BAHL and GOTTSCHALK, 1988). A strain converting these polymers directly to solvents, e.g., butanol, would be highly desirable.

Whey, on the other hand, seems to be an ideal alternative substrate for acetone–butanol fermentation and will be considered in more detail (MADDOX et al., 1994). Whey is a by-product during the manufacture of cheese or casein and represents a major waste disposal problem to the dairy industry if it is not further used, e. g., for the manufacture of lactose. The lactose content of whey is 4–5% (w/v) and *C. acetobutylicum* is capable of fermenting lactose directly. Depending on the strain, the metabolism of lactose involves either a $\beta$-galactosidase, a phospho-$\beta$-galactosidase, or both enzymes (YU et al., 1987; HANCOCK et al., 1991). The relatively low sugar content is unsuitable for many other fermentation processes without prior concentration, but almost optimal for the acetone–butanol fermentation. The amount of sugar that can be utilized by *C. acetobutylicum* is limited by the product toxicity, especially of butanol. *C. acetobutylicum* yields the maximum possible solvent concentration with about 6% (w/v) initial fermentable carbohydrate in the medium. Furthermore, it should be mentioned that butanol/acetone ratios after fermentation of whey are higher (e.g., 10:1) as compared to those obtained from starch or molasses (2:1). For an industrial process this preponderance of butanol is useful not only from the viewpoint of the desired product, but it would also simplify the product recovery process. Although not all factors responsible for the shift of the butanol/acetone ratio are known, it was shown that iron limitation had the greatest effect on this ratio (BAHL et al., 1986). On the other hand, whey is a relatively poor medium. Reduced productivities compared to molasses as substrate and incomplete utilization of lactose are the major problems (MADDOX, 1980; WELSH and VELIKY, 1984; ENNIS and MADDOX, 1985; LINDEN et al., 1986). However, an optimized fermentation process for the production of butanol from whey has been described recently (MADDOX et al., 1994). The improved process includes a fluidized bed reactor of bonechar-immobilized cells coupled with pervaporation to remove and concentrate the solvents. With respect to economics the use of such a process would result in a product price of $ 0.62 (U.S.) per liter for a plant capacity of 900 $m^3$ whey permeate per day based on a whey permeate price of $ 116 (U.S.). In the case that whey permeate can be obtained at zero cost, the price drops to $ 0.21 (U.S.). Thus the acetone–butanol fermentation may be one economically viable way of producing a useful product from whey.

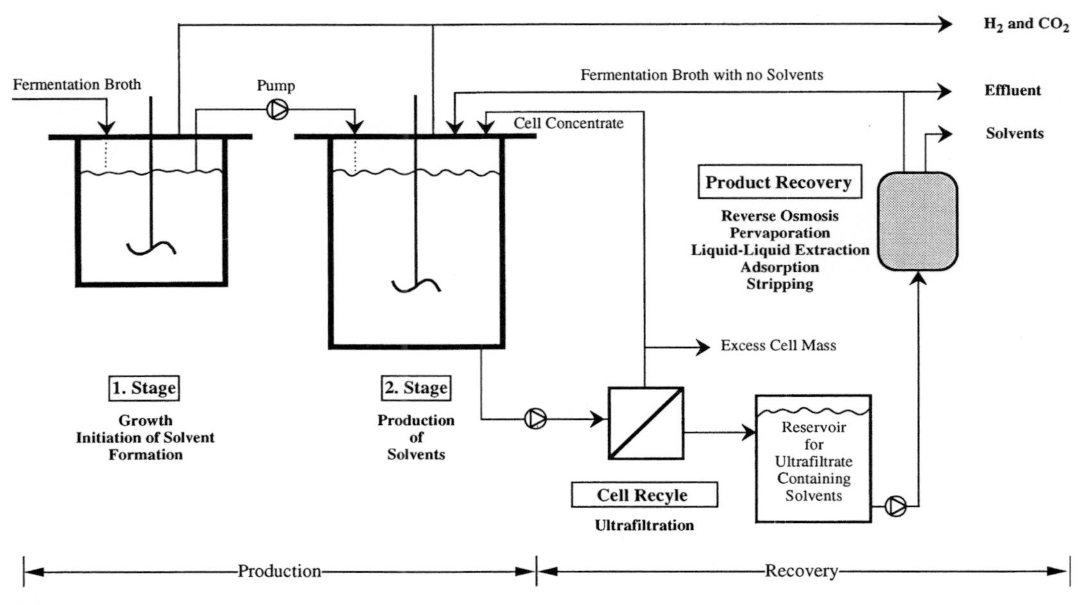


**Fig. 5.** Flow diagram of a continous acetone–butanol fermentation with cell recycle and integrated product recovery.

A combination of the different improved technologies described above will be necessary for an optimized acetone–butanol fermentation. A promising example was given by FRIEDL et al. (1991). A stable high solvent productivity of 3.5 g $L^{-1}$ $h^{-1}$ was obtained using immobilized cells of *C. acetobutylicum* coupled with product removal by pervaporation. A lactose utilization value of 97.9% was observed at a concentration of 130 g lactose $L^{-1}$ in the feed solution, and the solvent yield was remarkedly high (0.39 g solvents $g^{-1}$ lactose). In Fig. 5, a possible flow diagram of an optimized acetone–butanol fermentation is shown taking the recent developments of fermentation technology into account. In most cases the new developments were tested on a laboratory scale. Thus, it will be necessary to determine the economic viability of such an integrated acetone–butanol fermentation process as shown in Fig. 5 on a pilot and production scale.

# 5 Conclusions

Acetone–butanol fermentation was applied at a large industrial scale for about 40 years. After the last fermentation plants were closed, considerable progress was made regarding the physiology, biochemistry and genetics of *Clostridium acetobutylicum.* Furthermore, recent developments in technology of continuous fermentation using free or immobilized cells are promising and the integration of sophisticated product removal techniques into these highly productive systems can overcome the severe problem of product inhibition by butanol. Thus, it is possible now to achieve high fermenter productivity, high substrate utilization, high solvent yield, and a high solvent removal rate using continuous acetone–butanol fermentation with an integrated product removal. In addition, based on the progress made in recent years with respect to basic research on *C. acetobutylicum* it seems to be possible now to construct strains with improved features for a commercial application. In general, the results obtained to date suggest that an economically viable acetone–butanol fermentation is at least technically possible and could be reintroduced on an industrial scale.

Acknowledgement

Work in the authors' laboratories has been supported by grants from the Deutsche Forschungsgemeinschaft and the Bundesminister für Forschung und Technologie.

# 6 References

AFSCHAR, A. S., BIEBL, H., SCHALLER, K., SCHÜGERL, K. (1985), Production of acetone and butanol by *Clostridium acetobutylicum* in continuous culture with cell recycle, *Appl. Microbiol. Biotechnol.* **22**, 394–398.

ALLCOCK, E. R., WOODS, D. R. (1981), Carboxymethyl cellulase and cellobiase production by *Clostridium acetobutylicum* in an industrial fermentation medium, *Appl. Environ. Microbiol.* **41**, 539–541.

ANDERSCH, W., BAHL, H., GOTTSCHALK, G. (1982), Acetone–butanol production by *Clostridium acetobutylicum* in an ammonium-limited chemostat at low pH values, *Biotechnol. Lett.* **4**, 29–32.

ANDERSCH, W., BAHL, H., GOTTSCHALK, G. (1983), Level of enzymes involved in acetate, butyrate, acetone and butanol formation by *Clostridium acetobutylicum, Appl. Microbiol. Biotechnol.* **18**, 327–332.

ANDREESEN, J. R., GOTTSCHALK, G. (1969), The occurrence of a modified Entner-Doudoroff pathway in *Clostridium aceticum, Arch. Microbiol.* **69**, 160–170.

ANNOUS, B. A., BLASCHEK, H. P. (1991), Isolation and characterization of *Clostridium acetobutylicum* mutants with enhanced amylolytic activity, *Appl. Environ. Microbiol.* **57**, 2544–2548.

ARROYO, R. (1938), Art of producing butanol and acetone by fermentation of molasses, *U. S. Patent* 2113471.

ARZBERGER, C. F. (1936), Process for the production of butyl alcohol by fermentation, *U. S. Patent* 2050219.

ARZBERGER, C. F. (1938), Process for the production of butyl alcohol by fermentation, *U. S. Patent* 2139108.

AWANG, G. M., JONES, G. A., INGLEDEW, W. M. (1988), The acetone–butanol ethanol fermentation, *CRC Crit. Rev. Microbiol.* **15** (Suppl. 1), S33–S67.

AZEDDOUG, H., HUBERT, J., REYSSET, G. (1992), Stable inheritance of shuttle vectors based on plasmid pIM13 in a mutant strain of *Clostridium acetobutylicum, J. Gen. Microbiol.* **138**, 1371–1378.

BAHL, H. (1993), Heat shock response and onset of solvent formation in *Clostridium acetobutylicum,* in: *The Clostridia and Biotechnology* (WOODS, D. R., Ed.), pp. 247–259, Stoneham: Butterworth-Heinemann.

BAHL, H., DÜRRE, P. (1993), Clostridia, in: *Biotechnology*, 2nd Edn., Vol. 1 (REHM, H.-J., REED, G., Eds.), pp. 285–323, Weinheim: VCH.

BAHL, H., GOTTSCHALK, G. (1984), Parameters affecting solvent production by *Clostridium acetobutylicum* in continuous culture, *Biotechnol. Bioeng. Symp.* **14**, 215–223.

BAHL, H., GOTTSCHALK, G. (1988), Microbial production of butanol/acetone, in: *Biotechnology* 1st Edn., Vol. 6b (REHM, H.-J., REED, G., Eds.), pp. 1–30, Weinheim: VCH.

BAHL, H., ANDERSCH, W., BRAUN, K., GOTTSCHALK, G. (1982a), Effect of pH and butyrate concentration on the production of acetone and butanol by *Clostridium acetobutylicum* grown in continuous culture, *Eur. J. Appl. Microbiol. Biotechnol.* **14**, 17–20.

BAHL, H., ANDERSCH, W., GOTTSCHALK, G. (1982b), Continuous production of acetone and butanol by *Clostridium acetobutylicum* in a two-stage phosphate limited chemostat, *Eur. J. Appl. Microbiol. Biotechnol.* **15**, 201–205.

BAHL, H., GOTTWALD, M., KUHN, A., RALE, V., ANDERSCH, W., GOTTSCHALK, G. (1986), Nutritional factors affecting the ratio of solvents produced by *Clostridium acetobutylicum, Appl. Environ. Microbiol.* **52**, 169–172.

BALLONGUE, J., AMINE, J., MASION, E., PETITDEMANGE, H., GAY, R. (1985), Induction of acetoacetate decarboxylase in *Clostridium acetobutylicum, FEMS Microbiol. Lett.* **29**, 273–277.

BALLONGUE, J., AMINE, J., MASION, E., PETITDEMANGE, H., GAY, R. (1986), Rôle de l'acétate et du butyrate dans l'induction de la NADH:rubrédoxine oxydoréductase chez *Clostridium acetobutylicum, Biochimic* **68**, 575–580.

BEESCH, S. C. (1948), Process for fermenting carbohydrates to butanol and acetone, *U. S. Patent* 2439791.

BEESCH, S. C. (1952), Acetone–butanol fermentation of sugars, *Eng. Proc. Dev.* **44**, 1677–1682.

BEHRENS, S., NARBERHAUS, F., BAHL, H. (1993), Cloning, nucleotide sequence and structural analysis of the *Clostridium acetobutylicum dnaJ* gene, *FEMS Microbiol. Lett.* **114**, 53–60.

BEIJERINCK, M. W. (1893), Ueber die Butylalkoholgährung und das Butylferment, *Verhand. Kon. Akad. Wetenschappen Amsterdam* 2. Sect., Teil 1, Nr. **10**, 1–51.

BERNDT, H., SCHLEGEL, H. G. (1975), Kinetics and properties of $\beta$-ketothiolase from *Clostridium pasteurianum, Arch. Microbiol.* **103**, 21–30.

BERTRAM, J., DÜRRE, P. (1989), Conjugal transfer and expression of streptococcal transposons in *Clostridium acetobutylicum, Arch. Microbiol.* **151,** 551–557.

BERTRAM, J., KUHN, A., DÜRRE, P. (1990), Tn*916*-induced mutants of *Clostridium acetobutylicum* defective in regulation of solvent formation, *Arch. Microbiol.* **153**, 373–377.

BOTKIN, S. (1892), Ueber einen *Bacillus butyricus, Z. Hyg. Infektionskrankh.* **11**, 421–434.

BOWLES, L. K., ELLEFSON, W. L. (1985), Effects of butanol on *Clostridium acetobutylicum, Appl. Environ. Microbiol.* **50**, 1165–1170.

BREDEMANN, G. (1909), *Bacillus amylobacter* A.M. et Bredemann in morphologischer, physiologischer und systematischer Beziehung. Mit besonderer Berücksichtigung des Stickstoffbindungsvermögens dieser Spezies, *Cbl. Bakteriol. Parasitenkd. Infektionskrankh.,* II. Abt. 23, 385–568.

BREZNAK, J. A., COSTILOW, R. N. (1994), Physicochemical factors in growth, in: *Methods for General and Molecular Bacteriology* (GERHARDT, P., MURRAY, R. G. E., WOOD, W. A., KRIEG, N. R., Eds.), pp. 137–154, Washington: Am. Soc. Microbiol.

BUCHNER, E., MEISENHEIMER, J. (1908), Über Buttersäuregärung, *Ber. Dtsch. Chem. Ges.* **41**, 1410–1419.

BURCHHARDT, G., DÜRRE, P. (1990), Isolation and characterization of DNase-deficient mutants of *Clostridium acetobutylicum, Curr. Microbiol.* **21**, 307–311.

CALAM, C. T. (1979), Isolation of *Clostridium acetobutylicum* strains producing butanol and acetone, in: *Proc. 4th Symp. "Technische Mikrobiologie"* (DELLWEG, H., Ed.), pp. 233–242, Berlin: Verlag Versuchs- und Lehranstalt für Spiritusfabrikation und Fermentationstechnologie im Institut für Gärungsgewerbe und Biotechnologie.

CARNARIUS, E. H., MCCUTCHAN, W. N. (1938), Process for the production of butyl alcohol by fermentation, *U. S. Patent* 2139111.

CARY, J. W., PETERSEN, D. J., PAPOUTSAKIS, E. T., BENNETT, G. N. (1988), Cloning and expression of *Clostridium acetobutylicum* phosphotransbutyrylase and butyrate kinase genes in *Escherichia coli, J. Bacteriol.* **170**, 4613–4618.

CARY, J. W., PETERSEN, D. J., PAPOUTSAKIS, E. T., BENNETT, G. N. (1990), Cloning and expression of *Clostridium acetobutylicum* ATCC 824 acetoacetyl-coenzyme A:acetate/butyrate:coenzyme A-transferase in *Escherichia coli, Appl. Environ. Microbiol.* **56**, 1576–1583.

CATO, E. P., GEORGE, W. L., FINEGOLD, S. M. (1986), Genus *Clostridium* Prazmowski 1880, $23^{AL}$, in: *Bergey's Manual of Systematic Bacteriology* (SNEATH, P. H. A., MAIR, N. S., SHARPE, M. E., HOLT, J. G., Eds.), Vol. II, pp. 1141–1200, Baltimore: Williams & Wilkins.

CHEN, J.-S. (1993), Properties of acid- and solvent-forming enzymes of clostridia, in: *The Clostridia and Biotechnology* (WOODS, D. R., Ed.), pp. 51–76, Stoneham: Butterworth-Heinemann.

CHEN, J.-S., HIU, S. F. (1986), Acetone–butanol–isopropanol production by *Clostridium beijerinckii* (synonym, *Clostridium butylicum*), *Biotechnol. Lett.* **8**, 371–376.

CHOJECKI, A., BLASCHEK, H. P. (1986), Effect of carbohydrate source on $\alpha$-amylase and glucoamylase formation by *Clostridium acetobutylicum* SA1, *J. Ind. Microbiol.* **1**, 63–67.

COLBY, G. D., CHEN, J.-S. (1992), Purification and properties of 3-hydroxybutyryl-coenzyme A dehydrogenase from *Clostridium beijerinckii ("Clostridium butylicum")* NRRL B593, *Appl. Environ. Microbiol.* **58**, 3297–3302.

CONTAG, P. R., WILLIAMS, M. G., ROGERS, P. (1990), Cloning of a lactate dehydrogenase gene from *Clostridium acetobutylicum* B643 and expression in *Escherichia coli, Appl. Environ. Microbiol.* **56**, 3760–3765.

CYNKIN, M. A., DELWICHE, E. A. (1958), Metabolism of pentoses by clostridia. I. Enzymes of ribose dissimilation in extracts of *Clostridium perfringens, J. Bacteriol.* **75**, 331–334.

CYNKIN, M. A., GIBBS, M. (1958), Metabolism of pentoses by clostridia. II. The fermentation of $C^{14}$-labeled pentoses by *Clostridium perfringens, Clostridium beijerinckii,* and *Clostridium butylicum, J. Bacteriol.* **75**, 335–338.

DABROCK, B., BAHL, H., GOTTSCHALK, G. (1992), Parameters affecting solvent production by *Clostridium pasteurianum, Appl. Environ. Microbiol.* **58**, 1233–1239.

DADGAR, A. M., FOUTCH, G. L. (1985), Evaluation of solvents for the recovery of *Clostridium* fermentation products by liquid–liquid extraction, *Biotechnol. Bioeng. Symp.* **15**, 611–620.

DAVIES, R., STEPHENSON, M. (1941), Studies on the acetone–butyl alcohol fermentation. I. Nutritional and other factors involved in the preparation of active suspensions of *Cl. acetobutylicum* (Weizmann), *Biochem. J.* **35**, 1320–1331.

DIEZ-GONZALEZ, F., HUNTER, J. B., RUSSELL, J. B. (1994), The "acetilactic" fermentation of *Clostridium acetobutylicum* P262, *Abstr. Gen.*

*Meet. Am. Soc. Microbiol.,* Las Vegas, May 23–27, O-52.

DOMBECK, K. M., INGRAM, L. O. (1984), Effects of ethanol on the *Escherichia coli* plasma membrane, *J. Bacteriol.* **157**, 233–239.

DUCLAUX, E. (1895), Sur la nutrition intra-cellulaire, *Ann. Inst. Pasteur* **9**, 811–839.

DÜRRE, P. (1993), Transposons in clostridia, in: *The Clostridia and Biotechnology* (WOODS, D. R., Ed.), pp. 227–246, Stoneham: Butterworth-Heinemann.

DÜRRE, P., KUHN, A., GOTTWALD, M., GOTTSCHALK, G. (1987), Enzymatic investigations on butanol dehydrogenase in extracts of *Clostridium acetobutylicum, Appl. Microbiol. Biotechnol.* **26**, 268–272.

DÜRRE, P., BAHL, H., GOTTSCHALK, G. (1992), Die Aceton-Butanol-Gärung: Grundlage für einen modernen biotechnologischen Prozeß? *Chem. Ing. Tech.* **64**, 491–498.

DYR, J., PROTIVA, J., PRAUS, R. (1958), Formation of neutral solvents in continuous fermentation by means of *Clostridium acetobutylicum,* in: *Continuous Cultivation of Microorganisms. A Symposium* (MALEK, I., Ed.), pp. 210–226, Prague: Cechoslovakian Academy of Sciences.

EMMERLING, O. (1897), Butylalkoholische Gährung, *Ber. Dtsch. Chem. Ges.* **30**, 451–453.

ENNIS, B. M., MADDOX, I. S. (1985), Use of *Clostridium acetobutylicum* P 262 for production of solvents from whey permeate, *Biotechnol. Lett.* **7**, 503–508.

ENNIS, B. M., GUTIERREZ, N. A., MADDOX, I. S. (1986a), The acetone–butanol–ethanol fermentation: A current assessment, *Process Biochem.* **21**, 131–147.

ENNIS, B. M., MARSHALL, C. T., MADDOX, I. S., PATERSON, A. H. J. (1986b), Continuous product recovery by *in situ* gas stripping/condensation during solvent production from whey permeate using *Clostridium acetobutylicum, Biotechnol. Lett.* **8**, 725–730.

ENSLEY, B., MCHUGH, J. J., BARTON, L. L. (1975), Effect of carbon sources on formation of $\alpha$-amylase and glucoamylase by *Clostridium acetobutylicum, J. Gen. Appl. Microbiol.* **21**, 51–59.

FERNBACH, A., STRANGE, E. H. (1911), Improvements in the manufacture of higher alcohols, *British Patent* 15204.

FERNBACH, A., STRANGE, E. H. (1912), Improvements connected with fermentation processes for the production of acetone, and higher alcohols, from starch, sugars, and other carbohydrate materials, *British Patent* 21073.

FICK, M., PIERROT, P., ENGASSER, J. M. (1985), Optimal conditions for long-term stability of acetone–butanol production by continuous cultures of *Clostridium acetobutylicum, Biotechnol. Lett.* **7**, 503–508.

FINN, R. K, NOWREY, J. E. (1959), A note on the stability of clostridia when held in continuous culture, *Appl. Microbiol.* **7**, 29–32.

FISCHER, R. J., HELMS, J., DÜRRE, P. (1993), Cloning, sequencing, and molecular analysis of the *sol* operon of *Clostridium acetobutylicum,* a chromosomal locus involved in solventogenesis, *J. Bacteriol.* **175**, 6959–6969.

FITZ, A. (1876), Ueber die Gährung des Glycerins, *Ber. Dtsch. Chem. Ges.* **9**, 1348–1352.

FITZ, A. (1877), Ueber Schizomyceten-Gährungen II [Glycerin, Mannit, Stärke, Dextrin], *Ber. Dtsch. Chem. Ges.* **10**, 276–283.

FITZ, A. (1878), Ueber Schizomyceten-Gährungen III, *Ber. Dtsch. Chem. Ges.* **11**, 42–55.

FITZ, A. (1882), Ueber Spaltpilzgährungen. VII. Mittheilung, *Ber. Dtsch. Chem. Ges.* **15**, 867–880.

FOND, O., PETITDEMANGE, E., PETITDEMANGE, H., ENGASSER, J. M. (1983), Cellulose fermentation by a coculture of a mesophilic *Clostridium* and *Clostridium acetobutylicum, Biotechnol. Bioeng. Symp.* **13**, 217–224.

FÖRBERG, C., HÄGGSTRÖM, L. (1985), Control of cell adhesion and activity during continuous production of acetone and butanol with adsorbed cells, *Enzyme Microb. Technol.* **7**, 230–234.

FÖRBERG, C., ENFORS, S. O., HÄGGSTRÖM, L. (1983), Control of immobilized, non-growing cells for continuous production of metabolites, *Eur. J. Appl. Microbiol. Biotechnol.* **17**, 143–147.

FREIBERG, G. W. (1925), Process for producing acetone and butyl alcohol, *U. S. Patent* 1537597.

FREIER, D., GOTTSCHALK, G. (1987), L(+)-lactate dehydrogenase of *Clostridium acetobutylicum* is activated by fructose-1,6-bisphosphate, *FEMS Microbiol. Lett.* **43**, 229–233.

FREIER-SCHRÖDER, D., WIEGEL, J., GOTTSCHALK, G. (1989), Butanol formation by *Clostridium thermosaccharolyticum* at neutral pH, *Biotechnol. Lett.* **11**, 831–836.

FRENCH, D., KNAPP, D. W. (1950), The maltase of *Clostridium acetobutylicum.* Its specificity range and mode of action, *J. Biol. Chem.* **187**, 463–467.

FRICK, C., SCHÜGERL, K. (1986), Continuous acetone–butanol production with free and immobilized *Clostridium acetobutylicum, Appl. Microbiol. Biotechnol.* **25**, 186–193.

FRIDOVICH, I. (1972), Acetoacetate decarboxylase, in: *The Enzymes* (BOYER, P. D., Ed.,), Vol. VI, pp. 255–270, New York: Academic Press.

FRIEDL, A., QURESHI, N., MADDOX, I. S. (1991), Continuous acetone–butanol–ethanol (ABE) fermentation using immobilized cells of *Clostridium acetobutylicum* in a packed bed reactor and integration with product removal by pervaporation, *Biotechnol. Bioeng.* **38**, 518–527.

FUNK, F. J. (1925), Butyl alcohol and acetone fermentation process, *U. S. Patent* 1538516.

GABRIEL, C. L. (1928), Butanol fermentation process, *Ind. Eng. Chem.* **28**, 1063–1067.

GARCIA III, A., IANNOTTI, E. L., FISHER, J. R. (1984), Reverse osmosis application for acetone–butanol fermentation, *Biotechnol. Bioeng. Symp.* **14**, 543–552.

GARCIA III, A., IANNOTTI, E. L., FISHER, J. R. (1986), Butanol fermentation liquor production and separation by reverse osmosis, *Biotechnol. Bioeng.* **28**, 785–791.

GAVARD, R., HAUTECOEUR, B., DESCOURTIEUX, H. (1957), Phosphotransbutyrylase de *Clostridium acetobutylilcum, C. R. Acad. Sci.* **244**, 2323–2326.

GEORGE, H. A., JOHNSON, J. L, MOORE, W. E. C., HOLDEMAN, L. V., CHEN, J. S. (1983), Acetone, isopropanol, and butanol production by *Clostridium bejierinckii* (syn. *Clostridium butylicum*) and *Clostridium aurantibutyricum, Appl. Environ. Microbiol.* **45**, 1160–1163.

GERISCHER, U., DÜRRE, P. (1990), Cloning, sequencing, and molecular analysis of the acetoacetate decarboxylase gene region from *Clostridium acetobutylicum, J. Bacteriol.* **172**, 6907–6918.

GERISCHER, U., DÜRRE, P. (1992), mRNA analysis of the *adc* gene region of *Clostridium acetobutylicum* during the shift to solventogenesis, *J. Bacteriol.* **174**, 426–433.

GOODLOVE, P. E., CUNNINGHAM, P. R., PARKER, J., CLARK, D. P. (1989), Cloning and sequence analysis of the fermentative alcohol-dehydrogenase-encoding gene of *Escherichia coli, Gene* **85**, 209–214.

GOTTSCHAL, J. C., MORRIS, J. G. (1981a), The induction of acetone and butanol production in cultures of *Clostridium acetobutylicum* by elevated concentrations of acetate and butyrate, *FEMS Microbiol. Lett.* **12**, 385–389.

GOTTSCHAL, J. C., MORRIS, J. G. (1981b), Nonproduction of acetone and butanol by *Clostridium acetobutylicum* during glucose- and ammonium-limitation in continuous culture, *Biotechnol. Lett.* **3**, 525–530.

GOTTSCHAL, J. C., MORRIS, J. G. (1982), Continuous production of acetone and butanol by *Clostridium acetobutylicum* growing in turbidostat culture, *Biotechnol. Lett.* **4**, 477–482.

GOTTSCHALK, G. (1986), *Bacterial Metabolism*, 2nd Edn., New York: Springer-Verlag.

GOTTWALD, M., GOTTSCHALK, G. (1985), The internal pH of *Clostridium acetobutylicum* and its effect on the shift from acid to solvent formation, *Arch. Microbiol.* **143**, 42–46.

GOTTWALD, M., HIPPE, H., GOTTSCHALK, G. (1984), Formation of *n*-butanol from D-glucose by strains of the *"Clostridium tetanomorphum"* group, *Appl. Environ. Microbiol.* **48**, 573–576.

GRETHLEIN, A. J., WORDEN, R. M., JAIN, M. K., DATTA, R. (1991), Evidence for production of *n*-butanol from carbon monoxide by *Butyribacterium methylotrophicum, J. Ferment. Bioeng.* **72**, 58–60.

GRIMBERT, L. (1893), Fermentation anaérobie produite par le *Bacillus orthobutylicus.* Ses variations sous certaines influences biologiques, *Ann. Inst. Pasteur* **7**, 353–402.

GROOT, W. J., SCHOUTENS, G. H., VAN BEELEN, P. N., VAN DEN OEVER, C. E., KOSSEN, N. W. F. (1984a), Increase of substrate conversion by pervaporation in the continuous butanol fermentation, *Biotechnol. Lett.* **6**, 789–792.

GROOT, W. J., VAN DEN OEVER, C. E., KOSSEN, N. W. F. (1984b), Pervaporation for simultaneous product recovery in the butanol/isopropanol batch fermentation, *Biotechnol. Lett.* **6**, 709–714.

GROOT, W. J., VAN DER LANS, R. G. J. M., LUYBEN, CH. A. M. (1992), Technologies for butanol recovery integrated with fermentations, *Process Biochem.* **27**, 61–75.

GRUBER, M. (1887), Eine Methode der Cultur anaërobischer Bacterien nebst Bemerkungen zur Morphologie der Buttersäuregährung, *Cbl. Bacteriol. Parasitenkd.* **1**, 367–372.

GRUPE, H., GOTTSCHALK, G. (1992), Physiological events in *Clostridium acetobutylicum* during the shift from acidogenesis to solventogenesis in continuous culture and presentation of a model for shift induction, *Appl. Environ. Microbiol.* **58**, 3896–3902.

HÄGGSTRÖM, L., ENFORS, S. O. (1982), Continuous production of butanol with immobilized cells of *Clostridium acetobutylicum, Appl. Biochem. Biotechnol.* **7**, 35–37.

HANCOCK, K. R., ROCKMAN, E., YOUNG, C. A., PEARCE, L., MADDOX, I. S., SCOTT, D. B. (1991), Expression and nucleotide sequence of the *Clostridium acetobutylicum* $\beta$-galactosidase gene cloned in *Escherichia coli, J. Bacteriol.* **173**, 3084–3095.

HANSON, A. M., RODGERS, N. E. (1946), Influence of iron concentration and attenuation on the metabolism of *Clostridium acetobutylicum, J. Bacteriol.* **51**, 568–569.

HARRIS, J., MULDER, R., KELL, D. B., WALTER, R. P., MORRIS, J. G. (1986), Solvent production by *Clostridium pasteurianum* in media of high sugar content, *Biotechnol. Lett.* **8**, 889–892.

HARTMANIS, M. G. N. (1987), Butyrate kinase from *Clostridium acetobutylicum, J. Biol. Chem.* **262**, 617–621.

HARTMANIS, M. G. N., GATENBECK, S. (1984), Intermediary metabolism in *Clostridium acetobutylicum:* Levels of enzymes involved in the formation of acetate and butyrate, *Appl. Environ. Microbiol.* **47**, 1277–1283.

HARTMANIS, M. G. N., KLASON, T., GATENBECK, S. (1984), Uptake and activation of acetate and butyrate in *Clostridium acetobutylicum, Appl. Microbiol. Biotechnol.* **20**, 66–71.

HASTINGS, J. J. H. (1978), Acetone–butyl alcohol fermentation, in: *Economic Microbiology*, Vol. 2. Primary Products of Metabolism (ROSE, A. H., Ed.), pp. 31–45, New York: Academic Press.

HERMANN, M., FAYOLLE, F., MARCHAL, R., PODVIN, L., SEBALD, M., VANDECASTEELE, J.-P. (1985), Isolation and characterization of butanol-resistant mutants of *Clostridium acetobutylicum, Appl. Environ. Microbiol.* **50**, 1238–1243.

HERRERO, A. A. (1983), End-product inhibition in anaerobic fermentations, *Trends Biotechnol.* **1**, 49–53.

HIGGINS, C. F. (1992), ABC transporters from microorganisms to man, *Annu. Rev. Cell Biol.* **8**, 67–113.

HIPPE, H., ANDREESEN, J. R., GOTTSCHALK, G. (1992), The genus *Clostridium* – nonmedical, in: *The Prokaryotes,* 2nd Edn. (BALOWS, A., TRÜPER, H. G., DWORKIN, M., HARDER, W., SCHLEIFER, K.-H., Eds.), Vol. II, pp. 1800–1866, New York: Springer-Verlag.

HIU, S. F., ZHU, C.-Z., YAN, R.-T., CHEN, J.-S. (1987), Butanol–ethanol dehydrogenase and butanol–ethanol-isopropanol dehydrogenase: Different alcohol dehydrogenases in two strains of *Clostridium beijerinckii (Clostridium butylicum), Appl. Environ. Microbiol.* **53,** 697–703.

HOCKENHULL, D. J., HERBERT, D. (1945), The amylase and maltase of *Clostridium acetobutylicum, Biochem. J.* **39**, 102–106.

HOLT, R. A., STEPHENS, G. M., MORRIS, J. G. (1984), Production of solvents by *Clostridium acetobutylicum* cultures maintained at neutral pH, *Appl. Environ. Microbiol.* **48**, 1166–1170.

HOLT, R. A., CAIRNS, A. J., MORRIS, J. G. (1988), Production of butanol by *Clostridium puniceum* in batch and continuous culture, *Appl. Microbiol. Biotechnol.* **27**, 319–324.

HONGO, M. (1960), Process for producing butanol by fermentation, *U. S. Patent* 2945786.

HÜSEMANN, M. H. W., PAPOUTSAKIS, E. T. (1989), Comparison between *in vivo* and *in vitro* enzyme activities in continuous and batch fermentations of *Clostridium acetobutylicum, Appl. Microbiol. Biotechnol.* **30**, 585–595.

HUTKINS, R. W., KASHKET, E. R. (1986), Phosphotransferase activity in *Clostridium acetobutylicum* from acidogenic to solventogenic phases of growth, *Appl. Environ. Microbiol.* **51**, 1121–1123.

INGRAM, L. O. (1986), Microbial tolerance to alcohols: Role of the cell membrane, *Trends Biotechnol.* **4**, 40–44.

IZSAK, A., FUNK, F. J. (1933), Process of producing butyl alcohol, *U. S. Patent* 1908361.

JANATI-IDRISSI, R., JUNELLES, A. M., EL KANOUNI, A., PETITDEMANGE, H., GAY, R. (1987), Sélection de mutants de *Clostridium acetobutylicum* défectifs dans la production d'acétone, *Ann. Inst. Pasteur/Microbiol.* **138**, 313–323.

JÖBSES, I. M. L., ROELS, J. A. (1983), Experience with solvent production by *Clostridium beijerinckii* in continuous culture, *Biotechnol. Bioeng.* **25**, 1187–1194.

JONES, D. T., WOODS, D. R. (1986), Acetone–butanol fermentation revisited, *Microbiol. Rev.* **50**, 484–524.

JUNGERMANN, K., THAUER, R. K., LEIMENSTOLL, G., DECKER, K. (1973), Function of reduced pyridine nucleotide-ferredoxin oxidoreductases in saccharolytic clostridia, *Biochim. Biophys. Acta* **305**, 268–280.

KASHKET, E. R., CAO, Z.-Y. (1993), Isolation of a degeneration-resistant mutant of *Clostridium acetobutylicum* NCIMB 8052, *Appl. Environ. Microbiol.* **59**, 4198–4202.

KATAGIRI, H., IMAI, K., SUGIMARI, T. (1960), On the metabolism of organic acids by *Clostridium acetobutylicum.* Part I. Formation of lactic acid and racemase, *Bull. Agr. Chem. Soc. Jpn.* **24**, 163–172.

KEIS, S., BENNETT, C., JONES, D. T. (1994), Differentiation of solvent producing clostridial strains by restriction analysis typing, ribotyping and phage typing, *Abstr. 7th Int. Symp. Genet. Ind. Microorg.,* Montreal, June 26–July 1, p. 201.

KEMPNER, W., KUBOWITZ, F. (1933), Wirkung des Lichtes auf die Kohlenoxydhemmung der Buttersäuregärung, *Biochem. Z.* **265**, 245–252.

KIM, A. Y., BLASCHEK, H. P. (1991), Isolation and characterization of a filamentous viruslike particle from *Clostridium acetobutylicum* NCIB 6444, *J. Bacteriol.* **173**, 530–535.

KIM, A. Y., BLASCHEK, H. P. (1993), Construction and characterization of a phage–plasmid hybrid (phagemid), pCAK1, containing the replicative form of viruslike particle CAK1 isolated from

*Clostridium acetobutylicum* NCIB 6444, *J. Bacteriol.* **175**, 3838–3843.

KIM, B. H., ZEIKUS, J. G. (1985), Importance of hydrogen metabolism in regulation of solventogenesis by *Clostridium acetobutylicum, Dev. Ind. Microbiol.* **26**, 549–556.

KIM, A. Y., ATTWOOD, G. T., HOLT, S. M., WHITE, B. A., BLASCHEK, H. P. (1994), Heterologous expression of endo-$\beta$-1,4-D-glucanase from *Clostridium cellulovorans* in *Clostridium acetobutylicum* ATCC 824 following transformation of the *engB* gene, *Appl. Environ. Microbiol.* **60**, 337–340.

KUBOWITZ, F. (1934), Über die Hemmung der Buttersäuregärung durch Kohlenoxyd, *Biochem. Z.* **274**, 285–298.

KUTZENOK, A., ASCHNER, M. (1952), Degenerative processes in a strain of *Clostridium butylicum, J. Bacteriol.* **64**, 829–836.

LAPAGE, S. P., SHELTON, J. E., MITCHELL, T. G., MACKENZIE, A. R. (1970), Culture collections and the preservation of bacteria, in: *Methods in Microbiology* (NORRIS, J. R., RIBBONS, D. W., Eds.), Vol. 3 A, pp. 125–228, London: Academic Press.

LARGIER, S. T., LONG, S., SANTANGELO, J. D., JONES, D. T., WOODS, D. R. (1985), Immobilized *Clostridium acetobutylicum* P 262 mutants for solvent production, *Appl. Environ. Microbiol.* **50**, 477–481.

LEE, S. F., FORSBERG, C. W., GIBBINS, L. N. (1985a), Xylanolytic activity of *Clostridium acetobutylicum, Appl. Environ. Microbiol.* **50**, 1068–1076.

LEE, S. F., FORSBERG, C. W., GIBBINS, L. N. (1985b), Cellulytic activity of *Clostridium acetobutylicum, Appl. Environ. Microbiol.* **50**, 220–228.

LEE, S. F., FORSBERG, C. W., RATTRAY, J. B. (1987), Purification and characterization of two endoxylanases from *Clostridium acetobutylicum* ATCC 824, *Appl. Environ. Microbiol.* **53**, 664–650.

LEE, S. Y., BENNETT, G. N., PAPOUTSAKIS, E. T. (1992), Construction of *Escherichia coli–Clostridium acetobutylicum* shuttle vectors and transformation of *Clostridium acetobutylicum* strains, *Biotechnol. Lett.* **14**, 427–432.

LEGG, D. A., STILES, H. R. (1937), Process of producing butyl alcohol, *U. S. Patent* 2089562.

LEMMEL, S. A. (1985), Mutagenesis in *Clostridium acetobutylicum, Biotechnol. Lett.* **7**, 711–716.

LEMMEL, S. A., DATTA, R., FRANKIEWICZ, J. R. (1986), Fermentation of xylan by *Clostridium acetobutylicum, Enzyme Microb. Technol.* **8**, 217–221.

LENZ, T. G., MOREIRA, A. R. (1980), Economic evaluation of the acetone–butanol fermentation, *Ind. Eng. Chem. Prod. Res. Dev.* **19,** 478–483.

LESKIW, B. K., BIBB, M. J., CHATER, K. F. (1991), The use of a rare codon specifically during development? *Mol. Microbiol.* **5**, 2861–2867.

LEUNG, J. C. Y., WANG, D. I. C. (1981), Production of acetone and butanol by *Clostridium acetobutylicum* in continuous culture using free and immobilized cells, *Proc. 2nd World Congr. Chem. Eng.* **1**, 348–352.

LIN, Y.-L., BLASCHEK, H. P. (1983), Butanol production by a butanol-tolerant strain of *Clostridium acetobutylicum* in extruded corn broth, *Appl. Environ. Microbiol.* **45**, 966–973.

LIN, Y.-L., BLASCHEK, H. P. (1984), Transformation of heat-treated *Clostridium acetobutylicum* protoplasts with pUB110 plasmid DNA, *Appl. Environ. Microbiol.* **48**, 737–742.

LINDEN, J. C., MOREIRA, A. R., LENZ, T. G. (1986), Acetone and butanol, in: *Comprehensive Biotechnology. The Principles of Biotechnology; Engineering Consideration* (COONEY, C. L., HUMPHREY, A. E., Eds.), pp. 915–931, Oxford: Pergamon Press.

LONG, S., JONES, D. T., WOODS, D. R. (1984a), The relationship between sporulation and solvent production in *Clostridium acetobutylicum* P262, *Biotechnol. Lett.* **6**, 529–534.

LONG, S., JONES, D. T., WOODS, D. R. (1984b), Initiation of solvent production, clostridial stage and endospore formation in *Clostridium acetobutylicum* P262, *Appl. Microbiol. Biotechnol.* **20**, 256–261.

LOUGHLIN, J. F. (1935), Production of butyl alcohol and acetone by fermentation, *U. S. Patent* 1992921.

MADDOX, I. S. (1980), Production of *n*-butanol from whey filtrate using *Clostridium acetobutylicum* NCIB 2951, *Biotechnol. Lett.* **2**, 493–498.

MADDOX, I. S. (1983), Use of silicalite for the adsorption of *n*-butanol from fermentation liquors, *Biotechnol. Lett.* **5**, 89–94.

MADDOX, I. S., MURRAY, A. E. (1983), Production of *n*-butanol by fermentation of wood hydrolysate, *Biotechnol. Lett.* **5**, 175–178.

MADDOX, I. S., QURESHI, N., GUTIERREZ, N. A. (1994), Utilization of whey by clostridia and process technology, in: *The Clostridia and Biotechnology* (WOODS, D. R., Ed.), pp. 343–369, Stoneham: Butterworth-Heinemann.

MARCHAL, R., BLANCHET, D., VANDECASTEELE, J. P. (1985), Industrial optimization of acetone–butanol fermentation: A study of the utilization of Jerusalem artichokes, *Appl. Microbiol. Biotechnol.* **23**, 92–98.

MARCZAK, R., PETITDEMANGE, H., ALIMI, F., BALLONGUE, R., GAY, R. (1983), Influence de la phase de croissance et de la composition du milieu sur le taux de biosynthèse de la NADH:rubrédoxine oxydoréductase chez *Clostridium acetolbutylicum, C. R. Acad. Sci.,* Ser. 3 **296**, 469–474.

MARCZAK, R., BALLONGUE, R., PETITDEMANGE, H., GAY, R. (1984), Regulation of the biosynthesis of NADH-rubredoxin oxidoreductase in *Clostridium acetobutylicum, Curr. Microbiol.* **10**, 165–168.

MARCZAK, R., BALLONGUE, J., PETITDEMANGE, H., GAY, R. (1985), Differential levels of ferredoxin and rubredoxin in *Clostridium acetobutylicum, Biochimie* **67**, 241–248.

MCCLUNG, L. S., MCCOY, E. (1935), Studies on anaerobic bacteria. VII. The serological relations of *Clostridium acetobutylicum, Cl. felsineum* and *Cl. roseum, Arch. Mikrobiol.* **6**, 239–249.

MCCOY, E. F. (1946), Production of butyl alcohol and acetone by fermentation, *U. S. Patent* 2398837.

MCCOY, E., MCCLUNG, L. S. (1935), Studies on anaerobic bacteria. VI. The nature and systematic position of a new chromogenic *Clostridium, Arch. Mikrobiol.* **6**, 230–238.

MCCOY, E., FRED, E. B., PETERSON, W. H., HASTINGS, E. G. (1926), A cultural study of the acetone butyl alcohol organism, *J. Infect. Dis.* **39**, 457–483.

MCNEIL, B., KRISTIANSEN, B. (1986), The acetone butanol fermentation, *Adv. Appl. Microbiol.* **31**, 61–92.

MEINECKE, B., BAHL, H., GOTTSCHALK, G. (1984), Selection of an asporogenous strain of *Clostridium acetobutylicum* in continuous culture under phosphate limitation, *Appl. Environ. Microbiol.* **48**, 1064–1065.

MEINECKE, B., BERTRAM, J., GOTTSCHALK, G. (1989), Purification and characterization of the pyruvate–ferredoxin oxidoreductase from *Clostridium acetobutylicum, Arch. Microbiol.* **152**, 244–250.

MERMELSTEIN, L. D., PAPOUTSAKIS, E. T. (1993), *In vivo* methylation in *Escherichia coli* by the *Bacillus subtilis* phage ϕ3T I methyltransferase to protect plasmids from restriction upon transformation of *Clostridium acetobutylicum* ATCC 824, *Appl. Environ. Microbiol.* **59**, 1077–1081.

MERMELSTEIN, L. D., WELKER, N. E., BENNETT, G. N., PAPOUTSAKIS, E. T. (1992), Expression of cloned homologous fermentative genes in *Clostridium acetobutylicum* ATCC 824, *Bio/Technology* **10**, 190–195.

MERMELSTEIN, L. D., PAPOUTSAKIS, E. T., PETERSEN, D. J., BENNETT, G. N. (1993), Metabolic engineering of *Clostridium acetobutylicum* ATCC 824 for increased solvent production by enhancement of acetone formation enzyme activities using a synthetic acetone operon, *Biotechnol. Bioeng.* **42**, 1053–1060.

MICHELS, J. (1990), Reinigung der NADP$^+$-abhängigen Butanol-Dehydrogenase aus *Clostridium acetobutylicum* über Triazinyl-Farbstoffe, *Thesis,* University of Göttingen.

MINTON, N. P., BREHM, J. K., SWINFIELD, T.-J., WHELAN, S. M., MAUCHLINE, M. L., BODSWORTH, N., OULTRAM, J. D. (1993), Clostridial cloning vectors, in: *The Clostridia and Biotechnology* (WOODS, D. R., Ed.), pp. 119–150, Stoneham: Butterworth-Heinemann.

MITCHELL, W. J. (1992), Carbohydrate assimilation by saccharolytic clostridia, *Res. Microbiol.* **143**, 245–250.

MITCHELL, W. J., SHAW, J. E., ANDREWS, L. (1991), Properties of the glucose phosphotransferase system of *Clostridium acetobutylicum* NCIB 8052, *Appl. Environ. Microbiol.* **57**, 2534–2539.

MONOT, F., ENGASSER, J. M. (1983), Production of acetone and butanol by batch and continuous culture of *Clostridium acetobutylicum* under nitrogen limitation, *Biotechnol. Lett.* **5**, 213–218.

MONOT, F., MARTIN, J. R., PETITDEMANGE, H., GAY, R. (1982), Acetone and butanol production by *Clostridium acetobutylicum* in a synthetic medium, *Appl. Environ. Microbiol.* **44**, 1318–1324.

MONOT, F., ENGASSER, J. M., PETITDEMANGE, H. (1983), Regulation of acetone butanol production in batch and continuous cultures of *Clostridium acetobutylicum, Biotechnol. Bioeng. Symp.* **13**, 207–216.

MOREIRA, A. R. (1983), Acetone–butanol fermentation, in: *Organic Chemicals from Biomass* (WISE, D. L., Ed.), pp. 385–406, Menlo Park: Benjamin/Cummins Publ. Co., Inc.

MOREIRA, A. R., ULMER, D. C., LINDEN, J. C. (1981), Butanol toxicity in the butylic fermentation, *Biotechnol. Bioeng. Symp.* **11**, 567–579.

MÜLLER, J. (1938a), Production of neutral solvents by fermentation, *U. S. Patent* 2132039.

MÜLLER, J. (1938b), Fermentation mash, *U. S. Patent* 2123078.

NAIR, R. V., BENNETT, G. N., PAPOUTSAKIS, E. T. (1994), Molecular characterization of an aldehyde/alcohol dehydrogenase gene from *Clostridium acetobutylicum* ATCC 824, *J. Bacteriol.* **176**, 871–885.

NAKAMURA, S., OKADO, I., ABE, T., NISHIDA, S. (1979), Taxonomy of *Clostridium tetani* and related species, *J. Gen. Microbiol.* **113**, 29–35.

NAKAMURA, T., IGARASHI, Y., KODAMA, T. (1990), Acetoacetyl-CoA transferase and thiolase reactions are catalyzed by a single enzyme in *Clostridium acetobutylicum* ATCC824, *Agr. Biol. Chem.* **54**, 276–268.

NARBERHAUS, F., BAHL, H. (1992), Cloning, sequencing, and molecular analysis of the *groESL* operon of *Clostridium acetobutylicum, J. Bacteriol.* **174**, 3282–3289.

NARBERHAUS, F., GIEBELER, K., BAHL, H. (1992), Molecular characterization of the *dnaK* gene region of *Clostridiuin acetobutylicum,* including *grpE, dnaJ,* and a new heat shock gene, *J. Bacteriol.* **174**, 3290–3299.

NELSON, D. L., WEBB, B. P. (1978), Acetone, in: *Kirk-Othmer Encyclopedia of Chemical Technology,* 3rd Edn. (MARK, H. F., OTHMER, D. F., OVERBERGER, C. G., SEABORG, G. T., GRAYSON, M., ECKROTH, D., Eds.), Vol. 1, pp. 179–192, New York: John Wiley & Sons.

OGATA, S., HONGO, M. (1979), Bacteriophages of the genus *Clostridium, Adv. Appl. Microbiol.* **25**, 241–273.

OSBURN, O. L., WERKMAN, C. H. (1935), Utilization of agricultural wastes. II. Influence of nitrogenous substrate on production of butyl and isopropyl alcohols by *Clostridium butylicum, Ind. Eng. Chem.* **27**, 416–419.

OULTRAM, J. D., YOUNG, M. (1985), Conjugal transfer of plasmid pAMb1 from *Streptococcus lactis* and *Bacillus subtilis* to *Clostridium acetobutylicum, FEMS Microbiol. Lett.* **27**, 129–134.

OULTRAM, J. D., DAVIES, A., YOUNG, M. (1987), Conjugal transfer of a small plasmid from *Bacillus subtilis* to *Clostridium acetobutylicum* by cointegrate formation with plasmid pAMβ1, *FEMS Microbiol. Lett.* **42**, 113–119.

OULTRAM, J. D., LOUGHLIN, M., SWINFIELD, T.-J., BREHM, J. K., THOMPSON, D. E., MINTON, N. P. (1988a), Introduction of plasmids into whole cells of *Clostridium acetobutylicum* by electroporation, *FEMS Microbiol. Lett.* **56**, 83–88.

OULTRAM, J. D., PECK, H., BREHM, J. K., THOMPSON, D. E., SWINFIELD, T. J., MINTON, N. P. (1988b), Introduction of genes for leucine biosynthesis from *Clostridium pasteurianum* into *C. acetobutylicum* by cointegrate conjugal transfer, *Mol. Gen. Genet.* **214**, 177–179.

OULTRAM, J. D., SCHIMMING, S., WALMSLEY, R., LOUGHLIN, M., SCHWARTZ, W., STAUDENBAUER, W. L., MINTON, N. P. (1990), Introduction of cellulase (*cel*) genes from *Clostridium thermocellum* into *Clostridium acetobutylicum, Abstr. 6th Int. Symp. Genet. Ind. Microorg.,* Strasbourg, August 12–18, C 54.

OULTRAM, J. D., BURR, I. A., ELMORE, M. J., MINTON, N. P. (1993), Cloning and sequence analysis of the genes encoding phosphotransbutyrylase and butyrate kinase from *Clostridium acetobutylicum* NCIMB 8052, *Gene* **131**, 107–112.

OXFORD, A. E., LAMPEN, J. O., PETERSON, W. H. (1940), 190. Growth factor and other nutritional requirements of the acetone butanol organism, *Cl. acetobutylicum, Biochem. J.* **34**, 1588–1597.

PALOSAARI, N. R, ROGERS, P. (1988), Purification and properties of the inducible coenzyme A-linked butyraldehyde dehydrogenase from *Clostridium acetobutylicum, J. Bacteriol.* **170**, 2971–2976.

PAPA, A. J. (1982), Isopropyl alcohol, in: *Kirk-Othmer Encyclopedia of Chemical Technology,* 3rd Edn. (MARK, H. F., OTHMER, D. F., OVERBERGER, C. G., SEABORG, G. T., GRAYSON, M., ECKROTH, D., Eds.), Vol. 19, pp. 198–220, New York: John Wiley & Sons.

PAQUET, V., CROUX, C., GOMA, G., SOUCAILLE, P. (1991), Purification and characterization of the extracellular $\alpha$-amylase from *Clostridium acetobutylicum, Appl. Environ. Microbiol.* **57**, 212-218.

PASTEUR, L. (1861a), Animacules infusoires vivant sans gaz oxygène libre et déterminant des fermentations, *C. R. Hebd. Séances Acad. Sci.* **52**, 344–347.

PASTEUR, L. (1861b), Expériences et vues nouvelles sur la nature des fermentations, *C. R. Hebd. Séances Acad. Sci.* **52**, 1260–1264.

PASTEUR, L. (1862), Quelques résultats nouveaux relatifs aux fermentations acétique et butyrique, *Bull. Soc. Chim. Paris* Mai 1862, 52–53.

PASTEUR, L. (1863), Recherches sur le putréfaction, *C. R. Hebd. Séances Acad. Sci.* **56**, 1189–1194.

PERDRIX, L. (1891), Sur les fermentations produites par un microbe anaérobie de l'eau, *Ann. Inst. Pasteur* **5**, 287–311.

PETERSEN, M. (1991), Untersuchungen zur Lipase-Sekretion von *Clostridium tetanomorphum* und Reinigung des lipolytisch aktiven Enzyme, *Thesis*, University of Göttingen.

PETERSEN, D. J., BENNETT, G. N. (1990), Purification of acetoacetate decarboxylase from *Clostridium acetobutylicum* ATCC 824 and cloning of the acetoacetate decarboxylase gene in *Escherichia coli, Appl. Environ. Microbiol.* **56**, 3491–3498.

PETERSEN, D. J., BENNETT, G. N. (1991), Cloning of the *Clostridium acetobutylicum* ATCC 824 acetyl coenzyme A acetyltransferase (thiolase; EC 2.3.1.9) gene, *Appl. Environ. Microbiol.* **57**, 2735–2741.

PETERSEN, D. J., WELCH, R. W., RUDOLPH, F. B., BENNETT, G. N. (1991), Molecular cloning of an alcohol (butanol) dehydrogenase gene cluster from *Clostridium acetobutylicum* ATCC 824, *J. Bacteriol.* **173**, 1831–1834.

PETERSEN, D. J., CARY, J. W., VANDERLEYDEN, J., BENNETT, G. N. (1993), Sequence and arrangement of genes encoding enzymes of the acetone-production pathway of *Clostridium acetobutylicum* ATCC 824, *Gene* **123**, 93–97.

PETITDEMANGE, H., CHERRIER, C., RAVAL, G., GAY, R. (1976), Regulation of the NADH and NADPH-ferredoxin oxidoreductases in clostridia of the butyric group, *Biochim. Biophys. Acta* **421**, 334–347.

PETITDEMANGE, H., CHERRIER, C., BENGONE, J. M., GAY, R. (1977), Etude des activités NADH et NADPH-ferrédoxine oxydoréductasiques chez *Clostridium acetobutylicum, Can. J. Microbiol.* **23**, 152–160.

PETITDEMANGE, H., MARCZAK, R, BLUSSON, H., GAY, R. (1979), Isolation and properties of reduced nicotinamide adenine dinucleotide-rubredoxin oxidoreductase of *Clostridium acetobutylicum, Biochem. Biophys. Res. Commun.* **4**, 1258–1265.

PICH, A., BAHL, H. (1991), Purification and characterization of the DNA-dependent RNA polymerase from *Clostridium acetobutylicum, J. Bacteriol.* **173**, 2120–2124.

PICH, A., NARBERHAUS, F., BAHL, H. (1990), Induction of heat shock proteins during initiation of solvent formation in *Clostridium acetobutylicum, Appl. Microbiol. Biotechnol.* **33**, 697–704.

PIERROT, P., FICK, M., ENGASSER, J. M. (1986), Continuous acetone–butanol fermentation with high productivity by cell ultrafiltration and cell recycling, *Biotechnol. Lett.* **8**, 253–256.

PRESCOTT, S. G., DUNN, C. G. (1959), *Industrial Microbiology,* 3rd Edn., New York: McGraw-Hill Book Co.

PRINGSHEIM, H. H. (1906a), Ueber den Ursprung des Fuselöls und eine Alkohole bildende Bakterienform, *Cbl. Bakteriol. Parasitenkd. Infektionskrankh.,* II. Abt. 15, 300–321.

PRINGSHEIM, H. (1906b), Ueber ein Stickstoff assimilierendes *Clostridium, Cbl. Bakteriol. Parasitenkd. Infektionskrankh.,* II. Abt. 15, 795–800.

QURESHI, N., MADDOX, I. S. (1987), Continuous solvent production from whey permeate using cells of *Clostridium acetobutylicum* immobilized by adsorption onto bonechar, *Enzyme Microb. Technol.* **9**, 668–671.

REID, S. J., ALLCOCK, E. R., JONES, D. T., WOODS, D. R. (1983), Transformation of *Clostridium acetobutylicum* protoplasts with bacteriophage DNA, *Appl. Environ. Microbiol.* **45**, 305–307.

REYSSET, G., SEBALD, M. (1985), Conjugal transfer of plasmid-mediated antibiotic resistance from streptococci to *Clostridium acetobutylicum, Ann. Inst. Pasteur/Microbiol.* **136B**, 275–282.

REYSSET, G., SEBALD, M. (1993), Transformation/electrotransformation of clostridia, in: *The Clostridia and Biotechnology* (WOODS, D. R., Ed.), pp. 151–156, Stoneham: Butterworth-Heinemann.

REYSSET, G., HUBERT, J., PODVIN, L., SEBALD, M. (1988), Transfection and transformation of *Clostridium acetobutylicum* strain NI-4081 protoplasts, *Biotechnol. Techniques* **2**, 199–204.

ROGERS, P. (1986), Genetics and biochemistry of *Clostridium* relevant to development of fermentation processes, *Adv. Appl. Microbiol.* **31**, 1–60.

ROGERS, P., GOTTSCHALK, G. (1993), Biochemistry and regulation of acid and solvent production in clostridia, in: *The Clostridia and Biotechnology* (WOODS, D. R., Ed.), pp. 25–50, Stoneham: Butterworth-Heinemann.

ROGERS, P., PALOSAARI, N. (1987), *Clostridium acetobutylicum* mutants that produce butyraldehyde and altered quantities of solvents, *Appl. Environ. Microbiol.* **53**, 2761–2766.

ROOS, J. W., MCLAUGHLIN, J., PAPOUTSAKIS, E. T. (1984), The effect of pH on nitrogen supply, cell lysis, and solvent production in fermentations of *Clostridium acetobutylicum, Biotechnol. Bioeng.* **27**, 681–694.

ROSS, D. (1961), The acetone–butanol fermentation, *Prog. Ind. Microbiol.* **3**, 73–85.

RUBBO, S. D., MAXWELL, M., FAIRBRIDGE, R. A., GILLESPIE, J. M. (1941), The bacteriology, growth factor requirements and fermentation reactions of *Clostridium acetobutylicum* (Weizmann), *Aust. J. Exp. Biol. Med.* **19**, 185–197.

SANTANGELO, J. D., JONES, D. T., WOODS, D. R. (1991), Metronidazole activation and isolation of *Clostridium acetobutylicum* electron transport genes, *J. Bacteriol.* **173**, 1088–1095.

SANTANGELO, J. D., DÜRRE, P., WOODS, D. R. (1994), Characterization and expression of the hydrogenase I gene from *Clostridium acetobutylicum* P262, *Abstr. 7th Int. Symp. Genet. Ind. Microorg.,* Montreal, June 26–July 1, p. 96.

SASS, C., WALTER, J., BENNETT, G. N. (1993), Isolation of mutants of *Clostridium acetobutylicum* ATCC 824 deficient in protease activity, *Curr. Microbiol.* **26**, 151–154.

SAUER, U., DÜRRE, P. (1992), Possible function of $tRNA^{Thr}_{ACG}$ in regulation of solvent formation in *Clostridium acetobutylicum, FEMS Microbiol. Lett.* **100**, 147–154.

SAUER, U., DÜRRE, P. (1993), Sequence and molecular characterization of a DNA region encoding a small heat shock protein of *Clostridium acetobutylicum, J. Bacteriol.* **175**, 3394–3400.

SAUER, U., TREUNER, A., SANTANGELO, J. D., BUCHHOLZ, M., DÜRRE, P. (1994), Sporulation and primary σ factor homologous genes in *Clostridium acetobutylicum, Abstr. 7th Int. Symp. Genet. Ind. Microorg.,* Montreal, June 26–July 1, p. 249.

SCHARDINGER, F. (1905), *Bacillus macerans,* ein Aceton bildender Rottebacillus, *Cbl. Bakteriol. Parasitenkd. Infektionskrankh.,* II. Abt. 14, 772–781.

SCHARDINGER, F. (1907), Verhalten von Weizen- und Roggenmehl zu Methylenblau und zu Stärkekleister, nebst einem Anhange über die Bildung höherer Alkohole durch hitzebeständige Mikroorganismen aus Weizenmehl, *Cbl. Bakteriol. Parasitenkd. Infektionskrankh.,* II. Abt. 18, 748–767.

SCHATTENFROH, A., GRASSBERGER, R. (1899), Weitere Mitteilungen über Buttersäuregärung, *Cbl. Bakteriol. Parasitenkd. Infektionskrankh.,* II. Abt. 5, 697–702.

SCHATTENFROH, A., GRASSBERGER, R. (1900), Ueber Buttersäuregährung. I. Abhandlung, *Arch. Hyg.* **37**, 54–103.

SCHLOTE, D., GOTTSCHALK, G. (1986), Effect of cell recycle on continuous butanol–acetone fermentation with *Clostridium acetobutylicum* under phosphate limitation, *Appl. Microbiol. Biotechnol.* **24**, 1–6.

SCOTT, D., HEDRICK, L. R. (1952), The amylase of *Clostridium acetobutylicum, J. Bacteriol.* **63**, 795–803.

SHERMAN, P. D. (1978), Butyl alcohols, in: *Kirk-Othmer Encyclopedia of Chemical Technology,* 3rd Edn. (MARK, H. F., OTHMER, D. F., OVERBERGER, C. G., SEABORG, G. T., GRAYSON, M., ECKROTH, D. Eds.), Vol. 4, pp. 338–345, New York: John Wiley & Sons.

SPIVEY, M. J. (1978), The acetone/butanol/ethanol fermentation, *Process Biochem.* **13**, 2–5.

STEPHENS, G. M., HOLT, R. A., GOTTSCHAL, J. C., MORRIS, J. G. (1985), Studies on the stability of solvent production by *Clostridium acetobutylicum* in continuous culture, *J. Appl. Bacteriol.* **59**, 597–605.

STRÄTZ, M., SAUER, U., KUHN, A., DÜRRE, P. (1994), Plasmid transfer into the homoacetogen *Acetobacterium woodii* by electroporation and conjugation, *Appl. Environ. Microbiol.* **60**, 1033–1037.

TAGAKI, W., WESTHEIMER, F. H. (1968), Acetoacetate decarboxylase. The molecular weight of the enzyme and subunits, *Biochemistry* **7**, 895–900.

TAYA, M., ISHII, S., KOBAYASHI, T. (1985), Monitoring and control for extractive fermentation of *Clostridium acetobutylicum, J. Ferment. Technol.* **63**, 181–187.

THAUER, R. K., JUNGERMANN, K., DECKER, K. (1977), Energy conservation in chemotrophic anaerobic bacteria, *Bacteriol. Rev.* **41**, 100–180.

THOMPSON, D. K., CHEN, J.-S. (1990), Purification and properties of an acetoacetyl coenzyme A-reacting phosphotransbutyrylase from *Clostridium beijerinckii ("Clostridium butylicum")* NRRL B593, *Appl. Environ. Microbiol.* **56**, 607–613.

TRUFFAUT, N., SEBALD, M. (1988), Study of plasmids isolated from saccharolytic clostridia: Construction of hybrid plasmids and transfer to *Escherichia coli* and *Bacillus subtilis, Biotechnol. Lett.* **10**, 1–6.

TRUFFAUT, N., HUBERT, J., REYSSET, G. (1989), Construction of shuttle vectors useful for transforming *Clostridium acetobutylicum, FEMS Microbiol. Lett.* **58**, 15–20.

TWAROG, R., WOLFE, R. S. (1962), Enzymatic phosphorylation of butyrate, *J. Biol. Chem.* **237**, 2474–2477.

ULLMANN, S., DÜRRE, P. (1994), Cloning and sequencing of the DNA gyrase genes of *Clostridium acetobutylicum* DSM 792, *Abstr. 7th Int. Symp. Genet. Ind. Microorg.,* Montreal, June 26–July 1, p. 23.

VALENTINE, R. C., WOLFE, R. S. (1960), Purification and role of phosphotransbutyrylase, *J. Biol. Chem.* **235**, 1948–1952.

VELDKAMP, H. (1965), Enrichment cultures of prokaryotic organisms, in: *Methods in Microbiology* (NORRIS, J. R., RIBBONS, D. W., Eds.), Vol. 3A, pp. 305–361, London: Academic Press.

VERHASSELT, P., PONCELET, F., VITS, K., VAN GOOL, A., VANDERLEYDEN, J. (1989), Cloning and expression of a *Clostridium acetobutylicum* α-amylase gene in *Escherichia coli, FEMS Microbiol. Lett.* **59**, 135–140.

VOGET, C. E., MIGNONE, C. F., ERTOLA, R. J. (1985), Butanol production from apple pomace, *Biotechnol. Lett.* **7**, 43–46.

VOLESKY, B., SZCZESNY, T. (1983), Bacterial conversion of pentose sugars to acetone and butanol, *Adv. Biochem. Eng. Biotechnol.* **27**, 101–117.

VOLLHERBST-SCHNECK, K., SANDS, J. A., MONTENECOURT, B. S. (1984), Effect of butanol on lipid composition and fluidity of *Clostridium acetobutylicum* ATCC 824, *Appl. Environ. Microbiol.* **47**, 193–194.

VON HUGO, H., GOTTSCHALK, G. (1974), Distribution of 1-phosphofructokinase and PEP:fructose phosphotransferase activity in clostridia, *FEBS Lett.* **46**, 106–108.

WALTER, K. A., BENNETT, G. N., PAPOUTSAKIS, E. T. (1992), Molecular characterization of two *Clostridium acetobutylicum* ATCC 824 butanol dehydrogenase isozyme genes, *J. Bacteriol.* **174**, 7149–7158.

WALTER, K. A., NAIR, R. V., CARY, J. W., BENNETT, G. N., PAPOUTSAKIS, E. T. (1993), Sequence and arrangement of two genes of the butyrate-synthesis pathway of *Clostridium acetobutylicum* ATCC 824, *Gene* **134**, 107–111.

WALTON, M. T. (1945), Process for the production of riboflavin by butyl alcohol producing bacteria, *U. S. Patent* 2368074.

WANG, D. I. C., COONEY, C. L., DEMAIN. A. L., DUNHILL, P. (1979), *Fermentation and Enzyme Technology,* New York: John Wiley & Sons.

WARREN, S., ZERNER, B., WESTHEIMER, F. H. (1966), Acetoacetate decarboxylase. Identification of lysine at the active site, *Biochemistry* **5**, 817–823.

WATERSON, R. M., CASTELLINO, F. J., HASS, G. M., HILL, R. L. (1972), Purification and characterization of crotonase from *Clostridium acetobutylicum, J. Biol. Chem.* **247**, 5266–5271.

WEIZMANN, C. (1915), Improvements in the bacterial fermentation of carbohydrates and in bacterial cultures for the same, *British Patent* 4845.

WEIZMANN, C. (1919), Production of acetone and alcohol by bacteriological processes, *U. S. Patent* 1315585.

WELCH, R. W., RUDOLPH, F. B., PAPOUTSAKIS, E. T. (1989), Purification and characterization of the NADH-dependent butanol dehydrogenase from *Clostridium acetobutylicum* (ATCC 824), *Arch. Biochem. Biophys.* **273**, 309–318.

WELCH, R. W., CLARK, S. W., BENNETT, G. N., RUDOLPH, F. B. (1992), Effects of rifampicin and chloramphenicol on product and enzyme levels of the acid- and solvent-producing pathways of *Clostridium acetobutylicum* (ATCC 824), *Enzyme Microb. Technol.* **14**, 277–283.

WELSH, F. W., VELIKY, I. A. (1984), The production of acetone-butanol from acid whey, *Biotechnol. Lett.* **6**, 61–64.

WESTHEIMER, F. H. (1969), Acetoacetate decarboxylase from *Clostridium acetobutylicum, Methods Enzymol.* **14**, 231–241.

WIESENBORN, D. P., RUDOLPH, F. B., PAPOUTSAKIS, E. T. (1988), Thiolase from *Clostridium acetobutylicum* ATCC 824 and its role in the synthesis of acids and solvents, *Appl. Environ. Microbiol.* **54**, 2717–2722.

WIESENBORN, D. P., RUDOLPH, F. B., PAPOUTSAKIS, E. T. (1989a), Phosphotransbutyrylase from *Clostridium acetobutylicum* ATCC 824 and its role in acidogenesis, *Appl. Environ. Microbiol.* **55**, 317–322.

WIESENBORN, D. P., RUDOLPH, F. B., PAPOUTSAKIS, E. T. (1989b), Coenzyme A transferase from *Clostridium acetobutylicum* ATCC 824 and its role in the uptake of acids, *Appl. Environ. Microbiol.* **55**, 323–329.

WILDE, E., HIPPE, H., TOSUNOGLU, N., SCHALLEHN, G., HERWIG, K., GOTTSCHALK, G. (1989), *Clostridium tetanomorphum* sp. nov., nom. rev., *Int. J. Syst. Bacteriol.* **39**, 127–134.

WILKINSON, S. R., YOUNG, M. (1994), Targeted integration of genes into the *Clostridium acetobutylicum* chromosome, *Microbiology* **140**, 89–95.

WILLIAMS, D. R., YOUNG, D. I., YOUNG, M. (1990), Conjugative plasmid transfer from *Escherichia coli* to *Clostridium acetobutylicum, J. Gen. Microbiol.* **136**, 819–826.

WINOGRADSKY, S. (1902), *Clostridium pastorianum,* seine Morphologie und seine Eigenschaften als Buttersäureferment, *Cbl. Bakteriol. Parasitenkd. Infektionskrankh.,* II. Abt. 9, 43–54, 107–112.

WONG, J., BENNETT, G. N. (1994), Studies of DNA supercoiling in *Clostridium acetobutylicum* ATCC824, *Abstr. Gen. Meet. Am. Soc. Microbiol.,* Las Vegas, May 23–27, O-40.

WOODRUFF, J. C., STILES, H. R., LEGG, D. A. (1937), Process of producing butyl alcohol, *U. S. Patent* 2089522.

WOOLLEY, R. C., MORRIS, J. G. (1990), Stability of solvent production by *Clostridium acetobutylicum* in continuous culture: strain differences, *J. Appl. Bacteriol.* **69**, 718–728.

WOOLLEY, R. C., PENNOCK, A., ASHTON, R. J., DAVIES, A., YOUNG, M. (1989), Transfer of Tn*1545* and Tn*916* to *Clostridium acetobutylicum, Plasmid* **22**, 169–174.

YAN, R.-T., CHEN, J.-S. (1990), Coenzyme A-acylating aldehyde dehydrogenase from *Clostridium beijerinckii* NRRL B592, *Appl. Environ. Microbiol.* **56**, 2591–2599.

YAN, R.-T., ZHU, C.-X., GOLEMBOSKI, C., CHEN, J.-S. (1988), Expression of solvent-forming enzymes and onset of solvent production in batch cultures of *Clostridium beijerinckii ("Clostridium butylicum"), Appl. Environ. Microbiol.* **54**, 642–648.

YAROVENKO, V. L. (1964), Principles of the continuous alcohol and butanol–acetone fermentation processes, in: *Continuous Cultivation of Microorganisms, 2nd Symp.* (MALEK, I., Ed.), pp. 205–217, Prague: Cechoslovakian Academy of Sciences.

YOON, K.-H., LEE, J.-K., KIM, B. H. (1991), Construction of a *Clostridium acetobutylicum–Escherichia coli* shuttle vector, *Biotechnol. Lett.* **13**, 1–6.

YOSHINO, S., YOSHINO, T., HARA, S., OGATA, S., HAYASHIDA, S. (1990), Construction of shuttle vector plasmid between *Clostridium acetobutylicum* and *Escherichia coli, Agr. Biol. Chem.* **54**, 437–441.

YOUNG, M. (1993), Development and exploitation of conjugative gene transfer in clostridia, in: *The Clostridia and Biotechnology* (WOODS, D. R., Ed.), pp. 99–117, Stoneham: Butterworth-Heinemann.

YOUNG, M., MINTON, N. P., STAUDENBAUER, W. L. (1989), Recent advances in the genetics of the clostridia, *FEMS Microbiol. Rev.* **63**, 301–326.

YOUNGLESON, J. S., SANTANGELO, J. D., JONES, D. T., WOODS, D. R. (1988), Cloning and expression of a *Clostridium acetobutylicum* alcohol dehydrogenase gene in *Escherichia coli, Appl. Environ. Microbiol.* **54**, 676–682.

YOUNGLESON, J. S., JONES, W. A., JONES, D. T., WOODS, D. R. (1989a), Molecular analysis and nucleotide sequence of the *adh1* gene encoding an NADPH-dependent butanol dehydrogenase in the Gram-positive anaerobe *Clostridium acetobutylicum, Gene* **78**, 355–364.

YOUNGLESON, J. S., JONES, D. T., WOODS, D. R. (1989b), Homology between hydroxybutyryl and hydroxyacyl coenzyme A dehydrogenase enzymes from *Clostridium acetobutylicum* fermentation and vertebrate fatty acid $\beta$-oxidation pathways, *J. Bacteriol.* **171**, 6800–6807.

YU, P.-L., PEARCE, L. E. (1986), Conjugal transfer of streptococcal antibiotic resistance plasmids into *Clostridium acetobutylicum, Biotechnol. Lett.* **8**, 469–474.

YU, E. K. C., SADDLER, J. N. (1983), Enhanced acetone–butanol fermentation by *Clostridium, acetobutylicum* grown on D-xylose in the presence of acetic or butyric acid, *FEMS Microbiol. Lett.* **18**, 103–107.

YU, E. K. C., CHAN, M. K.-H., SADDLER, J. N. (1985), Butanol production from cellulosic substrates by sequential co-culture of *Clostridium thermocellum* and *C. acetobutylicum, Biotechnol. Lett.* **7**, 509–514.

YU, P. L., SMART, J. B., ENNIS, B. M. (1987), Differential induction of $\beta$-galactosidase activities in the fermentation of whey permeate by *Clostridium acetobutylicum, Appl. Microbiol. Biotechnol.* **26**, 254–257.

ZILLIG, W., HOLZ, I., WUNDERL, S. (1991), *Hyperthermus butylicus* gen. nov., sp. nov., a hyperthermophilic, anaerobic, peptide-fermenting, facultatively $H_2S$-generating archaebacterium, *Int. J. Syst. Bacteriol.* **41**, 169–170.

ZOUTBERG, G. R., WILLEMSBERG, R., SMIT, G., TEIXEIRA DE MATTOS, M. J., NEIJSSEL, O. M. (1989a), Aggregate-formation by *Clostridium butyricum, Appl. Microbiol. Biotechnol.* **32**, 17–21.

ZOUTBERG, G. R., WILLEMSBERG, R., SMIT, G., TEIXEIRA DE MATTOS, M. J., NEIJSSEL, O. M. (1989b), Solvent production by an aggregate-forming variant of *Clostridium butyricum, Appl. Microbiol. Biotechnol.* **32**, 22–26.

# 7 Microbial Production of 2,3-Butanediol

IAN S. MADDOX

Palmerston North, New Zealand

# 1 Introduction

2,3-Butanediol has been known as a bacterial fermentation product since early this century. Although interest in the development of a commercial process began during the 1930s it has not yet been operated on this scale. Much of the work into the fermentation process prior to 1963 has been well reviewed (PRESCOTT and DUNN, 1959; LONG and PATRICK, 1963). During World War II, research was intensified in Canada and U.S.A. in anticipation of shortages of the strategic chemical 1,3-butadiene, which can be produced by dehydration of butanediol and used in the manufacture of synthetic rubber. This work was largely discontinued in the face of competition from synthetic 1,3-butadiene, obtained from petrochemical sources. Nevertheless, fundamental research studies continued into the physiology and biochemistry of the organisms capable of 2,3-butanediol production. Following the oil-shocks and price rises of the 1970s, interest was rekindled in the production of chemical feedstocks and fuels from renewable resources (biomass), and this has included butanediol. Several reviews are available describing the activities during this period (JANSEN and TSAO, 1983; VOLOCH et al., 1985; MAGEE and KOSARIC, 1987). An interesting advantage of butanediol over ethanol and butanol is that it is much less toxic to the producing microorganism, thus allowing higher product concentrations to be achieved in fermentation broths (YU and SADDLER, 1985). Unfortunately, there are few research reports describing the recovery of butanediol from fermentation broths, and this may hinder future commercialization of the process. Another problem may be that if butanediol is used as a chemical feedstock rather than as a fuel, the dehydration reactions to yield 1,3-butadiene will need to be developed as a commercial process (PALSSON et al., 1981).

Currently, research into the fermentation process is continuing in many parts of the world, using a variety of substrates and microorganisms, and it is the purpose of this chapter to summarize our existing knowledge.

# 2 Microorganisms

Several organisms are known to accumulate butanediol in reasonable quantities, but there is some variation in the stereoisomers produced. Fig. 1 shows the three isomeric forms of butanediol, and Tab. 1 summarizes the isomers produced by different organisms. Other bacteria which are referred to in the literature include *Klebsiella pneumoniae*, *K. aerogenes*, and *Aerobacter aerogenes*, but these are synonyms for *K. oxytoca* as are *Enterobacter aerogenes* and *E. cloacae*. These organisms are gram-negative, facultatively anaerobic rods, and are members of the family Enterobacteriaceae. They produce the L(+)- and *meso*-stereoisomers in the approximate ratio 1:9. *Aeromonas (Pseudomonas) hydrophila* is also a gram-negative, facultatively anaerobic rod, but is distinguished from members of the Enterobacteriaceae on the basis of being oxidase-positive and having polar flagella when grown in liquid media.

*Bacillus polymyxa* displays vigorous anaerobic growth in the presence of a fermentable carbohydrate, and may be mistaken for a member of the Enterobacteriaceae as the Gram reaction is usually negative. *B. subtilis* differs from *B. polymyxa* in that it produces a mixture of the D(−)- and *meso*-stereoisomers (approximate ratio 3:2), and glycerol rather than ethanol as a reaction by-product. *B. amyloliquefaciens* is similar to *B. subtilis*, but differs in producing fewer by-products (ALAM et al., 1990). *B. licheniformis*, although a different species, behaves similarly to *B. polymyxa* except for the production of

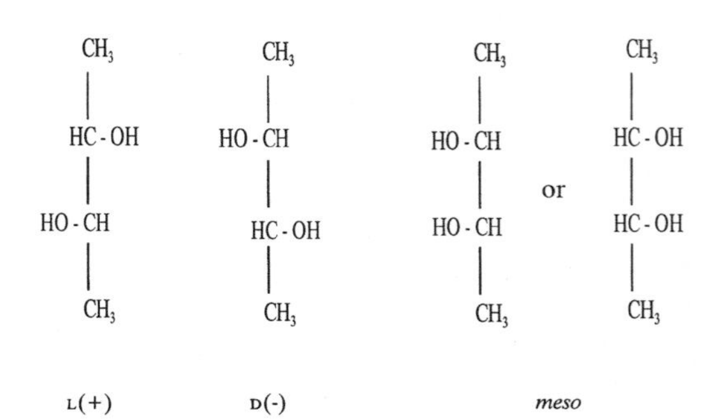


**Fig. 1.** The stereoisomers of 2,3-butanediol.

**Tab. 1.** Butanediol Isomers Produced by Various Bacteria

| Organism | L(+) | D(−) | *meso* |
|---|---|---|---|
| *Klebsiella oxytoca* | + | − | + |
| *Aeromonas hydrophila* | − | + | − |
| *Bacillus polymyxa* | − | + | − |
| *Bacillus subtilis* | − | + | + |
| *Bacillus amyloliquefaciens* | − | + | + |
| *Serratia marcescens* | − | − | + |

lactate as a by-product (RASPOET et al., 1991).

*K. oxytoca* and *B. polymyxa* are the organisms which have received most consideration in research studies, with the former receiving more attention in recent years. It has been claimed that *B. polymyxa* has the advantages of producing only the D(−)-isomer and of being able to ferment polymeric substrates, such as found in agricultural wastes, without any pretreatment. However, the isomeric form would have little relevance if the butanediol were used as either a fuel or chemical feedstock, and the most appropriate organism for any substrate would be case-specific. Attention must also be paid to the by-products produced and their relative amounts, as these have direct effects on product yields and purification procedures.

# 3 Biochemistry

Most studies have been performed using members of the Enterobacteriaceae, and Fig. 2 summarizes the mixed acid-2,3-butanediol pathway (KOSARIC et al., 1992). The genes involved in the conversion of pyruvate to 2,3-butanediol, i.e., those coding for $\alpha$-acetolactate synthase, $\alpha$-acetolactate decarboxylase, and acetoin reductase, have been isolated

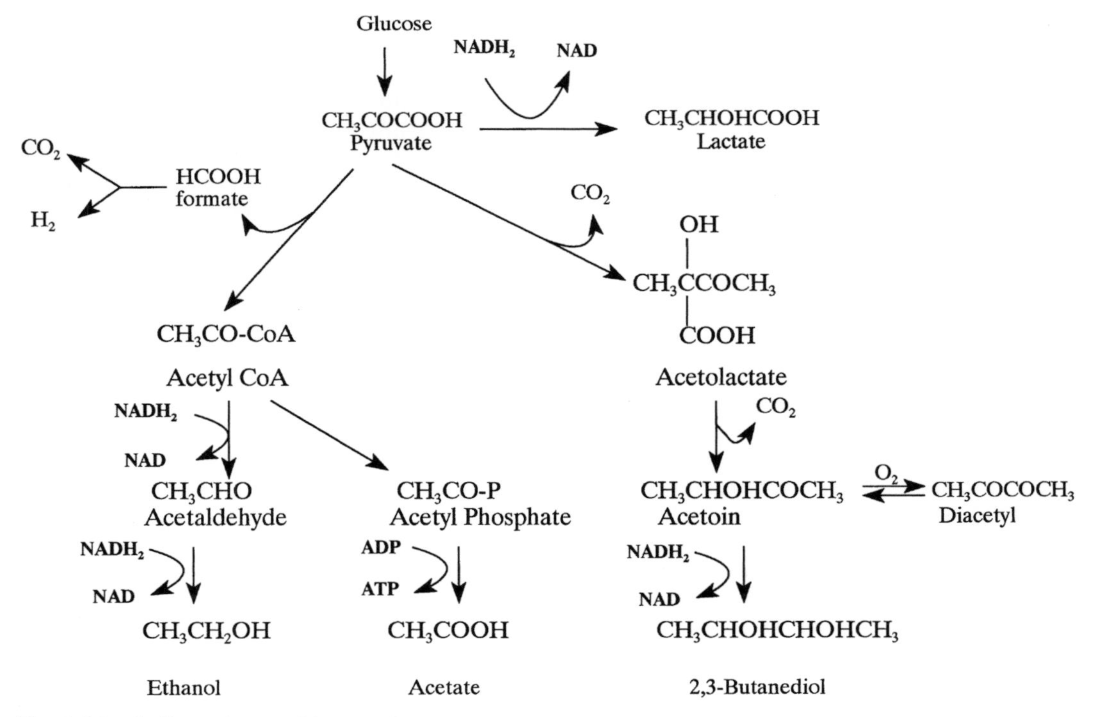


**Fig. 2.** Metabolic pathway of butanediol production from glucose.

and shown to be located in one operon (BLOMQVIST et al., 1993).

## 3.1 Production of Pyruvate

Production of pyruvate from glucose follows the normal glycolytic pathway, although there is evidence from theoretical studies that in some organisms the hexose monophosphate pathway may also be involved (PAPOUTSAKIS and MEYER, 1985). For pentose sugars, the pentose phosphate pathway operates to yield glyceraldehyde-3-phosphate which is then converted to pyruvate by glycolysis (JANSEN and TSAO, 1983).

## 3.2 Pyruvate to Acetolactate

The catabolic (pH 6) $\alpha$-acetolactate-forming enzyme (EC 4.1.3.18) catalyzes the condensation of two pyruvate molecules and a single decarboxylation to yield acetolactate. The size of the enzyme is 60 kDa (BLOMQVIST et al., 1993) and it displays an optimum at pH 6. A kinetic study of the enzyme from *Aerobacter aerogenes* has been described by STORMER (1968). The enzyme requires acetate for induction, and the acetate ion is also involved in its activation (JOHANSEN et al., 1975; STORMER, 1977).

## 3.3 Acetolactate to Acetoin

Acetolactate decarboxylase (EC 4.1.1.5) catalyzes the decarboxylation of acetolactate to acetoin. The enzyme from *A. aerogenes* has been purified and characterized by LOKEN and STORMER (1970), and has a pH optimum of 6.2–6.4. The stereochemical reaction product is invariably D(−)-acetoin, and this can be either oxidized non-enzymically to diacetyl by oxygen present in the medium, or reduced enzymically to butanediol. As with acetolactate formation, acetolactate decarboxylase requires acetate for induction.

## 3.4 Acetoin to Butanediol

The final reaction in the formation of 2,3-butanediol is catalyzed by the enzyme 2,3-butanediol dehydrogenase (EC 1.1.1.4). In *A. aerogenes* the reductions of diacetyl to acetoin (diacetyl reductase, EC 1.1.1.5) and of acetoin to butanediol are catalyzed by the same enzyme. However, whereas the former reduction is irreversible the latter is reversible (BRYN et al., 1971). As with the preceding enzymes in the biosynthetic pathway, butanediol dehydrogenase is induced by acetate. Conversely, for the oxidation of butanediol to acetoin, acetate acts as an inhibitor. This inhibition increases with decreasing pH, indicating that it is acetic acid which is the effector (LARSEN and STORMER, 1973).

Butanediol dehydrogenases from *B. polymyxa* and *Serratia marcescens* have also been purified and characterized (HOHN-BENTZ and RADLER, 1978).

The overall effect of acetate on butanediol formation from pyruvate is shown in Fig. 3. BLOMQVIST et al. (1993), during their work on the isolation of the three genes coding for those enzymes involved in the conversion of pyruvate to butanediol, concluded that the operon containing these genes is regulated at the transcriptional level and that induction is improved by limited-oxygen conditions. The cloning of these genes now opens the possibility of fully elucidating the regulation of butanediol production and its relationship to pH, anaerobic conditions and the regulation of amino acid biosynthesis (*via* the anabolic $\alpha$-acetolactate synthase involved in the biosynthesis of valine, leucine, and isoleucine).

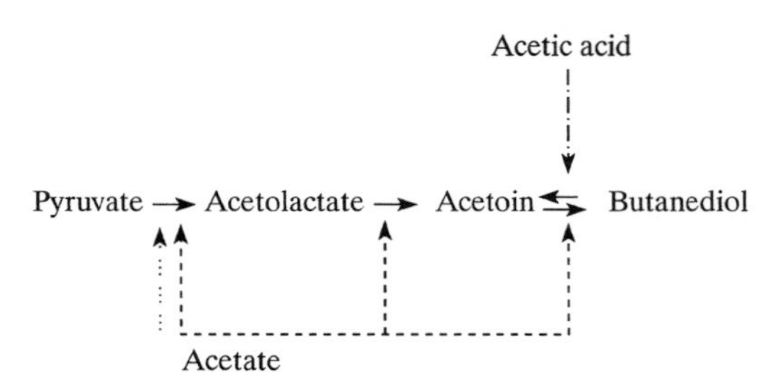


**Fig. 3.** The effect of acetate on butanediol formation from pyruvate. ----- induction; ··· ··· activation; -·-·-·- inhibition.

## 3.5 Mechanism of Formation of Different Stereoisomers

The D(−)-, L(+)-, and *meso*-isomers of butanediol are all known to be produced by different organisms, the ratio of isomers appearing to depend on the organism and the culture conditions. Three models have been postulated to account for the production of the different stereoisomers. In the first, developed by TAYLOR and JUNI (1960) and used by VOLOCH et al. (1983), the existence of an acetoin racemase is proposed (Fig. 4). The stereochemical reaction product of acetolactate decarboxylase is invariably D(−)-acetoin, and this can be reduced to D(−)-butanediol by the stereospecific D(−)-butanediol dehydrogenase. The stereospecific L(+)-butanediol dehydrogenase, on the other hand, converts D(−)-acetoin to *meso*-butanediol. To account for the production of L(+)-butanediol, the proposed acetoin racemase converts D(−)-acetoin to L(+)-acetoin which is then reduced to L(+)-butanediol and/or *meso*-butanediol by the L(+)- and/or D(−)-butanediol dehydrogenase, respectively. This model can account for the different product spectra of different organisms. Thus, e.g., *K. oxytoca* possesses the acetoin racemase and the L(+)-butanediol dehydrogenase, allowing the formation of L(+)- and *meso*-butanediol, but not D(−)-butanediol. Conversely, some *Bacillus* species possess the acetoin racemase and the D(−)-butanediol dehydrogenase, allowing the formation of D(−)- and *meso*-butanediol, but not L(+)-butanediol. Unfortunately, the flaw in this model is that the presence of acetoin racemase has never been demonstrated experimentally. Thus, HOHN-BENTZ and RADLER (1978) suggested two other possible reasons to account for the various spectra of butanediol isomers: (1) small amounts of L(+)-acetoin are produced by the stereospecific acetolactate decarboxylase; (2) L(+)-acetoin is formed from diacetyl.

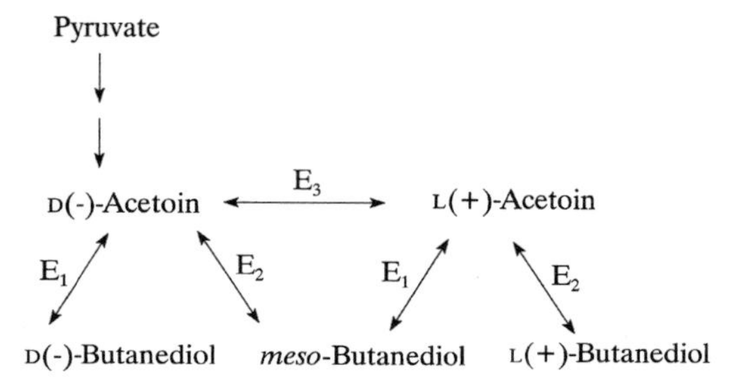


**Fig. 4.** Model proposed by TAYLOR and JUNI (1960) for production of different stereoisomers of butanediol. $E_1$, D(−)-dehydrogenase; $E_2$, L(+)-dehydrogenase; $E_3$, acetoin racemase.

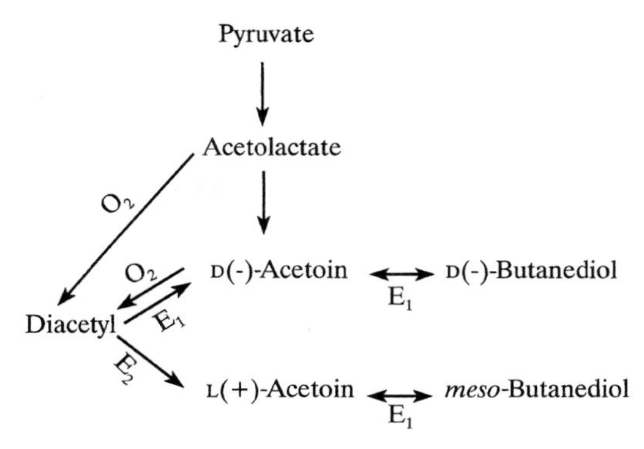


**Fig. 5.** Model proposed by UI et al. (1986) for production of different stereoisomers of butanediol by *B. polymyxa*. $E_1$, NADH-linked D(−)-butanediol dehydrogenase; $E_2$, NADPH-linked diacetyl reductase.

The second model was proposed by UI et al. (1986), for a *Bacillus* species, and is based on the experimentally-proved presence of the enzyme diacetyl reductase (Fig. 5). Thus, diacetyl is formed non-enzymically from acetolactate and/or D(−)-acetoin, and is then successively reduced to D(−)-acetoin and D(−)-butanediol by an NADH-linked D(−)-butanediol dehydrogenase. L(+)-acetoin is formed from diacetyl by the action of an NADPH-linked diacetyl reductase, but this enzyme cannot reduce acetoin to butanediol. Thus, the D(−)-butanediol dehydrogenase reduces L(+)-acetoin to *meso*-butanediol. In their experiments, UI et al. (1986) could not demonstrate the presence of an acetoin or butanediol racemase.

Finally, a model has been proposed by UI et al. (1984) for *K. pneumoniae* which, again, does not require the intervention of an acetoin racemase. In this model, three butanediol dehydrogenases are postulated so that D(−)-

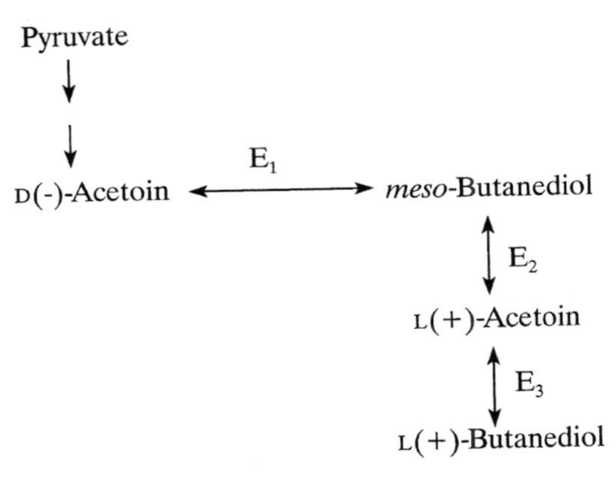


**Fig. 6.** Model proposed by UI et al. (1984) to account for production of different stereoisomers of butanediol by *K. pneumoniae*. $E_1$, *meso*-butanediol dehydrogenase (D(−)-acetoin-forming); $E_2$, *meso*-butanediol dehydrogenase (L(+)-acetoin-forming); $E_3$, L(+)-butanediol dehydrogenase (L(+)-acetoin-forming).

acetoin is converted successively to *meso*-butanediol, L(+)-acetoin and L(+)-butanediol (Fig. 6).

In summary, it is clear that for bacteria which produce only the D(−)-isomer, the existence of a sole D(−)-butanediol dehydrogenase is all that is necessary. But, to account for the production of the L(+)- and/or *meso*-isomers, because of the stereospecificity of the butanediol dehydrogenases, there must be a source of L(+)-acetoin. Since the presence of an acetoin racemase has never been demonstrated experimentally, the models of UI et al. (1984) and (1986) must be considered as the favored mechanisms.

## 3.6 Regulation and Physiological Significance of Butanediol Production

Most of the studies performed in this regard have been with members of the Enterobacteriaceae, hence it is this group of organisms that is considered here. However, it is likely that a similar situation pertains for species of *Bacillus* (DE MAS et al., 1988).

These organisms are facultative anaerobes which are able to obtain their energy for growth by either aerobic respiration or anaerobic fermentation. Butanediol is produced during oxygen-limited growth, by a fermentative pathway known as the mixed acid–butanediol pathway (Fig. 2). The major influence on butanediol production is the supply of oxygen and the dissolved oxygen concentration of the culture. As such, the mixed acid–butanediol pathway is a classic example of anaerobic fermentation in terms of maintaining an internal redox balance with respect to the pyridine nucleotide pool during glycolysis and biosynthesis. The second major influence on butanediol production is the pH of the culture, and this is manifest *via* the effect of acetate on the enzymes responsible for butanediol synthesis.

PIRT (1957), using *Aerobacter cloacae*, was the first to demonstrate the importance of oxygen on butanediol production. He showed that under anaerobic conditions cell synthesis and carbon dioxide production were at their minimum levels, and most of the carbon (sugar) was converted into a mixture of ethanol, formate, butanediol, acetoin, and acetate. A small supply of oxygen suppressed the formation of ethanol and formate, but still permitted the production of butanediol and acetoin, while increasing the proportion of carbon converted to acetate, cells, and carbon dioxide. As the oxygen supply was further increased, the yield of cells and carbon dioxide increased while butanediol and acetoin production were suppressed. This concept was developed further by HARRISON and PIRT (1967), who worked with *K. aerogenes* in continuous culture, and by JOHANSEN et al. (1975) who worked with *A. aerogenes*. In both cases, it was demonstrated that under aerobic conditions butanediol production was minimal. However, under anaerobic conditions butanediol, acetate, ethanol, and lactate were formed at the expense of biomass. The reason for the shift in metabolism from production of biomass to production of mixed acids–butanediol is so that the balance of NAD/NADH within the cell is maintained. During glycolysis there is a net conversion of NAD to NADH. Under aerobic conditions, regeneration of NAD occurs *via* respiration, while under anaerobic conditions, where respiration cannot occur, butanediol production serves the same purpose. Further, under aerobic conditions, the key enzyme acetolactate synthase is rapidly and irreversibly inactivated, thus preventing butanediol production in the presence of oxygen (KOSARIC et al., 1992). It should also be noted that because the reduction of acetoin to butanediol is a reversible reaction, the enzyme butanediol dehydrogenase can be used to regenerate NADH. Thus the butanediol pathway and the relative proportions of acetoin and butanediol serve to maintain the intracellular NAD/NADH balance in changing culture conditions (BLOMQVIST et al., 1993).

During their experiments in continuous culture, HARRISON and PIRT (1967) observed that under oxygen-limited conditions, butanediol production occurred at pH 6.0, but that at pH 7.4 acetate production occurred at the expense of butanediol. Similar results were reported by JOHANSEN et al. (1975). It is now known that the induction of the butanediol pathway at low pH values is not due to any change in the intracellular pH value. Rather, the accumulation of acidic products, e.g., acetate, in the medium causes a decrease in the extracellular pH value. The resulting transmembrane pH gradient causes the accumulation of acetate within the cytoplasm, and these high acetate concentrations induce the butanediol pathway (Fig. 3). Thus the role of the culture pH is to create a larger pH gradient such that induction of butanediol production can take place before the external acetate concentration becomes high enough to cause growth inhibition, i.e., butanediol is less toxic to the organism than is acetate (BOOTH, 1985).

The production of butanediol is usually accompanied by the simultaneous production of other metabolites such as ethanol, formate, acetoin, acetate and, possibly, lactate. TEXEIRA DE MATTOS and TEMPEST (1983) questioned why it was necessary for the organism to produce so many metabolic products as a means of maintaining the correct balance of pyridine nucleotides within the cell when the same balance could be maintained by the production of just lactate. The proposed solution was described in terms of a regulation of energy flux. Thus, although the formation of acetate is energetically more favorable than the formation of lactate (due to the production of one ATP per acetate formed), it requires the concomitant synthesis of a product such as ethanol or butanediol to maintain the redox balance. Under conditions where the energy demand, i.e., ATP requirement, is low, such as at low growth rates, butanediol production will be favored over that of acetate.

In summary, butanediol is a product of fermentative metabolism whose production serves to maintain the correct redox balance within the cell during growth under oxygen-limited conditions. In addition, its production is favored when the cell's requirement for energy (ATP) is low, and in this respect it contrasts with acetate production.

# 4 Fermentation Process

## 4.1 Theoretical Yields

The net equations for the reactions leading to butanediol from glucose and xylose are:

Glucose $\rightarrow$ 2 $CO_2$ + $NADH_2$ + 2 ATP + Butanediol

Xylose $\rightarrow$ 5/3 $CO_2$ + 5/6 $NADH_2$ + 5/3 ATP + 5/6 Butanediol

Of the total carbon contained in the sugar, 2/3 goes to butanediol, and 1/3 is lost as carbon dioxide. On a mass basis, the theoretical yield of butanediol from both glucose and xylose is 0.5. The molar yield from hexoses is 1.0 while that from pentoses is 0.83.

## 4.2 Substrates, Yields, and Productivities

*Klebsiella oxytoca* is able to ferment a range of sugars including glucose, mannose, galactose, xylose, arabinose, cellobiose, and lactose. *Bacillus polymyxa* can, in addition, ferment polymeric materials such as xylan, in-

**Tab. 2.** Summary of Some Batch Fermentation Data for Various Bacterial Strains and Substrates

| Substrate | Organism | Overall Butanediol Productivity [g/L h] | Overall Butanediol Yield [g/g sugar used] | Reference |
|---|---|---|---|---|
| Glucose | *A. aerogenes* NRRL B199 | 2.02 | 0.45 | SABLAYROLLES and GOMA (1984) |
| Glucose | *B. polymyxa* ATCC 12321 | 0.64 | 0.3 | DE MAS et al. (1988) |
| Xylose | *K. oxytoca* ATCC 8724 | 1.35 | 0.36 | JANSEN et al. (1984a) |
| Xylose | *B. polymyxa* NRCC 9035 | 0.1 | 0.24 | LAUBE et al. (1984) |
| Mannose | *K. pneumoniae* AU-1-d3 | 0.64 | 0.30 | TRAN and CHAMBERS (1986) |
| Lactose | *K. oxytoca* ATCC 8724 | 0.86 | 0.21 | RAMACHANDRAN et al. (1990) |
| Starch | *A. hydrophila* NCIB 9240 | 0.17 | 0.2 | WILLETTS (1984) |
| Starch | *B. polymyxa* DSM 356 | | 0.28 | AFSCHAR et al. (1993) |
| Starch, saccharified | *K. oxytoca* DSM 3539 | 0.66 | 0.5 | AFSCHAR et al. (1993) |
| Molasses, high-test | *K. oxytoca* DSM 3539 | 1.1 | 0.5 | AFSCHAR et al. (1991) |
| Molasses, high-test | *K. oxytoca* DSM 3539 | 1.23 | 0.5 | AFSCHAR et al. (1993) |
| Molasses, blackstrap | *K. oxytoca* DSM 3539 | 0.75 | 0.42 | AFSCHAR et al. (1993) |
| Whey | *B. polymyxa* ATCC 1232 | 0.02 | 0.16 | SPECKMAN and COLLINS (1982) |
| Whey permeate | *K. pneumoniae* NCIB 8017 | 0.08 | 0.46 | LEE and MADDOX (1984) |
| Whey | *K. pneumoniae* ATCC 13882 | 0.38 | | BARRETT et al. (1983) |
| Citrus waste | *A. aerogenes* | 1.1 | | LONG and PATRICK (1961) |
| Xylan | *B. polymyxa* NRCC 9035 | 0.02 | | LAUBE et al. (1984) |
| Wood hemicellulose hydrolyzate | *K. pneumoniae* ATCC 8724 | | 0.45 | YU et al. (1984a)<br>YU and SADDLER (1985) |
| Jerusalem artichoke | *B. polymyxa* ATCC 12321 | 0.79 | 0.4 | FAGES et al. (1986) |
| Agricultural residues | *K. pneumoniae* ATCC 8724 | | | YU et al. (1984b) |

ulin, and starch. Tab. 2 presents a summary of some of the work performed in recent years using free cells in traditional batch fermentation. A variety of bacterial strains and substrates have been used, and the fermentation conditions may not necessarily have been optimized. Hence it is difficult to directly compare the results described by different authors. Much work has been directed towards the fermentation of sugars present in wood hydrolyzates, particularly those derived from the hemicellulose component. Virtually all of these sugars can be fermented by both *K. oxytoca* and *B. polymyxa*. Given that neither the optimum fermentation conditions nor the most appropriate strains of organisms have necessarily been used, it appears on the present evidence that *K. oxytoca* is the preferred organism. Unfortunately, however, this organism possesses neither cellulase nor hemicellulase activity, so a variety of methods have been investigated as pretreatment techniques to render the sugars available from wood or agricultural residues. For example, YU et al. (1984c) have described the use of acid hydrolysis as a pretreatment technique. Unfortunately, this technique, and that of steam treatment, can cause the release of water-soluble lignin degradation products, such as vanillyl or syringyl derivatives, which are inhibitory to butanediol production by *K. oxytoca* (NISHIKAWA et al., 1988a, b; FRAZER and MCCASKEY, 1991). To some extent, the problem can be overcome by prolonged incubation of the cultures. Another approach to wood pretreatment has been that of simultaneous saccharification and fermentation. Culture filtrates from a strain of the cellulolytic fungus *Trichoderma harzianum* have been used to hydrolyze the substrates, while *K. pneumoniae* fermented the liberated sugars (YU et al., 1984b). The enzymatic hydrolysis was optimal at pH 5.0 and 50°C, but the combined hydrolysis and fermentation was most efficient at pH 6.5 and 30°C. The combined process resulted in butanediol levels that were 30-40% higher than could be obtained using separate hydrolysis and fermentation processes.

*Bacillus polymyxa*, the alternative organism to *K. pneumoniae*, has the advantage of being able to directly utilize polymers, e.g., xylan and starch, but the reactor productivities and yields which have been reported are generally inferior to those of the latter organism. An exception to this may be the use of Jerusalem artichokes (containing a mixture of inulin and sucrose) as a substrate for *B. polymyxa*, where very promising reactor productivities and yields have been recorded. As a substrate for butanediol production, starch has received little attention, probably because of its relatively high cost as a fermentation substrate when compared to agricultural wastes. Work described by WILLETS (1984) using *Aeromonas hydrophila*, and by AFSCHAR et al. (1993) using *B. polymyxa*, has demonstrated that prior hydrolysis is not necessary for these organisms, but the productivities and yields obtained were rather low. In contrast, starch which had undergone prior saccharification using a mixture of amylase and amyloglucosidase, proved to be an excellent substrate for *K. oxytoca* (AFSCHAR et al., 1993). High-test molasses has also been reported to be an excellent substrate for *K. oxytoca*, but blackstrap molasses, because of its higher salt content, appears to be less useful (AFSCHAR et al., 1991, 1993). Whey, a by-product of the dairy industry, contains lactose and has been studied as a substrate for butanediol production by several groups. Members of the Enterobacteriaceae appear to be more useful organisms than *B. polymyxa*, although one strain of *K. oxytoca* has been reported as producing butanediol only poorly from lactose (CHAMPLUVIER et al., 1989a). Prior hydrolysis of the lactose using a $\beta$-galactosidase enzyme confers an advantage over the unhydrolyzed substrate (LEE and MADDOX, 1984; CHAMPLUVIER et al., 1989b).

## 4.3 Factors Affecting the Fermentation

### 4.3.1 Temperature

In general, the optimum temperature for growth of butanediol-producing bacteria is in the range 30-37°C, but few definitive studies have been reported on the effect of temperature on butanediol production. With *K. pneumoniae*, an increase in temperature from 33°C to 37°C has been reported to reduce butanediol production by 66%, but little effect was observed over the same temperature range with *Enterobacter aerogenes* (BARRETT et al., 1983). MARTINEZ and SPECKMAN (1988) have reported that for *B. polymyxa* growing on whey the optimum temperature is 35°C, with little butanediol production at 40°C. Since different strains may possess different temperature optima, this should be checked for each strain and substrate being used.

## 4.3.2 pH Value

Despite the fact that the culture pH value is of major importance in butanediol production, relatively few studies have been performed where this parameter has been controlled. JANSEN et al. (1984a), using *K. oxytoca* growing on xylose in batch fermentation, used pH control to investigate the effect on both growth and butanediol production. The maximum specific growth rate occurred at pH 5.2, while there was no growth below pH 4.2. The butanediol yield was not strongly affected over the range pH 4.4 to 5.8, but appeared to reach a peak value between pH 5.2 and pH 5.6. HARRISON and PIRT (1967) used *K. aerogenes* in chemostat culture with pH control, and demonstrated that at pH 6.2 butanediol production was favored over acetate production, whereas at pH 7.4 the reverse was true. Similar results have subsequently been reported by SCHÜTZ et al. (1985) and by ZENG et al. (1990a) using *Enterobacter cloacae* and *E. aerogenes*, respectively, in continuous culture. A similar dependence of butanediol production on pH has been confirmed using *Bacillus licheniformis* (RASPOET et al., 1991) and *B. polymyxa* (MANKAD and NAUMAN, 1992). Thus, it is now well established that maximum butanediol productivity and yield occur in the range pH 5.0–5.8, while production of acetate is favored at higher pH values.

## 4.3.3 Oxygen

This is the most important environmental parameter affecting the fermentation process.

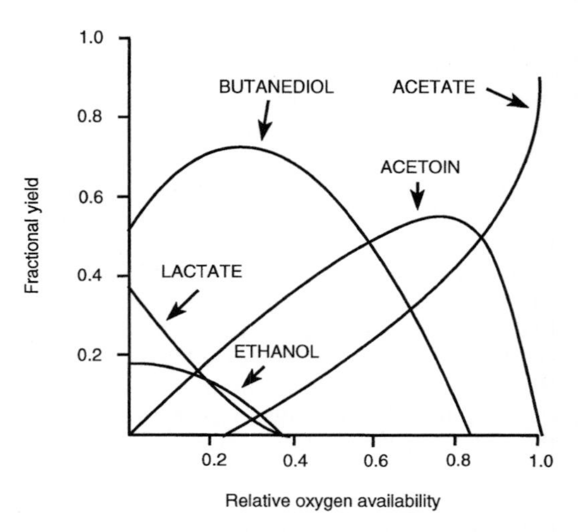


**Fig. 7.** The effect of relative oxygen availability on fractional product yields in *B. polymyxa* (DE MAS et al., 1988).

Although butanediol is a product of anaerobic fermentation, aeration is known to enhance its production. Essentially, this is for two reasons. First, since butanediol-producing bacteria are facultative anaerobes, they can grow in either the presence or absence of oxygen. Aerobic growth is more efficient, hence when cells are cultivated in the presence of oxygen, higher concentrations of biomass can be achieved. If the oxygen supply is subsequently withdrawn, fermentative metabolism ensues and butanediol production occurs. In this case, there is a direct relationship between the volumetric butanediol productivity and the biomass concentration, which was determined by the original oxygen supply. The second reason is that the oxygen supply influences the proportion of the various products of the mixed acid–butanediol pathway (Fig. 2). This has been illustrated by DE MAS et al. (1988) as shown in Fig. 7. Thus, when the supply of oxygen is high, biomass, acetate and, to a lesser extent, acetoin production are favored. At an intermediate oxygen supply, butanediol production is favored, while at a low oxygen supply, lactate and ethanol production predominate. The reason for these changes in product proportions is connected with energy supply and the maintenance of the correct balance of NAD/NADH within

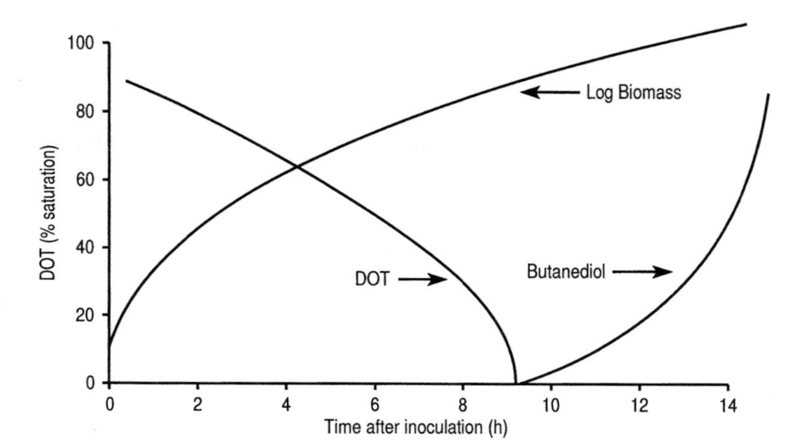


**Fig. 8.** Relationship between culture DOT value, butanediol and biomass during a typical fermentation process (based on JANSEN et al., 1984a).

the cell, as described earlier in this chapter. In addition, the situation is complicated by the inhibitory effects of some of the products, e.g., acetate, ethanol, and lactate, on cell growth and butanediol production. To optimize the production of butanediol, therefore, it is necessary to minimize the production of byproducts, which may be inhibitory, and this is achieved by control of the oxygen supply rate relative to the demand of the culture. In turn, the oxygen supply rate is controlled by the rates of aeration and agitation of the culture (QURESHI and CHERYAN, 1989a).

In a batch fermentation process there are typically two phases (Fig. 8; JANSEN et al., 1984a). In the first, the oxygen supply is sufficient to maintain a high value of dissolved oxygen tension (DOT) in the culture, thus the biomass increases exponentially, but no butanediol is produced. As the biomass concentration increases, so does the oxygen demand of the culture, and the DOT value falls to zero. After the oxygen supply becomes limiting (DOT $<$ 3% of saturation) the biomass increases linearly rather than exponentially, i.e., the growth rate decreases. During the second phase, the fate of the sugar depends on the oxygen supply rate (controlled by the aeration and agitation rates) and on the oxygen demand of the culture (controlled by the biomass concentration, the culture pH value, and the presence or absence of growth inhibitors). The aim in the present context is to optimize these parameters for butanediol production. Unfortunately, the exact fermentation conditions will be case-specific, and will need to be determined for a particular bacterial strain, medium composition and fermenter design and operation. However, an interesting approach to the optimization process has been described by KOSARIC et al. (1992) who used the redox potential, in conjunction with other parameters such as pH, to monitor culture activity and to maintain optimal growth conditions.

A recent report by SILVEIRA et al. (1993) has expressed the influence of oxygen in a slightly different way. Thus, when the DOT of the culture is greater than zero, cell growth is exponential and butanediol production is inversely proportional to the oxygen transfer coefficient ($K_La$). When the DOT is zero, growth becomes linear and the maximum specific butanediol production rate is a function of $K_La$. Thus, the maximum specific production rate is dependent on the $K_La$ of the particular fermentation run. Below its maximum, the specific production rate is a function of the specific oxygen uptake rate.

Similar studies have been performed on the relationship between oxygen supply and butanediol production in *B. polymyxa* (FAGES et al., 1986; DE MAS et al., 1988), and comparable conclusions have been drawn. In order to optimize the batch fermentation process, FAGES et al. (1986) devised a program for progressive lowering of the oxygen transfer rate ($K_La$) during the process from 41 $h^{-1}$ to 8 $h^{-1}$. However, the details of such an approach will be case-specific.

Since the original report of HARRISON and PIRT (1967) demonstrating the importance of oxygen supply in continuous culture, there have been few further reports describing this fermentation mode. However, ZENG et al. (1990b) showed that at a particular dilution rate there was an optimum oxygen uptake rate for butanediol production, and that this value increased with dilution rate.

## 4.3.4 Initial Sugar Concentration

The effect of the initial sugar concentration on the progress of a batch fermentation process will vary with the composition of the medium. For example, when using "technical" media, an increase in sugar concentration will simultaneously result in increases in other medium components, some of which may be inhibitory. Hence, the initial sugar concentration that can be used will be restricted. However, it is well established that for synthetic media, where no inhibitory compounds are present, initial sugar concentrations of up to 200 g/L can be fermented (JANSEN et al., 1984a; QURESHI and CHERYAN, 1989b; SABLAYROLLES and GOMA, 1984). The decrease in specific growth rate that occurs at sugar concentrations above 20 g/L can be explained by a decreasing water activity (Fig. 9). Butanediol productivity is much less influenced by the initial sugar concentration than is the growth rate, and maximum values

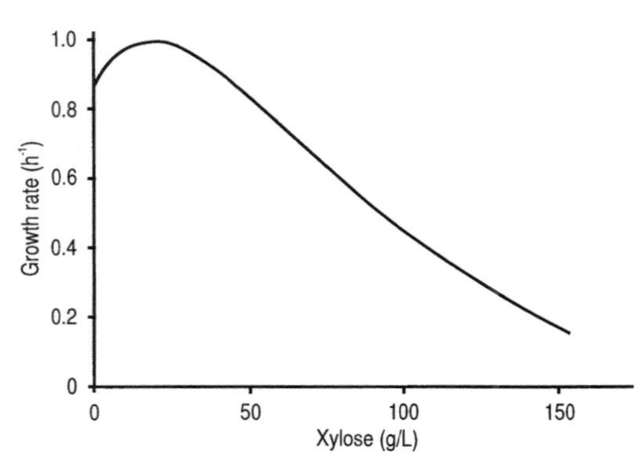


**Fig. 9.** Effect of initial sugar concentration on maximum specific growth rate of *K. oxytoca* at pH 5.2 (from JANSEN et al., 1984a).

occur at an initial concentration of approximately 100 g/L. Above this concentration, lower productivities may be related to changes in the oxygen demand of the culture due to increased biomass concentration, thereby lowering the specific oxygen supply per cell (ALAM et al., 1990; MARTINEZ and SPECKMAN, 1988). RAMACHANDRAN et al. (1990) observed that lactose, being a disaccharide, is less inhibitory on a weight basis than glucose, since reduction of water activity is a function of molarity. Using *B. polymyxa*, DE MAS et al. (1988) have shown that substrate inhibition of growth is significant only at concentrations exceeding 150 g/L.

## 4.3.5 Effect of Acetate and Other By-Products

The effects of acetate on the enzymes involved in butanediol biosynthesis were discussed in Sect. 3, and are shown in Fig. 3. With the Enterobacteriaceae, the presence of low levels of acetate (1–5 g/L) during batch fermentation is stimulatory to both sugar utilization and butanediol production. At higher concentrations, however, inhibition of growth occurs (YU and SADDLER, 1982; BARRETT et al., 1983). This inhibition is due to undissociated acetic acid rather than to the acetate ion, and so the effect is increased with decreasing culture pH value (FOND et al., 1985; ZENG et al., 1990a). Similar inhibitory effects have been observed with *A. hydrophila*, but the acetate concentrations involved are much lower (WILLETTS, 1984), and with *B. polymyxa* (DE MAS et al., 1988).

Lactate has been shown to be inhibitory to *K. oxytoca* at concentrations in the range 3–15 g/L (AFSCHAR et al., 1991; QURESHI and CHERYAN, 1989c), and to *B. polymyxa* at concentrations above 20 g/L (DE MAS et al., 1988). Growth inhibition of *E. aerogenes* by ethanol has been demonstrated at concentrations in the range 5-60 g/L (ZENG and DECKWER, 1991), and of *B. polymyxa* at similar concentrations (DE MAS et al., 1988). Finally, acetoin, at concentrations up to 20 g/L, is inhibitory to *B. polymyxa* (DE MAS et al., 1988).

During operation of processes for butanediol production, attention must be paid to the concentrations of the by-products since their inhibitory effects could influence the growth, and hence oxygen demand of the culture, with subsequent effects on butanediol production.

### 4.3.6 Product Inhibition

Butanediol is much less toxic to microorganisms that are the alcohols ethanol and butanol, and the effect on growth rate is much more marked than the effect on butanediol production rate (RAMACHANDRAN et al., 1990). Using *K. oxytoca* growing on xylose, JANSEN et al. (1984a) demonstrated that the organism failed to grow at butanediol concentrations exceeding 65 g/L. For *B. polymyxa*, the growth rate has been reported to decrease markedly above 60 g/L (DE MAS et al., 1988). However, it has been reported that there is little effect of butanediol on its own production rate even at concentrations exceeding 100 g/L (SABLAYROLLES and GOMA, 1984). Hence, in any fermentation process, product inhibition does not present a problem once the necessary biomass is established.

## 4.4 Modelling of the Fermentation

Mathematical models of batch fermentation data are useful for developing correlations between the operating conditions and the observed results. For the butanediol fermentation process, the most important operating condition is the oxygen transfer rate, followed by the initial sugar concentration (which helps determine the biomass concentration). SABLAYROLLES and GOMA (1984) performed a mathematical analysis of data obtained from batch fermentations involving *A. aerogenes* growing on glucose, and reported the following relationships:

$$\text{Biomass yield} = \frac{0.0218\,K_{\text{L}}a + 7.07}{S_0 + 15.9}$$

(g biomass formed/g glucose consumed)

$$\text{Butanediol yield} = 0.5 \cdot \frac{S_0}{0.0967\,S_0 + 2.95} \cdot \frac{0.015\,K_{\text{L}}a + 23.36}{K_{\text{L}}a + 142}$$

(g butanediol formed/g glucose consumed)

where $K_{\text{L}}a$ is the oxygen transfer coefficient when the fermentation is half complete, and $S_0$ is the initial sugar concentration.

JANSEN et al. (1984b), using *K. oxytoca*, applied the principles of bioenergetics to develop relationships for the butanediol yield and productivity based on the initial sugar concentration and oxygen supply rate (Figs. 10 and 11). These models demonstrate the interactive effects of these two operating parameters and confirm that the optimum conditions for the productivity do not necessarily coincide with those for the yield. It should be borne in mind, however, that these simulations are based on data obtained from experiments where the oxygen supply rate was maintained constant throughout the process. Programmed changes to this parameter are necessary to compensate for the differing oxygen requirements of biomass production and butanediol production, as demonstrated by FAGES et al. (1986).

More recently, MANKAD and NAUMAN (1992), using *B. polymyxa*, developed a predictive model that simulates product distribution at known oxygen transfer rates on the hypothesis that, in an energy-limited environment, the organism utilizes glucose and oxygen in the most efficient manner (the efficiency was measured in terms of ATP yields). The results confirmed that the product distribution is primarily influenced by an energetic limitation such as that imposed by limited oxygen availability.

## 4.5 Fed-Batch Fermentation

Fed-batch culture may be defined as a batch culture which is continuously fed with a growth-limiting substrate. Hence, both the growth rate and the limiting substrate utilization rate are controlled. Classical examples of fed-batch fermentation processes include

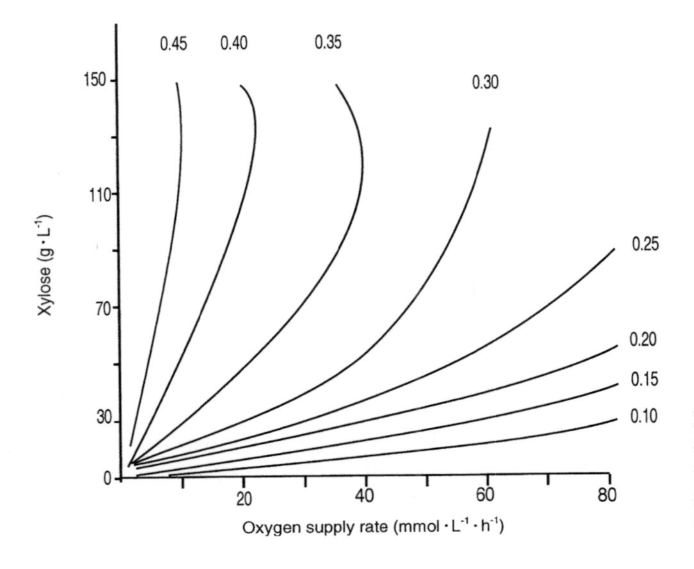


**Fig. 10.** Contour graph of the butanediol yield (g per g xylose) as a function of initial xylose concentration and oxygen supply rate, for batch fermentation (from JANSEN et al., 1984b).

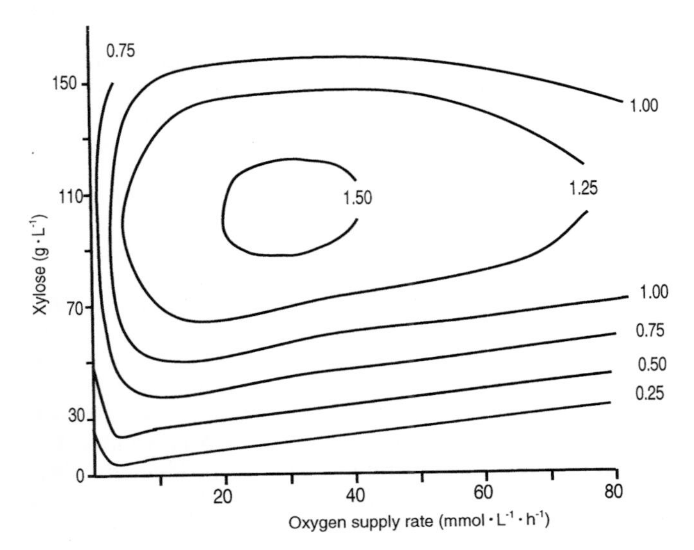


**Fig. 11.** Contour graph of the butanediol production rate (g/L h) as a function of initial xylose concentration and oxygen supply rate for a batch fermentation (from JANSEN et al., 1984b).

baker's yeast and penicillin production, where the growth rate and sugar utilization rate are controlled as a means of directing the metabolism of the organism. Other commercial fermentation processes, however, use fed-batch culture not as a means of directing metabolism, but as a means of increasing the total supply of sugar to the fermentation process. This is especially true in processes where a high initial sugar concentration is inhibitory to the growth rate, but the desired metabolic product has little inhibitory effect on its own production rate. Hence, the purpose of the fed-batch fermentation process in this situation is to (1) increase the final product concentration in the broth, which in turn aids the economics of product recovery, and (2) to improve the overall reactor productivity by extending the duration of the process and minimizing "downtime". On this basis, production of butanediol is an ideal candidate for the application of fed-batch culture. As stated earlier, butanediol has no inhibitory effect on its own production rate even at concentrations exceeding 100 g/L, whereas the specific growth rate of the organism is inhibited at sugar concentrations exceeding 20 g/L. On theoretical grounds, therefore, to maximize the growth rate during the "biomass-producing" phase of the process, the initial sugar concentration should not exceed 20 g/L, and it should be maintained by continuous feeding. Then, during the "butanediol-producing" phase, continuous feeding should be prolonged for as long as possible, i.e., until the butanediol production rate becomes too low. This is not a classical fed-batch fermentation process, however, since the growth rate is not being controlled by the feed-rate of a growth-limiting substrate, and the process should, more correctly, be termed "continuous feeding".

OLSON and JOHNSON (1948) demonstrated such a fed-batch process using *A. aerogenes* growing on a glucose-based synthetic medium. Initially, the glucose concentration was 100 g/L, and after this had decreased to approximately 30 g/L (fermentation 30 h, butanediol concentration 20–25 g/L) a continuous feed was commenced. The feed medium contained glucose (450 g/L), and nitrogen and phosphate in the same proportion to glucose as in the growth medium. The rate of feeding was such that the glucose concentration in the fermenter was maintained at 30 g/L, i.e., 4.5 g/L h. The results showed that the butanediol production rate was maintained for 108 h, allowing a butanediol concentration of 99 g/L, and a glucose utilization of 265 g/L, to be attained. The overall productivity of the fed-batch fermentation was 0.91 g/L h, compared with 0.66 g/L h for a typical batch fermentation.

Similar approaches have been taken by YU and SADDLER (1983), who provided the culture with a daily dose of glucose and yeast extract, and by QURESHI and CHERYAN (1989b) who added glucose in two steps. AFSCHAR et al. (1991) developed a pulsed substrate feeding strategy in which the glucose addition was controlled *via* the carbon dioxide content in the exhaust gas. In all of these examples, the desired aim of obtaining a high product concentration within the fermenter was achieved, thus demonstrating the usefulness of the fed-batch technique. In addition, improved productivities were observed because production was maintained for longer periods of time. However, little heed was paid to the oxygen supply, and it is likely that further improvements could be made if attention was paid to this parameter. DE MAS et al. (1988) recognized this concept during fed-batch culture of *B. polymyxa*. In this case, a constant value of the oxygen transfer rate was chosen to maximize butanediol production and to minimize acetate production. However, as the fermentation progressed, it became apparent that the oxygen demand decreased while the oxygen transfer rate remained constant. Thus, the oxygen availability increased, leading to increased levels of acetate. The authors recognized that to maximize butanediol production, the oxygen transfer rate should be progressively decreased during the course of the fermentation, to compensate for the decreased oxygen demand.

## 4.6 Intensified Fermentation Technologies

Most of the early studies into the butanediol fermentation process used traditional batch fermentation. More recently, however, it has been realized that to achieve improved reactor productivities, and so improve the economics of the process, it may be necessary to use intensified fermentation technologies such as continuous flow, immobilized cells, cell recycle, or combinations of these. Chemostat culture is not usually considered to be such a technology as it is more suited as a tool for understanding the kinetics and physiology of a process than as a means of improving reactor productivity (see, e.g., HARRISON and PIRT, 1967). Examples of intensified technologies that have been studied for butanediol production are shown in Tab. 3.

PIRT and CALLOW (1958) used continuous culture during a study into the effects of various environmental parameters on butanediol production by *A. aerogenes*. In addition to es-

**Tab. 3.** Summary of Data Obtained Using Some Intensified Fermentation Technologies

| Method | Substrate | Organism | Butanediol Productivity [g/L h] | Reference |
|---|---|---|---|---|
| Continuous flow/free cells | Sucrose | *A. aerogenes* NCIB 8017 | 4.6 | Pirt and Callow (1958) |
| Continuous flow/free cells | Glucose | *K. pneumoniae* NRRL B199 | 4.25 | Ramachandran and Goma (1987) |
| Continuous flow/free cells | Glucose | *E. aerogenes* DSM 30053 | 5.6 | Zeng et al. (1990b) |
| Continuous flow/free cells | Lactose | *E. cloacae* | 0.03 | Schütz et al. (1985) |
| Repeated batch | Molasses | *K. oxytoca* DSM 3539 | 2.4 | Afschar et al. (1991, 1993) |
| Batch/high cell density | Glucose | *K. oxytoca* NRRL B199 | 3.2 | Qureshi and Cheryan (1989b) |
| Continuous flow/cell recycle | Whey permeate | *B. polymyxa* | 1.04 | Shazer and Speckman (1984) |
| Continuous flow/cell recycle | Glucose | *K. pneumoniae* NRRL B199 | 9.84 | Ramachandran and Goma (1988) |
| Continuous flow/cell recycle | Glucose | *K. oxytoca* NRRL B199 | 1.40 | Qureshi and Cheryan (1989d) |
| Continuous flow/cell recycle | Glucose | *E. aerogenes* DSM 30053 | 14.6[a] | Zeng et al. (1991) |
| Continuous flow/carrageenan-immobilized cells | Glucose | *E. aerogenes* IAM 1133 | 0.75 | Chua et al. (1980) |
| Continuous flow/alginate-immobilized cells | Whey permeate | *K. pneumoniae* NCIB 8017 | 2.3 | Lee and Maddox (1986) |
| Batch/alginate-immobilized cells | Whey permeate | *B. polymyxa* ATCC 12321, mutant | 0.06 | Martinez and Speckman (1988) |
| Continuous flow/adsorption-immobilized cells | Whey permeate | *K. pneumoniae* NCIB 8017 | 11.7 | Maddox et al. (1988) |
| Continuous flow/adsorption-immobilized cells | Lactose | *K. oxytoca* NRRL B199 | 1.0 | Champluvier et al. (1989b) |

[a] Acetoin production is included in this value.

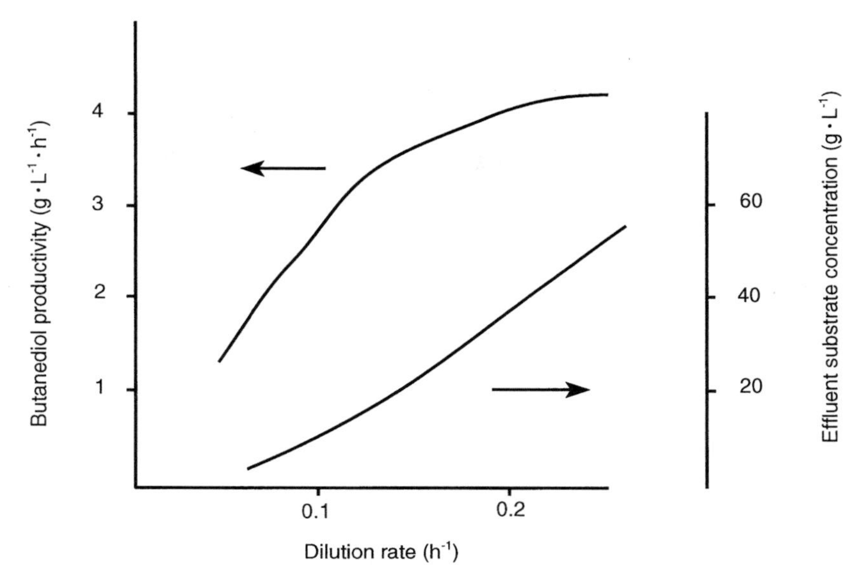


**Fig. 12.** Effect of dilution rate on butanediol productivity and effluent substrate concentration in continuous culture of *K. pneumoniae* (oxygen transfer rate, 23.6 mmol/L h; feed medium glucose concentration, 100 g/L), (from RAMACHANDRAN and GOMA, 1987).

tablishing some of the basic principles of the process, they achieved a reactor productivity well in excess of any reported in studies using batch fermentation. This was attributed to the fact that in batch processes there is a considerable delay between inoculation of the culture and attainment of maximum productivity, whereas in continuous culture the productivity can be held at the maximum for a considerable length of time. Nevertheless, the authors realized that in continuous culture, growth must be continuous, but the conditions required for maximum butanediol production are not necessarily identical with those for sufficient growth. On these grounds, they proposed a two-stage continuous process in which the first stage is optimized for biomass production, and the second for butanediol production.

Similar results have been reported by RAMACHANDRAN and GOMA (1987) (Tab. 3). In addition, they showed that as the dilution rate increased, so did the butanediol productivity until a plateau was reached at a dilution rate of approx. 0.2 $h^{-1}$ (Fig. 12). However, the increased productivity was achieved at the expense of an increased substrate concentration and decreased product concentration in the effluent stream. This study also confirmed that as the oxygen transfer rate increases, so does the proportion of product appearing as acetoin. ZENG et al. (1990b) showed that for a given dilution rate in continuous culture, there is an optimal oxygen uptake rate for butanediol and acetoin production, and that this optimal value increased with the dilution rate. Their overall findings indicated that growth and butanediol production are non-associated phenomena, and hence high reactor productivities should be achievable using cell recycle and cell immobilization techniques.

Results have been reported by QURESHI and CHERYAN (1989b) and AFSCHAR et al. (1991, 1993) using batch cultures in which a high cell (biomass) concentration has been artificially achieved, e.g., by concentrating the cells at the end of a batch process, and adding them to fresh medium. Although the reactor productivities obtained were superior to those from traditional batch culture, they were rather poorer than those from some other intensified fermentation technologies. Nevertheless, the approach of separating the two processes of biomass formation and butanediol production is sound, and, as stated above, is the rationale behind the use of cell recycle and cell immobilization techniques.

Cell recycle is a technique applied to the continuous culture of free cells, whereby the cells in the effluent stream are separated by, e.g., centrifugation or cross-flow microfiltration, and returned to the fermenter. In this way, a high concentration of biomass is achieved in the fermenter, and once the concentration is sufficiently high, the operating conditions can be optimized for butanediol production rather than for cell growth. SHAZER and SPECKMAN (1984) applied such a system to butanediol production from whey permeate using *B. polymyxa*. A stirred tank reactor was employed in which the bacterial cells were recycled *via* a hollow fiber ultrafiltration cartridge. A butanediol productivity of 1.04 g/L h was achieved, a value several times higher than that achieved in traditional batch fermentation. In the presence of acetate (4.5 g/L) the productivity was increased substantially to 8.2 g/L h. A similar system has been described by RAMACHANDRAN and GOMA (1988), but using *K. pneumoniae* growing on glucose (Tab. 3). They achieved a reactor productivity approaching 10 g/L h, and identified an optimum cell concentration (40 g/L) for maximum performance. Above this concentration, there is apparently severe oxygen limitation which, coupled with excess glucose availability, causes the organism to switch from butanediol to polysaccharide production. In the work of QURESHI and CHERYAN (1989d), poor butanediol productivity was observed although these authors used the same strain of organism and a similar cell recycle system to that of RAMACHANDRAN and GOMA (1988). The reason for this poor performance is not clear. ZENG et al. (1991) have reported that the achievable butanediol concentration of a cell recycle culture is limited by the accumulation of acetate, i.e., an inhibitory by-product. Hence, as with batch fermentation processes, due attention must be paid to the oxygen demand and supply rates in cell recycle systems, in order to take full advantage of the technology.

CHUA et al. (1980) were the first to apply a cell immobilization technique to butanediol production, and used *E. aerogenes* immobilized by entrapment in $\kappa$-carrageenan. (As with cell recycle techniques, cell immobilization lends itself to high reactor productivities by virtue of the high cell concentrations which can be attained in the reactor.) When the immobilized cells were used in continuous culture, the butanediol productivity was shown to be superior to that observed in batch culture. The authors made the significant observation that the incubation conditions for immobilized cells were not as sensitive as those for free cells. Unfortunately, this observation has not been studied further. Entrapment in alginate has been used to immobilize cells of *B. polymyxa* (MARTINEZ and SPECKMAN, 1988) and *K. pneumoniae* (LEE and MADDOX, 1986) for butanediol production from whey permeate. Productivities reported for the former organism were poor, but for *K. pneumoniae*, values up to 2.3 g/L h were observed when the immobilized cells were used in a packed bed reactor. To avoid gas hold-up and blockage, the reactor was operated at an angle of 10° to the horizontal and fitted on the inside with a cylindrical steel mesh. The system was completely stable during a seven week period of continuous operation.

Cell immobilization by adsorption is an alternative technique to entrapment, and has been used for species of *Klebsiella* by MADDOX et al. (1988) and CHAMPLUVIER et al. (1989b). The marked difference in reactor productivities obtained (Tab. 3) emphasizes the need for careful selection of organism and operating conditions. MADDOX et al. (1988) immobilized the cells on bonechar (a type of activated carbon), while CHAMPLUVIER et al. (1989b), unfortunately, used a strain of organism which could not directly utilize the substrate, so that the cells had to be co-immobilized, on glass wool, with dead cells of the yeast *Kluyveromyces lactis*.

WILLETTS (1985) has described a method of immobilization for *A. hydrophila* in which the cells were adsorbed onto a precipitate of titanium hydroxide. However, the method appeared to impose some metabolic limitation on butanediol production.

# 5 Product Recovery

Relatively few studies have been performed on this aspect of the process, probably because the fermentation has never been operated on a commercial scale. However, if butanediol production by fermentation is ever to be considered as a commercial proposition, it is essential that product recovery be optimized, not least because of its effect on the overall economics of the operation. There are two major difficulties involved in the recovery of butanediol, its high boiling point (180–184°C) and its high affinity for water. Recovery by distillation involves the removal of large quantities of water and, unless excess lime is added to the broth, leads to the formation of a thick tarry mass from which it is difficult to vaporize the butanediol (LONG and PATRICK, 1963; VOLOCH et al., 1985). Solvent extraction has been suggested as an alternative to distillation, possible solvents being ethyl acetate, diethyl ether and *n*-butanol. The butanediol could then be recovered as a bottom product in a subsequent distillation step, with the solvent being recycled. The use of dodecanol as the extractant has also been reported (EITEMAN and GAINER, 1989). In this case, *in situ* extraction has been compared with the use of an external column, the latter showing substantial advantages over the former due to reduced contact of the toxic extractant with the bacterial cells. In a patent, SRIDHAR (1988) has described the use of butanol or hexanol as the extractant, after removing a large part of the water by reverse osmosis. In contrast, FAGES (1986) has used a phosphoric acid ester (e.g., tertiary butyl phosphate) for extraction. These esters have boiling points higher than that of butanediol, very little solubility in water, and no tendency to form azeotropes with butanediol. The butanediol was extracted with the ester after addition to the broth of potassium carbonate, and then recovered by distillation.

A process whereby butanediol is salted out from the fermentation broth using water-free potassium carbonate has been described by AFSCHAR et al. (1993). After pre-cleaning of the broth by boiling followed by addition of a flocculating agent and solids removal, addition of potassium carbonate resulted in the formation of two phases. The upper phase, containing butanediol, contained 97% by weight of the diol. (This technique was also applied to the separation of 1,3-propanediol from a fermentation broth, but in this case it was necessary to add ethanol prior to adding the potassium carbonate.)

During the last decade, several energy-efficient membrane processes have been developed, principally for the ethanol fermentation. However, there are few reports of these techniques being applied to butanediol recovery. One exception is the application of pervaporation by DETTWILER et al. (1993). These authors have described a simulation model for the continuous production of butanediol using *B. subtilis*, integrated with separation by pervaporation. No doubt, other reports will be forthcoming in the future.

Another possible approach to product recovery is to convert the butanediol directly to methyl ethyl ketone while it is still in the fermentation broth. The ketone (boiling point 79.6°C) could then be recovered more easily than butanediol by either distillation or solvent extraction, and used in the manufacture of butadiene or as a fuel (for uses of butanediol, see Sect. 6). The dehydration of aqueous 2,3-butanediol to methyl ethyl ketone using a sulfuric acid catalyst has been investigated by EMERSON et al. (1982). The kinetics were first order with respect to butanediol, and ketone yields exceeding 90% were obtained. However, reaction rates and yields were lower when the dehydration was performed in a cell-free fermentation broth, probably due to the presence of residual sugar and salts. A similar approach has been described by TRAN and CHAMBERS (1987) using a solid acid catalyst consisting of sulfonic groups covalently bound to an inorganic matrix. Again, problems were encountered when the dehydration was performed in a fermentation broth, but these were reduced when the broth was treated with activated carbon prior to the dehydration.

# 6 Uses

At present, since there is no commercial production of butanediol, there are no commercial applications. Hence, this section presents potential rather than actual uses.

Butanediol has a potential use as a fuel (heating value 27.2 kJ/g; cf. ethanol 29.1 kJ/g) or as an octane booster for gasoline, but it is potentially more valuable as a chemical feedstock (YU and SADDLER, 1985). It can be successively dehydrated to yield methyl ethyl ketone and 1,3-butadiene. The former is an industrial solvent, and can also be dehydrogenated to yield high octane isomers suitable for high quality aviation fuels. Butadiene is used in the manufacture of synthetic rubber, and can also be further converted to styrene which is used in the manufacture of plastics and resins.

# 7 Conclusions and Future Developments

During the last decade there has been considerable research into the butanediol fermentation process, despite the fact that it has never been operated on the commercial scale. Much of the research has been aimed at optimizing the batch fermentation process, particularly by understanding the oxygen requirements, and in developing new fermentation technologies. Hence, the technologies now exist, at least on the laboratory scale, to achieve high reactor productivities. However, commercial application remains difficult due to competition from the petrochemical industry. In addition, fermentative production of butanediol still suffers from the difficulty of product recovery and the lack of large-scale technology to convert the diol to useful chemical feedstocks. Although there has been some recent progress in these areas, much remains to be done. Similarly, little attention has been paid to strain improvement of the butanediol-producing organisms, although the isolation of mutant strains of *B. polymyxa* with enhanced butanediol production has been described (MALLONEE and SPECKMAN, 1988). Perhaps the most significant recent development in this area is the cloning and characterization of the genes involved in butanediol biosynthesis (BLOMQVIST et al., 1993). This now makes possible the use of genetic engineering to improve butanediol production.

# 8 References

AFSCHAR, A. S., BELLGARDT, K. H., VAZ ROSSELL, C. E., CZOK, A., SCHALLER, K. (1991), The production of 2,3-butanediol by fermentation of high test molasses, *Appl. Microbiol. Biotechnol.* **34**, 582–585.

AFSCHAR, A. S., VAZ ROSSELL, C. E., JONAS, R., QUESADA CHANTO, A., SCHALLER, K. (1993), Microbial production and downstream processing of 2,3-butanediol, *J. Biotechnol.* **27**, 317–329.

ALAM, S., CAPIT, F., WEIGAND, W. A., HONG, J. (1990), Kinetics of 2,3-butanediol fermentation by *Bacillus amyloliquefaciens*: Effect of initial substrate concentration and aeration, *J. Chem. Tech. Biotechnol.* **47**, 71–84.

BARRETT, E. L., COLLINS, E. B., HALL, B. J., MATOI, S. H. (1983), Production of 2,3-butylene glycol from whey by *Klebsiella pneumoniae* and *Enterobacter aerogenes*, *J. Dairy Sci.* **66**, 2507–2514.

BLOMQVIST, K., NIKKOLA, M., LEHTOVAARA, P., SUIHKO, M.-L., AIRAKSINEN, U., STRABY, K., KNOWLES, J. K. C., PENTTILA, M. E. (1993), Characterization of the genes of the 2,3-butanediol operons from *Klebsiella terrigena* and *Enterobacter aerogenes*, *J. Bacteriol.* **175**, 1392–1404.

BOOTH, I. R. (1985), Regulation of cytoplasmic pH in bacteria, *Microbiol. Rev.* **49**, 359–378.

BRYN, K., HETLAND, O., STORMER, F. C. (1971), The reduction of diacetyl and acetoin in *Aerobacter aerogenes* : Evidence for one enzyme catalyzing both reactions, *Eur. J. Biochem.* **18**, 116–119.

CHAMPLUVIER, B., DECALLONNE, J., ROUXHET, P. G. (1989a), Influence of sugar source (lactose, glucose, galactose) on 2,3-butanediol production by *Klebsiella oxytoca* NRRL-B199, *Arch. Microbiol.* **152**, 411–414.

CHAMPLUVIER, B., FRANCART, B., ROUXHET, P. G. (1989b), Co-immobilization by adhesion of $\beta$-galactosidase in nonviable cells of *Kluyveromyces lactis* with *Klebsiella oxytoca*: Conversion of lactose into 2,3-butanediol, *Biotechnol. Bioeng.* **34**, 844–853.

CHUA, J. W., ERARSLAN, A., KINOSHITA, S., TAGUCHI, H. (1980), 2,3-butanediol production by immobilized *Enterobacter aerogenes* IAM 1133 with κ-carrageenan, *J. Ferment. Technol.* **58**, 123–127.

DE MAS, C., JANSEN, N. B., TSAO, G. T. (1988), Production of optically active 2,3-butanediol by *Bacillus polymyxa*, *Biotechnol. Bioeng.* **31**, 366–377.

DETTWILER, B., DUNN, I. J., HEINZLE, E., PRENOSIL, J. E. (1993), A simulation model for the continuous production of acetoin and butanediol using *Bacillus subtilis* with integrated pervaporation separation, *Biotechnol. Bioeng.* **41**, 791–800.

EITEMAN, M. A., GAINER, J. L. (1989), *In situ* extraction versus the use of an external column in fermentation, *Appl. Microbiol. Biotechnol.* **30**, 614–618.

EMERSON, R. R., FLICKINGER, M. C., TSAO, G. T. (1982), Kinetics of dehydration of aqueous 2,3-butanediol to methyl ethyl ketone, *Ind. Eng. Chem. Prod. Res. Dev.* **21**, 473–477.

FAGES, J. (1986), *French Patent* 2574784.

FAGES, J., MULARD, D., ROUQUET, J.-J., WILHELM, J.-L. (1986), 2,3-butanediol production from Jerusalem artichoke, *Helianthus tuberosus*, by *Bacillus polymyxa* ATCC 12321. Optimization of $K_La$ profile, *Appl. Microbiol. Biotechnol.* **25**, 197–202.

FOND, O., JANSEN, N. B., TSAO, G. T. (1985), A model of acetic acid and 2,3-butanediol inhibition of the growth and metabolism of *Klebsiella oxytoca*, *Biotechnol. Lett.* **7**, 727–732.

FRAZER, F. R., MCCASKEY, T. A. (1991), Effect of components of acid-hydrolysed hardwood on conversion of D-xylose to 2,3-butanediol by *Klebsiella pneumoniae*, *Enzyme Microb. Technol.* **13**, 110–115.

HARRISON, D. E. F., PIRT, S. J. (1967), The influence of dissolved oxygen concentration on the respiration and glucose metabolism of *Klebsiella aerogenes* during growth, *J. Gen. Microbiol.* **46**, 193–221.

HOHN-BENTZ, H., RADLER, F. (1978), Bacterial 2,3-butanediol dehydrogenases, *Arch. Microbiol.* **116**, 197–203.

JANSEN, N. B., TSAO, G. T. (1983), Bioconversion of pentoses to 2,3-butanediol by *Klebsiella pneumoniae*, *Adv. Biochem. Eng./Biotechnol.* **27**, 85–99.

JANSEN, N. B., FLICKINGER, M. C., TSAO, G. T. (1984a), Production of 2,3-butanediol from D-xylose by *Klebsiella oxytoca* ATCC 8724, *Biotechnol. Bioeng.* **26**, 362–369.

JANSEN, N. B., FLICKINGER, M. C., TSAO, G. T. (1984b), Application of bioenergetics to modelling the microbial conversion of D-xylose to 2,3-butanediol, *Biotechnol. Bioeng.* **26**, 573–582.

JOHANSEN, L., BRYN, K., STORMER, F. C. (1975), Physiological and biochemical role of the butanediol pathway in *Aerobacter (Enterobacter) aerogenes*, *J. Bacteriol.* **123**, 1124–1130.

KOSARIC, N., MAGEE, R. J., BLASZCZYK, R. (1992), Redox potential measurement for monitoring glucose and xylose conversion by *K. pneumoniae*, *Chem. Biochem. Eng. Q.* **6**, 145–152.

LARSEN, S. H., STORMER, F. C. (1973), Diacetyl (acetoin) reductase from *Aerobacter aerogenes:* Kinetic mechanism and regulation by acetate of the reversible reduction of acetoin to 2,3-butanediol, *Eur. J. Biochem.* **34**, 100–106.

LAUBE, V. M., GROLEAU, D., MARTIN, S. M. (1984), 2,3-butanediol production from xylose and other hemicellulosic components by *Bacillus polymyxa*, *Biotechnol. Lett.* **6**, 257–262.

LEE, H. K., MADDOX, I. S. (1984), Microbial production of 2,3-butanediol from whey permeate, *Biotechnol. Lett.* **6**, 815–818.

LEE, H. K., MADDOX, I. S. (1986), Continuous production of 2,3-butanediol from whey permeate using *Klebsiella pneumoniae* immobilized in calcium alginate, *Enzyme Microb. Technol.* **8**, 409–411.

LOKEN, J. P., STORMER, F. C. (1970), Acetolactate decarboxylase from *Aerobacter aerogenes*: Purification and properties, *Eur. J. Biochem.* **14**, 133–137.

LONG, S. K., PATRICK, R. (1961), Production of 2,3-butylene glycol from citrus wastes: I The *Aerobacter aerogenes* fermentation, *Appl. Microbiol.* **9**, 244–248.

LONG, S. K., PATRICK, R. (1963), The present status of the 2,3-butylene glycol fermentation, *Adv. Appl. Microbiol.* **5**, 135–155.

MADDOX, I. S., QURESHI, N., MCQUEEN, J. M. (1988), Continuous production of 2,3-butanediol from whey permeate using cells of *Klebsiella pneumoniae* immobilized onto bonechar, *N. Z. J. Dairy Sci. Technol.* **23**, 127–132.

MAGEE, R. J., KOSARIC, N. (1987), The microbial production of 2,3-butanediol, *Adv. Appl. Microbiol.* **32**, 89–161.

MALLONEE, D. H., SPECKMAN, R. A. (1988), Development of a mutant strain of *Bacillus polymyxa* showing enhanced production of 2,3-butanediol, *Appl. Environ. Microbiol.* **54**, 168–171.

MANKAD, T., NAUMAN, E. B. (1992), Effect of oxygen on steady-state product distribution in *Bacillus polymyxa* fermentations, *Biotechnol. Bioeng.* **40**, 413–426.

MARTINEZ, S. B., SPECKMAN, R. A. (1988), 2,3-Butanediol production from hydrolyzed whey permeate by immobilized cells of *Bacillus polymyxa*, *Appl. Biochem. Biotechnol.* **18**, 303–313.

NISHIKAWA, N. K., SUTCLIFFE, R., SADDLER, J. N. (1988a), The influence of lignin degradation products on xylose fermentation by *Klebsiella pneumoniae*, *Appl. Microbiol. Biotechnol.* **27**, 549–552.

NISHIKAWA, N. K., SUTCLIFFE, R., SADDLER, J. N. (1988b), The effect of wood-derived inhibitors on 2,3-butanediol production by *Klebsiella pneumoniae*, *Biotechnol. Bioeng.* **31**, 624–627.

OLSON, B. H., JOHNSON, M. J. (1948), The production of 2,3-butylene glycol by *Aerobacter aerogenes* 199, *J. Bacteriol.* **55**, 209–222.

PALSSON, B. O., FATHI-AFSHAR, S., RUDD, D. F., LIGHTFOOD, E. N. (1981), Biomass as a source of chemical feedstocks: An economic evaluation, *Science* **213**, 513–517.

PAPOUTSAKIS, E. T., MEYER, C. L. (1985), Equations and calculations of product yields and preferred pathways for butanediol and mixed-acid fermentations, *Biotechnol. Bioeng.* **27**, 50–66.

PIRT, S. J. (1957), The oxygen requirement of growing cultures of an *Aerobacter* species determined by means of the continuous culture technique, *J. Gen. Microbiol.* **16**, 59–75.

PIRT, S. J., CALLOW, D. S. (1958), Exocellular product formation by microorganisms in continuous culture. I. Production of 2,3-butanediol by *Aerobacter aerogenes* in a single stage process, *J. Appl. Bacteriol.* **21**, 188–205.

PRESCOTT, S. C., DUNN, C. G. (1959), *Industrial Microbiology*. New York: McGraw Hill.

QURESHI, N., CHERYAN, M. (1989a), Effect of aeration on 2,3-butanediol production from glucose by *Klebsiella oxytoca*, *J. Ferment. Bioeng.* **67**, 415–418.

QURESHI, N., CHERYAN, M. (1989b), Production of 2,3-butanediol by *Klebsiella oxytoca*, *Appl. Microbiol. Biotechnol.* **303**, 440–443.

QURESHI, N., CHERYAN, M. (1989c), Effect of lactic acid on growth and butanediol production by *Klebsiella oxytoca*, *J. Ind. Microbiol.* **4**, 453–456.

QURESHI, N., CHERYAN, M. (1989d), Production of 2,3-butanediol in a membrane recycle bioreactor, *Process Biochem.* **24**, 172–175.

RAMACHANDRAN, K. B., GOMA, G. (1987), Effect of oxygen supply and dilution rate on the production of 2,3-butanediol in continuous bioreactor by *Klebsiella pneumoniae*, *Enzyme Microb. Technol.* **9**, 107–111.

RAMACHANDRAN, K. B., GOMA, G. (1988), 2,3-Butanediol production from glucose by *Klebsiella pneumoniae* in a cell recycle system, *J. Biotechnol.* **9**, 39–46.

RAMACHANDRAN, K. B., HASHIM, M. A., FERNANDEZ, A. A. (1990), Kinetic study of 2,3-butanediol production by *Klebsiella oxytoca*, *J. Ferment. Bioeng.* **70**, 235–240.

RASPOET, D., POT, B., DE DEYN, D., DE VOS, P., KERSTERS, K., DE LEY, J. (1991), Differentiation between 2,3-butanediol producing *Bacillus licheniformis* and *B. polymyxa* strains by fermentation product profiles and whole-cell protein electrophoretic patterns, *Syst. Appl. Microbiol.* **14**, 1–7.

SABLAYROLLES, J. M., GOMA, G. (1984), Butanediol production by *Aerobacter aerogenes* NRRL B199: Effects of initial substrate concentration and aeration agitation, *Biotechnol. Bioeng.* **26**, 148–155.

SCHÜTZ, M., SCHONE, B., KLEMM, K. (1985), Butandiolbildung aus Laktose durch *Enterobacter cloacae* in kontinuierlicher Kultur, *Milchwissenschaft* **40**, 513–517.

SHAZER, W. H., SPECKMAN, R. A. (1984), Continuous fermentation of whey permeate by *Bacillus polymyxa*, *J. Dairy Sci.* **67** (Suppl. 1), 50–51.

SILVEIRA, M. M., SCHMIDELL, W., BERBERT, M. A. (1993), Effect of the air supply on the production of 2,3-butanediol by *Klebsiella pneumoniae* NRRL B199, *J. Biotechnol.* **31**, 93–102.

SPECKMAN, R. A., COLLINS, E. B. (1982), Microbial production of 2,3-butylene glycol from cheese whey, *Appl. Environ. Microbiol.* **43**, 1216–1218.

SRIDHAR, S. (1988), *German Patent* 3 623 827.

STORMER, F. C. (1968), Evidence for induction of the 2,3-butanediol-forming enzymes in *Aerobacter aerogenes*, *FEBS Lett.* **2**, 36–38.

STORMER, F. C. (1977), Evidence for regulation of *Aerobacter aerogenes* pH 6 acetolactate-forming enzyme by acetate ion, *Biochem. Biophys. Res. Commun.* **74**, 898–902.

TAYLOR, M. B., JUNI, E. (1960), Stereoisomeric specificities of 2,3-butanediol dehydrogenases, *Biochim. Biophys. Acta* **39**, 448–457.

TEIXEIRA DE MATTOS, M. J., TEMPEST, D. W. (1983), Metabolic and energetic aspects of the growth of *Klebsiella aerogenes* NCTC 418 on glucose in anaerobic chemostat culture, *Arch. Microbiol.* **134**, 80–85.

TRAN, A. V., CHAMBERS, R. P. (1986), Lignin and extractives derived inhibitors in the 2,3-butanediol fermentation of mannose-rich prehydrolysates, *Appl. Microbiol. Biotechnol.* **23**, 191–197.

TRAN, A. V., CHAMBERS, R. P. (1987), The dehydration of fermentative 2,3-butanediol into methyl ethyl ketone, *Biotechnol. Bioeng.* **29**, 343–351.

UI, S., MATSUYAMA, N., MASUDA, H., MURAKI, H. (1984), Mechanism for the formation of 2,3-butanediol sterioisomers in *Klebsiella pneumoniae*, *J. Ferment. Technol.* **62**, 551–559.

UI, S., MASUDA, T., MASUDA, H., MURAKI, H. (1986), Mechanism for the formation of 2,3-butanediol stereoisomers in *Bacillus polymyxa*, *J. Ferment. Technol.* **64**, 481–486.

VOLOCH, M., LADISCH, M. R., RODWELL, V. W., TSAO, G. T. (1983), Reduction of acetoin to 2,3-butanediol in *Klebsiella pneumoniae*, *Biotechnol. Bioeng.* **25**, 173–183.

VOLOCH, M., JANSEN, N. B., LADISCH, M. R., TSAO, G. T., NARAYAN, R., RODWELL, V. W. (1985), 2,3-butanediol, in: *Comprehensive Biotechnology* (BLANCH, H. W., DREW, S., WANG, D. I. C., Eds.), pp. 933–947, Oxford: Pergamon Press.

WILLETTS, A. (1984), Butane 2,3-diol production by *Aeromonas hydrophila* grown on starch, *Biotechnol. Lett.* **6**, 263–268.

WILLETTS, A. (1985), Butane 2,3-diol production by immobilized *Aeromonas hydrophila*, *Biotechnol. Lett.* **7**, 261–266.

YU, E. K. C., SADDLER, J. N. (1982), Enhanced production of 2,3-butanediol by *Klebsiella pneumoniae* grown on high sugar concentration in the presence of acetic acid, *Appl. Environ. Microbiol.* **44**, 777–784.

YU, E. K. C., SADDLER, J. N. (1983), Fed-batch approach to production of 2,3-butanediol by *Klebsiella pneumoniae* grown on high substrate concentrations, *Appl. Environ. Microbiol.* **46**, 630–635.

YU, E. K. C., SADDLER, J. N. (1985), Biomass conversion to butanediol by simultaneous saccharification and fermentation, *Trends Biotechnol.* **3**, 100–104.

YU, E. K. C., LEVITIN, N., SADDLER, J. N. (1984a), Utilization of wood hemicellulose hydrolysates by microorganisms for the production of liquid fuels and chemicals, *Dev. Ind. Microbiol.* **25**, 613–620.

YU, E. K. C., DESCHATELETS, L., SADDLER, J. N. (1984b), The combined enzymatic hydrolysis and fermentation of hemicellulose to 2,3-butanediol, *Appl. Microbiol. Biotechnol.* **19**, 365–372.

YU, E. K. C., DESCHATELETS, L., LEVITIN, N., SADDLER, J. N. (1984c), Production of 2,3-butanediol from HF-hydrolyzed aspen wood, *Biotechnol. Lett.* **6**, 611–614.

ZENG, A.-P., DECKWER, W.-D. (1991), A model for multiproduct-inhibited growth of *Enterobacter aerogenes* in 2,3-butanediol fermentation, *Appl. Microbiol. Biotechnol.* **35**, 1–3.

ZENG, A.-P., BIEBL, H., DECKWER, W.-D. (1990a), Effect of pH and acetic acid on growth and 2,3-butanediol production of *Enterobacter aerogenes* in continuous culture, *Appl. Microbiol. Biotechnol.* **33**, 485–489.

ZENG, A.-P., BIEBL, H., DECKWER, W.-D. (1990b), 2,3-butanediol production by *Enterobacter aerogenes* in continuous culture: Role of oxygen supply, *Appl. Microbiol. Biotechnol.* **33**, 264–268.

ZENG, A.-P., BIEBL, H., DECKWER, W.-D. (1991), Production of 2,3-butanediol in a membrane bioreactor with cell recycle, *Appl. Microbiol. Biotechnol.* **34**, 463–468.

# 8 Lactic Acid

JÁN S. KAŠČÁK

Bratislava, Slovakia

JIŘÍ KOMÍNEK
MAX ROEHR

Wien, Austria

# 1 Introduction

The principles of industrial lactic acid fermentation may be traced back to the year 1857 when LOUIS PASTEUR published his famous *"Mémoire sur la fermentation appelée lactique"*. Since then, and especially with the isolation of various lactic acid bacteria, it became evident that the use of these organisms is strongly associated with the cultural development of mankind, e.g., the preparation of fermented foods from milk, vegetables, cereals, and meat as well as the preservation of such materials. A survey of the various aspects of fermented foods is given in Vol. 9 of this treatise (REED and NAGODAWITHANA, 1995). The present chapter refers to the industrial production of more or less pure lactic acid as a commodity.

Lactic acid bacteria are an extensively studied and thus well-defined group of facultative anaerobic gram-positive bacteria. Their biology and biochemistry is documented in many excellent reviews (BALOW et al., 1991). The reader is especially referred to the contribution of TEUBER (1993) in Vol. 1 of *Biotechnology*.

Therefore, only properties of lactic acid bacteria that are relevant for understanding the peculiarities of industrial processes are considered in the following.

# 2 Biology and Biochemistry of Industrial Lactic Acid Fermentation

(1) Homofermentative and heterofermentative bacteria: Lactic acid bacteria either utilize the well-known EMP pathway of glucose metabolism to produce lactic acid as the only (main) end product, or they use pathways of pentose metabolism resulting in lactic acid plus other products such as acetic acid, ethanol, and $CO_2$. In 1919, ORLA-JENSEN proposed the designations "homofermentative" and "heterofermentative for the respective organisms. Whereas heterofermentative bacteria are involved in most of the typical fermentations leading to food or animal feed, homofermentative bacteria are producers of industrial lactic acid. It is known, however, that the homofermentative behavior of certain strains may change to heterofermentative if the concentration of the main carbon source is decreased to growth-limiting levels (DE VRIES et al., 1970).

(2) Mesophilic and thermophilic bacteria: The temperature range for optimal growth of mesophilic lactic acid bacteria is 28–45°C and that of thermophilic lactic acid bacteria is 45–62°C, respectively. In dairy science special traditional designations have been used to characterize the temperature relations of these bacteria: Thermobacterium, Streptobacterium and Betabacterium. A compilation of the most common lactic acid bacteria according to the above-mentioned criteria taken from TEUBER (1993) is presented in Fig. 1. As will be seen below, organisms operating at higher temperatures are preferred because of reduced risk of contamination during fermentation.

Besides these data, several other properties should be mentioned which are helpful in designing industrial fermentation processes:

(3) Lactic acid bacteria – despite being able to produce acids as the main metabolic products – are rather sensitive to acids. Thus processes aiming for high consumption of carbon sources to produce high concentrations of lactic acid have to be conducted by neutralizing the acid formed, i.e., by keeping the pH between 5.5 and 6.0. Thus devices for mixing the fermentation medium have to be provided. According to BUCHTA (1974), lactic acid fermentation is strongly inhibited at pH 5 and ceases at pH values below 4.5.

(4) Lactic acid bacteria are facultative anaerobic organisms. Therefore, in practice low oxygen tensions should be maintained but exclusion of oxygen (air) is not an absolute requirement.

(5) Lactic acid bacteria display complex nutrient requirements. These comprise many of the known vitamins, amino acids, and even small peptides. Based on the respective functional relationship between the requirement for one specific substance and the resulting

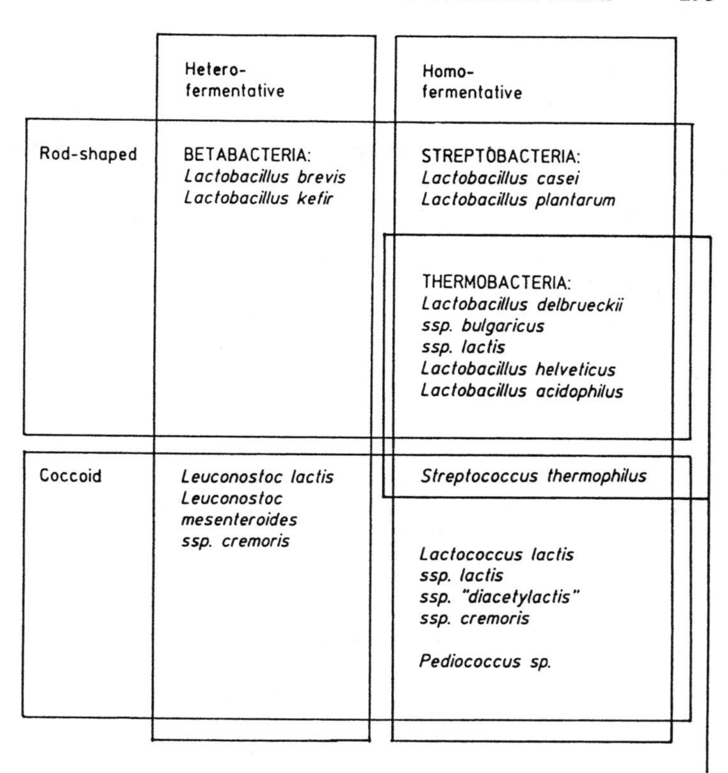


**Fig. 1.** Functional classification of lactic acid bacteria (TEUBER, 1993; with permission from VCH).

growth or (lactic) acid production, specific and highly sensitive assays for vitamins and amino acids were developed before other microassay procedures became available (BARTON-WRIGHT, 1952). In the technical process, this indicates the necessity of including complex and sometimes expensive nutrients in the fermentation media. The presence of such complex nutrients in turn may contaminate the end product and make even more expensive purification steps necessary.

(6) In general, lactic acid bacteria utilize common carbon sources for growth and acid production, such as glucose and fructose, lactose (of course), maltose, and sucrose. Starch cannot be utilized but there are several reports that certain members of lactic acid bacteria can use liquefied starch for lactic acid production (see below).

(7) Lactic acid bacteria differ in their ability to produce D-(−), L-(+), and DL-lactic acid, depending on the presence of respective lactate dehydrogenases and racemases. For details concerning this question, the reader is referred to GARVIE (1980) and TEUBER (1993).

# 3 Production Strains

There is a wealth of information concerning the genetics of lactic acid bacteria and possible applications of genetic engineering in this field. For recent developments see DEVOS and VAUGHAN (1944), FITZSIMONS et al. (1994), and WEI et al. (1995). Although all prerequisites for developing a recombinant DNA system have been thoroughly investigated, industrial production of lactic acid apparently still relies on producer strains that are selected empirically.

The choice of organisms to be applied and selected for higher productivities largely depends on the raw material.

Using glucose as raw material any homofermentative member of the genus *Lactobacillus* may be used. The preferred organism, however, is *Lactobacillus delbrueckii,* now designated as *Lactobacillus delbrueckii* ssp. *delbrueckii* (TEUBER, 1993; HAMMES et al., 1991). This organism can also utilize sucrose and is thus suitable for the fermentations of sucrose or molasses, but it cannot utilize lactose. In contrast, *L. delbrueckii* ssp. *bulgaricus,* the former *L. bulgaricus,* utilizes lactose but not sucrose and has been used for the conversion of lactose, i.e., whey. Interestingly several strains of this organism are unable to utilize galactose. *L. helveticus,* on the other hand, utilizes lactose and galactose but not sucrose, and thus represents another useful organism for the conversion of whey. *L. delbrueckii* ssp. *lactis,* the former *L. lactis,* may be used in fermentations of glucose, sucrose, and lactose. With respect to starch, which cannot be fermented by lactic acid bacteria and has to be hydrolyzed prior to fermentation, the choice of suitable organisms depends on the way of starch hydrolysis: If $\alpha$- and $\beta$-amylases (barley malt) are used, as in the traditional saccharifying procedures for, e.g. ethanol production, the resulting carbon source will be mainly maltose. Organisms capable of fermenting maltose are *L. delbrueckii* ssp. *lactis,* some strains of *L. delbrueckii* ssp. *delbrueckii* and several others (TEUBER, 1993). In case of saccharification with $\alpha$-amylases and glucoamylase, resulting in production of mainly glucose and glucose oligosaccharides, suitable strains will be those described above. One exception should be mentioned: *Lactobacillus amylophilus* (NAKAMURA and CROWELL, 1979) and *Lactobacillus amylovorus* (NAKAMURA, 1981) have been described to actively ferment starch, and this has lead to alternative processes of industrial lactic acid production (CHENG et al., 1991; ZHANG and CHERYAN, 1994).

Members of the lactococci are rarely used in commercial lactic acid fermentation.

Irrespective of the raw material used, it is desirable to select strains with high productivities. As mentioned above, this selection is traditionally performed by simple trial and error methods. Since productivity is strongly dependent on the hydrogen concentration of the fermentation liquid, it might be expected that adaptation to higher concentrations of lactic acid would increase the performance of a given strain. Increased yields have in fact been claimed following this procedure (ROBISON, 1988). Application of chemical mutagens (e.g., ethyl methane sulfonate) has been reported to yield mutants of *L. delbrueckii* (ATCC 9649) with improved tolerance to higher acid concentrations and higher productivities (DEMIRCI and POMETTO, 1992).

One problem often encountered in the dairy industry is the occurrence of various bacteriophages as serious contaminants. Little is known about phage infections in industrial lactic acid fermentations. While many of the known phages thrive on coccoid lactic acid bacteria, several phages of lactobacilli are also known (ACCOLAS and SPILLMANN, 1979). In one case there is evidence that virulent phages were derived from prophages which had remained in that stage for longer periods of time (SHIMIZU-KADOTA et al., 1983; SHIMIZU-KADOTA and TSUCHIDA, 1984). It was concluded that elimination of phages would be possible by curing of the prophages in the lysogenic strain. Other authors (e.g., TEUBER and LEMKE, 1983), however, do not agree that temperate phages are of great importance in phage infections.

In the dairy industry several means of counteracting phage contaminations and of selecting phage resistant strains have been devised (MATA-GILSINGER, 1985; TEUBER, 1993, and Vol. 9 of this treatise, REED and NAGODAWITHANA, 1995).

In order to create reproducible conditions, it is recommended to maintain cultures of producer organisms on standard culture media, e.g., the so-called MRS medium which is available from several sources (Merck 10660 0500, 10661 0500; Oxoid CM 395; Difco 0881). Inocula for seeding the main fermenter are prepared by transfers to increasing volumes of media with the chosen raw material as carbon source. Care should be taken in each step to maintain the pH above 5.5, e.g., by adding small amounts of calcium carbonate.

# 4 Industrial Production

First attempts of commercial lactic acid production were made in 1881 by AVERY in Littleton, USA. The aim was to substitute the imported tartrates then being used in baking powders by calcium lactate. Although not very prosperous, this probably represents the first industrial establishment in biotechnology. In 1895, a successful plant for lactic acid production was established by A. Boehringer in Ingelheim, Germany (BUCHTA, 1983).

Fermentative production of lactic acid as a commodity has always been in competition with chemical synthesis. Chemical synthesis on an industrial scale may be performed by oxidation of propene by nitric acid and/or nitrogen peroxide in the presence of oxygen, yielding $\alpha$-nitratopropionic acid which can be hydrolyzed to lactic and nitric acid (PLATZ et al., 1972; BOICHARD et al., 1973). Another way of chemical synthesis is hydrolysis of lactic acid nitrile (VOGT et al., 1966), which can be obtained by reacting acetaldehyde with HCN. For further processes see, e.g., SCHWALL (1979).

Chemical synthesis and fermentative production of lactic acid both have to cope with the problem of purification to obtain qualities that are required to meet the standards in different applications. These steps largely determine the price of the product.

Attempts to improve the fermentative process by application of entirely different organisms as potential lactic acid producers will be dealt with in Sect. 4.3.

## 4.1 Raw Materials

Besides their availability at reasonable prices, raw materials should meet certain purity requirements since these are decisive for the necessary purification procedures of the lactic acid produced. The choice of raw materials largely depends on the intended application and the respective costs of product purification. The relatively low selling price for lactic acid requires careful evaluation of these two factors.

Various readily available mono- and disaccharide materials are traditional substrates for lactic acid manufacturing:

- glucose (dextrose) and glucose syrups of varying qualities as end products of starch conversion processes applying $\alpha$-amylases and glucoamylases (see Chapter 1b, this volume);
- maltose as a product of specific enzymatic starch conversion applying $\alpha$-amylases and $\beta$-amylases from barley malt or other sources (see Chapter 1b, this volume);
- sucrose as end product, intermediate product (syrups, juices) and by-product (molasses) of beet and cane sugar production (see Chapter 1a, this volume);
- lactose as a constituent of whey as the natural substrate of most lactic acid bacteria.

As mentioned above, starch cannot be utilized by the common lactic acid bacteria, except *L. amylophilus* and *L. amylovorus.* Starch hydrolysis can be performed by either chemical or enzymatic methods. Chemical starch conversion has been abandoned for some time. Enzymatic saccharification of starch originally was performed similar to ethanol production using barley malt as the saccharifying agent (see RAUCH et al., 1960). Since maltose is the main product of this process, only a limited number of organisms is suitable for the fermentation process (see above). Analogous to the ethanol producing industry, starch liquefaction and saccharification is achieved at present by the combined action of different $\alpha$-amylases of bacterial origin and of fungal glucoamylases. Thus the preparation of sugar sources from various starch raw materials does not differ from procedures in other related fields. Hydrol, a by-product of dextrose production, has also been used as raw material.

Sucrose-derived raw materials are also suitable as raw materials for lactic acid fermentation and apparently they were the first substrates to be used in industrial fermentations. Most of the known homofermentative lactobacilli can utilize sucrose, with *L. bulgaricus* and *L. helveticus* being notable exceptions. The use of cheap molasses has been of great interest. It may be said that molasses, after appropriate treatments, are suitable carbon substrates, but lower yields have to be taken into account and more laborious purification procedures are necessary. The best treatment seems to be acid clarification at elevated temperatures.

Lactose, a natural constituent of milk and whey, poses a special problem because of its low concentration in the most readily available whey. In the past, several processes were developed for the processing of whey. Despite the low price of whey these processes could not compete with others based on the commonly used raw materials. Moreover, as with molasses, more expensive purification procedures have impeded successful utilization of whey as a most abundant raw material.

The fact that whey represents a considerable environmental load has stimulated the development of modern technologies of whey concentration and fractionation, such as ultrafiltration and electrodialysis. These, in turn, have exerted a strong stimulus on fermentation research laboratories as may be seen from recent literature (ROY et al., 1986; BOYAVAL et al., 1987; HOFMAN, 1988; LEH and CHARLES, 1989a, b; KULOZIK et al., 1992; CHIARINI et al., 1992; BOERGARDTS et al., 1994).

## 4.2 Fermentation Processes

Several procedures to ferment all these raw materials to lactic acid of varying purity have been elaborated in the course of about one century, as outlined below. Details in the older literature may be found in: SCHOPMEYER, 1954; INSKEEP et al., 1952; PRESCOTT and DUNN, 1959; RAUCH et al., 1960; REHM, 1967. For more recent literature the reader is referred to MIALL (1978); VICKROY (1985); CHAHAL (1990).

### 4.2.1 Classical Calcium Lactate Processes

Considering most of the information described above, a basic protocol for manufacturing lactic acid in a classical way is described in Fig. 2.

The sugar raw material (glucose, sucrose) is brought to a sugar concentration of 120–180 g $L^{-1}$. Complex nitrogen sources are added, e.g., mixtures of inorganic N-compounds such as ammonia, ammonia water, ammonium sulfate, or ammonium phosphates with complex organic materials such as corn steep liquor, yeast extracts, peptones and other protein digests, malt sprouts, etc. yielding N-concentrations between 1 and 10 g $L^{-1}$. If not included as constituents of other nutrient components, mineral components (macro elements) are provided by addition of phosphates and salts (e.g., sulfates of magnesium, manganese and iron). In general, only the complex components (yeast extract, peptone, etc.) determine the resulting cell concentration (MONTEAGUDO et al., 1993). Sterilization of medium constituents is advisable.

Fermentation is performed in reactor volumes of several to >100 $m^3$. One important factor is the material the fermenters for lactic acid production are made of since lactic acid is known to be highly corrosive. Corrosion of the vessels has to be prevented not only in order to protect the reactor but also to avoid contamination of the fermentation fluid by soluble compounds (heavy metals, etc.) which would complicate further purification steps.

In the past, smaller reactors were made of wood or concrete. Nowadays, concrete – with suitable coatings, if necessary – and stainless steel serve this purpose. Gentle agitation is performed by conventional stirrers. As the fermentation is conducted at temperatures >45°C, heating has to be provided in the first stages, and cooling as soon as the heat is generated by the fermentation itself.

Sterile calcium carbonate, preferably as powdered chalk, is added either at the beginning or in increments during the fermentation to keep the concentration of free lactic acid as low as possible. As mentioned above, pH values should be maintained between 5.5 and 6.0.

Inoculation is performed by adding a volume of about 10% of a seed culture obtained from precultures of increasing volumes started with a pure culture (see, e.g., RAUCH et al., 1960), or simply by using a small portion of the previous batch. The start of active fermentation (ca. 6 h following inoculation)

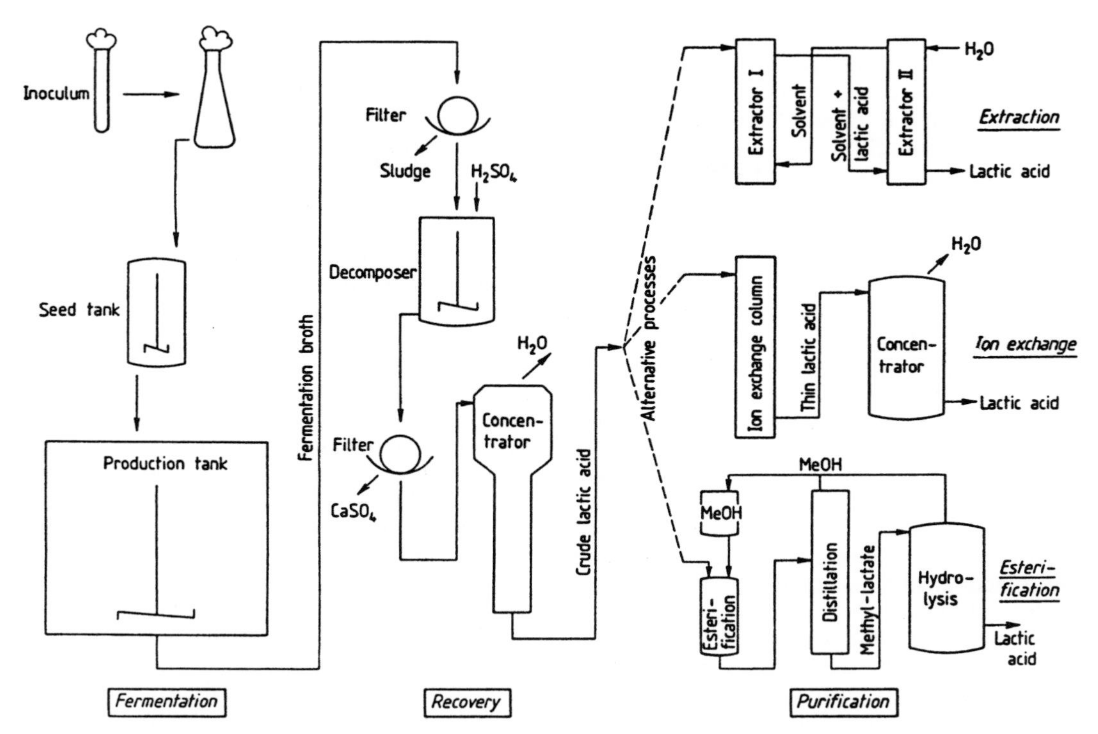


**Fig. 2.** Industrial production of lactic acid (BUCHTA, 1983; with permission from VCH).

can be recognized by the evolution of $CO_2$ from the added calcium carbonate, which also serves to prevent air from entering the system. Thus evolution of $CO_2$ from respiration processes may be fairly excluded. The application of milk of lime for neutralization does not afford these advantages.

Maintaining relatively high temperatures of up to 50°C for *L. delbrueckii* or similar strains reduces the probability of contaminations by, e.g., butyric acid forming anaerobic bacteria.

Active fermentation is completed after 2–6 days, depending on the concentration of the carbon source applied. In calcium lactate fermentations the upper limits of sugar concentration are said to be determined by the solubility of the resulting calcium lactate, which at higher concentrations tends to precipitate from the fermentation broth. It has been claimed, however, that the application of higher sugar concentrations (e.g., 260 g $L^{-1}$) would be feasible in a fermentation with a certain CaO dosage to adjust the pH to 6.3 causing continuous precipitation of calcium lactate. This would shift the equilibrium of the reaction to the direction of product formation. 99.6% conversion over 3 days were reported with this protocol (MASLOWSKI, 1988).

Usually, conversion yields of 85–95% of the theoretical accessible values are reported. Amounts of up to 2% of acetic acid and propionic acid as by-products may be due to temporary switches to heterolactic phases of fermentation by deviations from optimum conditions of pH or substrate concentrations due to, e.g., incomplete mixing.

### 4.2.2 Recovery and Purification Procedures

A detailed description of product recovery and purification following the classical calcium lactate process is given by RAUCH et al.

(1960), depending on the purity of the raw materials used. Using pure sugars (dextrose, refined beet or cane sugar), the fermentation liquid is heated to dissolve all calcium lactate and is subsequently treated with stoichiometric amounts of sulfuric acid (78%). The resulting calcium sulfate is separated by filtration with thorough washing of the filter cake. The combined liquids are evaporated (vacuum). Residual amounts of gypsum precipitating in the concentrated lactic acid solution are filtered off together with coloring substances, which may be adsorbed to activated carbon. Impurities due to heavy metal ions are removed by treatment with hexacyanoferrate. Further purification is achieved by passing the solution through ion exchange columns and by treatment with hydrogen peroxide or potassium permanganate. Concentration of the resulting solution yields rather pure lactic acid (food quality) with a concentration of 80%.

Fermentation liquids derived from the fermentation of raw materials of lower quality require more expensive purification steps. These mainly comprise prepurification by filtration of the hot calcium lactate with the addition of filter aids and subsequent crystallization of calcium lactate, followed by decomposition with sulfuric acid and additional treatments with ion exchange resins.

Further purification is achieved by repeating precipitation and decomposition of calcium lactate. Improvements have been reported when using zinc lactate or magnesium lactate instead. Purifications with magnesium lactate have been recommended in fermentations using crude carbohydrate sources such as molasses because of improved crystallization properties (BODE, 1969).

As may be calculated, the amount of calcium sulfate formed and to be disposed of is approximately the same as that of the lactic acid produced. Problems of waste gypsum disposal, therefore, become increasingly important with higher yields.

Solvent extraction of lactic acid as another purification procedure has been proposed. Several extractants have been described: isopropyl ether (JENEMANN, 1931); $\alpha,\omega$-diamino-oligoethers (MIESIAC et al., 1992); isobutanol (BERNING et al., 1985); trialkyl tertiary amines in an organic solvent (BAILEY et al., 1987); di-*n*-octylamine in hexane (HANO et al., 1993). As a novel improvement, the application of liquid membranes should be mentioned (CHAUDHURI and PYLE, 1992a, b; SCHOLLER et al., 1993; LAZAROVA and PEEVA, 1994). In the past, isopropyl ether has been used in several plants but its application is considered too hazardous.

Another method of lactic acid purification is esterification with, e.g., methanol and distillation of the volatile ester as a most effective separation step to yield pure products (e.g., SCHOPMEYER, 1943; Bowmans Chem. Ltd., 1962; SEPITKA and GAERTNER, 1962). Esterification is performed in countercurrent operation with concomitant separation of the volatile ester and subsequent deesterification with water in a second stage to yield free lactic acid and methanol to be reintroduced into the system. This method is said to yield lactic acid of optimum purity without any waste product to be disposed of. Disadvantages of this method are the expensive equipment and technical problems in handling a fluid with higher contents of inorganic compounds.

More recently, ion exchange, hitherto only used in later purification steps, has been proposed for primary separation of lactic acid from fermentation liquids (KANAZAWA, 1977; OBARA, 1989; KULPRATHIPANJA and OROSKAR, 1991; MANTOVANI et al., 1992; MATSUDA and YOSHIDA, 1992; EVANGELISTA et al., 1994). It is claimed that this method makes the production of food-grade, heat-stable (see below) lactic acid possible without the problem of waste disposal as in the calcium and sulfuric acid procedure.

Apparently, only extensive economic evaluations can decide which of the available purification procedures matches the requirements regarding product quality and production costs at best.

## 4.2.3 Process Variants

### 4.2.3.1 Ammonium Lactate Process

Maintenance of pH 5.5–6 during fermentation may be also accomplished by the addi-

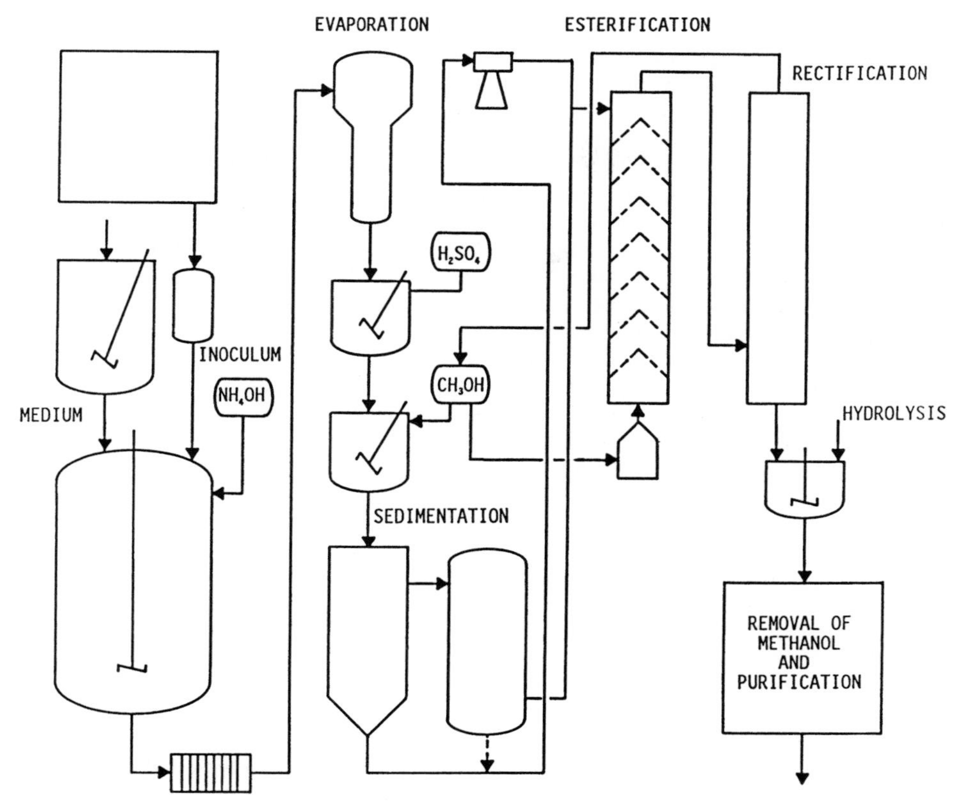


**Fig. 3.** Industrial production of lactic acid using the ammonium lactate process.

tion of sodium carbonate or hydroxide, but isolation of the resulting sodium lactate presents considerable problems. Several workers have tried to neutralize the lactic acid formed by adding ammonia (POZNANSKI et al., 1976; STIEBER et al., 1977; JARRY and LAMONERIE, 1988). The initial idea probably was to produce ammonium lactate as a combined animal feed. It was realized then that concentrated solutions of ammonium lactate may be esterified advantageously to recover lactic acid as lactic ester while liberating ammonia or leaving behind inorganic ammonium salts with useful properties (e.g., WALKUP et al., 1991; SEPITKA and SEPITKOVA, 1992; COCKREM and JOHNSON, 1993; KUMAGAI et al., 1994). In the patent claims of WALKUP et al. (1991), the ammonium lactate formed is reacted with an alcohol to yield the respective ester, which can either be separated and used to produce low-cost, purified lactic acid, or converted to low-cost, purified acrylic acid. The patent of COCKREM and JOHNSON (1993) includes a list of different alcohols to be applied. In the process described by KUMAGAI et al. (1994), esterification of ammonium lactate is performed at elevated temperatures effecting liberation and recovery of ammonia in the course of esterification. The process patented by SEPITKA and SEPITKOVA (1992) as outlined in Fig. 3 involves fermentation of sugar solutions (200 $g L^{-1}$, e.g., diluted syrups) by *L. delbrueckii*. The pH value is kept close to 6.0 by continuous addition of ammonia water. The fermented liquor is evaporated to a (calculated) lactic acid concentration of 50–60%, treated with appropriate amounts of concentrated sulfuric acid to liberate lactic acid, and mixed with an equal volume of methanol. Under these conditions ammonium sulfate is precipitated almost quantitatively. It is recovered by settling and subsequent centrifugation and may be used as liquid fertilizer. The clear supernatant containing small amounts of sulfuric acid is introduced into an esterification column consisting of conical sieve plates, where sulfuric acid catalyzed esterification takes place in a thin film in presence of countercurrent methanol vapors which are introduced to the system at the bottom of the column. Unreacted methanol and methyl lactate are separated in a rectification column and methyl lactate is reacted with water to yield pure lactic acid. The temperature during the entire process does not exceed 100°C. Lactic acid yield in the entire process is 95%, i.e., 90% based on sugar input. The colorless product thus obtained and finally concentrated shows the following characteristics: concentration: 80±0.8%; specific gravity: 1.196; ash: 0.0%; sulfate: none; calcium: none; chloride: traces to none; iron: traces to none. The patent also claims a continuous fermentation process as part of the technology.

#### 4.2.3.2 Continuous Fermentation Processes

In continuous fermentation processes considerably higher productivities are achieved, and thus they have been performed in various forms. Early studies comprised experiments using cell suspensions (CHILDS and WELSBY, 1966), eventually with cell recycling (VICK et al., 1983). In fermenter systems with high flow rates, productivities went up to about 50 g $L^{-1}$ $h^{-1}$ (RICHTER et al., 1987). Such high productivities, of course, were considered to be interesting for the conversion of whey, especially in places with high accumulation of whey. Several authors have reported continuous fermentations of whey permeates with high productivities (BOYAVAL et al., 1987; MEHAIA and CHERYAN, 1987a; AESCHLIMANN and VON STOCKAR, 1990; KRISCHKE et al., 1991; KULOZIK et al., 1992; BOERGARDTS et al., 1994). Some aspects of integrated processes have been discussed in a textbook published recently (CHMIEL and PAULSEN, 1991). In this context, it has been mentioned that in order to achieve high productivities considerable residual concentrations of substrates often have to be taken into account.

Combinations of these processes with electrodialysis have also been described (HONGO et al., 1986; CZYTKO et al., 1987; YAO and TODA, 1990).

#### 4.2.3.3 Application of Immobilized Cells

An appreciable amount of work has been devoted to studies of lactic acid production with immobilized cell systems (LINKO et al., 1984; MEHAIA and CHERYAN, 1987b; BOYAVAL and GOULET, 1988; BASSI et al., 1991), but industrial applications have not been realized so far (VENKATESH et al., 1993).

### 4.3 Organisms Other than Lactic Acid Bacteria

A limited number of other microorganisms is capable to produce larger amounts of lactic acid from common carbon sources. This is best known for *Rhizopus*. In his famous treatise on *"Chemical Activities of Fungi"* FOSTER (1949) recommended the use of *Rhizopus* for commercial lactic acid production (see WARD et al., 1936; 1938a, b; PRESCOTT and DUNN, 1959) because this organism is able to convert several sugars with appreciable yields in synthetic fermentation media. Thus downstream processing was considered to be easier than fermentations by the more fastidious lactic acid bacteria. So far, although being patented (SNELL and LOWERY, 1964), commercial use of this potential technique has not been accomplished. Recently, a process has been patented in which conversion of starch by *Rhizopus oryzae* in a single-step fermentation process with high lactic acid yields is claimed (HANG, 1990).

Several reports have been published on a group of spore-forming bacteria which are long known as lactic acid producing:

- *Sporolactobacillus inulinus* (e.g., ATCC 14897 or 15538). This organism specifically produces D-(−)lactic acid (KITAHARA, 1966; NAGAI et al., 1986; KOSAKI and KAWAI, 1986; KOBAYASHI and TANAKA, 1988);
- *Bacillus coagulans* (10 strains listed, e.g., in DSM, 1989). Lactic acid production by this organism (BLUMENSTOCK, 1984) has been patented by KIRKOVITS and EDLAUER (1990).

Owing to their common properties both organisms have been studied extensively, especially with regard to their phylogenetic relationships (SUZUKI and YAMASATO, 1994).

## 4.4 Product Quality and Applications, Production Figures

Lactic acid, a highly hygroscopic, syrupy liquid is commercially available at different grades (qualities):

*Technical-grade lactic acid* with light yellow to brown color of varying concentrations (20–80%) should be iron-free and is used in deliming hides, in the textiles industry, and for manufacturing esters for use as solvents and plasticizers.

Almost 85% of lactic acid produced is used in the food industry as an acidulant and as a food preservative. In different countries quality requirements are embodied in the respective food codices. *Food-grade lactic acid* should be colorless and almost odorless and should contain a minimum of 80% lactic acid. Because of its pleasant sour taste it is used as an additive in the manufacture of beverages, essences, fruit juices and syrups, as an acidulant in jams, jellies and confectionery, in the canning industry and in bakeries to produce sour flours and doughs, respectively.

*Pharmacopoeia-grade lactic acid* with 90% lactic acid content and <0.1% ash should also be colorless and odorless and is used for treatments of the intestine, in hygienic preparations as well as for the manufacture of pure pharmaceutical and other derivatives of lactic acid, e.g., calcium lactate.

*Plastic-grade lactic acid* should be absolutely colorless and for some applications it should exhibit the lowest ash content (<0.01%). It is used in the manufacture of various lacquers, varnishes, and impregnating agents as well as polymers. A more recent field of application is the production of biodegradable plastics utilizing, e.g., polylactic acid which is formed spontaneously in concentrated solutions, especially upon heating.

An important quality criterion of highly pure lactic acid is the so-called heat stability, i.e., no color development upon heating an 88% lactic acid solution to 180°C.

Physical properties and details regarding applications in the chemical industry may be found in the respective chemical encyclopedias (e.g., Ullmann, Kirk-Othmer).

The market volume of lactic acid production has changed little in the last decade and may be estimated as being ca. 50000 t $a^{-1}$, about 2/3 of which are produced by fermentation.

# 5 References

ACCOLAS, J. P., SPILLMANN, H. (1979), Morphology of bacteriophages of *Lactobacillus bulgaricus, L. lactis* and *L. helveticus, J. Appl. Bacteriol.* **47**, 309–319.

AESCHLIMANN, A., VON STOCKAR, U. (1990), The effect of yeast extract supplementation on the production of lactic acid from whey permeate by *Lactobacillus helveticus, Appl. Microbiol. Biotechnol.* **32**, 398–402.

BAILEY, R. B., JOSHI, D. K., MICHAELS, S. L., WISDOM, R. A. (1987), *U.S. Patent* 4698303.

BALOW, A., TRÜPER, H. G., DWORKIN, M., HARDER, W., SCHLEIFER, K. H. (Eds.) (1991), *The Prokaryotes,* Vols. 1 and 2, 2nd Edn., New York: Springer-Verlag.

BARTON-WRIGHT, E. C. (1952), *The Microbiological Assay of the Vitamin B-Complex and Amino Acids,* London: Pitman & Sons.

BASSI, A. S., ROHANI, S., MACDONALD, D. G. (1991), Fermentation of cheese whey in an immobilized-cell fluidized-bed reactor, *Chem. Eng. Commun.* **103**, 119–129.

BERNING, W., SIEGEL, H., HEIDA, B., WICKENHÄUSER, G. (1985), *Ger. Patent* 3415141.

BLUMENSTOCK, J. (1984), *Bacillus coagulans* Hammer 1915 und andere thermophile oder mesophile, säuretolerante *Bacillus*-Arten – eine taxonomische Untersuchung. *Dissertation*, University of Göttingen.

BODE, H. E. (1969), *U.S. Patent* 3429777.

BOERGARDTS, P., KRISCHKE, W., CHMIEL, H., TROESCH, W. (1994), Development of an integrated process for the production of lactic acid from whey permeate, *Prog. Biotechnol.* **9** (ECB6: Proc. 6th Eur. Congr. Biotechnol. 1993, Pt 2), 905–908.

BOICHARD, J., BROSSARD, B. P., GAY, M. L. M. J., JANIN, R. M. C. (1973), *U.S. Patent* 3715394.

Bowmans Chem. Ltd. (1962) *Brit. Patent* 907322.

BOYAVAL, P., GOULET, J. (1988), Optimal conditions for production of lactic acid from cheese whey permeate by calcium alginate-entrapped *Lactobacillus helveticus, Enzyme Microb. Technol.* **10**, 725–728.

BOYAVAL, P., CORRE, C., TERRE, S. (1987), Continuous lactic acid fermentation with concentrated product recovery by ultrafiltration and electrodialysis, *Biotechnol. Lett.* **9**, 207–212.

BUCHTA, K. (1974), Die biotechnologische Gewinnung von organischen Säuren, *Chem. Ztg.* **98**, 533–538.

BUCHTA, K. (1983), Lactic acid, in: *Biotechnology,* 1st Edn. Vol. 3 (REHM, H.-G., REED, G., Eds.), pp. 409–417. Weinheim: Verlag Chemie.

CHAHAL, S. P. (1990), Lactic acid, in: *Ullman's Encyclopedia of Industrial Chemistry,* 5th Edn., Vol. A 15 (ELVERS, B., HAWKINS, S., SCHULZ, G., Eds.), pp. 97–105. Weinheim: VCH.

CHAUDHURI, J. B., PYLE, D. L. (1992a), Emulsion liquid membrane extraction of organic acids, I. A theoretical model for lactic acid extraction with emulsion swelling, *Chem. Eng. Sci.* **47**, 41–48.

CHAUDHURI, J. B., PYLE, D. L. (1992b), Emulsion liquid membrane extraction of organic acids, II. Experimental, *Chem. Eng. Sci.* **47**, 49–56.

CHENG, P., MUELLER, R. E., JAEGER, S., BAJPAI, R., JANNOTTI, E. L. (1991), Lactic acid production from enzyme-thinned corn starch using *Lactobacillus amylovorus, J. Ind. Microbiol.* **7**, 27–34.

CHIARINI, L., MARA, L., TABACCHIONI, S. (1992), Influence of growth supplements on lactic acid production in whey ultrafiltrate by *Lactobacillus helveticus, Appl. Microbiol. Biotechnol.* **36**, 461–464.

CHILDS, C. G., WELSBY, B. (1966), Continuous lactic fermentation, *Process Biochem.* **1**, 441.

CHMIEL, H., PAULSEN, J. (1991), Die Produktion von Milchsäure aus Molkepermeat, in: *Bioprozeßtechnik,* Vol. 2 (CHMIEL, H., Ed.), pp. 272–278. Stuttgart: Gustav Fischer Verlag.

COCKREM, M. C. M., JOHNSON, P. D. (1993), *World Patent* 9300440 A1.

CZYTKO, M., ISHII, K., KAWAI, K. (1987), Continuous glucose fermentation for lactic acid production: recovery of acid by electrodialysis, *Chem. Ing. Techn.* **59**, 952–954.

DE VRIES, W., KAPTEIJN, W. M. C., VAN DER BEEK, E. G., STOUTHAMER, A. H. (1970), Molar growth yields and fermentation balances of *Lactobacillus casei* L3 in batch cultures and in continuous cultures, *J. Gen. Microbiol.* **63**, 333–345.

DEMIRCI, A., POMETTO, A. (1992), Enhanced production of D(–)-lactic acid by mutants of *Lactobacillus delbrueckii* ATCC 9649, *J. Ind. Microbiol.* **11**, 23–28.

DEVOS, W. M., VAUGHAN, E. E. (1994), Genetics of lactose utilization in lactic acid bacteria, *FEMS Microbiol. Rev.* **15**, 217–237.

DSM (1989), Deutsche Sammlung von Mikroorganismen, Catalogue of Strains 1989, 4th Edn., Braunschweig: DSM.

EVANGELISTA, R. L., MANGOLD, A. J., NIKOLOV, Z. L. (1994), Recovery of lactic acid by sorption – Resin evaluation, *Appl. Biochem. Biotechnol.* **45**, 131–144.

FITZSIMONS, A., HOLS, P., JORE, J., LEER, R. J., O'CONNELL, M., DELCOUR, J. (1994), Development of an amylolytic *Lactobacillus plantarum* silage strain expressing the *Lactobacillus amylovorus alpha*-amylase gene, *Appl. Environ. Microbiol.* **60**, 3529–3535.

FOSTER, J. W. (1949), *Chemical Activities of Fungi.* New York: Academic Press.

GARVIE, E. I. (1980), Bacterial lactate dehydrogenases, *Microbiol. Res.* **44**, 106–139.

HAMMES, W. P., WEISS, N., HOLZAPFEL, W. (1991), The genera *Lactobacillus* and *Carnobacterium,* in: *The Prokaryotes,* 2nd Edn., Vol. 2 (BALOW, A., TRÜPER, H. G., DWORKIN, M., HARDER, W., SCHLEIFER, K. H., Eds.), pp. 1535–1594. New York: Springer-Verlag.

HANG, Y. D. (1990), *U.S. Patent* 4963486 A.

HANO, T., MATSUMOTO, M., UENOYAMA, S., OHTAKE, T., KAWANO, Y., MIURA, S. (1993), Separation of lactic acid from fermented broth by solvent extraction, *Bioseparation* **3**, 321–326.

HOFMAN, M. (1988), *Eur. Patent* 265409.

HONGO, M., NOMURA, Y., IWAHARA, M. (1986), Novel method of lactic acid production by electrodialysis fermentation, *Appl. Environ. Microbiol.* **51,** 314–319.

INSKEEP, G. C., TAYLOR, G. G., BREITZKE, W. C. (1952), Lactic acid from corn sugar, *Ind. Eng. Chem.* **44,** 1955–1966.

JARRY, A., LAMONERIE, H. (1988), Microbial manufacture of lactic acid in the presence of fish protein hydrolyzate, *Eur. Pat. Appl.* 266258.

JENEMANN, J. A. (1931), *U.S. Patent* 1906068.

KANAZAWA, J. (1977), *Jpn. Patent* 52105284.

KIRKOVITS, A. E., EDLAUER, H. (1990), *Ger. Patent* 4000942 A1.

KITAHARA, K. (1966), *U.S. Patent* 3262862.

KOBAYASHI, T., TANAKA, M. (1988), Production of D-lactic acid with bioreactor, *Biol. Ind.* **5**, 800–806.

KOSAKI, M., KAWAI, K. (1986), *Eur. Patent* 190770 A2.

KRISCHKE, W., SCHROEDER, M., TROESCH, W. (1991), Continuous production of L-lactic acid from whey permeate by immobilized *Lactobacillus casei* ssp. *casei, Appl. Microbiol. Biotechnol.* **34**, 573–578.

KULOZIK, U., HAMMELEHLE, B., PFEIFER, J., KESSLER, H. G. (1992), High reaction rate continuous bioconversion process in a tubular reactor with narrow residence time distributions for the production of lactic acid, *J. Biotechnol.* **22**, 107–116.

KULPRATHIPANJA, S., OROSKAR, A. R. (1991), *U.S. Patent* 5068418.

KUMAGAI, A., YAGUCHI, M., ARIMURA, T., MIURA, S. (1994), *Eur. Patent* 614983 A2.

LAZAROVA, Z., PEEVA, L. (1994), Facilitated transport of lactic acid in a stirred transfer cell, *Biotechnol. Bioeng.* **43**, 907–912.

LEH, M. B., CHARLES, M. (1989a), Lactic acid production by batch fermentation of whey permeate: a mathematical model, *J. Ind. Microbiol.* **4**, 65–70.

LEH, M. B., CHARLES, M. (1989b), The effect of whey protein hydrolyzate on the lactic acid fermentation, *J. Ind. Microbiol.* **4**, 71–75.

LINKO, P., STENROOS, S. L., LINKO, Y. Y., KOISTINEN, T., HARJU, M., HEIKONEN, M. (1984), Applications of immobilized lactic acid bacteria, *Ann. N. Y. Acad. Sci.* **434**, 406–417.

MANTOVANI, G., VACCARI, G., CAMPI, A. L. (1992), *Eur. Patent* 517242 A2.

MASLOWSKI, B. (1988), *Pol. Patent* 144390 B2.

MATA-GILSINGER, M. (1985), Presence of phages in starter cultures, in: *Symposium on the Importance of Lactic Acid Fermentation in the Food Industry,* Mexico City, 1984. UNIDO Doc. ID/WG. 431/4.

MATSUDA, K., YOSHIDA, Y. (1992), *Jpn. Patent* 04320691 A2.

MEHAIA, M. A., CHERYAN, M. (1987a), Production of lactic acid from sweet whey permeate concentrates, *Process Biochem.* **22**, 185–188.

MEHAIA, M. A., CHERYAN, M. (1987b), Immobilization of *Lactobacillus bulgaricus* in a hollow-fiber bioreactor for production of lactic acid from acid whey permeate, *Appl. Biochem. Biotechnol.* **14**, 21–27.

MIALL, L. M. (1978), Lactic acid, in: *Economic Microbiology,* Vol. 2 (ROSE, A. H., Ed.), pp. 95–98. London: Academic Press.

MIESIAC, I., SZYMANOWSKI, J., SCHUEGERL, K., SOBCZYNSKA, A. (1992), Interfacial activity of lactic acid extractants containing oligo-oxyethylene fragments, *Colloids Surf.* **64**, 119–123.

MONTEAGUDO, J. M., RINCON, J., RODRIGUEZ, L., FUERTES, J., MOYA, A. (1993), Determination of the best nutrient medium for the production of L-lactic acid from beet molasses: a statistical approach, *Acta Biotechnol.* **13**, 103–110.

NAGAI, S., OZAKI, M., FUKUNISHI, K., YAMAZAKI, K. (1986), *Jpn. Patent* 61058588 A2.

NAKAMURA, L. K. (1981), *Lactobacillus amylovorus*, a new starch-hydrolyzing species from cattle waste-corn fermentations, *Int. J. Syst. Bacteriol.* **31**, 56–63.

NAKAMURA, L. K. CROWELL, C. D. (1979), *Lactobacillus amylophilus*, a new starch-hydrolyzing species from swine waste-corn fermentation, *Dev. Ind. Microbiol.* **20**, 531–540.

OBARA, H. (1989), *Jpn. Patent* 01091788.

ORLA-JENSEN, S. (1919), *The Lactic Acid Bacteria.* Kobenhaven: Det Kgl. Danske Vid. Selskab. Skrifter.

PASTEUR, L. (1857), Mémoire sur la fermentation appelée lactique, *Compt. Rend.* **45**, 913ff.

PLATZ, R., NOHE, H., DOCKNER, T. (1972), *U.S. Patent* 3642889.

POZNANSKI, S., KORNACKI, K., SMIETANA, Z., RYMASZEWSKI, J., SURAZYNSKI, A., JAKUBOWSKI, J., CHOJNOWSKI, W. (1976), *Pol. Patent* 86492.

PRESCOTT, S. C., DUNN, C. G. (1959), *Industrial Microbiology,* 3rd Edn. New York: McGraw-Hill Book Company, Inc.

RAUCH, M., GERNER, F., WECKER, A. (1960), Milchsäure, in: Ullmanns Encyklopädie der technischen Chemie, 3rd Edn., Vol. 12 (FOERST, W., Ed.), pp. 525–537. Weinheim: Verlag Chemie.

REED, G., NAGODAWITHANA, T. W. (Eds.) (1995), *Biotechnology,* 2nd Edn., Vol. 9. Weinheim: VCH.

REHM, H. J. (1967), *Industrielle Mikrobiologie.* Berlin: Springer-Verlag.

RICHTER, K., BECKER, U., MEYER, D. (1987), Continuous fermentation in high flow rate fermentor systems, *Acta Biotechnol.* **7**, 237–245.

ROBISON, P. D. (1988), *U.S. Patent* 4749652.

ROY, D., GOULET, J., LEDUY, A. (1986), Batch fermentation of whey ultrafiltrate by *Lactobacillus helveticus* for lactic acid production, *Appl. Microbiol. Biotechnol.* **24**, 206–213.

SCHOLLER, C., CHAUDHURI, J. B., PYLE, D. L. (1993), Emulsion liquid membrane extraction of lactic acid from aqueous solutions and fermentation broth, *Biotechnol. Bioeng.* **42**, 50–58.

SCHOPMEYER, H. H. (1943), *U.S. Patent* 2350370.

SCHOPMEYER, H. H. (1954), Lactic acid, in: *Industrial Fermentations,* Vol. 1 (UNDERKOFLER, L. A., HICKEY, R. J., Eds.), pp. 391–419. New York: Chemical Publishing Co., Inc.

SCHWALL, H. (1979), Milchsäure, in: Ullmanns Encyklopädie der technischen Chemie, 4th Edn., Vol. 17 (BARTHOLOMÉ, E., BIEKERT, E., HELLMANN, H., LEY, H., WEIGERT, M., Eds.), pp. 1–7. Weinheim: Verlag Chemie.

SEPITKA, A., GAERTNER, M. (1962), *Czech. Patent* 104398.

SEPITKA, A., SEPITKOVA, J. (1992), *Czech. Patent* 276959.

SHIMIZU-KADOTA, M., TSUCHIDA, N. (1984), Physical mapping of the virion and the prophage DNAs of temperate *Lactobacillus* phage FSW, *J. Gen. Microbiol.* **130**, 423–430.

SHIMIZU-KADOTA, M., SAKURAI, T., TSUCHIDA, N. (1983), Prophage origin of a virulent phage appearing on fermentations of *Lactobacillus casei* S-1, *Appl. Environ. Microbiol.* **45**, 669–674.

SNELL, R. L., LOWERY, C. E. (1964), *U.S. Patent* 3125494.

STIEBER, R. W., COULMAN, G., GERHARDT, P. (1977), Dialysis continuous process for ammonium lactate fermentation of whey: experimental tests, *Appl. Environ. Microbiol.* **34**, 733–739.

SUZUKI, T., YAMASATO, K. (1994), Phylogeny of spore-forming lactic acid bacteria based on 16S rRNA gene sequences, *FEMS Microbiol. Lett.* **115**, 13–17.

TEUBER, M. (1993), Lactic acid bacteria, in: *Biotechnology,* Vol. 1, 2nd Edn. (REHM, H.-J., REED, G., Eds.), pp. 325–366. Weinheim: VCH.

TEUBER, M., LEMKE, J. (1983), The bacteriophage of lactic acid bacteria with emphasis on genetic aspects of group N streptococci, *Antonie van Leeuwenhoek* **49**, 283–295.

VENKATESH, K. V., OKOS, M. R., WANKAT, P. C. (1993), Design of an immobilized cell reactor/separator for lactic acid production, in: *Proc. 9th Natl. Conv. Chem. Eng. Int. Symp.* "Importance Biotechnol. Coming Decades" (AYYANNA, C., Ed.), pp. 156–160. New Delhi: Tata McGraw-Hill.

VICK, R. T. B., MANDEL, D. K., DEA, D. K., BLANCH, H. W., WILKE, C. R. (1983), The application of cell recycle to continuous fermentative lactic acid production, *Biotechnol. Lett.* **5**, 665–670.

VICKROY, T. B. (1985), Lactic acid, in: *Comprehensive Biotechnology,* Vol. 3 (BLANCH, H. W., DREW, S., WANG, D. I. C., Eds.), pp. 761–776. Oxford: Pergamon Press.

VOGT, W., ERPENBACH, H., JOEST, H., STRIE, L. (1966), *U.S. Patent* 3284495.

WALKUP, P., ROHRMANN, C., HALLEN, R., EAKIN, D. E. (1991), *World Patent* 9111527 A2.

WARD, G. E., LOCKWOOD, L. B., MAY, O. E., HERRICK, H. T. (1936), Biochemical studies in the genus *Rhizopus.* I. The production of dextro-lactic acid, *J. Am. Chem. Soc.* **58**, 1286–1288.

WARD, G. E., LOCKWOOD, L. B., TABENKIN, B., WELLS, P. A. (1938a), Rapid fermentation process for dextrolactic acid, *Ind. Eng. Chem.* **30**, 1233–1235.

WARD, G. E., LOCKWOOD, L. B., MAY, O. E. (1938b), *U.S. Patent* 2132712.

WEI, M. Q., RUSH, C. M., NORMAN, J. M., HAFNER, L. M., EPPING, R. J., TIMMS, P. (1995), An improved method for the transformation of *Lactobacillus* strains using electroporation, *J. Microbiol. Meth.* **21**, 97–109.

YAO, P., TODA, K. (1990), Lactic acid production in electrodialysis culture, *J. Gen. Appl. Microbiol.* **36**, 111–125.

ZHANG, D. X., CHERYAN, M. (1994), Starch to lactic acid in a continuous membrane reactor, *Process Biochem.* **29**, 145–150.

# 9 Citric Acid

Max Roehr
Christian P. Kubicek
Jiří Komínek

Wien, Austria

# 1 Introduction and Historical View

In his fascinating book on the *"Chemical Activities of Fungi"* FOSTER (1949) stated that "probably more study has been devoted to citric acid fermentation than to any other function in mold metabolism, yet, there is probably less fundamental knowledge regarding it than there is about most processes treated in this book – notwithstanding the fact that historically it was the first fungal process of industrial potentialities to be discovered" (abridged quotation). Indeed, since this was quoted, the number of articles on this subject again has increased considerably, and it may be hoped that this could contribute adequately to our knowledge of microbial biochemistry and technology.

Citric acid was first observed as a fungal product by WEHMER in 1891–93 (WEHMER, 1896, 1907), as a by-product of oxalic acid produced by a culture of *Penicillium glaucum* on sugar media. In 1993, he isolated two novel fungal strains with the ability to accumulate citric acid which he designated as a new genus, *Citromyces,* now to be regarded as monoverticillate Penicillia (WEHMER, 1903).

Attempts to transfer this finding into industrial practice by Fabrique de Produits Chimiques de Thann et de Mulhouse failed. This was mainly due to the fact that fermentations had to be run for several weeks to attain appreciable yields; thus, severe contaminations were almost inevitable, especially since WEHMER relied on the then prevailing believe that in order to achieve and maintain substantial acid production, the acid would have to be neutralized as formed.

In 1913 ZAHORSKY obtained a patent for citric acid production using *Sterigmatocystis nigra* – a synonym for *Aspergillus niger.* It was the work of CURRIE, however, that opened up the way for successful industrial production of citric acid. In 1916 (THOM and CURRIE, 1916; CURRIE, 1916), as a coworker of Dr. Ch. THOM, head of the microbiological laboratory of the Bureau of Chemistry in Washington, DC, J. N. CURRIE first observed that numerous strains of *Aspergillus niger* collected in this laboratory produced appreciable amounts of citric acid under certain conditions. The most important finding was (CURRIE, 1917) that *Aspergillus niger* grew well at pH values around 2.5–3.5 and that citric acid was produced abundantly at these and even lower pH values. Most striking, moreover, was the finding that at such low pH values yields of <60% could be achieved in simple media requiring only 1–2 weeks time with little risk of contamination.

Up to this time, citric acid amounting to about 10,000 t $a^{-1}$ was produced by pressing citrus fruits (mainly lemons), which contain about 7–9% citric acid, and precipitating the acid as calcium salt. About 30–40 t of lemons (corresponding to an area of about 2.7 ha) were required for 1 t of citric acid. Italy was the supplier of more than 80% of world demand.

Several companies in different countries were active in processing imported calcium citrate to citric acid, mainly for applications in the preparation of beverages.

Fiscal restrictions by both Italy as the exporting country as well as by the main importing countries and general shortages at the end of World War I apparently stimulated interest in other, more reliable ways of citric acid production.

Hence, the pioneering work of CURRIE became most instrumental for the development of an industry for citric acid manufacture by fermentation. CURRIE's finding that *Aspergillus niger* could grow and accumulate citric acid at pH values <2 meant that "fermentation can be started off at a hydrogen ion concentration that will greatly reduce the chances of infection". Furthermore, he found that high sugar concentrations were most favorable for higher production rates and he observed that optimum yields were only attainable under conditions of restricted growth. This, as will be shown in the following, turned out to be the most effective prerequisite for successful operation of modern citric acid plants.

In 1919 another attempt at commercially producing citric acid by fermentation (Societé des Produits Organiques de Tirlemont, Belgium) had failed, but in 1923, Chas. Pfizer & Co. in New York, usinq CURRIE's know-how, successfully went into production – and are still the greatest producer in the world. Subsequently, in England J. and E. Sturge, formerly having processed calcium citrate, started production with a similar process based on patents granted to FERNBACH et al. (1927, 1932).

In 1928 a plant was built in Kaznějov near Plzen in the Czech Republic where beet molasses as a cheap carbon source was used with success (Montan- und Industrialwerke, vorm. J. D. Stark, 1924; SZÜCS, 1925). This material, however, presented difficulties due to its content of detrimental substances, of which metal ions of the transition metals group were found to be most crucial. According to LEOPOLD and VALTR (1964) in this plant hexacyanoferrate (HCF) was first used to precipitate or complex unwanted metals in the medium. Apparently kept secret, this was published as late as in 1938 in a patent by MEZZADROLI (1938).

Several other plants were erected subsequently, e.g., in the Soviet Union, in Germany (Joh. A. Benckiser GmbH), in Belgium (La Citrique Belge) and in several other countries. By 1933, according to PERQUIN (1938) world production of citric acid amounted to 10,400 t, with only 1,800 t produced from lemons in Italy, whereas the rest was already manufactured by fermentation (Europe 5,100 t, USA 3,500 t).

Concomitantly considerable efforts were made towards the elucidation of the biochemical mechanisms of citric acid production. Accounts of this early work may be found in several reviews, e.g., BERNHAUER (1940), FOSTER (1949), PRESCOTT and DUNN (1959) or JOHNSON (1954). It is interesting to note how many theories were put forward until it was clear that citric acid biosynthesis proceeds via the normal route of carbohydrate breakdown then established and the fundamental reaction of the tricarboxylic acid cycle – condensation of acetyl-CoA and oxaloacetate. The regulatory mechanisms, however, leading to the high amount of citric acid synthesized and readily excreted, were only studied beginning with the 1970s and they are still being investigated by several groups.

Originally, citric acid fermentations were carried out in surface cultures as devised by CURRIE (1917). The introduction of submerged fermentation presented considerable problems, mainly due to the fact that under such conditions fermentation systems and especially production strains developed much more sensitivity to factors hitherto considered as less crucial. Thus, the effect of transition metal ions was more pronounced, and it was necessary to consider raw materials much more carefully. Although preliminary work dates back to the 1930s (PERQUIN, 1938) and 1940s (SZÜCS, 1944), it was only during the 1950s that the basic principles of scaling-up this complex type of fermentation were elaborated. Only with this fermentation it was shown that special strategies have to be developed according to the type of raw material.

SZÜCS (1944) had shown that for successful operation the process had to be divided into two stages, the first supporting growth of the mycelium, and the second restricting growth according to the ideas of CURRIE (1917). This was achieved by incorporating sufficient phosphorus ($\sim 70$ mg $L^{-1}$) into the growth medium and by preparing a phosphorus-free medium for the production stage.

This was amended by SHU and JOHNSON (1947, 1948a, b) designing a fermentation medium that provided growth and production phases in a single system. These authors also showed that a certain balance between manganese, zinc, and phosphate was decisive for optimum fermentation conditions.

For commercial fermentations this was difficult to implement since they required raw materials with ill-defined composition, such as molasses. A tremendous amount of work was thus dedicated to elaborate optimum conditions for the utilization of cheap materials like cane molasses (SHU, 1947; KAROW and WAKSMAN, 1947; WOODWARD et al., 1949; MOYER, 1953), beet molasses (CLEMENT, 1952; MARTIN and WATERS, 1952; STEEL et al., 1955), starch (MOYER, 1953a, b), and starch hydrolyzates (SCHWEIGER and SNELL, 1949; SNELL and SCHWEIGER, 1949).

Various processes for treating and purifying molasses, especially for the removal of trace metals, were developed. Most of them comprise treatment with HCF as mentioned above, but the application of ion exchange resins was also suggested (PERLMAN et al.,

1945; WOODWARD et al., 1949). In the course of such experiments it was discerned that HCF not only served as a complexant to detrimental metal ions, but had to be added in small amounts to bring about that certain growth restriction accompanying the switch to abundant acid accumulation.

A major breakthrough was achieved in the USA by workers of Miles Laboratories, Inc. In the course of developing their technology, the respective authors first introduced the use of starch hydrolyzates as comparably cheap raw material (see above). When applying these rather pure raw materials, they discovered that the detrimental action of traces of iron could be counteracted by the addition of certain amounts of copper ions (SCHWEIGER, 1959, 1961). Moreover, as with HCF, it was found that a small excess of copper ions was beneficial to achieve high yields of citric acid. Details of these technologies are described below.

The present situation: Several companies use the advanced technology of submerged fermentation and are responsible for about 80% of world citric acid production. Only a few other companies still rely on the old surface (and solid-state) technologies.

In the 1960s new processes emerged in the course of extensive studies on the utilization of *n*-alkanes as carbon sources for the production of microbial biomass (single cell protein, SCP) and other products such as amino acids, vitamins, organic acids, etc. Originally a domain of Japanese workers, the use of complex mutants of mainly coryneform bacteria such as *Corynebacterium, Arthrobacter,* and *Brevibacterium* soon became an established technology with characteristic features of intellectual achievements. In 1966/67 several of such mutants were found to produce citric acid from *n*-paraffins ($C_9$–$C_{30}$) and related substrates. Although patented (TANAKA et al., 1968) this was not exploited at an industrial scale. Subsequently, however, it was discovered that certain yeasts, mainly *Candida* and related forms, which had been found to utilize *n*-alkanes as carbon source for biomass production, were able to produce appreciable amounts of citric acid (together with isocitric acid) from *n*-paraffins (FUKUI and TANAKA, 1980). First attempts at isolating citric acid as a by-product of SCP manufacture were soon superseded by investigations to make citric acid the main product of the utilization of such carbon sources. This was also due to the fact that interest in producing SCP from hydrocarbons had declined, as the product could hardly be applied as animal feed because of problems with animal health authorities. Although citric acid yields of almost 100% from $C_9$–$C_{20}$ alkanes were reported, industrial implementation (cf. FERRARA et al., 1977; STOTTMEISTER et al., 1982) was blocked by several instances: Citric acid turned out to be accompanied by appreciable amounts of isocitric acid, which was not acceptable for many applications, e.g., in the food industry; selection of strains and optimization of process parameters could improve the situation considerably. In the meantime it had become obvious that alkanes could no longer be considered a cheap substrate, which ended the economic expectations of citric acid production from such feedstocks.

Notwithstanding these facts, research on the application of yeasts (or bacteria) has produced several interesting side effects. Especially in the first period of work on citric acid as a by-product of single cell protein manufacture, it was necessary to elaborate isolation procedures for dilute citric acid solutions. Solvent extraction methods were developed which now can be applied advantageously in conventional fermentation practice.

On the other hand it was observed that there are yeast strains which can also produce citric acid when grown on glucose or similar carbon sources (TABUCHI et al., 1969; ABE et al., 1970). These findings were extensively studied and developed further in the former German Democratic Republic. The events connected with the German reunion have probably impeded further attempts towards an implementation of another novel technology at a larger scale.

# 2 Biological Fundamentals – Regulation of Citric Acid Accumulation

## 2.1 Type and Concentration of Sugar to Trigger Citric Acid Accumulation via Bypassing Glycolytic Regulation

Various parameters influence the efficacy of citric acid accumulation by *A. niger*. Among these, the type and concentration of the sugar used has a most pronounced effect on acid production. While the optimal concentrations of trace metals, phosphate, and nitrogen are interrelated, the concentration and type of the sugar used are the only parameters which cannot be influenced by appropriate manipulation of the others. The suitability of various carbon sources for citric acid production by *A. niger* has been discussed by XU et al. (1989b): only carbon sources such as glucose or sucrose which are rapidly metabolized by the fungus, result in both high yields and high rates of acid accumulation. The biochemical basis of this effect appears to be a bypass of glycolytic regulation: at least three enzymes of the hexose bisphosphate pathway in Aspergilli participate in its control to a major extent, i.e., hexokinase (EC 2.7.1.1), 6-phosphofructokinase (EC 2.7.1.11; PFK 1), and pyruvate kinase (EC 2.7.1.40) (SMITH and NG, 1972; KUBICEK, 1988b). It is noted that glycolytic regulation is mainly caused by fine control mechanisms, as increased activities of some glycolytic enzymes are not leading to increased citric acid production. Gene amplification of, e.g., *pki*A (encoding pyruvate kinase) did not lead to improved producer strains (L. DE GRAAFF, J. VISSER, S. CHOOJUN, M. ROEHR, and C. P. KUBICEK, unpublished results). Similarly, glycolytic regulation in *Saccharomyces cerevisiae* is not affected by amplification of selected genes (HEINISCH, 1986).

PFK 1 from *A. niger* has been most thoroughly characterized and is known to be susceptible to control by various metabolites, i.e., activation by Fru-2,6-$P_2$, AMP, and $NH_4^+$ ions, and inhibition by phosphoenol pyruvate, citrate, and ATP (HABISON et al., 1983; ARTS et al., 1987). There is also indirect evidence for its regulation by a c-AMP dependent protein kinase (LEGISA and BENCINA, 1994). Activation by Fru-2,6-$P_2$ links PFK 1 activity to activity and regulation of 6-phosphofructo-2-kinase (EC 2.7.1.105; PFK 2). It is intriguing that the latter enzyme is rather poorly regulated in *A. niger*, and the conclusion has been drawn that activity of this enzyme is mainly modulated by the availability of its substrate, Fru-6-P (HARMSEN et al., 1992). The conversion of Fru-6-P to Fru-1,6-$P_2$ has, therefore, been considered as a potential major point of regulation of glycolytic flux.

It is evident from the data reported above that citric acid accumulation can only occur under conditions where PFK 1 is unable to control the glycolytic flux efficiently. The antagonizing role of $NH_4^+$ – accumulating under manganese deficient conditions (KUBICEK et al., 1979a) – on citrate inhibition of PFK 1 (HABISON et al., 1979, 1983) is one parameter involved in this process. Induction of citric acid accumulation by high concentrations of appropriate sugars, however, involves another mechanism: the presence of high concentrations of sugars triggers a rise of the intracellular concentration of Fru-2,6-$P_2$ (KUBICEK-PRANZ et al., 1990). This is only observed with carbon sources permitting high yields of citric acid. Since the intracellular concentration of Fru-6-P was also high under these conditions, it is tempting to speculate that the effect of the type of carbon source and of concentration on citric acid accumulation is due to their different effects on the Fru-6-P steady-state concentrations, which in turn influence the formation of the glycolytic activator Fru-2,6-$P_2$.

This hypothesis – if correct – implies, however, that the metabolic steps prior to Fru-6-P, i.e., hexokinase (phosphoglucose isomerase catalyzes a rapid equilibrium and is unlikely to be of any regulatory importance) may participate in the regulation of glycolysis and citric acid accumulation under certain conditions. In fact, using steady-state sensitivity analysis, TORRES (1994a, b) has recently calculated that either transport or phosphoryla-

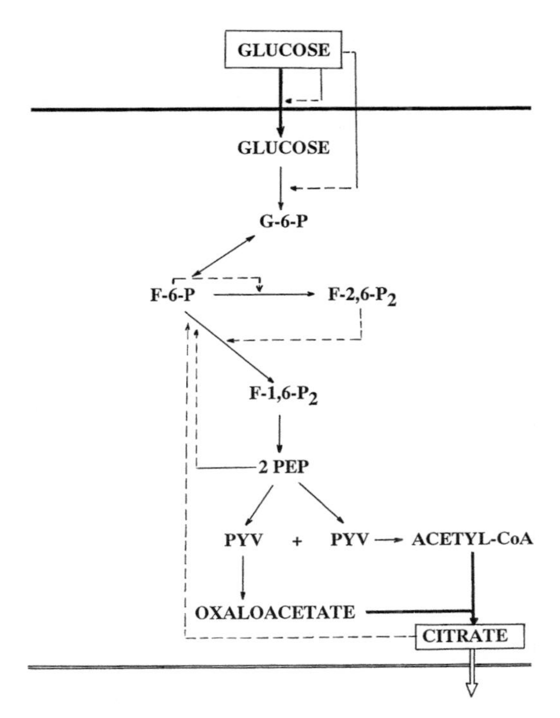


**Fig. 1.** Proposed mechanism of regulation of citric acid accumulation by elevated sugar concentrations.

tion of glucose have a major impact on the rate of citric acid accumulation. The biochemistry of hexose phosphorylation in *A. niger* has, therefore, been investigated (STEINBÖCK et al., 1993): unlike *S. cerevisiae* (LOBO and MAITRA, 1977), but consistent with *Kluyveromyces lactis* (PRIOR et al., 1993), *A. niger* contains only a single hexokinase. The enzyme is formed in increased amounts at high glucose or sucrose concentrations. Its kinetic characteristics are close to the mammalian enzyme, yet it is not inhibited by glucose-6-phosphate. 2-desoxyglucose-resistant mutants of *A. niger,* whose citric acid production had dropped down considerably (FIEDUREK et al., 1988; KIRIMURA et al., 1992), exhibited reduced hexokinase activity. However, a plot of hexokinase activity against citrate accumulation in these mutants suggests that hexokinase does not limit their metabolic flux. These findings suggest that the reduced hexokinase activity may rather be a consequence than the cause of the low rates of carbon flow in the 2-desoxyglucose-resistant mutants. Therefore, other factors controlling the efficacy of early steps in glucose metabolism are currently investigated in the authors' laboratory.

Based on this present information, a mechanism by which increased sugar concentrations may "trigger" high acid accumulation is proposed in Fig. 1.

## 2.2 Pyruvate Carboxylase and the Role of Metabolite Compartmentation

Independent of its regulation, the end product of aerobic glycolysis is pyruvate. In "normal" cells, the further catabolic fate of pyruvate is to enter the mitochondrion and to be used in the formation of acetyl-CoA. *A. niger* and several other fungi, however, possess a pyruvate carboxylase (EC 6.4.1.1), which forms oxaloacetate from pyruvate and carbon dioxide (OSMANI and SCRUTTON, 1983; BERCOVITZ et al., 1990; JAKLITSCH et al., 1991). This enzyme is also induced in the presence of high carbohydrate concentrations (FEIR and SUZUKI, 1969; HOSSAIN et al., 1984). The involvement of pyruvate carboxylase in the formation of citric acid has been proven by tracer experiments (CLELAND and JOHNSON, 1954) and by exit gas measurements from fermentations (KUBICEK et al., 1979b).

An essential difference of fungal pyruvate carboxylase from that of most other eukaryotic organisms is its localization in the cell: whereas in most eukaryotic organisms this enzyme is contained in the mitochondria (OSMANI and SCRUTTON, 1983; PUROHIT and RATLEDGE, 1988), in *Aspergillus* spp. it appears to be exclusively located in the cytosolic fraction (JAKLITSCH et al., 1991). This may be an essential advantage to acid accumulation: oxaloacetate cannot be transported by the mitochondria and, therefore, needs first to be reduced to malate. Since the tricarboxylate carrier uses dicarboxylic acids as partner substrates, this results in a stoichiometric relationship of efflux of citrate and the formation of malate. The activity of a cytosolic pyruvate carboxylase, therefore, links glycolysis directly to citrate efflux from the mitochondria.

## 2.3 The Activity of the Tricarboxylic Acid Cycle and Citric Acid Accumulation

Most of the previous work on the accumulation of citric acid concentrated on the role of the inhibition of a step in the tricarboxylic acid cycle, and conflicting reports were obtained. Whereas numerous workers claimed the necessity to inhibit a step in the cycle for the accumulation of organic acids, particularly citrate, other workers – using isolated mitochondria – provided evidence for the presence of an intact cycle (for review, see ROEHR et al. 1992; KUBICEK and ROEHR, 1986; KUBICEK et al., 1994). It is unknown whether or not a rise in the mitochondrial citrate concentration is necessary for citric acid accumulation, but it may occur as a consequence of the regulatory properties of some enzymes of the tricarboxylic acid cycle. While there is evidence now that aconitase is not inhibited under conditions of citrate overflow (KUBICEK and ROEHR, 1985), this cannot be ruled out for some other enzymes of the tricarboxylic acid cycle, since inhibition of NADP-specific isocitrate dehydrogenase by citrate (MATTEY, 1977; LEGISA and MATTEY, 1986), inhibition of NAD-specific isocitrate dehydrogenase by the "catabolic reduction charge" [NADH/(NAD+NADH)] ratio (KUBICEK, 1988a) and inhibition of 2-oxoglutarate dehydrogenase by oxaloacetate, *cis*-aconitate and NADH (MEIXNER-MONORI et al., 1985) have been reported. However, the importance of these inhibitions, if they occur *in vivo,* is unclear: Since citrate efflux is dependent on malate influx, one may envisage a mechanism by which the accumulation of citrate is solely caused by increased transport into the cytoplasm. Unfortunately, the properties of the tricarboxylate carrier of *A. niger* mitochondria have not yet been studied. Such an investigation is highly desirable since it may help to explain why isocitrate and *cis*-aconitate – despite of the equilibrium of aconitase – are not accumulated by *A. niger.*

It is also noted that steady-state sensitivity analysis showed the importance of mitochondrial citrate efflux on the rate of citrate accumulation (TORRES, 1994b).

## 2.4 Regulation of Product Formation by Environmental Conditions

Besides the sugar concentration and the type of sugar, as outlined above, as decisive factors for citrate accumulation, other environmental parameters considerably influence the amount of citric acid formed as well as the formation of other by-product metabolites such as gluconic and oxalic acid.

Environmental conditions favoring citric acid accumulation are:

- high sugar concentrations ($120–250$ g $L^{-1}$),
- deficiency in trace metals, especially Mn and Fe,
- high dissolved oxygen tension ($>140$ mbar),
- working pH between 1.6 and 2.2,
- phosphate concentration adjusted to minimum growth requirements,
- elevated ammonium ion concentrations.

The biochemical basis of their effects will be briefly discussed below.

### 2.4.1 Trace Metal Ions

The availability of trace metals, particularly $Mn^{2+}$ but also $Fe^{3+}$ and $Zn^{2+}$, has a significant impact on citric acid accumulation by *A. niger,* and a stringent deficiency particularly in manganese ions is obligatory for achieving high yields. Different explanations have been offered regarding the biochemical mechanism of action of trace metals (for review, see ROEHR et al., 1992). The role of manganese deficiency may reside in its impairment of macromolecular synthesis (KUBICEK et al., 1979a; HOCKERTZ et al., 1989), thereby inducing increased protein degradation (KUBICEK et al., 1979a; MA et al., 1985). As a consequence, elevated mycelial concentrations of $NH_4^+$ are accumulated which activate PFK 1, antagonize its inhibition by citrate, and hence accelerate glycolysis (KUBICEK et al., 1979a). Isolation of mutants of *A. niger* in which PFK 1 was partially citrate-insensitive, and which simultaneously were less sensitive to interference of accumulation of citric acid by $Mn^{2+}$ is in favor of this hypothesis (SCHREFERL et al., 1986). The reason for this impairment of macromolecular synthesis has been assumed to be an inhibition of ribonucleotide reductase, leading to impairment of DNA synthesis because of shortage of monomeric precursors of DNA replication (HOCKERTZ et al., 1988, 1989). However, several other effects of manganese deficiency have been noted, as on lipid composition, cell wall biosynthesis, and hyphal morphology, the relationship of which to disturbed DNA replication is unclear (for review, see KUBICEK and ROEHR, 1986). The latter effect is particularly striking and has been used as a criterion to predict the success of citric acid fermentation: Manganese-deficient grown mycelia are strongly vacuolated, highly branched, contain strongly enthickened cell-walls, and exhibit a bulbous appearance. This phenomenon may be related to a loss of orientation of apical growth by a disorientation of vesicle transport.

The well-known (and technically important, see below) ability of $Cu^{2+}$ ions to antagonize the deleterious effect of $Mn^{2+}$ appears to be due to inhibition of $Mn^{2+}$ uptake, which occurs by a specific, high-affinity transport system (SEEHAUS et al., 1990). The influence of other metal ions on the accumulation of organic acids by Aspergilli is even less clear: some workers have attributed a particularly strong influence to $Fe^{3+}$, which is not supported by others (for review, see KUBICEK and ROEHR, 1986). In view of the high concentrations of $Fe^{3+}$ needed for inhibition of citrate accumulation, the authors postulated that the manganese impurities may actually account for the observed effect (KUBICEK and ROEHR, 1986).

### 2.4.2 Dissolved Oxygen Tension and Aeration

Formation of all organic acids is increased under conditions of strong aeration. The necessity of high oxygen uptake is evident from the metabolic balance of citrate formation and the high sugar concentration used. It appears, however, that the capacity of the respiratory chain of *A. niger* is unable to deal with the high rate of formation of NADH. Moreover, the assembly of complex I from the respiratory chain (NADH:ubiquinone reductase) becomes disturbed under citric acid accumulating conditions (WALLRATH et al., 1991; SCHMIDT et al., 1992). This situation is compensated for by the induction of an alternative respiratory pathway, in which NADH reoxidation is not coupled to ATP-formation (ZEHENTGRUBER et al., 1980; KUBICEK et al. 1980; KIRIMURA et al., 1987). A yet unidentified component of this pathway must be stringently dependent on the maintenance of a high oxygen tension, since even short interruptions have been shown to impair its activity (KUBICEK et al., 1980). It is tempting to speculate that the importance of this mechanism is to aid in avoiding overproduction of ATP (see also below), which cannot be spent in biosynthetic reactions.

### 2.4.3 pH

The accumulation of citric acid by *A. niger* is markedly influenced by the pH, i.e., citric acid appears only in significant amounts at pH values $<2.5$. At higher pH values, oxalic and gluconic acid are formed as by-products. The external pH has only a small influence on the cytosolic pH (LEGISA and KIDRIC, 1989). It is unlikely that these small changes cause inactivation of intracellular enzymes, but they may influence fine regulation, as some enzymes have been proven to differ in their kinetic and regulatory properties when the pH changes only within a single unit (BANUELOS et al., 1977).

While the formation of gluconic acid at higher extracellular pH values is easily explainable by the effect of pH on the stability of the extracellular glucose oxidase (see Chapter 10 of this volume), the accumulation of oxalic acid still deserves explanation. Biochemical studies using $^{14}C$-labeling have shown that oxalate originates from oxaloacetate, and oxaloacetate hydrolase has been demonstrated to be located solely in the cytoplasm (KUBICEK et al., 1988). The enzyme, therefore, competes with malate dehydrogenase for the product of the pyruvate carboxylase reaction, and the relative activities of the two enzymes *in vivo* determine which acid is ultimately formed. Oxaloacetate hydrolase is specifically induced by raising the pH of the medium to at least 5 in the presence of nitrogen and phosphate (KUBICEK et al., 1988). However, given the fact that the activity of the cytosolic malate dehydrogenase isoenzyme depends on the ratio of cytosolic NADH/NAD, oxalic acid formation may also be influenced by the efficacy of NADH reoxidation.

Yet another perspective of the effect of pH has recently been highlighted (ROEHR et al., 1992): Considering the surplus of 1 ATP and 3 NADH arising from the accumulation of citric acid under conditions of equimolar conversion of glucose to citrate in the late stages of citric acid fermentation (ROEHR et al. 1981), it is likely that this may be used – or even be necessary – to maintain the pH gradient between the cytosol and the extracellular medium. MATTEY et al. (1988) have established the involvement of a plasma membrane-bound ATPase in the maintenance of the pH gradient in citric acid producing *A. niger*. The authors favor the idea that the low pH may be required for citric acid biosynthesis as a valve to spill over the surplus of ATP, which otherwise would lead to a metabolic imbalance and would stop acidogenesis. The induction of oxalic acid biosynthesis at higher extracellular pH values would be consistent with this assumption, since oxalic acid formation produces neither a surplus of ATP nor NADH. This hypothesis, however, still lacks experimental proof.

### 2.4.4 Effect of Phosphate and Nitrogen

It is evident from the literature that the concentrations of both phosphate and nitrogen are of influence in media optimally designed for organic acid production by Aspergilli. But for the accumulation of citric acid in batch cultures the balance of nitrogen, phosphate and trace metals appears to be important (SHU and JOHNSON, 1948b). Little is known regarding the biochemical basis of these observations. DAWSON et al. (1989), using fed-batch fermentation as experimental approach, claimed evidence for the regulation of citric acid production by nitrogen catabolite repression. However, in view of the molecular genetic knowledge available on this type of wide-domain control in *A. nidulans* (reviewed by CADDICK, 1992), this claim requires genetic confirmation.

# 3 Production Procedures

Since the publication of the chapter on citric acid in the 1st Edition of *Biotechnology* (ROEHR et al., 1983), the technology of citric acid production has been reviewed by several authors, e.g., BIGELIS (1985), MILSOM and MEERS (1985), MILSOM (1987), VERHOFF (1986), KUBICEK and ROEHR (1986), GOLDBERG et al. (1991), ROEHR et al. (1992), BIGELIS and ARORA (1992, SMITH (1994). NOYES (1969) provides a still useful and comprehensive account of relevant U.S. patent literature covering the period of 1944–1968 (see also LAWRENCE, 1974).

## 3.1 Production Strains

As citric acid manufacturers generally keep their strains and methods of strain selection secret, information on the origins of strains and specific selection procedures is hardly available.

Basically, selection of strains begins with isolation from natural habitats according to common microbiological methods. Several of such strains have been incorporated in governmental or industrial strain collections. It is known to most microbiologists that *Aspergillus niger* can be enriched from soil by soaking samples in a nutrient medium containing about 20% tannin as the main carbon source and incubating for about 2–3 weeks (RIPPEL-BALDES, 1940). Since *A. niger* possesses the enzymes necessary for tannin utilization – a property that has also been used industrially in the past – enrichment cultures with this substrate usually result in rather pure cultures of wild-type *A. niger.*

After applying the usual pure-culture techniques, production strain selection may be performed according to the following principles:

The basic assumption is that in a population of numerous fungal spores there is a distribution with respect to certain properties. Thus only a small number of spores are potent producer cells to be isolated. According to WENDEL (1957) there are two principal methods of selecting portions from a given population, the *single-spore method* and the *passage method.*

**The single-spore technique.** A great number of single cultures are prepared from suitable dilutions of spore suspensions. Spores from these cultures are used to inoculate, e.g., flasks containing standardized media with the respective carbon source, molasses in many cases. Such assays may be performed as surface or shake-flask cultures with titration of the acid formed after a given period of time. This procedure is only random for sure and most laborious considering that one conidial head of *A. niger* will yield several 10,000 spores. However, subculturing and reassaying of superior cultures obtained in this way may increase the success of the procedure considerably. In many cases high-yielding strains are obtained after a few cycles. Similar methods are applied to counteract the phenomenon of "degeneration", i.e., the slow change of distribution of the acid producing capability with storage time as well as with consecutive spore propagations and transfers of stock cultures.

Several attempts to facilitate this technique of strain selection have been described. The method of JAMES et al. (1956), improving earlier trials of DAVIS (1948) and QUILICO et al. (1949), consists of growing a certain number of single-spore cultures in a petri dish applying filter paper disks soaked with the medium and incorporating a pH indicator to detect the acid zones formed upon mycelial development. Dependent on the inorganic nutrients applied, however, this method may also register acidification by the utilization of physiologically acid salts (e.g., ammonium sulfate), leading to erroneous results as apparently obtained by BONATELLI and AZEVEDO (1982). Misinterpretation may be avoided either by carefully choosing suitable combinations of salts and indicators (AMIRIMANI and ROEHR, unpublished) or by using a specific reagent (*p*-dimethylaminobenzaldehyde) to detect citric acid formation (ROEHR et al., 1979).

Examples of strains obtained by the application of single-spore techniques are the so-called Wisconsin strain 72 (ATCC 1015), parent strain of Wisconsin 72-4 (ATCC 11414), the latter being a simple subculture of the former, used by the Wisconsin group (PERLMAN, 1949; SHU and JOHNSON, 1948a, b) and by Canadian workers (MARTIN and WATERS, 1952; CLARK et al., 1966). In Europe, the first industrial strains came from Vienna (SZÜCS, 1925; BERNHAUER, 1928) and from Prague (BERNHAUER, 1928). These strains found applications in the plants of Kaznèjov near Plzen, Czech Republic, in that of Benckiser GmbH, Ladenburg, Germany, and that of Citrique Belge, as well as in Russia and the USA (ATCC 10577).

**The passage method.** It basically consists of plating a spore suspension on solid media containing varying concentrations of a substance affecting the population with respect to germination time or growth rate. This may result in the selection of certain parts of the population displaying distinct properties, either based on scientific principles or being purely empirical. Examples are: resistance to low pH (CURRIE, 1917; BERNHAUER, 1929), to elevated concentrations of citric acid (Montan- und Industrialwerke, vorm. J. D. Starck, 1933; BERNHAUER, 1934; LEOPOLD, 1958), to elevated temperatures (LEOPOLD and VALTR, 1970), to deoxyglucose (KIRIMURA et al., 1992), adaptation to elevated sugar concentrations (DOELGER and PRESCOTT,

1934; SCHREFERL-KUNAR et al., 1989; XU et al., 1989a), selection and adaptation to a certain carbon source designated as raw material (PELECHOVA et al., 1990).

These procedures are often implemented in combination with single-spore techniques to improve strains and to provide different strains for specific applications.

As pointed out above, it is difficult to maintain the desired properties of production strains over longer periods of time without the risk of serious losses of viability and acid producing capabilities.

Various methods have been described for strain maintenance since no generally applicable procedure can be recommended. One method frequently applied in practice includes cultivation on a suitable sporulation medium, harvesting the spores aseptically in a dry state (e.g., by admixing inert powdery materials), and carefully drying the spores with dry air or by placing them into a desiccator with suitable drying agents. Finally, the spores are kept in sealed glass vessels, eventually with admixed sterile soil, quartz sand, or iron-free activated carbon, at low temperatures or at room temperature. Preservation at extremely low temperatures has also been recommended (e.g., SIMONYAN et al., 1988) but except patent literature little information is available.

Stability over periods of several years usually is attainable with these procedures, but frequent assaying of samples by spore propagation and small-scale fermentation trials is recommended.

Mutagenic treatment as a means of strain improvement has been described by several authors. Unfortunately, rather low producing strains were used, thus a great deal of published work is of little practical value. UV irradiation is the simplest and, therefore, the preferred method. Spores are irradiated either in suspension ($10^7$–$10^8$ $mL^{-1}$) or as agar cultures using ordinary germicidal lamps. According to GARDNER et al. (1956), UV mutants superior to the parent strain (Wisconsin 72-4) were growth-restricted multistep mutants. ILCZUK (1968, 1970), on the other hand, presented evidence that growth restriction due to auxotrophic mutations was the reason for elevated acid producing capabilities. Several authors have described correlations between elevated citric acid production and mutagen-induced morphological changes in *A. niger*.

Another possible method of strain improvement could be the parasexual cycle, as first described for *A. niger* by a PONTECORVO et al. (1953). So far, however, heterokaryosis as well as di(poly)ploidy have contributed little to the improvement of industrial strains (CHANG and TERRY, 1953; CIEGLER and RAPER, 1957; ILCZUK, 1971a). According to reports of ILCZUK (1971b) and DAS and ROY (1978), diploids display higher citric acid yields as their parent haploids, but tend to be less stable (BONATELLI et al., 1983).

Protoplast fusion, as described for *Aspergillus* including *A. niger* by the group of FERENCZY and KEVEI in Szeged, Hungary (cf. FERENCZY et al., 1975a, b; FERENCZY and FARKAS, 1980; FERENCZY, 1984) and by the group of PEBERDY in Nottingham, UK (cf. ANNÉ and PEBERDY, 1975, 1976; PEBERDY and FERENCZY, 1985; PEBERDY, 1987; POWELL et al., 1993) appears to be a promising tool (see also GADAU and LINGG, 1992) to extend the range of genetic manipulation of *A. niger* with respect to citric acid production. Some recent publications are cited which could stimulate further work:

KIRIMURA et al. (1986, 1988a) have studied protoplast fusion of production strains and were able to obtain fusants with acid production capacities exceeding those of the parent strains in solid state fermentation, but not in submerged operation. Using the well-known polyethylene glycol method of fusion, fusion frequencies of about 5% could be attained (see also MARTINKOVÁ et al., 1990). Haploidization of fusant diploids (KIRIMURA et al., 1988b; OGAWA et al., 1989) yielded strains that were also superior in submerged fermentation.

An enormous amount of work has been devoted to the study of the molecular genetics of Aspergilli including *A. niger* since the early 1980s. Main achievements to be expected could be: the development of novel strains with useful properties such as resistance to detrimental constituents of fermentation raw materials, capability of utilizing unconventional raw materials (starch, cellulose, pectin containing and other waste materials, e.g., whey, etc.), improved fermentation performance. Further goals would be to gain a more thorough and detailed understanding of the regulatory mechanisms of citric acid accumulation and excretion, which might in turn aid in strain improvement, e.g., by combining the capabilities of yeasts as unicellular organisms with those of *A. niger*. Accounts of such work are found in recent comprehensive publications, e.g., BENNETT and LASURE (1985, 1991), NEVALEINEN and PENTTILÄ (1989), BENNETT and KLICH (1992), POWELL et al. (1994), MARTINELLI and KINGHORN (1994).

## 3.2 Spore Propagation

Large-scale spore propagation is one of the most important and crucial operations in successfully running a citric acid plant. It must be considered that, e.g., in a 20,000 t $a^{-1}$ plant 30–300 kg of stable and excellent spores are required. In contrast to many other fermentation processes with several preculture stages inoculation of a citric acid fermentation normally comprises direct inoculation of production vessels with spores or only one stage for preforming mycelia to seed the main fermenter.

These large amounts of spores are produced by growing the respective production strains on heavy agar media in shallow trays in compact chambers under conditions of controlled aeration and humidity. Highest levels of sterility have to be maintained, which is done by steaming certain parts of the system and applying high-quality disinfectants. Sporulation media usually contain carbon and nitrogen sources in proportions that support rich sporulation, e.g., molasses and malt extract, peptones, etc. Care has to be taken to exclude detrimental metal ions from these media. Inoculation of the sterile sporulation media is performed by blowing in spores from a preserve (e.g., 10 mg of spores per $m^2$). Sporulation proceeds at 30°C in a laminar (below 0.5 m $s^{-1}$) stream of sterile air, providing about 70–90% humidity, for periods up to 6 d resulting in a velvety, undulated lawn of a densely sporulating mycelium. By successively applying dry air, the spores are dried to enhance stability upon storage. Yields of dry spores are in the range of 60–120 g $m^{-2}$. The dried spores are harvested carefully by a cyclone and stored at a cool dry place. Although information and experience is scattered, high acid producing capabilities in such mass spore preserves may be maintained for periods of several years. It is necessary, however, always to check the potency of a mass spore preserve before going into plant operation.

Several other means of spore propagation have been described, e.g., the use of bran (FOSTER, 1949) or grainy materials such as maize grits (SCHWEIGER, 1965) or other supports soaked with nutrients. Liquid sporulation media have also been recommended.

## 3.3 Raw Materials

Various carbohydrate materials (see also Chapters 1a and 1b of this volume) may be used in citric acid production. From a practical point of view they may be classified into two groups:

(1) Raw materials with a low ash content from which the cations can be removed by standard procedures. Examples of these are various qualities of cane or beet sugar and more or less refined starch hydrolyzates such as dextrose syrups and crystallized dextrose. Since these are mainly used in submerged fermentation their treatment will be described in the respective section (3.6).

(2) Raw materials with a high ash content and high amounts of other non-sugar substances, such as cane and beet molasses, the respective sugar juices including so-called high-test molasses, and crude unfiltered starch hydrolyzates.

Besides these rather cheap raw materials, even cheaper waste materials such as whey, sweet potato effluents, cassava meal, wheat flour, wheat bran, or sugar cane pressmud have been proposed as feedstocks. The utilization of whey or whey permeates has been studied especially in New Zealand (HOSSAIN et al., 1983; MADDOX and ARCHER, 1984; cf.

also SOMKUTI and BENCIVENGO, 1981). In most cases other raw materials are only of local significance and rarely available in consistent quality.

As pointed out earlier, an enormous amount of labor (concomitant with a huge amount of respective publications) has been invested in studies on the composition, suitability for citric acid manufacture (including respective predictabilities), and special treatments of molasses.

The composition of molasses depends on various factors, e.g., the kind of beets or cane, methods of cultivation (fertilizers, pesticides), conditions of storage, production procedures, methods of handling (e.g., transport, temperature variations), preceding applications, etc. A number of authors have tried to fractionate certain molasses to characterize components harmful to citric acid fermentation (e.g., ZAKOWSKA and JALOCHA, 1984; DRURI and ZAKOWSKA, 1986, 1987; ZAKOWSKA and GABARA, 1991). In spite of this and the existence of various methods of analyses, it is still necessary to evaluate the suitability of a special brand of molasses with respect to yield, productivity and adjustments for operation by appropriate trials on the laboratory or pilot plant scale.

Cane molasses (blackstrap molasses) with a high content of ash (7–10%) and organic non-sugars (20–25%) is considered less suitable for citric acid manufacture.

Beet molasses, a preferred material in Europe, should contain about 80% solids, with 50% sucrose and about 25–30% non-sugars. More recently, however, due to efforts of the beet sugar manufacturing industry to increase sugar yields from the respective juices by novel processes, molasses have come onto the market (e.g., the so-called Quentin molasses) with sugar contents of considerably less than 50% and hence higher contents of non-sugars rendering these less suitable for citric acid manufacture. More novel methods of sugar production and refining even remove almost all sugar from the respective raw materials without producing sugar containing by-products.

A noteworthy concern, hitherto considered less significant, is the fact that with the application of cheap raw materials such as molasses, enormous loads of inorganic substances are introduced into the waste water streams (cf., e.g., WELLER, 1981). In some cases this already excludes the use of such raw materials in certain areas demanding considerable changes of technology.

Concerning sources of other nutrients, it is interesting to note that, in principle, the basic design of CURRIE (1917) has never been modified substantially. CURRIE's medium contained 2–2.5 g $L^{-1}$ $NH_4NO_3$, 0.75–1.0 g $L^{-1}$ $KH_2PO_4$, 0.20–0.25 g $L^{-1}$ $MgSO_4 \cdot 7H_2O$. In modern industrial protocols, cheaper N and P sources such as gaseous ammonia and ammonium hydroxide and phosphoric acid, respectively, may be preferred. Trace metals, especially zinc, iron, and manganese were introduced as contaminants of nutrient chemicals and by the process water. In highly purified media these have to be added in specified amounts (see below).

## 3.4 Koji Process

The simplest process for citric acid production has been developed in Japan (cf. YAMADA, 1965). It largely resembles a solid-state fermentation, based on the know-how of the traditional Koji process. Raw materials are starch containing pulps from starch manufacture, e.g., fibrous residues of sweet potatoes, rice or wheat bran and similar substrates. These are produced abundantly in Japan – e.g., about 60,000 t of residues from sweet potato processing equivalent to 30,000–40,000 t of starch – which are priced similarly to molasses.

The starchy material is soaked with water to a water content of 65–70% and treated with steam to liquefy the starch. The resulting almost sterile paste is placed in trays or on the floor of fermentation compartments and inoculated by spreading spores of *A. niger* over this material. In case of less fibrous materials, bran or sugar cane bagasse may be admixed. Trays very often can only hold about 10 L (e.g., 0.6 by 0.4 by 0.05 m), thus thousands of these are required – and have to be handled – even for small production capacities. At 30 °C and pH values around 5.5 saccharification is brought about by the fungal amylases, followed by conversion to citric acid, with a marked decrease of pH to about 2. After 1–2 weeks the acid containing mass is crushed and extracted with warm water in simple countercurrent extractors. The resulting solution containing about 4–5% citric acid is further processed according to standard methods.

The successive saccharification and acidification to citric acid by a single strain of *A. niger* requires special strains with doubled peak capabilities. In this context it is noteworthy to mention that such strains have been selected and investigated (e.g., JIN et al., 1993) – also for submerged fermentations (RUGSASEEL et al., 1993). In the light of the complexity of this technique, the advantages of such a combined process appear rather questionable. In addition, these strains should tolerate inevitably higher concentrations of heavy metals in many waste materials. Special strains have been described to serve this purpose. On the other hand, it has been reported that the influence of metal ions is hardly critical in these types of fermentation processes (cf. SHANKARANAND and LONSANE, 1994).

Similar processes were described (cf. CAHN, 1935) for using materials such as cane sugar and bagasse (LAKSHMINARAYANA et al., 1975) and cane molasses (CHAUDHARY et al., 1978) in India.

More recent developments comprise the use of other waste materials such as pineapple wastes (YO, 1975; USAMI and FUKUTOMI, 1977), grape residues (cf. HANG, 1988), and apple pomace (XU and HANG, 1988; HANG and WOODAMS, 1987).

It is obvious that such processes are primarily appropriate to satisfy smaller demands for citric acid in decentralized economic systems. Production figures are hardly available, but overall production may be estimated to be several thousand tons only.

## 3.5 Surface Process

In the classical process for manufacturing citric acid the culture solution is held in shallow pans and the fungus develops as a mycelial mat on the surface of the medium (Fig. 2). This should create a rather large contact area between liquid phase, mycelium and air supplying the oxygen required. Mixing of the liquid phase in the course of the process largely occurs through temperature and concentration gradients along the liquid depth.

According to this, the system consists of fermentation rooms in which a large number of trays are mounted one over the other in stable racks. Trays are usually made of high-purity aluminum or of special qualities of stainless steel. Tray sizes vary from 2 m × 2.5 m × 0.25 m to 2.5 m × 4 m × 0.25 m with usable liquid depths of 0.1–0.18 m. By this liquid volumes with weights of 0.5–2 t per tray must be supported by the rack construction and have to be handled following fermentation. Provision is made for continuous filling by appropriate overflow devices.

The fermentation chambers are provided with an effective air circulation system which mainly serves the purpose of temperature regulation and only to a lesser extent to oxygen supply and humidity control. Air is introduced into the chamber as an almost laminar stream through inlets evenly placed at all levels of the chamber. Wall and floor covering materials should be washable and resistant to disinfectants as well as acids, doors should be tightly closed. The fermentation chambers should be cleaned aseptically by the use of suitable liquid (spray) and gaseous disinfectants (e.g., sulfur dioxide or formaldehyde) and by frequent cleaning of walls and floors with liquid disinfectants (caustic soda or formaldehyde solutions).

According to the authors' experience, however, this is not always obeyed sufficiently by personnel entering the chambers. Contamination is mainly caused by Penicillia, other Aspergilli, yeasts, and lactic acid bacteria.

Molasses which is generally employed in the surface process is treated as follows:

The raw material is diluted to a sugar concentration of 120–180 kg m$^{-3}$, nutrients are added, and the solution is acidified, e.g., with phosphoric acid to a pH of 6.0–6.5 and heated to about 90 °C for 15–45 min. Subsequently, potassium hexacyanoferrate (HCF) is added to the hot solution. As mentioned above, HCF has a two-fold effect: precipitating or complexing harmful metal ions (Fe, Mn) and

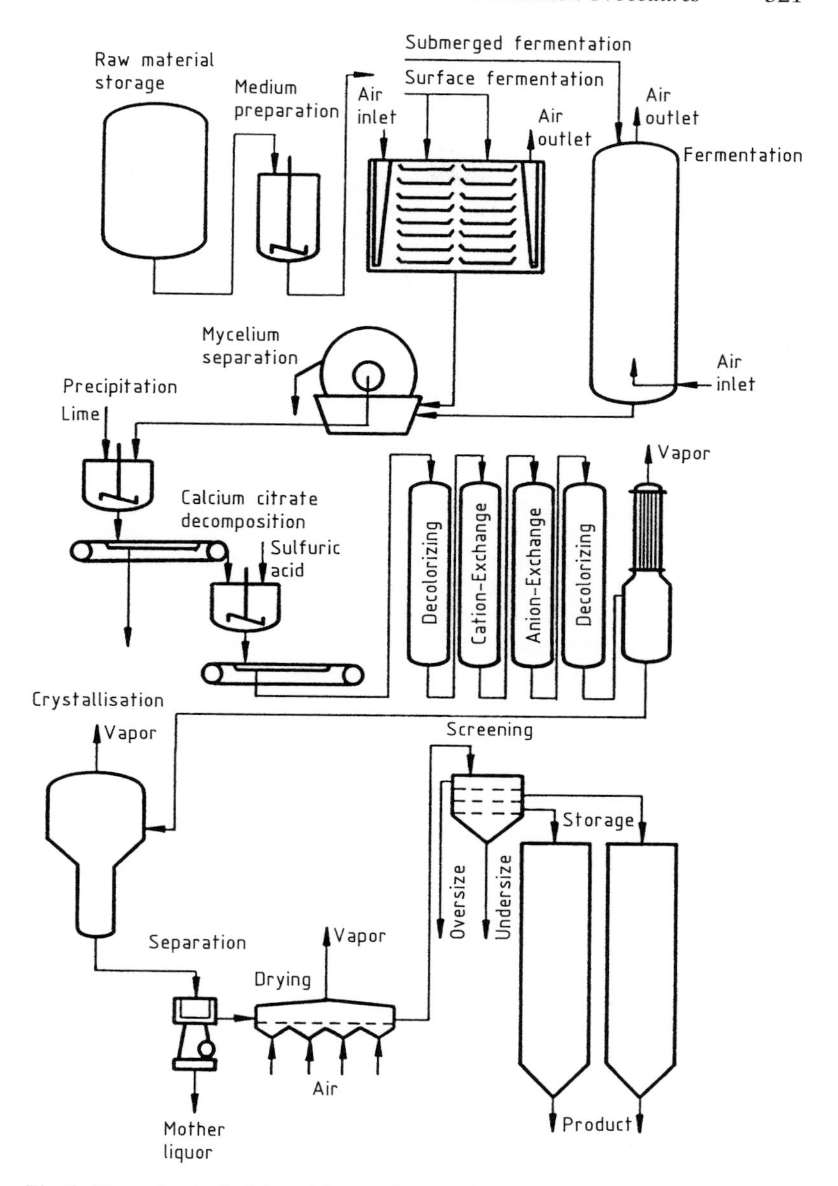


**Fig. 2.** Flow-sheet of citric acid manufacture by surface or submerged process (from ROEHR et al., 1992, with permission).

acting as a metabolic inhibitor restricting growth and promoting acid production (cf. MARTIN, 1955; CLARK, 1964). The amount of HCF needed for the complexing action in order to achieve the necessary excess of HCF as growth inhibitor has thus to be determined. Since this so-called HCF demand depends on the quality and brand of molasses available, it is preferred practice to determine the HCF demand by small-scale trials with varying additions of HCF. The amount of "free" HCF upon treatment with a sufficient amount of

HCF may be determined by the method of MARIER and CLARK (1960). HPLC methods have also been elaborated (SELAHZADEH and ROEHR, unpublished results). The HCF excess varies between 5–100 mg $L^{-1}$ for beet molasses and 5–200 mg $L^{-1}$ for cane molasses (cf., e.g., CLARK, 1964). The exact amount has to be determined by small-scale fermentation experiments according to the strain used.

The molasses solution is stored hot or is immediately used for charging the trays while cooling to approximately 40 °C. Beginning at this stage it is advisable to blow humid air through the system to reduce condensation of water. The rate of cooling is decisive for the fermentation yield (SCHMITZ, 1976), rates of less than 1 °C per hour being least detrimental.

Inoculation is performed in different ways: The necessary number of conidia (about 100–150 mg $m^{-2}$) may be introduced into the system either as a suspension or by mixing dry spores with the air stream blown over the trays. Germination requires 1–2 d during which a decrease in temperature may be counteracted by increasing the temperature of the moist air stream. During this and the preceding cooling step, avoidance of contamination is crucial (cf. MESSING and SCHMITZ, 1976). MESSING and WAMSER (1975) have, therefore, suggested to use specifically sterilized and conditioned air (about 1–2% of total) at this stage and to apply commonly filtered unconditioned air during the main fermentation period.

Following spore germination a wrinkled, multiply folded mycelial mat is formed with a large specific surface to promote the necessary metabolic activities of citric acid production. Care has to be taken that this rather dense mat does not become submerged since this would seriously impair acid production. With the beginning of mycelium formation, citric acid formation accompanied by a drop of pH to values $<2.0$ sets in, and the process may be completed within 6–8 d. Mixing of the fermentation liquid is brought about by temperature differences which may amount to 3 °C along the liquid depth.

During active fermentation citric acid is produced at an average rate of about 0.5 kg CAM $m^{-3}$ $h^{-1}$ (CAM denotes citric acid monohydrate, F.W. 210.14, as largely sold, e.g., in Europe; CAA denotes citric acid anhydrate, F.W. 192.13, sold, e.g., in the USA. The conversion factor is 1.094). At the same time, heat in the range of 600–1,100 kJ $m^{-2}$ $h^{-1}$ has to be removed from the system by air. Owing to the small heat capacity of air this entails an aeration rate of at least 4 volumes of air per volume of medium per minute (4 vvm). It may thus be calculated that aeration in the surface process mainly serves the purpose of cooling. On the other hand this calorifering process comprises the evaporation of substantial amounts of water from the medium, ca. 30–40%. Thus, with a fermentation yield of 75% citric acid based on an initial sugar concentration of 160 kg $m^{-3}$, the fermented liquid may have a citric acid concentration of 200–250 kg $m^{-3}$ (instead of 120).

At the end of fermentation, the trays are removed from the fermentation chambers and the contents are transferred to the purification and recovery plant. Emptying of trays may be performed by placing them into large rotary racks which are turned by 180° to release the contents into transport channels, and the trays can be prepared for the next run.

## 3.6 Submerged Process

Citric acid manufacture with the use of submerged fermentation (see Fig. 2) is constantly increasing in spite of being one of the more sophisticated technologies. It has been estimated that about 80% of world production is manufactured by submerged fermentation. Several advantages such as higher yields and productivities as well as lower labor costs have justified the fact that various operational parameters had to be reinvestigated in order to implement this technology successfully. The organism is extremely sensitive to traces of transition metal ions, especially iron and manganese (CLARK et al., 1966). Critical concentrations of iron and manganese are 1 mg $L^{-1}$ (cf. SNELL and SCHWEIGER, 1949) and 2 μg $L^{-1}$, respectively. This must be taken into consideration when choosing the material of construction of the fermenter and peripheral installations as well as when even planning the water supply of the plant. It is also necessary to carefully select the various chemicals used as nutrients and to set up larger stocks of suitable lots of these chemicals. Appropriate treatment of the carbon substrate is indispensable and will be treated in detail below. Maintaining the oxygen concentration above 25% of saturation is also required and interruptions in oxygen supply (KUBICEK et al., 1980) may be quite harmful.

Submerged citric acid fermentation may be performed using conventional stirred reactors as well as tower fermenters. Increasingly, the latter is the preferred type of bioreactor owing to several advantages over the stirred reactor: its lower price, the possibility to build larger reactors, operation without large rotating units with less risk of contamination and less heat production in the reactor, better conditions for working with suspended solids. Stirred fermenters have capacities of 50 to several 100 $m^3$, whereas tower fermenters may be even larger – up to 1,000 $m^3$.

Owing to the sensitivity of *A. niger* to traces of heavy metal ions that could be dissolved from the materials of construction and due to the rather corrosive action of citric acid it is advisable to use a particularly acid resistant steel for all parts of the fermentation system. The best material is stainless AISI 316 Ti (1.4571, DIN X10CrNiMoTi1810) or similar specifications. In addition, the fermenters and peripheral vessels should be passivated by common procedures, e.g., by filling with diluted solutions of sulfuric and nitric acid and aerating for about 20 h.

Fermenters for citric acid fermentation do not have to be built as pressure vessels since "sterilization" is usually performed by simply steaming the previously cleaned reactors without applying pressure (0.2 bar gauge only). Cooling can be done by an external water film over the entire outside wall of the fermenter.

Aeration is accomplished by the common dispersion devices of the fungal fermentation industry. Tower fermenters usually have a ratio of height to diameter of 4:1 to 6:1 thus increasing the solubility of oxygen by the increased hydrostatic pressure. In general, it is relatively easy to maintain oxygen concentrations above 25–50% of saturation throughout the process. Mixing may be less effective in tower fermenters, but seldom presents serious problems. Draught tubes may be built in, but their effect is limited when fed-batch operation is carried out.

The fermenters may be inoculated in different ways. Spore suspensions containing Tweens to facilitate even dispersion are made to provide final spore concentrations of $5–25 \times 10^6$ spores per liter. Preincubation for 6–8 h may shorten the first stage of fermentation. Alternatively, mycelium may be pregrown in a seed fermenter (TVEIT, 1963). The preferred type of production mycelium is in the form of pellets or flocks of uniform size.

As mentioned above, the kind of raw material used as carbon source largely determines the way of operating the process. In the following, some examples may illustrate this and the necessary operational details.

### 3.6.1 Molasses Batch and Fed-Batch Fermentations

Although first experiments to implement the method of submerged fermentation were done with pure sugars, processes using molasses as the cheapest raw material are the most studied of all citric acid manufacturing operations. The disadvantage of molasses is its high content of non-sugar substances which precludes the application of higher sugar concentrations. As with ethanol production, fermentation speed is drastically reduced under conditions of elevated concentrations. Hence, presently batch fermentations are only used in cases with less concentrated raw materials available. The preferred method is fed-batch operation. Proceeding on the assumption of a possible molasses concentration of 250 kg $m^{-3}$ in a batch process, corresponding to a sugar concentration of about 120 kg $m^{-3}$, fed-batch processes may be run with an average sugar concentration of 150 kg $m^{-3}$: e.g., starting with half a fermenter of pretreated molasses solution (with excess HCF, see above) with a sugar concentration of 100 kg $m^{-3}$ and subsequently feeding molasses containing 200 kg sugar $m^{-3}$, citric acid concentrations of 130 kg CAM $m^{-3}$ may be obtained within 6 d, corresponding to a productivity of 0.9 kg CAM $m^{-3} \cdot h$. Obviously, such procedures may be varied within certain limits. According to HUSTEDE and RUDY (1976) the necessary amount of HCF may advantageously be added in certain increments, i.e., treating the molasses with less than the estimated amount of HCF and making further additions at different stages of the fungal development.

### 3.6.2 Pure Sugar Batch Fermentations

As mentioned above, processes with the use of pure sugars were the first to be exploited but these were seldom implemented due to the higher prices of raw materials. Nevertheless, the exploitation of such procedures gave rise to interesting process variants being most important today.

One variation of the theme of pretreating molasses as described above was the introduction of ion exchange to prepare fermentable molasses sugar solutions (PERLMAN et al., 1945; WOODWARD et al., 1949). Apart from the fact that similar procedures are presently used by the sugar manufacturing industry, it was found that relatively pure sugar raw materials such as cane or beet sugar as well as starch hydrolyzates could be freed from harmful metal ions by treatment with industrial cation exchange resins. According to the above-mentioned patent (see also SNELL and SCHWEIGER, 1949), no addition of inhibiting substances is necessary if the concentration of iron is held $<1$ mg $L^{-1}$; in other cases it has been found that only small amounts of inhibitory HCF (ca. 1–10 mg $L^{-1}$) are sufficient to trigger transition from growth of the fungus to citric acid production. Suitable qualities of raw materials comprise:

- Sucrose: white beet or cane sugar, EU category II, with minimum 99.3% sucrose, maximum 0.03–0.05% water, maximum 0.03% ash (conductometric determination).
- Glucose: crystalline monohydrate, pharmacopoeia quality (as a guideline), with minimum 91% glucose, maximum 8–10% water, or anhydrate of similar quality, or glucose syrups of 90–97 DE.

As will be reported below, even less purified sugars, e.g., crude maize grits hydrolyzates (cf. SWARTHOUT, 1966) may be used under certain conditions.

In a typical process, 300 kg $m^{-3}$ sugar solutions are prepared at 40 °C, the pH is adjusted to 1.6–1.8 with sulfuric acid, and the solutions are applied to a strong cation exchanger. Ion exchange should reduce the content of iron and manganese ions to levels $<200$ $\mu g$ $L^{-1}$ and 5 $\mu g$ $L^{-1}$, respectively. Subsequently the decationized solutions are filtered, pasteurized, and transferred to the fermenter where they are diluted with decationized water. It is necessary to make sure that the process water used has to be checked for trace metals before its use in such fermentations. Initial concentrations of sucrose solutions may be 160–250 kg $m^{-3}$, whereas in many cases glucose concentrations should be about 120–160 kg $m^{-3}$. Care should be taken to allow as little inversion of sucrose as possible, since the presence of invert sugar may result in prolonged fermentation times. After addition of nutrient salts and HCF (see above), the pH is adjusted to an initial value of 2.5–3.0 by the addition of caustic soda or more advantageously by adding gaseous ammonia or ammonium hydroxide as part of the nutrient supply. Now the system is inoculated with spores or pregrown mycelium as described above. The amount of, e.g., spores may vary within a range of $5–25 \cdot 10^6$ spores per liter. Until germination of spores, aeration is maintained at a level of about 0.1 vvm. During active fermentation aeration should be performed such that oxygen saturation of ca. 50% is maintained. During the period of initial mycelial growth and pellet formation the morphology of the mycelium is characterized by abnormally short, forked, bulby ("knobby") hyphae with areas of swollen cells showing more granulations and vacuoles (Fig. 3a). This characteristic morphology of restricted growth, a prerequisite for abundant citric acid production, is brought about by the inhibitory action of HCF and the reduction in concentration of iron and manganese. If the raw materials are not treated properly, or the purity

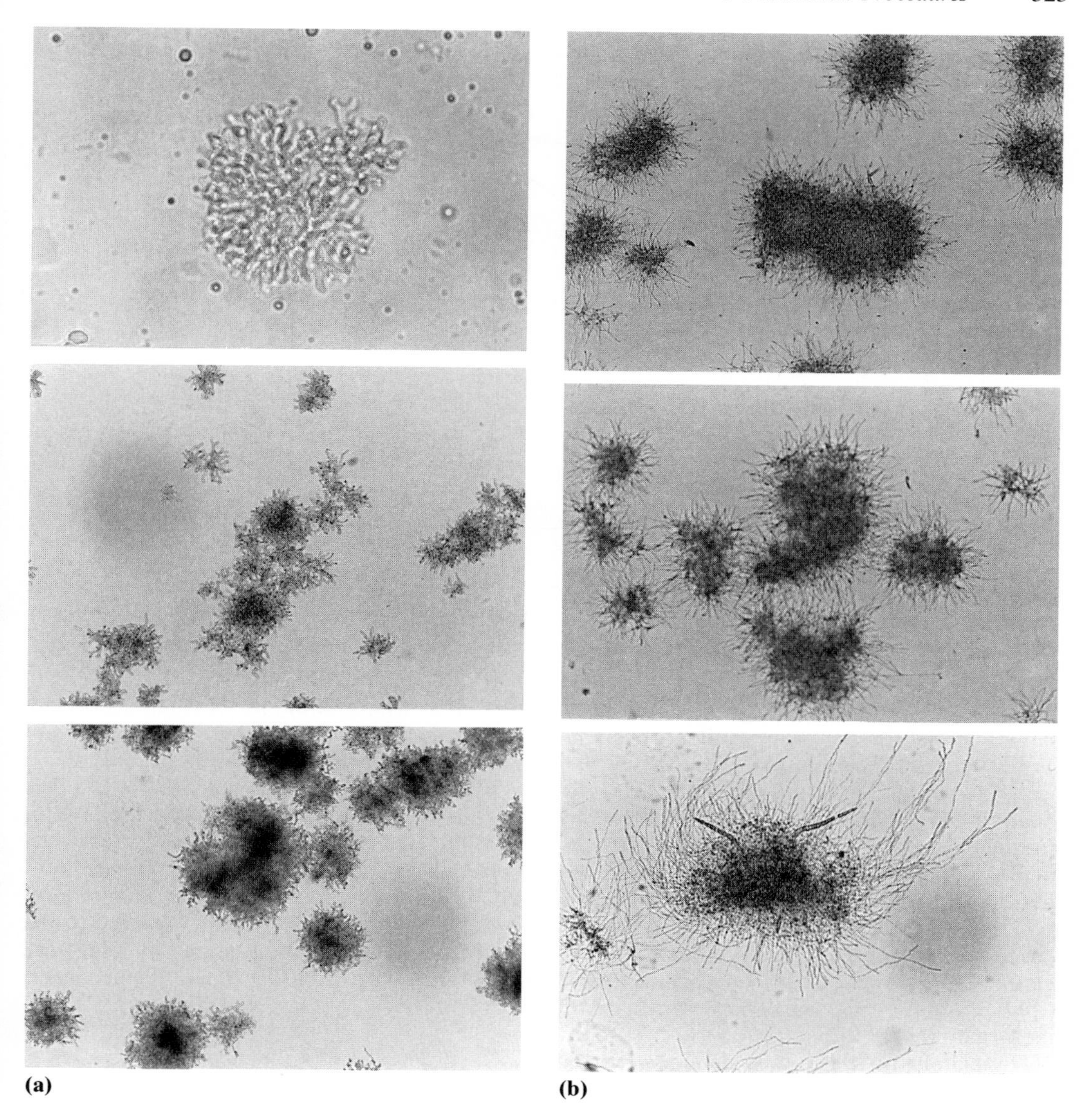

**(a)** **(b)**

**Fig. 3.** Morphology of production mycelium of *A. niger* under conditions of low (a) and elevated (b) concentrations of Mn and Fe.

of nutrient chemicals is insufficient, or if the material of construction gives rise to elevated concentrations of these ions, unrestricted growth will set in. This is characterized by the appearance of long, unbranched hyphae forming spider-like aggregations of hyphal threads (Fig. 3b). When only first tendencies to form such morphology are observed under the microscope, it is sometimes possible to counteract this phenomenon by proper additions of inhibitor.

The formation of pellets or flocks of this special kind of mycelium requires about 2–3 d and is accompanied by a drop in pH to values

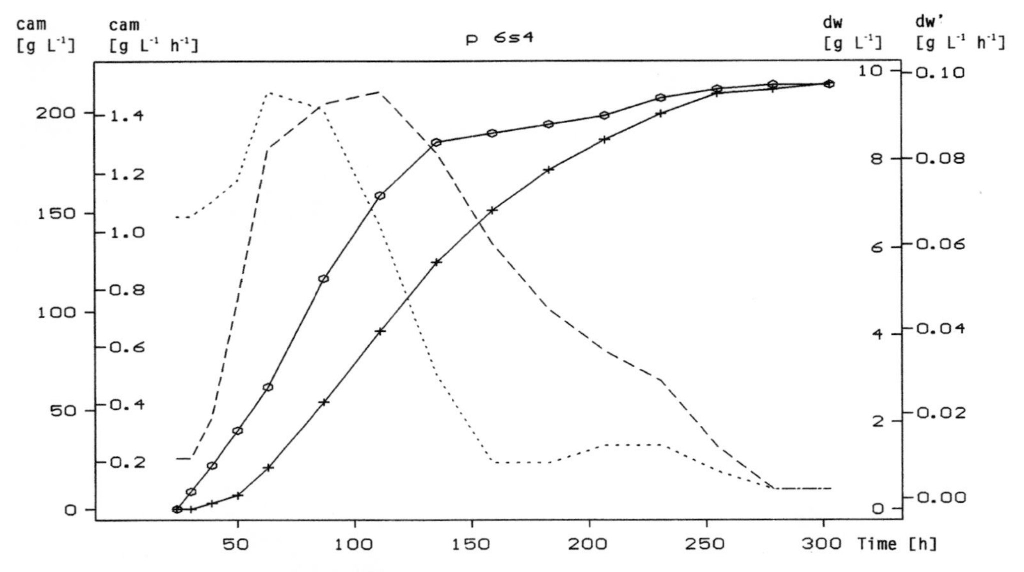


**Fig. 4.** Time-course of a typical industrial citric acid fermentation showing: + citric acid monohydrate (cam) and O mycelium dry weight (dw) [g $L^{-1}$]; ——— citric acid volumetric productivity (cam) and - - - specific growth rate (dw') [g $L^{-1}$ $h^{-1}$].

of 1.6–1.8. In many systems the pH is maintained at values >2 during the process to avoid the restraining action of lower pH values. Following the second day citric acid production begins and should be monitored by frequent titration. As mentioned above, accidental cessation of aeration, even for shorter periods of time, may slow down the fermentation considerably. It has been found that recovery of the culture may be attained by elevating the pH to about 2.0–4.0 as soon as possible (BATTI, 1966) until the culture regains its former capability.

Dependent on the initial sugar concentration, acid production ends between the 9th and 12th day of fermentation. Productivities are often >1 kg $m^{-3}$ $h^{-1}$ with yields of CAM based on sugar consumed exceeding 90%. A typical fermentation diagram is shown in Fig. 4.

## 3.6.3 Mixed-Type Fermentations

As may be derived from the description of these processes, pure sugar fermentations with their increased sensitivity to impurities as well as to the influence of smaller doses of inhibitors have lead to the evolution of another variant of manufacturing process:

This simply comprises a fed-batch process started by inoculating a molasses medium of medium strength (60–100 kg $m^{-3}$ sugar), accordingly inhibiting with higher excess of HCF. Subsequently, pure sugar solutions of high strength (ca. 300 kg $m^{-3}$) are fed to the fermenter as described above. Yields and rates are comparable to those of pure sugar fermentations.

## 3.6.4 Fermentation of Relatively "Impure" Raw Materials

The utilization of relatively "impure" raw materials such as crude, unfiltered starch hydrolyzates marks another significant improvement in industrial citric acid production.

In 1961, SCHWEIGER (1961), apparently while trying to stop contaminations (cf. SCHWEIGER, 1959), observed that the deleterious effect of iron impurities in the fermentation fluid could be counteracted by supplying copper(II)ions in a certain proportion to the iron concentration (cf. Tab. 2 in ROEHR et al., 1983a, p. 426). In the meantime, experience in various locations has shown that this relation only holds for low iron and corresponding copper concentrations as described above. Thus addition of low doses of copper ions (ca. 1–3 mg $L^{-1}$) invokes the same morphological changes as the respective additions of HCF. If iron is not removed to the level indicated above, the required higher additions of copper result in impaired fermentations with lower yields. In the light of results of several other workers it may be assumed that the proper target of copper action is manganese, the much more pronounced deleterious action of which has been studied extensively since then (see Sect. 2.4.1).

The use of less refined raw materials and the addition of copper ions to check the deleterious action of iron/manganese essentially characterizes the famous Miles technology and variants of it which may be considered the most advanced technologies at present. It goes without saying that these technologies are equally applicable to the conversion of purer sucrose-based raw materials or respective mixtures.

Information on control systems in modern citric acid plants is scarce. Two examples of such systems, however, have been reported (POWELL, 1973; SIEBERT and HUSTEDE, 1982).

Attempts to transfer novel techniques such as the application of immobilized cells and that of continuous operation into industrial practice have not been successful so far. Continuous culture experiments without cell immobilization (e.g., KRISTIANSEN and SINCLAIR, 1979; KRISTIANSEN and CHARLEY, 1981), applying glucose (50 g $L^{-1}$) resulted in citric acid concentrations between 2 and 6 g $L^{-1}$ with productivities of about 0.4 g $L^{-1}$ $h^{-1}$ at dilution rates around 0.075 $h^{-1}$, but measuring residual sugar concentrations of 40 g $L^{-1}$.

Several groups have studied different ways of immobilization of *A. niger*: adsorption to glass carriers (HEINRICH and REHM, 1982), adsorption to polyurethane foam (LEE et al., 1989), entrapment in calcium alginate (VAIJA et al., 1982; VAIJA and LINKO, 1986; EIKMEIER and REHM, 1984, 1987a, b; HONECKER et al., 1989), entrapment in polyacrylamide gels (HORITSU et al., 1985; GARG and SHARMA, 1992), inclusion in a polypropylene hollow fiber system (CHUNG and CHANG, 1988), entrapment in agar gel or $\kappa$-carrageenan beads (BORGLUM and MARSHALL, 1984), immobilization in porous cellulose carriers (FUJII et al., 1994). The main drawbacks so far encountered apparently are the low dilution rates necessary to achieve acceptable residual sugar concentrations, still low productivities, and product concentrations under conditions of industrial operation. In addition, attainment of optimum conditions seem to require even more sophisticated strategies than the established art.

# 4 Product Recovery and Purification

Mycelial mats ("Pilzdecken") from the surface process are easily separated from the fermentation broth. The latter is drained off, the remaining mats are disintegrated carefully and flushed onto a washing vessel. Thorough washing of the resulting mass is necessary as the spongy material retains about 15% of the citric acid produced in the fermentation. The washing water is added to the fermentation broth. The washed mycelium is dehydrated by filtration and subsequently dried to yield a protein-rich feedstuff. In an amount of 150–200 kg $t^{-1}$ of citric acid produced it represents a valuable fermentation by-product.

In the submerged process removal of mycelial material from the liquid is far more difficult. Usually this may be accomplished by a belt discharge filter or by a belt filter. Frequently impaired filtration may be encountered due to slimy mycelial excretions formed during fermentation. In such cases filter aids have to be employed. If the mycelium is designated for use as animal feedstuff, filter aids should be digestible, e.g., cellulosic waste materials such as disintegrated crop residues. An alternative filter aid is gypsum which accumulates in the main step of citric acid recovery and purification (cf. SALZBRUNN et al., 1989).

Further filtration steps are usually combined with the precipitation of oxalate which has to be removed owing to its toxicity. This is performed by adding milk of lime in slight excess at pH values <3.0. Calcium oxalate thus formed is filtered off using filter aids as indicated above. An elegant variant is the use of gypsum (SALZBRUNN et al., 1989) which reacts with oxalate to form calcium oxalate. In contrast to milk of lime, however, gypsum may be added in excess without the risk of unwanted precipitation of calcium citrate. Oxalate removal may of course be combined with mycelium filtration.

The fermentation broth also contains the bulk of unwanted substances present in the raw materials as well as substances resulting from the decomposition of fungal biomass.

Direct crystallization of citric acid from the fermented liquid, therefore, is almost impossible or requires rather laborious purification steps (see, however, FELMAN et al., 1995b). On the other hand, several procedures have been described dealing with the recovery of alkali salts of citric acid by direct crystallization from fermentation broths following neutralization with the respective alkaline alkali metal compounds equivalent to obtain mono-, di-, or trialkali citrates and evaporation to about 400 g $L^{-1}$, calculated as citric acid (SCHULZ, 1959, 1963; TSUDA et al., 1975; FELMAN et al., 1994, 1995a).

Recovery of citric acid from fermentation broths is generally accomplished by three basic procedures: (1) precipitation, (2) extraction, (3) adsorption and absorption (mainly using ion exchange resins).

Precipitation, the classical method, is performed by the addition of calcium oxide/hydrate (milk of lime) to form the slightly soluble tricalcium citrate tetrahydrate. Successful operation of the precipitation depends on several factors (SCHMITZ, 1977; SCHMITZ and WÖHLK, 1980): citric acid concentration, temperature, pH during precipitation, rate of lime addition. If precipitation is properly done, impurities remain largely in the residual solution and may be removed by washing the filtered precipitate. Upon addition of milk of lime, soluble monocalcium citrate is formed, followed by dicalcium citrate and tricalcium citrate, the latter being precipitated at a higher rate. In the course of lime addition, varying degrees of supersaturation occur. Frequently this results in the formation of amorphous to microcrystalline precipitates with a tendency to form dense aggregations and containing higher amounts of enclosed and adhering impurities. This, on the other hand, lowers filtration rates, and the filter cakes cannot be washed properly. In order to attain larger crystals of higher purity and better performance upon filtration and washing, milk of lime containing about 180–250 kg CaO $m^{-3}$ is added at empirically determined rates at temperatures around 90 °C and at pH values below but close to 7. SCHMITZ and WÖHLK (1980) have designed a process for continuous precipitation of tricalcium citrate. Similar protocols have been elaborated by SALZBRUNN et al. (1991) to obtain almost entirely macrocrystalline calcium citrate and yielding end products of considerably higher quality.

Alternatively, cheaper calcium carbonate may be applied as precipitating agent (SALZBRUNN et al., 1991) if foaming is counteracted by mixing with a vibrating mixer.

After filtration and washing, the moist calcium citrate is reacted at about 50 °C with sulfuric acid of medium strength (60–70%) maintaining a slight excess (1–2 g $L^{-1}$) to secure complete reaction. Calcium sulfate (gypsum) and free citric acid are separated by filtration on drum or belt filters. The filtrate is treated with activated carbon to remove impurities such as colored organic substances, and passed over ion exchange resins to remove residual calcium sulfate and metal ions. Strong cation exchange resins (e.g., Dowex-50 or Lewatit 100, $H^+$ form) and anion exchangers of medium strength (e. g., Dowex-2, Duolite A 2) are used. Treatment of the crude solutions with hydrogen peroxide before passing through the ion exchange columns has been claimed by Miles Laboratories (1961) to be effective, especially to remove readily carbonizable substances (see below). Subsequently, the purified citric acid solution is evaporated in a multi-stage evaporator at temperatures <40 °C to avoid caramelization.

Crystallization is performed in vacuum crystallizers. Citric acid monohydrate (CAM),

**Tab. 1.** Solvent Extraction of Citric Acid from Fermentation Broths

| Inventors | Extractants |
|---|---|
| Chemische Fabrik J. A. Benckiser (1932) | *n*-butanol |
| COLIN (1960, 1968) | isobutanol |
| COLIN and MOUNDLIC (1967) | tri-isobutylphosphate + diluents such as: octane, benzene, kerosene or *n*-butylacetate, methyl isobutylketone, isobutanol |
| IMI-TAMI (1972) | various amines |
| BANIEL et al. (1974) | secondary or tertiary amine + organic solvent |
| RIEGER and KIOUSTELIDIS (1975) | tridecylamine or triisononylamine + water-insoluble alcohol or ketone |
| Food and Drug Administration (1975) | tridodecylamine + *n*-octanol + synthetic isoparaffine petroleum hydrocarbons |
| ALTER and BLUMBERG (1981) | N,N-diethyl dodecanamide |
| BANIEL (1982) | trilaurylamine + oleic acid + aliphatic naphtha |
| KAWANO (1987) | trilaurylamine + xylene |
| JIANG and SU (1987) | trialkylphosphine oxides |
| YI et al. (1987) | dialkylamide + butylacetate |
| BAUER et al. (1988, 1989) | amines + nonpolar solvents |
| BANIEL and GONEN (1991) | amines |
| BIZEK et al. (1992a, b) | trialkylamine + octanol/*n*-heptane |
| HARTL und MARR (1993) | amines + nonpolar solvents |
| BEMISH et al. (1993) | trilaurylamine + octanol + cyclohexanone |
| PROCHAZKA et al. (1994) | trialkylamine + octanol/*n*-heptane |
| LEHNHARDT et al. (1995) | oxygenated solvents ($C_4-C_{12}$) having $\geq 1$ functional group |

the product mainly sold in Europe, is formed at temperatures below the transition temperature of 36.5 °C. Above this temperature citric acid anhydrate (CAA) is formed which is traded in the USA and in Asia. Drying of the product is usually performed in a fluidized bed.

A serious drawback of liming to recover citric acid is the fact that large amounts of gypsum, about one ton per ton of CAM, are formed that in most cases are disposed of in waste deposits. In view of increasing efforts towards environmental protection this may cause considerable problems.

Several authors (e.g., AYERS, 1957) have suggested to precipitate calcium hydrogen citrate (dicalcium acid citrate), with the aim of decreasing the amount of milk of lime by about one third, according to the following reaction:

$$Ca_3(Cit)_2 + CitH \rightleftharpoons 3\,CaHCit$$

It has been claimed that the equilibrium between tricalcium citrate + citric acid and calcium hydrogen citrate can be shifted almost completely to the right by increasing the temperature to values >40 °C, preferentially to 80–95 °C, and holding it for about 24 h. Thus by treating 2/3 of a given volume of citric acid mash with milk of lime to obtain tricalcium citrate, this precipitate may subsequently be reacted with the remaining 1/3 to yield insoluble calcium hydrogen citrate. A similar procedure has been claimed by SHISHIKURA (1994).

It has been proposed that this may also be accomplished by reacting equimolar mixtures of citric acid and calcium hydroxide, calcium oxide or calcium carbonate at the temperatures mentioned above, but the stoichiometry of such reactions is rather complex.

An alternative way of citric acid isolation and purification is solvent extraction. A tremendous amount of information including many patents has been published during the last three decades. A selection of patents is given in Tab. 1.

As may be seen, various solvent combinations have been exploited, of which aliphatic alcohols and ketones, amines and phosphines with hydrocarbons are predominant. Little is known about comparative evaluations on a larger scale. It may be stated, however, that alcohols (such as butanols) are too much water-miscible, thus requiring energy-consuming steps for recovery. Arguments concerning the question whether the concentration of reextracted citric acid solutions is comparable to that before extraction are yet to be settled. Noteworthy is the report of ZHANG and CHENG (1987), who compared citric acid obtained from trialkylphosphine oxide and from tributylphosphate extraction with that obtained from common purification by calcium precipitation with respect to toxic effects in food applications: no such effects were detected but it was found that the tributylphosphate sample showed significant teratogenic action on rats. According to MILSOM (1987) the extractant apparently used by Miles Laboratories, a mixture of a secondary or tertiary amine having at least 20 C-atoms and a diluting organic solvent (BANIEL et al., 1974; BANIEL, 1982), has received approval by the U.S. Food and Drug Administration.

Recovery and purification by ion-exchange procedures is gaining importance since purer raw materials are used in citric acid manufacture. During the last decade, a variety of processes has been disclosed. Several anion-exchange resins are commercially available, the preparation of others is reported in the technical literature. Examples of more recent patents are the following:

HELLMIG et al. (1983), HIRASHIMA et al. (1985), KULPRATHIPANJA (1988, 1989), KULPRATHIPANJA et al. (1989), KULPRATHIPANJA and STRONG (1990), KULPRATHIPANJA and OROSKAR (1991), DUFLOT and LELEV (1989), MATSUDA and YOSHIDA (1991), EDLAUER et al. (1990), SCHELLENBERGER et al. (1991), MCQUIGG et al. (1992).

Possible drawbacks of this technique may be seen in the fact that elution of citric acid from the ion exchanger may cause considerable dilution of the resulting citric acid solutions. Besides this, the regeneration of large amounts of ion-exchange resins may produce additional sources of environmental pollution.

More sophisticated developments may be expected with the application of liquid membranes (cf. EYAL and BRESSLER, 1993).

# 5 Citric Acid from Other Substances and Organisms

Until the early 1970s *A. niger* was the organism almost exclusively used for the manufacture of citric acid, although it was by no means clear why only this organism or close relatives should be capable of citric acid production. Such relatives are: *A. awamori, A. carbonarius, A. fonsecaeus, A. foetidus, A. phoenicis,* which all belong to the section Nigri Gams et al. (*A. niger* group Thom and Church 1926) as discussed by SAMSON (1992). One noteworthy exception is *A. wentii,* belonging to the section Wentii Gams et al. (*A. wentii* group Thom and Raper 1945). *A. wentii* has been patented in a well documented production process (WAKSMAN and KAROW, 1946; KAROW and WAKSMAN, 1947) but never been used in industrial practice. PERLMAN and SIH (1960) have further listed *A. clavatus, A. fumaricus, A. luchensis, A. saitoi, A. usamii.* Similarly, the potential of other filamentous fungi has not been utilized, but some patents should be mentioned: KINOSHITA et al. (1961a) have investigated the use of *Penicillium janthinellum* var. *kuensanii* (ATCC 13154) and *P. restrictum* var. *kuensanii* (ATCC 13155). Yields >80% on sugar consumed (blackstrap molasses) after 7 d of fermentation in, e.g., a 100 L fermenter were claimed. In this patent, the authors also made reference to other Penicillia listed in the *"Manual of the Penicillia"* of RAPER and THOM (1949) as being able to produce citric acid. In another patent KINOSHITA et al. (1961b) described the use of several strains of *Trichoderma viride,* e.g., strain ATCC 13233, for citric acid fermentation. In media with about 100 g $L^{-1}$ of crude potato or sweet potato materials as carbon substrate, yields of 75% on starch added and 85% on starch consumed, respectively, within 7 d were reported.

Thus the advantages of such a process as claimed by the inventors are the amylase and the cellulase content of the organism, apparently no sensitivity to heavy metals, and no appreciable formation of by-products. Nevertheless, these processes never got into industrial operation.

More promising organisms, however, have been found within the group of yeasts, esp. the genus *Candida* and related genera. The knowledge dates back to investigations on the abilities of yeasts to grow on *n*-alkanes as carbon sources and to produce various valuable substances such as citric acid as by-products. According to YAMADA (1977) citric acid production from straight-chain *n*-paraffins has been studied since 1960 by the Japanese industry. Various fractions of straight-chain paraffins (about $C_9$ to $C_{20}$) were the preferred substrates. Other factors of importance were reported to be pH and the concentration of iron ions. The pH value should be kept above 5; lowering pH results in the production of polyols such as erythritol and arabitol (TABUCHI and HARA, 1973). Iron (e.g., 400 mg $FeSO_4 \cdot 7H_2O$ $L^{-1}$) considerably decreased the productivity of *Saccharomycopsis (Candida) lipolytica* (MARCHAL et al., 1977a), probably by increasing activity of aconitate hydratase (TABUCHI et al., 1973). MARCHAL et al. (1977b) and BEHRENS et al. (1978) described the fermentation as being biphasic, citric acid beginning to accumulate after exhaustion of the nitrogen source in the medium. Maintaining a certain ratio of phosphorus to carbon of $0.1 \cdot 10^{-3}$–$2.0 \cdot 10^{-3}$:1 throughout the fermentation was claimed to guarantee constant high yields of citric acid (FUKUDA et al., 1982).

The list of more important citric acid producing yeasts comprises:

*Candida brumptii,* new designation:
*C. catenulata*
*Candida guilliermondii*
*Candida intermedia*
*Candida lipolytica,* new designation:
*Yarrowia lipolytica*
*Candida oleophila*
*Candida parapsilosis*
*Candida tropicalis*
*Candida zeylanoides*
*Rhodotorula*
*Saccharomycopsis lipolytica,* new designation:
*Yarrowia lipolytica*

Several strains of these are deposited in international collections of microorganisms.

One disadvantage of the use of yeasts was the formation of significant amounts of isocitric acid, up to 50% of total acid production. This was counteracted by selecting mutant strains, mainly with lowered activities of aconitate hydratase. Monofluoroacetate was the most widely used selecting agent (AKIYAMA et al., 1972, 1973a, b). On the other hand, formation of isocitric acid can also be repressed by the addition of surfactants (MATSUMOTO et al., 1983) or by the introduction of oxygen (MATSUMOTO and ICHIKAWA, 1986).

Industrial know-how, mainly acquired in Japan by several companies (e.g., Takeda Chemical Industries, Hitachi Chemicals, Kyowa Hakko Kogyo, Ajinomoto Co., Inc., etc.), was exported to the USA (Pfizer) as well as to Europe (Liquichimica) in the early 1970s. These activities are reflected by the great number of patents as documented in the 1st Edition of *"Biotechnology"* (ROEHR et al., 1983a). The world oil crisis of 1973/74 almost entirely ended the exploitation of *n*-alkanes as feedstock for the industrial production of citric acid. In the meantime, however, it had been discovered that certain yeasts could also produce citric acid from carbohydrates, especially with glucose as carbon source (TABUCHI and ABE, 1968; TABUCHI et al., 1969; IIZUKA et al., 1969; SHIMIZU et al., 1970; ABE et al., 1970). Citric acid production on molasses media was also described (MIALL and PARKER, 1975). Since many of the producer organisms displayed low fructose assimilation, molasses or invert sugar mixtures could also be used for the simultaneous production of citric acid and fructose (UCHIO et al., 1979). Extensive investigations following the findings with glucose fermentations revealed that citric acid productivity on glucose was as high as on alkanes (BEHRENS et al., 1978). This line was followed mainly by working groups in the former German Democratic Republic. Some more recent developments include: varying oxygen supply (STOTTMEISTER et al., 1986), preadaptation of cells to high-methionine media (BARTH and KREBS, 1985), or preculturing cells on alkanes and subsequent transfer to glucose media (WEISSBRODT et al., 1987a). With the latter procedure substantial improvements of yields and productivities have been claimed. Besides glucose, ethanol has been proposed as feedstock offering advantages in product purification.

As with *A. niger,* several attempts have been made to exploit the potential of novel techniques such as cell immobilization and continuous operation in this art. Several patents describe the more recent achievements, e.g., application of semicontinuous culture (WEISSBRODT et al., 1987b) or immobilization of cells (KAISER et al., 1990).

# 6 Process Kinetics

***Aspergillus niger.*** According to the classical scheme of GADEN (1955) citric acid fermentation was specified as a "type II" fermentation. According to this scheme, product formation apparently arises from energy metabolism, but somehow indirectly. Based on rather scarce data of SHU and JOHNSON (1948a) with only 7 data points over a fermentation period of 280 h (!) it was concluded that growth rate displayed two maxima whereas product formation showed only one rate maximum. Based on this concept further model studies were carried out by KHAN and GHOSE (1973) and CHMIEL (1975a, b). KRISTIANSEN and SINCLAIR (1979) studied *Aspergillus foetidus* in continuous culture and found that citric acid production could be described by the model of LUEDEKING and PIRET (1959). This was modified and extended by ROEHR et al. (1981) monitoring a pilot-size citric acid fermentation with high data point frequency. It was observed that the process is characterized by the following features: In the first phase (trophophase) there is significant product formation depending on growth rate. In the second phase (idiophase) in which almost no growth can be observed (but is simulated by accumulation of storage compounds by the mycelium), product formation is maximized and depends on biomass concentration. A modification of the model of LUEDEKING and PIRET (cf. BROWN and VAS, 1973), considering a lag time needed for a hyphal cell to enter the stage of citric acid production, was suggested. Thus the relation between citric acid formation and growth would be:

$$(\mathrm{d}p/\mathrm{d}t)_{\mathrm{t}}=k_1\,(\mathrm{d}x/\mathrm{d}t)_{\mathrm{t}-\mathrm{t}^*}+k_2\,(x)_{\mathrm{t}-\mathrm{t}^*}. \quad (1)$$

In Eq. (1), $p$ denotes citric acid concentration (g $\mathrm{L}^{-1}$), $x$ denotes biomass concentration (g d wt $\mathrm{L}^{-1}$) and $t^*$ is the time needed by a certain hyphal cell to enter the stage of citric acid production. Experimentally, $t^*$ has been determined to be 15 h. This is within the range of the doubling time of active biomass (under special environmental conditions) suggesting that citric acid is formed mainly by the subapical part of the mycelium.

$k_1$ and $k_2$ were determined as 1.9 and 0.09, respectively.

Since $\mathrm{d}p/\mathrm{d}t$ may be expressed as

$$\mathrm{d}p/\mathrm{d}t=\mathrm{d}p/\mathrm{d}s\cdot\mathrm{d}s/\mathrm{d}x\cdot\mathrm{d}x/\mathrm{d}t=k_1\,\mathrm{d}x/\mathrm{d}t \quad (2)$$

where

$$k_1=\mathrm{d}p/\mathrm{d}s\cdot\mathrm{d}s/\mathrm{d}x=Y_{\mathrm{p/s}}\cdot Y_{\mathrm{x/s}}^{-1}. \quad (3)$$

$k_1$ thus represents the quotient of product yield coefficient $Y_{\mathrm{p/s}}$ and biomass yield coefficient $Y_{\mathrm{x/s}}$. With average values of 0.8 for the former and 0.4 (under limiting conditions) for the latter, the theoretical value is about 2 which agrees well with the experimental value of 1.9.

Based on these findings, citric acid as "type II" fermentation may be better characterized as rather displaying one growth rate maximum but two maxima of production rate.

**Yeasts.** Kinetic studies have been reported by MARCHAL et al. (1977a), BEHRENS et al. (1978), BRIFFAUD and ENGASSER (1979) and MORESI et al. (1980). It is agreed that citric acid is only produced after growth has ended due to exhaustion of the nitrogen source in the medium (Fig. 5). Thus citric acid production by yeasts may be considered as a relatively simple idiophase event. According to

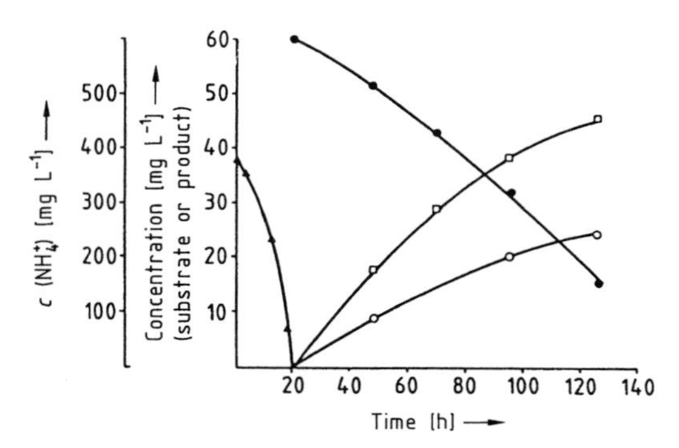


**Fig. 5.** Kinetics of citric acid production from glucose by alkane grown yeast (from BEHRENS et al., 1978, with permission).
▲ $NH_4^+$-nitrogen, ● glucose, □ total acids, ○ isocitric acid.

BEHRENS et al. (1978) there are only small differences between specific rates of citric acid production on $n$-alkanes (0.18 $h^{-1}$) and on glucose (0.19 $h^{-1}$).

# 7 Process Ecology

Recovery of citric acid from the fermentation broth is accompanied by the accumulation of considerable amounts of by-products. About 2.5 t of waste material (calculated on a dry weight basis) per ton of CAM are formed. The main component is gypsum (about 60% of total), followed by soluble products in the spent fermentation broth, and the waste mycelium. Increasing ecological constraints demand adequate steps towards eliminating environmental loads and possible utilization of waste materials.

**Gypsum.** As has been mentioned above, in many cases depositing of waste gypsum can no longer be tolerated. Several purification procedures have, therefore, been elaborated to achieve removal of accompanying substances such as residual mycelium (cf., e.g., WEGNER and KELLERWESSEL, 1977), Berlin blue (FOERSTER and POTENCSIK, 1971; Donau Chemie AG, 1993), etc. Purified gypsum, accumulating finely dispersed, can be made an ideal construction material, e.g., as gypsum board, yet cheaper than gypsum from natural sources. Purified gypsum may also be used as a pigment in coatings for paper or it can be converted into $\alpha$-gypsum binders (SATAVA et al., 1977; WENDLER, 1988) or into so-called sulfomineral cements (ATAKUZIEV et al., 1980; UNROD et al., 1992).

At present, there are indications that cost-effective procedures to produce purified gypsum from waste gypsum derived from various sources will be achieved; and this should in turn influence the conditions and economy of operations leading to waste gypsum.

Alternatively, several authors (e.g., KACZMAROWICZ, 1987) have suggested to consider the well-known reaction of calcium sulfate with carbon dioxide and ammonia resulting in the formation of calcium carbonate and ammonium sulfate which both could find diverse applications. Similarly, reaction with sodium carbonate (HUANG et al., 1990) may yield calcium carbonate and sodium sulfate.

**Mycelium.** The waste mycelium from both the surface and the submerged process may be dried and used as an animal feedstuff or alternatively as a supplement to fertilizers.

**Waste Fluids.** The waste fluid from the calcium citrate precipitation may be evaporated to yield a product ("vinasse") which can be used as an animal feedstuff. Due to its high content of inorganic substances, especially potassium, applications are rather limited. Due to technical problems and the high energy input such processes are less accepted.

Nowadays, in larger plants the bulk of liquid waste streams is preferably purified by the installation of larger waste water treatment plants. Additional anaerobic stages to common activated sludge systems have been suggested, e.g., by KOEPP et al. (1985), KOEPP-BANK (1986), and DERYCKE and PIPYN (1990). Examples of combined systems for large facilities have recently been reported for, e.g., the plant of Miles, Inc. in Elkhard, USA (COOPER and FOSTER, 1990) and for A. G. Jungbunzlauer Spiritus und Chemische Fabrik in Pernhofen, Austria (KROISS, 1987; SVARDAL et al., 1993). In both cases the effectiveness of anaerobic pretreatment stages is stressed; detailed descriptions of processes demonstrate the economic feasibility of these installations.

# 8 Utilization of Citric Acid

Citric acid is mainly used in the food industry (60%) because of its pleasant acid taste and its high solubility in water. It is worldwide accepted as "GRAS" (generally regarded as safe), approved by the Joint FAO/WHO Expert Committee on Food Additives without limitations as an additive in beverages, jams, all kinds of sweets, ice-cream, etc. (see, e.g., RUDY 1967; LAWRENCE, 1974; SCHULZ and RAUCH, 1975; BOUCHARD and MERRITT, 1979). GOLDBERG et al. (1991) have provided an excellent compilation of properties and applications in the food industry. They have also included a highly interesting comparative overview regarding citric, fumaric, and malic acids based on earlier data of GARDNER (1966). Certain citric acid esters are also used in the food industry as emulsifiers.

Other applications are in the cosmetics and the pharmaceutical industry (10%), where purity requirements are strictly regulated by the respective pharmacopoeias (see Tab. 2). Food and beverages industries generally follow the same regulations.

Various citric acid salts are used in certain areas such as food and pharmaceutical industries either as salts of different metals, e.g., iron or calcium, or as buffers. Trisodium citrate and citric acid, e.g., may be mixed to yield buffer salts for almost any acidity and pH value. Aluminum citrate has been proposed as a crosslinking agent for CM-cellulose or partially hydrolyzed polyacrylamide in enhanced oil recovery (FODOR and COBB, 1986).

Increasing amounts of citric acid (up to 30%) are applied in other industrial areas. An important property of citric acid is its chelating (sequestering) action. This can be utilized in various household and industrial cleaners and detergents. For this application mainly sodium citrates are used, the production of which has been described above. Alternatively technical-grade citric acid may be neutralized to yield the respective salts. High production figures were expected when it was planned to design novel washing agents; the detergent industry, however, at least in part, has found other, cheaper ways of substitution. It is remarkable to note that this could hardly slow down production of citric acid. In the 1970s further sequestering agents based on citric acid were invented (Citrique Belge and UCB) which were produced by thermal decomposition of calcium citrate. These yielded unsaturated polycarboxylic acids which were sulphonated to yield sulphopolycarboxylic acids (SPC). Esters of SPC are excellent detergents, non-toxic and almost completely biodegradable. The manufacture of citric acid esters for the plastics industry should also be mentioned. An interesting use of citric acid as an absorbent in the removal of $SO_2$ from waste gases has been proposed (ROSENBAUM et al., 1973; TRONDHEIM, 1975)). Citric acid can be used together with tartaric acid in set-

**Tab. 2.** Specifications of Citric Acid for Pharmaceutical, Food, and Related Uses

| | | DAB | BP | USP |
|---|---|---|---|---|
| Citric acid (anhydrous) [%] | | 99.5 | 99.5 | 99.5 |
| Ash [18%] | max. | 0.1 | 0.1 | 0.05 |
| Heavy metals (as Pb) [mg kg$^{-1}$] | max. | 10 | 10 | 10 |
| Arsenic [mg kg$^{-1}$] | max | 2 | 1 | 3 |
| Sulfate [%] | max. | 0.02 | 0.03 | – |
| Oxalate | | not detectable | | |
| Readily carbonizable substances | | below limits in specified procedures | | |

DAB: Deutsches Arzneibuch
BP: British Pharmacopoeia
USP: U.S. Pharmacopoeia

**Tab. 3.** Main Producers of Citric Acid

| Company | Location | Total Capacity [t a$^{-1}$] (approx.) |
|---|---|---|
| Miles Inc.[a] | USA, Brazil, Mexico, Columbia | 120,000 |
| Chas. Pfizer[b] | USA, Ireland | 115,000 |
| A.G. Jungbunzlauer Chemische Fabrik | Austria, Germany, France, Indonesia | >100,000 |
| Citrique Belge | Belgium | 60,000 |
| Biacor[c] | Italy | 28,000 |
| John Sturge[a] | UK | 25,000 |
| Cargill | USA | 25,000 |
| AKTIVA | Czech Republic | 15,000 |
| Gadot | Israel | 12,000 |
| + Smaller plants in | | |
| | China | 50,000 |
| | Indonesia | 25,000 |
| | Russia | 22,000 |
| | India | 10,000 |
| | Slovakia | 4,500 |
| | Turkey | 4,500 |
| | Thailand | 4,000 |

[a] Bayer USA subsidiary Haarmann & Reimer Corp.
[b] ADM (Archer Daniels Midland), USA
[c] Gruppo Ferruzzi, Italy

ting retarder compositions for technical gypsum (KRETZSCHMANN et al., 1975). Similar applications are additives in building materials such as cement influencing the retardation and formation of specific hydrates of such materials (cf. POELLMANN et al., 1990). For further industrial uses of citric acid and citrates the reader is referred to the relevant encyclopedias *(Ullmann's Encyclopedia of Industrial Chemistry, Kirk-Othmer Encyclopedia of Chemical Technology).*

# 9 Production Figures

Citric acid is the major organic acid produced by fermentation and the second of all fermentation commodities following industrial ethanol.

Present world production capacity may be estimated as being around 700,000 t a$^{-1}$. Major producers are listed in Tab. 3.

# 10 References


ABE, M., TABUCHI, T., TAHARA, Y. (1970), Studies on organic acid fermentation in yeasts, *J. Agr. Chem. Soc. Jpn.* **44**, 493–498.

AKIYAMA, S., SUZUKI, T., SUMINO, Y., NAKAO, Y., FUKUDA, H. (1972), Production of citric acid from *n*-paraffins by fluoroacetate sensitive mutant strains of *Candida lipolytica, Agr. Biol. Chem.* **36**, 339–341.

AKIYAMA, S., SUZUKI, T., SUMINO, Y., NAKAO, Y., FUKUDA, H. (1973a), Production of citric acid from *n*-paraffins by fluoroacetate-sensitive mutants. I. Induction and citric acid productivity of fluoroacetate-sensitive mutant strains of *Candida lipolytica, Agr. Biol. Chem.* **37**, 879–884.

AKIYAMA, S., SUZUKI, T., SUMINO, Y., NAKAO, Y., FUKUDA, H. (1973b), Production of citric acid from *n*-paraffins by fluoroacetate-sensitive mutants. II. Relation between aconitate hydratase activity and citric acid productivity in fluoroacetate-sensitive mutant strains of *Candida lipolytica, Agr. Biol. Chem.* **37**, 885–888.

ALTER, J. E., BLUMBERG, R. (1981), *U.S. Patent* 4251671.

ANNÉ, J., PEBERDY, J. F. (1975), Conditions for induced fusion of fungal protoplasts in polyethylene glycol solutions, *Arch. Microbiol.* **105**, 201–205.

ANNÉ, J., PEBERDY, J. F. (1976), Induced fusion of fungal protoplasts following treatment with polyethyleneglycol, *J. Gen. Microbiol.* **92**, 413–417.

ARTS, E., KUBICEK, C. P., ROEHR, M. (1987), Regulation of phosphofructokinase from *Aspergillus niger:* effect of fructose-2,6-bisphosphate on the action of citrate, amonium ions and AMP, *J. Gen. Microbiol.* **133**, 1195–1199.

ATAKUZIEV, T. A., KHASANOV, R. S., MIRZAEV, F. M., RAZDORSKIKH, L. M. (1980), Citric acid production waste as the main component of sulfomineral cement, *Dokl. Akad. Nauk UzSSR* **(3)**, 48–50.

AYERS, R., Jr. (1957), *U.S. Patent* 2810755.

BANIEL, A. M. (1982), *Eur. Patent* 49429.

BANIEL, A. M., GONEN, D. (1991), *U.S. Patent* 4994609.

BANIEL, A. M., BLUMBERG, R., HAJDU, K. (1974), *Ger. Patent* 2329480.

BANUELOS, M., GANCEDO, C., GANCEDO, J. M. (1977), Activation by phosphate of yeast phosphofructokinase, *J. Biol. Chem.* **252**, 6394–6398.

BARTH, G., KREBS, D. (1985), *Ger. Patent* (GDR) 227448.

BATTI, M. A. (1966), *U.S. Patent* 3290227.

BAUER, U., MARR, R., RUECKL, W., SIEBENHOFER, M. (1988), Extraction of citric acid from aqueous solutions, *Chem. Biochem. Eng. Q.* **2**, 230–232.

BAUER, U., MARR, R., RUECKL, W., SIEBENHOFER, M. (1989), Reactive extraction of citric acid from an aqueous fermentation broth, *Ber. Bunsen-Ges. Phys. Chem.* **93**, 980–984.

BEHRENS, U., WEISSBRODT, E., LEHMANN, W. (1978), Zur Kinetik der Citronensäurebildung bei *Candida lipolytica, Z. Allg. Mikrobiol.* **18**, 549–558.

BEMISH, T. A., CHIANG, J. P., PATWARDHAN, B. H., SOLOW, D. J. (1993), *U.S. Patent* 5237098.

BENNETT, J. W., KLICH, M. A. (Eds.) (1992), *Aspergillus: Biology and Industrial Applications.* Stoneham: Butterworth-Heinemann.

BENNETT, J. W., LASURE, L. L. (Eds.) (1985), *Gene Manipulations in Fungi.* Orlando, FA: Academic Press.

BENNETT, J. W., LASURE, L. L. (1991), *More Gene Manipulations in Fungi.* San Diego, CA: Academic Press.

BERCOVITZ, A., PELEG, Y., BATTAT, E., ROKEM, J. S., GOLDBERG, I. (1990), Localization of pyruvate carboxylase in organic acid producing *Aspergillus* strains, *Appl. Environ. Microbiol.* **56**, 1594–1597.

BERNHAUER, K. (1928), Über die Charakterisierung der Stämme von *Aspergillus niger* aufgrund ihres biochemischen Verhaltens. 1. Vergleichende Untersuchungen über die Säurebildung durch verschiedene Pilzstämme, *Biochem. Z.* **197**, 278–286.

BERNHAUER, K. (1929), Über die Charakterisierung der Stämme von *Aspergillus niger* aufgrund ihres biochemischen Verhaltens. 2. Die Bedeutung saurer Substrate für die Charakterisierung und Züchtung der Pilzstämme, *Biochem. Z.* **205**, 240–244.

BERNHAUER, K. (1934), Die Citronensäuregärung, *Ergebn. Enzymforsch.* **3**, 185–277.

BERNHAUER, K. (1940). Oxydative Gärungen, in: *Handbuch der Enzymologie* (NORD, F. F., WEIDENHAGEN, R., Eds.). Leipzig: Akademische Verlagsgesellschaft.

BIGELIS, R. (1985), Primary metabolism and industrial fermentations, in: *Gene Manipulations in Fungi* (BENNETT, J. W., LASURE, L. L., Eds.), pp. 357–401. Orlando, FA: Academic Press.

BIGELIS, R., ARORA, D. K. (1992), Organic acids of fungi, in: *Handbook of Applied Mycology,* Vol. 4. (ARORA, D. K., ELANDER, R. P., MUKERJI, K. G., Eds), pp. 357–376. New York: Marcel Dekker.

BIZEK, V., HORACEK, J., RERICHA, R., KOUSOVA, M. (1992), Amine extraction of hydroxycarboxylic acids. 1. Extraction of citric acid with 1-octanol/*n*-heptane solutions of trialkylamine, *Ind. Eng. Chem. Res.* **31**, 1554–1562; 2632.

BONATELLI, R., Jr., AZEVEDO, J. L. (1982), Improved reproducibility of citric acid production in *Aspergillus niger, Biotechnol. Lett.* **4**, 761–766.

BONATELLI, R., Jr., AZEVEDO, J. L., VALENT, G. U. (1983), Parasexuality in a citric acid producing strain of *Aspergillus niger, Rev. Brasil. Genet.* **VI**, (3), 399–405.

BORGLUM, G. B., MARSHALL, J. J. (1984), The potential of immobilized biocatalysts for production of industrial chemicals, *Appl. Biochem. Biotechnol.* **9**, 117–130.

BOUCHARD, E. F., MERRITT, E. G. (1979), Citric acid, in: *Kirk-Othmer's Encyclopedia of Chemical Technology,* Vol. 6, (GRAYSON, M., ECKROTH, D., Eds.), 3rd Edn., pp. 150–179. New York: Wiley.

BRIFFAUD, J., ENGASSER, M. (1979), Citric acid production from glucose. I. Growth and excretion kinetics in a stirred fermentor, *Biotechnol. Bioeng.* **21**, 2083–2092.

BROWN, D. E., VAS, R. C. (1973), Maturity and product formation in cultures of microorganisms, *Biotechnol. Bioeng.* **15**, 321–330.

CADDICK, M. X. (1992), Characterization of a major *Aspergillus* regulatory gene *are*A, in: *Molecular Biology of Filamentous Fungi* (STAHL, U., TUDZYNSKI, P., Eds.), pp. 141–151. Weinheim: VCH.

CAHN, F. J. (1935), Citric acid fermentation on solid materials, *Ind. Eng. Chem.* **27**, 201.

CHANG, L. T., TERRY, C. A. (1953), Intergenic complementation of glucoamylase and citric acid production in two species of *Aspergillus, Appl. Microbiol.* **25**, 890–895.

CHAUDHARY, K., ETHIRAJ, S., LAKSHMINARAYANA, K., TAURO, P. (1978), Citric acid production from Indian cane molasses by *Aspergillus niger* under solid state fermentation conditions, *J. Ferment. Technol.* **56**, 554–557.

Chemische Fabrik J. A. Benckiser (1932), *Ger. Patent* 555810.

CHMIEL, A. (1975a), Kinetic studies on citric acid production by *Aspergillus niger.* I. Phases of mycelium growth and product formation, *Acta Microbiol. Pol. Ser. B* **7**, 185–193.

CHMIEL, A. (1975b), Kinetic studies on citric acid production by *Aspergillus niger.* II. The two-stage process, *Acta Microbiol. Pol. Ser. B* **7**, 237–242.

CHUNG, B. H., CHANG, H. N. (1988), Aerobic fungal cell immobilization in a dual hollow-fiber bioreactor: continuous production of citric acid, *Biotechnol. Bioeng.* **32**, 205–212.

CIEGLER, A., RAPER, K. B. (1957), Application of heterokaryons of *Aspergillus* to commercial type fermentations, *Appl. Microbiol.* **5**, 106–110.

CLARK, D. S. (1964), *U.S. Patent* 3118821.

CLARK, D. S., ITO, K., HORITSU, H. (1966), Effect of manganese and other heavy metals on submerged citric acid fermentation of molasses, *Biotechnol. Bioeng.* **8**, 465–471.

CLELAND, W. W., JOHNSON, M. J. (1954), Tracer experiments on the mechanism of citric acid formation by *Aspergillus niger, J. Biol. Chem.* **208**, 679–692.

CLEMENT, M. T. (1952), Citric acid fermentation of beet molasses by *Aspergillus niger* in submerged culture, *Can. J. Technol.* **30**, 82–88.

COLIN, P. (1960), *French Patent* 1211066.

COLIN, P. (1968), *French Patent* 1548328.

COLIN, P., MOUNDLIC, J. (1967), *French Patent* 1494958.

COOPER, J. I., FOSTER, D. E. (1990), Anaerobic digestion of 400000 gallons/day of fermentation wastes in fixed-bed, downflow digesters. Waste characterization studies to commercial operation, *Energy Biomass Wastes* **13**, 1061–1077.

CURRIE, J. N. (1916), On the citric acid production of *Aspergillus niger, Science* **44**, 215.

CURRIE, J. N. (1917), The citric acid fermentation of *Aspergillus niger, J. Biol. Chem.* **31**, 15–37.

DAS, A., ROY, P. (1978), Improved production of citric acid by a diploid strain of *Aspergillus niger, Can. J. Microbiol.* **24**, 622–625.

DAVIS, H. (1948), Detection and occurrence of acid-producing fungi, *MSc Thesis,* University of Texas, Austin, TX.

DAWSON, M. W., MADDOX, I. S., BROOKS, J. D. (1989), Evidence for nitrogen catabolite repression during citric acid production by *Aspergillus niger* under phosphate limited growth conditions, *Biotechnol. Bioeng.* **33**, 1500–1504.

DERYCKE, D., PIPYN, P. (1990), Anaerobic digestion, ammonia stripping/recovery and (de)nitrification of a citric acid factory effluent, *Meded. Fac. Landbouwwet. Rijksuniv. Gent* **55**, 1481–1483.

DOELGER, W. P., PRESCOTT, S. C. (1934), Citric acid fermentation, *Ind. Eng. Chem.* **26**, 1142–1149.

Donau Chemie AG (1993), *Austrian Patent* 396464 B.

DRURI, M. ZAKOWSKA, Z. (1986), Isolation from beet molasses of fractions toxic to *Aspergillus niger* by extraction with organic solvents, *Przem. Ferment. Owocowo-Warzywny* **30**, 14–16.

DRURI, M., ZAKOWSKA, Z. (1987), Untersuchung von Bestandteilen der Zuckerrübenmelasse, die auf *Aspergillus niger* bei der Citronensäureproduktion toxisch wirken, *Acta Biotechnol.* **7**, 275–278.

DUFLOT, P., LELEU, J. B. (1989), *Eur. Patent* 346196.

EDLAUER, R., KIRKOVITS, A. E., WESTERMAYER, R., STOJAN, O. (1990), *Eur. Patent* 377430.

EIKMEIER H., REHM, H. J. (1984), Production of citric acid with immobilized *Aspergillus niger, 3rd, Eur. Congr. Biotechnol.,* Vol. 2, pp. 179–184. Weinheim: Verlag Chemie.

EIKMEIER H., REHM, H. J. (1987a), Semicontinuous and continuous production of citric acid with immobilized cells of *Aspergillus niger, Z. Naturforsch. C: Biosci.* **42**, 408–413.

EIKMEIER H., REHM, H. J. (1987b), Stability of calcium alginate during citric acid production of immobilized *Aspergillus niger, Appl. Microbiol. Biotechnol.* **26**, 105–111.

EYAL, A. M., BRESSLER, E. (1993), Industrial separation of carboxylic and amino acids by liquid membranes: applicability, process considerations, and potential advantages, *Biotechnol. Bioeng.* **41**, 287–295 (mini-review).

FEIR, H. A., SUZUKI, I. (1969), Pyruvate carboxylase from *Aspergillus niger:* kinetic study of a biotin-containing carboxylase, *Can. J. Biochem.* **47**, 697–710.

FELMAN, S. W., PATEL, C., PATWARDHAN, B. H., SOLOW, D. J. (1994), *U.S. Patent* 5,352.825

FELMAN, S. W., PATEL, C., PATWARDHAN, B. H., SOLOW, D. J. (1995a), *Eur. Pat. Appl.* 633239.

FELMAN, S. W., PATEL, C., PATWARDHAN, B. H., SOLOW, D. J. (1995b), *Eur. Pat. Appl.* 633240.

FERENCZY, L. (1984), in: *Cell Fusion: Gene Transfer and Transformation* (BEERS, R. F., Jr., BASSETT, E. G., Eds.), pp. 145–169. New York: Raven Press.

FERENCZY, L., FARKAS, G. L. (Eds.) (1980), *Advances in Protoplast Research,* Budapest: Akademiai Kiado; Oxford: Pergamon Press.

FERENCZY, L., KEVEI, F., SZEGEDI, M. (1975a), High-frequency fusion of fungal protoplasts, *Experientia* **31**, 1028–1029.

FERENCZY, L., KEVEI, F., SZEGEDI, M. (1975b), Fusion of fungal protoplasts, in: *Abstracts, 4th Int. Symp. Yeasts and Other Protoplasts,* Nottingham, p. 29.

FERNBACH, A., YUILL, J. L., Rowntree & Co, Ltd. (1927), *Brit. Patents* 266414; 266415.

FERNBACH, A., YUILL, J. L., MCLELLAN, B. G., Rowntree & Co. (1932), *Ger. Patent* 461356.

FERRARA, L., DE CESARI, L., SALVINI, E. (1977), *n*-Paraffins: a new feedstock for the production of citric acid and its derivatives, *Chim. Ind.* **59**, 202–206.

FIEDUREK, J., SZCZODRAK, J., ILCZUK, Z. (1988), Citric acid synthesis by *Aspergillus niger* mutants resistant to 2-desoxyglucose, *Acta Microbiol. Pol.* **36,** 303–307.

FODOR, L. M., COBB, R. L. (1986), *U.S. Patent* 4601340.

FOERSTER, H. J., POTENCSIK, I. (1971), *Ger. Patent Appl.* 2013756.

Food and Drug Administration, Washington (1975), Food additives. Solvent extraction process for citric acid, *Fed. Regist.* **40,** 49080–49082.

FOSTER, J. W. (1949), *Chemical Activities of Fungi.* New York: Academic Press.

FOSTER, J. W., DAVIS, H. (1949), *Bull. Torrey Cl.* **76**, 174.

FUJII, N., YASUDA, K., SAKAKIBARA, M. (1994), Effect of volume ratio of cellulose carriers and time interval of repeated batch culture on citric acid productivity by immobilized *Aspergillus niger, J. Ferment. Bioeng.* **78**, 389–393.

FUKUDA, H., SUZUKI, T., AKIYAMA, S., SUMINO, Y. (1982), *Ger. Patent* 2108094 C2.

FUKUI, S., TANAKA, A. (1980), Production of useful compounds from alkane media in Japan, *Adv. Biochem. Eng.* **17**, 1–35.

GADAU, M. E., LINGG, A. J. (1992), Protoplast fusion in fungi, in: *Handbook of Applied Mycology,* Vol. 4. (ARORA, D. K., ELANDER, R. P., MUKERJI, K. G., Eds.), pp. 101–128. New York: Marcel Dekker.

GADEN, E. L. (1955), Fermentation kinetics and productivity, *Chem. Ind. (N.Y.)* **12**, 154–159.

GARDNER, W. H. (1966), *Food acidulants,* New York: Allied Chemical Corporation.

GARDNER, J. F., JAMES, L. V., RUBBO, S. D. (1956), Production of citric acid by mutants of *Aspergillus niger, J. Gen. Microbiol.* **14**, 228–237.

GARG, K., SHARMA, C. B. (1992), Repeated batch production of citric acid from sugar cane molasses using recycled solid-state surface culture of *Aspergillus niger, J. Gen. Appl. Microbiol.* **38**, 605–615.

GOLDBERG, I., PELEG, Y., ROKEM, J. S. (1991), Citric, fumaric, and malic acids, in: *Biotechnology and Food Ingredients* (GOLDBERG, I., WILLIAMS, R. A., Eds.), pp. 349–374. New York: Van Nostrand Reinhold.

HABISON, A., KUBICEK, C. P., ROEHR, M. (1979), Phosphofructokinase as a regulatory enzyme in citric acid producing *Aspergillus niger, FEMS Microbiol. Lett.* **5**, 39–41.

HABISON, A., KUBICEK, C. P., ROEHR, M. (1983), Partial purification and regulatory properties of phosphofructokinase from *Aspergillus niger, Biochem. J.* **209**, 669–676.

HANG, Y. D. (1988), *U.S. Patent* 4,791.058 A.

HANG, Y. D., WOODAMS, E. E. (1987), Apple pomace: a potential substrate for citric acid production by *Aspergills niger, Biotechnol. Lett.* **9**, 183–186.

HARMSEN, H. J. M., KUBICEK-PRANZ, E. M., ROEHR, M., VISSER, J., KUBICEK, C. P. (1992), Regulation of 6-phosphofructo-2-kinase from the citric acid accumulating fungus *Aspergillus niger, Appl. Microbiol. Biotechnol.* **37**, 784–788.

HARTL, J., MARR, R. (1993), Extraction processes for bioproduct separation, *Sep. Sci. Technol.* **28**, 805–819.

HEINISCH, J. (1986), Isolation and characterization of the two structural genes coding for phosphofructokinase in yeast, *Mol. Gen. Genet.* **202**, 75–83.

HEINRICH, M., REHM, H. J. (1982), Formation of gluconic acid at low pH-values by free and immobilized *Aspergillus niger* cells during citric acid fermentation, *Eur. J. Appl. Microbiol. Biotechnol.* **15**, 88–92.

HELLMIG, R., BEHRENS, U., BOEHM, R., GOEBEL, R., STRIEGLER, J., WEISSBRODT, E. (1983), *Ger. Patent* (GDR) 203533.

HIRASHIMA, T., IKUTA, N., KUROHARA, T. (1985), *Jpn. Patent* 60193942.

HOCKERTZ, S., SCHMID, J., AULING, G. (1988), A specific transport system for manganese in the filamentous fungus *Aspergillus niger, J. Gen. Microbiol.* **133**, 2515–3519.

HOCKERTZ, S., PLÖNZIG, J., AULING, G. (1989), Impairment of DNA formation is an early event in *Aspergillus niger* under manganese starvation, *Appl. Microbiol. Biotechnol.* **25**, 590–593.

HONECKER, S., BISPING, B., YANG, Z., REHM, H. J. (1989), Influence of sucrose concentration and phosphate limitation on citric acid production by immobilized cells of *Aspergillus niger, Appl. Microbiol. Biotechnol.* **31**, 17–24.

HORITSU, H., ADACHI, S., TAKAHASHI, Y., KAWAI, K., KAWANO, Y. (1985), Production of citric acid by *Aspergillus niger* immobilized in polyacrylamide gels, *Appl. Microbiol. Biotechnol.* **22**, 8–12.

HOSSAIN, M., BROOKS, J. D., MADDOX, I. S. (1983), Production of citric acid from whey permeate by fermentation using *Aspergillus niger, N. Z. J. Dairy Sci. Technol.* **18**, 161–168.

HOSSAIN, M., BROOKS, J. D., MADDOX, I. S. (1984), The effect of sugar source on citric acid production by *Aspergillus niger, Appl. Microbiol. Biotechnol.* **19**, 383–391.

HUANG, Y., ZHAO, Y., SU, S. , GAO, L., ZHANG, W. (1990), Comprehensive utilization of calcium sulfate waste residues from production of citric acid, *Huaxue Shijie* **31**, 570–571.

HUSTEDE, H., RUDY, H. (1976), *U.S. Patent* 3941656.

IIZUKA, H., SHIMIZU, J., ISHII, K., NAKAJIMA, Y. (1969), *Ger. Patent* 1812710.

ILCZUK, Z. (1968), Genetics of citric acid producing strains of *Aspergillus niger,* I. Citric acid synthesis by morphological mutants induced with UV, *Acta Microbiol. Pol.* **17**, 331–336.

ILCZUK, Z. (1970), Genetik der Citronensäure erzeugenden Stämme von *Aspergillus niger.* 2. Citronensäuresynthese von mittels UV-Strahlen induzierten auxotrophen Mutanten von *A. niger, Die Nahrung* **14**, 97–105.

ILCZUK, Z. (1971a), Genetik der Citronensäure erzeugenden Stämme von *Aspergillus niger.* 3. Citronensäuresynthese durch erzwungene Heterokaryen zwischen auxotrophen Mutanten von *A. niger, Die Nahrung* **15**, 251–262.

ILCZUK, Z. (1971b), Genetik der Citronensäure erzeugenden Stämme von *Aspergillus niger.* 4. Citronensäuresynthese durch heterozygotische Diploide von *A. niger, Die Nahrung* **15**, 381–388.

IMI (TAMI) Inst. f. Research & Development, Haifa) (1972), *Isr. Patent* 39710.

JAKLITSCH, W. M., KUBICEK, C. P., SCRUTTON, M. C. (1991), Intracellular location of enzymes involved in citrate accumulation in *Aspergillus niger, Can. J. Microbiol.* **37**, 823–827.

JAMES, L. V., RUBBO, J. F., GARDNER, S. D. (1956), Isolation of high acid-yielding mutants of *Aspergillus niger* by a paper culture selection technique, *J. Gen. Microbiol.* **14**, 223–227.

JIANG, Y. M., SU, Y. F. (1987), Study on synthetic methods of trialkylphosphine oxides and their extraction behavior toward some acids, *Sep. Sci. Technol.* **22**, 315–323.

JIN, Q., LAN, Q., ZHU, B. (1993), The breeding of thermotolerant citric acid overproducing strains from *Aspergillus niger, Weishengwu Xuebao* **33**, 204–209.

JOHNSON, M. J. (1954), in: *Industrial Fermentations* Vol. 1 (UNDERKOFLER, L. A., HICKEY, R. J., Eds.), pp. 420–445. New York: Chem. Publ. Co.

KACZMAROWICZ, G. (1987), *Pol. Patent* 136823 B1.

KAISER, M., NOELTE, D., WEIZENBECK, E., MAY, U., KREIBICH, G., BEHRENS, U., STOTTMEISTER, U., WEISSBRODT, E., SCHMIDT, J., et al. (1990), *Ger. Patent* (GDR) 275480.

KAROW, E. O., WAKSMAN, S. A. (1947), Production of citric acid in submerged culture, *Ind. Eng. Chem.* **39**, 821–825.

KAWANO, T. (1987), *Jpn. Patent* 62062896.

KHAN, A. H., GHOSE, T. K. (1973), Kinetics of citric acid fermentation by *Aspergillus niger, J. Ferment. Technol.* **51**, 734–741.

KINOSHITA S., TANAKA, K., AKITA, S. (1961a), *U.S. Patent* 2973303.

KINOSHITA, S., TERADA, O., OISHI, K. (1961b), *U.S. Patent* 2993838.

KIRIMURA, K., YAGUCHI, T., USAMI, S. (1986), Intraspecific protoplast fusion of citric acid-producing strains of *Aspergillus niger, J. Ferment. Technol.* **64**, 473–479.

KIRIMURA, K., HIROWATARI, Y., USAMI, S. (1987), Alterations of respiratory systems in *Aspergillus niger* under the conditions of citric acid fermentation, *Agric. Biol. Chem.* **51**, 1299–1303.

KIRIMURA, K., NAKAJIMA, I., LEE, S. P., KAWABE, S., USAMI, S. (1988a), Citric acid production by the diploid strains of *Aspergillus niger* obtained by protoplast fusion, *Appl. Microbiol. Biotechnol.* **27**, 504–506.

KIRIMURA, K., LEE, S. P., NAKAJIMA, I., KAWABE, S., USAMI, S. (1988b), Improvement in citric acid production by haploidization of *Aspergillus niger* diploid strains, *J. Ferment. Technol.* **66**, 375–382.

KIRIMURA, K., SARANGBIN, S., RUGSASEEL, S., USAMI, S. (1992), Citric acid production by 2-desoxyglucose resistant mutant strains of *Aspergillus niger, Appl. Microbiol. Biotechnol.* **36**, 573–577.

KOEPP, H. J., SCHOBERTH, S. M., SAHM, H. (1985), Anaerobic treatment of wastewater from citric acid production, *Conserv. Recycl.* **8**, 211–219.

KOEPP-BANK, H. J. (1986), Microbial anaerobic treatment of wastewater from citric acid production, *Ber. Kernforschungsanlage Jülich,* 153 p.

KRETZSCHMANN, G., HUTT, G., SCHOLZ, D. (1975), *Ger. Patent* 2331670.

KRISTIANSEN, B., CHARLEY, R. (1981), *Adv. Biotechnol.* (Proc. 6th Int. Ferm. Symp.) Vol. 1, pp. 221–227 (MOO-YOUNG, M., ROBINSON, C. W., VEZINA, C., Eds.). Toronto: Pergamon.

KRISTIANSEN, B., SINCLAIR, C. G. (1979), Production of citric acid continuous culture, *Biotechnol. Bioeng.* **21**, 297–315.

KROISS, H. (1987), Aerobic and anaerobic treatment of industrial wastewater, *Chem.-Tech.* **16**, 98, 100.

KUBICEK, C. P. (1988a), Regulatory aspects of the tricarboxylic acid cycle in filamentous fungi – a review, *Trans. Br. Mycol. Soc.* **90**, 339–349.

KUBICEK, C. P. (1988b), The role of the citric acid cycle in fungal organic acid fermentations, *Biochem. Soc. Symp.* **54**, 113–126.

KUBICEK, C. P., ROEHR, M. (1985), Aconitase and citric acid fermentation by *Aspergillus niger, Appl. Environ. Microbiol.* **50**, 1336–1338.

KUBICEK, C. P., ROEHR, M. (1986), Citric Acid Fermentation, *CRC Crit. Rev. Biotechnol.* **3**, 331–373.

KUBICEK, C. P., HAMPEL, W. A., ROEHR, M. (1979a), Manganese deficiency leads to elevated amino acid pools in citric acid producing *Aspergillus niger, Arch. Microbiol.* **123**, 73–79.

KUBICEK, C. P., ZEHENTGRUBER, O., ROEHR, M. (1979b), An indirect method for studying the fine control of citric acid formation by *Aspergillus niger, Biotechnol. Lett.* **1**, 57–62.

KUBICEK, C. P., ZEHENTGRUBER, O., EL-KALAK, H., ROEHR, M. (1980), Regulation of citric acid production by oxygen: effect of dissolved oxygen tension on adenylate levels and respiration in *Aspergillus niger, Eur. J. Appl. Microbiol. Biotechnol.* **9**, 101–116.

KUBICEK, C. P., SCHREFERL-KUNAR, G., WOEHRER, W., ROEHR, M. (1988), Evidence for a cytoplasmic pathway of oxalate biosynthesis in *Aspergillus niger, Appl. Environ. Microbiol.* **54**, 633–637.

KUBICEK, C. P., WITTEVEEN, C. F. B., VISSER, J. (1994), Regulation of organic acid production by Aspergilli. In: *The Genus Aspergillus. From Taxonomy and Genetics to Industrial Application* (POWELL, K. A., RENWICK, A., PEBERDY, J. F., Eds.), pp 135–144. New York, London: Plenum Press.

KUBICEK-PRANZ, E. M., MOZELT, M., ROEHR, M., KUBICEK, C. P. (1990), Changes in the concentration of fructose-2,6-bisphosphate in *Aspergillus niger* during stimulation of acidogenesis by elevated sucrose concentrations, *Biochim. Biophys. Acta* **1033**, 250–255.

KULPRATHIPANJA, S. (1988), *U.S. Patent* 4720579.

KULPRATHIPANJA, S. (1989), *Eur. Patent* 324210.

KULPRATHIPANJA, S., OROSKAR, A. R. (1991), *U.S. Patent* 5068419.

KULPRATHIPANJA, S., STRONG, S. A. (1990), *U.S. Patent* 4924027.

KULPRATHIPANJA, S., OROSKAR, A. R., PRIEGNITZ, J. W. (1989), *U.S. Patent* 4851573.

LAKSHMINARAYANA, K., CHAUDHARY, K., ETHIRAJ, S., TAURO, P. (1975), Solid state fermentation method for citric acid production using sugar cane bagasse, *Biotechnol. Bioeng.* **17**, 291–293.

LAWRENCE, A. A. (1974), *Food Acid Manufacture. Recent Developments.* Park Ridge, NJ: Noyes Data Corporation.

LEE, Y. H., LEE, C. W., CHANG, H. N. (1989), Citric acid production by *Aspergillus niger* immobilized on polyurethane foam, *Appl. Microbiol. Biotechnol.* **30**, 141–143.

LEGISA, M., BENCINA, M. (1994), Evidence for the activation of 6-phosphofructo-1-kinase by cAMP-dependent protein kinase in *Aspergillus niger, FEMS Microbiol. Lett.* **118**, 327–334.

LEGISA, M. KIDRIC, J. (1989), Initiation of citric acid accumulation in the early stages of *Aspergillus niger* growth, *Appl. Microbiol. Biotechnol.* **31**, 453–457.

LEGISA, M. MATTEY, M. (1986), Glycerol as an initiator of citric acid accumulation in *Aspergillus niger, Enzyme Microb. Technol.* **8**, 607–609.

LEHNHARDT, W. F., SCHANEFELT, R. V., NAPIER, L. L. (1995), *World Patent* 9503268.

LEOPOLD, H. (1958), Versuche zur Erhöhung der Citronensäureproduktion eines aktiven Stammes von *Aspergillus niger, Die Nahrung* **2**, 140–155.

LEOPOLD, H., VALTR, Z. (1964), Zur Wirkung des Kaliumferrocyanids bei der Herstellung von Melasselösungen für die Citronensäuregärung, *Die Nahrung* **8**, 37–48.

LEOPOLD, H., VALTR, Z. (1970), Zur Frage der Qualitätsverbesserung der Konidien von *Aspergillus niger*-Stämmen bei der Citronensäuregärung, *Z. Allg. Mikrobiol.* **10**, 121–127.

LOBO, Z., MAITRA, P. K. (1977), Genetics of yeast hexokinase, *Genetics* **86**, 727–744.

LOCKWOOD, L. B. (1979), Citric acid, in: *Microbial Technology*, Vol. 1 (PEPPLER, H. J., Ed.), 2nd Edn., pp. 356–387. New York: Academic Press.

LUEDEKING, R., PIRET, E. L. (1959), A kinetic study of the lactic acid fermentation. Batch process at controlled pH, *J. Biochem. Microbiol. Technol. Eng.* **1**, 393–412.

MA, H., KUBICEK, C. P., ROEHR, M. (1985), Metabolic effects of manganese deficiency in *Aspergillus niger*: evidence for increased protein degradation, *Arch. Microbiol.* **141**, 266–268.

MADDOX, I. S., ARCHER, R. H. (1984), *Food Technol. N. Z.* **19**, 24–26.

MARCHAL, R., CHAUDE, O., METCHE, M. (1977a), Production of citric acid from *n*-paraffins by *Saccharomycopsis lipolytica*: kinetics and balance of the fermentation, *Eur. J. Appl. Microbiol.* **4**, 111–123.

MARCHAL, R., VANDECASTEELE, J. P., METCHE, M. (1977b), Regulation of the central metabolism in relation to citric acid production in *Saccharomycopsis lipolytica, Arch. Microbiol.* **113**, 99–104.

MARIER, J. R., CLARK, D. S. (1960), An improved colorimetric method for determining ferrocyanide ion, and its application to molasses, *Analyst* **85,** 574–578.

MARTIN, S. M. (1955), Effect of ferrocyanide on growth and acid production of *Aspergillus niger, Can. J. Microbiol.* **1,** 644–652.

MARTIN, S. M. WATERS, W. R. (1952), Production of citric acid by submerged fermentation, *Ind. Eng. Chem.* **44,** 2229–2239.

MARTINELLI, S. D., KINGHORN, J. R. (Eds.) (1994), *Aspergillus: 50 Years on.* Amsterdam, London, New York, Tokyo: Elsevier.

MARTINKOVÁ, L., MUSILKOVÁ, M., UJCOVÁ, E., MACHEK, F., SEICHERT, L. (1990), Protoplast fusion in *Aspergillus niger* strains accumulating citric acid, *Folia Microbiol. (Prague)* **35**, 143–148.

MATSUDA, K., YOSHIDA, J. (1991), *Jpn. Patent* 03183487.

MATSUMOTO, T., ICHIKAWA, Y. (1986), *Jpn. Patent* 61219391.

MATSUMOTO, T., FUJIMAKI, A., NAGATA, T. (1983), *U.S. Patent* 4411998.

MATTEY, M. (1977), Citrate regulation of citric acid production in *Aspergillus niger, FEMS Microbiol. Lett.* **2**, 71–74.

MATTEY, M., LEGISA, M., LOWE, S. (1988), Effect of lectins and inhibitors on membrane transport in *Aspergillus niger, Biochem. Soc. Trans.* **16**, 969–970.

MCQUIGG, D., MARSTON, C. K., FITZPATRICK, G., CROWE, E., VORHIES, S. (1992), *World Patent* 9216490.

MEIXNER-MONORI, B., KUBICEK, C. P., HABISON, A., KUBICEK-PRANZ, E. M., ROEHR, M. (1985), Presence and regulation of $\alpha$-ketoglutarate dehydrogenase multienzyme complex in the filamentous fungus *Aspergillus niger, J. Bacteriol.* **161,** 265–271.

MESSING, T., SCHMITZ, R. (1976), Citronensäure aus Saccharose – ein technisch genutzter mikrobieller Prozeß auf der Basis von Melasse, *Chem. Exp. Didakt.* **2,** 309–316.

MESSING, T., WAMSER, K. (1975), *Ger. Patent Appl.* 2543307.

MEZZADROLI, G. (1938), *French Patent* 833631.

MIALL, L. M., PARKER, G. F. (1975), *Ger. Patent* 2429224.

Miles Laboratories, Inc. (1961), *Brit. Patent* 905817.

MILSOM, P. E. (1987), Organic acids by fermentation, especially citric acid, in: *Food Biotechnology,* Vol. 1 (KING, R. D., CHEETHAM, P. S. J., Eds.), pp. 273–307. London: Elsevier.

MILSOM, P. E., MEERS, J. L. (1985), Citric acid, in: *Comprehensive Biotechnology,* Vol. 3 (BLANCH, H. W., DREW, S., WANG, D. I. C., Eds.), pp. 665–680. Oxford: Pergamon Press.

Montan- und Industrialwerke (vorm. J. D. Starck) (1924), *Ger. Patent* 461356.

Montan- und Industrialwerke (vorm. J. D. Starck) (1933), *Ger. Patent* 578820.

MORESI, M., CIMARELLI, D., GASPARRINI, G., LIUZZO, G., MARINELLI, R. (1980), Kinetics of citric acid fermentation from *n*-paraffins by yeasts, *J. Chem. Tech. Biotechnol.* **30**, 266–277.

MOYER, A. J. (1953a), Effect of alcohols on the mycological production of citric acid in surface and submerged culture. 1. Nature of the alcohol effect, *Appl. Microbiol.* **1**, 1–7.

MOYER, A. J. (1953b), Effect of alcohols on the mycological production of citric acid in surface and submerged culture. 2. Fermentation of crude carbohydrates, *Appl. Microbiol.* **1**, 8–13.

NEVALEINEN, H., PENTTILÄ, M., (Eds.) (1989), *Molecular Biology of Filamentous Fungi, Foundation for Biotechnical and Industrial Fermentation Research,* Vol. 6. Helsinki: Foundation for Biotechnical and Industrial Fermentation Research.

NOYES, R. (1969), *Citric Acid Production Processes.* Park Ridge, NJ: Noyes Development Corporation.

OGAWA, K., TSUCHIMOCHI, M., TANIGUCHI, K., NAKATSU, S. (1989), Interspecific hybridization of *Aspergillus usamii* mut. *shirousamii* and *Aspergillus niger* by protoplast fusion, *Agric. Biol. Chem.* **53**, 2873–2880.

OSMANI, S. A., SCRUTTON, M. C. (1983), The subcellular location of pyruvatecarboxylase and some other enzymes in *Aspergillus nidulans, Eur. J. Biochem.* **133**, 551–560.

PEBERDY, J. F. (1987), Developments in protoplast fusion in fungi, *Microbiol. Sci.* **4**, 108–114.

PEBERDY, J. F., FERENCZY, L. (Eds.) (1985), *Fungal Protoplasts: Applications in Biochemistry and Genetics.* New York: Marcel Dekker.

PELECHOVA, J., PETROVA, L., UJCOVA, E., MARTINKOVA, L. (1990), Selection of a hyperproducing strain of *Aspergillus niger* for biosynthesis of citric acid on unusual carbon substrates, *Folia Microbiol. (Prague)* **35**, 138–142.

PERLMAN, D. (1949), Mycological production of citric acid – the submerged culture method, *Econ. Bot.* **3**, 360–374.

PERLMAN, D., SIH, C. J. (1960), Fungal synthesis of citric, fumaric, and itaconic acid, *Progr. Ind. Microbiol.* **2**, 167–194.

PERLMAN, D., KITA, D. A., PETERSON, W. H. (1945), Production of citric acid from cane molasses, *Arch. Biochem. Biophys.* **11**, 123–129.

PERQUIN, L. H. C. (1938), Bijdrage tot de kennis der oxydatieve dissimilatie van *Aspergillus niger* van tieghem, *Dissertation,* Technical University, Delft.

POELLMANN, H., MICHAUX, M., NELSON, E. B. (1990), Study on the influence of citric acid on retardation and formation of new hydrates in cement, *Proc. 12th Int. Conf. Cem. Microsc.,* pp. 303–322.

PONTECORVO, G., ROPER, J. A., FORBES, E. (1953), Genetic recombination without sexual reproduction in *Aspergillus niger, J. Gen. Microbiol.* **8**, 198–210.

POWELL, C. (1973), Citric acid process is safeguarded by complex control system, *Proc Eng.,* January, 87–89.

POWELL, K. A., RENWICK, A., PEBERDY, J. (Eds.) (1994), *The Genus Aspergillus from Taxonomy and Genetics to Industrial Application.* FEMS Symposium No. 69. New York: Plenum Publishing Corp.

PRESCOTT, S. C., DUNN, C. G. (1959), *Industrial Microbiology,* 3rd Edn., pp. 533–618; 642–646. New York: McGraw Hill.

PRIOR, C., MAMESSIER, P., FUKUHARA, H., CHEN, X. J., WESOLOWSKI-LOUVEL, M. (1993), The hexokinase gene is required for transcriptional regulation of the glucose transporter gene RAGI in *Kluyveromyces lactis, Mol. Cell. Biol.* **13**, 3882–3889.

PROCHAZKA, J., HEYBERGER, A., BIZEK, V., KOUSOVA, M., VOLAUFOVA, E. (1994), Amine extraction of hydroxycarboxylic acids. 2. Comparison of equilibria for lactic, malic, and citric acids, *Ind. Eng. Chem. Res.* **33**, 1565–1573.

PUROHIT, H. J., RATLEDGE, C. (1988), Mitochondrial location of pyruvate carboxylase in *Aspergillus niger, FEMS Microbiol. Lett.* **55**, 129–132.

QUILICO, A., PANIZZI, L., VISCONTI, N. (1949), *R. C. Accad. Lincei* **6**, 40.

RAPER, K. B., THOM, C. (1949), *A Manual of the Penicillia.* Baltimore, MD: Williams and Wilkins.

RIEGER, M., KIOUSTELIDIS, J. (1975), *Ger. Patent* 2355059.

RIPPEL-BALDES, A. (1940), Über die Verbreitung von *Aspergillus niger*, insbesondere in Deutschland, *Arch. Mikrobiol.* **11**, 1–32.

ROEHR, M., KUBICEK, C. P. (1981), Regulatory aspects of citric acid fermentation by *Aspergillus niger, Proc. Biochem.* **16**, 34–38.

ROEHR, M., STADLER, P. J., SALZBRUNN, W., KUBICEK, C. P. (1979), An improved method for characterization of citrate production by *Aspergillus niger, Biotechnol. Lett.* **1**, 281–286.

ROEHR, M., ZEHENTGRUBER, O., KUBICEK, C. P. (1981), Kinetics of biomass formation and citric acid production by *Aspergillus niger* on pilot plant scale, *Biotechnol. Bioeng.* **23**, 2433-2445.

ROEHR, M., KUBICEK, C. P., KOMINEK, J. (1983), Citric acid, in: *Biotechnlogy,* 1st Edn., Vol. 3 (REHM, H.-J., REED, G., Eds.), pp. 419–454. Weinheim: Verlag Chemie.

ROEHR, M., KUBICEK, C. P., KOMINEK, J. (1992), Industrial acids and other small molecules, in: *Aspergillus: Biology and Industrial Applications* (BENNETT, J. W., KLICH, M. A., Eds.), pp. 91–131. Stoneham: Butterworth-Heinemann.

ROSENBAUM, J. B. , MCKINNEY, W. A., BEARD, H. R., CROCKER, L., NISSEN, W. I. (1973), Sulfur dioxide emission control by hydrogen sulfide reaction in aqueous solution, *Report of Investigations 7774,* U.S. Department of the Interior, Bureau of Mines, Washington, DC: U.S. Government Printing Office.

RUDY, H. (1967), *Fruchtsäuren.* Heidelberg: Hüthig-Verlag.

RUGSASEEL, S., KIRIMURA, K., USAMI, S. (1993), Selection of mutants of *Aspergillus niger* showing enhanced productivity of citric acid from starch in shaking culture, *J. Ferment. Bioeng.* **75**, 226–228.

SALZBRUNN, W., KOMINEK, J., ROEHR, M. (1989), *Eur. Patent* 221880 B1.

SALZBRUNN, W., KOMINEK, J., ROEHR, M. (1991), *Eur. Patent* 221888 B1.

SAMSON, R. A. (1992), Current taxonomic schemes of the genus *Aspergillus* and its teleomorphs, in: *Aspergillus: Biology and Industrial Applications* (BENNETT, J. W., KLICH, M. A., Eds.), pp. 355–390. Stoneham: Butterworth-Heinemann.

SATAVA, V., VEPREK, O., SKVARA, F. (1977), Possibilities of utilizing waste gypsum from the chemical industry, *Sb. Vys. Sk. Chem. Technol. Praze, Chem. Technol. Silik.* **L7**, 101–148.

SCHELLENBERGER, A., DOEPFER, K. P., FOERSTER, M., HOFFMANN, T., STEINMETZER, K., SCHAAF, R., MUELLER, U., KLUGE, H. (1991), *Ger. Patent* (GDR) 285792.

SCHMIDT, M., WALLRATH, J., DÖRNER, A., WEISS, H. (1992), Disturbed assembly of the respiratory chain NADH:ubiquinone reductase (complex I), in citric acid accumulating *Aspergillus niger* strain B60, *Appl. Microbiol. Biotechnol.* **36**, 667–672.

SCHMITZ, R. (1976), Modifiziertes Kammersystem für die Fermentation kohlenhydrathaltiger Lösungen zu Citronensäure nach dem Oberflächenverfahren, *Chem.-Tech.* **5**, 271–272.

SCHMITZ, R. (1977), Problemlösungen bei der Aufarbeitung von Fermentationslösungen zu reiner Citronensäure, *Chem.-Tech.* **6**, 255–259.

SCHMITZ, R., WÖHLK, W. (1980), Die Fällungskristallisation am Beispiel der Tricalciumcitrat-Fällung im Aufbereitungsprozeß der Citronensäureherstellung, *Chemie-Technik* **9**, 345–348.

SCHREFERL, G., KUBICEK, C. P., ROEHR, M. (1986), Inhibition of citric acid accumulation by manganese ions in *Aspergillus niger* mutants with reduced citrate control of phosphofructokinase, *J. Bacteriol.* **165**, 1019–1022.

SCHREFERL-KUNAR, G., GROTZ, M., ROEHR, M., KUBICEK, C. P. (1989), Increased citric acid production by mutants of *Aspergillus niger* with increased glycolytic capacity, *FEMS Microbiol. Lett.* **59**, 297–300.

SCHULZ, G. (1959), *Ger. Patent* 1055483.

SCHULZ, G. (1963), *U.S. Patent* 3086928.

SCHULZ, G., RAUCH, J. (1975), Citronensaure, in: *Ullmanns Encyklopädie der technischen Chemie,* 4th Edn., Vol. 9 (BARTHOLOMÉ, E., BIEKERT, E., HELLMANN, H., LEY, H., WEIGERT, M., Eds.), pp. 624–636. Weinheim: Verlag Chemie.

SCHWEIGER, L. B. (1959), *Swiss Patent* 342191.

SCHWEIGER, L. B. (1961), *U.S. Patent* 2970084.

SCHWEIGER, L. B. (1965), *Ger. Patent* 1198311.

SCHWEIGER, L. B., SNELL, R. L. (1949), *U.S. Patent* 2476159.

SEEHAUS, C., PILZ, F., AULING, G. (1990), High affinity manganese transport system by filamentous fungi – e.g. *Aspergillus nidulans, Aspergillus niger*; potential application to e.g. citric acid production, *Zbl. Bakt.* **272**, 357–358 (abstract).

SHANKARANAND, V. S., LONSANE, B. K. (1994), Idiosyncrasies of solid-state fermentation systems in the biosynthesis of metabolites by some bacterial and fungal cultures, *Proc. Biochem.* **29**, 29–37.

SHIMIZU, J., ISHII, K., NAKAJIMA, Y., IIZUKA, H. (1970), *Jpn. Patent* 45039035.

SHISHIKURA, A. (1944), *Jpn. Patent* 06048979.

SHU, P. (1947), *Ph. D. Thesis,* University of Wisconsin, Madison.

SHU, P., JOHNSON, M. J. (1947), Effect of the composition of the sporulation medium on citric acid production by *Aspergillus niger* in submerged culture, *J. Bacteriol.* **54**, 161–167.

SHU, P., JOHNSON, M. J. (1948a), Citric acid production by submerged fermentation with *Aspergillus niger, Ind. Eng. Chem.* **40**, 1202–1205.

SHU, P., JOHNSON, M. J. (1948b), The interdependence of medium constituents in citric acid production by submerged fermentation, *J. Bacteriol.* **56**, 577–585.

SIEBERT, D., HUSTEDE, H. (1982), Citronensäure-Fermentation – biotechnologische Probleme und Möglichkeiten der Rechnersteuerung, *Chem.-Ing.-Tech.* **54**, 659–669.

SIMONYAN, M. G., RESHETNIKOVA, I. A., GORLENKO, M. V. (1988), Metabolite synthesis by *Aspergillus niger* v. Tiegh. after its storage in different conditions during growth on polymers, *Mikol. Fitopatol.* **22**, 240–243 (from CA 109:146227).

SMITH, J. E. (Ed.) (1994), *Aspergillus. Biotechnology Handbooks*, Vol. 7. New York: Plenum Publishing Corporation.

SMITH, J. E., NG, W. S. (1972), Fluorometric determination of glycolytic intermediates and adenylates during sequential changes in replacement culture of *Aspergillus niger, Can. J. Microbiol.* **18**, 1657–1664.

SNELL, R. L., SCHWEIGER, L. B. (1949), *U.S. Patent* 2492667.

SOMKUTI, G. A., BENCIVENGO, M. M. (1981), Citric acid fermentation in whey permeate, *Dev. Ind. Microbiol.* **22**, 557–563.

STEEL, R., LENTZ, C. P., MARTIN, S. M. (1955), Submerged citric acid fermentation of sugar beet molasses: increase in scale, *Can. J. Microbiol.* **1**, 299–311.

STEINBÖCK, F., CHOOJUN, S., HELD, I., ROEHR, M., KUBICEK, C. P. (1993), Characterization and regulatory properties of a single hexokinase from the citric acid accumulating fungus *Aspergillus niger, Biochim. Biophys. Acta* **1200**, 215–223.

STOTTMEISTER, U., BEHRENS, U., WEISSBRODT, E., BARTH, G., FRANKE-RINKER, D., SCHULZE, E. (1982), Use of paraffins and other noncarbohydrate carbon sources for microbial citric acid synthesis, *Z. Allg. Mikrobiol.* **22**, 399–424.

STOTTMEISTER, U., BEHRENS, U., WEISSBRODT, E., WEIZENBECK, E., DUERESCH, R., KAISER,

M., NOELTE, D., RICHTER, H. P., SCHMIDT, J., et al. (1986), *Ger. Patent* (GDR) 239610.

SVARDAL, K., GOETZENDORFER, K., NOWAK, O., KROISS, H. (1993), Treatment of citric acid wastewater for high quality effluent on the anaerobic-aerobic route, *Water Sci. Technol.* **28**, 177–186.

SWARTHOUT, E. J. (1966), *U.S. Patent* 3285831.

SZÜCS, J. (1925), *Austrian Patent* 101009.

SZÜCS, J. (1944), *U.S. Patent* 2353771.

TABUCHI, T., ABE, M. (1968), *Jpn. Patent* 43020707.

TABUCHI, T., HARA, S. (1973), Organic acid fermentation by yeasts. VII. Conversion of citrate fermentation to polyol fermentation in *Candida lipolytica, J. Agr. Chem. Soc. Jpn.* **47**, 485–490.

TABUCHI, T., TANAKA, M., ABE, M. (1969), Production of citric acid by *Candida lipolytica* No. 228, *J. Agr. Chem. Soc. Jpn.* **43**, 154–158.

TABUCHI, T., TAHARA, Y., TANAKA, M., YANAGIUCHI, S. (1973), Organic acid fermentation by yeasts. IX. Preliminary experiments on the mechanism of citrate fermentation in yeasts. *J. Agr. Chem. Soc. Jpn.* **47**, 617–622.

TANAKA, K., KIMURA, K, YAMAGUCHI, K. (1968), *Jpn. Patent* 43013677.

THOM, C., CURRIE, J. N. (1916), *Aspergillus niger* group, *J. Agric. Res.* **VII**, 1–15.

TORRES, N. V. (1994a), Modelling approach to control carbohydrate metabolism during citric acid accumulation by *Aspergillus niger*: I: Model definition and stability of the steady state, *Biotechnol. Bioeng.,* **44**, 104–111.

TORRES, N. V. (1994b), Modelling approach to control carbohydrate metabolism during citric acid accumulation by *Aspergillus niger*: II. Sensitivity analysis, *Biotechnol. Bioeng.* **44**, 112–118.

TRONDHEIM, O. E. (1975), *U.S. Patent* 3886069.

TSUDA, M., FUJIWARA, Y., SHIRAISHI, T. (1975), *U.S. Patent* 3904684.

TVEIT, M. T. (1963), *U.S. Patent* 3105015.

UCHIO, R., KIYOMI, N., TAKINAMI, K. (1979), *Jpn. Patent* 54023193.

UNROD, V. I., DEJKALO, A. A., LYZHENKOVA, I. I., FEDORYAK, V. N. (1992), from: *Izobreteniya 1992,* **(13)**, 86–87, *S.U. Patent* 1724618 A1.

USAMI, S. FUKUTOMI, N. (1977), Citric acid production by a solid fermentation method using sugar cane bagasse and concentrated liquor of pineapple waste, *Hakko Kogaku Kaishi* **55**, 44–50.

VAIJA, J., LINKO, P. (1986), Continuous citric acid production by immobilized *Aspergillus niger.* Reactor performance and fermentation kinetics, *J. Mol. Catal.* **38**, 237–253.

VAIJA, J., LINKO, Y. Y., LINKO, P. (1982), Citric acid production with alginate bead entrapped *Aspergillus niger* ATCC 9142, *Appl. Microbiol. Biotechnol.* **7**, 51–54.

VERHOFF, F. H. (1986), Citric acid, in: *Ullmann's Encyclopedia of Industrial Chemistry,* 5th Edn., Vol. A7 (GERHARTZ, W., Ed.), pp. 103–108. Weinheim: VCH.

WAKSMAN, S. A., KAROW, E. O. (1946), *U.S. Patent* 2394031.

WALLRATH, J., SCHMIDT, M., WEISS, H. (1991), Concomitant loss of respiratory chain NADH:ubiquinone reductase (complex I), and citric acid accumulation by *Aspergillus niger, Appl. Microbiol. Biotechnol.* **36**, 76–81.

WEGNER, H., KELLERWESSEL, H. (1977), *Ger. Patent* 2618608.

WEHMER, C. (1896), in: *Technische Mykologie* (LAFAR, F., Ed.). Jena: Fischer.
(*Technical Mycology,* translation by SALTER, C. T. C., 1898–1910. London: C. Griffith.)

WEHMER, C. (1903), *Beiträge zur Kenntnis einheimischer Pilze. I. Zwei neue Schimmelpilze als Erreger einer Citronensäure-Gärung.* Hannover/Leipzig: Halm.

WEHMER, C. (1907), in: *Handbuch der Technischen Mykologie,* Vol. 4 (LAFAR, F., Ed.), pp. 192–238, 239–270. Jena: Fischer.

WEISSBRODT, E., BEHRENS, U., STOTTMEISTER, U., DUERESCH, R., RICHTER, H. P., SCHMIDT, J. (1987a), *Ger. Patent* (GDR) 248376.

WEISSBRODT, E., BEHRENS, U., STOTTMEISTER, U., DUERESCH, R., RICHTER, H. P., SCHMIDT, J. (1987b), *Ger. Patent* (GDR) 248377.

WELLER, G. (1981), Problems with determination of [wastewater discharge] minimum requirements in the molasses-utilizing industry, *Münch. Beitr. Abwasser-, Fisch-. Flussbiol.* **33** *(Allg. Anerkannte Regeln Techn. Mindestanforderungen Gewässerschutz),*141–154.

WENDEL, K. (1957), Zur Frage der Stoffwechselsteigerung und Säuerung bei *Aspergillus niger* in Abhängigkeit von Züchtungs- und Substratbedingungen. I. Verfahren zur Erhaltung und Steigerung der Gärleistung säurebildender Pilze, insbesondere der Citronensäuregärstämme, *Zentralbl. Bakt. 2 Abt.* **110**, 312–318.

WENDLER, L. (1988), Unused possibilities in the manufacture of high-quality plaster of Paris, *Sklar Keram.* **38**, 35–36.

WOODWARD, J. C., SNELL, R. L., NICHOLLS, R. S. (1949), *U.S. Patent* 2492673.

XU, W. Q., HANG, Y. D. (1988), Roller culture technique for citric acid production by *Aspergillus niger, Proc. Biochem.* **23**, 117–118.

XU, D. B., MADRID, C. P., ROEHR, M., KUBICEK, C. P. (1989a), The influence of type and concentration of the carbon source on production of citric acid by *Aspergillus niger, Appl. Microbiol. Biotechnol.* **30**, 553–558.

XU, D. B., ROEHR, M., KUBICEK, C. P. (1989b), *Aspergillus niger* cyclic AMP levels are not influenced by manganese deficiency and do not correlate with citric acid accumulation, *Appl. Microbiol. Biotechnol.* **32**, 124–128.

YAMADA, K. (1965), "Science in Japan", 401, cited by LOCKWOOD, L. B. (1979).

YAMADA, K. (1977), Recent advances in industrial fermentation in Japan, *Biotechnol. Bioeng.* **19**, 1563–1622.

YI, M., PEN, Q., CHEN, D., PEN, L., ZHANG, M., WEN, R., MOU, X., WANG, W. (1987), Extraction of citric acid by N,N-disubstituted alkyl amides from fermentation aqueous solution, *Beijing Daxue Xuebao, Ziran Kexueban* **4**, 30–37. (from CA 108:110761).

YO, K. (1975), *Jpn. Patent* 50154488.

ZAHORSKY, B. (1913), *U.S. Patent* 1065358.

ZAKOWSKA, Z., GABARA, B. (1991), Changes in development and ultrastructure of *Aspergillus niger* mycelium in the presence of toxic substances from molasses in citric acid fermentation medium, *Acta Microbiol. Pol.* **40**, 243–254.

ZAKOWSKA, Z., JALOCHA, A. (1984), Fraktionierung von Inhaltsstoffen der Zuckerrübenmelasse, die toxisch für die Citronensäureproduktion mit *Aspergillus niger* sind, *Acta Biotechnol.* **4**, 171–178.

ZEHENTGRUBER, O., KUBICEK, C. P., ROEHR, M. (1980), Alternative respiration of *Aspergillus niger, FEMS Microbiol. Lett.* **8**, 71–74.

ZHANG, Y., CHENG, Z. (1987), Teratogenic effect of citric acid in rats, *Shanghai Yike Daxue Xuebao* **14**, 195–198 (from CA 107:216314).

# 10 Gluconic Acid

MAX ROEHR
CHRISTIAN P. KUBICEK
JIŘÍ KOMÍNEK

Wien, Austria

# 1 Introduction and Historical View

Gluconic acid was discovered in 1870 by HLASIWETZ and HABERMANN upon oxidation of glucose by chlorine. Gluconic acid production by microorganisms was observed in 1880 by BOUTROUX in the course of studies on lactic acid fermentation and verified by several authors as the action of acetic acid bacteria (see LAFAR, 1914).

In 1922, MOLLIARD detected gluconic acid in cultures of *Sterigmatocystis nigra (Aspergillus niger)*. Subsequently, production of gluconic acid was observed with a great number of microorganisms, especially filamentous fungi (e.g., *Penicillium*) and oxidative bacteria such as *Pseudomonas* and related genera, *Acetobacter, Gluconobacter, Zymomonas* etc.

Systematic studies by BERNHAUER (1924, 1926; cf. BERNHAUER, 1940) revealed that *Aspergillus niger* can convert glucose to gluconic acid with high yields if the acid formed is neutralized, e.g., by the addition of calcium carbonate. Although BERNHAUER's findings were patented (BERNHAUER and SCHULHOF, 1932), industrial production was first established in the USA following pilot plant studies in the laboratories of the U.S. Department of Agriculture (cf. FOSTER, 1949; UNDERKOFLER, 1954; PRESCOTT and DUNN, 1959; WARD, 1967, for historical and MIALL, 1978 for more recent details). According to these sources, MAY et al. (1927), when screening 172 fungal strains in search of potential tartaric acid producers, found active gluconic acid production mainly by strains of the *Penicillium luteum–purpurogenum* group. A semi-plant scale surface process was established (MAY et al., 1929) which gave rather low yields (about 57% of the theoretical values) and low productivities (about 2.3 g $L^{-1}$ $h^{-1}$). Subsequently, after studying the influence of increased aeration (MAY et al., 1934), a submerged process was developed comprising a revolving drum (9–12 rev $min^{-1}$) equipped with internal baffles and enabling the introduction of air under pressure (HERRICK et al., 1935). As a commercial reactor, this type of fermenter apparently was a deadlock, but it was suitable for studying most of the essential fermentation parameters (WELLS et al., 1937). With a new strain of *Aspergillus niger* (MOYER et al., 1937), high aeration rates (about 0.375 vvm) and addition of sufficient amounts of calcium carbonate, 15–20% glucose solutions could be converted with yields in excess of 95%, based on sugar present, in a 24 h fermentation period (GASTROCK et al., 1938). The concentration of glucose was mainly limited by the solubility of the calcium gluconate formed (about 40 g $L^{-1}$); fortunately calcium gluconate was found to be able to form supersaturated solutions. The semi-plant scale reactor of 1.75 $m^3$ could be charged with about 33–42% of its volume. Repeated use of the mycelium (more than ten times) was found to be feasible (PORGES et al., 1940, 1941). It was also discovered that the concentration of glucose could be increased up to 350 g $L^{-1}$ by the addition of boron compounds as complexing agents (MOYER et al., 1940; MOYER, 1944). The use of 0.5–2.5 g $L^{-1}$ borates in solutions of 200–350 g $L^{-1}$ glucose prevents precipitation of calcium gluconate thus yielding calcium borogluconates; detrimental effects upon fungal growth were minimized by selection of resistant strains and by adding these compounds only during the later phases of fungal growth. Calcium borogluconate thus formed was even recommended as a special curative agent in the treatment of milk fever in cows. However, the finding that borogluconate is harmful to animal blood vessels was reason enough to disregard this process variant.

In 1933, CURRIE et al. (Ch. Pfizer & Co.) had developed an advanced submerged process for gluconic acid production which apparently was not recognized by other workers in the field. Only in the early 1950s, when the sequestering properties of sodium gluconate were appreciated, it was at the laboratories of the U.S.D.A. (Peoria) that a modified process utilizing sodium hydroxide was developed (BLOM et al., 1952) which formed the basis of modern sodium gluconate production processes (HUSTEDE et al., 1989).

In the meantime, several attempts have also been described to transfer bacterial gluconic acid production into commercial practice. Several aspects of these and more recent developments are treated below.

# 2 Biological Fundamentals

## 2.1 Fungal Gluconate Production, Regulation of Gluconate Accumulation

The biochemical processes involved in gluconic acid formation by *A. niger* are rather straightforward, as shown in Fig. 1. The initial step of the pathway is the oxidation of glucose to glucono-δ-lactone which is catalyzed by flavoprotein glucose oxidase (EC 1.1.3.4) with concomitant reduction of oxygen to hydrogen peroxide. Glucono-δ-lactone is hydrolyzed either by a lactonase (EC 3.1.1.17) or hydrolyzes spontaneously. Hydrogen peroxide is detoxified by catalases (EC 1.11.1.6). The physiological function of glucose oxidase is not quite clear. Most likely is its contribution to the competitiveness of the organism by removal of glucose, by concomitant acidification of the environment and/or by the formation of hydrogen peroxide to which *A. niger* itself is very resistant. Biocontrol of the phytopathogenic fungus *Verticillium dahliae* by *Talaromyces flavus* has been ascribed to hydrogen peroxide (KIM et al., 1988). It has been suggested that in some white-rot fungi like *Phanerochaete chrysosporium* glucose oxidase contributes to the supply of hydrogen peroxide required for lignin degradation by lignin peroxidases (RAMASAMY et al., 1985; KELLEY and REDDY, 1986).

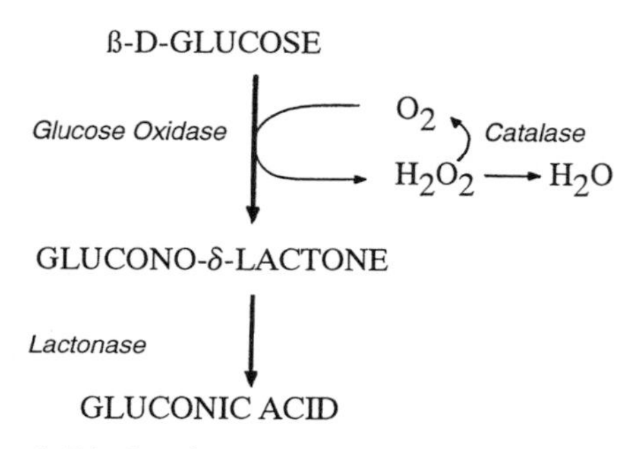

**Fig. 1.** Biochemistry of gluconate formation.

### 2.1.1 Localization of the Enzymes Involved in Gluconate Formation

For a long time it was assumed that in *A. niger* glucose oxidase is located intracellularly (PAZUR, 1966). This was supported by cytochemical and ultrastructural studies by VAN DIJKEN and VEENHUIS (1980). MISCHAK et al. (1985), however, showed that under manganese-deficient growth conditions glucose oxidase activity was found almost quantitatively in the medium. According to their explanation a cell-wall located glucose oxidase enters the culture fluid because of an altered cell wall composition resulting from manganese deficiency. There are also several other arguments in favor of an extracellular localization of glucose oxidase: the enzyme is strongly glycosylated which has never been observed for any peroxisomal protein; the $K_m$ value for glucose is high (15 mM), and hence the enzyme is unlikely to be active *in vivo* if located intracellularly. Finally, the aa-sequence of glucose oxidase from *A. niger* as deduced from its gene sequence (KRIECHBAUM et al., 1989; FREDERICK et al., 1990; WHITTINGTON et al., 1990) is preceded by a 22 amino acid N-terminal prepro-sequence bearing a putative signal peptidase processing site and a monobasic processing site (FREDERICK et al., 1990). These are all typical features of a secretory protein. WITTEVEEN et al. (1992) have used immuno-electron microscopical methods to definitely demonstrate that glucose oxidase is indeed localized in the fungal cell wall. These authors also demonstrated the presence of 4 different catalases in *A. niger*. Two of these catalases (CAT I, localized intracellularly; CAT II, occurring in the cell wall) are formed constitutively. Under conditions inducing glucose oxidase and gluconic acid formation two further catalases (the intracellular CAT III and the extracellular CAT IV) appear, thus protecting the fungal cell against hydrogen peroxide by a multiple barrier. The gene encoding one of these catalases has recently been cloned and characterized (FOWLER et al., 1993). Lactonase is also secreted and found to be distributed in equal amounts over culture fluid and cell wall (WITTEVEEN et al., 1992). The cell-wall bound localization of glucose oxidase and other auxiliary enzymes for gluconic acid formation is the reason why the mycelium can be successfully reused for subsequent fermentations (REUSS et al., 1986).

### 2.1.2 Recombinant Glucose Oxidase Production

Glucose oxidase overexpression has also been achieved by transforming *A. niger* with the glucose oxidase gene (WHITTINGTON et al., 1990; WITTEVEEN et al., 1993). The three- to four-fold increase in glucose oxidase activity which was observed when using optimal induction conditions (WITTEVEEN et al., 1993), was also observed in amplifications of genes encoding other *Aspergillus* secretory proteins (WARD, 1991). The glucose oxidase gene was introduced and expressed in yeast under control of the regulated *ADH2–GAPDH* hybrid promoter (FREDERICK et al., 1990) and in *Aspergillus nidulans* under its own promoter (WHITTINGTON et al., 1990; WITTEVEEN et al., 1993). In yeast the enzyme is secreted and it becomes overglycosylated (MW 350–400 kDa) whereas the MWs (150–180 kDa) in *A. niger* and *A. nidulans* are indistinguishable. However, WITTEVEEN et al. (1993) observed host-dependent differences in the isoelectric focusing patterns of the two glucose oxidases which are probably the result of slight differences in glycosylation.

### 2.1.3 Induction of Glucose Oxidase of *Aspergillus niger*

Because of the high $K_m$ for oxygen and glucose of glucose oxidase (GIBSON et al., 1964) both the dissolved oxygen level as well as the available glucose concentration are key parameters in the process kinetics of gluconate formation (REUSS et al., 1986). Both conditions also induce the enzyme system involved in gluconic acid production. Isolation of glucose oxidase overproducing strains has been achieved in several laboratories, but their biochemical alterations have not been analyzed (KUNDU and DAS, 1985; FIEDUREK et al., 1986; MARKWELL et al. 1989). WITTEVEEN et al. (1990), using a very sensitive colony screening method, isolated a variety of glucose oxidase-overproducing mutants as well as a single glucose oxidase-negative mutant. These *gox* mutations have been genetically localized and belong to 9 different complementation groups (SWART et al., 1990): Three *gox* loci (*gox*B, *gox*F, *gox*C) belong to linkage group II, one (*gox*I) to linkage group III, two (*gox*D, *gox*G) to linkage group V, two (*gox*A, *gox*E) to linkage group VII, whereas *gox*H could not be assigned. All *gox* mutations were found to be recessive. Based on these findings, SWART et al. (1990) outlined a strategy to construct improved strains which carry up to 4 different *gox* overproducing mutations. WITTEVEEN et al. (1993) investigated the induction characteristics of glucose oxidase, catalase and lactonase activities in various *gox* mutants. Analysis of the overproducing mutants revealed that the induction of glucose oxidase by the high oxygen tension and the high glucose concentrations both involve different mechanisms. The *gox*C17 mutant – in which glucose oxidase activity is absent – can be complemented by the glucose oxidase structural gene. None of the three enzyme activities are induced in this mutant. Induction of lactonase and catalase, therefore, requires an active glucose oxidase. This is in agreement with the findings that all three enzymes are induced in the wild-type strain by hydrogen peroxide even in the absence of glucose. The *gox*B and the *gox*E glucose oxidase overproducing mutants most likely concern regulatory genes: in *gox*B, all three enzymes are induced regardless of the carbon source or of the level of oxygenation. *Gox*E mutants are very similar to those in *gox*B except that their glucose oxidase formation still depends on high levels of aeration. This indicates that in gluconate biosynthesis at least two different control systems operate which have partially overlapping targets.

## 2.2 Bacterial Gluconate Production

As mentioned in the introductory section of this chapter, many acetic acid bacteria, besides being able to oxidize ethanol, are also capable of oxidizing D-glucose to D-gluconic acid. This property, first detected by BROWN (1886) and several other authors (cf. LAFAR, 1914), has also been used as a taxonomic criterion to distinguish certain groups of acetic acid bacteria.

In 1924, KLUYVER and DE LEEUW described a special type of *Acetobacter, A. suboxydans,* frequently occurring in Dutch beers. This species, while having low acetogenic potency, was able to oxidize glucose almost completely to gluconic acid. When calcium carbonate was added, rather high yields of calcium 5-oxo-gluconate were obtained in addition to calcium gluconate. In 1929, HERMANN isolated another gluconic acid-producing strain from "kombucha", the Asian tea fungus, which he designated as *Bacterium gluconicum.* The significant differences in producing either acetic acid from ethanol or gluconic acid from glucose as the main oxidation products prompted ASAI in 1935 to propose the introduction of a new genus, *Gluconobacter,* for the latter type of organism (ASAI, 1968). This genus, together with the established genus *Acetobacter,* now comprises the family of Acetobacteraceae, which is generally accepted and also supported by molecular biological studies (SIEVERS et al., 1994a). *A. suboxydans,* an organism which has been investigated in numerous studies on the oxidative behavior of the so-called non-overoxidizing acetic acid bacteria (especially the frequently used strain ATCC 621) is now listed as *Gluconobacter oxydans.*

The metabolic pathways as well as the respective enzymes responsible for the oxidation of D-glucose to D-gluconic acid and oxogluconic acids have been extensively reviewed by MATSUSHITA et al. (1994). D-Glucose is oxidized to D-gluconic acid mainly by a membrane-bound D-glucose dehydrogenase which is a 87 kDa protein with tightly (noncovalently) bound pyrroloquinoline quinone (PQQ) (AMEYAMA et al., 1981a, b). In contrast to this, a cytoplasmic $NADP^+$-dependent dehydrogenase is not directly involved in this process (PRONK et al., 1989). Apparently this enzyme mainly is active in a pathway alternative to the common phosphotransferase system of glucose assimilation (MIDGLEY and DAWES, 1973).

Gluconate dehydrogenase, catalyzing the production of 2-oxo-D-gluconate, and 2-oxo-D-gluconate dehydrogenase, catalyzing the formation of 2,5-dioxo-D-gluconate, were isolated from *Gluconobacter dioxyacetonicus* (SHINAGAWA et al., 1984) and *G. melanogenus* (SHINAGAWA et al., 1981), respectively. They were characterized as flavoproteins, with cytochrome c as a second and an unknown protein as a third subunit. As mentioned earlier, *G. oxydans* can also produce 5-oxo-D-gluconate but the respective enzyme has not been characterized (STADLER-SZÖKE et al., 1980; SHINAGAWA et al., 1983).

The respiratory system of ketogenic acetic acid bacteria thus comprises quinoproteins and flavoproteins, donating electrons to ubiquinone with subsequent electron transfer to cytochromes including a cyanide-sensitive energy-generating cytochrome o and an alternative cyanide-insensitive bypass with cytochrome c.

D-Gluconate and 2-oxo-D-gluconate thus formed are believed to be transported into the cytoplasm, where the latter may be reduced to D-gluconate by an NADP-dependent 2-oxo-D-gluconate reductase. Subsequently, the cytoplasmic D-gluconate is further metabolized through the pentose phosphate pathway (ASAI, 1968). According to OLIJVE (1978), enzyme synthesis of the pentose phosphate pathway is almost completely repressed if the glucose concentration exceeds 5–15 mM and the pH of the medium is kept below 3.5–4.0. Maintaining the pH at lower levels, in addition, has been shown to suppress the unwanted formation of oxo-gluconates in *Gluconobacter* fermentations (BERNHAUER and KNOBLOCH, 1938; STADLER-SZÖKE et al., 1980). This evidently forms the basis of gluconic acid production by *Gluconobacter* at an industrial scale with sufficiently high yields.

The $NAD(P)^+$-independent quinoprotein GDH occurs in a variety of members of the genus *Acetobacter* but also in related bacteria such as *Zymomonas,* several pseudomonads, and in some enterobacteria.

Similar (but not identical) enzymes are also found in a newly described methylotrophic member of the acetic acid bacteria, *Acetobacter methanolicus* (UHLIG et al., 1986; SIEVERS et al., 1994b). Its capability of gluconic acid production will be discussed below. The organism is unique in its ability to grow on methanol and glycerol or glucose as carbon and energy sources and to oxidize these substrates with high activities.

An interesting enzyme has been found in the course of studies on the production of sorbitol by *Zymomonas mobilis* grown on sucrose or mixtures of glucose and fructose (VIIKARI, 1984a, b; BARROW et al., 1984; BRINGER-MEYER et al., 1985). It was shown that reduction of fructose to sorbitol was coupled with the oxidation of glucose to gluconate (LEIGH et al., 1984). The respective enzyme, designated as D-glucose-1:D-fructose-2-oxidoreductase, has been isolated and characterized by ZACHARIOU and SCOPES (1986). Gluconate, after phosphorylation, is further metabolized by entering the ED-pathway. SCOPES et al. (1986), however, have described a procedure using toluene-treated cells in highly concentrated mixtures of glucose and fructose (up to 60%), that permits an almost quantitative conversion to gluconic acid and sorbitol as a potentially useful industrial process (see below).

# 3 Industrial Production Procedures

Several processes for the production of gluconic acid by chemical means have been elaborated (SCHULZ, 1983) comprising simple chemical, electrochemical, or catalytic oxidation of glucose. The disadvantages of these processes are their lower yields due to inevitable side-reactions. Thus the specificity of the biochemical reaction has made fermentation the preferred procedure. Several procedures for the fermentative production of gluconic acid have thus been reported, as outlined below.

## 3.1 *Aspergillus niger* Fermentations

The following conditions have been found to be mandatory for successful operation of gluconate fermentations with molds, especially with *A. niger:*

(1) High concentrations of glucose: 110–250 g $L^{-1}$.
(2) Low concentrations of the nitrogen source: about 20 mM nitrogen.
(3) Low phosphorus concentrations.
(4) Sufficient concentrations of trace elements, which are generally supplied by the mineral salts used and/or, e.g., by corn steep liquor. As shown by BERNHAUER (1928) this could be restored by the addition of 0.7–13 mM manganese in mycelia that had lost their ability to produce gluconic acid.
(5) The pH value of the fermentation medium throughout the process is to be held in the range of 4.5–6.5; below pH 3, glucose oxidase is inactivated triggering metabolism to increased glycolysis with concomitant formation of other acids.
(6) High aeration rates – preferably by applying elevated air pressure (up to 4 bar).

### 3.1.1 Calcium Gluconate Fermentation

This is the oldest process based on the work of BERNHAUER in Prague and the work at the laboratories of the U.S. Department of Agriculture during the 1930s and 1940s.

Inoculum for the production stage may be either a conidial suspension from a suitable spore propagation medium (MOYER et al., 1937), or precultured mycelium from a liquid culture in a medium lower in glucose and higher in nitrogen and phosphorus than the production medium. In the case of pregrown mycelium, inocula in amounts of 5–30% of the final batch volume are usually employed.

The production medium may contain glucose of medium purity (e.g., corn sugar) in concentrations up to 150 g $L^{-1}$. As mentioned above, limitation to this rather low concentration is due to the low solubility of calcium gluconate. A typical medium composition as described in the literature and checked by the present authors is as follows:

- 110–150 g $L^{-1}$ glucose (e.g., corn sugar)
- 0.388 g $L^{-1}$ $(NH_4)_2HPO_4$
- 0.188 g $L^{-1}$ $KH_2PO_4$
- 0.156 g $L^{-1}$ $MgSO_4 \cdot 7H_2O$
- 26 g $L^{-1}$ $CaCO_3$ (separately sterilized)

The fermentation process is usually carried out in conventional bioreactors with intensive aeration (0.25–1 vvm), preferably applying elevated air pressures of up to 3 bar.

Calcium carbonate is added as a steam-sterilized slurry in increments according to the course of acid production keeping the pH of the fermentation broth >3.5. In order to avoid precipitation of calcium gluconate the total amount of calcium gluconate should be only about 2/3 of the stoichiometric requirement.

Fermentations with almost quantitative yields are usually completed within less than 24 h, corresponding to a productivity of about 6 g $L^{-1}$ $h^{-1}$. Hence, sterility requirements are not considered too crucial unless the mycelium is reused. As has been pointed out, this may be done for several times with the advantage of reducing the respective lag periods. As the mycelium tends to rise to the top after aeration and agitation have been stopped, this may be accomplished by simply withdrawing the fermentation broth from the bottom and substituting with fresh medium. Filtration of the mycelium has likewise been proposed (Porges et al., 1940, 1941). This may be optimized by monitoring the glucose oxidase activity of the mycelium (Hatcher, 1972). It is claimed, in addition, that after several rounds maximum enzyme activity may be recovered by proper additions of nutrients to the next production medium.

### 3.1.2 Sodium Gluconate Fermentation

As mentioned above, knowledge of the special sequestering properties of sodium gluconate has now made this type of fermentation the preferred one. Blom et al. (1952) have outlined such a process as follows (see also Milsom and Meers, 1985): Conidia are propagated in a liquid sporulation medium containing

- 50.0 g $L^{-1}$ glucose
- 0.6 g $L^{-1}$ $(NH_4)_2HPO_4$
- 0.15 g $L^{-1}$ $KH_2PO_4$
- 0.2 g $L^{-1}$ KCl
- 0.12 g $L^{-1}$ $MgSO_4 \cdot 7H_2O$
- 0.01 g $L^{-1}$ Fe-tartrate
- 30 mL potato extract
- 30 mL beer
- 1.5 g $L^{-1}$ agar

These conidia may be used to inoculate the production medium (see below). Alternatively, precultures are made in a medium containing the same ingredients as the production medium with the exception of a lower glucose concentration (60 g $L^{-1}$). After 24–48 h of cultivation with aeration (0.2 vvm), a mycelial suspension is obtained sufficient to inoculate a sterile production medium composed of

- 240–380 g $L^{-1}$ glucose
- 3.7 g $L^{-1}$ corn steep liquor
- 0.42 g $L^{-1}$ $(NH_4)_2HPO_4$
- 0.20 g $L^{-1}$ $KH_2PO_4$
- 0.11 g $L^{-1}$ urea
- 0.17 g $L^{-1}$ $MgSO_4 \cdot 7H_2O$
- ca. 0.17 mL $H_2SO_4$ to pH 4.5

After sterilization, the pH of the medium is adjusted to 6.0–6.5 before inoculation and held close to 6.5 throughout the fermentation by adding sodium hydroxide. Aeration rate should be set at about 1 vvm, and temperature maintained at 33–34 °C. Again the application of elevated pressure was shown to be advantageous. Thus tower types of fermenters might be superior to conventional stirred reactors (Traeger et al., 1989). Foaming is controlled by common agents. Under these conditions, productivities (without consideration of lag times) of more than 13 kg $m^{-3}$ $h^{-1}$ were reported. Lag times may be eliminated by reuse of the mycelium. Such mycelium should be used immediately after separation or held suspended with aeration until use.

Studies with the aim of obtaining higher levels of glucose oxidase in the mycelium have shown that this may be achieved by starting with lower concentrations of glucose in the stage of mycelial growth and gradually increasing this concentration in the production stage. In addition, according to a patent held by ZIFFER et al. (1971), the total amount of glucose is increased to as much as 600 g $L^{-1}$ by neutralizing only half the gluconic acid formed. This apparently means that compromises have to be made with respect to optimum pH, at least in the later phase of the process.

On the other hand, continuous processes have been developed, e.g., by Fujisawa Pharmaceutical Group (YAMADA, 1977) claiming the conversion of glucose solutions of 350 g $L^{-1}$ with yields of 95% and with significantly increased productivities.

### 3.1.3 Combined Processes – Immobilized Cells and Enzymes

Besides using glucose or starch hydrolyzates as raw materials for gluconic acid production, the application of sucrose or mixtures of glucose and fructose such as inverted sucrose or syrups derived from glucose isomerization processes may offer interesting process variants. Thus ICHIKAWA et al. (1975) patented a process for fructose production by shaking sucrose solutions with cells or enzymes of invertase-producing *A. niger*. After separating gluconic acid as calcium gluconate, crystalline fructose could be recovered from the mother liquor. Similar processes have been designed for invert sugar or liquid isomerose sugars (SATO et al., 1967).

The possibility of frequent reuse of mycelium as mentioned above clearly suggests the application of immobilization procedures for cells as well as for the respective enzymes:

*A. niger* mycelia (pellets) have been immobilized by flocculation with polyelectrolytes (LEE and LONG, 1974), by covalently binding to glycidyl ester copolymers (NELSON, 1976), by attaching to glass rings (HEINRICH and REHM, 1982) or to other porous materials such as a nonwoven felt (SAKURAI and TAKAHASHI, 1989a, b; SAKURAI et al., 1989) or polyurethane foam (VASSILEV et al., 1993).

In several studies it has been shown that glucose oxidase is rather sensitive to the hydrogen peroxide formed in the course of glucose oxidation. In a system of free mycelium, hydrogen peroxide is readily decomposed by fungal catalase, and the oxygen formed contributes to the necessary oxygen supply. In several immobilized systems, however, losses of catalase may occur and the inclusion of catalase or peroxidase will prove beneficial to maintain stability and productivity (BUCHHOLZ and GOEDELMANN 1978; HARTMEIER and DOEPPNER, 1983, 1984).

Apparently, these and other problems suggest the establishment of systems entirely consisting of enzymes, i.e., glucose oxidase and catalase. Various methods of immobilization of glucose oxidase and coimmobilization with catalase or peroxidase have been described: BUCHHOLZ and GOEDELMANN (1978), HARTMEIER and TEGGE (1979), COPPENS (1980), BELHAJ et al. (1983), KIRSTEIN and SCHEELER (1987).

By the application of glucose oxidizing enzymes other than glucose oxidase, namely glucose dehydrogenases, hydrogen peroxide problems can be circumvented. Systems incorporating bacterial or other glucose dehydrogenases have been exploited, but in these reactions it is necessary to regenerate the free coenzymes, i.e., NAD or NADP. This problem can be solved by coupling glucose dehydrogenation to another redox reaction working in the opposite direction. An interesting example is the system

D-Glucose + $NAD^+$
→ D-Gluconolactone + NADH + $H^+$

coupled to

D-Fructose + NADH + $H^+$
→ D-Mannitol + $NAD^+$

Thus a mixture of glucose and fructose, obtained either by inverting sucrose or by isomerizing commercial dextroses, could be converted to gluconolactone and mannitol on a commercial basis, provided that the enzymes, and particularly the coenzymes, could be retained in the system for longer periods of time.

This has been verified in different membrane reactors with the respective enzymes, glucose dehydrogenase and mannitol dehydrogenase, from different sources (KULBE et al., 1985, 1986). NAD, the coenzyme to be recycled, was retained in the membrane system by coupling to a polymer, e.g., polyethylene glycol or polyethylene imine. HOWALDT (1988) then found that an even better means of retaining NAD in an active state is by employing a negatively charged ultrafiltration membrane. This system has been described in detail by HOWALDT et al. (1988, 1990) and by SCHMIDT et al. (1990).

## 3.2 *Aureobasidium pullulans* Fermentation

In the 1960s, gluconic acid production by *Aureobasidium (Pullularia) pullulans* has been described by several authors (SASAKI and TAKAO, 1970; NAKAMURA et al., 1977; SU et al., 1977) as an alternative to the traditional gluconic acid production methods. Yields of sodium gluconate or glucono-$\delta$-lactone from glucose (concentrations of 150–200 g $L^{-1}$) or starch hydrolyzates were 97% and 85%, respectively, and productivities above 6 g $L^{-1}$ $h^{-1}$ were obtained. In the course of an extensive screening program, ANASTASSIADIS et al. (1992, 1994) have isolated about 50 producer strains of *A. pullulans.* In a fed-batch process more than 400 g $L^{-1}$ gluconic acid were produced, and more than 260 g $L^{-1}$ were obtained in a continuous process with residence times of less than 20 h. Maximum productivities in the continuous process exceeded 19 g $L^{-1}$ $h^{-1}$.

## 3.3 Bacterial Gluconate Fermentations

Various methods of gluconic acid production by bacterial fermentations have been described so far but only few of them have been used at a larger scale although, in various respects, several of the processes discussed below are attractive.

### 3.3.1 Fermentations Using *Gluconobacter* and *Acetobacter*

Fermentation processes using *Gluconobacter oxydans* (*Acetobacter suboxydans*) were first patented by VERHAVE (1930), HERMANN (1933), and CURRIE and CARTER (1933). These authors recommended procedures similar to the quick vinegar process. As early as in 1933, however, CURRIE and FINLAY obtained a patent for a submerged fermentation process.

In the Netherlands, processes have been dealt with by Noury and van der Lande (1962) and by MEIBERG and SPA (1983). In all of these, it was shown that the decisive factors for optimum performance of bacterial gluconate production processes are high glucose concentrations, lowered pH, and high oxygen concentrations (OOSTERHUIS et al., 1983; TRAEGER and ONKEN, 1988). In a more recent study the performance of a *Gluconobacter* and an *A. niger* fermentation system has been compared with respect to varying oxygen concentrations in an airlift reactor (TRAEGER et al., 1992).

Several other publications were devoted rather to investigations of the kinetics of the process than to studies of commercial processes (KONO and ASAI, 1968; NYESTE et al., 1980).

More recent investigations are focussed on the application of immobilized cells in gluconic acid production processes (TRAMPER et al., 1983; SEISKARI et al., 1984, 1985; HARTMEIER and HEINRICHS, 1986; SHIRAISHI et al., 1989a, b).

Beginning in about 1985, a research group around W. BABEL in Leipzig reported on the production of gluconic acid from glucose by acidophilic methylotrophic bacteria, soon identified as strains of a novel acetic acid bacterium, *Acetobacter methanolicus* (UHLIG et al., 1986). The possibility to use this organism in a commercial process is documented in several patents (BABEL et al., 1985, 1986a, b, 1987, 1991; DUERESCH et al., 1988; ISKE et al., 1991; see also POEHLAND et al., 1993). According to these sources, glucose concentrations of 150–250 g $L^{-1}$ in fed-batch operation were processed with yields of up to 100% (dependent on glucose concentrations) and productivities >30 g $L^{-1}$ $h^{-1}$.

### 3.3.2 Related Processes with *Zymomonas mobilis*

As mentioned above, the presence of a glucose–fructose oxidoreductase in *Z. mobilis* may be utilized to convert mixtures of glucose and fructose, or sucrose (in presence of β-fructosidase) to sorbitol and gluconic acid, provided that gluconic acid formed is not further metabolized by the common ethanol producing mechanism of the organism. This is accomplished by treating the cells of *Z. mobilis* in a way that brings about leakage of the cells with consecutive loss of essential coenzymes such as ADP and ATP as well as NAD(P). Several procedures of cell treatment have been described together with attempts at commercial solutions:

- Permeabilization of cells by treating with toluene (ROGERS and CHUN, 1987; SCOPES et al., 1988; PATERSON et al., 1988; CHUN and ROGERS, 1988);
- freezing of cells at −20°C (BRINGER-MEYER and SAHM, 1989);
- drying of cells (ICHIKAWA et al., 1989);
- application of cationic detergents (REHR and SAHM, 1991a, b; REHR et al., 1991).

Basic protocols comprise treatment of cells as indicated above, applying mixtures of up to 600 g $L^{-1}$ glucose and fructose (available from various sources). By keeping the pH at optimum values of 6.5 and the temperature at 39°C, almost 100% yields of both sorbitol and gluconic acid may be achieved within several hours (REHR and SAHM, 1990; Forschungszentrum Jülich, 1990). Attempts at improving these processes by cell immobilization have also been reported (REHR and SAHM, 1991b; RO and KIM, 1991; KIM and KIM, 1992; JANG et al., 1992).

# 4 Product Recovery and Processing

Methods of product recovery are determined by the kind of fermentation and the type of product required. Crude calcium gluconate broths are filtered and neutralized with the appropriate amount of milk of lime or calcium carbonate with heating and stirring. Subsequently the liquid is concentrated to a hot supersaturated solution of about 20% calcium gluconate. Upon cooling to about 20°C, crystallization sets in or may be promoted by seeding and/or adding small amounts of water-miscible solvents. Calcium gluconate crystals precipitate as small needles forming characteristic cauliflower-like aggregations. These are separated and washed several times with cold water. Calcium gluconate thus obtained can be dried at temperatures of ca. 80°C or processed further in aqueous suspensions. Mother liquors should be treated with activated carbon before being reintroduced into the purification procedure.

By treating suspensions of calcium gluconate with stoichiometric amounts of sulfuric acid, free gluconic acid can be obtained. After repeating this step, in general the clear liquid is concentrated to a 50% solution of gluconic acid.

Gluconic acid is readily soluble in water. In aqueous solutions an equilibrium mixture of gluconic acid, 1,5- or δ-gluconolactone and 1,4- or γ-gluconolactone is present, the ratios depending on concentration and temperature. Thus, from oversaturated solutions at 0–30°C, crystals of gluconic acid will precipitate, whereas at temperatures between 30–70°C crystals of the δ-lactone, and >70°C those of the γ-lactone are precipitated.

In general, products of commerce are purified solutions of 50% gluconic acid, or the δ-lactone, which is produced by evaporating gluconic acid solutions at 35–60 °C to a concentration of 65–85%. Upon keeping the concentrated solution at the specified temperature crystallization sets in spontaneously or may be promoted by seeding or adding solvents (e.g., butanol).

Sodium gluconate from sodium gluconate fermentations is usually recovered by simply concentrating the filtered fermentation broths to about 45% solids (BLOM et al., 1952), followed by addition of sodium hydroxide to pH 7.5 and drum drying. Only if necessary, additional purification steps are included, e.g., treatments with adsorbents.

Owing to the fact that sodium gluconate has become the main product of commercial gluconate fermentations, free gluconic acid as well as the δ-lactone are now predominantly prepared from pure sodium gluconate solutions using ion exchange as separation technique.

Gluconic acid salts are prepared by reacting gluconic acid with the respective cations applying common procedures.

# 5 Utilization of Gluconic Acid and Gluconates

Due to some outstanding properties – extremely low toxicity, low corrosivity and the capability of forming water-soluble complexes with divalent and trivalent metal ions – gluconic acid has found applications, e.g., in the dairy industry to prevent the deposition of milkstone or to remove it. The remarkable non-corrosiveness of gluconic acid may generally be utilized in gentle metal cleaning operations, e.g., in cleaning aluminum cans and other equipment. According to German law gluconic acid is considered as food and thus not as an additive. In beverages it prevents cloudiness and scaling by calcium compounds. In various foods it produces and improves a mild sour taste and complexes traces of heavy metals.

Gluconic acid δ-lactone may be used similar to gluconic acid. It is preferred when slow acidulating action is desired, e.g., in baking aids or in the production of cured meat products, especially in the processing of ground meat for sausage manufacture. According to YAMADA (1977) several thousand tons of gluconolactone are used in Japan for coagulating soybean protein in the manufacture of tofu.

The main product of commerce, however, is the sodium salt. This is due to the outstanding property of forming stable complexes with various metal ions, especially in alkaline (i.e., sodium hydroxide or carbonate) solutions. These may be used, e.g., to scale-off oxides of heavy metals from metal surfaces, to remove zinc from metal surfaces, or in removing paints and lacquers from various objects. The sequestering action on calcium and similar ions may be used in alkaline glass washing preparations or in the textile industry to prevent iron deposition. Sodium gluconate is also recommended as an additive to concrete acting as a plastifier and retarding the setting process.

Several other salts of gluconic acid have gained importance: Gluconates of calcium and iron are the preferred carriers used in calcium and iron therapy because of the extremely low toxicity of gluconate and its favorable biodegradability. Various gluconates together with gluconic acid are used in the tanning and textile industry.

# 6 Production Figures

Total world production of gluconic acid and gluconates is estimated to be 50,000–60,000 t. More than 80% of this production is sold as sodium gluconate.

# 7 References


AMEYAMA, M., MATSUSHITA, K., OHNO, Y., SHINAGAWA, E., ADACHI, O. (1981a), Existence of a novel prosthetic group, PQQ, in membrane-bound, electron transport chain-linked, primary dehydrogenases of oxidative bacteria, *FEBS Lett.* **130**, 179–183.

AMEYAMA, M., SHINAGAWA, E., MATSUSHITA, K., ADACHI, O. (1981b), D-Glucose dehydrogenase of *Gluconobacter suboxydans:* Solubilization, purification and characterization, *Agric. Biol. Chem.* **45**, 851–861.

ANASTASSIADIS, S., AIVASIDIS, A., WANDREY, C. (1992), Continuous gluconic acid fermentation with yeast like molds, *DECHEMA Biotechnol. Conf.* **5** (Pt B, Bioprocess Engineering), 537–540.

ANASTASSIADIS, S., AIVASIDIS, A., WANDREY, C. (1994), *Ger. Patent Appl.* 4317488.

ASAI, T. (1968), *Acetic Acid Bacteria. Classification and Biochemical Activities,* Tokyo: Tokyo University Press.

BABEL, W., MIETHE, D., ISKE, U., SATTLER, K., RICHTER, H. P., SCHMIDT, J. , DUERESCH, R. (1985), *Ger. Patent* (GDR) 218387.

BABEL, W., MIETHE, D., ISKE, U., SATTLER, K., RICHTER, H. P., SCHMIDT, J., DUERESCH, R. (1986a), *Ger. Patent* (GDR) 236754.

BABEL, W., MUELLER, R. KLEBER, H. P., UHLIG, H. (1986b), *Ger. Patent* (GDR) 238067.

BABEL, W., ISKE, U., JECHOREK, M., MIETHE, D. (1987), *Ger. Patent* (GDR) 251571.

BABEL, W., LOFFHAGEN, N., MIETHE, D. et al. (1991), *Ger. Patent* (GDR) 293135.

BARROW, K. D., COLLINS, J. G., LEIGH, D. A., ROGERS, P. L., WARR, R. G. (1984), Sorbitol production by *Zymomonas mobilis, Appl. Microbiol. Biotechnol.* **20**, 225–232.

BELHAJ, S., GELLF, G., THOMAS, D., ZOULALIAN, A., BESSON, G. (1983), Development of an immobilized enzyme reactor for enzymic oxidation, *Bio-Sciences* **2**, 187–190.

BERNHAUER, K. (1924), Zum Problem der Säurebildung durch *Aspergillus niger, Biochem. Z.* **153**, 517.

BERNHAUER, K. (1926), Über die Säurebildung durch *Aspergillus niger, Biochem. Z.* **172**, 313.

BERNHAUER, K. (1928), Beiträge zur Enzymchemie der durch *Aspergillus niger* bewirkten Säurebildungsvorgänge, *Z. Physiol. Chem.* **177**, 86–106.

BERNHAUER, K. (1940), Oxydative Gärungen, in: *Handbuch der Enzymologie* (NORD, F. F., WEIDENHAGEN, R., Eds.), pp. 1036–1120. Leipzig: Akademische Verlagsgesellschaft.

BERNHAUER, K., KNOBLOCH, H. (1938), *Naturwissenschaften* **26**, 819.

BERNHAUER, K., SCHULHOF, L. (1932), *U.S. Patent* 1849053.

BLOM, R. H., PFEIFER, V. F., MOYER, A. J., TRAUFLER, D. H., CONWAY, H. F., CROCKER, C. K., FARISON, R. E., HANNIBAL, D. V. (1952), Sodium gluconate production, *Ind. Eng. Chem.* **44**, 435–440.

BOUTROUX, L. (1880), Sur une fermentation nouvelle du glucose, *C. R. Acad. Sci.* **91**, 236.

BRINGER-MEYER, S., SAHM, H. (1989), *Ger. Patent* 3841702.

BRINGER-MEYER, S., SCOLLAR, M., SAHM, H. (1985), *Zymomonas mobilis* mutants blocked in fructose utilization, *Appl. Microbiol. Biotechnol.* **23**, 134–139.

BROWN, A. J. (1886), The chemical action of pure cultivations of *Bacterium aceti, J. Chem. Soc. Transact.* (Lond.), 172.

BUCHHOLZ, K., GÖDELMANN, B. (1978), Macrokinetics and operational stability of immobilized glucose oxidase and catalase, *Biotechnol. Bioeng.* **20**, 1201–1220.

CHUN, U. H., ROGERS, P. L. (1988), The simultaneous production of sorbitol from fructose and gluconic acid from glucose using an oxidoreductase of *Zymomonas mobilis, Appl. Microbiol. Biotechnol.* **29**, 19–24.

COPPENS, G. (1980), *Eur. Patent* 800806.

CURRIE, J. N., CARTER, R. H. (1933), *U.S. Patent* 1896811.

CURRIE, J. N., FINLAY, A. (1933), *U.S. Patent* 1908225.

CURRIE, J. N., KANE, J. H., FINLAY, A. (1933), *U.S. Patent* 1893819.

DUERESCH, R., SCHMIDT, J., RICHTER, H. P., BABEL, W., ISKE, U., JECHOREK, M., MIETHE, D., KOCHMANN, W. (1988), *Ger. Patent* (GDR) 259969.

FIEDUREK, J., ROGALSKI, J., ILCZUK, Z., LEONOWICZ, A. (1986), Screening and mutagenesis of moulds for the improvement of glucose oxidase production, *Enzyme Microb. Technol.* **8**, 734–736.

Forschungszentrum Jülich GmbH (1990), *Report on activities of the Institut für Biotechnologie and Institut für Enzymtechnologie.* Jülich: Forschungszentrum Jülich GmbH.

FOSTER, J. W. (1949), Gluconic and other sugar acids, in: *Chemical Acivities of Fungi,* pp. 446–467. New York: Academic Press.

FOWLER, T., REY, M. W., VÄHÄ-VAHE, P., POWER, S. D., BERKA, R. M. (1993), The *cat*R gene encoding a catalase from *Aspergillus niger:* Primary structure and elevated expression through increased copy number and use of strong promoter, *Mol. Microbiol.* **9**, 989–998.

FREDERICK, K. R., TUNG, J., EMERICK, R. S., MASIARZ, F. R., CHAMBERLAIN, S. H., VASAVADA, A., ROSENBERG, S., CHAKRABORTY, S., SCHOPFER, L. M., MASSEY, V. (1990), Glucose oxidase from *Aspergillus niger.* Cloning, gene sequence, secretion from *Saccharomyces cerevisiae* and kinetic analysis of a yeast-derived enzyme, *J. Biol. Chem.* **265**, 3793–3802.

GASTROCK, E. A., PORGES, N., WELLS, P. A., MOYER, A. J. (1938), Gluconic acid production on pilot-plant scale ... Effect of variables on production by submerged mold growths, *Ind. Eng. Chem.* **30**, 782–789.

GIBSON, Q. H., SWOBODA, B. E. P., MASSEY, B. (1964), Kinetics and mechanism of action of glucose oxidase, *J. Biol. Chem.* **239**, 3927–3924.

HARTMEIER, W., DOEPPNER, T. (1983), Preparation and properties of mycelium bound glucose oxidase co-immobilizied with excess catalase, *Biotechnol. Lett.* **5**, 743–748.

HARTMEIER, W., DOEPPNER, T. (1984), *Ger. Patent* 3301992.

HARTMEIER, W., HEINRICHS, A. (1986), Membrane-enclosed alginate beads containing *Gluconobacter* cells and molecular dispersed catalase, *Biotechnol. Lett.* **8**, 565–570.

HARTMEIER, W., TEGGE, G. (1979), Versuche zur Glucoseoxidation in Glucose-Fructose-Gemischen mittels fixierter Glucoseoxidase und Katalase, *Starch/Stärke* **31**, 348–353.

HATCHER, H. J. (1972), *U.S. Patent* 3669840.

HEINRICH, M., REHM, H. J. (1982), Formation of gluconic acid at low pH values by free and immobilized *Aspergillus niger* cells during citric acid fermentation, *Eur. J. Appl. Microbiol. Biotechnol.* **15**, 88–92.

HERMANN, S. (1929), *Bacterium gluconicum,* ein in der sogenannten Kombucha (japanischer oder indischer Teepilz) vorkommender Spaltpilz, *Biochem. Z.* **205**, 297–305.

HERMANN, S. (1933), *Austrian Patent* 133139.

HERRICK, H. T., HELLBACH, R., MAY, O. E. (1935), Apparatus for the application of submerged mold fermentations under pressure, *Ind. Eng. Chem.* **27**, 681–683.

HLASIWETZ, H., HABERMANN, J. (1870), Zur Kenntniss einiger Zuckerarten (Glucose, Rohrzucker, Levulose, Sorbin, Phloroglucin), *Liebigs Ann. Chem.* **155**, 123.

HOWALDT, M. (1988), *Thesis,* University of Stuttgart.

HOWALDT, M., GOTTLOB, A., KULBE, K. D., CHMIEL, H. (1988), Simultaneous conversion of glucose/fructose mixtures in a membrane reactor, *Ann. N.Y. Acad. Sci.* **542**, 400–405.

HOWALDT, M., KULBE, K. D., CHMIEL, H. (1990), A continuous enzyme membrane reactor retaining the native nicotinamide cofactor NAD(H), *Ann. N.Y. Acad. Sci.* **589**, 253–260.

HUSTEDE, H., HABERSTROH, H. J., SCHINZIG, E. (1989), Gluconic acid, in: *Ullmann's Encyclopedia of Industrial Chemistry,* 5th Edn., Vol. A12 (ELVERS, B., HAWKINS, S., RAVENSCROFT, M., ROUNSAVILLE, J. F., SCHULZ, G., Eds.), pp. 449–456. Weinheim: VCH.

ICHIKAWA, Y., SATO, S., HIROSE, S. (1975), *Jpn. Patent* 50154484.

ICHIKAWA, Y., KITAMOTO, Y., KATO, N., MORI, N. (1989), *Eur. Patent* 322723.

ISKE, U., JECHOREK, M., BABEL, W. et al. (1991), *Ger. Patent* (GDR) 295190.

JANG, K. H., PARK, C. J., CHUN, U. H. (1992), Improvement of oxidoreductase stability of cethyltrimethylammonium bromide permeabilized cells of *Zymomonas mobilis* through glutaraldehyde crosslinking, *Biotechnol. Lett.* **14**, 311–316.

KELLEY, R. L., REDDY, C. A. (1986), Purification and characterization of glucose oxidase from ligninolytic cultures of *Phanerochaete chrysosporium, J. Bacteriol.* **166**, 269–274.

KIM, D. M., KIM, H. S. (1992), Continuous production of gluconic acid and sorbitol from Jerusalem artichoke and glucose using an oxidoreductase of *Zymomonas mobilis* and inulinase, *Biotechnol. Bioeng.* **39**, 336–342.

KIM, K. K., FRAVEL, D. R., PAPAVIZAS, G. C. (1988), Identification of a metabolite produced by *Talaromyces flavus* as glucose oxidase and its role in the biocontrol of *Verticillium dahliae, Phytopathol.* **78**, 488–492.

KIRSTEIN, D., SCHELLER, F. (1987), *Ger. Patent* (GDR) 243714.

KLUYVER, A. J., DE LEEUW, F. J. G. (1924), *Acetobacter suboxydans,* een merkwaardige azijnbacterie, *Tijtschr. Vgl. Geneesk.* **10**, 170–182.

KONO, T., ASAI, T. (1968), Studies on fermentation kinetics. Kinetic analysis of gluconic acid fermentation, *Hakko Kogaku Zasshi* **46**, 398–405.

KRIECHBAUM, M., HEILMAN, H. J., WIENTJES, F. D., HAHN, M., JANY, K. D., GASSEN, H. G. (1989), Cloning and DNA sequence analysis of the glucose oxidase gene from *Aspergillus niger* NRRL-3, *FEBS Lett.* **255**, 63–66.

KULBE, K. D., SCHWAB, U., CHMIEL, H., STRATHMANN, H. (1985), *Ger. Patent* 3326546.

KULBE, K. D., SCHWAB, U., HOWALDT, M., KIMMERLE, K. (1986), Use of membrane processes in the simultaneous preparation of mannitol and gluconic acid from sucrose by conjugated $NAD^+$-related dehydrogenase, *GBF Monogr. Ser.* **9** (Tech. Membr. Biotechnol.), 189–200.

KUNDU, P. N., DAS, A. (1985), A note on crossing experiments with *Aspergillus niger* for the production of sodium gluconate, *J. Appl. Bacteriol.* **59**, 1–5.

LAFAR, F. (1914), Die Essigsäuregärung, in: *Handbuch der Technischen Mykologie,* Vol. 5 (LAFAR, F., Ed.), pp. 539–632. Jena: Verlag von Gustav Fischer.

LEE, C., LONG, M. (1974), *U.S. Patent* 3821086.

LEIGH, D., SCOPES, R. K., ROGERS, P. L. (1984), A proposed pathway for sorbitol production by *Zymomonas mobilis, Appl. Microbiol. Biotechnol.* **20**, 413–415.

MARKWELL, J., FRAKES, L. G., BROTT, E. C., OSTERMAN, J., WAGNER, F. W. (1989), *Aspergillus niger* mutants with increased glucose oxidase production, *Appl. Microbiol. Biotechnol.* **30**, 166–169.

MATSUSHITA, K., TOYAMA, H., ADACHI, O. (1994), Respiratory chains and bioenergetics of acetic acid bacteria, *Adv. Microb. Physiol.* **36**, 247–301.

MAY, O. E., HERRICK, H. T., THOM, C., CHURCH, M. B. (1927), The production of gluconic acid by the *Penicillium luteum–purpurogenum* group, *J. Biol. Chem.* **75**, 417–422.

MAY, O. E., HERRICK, H. T., MOYER, A. J., HELLBACH, R. (1929), Semi-plant scale production of gluconic acid by mold fermentation, *Ind. Eng. Chem.* **21,** 1198–1203.

MAY, O. E., HERRICK, H. T., MOYER, A. J., WELLS, P. A. (1934), Gluconic acid production by submerged mold growths under increased air pressure, *Ind. Eng. Chem.* **26,** 575–578.

MEIBERG, J. B. M., SPA, H. A. (1983), Microbial production of gluconic acid and gluconates, *Antonie van Leeuwenhoek* **49,** 89–90.

MIALL, L. M. (1978), Gluconic acid, in: *Economic Microbiology,* Vol. 2 (ROSE, A. H., Ed.), pp. 99–105. London–New York–San Francisco: Academic Press.

MIDGLEY, M., DAWES, E. A. (1973), The regulation of transport of glucose and methyl $\alpha$-glucoside in *Pseudomonas aeruginosa*, *Biochem. J.* **132**, 141.

MILSOM, P. E., MEERS, J. L. (1985), Gluconic and itaconic acid, in: *Comprehensive Biotechnology,* Vol. 3 (BLANCH, H. W., DREW, S., WANG, D. I. C., Eds.), pp. 665–680. Oxford: Pergamon Press.

MISCHAK, H., KUBICEK, C. P., ROEHR, M. (1985), Formation and location of glucose oxidase in citric acid producing mycelia of *Aspergillus niger, Appl. Microbiol. Biotechnol.* **21**, 27–31.

MOLLIARD, M. (1922), Sur une nouvelle fermentation acide produit par le *Sterigmatocystis nigra, C. R. Acad. Sci.* **174**, 881.

MOYER, A. J. (1944), *U.S. Patent* 2351500.

MOYER, A. J., WELLS, P. A., STUBBS, J. J., HERRICK, H. T., MAY, O. E. (1937), Gluconic acid production. Development of inoculum and composition of fermentation solution for gluconic acid production by submerged mold growths under increased air pressure, *Ind. Eng. Chem.* **29**, 777–781.

MOYER, A. J., UMBARGER, E. J. , STUBBS, J. J. (1940), Fermentation of concentrated solutions of glucose to gluconic acid, *Ind. Eng. Chem.* **32,** 1379–1383.

NAKAMURA, H., ISHIHARA, K., HASHIMOTO, Y. (1977), *Jpn. Patent* 52010487.

NELSON, R. P. (1976), *U.S. Patent* 3957580.

Noury and van der Lande, N. V. (1962), *Brit. Patent* 902609.

NYESTE, L., SEVELLA, B., SZIGETI, L., SZÖKE, A., HOLLÓ, J. (1980), Modeling and off line optimization of batch gluconic acid fermentation, *Eur. J. Appl. Microbiol. Biotechnol.* **10**, 87–94.

OLIJVE, W. (1978), *Ph. D. Thesis,* University of Groningen.

OOSTERHUIS, N. M. G., GROESBEEK, N. M., OLIVIER, A. P. C., KOSSEN, N. W. F. (1983), Scale-down aspects of the gluconic acid fermentation, *Biotechnol. Lett.* **5**, 141–146.

PATERSON, S. L., FANE, A. G., FELL, C. J. D., CHUN, U. H., ROGERS, P. L. (1988), Sorbitol and gluconate production in a hollow fiber membrane reactor by immobilized *Zymomonas mobilis, Biocatalysis* **1**, 217–229.

PAZUR, J. H. (1966), Glucose oxidase from *Aspergillus niger, Methods Enzymol.* **9**, 82–86.

POEHLAND, H. D., SCHIERZ, V., SCHUMANN, R. (1993), Optimization of gluconic acid synthesis by removing limitations and inhibitions, *Acta Biotechnol.* **13**, 257–268.

PORGES, N., CLARK, T. F., GASTROCK, E. A. (1940), Gluconic acid production. Repeated use of submerged *Aspergillus niger* for semicontinuous production, *Ind. Eng. Chem.* **32**, 107–111.

PORGES, N., CLARK, T. F., ARONOVSKY, S. I. (1941), Gluconic acid production. Repeated recovery and re-use of submerged *Aspergillus niger* by filtration, *Ind. Eng. Chem.* **33**, 1065–1067.

PRESCOTT, S. C., DUNN, C. G. (1959), *Industrial Microbiology*, 3rd Edn. New York: McGraw Hill.

PRONK, J. T., LEVERING, P. R., OLIJVE, W., VAN DIJKEN, J. P. (1989), Role of NADP-dependent and quinoprotein glucose dehydrogenase in gluconic acid production by *Gluconobacter oxydans, Enzyme Microb. Technol.* **11**, 160–164.

RAMASAMY, K., KELLEY, R. L., REDDY, C. A. (1985), Lack of lignin degradation by glucose oxidase negative mutants of *Phanerochaete chry-*

*sosporium, Biochem. Biophys. Res. Comm.* **131**, 436–441.

REHR, B., SAHM, H. (1990), Produktion von Sorbitol und Gluconsäure durch das Ethanol produzierende Bakterium *Zymomonas mobilis, Forum Mikrobiologie* **13**, 166–169.

REHR, B., SAHM, H. (1991a), *Ger. Patent* 3936757.

REHR, B., SAHM, H. (1991b), *Ger. Patent* 4017103.

REHR, B., WILHELM, C., SAHM, H. (1991), Production of sorbitol and gluconic acid by permeabilized cells of *Zymomonas mobilis, Appl. Microbiol. Biotechnol.* **35**, 144–148.

REUSS, M., FRÖHLICH, S., KRAMER, B., MESSERSCHMIDT, K., POMMERENING, G. (1986), Coupling of microbial kinetics and oxygen transfer for analysis and optimization of gluconic acid production by *Aspergillus niger, Bioproc. Eng.* **1**, 79–91.

RO, H. S., KIM, H. S. (1991), Continuous production of gluconic acid and sorbitol from sucrose using invertase and an oxidoreductase of *Zymomonas mobilis, Enzyme Microb. Technol.* **13**, 920–924.

ROGERS, P. L., CHUN, U. H. (1987), Novel enzymatic process for sorbitol and gluconate production, *Aust. J. Biotechnol.* **1**, 51–54.

SAKURAI, H., TAKAHASHI, J. (1989a), *Jpn. Patent* 01095781.

SAKURAI, H., TAKAHASHI, J. (1989b), *Jpn. Patent* 01165380.

SAKURAI, H., LEE, H. W., SATO, S., MUKATAKA, S., TAKAHASHI, J. (1989), Gluconic acid production at high concentrations by *Aspergillus niger* immobilized on a nonwoven fabric, *J. Ferment. Bioeng.* **67**, 404–408.

SASAKI, Y., TAKAO, S. (1970), Gluconic acid fermentation by *Pullularia pullulans*. II. Acid production from various carbon sources, *Hakko Kogaku Zasshi* **48**, 368–373.

SATO, T., TUMURA, N., SHIMIZU, K. (1967), Utilization of isomerized liquid sugar. II. Preparation of gluconic acid and fructose from isomerized sugar, *Nippon Shokuhin Kogyo Gakkaishi* **14**, 508–509 (CA 68:113323).

SCHMIDT, K. CHMIEL, H., KULBE, K. D. (1990), Umwandlung nachwachsender Rohstoffe, *Chem. Anlagen Verfahren* High Tech Dezember, 26–30.

SCHULZ, G. (1983), Gluconsäure, in: *Ullmanns Encyklopädie der technischen Chemie,* 4th Edn., Vol. 24 (BARTHOLOMÉ, E., BIEKERT, E., HELLMANN, H., LEY, H., WEIGERT, M., Eds.), pp. 783–793. Weinheim: Verlag Chemie.

SCOPES, R. K., LEIGH, D., ROGERS, P. L. (1986), Sorbitol and gluconate production by *Zymomonas mobilis, Proc. 7th Australian Biotechnol. Conf.,* Melbourne, pp. 410–413.

SCOPES, R. K., ROGERS, P. L., LEIGH, D. A. (1988), *U.S. Patent* 4755467.

SEISKARI, P., LINKO, Y. Y., LINKO, P. (1984), Continuous production of gluconic acid by immobilized *Gluconobacter* organisms, *Eur. Congr. Biotechnol.,* 3rd Edn., Vol. 1, pp. 339–344. Weinheim: Verlag Chemie.

SEISKARI, P., LINKO, Y. Y., LINKO, P. (1985), Continuous production of gluconic acid by immobilized *Gluconobacter oxydans* cell bioreactor, *Appl. Microbiol. Biotechnol.* **21**, 356–360.

SHINAGAWA, E., MATSUSHITA, K., ADACHI, O., AMEYAMA, M. (1981), Purification and characterization of 2-keto-D-gluconate dehydrogenase from *Gluconobacter melanogenus, Agric Biol. Chem.* **45**, 1079–1085.

SHINAGAWA, E., MATSUSHITA, K., ADACHI, O., AMEYAMA, M. (1983), Selective production of 5-keto-D-gluconate by *Gluconobacter* strain, *J. Ferment. Technol.* **61**, 359.

SHINAGAWA, E., MATSUSHITA, K., ADACHI, O., AMEYAMA, M. (1984), D-Gluconate dehydrogenase, 2-keto-D-gluconate yielding, from *Gluconobacter dioxyacetonicus:* Purification and characterization, *Agric. Biol. Chem.* **48**, 1517–1522.

SHIRAISHI, F., KAWAKAMI, K., KONO, S., TAMURA, A., TSURUTA, S., KUSUNOKI, K. (1989a), Characterization of production of free gluconic acid by *Gluconobacter suboxydans* adsorbed on ceramic honeycomb monolith, *Biotechnol. Bioeng.* **33**, 1413–1418.

SHIRAISHI, F., KAWAKAMI, K., TAMURA, A., TSURUTA, S., KUSUNOKI, K. (1989b), Continuous production of free gluconic acid by *Gluconobacter suboxydans* IFO 3290 immobilized by adsorption on ceramic honeycomb monolith: effect of reactor configuration on further oxidation of gluconic acid to ketogluconic acid, *Appl. Microbiol. Biotechnol.* **31**, 445–447.

SIEVERS, M., LUDWIG, W., TEUBER, M. (1994a), Phylogenetic positioning of *Acetobacter, Gluconobacter, Rhodopila* and *Acidiphilium* species as a branch of acidophilic bacteria in the *alpha*-subclass of Proteobacteria based on 16S ribosomal RNA sequences, *Syst. Appl. Microbiol.* **17**, 189–196.

SIEVERS, M., LUDWIG, W., TEUBER, M. (1994b), Revival of the species *Acetobacter methanolicus* (ex: UHLIG et al., 1986) nom. rev., *Syst. Appl. Microbiol.* **17**, 352–354.

STADLER-SZÖKE, A., NYESTE, L., HOLLÓ, J. (1980), Studies on the factors affecting gluconic acid and 5-ketogluconic acid formation by *Acetobacter, Acta Aliment.* **9**, 155-172.

SU, Y., LIU, W., JANG, L. (1977), Studies on microbial production of sodium gluconate and glucono-delta-lactone from starch, *Proc. Natl. Sci. Counc. Taiwan*, Part 2, **10**, 143–160.

SWART, K., VAN DEN VONDERVOORT, P. J. I., WITTEVEEN, C. F. B., VISSER, J. (1990), Genetic localization of a series of genes affecting glucose oxidase levels in *Aspergillus niger, Curr. Genet.* **18**, 435–439.

TRAEGER M., ONKEN, U. (1988), Influence of high oxygen partial pressure on the production of gluconic acid by *Gluconobacter oxydans, DECHEMA Biotechnol. Conf.* 1987, pp. 199–201.

TRAEGER, M., QAZI, G. N., ONKEN, U., CHOPRA, C. L. (1989), Comparison of airlift and stirred reactors for fermentation with *Aspergillus niger, J. Ferment. Bioeng.* **68**(2), 112–116.

TRAEGER M., QAZI, G. N., GULAM, N., BUSE, R., ONKEN, U. (1992), Comparison of direct glucose oxidation by *Gluconobacter oxydans* ssp. *suboxydans* and *Aspergillus niger* in a pilot scale airlift reactor, *J. Ferment. Bioeng.* **74**, 274–281.

TRAMPER, J., LUYBEN, K. C. A. M., VAN DEN TWEEL, W. J. J. (1983), Kinetic aspects of glucose oxidation by *Gluconobacter oxydans* cells immobilized in calcium alginate, *Eur. J. Appl. Microbiol. Biotechnol.* **17**, 13–18.

UHLIG, H., KAHRBAUM, K., STEUDEL, A. (1986), *Acetobacter methanolicus* sp. nov., an acidophilic facultatively methylotrophic bacterium, *Int. J. Syst. Bacteriol.* **36**, 317–322.

UNDERKOFLER, L. A. (1954), Gluconic acid, in: *Industrial Fermentations,* Vol. 1 (UNDERKOFLER, L. A., HICKEY, R. J., Eds.), pp. 446–469. New York: Chemical Publishing Company.

VAN DIJKEN, J. P., VEENHUIS, M. (1980), Cytochemical localization of glucose oxidase in peroxisomes of *Aspergillus niger, Eur. J. Appl. Microbiol. Biotechnol.* **9**, 275–283.

VASSILEV, N. B., VASSILEVA, M. C., SPASSOVA, D. I. (1993), Production of gluconic acid by *Aspergillus niger* immobilized on polyurethane foam, *Appl. Microbiol. Biotechnol.* **39**, 285–288.

VERHAVE, T. H. (1930), *Ger. Patent* 563758.

VIIKARI, L. (1984a), Formation of levan and sorbitol from sucrose by *Zymomonas mobilis, Appl. Microbiol. Biotechnol.* **19**, 252–255.

VIIKARI, L. (1984b), Formation of sorbitol by *Zymomonas mobilis, Appl. Microbiol. Biotechnol.* **20**, 118–123.

WARD, G. E. (1967), Production of gluconic acid, glucose oxidase, fructose, and sorbose, in: *Microbial Technology,* 1st Edn. (PEPPLER, H. J., Ed.), pp. 200–221. New York: Van Nostrand Reinhold.

WARD, M. (1991), *Aspergillus nidulans* and other filamentous fungi as genetic systems. in: *Modern Microbial Genetics,* pp. 455–496. New York: Wiley-Liss, Inc.

WELLS, P. A., MOYER, A. J., STUBBS, J. J., HERRICK, H. T., MAY, O. E. (1937), Gluconic acid production. Effect of pressure, air flow, and agitation on gluconic acid production by submerged mold growths, *Ind. Eng. Chem.* **29**, 653–656.

WHITTINGTON, H., KERRY-WILLIAMS, S., BIDGOOD, K., DODSWORTH, N., PEBERDY, J. F., DOBSON, M., HINCHLIFFE, E., BALLANCE, D. J. (1990), Expression of the glucose oxidase gene in *A. niger, A. nidulans* and *Saccharomyces cerevisiae, Curr. Genet.* **18**, 531–536.

WITTEVEEN, C. F. B., VAN DEN VONDERVOORT, P., SWART, K., VISSER, J. (1990), Glucose oxidase overproducing and negative mutants of *Aspergillus niger, Appl. Microbiol. Biotechnol.* **33**, 683–686.

WITTEVEEN, C. F. B., VEENHUIS, M., VISSER, J. (1992), Location of glucose oxidase and catalase activities in *Aspergillus niger, Appl. Environ. Microbiol.* **58**, 1190–1194.

WITTEVEEN, C. F. B., VAN DEN VONDERVOORT, P. J. I., VAN DEN BROECK, H. C., VAN ENGELENBURG, F. A. C., DE GRAAFF, L. H., HILLEBRAND, M. H. B. C., SCHAAP, P. J., VISSER, J. (1993), The induction of glucose oxidase, catalase and lactonase in *Aspergillus niger, Curr. Genet.* **24**, 408–416.

YAMADA, K. (1977), Recent advances in industrial fermentation in Japan, *Biotechnol. Bioeng.* **19**, 1563–1622.

ZACHARIOU, M., SCOPES, R. K. (1986), Glucose–fructose oxidoreductase, a new enzyme isolated from *Zymomonas mobilis* that is responsible for sorbitol production, *J. Bacteriol.* **167**, 863–869.

ZIFFER, J., GAFFNEY, A. S., ROTHENBERG, S., CAIRNEY, T. J. (1971), *Brit. Patent* 1249347.

# 11 Further Organic Acids

Max Roehr
Christian P. Kubicek

Wien, Austria

# 1 Introduction

In addition to citric acid and gluconic acid which are produced in large amounts, several other acids demand adequate attention. Among these lactic acid and acetic acid are treated in Chapters 8 and 12 of this volume. The following chapter is mainly devoted to itaconic acid and the so-called fruit acids, i.e., malic, succinic, and tartaric acid. Due to its close chemical and biological relationship to fruit acids, fumaric acid is also included. Furthermore, data are presented on acids somehow related to mechanisms of direct oxidation of glucose such as kojic acid. All of these have in common that they principally can be produced by fermentation processes.

For more recent reviews with regard to the production of these acids the reader is referred to BUCHTA (1983), BIGELIS (1985), GOLDBERG et al. (1991), and BIGELIS and ARORA (1992).

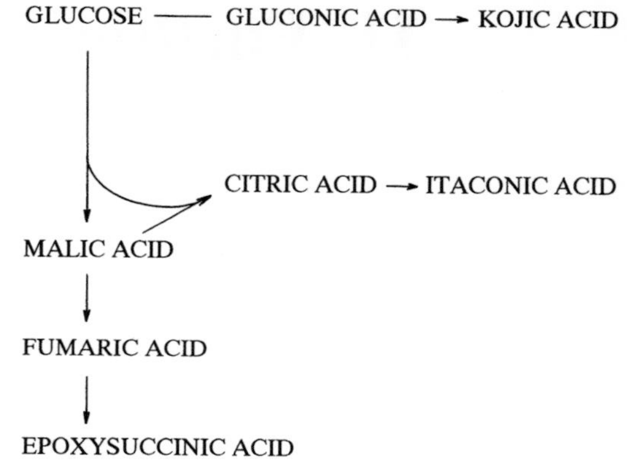


**Fig. 1.** Biosynthetic routes.

# 2 Biological Fundamentals, Biochemical Pathways of Organic Acid Production and Their Regulation

In the attempt to explain the accumulation of organic acids by filamentous fungi on a biochemical basis, it is appropriate to consider their metabolic relationship first. It is evident from Fig. 1 that itaconate, malate, fumarate, and epoxysuccinate are obviously derived from reactions close to the tricarboxylic acid cycle, whereas kojic acid seems to be formed by certain steps in carbohydrate oxidation (ROEHR et al., 1992).

Biosynthesis of the latter has not yet been explained sufficiently: tracer studies using [1-$^{14}$C]-glucose revealed that most of the kojic acid is formed from glucose without previous breakage of the carbon chain (ARNSTEIN and BENTLEY, 1953), and a small amount via the pentosephosphate pathway (ARNSTEIN and BENTLEY, 1956). Later investigations by KITADA and FUKIMBARA (1971) using [1-$^{14}$C]-, [6-$^{14}$C]-, [3-$^{3}$H]-, and [5-$^{3}$H]-labeled glucose showed that the major pathway of kojic acid formation in a submerged culture of *Aspergillus oryzae* is a direct conversion of glucose without splitting of the carbon chain. BAJPAI et al. (1981), on the basis of enzyme activity measurement, postulated a pathway involving glucose dehydrogenase and gluconate dehydrogenase for kojic acid biosynthesis. Glucose dehydrogenase has been preliminarily characterized (GIBAK, 1967). However, direct evidence for the involvement of this pathway in kojic acid biosynthesis is not yet available.

Complete glycolysis, on the other hand, is a prerequisite for the formation of the other di- and tricarboxylic acids. In most eukaryotes, pyruvate is converted to acetyl-CoA in the mitochondrion. However, in organic acid accumulating fungi there is also a pyruvate carboxylase present which – in contrast to most other eukaryotic organisms – is mainly located in the cytosol and forms oxaloacetate from pyruvate and carbon dioxide. Its involvement in the formation of *cis*-itaconic, malic, and fumaric acid as well as citric and oxalic acid (see Chapter 9 of this volume) has been proven by tracer experiments (CLELAND and JOHNSON, 1954; BENTLEY and THIESSEN, 1957a; WINSKILL, 1983; KUBICEK, 1988; PELEG et al., 1988). The enzyme purified from *Aspergillus niger* (FEIR and SUZUKI, 1969; WONGCHAI and JEFFERSON, 1974) and *A. nidulans* (OSMANI and SCRUTTON, 1983) is a tetramer of 125 kDa subunits, and is allosterically regulated by acetyl-CoA, 2-oxoglutarate, and aspartate (OSMANI and SCRUTTON, 1983, 1985). A general role of pyruvate carboxylase in the accumulation of organic acids related to the tricarboxylic acid cycle has been proposed (KUBICEK, 1988). Consistent with this hypothesis, BERCOVITZ et al. (1990) found a cytosolic pyruvate carboxylase in all *Aspergillus* species producing malic or citric acid. In addition, however, a mitochondrially located pyruvate carboxylase isoenzyme has also been found, which was present in some strains during the growth phase. There is some evidence in *Saccharomyces cerevisiae* that the cytosolic and the mitochondrial isoenzyme can be derived from the same gene product (PELEG et al., 1990). Therefore, the conditions for induction of the cytosolic enzyme and its regulation still deserve clarification.

Most investigations dealing with the biochemistry of accumulation of di- and tricarboxylic acids have been concentrated on the role of the tricarboxylic acid cycle in their biosynthesis. Conflicting reports have been published, however (for review, see ROEHR et al., 1992). Evidence for a non-mitochondrial pathway has been obtained for malic acid formation. By applying $^{13}$C-NMR (PELEG et al., 1989) it was found that [1-$^{13}$C]-glucose was predominantly incorporated into C-3 (—CH2—) of malate, which is consistent with its formation via oxaloacetate. Acid production correlated with a *de novo* induction of the cytosolic isoenzyme of malate dehydrogenase. This suggests a cytosolic pathway for malic acid biosynthesis which is self-contained in ATP and NADH generation (Fig. 2). A similar pathway has been established for fumaric acid accumulation by *Rhizopus arrhizus* (KENEALY et al., 1986; Fig. 2), which contains a cytosolic isoenzyme of fumarase (PELEG et al., 1989). In analogy to malic acid accumulation, the cytosolic fumarase was specifically induced under conditions of fumaric acid production. These findings stress the importance of the cytosolic reductive pathway in organic acid accumulation.

Another interesting feature of formation of malic and fumaric acid is that they need an

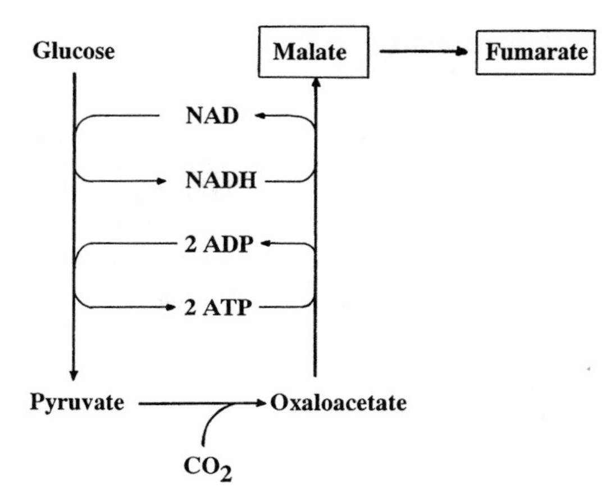


**Fig. 2.** Biosynthetic routes to malate and fumarate.

exogenous source of carbon dioxide for the pyruvate carboxylic reaction for high molar yields from glucose. Unfortunately, yields in these fermentations are not always consistently defined. It should be noted here that – in view of the biochemical pathway outlined above – a 100% conversion of glucose (C6) to fumaric acid (C4) would mean that 2 moles of fumaric acid are formed from 1 mol glucose (i.e., 64% w/w). Yields in industrial production in fact come close to this value, suggesting that the additional 2 carbons have to be derived from a source other than glucose. Obviously, this is provided by $CaCO_3$ added for pH control. Both the lack of ATP overproduction and this requirement for exogenous $CO_2$ may, therefore, at least in part, explain why these acids are formed at higher pH values than itaconic or citric acid (ROEHR et al., 1992).

A further, less common acid accumulated by *Aspergillus fumigatus* is 1-*trans*-2,3-epoxysuccinic acid. Its biosynthesis involves incorporation of molecular oxygen into fumarate (WILKOFF and MARTIN, 1963; AIDA and FOSTER, 1962). Based on the findings for malic and fumaric acid, it is tempting to speculate that the oxygen incorporating enzyme, the nature of which has not yet been identified, is also located in the cytoplasm.

The chemical relationship of itaconic acid to *cis*-aconitate, on the other hand, suggests an obvious linkage of its biosynthesis to the reactions catalyzed by citrate synthase and aconitase, which was confirmed by the demonstration of a *cis*-aconitate decarboxylase in *A. terreus* (BENTLEY and THIESSEN, 1957b). More recent studies have provided evidence that this enzyme is specifically induced under conditions of itaconic acid production (JAKLITSCH et al., 1991). Although controversial biochemical pathways of *cis*-itaconate biosynthesis have been proposed (for review, see ROEHR et al., 1992), the involvement of the tricarboxylic acid cycle-related pathway in *cis*-itaconate production is now well established from $^{14}C$-tracer studies (BENTLEY and THIESSEN, 1957b; WINSKILL, 1983). *Cis*-itaconate, on the other hand, clearly requires at least one mitochondrial step, i.e., citrate synthase, which is located exclusively in the mitochondria (JAKLITSCH et al., 1991). In this regards it closely resembles the biosynthesis of citric acid. Export of citrate from mitochondria has been discussed to be triggered by the cytosolic increase of its counter-ion malate (KUBICEK, 1988) or by a rise in the intramitochondrial concentration of citrate as a consequence of the metabolic properties of some tricarboxylic acid-cycle enzymes (see Chapter 9 of this volume). It is unlikely, however, that citrate transport from the mitochondria is involved in *cis*-aconitate formation, since aconitate hydratase is exclusively mitochondrial in *A. terreus,* whereas *cis*-aconitate decarboxylase is located in the cytoplasm (JAKLITSCH et al., 1991). This indicates that in this species of *Aspergillus cis*-aconitate apparently leaves the mitochondrion, which may explain why little citric acid is produced simultaneously along with itaconate. A comparative analysis of the tricarboxylate carriers of *A. niger* and *A. terreus* would be expected to reveal biochemical proof for this interpretation.

The typical feature of by-production of itatartaric acid, e.g., by *Aspergillus itaconicus* (JAKUBOWSKA, 1977), citramalic acid and $\beta$-hydroxyparaconic acid is only poorly explained. Since no label of $^{14}C$-itaconate is incorporated into one of these acids, the conclusion is favored that they are produced via separate pathways. Particularly citramalate is a common metabolite in several microorganisms, and can be formed by a condensation of acetyl-CoA and pyruvate (ROEHR, 1963).

# 3 Production Procedures

## 3.1 Itaconic Acid

There has been continued interest in biological ways of producing compounds with double bonds suitable for the manufacture of various polymers. Itaconic acid and, to a lesser extent, fumaric acid occupy such positions.

Itaconic acid was originally known as a product of pyrolytic distillation of citric acid. In 1929, KINOSHITA (1929, 1931) observed the formation of itaconic acid by an osmophilic strain of a green *Aspergillus* species, which he had isolated from dried salted plums. KINOSHITA named this organism *A. itaconicus.* THOM and RAPER (1945), in their well-known manual, considered *A. itaconicus* as morphologically identical with *A. varians.* At the same time it was found (MOYER and COGHILL, 1945) that KINOSHITA's strain had almost completely lost its potency. In the meantime, however, in one of their famous "studies on the biochemistry of microorganisms" from RAISTRICK's laboratory, CALAM et al. (1939) had found that one out of several strains of *A. terreus* produced considerable quantities of itaconic acid. Extensive screening work at the Northern Regional Research Laboratory (NRRL), Peoria, USA (MOYER and COGHILL, 1945; LOCKWOOD and REEVES, 1945) resulted in the isolation of a strain (NRRL 1960 = ATCC 10020) which was further investigated on the pilot plant scale. Literature on these earlier investigations, describing also transfer from surface to submerged operation, may be found, e.g., in the reviews of FOSTER (1949), LOCKWOOD (1954, 1975, 1979) and MIALL (1978). In 1955, industrial production in submerged fermentation was started by Chas. Pfizer & Co. Inc. in their Brooklyn plant (ANONYMOUS, 1955). According to MIALL (1978), further plants were established in England, France, Russia, and Japan.

Miles Laboratories (BATTI and SCHWEIGER, 1963) and VON FRIES (1966) took out patents for processes rather similar to the art of citric acid manufacture.

A Japanese group (KOBAYASHI, 1967; KOBAYASHI et al., 1972) described complete protocols for a process of itaconic acid manufacture, also using *A. terreus,* namely a strain K 26 = ATCC 32359, derived from ATCC 10020.

The process of Pfizer & Co. (Pfizer Inc., 1948; NUBEL and RATAJAK, 1964) utilizes cheap raw materials such as molasses (10–30% of total) with additions of glucose or high-test cane molasses giving final sugar concentrations of 100, 150, or 180 g $L^{-1}$. With molasses, only supplementation of zinc, magnesium, and copper is required. It has been found, however, that the fermentation is rather sensitive to volatile acids which may be present in molasses. Procedures for their removal have been elaborated. Fermentations are carried out at 39–42°C, giving yields of 47% (equal to 62% of theory) within 3 d (NUBEL and RATAJAK, 1964). The resulting itaconic acid is partly neutralized by adding milk of lime.

The Miles process (BATTI and SCHWEIGER, 1963), similar to experiences with citric acid production, utilizes decationized molasses to make up media with sugar concentrations of about 125 g $L^{-1}$, with the addition of copper (0.5–200 mg $L^{-1}$) and zinc ions. Yields of 58% based on sugar concentration within 7.5 d have been claimed. According to BATTI (1964), the pH should be kept >3.0 to avoid formation of unwanted by-products, especially itatartaric acid. The occurrence of this acid as a fermentation product of a mutant of *A. terreus* had been reported earlier by STODOLA et al. (1945). According to BATTI and SCHWEIGER (1963), the addition of calcium ions is beneficial for higher yields of itaconic acid. RIVIERE et al. (1977) have shown that calcium ions inhibit itaconic acid oxidase, which catalyzes the formation of itatartaric acid. RYCHTERA (1977) observed maximum itaconic acid production rates at pH values between 2.8 and 3.1 and decreasing rates above these values due to changes in pellet morphology (RYCHTERA and WASE, 1981).

The procedure of VON FRIES (1966) equally resembles citric acid manufacturing practice: crude molasses solutions (160 g $L^{-1}$ sugar) are heat-treated according to common procedures. VON FRIES (1966) suggested alkaline treatment, now largely abandoned. Hexacyanoferrate (HCF) is added to the hot solution in amounts sufficient to obtain an excess of 5 mg $L^{-1}$ HCF. One tenth of this solution is diluted to a sugar concentration of 37 g $L^{-1}$ and used as the initial volume. After addition of nutrient salts, especially zinc, HCF is added again to achieve an excess of about 30 mg $L^{-1}$. The fermenter is inoculated with spores of *A. terreus* ATCC 10020 to a density of $10^6$–$10^7$ spores per liter. After propagation for 2 d, the remaining amount of molasses solution is fed to the fermenter in increments to maintain sugar concentrations of 15–20 g $L^{-1}$. After a fermentation period of 5 d, yields of 60–65% equal to about 85% of theoretical values can be realized. Crude starch hydrolyzates are decationized before being fermented similarly with 2-fold addition of HCF.

Recovery and purification of itaconic acid is usually performed by separating the fungal biomass by filtration followed by evaporation of the liquid to less than one tenth of its volume. Upon cooling, the bulk of product may be separated as a crude crystalline mass. Purification is accomplished by treating the dissolved mass with activated carbon and recrystallization. Mother liquors in some cases are refed to the fermentation.

Several attempts have been made to improve the process economically. The Japanese group mentioned above studied the process thoroughly with these aims in mind. The authors relied on *A. terreus,* using strain K26, derived from ATCC 10020 which apparently proved to be superior to *A. itaconicus* (KOBAYASHI, 1967) although the latter had been reisolated and reselected by KINOSHITA and TANAKA (1961). The group of KOBAYASHI designed a two-stage continuous process considering the fact that the itaconic acid production rate was maximal at mycelium concentrations of about 20 g $L^{-1}$ (d.w.). High productivities, however, go along with lower concentrations of itaconic acid in the output; it was thus suggested to design novel methods of product recovery. The authors designed and investigated a procedure comprising electrodialysis using ion-exchange membranes, which is comprehensively documented (KOBAYASHI, 1967; KOBAYASHI et al., 1972, 1973, 1980).

High productivities (0.937 g $L^{-1}$ $h^{-1}$) and yields >80% have been claimed by CROS and SCHNEIDER (1989). More recently, attempts have been reported to study the use of xylose (KAUTOLA et al., 1985; KAUTOLA, 1990) in order to utilize cheap wood hydrolyzates as carbon source.

Several workers have tried to immobilize the fungus in order to improve the performance of fermentation systems (HORITSU et al., 1983; JU and WANG, 1986; KAUTOLA et al., 1990, 1991; VASSILEV et al., 1992; OKABE et al., 1993). Others have reported itaconic acid production by novel producer organisms such as yeasts (TABUCHI et al., 1981) or *Ustilago* (TABUCHI, 1991).

The main potential of utilizing itaconic acid according to MIALL (1978) is in the manufacture of styrene butadiene copolymers and for lattices and emulsions. Synthetic fiber manufacturing makes use of acrylnitrile copolymers. Obviously, all these materials have to compete with respective similar petrochemical products.

## 3.2 Fruit Acids

Fruit acids are typical constituents of fruits and other plant materials which are responsible for the characteristic acid flavor components of such materials. The whole subject has been most comprehensively treated by GARDNER (1966) and RUDY (1967).

The discovery and study of these acids dates back to the 18th century and is closely related to the name of C. W. SCHEELE, who discovered tartaric acid in 1770, citric acid in 1784, and malic acid in 1785.

First of all, one obvious question should be discussed: What are the reasons for producing these various kinds of acids in quite different amounts? The following remarks may be helpful in providing some answers.

Tartaric acid as the typical constituent of wine is deposited as scarcely soluble potassium salt (cremor tartari) in the course of wine fermentation. At the turn of the century, it was the first fruit acid to be produced professionally in wine producing countries from the readily available residues of wine making. Together with citric acid, which was produced mainly in Italy from citrus fruits via precipitation of the calcium salt, a rather stable and flourishing market of fruit acids was established for the manufacture of beverages and for drug uses. With the development of the citric acid fermentation, production of citric acid increased from several ten thousand tons to several hundred thousand tons per year, leaving tartaric acid as a product for special, in part traditional, applications.

Fumaric acid, on the other hand, is available by a rather simple chemical process. It is cheaper, therefore, but its use is limited because of its rather low solubility in water. It may serve, however, as a feedstock for the production of malic acid by an enzymatic procedure or of succinic acid by chemical means.

Malic acid, according to its name, is found in apples, grapes and other fruits and displays some kind of tart acid flavor as compared to other fruit acids. It would, therefore, find more specific applications and wider uses, if cheaper ways of production could be made available.

However, decisions as which acid is used in whatever field of application are not only determined by economic considerations but also by other judgements such as (especially in food, medical, and cosmetics applications) sensory properties or even traditional aspects. The acid taste of the different acids is not only related to the hydrogen concentration (dissociation) exhibited at different concentrations (RUDY, 1967) but also a matter of taste, regarding tartaric acid as the sharpest and, e.g., lactic acid as the most bland. Gradual differences between, e.g., citric, malic, and fumaric acid are reflected best by the fact that mixtures of these or with other food components exhibit special variations in taste and acceptability that can be discriminated by sensorial means. GOLDBERG et al. (1991) have provided an excellent comparative compilation of data and properties regarding citric, fumaric, and malic acid based on earlier data of GARDNER (1966).

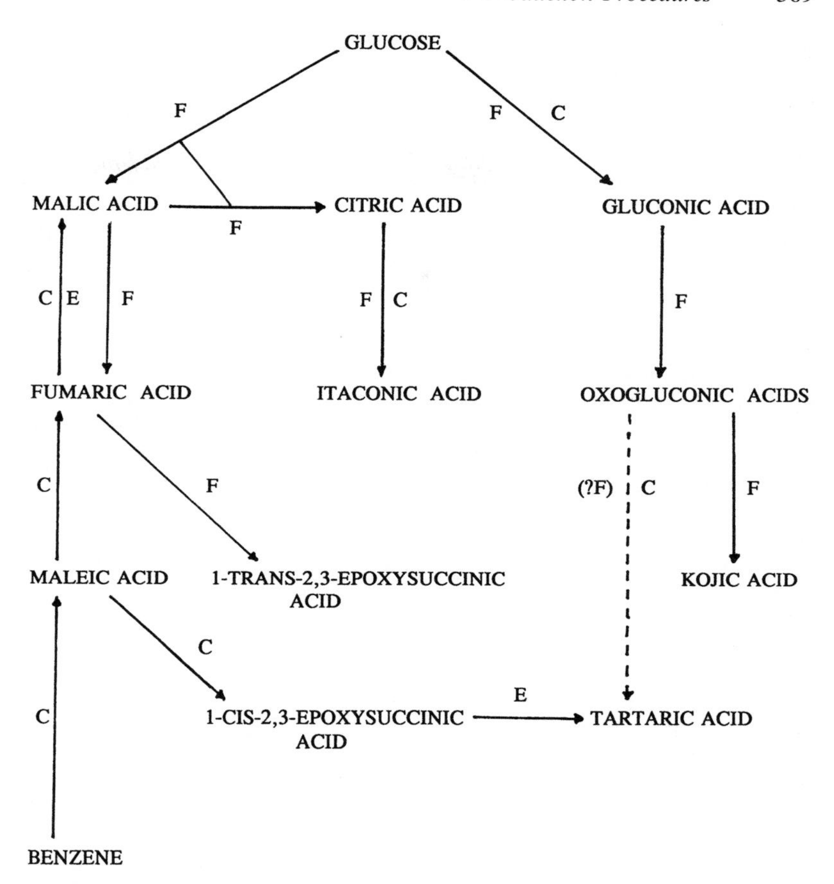


**Fig. 3.** Fruit acids – ways of production; (C) chemical, (E) enzymatic, (F) fermentation process.

Fig. 3 shows the relations of fruit acids with respect to possible production routes.

## 3.2.1 Fumaric Acid

Fumaric acid is presently produced in large amounts by a chemical route, namely catalytic vapor-phase oxidation of benzene or $C_4$ hydrocarbons to maleic acid (and maleic acid anhydride) followed by isomerization to fumaric acid, e.g., by boiling the aqueous solution with hydrochloric acid or other catalysts. As fumaric acid, in contrast to maleic acid, is scarcely soluble in water, it can be recovered almost quantitatively. Several process variants have been described (see IRWIN et al., 1967; LAWRENCE, 1974; LOHBECK and HAFERKORN, 1978; LOHBECK et al., 1990). Maleic acid formed as a by-product of phthalic acid anhydride manufacture may also be used as a raw material. Since maleic acid is rather toxic, fumaric acid has to be purified appropriately for food applications. Specifications for food applications are: fumaric acid $>99.5\%$; arsenic $<3$ mg kg$^{-1}$; heavy metals (determined as iron) $<10$ mg kg$^{-1}$; maleic acid $<0.1\%$; ash $<0.1\%$; water $<0.5\%$. Thus biological routes, although seemingly more expensive, might be advantageous under certain conditions (GOLDBERG et al., 1991).

Relatively small amounts of fumaric acid occur in plant materials or in higher fungi, e.g., in species of *Fumaria,* hence the name of this acid. Following the line of fungal origin, in 1911 EHRLICH discovered production of fumaric acid upon cultivation of species of *Rhizopus*, especially *R. nigricans*. This finding was extended in studies of *R. oryzae* (LOCKWOOD et al., 1936) and by FOSTER and WAKSMAN (1939a, b). In this context it should be mentioned that many strains of *Rhizopus* produce lactic acid (and ethanol) under aerobic conditions (FOSTER, 1949). It appears that the capability to produce fumaric acid in larger amounts is mainly strain-dependent and to a lesser extent determined by adjustments of media composition. It has been shown, however, that formation of lactic acid and ethanol can be suppressed by the addition of specific inhibitors such as *p*-benzoquinone (1 g $L^{-1}$) (MIKSCH et al., 1950).

Utilizing the early know-how of citric acid fermentation, attempts were made to accomplish commercial production of fumaric acid by fermentation with members of the genus *Rhizopus*. Accounts of earlier work may, e.g., be found in the following sources: FOSTER and WAKSMAN, 1939a, b; FOSTER, 1949, 1954; BERNHAUER, 1950; RHODES et al., 1959, 1962; PERLMAN and SIH, 1960 (see also the excellent review of MIALL, 1978). First patents date from 1943 (WAKSMAN, 1943; KANE et al., 1943). Fumaric acid fermentation was the first fermentation process which served as a model for implementing and scaling up the technique of submerged fermentation with molds as producing organisms. It has been reported that it was the know-how of this fermentation that permitted almost instantaneous large-scale production of penicillin during World War II. Main features of this fermentation are summarized as follows (RHODES et al., 1959, 1962; GOLDBERG and STIEGLITZ, 1985, 1986).

Glucose or, with strains possessing amylase, starch was used as carbon source. Only few *Rhizopus* strains produce sucrose inverting enzymes, exceptions being, e.g., *R. arrhizus* (now *R. oryzae*) NRRL 2582. Nitrate cannot be utilized as nitrogen source. The proportion of growth and fumaric acid formation largely depends on the N/C ratio and the amount of zinc added (ca. 20 mg $L^{-1}$), which has to be optimized. Fumaric acid production starts at the end of the growth period and requires neutralization of the acid as formed. This was mainly done by adding calcium or magnesium carbonate which also provided the beneficial above-mentioned excess of carbon dioxide, resulting, however, in heavy foaming and the formation of hardly soluble, gel-forming fumarates in the broth. The use of sodium carbonate (RAUCH et al., 1950) apparently led to problems in maintaining pH values slightly below neutrality. As mentioned in the introduction of Chapter 9, maintaining sterility in a fermentation conducted at pH values near neutrality was a distinct challenge in the evolution of fungal biotechnology.

Recently, an Italian group has studied optimum conditions of fumaric acid production with various raw materials using statistical design methods (MORESI et al., 1991, 1992). As with other mold fermentations, the employment of immobilized *R. nigricans* has been suggested (LINKO et al., 1990; BUZZINI et al., 1993).

Due to its low solubility in water (0.6 g per 100 g at 25 °C), applications of fumaric acid in the food industry, especially as an acidulant in beverages, are rather limited. Its low hygroscopicity, on the other hand, offers certain advantages. Various mixed solid compositions have therefore been designed, e.g., for jellies or dry beverage mixes. Several trials have been made to improve cold-water solubility (LAWRENCE, 1974). Other uses are in the manufacture of alkyd or polyester resins and related materials. Fumaric acid can serve in manufacturing aspartic acid (for, e.g., aspartame production) using aspartase (KOYAMA et al., 1986). An interesting use, if economically feasible, would be the manufacture of "natural" malic acid, as described below.

**1-*trans*-2,3-epoxysuccinic acid** may be regarded as an oxygenation product of fumaric acid. Its formation by molds has been reported for *Paecilomyces* (SAKAGUCHI et al., 1939) and for *Aspergillus fumigatus* (T. H. ANDERSON, 1955, quoted by MIALL, 1978; MARTIN and FOSTER, 1955). MARTIN and FOSTER investigated conversion to *meso*-tartaric acid by a crude bacterial enzyme preparation ("*trans*-succinepoxide hydrolase"). Later this enzyme was isolated from *Pseudomonas putida* (ALLEN and JAKOBY, 1969). Unfortunately, there is only demand for

L(+)-tartaric acid. More recently, an improved method of epoxysuccinic acid production using *Aspergillus clavatus* has been patented (YAMAGUCHI and NOGAMI, 1987), specifying epoxysuccinic acid as a $\beta$-lactam antibiotic precursor.

### 3.2.2 Malic Acid

Malic acid is a fruit acid with good solubility in water, characteristic tart taste and special blending and flavor fixing properties. Thus it could find increasing applications in various fields of food technology. This becomes also evident when considering the amount of work including patents published during the last decades. However, in contrast to other fruit acids, malic acid is rather hygroscopic – a disadvantage in various fields of application.

Presently, malic acid is produced mainly by chemical procedures, e.g., by heating aqueous solutions of maleic/fumaric acid mixtures (resulting from the production of maleic/fumaric acid from benzene, see above) at temperatures of 180–220 °C, pressures of 14–18 bar in special corrosion resistant reactors for several hours (AHLGREN, 1968; WINSTROM and FRINK, 1968; MILTENBERGER, 1989). The resulting racemic mixture of D-(+)- and L-(−)-malic acid can be purified by ion-exchange and/or recrystallization procedures. About 30,000 t of malic acid are produced annually.

Although U.S. and European regulations permit the use of D,L-malic acid as a food additive, procedures for the production of the "natural" form of malic acid, L-(−)- or *S*-(−)-malic acid, are receiving increasing interest.

Two principal routes of L-malic acid manufacture have been developed during the last three decades:

(1) fermentation of sugars or other carbon sources following the pathways indicated above,
(2) conversion of fumaric acid catalyzed by fumarate hydratase (EC 4.2.1.2).

As mentioned above, malic acid fermentation may be regarded as an abridged version of fumaric acid fermentation, in which the last step is blocked. Microorganisms exhibiting such properties have been described by ABE et al. (1962), namely several strains of the *Aspergillus flavus* group (SAMSON, 1992): *A. parasiticus* ATCC 13696, *A. flavus* ATCC 13697, and *A. oryzae* QM 82i.

Fermentation conditions are similar to those reported for the production of fumaric acid. Thus media are made up to produce a certain growth phase followed by a production phase in which most of the carbon source (preferentially glucose) is converted to a mixture of malic acid with smaller portions of succinic acid and fumaric acid. Calcium carbonate again serves as neutralizing agent.

Other organisms such as *Aureobasidium pullulans* (YONEMITSU et al., 1976), *Schizophyllum commune* (KUMAGAI and TOTSUKA, 1988), or *Ustilago* and *Tolyposporium* as claimed by TABUCHI and NAKAHARA (1991) have been described as malic acid producers.

Besides the use of *n*-paraffins as unconventional raw materials (SATO et al., 1977), processes involving fermentation of ethanol and calcium carbonate (TACHIBANA and MURAKAMI, 1973) or calcium acetate (CAMPBELL et al., 1987) with acceptable yields have also been described.

With the *A. flavus* strain, malic acid yields of about 50% of the theoretical estimation were claimed with originally rather low productivities ($0.1 \text{ g L}^{-1} \text{ h}^{-1}$). More recent studies, however, using the same strain (PELEG et al., 1988) resulted in similar yields but higher productivities ($0.3 \text{ g L}^{-1} \text{ h}^{-1}$). In 1991, BATTAT et al. (1991) optimized the process mainly by increasing agitation rate and iron concentration (to $12 \text{ mg L}^{-1}$) and by lowering the N and P concentrations (to $271 \text{ mg L}^{-1}$ and 1.5 mM, respectively) thus obtaining yields ($113 \text{ g L}^{-1}$ from $120 \text{ g L}^{-1}$ glucose utilized) exceeding 60% of theoretical values or 120% on a molar basis. A productivity of 0.59 $\text{g L}^{-1} \text{ h}^{-1}$ was achieved.

As the authors claim, fermentation of less expensive carbohydrate raw materials could be competitive to other acid producing fermentation processes (see also GOLDBERG et al., 1991). The fact that, e.g., 1 mol glucose theoretically yields 268 g L-(−)-malic acid as compared to only 192 g citric acid should support this opinion. As a disadvantage it should be noted that the use of members of the *A.*

*flavus* group, especially that of *A. flavus* itself, is somehow critical due to the known tendency to form mycotoxins.

It appears, however, that the second route, enzymatic conversion of synthetic fumaric acid to L-(−)-malic acid, will be the most promising way to obtain this acid on an industrial scale. This is also reflected by the great number of patents taken out during the last years.

Fumarase (fumarate hydratase, EC 4.2.1.2) has been found in a large number of microorganisms, but could also be obtained from various plant and animal sources. Maleate may also serve as a substrate in these processes.

In an early patent (KITAHARA, 1961), lactic acid bacteria (e.g., *Lactobacillus brevis* or *L. delbrueckii*) or *Escherichia coli* were proposed as enzyme sources in a process converting solid calcium fumarate in suspension with precultured *L. brevis* for periods of 16 h to 3 d. The resulting precipitate contained almost pure calcium (L)-malate presenting a yield of 90%.

Other organisms such as *Pullularia pullulans* (HANDA et al., 1968), *Paracolobactrum aerogenoides* (DEGEN et al., 1974; Snam Progetti S.p.A., 1975), yeasts (YASUDO et al., 1979; YANG, 1992), *Aspergillus wentii* (SCHINDLER, 1984; FURITSETSU, 1986), *Thermus* or *Bacillus* (Kyowa Hakko Kogyo, 1981; KIMURA et al., 1988) have been described as potent fumarase containing malic acid producers. Most of the patents deal with organisms belonging to the coryneform group (*Micrococcus, Arthrobacter, Brevibacterium*) of bacteria (OZAWA and WATANABE, 1969a; NAKAYAMA and KOBAYASHI, 1991), especially *Brevibacterium flavum* and *B. ammoniagenes* (OZAWA and WATANABE, 1968, 1969b, c; YAMAGATA et al., 1987, 1988).

In most of these processes, fumarate (or eventually maleate) in concentrations of about 1 M is treated with cells or isolated enzyme preparations for periods of 0.5 to several hours with average yields up to 80% on a weight basis. Inclusion of detergents has been found beneficial. Once again, immobilization procedures are claimed to be more effective and economical (CHIBATA et al., 1977, 1987; TAKATA and TOSA, 1993). Recently, the *FUM1* gene of *S. cerevisiae,* encoding the cytosolic as well as the mitochondrial fumarate hydratase, has been cloned downstream of a strong *GAL10* promoter resulting in overexpression of this gene (PELEG et al., 1990). Fumarate hydratase activity of such cells was 80 mmol $h^{-1}$ $g^{-1}$ wet cell weight, permitting conversion yields of 80–90% with productivities of about 2 g $L^{-1}$ $h^{-1}$. Excellent results were obtained with agarose-immobilized cells (NEUFELD et al., 1991).

Novel purification procedures have also been reported, e.g., electrodialysis (SEIPENBUSCH, 1986; SRIDHAR, 1987).

It is in fact unpredictable whether L-malic acid from these processes will be able to compete with synthetic D,L-malic acid and fumaric acid or with citric acid.

### 3.2.3 Succinic Acid

Succinic acid is manufactured entirely by chemical means, i.e., from maleic or fumaric acid. Several biotechnical routes of production are known but cannot compete. No specific applications apart from those of other fruit acids are known so far.

### 3.2.4 Kojic Acid

The production of kojic acid, a 2-hydroxymethyl-5-hydroxy-γ-pyrone, has attracted workers especially in the Far East for many years although practical uses were only limited. Kojic acid was detected as a fermentation by-product of *A. oryzae* on "koji" (SAITO, 1907) and named accordingly by YABUTA (1924). A great many of members of the *A. flavus* group (SAMSON, 1992) are able to produce this acid under certain conditions. Earlier work on kojic acid fermentation has been summarized by FOSTER (1949) or PRESCOTT and DUNN (1959). As this group has already been described above as producers of acids of the tricarboxylic acid cycle, fumaric acid and malic acid, strain differences besides environmental conditions must be responsible for differences of the metabolic spectrum. According to MIALL (1978), differences in fermentation conditions could be the sugar concentrations and aeration rates. MIALL observed that a culture of *A. flavus* produced kojic acid on sucrose in surface culture, and citric acid on molasses in submerged culture. Similar differences had been reported for aflatoxin and kojic acid production by *A. flavus* (BASAPPA et al., 1970). Thus organisms with impaired activities of parts of the tricarboxylic acid cycle produce considerable quantities of kojic acid upon cultivation on elevated concentrations of mono- or disaccharides and low phosphate concentrations.

KITADA et al. (1967) achieved kojic acid yields of 54% in cultures containing 100 g $L^{-1}$ glucose, 1 g $L^{-1}$ $KH_2PO_4$, 0.5 g $L^{-1}$ $MgSO_4 \cdot 7H_2O$ and several grams per liter of natural N-sources (bonito extract or peptone). In larger-scale processes, e.g., 700 L fermenters, up to 30 g $L^{-1}$ of kojic acid were obtained (KITADA et al., 1971). In general, productivities are rather low. Recently, KWAK and RHEE (1992) have described experiments comparing free suspended cells of *A. oryzae* (NRRL 484) and immobilized cells entrapped in alginate gel beads. In the free mycelial fermentation, production of kojic acid started after depletion of the nitrogen source (0.75 g $L^{-1}$ ammonium sulfate, 0.5 g $L^{-1}$ yeast extract) and yielded 23 g $L^{-1}$ kojic acid from 100 g $L^{-1}$ glucose within 14 d. When using immobilized cells, applying a volume ratio of immobilized particles to fermentation broth of 1:3, kojic acid concentrations of 83 g $L^{-1}$ were attained within 6 d with an almost linear time course. When feeding further glucose portions to an immobilized system in the course of a fermentation, kojic acid began to crystallize upon reaching a concentration of about 83 g $L^{-1}$.

Kojic acid is recovered from fermentation fluids by evaporation and crystallization. It is said that purification is crucial as even traces of iron-III-ions (0.1 mg $L^{-1}$) give rise to the formation of a deeply red colored complex. It is interesting to note that in the protocol of KWAK and RHEE (1992) the nutrient medium for growth and kojic acid formation contains 20 g $L^{-1}$ $FeSO_4 \cdot 7H_2O$ and the aeration rate is 1.5 vvm.

Potential uses of kojic acid in the earlier literature were given as an analytical reagent (iron), as an insecticide or antibiotic, and as an intermediate in the production of maltol (a flavor enhancing agent) via comenic acid. According to recent patents, its inhibitory action on tyrosinase (melanin formation; OHYAMA and MISHIMA, 1990) can be used in developing cosmetics with skin-lightening properties (SARUNO and IZUMI, 1978; HIGA, 1988; OYAMA, 1989). NAKAGAWA et al. (1993) have described the production of a less toxic dimer of kojic acid by *Pseudonocardia spinosa* exhibiting equal inhibitory action on melanin formation.

### 3.2.5 Tartaric Acid

L-(+)-tartaric acid, the natural form of tartaric acids, is manufactured traditionally from residues of wine making. These raw materials may roughly be divided into two groups, according to their tartaric acid content:

(1) Low-grade materials (dregs, wine lees, "Weinhefegeläger", "Weinhefe"; HABERSTROH and HUSTEDE, 1983) contain only 15–25% tartaric acid as potassium tartrate and smaller amounts of calcium tartrate, and the bulk of yeast cell and grapes debris. It may be enriched to 40–45% tartaric acid by flotation and centrifugation of the lighter particles.

(2) High-grade materials (cream of tartar; "Faßweinstein") contains 60–70% tartaric acid, mainly as potassium tartrate.

The usual purification procedure of higher-grade material consists of carefully heating to 140–150 °C to facilitate suspending in water. The suspension is neutralized at 60–70 °C with milk of lime with added calcium sulfate. Calcium tartrate is formed which is filtered off, treated with sufficient sulfuric acid and, after filtration of calcium sulfate, evaporated and crystallized. Various other purification procedures have been suggested (RUDY, 1967; LAWRENCE, 1974; HABERSTROH and HUSTEDE, 1983).

L-(+)-tartaric acid is the most expensive fruit acid. Although there is hardly any application that could not be replaced by another of the fruit acids, demand for tartaric acid remains constant at least. In several countries, production of tartaric acid has been suspended due to the competition by citric acid.

Attempts have been made to explore alternative ways of producing tartaric acid. In 1924, KLUYVER and DE LEEUW (1924) described a novel acetic acid bacterium, *Acetobacter suboxydans,* and its ability to oxidize glucose to gluconic acid and further to oxogluconic acids. PASTERNACK and BROWN (1940) and LOCKWOOD and NELSON (1951) disclosed the possibility of producing tartaric acid by an oxidative conversion of D-5-oxogluconic acid in the presence of a metallic catalyst, with vanadium compounds, e.g., $V_2O_5$, being most effective. KAMLET (1943) patented a process in which a culture of *A. suboxydans* strain ATCC 621 (now *Gluconobacter oxydans*) was claimed to produce tartaric acid by oxidizing glucose in presence of a vanadium compound as catalyst. Apparently, this was interpreted in terms of the existence of bacterial enzymes being able to convert 5-oxogluconic acid to tartaric acid. In 1971, YAMADA et al. (1971a) and KODAMA et al. (1971) investigated more than 9,000 bacterial isolates, of which 8 strains, mainly belonging to the genus *Gluconobacter,* were found to produce tartaric acid from glucose. The process was also patented (YAMADA et al., 1971b, 1972; MINOTA et al., 1972). Upon inspection of the Japanese publication (KODAMA et al., 1971) it may be seen, however, that apparently no tartaric acid formation occurred unless a vanadium compound was added. KRUMPHANCL et al. (1968) had already shown that no tartaric acid was formed under varied conditions of cultivating *Acetobacter suboxydans* and that tartaric acid had to be produced by a vanadium catalyzed chemical process. This question has finally been settled by KLASEN et al. (1992) who demonstrated that *Gluconobacter oxydans,* including the classic strain ATCC 621, does not possess the necessary enzymes for converting 5-oxogluconic acid to tartaric acid.

It has been mentioned above that *meso*-tartaric acid can be produced enzymatically by hydration of *trans*-epoxysuccinic acid. Thus the availability of *cis*-epoxysuccinic acid would open an interesting route to L-(+)-tartaric acid (Fig. 3). *cis*-Epoxysuccinic acid may be obtained by catalytic oxidation of maleic acid. Several patents were taken out during the last two decades claiming the conversion of *cis*-epoxysuccinic acid to L-(+)-tartaric acid using cells or enzyme preparations of various microorganisms which were often named according to their respective ability:

*Alcaligenes levotartaricus* (SATO and YAUCHI, 1975); *Acinetobacter tartarogenes* (KAMATANI et al., 1976a); *Pseudomonas, Agrobacterium, Rhizobium* (KAMATANI et al., 1976a, b); *Nocardia tartaricans* (MIURA et al., 1976; YUTANI et al., 1977). Apparently, industrial realization has not been successful so far.

L-(+)-tartaric acid is used mainly in effervescent powders or beverages, in baking and jelly powders, and in some pharmaceuticals. Pyrotechnical applications have also been claimed.

As may be seen from the foregoing, conventional production from residues of wine making still prevails. Therefore, production figures are rather uncertain and may be estimated between 50,000 and 70,000 t $a^{-1}$.

# 4 References


ABE, S., FURUYA, A., SAITO, T., TAKAYAMA, K. (1962), *U.S. Patent* 3063910.

AHLGREN, C. R. (1968), *U.S. Patent* 3379756.

AIDA, K., FOSTER, J. W. (1962), Incorporation of molecular oxygen into *trans*-L-epoxysuccinic acid, *Nature* **196**, 672.

ALLEN, R. H., JAKOBY, W. B. (1969), Tartaric acid metabolism. IX. Synthesis with tartrate epoxidase, *J. Biol. Chem.* **244**, 2078–2084.

Anonymous (1955), Fermentation opens new chemical doors, *Chem. Eng.* (June), 116–118.

ARNSTEIN, H. R. V., BENTLEY, R. (1953), The biosynthesis of kojic acid. 1. Production from-[1-$^{14}$C]- and [3:4-$^{14}C_2$]-glucose and [2-$^{14}$C]-1:3-dihydroxyacetone, *Biochem. J.* **54**, 493–508.

ARNSTEIN, H. R. V., BENTLEY, R. (1956), The biosynthesis of kojic acid. 4. Production from pentoses and methylpentoses, *Biochem. J.* **62**, 403–411.

BAJPAI, P., AGRAWAL, P. K., VISHWANATHAN, L. (1981), Enzymes relevant to kojic acid biosynthesis in *Aspergillus flavus, J. Gen. Microbiol.* **127**, 131–136.

BASAPPA, S. C., SREENIVASAMURTHY, V., PARPIA, H. (1970), Aflatoxin and kojic acid production by resting cells of *Aspergillus flavus, J. Gen. Microbiol.* **61**, 81–86.

BATTAT, E., PELEG, Y., BERCOVITZ, A., ROKEM, J. S., GOLDBERG, I. (1991), Optimization of L-malic acid production by *Aspergillus flavus* in a stirred fermentor, *Biotechnol. Bioeng.* **37**, 1108–1116.

BATTI, M. A. (1964), *U.S. Patent* 3162582.

BATTI, M., SCHWEIGER, L. (1963), *U.S. Patent* 3078217.

BENTLEY, R., THIESSEN, C. P. (1957a), Biosynthesis of itaconic acid in *Aspergillus terreus.* I. Tracer studies with $^{14}$C-labelled substrates, *J. Biol. Chem.* **226**, 673–687.

BENTLEY, R., THIESSEN, C. P. (1957b), Biosynthesis of itaconic acid in *Aspergillus terreus.* III. The properties and reaction mechanism of *cis*-aconitic decarboxylase, *J. Biol. Chem.* **226**, 689–702.

BERCOVITZ, A., PELEG, Y., BATTAT, E., ROKEM, J. S., GOLDBERG, I. (1990), Localization of pyruvate carboxylase in organic acid-producing *Aspergillus* strains, *Appl. Environ. Microbiol.* **56**, 1594–1597.

BERNHAUER, K. (1950), Fortschritte der mikrobiologischen Chemie in Wissenschaft und Technik, in: *Ergebn. Enzymforsch.* **11**, 151.

BIGELIS, R. (1985), Primary metabolism and industrial fermentations, in: *Gene Manipulations in Fungi* (BENNETT, J. W., LASURE, L. L., Eds.), pp. 357–401. Orlando, FL: Academic Press.

BIGELIS, R., ARORA, D. K. (1992), Organic acids of fungi, in: *Handbook of Applied Mycology*, Vol. 4 (ARORA, D. K., ELANDER, R. P., MUKERJI, K. G., Eds.), pp. 357–376. New York: Marcel Dekker.

BUCHTA, K. (1983), Organic acids of minor importance, in: *Biotechnology*, 1st Edn., Vol. 3 (REHM, H.-J., REED, G., Eds.), pp. 467–478. Weinheim: Verlag Chemie.

BUZZINI, P., GOBBETTI, M., ROSSI, J. (1993), Different unconventional supports for the immobilization of *Rhizopus arrhizus* cells to produce fumaric acid on grape must, *Ann. Microbiol. Enzymol.* **43** (Pt. 1), 53–60.

CALAM, C. T., OXFORD, A. E., RAISTRICK, H. (1939), The biochemistry of microorganisms LXIII. Itaconic acid, a metabolic product of a strain of *Aspergillus terreus* Thom., *Biochem. J.* **33**, 1488–1495.

CAMPBELL, S. M., TODD, J. R., ANDERSON, J. G. (1987), Production of L-malic acid by *Paecilomyces varioti, Biotechnol. Lett.* **9**, 393–398.

CHIBATA, I., TOSA, T., SATO, T., YAMAMOTO, K. (1977), *Ger. Patent* 2629447.

CHIBATA, I., TOSA, T., YAMAMOTO, K., TAKATA, I. (1987), Production of L-malic acid by immobilized microbial cells, *Methods Enzymol.* **136** (Immobilized Enzymes Cells, Pt. C), 455–463.

CLELAND, W. W., JOHNSON, M. J. (1954), Tracer experiments on the mechanism of citric acid formation by *Aspergillus niger, J. Biol. Chem.* **208**, 679–687.

CROS, P., SCHNEIDER, D. (1989), *Eur. Patent* 341112.

DEGEN, L., ODO, N., OLIVIERI, R. (1974), *Ger. Patent* 2363285.

EHRLICH, F. (1911), Über die Bildung von Fumarsäure durch Schimmelpilze, *Ber. Dtsch. Chem. Ges.* **44**, 3737–3742.

FEIR, H. A., SUZUKI, I. (1969), Pyruvate carboxylase of *Aspergillus niger*: kinetic study of a biotin-containing enzyme, *Can. J. Biochem.* **47**, 697–710.

FOSTER, J. W. (1949), *Chemical Activities of Fungi.* New York: Academic Press.

FOSTER, J. W. (1954), Fumaric Acid, in: *Industrial Fermentations*, Vol. 2 (UNDERKOFLER, L. A., HICKEY, R. J., Eds.), pp. 470–487. New York: Chemical Publishing Co., Inc.

FOSTER, J. W., WAKSMAN, S. A. (1939a), The production of fumaric acid by molds belonging to the genus *Rhizopus, J. Am. Chem. Soc.* **61**, 127–135.

FOSTER, J. W., WAKSMAN, S. A. (1939b), The specific effect of zinc and other heavy metals on growth and fumaric-acid production by *Rhizopus, J. Bacteriol.* **37**, 599–617.

FRIES, H., VON (1966), *Ger. Patent* 1219430.

FURITSETSU, S. (1986), *Jpn. Patent* 61078391.

GARDNER, W. H. (1966), *Food Acidulants.* New York: Allied Chemical Corporation.

GIBAK, T. (1967), Studies on glucose dehydrogenase of *Aspergillus oryzae*. III. General enzymatic properties, *Biochim. Biophys. Acta* **146**, 317–327.

GOLDBERG, I., STIEGLITZ, B. (1985), Improved rate of fumaric acid production by Tweens and vegetable oils in *Rhizopus arrhizus, Biotechnol. Bioeng.* **27**, 1067–1069.

GOLDBERG, I., STIEGLITZ, B. (1986), *U.S. Patent* 4564594.

GOLDBERG, I., PELEG, Y., ROKEM, J. S. (1991), Citric, fumaric and malic acids, in: *Biotechnology and Food Ingredients*, (GOLDBERG, I., WILLIAMS, R. A., Eds.), pp. 349–374. New York: Van Nostrand Reinhold.

HABERSTROH, H. J., HUSTEDE, H. (1983), Weinsäure, in: *Ullmans Encyklopädie der technischen Chemie*, 4th Edn., Vol. 24 (BARTHOLOMÉ, E., BIEKERT, E., HELLMANN, H., LEY, H., WEIGERT, M., Eds.), pp. 431–437. Weinheim: Verlag Chemie.

HANDA, M., SUGAWARA, Y., NISHIMURA, S. (1968), *Jpn. Patent* 43020705.

HIGA, Y. (1988), *Jpn. Patent* 63008310.

HORITSU, H., TAKAHASHI, Y., TSUDA, J., KAWAI, K., KAWANO, Y. (1983), Production of itaconic acid by *Aspergillus terreus* immobilized in polyacrylamide gels, *Eur. J. Appl. Microbiol. Biotechnol.* **18**, 358–360.

IRWIN, W. E., LOCKWOOD, L. B., ZIENTY, M. F. (1967), Malic acid, in: *Kirk-Othmer Encyclopedia of Chemical Technology*, 2nd Edn., Vol. 12 (STANDEN, A., MARK, H. F., MCKETTA, J. J., OTHMER, F., Eds.), pp. 837–849. New York: John Wiley and Sons.

JAKLITSCH, W. M., KUBICEK, C. P., SCRUTTON, M. C. (1991), The subcellular organization of itaconate biosynthesis in *Aspergillus terreus, J. Gen. Microbiol.* **137**, 533–539.

JAKUBOWSKA, J. (1977), Itaconic and itatartaric acid biosynthesis, in: *Genetics and Physiology of Aspergillus* (SMITH, J. E., PATEMAN, J. A., Eds.), pp. 427–451. London: Academic Press.

JU, N., WANG, S. (1986), Continuous production of itaconic acid by *Aspergillus terreus* immobilized in a porous disk bioreactor, *Appl. Microbiol. Biotechnol.* **23**, 311–314.

KAMATANI, Y., OKAZAKI, H., IMAI, K., FUJITA, N., YAMAZAKI, Y., OGINO, K. (1976a), *Ger. Patent* 2600589.

KAMATANI, Y., OKAZAKI, H., IMAI, K., FUJITA, N., YAMAZAKI, Y., OGINO, K. (1976b), *Ger. Patent* 2600682.

KAMLET, J. (1943), *U.S. Patent* 2314831.

KANE, J. H., FINLAY, A., AMANN, P. F. (1943), *U.S. Patent* 2327191.

KAUTOLA, H. (1990), Itaconic acid production from xylose in repeated-batch and continuous bioreactors, *Appl. Microbiol. Biotechnol.* **33**, 7–11.

KAUTOLA, H., VAHVASELKA, M., LINKO, Y. Y., LINKO, P. (1985), Itaconic acid production by immobilized *Aspergillus terreus* from xylose and glucose, *Biotechnol. Lett.* **7**, 167–172.

KAUTOLA, H., VASSILEV, N., LINKO, Y. Y. (1990), Continuous itaconic acid production by immobilized biocatalysts, *J. Biotechnol.* **13**, 315–323.

KAUTOLA, H., RYMOWICZ, W., LINKO, Y. Y., LINKO, P. (1991), Itaconic acid production by immobilized *Aspergillus terreus* with varied metal additions, *Appl. Microbiol. Biotechnol.* **35**, 154–158.

KENEALY, W., ZAADY, E., DU PREEZ, J. C., STIEGLITZ, B., GOLDBERG, I. (1986), Biochemical aspects of fumaric acid accumulation by *Rhizopus arrhizus, Appl. Environ. Microbiol.* **52**, 128–133.

KIMURA, K., TAKAYAMA, K., YASUDO, A., KAWAMOTO, T., KAWAMORI, M., MASUNAGA, I. (1988), *Jpn. Patent* 63022193.

KINOSHITA, K. (1929), Formation of itaconic acid and mannitol by a new filamentous fungus, *J. Chem. Soc. Jpn.* **50**, 583–593.

KINOSHITA, K. (1931), Über die Produktion von Itaconsäure und Mannit durch einen neuen Schimmelpilz, *Aspergillus itaconicus, Acta Phytochim. Jpn.* **5**, 271–287.

KINOSHITA, S., TANAKA, K. (1961), *Brit. Patent* 878152.

KITADA, M., FUKIMBARA, T. (1971), Kojic acid fermentation. VII. Mechanism of the conversion of glucose to kojic acid, *Hakko Kogaku Zasshi* **49**, 847–851.

KITADA, M., TERUI, G., KITADA, M., FUKIMBARA, T. (1967), Kojic acid fermentation. I. Cultural condition in submerged culture, *Hakko Kogaku Zasshi* **45**, 1101–1107.

KITADA, M., KANAEDA, J., MIYAZAKI, K., FUKIMBARA, T. (1971), Kojic acid fermentation. VI. Production and recovery of kojic acid on an industrial scale, *Hakko Kogaku Zasshi* **49**, 343–349.

KITAHARA, K. (1961), *U.S. Patent* 2972566.

KLASEN, R., BRINGER-MEYER, S., SAHM, H. (1992), Incapability of *Gluconobacter oxydans* to produce tartaric acid, *Biotechnol. Bioeng.* **40**, 183–186.

KLUYVER, A. J., DE LEEUW, F. J. G. (1924), *Acetobacter suboxydans* een merkwaardige azijnbacterie, *Tidschr. Vgl. Geneesk.* **10**, (afl. 2–3), 170.

KOBAYASHI, T. (1967), Itaconic acid fermentation, *Process Biochem.* **2**, 61–65.

KOBAYASHI, T., NAKAMURA, I., NAKAGAWA, M. (1972), in: *Fermentation Technology Today, Proc. 4th Int. Ferment. Symp.* (TERUI, G., Ed.), pp. 215–221. Osaka: Soc. Ferment. Technol.

KOBAYASHI, T., NAKAMURA, I., NAKAGAWA, M. (1973), *Jpn. Patent* 48092584.

KOBAYASHI, T., NAKAMURA, I., NAKAGAWA, M. (1980), *Jpn. Patent* 55007095.

KODAMA, T., KOTERA, U., YAMADA, K. (1971), Microbial formation of tartaric acid from glucose. II. Fundamental culture conditions, *J. Ferment. Technol.* **49**, 93–97.

KOYAMA, Y., MTSUISHI, T., AKASHI, K. (1986), *Jpn. Patent* 61289890.

KRUMPHANCL, V., DYR, J., HONZOVA, H., PARDON, J. (1968), Tartaric acid by fermentation, *Sb. Vys. Sk. Chem.-Technol. Praze, Potraviny* **E21**, 19–24.

KUBICEK, C. P. (1988), The role of the citric acid cycle in fungal organic acid fermentations, *Biochem. Soc. Symp.* **54**, 113–126.

KUBICEK, C. P., SCHREFERL-KUNAR, G., WOEHRER, W., ROEHR, M. (1988), Evidence for a cytoplasmic pathway of oxalate biosynthesis in *As-*

*pergillus niger, Appl. Environ. Microb.* **54**, 633–637.

KUMAGAI, C., TOTSUKA, A. (1988), *Jpn. Patent* 63091087.

KWAK, M. Y., RHEE, J. S. (1992), Cultivation characteristics of immobilized *Aspergillus oryzae* for kojic acid production, *Biotechnol. Bioeng.* **39**, 903–906.

Kyowa Hakko Kogyo Co. Ltd. (1981), *Jpn. Patent* 56042589.

LAWRENCE, A. A. (1974), *Food Acid Manufacture. Recent Developments.* Park Ridge, NJ: Noyes Data Corporation.

LINKO, Y. Y., KAUTOLA, H., LINKO, P. (1990), *Fin. Patent* 82484.

LOCKWOOD, L. B. (1954), Itaconic Acid, in: *Industrial Fermentations*, Vol. 1 (UNDERKOFLER, L. A., HICKEY, R. J., Eds.), pp. 488–497. New York: Chemical Publishing Company.

LOCKWOOD, L. B. (1975), Organic acid production, in: *The Filamentous Fungi*, Vol. 1 (SMITH, J. E., BERRY, D. R., Eds.), pp. 142–154. London: Edward Arnold.

LOCKWOOD, L. B. (1979), Production of organic acids by fermentation, in: *Microbial Technology*, 2nd Edn., Vol. 1 (PEPPLER, H. J., PERLMAN, D., Eds.), pp. 367–387. New York: Academic Press.

LOCKWOOD, L. B., NELSON, G. E. N (1951), *U.S. Patent* 2559650.

LOCKWOOD, L. B., REEVES, M. D. (1945), Some factors affecting the production of itaconic acid by *Aspergillus terreus, Arch. Biochem.* **6**, 455–469.

LOCKWOOD, L. B., WARD, G. E., MAY, O. E. (1936), The physiology of *Rhizopus oryzae, J. Agr. Res.* **53**, 849–857.

LOHBECK, K., HAFERKORN, H. (1978), Maleinsäure mit Fumar-, Citracon- und Mesaconsäure, in: *Ullmanns Encyklopädie der technischen Chemie*, 4th Edn., Vol. 16 (BARTHOLOMÉ, E., BIEKERT, E., HELLMANN, H., LEY, H., WEIGERT, M., Eds.), pp. 407–414. Weinheim: Verlag Chemie.

LOHBECK, K., HAFERKORN, H., FUHRMANN, W., FEDTKE, N. (1990), Maleic and fumaric acids, in: *Ullmann's Encyclopedia of Industrial Chemistry*, 5th Edn., Vol. 1 16 (ELVERS, B., HAWKINS, S., SCHULZ, G., Eds.), pp. 53–62. Weinheim: VCH.

MARTIN, W. R., FOSTER, J. W. (1955), Production of *trans*-L-epoxysuccinic acid by fungi and its microbiological conversion to *meso*-tartaric acid, *J. Bacteriol.* **70**, 405–414.

MIALL, L. M. (1978), Organic acids, in: *Economic Microbiology*, Vol. 2 (ROSE, A. H., Ed.), pp. 48–119. London: Academic Press.

MIKSCH, J. N., RAUCH, J., MIELKE-MIKSCH, R., BERNHAUER, K. (1950), Über die Säurebildung durch Rhizopusarten. 5. Einige Beobachtungen über die Beeinflussung der Säurebildung in der Submerskultur durch Schwermetallionen und Gärungshemmstoffe, *Biochem. Z.* **320**, 398–401.

MILTENBERGER, K. H. (1989), Aliphatic hydroxycarboxylic acids, in: *Ullmann's Encyclopedia of Industrial Chemistry* 5th Edn., Vol. A 13 (GRAYSON, M., Ed.), pp. 507–517. Weinheim: VCH.

MINOTA, Y., KODAMA, T., KOTERA, U., YAMADA, K. (1972), *Jpn. Patent* 47033154.

MIURA, Y., YUTANI, K., TAKESUE, H., FUJII, K., IZUMI, Y. (1976), *Ger. Patent* 2605921.

MORESI, M., PARENTE, E., PETRUCCIOLI, M., FEDERICI, F. (1991), Optimization of fumaric acid production from potato flour by *Rhizopus arrhizus, Appl. Microbiol. Biotechnol.* **36**, 35–39.

MORESI, M., PARENTE, E., PETRUCCIOLI, M., FEDERICI, F. (1992), Fumaric acid production from hydrolyzates of strach-based substrates, *J. Chem. Tehnol. Biotechnol.* **54**, 283–290.

MOYER, A. J., COGHILL, R. D. (1945), The laboratory-scale production of itaconic acid by *Aspergillus terreus, Arch. Biochem.* **7**, 167–183.

NAKAGAWA, S., NAKANISHI, O., FURUYA, T., OKUMYA, T., SUGIHARA, R. (1993), *Jpn. Patent* 05310727.

NAKAYAMA, K., KOBAYASHI, Y. (1991), *Jpn. Patent* 03053888.

NEUFELD, R. J., PELEG, Y., ROKEM, J. S., PINES, O., GOLDBERG, I. (1991), L-Malic acid formation by immobilized *Saccharomyces cerevisiae* amplified for fumarase, *Enzyme Microb. Technol.* **13**, 991–996.

NUBEL, R. C., RATAJAK, E. J. (1964), *Brit. Pat.* 950570.

OHYAMA, Y., MISHIMA, Y. (1990), *Fragrance J.* **6**, 53–58.

OKABE, M., OHTA, N., PARK, Y. S. (1993), Itaconic acid production in an air-lift bioreactor using a modified draft tube, *J. Ferment. Bioeng.* **76**, 117–122.

OSMANI, S. A., SCRUTTON, M. C. (1983), The subcellular localization of pyruvate carboxylase and of some other enzymes in *Aspergillus nidulans, Eur. J. Biochem.* **133**, 551–560.

OSMANI, S. A., SCRUTTON, M. C. (1985), The subcellular localization and regulatory properties of pyruvate carboxylase from *Rhizopus arrhizus, Eur. J. Biochem.* **147**, 119–128.

OYAMA, Y. (1989), *Eur. Patent* 88-115543.

OZAWA, T., WATANABE, S. (1968), *Jpn. Patent* 43028948.

OZAWA, T., WATANABE, S. (1969a), *Jpn. Patent* 44014786.

OZAWA, T., WATANABE, S. (1969b), *Jpn. Patent* 44001191.

OZAWA, T., WATANABE, S. (1969c), *Jpn. Patent* 44001194.

PASTERNACK, R., BROWN, E. V. (1940), *U.S. Patent* 2197021.

PELEG, Y., STIEGLITZ, B., GOLDBERG, I. (1988), Malic acid accumulation by *Aspergillus flavus*. I. Biochemical aspects of acid biosynthesis, *Appl. Microbiol. Biotechnol.* **28**, 69–75.

PELEG, Y., BARAK, A., SCRUTTON, M. C., GOLDBERG, I. (1989), Isoenzyme pattern and sub-cellular localization of enzymes involved in fumaric acid accumulation by *Rhizopus oryzae*, *Appl. Microbiol. Biotechnol.* **30**, 176–183.

PELEG, Y., ROKEM, J. S., GOLDBERG, I., PINES, O. (1990), Inducible overexpression of the FUM1 gene in *Saccharomyces cerevisiae*: localization of fumarase and efficient fumaric acid bioconversion to L-malic acid, *Appl. Environ. Microbiol.* **56**, 2777–2783.

PERLMAN, D., SIH, C. J. (1960), Fungal synthesis of citric, fumaric, and itaconic acids, in: *Progress in Industrial Microbiology*, Vol. II (HOCKENHULL, D. J. D., Ed.), pp. 167–194. London: Heywood & Company, Ltd.

Pfizer Inc. (1948), *Brit. Patent* 602866.

PRESCOTT, S. C., DUNN, C. G. (1959), *Industrial Microbiology*, 3rd Edn. New York: McGraw-Hill.

RAUCH, J., MIELKE-MIKSCH, R., BERNHAUER, K. (1950), Über die Säurebildung durch Rhizopusarten. 3. Vergleichende Prüfung verschiedener Fumarsäurebildner in der Oberflächenkultur, *Biochem. Z.* **320**, 384–389.

RHODES, R. A., MOYER, A. J., SMITH, M. L., KELLEY, S. E. (1959), Production of fumaric acid by *Rhizopus arrhizus*, *Appl. Microbiol.* **7**, 74–80.

RHODES, R. A., LAGODA, A. A., MISENHEIMER, T. J., SMITH, M. L., ANDERSON, R. F., JACKSON, R. W. (1962), Production of fumaric acid in 20-liter fermentors, *Appl. Microbiol.* **10**, 9–15.

RIVIERE, J., MOSS, M., SMITH, J. M. (1977), *Industrial Applications of Microbiology*, pp. 159–161. London: Surrey University Press.

ROEHR, M. (1963), Zur Biosynthese der Glutaminsäure durch Essigsäurebakterien der Suboxydans-Gruppe, *Zentralbl. Bakteriol.* 2. Abt. **117**, 129–144.

ROEHR, M., KUBICEK, C. P., KOMINEK, J. (1992), Industrial acids and other small molecules, in: *Aspergillus: Biology and Industrial Application* (BENNETT, J. W., KLICH, M. A., Eds.), pp. 91–131. Stoneham: Butterworth-Heinemann.

RUDY, H. (1967), *Fruchtsäuren, Wissenschaft und Technik*, Heidelberg: Dr. Alfred Hüthig Verlag.

RYCHTERA, M. (1977), Effect of the acidity of the fermentation medium on the formation of itaconic acid during the culturing of *Aspergillus terreus*, *Kvasny Prum.* **23**, 154, 159–162.

RYCHTERA, M., WASE, D. A. (1981), The growth of *Aspergillus terreus* and the production of itaconic acid in batch and continuous cultures. The influence of pH, *J. Chem. Tech. Biotechnol.* **31**, 509–521.

SAITO, K. (1970), Über die Säurebildung von *Aspergillus oryzae*, *Bot. Mag.*, Tokyo **21**, 7–11.

SAKAGUCHI, K., INOUE, T., TADA, S. (1939), On the production of an ethylene-$\alpha$-$\beta$-di-carboxylic acid by molds, *Zentralbl. Bakteriol.* 2. Abt. **100**, 302–307.

SAMSON, R. A. (1992), Current taxonomic schemes of the genus *Aspergillus* and its teleomorphs, in: *Aspergillus: Biology and Industrial Applications* (BENNETT, J. W., KLICH, M. A., Eds.), pp. 355–390. Stoneham: Butterworth-Heinemann.

SARUNO, R., IZUMI, T. (1978), *Jpn. Patent* 53018739.

SATO, H., YAUCHI, A. (1975), *Jpn. Patent* 50145586.

SATO, S., NAKAHARA, T., MINODA, Y. (1977), Enzymatic studies on L-malic acid production from *n*-paraffins by *Candida brumptii* IFO-0731, *Agr. Biol. Chem.* **41**, 1903–1907.

SCHINDLER, F. (1984), *Ger. Patent* 3310849.

SEIPENBUSCH, R. (1986), *Ger. Patent* 3434918.

Snam Progetti S.p.A. (1975), *Jpn. Patent* 50155681.

SRIDHAR, S. (1987), *Ger. Patent* 3542861.

STODOLA, F. H., FRIEDKIN, M., MOYER, A. J., COGHILL, R. D. (1945), Itatartaric acid, a metabolic product of an ultraviolett-induced mutant of *Aspergillus terreus*, *J. Biol. Chem.* **161**, 739–742.

TABUCHI, T. (1991), *Jpn. Patent* 03035785.

TABUCHI, T., NAKAHARA, T. (1991), *Jpn. Patent* 03180187.

TABUCHI, T., SUGISAWA, T., ISHIDORI, T., NAKAHARA, T., SUGIYAMA, J. (1981), Itaconic acid fermentation by a yeast belonging to the genus *Candida*, *Agric. Biol. Chem.* **45**, 475–479.

TACHIBANA, S., MURAKAMI, T. (1973), L-Malate production from ethanol and calcium carbonate by *Schizophyllum commune*, *J. Ferment. Technol.* **51**, 858–864.

TAKATA, I., TOSA, T. (1993), Production of L-malic acid, *Bioprocess Technol.* **16**, 53–65.

THOM, C., RAPER, K. B. (1945), *A Manual of the Aspergilli*. Baltimore, MD: Williams & Wilkins.

VASSILEV, N., KAUTOLA, H., LINKO, Y. Y. (1992), Immobilized *Aspergillus terreus* in itaconic acid production from glucose, *Biotechnol. Lett.* **14**, 201–206.

WAKSMAN, S. A. (1943), *U.S. Patent* 2326986.

WILKOFF, L. J., MARTIN, W. R. (1963), Studies on the biosynthesis of *trans*-L-epoxysuccinic acid by *Aspergillus fumigatus, J. Biol. Chem.* **238**, 843–847.

WINSKILL, N. (1983), Tricarboxylic acid cycle activity in relation to itaconic acid biosynthesis by *Aspergillus terreus, J. Gen. Microbiol.* **129**, 2877–2883.

WINSTROM, L. O., FRINK, J. W. (1968), *U.S. Patent* 3379757.

WONGCHAI, W., JEFFERSON, W. E., JR. (1974), Pyruvate carboxylase of *Aspergillus niger:* purification and some properties, *Fed. Proc.* **33**, 1378 (conference abstract).

YABUTA, T. (1924), The construction of kojic acid, a δ-pyrone derivative formed by *Aspergillus oryzae* from carbohydrates, *J. Agr. Chem. Soc. Jpn.* **1**, 1–15.

YAMADA, K., KODAMA, T., OBATA, T., TAKAHASHI, N. (1971a), Microbial formation of tartaric acid from glucose, *J. Ferment. Technol.* **49**, 85–92.

YAMADA, K., MINODA, Y., KODAMA, T., KOTERA, U. (1971b), *U.S. Patent* 3585109.

YAMADA, K., KODAMA, T., KOTERA, U. (1972), *Jpn. Patent* 47029581.

YAMAGATA, H., NARA, S., TERASAWA, M., YUGAWA, H. (1987), *Jpn. Patent* 62083893.

YAMAGATA, H., SATO, Y., TERASAWA, M., YUGAWA, H. (1988), *Jpn. Patent* 63160590.

YAMAGUCHI, T., NOGAMI, I. (1987), *Eur. Patent* 236760.

YANG, L. (1992), *Chin. Patent* 1067451.

YASUDO, A., KAWAMOTO, T., YOSHINO, M., KIMURA, K. (1979), *Jpn. Patent* 54076894.

YONEMITSU, E., TSURUMI, Y., MATSUZAKI, T. (1976), *Jpn. Patent* 51079783.

YUTANI, K., TAKESUE, H., MIURA, Y., NOMA, Y. (1977), *Jpn. Patent* 52047987.

# 12 Acetic Acid

HEINRICH EBNER

Linz, Austria

SYLVIA SELLMER
HEINRICH FOLLMANN

Bonn, Germany

# 1 Introduction

Worldwide acetic acid is called "vinegar", if obtained by oxidative fermentation of ethanol-containing solutions by acetic acid bacteria. Although "vinegar" does not entirely exclude "diluted chemically produced acetic acid" in every country of the world, this term is used here to define acetic acid which is produced from ethanol by primary microbial metabolism, the so-called "acetic acid fermentation" or "vinegar fermentation".

In the future, acetic acid may possibly also be produced by anaerobic fermentation of glucose by *Clostridium thermoaceticum.* At least theoretically, this metabolic pathway should achieve higher yields of acetic acid per unit of glucose. The basic idea for carrying out laboratory research is the potential use of renewable resources as raw materials for chemical feedstocks, but it is not suitable for food production.

# 2 Bases of Acetic Acid Fermentation

Acetic acid fermentation (acetous fermentation) is an oxidative fermentation by which solutions of ethanol are oxidized to acetic acid and water by acetic acid bacteria using atmospheric oxygen. The oxidation proceeds according to the basic equation

$$C_2H_5OH + O_2 \rightarrow CH_3COOH + H_2O$$
$$G_o = -455 \text{ kJ mol}^{-1}$$

The alcohol-containing solution is called "mash". Its alcohol concentration is given in percent per volume. Usually, it also contains some acetic acid, expressed in grams of acetic acid per 100 mL (% w/v). The sum of ethanol (vol%) and acetic acid (g per 100 mL) is called "total concentration" because the sum of these rather incommensurable values gives the maximal concentration of acetic acid that can be obtained by complete fermentation. For detailed reasons of this somewhat strange but commonly used calculation see EBNER and FOLLMANN (1983). The quotient of the total vinegar concentration produced over the total mash concentration indicates the concentration yield.

# 3 Raw Materials for Acetic Acid Fermentation

All mashes must contain ethanol and water and provide the acetic acid bacteria with nutrients.

## 3.1 Alcohol

By far the largest percentage of vinegar is alcohol vinegar, which is produced from diluted purified ethanol. Other common names for the same product are "white vinegar" or "spirit vinegar". It is customary in almost all countries to denature the ethanol serving as raw material for the vinegar industry. In most European countries denaturation is carried out with alcohol vinegar. In the U.S. this is usually done with ethyl acetate, which is split into ethanol and acetic acid during fermentation.

Mashes obtained by alcoholic fermentation of natural sugar-containing liquids also serve as raw materials. The designation of vinegar is according to the particular raw material used: Wine vinegar, e.g., is produced by acetic acid fermentation of grape wine; cider vinegar is produced from fermented apple juice; malt vinegar is the undestilled product of alcoholic and subsequent acetous fermentation of an infusion of barley or other cereals, the starch of which was converted by malting; rice vinegar is made from saccharified rice starch, followed by alcoholic and acetous fermentation.

## 3.2 Water for Processing

The water used for the preparation of mashes must be clear, colorless, odorless, bacteriologically clean, and without sediments or suspended particles. It must also be free of chlorine, ozone, and other chemical compounds damaging bacteria.

### 3.3 Nutrients

Most natural raw materials do not require the addition of extra nutrients. However, apple cider is usually low in nitrogenous compounds, but this can be corrected by the addition of ammonium phosphate. Some grape wines also require the addition of ammonium phosphate for an optimal fermentation. In rare cases all nutrients listed below must be added for the production of alcohol vinegar.

For the production of alcohol vinegar, a mixture of the required nutrients was developed. When producing vinegar with up to 15% acetic acid, the acetic acid bacteria definitely require glucose and potassium, sodium, magnesium, calcium, ammonium as ammonium phosphate, sulfate, and chloride. The following trace minerals are needed: iron, manganese, cobalt, copper, molybdenum, vanadium, zinc. These additions are sufficient for an optimal acetic acid fermentation.

Commercially available mixtures contain supplements such as dried yeast extract in order to restart a fermentation more quickly if stopped by a disturbance such as, e.g., a power failure. Principally, nutrients should be added sparingly in order to exert a selection pressure directing to a low requirement for nutrients.

## 4 Production by Acetic Acid Fermentation

### 4.1 Microorganisms

#### 4.1.1 Overview and Basic Problems of Classification

The microorganisms oxidizing ethanol to acetic acid are commonly called acetic acid bacteria. This special primary microbial metabolism at low pH of the surrounding medium differentiates them from all other bacteria. Acetic acid bacteria are polymorphous. Cells are gram-negative, ellipsoidal to rod shaped, straight or slightly curved, 0.6–0.8 μm long, occurring singly, in pairs or in chains. There are non-motile and motile forms with polar or peritrichous flagella. They are obligately aerobic, some produce pigments, some cellulose.

Many attempts to classify acetic acid bacteria were carried out until 1980. The proposed classifications were summarized by EBNER and FOLLMANN (1983). At that time, a major contribution to the exact taxonomy of acetic acid bacteria were the DNA–rRNA hybridization studies of GILLIS and DE LEY (1980). A first conclusion from this work was that the two groups *Gluconobacter* and *Acetobacter,* as classified by DE LEY (1961), were closely related groups, justifying their union in the family of Acetobacteraceae where they are clearly distinguishable as a separate branch in the superfamily IV (the $\alpha$-subclass; STACKEBRANDT et al., 1988). Although DE LEY et al. (1984) give a set of 32 features to differentiate the genera *Gluconobacter, Acetobacter,* and *Frateuria,* not all of them are necessary for a satisfactory identification.

The feature of acetate overoxidation identifies the strains used in "high acid" fermentations as members of the genus *Acetobacter,* and thus separates them from the genera *Gluconobacter,* and *Frateuria.* SWINGS (1992) divided *Acetobacter* into 7 species. GILLIS et al. (1989) described a species capable of fixing $N_2$ microaerobically, called *Acetobacter diazotrophicus.*

#### 4.1.2 Required Properties of Industrially Used Strains

Considering the interest of many microbiologists in acetic acid bacteria it is surprising that the available information about these bacteria has so little effect on vinegar production. "Working" strains were isolated, but lost again, demonstrating the difficulties to isolate and grow *Acetobacter* strains from industrial vinegar fermentations on solid media.

A considerable progress was achieved by propagating bacteria from fermenters in Japan on a special double layer agar. The bacteria isolated on this medium have been described as *Acetobacter polyoxogenes* by ENTANI et al. (1985). However, *A. polyoxogenes* was not available from the Japan Collection of Microorganisms due to problems of propagation and preservation.

A new species in the genus *Acetobacter* for which SIEVERS et al. (1992) proposed the name *Acetobacter europaeus,* has been isolated and characterized in pure culture from vinegar fermentations at high acidity in Germany and Switzerland. All investigated strains isolated from submerged fermenters and trickle generators had very low (0–22%) DNA–DNA similarities with the traditional strains of *Acetobacter* and *Gluconobacter.* The phenotypical differentiation on the species level of the genus *Acetobacter* is rather difficult, since different strains of a single species do not necessarily utilize the same carbon source. A useful and significant criterion for the distinction of *Acetobacter europaeus* from the other *Acetobacter* species is the strong tolerance to acetic acid of 4–8% in AE-Agar and the absolute requirement of acetic acid for growth.

Independent of all these problems and difficulties, the first and foremost interest of vinegar industry is to use a strain of acetic acid bacteria that tolerates high concentrations of acetic acid and total concentration, requires small amounts of nutrients, does not overoxidize the acetic acid formed, and yields high production rates. At present, it also seems to be necessary to make this strain phage-resistant (SCHOCHER et al., 1979).

## 4.1.3 Industrially Used Strains

The vinegar industry has always worked with acetic acid bacteria that, in most cases, were not derived from pure cultures. The striking fact that common microorganisms used on a large scale for industrial vinegar production have not been properly described and characterized in taxonomic terms is caused by the difficulty to cultivate them on semisolid media.

Industrial submerged vinegar fermentations are started by inoculation with "inoculation vinegar", i.e., microbiologically undefined remains from previous fermentations. Acetic acid fermentation can be carried out for years without interruption or decrease in efficiency or yield, if suitable conditions are chosen to allow a permanent selection of strains tolerating high acidity at a minimum of nutrient concentration.

From industrial practice, it has long been known that the properties of a newly isolated strain of acetic acid bacteria may change from the very first moment of cultivation in Petri dishes and, that this strain, if cultivated over a number of generations, may show other properties, especially as far as the adaptation to certain concentrations of acetic acid is concerned, but also regarding its phenotypic features.

But, investigations of the last five years using plasmid profile analysis for characterization and identification of acetic acid bacteria derived from vinegar fermentations could not verify this variability.

### 4.1.3.1 Plasmid Profile Analysis for Characterization of *Acetobacter* Strains

Since DAVIES et al. (1981) used plasmid profile analysis for the identification of lactic acid bacteria, this method has become a powerful tool to identify and characterize strains of all genera. Plasmids are circular, extrachromosomal DNA molecules, which can be extracted from bacteria by a sophisticated technique. The plasmid profile of a strain is very specific. Plasmids can determine certain properties of bacteria, i.e., phage resistance or citrate metabolism.

The potential of this technique for vinegar fermentations was investigated by INOUE et al. (1985), FUKAYA et al. (1985a), and TEUBER et al. (1987b). The majority of the acetic acid bacteria contain 1–8 plasmids from 1 to >17 MDa. SIEVERS et al. (1990) applied plasmid profile analysis to characterize the *Acetobacter europaeus* strains and the microflora of industrial vinegar production plants. It was shown that the microflora in submerged fermentations consists of one predominant strain, whereas in trickling generators the microflora is quite heterogenous, containing a mixture of several *Acetobacter europaeus* strains with quite different plasmid profiles. This is probably the reason for the susceptibility of submerged fermentations to bacteriophage attack. In conclusion, the determination of plasmid profiles from submerged vinegar fermenters represents a new and powerful technique to characterize the microflora of such fermentations and to answer questions that could not have been approached until now, e.g., stability, origin, and identity of the prevailing microflora.

Investigations of the last years have shown that the plasmid profiles of these strains are able to rest stable over a period of at least five years and are not changing their phenotypic features during this time.

Fermentations at high total concentrations are carried out continuously in the laboratories of Heinrich Frings GmbH & Co. KG, Bonn, Germany, to increase the tolerance of the strains against acetic acid. These cultures are supplied to vinegar factories in all parts of the world.

In 1993 Frings established a Phage and Plasmid Laboratory for further investigations.

### 4.1.4 Genetic Improvement of *Acetobacter* Strains

To improve strains of *Acetobacter* genetically, recombinant DNA techniques are considered to be useful. Host–vector systems and an efficient transformation method for *Acetobacter* have been developed. The preparation of spheroplasts and the development of a spheroplast fusion method for *Acetobacter* have been described by FUKAYA et al. (1989a). A strain growing at 37°C and 4% acetic acid and a strain unable to grow at 35°C and 5% acetic acid were spheroplasted and fused. Fusion products were recieved growing at 37°C on agar plates containing 5% acetic acid, on which both parent strains were unable to grow.

Genetic transformation systems have been described for *Acetobacter xylinum* (VALLA et al., 1983), *Acetobacter aceti* (OKUMURA et al., 1985), *Acetobacter* (FUKAYA et al., 1985a, b, c), and *Gluconobacter suboxydans* (FUKAYA et al., 1985d).

### 4.1.5 Bacteriophages

*Acetobacter*-specific bacteriophages demonstrated by electron microscopy in submerged and trickling fermenters are discussed to be responsible for fermentation problems in industrial vinegar fermenters (TEUBER et al., 1987a, STAMM et al., 1989; SELLMER et al., 1992; DEFIVES et al., 1990). In trickling generators, fermentation is sometimes slowed down but hardly stops completely. In submerged fermentations, complete loss of productivity has been observed. Only 3 morphologically different *Acetobacter* phage types were described until 1992. SELLMER et al. (1992) reported a great variety in phage types regarding the dimensions of the phage heads and tails. All phages isolated from vinegar fermentations in Germany, Austria, and Denmark showed long contractile tails and isometric heads belonging to Bradley's group A (BRADLEY, 1967) and the family of Myoviridae (ACKERMANN, 1987), respectively. Although not all phages described have actually been shown to be infective for defined *Acetobacter* strains due to a lack of a suitable technique, their occurrence in high numbers ($10^6 - 10^9$ mL$^{-1}$) in disturbed fermentations suggests that bacteriophages may actually be the source of fermentation problems.

## 4.2 Biochemistry

During acetic acid fermentation ethanol is almost quantitatively oxidized to acetic acid. Concentration yields between 95% and 98% are common, and the remainder is mainly lost in the effluent gas. At the same time a suitable C6-carbon source (preferably glucose) is oxidized. End products of this oxidation are $CO_2$ and $H_2O$.

### 4.2.1 Ethanol

Ethanol oxidation is performed by two sequential reactions catalyzed by alcohol dehydrogenase (ADH) and aldehyde dehydrogenase (ALDH) located at the outer surface of the cytoplasmic membrane. Their function is linked to the respiratory chain of the organism (AMEYAMA and ADACHI, 1982a, b).

Alcohol dehydrogenase has been purified from several acetic acid bacteria and is shown to contain pyrroloquinoline quinone (PQQ) as well as heme C as prosthetic groups (ADACHI et al., 1978a, b; MURAOKA et al., 1982). Extensive purification and characterization of the enzymes involved in ethanol oxidation of *Acetobacter aceti, A. rancens,* and *G. suboxydans* have been reported by ADACHI et al. (1978a, b, 1980), AMEYAMA et al. (1981), HOMMEL and KLEBER (1984), and MURAOKA et al. (1981, 1982). It has been shown that ADH of acetic acid bacteria donates electrons directly to ubichinone in the cytoplasmic membranes. Thus, the ethanol oxidase respiratory chain of acetic acid bacteria is constituted of only three membranous respiratory components – alcohol dehydrogenase, ubichinone, and terminal oxidase (MATSUSHITA et al., 1992).

AMEYAMA et al. (1981) confirmed that alcohol and further aldehyde dehydrogenase of acetic acid bacteria had the same prosthetic group, PQQ, as their glucose dehydrogenase. These quinoproteins, glucose, alcohol and aldehyde dehydrogenases are involved in sugar oxidation that links them to an electron transport system in the membrane of oxidative bacteria. FUKAYA et al. (1989b) and TAYAMA et al. (1989) purified ADH and ALDH from *A. polyoxogenes* sp. nov. This strain originated from a vinegar factory showing higher acid productivity and higher tolerance to acetic acid than strains isolated from natural sources. In contrast to the ADH and ALDH obtained from *A. aceti* and *G. suboxydans,* the enzymes of *A. polyoxogenes* were quite stable and this high stability seems at least to be partially involved in the ability of *A. polyoxogenes* to produce high concentrations of acetic acid.

### 4.2.2 Sugar

For the breakdown of sugars, *Acetobacter* is equipped with the hexose monophosphate (HMP) pathway and the TCA cycle (ASAI, 1968) whereas glycolysis is either absent or very weak. The enzymes of the Entner-Doudoroff pathway are present in *Gluconobacter* and in *Acetobacter xylinum* (KERSTERS and DE LEY, 1968). *Gluconobacter* and *Acetobacter* are known for their direct oxidative capacity on sugars, alcohols, and steroids (DE LEY and KERSTERS, 1964). For further details, see EBNER and FOLLMANN (1983). The biochemistry of the ketogenic activities of acetic acid bacteria have been reviewed by KULHANEK (1984).

### 4.2.3 Acetate

Strains of *Acetobacter* are able to oxidize acetate as well as lactate either in the presence or in the absence of ethanol by using enzymes of the tricarboxylic cycle.

### 4.2.4 Carbon Dioxide

It has been known since the work of RAZUMOVSKAYA and BELOUSOVA (1952) that *Acetobacter* needs $CO_2$ for growth. $CO_2$ is incorporated into cell substances, where approximately 0.1% of the cell carbon is derived from $CO_2$. A very small but measurable portion of acetic acid is derived from $CO_2$ metabolism (HROMATKA and GSUR, 1962).

### 4.2.5 Nitrogen

If an organic carbon source is present a number of acetic acid bacteria can use ammonium as sole nitrogen source. Some strains need the presence of amino acids, others need cofactors such as vitamins or purines (BROWN and RAINBOW, 1956). No essential amino acids are known.

### 4.2.6 Growth Factors

Growth factor requirement depends on the carbon source supplied (RAGHAVENDRA RAO and STOKES, 1953). Some strains require *p*-aminobenzoic acid, niacin, thiamin, or pantothenic acid as growth factors (GOSSELE et al., 1980). AMEYAMA and KONDO (1966) found that some *Acetobacter* strains did not need vitamins in the presence of glucose.

But the combined addition of glutathione with sodium glutamate had a cumulative effect on the growth of *A. aceti,* as shown by ADACHI et al. (1990).

PQQ, the novel prosthetic group in acetic acid bacteria, also shows an interesting growth-stimulating effect for various microorganisms (AMEYAMA et al., 1984, 1985; ADACHI et al., 1990). One of two PQQ effects was the stimulation of both the growth rate and the total cell yield with only a trace amount of PQQ as an essential growth factor. The second effect of PQQ observed universally was a marked reduction of the lag phase in microbial growth but no increase of the growth rate at the subsequent exponential phase or in the total cell yield at the stationary phase. PQQ was produced in culture media by most strains of microorganisms.

## 4.3 Physiology

### 4.3.1 Oxygen Demand and Total Concentration

During commercial use of many aerobic microorganisms it is common practice to let the inoculum remain in a fermenter for some time up to several days without aeration before inoculation into the main fermenter. However, application of this procedure is impossible in the case of acetic acid fermentation.

Extensive tests have been conducted by HROMATKA and EBNER (1951) and HROMATKA and EXNER (1962) concerning the damage of acetic acid bacteria due to an interruption of aeration as a function of total concentration, acetic acid concentration, rate of fermentation, and length of interruption of aeration. The longer the interruption of aeration and the higher the total concentration of the mash, the more marked is the effect. At a total concentration of 5% an interruption of aeration for 2–8 min leads to the same damage as an interruption for 15–60 s at total concentrations of 11–12%. At constant total concentrations damage increases with increasing concentration of acetic acid and with increasing fermentation rate. The specific production of cell material is extremely low. For more details see EBNER and FOLLMANN (1983).

Experiments conducted by MURAOKA et al. (1983) confirmed these results. They found that there was a sharp drop in the oxidation rate of ethanol as well as in the activity of enzymes involved (ADH and ALDH) after an interruption of oxygen supply, with an acetic acid concentration of 6%. ADH activity in the membrane fraction was highly stable in the culture broth containing 4% or less acetic acid. But the purified enzyme was very unstable even in the culture broth of 2% total concentration. It was suggested from this fact that the alcohol dehydrogenase has a membrane component contributing to its own stability at acetic acid concentrations of less than 4%, which becomes gradually liable to damage with further increasing acidity. But, the relationship between the alcohol dehydrogenase system and oxygen has not yet been elucidated at all.

Under conditions of industrial acetic acid production the microorganisms undergo considerable stress, caused by a high concentration of acetic acid and limited oxygen supply. A high utilization of oxygen up to 70% can be achieved in industrial production. A sparing use of air is very important because of the volatility of ethanol and acetic acid. However, the use of pure oxygen or highly oxygen-enriched air may easily damage the bacteria (HROMATKA and EBNER, 1951).

HITSCHMANN and STOCKINGER (1985) found that at a total concentration of 13%, the ATP pool and the growth rate show a reverse behavior. The energy charge (average value of intracellular ATP, ADP, and AMP) showed a rather high value of 0.84 at the beginning of the fermentation. During fermentation, while the acetic acid concentration increased from 9 to 12%, the energy charge decreased to 0.7. After interruption of aeration for 45 s, the energy charge dropped to 0.58 and after several weeks of storage the charge of the inoculum was only 0.50.

### 4.3.2 Lack of Ethanol

Acetic acid bacteria are also damaged if a vinegar fermentation is carried out to the point where all of the ethanol has been oxidized and the addition of fresh ethanol-containing mash is delayed. This is analogous to the damage resulting from an interruption of oxygen supply and also depends primarily on the total concentration and the duration of ethanol lack.

### 4.3.3 Specific Growth Rate

HROMATKA et al. (1953) were the first to calculate the specific growth rate of acetic acid bacteria. In semicontinuous fermentations there was no decrease of the specific growth rate at increasing acetic acid concentrations, but at increasing total concentrations the specific growth rate decreased from 0.49 $h^{-1}$ at 5% to 0.16 $h^{-1}$ at 11%.

In continuous culture at a total concentration of 12% the specific growth rate decreased from 0.027 $h^{-1}$ at 4.5% ethanol to 0.006 $h^{-1}$ at 1% ethanol. It should be mentioned that the specific growth rate obtained in continuous fermentations with constant concentrations of acetic acid and ethanol are less favorable than those obtained under semicontinuous conditions.

VERA and WANG (1977) carried out batch fermentations with *Acetobacter suboxydans* and showed that at a total concentration of 7.5%, a specific growth rate of 0.30 $h^{-1}$ is in good agreement with the results found by HROMATKA et al. (1953).

### 4.3.4 Specific Product Formation

HROMATKA and EBNER (1949) found a specific product formation of 21 g acetic acid per gram of bacterial dry substance per hour in semicontinuous tests at a total concentration of 10%. This very high value was independent of the acetic acid concentration of 4.5–7.2%. MORI and TERUI (1972) found a strong dependence between specific product formation and acetic acid concentration in continuous fermentations at a total concentration of 7%. The maximum specific product formation was 15 g $g^{-1}$ $h^{-1}$. VERA and WANG (1977) found that the specific product formation follows a growth-associated mechanism as long as the acetic acid concentration is below 3%. At higher acetic acid concentrations up to 7% an additional non-growth associated term must be considered. The highest value obtained with cell recycle in continuous culture of *A. suboxydans* was 11.5 g $g^{-1}$ $h^{-1}$.

### 4.3.5 Changes in Concentration and Temperature

Charging of a fermentation must be done with mash of the same total concentration and under rapid mixing because a locally formed strong concentration gradient damages the bacteria. The next cycle starts without any lag-phase. At the same time temperature has to be kept constant because a repeated change in temperature causes a similar result (EBNER and FOLLMANN 1983).

### 4.3.6 Overoxidation

Overoxidation is the undesirable oxidation of acetic acid to $CO_2$ and $H_2O$ that has to be prevented by avoiding lack of ethanol and keeping the total concentration at a high level. In trickle fermenters this is much more difficult than in submerged fermenters. For more details, see EBNER and FOLLMANN (1983).

## 4.4 Industrial Processes

### 4.4.1 Submerged Vinegar Fermentation

#### 4.4.1.1 Semicontinuous and Continuous Fermentations

For the production of vinegar with more than 12% and up to 15% acetic acid, as reported by EBNER (1985), the process has to be carried out in a semicontinuous manner, each fermentation cycle taking about the same time as the preceding and following cycles. The starting concentration of each cycle is 7–10% acetic acid and about 5% ethanol. When an alcohol concentration between 0.05% and 0.3% has been achieved in the fermentation liquid, a quantity of vinegar is discharged from the fermenter. Refilling with new mash of 0–2% acetic acid and 12–15% alcohol leads to the starting concentrations for the new cycle mentioned above. Discharging must be carried out quickly in order to avoid complete alcohol depletion. Charging must be done slowly under constant fermentation temperature and rapid mixing.

Since the middle of 1994 it is possible to produce vinegar of up to 19% acetic acid in a modified single-stage process. Starting with similar concentrations as mentioned above, alcohol is slowly added at a constant alcohol level of 2–3% in the fermentation liquid creating a corresponding increase of the total concentration. Addition of alcohol is stopped when the desired total concentration has been reached. When alcohol approaches zero a part of the fermentation liquid is discharged and replaced by mash with lower total concentration to bring acetic acid concentration and total concentration back to starting conditions, thereby enabling the bacteria to multiply faster. Later, total concentration is increased again by adding alcohol.

Continuous fermentation is only possible up to a maximum of 9–10% acetic acid because the specific growth rate of the bacteria decreases with decreasing ethanol concentration. To obtain high yields the fermentation must be carried out at a low alcohol concentration.

#### 4.4.1.2 Two-Stage Processes

In the canning industry vinegar with a very high percentage of acetic acid is in demand. The vinegar industry is interested in producing vinegar of high acidity in order to save storage and transport costs. Therefore, as reported by EBNER and ENENKEL (1978) a two-stage process for the production of vinegar with more than 15% acetic acid was developed. During a submerged fermentation with a total concentration below 15% in a first fermenter, alcohol is added slowly to increase the total concentration up to about 18.5%. After the acetic acid concentration has reached 15%, about 30% of the fermenting liquid are transferred into a second fermenter. The first fermenter is resupplied with new mash of lower total concentration. In the second fermenter the fermentation continues until the alcohol is almost depleted. The whole quantity of the finished vinegar is discharged. The fermentation liquid in the first fermenter is supplied with alcohol at the appropriate time and later divided again.

Since the end of 1993, vinegar of more than 20% acetic acid can be produced using this process. An automatic control for this process was described by ENENKEL (1988).

In 1981, a similar two-stage process for the production of vinegar with more than 20% acetic acid was described by KUNIMATSU et al. (1981, 1982). The second fermentation stage is carried out at a reduced temperature of only 18–24 °C, while in the first fermentation stage the normal temperature range of 27–32 °C is used.

MASELLI and HORWATH (1984) claimed a similar two-stage process with semicontinuous fermentation in at least two fermenters and membrane filtration of the mash from the semicontinuous stage after combining it with an oxygen-containing gas. After adding ethanol and nutrients the concentrated mash containing the bacteria is subjected to batch fermentation at a substantially lower temperature until 18–20% acetic acid are reached. The membrane-filtered vinegar is used as a second end product.

### 4.4.1.3 The Frings Acetator®

Frings Acetator® is the most common equipment for the production of all kinds of vinegar. At the end of 1993 more than 600 acetators, continuously improved, enlarged and automated, were in operation all over the world with a total production of $1354 \cdot 10^6$ L $a^{-1}$ of vinegar with 10% acetic acid. Energy consumption is about 400 W $L^{-1}$ of fermented ethanol with decreasing tendency, yield is up to 98%.

Commercial sizes of acetators are for the fermentation of up to 3,600 L of pure ethanol in 24 h.

The different processes demand excellent performance of the fermenter, especially of its aeration system.

The Frings aerator, as reported by EBNER and ENENKEL (1974), is self-aspirating, i.e., no compressed air is needed. The rotor is installed at the shaft of a motor mounted under the fermenter, connected with an air suction pipe and surrounded by a stator. It sucks air and pumps liquid, thereby creating an air–liquid emulsion that is ejected radially outward through the stator at a given speed. This speed must be chosen adequately so that the turbulence of the stream causes a uniform distribution of the air over the whole cross section of the fermenter.

As foam formation cannot always be avoided the acetator is equipped with a mechanical defoamer. From 1966 until 1991, the equipment described by EBNER (1966) was used. The rotor, turning in a spiral housing, separates liquid and gas and returns the liquid into the lower part of the fermenter. Only during discharging of the finished vinegar a part of this returned liquid is pumped out of the fermenter. As this liquid causes foam formation, the defoamer must operate most of the time of a fermentation cycle. In an improved version described by WITTLER (1991) the separated liquid portion is not pumped back into the fermenter, but into a small collection vessel to be mixed with the vinegar end product later. Lysis of harmed vinegar bacteria releases foam causing surface-active substances. By means of the total elimination of the separated liquid, further foam formation is avoided. Minimized foam formation with a shorter operating time of the defoamer and consequently a lower power uptake and a negligibly small volume of the separated liquid are the result.

The Frings Alkograph®, described by EBNER and ENENKEL (1966), an automatic instrument for measuring the amount of ethanol in the fermenting liquid has been in use since 1965. Small amounts of liquid flow through the analyzer continuously, at first through a heating vessel and then through three boiling vessels. The boiling temperature of the incoming liquid is measured in the first boiling vessel. While alcohol is distilled off continuously from the second and the third boiling vessel, the higher boiling point of the liquid from which ethanol has been removed is measured in the third boiling vessel. The difference in temperature corresponds to the ethanol content and is recorded automatically. As the flow through the vessels takes some time, there is a delay of about 15 min between the beginning of the inflow of a sample and the appearance of the correct value on the recorder. As alcohol concentration is decreasing slowly during fermentation this delay has no disadvantage on fermentation control.

To carry out the single-stage semicontinuous process at a defined alcohol content a contact in the alkograph activates the vinegar discharge pump. As soon as a preset level has been reached the mash pump starts adding fresh mash. This pump is controlled by the fermentation temperature in order to refill under constant temperature. The pump is stopped when the desired level is reached and an automatic cooling system is activated. A fermentation cycle takes 24–48 h.

The Frings Alkograph® has worked with high reliability for about 30 years in 231 acetators all over the world. New biosensors together with modern membranes permitted the development of a device to measure alcohol without any time delay: the Frings Alkosens®. HEIDER et al. (1983) described a device for measuring volatile constituents of the fermentation medium. A membrane serves to filter these volatile constituents from the fermentation liquid into a carrier medium, and they are recorded by an electronic sensor. As the measurement relies on diffusion, ambient temperature has to be kept always at the same level. Thus the alcohol probe is situated in the fermentation broth. The alcohol value displayed at the analyzer is influenced by the concentration of acetic acid and, therefore, new calibrations are needed at different total concentration levels. This disadvantage has been overcome recently by the new system Frings Alkosens II®, which incorporates a multi-layer membrane. New sensor technologies and more sophisticated computer equipment permit a fully automated process control of continuous, semicontinuous as well as single-stage or two-stage high strength fermentations – preparation of mash included.

### 4.4.1.4 Further Trends in Submerged Processes

HEKMAT and VORTMEYER (1992) applied a computerized experimental laboratory system to submerged acetic acid fermentation with industrial *Acetobacter* strains at about 10% of final acidity using automated semicontinuous cycles over a period of one year. Based on these data a mathematical model was developed to predict the time course of a cycle, even in the case when at the beginning of the cycle a lag phase occurred due to a lack of substrate at the end of the foregoing cycle. The parameters substrate, product, and biomass, however, were not sufficient to characterize the physiological status of the culture, since the behavior of the microorganisms also depends on their past. RNA concentration per unit of cell mass was found to be an internal key parameter that showed a reasonable agreement of the experimental data and the mathematical model.

MORI (1989) summarized the possibilities of on-line measurement of alcohol concentration in vinegar production. In addition to the Frings Alkograph® the Alcolyzer®, a Japanese instrument based on the same principle, immobilized microbial sensors, immobilized enzyme sensors, and semiconductor alcohol sensors have been described.

NOMURA et al. (1988, 1989) and NOMURA (1992) reported experiments for the continuous production of acetic acid by submerged fermentation with continuous removal of acetic acid by electrodialysis, continuous supply of ethanol and concentrated nutrient solution, and control of ethanol and pH. The reason for the continuous removal of acetic acid from the reaction mixture was to alleviate the inhibitory effect of the produced acetic acid. Ethanol concentration was kept constant at 1%, acetic acid concentration at ca. 2%. Without dialysis and pH control, acetic acid production already stopped after 2 d at ca. 4%, while with dialysis it lasted 30 d.

### 4.4.1.5 Other Submerged Fermentation Processes

Other processes of little practical significance and abandoned processes are described by EBNER and FOLLMANN (1983).

### 4.4.1.6 Continuous Culture and Cell Recycle

VERA and WANG (1977) and BÄNSCH and BAUER (1991) invested much work to increase the productivity of acetic acid fermentation by continuous culture and cell recycle.

It must indeed be admitted that at first sight the productivity in an acetator does not seem to be very high. The Acetator 3600®, e.g., produces during 24 h 4,000 kg acetic acid with a concentration of up to 15% in a fermenting volume of 100 $m^3$, i.e., 1.6 g $L^{-1}$ $h^{-1}$ acetic acid in average or about 2.4 g $L^{-1}$ $h^{-1}$ at the maximum fermentation rate. The aeration rate is 0.1 vvm only, the nutrient consumption is the same as described in Sect. 4.3.

Kinetic studies carried out by BÄNSCH and BAUER (1991) clarified the dependence of the bacterial growth rate on ethanol concentration, on acetic acid concentration, and on concentration of dissolved oxygen. The optimum ethanol concentration is 23 g $L^{-1}$, the optimum acetic acid concentration 36 g $L^{-1}$, and the optimum oxygen concentration 2.5–3 mg $L^{-1}$. The productivity of acetic acid formation mainly depends on the growth rate but also on the cell concentration. Therefore, the maximum productivity of 6–6.6 g $L^{-1}$ $h^{-1}$

acetic acid can be obtained at 32 g $L^{-1}$ ethanol and 60–66 g $L^{-1}$ acetic acid.

NANBA et al. (1984) studied the synergistic effects of acetic acid and ethanol on the growth of *Acetobacter* sp. at low concentrations with similar results.

At first sight, the potential productivity is about 4 times higher than presently achieveable and could even be further increased by recycling of the bacteria. However, this high degree of productivity can only be achieved at a narrow concentration range. A slight change of ethanol, acetic acid or oxygen concentration results in an immediate return to lower values.

For technical applications a second fermenter in series to the fermenter operating at optimal fermentation conditions would be necessary to complete the fermentation. To keep a high oxygen concentration in the fermentation liquid at the higher fermentation rate, a high aeration rate would be required with recovery of ethanol and acetic acid from the exhaust gas to obtain high yields, and probably addition of oxygen would be necessary. Higher nutrient consumption as well as increasing foam formation would result. The cell recycle system would need a continuous microfiltration avoiding a lack of oxygen and require an automatic control of cell concentration. But all the expenditure in process control only permits the production of the same amount of vinegar, containing not more than 10% acetic acid, in a smaller fermenter. Apparently, the demand for high productivity does not necessarily indicate the optimal developments for future production.

PARK et al. (1989a, c) tried to enhance continuous acetic acid production in a laboratory fermenter equipped with a hollow fiber filter module by increasing the concentration of dissolved oxygen. The maximum acetic acid concentration achieved was 5%. The highest productivity obtained with yeast extract as nutrient was 107 g $L^{-1}$ $h^{-1}$. The concentration of dissolved oxygen was already lower than the critical concentration of 1 ppm, although pure oxygen was supplied at an aeration rate of 1 vvm. Furthermore, it was difficult to maintain a high concentration of viable cells in the cell recycle culture, because the cells were inactivated by the lethal conditions of acetic acid concentration, oxygen deficiency, and nutritional limitation.

The effect of dissolved oxygen and acetic acid concentration on acetic acid production in continuous culture has extensively been investigated by PARK et al. (1989b). The optimum concentration of dissolved oxygen was found to be about 2 ppm for acetic acid production at low acidity. The specific acetic acid production rate, however, was diminished to complete inhibition when the dissolved oxygen concentration was higher than 2 ppm at an acetic acid concentration of more than 4.5%. This is in good agreement with HROMATKA and EBNER (1951), who found that the use of pure oxygen during fermentation can completely inhibit acetic acid formation.

As viability of the bacteria is important for high productivity the factors affecting viability have been investigated by PARK and TODA (1990). To study the effect of cell bleeding (withdrawing at constant flow rate) serial experiments were performed in a bioreactor with cell recycle and were supported by mathematical analysis. Acetic acid concentrations were up to 5.6%. The maximum value of productivity was 123 g $L^{-1}$ $h^{-1}$. The bleed ratio and dilution rate at which both concentration of acetic acid and viable cells are relatively high can be predicted now.

The authors expect these results to be useful for designing a two-stage culture for acetic acid production using the active cells in the bleed from the first stage in a second-stage vessel beyond a lethal acetic acid concentration.

How far these certainly interesting results will influence the construction of industrial equipment in Japan remains to be seen. The costs of such complicated fermenters may slow down or even prevent a fast conversion to industrial scale, although in Japan the demand for vinegar with 5% acetic acid is relatively high. However, the above mentioned results will certainly have no influence on the production of vinegar with more than 10% acetic acid. In a review of the Japanese developmental work FUKAYA et al. (1992) point out that the problem of scale-up remains for further investigation.

#### 4.4.1.7 Semicontinuous Culture with Cell Recycle

PARK et al. (1991a) investigated semicontinuous cultures with cell recycle and found an increase in productivity compared to semicontinuous culture without cell recycle at an acidity of up to 6%. While the concentration of total cells increased in every cycle, the viable cells reached a much lower *quasi* steady-state value, which was explained by product inhibition. The productivity was 2.9 g $L^{-1}$ $h^{-1}$ without and 5.0 g $L^{-1}$ $h^{-1}$ with cell recycle at 6% final acetic acid concentration, but 14 g $L^{-1}$ $h^{-1}$ at 3% final acetic acid concentration.

In another study PARK et al. (1991b) reported semicontinuous fermentations in which the ethanol concentration was kept for a certain time between 20 and 30 g $L^{-1}$ by adding ethanol-rich medium. At the end of the cycle ethanol-poor medium was added. Cells were recycled by membrane filtration. The goal was to obtain high concentrations of acetic acid; the highest amount reached was 9%. It was found that the *Acetobacter* cells lost their viability at about 6% acetic acid, but these non-viable cells were still effective for ethanol oxidation. Already 1976, this interesting feature of *Acetobacter* cells had been reported by EBNER.

### 4.4.2 Surface Fermentation

#### 4.4.2.1 Old Surface Fermentation Processes

CONNER and ALLGEIER (1976) published detailed descriptions of the history of vinegar fermentation. Today, the equipment used in the famous "Orleans Process" for the production of wine vinegar can only be seen in museums.

#### 4.4.2.2 New Experiments with Surface Fermentation

A new approach for surface fermentation was made by TODA et al. (1989, 1990). The authors used a bioreactor with a shallow, horizontal flow of medium under a bacterial film with a surface of a few hundred $cm^2$. Liquid depth was up to 10 mm and the acetic acid concentration of the effluent was 5.76%. The oxygen absorption rate through the microbial film was found to be very high; a kinetic study of acetic acid production was published by PARK et al. (1990).

Kewpie Jozo Kabushiki Kaisa (1970) described a surface fermentation process in a system of serial intercommunicating fermentation vessels with flow-regulating devices that enforce the flow to approach plug flow conditions. The liquid depth under the layer of acetic acid bacteria was about 50 mm. The inflowing liquid had 2% acidity and contained 3.5% alcohol. The acidity of the effluent stabilized at 5% acetic acid. The residence time was about 21 h.

HIGASHIDE et al. (1992) reported continuous surface fermentations in a single vessel with a working volume of 16 L. Under optimum conditions with a liquid depth of 100 mm, 2–3% acidity and 20–30 g $L^{-1}$ alcohol concentration of the inflowing liquid, vinegar with 4.5% acetic acid was produced during 80 d. The flavor of this vinegar was claimed to be better than that of vinegar produced by other processes.

For the production of rice vinegar with good quality surface fermentation develops to an automatic fermentation process at lower costs in Japan.

### 4.4.3 Trickling Processes

The old semicontinuous generator process used beech wood shavings, birch twigs, or corn cobs as carrier material for the vinegar bacteria. Large wooden tanks with up to 100 $m^3$ shavings are still in use. The fermenting liquid is collected in the lower part of the tank and continuously pumped under cooling over the carrier column. Air is aspirated or blown through the column by ventilators. A number of generators older than 40 years are still in operation.

The inhomogeneity of the carrier column in general makes a sufficiently homogenous distribution of air and trickling liquid impossible. Variations in temperature within the column can hardly be avoided. Mashes containing a high amount of nutrients that are low in total concentration form slimy deposits in the carrier material which may clog the column.

Nevertheless, in 1993, Carl Kühne KG proposed the use of a silicate carrier material with a large specific inner surface and an open pore structure.

### 4.4.4 Experiments with Immobilized Acetic Acid Bacteria

Many Japanese scientists investigated the fermentation of rice or fruit vinegar with the usual concentration of about 5% acetic acid in continuous fermentation of high productivity with immobilized acetic acid bacteria.

MORI et al. (1989) reported a 460 days run in an air lift bioreactor with acetic acid bacteria immobilized on $\kappa$-carrageenan gel beads, where living cells were newly released into the reactor. These newly released bacteria showed extremely high growth rate and respiratory activity which they retained for a few generations after leakage. The fact that the bacteria entrapped in $\kappa$-carrageenan gel and treated in a fluidized-bed reactor were gradually released from the gel beads had already been published before by OSUGA et al. (1984).

OKUHARA (1985) and NANBA et al. (1985) proposed polypropylene fibers as carrier material for *Acetobacter* producing rice vinegar of 3% acetic acid. KONDO et al. (1988) used ceramic honeycomb-monolith as carrier in a fixed-bed reactor and achieved a maximum of 4% acetic acid. SAEKI (1990a) investigated calcium alginate beads in a fluidized-bed reactor and obtained 3,5% acetic acid. The same author (SAEKI, 1990b, 1991) used this reactor in combination with an ethanol fermenter to produce rice vinegar of up to 6,5% acetic acid from a saccharified rice medium.

TAKADA and HIRAMITSU (1991) tested a two column bioreactor of porous ceramics to produce vinegar of an orange wine with up to 8,4% acetic acid. YAMASHITA et al. (1991) fixed the cells on woven cotton fabric to produce vinegar of kiwi fruit and persimmon of 4,5% acetic acid. Because of the disadvantages of all known carriers SAEKI et al. (1991) tried a new ceramic carrier called "Aphrocell" with a continuous pore structure, but the oxygen supply to this structure was difficult. None of the tested carriers proved to be ideal.

SUN et al. (1990) carried out experimental and theoretical studies on the continuous acetic acid production by immobilized cells in a three-phase fluidized-bed bioreactor at very low acetic acid concentrations. The influence of gel-entrapped and suspended cells, gel size, $k_L a$, solid holdup on productivity was studied by theoretical calculations and experiments.

MORI (1985, 1992) and MORI et al. (1992) stated that immobilized acetic acid bacteria are highly oxygen-dependent. Therefore, it is advantageous to supply oxygen to a closed and pressurized culture. Feeding medium at a higher dilution rate gives a higher production rate. As acetic acid inhibits the bacteria, it is considered that multiple reactors in series would be the best solution to obtain an increasing acidity, thereby keeping the dilution rate higher and creating the prolonged residence time necessary. An output acidity of 5.5% was obtained in two fluidized-bed type tabletop reactors in series. In scaling-up to a pilot reactor carrageenan gel beads were not used any longer because many of them had broken. Porous chitosan beads were selected with a diameter of 0.3–3 mm, that gave a higher productivity up to 9 g $L^{-1}$ $h^{-1}$. The microbial layer, however, was easily peeled off from the chitopearls. The authors state that more operational know-how is needed for the practical application of this technology.

### 4.4.5 Environmental Protection

Today, it is possible to produce vinegar without any environmental pollution.

The exhaust gas contains alcohol, acetic acid, and ethyl acetate according to their vapor tension at a fermentation temperature of about 30 °C. To obtain high yields the aeration rate is chosen as low as possible. It is furthermore common practice today to cool the exhaust gas down with cooling water as far as possible, and in order to absorb the rest of the volatiles it can be scrubbed with water that is recycled and used for mash preparation later.

The raw vinegar contains bacteria that have to be filtered off. Today most modern factories use cross-flow filtration without adding any filter aid. The very small quantity of remaining solids of about 100 g per 1,000 L of vinegar formed by acetic acid bacteria are degraded during waste water treatment.

## 4.5 Production Volumes

### 4.5.1 European Union

In 1992, the European Union produced $325 \cdot 10^6$ L of alcohol vinegar, $143 \cdot 10^6$ L of wine vinegar, and $45 \cdot 10^6$ L of other kinds of vinegar (all with 10% acetic acid).

### 4.5.2 United States of America

Years ago, the U.S. vinegar industry ceased to control and publish vinegar production figures. Therefore, older figures must be quoted here. In 1987, production volumes were $581 \cdot 10^6$ L of distilled vinegar, $54 \cdot 10^6$ L of cider vinegar, and $126 \cdot 10^6$ L of other types of vinegar – all of different acetic acid concentrations.

### 4.5.3 Japan

In 1992, Japan produced $56 \cdot 10^6$ L of rice vinegar, $146 \cdot 10^6$ L of other grain vinegars, and $161 \cdot 10^6$ L of other brewed vinegar, i.e., a total of $394 \cdot 10^6$ L at concentrations of 4–5% acetic acid.

### 4.5.4 World

From statistics and personal knowledge of the situation in many countries, the world production (excluding the USSR and China) of vinegar of 10% acetic acid is assumed to be about $1{,}900 \cdot 10^6$ L $a^{-1}$ or 190,000 t of pure acetic acid.

# 5 Experiments with Anaerobic Fermentation of Glucose

A novel route to the production of acetic acid by direct anaerobic fermentation of glucose has been under investigation since 1978 using an anaerobic and thermophilic microorganism, *Clostridium thermoaceticum.* While during anaerobic fermentation of glucose to ethanol and further aerobic fermentation to acetic acid, a maximum yield of 0.67 g acetic acid per g glucose is only achieveable. An – at least theoretical – maximum yield is 1 g $g^{-1}$ if the acid is produced directly by *Clostridium thermoaceticum.*

WANG et al. (1978) carried out batch fermentations in a 3 L fermenter at 58°C with anaerobiosis maintained by an overlay of $CO_2$. Maximum final acetic acid concentration was 30 g $L^{-1}$. This yield could only be maintained at a constant pH of 7.0 by adding NaOH. Total inhibition of growth resulted at a sodium acetate concentration of 40 g $L^{-1}$. The yield was 0.85 g acetic acid per g glucose.

SCHWARTZ and KELLER (1982) compared different strains of *Clostridium thermoaceticum* with regard to growth rate, tolerance to acetate, efficiency of converting glucose to acetic acid, and to cell mass. An interesting discovery was the effect of the redox potential on the growth rate. At pH 7 a redox potential of about $-360$ mV was needed before growth occurred. At pH 6 the potential had to be about $-300$ mV and at pH 5 $-240$ mV. The cells are able to reduce the redox potential, but the nature of the reducing agent is unknown. At pH 7 the maximum acetic acid concentration obtained was 20 g $L^{-1}$.

WANG and WANG (1983) reported continuous culture experiments for the production of acetic acid by immobilized whole cells of *Clostridium thermoaceticum* in a fermenter with 500 mL working volume. Carrageenan gels were used for immobilization. High cell loading (60 g $L^{-1}$ of gel) can be achieved inside the gel. Cell concentration is significantly higher than in free-cell fermentations. Consequently, dilution rates that are much higher than the maximum growth rate could be attained, resulting in high overall productivity. The highest volumetric acetic acid productivity achieved was 6.9 g $L^{-1}$ $h^{-1}$ at a dilution rate of 0.40 $h^{-1}$, an acetic acid concentration of 19 g $L^{-1}$, and a pH adjusted to 6.9.

WANG and WANG (1984) carried out pH controlled batch fermentations in a 5 L fermenter. The initial glucose concentration was 18 g $L^{-1}$ and a concentrated glucose solution was fed periodically maintaining the dissolved glucose concentration between 5 and 15 g $L^{-1}$. At pH 6.9, 56 g $L^{-1}$ acetic acid were obtained. At lower and higher pH values growth of *Clostridium thermoaceticum* was inhibited. At pH values below 6.0 undissociated acetic acid is responsible for growth inhibition, and at pH values above 6.0 ionized acetate is responsible for the same effect.

SUGAYA et al. (1986) carried out batch fermentations in 16 L fermenters, and continuous fermentations in 0.3 L chemostats. Acetic acid production was found to be both growth associated and non-growth associated. The obtained yield was 0.67 g $g^{-1}$ glucose only. Reactions with phosphate may be a major reason for glucose loss together with Maillard condensation reactions.

Although an acceptable concentration of acetic acid can already be obtained it has to be considered that the way to a potential industrial use is still very long and uncertain. Acetic acid produced out of crude oil is very cheap. So far, the yields from glucose are still too low. Nutrients have to be minimized. Acetic acid must be set free by another acid, then extracted and concentrated which is a costly procedure. A plant for the production of 2 t $d^{-1}$ acetic acid and vinegar from molasses via ethanol was established in Turkey in 1962 (EBNER and FOLLMANN, 1983). It had to be closed down in 1992 because prices could not be maintained even at larger production volumes.

# 6 References


ACKERMANN, H. W. (1987), Bacteriophage taxonomy, *Microbiol. Sci.* **4**, 214–218.

ADACHI, O., TAYAMA, K., SHINAGAWA, E., MATSUSHITA, K., AMAYAMA, M. (1978a), Purification and characterization of particulate alcohol dehydrogenase from *Gluconobacter suboxydans, Agric. Biol. Chem.* **42**, 2045–2056.

ADACHI, O., MIYAGAWA, E., SHINAGAWA, E., MATSUSHITA, K., AMAYAMA, M. (1978b), Purification and properties of particulate alcohol dehydrogenase from *Acetobacter aceti, Agric. Biol. Chem.* **42**, 2331–2340.

ADACHI, O., TAYAMA, K., SHINAGAWA, E., MATSUSHITA, K., AMAYAMA, M. (1980), Purification and characterization of membrane-bound aldehyde dehydrogenase from *Gluconobacter suboxydans, Agric. Biol. Chem.* **44**, 503–515.

ADACHI, O., OKAMOTO, K., MATSUSHITA, K., SHINAGAWA, E., AMEYAMA, M. (1990), An ideal basal medium for assaying stimulating activity of pyrroloquinoline quinone with *Acetobacter aceti, Agric. Biol. Chem.* **54**, 2751–2752.

AMEYAMA, M., ADACHI, O. (1982a), Alcohol dehydrogenase from acetic acid bacteria, membrane bound, in: *Methods in Enzymology,* (WOOD, W. A., Ed.) Vol. 89, pp. 450–457. New York: Academic Press.

AMEYAMA, M., ADACHI, O. (1982b), Alcohol dehydrogenase from acetic acid bacteria, membrane bound, in: *Methods in Enzymology* (WOOD, W. A., Ed.) Vol. 89, pp. 491–497. New York: Academic Press.

AMEYAMA, M., KONDO, K. (1966), Carbohydrate metabolism by the acetic acid bacteria. Part V. On the vitamin requirements for the growth, *Agric. Biol. Chem.* **30**, 203–211.

AMEYAMA, M., MATSUSHITA, K., OHNO, Y., SHINAGAWA, E., ADACHI, O. (1981), Existence of a novel prosthetic group, PQQ, in membrane-bound, electron transport chain-linked, primary dehydrogenases of oxidative bacteria, *FEBS Lett.* **130** (2), 179–183.

AMEYAMA, M., SHINAGAWA, K., MATSUSHITA, K., ADACHI, O. (1984), Growth stimulating of microorganisms by pyrroloquinoline quinone, *Agric. Biol. Chem.* **48**, 2909–2911.

AMEYAMA, M., SHINAGAWA, E., MATSUSHITA, K., ADACHI, O. (1985), Growth stimulating activity for microorganisms in naturally occurring substances and partial characterization of the substance for the activity as pyrroloquinoline quinone, *Agric. Biol. Chem.* **49**, 699–709.

ASAI, T. (1968), *Acetic Acid Bacteria.* Tokyo: Univ. of Tokyo Press; Baltimore, MD: Univ. Park Press.

BÄNSCH, J. J., BAUER, W. (1991), Substrate- and product-dependent kinetic of the continuous aerobic vinegar fermentation, *BioEngineering* **7** (3), 26–36.

BRADLEY, D. E. (1967), Ultrastructure of bacteriophages and bacteriocins, *Bacteriol. Rev.* **31**, 230–314.

BROWN, G. D., RAINBOW, C. (1956), Nutritional patterns in acetic acid bacteria, *J. Gen. Microbiol.* **15**, 61.

CONNER, H. A., ALLGEIER, R. J. (1976), Vinegar: Its history and development, *Adv. Appl. Microbiol.* **20**, 81–133.

DAVIES, F. L., UNDERWOOD, H. M., GASSON, M. J. (1981), The value of plasmid profiles for strain identification in lactic streptococci and the relationship between *Streptococcus lactis* 712, ML3 and C2, *J. Appl. Bacteriol.* **51**, 325–337.

DEFIVES, C., OCHIN, D., HORNEZ, J. P., WERQUIN, M. (1990), Accidents of fabrication of vinegar caused by a bacteriophage, *Microbiol. Aliments Nutr.* **8**, 77–79.

DE LEY, J. (1961), Comparative carbohydrate metabolism and a proposal for a phylogenetic relationship of the acetic acid bacteria, *J. Gen. Microbiol.* **24**, 31.

DE LEY, J., KERSTERS, K. (1964), Oxidation of aliphatic glycols by acetic acid bacteria, *Bacteriol. Rev.* **28**, 83–95.

DE LEY, J., SWINGS, J., GOSSELE, F. (1984), Genus *Acetobacter,* in: *Bergey's Manual of Systematic Bacteriology,* Vol. 1, (KREIG, N. R., HOLT, J. G., Eds.), pp. 268–274. Baltimore, MD: Williams and Wilkins.

EBNER, H. (1966), Apparatus for separating gas from foam, *U.S. Patent* 3262252.

EBNER, H. (1976), Essig, in: *Ullmanns Enzyklopädie der technischen Chemie*, 4. Aufl., Bd. 11, pp. 41–55. Weinheim: Verlag Chemie.

EBNER, H. (1985), Process for the production of vinegar with more than 12 gms/100 ml acetic acid, *U.S. Patent* 4503078.

EBNER, H., ENENKEL, A. (1966), Process and apparatus for the analysis of mixtures of liquids, *U.S. Patent* 3290924.

EBNER, H., ENENKEL, A. (1974), Device for aerating liquids, *U.S. Patent* 3813086.

EBNER, H., ENENKEL, A. (1978), Two stage process for the production of vinegar with high acetic acid concentration, *U.S. Patent* 4076844.

EBNER, H., FOLLMANN, H. (1983), Acetic Acid, in: *Biotechnology,* 1st Edn., Vol. 3 (REHM, H. J., REED, G., Eds.), pp. 387–407. Weinheim: Verlag Chemie.

ENENKEL, A. (1988), Control arrangement for a vinegar fermentation process, *U.S. Patent* 4773315.

ENTANI, E., OHMORI, S., MASAI, H., SUZUKI, K.-J. (1985), *Acetobacter polyoxogenes* sp. nov., a new species of an acetic acid bacterium useful for producing vinegar with high acidity, *J. Gen. Appl. Microbiol.* **31**, 475–490.

FUKAYA, M., IWATA, T., ENTANI, E., MASAI, H., UOZUMI, T., BEPPU, T. (1985a), Distribution and characterization of plasmids in acetic acid bacteria, *Agric. Biol. Chem.* **49**, 1349–1355.

FUKAYA, M., OKUMURA, H., MASAI, H., UOZUMI, T., BEPPU, T. (1985b), Construction of new shuttle vectors for *Acetobacter, Agric. Biol. Chem.* **49**, 2083–2090.

FUKAYA, M., TAYAMA, K., OKUMURA, H., MASAI, H., UOZUMI, T., BEPPU, T. (1985c), Improved transformation method for *Acetobacter* with plasmid DNA, *Agric. Biol. Chem.* **49**, 2091–2097.

FUKAYA, M., OKUMURA, H., MASAI, H., UOZUMI, T., BEPPU, T. (1985d), Development of host vector system for *Gluconobacter suboxydans, Agric. Biol. Chem.* **49**, 2407–2411.

FUKAYA, M., TAGAMI, H., TAYAMA, K., OKUMURA, H., KAWAMURA, Y., BEPPU, T. (1989a), Spheroplast fusion of *Acetobacter aceti* and its application to the breeding of strains for vinegar production, *Agric. Biol. Chem.* **53**, 2435–2440.

FUKAYA, M., TAYAMA, K., TAMAKI, T., TAGAMI, H., OKUMURA, H., KAWAMURA, Y., BEPPU, T. (1989b), Cloning of membrane-bound aldehyde dehydrogenase gene of *Acetobacter polyoxogenes* and improvement of acetic acid production by use of the cloned gene, *Appl. Environ. Microbiol.* **55**, 171–176.

FUKAYA, M., PARK, Y. S., TODA, K. (1992), Improvement of acetic acid fermentation by molecular breeding and process development, *J. Appl. Bacteriol.* **73**, 447–454.

GILLIS, M., DE LEY, J. (1980), Intra- and intergeneric similarities of the ribosomal nucleic acid cistrons of *Acetobacter* and *Gluconobacter, Int. J. Syst. Bacteriol.* **30**, 7–27.

GILLIS, M., KERSTERS, K., HOSTE, B., JANSSENS, D., KROPPENSTEDT, R. M., STEPHAN, M. P., TEIXEIRA, K. R. S., DÖBEREINER, J., DE LEY, J. (1989), *Acetobacter diazotrophicus* sp. nov., a nitrogen-fixing acetic acid bacterium associated with sugarcane, *Int. J. Syst. Bacteriol.* **39**, 361–364.

GOSSELE, F., SWINGS, J., DE LEY, J. (1980), Intra- and intergeneric similarities of the ribosomal ribonucleic acid cistrons of *Acetobacter* and *Gluconobacter, Zentralbl. Bakteriol. Parasitenkd. Infektionskrankh. Hyg. I. Abt. Orig.* **C1**, 348–350.

HEIDER, M., HOFSCHNEIDER, J., SCHALLENBERGER, W., NODES, F., KEMPE, E. (1983), Measurement of volatile constituents of a culture medium, *U.S. Patent* 4404284.

HEKMAT, D., VORTMEYER, D. (1992), Measurement, control, and modeling of submerged acetic acid fermentation, *J. Ferment. Bioeng.* **73** (1), 26–30.

HIGASHIDE, T., SHINADA, C., KAWAMURA, Y., HISAMATSU, M., YAMADA, T. (1992), Vinegar making. (I) Improvement of surface fermentation by continuous methods, *Bull. Fac. Biores. Mie Univ. Tsu* **8**, 43–49.

HITSCHMANN, A., STOCKINGER, H. (1985), Oxygen deficiency and its effect on the adenylate system in *Acetobacter* in the submerse acetic fermentation, *Appl. Microbiol. Biotechnol.* **22**, 46–49.

HOMMEL, R., KLEBER, H. P. (1984), Purification and properties of a membrane-bound aldehyde dehydrogenase involved in the oxidation of alkanes by *Acetobacter rancens* CCM1774, *Proc. 3rd Eur. Congr. Biotechnol.* **I**, 133–137.

HROMATKA, O., EBNER, H. (1949), Investigations of the vinegar fermentation. I. Trickling method and submerged method, *Enzymologia* **13**, 369–387.

HROMATKA, O., EBNER, H. (1951), Investigations of the vinegar fermentation. III. The influence of aeration on submerged fermentation, *Enzymologia* **15**, 57–69.

HROMATKA, O., EXNER, W. (1962), Investigations of the vinegar fermentation. VIII. Further knowledge on interruption of aeration, *Enzymologia* **25**, 37–51.

HROMATKA, O., GSUR, H. (1962), Investigations of the vinegar fermentation. XII. The influence of radioactive $^{14}CO_2$ on *Acetobacter, Enzymologia* **25**, 81–86.

HROMATKA, O., KASTNER, G., EBNER, H. (1953), Investigations of the vinegar fermentation. V. The influence of temperature and total concentration on the submerged fermentation, *Enzymologia* **15**, 337–350.

INOUE, T., FUKUDA, M., YANO, K. (1985), Efficient introduction of vector plasmids into acetic acid bacteria, *J. Ferment. Technol.* **63** (1), 1–4.

KERSTERS, K., DE LEY, J. (1968), The occurrence of the Entner-Doudoroff pathway in bacteria, *Antonie van Leeuwenhoek J. Microbiol. Ser.* **34**, 393–408.

Kewpie Jozo Kabushiki Kaisha (1970), Process for continuous production of vinegar by surface fermentation, *British Patent* 1305868.

KONDO, M., SUZUKI, Y., KATO, H. (1988), Vinegar production by *Acetobacter* cells immobilized on ceramic honeycomb-monolith, *Hakko Kogaku Kaishi* **66**, 393–399.

KÜHNE, C. KG (1993), Process and fixed-bed reactor for the production of vinegar, *Eur. Patent* 0523524 A1.

KULHANEK, M. (1984), Ketofermentations, in: *Modern Biotechnology,* Vol. 2 (KRUMPKANZL, V., REHACEK, Z., Eds.), pp. 614–676. Prague: Institute of Microbiology, Czechoslovak Academy of Sciences.

KUNIMATSU, Y., OKUMURA, H., MASAI, H., YAMADA, K., YAMADA, M. (1981), Production of vinegar with high acetic acid concentration, *U.S. Patent* 4282257.

KUNIMATSU, Y., OKUMURA, H., MASAI, H., YAMADA, K., YAMADA, M. (1982), Process for the production of vinegar with high acetic acid concentration, *U.S. Patent* 4364960.

MASELLI, J. A., HORWATH, R. O. (1984), Combination of semi-continuous and batch process for preparation of vinegar, *U.S. Patent* 4456622.

MATSUSHITA, K., TAKAKI, Y., SHINAGAWA, E., AMEYAMA, M., ADACHI, O. (1992), Ethanol oxidase respiratory chain of acetic acid bacteria. Reactivity with ubiquinone of pyrroloqinoline quinone-dependent alcohol dehydrogenases purified from *Acetobacter aceti* and *Gluconobacter suboxydans, Biosci. Biotech. Biochem.* **56**, 303–310.

MORI, A. (1985), Production of vinegar by immobilized cells, *Proc. Biochem.* **20**, 67–74.

MORI, A. (1989), Technical trends in the on-line measurement and control of alcohol concentrations in bioreactors for vinegar production, *Instrum. Control Engin. Jap., Kogyo Gijutsu-sha* 18–24.

MORI, A. (1992), in: *Industrial Application of Immobilized Biocatalysts* (TANAKA, A., TOSA, T., KOBAYASHI, T., Eds.), pp. 291–313. New York–Basel–Hong Kong: Marcel Dekker, Inc.

MORI, A., TERUI, G. (1972), Kinetic studies on submerged acetic acid fermentation. II. Process kinetics, *J. Ferment. Technol.* **50**, 70–78.

MORI, A., MATSUMOTO, N., IMAI, C. (1989), Growth behavior of immobilized acetic acid bacteria, *Biotechnol. Lett.* **11** (3), 183–188.

MORI, A., TANAKA, S., MATSUMOTO, N., IMAI, C. (1992), Vinegar production in a bioreactor with chitosan beads as support of immobilized bacteria, in: *Biochemical Engineering for 2001,* (FURUSAKI, S., ENDO, I., MATSUNO, R., Eds.), pp. 441–443. Tokyo–Berlin–Heidelberg–New York: Springer-Verlag.

MURAOKA, H., WATABE, Y., OGASAWARA, N., TAKAHASHI, H. (1981), Purification and properties of coenzyme-independent aldehyde dehydrogenase from the membrane fraction of *Acetobacter aceti, J. Ferment. Technol.* **59**, 247–255.

MURAOKA, H., WATABE, Y., OGASAWARA, N. (1982), Effect of oxygen deficiency on acid production and morphology of bacterial cells in submerged acetic acid fermentation by *Acetobacter aceti, J. Ferment. Technol.* **60**, 171–180.

MURAOKA, H., WATABE, Y., OGASAWARA, N., TAKAHASHI, H. (1983), Trigger of damage by oxygen deficiency to the acid production system during submerged acetic fermentation with *Acetobacter aceti, J. Ferment. Technol.* **61**, 89–93.

NANBA, A., TAMURA, A., NAGAI, S. (1984), Synergistic effects of acetic acid and ethanol on the growth of *Acetobacter* sp., *J. Ferment. Technol.* **62**, 501–505.

NANBA, A., KIMURA, K., NAGAI, S. (1985), Vinegar production by *Acetobacter rancens* cells fixed on a hollow fiber module, *J. Ferment. Technol.* **63**, 175–179.

NOMURA, Y. (1992), An approach to high-speed batch culture by built-in electrodialysis culture, *Hakko Kogaku Kaishi* **70**, 205–216.

NOMURA, Y., IWAHARA, M., HONGO, M. (1988), Acetic acid production by an electrodialysis fermentation method with a computerized control system, *Appl. Environ. Microbiol.* **54**, 137–142.

NOMURA, Y., IWAHARA, M., HONGO, M. (1989), Continuous production of acetic acid by electrodialysis bioprocess with a computerized control of fed batch culture, *J. Biotechnol.* **12**, 317–326.

OKUHARA, A. (1985), Vinegar production with *Acetobacter* grown on a fibrous support, *J. Ferment. Technol.* **63**, 57–60.

OKUMURA, H., UOZUMI, T., BEPPU, T. (1985), Construction of plasmid vectors and a genetic transformation system for *Acetobacter aceti, Agric. Biol. Chem.* **49**, 1011–1017.

OSUGA, J., MORI, A., KATO, J. (1984), Acetic acid production by immobilized *Acetobacter aceti* cells entrapped in a κ-carrageenan gel, *J. Ferment. Technol.* **62**, 139–148.

PARK, Y. S., TODA, K. (1990), Simulation study on bleed effect in cell recycle culture of *Acetobacter aceti, J. Gen. Appl. Microbiol.* **36**, 221–233.

PARK, Y. S., OHTAKE, H., TODA, K., FUKAYA, M., OKUMURA, H., KAWAMURA, Y. (1989a), Acetic acid production using a fermenter equipped with a hollow fiber module, *Biotechnol. Bioeng.* **33**, 918–923.

PARK, Y. S., OHTAKE, H., FUKAYA, M., OKUMURA, H., KAWAMURA, Y., TODA, K. (1989b), Effects of dissolved oxygen and acetic acid concentrations on acetic acid production in continuous culture of *Acetobacter aceti, J. Ferment. Bioeng.* **68**, 96–101.

PARK, Y. S., OHTAKE, H., FUKAYA, M., OKUMURA, H., KAWAMURA, Y., TODA, K. (1989c), Enhancement of acetic acid production in a high cell-density culture of *Acetobacter aceti, J. Ferment. Bioeng.* **68**, 315–319.

PARK, Y. S., OHTAKE, H. TODA, K. (1990), A kinetic study of acetic acid production by liquid-surface cultures of *Acetobacter aceti, Appl. Microbiol. Biotechnol.* **33**, 259–263.

PARK, Y. S., FUKAYA, M., OKUMURA, H., KAWAMURA, Y., TODA, K. (1991a), Production of acetic acid by a repeated batch culture with cell recycle of *Acetobacter aceti, Biotechnol. Lett.* **13**, 271–276.

PARK, Y. S., TODA, K., FUKAYA, M., OKUMURA, H., KAWAMURA, Y. (1991b), Production of a high concentration acetic acid by *Acetobacter aceti* using a repeated fed-batch culture with cell recycling, *Appl. Microbiol. Biotechnol.* **35**, 149–153.

RAGHAVENDRA RAO, M. R., STOKES, J. L. (1953), Utilization of ethanol by acetic acid bacteria, *J. Bacteriol.* **66**, 634–638.

RAZUMOVSKAYA, Z. G., BELOUSOVA, T. Z. (1952), Relations of *Acetobacter* to carbon dioxide in vinegar production, *Mikrobiologiya* **21**, 403–407.

SAEKI, A. (1990a), Studies on acetic acid fermentation. II. Vinegar production using immobilized *Acetobacter aceti* cells entrapped in calcium alginate gel beads, *Nippon Shokuhin Kogyo Gakkaishi* **37**, 191–198.

SAEKI, A. (1990b), Studies on acetic acid fermentation. III. Continuous production of vinegar with immobilized *Saccharomycodes ludwigii* cells and immobilized *Acetobacter aceti* cells entrapped in calcium alginate gel beads, *Nippon Shokuhin Kogyo Gakkaishi* **37**, 722–725.

SAEKI, A. (1991), Studies on acetic acid fermentation. V. Continuous vinegar production using twin bioreactors made from ethanol fermentor and acetic acid fermentor, *Nippon Shokuhin Kogyo Gakkaishi* **38**, 891–896.

SCHOCHER, A. J., KUHN, H., SCHINDLER, B., PALLERONI, N. J., DESPREAUX, C. W., BOUBLIK, M., MILLER, P. A. (1979), *Acetobacter* bacteriophage A-1, *Arch. Microbiol.* **121**, 193–197.

SCHWARTZ, R. D., KELLER, F. A., Jr. (1982), Acetic acid production by *Clostridium thermoaceticum* in pH-controlled batch fermentations at acidic pH, *Appl. Environ. Microbiol.* **43**, 1385–1392.

SELLMER, S., SIEVERS, M., TEUBER, M. (1992), Morphology, virulence and epidemiology of bacteriophage particles isolated from industrial vinegar fermentations, *System. Appl. Microbiol.* **15**, 610–616.

SIEVERS, M., ANDERESEN A., TEUBER, M. (1990), Plasmid profiles as tools to characterize the microflora of industrial vinegar fermenters, *Food Biotechnol.* **4**, 555.

SIEVERS, M., SELLMER, S., TEUBER, M. (1992), *Acetobacter europaeus* sp. nov., a main component of industrial vinegar fermenters in Central Europe, *Syst. Appl. Microbiol.* **15**, 386–392.

STACKEBRANDT, E., MURRAY, R. G. E., TRÜPER, H. G. (1988), Proteobacteria classis nov. a name for the phylogenetic taxon that includes the purple bacteria and their relatives, *Int. J. Syst. Bacteriol.* **38**, 321–325.

STAMM, W. W., KITTELMANN, M., FOLLMANN, H., TRÜPER, H. G. (1989), The occurrence of bacterophages in spirit vinegar fermentations, *Appl. Microbiol. Biotechnol.* **30**, 41–46.

SUEKI, M., KOBAYASHI, N., SUZUKI, A. (1991), Continuous acetic acid production by the bioreactor system loading a new ceramic carrier for microbial attachment, *Biotechnol. Lett.* **13**, 185–190.

SUGAYA, K., TUSE, D., JONES, J. L. (1986), Production of acetic acid by *Clostridium thermoaceticum* in batch and continuous fermentations, *Biotechnol. Bioeng.* **28**, 678–683.

SUN, Y., FURUSAKI, S. (1990), Continuous production of acetic acid using immobilized *Acetobacter aceti* in a three-phase fluidized bed bioreactor, *J. Ferment. Bioeng.* **69**, 102–110.

SWINGS, J. (1992), The genera *Acetobacter* and *Gluconobacter,* in: *The Prokaryotes,* 2nd Edn., Vol. 3 (BALOWS, A., TRÜPER, H. G., DWORKIN, M., HARDER, W., SCHLEIFER, K.-H., Eds.), pp. 2268–2286, New York: Springer-Verlag.

TAKADA, M., HIRAMITSU, T. (1991), Continuous production of vinegar using bioreactor with supports of porous ceramics, *Nippon Shokuhin Kogyo Gakkaishi* **38**, 967–971.

TAYAMA, K., FUKAYA, M., OKUMURA, H., KAWAMURA, Y., BEPPU, T. (1989), Purification and characterization of membrane-bound alcohol dehydrogenase from *Acetobacter polyoxogenes* sp. nov., *Appl. Microbiol. Biotechnol.* **32**, 181–185.

TEUBER, M., ANDRESEN, A., SIEVERS, M. (1987a), Bacteriophage problems in vinegar fermentations, *Biotechnol. Lett.* **9** (1), 37–38.

TEUBER, M., SIEVERS, M., ANDRESEN, A. (1987b), Characterization of the microflora of high acid submerged vinegar fermenters by distinct plasmid profiles, *Biotechnol. Lett.* **9**, 265–268.

TODA, K., PARK, Y. S., ASAKURA, T., CHENG, C. Y., OHTAKE, H. (1989), High-rate acetic acid production in a shallow flow bioreactor, *Appl. Microbiol. Biotechnol.* **30**, 559–563.

TODA, K., MIYAKE, M., OHTAKE, H. (1990), Enhanced oxygen absorption in shallow-flow surface culture, *J. Ferment. Bioeng.* **70**, 114–118.

VALLA, S., COUCHERON, D. H., KJOSBAKKEN, J. (1983), *Acetobacter xylinum* contains several plasmids: evidence for their involvement in cellulose formation, *Arch. Microbiol.* **134**, 9–11.

VERA, F. M., WANG, D. I. C. (1977), Increasing productivity in the acetic acid fermentation by continuous culture and cell recycle, *174th Am. Chem. Soc. Meet.,* Chicago.

WANG, G., WANG, D. I. C. (1983), Production of acetic acid by immobilized whole cells of *Clostridium thermoaceticum, Appl. Biochem. Biotechnol.* **8**, 491–503.

WANG, G., WANG, D. I. C. (1984), Elucidation of growth inhibition and acetic acid production by *Clostridium thermoaceticum, Appl. Environ. Microbiol.* **47** (2), 294–298.

WANG, D. I. C., FLEISCHAKER, R. J., WANG, G. (1978), A novel route to the production of acetic acid by fermentation, *Am. Inst. Chem. Eng. Symp. Ser.* **182**, 105–110.

WITTLER, R. (1991), Method and apparatus for controlling foam in a vinegar fermentation process, *U.S. Patent* 4997660.

YAMASHITA, S., OHTA, S., SUENAGA, H. (1991), Production of kiwi fruit and persimmon vinegar using *Acetobacter aceti* cells fixed on woven cotton fabrics, *Nippon Shokuhin Kogyo Gakkaishi* **38**, 608–613.

# 13 PHB and Other Polyhydroxyalkanoic Acids

ALEXANDER STEINBÜCHEL

Münster, Federal Republic of Germany

# 1 Introduction

The discovery of the storage polyester poly(3-hydroxybutyric acid), poly(3HB), in *Bacillus megaterium* is now celebrating its 70th anniversary (LEMOIGNE, 1926). After having detected poly(3HB) also in a wide range of other bacteria (STEINBÜCHEL, 1991a), this and related biosynthetic polyesters have experienced a dramatic development during the last 6 years (Tab. 1).

(1) Poly(3HB) was detected also in eukaryotic organisms and even in humans. (2) More than 90 different hydroxyalkanoic acids are now known as constituents of bacterial PHA; this variety of constituents and the possibility of having different combinations allow the biosynthesis of a large number of different polyesters. (3) The biosynthetic routes for many of these polyesters from related or from unrelated substrates have been investigated in much detail. (4) From many bacteria the genes for PHA biosynthesis have been cloned and analyzed at a molecular level. (5) Zeneca Bioproducts (formerly ICI BioProducts) in the United Kingdom have commercialized the biotechnological production of PHA, first consumer products were launched and many new applications are now being considered. (6) The availability of the bacterial genes provides the perspective for production of PHA in transgenic plants. (7) Like biosynthesis, also the degradation of PHA is understood

**Tab. 1.** Landmarks of PHA Research

| Year | Event | Reference |
|---|---|---|
| 1926 | Discovery of poly(3HB) in *Bacillus megaterium* | LEMOIGNE (1926) |
| 1961 | Isolation of *Alcaligenes eutrophus* as a reliable producer of poly(3HB) | SCHLEGEL et al. (1961) |
| 1962 | Detection of thermoplasticity of poly(3HB) | BAPTIST (1962) |
| 1974 | Detection of PHA others than poly(3HB) in environmental samples | WALLEN and ROWEDDER (1974) |
| 1981 | Patenting of a process for biotechnological production of poly(3HB-*co*-3HV) | HOLMES et al. (1981) |
| 1982 | Marketing of poly(3HB-*co*-3HV) under the trade name Biopol™ by ICI | HOWELLS (1982) |
| 1983 | Discovery of poly(3HO) in *Pseudomonas oleovorans* | DE SMET et al. (1983) |
| 1988 | Cloning of the poly(3HB)-biosynthesis pathway of *Alcaligenes eutrophus* | SLATER et al. (1988)<br>SCHUBERT et al. (1988) |
| Since 1988 | Detection of an increasing number of new constituents of PHA from axenic bacterial cultures | Most contributions were made by the laboratories of B. WITHOLT (Groningen), Y. DOI (Tokyo), R. C. FULLER and R. W. LENZ (Amherst), and A. STEINBÜCHEL (Göttingen), see text for details |
| 1990 | Launching of the first commercial product manufactured from PHA | ANONYMOUS (1990) |
| Since 1990 | Detection of bacteria capable for PHA biosynthesis from unrelated substrates | HAYWOOD et al. (1990, 1991), TIMM and STEINBÜCHEL (1990), STEINBÜCHEL and PIEPER (1992), RODRIGUES et al. (in press) |
| 1992 | Expression of poly(3HB)-biosynthesis genes in transgenic plants | POIRIER et al. (1992a, b) |
| 1994 | Vulcanization of $PHA_{MCL}$ to a natural rubber | DE KONING et al. (1994) |

PHA, polyhydroxyalkanoic acid; 3HB, 3-hydroxybutyric acid; 3HV, 3-hydroxyvaleric acid; 3HO, 3-hydroxyoctanoic acid; ICI, Imperial Chemical Industries.

much better now. Besides nucleic acids, proteins, polysaccharides, polyphosphates, polyisoprenoids PHA was recently referred to as an additional class of physiologically relevant biopolymers (MÜLLER and SEEBACH, 1993). Together with lignin, which represents one of the most abundant biopolymers, PHA therefore represents the seventh class of physiologically relevant biopolymers.

Various aspects of PHA such as the physiology and metabolism (ANDERSON and DAWES, 1990; STEINBÜCHEL, 1991a), the molecular genetics (STEINBÜCHEL et al., 1992), the properties and applications (DOI, 1990; BABEL et al., 1990; STEINBÜCHEL, 1991b; HOCKING and MARCHESSAULT, 1994) and others (MÜLLER and SEEBACH, 1993) have been reviewed very recently. In these review articles all previous review articles are mentioned; therefore, they are not cited here. This chapter will focus on the diversity of microbial PHA, on its analysis and properties and on various aspects of the production of these polyesters.

# 2 Occurrence and Variability of PHA

## 2.1 Constituents of Biosynthetic PHA

Approximately 90 different hydroxyalkanoic acids are now known to occur as constituents of biosynthetic PHA in addition to 3-hydroxybutyric acid. Most of these new constituents have been detected only during the last few years. The detection of all these constituents did not only reveal a very complex class of biopolymers, it also indicated that the enzymes involved in the synthesis are in general highly unspecific. It is interesting to know that most of these constituents were found in the PHA of only two bacteria: *Alcaligenes eutrophus* and *Pseudomonas oleovorans*.

### 2.1.1 Composition of PHA from Axenic Cultures

Besides 3-hydroxybutyric acid all aliphatic 3-hydroxyalkanoic acids with a carbon chain length ranging from 3–14 carbon atoms, i.e., from 3-hydroxypropionic acid to 3-hydroxytetradecanoic acid, were identified as constituents of biosynthetic PHA (Fig. 1). Unsaturated 3-hydroxyalkanoic acids also occurred in bacterial PHA. With the exception of 3-hydroxy(2)-butenoic acid found in a polyester synthesized by *Nocardia* sp. (DAVIS, 1964), the double bond occurs in the side chain. 3-Hydroxyalkenoic acids with one and even two double bonds were detected (Fig. 2). Furthermore, 3-hydroxyalkanoic acids with a branched *R*-pendant group were found. Methyl groups were detected at different positions of the side chains and at side chains of varying chain length (Fig. 3). In addition, 3-hydroxyalkanoic acids with a methyl group at the $\alpha$-carbon atom, e.g., in that part of the molecule contributing to the polyester backbone, also occurred (Fig. 3). Branches consisting of ethyl groups or other alkyl groups have not yet been found. The *R*-pendant group of biosynthetic poly(3HA) was also substituted with other atoms. Not only halogens such as bromine, chlorine or fluorine, but also cyano groups were detected. These substituents have been only detected at the $\omega$-position and only at 3-hydroxyalkanoic acids, yet (Fig. 4). Formyl, acetyl, propionyl and benzyl esters did also occur as substituents at the $\omega$-position of various 3-hydroxyalkanoic acids (Fig. 5).

Other 3-hydroxyalkanoic acids, which were identified as constituents of biosynthetic PHA, were 3-hydroxyphenylvaleric acid, 3-hydroxycyclohexylvaleric acid and 3,12-dihydroxydodecanoic acid (Fig. 6).

In addition to 3-hydroxyalkanoic acids, various 4-hydroxyalkanoic acids and also 5-hydroxyalkanoic acids were detected in biosynthetic PHA. These were 4-hydroxybutyric acid, 4-hydroxyvaleric acid and 4-hydroxyhexanoic acid as well as 5-hydroxyvaleric acid (Fig. 6). In contrast to 3-hydroxyalkanoic acids, substituted, branched or unsaturated non 3-hydroxyalkanoic acids have not yet

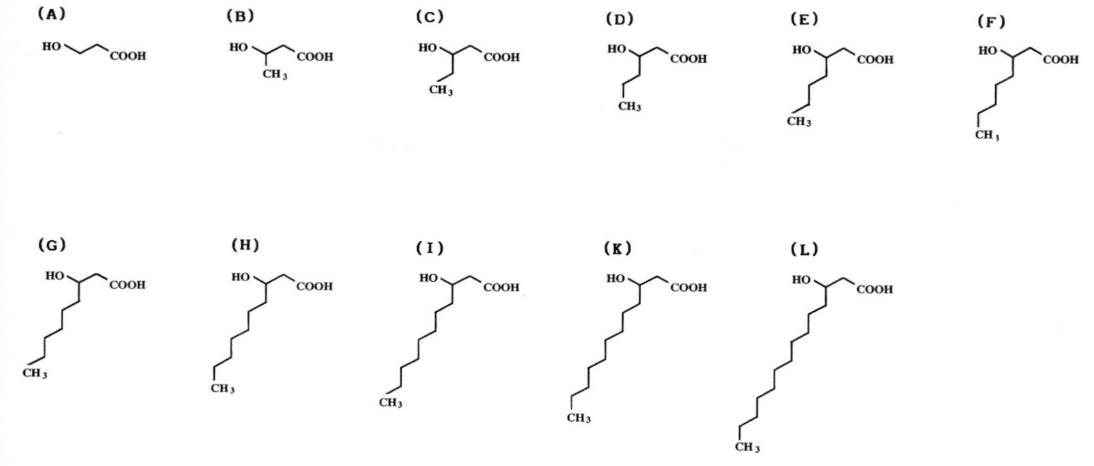


**Fig. 1.** 3-Hydroxyalkanoic acid constituents of biosynthetic PHA. (A) 3-hydroxypropionic acid (DOI et al., 1990c); (B) 3-hydroxybutyric acid (LEMOIGNE, 1926); (C) 3-hydroxyvaleric acid (HOLMES et al., 1981); (D) 3-hydroxyhexanoic acid (LAGEVEEN et al., 1988); (E) 3-hydroxyheptanoic acid (LAGEVEEN et al., 1988); (F) 3-hydroxyoctanoic acid (DE SMET et al., 1983); (G) 3-hydroxynonanoic acid (LAGEVEEN et al., 1988); (H) 3-hydroxydecanoic acid (LAGEVEEN et al., 1988); (I) 3-hydroxyundecanoic acid (LAGEVEEN et al., 1988); (K) 3-hydroxydodecanoic acid (LAGEVEEN et al., 1988); (L) 3-hydroxytetradecanoic acid (LEE et al., 1995). The first report on the occurrence of the respective hydroxyalkanoic acid in biosynthetic PHA is provided in brackets.

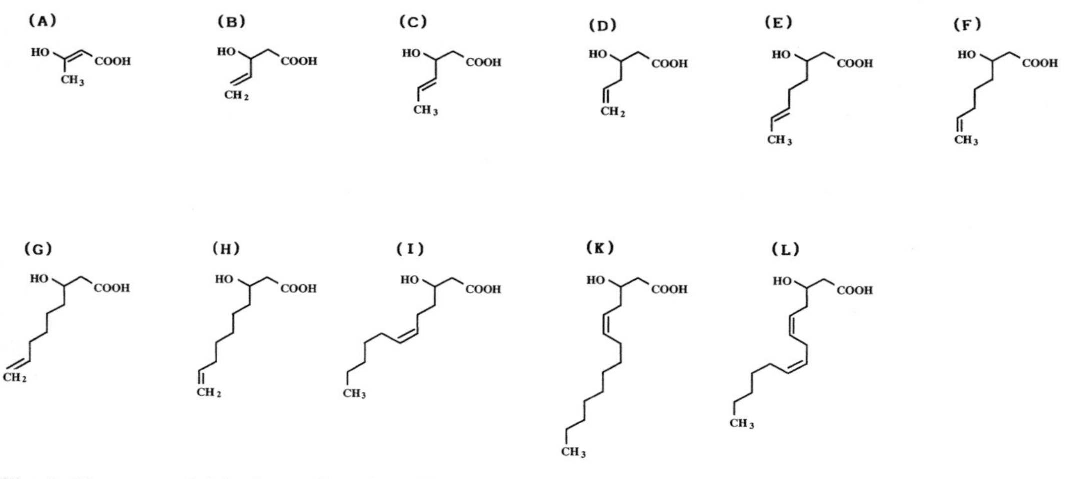


**Fig. 2.** Unsaturated 3-hydroxyalkanoic acid constituents of biosynthetic PHA. (A) 3-hydroxy-2-butenoic acid (DAVIS, 1964); (B) 3-hydroxy-4-pentenoic acid (LENZ et al., 1990); (C) 3-hydroxy-4-hexenoic acid (FRITZSCHE et al., 1990a); (D) 3-hydroxy-5-hexenoic acid (FRITZSCHE et al., 1990a); (E) 3-hydroxy-6-octenoic acid (FRITZSCHE et al., 1990a); (F) 3-hydroxy-7-octenoic acid (FRITZSCHE et al., 1990a); (G) 3-hydroxy-8-nonenoic acid (LAGEVEEN et al., 1988); (H) 3-hydroxy-9-decenoic acid (LAGEVEEN et al., 1988); (I) 3-hydroxy-5-dodecenoic acid (EGGINK et al., 1990); (K) 3-hydroxy-5-tetradecenoic acid (EGGINK et al., 1990); (L) 3-hydroxy-5,8-tetradecenoic acid (EGGINK et al., 1990). The first report on the occurrence of the respective hydroxyalkanoic acid in biosynthetic PHA is provided in brackets.

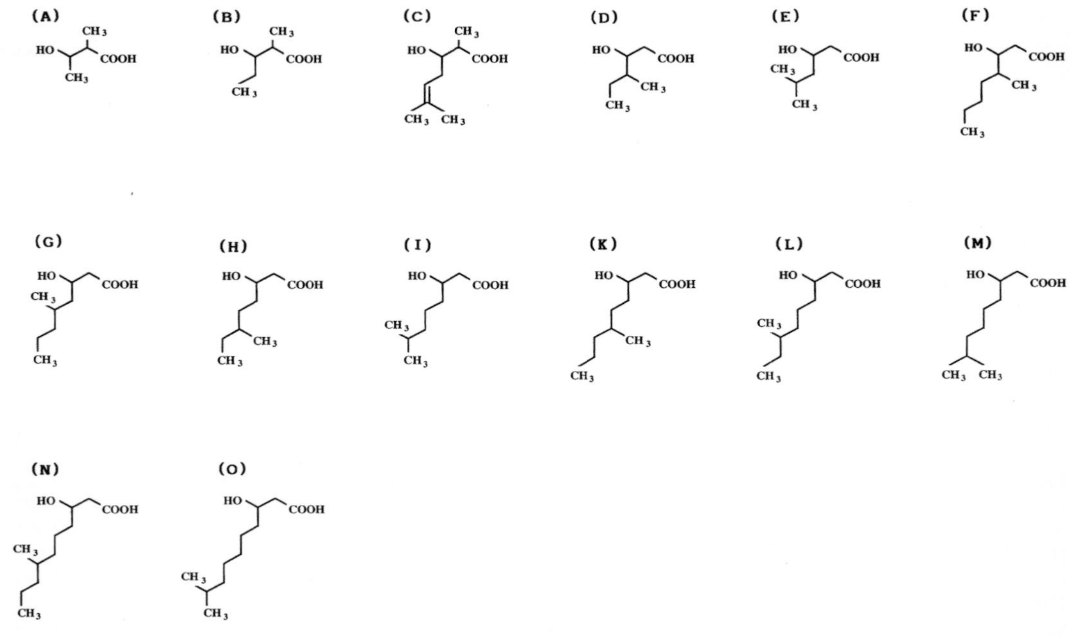


**Fig. 3.** Branched 3-hydroxyalkanoic acid constituents of biosynthetic PHA. (A) 3-hydroxy-2-methylbutyric acid (SATOH et al., 1992); (B) 3-hydroxy-2-methylvaleric acid (SATOH et al., 1992); (C) 3-hydroxy-2,6-dimethyl-5-heptenoic acid (HAZER et al., 1993); (D) 3-hydroxy-4-methylhexanoic acid (FRITZSCHE et al., 1990b); (E) 3-hydroxy-5-methylhexanoic acid (FRITZSCHE et al. 1990b); (F) 3-hydroxy-4-methyloctanoic acid (FRITZSCHE et al., 1990b); (G) 3-hydroxy-5-methyloctanoic acid (FRITZSCHE et al., 1990b); (H) 3-hydroxy-6-methyloctanoic acid (FRITZSCHE et al., 1990b); (I) 3-hydroxy-7-methyloctanoic acid (FRITZSCHE et al., 1990 b); (K) 3-hydroxy-6-methylnonanoic acid (HAZER et al., 1993); (L) 3-hydroxy-7-methylnonanoic acid (HAZER et al., 1993); (M) 3-hydroxy-8-methylnonanoic acid (HAZER et al., 1993); (N) 3-hydroxy-7-methyldecanoic acid (HAZER et al., 1993); (O) 3-hydroxy-9-methyldecanoic acid (HAZER et al., 1993). The first report on the occurrence of the respective hydroxyalkanoic acid in biosynthetic PHA is provided in brackets.

been found in biosynthetic PHA. Biosynthetic polyesters consisting of 2-hydroxyalkanoic acids such as, e.g., lactic acid, or of 6-hydroxyalkanoic acids, such as 6-hydroxyhexanoic acid – the building block of poly($\varepsilon$-caprolactone) – have also not yet been detected. Enzymatic analysis and studies on the *in vitro* biosynthesis of PHA provided some evidence that D(−)-lactyl-CoA is also a substrate for PHA synthases. However, it is a rather poor substrate and only very few PHA synthases seem to accept the coenzyme A thioester of this $\alpha$-hydroxyalkanoic acid (VALENTIN and STEINBÜCHEL, 1994).

Some fungi such as *Physarum polycephalum*, *Penicillium cyclopium* and *Aureobasidium* sp. are able to synthesize poly(L-malic acid) (see HOLLER et al., 1992, for a recent review; NAGATA et al., 1993). In biosynthetic poly(L-malic acid) the hydroxyl group at the 3′carbon atom is involved in the ester bond; therefore, the backbone of this polyester is identical with those of other poly(3HA) (Fig. 6). However, due to the carboxyl groups this polyester has quite different properties. Since it has a quite different physiological function and since relatively little is known about its biosynthesis, polymalic acid will not be considered further in this review.

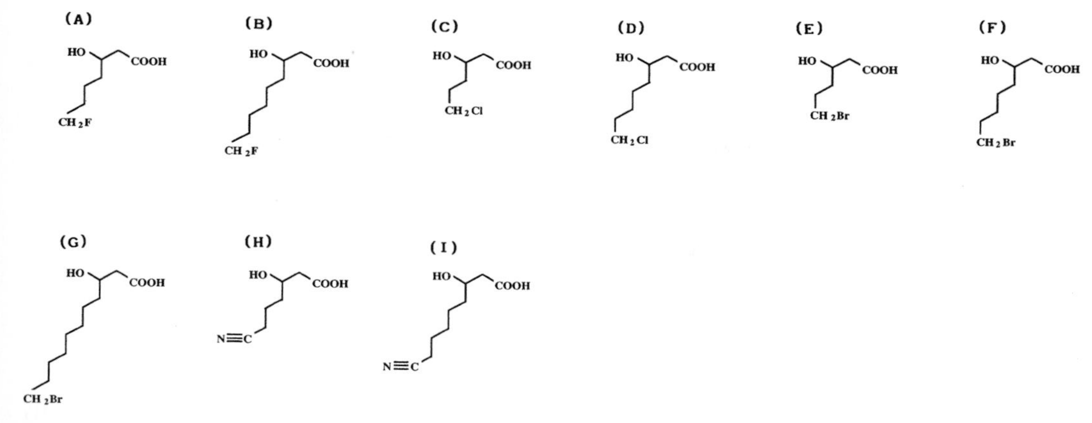


**Fig. 4.** 3-Hydroxyalkanoic acid constituents of biosynthetic PHA substituted with halogen atoms or other functional groups. (A) 3-hydroxy-7-fluoroheptanoic acid (ABE et al., 1990); (B) 3-hydroxy-9-fluorononanoic acid (ABE et al., 1990); (C) 3-hydroxy-6-chlorohexanoic acid (LENZ et al., 1990); (D) 3-hydroxy-8-chlorooctanoic acid (DOI and ABE, 1990); (E) 3-hydroxy-6-bromohexanoic acid (LENZ et al., 1990); (F) 3-hydroxy-8-bromooctanoic acid (LENZ et al., 1990); (G) 3-hydroxy-11-bromoundecanoic acid (LENZ et al., 1990); (H) 7-cyano-3-hydroxyheptanoic acid (LENZ et al., 1992); (I) 9-cyano-3-hydroxynonanoic acid (LENZ et al., 1992). The first report on the occurrence of the respective hydroxyalkanoic acid in biosynthetic PHA is provided in brackets.

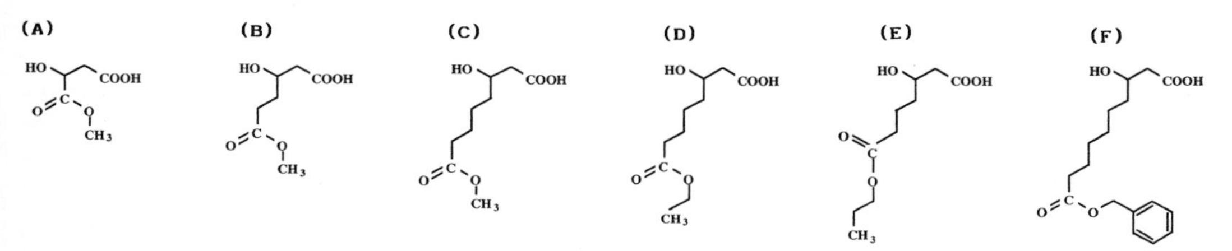


**Fig. 5.** 3-Hydroxyalkanoic acid constituents of biosynthetic PHA substituted with ester groups. (A) 3-hydroxysuccinic acid methyl ester; (B) 3-hydroxyadipic acid methyl ester; (C) 3-hydroxysuberic acid methyl ester; (D) 3-hydroxysuberic acid ethyl ester; (E) 3-hydroxypimelic acid propyl ester; (F) 3-hydroxysebacic acid-benzyl ester (A–F: C. SCHOLZ, personal communication). The first report on the occurrence of the respective hydroxyalkanoic acid in biosynthetic PHA is provided in brackets.

## 2.1.2 Composition of PHA Isolated from the Environment

The diversity of biosynthetic PHA was already indicated many years before the discovery of new PHAs in axenic bacterial cultures when samples taken from the environment were chemically analyzed. WALLEN and ROHWEDDER reported already in 1974 that activated sewage sludge from a domestic plant contained up to 1.3% (wt/wt) PHA consisting of 3-hydroxybutyric acid, 3-hydroxyvaleric acid as well as 3-hydroxyhexanoic acid and most probably 3-hydroxyheptanoic acid. Later, FINDLAY and WHITE (1983) reported on the occurrence of at least 10 different hydroxyalkanoic acids in PHA isolated from marine sediments. In addition to 3-hydroxybutyric acid straight and branched hydroxyalkanoic acids with 5, 6, 7 and 8 carbon atoms were also detected among the constituents. Occurrence of PHA other than poly(3HB) in sewage sludge was also detected by ODHAM

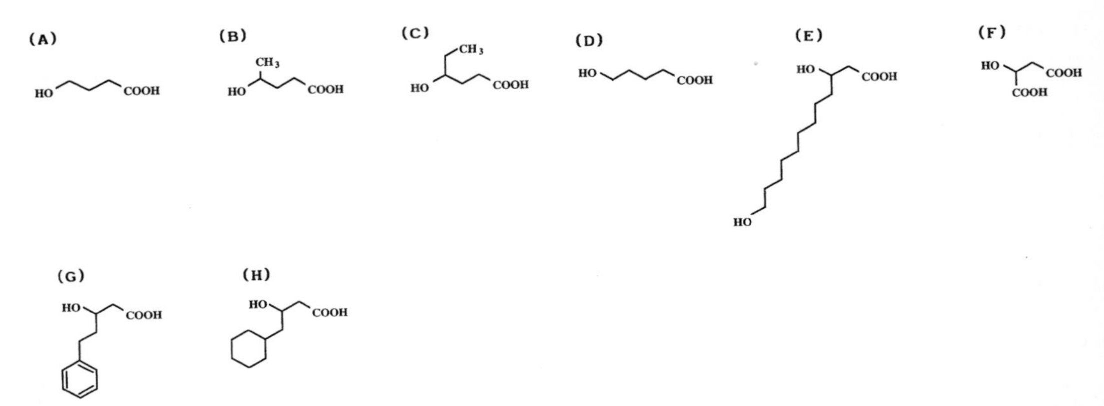


**Fig. 6.** Miscellaneous 3-hydroxyalkanoic acids. (A) 4-hydroxybutyric acid (KUNIOKA et al., 1988); (B) 4-hydroxyvaleric acid (VALENTIN et al., 1992); (C) 4-hydroxyhexanoic acid (VALENTIN et al., 1995); (D) 5-hydroxyvaleric acid (DOI et al., 1987b); (E) 3,12-dihydroxydodecanoic acid (KIM et al., 1992); (F) malic acid (FISCHER et al., 1989); (G) 3-hydroxy-5-phenylvaleric acid (FRITZSCHE et al., 1990c; KIM et al., 1991); (H) 3-hydroxy-4-cyclohexylbutyric acid (LENZ et al., 1992). The first report on the occurrence of the respective hydroxyalkanoic acid in biosynthetic PHA is provided in brackets.

et al. (1986). Recently, SATOH et al. (1992) isolated PHA from sewage sludge consisting of 3-hydroxy-2-methylbutyric acid and 3-hydroxy-2-methylvaleric acid (Fig. 3) in addition to 3-hydroxybutyric acid and 3-hydroxyvaleric acid. Although the origin of these PHA could not be analyzed since the polyesters were taken from complex samples, it is, however, very likely that they were derived from biosynthesis.

## 2.2 Prokaryotic Inclusions

Many bacteria are able to synthesize and accumulate PHA. The capability for biosynthesis of these polyesters occurs in all major physiological and taxonomic groups of eubacteria, and even some archaebacteria are able to synthesize PHA. Only relatively few groups, such as, e.g., the methanogenic bacteria and lactic acid bacteria, are obviously unable to synthesize PHA. In most bacteria PHA are deposited as inclusions in the cell, and the polyester can contribute considerably to the dry matter of the cell (Fig. 7). PHA inclusions exclusively occur in bacteria. However, they are only one type of inclusion among many others (listed in Tab. 2) that are common in bacteria (SHIVELY, 1974; MAYER, 1986).

## 2.3 Poly(3HB)–$Ca^{2+}$–Polyphosphate Complexes in Prokaryotes

In addition to the occurrence of PHA in inclusions other forms of these polyesters were detected (Tab. 3). There were early reports on the detection of PHA also in *Escherichia coli*. The occurrence of PHA in this bacterium was linked with the late exponential and the early stationary growth phase (REUSCH et al., 1986). PHA inclusions have never been reported to occur in *E. coli* harboring no foreign genes for PHA biosynthesis. Instead, complexes consisting of poly(3HB), calcium ions and polyphosphate, poly($P_i$), molecules at molar ratios of 1:0.5 were isolated from *E. coli* (REUSCH and SADOFF, 1988). The authors proposed that these molecules form a helical complex which is located in the cytoplasmic membrane. According to this model an outer cylinder of helically wounded poly(3HB) consisting of 130–170 building blocks with the nonpolar methyl and methylene groups at the outer side and an inner cylinder of helically wounded poly(Pi) consisting of 130–150 building blocks are separated by a layer of calcium ions which forms ionic bonds to the ester carbonyl oxygens and the

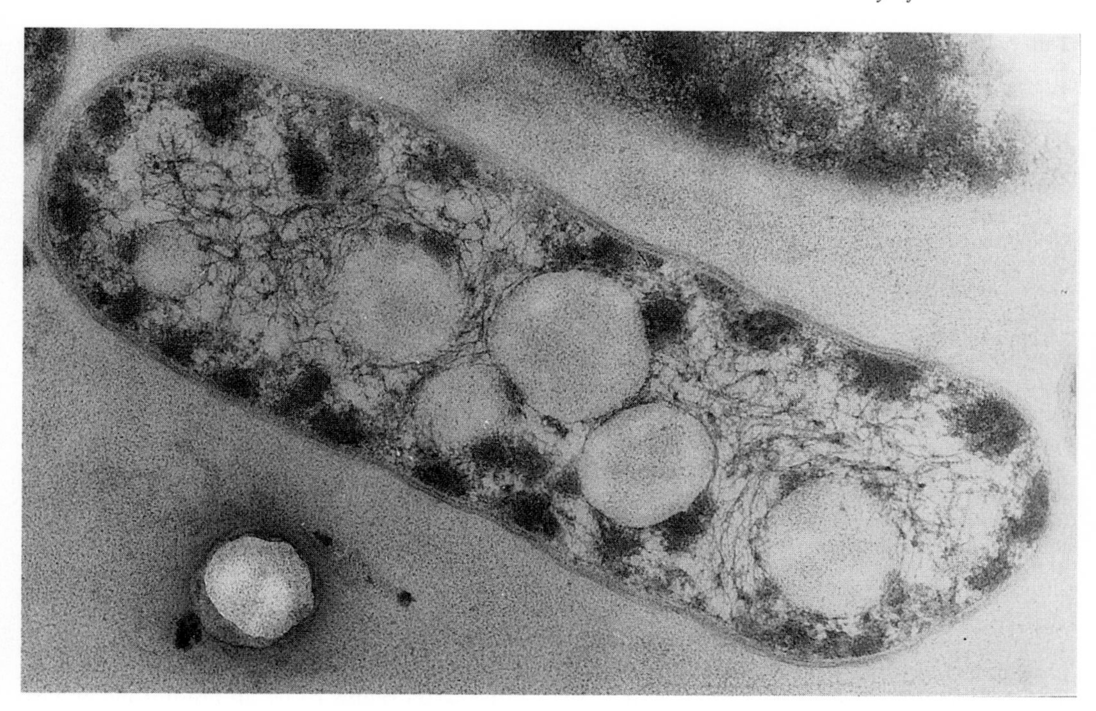

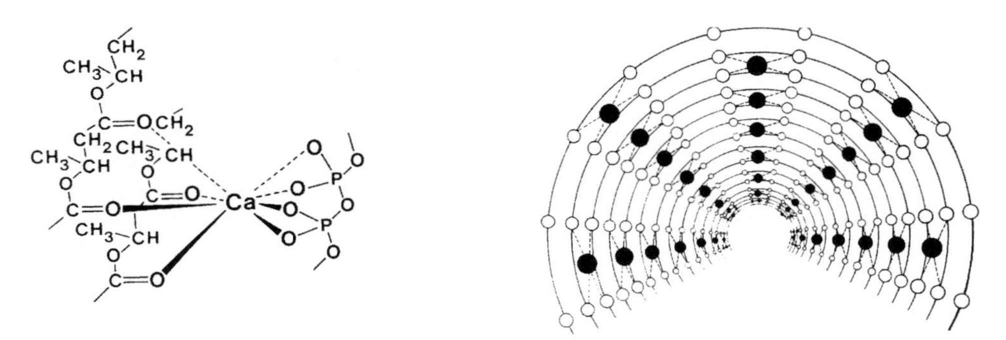

**Fig. 7.** Occurrence of PHA in bacteria. Top: thin section of cells of *A. eutrophus* having accumulated poly(3HB) (original micrograph by F. MAYER). Bottom: proposed structure of a channel consisting of poly(3HB)–$Ca^{2+}$–polyphosphate complex spanning the cytoplasmic membrane (reproduced from REUSCH and SADOFF, 1988, by courtesy of R. REUSCH).

**Tab. 2.** Bacterial Intracellular Inclusions

| Type of Inclusion | Chemical Components |
|---|---|
| Bacteriophages | Protein, nucleic acids |
| Calcium carbonate granules | $CaCO_3$ |
| Carboxysomes | Ribulosebisphosphate carboxylase and other proteins |
| Chlorosomes | Protein, bacteriochlorophyll, phospholipids |
| Cyanophycin granules | Protein (aspartate, arginine) |
| Endospores | Complex composition |
| Gas vesicles, gas vacuoles | Protein, gas |
| Lipid granules | Triacylglycerols, wax esters |
| Magnetosomes | Magnetite ($Fe_3O_4$), Greigite ($Fe_3S_4$), Pyrite ($FeS_2$) |
| Polyglycoside grana | Glycogen, starch |
| Polyhydroxyalkanoic acid granules | Polyesters of hydroxyalkanoic acids |
| Polyphosphate granules | Polyesters of inorganic phosphate |
| Phycobilisomes | Protein, phycocyanin, phycoerythrin |
| Protein crystals | Protein (e.g., BT toxin) |
| R-Bodies | Protein |
| Recombinant proteins | Protein |
| Rhapidosomes, microtubuli, etc. | Protein |
| Sulfur granules | Protein, orthorhombic sulfur |

phosphoryl oxygens (Fig. 7). A length of 45 Å and a diameter of 24 Å were calculated for this complex.

Since transformation efficiencies correlated with the concentration of poly(3HB) in the membrane, it was speculated whether this complex is contributing to the genetical competence of *E. coli* and whether foreign DNA or calcium or phosphate ions are taken up by these tubes (REUSCH et al., 1986, 1987). The detection of short molecules of single-stranded DNA ($24 \pm 4$ nucleotides) in the complexes was discussed as a support of this assumption (REUSCH, 1992). Poly(3HB) was also detected in the cytoplasmic membranes of genetically competent cells of other bacteria such as *Azotobacter vinelandii*, *Bacillus subtilis* and *Haemophilus influenzae* (REUSCH and SADOFF, 1983). However, there were also critical remarks on the validity of this model (MÜLLER and SEEBACH, 1993).

## 2.4 Occurrence of PHA in Eukaryotes

Due to the studies performed in the laboratory of R. REUSCH, PHA are now also known to occur in eukaryotic organisms. Therefore, poly(3HB) seems to be an ubiquitous polyester (REUSCH, 1992). Unbound poly(3HB) and poly(3HB)–$Ca^{2+}$–poly($P_i$) complexes were detected in *Saccharomyces cerevisiae*, in various plant tissues such as, e.g., spinach leaves, broccoli flowers or carrot roots, and in organs of animals such as, e.g., chicken liver, beef heart or pork kidney. However, poly(3HB) concentrations were one or two orders of magnitude lower than in competent cells of *E. coli*, and the polyesters isolated from the complexes of eukaryotes were larger than those of prokaryotes, with poly(3HB) and the poly($P_i$) molecules consisting of 120–200 or 170–220 building blocks, respectively (REUSCH, 1989). Recently, poly(3HB) has also been detected in human blood plasma at concentrations ranging from 0.60 to 18.2 mg $L^{-1}$. Most of the polyesterwas bound to albumins and various lipoproteins (REUSCH and SPARROW, 1992).

# 3 Biosynthesis of PHA

In most bacteria that have been investigated so far, PHA are accumulated if a car-

**Tab. 3.** Occurrence of Polyhydroxyalkanoic Acids

| Organism | Type of PHA | Occurrence | Degree of Polymerization | Amount | Function |
|---|---|---|---|---|---|
| Bacteria | | | | | |
| *Alcaligenes eutrophus*[a] | Poly(3HB) and other PHA | Granules | up to 30,000 | up to 96% of CDM | Storage of energy<br>Storage of carbon<br>Sink of reducing power |
| *Escherichia coli*[b] | Only poly(3HB) | Poly(3HB)–$Ca^{2+}$–Poly($P_i$)–complex and soluble forms | 150–180 | 156 μg $g^{-1}$ CDM | Uptake of DNA or $Ca^{2+}$ ions ? |
| Fungi | | | | | |
| *Saccharomyces cerevisiae*[c] | Only poly(3HB) | Poly(3HB)–$Ca^{2+}$–Poly($P_i$)–complex and soluble forms | 120–200 | 0,26 μg $g^{-1}$ CWW | not known |
| *Physarum polycephalum*[d] | Only polymalic acid[f] | plasmodial stage soluble forms or associated to proteins | approx. 750 | 100–150 mg $mL^{-1}$ nucleus<br>0.1–0.2 mg $mL^{-1}$ cytoplasm | Binding of proteins to DNA? |
| Plants[c] | Only poly(3HB)[g] | Poly(3HB)–$Ca^{2+}$–Poly($P_i$)–complex and soluble forms | 120–200 | 0.25–1.6 μg $g^{-1}$ CWW | not known |
| Animals[c] | Only poly(3HB)[h] | Poly(3HB)–$Ca^{2+}$–Poly($P_i$)–complex and soluble forms | 120–200 | 0.15–9.2 μg $g^{-1}$ CWW | not known |
| Humans[e] | Only poly(3HB) | Blood plasma, associated to albumin and to other lipoproteins | not known | 0.60–18 mg $L^{-1}$ plasma | not known |

[a] For an extended list of PHA accumulating bacteria see STEINBÜCHEL (1991a);
[b] for details see REUSCH and SADOFF (1988);
[c] for details see REUSCH (1989);
[d] for details see HOLLER et al. (1992);
[e] for details see REUSCH and SPARROW (1992);
[f] polymalic acid was also detected in *Penicillium cyclopium*;
[g] poly(3HB) was detected in spinach (leaves), celery (leaves and stems), cauliflower (leaves and stems), and carrots (roots);
[h] poly(3HB) was detected in clams, chicken (liver, legs), turkey (liver, heart muscle), beef (liver, brain), sheep (heart, legs), pig (heart muscle) and calf (kidney).

CDM, cellular dry matter; CWW, cellular wet weight; Poly(3HB), poly(3-hydroxybutyric acid); PHA, polyhydroxyalkanoic acids; Poly($P_i$), polyphosphates.

bon source is provided in excess and if at the same time growth is limited by one other nutrient, resulting in imbalanced growth. As demonstrated in *A. eutrophus*, the limiting nutrient may be the source of nitrogen, sulfur, phosphate, iron, magnesium, potassium (STEINBÜCHEL and SCHLEGEL, 1989). Furthermore, in many aerobic bacteria such as *Azotobacter beijerinckii* and *A. eutrophus* PHA accumulation is induced during oxygen limitation (SENIOR and DAWES, 1973; WARD et al., 1977; MORINAGA et al., 1978; VOLLBRECHT et al., 1979; VOLLBRECHT and EL NAWAWY, 1980; SONNLEITNER et al., 1979; SCHLEGEL and STEINBÜCHEL, 1981; STEINBÜCHEL et al., 1983; STEINBÜCHEL and SCHLEGEL, 1989).

## 3.1 Biosynthetic Pathways in Prokaryotes

Three different stages of PHA biosynthesis can be distinguished:

(1) The carbon source has to be taken up by the bacterial cell. This step is usually catalyzed by specific transport proteins for the respective carbon source. Alternatively, the carbon source may enter the cell by passive diffusion.
(2) The carbon source is converted into a coenzyme A thioester of a hydroxyalkanoic acid which can be used as a substrate by PHA synthase. These reactions represent the complexicity of the bacterial metabolism, and they may vary considerably according to the respective bacterium (Fig. 8). The reactions comprising this second stage are mostly part of various catabolic or anabolic pathways; at present, they are only partially understood. Only in some bacteria, clear evidence was obtained for a separate set of enzymes present in addition to the catabolic and anabolic enzymes and expressed for poly(3HB) synthesis.
(3) The polyester is formed by an enzyme referred to as PHA synthase. This enzyme represents the key enzyme of PHA biosynthesis and it is tightly associated with the granules, as revealed, e.g., by enzymatic (HAYWOOD et al., 1989a) and by immunocytochemical studies (GERNGROSS et al., 1993; LIEBERGESELL et al., 1994) executed with *A. eutrophus* and *Chromatium vinosum*.

### 3.1.1 Routes to Hydroxyacyl Coenzyme A Thioesters

Central intermediates of the metabolism can be converted to a hydroxyacyl coenzyme A thioester by four major routes as recently outlined by STEINBÜCHEL (1991a). Fig. 8 provides an overview of these pathways. In addition, many other routes exist when special precursor subtrates are provided as a carbon source during the PHA accumulation phase. For example, hydroxyalkanoic acids such as 3-hydroxypropionic acid or 4-hydroxybutyric acid are very often incorporated into PHA, if these fatty acids are provided as carbon source. This indicates that they are directly used for PHA biosynthesis via the corresponding coenzyme A thioesters.

(1) If *A. eutrophus* is cultivated on carbon sources such as carbohydrates, pyruvate or acetate, biosynthesis of poly(3HB) starts from acetyl-CoA (GOTTSCHALK 1964a, b). Two enzymes are required for the synthesis of D(−)-3-hydroxybutyryl-CoA (Fig. 8, upper central part). A biosynthetic 3-ketothiolase catalyzes a biologiocal Claison condensation of two acetyl-CoA moieties resulting in the formation of a carbon–carbon bond and in the release of one coenzyme A. The resulting acetoacetyl-CoA is then stereoselectively reduced to D(−)-3-hydroxybutyryl-CoA by an NADPH-dependent acetoacetyl-CoA reductase (for a recent review, see STEINBÜCHEL and SCHLEGEL, 1991). This pathway is most probably present in the majority of poly(3HB) accumulating bacteria.
(2) In *Rhodospirillum rubrum* a modification of this pathway was described (Fig.

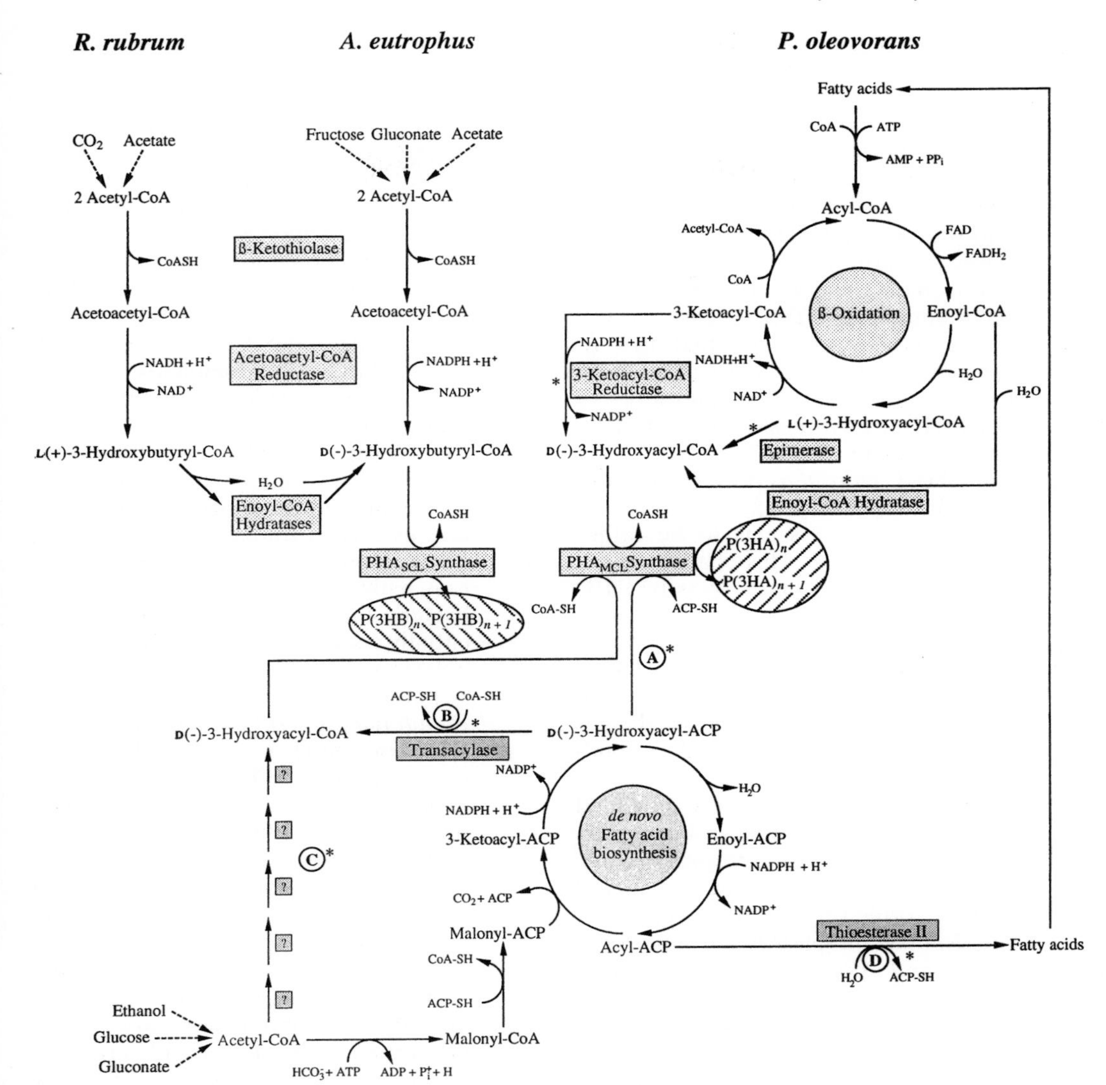


**Fig. 8.** Pathways of PHA biosynthesis. Four basic pathways for the synthesis of poly(3HB) in *R. rubrum* and *A. eutrophus* or for the synthesis of $PHA_{MCL}$ in *P. oleovorans* and *P. aeruginosa* are outlined. Asterisks indicate uncertain pathways or reactions that are theoretically possible but have not been confirmed; 3HB, 3-hydroxybutyric acid; PHA, polyhydroxyalkanoic acids; MCL, medium carbon chain length; SCL, short carbon chain length; ACP, acyl carrier protein (original figure by A. TIMM).

8, upper left part). Acetoacetyl-CoA is reduced to L(+)-3-hydroxybutyryl-CoA by an NADH-dependent aceto-acetyl-CoA reductase, and two enoyl-CoA hydratases convert the L(+)-isomer to the D(−)-isomer (MOSKOWITZ and MERRICK, 1969). Evidence for this pathway as the solely poly(3HB) biosynthesis pathway in other bacteria has not been obtained yet. However, analysis of mutants and theoretical considerations provide evidence that this *R. rubrum* pathway may exist in addition to the *A. eutrophus* pathway of poly(3HB) biosynthesis (STEINBÜCHEL and SCHLEGEL, 1991).

(3) Like the *A. eutrophus* poly(3HB) pathway, the third major pathway (Fig. 8, upper right part) seems to be rather widespread, too. This pathway accounts for the synthesis of poly(3HB) and of poly($3HA_{MCL}$) when the bacteria are cultivated on medium- or long-chain fatty acids such as, e.g., octanoic acid, or on substrates which are converted to fatty acids, such as alkanes and alkanols. Even- as well as odd-numbered fatty acids are activated by a thiokinase and are normally degraded via β-oxidation resulting in the formation of acetyl-CoA or propionyl-CoA, respectively. Subsequently, poly(3HB) may be formed by the *A. eutrophus* or the *R. rubrum* pathway from acetyl-CoA, and poly(3HB-*co*-3HV) from acetyl-CoA plus propionyl-CoA. In addition, intermediates of the β-oxidation pathway can also be directed to PHA biosynthesis (DE SMET et al., 1983; LAGEVEEN et al., 1988). Any coenzyme thioester of a 3-ketoalkanoic acid comprising four or more carbon atoms may be reduced to the corresponding D(−)-3-hydroxyacyl-CoA thioester by a 3-ketoacyl-CoA reductase. Furthermore, the L(+)-3-hydroxyacyl CoA thioester or the enoyl-CoA thioester intermediates of the β-oxidation cycle may be converted to the corresponding D(−)-3-hydroxyacyl-CoA thioesters thus providing a substrate for the PHA synthase. It is currently unknown which of these possible three reactions occurs during PHA biosynthesis from fatty acids.

PHA biosynthesis from intermediates of the β-oxidation pathway has been studied in most detail in *P. oleovorans*. The pathway seems to be widespread among pseudomonads belonging to the rRNA homology group I (BRANDL et al., 1988; HAYWOOD et al., 1989b, 1990; GROSS et al., 1989: HUISMAN et al., 1989; TIMM and STEINBÜCHEL, 1990). Since these pseudomonads possess PHA synthases which preferentially accept coenzyme A thioesters of $HA_{MCL}$ as substrates, these bacteria accumulate poly($HA_{MCL}$) such as, e.g., poly(3HO) if they are cultivated on, e.g., octanoic acid as carbon source (DE SMET et al., 1983). Modifications of this *P. oleovorans* pathway seem to be also active in bacteria that, like *A. eutrophus*, preferentially accumulate poly(3HB) or other poly($HA_{SCL}$).

(4) A pathway, which is referred to as the *Pseudomonas aeruginosa* pathway (Fig. 8, bottom part), is responsible for the formation of poly($HA_{MCL}$) from unrelated substrates such as, e.g., gluconate or acetate (HAYWOOD et al., 1990; TIMM and STEINBÜCHEL, 1990). The labeling pattern of the constituents in PHA as revealed by $^{13}C$ nuclear magnetic resonance spectroscopy obtained from $^{13}C$-labeled carbon sources, the occurrence of unsaturated constituents, and PHA accumulation experiments performed in the presence of cerulenin, which is an inhibitor of *de novo* fatty acid biosynthesis, demonstrated the involvement of *de novo* fatty acid biosynthesis in the synthesis of poly($3HA_{MCL}$) from, e.g., acetate (EGGINK et al., 1992; HUIJBERTS et al., 1992, 1994a; SAITO and DOI, 1993). This pathway seems to occur solely in pseudomonads belonging to the rRNA homology group I; interestingly *P. oleovorans* does not possess this pathway. Most other pseudomonads obviously use both, the *P. oleovorans* as well as the *P. aeruginosa* pathway, and at least in *P. putida* clear

evidence was provided that $\beta$-oxidation and *de novo* fatty acid biosynthesis contribute independently and in parallel to the provision of precursors for PHA biosynthesis (HUIJBERTS et al., 1994a). In addition, the authors found evidence that precursors of the PHA constituents are also obtained by the elongation of fatty acid derivatives by the addition of $C_2$ carbon units from acetyl-CoA. Presumably the enzyme $\beta$-ketothiolase is involved in these reactions (HUIJBERTS et al., 1994a).

The *P. aeruginosa* pathways of PHA biosynthesis represents only one major relatively widespread pathway for the synthesis of PHA from unrelated substrates. Several other minor pathways, that have not been investigated in much detail and that occur only in relatively few bacteria, have been detected recently. One of these pathways allows the synthesis of poly(3HB-*co*-3HV) from single carbon sources in the gram-positive *Rhodococcus ruber* (HAYWOOD et al., 1991); another example is a pathway for synthesis of this copolyester in a mutant of *A. eutrophus*, which is affected in the regulation of branched-chain amino acid metabolism (STEINBÜCHEL and PIEPER, 1992). In newly isolated species of the genus *Burkholderia* biosynthesis and accumulation of a copolyester of 3-hydroxybutyric acid and 3-hydroxy-4-pentenoic acid by a hitherto unknown pathway has been described (RODRIGUES et al., in press).

### 3.1.2 PHA Synthases

During the last 6 years the PHA synthase structural genes have been cloned from approximately 20 different bacteria, and from 12 of them the nucleotide sequences are available (STEINBÜCHEL et al., 1992, and references cited therein; HUSTEDE and STEINBÜCHEL, 1993; SCHEMBRI et al., 1994; S. RAHALKAR, M. LIEBERGESELL and A. STEINBÜCHEL, unpublished results). Comparison of the protein primary structures revealed that bacterial PHA synthases represent a group of highly homologous enzymes with 23 amino acid residues that are strictly conserved in 10 different PHA synthases, and with 270 residues representing 45% of all residues that are conserved in at least 50% of the known PHA synthases (STEINBÜCHEL et al., 1992). At present, 4 different types of PHA synthases can be distinguished due to their structures and the substrate range (Fig. 9). This classification was made according to the results of physiological studies as well as to the analysis of the primary structures of the PHA synthases derived from nucleic acid sequence data. Data obtained from biochemical characterization of the isolated PHA synthase proteins are still scarce at present.

The PHA synthase of *A. eutrophus* H16 has been studied in most detail and in several laboratories (Fig. 9; STEINBÜCHEL and SCHLEGEL, 1991). It is the prototype of class I PHA synthases and presumably widespread among bacteria. PHA synthases with a similar primary structure were detected in *Methylobacterium extorquens* (38.1% amino acid identity), *Rhodococcus ruber* (38.4%), and *Rhodobacter sphaeroides* (36.8%). This class of PHA synthase is characterized by a preference for coenzyme A thioesters of $HA_{SCL}$. The *A. eutrophus* enzyme accepts as substrates 3-hydroxypropionyl-CoA, 3-hydroxybutyryl-CoA, 3-hydroxyvaleryl-CoA, 4-hydroxybutyryl-CoA, 4-hydroxyvaleryl-CoA and 5-hydroxyvaleryl-CoA. It is a large enzyme with subunits exhibiting a relative molecular mass of 63,900. The quaternary structure of the *A. eutrophus* enzyme is not known. The enzyme is bound to the granules (HAYWOOD et al., 1989a) and located at their surface (GERNGROSS et al., 1993). From kinetic studies it was calculated that approximately 18,000 PHA synthase molecules per cell are present in *A. eutrophus* (KAWAGUCHI and DOI, 1992).

Alignments of the primary structures of ten different PHA synthases provided evidence for the presence of one highly conserved cysteine (STEINBÜCHEL et al., 1992). Site-directed mutagenesis confirmed that only the cystein residue $Cys_{319}$ is required for an active PHA synthase in *A. eutrophus* (GERNGROSS et al., 1994). Recently, evidence has been obtained that this enzymes requires post-translational modification by phosphopantetheine to provide a second thiol group per PHA synth-

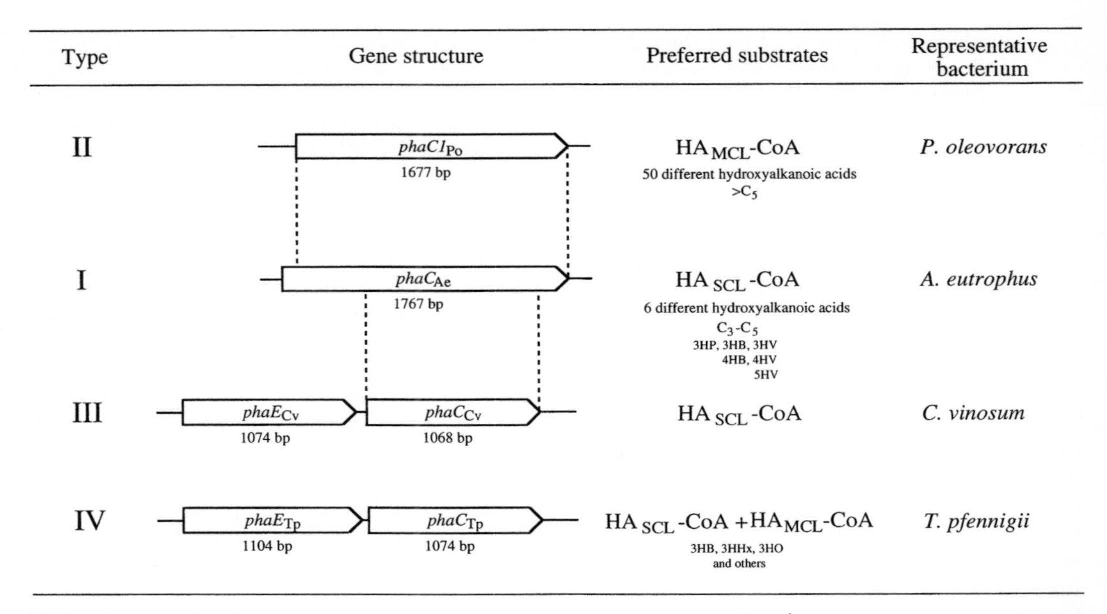


**Fig. 9.** Classes of PHA synthases. The gene structure and the preferred substrates of four different classes of PHA synthases and a representative bacterium harboring the respective are shown. HA, hydroxyalkanoic acid; SCL, short carbon chain length; MCL, medium carbon chain length; 3HP, 3-hydroxypropionic acid; 3HB, 3-hydroxybutyric acid; 3HV, 3-hydroxyvaleric acvid; 3HHx, 3-hydroxyhexanoic acid; 3HO, 3-hydroxyoctanoic acid; 4HB, 4-hydroxybutyric acid; 4HV, 4-hydroxyvaleric acid; 5HV, 5-hydroxyvaleric acid (adapted from STEINBÜCHEL et al., 1992).

ase subunit (GERNGROSS et al., 1994) which could support the mechanistic model for the polymerization process depending on the presence of two thiol groups (GRIEBEL et al., 1968; KAWAGUCHI and DOI, 1992; GERNGROSS et al., 1994). The putative mechanism, as it was proposed repeatedly, is shown in Fig. 10.

In an initiation or priming process, which is not understood, the 3-hydroxybutyric acid moiety from 3-hydroxybutyryl-CoA is transferred to one of the thiol groups, coupled with the release of coenzyme A. Another monomer is then added to the second thiol group, and its 3′-hydroxy group attacks the thiol ester of the primer to form a covalently bound dimer. After thiol transesterification of this dimer to the first thiol group, the second thiol group can accept an additional 3-hydroxybutyric acid moiety, and by repeated propagation steps the covalently bound oligomer/polymer is growing by one unit during each cycle. This propagation seems to be the rate limiting step of PHA biosynthesis, and from kinetic studies rate constants between only 1.2 and 3.1 propagations per s were calculated (KAWAGUCHI and DOI, 1992). The oligomer/polymer is released from the enzyme if a chain transfer with water occurs; this transfer terminates polyester biosynthesis. However, for this reaction rather low rate constants between 0.28 and 1.50 terminations per h (!) were calculated (KAWAGUCHI and

**Fig. 10.** Putative mechanism of the PHA synthase of *A. eutrophus*. The hypothetical model of the catalytic mechanism of the PHA synthase shows the initiation step, two propagation steps and the termination step for the formation of poly(3HB) from 3-hydroxybutyryl-coenzyme A. One sulfhydryl group at the protein represents the sulfhydryl group of Cys-319, the second may represent the sulfhydryl group of phosphopantetheine covalently linked to the PHA synthase protein in *A. eutrophus* (adapted from GERNGROSS et al., 1994).

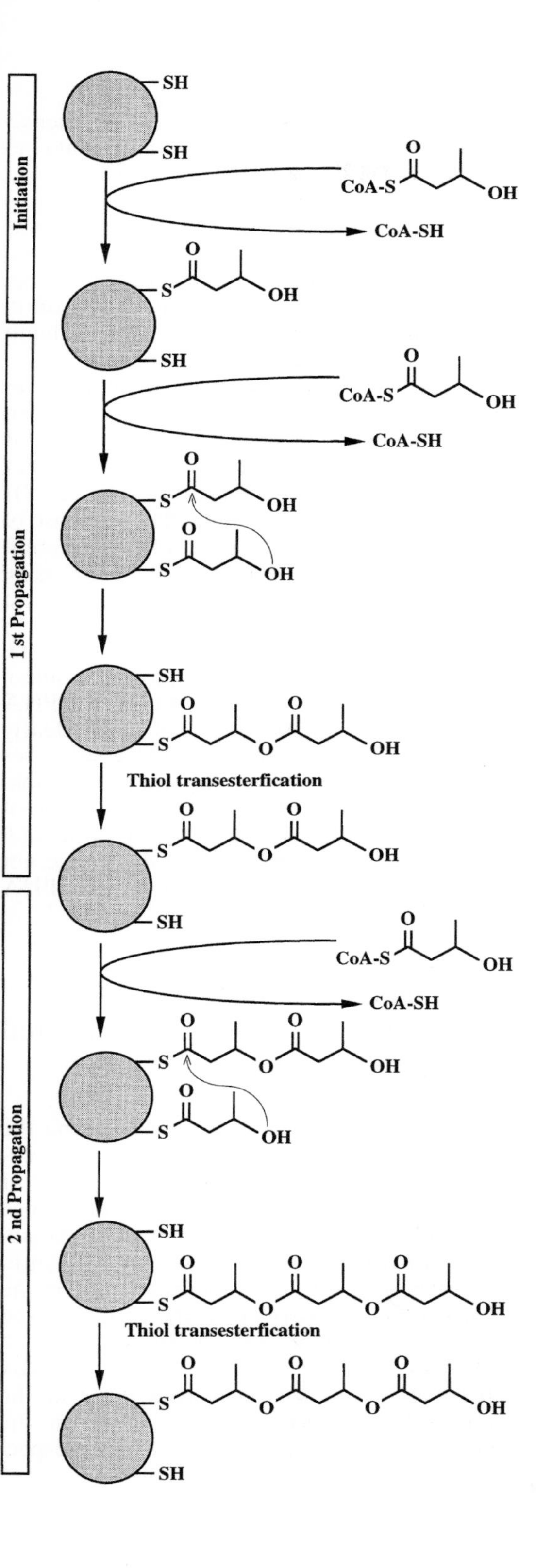
Initiation
1 st Propagation
2 nd Propagation
SH
CoA-S
OH
CoA-SH
Thiol transesterfication
Termination
$H_2O$
HO

DOI, 1992). This may explain the rather high degree of polymerization and also the rather low polydispersities ($M_W/M_N$) of poly(3HB) in *A. eutrophus*.

Class II PHA synthases are represented by the enzymes of *P. oleovorans*. The relative molecular mass of the PHA synthase No. 1 of *P. oleovorans* is 62,400. Its molecular structure is rather similar to that of the *A. eutrophus* PHA synthase (37.4% amino acid identity), except for approximately 26 amino acids which are lacking at the N-terminus of the enzyme of *P. oleovorans* (Fig. 9). However, the substrate range of both enzymes is quite different. The enzyme is almost inactive with 3-hydroxybutyryl-CoA, and physiological data provide evidence that it is not active with the coenzyme A thioesters of 4- or 5-hydroxyalkanoic acids.

The PHA synthase of *Chromatium vinosum* represents the class III PHA synthase that also occurs in *Thiocystis violacea* (LIEBERGESELL and STEINBÜCHEL, 1992, 1993). Whereas the substrate range seems to be rather similar to that of class I PHA synthases, the structure of the enzyme is different (Fig. 9). In contrast to class I and class II PHA synthases, this PHA synthase consists of two different types of subunits encoded by $phaE_{Cv}$ and $phaC_{Cv}$. Both genes constitute one single operon. Since the molecular weights of the complex and of the subunits were 400,000, 40,500 or 39,500 Da, respectively, the complex most probably consists of 10 subunits. Immunological analyses indicated that the $PhaC_{Cv}$ protein is the minor component contributing 2–4 subunits to this complex (LIEBERGESELL et al., 1994). Although the molecular weight of $PhaC_{Cv}$ is almost 40% less as compared to the *A. eutrophus* PHA synthase, due to the lack of approximately 175 amino acids at the N-terminus and of approximately 55 amino acids at the C-terminus, the homologies of $Pha_{Cv}$ to the central regions of these PHA synthases encoded by only one single gene (class I and class II PHA synthases) are obvious (STEINBÜCHEL et al., 1992). Analysis of clones that were completely or partially deleted for $phaE_{Cv}$, but possessing intact $phaC_{Cv}$, did not express active PHA synthase (LIEBERGESELL et al., 1992). Furthermore, after $PhaE_{Cv}$ was separated from $PhaC_{Cv}$ by chromatography on Phenyl-Sepharose, PHA synthase activity could no longer be measured (LIEBERGESELL et al., 1994). Therefore, the expression of active PHA synthase requires both, *phaC* plus *phaE*. The PhaE protein may have an essential function for the polymerization step, which the larger class I PHA synthases fulfill by themselves and the smaller PhaC subunit of the class III PHA synthases cannot.

Recently, a new type of PHA synthase was characterized from the sulfur purple bacterium *Thiocapsa pfennigii*. It is referred to as class IV PHA synthase and its molecular structure is similar to those of both class III PHA synthases (Fig. 9). The PHA synthase of *T. pfennigii* is also composed of two different subunits, and the *phaC* and the *phaE* gene products exhibit 52.8 and 85.2% amino acid identity to the corresponding proteins of *C. vinosum* (S. RAHALKAR, M. LIEBERGESELL and A. STEINBÜCHEL, unpublished results). From the substrate range, however, the PHA synthase of *T. pfennigii* seems to be clearly distinguished from all other PHA synthases investigated so far, since this enzyme is capable to synthesize copolyesters consisting of $3HA_{SCL}$ and $3HA_{MCL}$ or of various 4- as well as of 5-hydroxyalkanoic acids which have not been detected hitherto in biosynthetic PHA (see below).

### 3.1.3 Factors Affecting the Composition of PHA Accumulated in the Cell

The most significant factor influencing the composition of PHA synthesized and accumulated in the cell is the bacterial species or strain used. It determines the substrate range of PHA synthases and the peripheric metabolism providing the substrate for PHA synthase. The second most relevant factor is the carbon source provided to the cell, followed by environmental factors. As shown in Tab. 4, the same strain of *P. putida* synthesizes PHA from gluconate varying considerably in the composition of $HA_{MCL}$, depending on the cultivation temperature of 15 or 30 °C, respectively (HUIJBERTS et al., 1992).

**Tab. 4.** Influence of Temperature on the Composition of PHA Synthesized by *P. putida* KT2442 from Gluconate (data compiled from HUIJBERTS et al., 1992)

| PHA Content/Composition | Temperature | |
|---|---|---|
| | 15°C | 30°C |
| PHA content of cells | 20.5% of CDM | 17.0 % of CDM |
| 3-Hydroxyhexanoic acid | 0.4 mol% | <0.1 mol% |
| 3-Hydroxyoctanoic acid | 8.4 mol% | 6.9 mol% |
| 3-Hydroxydecanoic acid | 64.5 mol% | 74.3 mol% |
| 3-Hydroxy-5-*cis*-dodecanoic acid | 16.7 mol% | 8.8 mol% |
| 3-Hydroxydodecanoic acid | 6.0 mol% | 7.7 mol% |
| 3-Hydroxy-7-*cis*-tetradecanoic acid | 4.0 mol% | 1.6 mol% |
| 3-Hydroxytetradecanoic acid | <0.1 mol% | <0.1 mol% |

CDM: cellular dry matter.

### 3.1.4 PHA in Recombinant Bacteria

At the very beginning of the molecular analysis of the poly(3HB) biosynthesis pathway of *A. eutrophus* and of other bacteria it was demonstrated that this pathway could be expressed in a functionally active form in *E. coli* (see below, Sect. 8.1). Like in *A. eutrophus*, the PHA synthase structural genes often occur together with structural genes for other proteins relevant for PHA biosynthesis such as for $\beta$-ketothiolase and acetoacetyl-CoA reductase (PEOPLES and SINSKEY, 1989a, b) or for phasins (see below) (Fig. 11). Recently, the PHA biosynthesis genes of *A. eutrophus* were also transferred to various pseudomonads of the rRNA homology group I. Enzymatic analysis of recombinant pseudomonads clearly demonstrated that the genes were expressed, and detection of poly(3HB) in gluconate-grown cells showed a functionally active poly(3HB) biosynthesis pathway (STEINBÜCHEL and SCHUBERT, 1989). When cultivating the recombinant strain of *P. oleovorans* ATCC 29347 in the presence of excess octanoic acid, PHA consisting of 3-hydroxybutyric and 3-hydroxyoctanoic acid as main constituents, plus 3-hydroxyhexanoic acid as a minor constituent, was accumulated (TIMM et al., 1990). Detailed analyses of the polyester accumulated as well as of the accumulation and extraction kinetics revealed that a blend of poly(3HB) and poly(3HHx-*co*-3HO) was synthesized rather than poly(3HB-*co*-3HHx-*co*-3HO). This was confirmed by the analysis of partial methanolysis products and by the differential precipitation and subsequent separation of poly(3HB) and poly(3HHx-*co*-3HO) with petrol ether from a solution of the polyester in chloroform solution (TIMM et al., 1990). Furthermore, the cells did not only synthesize two different polyesters, they even deposited poly(3HB) and poly(3HHx-*co*-3HO) in separate granules which in freeze-fracture electron microscopy could be distinguished by the occurrence of needle-like or mushroom-like deformations, respectively (Fig. 12). Both types of granules could be separated in sucrose density gradients (PREUSTING et al., 1993a).

The unusual substrate specificity of the *T. pfennigii* PHA synthase could be only demonstrated by heterologous expression of this enzyme in the PHA-negative mutant GpP104 of *P. putida*. Cells of the recombinant strain cultivated on octanoic acid accumulated a copolyester consisting of almost equimolar amounts of 3-hydroxybutyric acid and 3-hydroxyhexanoic acid plus small amounts of 3-hydroxyoctanoic acid (LIEBERGESELL et al., 1993). This enzyme is also able to use coenzyme A thioesters of non-3-hydroxyalkanoic acids as substrates, as demonstrated by the incorporation of 4-hydroxyhexanoic acid, 4-hydroxyheptanoic acid, 4-hydroxyoctanoic acid, and 5-hydroxyhexanoic acid into the polyes-

***Alcaligenes eutrophus***

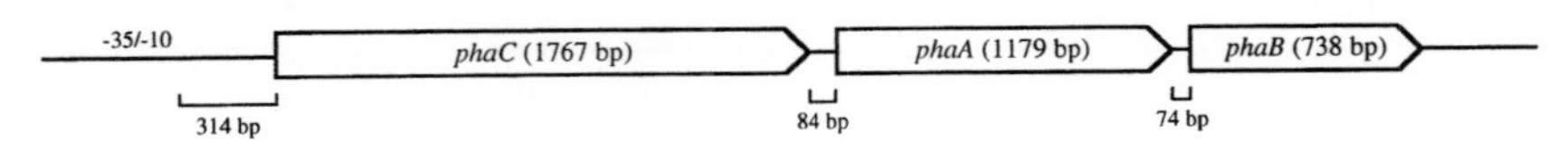


***Zoogloea ramigera***

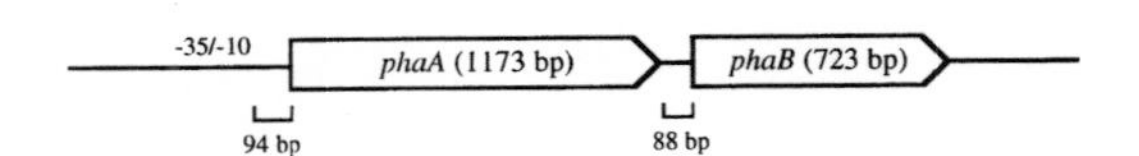


***Methylobacterium extorquens***

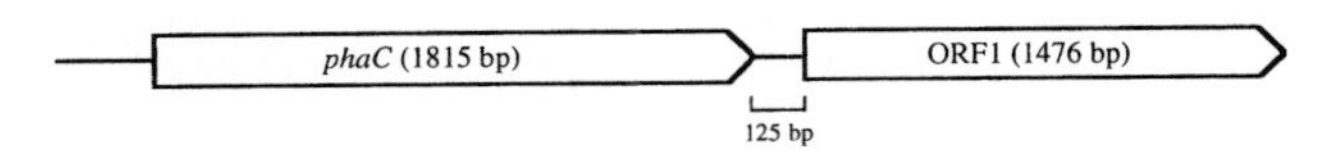


***Pseudomonas oleovorans***

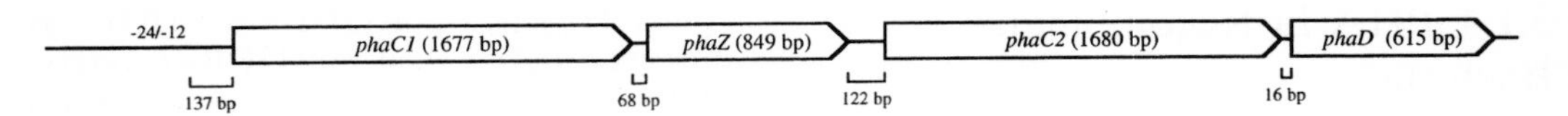


***Pseudomonas aeruginosa***

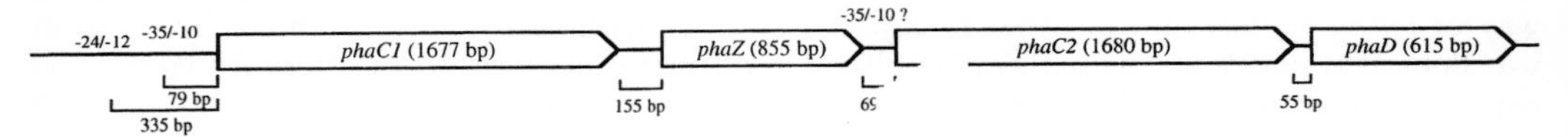


***Rhodococcus ruber***

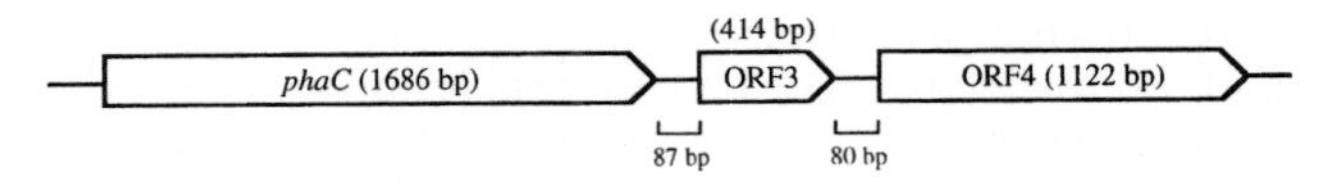


***Chromatium vinosum***

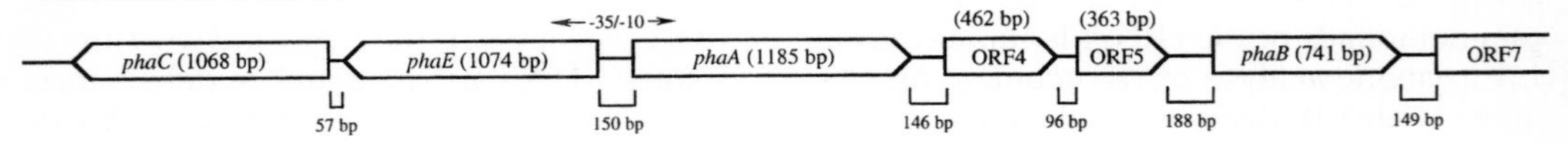


***Thiocystis violacea***

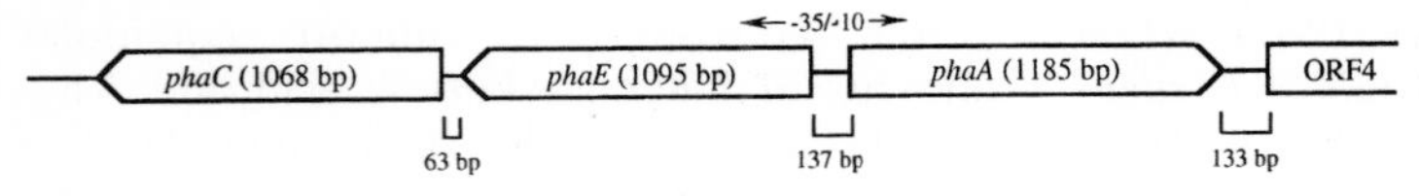


***Thiocapsa pfennigii***

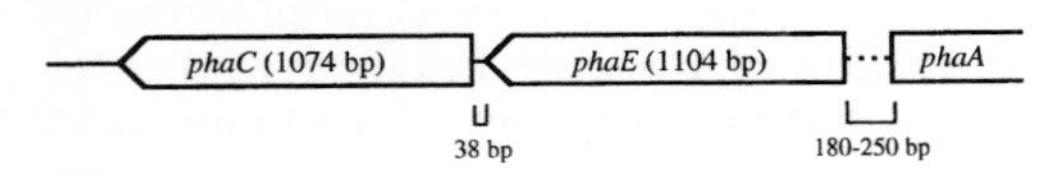

ter, when the respective organic acid was provided to the recombinant mutant strain as carbon source (VALENTIN et al., 1994; H. E. VALENTIN and A. STEINBÜCHEL, unpublished results). This example clearly demonstrated the importance of the physiological background in which a PHA synthase is expressed. *T. pfennigii* itself was only able to synthesize poly(3HB), even if precursor substrates were provided (LIEBERGESELL et al., 1991).

## 3.2 Regulation of PHA Biosynthesis

No evidence has been obtained yet that PHA biosynthesis is regulated at the transcriptional or the translational level. These enzymes seem to be constitutively expressed (STEINBÜCHEL and SCHLEGEL, 1991). Instead, PHA biosynthesis is regulated at the enzymatic level and by the availibity of suitable substrates for the PHA biosynthesis routes resulting from an imbalanced supply of nutrients. As pointed out above, PHA biosynthesis and accumulation occurs in most bacteria if a carbon source is provided in excess and if another nutrient is lacking. Regulation of β-ketothiolase by coenzyme A and acetyl-coenzyme A reductase (OEDING and SCHLEGEL, 1973; HAYWOOD et al., 1988a) and the availibility of reducing power for the acetoacetyl-coenzyme A (HAYWOOD et al., 1988b) are of great importance. The competition of the reductase with other enzyme systems for reducing equivalence has also been demonstrated recently by an increase of photoproduction of molecular hydrogen from acetate by PHA-negative mutants of *Rhodobacter sphaeroides* (HUSTEDE et al., 1993). There are, however, bacteria – such as, e.g., *Alcaligenes latus* – that acumulate significant amounts of PHA during the exponential growth phase; the limiting factor has not been identified yet. In addition, unrestricted provision of acetyl-CoA by the pyruvate dehydrogenase complex is important for maximum biosynthesis of poly(3HB), at least in *A. eutrophus* (PRIES et al., 1992).

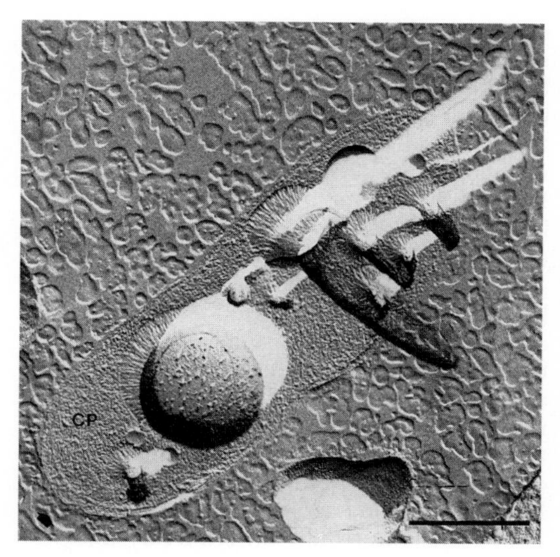


**Fig. 12.** Occurrence of poly(3HB) and poly(3HHx-*co*-3HO) granules in a recombinant *P. oleovorans*. Freeze-fracture electron micrograph of octanoate-grown cells of a recombinant strain of *P. oleovorans* GPo1 harboring plasmid pVK101–PP1 and expressing the poly(3HB)-biosynthesis genes of *A. eutrophus*. Inclusion with mushroom-type deformations represent poly(3HO) granules wherease inclusions with needle-type deformations represent poly(3HB) granules (bar=0.5 μm) (taken from PREUSTING et al., 1993c; reproduced with permission of Plenum Press).

**Fig. 11.** Molecular organization of PHA synthase loci. Designation of genes: *phaA*, β-ketothiolase; *phaB*, acetoacetyl-CoA reductase; *phaC*, PHA synthase; *phaE*, second component of PHA synthase complex in sulfur purple photosynthetic bacteria; *phaZ*, PHA depolymerase; ORF3, GA14-protein (a phasin) from *R. ruber*. The function of all other genes and open reading frames is unknown. The length of genes, open reading frames, and intergenic regions is provided in kilobase pairs. Positions of putative promoters (−35/−10 and −24/−12) are indicated (adapted and extended from STEINBÜCHEL et al., 1992).

## 3.3 Biosynthesis in Eukaryotes

Nothing is known about the biosynthesis routes of PHA or about the enzymes and genes required for PHA biosynthesis in eukaryotic organisms. Reports on the measurement of PHA synthase activity, on the identification of the corresponding genes by hybridization or by use of the polymerase chain reaction, and on tracer experiments using radioisotopes to reveal the biosynthesis pathway of PHA in eukaryotic organisms are not available. Since PHA has been detected in yeasts and in plants, these organisms must possess enzymes catalyzing the synthesis of the polyester. However, since the amounts of PHA detected in these organisms were rather low, the activities of the key enzyme PHA synthase may be by orders of magnitudes less than in PHA accumulating bacteria and therefore difficult to detect. Alternatively, the synthesis of poly(3HB) in these organisms may also result from the side activity of an unspecific enzyme which has a quite different physiological function. The situation is even much more complex in animals and humans since these organisms can take up PHA concommitantly with the diet. Therefore, it has clearly to be shown whether poly(3HB) was derived from own biosynthesis capabilities or from bacteria present in the diet.

## 3.4 Function of PHA

Some possible functions of PHA in eukaryotes have already been mentioned above. The function as a carbon and/or energy storage compound is probably most important for prokaryotes. In most bacteria accumulation of PHA is triggered if a carbon source is provided in excess and if the proliferation of the cells is impaired due to the lack of another nutrient which is essential for growth. The lacking nutrient may be the source of nitrogen, sulfur, phosphorous, iron, magnesium or potassium (e.g., STEINBÜCHEL and SCHLEGEL, 1989). This seems to be the standard situation for the occurrence of PHA inclusions in most bacteria. PHA can occur in large quantities in the bacterial cell, it is osmotically inert and does not exert any known deterious effect on the cells. Therefore, and since it is readily mobilized by the PHA degrading enzyme system in the cells, it can be beneficial to the bacterial cells in the absence of an exogenous carbon source because there is no need to degrade other essential macromolecules such as proteins or nucleic acids (HIPPE, 1967). Cells of *P. aeruginosa*, e.g., degraded the accumulated poly($HA_{MCL}$) at a rate comparable to that for the accumulation as soon as gluconate was depleted from the medium (TIMM and STEINBÜCHEL, 1990). Octane-grown cells of *P. oleovorans* mobilized the accumulated poly($HA_{MCL}$) also at a rather high rate (LAGEVEEN et al., 1988). In chemostat cultures during a 4 h period in the presence of thiosulfate and acetate, *Thiobacillus* strain S accumulated significant quantities of poly(3HB) during heterotrophic growth on acetate. The polyester was utilized as a supplementary carbon source during the initial stages of chemolithoautotrophic growth on thiosulfate when the carbon dioxide fixing capacity was low (KUENEN and ROBERTSON, 1984). As repeatedly shown, PHA contributes to the survival of the bacterial cell in the absence of an exogeneous carbon source (DAWES, 1986).

There are many different physiological functions discussed for PHA in bacteria. In some species of *Bacillus* and *Azotobacter*, poly(3HB) serves as a carbon and energy source for endospore formation (SLEPECKY and LAW, 1961; THOMPSON and NAKATA, 1973) or for encystment (LIN and SADOFF, 1968; STEVENSON and SOCOLOFSKY, 1966, 1973; REUSCH and SADOFF, 1981), respectively.

In many bacteria such as *Azotobacter* sp., *Bradyrhizobium japonicum*, *A. eutrophus* and some others, another important function has been assigned to the synthesis of poly(3HB) and other PHA. In these bacteria poly(3HB) synthesis is also triggered by oxygen deficiency (SENIOR and DAWES, 1973; WARD et al., 1977; MORINAGA et al., 1978; VOLLBRECHT et al., 1979; STEINBÜCHEL and SCHLEGEL, 1989). In some bacteria lack of oxygen is even a much better trigger for PHA biosynthesis than, e.g., lack of the nitrogen source. Due to the involvement of the acetoacetyl-CoA reductase, poly(3HB) biosynthesis starting from acetyl-CoA or acetoacetyl-CoA provides a sink for reducing equivalents, and the polyester can be regarded as a product of fermentative metabolism. Although much more reducing equivalents can be deposited within one single cell than a single cell is able to synthesize and excrete (the more typical fermentation products such as, e.g., ethanol or lactic acid), poly(3HB) provides several advantages since it is osmotically inert, non-toxic and not available to other bacteria in mixed cultures occurring in natural environments (STEINBÜCHEL, 1991a, b). Another advantage is that in *A. eutrophus* and most probably in many other bacteria the biosynthesis of poly(3HB) can start immediately and does not require the synthesis of new proteins.

STAM et al. (1986) suggested an additional interesting physiological function of PHA in bacteria. Poly(3HB) degradation releasing reducing equivalents in the nitrogen fixing *Rhizobium* ORS571 will protect the nitrogenase from oxygen damage due to the concommittant removal of oxygen in the nodules. This putative function could be supported by findings in *Azotobacter beijerinckii* (SENIOR and DAWES, 1971).

# 4 PHA Granules

## 4.1 Morphology and Physical State of the Polyester

PHA granules isolated from *Bacillus megaterium* contain approximately 97.7% PHA, 1.87% proteins, and 0.46% lipids and phospholipids (GRIEBEL et al., 1968). Although the detailed structural organization of the granules is not known at present, it is generally assumed that a PHA core is surrounded by a layer or "membrane" consisting of proteins and lipids (MAYER, 1992).

NMR spectroscopy of various bacteria has clearly shown that in the cells poly(3HB) occurs in the metastable amorphous state and that it is stored as a mobile elastomer (BARNARD and SANDERS, 1988, 1989; KAWAGUCHI and DOI, 1990). At first, plasticizers were assumed to be responsible for this, and water was considered as a candidate (BARNARD and SANDERS, 1989). Recently, it has been suggested that this simply results from the slow nucleation kinetics of small PHA particles. For a typical poly(3HB) granule exhibiting a diameter of 0.5 mm, the time to nucleate for crystallization is approximately 30,000 years (DE KONING and MAXWELL, 1993). A much shorter crystallization period can be induced by compaction of the granules by centrifugation, perturbation by sonication, damage of the surrounding membrane by exposure of the granules to solvents, sodium hydrochlorite or to hydrolytic enzymes, or by freezing and thawing cycles (BARNARD and SANDERS, 1989; DE KONING and MAXWELL, 1993). This is due to the formation of much larger PHA particles or to exogenous substances which get in excess to the PHA core of the granules to act as heterogeneous nucleation sites for crystallization (DE KONING and MAXWELL, 1993). Granules that never had been dried and that were obtained by the surfactant-hypochlorite treatment (RAMSAY et al., 1990), showed a crystalline shell at the surface and an amorphous core (LAUZIER et al., 1992).

## 4.2 Isolation of Native Granules

Depending on the chemical structure of the polyesters, the specific buoyant densities of PHA granules vary. For poly(3HB) granules buoyant densities of 1.18–1.24 g cm$^{-3}$ were reported (NICKERSON, 1982; BAUER and OWEN, 1988; HOROWITZ et al., 1993), whereas the buoyant density of poly(3HO) granules is only approximately 1.05 g cm$^{-3}$ (PREUSTING et al., 1993c). In addition to the chemical composition the buoyant density of PHA also depends on the physical state of the polyester (HOROWITZ et al., 1993). Therefore, differential and density gradient centrifugation were employed repeatedly to isolate native intact granules from various bacteria and separate them from the other cell constituents (Tab. 5). Different materials such as glycerol, sucrose, or sodium bromide were taken for the preparation of the gradients. The granules were used to study the biochemical structure of the inclusions or to investigate the *in vitro* PHA biosynthesis.

Large amounts of PHA in the cells affect the buoyant density of the whole cell. This was repeatedly utilized to separate PHA ac-

**Tab. 5.** Isolation of PHA Granules by Sedimentation or by Centrifugation in Density Gradients

| Bacterium | Gradient | | Centrifugation | Reference |
|---|---|---|---|---|
| *Acinetobacter calcoaceticus* | Sucrose | Continous, linear gradient: 1M to 2M one repeat | 4 h, 110,000·g | SCHEMBRI et al. (1994) |
| *Alcaligenes eutrophus* | Glycerol | Cushion: 100% | 12 min; 5,000·g | HIPPE and SCHLEGEL (1967) |
| | none | Sedimentation | 10 min; 5,000·g | KAWAGUCHI and DOI (1990) |
| | none | Sedimentation | 10 min; 10,000·g | GERNGROSS et al. (1993) |
| | Glycerol | Cushion: 100% two repeats | 20 min; 10,000·g | HARRISON et al. (1992) |
| | (1) Glycerol | Discontinous gradient: 44 + 88% | 0.5 h, 110,000·g | WIECZOREK et al. (1995) |
| | (2) Sucrose | Continous linear gradient: 1 to 2 M | | |
| *Azotobacter beijerinckii* | Glycerol | Cushion: 100% four repeats | 20 min; 4,100·g | RITCHIE and DAWES (1969) |
| *Bacillus cereus* | none | Sedimentation | 15 min; 2,700·g | LUNDGREN et al. (1964) |
| *Bacillus megaterium* | Glycerol | Cushion: 100% two repeats | 20 min; 9,000·g | MERRICK and DOUDOROFF (1964)<br>LUNDGREN et al. (1964) |
| | Glycerol | Cushion: 100% two repeats | 30 min; 1,600·g | GRIEBEL et al. (1968)<br>GRIEBEL and MERRICK (1971)<br>ELLAR et al. (1968) |
| *Bacillus thuringiensis* | NaBr | Discontinous gradient 5 + 10 + 15 + 20%, + linear 25 to 40% + cushion 42% | Zonal rotor No details for acceleration provided | NICKERSON (1982) |
| *Chromatium vinosum* | Sucrose | Continous, linear gradient: 1 M to 2 M | 4 h; 110,000·g | LIEBERGESELL et al. (1992) |
| | (1) Glycerol | Cushion: 50% | 1 h; 100,000·g | LIEBERGESELL et al. (1994) |
| | (2) Sucrose | Continous, linear gradient: 1 M to 2 M | 1 h; 197,000·g | |
| *Methylobacterium* AM1 | Glycerol | Cushion: 100% | 15 min; 9,000·g | BARNARD and SANDERS (1988) |
| *Pseudomonas oleovorans* | Sucrose | Cushion: 20% | 2.5 h; 140,000·g | FULLER et al. (1992) |
| | Sucrose | Discontinous gradient 15 + 20 + 25 + ... + 45 + 50 + 55% | 68 h; 110,000·g | PREUSTING et al. (1993c) |
| *Rhodococcus ruber* | (1) Glycerol | Cushion: 50% | 1 h; 100,000·g | PIEPER and STEINBÜCHEL (1992) |
| | (2) Sucrose | Discontinous gradient 25, 30, 35 and 40% | 1 h; 197,000·g | PIEPER-FÜRST et al. (1994) |
| *Zoogloea ramigera* | (1) none | Sedimentation | 60 min; 100,000·g | |
| | (2) Sucrose | Continous, linear gradient: 1.2 to 1.6 M | 16 h; 64,000·g | FUKUI et al. (1976) |

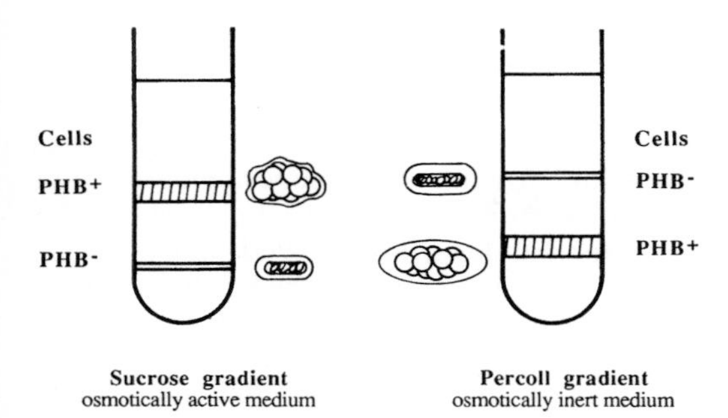


**Fig. 13.** Separation of PHB$^+$ from PHA$^-$ cells in density gradients by centrifugation. An identical number of cells of *A. eutrophus* strains H16 (phenotype PHB$^+$) and H16-PHB$^-$4 (phenotype PHB$^-$) are separated by centrifugation in sucrose (left side) or in Percoll (right side) density gradients (adapted from PEDROS-ALIO et al., 1985).

cumulating cells from cells that did not accumulate PHA for the enrichment of mutants unable to synthesize and/or accumulate PHA (SCHLEGEL et al., 1970). Depending on the osmolarity of the medium, PHA$^+$ cells had a higher or a lower buoyant density than PHA$^-$ cells (Fig. 13; PEDROS-ALIO et al., 1985).

## 4.3 Granule-Associated Proteins

There are several proteins bound to the PHA granules, among them catalytically active proteins such as PHA synthases and PHA depolymerases as well as other proteins which may have a structural function. PHA synthase protein has been found associated with the PHA granules in *Acinetobacter* sp. (SCHEMBRI et al., 1994), *Alcaligenes eutrophus* (HAYWOOD et al., 1989a; GERNGROSS et al., 1993), *Azotobacter beijerinckii* (RITCHIE and DAWES, 1969), *Bacillus megaterium* (GRIEBEL et al., 1968), *Chromatium vinosum* (LIEBERGESELL et al., 1992), *Methylobacterium extorquens* (VALENTIN and STEINBÜCHEL, 1993), *Rhodococcus ruber* (PIEPER and STEINBÜCHEL, 1992), *Rhodospirillum rubrum* (MERRICK and DOUDOROFF, 1961), and *Zoogloea ramigera* (FUKUI et al., 1976; TOMITA et al., 1983). In *A. eutrophus* and in *C. vinosum*, immunoelectron microscopy localized the PHA synthases at the periphery of the PHA granules (GERNGROSS et al., 1993; LIEBERGESELL et al., 1994). As demonstrated in *A. eutrophus* the PHA synthase was predominantly soluble during non-restricted growth and became granule-associated during accumulation of poly(3HB) in the stationary growth phase (HAYWOOD et al., 1989a). Soluble and granule-associated forms of PHA synthase were also detected in *Z. ramigera* (FUKUI et al., 1976).

PHA depolymerases responsible for the mobilization of the storage polyester and initiating PHA degradation in the cells were also found repeatedly associated to the PHA granules, e.g., in *A. eutrophus* (HIPPE and SCHLEGEL, 1967), *P. oleovorans* (HUISMAN, 1991; FOSTER et al., 1994), *Rhodospirillum rubrum* (MERRICK and DOUDOROFF, 1961, 1964), and *Z. ramigera* (TOMITA et al., 1983; SAITO et al., 1992).

In addition to these catalytically active proteins other proteins were found to be associated to the granules. These proteins were detected when isolated granules were treated with detergents and when the solubilized proteins were separated under denaturating conditions in SDS–polyacrylamide gels. At least one or two additional proteins were identified in the granule preparations of *Acinetobacter* sp. (SCHEMBRI et al., 1994), *A. eutrophus* (WIECZOREK et al., 1995), *C. vinosum* (LIEBERGESELL et al., 1992), *P. oleovorans* (FULLER et al., 1992), *R. ruber* (PIEPER and STEINBÜCHEL, 1992; PIEPER-FÜRST et al., 1994), and in many other bacteria (unpublished results from the laboratory of A. STEINBÜCHEL). In Fig. 14 the localization of the 14 kDa granule-associated protein GA14 in cells of *R. ruber* is shown, the structural gene of which is located downstream of the PHA synthase structural gene. In most cases, these additional proteins represented the major protein fraction of the granule. In contrast, the PHA synthase did never represent the most dominant protein band in isolated PHA granules, except in those isolated from cells of *Acinetobacter* sp. (SCHEMBRI et al., 1994). It was assumed (PIEPER-FÜRST et al., 1994) that these proteins might have a function similar to that oleosins on the surface of oil seed bodies in plants (HUANG, 1992;

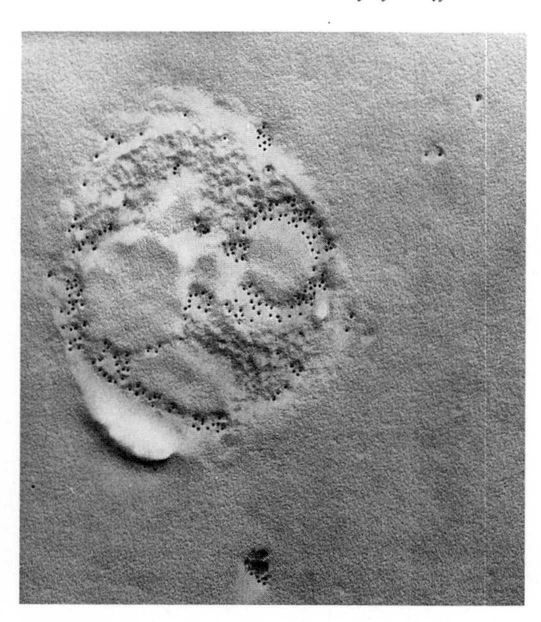

**Fig. 14.** Immunoelectron microscopic localization of the GA14 protein in *R. ruber*. Sections of the cells were labeled with anti-GA14 protein antibodies and an immunoglobulin G–gold complex and shadowed with platinum-carbon to give a three-dimensional impression, as described by PIEPER-FÜRST et al. (1994) (original micrograph by M. H. MADKOUR and F. MAYER).

MURPHY, 1993). Therefore, the designation of "phasins" was proposed for these proteins (STEINBÜCHEL et al., 1995). Together with phospholipids phasins most probably represent the main constituents of the membrane surrounding the polyester core of PHA granules.

In this context it is interesting that upon centrifugation of crude extracts obtained from cells of *Bacillus thuringiensis* in NaBr gradients, poly(3HB) granules sedimented in two separate bands: one band of granules with bound protein and another band of granules without bound protein, the latter of which had most probably lost their membranes (NICKERSON, 1982).

# 5 Degradation of PHA

PHA can be degraded by different mechanisms and for different scopes. Degradation may occur inside the bacterial cell where the polyester has been synthesized and accumulated or outside the bacterial cell by other bacteria, fungi or higher organisms. They use the polyester as a carbon source after release from a dying and lysing bacterial cell or after entering the ecosystem due to a technical or medical application of PHA.

## 5.1 Extracellular Biodegradation

Extracellular degradation of PHA has been investigated in much more detail than intracellular degradation. In general, degradation seems to follow a hydrolytic mechanism, and the enzymes catalyzing this reaction are referred to as PHA depolymerases. On PHA containing agar media, PHA-degrading microorganisms were made visible by the formation of clearing zones (halos) around the colonies (Fig. 15). These halos result from the hydrolysis of the water insoluble PHA into water soluble cleavage products by the excreted PHA

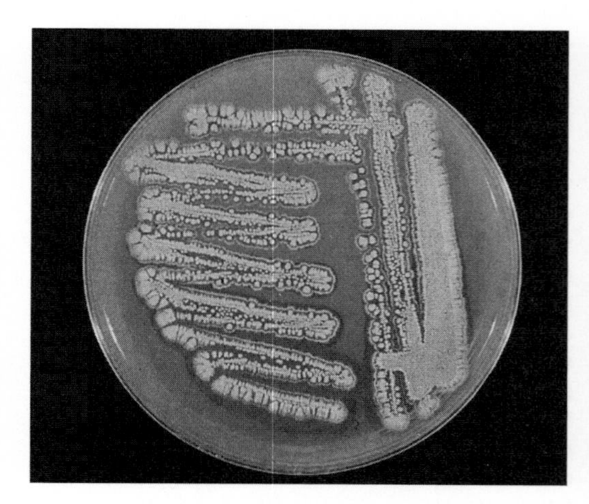

**Fig. 15.** Growth of PHA degrading bacteria and formation of clearing zones. A bacterial isolate was cultivated on mineral agar plates containing a suspension of poly(3HB) as the sole carbon source, as described by JENDROSSEK et al. (1993a).

depolymerases. Bacteria as well as fungi have been found to be capable of PHA hydrolysis. Since the first demonstration of biodegradation of poly(3HB) by an exoenzyme more than 30 years ago (CHOWDHURY, 1963), several aerobic poly(3HB) degrading bacteria were isolated (e.g., DELAFIELD et al., 1965a, b; JENDROSSEK et al., 1993a), and the hydrolysis of PHB was mostly studied at a physiological level. Meanwhile it has been shown that poly(3HB) is also degraded by anaerobic bacteria such as *Ilyobacter delafieldii* (JANSSEN and HARFOOT, 1990; JANSSEN and SCHINK, 1993) or by eukaryotic organisms such as *Penicillium simplicissimum*, *P. funiculosum*, *Eupenicillium* sp., and *Aspergillus fumigatus* (MCLELLAN and HALLING, 1988; BRUCATO and WONG, 1991; MERGAERT et al., 1993). It has also been shown that poly(3HV) (MÜLLER and JENDROSSEK, 1993), the copolyester poly(3HB-*co*-3HV), and copolyesters containing 4HB (MUKAI et al., 1992; DOI et al., 1992) as well as $PHA_{MCL}$ (SCHIRMER et al., 1993) and some other PHA are degraded by axenic cultures. Even poly($3HA_{MCL}$), which had been crosslinked by electron beam irradiation, was degraded by the extracellular $PHA_{MCL}$ depolymerase of *Pseudomonas fluorescens* GK13 (DE KONING et al., 1994). In addition to PHA depolymerases, related enzymes of higher organisms seem to be capable to hydrolyze these polyesters, too.

It was repeatedly demonstrated that the degree of crystallinity significantly affects the degradation rate of PHA, and KUMAGAI et al. (1992) concluded that poly(3HB) molecules in the amorphous state are more easily hydrolyzed than poly(3HB) molecules in the crystalline state. Chemosynthetic atactic and syndiotactic poly(3HB) consisting of both, *R*-3HB and *S*-3HB isomers, were also degraded by either purified poly(3HB) depolymerase of *Alcaligenes faecalis* T1 (KUMAGAI and DOI, 1992) or by cells of *Pseudomonas lemoignei* or *A. fumigatus* (HOCKING et al., 1994; JESUDASON et al., 1993).

### 5.1.1 Biochemistry and Molecular Genetics of Extracellular PHA Depolymerases

Meanwhile, PHA depolymerases have been isolated from several organisms, and they were characterized regarding their biochemical and molecular properties; from some bacteria also the genes were cloned. $PHA_{SCL}$ depolymerases from *Comamonas* sp. (JENDROSSEK et al., 1993a, b), *Comamonas testosteroni* (MUKAI et al., 1992), and *A. faecalis* (SAITO et al., 1989, 1993) were investigated in detail; even from the eukaryote *P. funiculosum* a $PHA_{SCL}$ depolymerase was isolated (BRUCATO and WONG, 1991).

*Pseudomonas lemoignei*, which is highly specialized for the utilization of poly(3HB) and unable to utilize other substrates, except some fatty acids, as sole carbon sources for growth, synthesizes more than one PHA depolymerase. LUSTY and DOUDOROFF (1966) detected two different $PHA_{SCL}$ depolymerases in poly(3HB) or succinate-grown cells. Later, NAKAYAMA et al. (1985) separated 4 different isoenzymes from *P. lemoignei*. Recently, MÜLLER and JENDROSSEK (1993) have detected an additional $PHA_{SCL}$ depolymerase, which was induced during the cultivation on poly(3HV) or on valeric acid. Although the activities with poly(3HB) were still high, the formation of this enzyme was not induced during cultivation on this polyester. Therefore, and since the enzyme exhibits a comparably high specificity for cleavage of poly(3HV) or poly(3HB-*co*-3HV), this enzyme was referred to as poly(3HV) depolymerase. Molecular analysis identified 5 different structural genes for $PHA_{SCL}$ depolymerases in *P. lemoignei* localized at five different regions in the genome (JENDROSSEK et al., 1993b; BRIESE et al., 1994b).

The $PHA_{MCL}$ depolymerase of *P. fluorescens* GK13, which is a representative of 26 isolates of bacteria capable to utilize poly(3HO) as sole carbon source for growth, was investigated in most detail. The enzyme was purified and biochemically characterized (SCHIRMER and JENDROSSEK, 1993). Subsequently, the gene was identified with a specific oligonucleotide, the sequence of which was designed according to the amino acid sequence of the N-terminus and the codon usage of this bacterium. The cloned gene could be expressed in a functionally active form in *E. coli* resulting in clearing zones around the colonies grown on LB-poly(3HO) agar plates (SCHIRMER and JENDROSSEK, 1994). No details are known for the $PHA_{MCL}$ depolymerase of a gram-positive isolate *Streptomyces* K10 (SCHIRMER et al., 1993).

The relative molecular masses of the $PHA_{SCL}$ depolymerases varied between 36,000 and 59,000, and the enzymes seem to exist as monomers. Only the $PHA_{MCL}$ depolymerases of *P. fluorescens* GK13 seems to have a dimeric quarternary structure consisting of two identical subunits of $M_r$ 25,000.

One striking feature of the 4 $PHA_{SCL}$ depolymerases of *P. lemoignei* are regions consisting of clusters of threonine and of other small and uncharged amino acids such as glycine and alanine located close to the C-terminus. Such threonine-rich regions were also detected in other extracellular hydrolases such as cellulases, xylanases, and glucanases; it was suggested that these regions are involved in the binding of the enzyme to the polyester (JENDROSSEK et al., 1993b, and references cited therein). A corresponding region was not identified in the $PHA_{MCL}$ depolymerase of *P. fluorescens* GK13 (SCHIRMER and JENDROSSEK, 1994) or in poly(3HB) depolymerases from other sources. These PHA depolymerases possess a region at the C-terminus homologous to the type III sequence of fibronectin. Similar regions were also detected in other bacterial extracellular hydrolases such as chitinases or cellulases (SAITO et al., 1993; JENDROSSEK et al., 1995a). The PHA depolymerases from *Penicillium funiculosum* was inhibited by diisopropyl fluorophosphate indicating that this enzyme is a serine esterase (BRUCATO and WONG, 1991).

All PHA depolymerases investigated so far had the sequence "Gly-Leu-Ser-Xaa-Gly" at the N-terminus, which corresponds to the "lipase box" found in lipases and esterases (JENDROSSEK et al., 1995b). Comparison of the primary structures of PHA depolymerases from various bacteria and biochemical studies revealed that these enzymes possess a catalytic triad consisting of serine (from the "lipase box") plus histidine and aspartate. Interestingly, identical triads were also identified as active sites of bacterial lipases (JAEGER et al., 1994). Some PHA depolymerases were glycosylated with glucose and N-acetyl glucosamine (BRIESE et al., 1994b; JENDROSSEK et al., 1994a). DOI and collaborators isolated a $M_r$ 35,000 protein which inhibited the inducible poly(3HB) depolymerases from *P. lemoignei* and other sources; the authors suggested that this inhibitor protein reversibly binds to the serine residues at the active site of the enzymes (MUKAI et al., 1992).

## 5.1.2 Degradation in the Environment

It was shown that PHA degradation occurs in a large variety of complex environments. These studies included oxic as well as anoxic environments. In addition, PHA degrading organisms frequently occur in the environment. Approximately 55% of terrestrial fungal isolates were able to hydrolyze poly(3HB) and/or poly(3HB-*co*-3HV) as indicated by the formation of halos around the colonies on PHA-agar plates (MATAVULJ and MOLITORIS, 1992). The numbers of poly(3HB) degrading bacteria were counted in various environmental samples, e.g., landfill leachates, sewage sludge compost, sewage sludge supernatant, forest soils, farm soil, paddy soil, weed field soil, road side sand and pond sediments. Comparison of the numbers of colonies forming clearing zones on yeast extract–poly(3HB) or on nutrient broth–poly(3HB) agar plates to colonies without clearing zones, demonstrated that 0.2–11.4% of all cultivatable aerobic microorganisms were able to hydrolyze poly(3HB). Interestingly, the number of poly(3HB) hydrolyzing bacteria often paralleled the number of bacteria able to hydrolyze the chemosynthetic polyester poly($\varepsilon$-caprolactone) (NISHIDA and TOKIWA, 1993). In a recent study, 295 different microorganisms were isolated from samples buried for biodegradation test studies and taxonomically characterized. 105 of these strains were gram-negative, mostly belonging to *Acidovorax facilis* and *Variovorax paradoxus*, 36 belonged to the genus *Bacillus*, 68 to *Streptomyces*, and 86 were fungi, mainly of the genus *Penicillium* or strains of *Aspergillus fumigatus* (MERGAERT et al., 1993). Other authors determined the total numbers of aerobic poly(3HB) degrading bacteria in different ecosystems and detected $1.2 \cdot 10^5$ bacteria per mL sewage sludge, $3.8 \cdot 10^3$ per mL sludge of fresh-water lake, $9.2 \cdot 10^5$ per g of garden soil, $1.3 \cdot 10^6$ per g of field-soil and $4.3 \cdot 10^6$ per g of compost (BRIESE et al., 1994a).

Several studies investigated the general biodegradability of various PHA in different ecosystems under *in situ* conditions such as in activated aerobic sewage sludge (BRIESE et al., 1994a; GILMORE et al., 1993) or in aquatic ecosystems such as in lake sediments, lake water or marine environments (MERGAERT et al., 1992; BRANDL and PÜCHNER, 1992; DOI et al., 1992), in soil (MERGAERT et al., 1993) or in anaerobic sewage sludge (BUDWILL et al., 1992). Most of these studies provided clear evidence that the biosynthetic PHA are readily degraded in ecosystems under natural conditions and also in compost heaps. It was, however, interesting that PHA samples mainly consisting of 3-hydroxyoctanoic acid obviously were not degraded over a period as long as 4 months during incubation in aerobic activated sewage sludge (GILMORE et al., 1993), although poly(3HO) degrading bacteria were readily enriched from various environments (SCHIRMER et al., 1993).

## 5.2 Intracellular Degradation

Relatively little ist known about the intracellular degradation of PHA. In some studies the degradation of poly(3HB) with isolated granules from *B. megaterium* and *R. rubrum* (MERRICK and DOUDOROFF, 1964), from *A. eutrophus* (HIPPE and SCHLEGEL, 1967; PRIES et al., 1991), and *Z. ramigera* (SAITO et al., 1992) was investigated. From *A. eutrophus* mutants were isolated exhibiting an accelerated rate of PHA degradation (PRIES et al., 1991), and some contrary results were obtained whether or not PHA synthesis and degradation can occur in parallel (DOI et al., 1990a, c; HAYWOOD et al., 1991). However, the poly(3HB) depolymerases have not been purified from these bacteria nor have the corresponding genes been cloned.

In contrast, there is some evidence for the identification of the genes for the intracellular $PHA_{MCL}$ depolymerase; they were localized between the structural genes of two $PHA_{MCL}$ synthases in *P. aeruginosa* and *P. oleovorans* (HUISMAN et al., 1991; TIMM and STEINBÜCHEL, 1992). These bacteria degrade the accumulated polyester at a rather high rate which is in the same order of magnitude than the rate of synthesis (TIMM and STEINBÜCHEL, 1990; see also LAGEVEEN et al., 1988). With the $PHA_{MCL}$ granules of *P. oleovorans* degradation studies were done as well. These experiments showed that Triton X-100 and phenylmethylsulfonyl fluoride inhibited the depolymerase activity, indicating that the enzyme most probably possessed a serine residue in its catalytically active site (FOSTER et al., 1994). Interestingly, mutants of *P. oleovorans* defective in the utilization of intracellularly stored PHA accumulated the same amount of PHA with the same molecular weight than the wild type (HUISMAN et al., 1992).

## 5.3. Degradation by Other Enzymes

Whereas none of the extracellular PHA depolymerases investigated exhibited lipase activity as revealed by measuring hydrolysis of triolein, several lipases from bacteria and fungi hydrolyzed PHA as revealed by measuring the weight loss of PHA films (MUKAI et al., 1993) or by a photometric assay of PHA granule suspensions (JAEGER et al., 1995). However, both laboratories showed that only PHA consisting of $\omega$-hydroxyalkanoic acids were hydrolyzed by the lipases; PHA containing hydroxyalkanoic acids with side chains attached to the polyester backbone such as, e.g., poly(3HB) were not or only marginally hydrolyzed. Hydrolysis of PHA by side activities of non-specific enzymes is relevant for the consideration of medical applications of PHA since, e.g., animal cells do most probably not possess PHA depolymerases, and for the consideration of technical applications.

## 5.4 Abiotic Degradation

Degradation of PHA by spontaneous hydrolysis has also been described. This has been investigated in detail only for poly(3HB) and poly(3HB-*co*-3HV). Hydrolysis occurred at rather low rates and preferentially at pH values in the acidic or in the alkaline range (HOLLAND et al., 1987; MILLER and WILLIAMS, 1987; DOI, 1990); rising temperatures also increased the rate of hydrolysis. In addition, thermal degradation of PHA occurs (GRASSIE et al., 1984), and for many PHA the temperature of significant decomposition is not far above the melting point, the mechanisms of thermal or hydrolytic decomposition of PHA have been compiled recently by HOCKING and MARCHESSAULT (1994).

# 6 Properties of PHA

## 6.1 General Properties

Biosynthetic PHA share different properties which make them attractive for several applications (Tab. 6). These polyesters are thermoplastic and/or elastomeric and exhibit a degree of polymerization as high as 30,000. They are generally lipophilic substances and not soluble in water. The only water soluble biosynthetic polyester is polymalic acid (PMA). According to investigations of the overwhelming number of novel PHA, these polymers are biodegradable; there is no evidence for one of these novel polyesters that they are not. All biosynthetic PHA are optically active polymers since they are composed of chiral hydroxyalkanoic acids. All constituents of bacterial PHA have the *R*-stereochemical configuration, and they are, therefore, fully isotactic. In contrast, the chiral centers of fungal PMA posses the *S*-stereochemical configuration. Only PHA consisting exclusively of $\omega$-hydroxyalkanoic acids such as, e.g., the homopolyester poly(4HB), are nonchiral. However, poly(4HB) is the only known representative of a biosynthetic poly($\omega$HA) homopolyester. All other polyesters with $\omega$HA also contain other hydroxyalkanoic acids, mostly 3HB (see below). Poly(3HB) was shown to be non-toxic and biocompatible. Poly(3HB) and some other PHA exhibit piezoelectricity.

**Tab. 6.** Properties of PHA

| Properties |
|---|
| Thermoplastic and/or elastomeric |
| High degree of polymerization |
| Insoluble in water |
| Biodegradable |
| Optically active[a] |
| Non-toxic[b] |
| Biocompatible[b] |
| Piezoelectric[b] |

[a] not valid for PHA consisting solely of $\omega$-hydroxy fatty acids
[b] only shown for poly(3HB)

## 6.2 Primary and Secondary Structure

Biosynthetic PHA occur as homopolyesters, copolyesters, and as blends. The 90 different constituents shown in Sect. 2.1 would theoretically allow the synthesis of 90 different homopolyesters, of more than 2,000 different dicopolyesters, and of more than 100,000 different tercopolyesters. In addition, for any given combination of constituents in a copolyester the molar fraction of each comonomer can vary. In practice, however, only very few biosynthetic homopolyesters exist. These are poly(3HB) and poly(3HV) (STEINBÜCHEL et al., 1993; and references cited therein), poly(4HB) (NAKAMURA et al., 1992; STEINBÜCHEL et al., 1994), poly(3-hydroxy-5-phenylvaleric acid) (KIM et al., 1991), and homopolyesters of 3-hydroxyhexanoic, 3-hydroxyheptanoic, 3-hydroxyoctanoic, or 3-hydroxynonanoic acid (LAGEVEEN et al., 1988; ANDERSON et al., 1990). Even if precursor substrates such as hydroxyalkanoic acids are used that are converted by a thiokinase to the corresponding coenzyme A thioester in one step, the PHA synthase competes for this thioester with catabolic as well as with anabolic enzymes. The latter enzymes convert this particular thioester to other hydroxyacyl coenzyme A thioesters such as, e.g., 3-hydroxybutyryl-CoA resultig in the synthesis of other potential substrates of the respective PHA synthase that are subsequently incorporated into the polyester thus resulting in the formation of a copolyester.

Among the different types of copolyesters that are theoretically possible, only random copolyesters were detected so far. In order to study the distribution of the comonomers in copolyesters, various copolyesters were subjected to partial methanolysis or to partial pyrolysis resulting in the formation of oligomers which were then separated by high-performance liquid chromatography (HPLC) and analyzed by fast atom bombardment mass spectrometry (FAB-MS). The analysis of poly(3HB-*co*-3HV) isolated from *A. eutrophus* (BALLISTRERI et al., 1989) and of various $PHA_{MCL}$ isolated from cells of *P. oleovorans* grown on different single alkanoic acids, e.g., nonanoic acid, or on mixtures of two different alkanoic acids, e.g., nonanoic acid plus 10-undecenoic acid (BALLISTRERI et al., 1990, 1992), clearly demonstrated that the bacteria synthesized random copolyesters. Biosynthetic alternating, block or graft copolyesters have not been detected yet. Due to the sluggishness of the metabolism in the whole cell, it is, however, not very likely that it will be possible to obtain such copolyesters by a fermentation process.

From certain substrates some bacteria accumulate blends of two different polyesters. One noticeable example is the formation of a blend of poly(3HB) and poly(3HHx-*co*-3HO) from octanoate by a recombinant strain of *P. oleovorans* expressing the PHA biosynthesis genes of *A. eutrophus* (TIMM et al., 1990). Both polyesters were even deposited in separate granules (PREUSTING et al., 1993a). The recently isolated wild type *Pseudomonas* strain GP4BH1 synthesized poly(3HB) plus poly($3HA_{MCL}$) from gluconate as well as from octanoic acid, rather than a copolyester of 3-hydroxybutyric acid plus medium chain length 3-hydroxyalkanoic acids (STEINBÜCHEL and WIESE, 1992). In cells of *P. oleovorans* grown on 5-phenylvaleric acid plus nonanoic acid, a blend of poly(3-hydroxy-5-phenylvaleric acid) and poly($HA_{MCL}$) was found in the same granule (LENZ et al., 1992).

## 6.3 Physical Properties

It would be beyond the scope of this chapter to review all reports on the physical properties of PHA or on the material properties of specimens manufactured from PHA. Comprehensive overviews on various aspects of the physical properties have been published recently (e.g., BARHAM et al., 1992; DOI, 1990; HOCKING and MARCHESSAULT, 1994). PHA are thermoplastic polymers which become highly viscous and mouldable at temperatures close to or above their melting points ($T_m$). The $T_m$ values and most other properties such as the glass transition temperature ($T_g$) or the crystallinity depend on the composition of the PHA, i.e., on the structures of the constituents, and on the relative amounts of comonomers in copolyesters. Regarding $T_m$, $T_g$ and crystallinity poly(3HB) resembles polypropylene (HOCKING and MARCHESSAULT, 1994). Poly(3HB) exhibits the highest $T_m$ (175 °C) of biosynthetic PHA. With an increasing length of the *R*-pendant group for other PHA or with an increasing molar fraction of 3HV in poly(3HB-*co*-3HV) copolyesters $T_m$ decreases. $T_m$ values as low as 39 °C have been reported for a copolyester of 3HV and 3-hydroxyheptanoic acid (PREUSTING et al., 1990).

With an increasing length of the *R*-pendant group and with increasing numbers of different HA in copolyesters the elasticity of PHA increases. Whereas poly(3HB) gives a rather brittle material, poly(3HB-*co*-3HV) is much less brittle, and poly(3HO) has a rubber-like consistency. Treatment of PHA consisting of unsaturated $HA_{MCL}$ by radiation yielded a complex cross-linked polyester which resembled a true rubber; this polyester exhibited constant mechanical properties over a wide temperature range from $T_g$ up to the degradation temperature (DE KONING et al., 1994).

## 6.4 Solution Properties

In Volume 6b of the First Edition of *"Biotechnology"* a large list of solvents in which poly(3HB) is highly soluble, fairly soluble or insoluble has been provided (LAFFERTY et al., 1988). In chlorinated hydrocarbons such as chloroform, methylene chloride, 1,2-dichloroethane or 1,1,2-trichloroethane and in propylene carbonate, poly(3HB) and some other PHA that have been investigated are soluble at a relatively high concentration. It should be emphasized that the solubilities are strongly temperature-dependent. However, the solubility of PHA in the solvents varies according to their chemical composition. Whereas, e.g., poly(3HB) is insoluble in petrol ether, poly(3HO) is soluble in this organic solvent (TIMM et al., 1990).

## 6.5 Piezoelectric Properties

Like many other biopolymers such as polysaccharides, proteins or nucleic acids, poly(3HB) possesses the property of piezoelectricity. The piezoelectric properties depend on the degree of crystallinity and on the temperature. A deformation by shear forces will produce a polarization in the poly(3HB) crystals resulting in the generation of a charge at the surface. On the other side application of voltage will cause deformations of the polyester. Since it was shown that other chemosynthetic piezoelectric polymers such as polyvinylidone fluoride stimulated bone growth, this property of poly(3HB) may be important for some medical applications (see below) of this polyester. Except for poly(3HB-*co*-3HV), the piezoelectric properties of other PHA have not been investigated so far (ANDO and FUKUDA, 1984; FUKUDA and ANDO, 1986).

# 7 General Methods for Detection and Analysis of PHA

Several methods have been developed to examine the bacterial cells for the presence of PHA. In the past, when poly(3HB) was the only known example of these biosynthetic polyesters, more general methods that did not distinguish between different constituents were applied. Due to the large variety of different constituents that are incorporated into PHA and that are also synthesized from non-related substrates, now more sophisticated methods must be applied for an accurate quantitative and qualitative analysis of PHA. Meanwile, methods are available for the analysis of the polyester composition in whole cells, without separation of the residual biomass from the polyester. It is, however, highly recommended also to analyze the isolated polyester in order to confirm the composition of a new polyester. PHA isolation can be routinely done by extracting dried cell material with chloroform and subsequent repeated precipitation of the extracted polyester by pouring the solution to an excess of cold ethanol. The precipitate is easily separated from the solvent, and small amounts of remaining contaminating material may be removed by washing the isolated polyester with a different solvent.

## 7.1 Staining Reactions

Due to the hydrophobic character of PHA, staining of cells or of colonies with lipophilic dyes has been applied for the identification of PHA. Sudan black B was introduced for this purpose very early (BURDON, 1946). Although this dye is not selective and also stains other lipid inclusions, it is now most widely used (SMIBERT and KRIEG, 1981) in combination with the light microscopic investigation of cells or colonies. The cells are exposed to a solution of 0.02% Sudan black in 96% ethanol. Excess stain, which has not been absorbed selectively, is removed by washing with a solution of 96% ethanol (SCHLEGEL et al., 1970). This staining technique is particularly suitable to distinguish between colonies of a $PHA^+$ wild type and a $PHA^-$ mutant derived thereof (Fig. 16). Other stains suitable for PHA are the fluorescent dyes Nile Blue A (OSTLE and HOLT, 1982), Nile Red (SCHEPER, 1991), and Phosphin 3R (BONARTSEVA, 1985). Nile Blue A seems to be rather specific and does not affect glycogen or polyphosphate. Colonies of poly(3HB) containing cells grown in the presence of 40 mg $L^{-1}$ Phosphin 3R exhibit a bright fluorescence at a wavelength of 460 nm. A 1% aque-

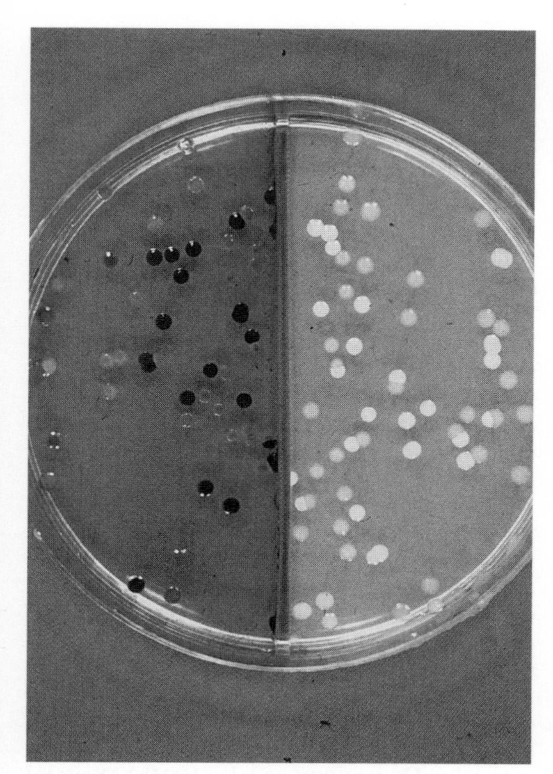

**Fig. 16.** Colonies of PHA$^+$ wild type and a PHA$^-$ mutant of *A. eutrophus* H16 on fructose mineral agar plates containing 0.005% (wt/vol) ammonium chloride. Colonies of the wild type H16 accumulate poly(3HB) and appear opaque whereas colonies of the PHA-negative mutant H16-PHB$^-$4 appear translucent (left side). Only colonies of the wild type stained with Sudan black (right side) (original photograph by V. OEDING, reproduced with permission of Blackwell Scientific Publications).

ous solution of Nile Blue can also be applied to stain dried cells on a slide in combination with fluorescence microscopy, after excess stain has been removed by washing with 8% acetic acid (OSTLE and HOLT, 1982). The lipophilic oxazon derivative, that spontaneously is formed by oxidation of the hydrophilic Nile Blue A and is always present in Nile Blue A solutions, is most probably responsible for staining. The optimum excitation wavelength of Nile Red stained cells of *A. eutrophus* is 565 nm, the optimum emission wavelength 588 nm. Cells of various bacteria can be stained with Nile Red, and this dye was also used to determine by flow cytometric analysis the amount of poly(3HB) in cells of, e.g., a population of *A. eutrophus* (SRIENC et al., 1984; SCHEPER, 1991).

## 7.2 Determination of Composition by Gravimetric and Turbidometric Methods

For a long time, poly(3HB) has been determined gravimetrically by assessment of the mass of the material precipitating in the presence of diethylether or acetone from a chloroform solution, that was obtained by extraction of lyophilized cell material (LEMOIGNE, 1926; SCHLEGEL et al., 1961; SCHLEGEL and GOTTSCHALK, 1962). A gravimetric method was introduced by WILLIAMSON and WILKINSON (1958) based on the selective removal of all other cell constituents by exposing the cell to sodium hypochlorite, a strongly oxidizing agent. The remaining material was dried, and the mass was determined. Alternatively, the residual turbidity of the solution due to the presence of PHA granules was assayed.

## 7.3 Determination of Composition by Spectroscopic Methods

With an UV-spectrometric method (SLEPECKY and LAW, 1960; LAW and SLEPECKY, 1961) utilizing a bathochromic effect on the absorption of the double bond of crotonic acid at 235 nm, that was formed upon heating of the polyester in the presence of concentrated sulfuric acid can be assayed. With an IR spectrophotometric method the absorption of the carbonyl group of the ester bond at 1.740 $cm^{-1}$ is determined (JÜTTNER et al., 1975). The development of this method was based on the earlier observations that the IR spectra of bacterial cell populations depend on the species and the nutritional state of the cells (BLACKWOOD and EPP, 1957, and references cited therein; HAYNES et al., 1958; NORRIS and GREENSTREET, 1958). For poly(3HB) and at 1.740 $cm^{-1}$, an absorption coefficient ($\varepsilon$) of 4.452 $cm^2\ mg^{-1}$ was determined. IR spectrophotometry also provided some hints for the presence of double bonds in the side chains of the polyester as well as on their lengths. Due to the C—H bonds polyesters also absorb at 3.000 $cm^{-1}$. The ratio of the absorption at 3.000 $cm^{-1}$ and at 1.740 $cm^{-1}$ increases with the length of the *R*-pendant group. PHA, which contain 3-hydroxyoctenoic acid or 3-hydroxy-2-butenoic acid, exhibit additional absorption peaks at 920, 1.640, and 3.080 $cm^{-1}$ (LAGEVEEN et al., 1988) or, due to the C=C bonds, at 1.580 $cm^{-1}$ (DAVIS, 1964), respectively. Fourier transform-infrared spectrometry allowed the analysis of PHA in complex environmental samples (NICHOLS et al., 1985). Other functionalized groups of the new PHA detected recently could be identified by IR spectrometry as well. Today, however, this method is not very often applied for the analysis of PHA.

Instead, $^{13}C$- and $^{1}H$-NMR spectrometry as well as two-dimensional Cosy NMR spectrometry are much more widely used to confirm the composition of new PHA in bacteria. In *P. putida* and in *P. aeruginosa* NMR spectroscopy has been successfully applied for the analysis of poly($3HA_{MCL}$) (HUIJBERTS et al., 1994a; SAITO and DOI, 1993), in *R. ruber* for poly(3HB-*co*-3HV) (WILLIAMS et al., 1994), in *A. eutrophus* for poly(3HB), poly(3HB-*co*-3HV), poly(3HB-*co*-4HB), poly(3HB-*co*-3HV-*co*-4HV), and poly(4HB) (DOI et al. 1987a; YOSHIE et al., 1992; VALENTIN et al., 1992; KUNIOKA et al., 1988; STEINBÜCHEL et al., 1994), and in recombinant strains of *P. putida* for copolyesters containing 4HHx (VALENTIN et al., 1994).

$^{1}H$-, $^{2}H$- and $^{13}C$-NMR spectrometry was also successfully applied to investigate the biosynthesis pathways of PHA in various bacteria. This was done to analyze the biosynthesis pathways for poly($3HA_{MCL}$) from, e.g., [1-$^{13}C$]- or [2-$^{13}C$]-labeled acetate or from [1-$^{13}C$]-labeled propionate in *P. putida* (SAITO and DOI, 1993; HUIJBERTS et al., 1994a), poly(3HB-*co*-4HB) from [1-$^{13}C$]- or [2-$^{13}C$]-labeled 4-hydroxybutyric acid in *A. eutrophus* (VALENTIN et al., 1995), poly(3HB) from [$^{1}H$]- or [$^{2}H$]-labeled acetate in *A. eutrophus* (YOSHIE et al., 1992), or poly(3HB-*co*-3HV) from [2,3-$^{13}C$]- or [1,4-$^{13}C$]-labeled succinate in *R. ruber* (WILLIAMS et al., 1994).

## 7.4 Determination of Composition by Chromatographic Methods

For routine analysis of PHA gas–liquid chromatography is now most widely used. This method is a good compromise because a large number of samples can be analyzed at reasonable costs, the identification and quantification of a large number of different hydroxyalkanoic acid constituents is possible, even in crude cell material or in activated sludge (APOSTILIDES and POTGIETER, 1981), and it is rather sensitive allowing the detection of even small amounts of PHA in cells. Gas chromatographic analysis was first introduced by BRAUNEGG et al. (1978). It is based on the conversion of the hydroxyalkanoic acid constituents of PHA into the corresponding methyl esters by acid or alkaline methanolysis. Recently, optimal conditions to quantitatively convert PHA into methyl esters by acid methanolysis have been investigated by HUIJBERTS et al. (1994b). Several modifications of this method such as ethanolysis (FINDLAY and WHITE, 1983), propanolysis (RIIS and MAI, 1981), the use of capillary columns (FINDLAY and WHITE, 1983; BRANDL et al., 1988) were applied later. It is also possible to analyze 4-hydroxyalkanoic acid methyl esters by this method (VALENTIN et al., 1992, 1994; STEINBÜCHEL et al., 1994). FINDLAY and WHITE (1983) coupled gas chromatographic separation of the ethyl esters of hydroxyalkanoic acids with mass spectrometric detection of the esters and its derivatives. GC-MS analysis of PHA was subsequently done in several laboratories. Analysis of cleavage products of PHA obtained from partial methanolysis or from partial pyrolysis by fast atom bombardment mass spectrometry allowed the determination of the distribution of comonomers in various copolyesters (BALLISTRERI et al., 1989, 1990, 1992).

## 7.5 Pyrolysis of PHA

Pyrolysis of PHA at controlled temperatures and under vacuum or under nitrogen gas has been described early (MORIKAWA and MARCHESSAULT, 1981; HAUTECOEUR et al., 1972). From poly(3HB) oligomers such as 3HB dimers or 3HB tetramers, unsaturated monomeric acids like crotonic acid or isocrotonic acid as well as carbon dioxide, propene, ketene, acetaldehyde or $\beta$-butyrolactone were released, depending on the conditions applied (GRASSIE et al., 1984). Thermal degradation of PHA and separation plus analysis of the pyrolysis products or its derivatives by gas chromatography or by GC-MS has been repeatedly done (e.g., HELLEUR, 1988).

## 7.6 Determination and Separation by HPLC

Oligomers of 3HB obtained from poly(3HB) by alkaline depolymerization or by methanolysis in the presence of BF3 were separated on a Lichrosorb RP8 column and quantified by UV detection (COULOMBE et al., 1978). After hydrolysis of poly(3HB) from *Rhizobium japonicum* in the presence of concentrated sulfuric acid the monomers and crotonic acid were separated on an Aminex HPX-87H column by ion exclusion high-pressure liquid chromatography and quantified by UV detection (KARR et al., 1983). A Micropack-MCH-N-Cap column was used to separate 3HB oligomers obtained by high-pressure liquid chromatography by BALLISTRERI et al. (1989).

# 8 Technical Production of PHA

Commercialization of PHA requires as a prerequisite production by industry at a large scale. Since approximately 15 years chemical companies in England and Austria try to establish the respective processes. Recently, Zeneca BioProducts (formerly ICI BioProducts) succeeded in establishing a process for fermentative production of PHA, and the company intends to scale up this process further (HOLMES, 1985; BYROM, 1987, 1990, 1992). Since some of these polyesters can be produced from renewable resources, production of PHA may offer additional ecological advantages as compared to thermoplastics and elastomers produced from fossil carbon sources (SCHLEGEL, 1992; EGGERSDORFER et al., 1992).

## 8.1 Fermentation Processes

Many bacteria have been screened for their suitability to produce PHA. However, there are only few to be used for biotechnological production of PHA at a large scale (Tab. 7). The suitability of a bacterium for PHA production depends on many different factors such as stability and safety of the organism, growth and accumulation rates, achievable cell densities and PHA contents, extractability of PHA, molecular weights of accumulated PHA, range of utilizable carbon sources, costs for the carbon source and the other components of the medium, and occurrence of byproducts (BYROM, 1992). The theoretical yield of PHA reflects the carbon source used for its production and also the metabolic pathways by which these carbon sources are assimilated. For the production of poly(3HB) theoretical yields between 0.32 and 1.16 $g\ g^{-1}$ were calculated for different carbon sources and for different metabolic backgrounds (YAMANE, 1992). The prices for technical grades of sucrose and methanol on the world market are most attractive at present (YAMANE, 1992); this makes these carbon sources the most likely candidates for technical production. For some bacteria promising processes have been developed in the last years. However, for those processes currrently used by industry relatively little detailed information is available for the public.

*Alcaligenes eutrophus*

In the 1960s *A. eutrophus* was already found to be a good producer of poly(3HB) (SCHLEGEL et al., 1961; WILDE, 1962). After

**Tab. 7.** Biotechnological Production of PHA

| Microorganism | Substrate | Polymer | Cell Density [g $L^{-1}$] | PHA Content [%] | Reference |
|---|---|---|---|---|---|
| *Alcaligenes eutrophus* | Glucose + propionate | Poly(3HB-*co*-3HV) | >100 | 70–80 | BYROM (1990) |
| *Alcaligenes latus* | Sucrose | Poly(3HB) | >100 | 75–80 | HÄNGGI (1990) |
| *Aureobasidium* sp. A-91 | Glucose + succinate corn steep liquor | Poly(β-L-malic acid) | – [a] | 61 g $L^{-1}$ [b] | NAGATA et al. (1993) |
| *Azotobacter vinelandii* UWD | Molasses + valerate | Poly(3HB-*co*-3HV) | 31 | 71.0 | PAGE et al. (1992) |
| *Chromobacterium violaceum* | Valerate | Poly(3HV) | 41 | 65–70 | STEINBÜCHEL and SCHMACK (in press) |
| *Escherichia coli* XL1-Blue (pSYL104) | LB medium + glucose | Poly(3HB) | 117 | 76.0 | KIM et al. (1992) LEE et al. (1994a) |
| *Escherichia coli* W (pSYL104) | Sucrose | Poly(3HB) | 125 | 27.5 | LEE and CHANG (1993) |
| *Klebsiella aerogenes* (pJM9131) | Molasses | Poly(3HB) | 38 | 70.0 | ZHANG et al. (1994) |
| *Protomonas extorquens* | Methanol | Poly(3HB) | 190 | 60.0 | SUZUKI et al. (1986c) |
| *Pseudomonas* strain K | Methanol | Poly(3HB) | 233 | 63.9 | SUZUKI et al. (1986b) |
| *Pseudomonas oleovorans* | Octane | Poly(3HO-*co*-3HHx) | 37 | 30–35 | PREUSTING et al. (1993c) |

Abbreviations and symbols: 3HB, 3-hydroxybutyric acid; 3HV, 3-hydroxyvaleric acid; 3HHx, 3-hydroxyhexanoic acid; 3HO, 3-hydroxyoctanoic acid;
[a] data not provided;
[b] the value provides the content of poly(β-L-malic acid) in the fermentation broth in g $L^{-1}$, since the polyester is excreted into the medium

having screened a large number of candidates, Zeneca BioProducts selected *A. eutrophus* as the production organism (BYROM, 1992). In the production plant of Zeneca BioProducts at Billigham a copolyester of 3-hydroxybutyric and 3-hydroxyvaleric acid, poly(3HB-*co*-3HV), is produced biotechnologically with a capacity of approximately 300 t per year. The process utilizes a mutant of *A. eutrophus*, and the fermentation is done in a mineral salt medium, with glucose and propionic acid as the sole carbon sources. The mutant was derived from strain H16 and is able to utilize glucose as a carbon source for growth (KÖNIG et al., 1969; SCHLEGEL and GOTTSCHALK, 1965). In addition, the mutant utilizes propionic acid much more efficiently than the wild type. The process is run as a two-step fed-batch fermentation process. During the first step, the cells are grown to high cell densities, and glucose is the only carbon source provided. During the second step the phosphate source is limited and glucose plus propionic acid are fed. Each phase lasts for approximately 48 h. Finally, cell densities higher than 100 g cell dry matter per L are obtained, and poly(3HB-*co*-3HV) contributes to approximately 60–70% of the cellular dry matter (BYROM, 1990). A non-solvent process was developed by Zeneca to recover the polyester from the cells (see below).

*Alcaligenes latus*

A different industrial process for the production of PHA has been developed at btF (Biotechnologische Forschungsgesellschaft) in Linz (Austria). This process employs strain btF-96 of *A. latus* DSM 1124 (HRABAK, 1992; HÄNGGI, 1990). *A. latus* is a nitrogen fixing bacterium that accumulates the homopolyester poly(3HB) during normal cell growth up to 80% of the cellular dry matter. Since accumulation of poly(3HB) is growth-associated, production of poly(3HB) is achieved in a one-step fed-batch fermentation process. Biomasses as high as 60 g $L^{-1}$ are easily achieved. Production is done in a 15 $m^3$ fermenter in a mineral salt medium with sucrose as the sole carbon source. The pH value is controlled by the addition of ammonia, that also provides the nitrogen source. The capacity of the plant exceeds 1 t of polyester per week (HRABAK, 1992).

The cells are harvested by centrifugation and washed with tap water. A single extraction step and a single precipitation step are employed to recover the polyester from the wet cells. Poly(3HB) is extracted with methylene chloride from the suspended cells. The solvent containing the polyester is decanted from the cell debris, and the polyester is precipitated by the addition of water. After drying of the precipitated polymer a polyester with a purity of 99% is recovered (HRABAK, 1992). The process also includes the recovery of the solvent. In 1993, however, the company stopped the production of poly(3HB), and the process was obviously not further developed.

*Azotobacter vinelandii*

*Azotobacter* species were the first organisms grown by ICI for PHA production. However, due to the production of large amounts of exopolysaccharides, these bacteria were not considered further (BYROM, 1992). An additional, very promising process that is being developed for the production of poly(3HB), employs the mutant UWD of the nitrogen fixing bacterium *A. vinelandii* (PAGE, 1992b). Mutant UWD was isolated from frequently occurring sectored colonies of rifampicin resistant colonies of transformants of *A. vinelandii* obtained with *A. vinelandii* 113 DNA. The distinct white rays within the pinkish colonies resulted from the formation of abundant amounts of poly(3HB) in the cells (PAGE and KNOSP, 1989). Whereas poly(3HB) synthesis was dependent on $O_2$ limitation and amounted up to 25% of the cellular dry matter in the wild type, synthesis of the polyester occurred in the mutant UWD during unrestricted growth in the exponential growth phase. This mutant accumulated poly(3HB) up to 65 or 75% of the cellular dry matter depending on whether ammonium or $N_2$ was used as nitrogen source (PAGE and KNOSP, 1989; MANCHAK and PAGE, 1994). Oxygen deficiency, however, still stimulated the conversion of the carbon source into the polyester from 0.25–0.33 mg poly(3HB) per mg glucose consumed. This is related to the observation made in the laboratory of E. A. DAWES that in its close relative *A. beijerinckii* oxygen deficiency induces poly(3HB) synthesis better than nitrogen deficiency (SENIOR and DAWES, 1973; WARD et al., 1977).

In mutant UWD, the activity of NADH-oxidase amounted to only 6% of the activity in the parent strain (PAGE and KNOSP, 1989). Later, PAGE and collaborators demonstrated that poly(3HB) was also synthesized from sucrose. The authors showed that the yield of poly(3HB) formation was comparable or even better than with refined carbon sources in complex unrefined media such as beet molasses or malt extract (PAGE, 1990, 1992a; PAGE et al., 1992). The addition of relatively small amounts ($\leq$0.2%) of complex nitrogen sources such as fish peptone, proteose peptone or yeast extract further promoted the amount of poly(3HB) produced (PAGE and CORNISH, 1993). Upon addition of 0.1% of fish peptone, a yield of 0.65 g poly(3HB) per g glucose was obtained, and the polyester was deposited in large, pleomorphic and osmotically sensitive cells. Later it was shown that mutant UWD was also able to synthesize poly(3HB-*co*-3HV), and that 3HV amounted to a molar fraction of up to a maximum of 52% if valeric acid or heptanoic acid were provided as precursor substrates in addition to glucose (PAGE et al., 1992; PAGE, 1992b). In fed-batch beet molasses cultures with 10–26 mM valeric acid up to 31 g $L^{-1}$ cell mass containing up to 71% poly(3HB-*co*-3HV) were obtained with yields of up to 0.38 g PHA per g glucose equivalents.

Whether this process will finally be of commercial relevance and whether it will be superior to the presently established process for producing PHA by *A. eutrophus* (see above) or not, will depend on the improvement of the process allowing the cells of *A. vinelandii* to grow at much higher densities. An increase of the cell densities by almost one order of magnitude will be required.

### *Pseudomonas oleovorans*

WITHOLT and collaborators developed two-liquid phase batch, fed-batch and continuous fermentation processes to produce poly($HA_{MCL}$) from octane as carbon and energy source (WITHOLT et al., 1990; PREUSTING et al., 1991). They demonstrated that efficient poly($HA_{MCL}$) production depended on an optimum supply of the cells with oxygen. Taking this into account by optimizing the air flow and the stirrer speed and by decreasing the temperature from 30 °C to 18 °C, cell densities as high as 35 g $L^{-1}$ were obtained (PREUSTING et al., 1993a). In a two-phase fed-batch process cell densities as high as 37.1 g $L^{-1}$ with a polymer content of 33% were obtained within 38 h with a feed of the limiting nutrient ammonium. The volumetric productivity was 0.25 g PHA per L per h (PREUSTING et al., 1993a).

During continuous fermentation in a two-phase medium with limiting amounts of ammonium and at a steady state cell density of 2–3 g $L^{-1}$, PHA could be maximally produced at a dilution rate of 0.20 $h^{-1}$ with a volumetric productivity of 0.17 g $L^{-1}$ $h^{-1}$ PHA. When doubling the concentration of the limiting nutrient in the medium feed the steady state cell density and the volumetric productivity were raised to 5.4 g $L^{-1}$ and 0.28 g $L^{-1}$ $h^{-1}$ PHA, respectively. By a subsequently increased air flow both values were raised to 11.6 g cells per L and 0.58 g PHA per L and h, respectively (PREUSTING et al., 1993b).

### *Chromobacterium violaceum*

A recent study described the formation of poly(3HV) homopolyester by strains of *C. violaceum*. This purple-colored bacterium accumulated the homopolyester during cultivation on valeric acid as sole carbon source (STEINBÜCHEL et al., 1993). Meanwhile, a fed-batch process has been developed that allows the cultivation of *C. violaceum* DSM 30191 to densities as high as 41 g $L^{-1}$ cellular dry matter and that provided cells consisting of approximately 65% poly(3HV). Some kilograms of this homopolyester could be obtained from a 250 L bioreactor allowing the manufacturing of mold extruded materials and of fibers (STEINBÜCHEL and SCHMACK, in press). Although some other bacteria synthesized poly(3HV) too, these bacteria were not suitable for mass production of cells and/or of PHA for several reasons (STEINBÜCHEL et al., 1993, and references cited therein).

Methylotrophic bacteria

Deposition of large amounts of poly(3HB) in methylotrophic bacteria was first described in *Methylomonas methanica* (KALLIO and HARRINGTON, 1960). A large number of different methylotrophic bacteria was screened for their capability to produce poly(3HB) by WHITTENBURY et al., (1970), and some of these bacteria such as *Methylocystis parvus* (ASENJO and SUK, 1986) and *Methylosinus trichosporium* (WEAVER et al., 1975; THOMPSON et al., 1976) were investigated in some more detail regarding the accumulation of this polyester. These studies showed that only methylotrophic bacteria using the Calvin-Benson-Bassham pathway or the serine pathway for the assimilation of C1 compounds are able to synthesize and accumulate PHA (BABEL, 1992). Obligately methanol assimilating bacteria based on the ribulose monophosphate (RMP) pathway are obviously unable to accumulate PHA (BABEL, 1992; TROTSENKO et al., 1992). This was confirmed by showing that four representatives of the RMP bacteria, *Methylophilus methylotrophus*, two strains of *Methylobacillus glycogenes*, and *Acetobacter methanolicus*, failed to accumulate even traces of poly(3HB), lacked active poly(3HB) synthase as well as the genetic information for the formation of PHA synthase, as revealed by enzymatic studies and hybridization experiments (FÖLLNER et al., 1993). However, recombinant strains of these bacteria harboring and expressing the PHA-biosynthesis genes of *A. eutrophus* (SCHUBERT et al., 1988) or *C. vinosum* (LIEBERGESELL and STEINBÜCHEL, 1992) were able to synthesize poly(3HB) from methanol (FÖLLNER et al., 1993). This clearly demonstrated that physiological constraints do obviously not prevent these bacteria from synthesizing and accumulating poly(3HB).

Since methylotrophic bacteria of the genus *Methylobacterium* were seriously considered for the industrial production of single cell protein, much experience in methanol fermentation technology is already available. Since methanol is a cheap substrate (YAMANE, 1992), these bacteria were promising candidates for the production of poly(3HB) (BYROM, 1992). However, due to difficulties in the extraction of the polyesters from the cells, due to the low molecular weight of the accumulated polyester, and due to other problems *Methylobacterium* was not considered further as a candidate for production by ICI (BYROM, 1992). As a result of the screening of a large number of methylotrophic bacteria for the production of poly(3HB) from methanol, YAMANE and collaborators detected *Pseudomonas* sp. K. as a suitable candidate (SUZUKI et al., 1986a). *Pseudomonas* sp. K. could be cultivated in a fed-batch process with nitrogen deficiency to cell densities as high as 233 g $L^{-1}$ cell dry matter and with a polymer content of 64%; the yield was 0.20 g $g^{-1}$ poly(3HB) methanol (SUZUKI et al., 1986b). *Protomonas extorquens* could also be grown to cell densities as high as 190 g $L^{-1}$ cellular dry matter with a maximum poly(3HB) content of 60% (SUZUKI et al., 1986c).

Recently, it has been shown that *M. rhodesianum* and *M. extorquens* are also able to synthesize poly(3HB-*co*-3HV) if propionic acid or valeric acid are provided as a precursor substrate in addition to methanol (BABEL, 1992). YAMANE and collaborators achieved synthesis of poly(3HB-*co*-3HV) with a high 3HV content from a mixture of methanol plus *n*-amyl alcohol with various methylotrophic bacteria (UEDA et al., 1992a). *Paracoccus denitrificans* and *M. extorquens* seemed to be the most suitable candidates; the poly(3HB-*co*-3HV) consisted of 91.5 or 38.2 mol% 3HV, respectively. The conversion of *n*-amyl alcohol to 3HV initiated by the oxidation of *n*-amyl alcohol by methanol dehydrogenase was very efficient. The yield of 0.968 g PHA per g *n*-amyl alcohol obtained with *P. denitrificans* was very close to the theoretical yield of 1.14 g $g^{-1}$ amyl alcohol calculated on the basis of chemical stoichiometry (UEDA et al., 1992b).

Recombinant *E. coli*

At the very beginning of the analysis of poly(3HB) biosynthesis genes it was detected that recombinant strains of *E. coli* expressing the PHA operon of *A. eutrophus* H16 (*pha-CAB*$_{Ae}$), accumulated poly(3HB) if the cells were cultivated in the presence of an excess of glucose (SLATER et al., 1988; SCHUBERT et al., 1988; PEOPLES and SINSKEY, 1989c). Formation of poly(3HB) occurred only when all three PHA biosynthesis genes of *A. eutrophus* were expressed. Expression of the PHA synthase only did not confer poly(3HB) biosynthesis to *E. coli*. Biosynthesis and accumulation of poly(3HB) depended strongly on the coexpression of $\beta$-ketothiolase and NADPH-linked acetoacetyl-CoA reductase. Meanwhile it could also be demonstrated for PHA synthases from other sources that a functional active poly(3HB) synthesis pathway is only established in recombinant *E. coli* if $\beta$-ketothiolase plus acetoacetyl-CoA reductase are expressed, such as in strains harboring the PHA locus of *C. vinosum* (LIEBERGESELL and STEINBÜCHEL, 1992). Genomic fragments, that did not comprise the structural genes for a $\beta$-ketothiolase plus acetoacetyl-CoA reductase in addition to the PHA synthase structural gene, such as those of *Thiocystis violacea* (LIEBERGESELL and STEINBÜCHEL, 1993), *Ectothiorhodospira shaposnikovii*, *Lamprocystis roseopersicina* and *Thiocapsa pfennigii* (LIEBERGESELL et al., 1993), *Rhodococcus ruber* (PIEPER and STEINBÜCHEL, 1992), *Methylobacterium extorquens* (VALENTIN and STEINBÜCHEL, 1993), or *Rhodobacter sphaeroides* and *Rhodospirillum rubrum* (HUSTEDE et al., 1992), or that which harbored the poly(HA$_{MCL}$) synthases of various pseudomonads belonging to the rRNA homology group I (HUISMAN et al., 1991; TIMM and STEINBÜCHEL, 1992; TIMM et al., 1994), did not confer the ability to synthesize and accumulate poly(3HB) or other PHA from carbohydrates or from fatty acids to *E. coli*.

DENNIS and collaborators constructed plasmid p4A, which harbored the *A. eutrophus phaCAB*$_{Ae}$ operon approximately 400 bp downstream of and colinear to the *lac* promotor, and they introduced this plasmid into various strains of *E. coli* (JANES et al., 1990). If recombinant strains of *E. coli* DH5α harboring this plasmid were cultivated in complex Luria Bertani broth or M9 mineral salts medium, each supplemented with 1% glucose, the cells accumulated poly(3HB) up to approximately 85% of the cellular dry matter and with a yield of approximately 0.5 g poly(3HB) per g glucose. Cells of *E. coli* HMS174 with plasmid p4A accumulated poly(3HB) even up to more than 90% if when cultivated in a mineral salts medium supplemented with whey. With M9 mineral salts medium plus 8% whey as much as 11.7 g $L^{-1}$ poly(3HB) were obtained (JANES et al., 1990).

DENNIS and collaborators extended their studies on the production of PHA in *E. coli*, and they accomplished production of poly(3HB-*co*-3HV) in a recombinant *E. coli fadR atoC* (Con) mutant harboring p4A or a derivative plasmid in M9 mineral salts medium, that were supplemented with glucose plus propionic acid. 3HV contributed up to 39 mol% to the accumulated polyester (SLATER et al., 1992). This mutant expressed the enzymes required for the catabolism of fatty acids of short chain length constitutively. A *fadR* mutant of *Klebsiella oxytoca* was also able to produce poly(3HB-*co*-3HV) from propionic acid, the molar fraction of 3HV was as high as 59%. The authors also introduced the *A. eutrophus* PHA biosynthesis genes into sucrose utilizing strains of *E. coli*, *K. aerogenes* and *K. oxytoca*. In a 10 L fed-batch culture, they were able to obtain approximately 38 g $L^{-1}$ of cells containing 70% poly(3HB) from molasses as the sole carbon source (ZHANG et al., 1994) within 32 h. Recently, production of poly(3HB) in a fed-batch process with a recombinant strain of *E. coli* XL1-Blue expressing the *A. eutrophus* PHA biosynthesis genes from pBluescript KS$^-$-SE52 using LB medium supplemented with glucose and with glucose feeding has been investigated. The strain could be cultivated to cell densities as high as 117 g $L^{-1}$ containing approximately 76% poly(3HB) within 42 h. Therefore, 88 g $L^{-1}$ poly(3HB) were produced, and the productivity was as high as 2.1 g $L^{-1}$ $h^{-1}$ poly(3HB) (KIM et al., 1992). With a recombinant strain of *E. coli* W expressing the PHA biosynthesis genes of *A. eutrophus* from pSYL104, which is a derivative of plasmid pBluescript KS$^-$-SE52 harboring the *parB* locus of plasmid R1, cell densities as high as 124.6 g $L^{-1}$ were obtained in a fed-batch culture from sucrose as the sole carbon source within 48 h. However, poly(3HB) only contributed 27.5% to the cellular dry matter (LEE and CHANG, 1993; LEE et al., 1994b).

## 8.2 Production in Plants

The availibility of PHA biosynthesis genes from bacteria initiated research on the transfer and expression of these genes in plants in several laboratories. For the first time, poly(3HB) biosynthesis was successfully conferred to *Arabidopsis thaliana* (Fig. 17). However, the transgenic plant expressing the *A. eutrophus* $phaC_{Ae}$ and $phaB_{Ae}$ genes accumulated only small amounts of the polyester in the cytoplasm, in the nucleus and in the vacuole (POIRIER et al., 1992a). Since then the production of PHA in plants appears to be a promising approach for the large-scale production of the polyesters, and many other plants such as oilseed rape are now also being considered for the expression of PHA biosynthesis genes (e.g., SMITH et al., 1994). In the future, it will be necessary to efficiently divert the carbon flow from starch, sucrose or lipid synthesis to PHA biosynthesis and express the genes in the appropriate compartment of the cell (POIRIER et al., 1992b). Recently, it has been shown that transgenic *A. thaliana* expressing the PHA biosynthesis pathway in chloroplasts by targeting all three *A. eutrophus* enzymes upon modification of the enzymes by the addition of the pea chloroplast transit peptide, accumulated poly(3HB) up to 14% of the dry weight of the plant (NAWRATH et al., 1994).

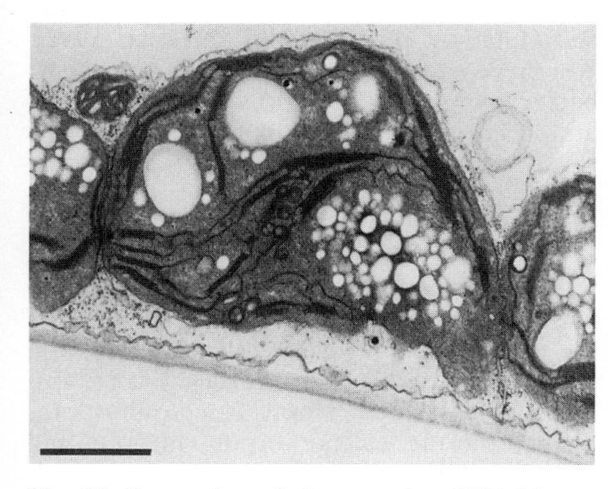

**Fig. 17.** Expression of *A. eutrophus* PHA-biosynthesis genes in *Arabidopsis thaliana.* Transmission electron micrograph of a thin section from a poly(3HB) producing transgenic cell of *A. thaliana* expressing $\beta$-ketothiolase, NADPH-dependent acetoacetyl-CoA reductase and PHA synthase from *A. eutrophus* and showing agglomeration of poly(3HB) granules in the chloroplasts (bar = 1 μm; original micrograph by Y. POIRIER).

## 8.3 Recovery Processes for PHA

Several methods are currently used to isolate PHA from the bacterial cells and to separate the released polyester from the other cell constituents. Routine laboratory processes have to be distinguished from processes established for technical processes. Together with reliable processes for large-scale fermentation or for *in vitro* production of PHA employing purified PHA synthases, this step is crucial with respect to the time and the cost required for the whole production process. Large-scale processes for the isolation of PHA from transgenic plants have not been established yet, since suitable plant material was not available. However, solvent extraction was successfully used to isolate poly(3HB) at a small scale from transgenic *A. thaliana* that had accumulated poly(3HB) (POIRIER et al., 1992a; NAWRATH et al., 1994).

Most processes for the isolation of PHA require the separation of bacterial biomass from the fermentation broth. In most cases, this is achieved either by centrifugation or by flocculation. Subsequently, the remaining water is removed from the cells either by lyophilization or by spray-drying, and PHA is selectively solubilized from the biomass by solvent extraction. This method is used at a technical scale for large batches as well as at a laboratory scale for small samples. For the latter, solvent extraction offers the advantage of being applicable to many different microorganisms, although PHA are usually extracted much faster from cells of gram-negative than of gram-positive bacteria. This supports, e.g., the practicability of extended screening programs under comparable conditions. One disadvantage of solvent extraction is that large amounts of solvents are required since PHA solutions became rather viscous even at low concentrations of the polyester.

Three different types of solvents are used for the extraction of PHA from the biomass. (1) Most widespread are chlorinated hydrocarbons, such as chloroform, methylene chloride or 1,2-dichloroethane in which all biosynthetic PHA are soluble, except polymalic acid (WALKER et al., 1982; VAN LAUTEM and GILAIN, 1986; BARHAM and SELWOOD, 1982; HOLMES et al., 1981). The btF process for the production of poly(3HB) by *A. latus* involves, e.g., extraction with methylene chloride. BERGER (1990) investigated the pretreatment of the cells in order to make the polyester more accessible to the solvent. This pretreatment was done either by washing the cells in a hydrophilic solvent like acetone or methanol or by physical treatment of the cells such as grinding, wet-milling, French Press treatment or freezing and thawing cycles. RAMSAY (1994) showed that pretreatment with acetone increased the purity to approximately 96%, whereas it was only 90–93% without pretreatment; in addition, the recovery of the polyester increased from 50 or 55% to 69 or 70%. (2) Cyclic carbonates such as ethylene carbonate or 1,2-propylene carbonate were also applied. These solvents offer the advantage of PHA solubility at rather high concentrations and at elevated temperatures, and solubility is reduced at lower temperatures (LAFFERTY and HEINZLE, 1978). (3) Finally, azeotropic mixtures of, e.g., 1,1,2-trichloroethane with water or chloroform with methanol, ethanol, acetone or hexane were also used (VAN LAUTEM and GILAIN, 1986; STAGEMAN, 1985).

PAGE and collaborators developed a simple procedure for the release of poly(3HB) from *A. vinelandii* that utilizes the fragility of the cells grown on a glucose medium supplemented with fish peptone (PAGE and CORNISH, 1993). Treatment of the cells suspended in distilled water and stirred with 1 N aqueous $NH_3$ at pH 11.4 and at 45 °C for 10 min and sedimentation of the polyester by centrifugation yielded a final product consisting of 94% poly(3HB), 2% protein, and 4% nonprotein biomass (PAGE and CORNISH, 1993). This procedure was also possible with initial cell masses as high as 200 g $L^{-1}$, and the extract could be recycled and used as a nitrogen source for subsequent fermentations, although it could not fully substitute fish peptone (PAGE and CORNISH, 1993).

Other processes for the release of PHA from cells are biological measures such as expression of protein E-mediated lysis at the end of the accumulation phase (KALOUSEK et al., 1990). However, successful applications of these processes depended on the complete repression of the formation of this protein during the growth and accumulation phases, which seemed to be a problem. FIDLER and DENNIS (1992) described a system in which a bacteriophage lysis gene under the control of a temperature-dependent promoter was derepressed by heating the culture. The lysis protein was synthesized and the cells lyzed; however, lysis occurred only during the exponential growth phase before maximal poly(3HB) accumulation (FIDLER and DENNIS, 1992). Another system utilized the lysozyme gene of bacteriophage T7, which was cloned on a separate plasmid in a recombinant *E. coli*. This lysozyme was also expressed during the growth phase, but it could penetrate the cytoplasmic membrane and disrupt the integrity of the peptidoglycan layer only after the cells were treated with the chelating agent ethylene diamine tetraacetic acid at the end of the accumulation phase. Subsequently, the poly(3HB) granules were released at an efficiency of 99%, and empty *E. coli* ghosts remained in the culture broth (FIDLER and DENNIS, 1992).

Enzymatic treatment of bacterial cells has been used for a long time to release intact PHA granules for studying PHA synthase or PHA depolymerase activity. Most often, only lysozyme or lysozyme in combination with other enzymes was used (GRIEBEL et al., 1968; LIEBERGESELL et al., 1993). With isolated granules from cells of *Chromatium vino-*

*sum* it was demonstrated that lysozyme exhibits an extraordinarily high affinity to the granules (LIEBERGESELL et al., 1994). For the commercial production of PHA by *A. eutrophus* a process has been described in which the cells were first exposed to a temperature of 80 °C and subsequently treated with a cocktail of various hydrolytic enzymes consisting of lysozyme, phospholipase, lecithinase, the proteinase alcalase and others (HOLMES and LIM, 1985). Most of the cellular components were hydrolyzed by these enzymes, whereas PHA remained intact. Zeneca use such a non-solvent process for the isolation of the Biopol material (BYROM, 1990).

Chemical methods are also suitable for the solubilization of non-PHA components of cells. Sodium hypochlorite has been most widely used for this purpose (WILLIAMSON and WILKINSON, 1958). However, this strong oxidizing agent has to be applied carefully, since it also partially hydrolyzed poly(3HB) if used at a too high concentration, at a rather alkaline pH, or for an extended period of time (BERGER et al., 1989). Only if this was taken into account, the relative high molecular masses of the polyesters were retained. The exposure time of the granules to sodium hypochlorite could be reduced by pretreating the biomass with a surfactant (RAMSAY et al., 1990). Many low molecular weights of bacterial polyesters that have been reported in literature most probably resulted from the non-appropriate application of sodium hypochlorite.

There are of course many physical methods to disrupt the bacterial cell and to release the PHA granules. These methods are widely established in microbiological laboratories and are usually employed at a small scale (SCHNAITMAN, 1981). ICI developed a special process for the release of PHA from *A. eutrophus* at the beginning of the 1980s. The biomass was heated in an autoclave to 220 °C under nitrogen and was injected into a tank of cold water through a fire-jet resulting in cell brackage and in the release of the granules (HOLMES and JONES, 1982). However, at present this process or other physical methods are not applied for the large-scale production of PHA. The reasons may be the high equipment cost and the circumstance that not very much of the other cell material will be removed.

# 9 Applications for PHA

PHA share several properties (see Tab. 6) that make them attractive candidates for various applications. Since, e.g., poly(3HB) and poly(3HB-*co*-3HV) can be processed on conventional equipment for polyolefin or other plastics for, e.g., injection molding, extrusion blow molding or fiber-spray gun molding (Zeneca, 1993), PHA can be manufactured to many different materials for a wide range of applications (HOLMES, 1985). Meanwhile, and since the time when the review on microbial polyesters for the first edition of this handbook of biotechnology was written (LAFFERTY et al., 1988), technical applications do not only exist in theory. First products came onto the market, several new applications were already tested, and many new potential applications are now being considered. Tab. 8 provides a short summary of applications.

## 9.1 Packaging Industry

Poly(3HB) and poly(3HB-*co*-3HV) are most well known to the public as potential substitutes of conventional, non-biodegradable plastics used for packaging (Tab. 8). In May 1990, the German hair care company "Wella" launched the first commercial product of PHA. It was a shampoo bottle manufactured from Biopol material produced by Zeneca. These shampoo bottles were available for some time on a limited market in Germany. Meanwhile, bottles for other cosmetics also came onto the market in Japan, and many other products were manufactured (Fig. 18).

However, with these polyesters not only containers or bags can be produced. They can be also used as coatings for, e.g., paper cups or paper trays for food service or food packaging. The PHA component provides a material that protects the paper or cardboard

**Tab. 8.** Applications for PHB and Other PHA

**Packaging Industry**
- Bottles
- Shopping bags
- Films
- Packaging flakes
- Six pack rings
- Paper coating

**Agriculture**
- Controlled-release coating for, e.g., pesticides
- Mulch films
- Irrigation tubes

**Fishing Industry**
- Fishing nets

**Medicine**
- Controlled drug release (e.g., hormones, cytostatica)
- Surgical sutures
- Surgical swabs
- Osteosynthesis materials (e.g., bone plates)
- Wound dressings
- Lubricant for surgeon's gloves
- Vascular grafts
- Blood vessels
- Disposable syringes
- Blister packs and strips
- Pill trays
- Infusion (monomeric 3HB)

**Foodstuff Industry**
- Flavor delivery systems
- Emulsifiers

**Tobacco Industry**
- Cigarette filter tips

**Chemical Industry**
- Synthesis of EPC
- Raw materials for paints

**Municipal Water Purification Plant**
- Denitrification

**Others**
- Golf tees
- Disposable rasors
- Diapers
- Geotextiles
- Toner for xerography
- Developer for photography
- Flower pots
- Pressure sensors
- Adhesives
- Separation media for enantiomers
- Disposable hygiene articles

against damages caused by the moisture of the packaged food or by the environment. Moisture resistance is achieved either by combining the hydrophilic material with a laminate of the polyester or with a latex suspension of PHA granules (LAUZIER et al., 1993; MARCHESSAULT et al., 1993). Biopol is currently under evaluation by the European Union and by the U.S. FDA for use in food contact applications (Zeneca, 1993).

## 9.2 Applications in Medicine and Pharmacy

Since the hydrolysis product of poly(3HB) belongs to the keton bodies occurring in the serum, which are normal intermediates of the metabolism of animals and humans (LEHNINGER, 1975), it was not assumed that this polyester is highly toxic to animal tissues. Also the polyester itself most probably will not cause any problems since poly(3HB) was found at low concentrations in the human serum (REUSCH and SPARROW, 1992). Therefore, a wide range of medical and pharmaceutical applications of PHA was proposed. The applications of PHA may avoid secondary surgeries or may allow the long-term release of pharmaceuticals. Some applications, such as the use of poly(3HB) for making absorbable prosthetic devices and surgical sutures, were already proposed at a very early stage (GRACE, 1966).

Poly(3HB) is slowly resorbable and highly compatible within the human body offering the potential as a biomaterial, and it is already being used to produce non-woven patches for pericardium repair following open-heart surgery. Other medical applications are being explored. The monomer (3HB) is a better nutrient than glucose and can be applied as an intraveneos supply during surgery.

Due to its biodegradability, PHA were considered as matrix material for the preparation of retard materials. This has been widely described in Volume 6b the First Edition of *"Biotechnology"* (LAFFERTY et al., 1988). Retard materials prepared from poly(3HB) containing the hormon analogon Buserelin were successfully applied to control ovulation in fe-

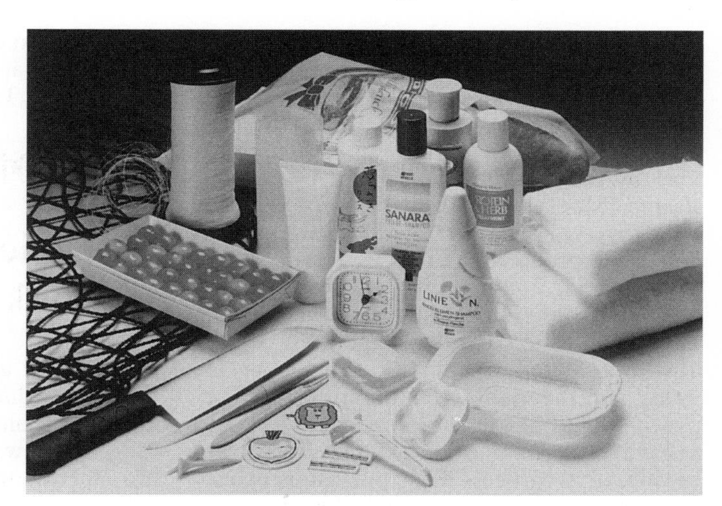


**Fig. 18.** Products manufactured from Biopol (original photograph by Zeneca Bioproducts).

male monkeys (FRASER et al., 1989). Whereas tablets were used in the latter application, microcapsules containing the cytostaticum aclarubicin for treatment of prostata carcinoma were also tested (KUBOTA et al., 1988). Feasibility of using PHA as a retard material will depend widely on the biodegradability and on the biocompatibility of the respective polyester. This has been studied in detail only with poly(3HB) and to some extent also with poly(3HB-*co*-3HV). However, the other new biosynthetic PHA described only recently have not been investigated for their use as retard materials. Since it has been shown that some chemosynthetic PHA were susceptible to hydrolytic cleavage by esterases and lipases (see above; MUKAI et al., 1993), the biodegradability of the new PHA in animal and human tissues should be investigated.

## 9.3 Applications in Agriculture

Several applications of PHA such as the use as mulch films, one-season irrigation tubes and retard materials for the controlled release of pesticides, herbicides or fertilizers are considered in agriculture. These applications may be quite interesting since they do not demand a highly pure polyester material as it is required for applications in medicine or food packaging.

## 9.4 Miscellaneous

Besides applications in packaging industry, agriculture, medicine or pharmacy many other applications of PHA are currently discussed. These applications range from rather different areas such as foodstuff industry, tobacco industry, municipal water treatment, xerography or horticulture etc. (see Tab. 8). Even marine fishing and seaweed nets have been manufactured from Biopol, that will, if accidentally lost at sea, sink to the seabed and finally fully biodegrade, minimizing the detrimental effects on marine life.

Furthermore, PHA may be an important source for the chemical industry to synthesize enantiomeric pure chemicals since most of the polyesters consist of chiral building blocks, as outlined above (SEEBACH, 1992; MÜLLER and SEEBACH, 1993, and references cited therein).

## 9.5 Combination with Other Materials and Derivatization

Due to the formation of hydrogen bonds by the ester group of PHA, these polyesters can be combined with other chemosynthetic polymers or with starch and cellulose. Often modification occurs when blends consisting of PHA and another polymer or chemical are heated. Poly(3HB) and polychloroprene were for example crosslinked by a radicalic reaction while heated (SIAPKAS, 1976). Recently, crosslinking of poly(3HB) and polycaprolactone by transesterification has been demonstrated (McCARTHY et al., 1993).

Poly($HA_{MCL}$) exhibit a rather low crystallinity of approximately only 25% (MARCHESSAULT et al., 1984). Furthermore, it crystallizes at a rather low rate (GAGNON et al., 1992). The glass transition temperature is below room temperature (PREUSTING et al., 1990), and the melting temperatures vary between 39 and 61 °C. Therefore, poly($HA_{MCL}$) behave as thermoplastic elastomers, but it will be difficult to process these polyesters by certain techniques, since they will lose their coherence and will soften at temperatures as low as 40 °C. Into poly($HA_{MCL}$), isolated from cells of *P. oleovorans* grown on a mixture of octane plus octene and therefore consisting of 3HHx, 3HO, 3-hydroxyhexenoic acid plus 3-hydroxyoctenoic acid, chemical crosslinks between the unsaturated *R*-pendant groups were introduced by electron beam irradiation. The resulting polymer represented a true rubber with constant properties over a temperature range from −20 °C to +170 °C (DE KONING et al., 1994). The biodegradability of this cross-linked polyester could be demonstrated.

# 10 Patent Applications

Considering only poly(3HB) and according to MÜLLER and SEEBACH (1993) an increasing number of patent applications have been made during the last years: 1980 (3 applications), 1981 (1), 1982 (7), 1983 (5), 1984 (5), 1985 (9), 1986 (10), 1987 (7), 1988 (20), 1989 (13), 1990 (14) and 1991 (29); applications for other PHA are not considered. There are many patent applications to protect biotechnological production processes of various PHA, the isolation of the polyesters from the cells, and their application in quite different areas. Many of these applications have been described in Sect. 9 and will, therefore, not be mentioned in detail again. Instead, some relevant patent applications known to the author are listed in Tab. 9.

# 11 Conclusions and Outlook

Basic and applied research during the last few years has considerably increased our knowledge on PHA and has provided a wide range of new biosynthetic polyesters. There are still many open academic questions regarding the biochemistry of the key enzyme PHA synthase and on the pathways resulting in the biosynthesis of PHA from unrelated substrates, which will keep the scientists busy for a long period. The results of biochemical and molecular genetic studies provide a solid basis to approach the establishment of transgenic plants suitable for the production of one or the other PHA, utilizing directly the radiation energy and the atmospheric carbon dioxide.

It is still very likely, that many novel constituents of biosynthetic PHA will be discovered. In the future, no clear border will exist between biosynthetic and chemosynthetic polyesters since many polyesters, that were thought to be accessible only by chemical synthesis, might be also accessible by biotechnical production. On the other side, for many biosynthetic polyesters more and more effective and reliable chemosynthetic processes will become available. Furthermore, research on the microbiology and biochemistry of biodegradation of these polyesters will contribute to understand and eventually effect biodegradation of PHA of both origins. Therefore, the struggle for biosynthesis and chemosynthesis will become very competitive although both will synergistically profit from each other. Whether or not PHA will have an economic future will mainly depend on the costs required for production. To some extent this will also depend on the decision on the waste management strategy used by the respective society.

**Tab. 9.** Relevant Patents for Production and Application of Polyhydroxyalkanoic Acids

| Main Subject | Patent Application | Assigned to | Year |
|---|---|---|---|
| *Production of PHA* | | | |
| Poly(3HB) production | EP 149744 | PCD | (1944) |
| Poly(3HB) and thermoplastic applications | US 3.036.959 | GRACE | (1962) |
| PHA | US 3,275,610 | Mobil Oil Corporation | (1966) |
| Poly(3HB) production | DE 2733203-A1 | Agroferm AG | (1978) |
| Propionate feeding | EP 0052459 | ICI | (1980) |
| Poly(3HB) production with *Alcaligenes* | EP 46344-A1 | ICI Plc | (1982) |
| Poly(3HB) production | EP 144017 | PCD | (1984) |
| Poly(3HB) production with *A. latus* | EP 144017-A1 | Chemie Linz AG | (1985) |
| Poly(3HB) production | JP 06251889-A2 | Mitsubishi Rayon Co. | (1985) |
| Poly(3HB-*co*-3HV) production from alcohols | EP 204442-A2 | ICI | (1986) |
| Continuous production of poly(3HB) | AT 379613-B1 | Chemie Linz AG | (1986) |
| Poly(3HB) production with pre-extraction solvent | DD 266116-A1 | AdW DDR | (1989) |
| Poly(3HB) – influence of phosphate | DD 266584-A1 | AdW DDR | (1989) |
| Poly(3HB-*co*-3HV) from bacteria | EP 288908-A2 | Mitsubishi Gas | (1988) |
| Poly(3HB-*co*-3HV) production | EP 396289-A1 | ICI | (1990) |
| Poly(3HB) production during log-phase by *A. vinelandii* | WO 9005190-A1 | University of Alberta | (1990) |
| | US 4,910,145 | ICI | (1990) |
| Poly(3HB-*co*-3HV) from unrelated substrates | EP 90304267.9 | ICI | (1990) |
| | | Chemical Corp. | |
| Poly(3HB-*co*-4HB) from *Alcaligenes* | JP 2234683-A2 | Mitsubishi Kasei Corp. | (1990) |
| Poly(3HB-*co*-3HV) from *Methylobacterium* | DD 290914-A5 | AdW DDR | (1991) |
| PHA plus UQ10 from *M. rhodesianum* | DD290915-A5 | AdW DDR | (1991) |
| PHA production with recombinant *E. coli* | WO 9118993-A1 | Center for Innovative Research | (1991) |
| Poly(3HB-*co*-4HB) from *Alcaligenes* | JP 03216193-A2 | Mitsubishi Kasei Corp. | (1991) |
| Poly(3HB-*co*-4HB) from *A. eutrophus* | JP 03292889-A2 | Mitsubishi Kasei Corp. | (1991) |
| Poly(3HA) production | AT 8902942-A1 | PCD | (1991) |
| PHA from *Alcaligenes* | EP 440165-A2 | Showa Denko K. K. | (1991) |
| 3HB and poly(3HB) from cultivation of activated sludge | JP 03143397-A2 | Taisei Corp. | (1991) |
| PHA production by fermentation | ES 2019169-A6 | Universidad de Alicante | (1991) |
| Poly(3HB) production during log-phase by *A. vinelandii* | WO 9118994-A1 | University of Alberta | (1991) |
| Poly(3HB-*co*-4HV) production | EP 466050-A1 | Showa Denko K. K. (1992) | |
| PHA biosynthesis genes | US 5229279 | Metabolix | (1993) |
| PHA biosynthesis genes | US 5245023 | Metabolix | (1993) |
| PHA biosynthesis genes | US 5250430 | Metabolix | (1993) |

**Tab. 9.** (continued)

| Main Subject | Patent Application | Assigned to | Year |
|---|---|---|---|
| *Isolation of PHA* | | | |
| Poly(3HB) isolation using cyclic carbonate solvent | BE 850332-A1 | Agroferm AG | (1977) |
| Separation of poly(3HB) from biomass | EP 14490-A1 | Solvay & Cie | (1980) |
| Physical breackage of cells for release of PHA | EP 46,335 | ICI | (1980) |
| Solvent extraction of aqueous cell suspension | US 4358583-A1 | ICI | (1982) |
| Poly(3HB) recovery by enzyme treatment | EP 145233-A1 | ICI | (1985) |
| Production of poly(3HB) | DE 3343551-A1 | Lentia GmbH | (1985) |
| Production of poly(3HB) with *A. latus* | DE 3343576-A1 | Lentia GmbH | (1985) |
| Poly(3HB) extraction from aqueous solutions by solvent | EP 168095-A1 | Solvay & Cie | (1986) |
| Separation and purification of poly(3HB) | JP 62205787-A2 | Mitsubishi Rayon Co. | (1987) |
| PHA from recombinant bacteria | WO 8900202-A2 | MIT | (1989) |
| Solvent extraction | DD 276304-A1 | AdW DDR | (1990) |
| 3HB esters from poly(3HB) by alcoholysis | JP 2086786-A1 | Nippon Synthetic Chemical Industries | (1990) |
| Lysis of bacteria for release of poly(3HA) | DE 4003827-A1 | PCD Deutschland GmbH | (1991) |
| PHA from recombinant bacteria and plants | WO 9100917-A2 | MIT | (1991) |
| Extraction with supercritical fluids | DD 294280-A5 | AdW DDR | (1991) |
| Poly(3HB-*co*-3HV) production | IN 167933-A1 | ICI | (1991) |
| Poly(3HB-*co*-3HV) production | EP 431883-A1 | ICI | (1991) |
| Poly(3HB-*co*-3HV) from aa-overproducing bacteria | WO 9118995-A1 | ICI | (1991) |
| PHA production | GB 9203915-A0 | ICI | (1992) |
| Copolymer production | US 5126255-A1 | ICI | (1992) |
| Isolation of PHA from microorganisms | DE 4036067-A1 | PCD Deutschland GmbH | (1992) |
| Layers from dispersion of PHA on bottles | DE 4040158-A1 | PCD Deutschland GmbH | (1992) |
| *Manufacturing/Modification* | | | |
| Poly(3HB) films by casting | US 3,182,036 | GRACE | (1965) |
| 3HB and oligomers from 3HB by hydrolysis | US 4365088-A1 | Solvay & Cie | (1982) |
| 3HB from poly(3HB) by hydrolysis | EP 320046-A1 | Solvay & Cie | (1989) |
| Preparation of esters of 3HB | EP 377260-A1 | Solvay & Cie | (1990) |
| Influence of nucleating agents | EP 400855-A1 | ICI | (1990) |
| Nucleating agents | WO 9119759-A1 | ICI | (1991) |
| Derivatives of PHA in the melt | EP 491171-A2 | PCD Deutschland GmbH | (1992) |
| Preparation of esters of 3HB | US 5107016-A1 | Solvay & Cie | (1992) |

| | | | |
|---|---|---|---|
| *Miscelleaneous on Production/Isolation* | | | |
| Plastic shaped body | US 5124371-A1 | Agency of Industrial Science and Technology | (1992) |
| *Applications of PHA* | | | |
| Retard materials | DE 3417576-A1 | Lentia GmbH | (1985) |
| Microporous membranes prepared from PHA | JP 60137402-A1 | Mitsubishi Rayon Co. Ltd. | (1985) |
| Surgical devices for bone fractures | WO 8607250-A1 | Material Consultant | (1986) |
| Microcapsules for retard materials | DE 3428372-A1 | Hoechst AG | (1986) |
| Microcapsules for retard materials | EP 315875-A1 | Hoechst AG | (1989) |
| Toner and developer from semicrystalline PHA | US 5004664-A1 | Xerox Corp. | (1991) |
| Cigarette filter tips | DE 4013304-A1 | BAT Cigarettenfabriken | (1992) |
| Cigarette filter tips | DE 4013293-C2 | BAT Cigarettenfabriken | (1992) |
| Flavor delivery systems from PHA | WO 9209210-A1 | Nutrasweet Co. | (1992) |
| Cream substitutes from PHA | WO 9209211-A1 | Nutrasweet Co. | (1992) |

*Abbreviatons:* AdW, Akademie der Wissenschaften; BAT, British American Tobacco; ICI, Imperial Chemical Industries; MIT, Massachusetts Institute of Technology; PCD, Petrochemie Danubia; PHA, polyhydroxyalkanoic acids; 3HB, 3-hydroxybutyric acid; 4HB, 4-hydroxybutyric acid; 3HV, 3-hydroxyvaleric acid; 4HV, 4-hydroxyvaleric acid.

Biosynthesis of PHA has to compete with chemical synthesis, and some PHA can also be synthesized chemically at reasonable efficiencies. Since the costs are comparably high, the production of PHA by chemical synthesis is not feasible at present. However, many interesting model compounds, such as homopolyesters of certain hydroxyalkanoic acids or PHA with a different stereoregularity, are only available by chemical synthesis at present. The most abundantly used procedure is the ring-opening polymerization of the corresponding lactones using specific catalysts. Recently, MARCHESSAULT and collaborators have described the synthesis of poly(3HHp) and poly(3HN) from 3-heptanolactone or 3-nonanolactone, respectively, using a trimethylaluminium–water catalyst (JESUDASON and MARCHESSAULT, 1994).

PHA will also have to compete with many other biodegradable polymers, obtained biotechnologically or from chemical synthesis. Recently, Showa Dencko launched a biodegradable polyester which is produced chemosynthetically from diols such as, e.g., ethylene glykol and from dicarboxylic acid such as, e.g., succinic acid (SHOWA Highpolymer Co., 1993). Another biosynthetic polymer which can be directly manufactured as a thermoplastic material is starch (WITTWER and TOMKA, 1984). In addition, many other biodegradable polymers that can be used as thermoplastics or elastomers are currently developed by the chemical industry (BARENBERG et al., 1990).

# 12 References

ABE, C., TAIMA, Y., NAKAMURA, Y., DOI, Y. (1990), New bacterial copolyester of 3-hydroxyalkanoates and 3-hydroxy-ω-fluoroalkanoates produced by *Pseudomonas oleovorans*, *Polym. Commun.* **31**, 404–406.

ANDERSON, A. J., DAWES, E. A. (1990), Occurrence, metabolism, metabolic role, and industrial uses of bacterial polyhydroxyalkanoates, *Microbiol. Rev.* **54**, 450–472.

ANDERSON, A. J., HAYWOOD, G. W., WILLIAMS, D. R., DAWES, E. A. (1990), The production of polyhydroxyalkanoates from unrelated carbon sources, in: *Novel Biodegradable Microbial Polymers* (DAWES, E. A., Ed.). Dordrecht: Kluwer Academic Publishers.

ANDO, Y., FUKUDA, E. (1984), Piezoelectric properties and molecular motion of poly(β-hydroxybutyrate) films, *J. Polym. Sci., Polym. Phys. Ed.* **22**, 1821–1834.

ANONYMOUS (1990), Biodegradable plastic hits the production line. *New Sci.,* May 5.

APOSTOLIDES, Z., POTGIETER, D. J. J. (1981), Determination of PHB (poly-β-hydroxybutyric acid) in activated sludge by a gas chromatographic method, *Eur. J. Appl. Microbiol. Biotechnol.* **13**, 62–63.

ASENJO, J. A., SUK, J. S. (1986), Microbial conversion of methane into poly-β-hydroxybutyrate (PHB): Growth and intracellular product accumulation in a type II methanotroph, *J. Ferment. Technol.* **64**, 271–278.

BABEL, W. (1992), Pecularities of methylotrophs concerning overflow metabolism, especially the synthesis of polyhydroxyalkanoates, *FEMS Microbiol. Rev.* **103**, 141–148.

BABEL, W., RIIS, V., HAINICH, E. (1990), Mikrobielle Thermoplaste: Biosynthese, Eigenschaft und Anwendung, *Plaste Kautsch.* **37**, 109–115.

BALLISTRERI, A., GAROOZO, D., GIUFFRIDA, M., IMPALLOMENI, G., MONTAUDO, G. (1989), Sequencing bacterial poly(β-hydroxybutyrate-*co*-β-hydroxyvalerate) by partial methanolysis, high-performance liquid chromatography fractionation, and fast atom bombardment mass spectrometry analysis, *Macromolecules* **22**, 2107–2111.

BALLISTRERI, A., MONTAUDO, G., IMPALLOMENI, G., LENZ, R. W., KIM, Y. B., FULLER, R. C. (1990), Sequence distribution of β-hydroxyalkanoate units with higher alkyl groups in bacterial copolyesters, *Macromolecules* **23**, 5059–5064.

BALLISTRERI, A., MONTAUDO, G., GIUFFRIDA, M., LENZ, R. W., LIM, Y. B., FULLER, R. C. (1992), Determination of sequence distribution in bacterial copolyesters containing higher alkyl and alkenyl pendant groups, *Macromolecules* **25**, 1845–1851.

BAPTIST, J. N. (1962), Process for preparing poly-β-hydroxybutyric acid, *US Patent Application* 3044942.

BARENBERG, S. H., BRASH, J. L., NARAYAN, R., REDPATH, A. E. (1990), *Degradable Materials: Perspectives, Issues and Opportunities,* 1st Edn. Boca Raton: CRC Press.

BARHAM, P. J. SELWOOD, A. (1982), *Eur. Patent Application* 58480.

BARHAM, P. J., BARKER, P., ORGAN, S. J. (1992), Physical properties of poly-hydroxybutyrate and copolymers of hydroxybutyrate and hydroxyvalerate, *FEMS Microbiol. Rev.* **103**, 289–298.

BARNARD, G. N., SANDERS, J. K. M. (1988), Observation of mobile poly(β-hydroxybutyrate) in the storage granules of *Methylobacterium* AM1 by *in vivo* $^{13}$C-NMR spectroscopy, *FEBS Lett.* **231**, 16–18.

BARNARD, G. N., SANDERS, J. K. M. (1989), The poly-β-hydroxybutyrate granule *in vivo:* A new insight based on NMR spectroscopy of whole cells, *J. Biol. Chem.* **264**, 3286–3281.

BAUER, H., OWEN, A. J. (1988), Some structural and mechanical properties of bacterially produced poly-β-hydroxybutyrate-co-β-hydroxyvalerate, *Colloid Polym. Sci.* **266**, 241–247.

BERGER, E. (1990), *M. Sc. A Thesis,* Ecole Polytechnique de Montréal, Montréal, Canada.

BERGER, E., RAMSAY, B. A., RAMSAY, J. A., CHAVARIE, C., BRANUEGG, G. (1989), PHB recovery by hypochlorite digestion of non-PHB biomass, *Biotechnol. Lett.* **3**, 227–232.

BLACKWOOD, A. C., EPP, A. (1957), Identification of β-hydroxybutyric acid in bacterial cells by infrared spectrometry, *J. Bacteriol.* **74**, 266–267.

BONARTSEVA, G. A. (1985), Activity of nodule bacteria in terms of poly(β-hydroxybutyrate) accumulation during colony staining with phosphin 3R. *Mikrobiologiya* **54**, 461–464.

BRANDL, H., PÜCHNER, P. (1992), Biodegradation of plastic bottles made from 'Biopol' in an aquatic ecosystem under *in situ* conditions, *Biodegradation* **2**, 237–243.

BRANDL, H., GROSS, R. A., LENZ, R. W., FULLER, R. C. (1988), *Pseudomonas oleovorans* as a source of poly(β-hydroxyalkanoates) for potential applications as biodegradable polyesters, *Appl. Environ. Microbiol.* **54**, 1977–1982.

BRAUNEGG, G., SONNLEITNER, B., LAFFERTY, R. M. (1978), A rapid gas chromatographic method for the determination of poly-β-hydroxybutyric acid in microbial biomass, *Eur. J. Appl. Microbiol. Biotechnol.* **6**, 29–37.

BRIESE, B. H., JENDROSSEK, D., SCHLEGEL, H. G. (1994a), Degradation of poly(3-hydroxybutyrate-*co*-3-hydroxyvalerate) by aerobic sewage sludge, *FEMS Microbiol. Lett.* **117**, 107–112.

BRIESE, B. H., SCHMIDT, B., JENDROSSEK, D. (1994b), *Pseudomonas lemoignei* has five depolymerase genes. A comparative study of bacterial and eukaryotic PHA depolymerases, *J. Environ. Polym. Degrad.* **2**, 75–87.

BRUCATO, C. L., WONG, S. S. (1991), Extracellular poly(3-hydroxybutyrate) depolymerase from *Penicillium funiculosum*: general characterization and active side studies, *Arch. Biochem. Biophys.* **290**, 497–502.

BUDWILL, K., FEDORAK, P. M., PAGE, W. J. (1992), Methanogenic degradation of poly(3-hydroxyalkanoates), *Appl. Environ. Microbiol.* **58**, 1398–1401.

BURDON, K. L. (1946), Fatty material in bacteria and fungi by staining dried fixed slide preparations, *J. Bacteriol.* **52**, 665–678.

BYROM, D. (1987), Polymer synthesis by microorganisms: technology and economics, *TIBTECH* **5**, 246–250.

BYROM, D. (1990), Industrial production of copolymer from *Alcaligenes eutrophus*, in: *Novel Biodegradable Microbial Polymers* (DAWES, E. A., Ed.), pp. 113–117. Dordrecht: Kluwer Academic Publishers.

BYROM, D. (1992), Production of poly-β-hydroxybutyrate: poly-β-hydroxyvalerate copolymers, *FEMS Microbiol. Rev.* **103**, 247–250.

CHOWDHURY, A. A. (1963), Poly-β-hydroxybuttersäureabbauende Bakterien und Exoenzym, *Arch. Mikrobiol.* **47**, 167–200.

COULOMBE, S., SCHAUWECKER, P., MARCHESSAULT, R. H., HAUTTECOEUR, R. (1978), High-pressure liquid chromatography for fractionating oligomers from degraded poly(β-hydroxybutyrate), *Macromolecules* **11**, 279–281.

DAVIS, J. B. (1964), Cellular lipids of a *Nocardia* grown on propane and *n*-butane, *Appl. Microbiol.* **12**, 301–304.

DAWES, E. A. (1986), *Microbial Energetics,* 1st Edn. Glasgow: Blackie.

DE KONING, G. J. M., MAXWELL, I. A. (1993), Biosynthesis of poly-(*R*)-3-hydroxy-alkanoate: An emulsion polymerization, *J. Environ. Polym. Degrad.* **1**, 223–226.

DE KONING, G. J. M., VAN BILSON, H. M. M., LEMSTRA, P. J., HAZELBERG, W., WITHOLT, B., SCHIRMER, A., D. JENDROSSEK, D. (1994), A biodegradable rubber by crosslinking polyhydroxyalkanoates from *Pseudomonas oleovorans* using electron beam irradiation, *Polymer* **35**, 2090–2097.

DE SMET, M. J., EGGINK, G., WITHOLT, B., KINGMA, J., WYNBERG, H. (1983), Characterization of intracellular inclusions formed by *Pseudomonas oleovorans* during growth on octane, *J. Bacteriol.* **154**, 870–878.

DELAFIELD, F. P., COOKSEY, K. E., DOUDOROFF, M. (1965a), β-hydroxybutyric dehydrogenase and dimer hydrolase of *Pseudomonas lemoignei, J. Biol. Chem.* **240**, 4023–4028.

DELAFIELD, F. P., DOUDOROFF, M., PALLERONI, N. J., LUSTY, C. J., CONTOPOULOS, R. (1965b), Decomposition of poly-β-hydroxybutyrate by pseudomonads, *J. Bacteriol.* **90**, 1455–1466.

DOI, Y. (1990), *Microbial Polyesters*. New York: VCH.

DOI, Y., ABE, C. (1990) Biosynthesis and characterization of a new bacterial copolyester of 3-hydroxyalkanoates and 3-hydroxy-$\omega$-chloroalkanoates, *Macromolecules* **23**, 3705–3707.

DOI, Y., KUNIOKA, M., NAKAMURA, Y., SOGA, K. (1987a), Biosynthesis of copolymer in *Alcaligenes eutrophus* H16 from $^{13}$C-labeled acetate and propionate, *Macromolecules* **20**, 2988–2991.

DOI, Y., TAMAKI, A., KUNIOKA, M., SOGA, K. (1987b), Biosynthesis of terpolyesters of 3-hydroxybutyrate, 3-hydroxyvalerate and 5-hydroxyvalerate from 5-chlorpentanoic and pentanoic acids, *Makromol. Chem. Rapid Commun.* **8**, 631—635.

DOI, Y., SEGAWA, A., KAWAGUCHI, Y., KUNIOKA, M. (1990a), Cyclic nature of poly(3-hydroxyalkanoate) metabolism in *Alcaligenes eutrophus*, *FEMS Microbiol. Lett.* **67**, 165–170.

DOI, Y., SEGAWARA, A., KUNIOKA, M. (1990b), Biosynthesis and characterization of poly-$\beta$-hydroxybutyrate-*co*-4-hydroxybutyrate in *Alcaligenes eutrophus, Int. J. Biol. Macromol.* **12**, 106–111.

DOI , Y., SEGAWA, A., NAKAMURA, S., KUNIOKA, M. T. (1990c), Production of biodegradable copolyesters by *Alcaligenes eutrophus*, in: *New Biosynthetic Biodegradable Polymers of Industrial Interest from Microorganisms* (DAWES, A. E., Ed.), pp. 37–48, Dordrecht: Kluwer Academic Publishers.

DOI, Y., KANESAWA, Y., TANAHASHI, N., KUMAGAI, Y. (1992), Biodegradation of microbial polyesters in the marine environment, *Polym. Degrad. Stab.* **36**, 173–177.

EGGERSDORFER, M., MEYER, J., ECKES, P. (1992), Use of renewable resources for non-food materials, *FEMS Microbiol. Rev.* **103**, 355–364.

EGGINK, G., VAN DER WAL, H., HUIJBERTS, G. (1990), Production of poly-3-hydroxyalkanoates by *P. putida* during growth on long-chain fatty acids, in: *New Biosynthetic Biodegradable Polymers of Industrial Interest from Microorganisms* (DAWES, E. A., Ed.), pp. 37–48. Dordrecht: Kluwer Academic Publishers.

EGGINK, G., DE WAARD, P., HUIJBERTS, G. N. M. (1992), The role of fatty acid biosynthesis and degradation in the supply of substrates for poly(3-hydroxy-alkanoate) formation in *Pseudomonas putida, FEMS Microbiol. Rev.* **103**, 159–164.

ELLAR, D., LUNDGREN, D. G., OKAMURA, K., MARCHESSAULT, R. H. (1968), Morphology of poly-$\beta$-hydroxybutyrate granules, *J. Mol. Biol.* **35**, 489–502.

FIDLER, S., DENNIS, D. (1992), Polyhydroxyalkanoate production in recombinant *Escherichia coli, FEMS Microbiol. Rev.* **103**, 231–236.

FINDLAY, R. H. D., WHITE, D. C. (1983), Polymeric $\beta$-hydroxyalkanoates from environmental samples and *Bacillus megaterium, Appl. Environ. Microbiol.* **45**, 71–78.

FISCHER, H., ERDMANN, S., HOLLER, E. (1989) An unusual polyanion from *Physarum polycephalum* that inhibits homologous DNA polymerase $\alpha$ *in vitro, Biochemistry* **28**, 5219–5226.

FÖLLNER, C. G., BABEL, W., VALENTIN, H., STEINBÜCHEL, A. (1993), Expression of polyhydroxyalkanoic acid-biosynthesis genes in methylotrophic bacteria relying on the ribulosemonophosphate pathway, *Appl. Microbiol. Biotechnol.* **40**, 284–291.

FOSTER, L. J. R., LENZ, R. W., FULLER, R. C. (1994), Quantitative determination of intracellular depolymerase activity in *Pseudomonas oleovorans* inclusions containing poly-3-hydroxyalkanoates with long alkyl substituents, *FEMS Microbiol. Lett.* **118**, 279–282.

FRASER, H. M., SANDOW, J., SEUDEL, H. R., NUNN, S. F. (1989), Controlled release of a GnRH agonist from a polyhydroxybutyric acid implant: Reversible suppression of the menstrual cycle in the macaque, *Acta Endocrinol. (Copenhagen)* **121**, 841–848.

FRITZSCHE, K., LENZ, R. W., FULLER, R. C. (1990a), Production of unsaturated polyesters by *Pseudomonas oleovorans, Int. J. Biol. Macromol.* **12**, 85–91.

FRITZSCHE, K., LENZ, R. W., FULLER, R. C. (1990b) Bacterial polyesters containing branched poly($\beta$-hydroxyalkanoate) units, *Int. J. Biol. Macromol.* **12**, 92–101.

FRIZSCHE, K., LENZ, R. W., FULLER, R. C. (1990c) An unusual bacterial polyester with a phenyl pendant group, *Macromol. Chem.* **191**, 1957–1965.

FUKADA, E., ANDO, Y. (1986), Piezoelectric properties of poly-$\beta$-hydroxy-butyrate and copolymers of $\beta$-hydroybutyrate and $\beta$-hydroxyvalerate, *Int. J. Biol. Macromol.* **8**, 361–366.

FUKUI, T., YOSHIMOTO, A., MATSUMOTO, M., HOSOKAWA, S., SAITO, T., NISHIKAWA, H., KENKICHI, T. (1976), Enzymatic synthesis of poly-$\beta$-hydroxybutyrate in *Zoogloea ramigera, Arch. Microbiol.* **110**, 149–156.

FULLER, R. C., O'DONNELL, J. P., SAULNIER, J., REDLINGER, T. E., FOSTER, J., LENZ, R. W. (1992), The supramolecular architecture of the polyhydroxy-alkanoate inclusions in *Pseudomonas oleovorans, FEMS Microbiol. Rev.* **103**, 279–288.

GAGNON, K. D., LENZ, R. W., FARRIS, R. J., FULLER, R. C. (1992), Crystallization behaviour and its influence on the mechanical properties of a thermoplastic elastomer produced by *Pseudomonas oleovorans*, *Macromolecules* **25**, 3723–3728.

GERNGROSS, T. U., REILLY, P., STUBBE, J., SINSKEY, A. J., PEOPLES, O. P. (1993), Immunocytochemical analysis of poly-$\beta$-hydroxybutyrate (PHB) synthase in *Alcaligenes eutrophus* H16: Localization of the synthase enzyme at the surface of PHB granules, *J. Bacteriol.* **175**, 5289–5293.

GERNGROSS, T. U., SNELL, K. D., PEOPLES, D. D., SINSKEY, A. J., CSUHAI, E., MASAMUNE, S., STUBBE, J. (1994), Overexpression and purification of the soluble polyhydroxyalkanoate synthase from *A. eutrophus*. Evidence for a required posttranslational modification for catalytic activity, *Biochemistry* **33**, 9311–9320.

GILMORE, D. F., ANTOUN, S., LENZ, R. W., FULLER, R. C. (1993), Degradation of poly($\beta$-hydroxyalkanoates) and polyolefin blends in a municipal wastewater treatment facility, *J. Environ. Polym. Dedgrad.* **1**, 269–274.

GOTTSCHALK, G. (1964a), Die Biosynthese der Poly-$\beta$-Hydroxybuttersäure durch Knallgasbakterien. I. Ermittlung der $^{14}C$-Verteilung in Poly-$\beta$-hydroxybuttersäure, *Arch. Mikrobiol.* **47**, 225–229.

GOTTSCHALK, G. (1964b), Die Biosynthese der Poly-$\beta$-Hydroxybuttersäure durch Knallgasbakterien. II. Verwertung organischer Säuren, *Arch. Mikrobiol.* **47**, 230–235.

GRACE, W. R. (1966), Absorbable prosthetic devices and surgical sutures, *British Patent* 1034123.

GRASSIE, N., MURRAY, E. J., HOLMES, P. A. (1984), The thermal degradation of poly((D)-$\beta$-hydroxybutyric acid). 1. Identification and quantitative analysis of products, *Polym. Degrad. Stab.* **6**, 47–61.

GRIEBEL, R. J., MERRICK, J. M. (1971), Metabolism of poly-$\beta$-hydroxybutyrate: Effect of mild alkaline extraction on native poly-$\beta$-hydroxybutyrate granules, *J. Bacteriol.* **108**, 782–789.

GRIEBEL, R. J., SMITH, Z., MERRICK, J. M. (1968), Metabolism of poly-$\beta$-hydroxybutyrate. I. Purification, composition, and properties of native poly-$\beta$-hydroxybutyrate granules from *Bacillus megaterium*, *Biochemistry* **7**, 3676–3681.

GROSS, R. A., DEMELLO, C., LENZ, R. W., BRANDL, H., FULLER, R. C. (1989), Biosynthesis and characterization of poly($\beta$-hydroxyalkanoates) produced by *Pseudomonas oleovorans*, *Macromolecules* **22**, 1106–1115.

HÄNGGI, U. J. (1990), Pilot scale production of PHB with *Alcaligenes latus*, in: *Novel Biodegradable Microbial Polymers* (DAWES, E. A., Ed.), pp. 65–70. Dordrecht: Kluwer Academic Publishers.

HARRISON, S. T. L., CHASE, H. A., AMOR, S. R., BONTHRONE, K. M., SANDERS, J. K. M. (1992), Plasticization of poly(hydroxybutyrate) *in vivo*, *Int. J. Biol. Macromol.* **14**, 50–56.

HAUTTECOEUR, B., JOLIVET, M., GAVARD, R. (1992), Controlled chemical depolymerization of the $\beta$-hydroxybutyric lipid from *Bacillus megaterium*, *C. R. Acad. Sci., Ser. D.* **274**, 2729–2732.

HAYNES, W. C., MELVIN, E. H., LOCKE, J. M., GLASS, C. A., SENTI, F. R. (1958), Certain factors affecting the infrared spectra of selected microorganisms, *Appl. Microbiol.* **6**, 298–304.

HAYWOOD, G. W., ANDERSON, A. J., CHU, L., DAWES, E. A. (1988a), Characterization of two 3-ketothiolases possessing differing substrate specificities in the polyhydroxyalkanoate synthesizing organism *Alcaligenes eutrophus*, *FEMS Microbiol. Lett.* **52**, 91–96.

HAYWOOD, G. W., ANDERSON, A. J., CHU, L., DAWES, E. A. (1988b), The role of NADH- and NADPH-linked acetoacetyl-CoA reductases in the poly-3-hydroxybutyrate synthesizing organism *Alcaligenes eutrophus*, *FEMS Microbiol. Lett.* **52**, 259–264.

HAYWOOD, G. W., ANDERSON, DAWES, E. A. (1989a), The importance of PHB-synthase substrate specificity in polyhydroxyalkanoate synthesis by *Alcaligenes eutrophus*, *FEMS Microbiol. Lett.* **57**, 1–6.

HAYWOOD, G. W., ANDERSON, A. J., DAWES, E. A. (1989b), A survey of the accumulation of novel polyhydroxyalkanoates by bacteria, *Biotechnol Lett.* **11**, 471–476.

HAYWOOD, G. W., ANDERSON, A. J., EWING, D. F., DAWES, E. A. (1990), Accumulation of a polyhydroxyalkanoate containing primarily 3-hydroxydecanoate from simple carbohydrate substrates by *Pseudomonas* sp. strain NCIMB 40135, *Appl. Environ. Microbiol.* **56**, 3354–3359.

HAYWOOD, G. W., ANDERSON, A. J., WILLIAMS, D. R., DAWES, E. A., EWING, D. F. (1991), Accumulation of a poly(hydroxyalkanoate) copolymer containing 3-hydroxyvalerate from simple carbohydrate substrates by *Rhodococcus* sp. NCIMB 40126, *Int. J. Biol. Macromol.* **13**, 83–88.

HAZER, B., LENZ, R. W., FULLER, R. C. (1993), Biosynthesis of some new polyesters from methyl-branched alkanoic acids by *P. oleovorans*. *Poster* presented at the *2nd National Meeting of the Bio/Environmentally Degradable Polymer Society,* Chicago, August 19–21.

HELLEUR, R. J. (1988), Pyrolysis-gas chromatography for the rapid characterization of bacterial poly-β-hydroxybutyrate-*co*-β-hydroxyvalerate, *Polym. Prepr.* **29**, 609–610.

HIPPE, H. (1967), Abbau und Wiederverwertung von Poly-β-Hydroxybuttersäure durch *Hydrogenomonas* H16, *Arch. Mikrobiol.* **56**, 248–288.

HIPPE, H., SCHLEGEL, H. G. (1967), Hydrolyse von PHBs durch intracelluläre Depolymerase von *Hydrogenomonas* H16, *Arch. Mikrobiol.* **56**, 278–299.

HOCKING, J. MARCHESSAULT, R. H. (1994), Biopolyesters, in: *Chemistry and Technology of Biodegradable Polymers* (GRIFFIN, G. F. L., Ed.), pp. 48–96. London: Chapman and Hall.

HOCKING, P. J., MARCHESSAULT, R. H., TIMMINS, M. R., SCHERER, T. M., LENZ, R. M., FULLER, R. C. (1994), Enzymatic degradability of isotactic versus syndiotactic poly(β-hydroxybutyrate), *Macromol. Rapid Commun.* **15**, 447–452.

HOLLAND, S. J., JOLLY, A. M., YASIN, M., TIGHE, B. J. (1987), Polymers for biodegradable medical devices. II. Hydroxybutyrate–hydroxyvalerate copolymers: hydrolytic degradation studies, *Biomaterials* **8**, 289–295.

HOLLER, E., ANGERER, B., ACHHAMMER, G., MILLER, S., WINDISCH, C. (1992), Biological and biosynthetic properties of poly-L-malate, *FEMS Microbiol. Rev.* **103**, 109–118.

HOLMES, P. A. (1985), Applications of PHB – a microbially produced biodegradable thermoplastic, *Phys. Technol.* **16**, 32–36.

HOLMES, P. A., JONES, E. (1982), Extraction of poly(β-hydroxy butyric acid), *Eur. Patent Application* EP 46335.

HOLMES, P. A., LIM, G. B. (1985), Separation of 3-hydroxybutyrate polymer from microorganism cells, *Eur. Patent Application* EP 145233.

HOLMES, P. A., WRIGHT, L. F., COLLINS, S. H. (1981), Beta-hydroxybutyrate polymers, *Eur. Patent Application* 0052459.

HOROWITZ, D. M., CLAUSS, J., HUNTER, B. K., SANDERS, J. K. M. (1993), Amorphous polymer granules, *Nature* **363**, 23.

HOWELLS, E. R. (1982), Single-cell protein and related technology, *Chem. Ind.* **15**, 508–511.

HRABAK, O. (1992), Industrial production of poly-β-hydroxybutyrate, *FEMS Microbiol. Rev.* **103**, 251–256.

HUANG, A. H. C. (1992), Oil bodies and oleosins in seeds, *Annu. Rev. Plant Physiol. Plant Mol. Biol.* **43**, 177–200.

HUIJBERTS, G. N. M., EGGINK, G., DE WAARD, P., HUISMAN, G. W., WITHOLT, B. (1992), *Pseudomonas putida* KT2442 cultivated on glucose accumulates poly(3-hydroxyalkanoates) consisting of saturated and unsaturated monomers, *Appl. Environ. Microbiol.* **58**, 536–544.

HUIJBERTS, G. N. M., DE RIJK, T. C., DE WAARD, P., EGGINK, G. (1994a), $^{13}$C Nuclear magnetic resonance studies of *Pseudomonas putida* fatty acid metabolic routes involved in poly(3-hydroxyalkanoate) synthesis, *J. Bacteriol.* **176**, 1661–1666.

HUIJBERTS, G. N. M., VAN DER WAL, H., WILKINSON, C., EGGINK, G. (1994b) Gas-chromatographic analysis of poly(3-hydroxyalkanoates) in bacteria, *Biotechnol. Techniques* **8**, 187–192.

HUISMAN, G. W. (1991), Poly(3-hydroxyalkanoates) from *Pseudomonas putida*: from DNA to plastic. *Ph. D. Thesis,* Rijksuniversiteit Groningen.

HUISMAN, G. W., DE LEEUW, O., EGGINK, G., WITHOLT, B. (1989), Synthesis of polyhydroxyalkanoates is a common feature of fluorescent pseudomonads, *Appl. Environ. Microbiol.* **55**, 1949–1954.

HUISMAN, G. W., WONINK, E., MEIMA, R., KAZEMIER, B., TERPSTRA, P., WITHOLT, B. (1991), Metabolism of poly(3-hydroxyalkanoates) (PHAs) by *Pseudomonas oleovorans*. Identification and sequences of genes and function of the encoded proteins in the synthesis and degradation of PHA, *J. Biol. Chem.* **266**, 2191–2198.

HUISMAN, G. W., WONINK, E., DE KONING, G., PREUSTING, H., WITHOLT, B. (1992), Synthesis of poly(3-hydroxyalkanoates) by mutant and recombinant strains of *Pseudomonas putida*, *Appl. Microbiol. Biotechnol.* **38**, 1–5.

HUSTEDE, E., STEINBÜCHEL, A. (1993), Characterization of the polyhydroxyalkanoate synthase gene locus of *Rhodobacter sphaeroides*, *Biotechnol. Lett.* **15**, 709–714.

HUSTEDE, E., STEINBÜCHEL, A., SCHLEGEL, H. G. (1992), Cloning of poly(3-hydroxybutyric acid) synthase genes of *Rhodobacter sphaeroides* and *Rhodospirillum rubrum* and heterologous expression in *Alcaligenes eutrophus*, *FEMS Microbiol. Lett.* **93**, 285–290.

HUSTEDE, E., STEINBÜCHEL, A., SCHLEGEL, H. G. (1993), Relationship between the photoproduction of hydrogen and the accumulation of PHB in non-sulphur purple bacteria, *Appl. Microbiol. Biotechnol.* **39**, 87–93.

JAEGER, K.-E., RANSAC, S., DIJKSTRA, B. W., COLSON, C., VAN HEUVEL, M., MISSET, O. (1994) Bacterial lipases, *FEMS Microbiol. Rev.* **15**, 29–63.

JAEGER, K.-E., STEINBÜCHEL, A., JENDROSSEK, D. (1995), Substrate specificities of bacterial polyhydroxyalkanoate depolymerases and lipases: Bacterial lipases hydrolyze poly(ω-hydroxyalkanoates), *Appl. Environ. Microbiol.* **61**, 3113–3118.

JANES, B., HOLLAR, J., DENNIS, D. (1990), Molecular characterization of the poly-$\beta$-hydroxybutyrate biosynthetic pathway of *Alcaligenes eutrophus* H16, in: *Novel Biodegradable Microbial Polymers* (DAWES, E. A., Ed.), pp. 175–190. Dordrecht: Kluwer Academic Publishers.

JANSSEN, P. H., HARFOOT, C. G. (1990), *Ilyobacter delafieldii* sp. nov., a metabolically restricted anaerobic bacterium fermenting PHB, *Arch. Microbiol.* **154**, 253–259.

JANSSEN, P. H., SCHINK, B. (1993), Pathway of anaerobic poly-$\beta$-hydroxybutyrate degradation by *Ilyobacter delafieldii*, *Biodegradation* **4**, 179–185.

JENDROSSEK, D., KNOKE, I., HABIBIAN, R., STEINBÜCHEL, A., SCHLEGEL, H. G. (1993a), Degradation of poly(3-hydroxybutyric acid), PHB, by bacteria and purification of a novel PHB depolymerase from *Comamonas* sp., *J. Environ Polym. Degrad.* **1**, 53–63.

JENDROSSEK, D., MÜLLER, B., SCHLEGEL, H. G. (1993b), Cloning and characterization of the PHA depolymerase gene locus, phaZ1, from *Pseudomonas lemoignei* and its gene product, *Eur. J. Biochem.* **218**, 701–710.

JENDROSSEK, D., FRISSE, A., BEHRENDES, A., ANDERMANN, M., KRATZIN, H. D., STANISLAWSKI, T., SCHLEGEL, H. G. (1995a), Biochemical and molecular characterization of the *Pseudomonas lemoignei* polyhydroxyalkanoate (PHA) depolymerase system, *J. Bacteriol.* **177**, 596–607.

JENDROSSEK, D., BACKHAUS, M., ANDERMANN, M. (1995b), Characterization of the *Comamonas* sp. poly(3-hydroxybutyrate) (PHB) depolymerase and of its structural gene, *Can. J. Microbiol.* **41**, 160–169.

JESUDASON, J. J., MARCHESSAULT, R. H. (1994), Synthetic poly[$(R,S)$-$\beta$-hydroxyalkanoates] with butyl and hexyl side chains, *Macromolecules* **27**, 2595–2602.

JESUDASON, J. J., MARCHESSAULT, R. H., SAITO, T. (1993), Enzymatic degradation of poly([$R,S$]$\beta$-hydroxybutyrate), *J. Environ. Polym. Degrad.* **1**, 89–98.

JÜTTNER, R. R., LAFFERTY, R. M., KNACKMUSS, H. J. (1975), A simple method for the determination of poly-$\beta$-hydroxybutyric acid in microbial biomass, *Eur. J. Appl. Microbiol.* **1**, 233–237.

KALLIO, R. E., HARRINGTON, A. A. (1960), Sudanophilic granules and lipids of *Pseudomonas methanica*, *J. Bacteriol.* **80**, 321–324.

KALOUSEK, S., DENNIS, D. E., LUBITZ, W. (1990), Release of poly-$\beta$-hydroxybutyrate granules from *Escherichia coli* by protein E-mediated lysis. *Poster* presented at the *International Symposium on Biodegradable Polymers*, Tokyo, October 28–31.

KARR, D. B., WATERS, J. K., EMERICH, D. W. (1983), Analysis of poly-$\beta$-hydroxybutyrate in *Rhizobium japonicum* bacteroides by ion-exclusion high pressure liquid chromatography and UV detection, *Appl. Environ. Microbiol.* **46**, 1339–1344.

KAWAGUCHI, Y., DOI, Y. (1990), Structure of native poly(3-hydroxybutyrate) granules characterized by X-ray diffraction, *FEMS Microbiol. Lett.* **70**, 151–156.

KAWAGUCHI, Y., DOI, Y. (1992), Kinetics and mechanism of synthesis and degradation of poly(3-hydroxybutyrate) in *Alcaligenes eutrophus*, *Macromolecules* **25**, 2324–2329.

KIM, Y. B., LENZ, R. W., FULLER, R. C. (1991) Preparation and characterization of poly($\beta$-hydroxyalkanoates) obtained from *Pseudomonas oleovorans* grown with mixtures of 5-phenylvaleric acid and *n*-alkanoic acids, *Macromolecules* **24**, 5256–5360.

KIM, B. S., LEE, S. Y., CHANG, H. N. (1992), Production of poly-$\beta$-hydroxybutyrate by fed-batch culture of recombinant *Escherichia coli*, *Biotechnol. Lett.* **14**, 811–816.

KÖNIG, C., SAMMLER, J., WILDE, E., SCHLEGEL, H. G. (1969), Konstitutive Glucose-6-phosphat-Dehydrogenase bei Glucose-verwertenden Mutanten von einem kryptischen Wildstamm, *Arch. Mikrobiol.* **67**, 51–57.

KUBOTA, M., NAKANO, M., JUNI, K. (1988), *Chem. Pharm. Bull.* **36**, 333–341.

KUENEN, J. G., ROBERTSON, L. A. (1984), Interactions between obligately and facultatively chemolithotrophic sulphur bacteria, in: *Continuous Cultures 8: Biotechnology, Medicine and the Environment* (DEAN, A. C. R., ELLWOOD, D. C., EVANS, C. G. T., Eds.), pp. 139–158. Chichester: Ellis Horwood Ltd. Publishers.

KUMAGAI, Y., DOI, Y. (1992) Physical properties and biodegradability of blends of isotactic and atactic poly(3-hydroxybutyrate), *Makromol. Chem., Rapid Commun.* **13**, 179–183.

KUMAGAI, Y., KANESAWA, Y., DOI, Y. (1992), Enzymatic degradation of microbial poly(3-hydroxybutyrate) films, *Makromol. Chem.* **193**, 53–57.

KUNIOKA, M., NAKAMURA, Y., DOI, Y. (1988), New bacterial copolyesters produced in *Alcaligenes eutrophus* from organic acids, *Polym. Commun.* **29**, 174–176.

LAFFERTY, R. M., HEINZLE, E. (1978), *US Patent Application* 4101533.

LAFFERTY, R. M., KORSATKO, B., KORSATKO, W. (1988), Microbial production of poly-$\beta$-hydroxybutyric acid, in: *Biotechnology,* 1st Edn., Vol. 6b

(REHM, H. J., REED, G., Eds.), pp. 136–176. Weinheim: VCH.

LAGEVEEN, R. G., HUISMAN, G. W., PREUSTING, H., KETELAAR, P., EGGINK, G., WITHOLT, B. (1988), Formation of polyesters by *Pseudomonas oleovorans*: effect of substrates on formation and composition of poly-(*R*)-3-hydroxyalkanoates and poly-(*R*)-3-hydroxyalkenoates, *Appl. Environ. Microbiol.* **54**, 2924–2932.

LAUZIER, C., REVOL, J.-F., MARCHESSAULT, R. H. (1992), Topotactic crystallization of isolated poly(β-hydroxybutyrate) granules from *Alcaligenes eutrophus, FEMS Microbiol. Rev.* **103**, 299–310.

LAUZIER, C. A., MONASTERIOS, C. J., SARACOVAN, I., MARCHESSAULT, R. H., RAMSAY, B. A. (1993), Film formation and paper coating with poly(β-hydroxyalkanoate), a biodegradable latex, *Tappi J.* **76**, 71–77.

LAW, J. H., SLEPECKY, R. A. (1961), Assay of poly-β-hydroxybutyric acid, *J. Bacteriol.* **82**, 33–36.

LEE, S. Y., CHANG, H. N. (1993), High cell density cultivation of *Escherichia coli* W using sucrose as a carbon source, *Biotechnol. Lett.* **15**, 971–974.

LEE, S. Y., YIM, K. S., CHANG, H. N., CHANG, Y. K. (1994a), Construction of plasmids, estimation of plasmid stability, and use of stable plasmids for the production of poly(3-hydroxybutyric acid) by recombinant *Escherichia coli, J. Biotechnol.* **42**, 901–909.

LEE, E. Y., LEE, K. M., CHANG, H. N., STEINBÜCHEL, A. (1994b), Comparison of *Echerichia coli* strains for synthesis and accumulation of poly(3-hydroxybutyric acid) and morphological changes, *Biotechnol. Bioeng.* **44**, 1337–1347.

LEE, S. Y., JENDROSSEK, D., SCHIRMER, A., CHOI, C. Y., STEINBÜCHEL, A. (1995), Biosynthesis of polyesters consisting of 3-hydroxybutyric acid and medium-chain-length 3-hydroxyalkanoic acids from 1,3-butanediol or from 3-hydroxybutyrate by *Pseudomonas* sp. A33, *Appl. Microbiol. Biotechnol.* **42**, 901–909.

LEHNINGER, A. L. (1975), *Biochemistry,* 2nd Edn., New York: Worth Publishers.

LEMOIGNE, M. (1926), Produits de deshydration et de polymerisation de l'acide β-oxybutyric, *Bull. Soc. Chim. Biol.* (Paris) **8**, 770–782.

LENZ, R. W., KIM, B. W., ULMER, H. W., FRITZSCHE, K., KNEE, E., FULLER, R. C. (1990), Functionalized poly-β-hydroxyalkanoates produced by bacteria, in: *Novel Biodegradable Microbial Polymers* (DAWES, E. A., Ed.), pp. 23–35. Dordrecht: Kluwer Academic Publishers.

LENZ, R. W., KIM, Y. B., FULLER, R. C. (1992), Production of unusual bacterial polyesters by *Pseudomonas oleovorans* through cometabolism, *FEMS Microbiol. Rev.* **103**, 207–214.

LIEBERGESELL, M., STEINBÜCHEL, A. (1992), Cloning and nucleotide sequences of genes relevant for biosynthesis of polyhydroxyalkanoic acid in *Chromatium vinosum* strain D, *Eur. J. Biochem.* **209**, 135–150.

LIEBERGESELL, M., STEINBÜCHEL, A. (1993), Cloning and molecular characterization of the poly(3-hydroxybutyric acid)-biosynthetic genes of *Thiocystis violacea, Appl. Microbiol. Biotechnol.* **38**, 493–501.

LIEBERGESELL, M., HUSTEDE, E., TIMM, A., STEINBÜCHEL, A., FULLER, R. C., LENZ, R. W., SCHLEGEL, H. G. (1991), Formation of poly(3-hydroxyalkanoic acids) by phototrophic and chemolithotrophic bacteria, *Arch. Microbiol.* **155**, 415–421.

LIEBERGESELL, M., SCHMIDT, B., STEINBÜCHEL, A. (1992), Isolation and identification of granule-associated proteins relevant for poly(3-hydroxybutyric acid) biosynthesis in *Chromatium vinosum* D, *FEMS Microbiol. Lett.* **99**, 227–232.

LIEBERGESELL, M., MAYER, F., STEINBÜCHEL, A. (1993), Analysis of poly-hydroxyalkanoic acid-biosynthesis genes of anoxygenic phototrophic bacteria reveals synthesis of a polyester exhibiting an unusual composition, *Appl. Microbiol. Biotechnol.* **40**, 292–300.

LIEBERGESELL, M., SONOMOTO, K., MADKOUR, M., MAYER, F., STEINBÜCHEL, A. (1994), Purification and characterization of the polyhydroxyalkanoic acid (PHA) synthase from *Chromatium vinosum* and localization of the enzyme at the surface of polyhydroxyalkanoic acid granules, *Eur. J. Biochem.* **226**, 71–80.

LIN, L. P., SADOFF, H. L. (1968), Encystment and polymer production by *Azotobacter vinelandii* in the presence of β-hydroxybutyrate, *J. Bacteriol.* **95**, 2336–2343.

LUNDGREN, D. G., PFISTER, R. M., MERRICK, J. M. (1964), Structure of poly-β-hydroxybutyric acid granules, *J. Gen. Microbiol.* **34**, 441–446.

LUSTY, C. J., DOUDOROFF, M. (1966), Poly-β-hydroxybutyrate depolymerase of *Pseudomonas lemoignei, Proc. Natl. Acad. Sci. USA* **56**, 960–965.

MANCHAK, J, PAGE, W. J. (1994), Control of polyhydroxyalkanoate synthesis in *Azotobacter vinelandii* strain UWD, *Microbiology* **140**, 953–963.

MARCHESSAULT, R. H., MORIKAWA, H., REVOL, J.-F., BLUHM, T. L. (1984), Physical properties of a naturally occurring polyester: poly(β-hydroxyvalerate)/poly(β-hydroxybutyrate), *Macromolecules* **17**, 1882–1884.

MARCHESSAULT, R. H., RIOUX, P., SARACOVAN, I. (1993), Direct electrostatic coating of paper, *Nord. Pulp Paper Res. J.* **1**, 211–216.

MATAVULJ, M., MOLITORIS, H. P. (1992), Fungal degradation of polyhydroxy-alkanoates and a semiquantitative assay for screening their degradation by terrestrial fungi, *FEMS Microbiol. Rev.* **103**, 323–332.

MAYER, F. (1986), *Cytology and Morphogenesis of Bacteria,* pp. 89–115, Berlin: Bornträger.

MAYER, F. (1992), Structural aspects of poly-β-hydroxybutyrate granules, *FEMS Microbiol. Rev.* **103**, 265–267.

MCCARTHY, S. P., GROSS, R. A., DUBEY, D. (1993), Reactive processing: P(3HB-HV)/PCL transesterification. *Poster* presented at the *2nd National Meeting of the Bio/Environmentally Degradable Polymer Society,* Chicago, August 19–21.

MCLELLAN, D. W., HALLING, P. J. (1988), Acid-tolerant poly(3-hydroxybutyrate) hydrolases from moulds, *FEMS Microbiol. Lett.* **52**, 215–218.

MERGAERT, J., ANDERSON, C., WOUTERS, A., SWINGS, J., KERSTERS, K. (1992), Biodegradation of polyhydroxyalkanoates, *FEMS Microbiol. Rev.* **103**, 317–322.

MERGAERT, J., WEBB, A., ANDERSON, C., WOUTERS, A., SWINGS, J. (1993), Microbial degradation of poly(3-hydroxybutyrate) and poly(3-hydroxybutyrate-*co*-3-hydroxyvalerate) in soils, *Appl. Environ. Microbiol.* **59**, 3233–3238.

MERRICK, J. M., DOUDOROFF, M. (1961), Enzymatic synthesis of poly-β-hydroxybutyric acid in bacteria, *Nature* **189**, 890–892.

MERRICK, J. M., DOUDOROFF, M. (1964), Depolymerization of poly-β-hydroxybutyrate by an intracellular enzyme, *J. Bacteriol.* **88**, 60–71.

MILLER, N. D., WILLIAMS, D. F. (1987), On the biodegradation of poly-β-hydroxybutyrate (PHB) homopolymer and poly-β-hydroxybutyrate-hydroxyvalerate copolymers, *Biomaterials* **8**, 129–137.

MORIKAWA, H., MARCHESSAULT, R. H. (1981), Pyrolysis of bacterial polyalkanoates, *Can. J. Chem.* **59**, 2306–2313.

MORINAGA, Y., YAMANAKA, S., ISHIZAKI, A., HIROSE, Y. (1978), Growth characteristics and cell composition of *Alcaligenes eutrophus* in chemostat culture, *Agric. Biol. Chem.* **42**, 439–444.

MOSKOWITZ, G. J., MERRICK, J. M. (1969), Metabolism of poly-β-hydroxybutyrate. Enzymatic synthesis of D-(−)-β-hydroxybutyryl coenzyme A by an enoyl hydrase from *Rhodospirillum rubrum*, *Biochemistry* **8**, 2748–2755.

MUKAI, K., YAMADA, K., DOI, Y. (1992), Extracellular poly(hydroxyalkanoate) depolymerase and their inhibitor from *Pseudomonas lemoignei*, *Int. J. Biol. Macromol.* **14**, 235–240.

MUKAI, K., DOI, Y., SEMA, Y., TOMITA, K. (1993), Substrate specificities in hydrolysis of polyhydroxyalkanoates by microbial esterases, *Biotechnol. Lett.* **15**, 601–604.

MÜLLER, B., JENDROSSEK, D. (1993), Purification and properties of a poly(3-hydroxyvalerate) depolymerase from *Pseudomonas lemoignei, Appl. Microbiol. Biotechnol.* **38**, 487–492.

MÜLLER, H. M., SEEBACH, D. (1993) Poly(hydroxyfettsäureester), eine fünfte Klasse von physiologisch bedeutsamen organischen Biopolymeren? *Angew. Chem.* **105**, 483–509.

MURPHY, D. J. (1993), Structure, function and biogenesis of storage lipid bodies and oleosins in plants, *Prog. Lipid Res.* **32**, 247–280.

NAGATA, N., NAKAHARA, T., TABUCHI, T. (1993), Fermentative production of Poly(β-L-malic acid), a polyelectrolytic biopolyester, by *Aureobasisium* sp. *Biosci. Biotech. Biochem.* **57**, 638–642.

NAKAMURA, S., DOIU, Y., SCANDOLA, M. (1992), Microbial synthesis and characterization of poly(3 - hydroxybutyrate - *co* - 4 - hydroxybutyrate), *Macromolecules* **25**, 4237–4241.

NAKAYAMA, K., SAITO, T., FUKUI, T., SHIRAKURA, Y., TOMITA, K. (1985), Purification and properties of extracellular poly(3-hydroxybutyrate) depolymerases from *Pseudomonas lemoignei*, *Biochim. Biophys. Acta* **827**, 63–72.

NAWRATH, C., POIRIER, Y., SOMERVILLE, C. (1994), Targeting of the polyhydroxybutyrate biosynthetic pathway to the plastids of *Arabidopsis thaliana* results in high levels of polymer accumulation, *Proc. Natl. Acad. Sci. USA* **91**, 1983–1989.

NICHOLS, P. D., HENSON, J. M., GUCKERT, J. B., NIVENS, D. E., WITTE, D. C. (1985), Fourier transform-infrared spectroscopic methods for microbial ecology: Analysis of bacteria, bacteria–polymer mixtures and biofilms, *J. Microbiol. Methods* **4**, 79–85.

NICKERSON, K. W. (1982), Purification of poly-β-hydroxybutyrate by density gradient centrifugation in sodium bromide, *Appl. Environ. Microbiol.* **43**, 1208–1209.

NISHIDA, H., TOKIWA, Y. (1993), Distribution of poly(β-hydroxybutyrate) and poly(ε-caprolactone) aerobic degrading microorganisms in different environments, *J. Environ. Polym. Degrad.* **1**, 227–233.

NORRIS, K. P., GREENSTREET, J. E. S. (1958), On the infrared absorption spectrum of *Bacillus megaterium*, *J. Gen. Microbiol.* **19**, 566–580.

ODHAM, G., TUNLID, A., WESTERDAHL, G., MARDEN, P. (1986), Combined determination of poly-β-hydroxyalkanoic and cellular fatty acids in starved marine bacteria and sewage sludge by gas-chromatography with flame ionization or mass spectrometry detection, *Appl. Environ. Microbiol.* **52**, 905–910.

OEDING, V., SCHLEGEL, H. G. (1973), β-Ketothiolase from *Hydrogenomonas eutropha* H16 and its significance in the regulation of poly-β-hydroxybutyrate metabolism, *Biochem. J.* **134**, 239–248.

OSTLE, A. G., HOLT, J. G. (1982), Nile blue A as a fluorescent stain for poly-β-hydroxybutyrate, *Appl. Environ. Microbiol.* **44**, 238–241.

PAGE, W. J. (1990), Production of poly-β-hydroxybutyrate by *Azotobacter vinelandii* strain UWD during growth on molasses and other complex carbon sources, *Appl. Microbiol. Biotechnol.* **31**, 329–333.

PAGE, W. J. (1992a), Production of poly-β-hydroxybutyrate by *Azotobacter vinelandii* UWD in media containing sugars and complex nitrogen sources, *Appl. Microbiol. Biotechnol.* **38**, 117–121.

PAGE, W. J. (1992b), Production of polyhydroxyalkanoates by *Azotobacter vinelandii* UWD in beet molasses culture, *FEMS Microbiol. Rev.* **103**, 149–158.

PAGE, W. J., CORNISH, A. (1993), Growth of *Azotobacter vinelandii* UWD in fish peptone medium and simplified extraction of poly-β-hydroxybutyrate, *Appl. Environ. Microbiol.* **59**, 4236–4244.

PAGE, W. J., KNOSP, O. (1989), Hyperproduction of poly-β-hydroxybutyrate during exponential growth of *Azotobacter vinelandii* UWD, *Appl. Environ. Microbiol.* **55**, 1334–1339.

PAGE, W. J., MANCHAK, J., RUDY, B. (1992), Formation of poly(hydroxybutyrate-*co*-hydroxyvalerate) by *Azotobacter vinelandii* UWD, *Appl. Environ. Microbiol.* **58**, 2866–2873.

PEDROS-ALIO, C., MAS, J., GUERRERO, R. (1985), The influence of poly-β-hydroxybutyrate accumulation on cell volume and buoyant density in *Alcaligenes eutrophus*, *Arch. Microbiol.* **143**, 178–184.

PEOPLES, O. P., SINSKEY, A. J. (1989a), Fine structural analysis of the *Zoogloea ramigera phbA-phbB* locus encoding β-ketothiolase and acetoacetyl-CoA reductase: nucleotide sequence of *phbB, Mol. Microbiol.* **3**, 349–357.

PEOPLES, P. O., SINSKEY, A. J. (1989b), Poly-β-hydroxybutyrate biosynthesis in *Alcaligenes eutrophus* H16. Characterization of the genes encoding β-ketothiolase and acetoacetyl-CoA reductase, *J. Biol. Chem.* **264**, 15293–15297.

PEOPLES, P. O., SINSKEY, A. J. (1989c), Poly-β-hydroxybutyrate (PHB) biosynthesis in *Alcaligenes eutrophus* H16. Identification and characterization of the PHB polymerase gene (*phaC*), *J. Biol. Chem.* **264**, 15298–15303.

PIEPER, U., STEINBÜCHEL, A. (1992), Identification, cloning and sequence analysis of the poly(3-hydroxyalkanoic acid) synthase gene of the Gram-positive bacterium *Rhodococcus ruber*, *FEMS Microbiol. Lett.* **96**, 73–80.

PIEPER-FÜRST, U., MADKOUR, M. H., MAYER, F., STEINBÜCHEL, A. (1994), Purification and characterization of a 14-kDa protein that is bound to the surface of polyhydroxyalkanoic acid granules in *Rhodococcus ruber*, *J. Bacteriol.* **176**, 4328–4337.

POIRIER, Y., DENNIS, D. E., KLOMPARENS, K., SOMERVILLE, C. (1992a), Polyhydroxybutyrate, a bidegradable thermoplastic, produced in transgenic plants, *Science* **256**, 520–523.

POIRIER, Y., DENNIS, D., KLOMPARENS, K., NAWRATH, C., SOMERVILLE, C. (1992b), Perspectives on the production of polyhydroxyalkanoates in plants, *FEMS Microbiol. Rev.* **103**, 237–246.

PREUSTING, H., NIJENHUIS, A., WITHOLT, B. (1990), Physical characteristics of poly(3-hydroxyalkanoates) and poly(3-hydroxyalkenoates) produced by *Pseudomonas oleovorans* grown on aliphatic hydrocarbons, *Macromolecules* **23**, 4220–4224.

PREUSTING, H., KINGMA, J., WITHOLT, B. (1991), Physiology and polyester formation of *Pseudomonas oleovorans* in continuous two-liquid phase cultures, *Enzyme Microb. Technol.* **13**, 770–780.

PREUSTING, H., VAN HOUTEN, R., HOEFS, A., VAN LANGENBERGHE, E. K., FAVRE-BULLE, O., WITHOLT, B. (1993a), High cell density cultivation of *Pseudomonas oleovorans*: growth and production of poly(3-hydroxyalkanoates) in two-liquid phase batch and fed-batch systems, *Biotechnol. Bioeng.* **41**, 550–556.

PREUSTING, H., HAZENBERG, W., WITHOLT, B. (1993b), Continuous production of poly(3-hydroxyalkanoates) by *Pseudomonas oleovorans* in a high cell density two-liquid phase chemostat, *Enzyme Microb. Technol.* **15**, 1–6.

PREUSTING, H., KINGMA, J., HUISMAN, G., STEINBÜCHEL, A., WITHOLT, B. (1993c), Formation of polyester blends by a recombinant strain of *Pseudomonas oleovorans*: different poly(3-hydroxyalkanoates) are stored in separate granules, *J. Environ. Polym. Degrad.* **1**, 45–53.

PRIES, A., PRIEFERT, H., KRÜGER, N., STEINBÜCHEL, A. (1991), Identification and characterization of two *Alcaligenes eutrophus* gene loci relevant to the phenotype poly(β-hydroxybutyric acid)-leaky which exhibit homology to *ptsH* and *ptsI* or *Escherichia coli*, *J. Bacteriol.* **173**, 5843–5853.

Pries, A., Hein, S., Steinbüchel, A. (1992), Identification of a lipoamide dehydrogenase gene as second locus affected in poly(3-hydroxyalkanoic acid)-leaky mutants of *Alcaligenes eutrophus*, *FEMS Microbiol. Lett.* **97**, 227–234.

Ramsay, J. A. (1994) PHA: its separation from microbial biomass and its biodegradation, in: *Proc. Symp. Physiology, Kinetics, Production and Use of Biopolymers* (Braunegg, G., Ed.), pp. 49–58, Schloss Seggau, Austria.

Ramsay, J. A., Berger, E., Ramsay, B. A., Chavarie, C. (1990), Recovery of polyhydroxyalkanoic acid granules by a surfactant-hypochlorite treatment, *Biotechnol. Lett.* **4**, 221–226.

Reusch, R. N. (1989), Poly-β-hydroxybutyrate/calcium polyphosphate complexes in eukaryotic membranes, *Proc. Soc. Exp. Med. Biol.* **191**, 377–381.

Reusch, R. N. (1992), Biological complexes of poly-β-hydroxybutyrate, *FEMS Microbiol. Rev.* **103**, 119–130.

Reusch, R. N., Sadoff, H. L. (1981) Lipid metabolism during encystment of *Azotobacter vinelandii*, *J. Bacteriol.* **145**, 889–895.

Reusch, R. N., Sadoff, H. L. (1983), D-(−)-poly-β-hydroxybutyrate in membranes of genetically competent bacteria, *J. Bacteriol.* **156**, 778–788.

Reusch, R. N., Sadoff, H. L. (1988), Putative structure and functions of a poly-β-hydroxybutyrate/calcium polyphosphate channel in bacterial plasma membranes, *Proc. Natl. Acad. Sci. USA* **85**, 4176–4180.

Reusch, R. N., Sparrow, A. (1992), Transport of poly-β-hydroxybutyrate in human plasma, *Biophys. Biochim. Acta* **1123**, 33–40.

Reusch, R. N., Hiske, T. W., Sadoff, H. L. (1986), Poly-β-hydroxybutyrate membrane structure and its relationship to genetic transformability in *Escherichia coli*, *J. Bacteriol.* **168**, 553–562.

Reusch, R., Hiske, T., Sadoff, H., Harris, R., Beveridge, T. (1987), Cellular incorporation of poly-β-hydroxbutyrate into plasma membranes of *Eschericha coli* and *Azotobacter vinelandii* alters native membrane structure, *Can. J. Microbiol.* **33**, 435–444.

Riis, V., Mai, V. (1981), Gas chromatographic determination of poly-β-hydroxybutyric acid in microbial biomass after hydrochloric acid propanolysis, *J. Chromatogr.* **445**, 285–287.

Ritchie, G. A. F., Dawes, E. A. (1969), The non-involvement of acyl-carrier protein in poly-β-hydroxybutyric acid biosynthesis in *Azotobacter beijerinckii*, *Biochem. J.* **112**, 803–805.

Rodrigues, M. F. A., Da Silva, L. F., Gomez, G. C., Valentin, H. E., Steinbüchel, A. (in press), Biosynthesis of poly(3-hydroxybutyric acid-*co*-3-hydroxy-4-pentenoic acid) from unrelated substrates by *Burkholderia* sp., *Appl. Microbiol. Biotechnol.*

Saito, Y., Doi, Y. (1993), Biosynthesis of poly(3-hydroxyalkanoates) in *Pseudomonas aeruginosa* AO-232 from $^{13}$C-labelled acetate and propionate, *Int. J. Biol. Macromol.* **15**, 287–292.

Saito, T., Suzuki, K., Yamamoto, J., Fukui, T., Miwa, K., Tomita, K., Nakanishi, S., Odani, S., Suzuki, J. I., Ishikawa, K. (1989) Cloning, nucleotide sequence, and expression in *Escherichia coli* of the gene for poly(3-hydroxybutyrate) depolymerase from *Alcaligenes faecalis*, *J. Bacteriol.* **171**, 184–189.

Saito, T., Saegusa, H., Miyata, Y., Fukui, T. (1992), Intracellular degradation of poly(3-hydroxybutyrate) granules of *Zoogloea ramigera* I-16-M, *FEMS Microbiol. Rev.* **103,** 333–338.

Saito, T., Iwata, A., Watanabe, T. (1993), Molecular structure of extracellular poly(3-hydroxybutyrate) depolymerase from *Alcaligenes faecalis* T1, *J. Environ. Polym. Degrad.* **1**, 99–105.

Satoh, H., Mino, T., Matsuo, T. (1992), Uptake of organic substrates and accumulation of polyhydroxyalkanoates linked with glycolysis of intracellular carbohydrates under anaerobic conditions in the biological excess phosphate removal processes, *Water Sci. Technol.* **26**, 933–942.

Schembri, M. A., Bayly, R. C., Davies, J. K. (1994), Cloning and analysis of the polyhydroxyalkanoic acid synthase gene from *Acinetobacter* sp.: Evidence that the gene is both plasmid and chromosomally located, *FEMS Microbiol. Lett.* **118**, 145–152.

Scheper, T. (1991), *Bioanalytik,* 1st Edn. Braunschweig: Vieweg.

Schirmer, A., Jendrossek, D. (1994), Molecular characterization of the extracellular poly(3-hydroxyoctanoic acid) [P(3OH)] depolymerase gene of *Pseudomonas fluorescens* GK13 and of its gene product, *J. Bacteriol.* **176**, 7065–7073.

Schirmer, A., Jendrossek, D., Schlegel, H. G. (1993), Degradation of poly(3-hydroxyoctanoic acid) [P(3HO)] by bacteria: purification and properties of a P(3HO) depolymerase from *Pseudomonas fluorescens* GK13, *Appl. Environ. Microbiol.* **59**, 1220–1227.

Schlegel, H. G. (1992), Past and present cycle of carbon on our planet, *FEMS Microbiol. Rev.* **103**, 347–354.

Schlegel, H. G., Gottschalk, G. (1962), Poly-β-hydroxybuttersäure, ihre Verbreitung, Funktion und Biosynthese, *Angew. Chem.* **74**, 342–346.

SCHLEGEL, H. G., GOTTSCHALK, G. (1965), Verwertung von Glucose durch eine Mutante von *Hydrogenomonas* H16, *Biochem. Z.* **342**, 249–259.

SCHLEGEL, H. G., STEINBÜCHEL, A. (1981), Die relative Respirationsrate (RRR), ein neuer Belüftungsparameter, in: *Fermentation* (LAFFERTY, R. M., Ed.), pp. 11–26. Wien: Springer-Verlag.

SCHLEGEL, H. G., GOTTSCHALK, G., VON BARTHA, R. (1961), Formation and utilization of poly-$\beta$-hydroxybutyric acid by Knallgas bacteria (*Hydrogenomonas*), *Nature* **191**, 463–465.

SCHLEGEL, H. G., LAFFERTY, R., KRAUSS, I. (1970), The isolation of mutants not accumulating poly-$\beta$-hydroxybutyric acid, *Arch. Mikrobiol.* **71**, 283–294.

SCHNAITMAN, C. A. (1981), Cell fractionation, in: *Manual of Methods for General Bacteriology* (GERHARDT, P., MURRAY, R. G. E., COSTILOW, R. N., NESTER, E. W., WOOD, W. A., KRIEG, N. R., PHILLIPS, G. B., Eds.), pp. 52–64, 1st Edn. Washington, DC: American Society for Microbiology.

SCHUBERT, P., STEINBÜCHEL, A., SCHLEGEL, H. G. (1988), Cloning of the *Alcaligenes eutrophus* gene for synthesis of poly-$\beta$-hydroxybutyric acid and synthesis of PHB in *Escherichia coli*, *J. Bacteriol.* **170**, 5837–5847.

SEEBACH, D. (1992), PHB in the hands of a chemist, *FEMS Microbiol. Rev.* **103**, 215.

SENIOR, P. J., DAWES, E. A. (1971), PHB-biosynthesis and regulation of glucose metabolism in *Azotobacter beijerinckii*, *Biochem. J.* **125**, 55–66.

SENIOR, P. J., DAWES, E. A. (1973), The regulation of poly-$\beta$-hydroxybutyrate metabolism in *Azotobacter beijerinckii*, *Biochem. J.* **134**, 225–238.

SHIVELY, J. M. (1974), Inclusion bodies of prokaryotes, *Annu. Rev. Microbiol.* **28**, 167–187.

Showa Highpolymer Co. (1993), Bionolle. Biodegradable aliphatic polyester. *Technical Data Sheet.*

SIAPKAS, S. (1976), Radikalische Vernetzung von Verschnitten aus PHB and Polychloropren. *Thesis,* Technical University, Graz.

SLATER, S. C., VOIGE, W. H., DENNIS, D. E. (1988), Cloning and expression in *Escherichia coli* of the *Alcaligenes eutrophus* H16 poly-$\beta$-hydroxybutyrate biosynthetic pathway, *J. Bacteriol.* **170**, 4431–4436.

SLATER, S., GALLAHER, T., DENNIS, D. (1992), Production of poly-(3-hydroxybutyrate-*co*-3-hydroxyvalerate) in a recombinant *Escherichia coli* strain, *Appl. Environ. Microbiol.* **58**, 1089–1094.

SLEPECKY, R. A., LAW, J. H. (1960), A rapid spectrophotometric assay of alpha, beta-unsaturated acids and $\beta$-hydroxy acids, *Anal. Chem.* **32**, 1697–1699.

SLEPECKY, R. A., LAW, J. M. (1961), Synthesis and degradation of poly-$\beta$-hydroxybutyric acid in connection with sporulation of *Bacillus megaterium*, *J. Bacteriol.* **82**, 37–42.

SMIBERT, R. M., KRIEG, N. R. (1981), General characterization, in: *Manual of Methods for General Bacteriology* (GERHARDT, P., MURRAY, R. G. E., COSTILOV, R. N., NESTER, E. W., WOOD, W. A., KRIEG, N. R., PHILLIPS, G. B., Eds.), pp. 409–443. Washington, DC: American Society for Microbiology.

SMITH, E., WHITE, K. A., HOLT, D., FENTEM, P. A., BRIGHT, S. W. J. (1994), Expression of polyhydroxybutyrate in oilseed rape. *Lecture* presented at the *Conference "Advances in Biopolymer Engineering"*, Florida, January 23–28.

SONNLEITNER, B., HEINZLE, E., BRAUNEGG, G., LAFFERTY, R. M. (1979), Format kinetics of poly-$\beta$-hydroxybutyric (PHB) production in *Alcaligenes eutrophus* H16 and *Mycoplana rubra* R14 with respect to the dissolved oxygen tension in ammonium limited batch cultures, *Eur. J. Appl. Microbiol. Biotechnol.* **7**, 1–10.

SRIENC, F., ARNOLD, B., BAILEY, J. E. (1984), Characterization of intracellular accumulation of poly-$\beta$-hydroxybutyrate (PHB) in individual cells of *Alcaligenes eutrophus* H16 by flow cytometry, *Biotechnol. Bioeng.* **26**, 982–987.

STAGEMAN, J. F. (1985), *US Patent Application* 4562245.

STAM, H., VAN VERSEVELD, H. W., DE VRIES, W., STOUTHAMER, A. W. (1986), Utilization of poly-$\beta$-hydroxybutyrate in free-living cultures of *Rhizobium* ORS571, *FEMS Microbiol. Lett.* **35**, 215–220.

STEINBÜCHEL, A. (1991a), Polyhydroxyalkanoic acids, in: *Biomaterials* (BYROM, D., Ed.), pp. 123–213. Basingstoke: MacMillan Publishers.

STEINBÜCHEL, A. (1991b), Polyhydroxyfettsäuren – thermoplastisch verformbare und biologisch abbaubare Polyester aus Bakterien, *Nachr. Chem. Tech. Lab.* **38**, 1112–1124.

STEINBÜCHEL, A., PIEPER, U. (1992), Production of copolyesters of 3-hydroxybutyric acid and 3-hydroxyvaleric acid by a mutant of *Alcaligenes eutrophus* from single unrelated carbon sources, *Appl. Microbiol. Biotechnol.* **37**, 1–6.

STEINBÜCHEL, A., SCHLEGEL, H. G. (1989), Excretion of pyruvate by mutants of *Alcaligenes eutrophus,* which are impaired in the accumulation of poly($\beta$-hydroxybutyric acid) (PHB), under conditions permissive for synthesis of PHB, *Appl. Microbiol. Biotechnol.* **31**, 168–175.

STEINBÜCHEL, A., SCHLEGEL, H. G. (1991), Genetics of poly($\beta$-hydroxyalkanoic acid) synthesis in *Alcaligenes eutrophus, Mol. Microbiol.* **5**, 535–542.

STEINBÜCHEL, A., SCHMACK, H. (in press), Large-scale production of poly(3-hydroxyvaleric acid) by fermentation of *Chromobacterium violaceum,* processing and characterization of the homopolyester, *J. Environ. Polym. Degrad.*

STEINBÜCHEL, A., SCHUBERT, P. (1989), Expression of the *Alcaligenes eutrophus* poly(β-hydroxybutyric acid)-synthetic pathway in *Pseudomonas* sp., *Arch. Microbiol.* **153**, 101–104.

STEINBÜCHEL, A., WIESE, S. (1992), A *Pseudomonas* strain accumulating polyesters of 3-hydroybutyric acid and medium-chain-length 3-hydroxyalkanoic acids, *Appl. Microbiol. Biotechnol.* **37**, 691–697.

STEINBÜCHEL, A., KUHN, M., NIEDRIG, M., SCHLEGEL, H. G. (1983), Fermentation enzymes in strictly aerobic bacteria: comparative studies on strains of the genus *Alcaligenes* and on *Nocardia opaca* and *Xanthobacter autotrophicus, J. Gen. Microbiol.* **129**, 2825–2835.

STEINBÜCHEL, A., HUSTEDE, E., LIEBERGESELL, M., TIMM, A., PIEPER, U., VALENTIN, H. (1992), Moleclular basis for biosynthesis and accumulation of polyhydroxyalkanoic acids in bacteria, *FEMS Microbiol. Rev.* **103**, 217–230.

STEINBÜCHEL, A., GEBDZI, E. M., MARCHESSAULT, R. H., TIMM, A. (1993), Synthesis and production of poly(3-hydroxyvaleric acid) by *Chromobacterium violaceum, Appl. Microbiol. Biotechnol.* **39**, 443–449.

STEINBÜCHEL, A., VALENTIN, H. E., SCHÖNEBAUM, A. (1994), Application of recombinant gene technology for production of polyhydroxyalkanoic acids: Biosynthesis of poly(4-hydroxybutyric acid) homopolyester, *J. Environ. Polym. Degrad.* **2**, 67–76.

STEINBÜCHEL, A., AERTS, K., BABEL, W., FÖLLNER, LIEBERGESELL, M., MADKOUR, M. H., MAYER, F., PIEPER-FÜRST, U., PRIES, A., VALENTIN, H. E., WIECZOREK, R. (1995), Considerations on the structure and biochemistry of bacterial polyhydroxyalkanoic acid inclusions, *Can. J. Microbiol.* **41**, 94–105.

STEVENSON, L. H., SOCOLOFSKY, M. D. (1966), Cyst formation and poly(β-hydroxybutyric acid) accumulation in *Azotobacter, J. Bacteriol.* **91**, 639–645.

STEVENSON, L. H., SOCOLOFSKY, M. D. (1973), Role of poly-β-hydroxybutyric acid in cyst formation by *Azotobacter, Antonie van Leeuwenhoek* **39**, 341–350.

SUZUKI, T., YAMANE, T., SHIMIZU, S. (1986a), Mass production of poly-β-hydroxybutyric acid by fully automatic fed-batch culture of methylotrophs, *Appl. Microbiol. Biotechnol.* **23**, 322–329.

SUZUKI, T., YAMANE, T., SHIMIZU, S. (1986b), Mass production of poly-β-hydroxybutyric acid by fed-batch culture with controlled carbon/nitrogen feeding, *Appl. Microbiol. Biotechnol.* **24**, 370–374.

SUZUKI, T., YAMANE, T., SHIMIZU, S. (1986c), Kinetics and effect of nitrogen source feeding on production of poly-β-hydroxybutyric acid by fed-batch culture, *Appl. Microbiol. Biotechnol.* **24**, 366–369.

THOMPSON, E. D., NAKATA, H. M. (1973), Characterization and partial purification of β-hydroxybutyrate dehydrogenase from sporulating cells of *Bacillus cereus* T, *Can. J. Microbiol.* **19**, 673–677.

THOMSON, A. W., O'NEILL, J. G., WILKINSON, J. F. (1976), Acetone production by methylobacteria, *Arch. Microbiol.* **109**, 243–246.

TIMM, A., STEINBÜCHEL, A. (1990), Formation of polyesters consisting of medium-chain-length 3-hydroxyalkanoic acids from gluconate by *Pseudomonas aeruginosa* and other fluorescent pseudomonads, *Appl. Environ. Microbiol.* **56**, 3360–3367.

TIMM, A., STEINBÜCHEL, A. (1992), Cloning and molecular characterization of polyhydroxyalkanoic acid gene locus of *Pseudomonas aeruginosa* PAO1, *Eur. J. Biochem.* **209**, 15–30.

TIMM, A., BYROM, D., STEINBÜCHEL, A. (1990), Formation of blends of various poly(3-hydroxyalkanoic acids) by a recombinant strain of *Pseudomonas oleovorans, Appl. Microbiol. Biotechnol.* **33**, 296–301.

TIMM, A., WIESE, S., STEINBÜCHEL, A. (1994), A general method for cloning of polyhydroxyalkanoic acid-synthase genes from pseudomonads belonging to the rRNA homology group I, *Appl. Microbiol. Biotechnol.* **40**, 669–675.

TOMITA, K., SAITO, T., FUKUI, T. (1983), Bacterial metabolism of poly-β-hydroxybutyrate, in: *Biochemistry of Metabolic Processes* (LENNON, D. L. F., STRATMAN, F. W., ZAHLTEN, R. N., Eds.), pp. 353-366. New York: Elsevier Science Publishing Inc.

TROTSENKO, Y. A., DORONINA, N. V., SOKOLOV, A. P., OSTAFIN, M. (1992), PHB synthesis by methane- and methanol-utilizing bacteria, in: *Proc. Int. Symp. Bacterial Polyhydroxyalkanoates* (SCHLEGEL, H., G., STEINBÜCHEL, A., Eds.), pp. 393–394. Göttingen: Goltze-Druck.

UEDA, S., MATSUMOTO, S., TAKAGI, A., YAMANE, T. (1992a), *N*-amyl alcohol as a substrate for the production of poly(3-hydroxy-*co*-3-hydroxyvalerate) by bacteria, *FEMS Microbiol. Lett.* **98**, 57–60.

UEDA, S., MATSUMOTO, S., TAKAGI, A., YAMANE, T. (1992b), Synthesis of poly(3-hydroxybutyrate-

*co*-3-hydroxyvalerate) from methanol and *n*-amyl alcohol by the methylotrophic bacteria *Paracoccus denitrificans* and *Methylobacterium extorquens*, *Appl. Environ. Microbiol.* **58**, 3574–3579.

Valentin, H., Steinbüchel, A. (1993), Cloning and characterization of the *Methylobacterium extorquens* polyhydroxyalkanoic acid synthase structural gene, *Appl. Microbiol. Biotechnol.* **39**, 309–317.

Valentin, H. E., Steinbüchel, A. (1994), Application of enzymatically synthesized short-chain-length hydroxy fatty acid coenzyme A thioesters for assay of polyhydroxyalkanoic acid synthases, *Appl. Microbiol. Biotechnol.* **40**, 699–709.

Valentin, H., Schönebaum, A., Steinbüchel, A. (1992), Identification of 4-hydroxyvaleric acid as a constituent in biosynthetic polyhydroxyalkanoic acids from bacteria, *Appl. Microbiol. Biotechnol.* **36**, 507–514.

Valentin, H. E., Lee, E. Y., Choi, C. Y., Steinbüchel, A. (1994), Identification of 4-hydroxyhexanoic acid as a new constituent of biosynthetic polyhydroxyalkanoic acids from bacteria, *App. Microbiol. Biotechnol*, **40**, 710–716.

Valentin, H. E., Zwingmann, G., Schönebaum, A., Steinbüchel, A. (1995), Metabolic pathway for biosynthesis of poly(3-hydroxybutyrate-*co*-4-hydroxybutyrate) from 4-hydroxybutyrate by *Alcaligenes eutrophus*, *Eur. J. Biochem.* **227**, 43–60.

Van Lautem, N., Gilain, J. (1986), Extraction of poly(β-hydroxybutyrates) from an aqueous suspension of microorganisms, *Eur. Patent Application* EP 168,095.

Vollbrecht, D., El Nawawy, M. A. (1980), Restricted oxygen supply and excretion of metabolites. I. *Pseudomonas* spec. and *Paracoccus denitrificans, Eur. J. Appl. Microbiol. Biotechnol.* **9**, 1–8.

Vollbrecht, D., Schlegel, H. G., Stoschek, G., Janczikowski, A. (1979), Excretion of metabolites by hydrogen bacteria. IV. Respiration rate-dependent formation of primary metabolites and of poly-3-hydroxybutanoate, *Eur. J. Appl. Microbiol. Biotechnol.* **7**, 267–276.

Wallen, L. L., Rohwedder, W. K. (1974), Poly-β-hydroxyalkanoate from activated sludge, *Environ. Sci. Technol.* **8**, 576–579.

Walker, J., Whitton, J. R., Alderson, B. (1982), *Eur. Patent Application* 46017.

Ward, A. C., Rowley, B. I., Dawes, E. A. (1977), Effect of oxygen and nitrogen limitation on poly-β-hydroxybutyrate biosynthesis in ammonium grown *Azotobacter beijerinckii*, *J. Gen. Microbiol.* **102**, 61–68.

Weaver, T. L., Patrick, M. A., Dugan, P. R. (1975), Whole-cell and membrane lipids of the methylotrophic bacterium *Methylosinus trichosporium, J. Bacteriol.* **124**, 602–605.

Whittenbury, R., Phillips, K. C., Wolikinson, J. F. (1970), Enrichment, isolation and some properties of methane-utilizing bacteria, *J. Gen. Microbiol.* **61**, 205–218.

Wieczorek, R., Pries, A., Steinbüchel, A., Mayer, F. (1995), Analysis of a 24-Kilodalton protein associated with the polyhydroxyalkanoic acid granules in *Alcaligenes eutrophus*, *J. Bacteriol.* **177**, 2425–2435.

Wilde, E. (1962), Untersuchungen über Wachstum und Speicherstoffsynthese von *Hydrogenomonas, Arch. Mikrobiol.* **43**, 109–137.

Williams, D. R., Anderson, A. J., Dawes, E. A., Ewing, D. F. (1994), Production of a copolyester of 3-hydroxybutyric acid and 3-hydroxyvaleric acid from succinic acid by *Rhodococcus ruber*. Biosynthetic considerations, *Appl. Microbiol. Biotechnol.* **40**, 717–723.

Williamson, D. H., Wilkinson, J. F. (1958), The isolation and estimation of poly-β-hydroxybutyrate inclusions of *Bacillus* species, *J. Gen. Microbiol.* **19**, 198–209.

Witholt, B., DeSmet, M.-J., Kingma, J., Van Beilen, J. B., Kok, M., Lageveen, R. G., Eggink, G. (1990), Bioconversions of aliphatic compounds by *Pseudomonas oleovorans* in multiphase bioreactors: background and economic potential, *Trends Biotechnol.* **8**, 46–52.

Wittwer, F., Tomka, I. (1984), Polymer composition for injection mouling, *Eur. Patent Application* 0118240 A2.

Yamane, T. (1992), Cultivation engineering of microbial bioplastics production, *FEMS Microbiol. Rev.* **103**, 257–264.

Yoshie, N., Goto, Y., Sakurai, M., Inoue, Y., Chujo, R., Doi, Y. (1992), Biosynthesis and NMR studies of deuterated poly(3-hydroxybutyrate) produced by *Alcaligenes eutrophus* H16, *Int. J. Biol. Macromol.* **14**, 81–86.

Zeneca (1993), Biopol Resin – Nature's plastic. The natural choice. *Brochure.* Zeneca BioProducts, Billingham, UK.

Zhang, H., Obias, V., Gonyer, K., Dennis, D. (1994), Production of polyhydroxyalkanoates in sucrose-utilizing recombinant *Escherichia coli* and *Klebsiella* strains, *Appl. Environ. Microbiol.* **60**, 1198–1205.

# 14a Amino Acids – Technical Production and Use

W. LEUCHTENBERGER

Halle-Künsebeck, Germany

# 1 Introduction

Amino acids are rather simple organic compounds that contain at least one amino group and one carboxylic function in their molecular structure.

Among the naturally occurring amino acids the protein-forming α-amino acids are most widely distributed and of considerable economical interest. They differ from glycine (aminoacetic acid), the simplest proteinogenic amino acid, in that they possess a side chain attached *alpha* to the carboxylic function. The side chain may be aliphatic, aromatic, or heterocyclic and may carry further functionality. In addition, the α-amino acids found in proteins are optically active and occur as L-enantiomers.

About 20 L-amino acids are the building blocks for the immense variety of peptides and proteins, the exciting biopolymers of life. Eight L-amino acids are essential for mammalians (Tab. 1).

On the other hand, α-amino acids with D-configuration were also found as metabolites, e.g., in cell walls of bacteria, in peptide antibiotics, and in plants.

# 2 Use and Market of Amino Acids

The use of amino acids is based on their nutritional value, taste, physiological activities, and chemical characteristics. The main fields of application are food (human nutrition), feed (animal nutrition), cosmetics and

**Tab. 1.** Protein-Forming L-Amino Acids

General structure: (R)–CH(NH$_2$)–COOH

| Name | IUPAC-Abbr. | Symbol | Group R- | M | Essential | Taste |
|---|---|---|---|---|---|---|
| L-Alanine | Ala | A | $CH_3-$ | 89.09 | no | sweet |
| L-Arginine | Arg | R | $H_2N-C(=NH)-NH-CH_2-CH_2-CH_2-$ | 174.20 | semi | bitter |
| L-Asparagine | Asn | N | $H_2N-C(=O)-CH_2-$ | 132.13 | no | weakly acidic |
| L-Aspartic acid | Asp | D | $HOOC-CH_2-$ | 133.10 | no | acidic |
| L-Cysteine | Cys | C | $HS-CH_2-$ | 121.16 | no | sulfurous |
| L-Glutamic acid | Glu | E | $HOOC-CH_2-CH_2-$ | 147.13 | no | acidic umami[a] |
| L-Glutamine | Gln | Q | $H_2N-C(=O)-CH_2-CH_2-$ | 146.15 | no | weakly sweet |
| Glycine | Gly | G | H | 75.07 | no | sweet |
| L-Histidine | His | H | imidazol-4-yl–$CH_2-$ (ring N, NH) | 155.16 | semi | bitter |

| | | | | | | |
|---|---|---|---|---|---|---|
| L-Isoleucine | Ile | I | $CH_3-CH_2-CH(CH_3)-$ | 131.18 | yes | bitter |
| L-Leucine | Leu | L | $CH_3-CH(CH_3)-CH_2-$ | 131.18 | yes | bitter |
| L-Lysine | Lys | K | $H_2N-CH_2-CH_2-CH_2-CH_2-$ | 146.19 | yes | sweet |
| L-Methionine | Met | M | $CH_3-S-CH_2-CH_2-$ | 149.21 | yes | bitter sulfurous |
| L-Phenylalanine | Phe | F | $C_6H_5-CH_2-$ | 165.19 | yes | bitter |
| L-Proline | Pro | P | ([b]) | 115.13 | no | sweet |
| L-Serine | Ser | S | $HO-CH_2-$ | 105.09 | no | sweet |
| L-Threonine | Thr | T | $CH_3-CH(OH)-$ | 119.12 | yes | sweet |
| L-Tryptophan | Trp | W | indol-3-yl$-CH_2-$ | 204.23 | yes | bitter |
| L-Tyrosine | Tyr | Y | $HO-C_6H_4-CH_2-$ | 181.19 | no | bitter |
| L-Valine | Val | V | $CH_3-CH(CH_3)-$ | 117.15 | yes | bitter |

[a] Umami is a basic flavor.
[b] L-Proline is a cyclic amino acid ("imino acid"); the amino group is integrated in the ring system. (pyrrolidine ring: H, COOH, NH)

medicine, and intermediates in the chemical industry. Although during the last decades dynamic developments in both areas, in the market (applications) and in industry (methods of production), have occurred, only few amino acids play an important role as bulk products. Except for methionine, which is manufactured and used in the racemic (D,L) form, all other industrially produced amino acids need the L-form when used as food and feed additives or as components in cosmetics and pharmaceuticals. Some antibiotics contain D-amino acids.

The whole market is estimated to amount to about 3 billion US $ in 1995, covering 38% for food, 54% for feed, and 8% for other applications such as medicine and cosmetics.

## 2.1 Use in Food

Amino acids are almost tasteless. However, they contribute to the flavor of food. They exhibit synergistic flavor-enhancing properties and are precursors of natural aromas. Therefore, L-amino acids and protein hydrolyzates are useful additives in the food industry.

*Monosodium L-glutamate (MSG)* has been recognized as a flavoring agent for seaweed ("konbu") and soy sauce since 1908 (IKEDA, 1908). MSG has a meaty taste and enhances the flavor of food, usually in combination with nucleotides (inosinate). Currently, the annual demand for MSG is more than 800,000 t world-wide. Thus MSG is the amino acid with the largest production capacity.

*Glycine* has a refreshing sweetish flavor and is, e.g., an important component in mussels and prawns. When added to vinegar, pickles, and mayonnaise it attenuates the sour taste and lends a note of sweetness to their aroma. Moreover, glycine is bacteriostatic in food and is also used as an antioxidant for emulsifiers.

Dipeptides and oligopeptides mostly reveal bitter flavors. One exception is the methyl ester of the dipeptide L-aspartyl-L-phenylalanine (aspartame), which is 150–200 times sweeter than sucrose and now commonly used as an artificial sweetener in low-calorie drinks. Aspartame is a striking example for a successful introduction of a new product in the food market based on L-amino acids (*L-aspartic acid, L-phenylalanine*). In 1994, about 10,000 t aspartame were produced world-wide.

*L-Cysteine* is used as a flour additive in the manufacture of baked goods and pasta. It improves the structure of baked products while permitting shorter kneading times and is effective as an anti-browning agent for food.

## 2.2 Use in Feed

The use of amino acids for the nutrition of monogastric animals is a significant driving force and still growing in the amino acid market. Half of the proteinogenic amino acids are essential for animals, such as pig and poultry. Whereas most natural feeds as fish meal, soya bean, wheat or maize protein are deficient in methionine, lysine and threonine, the requirements of those limiting amino acids in livestock, however, are comparatively high. When formulating a feed mix, the manufacturer may use either an excess of feed protein to make sufficient amounts of the limiting amino acids available or he may provide a minimum of natural protein and supplement it with appropriate quantities of *D,L-methionine, L-lysine* and *L-threonine.* The latter variant is more economic due to the relatively cheap production of those amino acids. Moreover, minimizing the total protein content in feedstuff helps to decrease enviromnental problems, such as manure formation in pig keeping.

For D,L-methionine and L-lysine (hydrochloride) a market of 300,000 t $a^{-1}$ and 250,000 t $a^{-1}$, respectively, has been established. Recently, L-threonine has successfully entered the feed market due to improved fermentation processes; the world-wide capacity is estimated to reach already 10,000 t $a^{-1}$.

*L-Tryptophan* is also a limiting amino acid, especially in corn-based feed. However, it is still too expensive to be used regularly for supplementation.

## 2.3 Pharmaceuticals

Parenteral nutrition with L-amino acid infusion solutions is a well-known component of clinical pre- and post-operative nutrition therapy. A standard infusion solution contains the eight classical essential amino acids, the semi-essential amino acids L-arginine and L-histidine, and several non-essential amino acids, generally glycine, L-alanine, L-proline, L-serine and L-glutamic acid.

In the pharmaceutical industry these amino acids are required in high quality (pyrogen free) at a rate of more than 5,000 t $a^{-1}$. Many therapeutic agents are derived from natural or non-natural amino acids. L-Arginine and its salt with L-aspartic or L-glutamic acid is effective in the treatment of hyperammonemia and hepatic disorders. Potassium–magnesium L-aspartate is used to assist recovery from fatigue, heart failure, and liver disease. L-Cysteine protects tissues from oxidation or inactivation and shows detoxifying effects; it also protects from radiation injury.

Some amino acid analogs are also used therapeutically, e.g., levodopa, (*S*)-3-(3,4-dihydroxyphenyl)alanine for treatment of Parkinson's disease, and oxitriptan, (*S*)-5-hydroxytryptophan as psychotherapeutic agent (antidepressant). Other attractive drugs based on L-proline derivatives and peptides are the angiotensin converting enzyme (ACE) inhibitors captopril, enalapril, and lisinopril. Finally, some D-amino acids are used in drugs: D-phenylglycine and D-4-hydroxyphenylglycine are components of the semi-synthetic antibiotics ampicillin and amoxycillin. D-Penicillamine is a strong chelating agent and helps to eliminate toxic metal ions, i.e., copper in Wilson's disease. It is also used in the treatment of severe rheumatoid arthritis.

## 2.4 Other Applications

Because of skin protecting properties amino acids as well as proteins are widely used in skin and hair cosmetics. Cysteine acts as a reducing agent in permanent wave preparations of hair.

Other interesting physicochemical properties such as thermal stability, low volatility, buffering capacity, chelating capability, but also low toxicity and easy biodegradation are the reason for a range of potential uses of amino acids and their derivatives in specific applications of the chemical industry.

Potential uses include N-acyl-amino acids as surfactants, detergents, and monomers of synthetic resins, polyamino acid esters as coatings for natural or synthetic materials (leather) and various amino acids as corrosion inhibitors, stabilizers of emulsions, and even as building blocks for agricultural products (pesticides, plant growth regulators).

# 3 Industrial Production of Amino Acids

## 3.1 General Methods

Four basic processes are suitable for the production of amino acids

- chemical synthesis,
- extraction,
- fermentation,
- enzymatic catalysis.

*Chemical synthesis* is applied to produce either the achiral amino acid glycine or an amino acid as a racemate that can be marketed as such, e.g., D,L-methionine. The manufacture of L-amino acids by this route affords further steps: resolution as usual after derivation, separation of the L-enantiomer, and recycling of the unwanted D-enantiomer. In some cases L-amino acids are produced directly from prochiral precursors by means of chiral catalysts (asymmetric synthesis).

The *extraction* process has the advantage to offer access to nearly all proteinogenic L-amino acids by isolation from protein hydrolyzates; starting materials are protein-rich products, such as keratin, feather, or blood meal.

The use of overproducing microbial strains in *fermentation* processes with sucrose or glucose as carbon source is currently the most economic access for the bulk amino acids such as monosodium L-glutamate, L-lysine hydrochloride, and L-threonine. Even the aromatic amino acids, L-phenylalanine and L-tryptophan – ten years ago only available by enzymatic processes – are now being produced by means of potent mutants of bacteria.

The forth method, *enzymatic catalysis,* uses whole cells or active cell components (enzymes) as such or, if possible, as immobilized biocatalysts in continuously operated reactors. The competitiveness of enzymatic processes depends on the availability and the price of the substrate, on the activity and stability of the involved enzyme(s), and on the simplicity of product recovery. In Tab. 2 the amino acids and their preferred production methods are listed, classified by size of production volume.

## 3.2 Production of Amino Acids Mainly Used in the Food Industry

### 3.2.1 L-Glutamic Acid

In 1957, a remarkable soil bacterium was discovered (KINOSHITA et al., 1957) which was able to excrete considerable amounts of L-glutamate. Fermentation processes using strains of this bacterium, later called *Corynebacterium glutamicum,* have been successfully commercialized not only for L-glutamic acid but also for the production of other economically important amino acids until today (KINOSITHA, 1972).

Numerous coryneform microorganisms have been isolated and found to be able to

**Tab. 2.** Production of Amino Acids, Methods and Volume

| Type | Amino Acid | Preferred Production Method |
|---|---|---|
| I | L-Alanine | enzymatic catalysis |
| I | L-Asparagine | extraction |
| I | L-Glutamine | fermentation, extraction |
| I | L-Histidine | fermentation, extraction |
| I | L-Isoleucine | fermentation, extraction |
| I | L-Leucine | fermentation, extraction |
| I | L-Methionine | enzymatic catalysis |
| I | L-Proline | fermentation, extraction |
| I | L-Serine | fermentation, extraction |
| I | L-Tyrosine | extraction |
| I | L-Valine | enzymatic catalysis, fermentation |
| II | L-Arginine | fermentation, extraction |
| II | L-Cysteine | reduction of L-cystine (extraction) |
| II | L-Tryptophan | fermentation |
| III | Glycine | chemical synthesis |
| III | L-Aspartic acid | enzymatic catalysis |
| III | L-Phenylalanine | fermentation |
| III | L-Threonine | fermentation |
| IV | D,L-Methionine | chemical synthesis |
| IV | L-Glutamic acid | fermentation |
| IV | L-Lysine | fermentation |

| | |
|---|---|
| Type I | 100–1000 t $a^{-1}$ |
| Type II | 1000–8000 t $a^{-1}$ |
| Type III | 8000–100000 t $a^{-1}$ |
| Type IV | 100000–800000 t $a^{-1}$ |

overproduce L-glutamic acid and other amino acids, e.g., *Brevibacterium flavum, B. lactofermentum,* and *Microbacterium ammoniaphilum.* Because of minor differences in the character of these bacteria (EIKMANNS et al., 1991) which are all gram-positive, non-spore forming, non-motile and requiring biotin for growth, the name of genus *Corynebacterium* was suggested for these coryneform bacteria (LIEBL et al., 1991).

For industrial production of L-glutamic acid, molasses (sucrose), starch hydrolyzates (glucose), and ammonium sulfate are generally used as carbon and nitrogen sources, respectively (HIROSE et al., 1985).

Key factors in controlling the fermentation are

- the presence of biotin in an optimal concentration that influences cell growth and the excretion of L-glutamate,
- a sufficient supply of oxygen to reduce the accumulation of by-products, such as lactic and succinic acid.

In biotin-rich fermentation media the addition of penicillin or cephalosporin C favors the overproduction of L-glutamic acid due to effects on the cell membrane. The supplementation of fatty acids also results in an increased permeability of the cells thus enhancing glutamate excretion.

In the past the mechanism of glutamate excretion was simply explained as a "leakage" or "overflow" phenomenon (KIKUCHI and NAKAO, 1986). Recently, KRÄMER (1993, 1994) discussed three mechanisms of amino acid secretion and reported that a specific carrier system exists, which is responsible for active glutamate transport in *C. glutamicum* (Fig. 1).

Generally, the intracellular accumulation of glutamate does not reach levels sufficient for feedback control in glutamate overproducers due to rapid excretion of glutamate. However, the regulatory mechanisms of L-glutamic acid biosynthesis have been studied intensively to obtain mutants with increased productivity (Fig. 2).

Two enzymes have been shown to play key roles in the biosynthesis of L-glutamic acid (SHIIO and UJIGAWA, 1980):

- Phosphenolpyruvate carboxylase (PEPC) catalyzes carboxylation of phosphenolpyruvate to yield oxaloacetate; it is inhibited by L-aspartic acid and repressed by both L-aspartic acid and L-glutamic acid.
- $\alpha$-Ketoglutarate dehydrogenase (KDH) converts $\alpha$-ketoglutarate to succinyl-CoA; in L-glutamate overproducing strains KDH limits further oxidation of $\alpha$-ketoglutarate to carbon dioxide and succinate thus favoring the formation of L-glutamic acid:

In L-glutamate overproducing strains the $k_m$ value of KDH for $\alpha$-ketoglutarate was nearly two magnitudes lower than that of L-glutamic acid dehydrogenase (GDH) which catalyzes the last step to L-glutamate. Consequently, $v_{max}$ of GDH was proved to be about 150 times higher than that of KDH. The *C. glutamicum gdh* gene has been isolated and characterized recently (BÖRMANN et al., 1992).

In acetate-containing media the formation of enzymes in the glyoxylate cycle is stimulated (isocitrate lyase, malate synthetase) and the metabolic switch of the TCA cycle to the glyoxylate cycle may be enhanced (SHIIO and OZAKI, 1986). A strain of *Microbacterium ammoniaphilum* cultured under biotin-deficient conditions produced 58% of L-glutamic acid formed from glucose via phosphoenolpyruvate, citrate, and $\alpha$-ketoglutarate, and the other 42% via the TCA or the glyoxylate cycle (WALKER et al., 1982).

Further strain development resulted in mutants either having sensitivity in cell permeability (MOMOSE and TAGAKI, 1978) or having the capability of increased carbon dioxide fix-

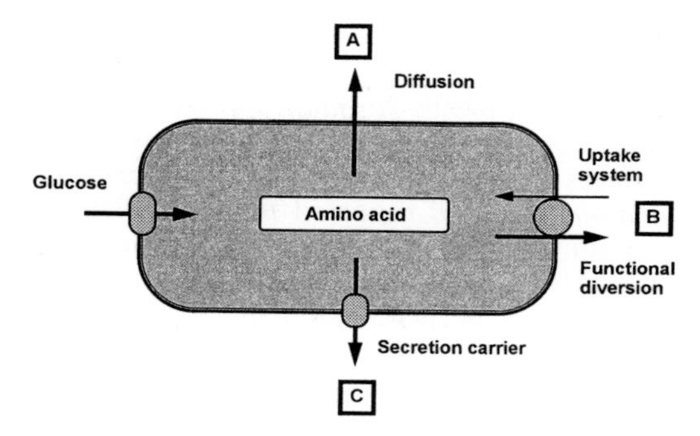


**Fig. 1.** Three models (A–C) explaining the mechanisms of amino acid secretion in *Corynebacterium glutamicum* (KRÄMER, 1993).

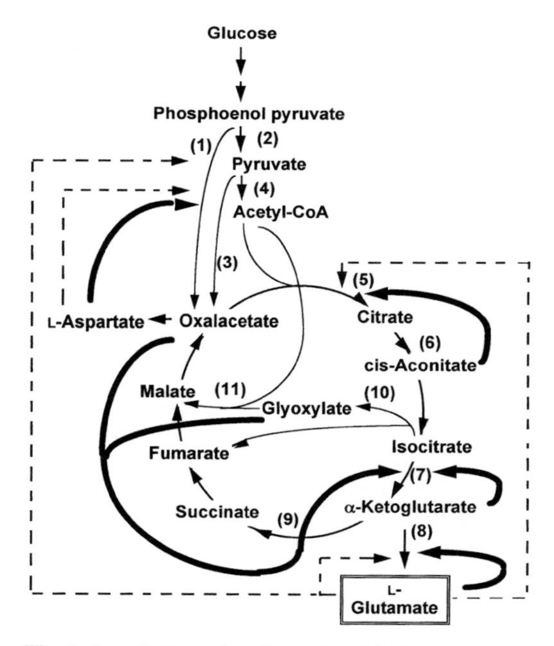


**Fig. 2.** Regulation of L-glutamic acid biosynthesis in *Corynebacterium glutamicum* (SHIIO and UJIGAWA, 1978).
→ feedback inhibition, - - -→ repression; enzyme (coding gene): 1 phosphoenolpyruvate carboxylase (*ppc*), 2 pyruvate kinase (*pyk*), 3 pyruvate carboxylase (*pyc*), 4 pyruvate dehydrogenase (*pdh*), 5 citrate synthase (*glt*A), 6 aconitase (*cit*B), 7 isocitrate dehydrogenase (*icd*), 8 L-glutamate dehydrogenase (*gdh*), 9 $\alpha$-ketoglutarate dehydrogenase (*ace*E), 10 isocitrate lyase (*ace*A), 11 malate syntethase (*ace*B).

ation (TOSAKA et al., 1981) or having a too low activity level in pyruvate dehydrogenase (ONO et al., 1980). Another approach was focussed on the development of thermophilic mutants. A strain of *Corynebacterium thermoaminogenes* is reported to accumulate L-glutamic acid at a temperature above 43 °C (YAMADA and SETO, 1988).

In large-scale production the formation of trehalose very often reduces the product yield. Trehalose consists of two $\alpha$-1,1 bond glucose molecules and is excreted by the bacteria as an osmoprotectant. A process has recently been developed and successfully industrialized in which trehalose formation is controlled and decreased by culturing the overproducing mutant in media with inverted sucrose from molasses (YOISHII et al., 1993a, b).

Today, wild-type isolates as well as mutants developed by classical breeding or even strains constructed by modern techniques using cell fusion or recombinant DNA methods (TSUCHIDA et al., 1981) are available for industrial L-glutamate production.

In addition, a new type of fermentation process was reported which uses a strain to overproduce L-glutamic acid and L-lysine simultaneously. The cultivation of an auxotrophic regulatory mutant of *B. lactofermentum* in a medium supplemented with polyoxyethylenesorbitan monopalmitate as surface-active agent resulted in the accumulation of 162 g $L^{-1}$ amino acids (105 g L-lysine·HCl + 57 g L-glutamic acid). This corresponds to a production rate of 3.8 g $L^{-1}$ $h^{-1}$ (L-lysine·HCl + L-glutamic acid) (SHIRATSUCHI et al., 1995).

The success of an economic production basically depends on the know-how in fermentation technology and in upstream and downstream processing and on the skill of employees in research, development, and production of the manufacturer.

A multi-step inoculation procedure followed by the fed-batch mode in the main fermentation up to the 500 $m^3$ scale is still the preferred technology for the production of L-glutamic acid. Alternative processes such as productions in airlift fermenters (WU and WU, 1992) or in continuously operated fermenters using immobilized cells (HENKEL et al., 1990; AMIN, 1994) or cell-recycling techniques (ISHIZAKI et al., 1993) are not competitive so far.

A scheme of production steps is given in Fig. 3. Critical operations are the steam-forced sterilization of the fermenter and batchwise or continuous sterilization of the culture medium to prevent contamination by foreign microbes. During the fermentation temperature, pH, dissolved oxygen and sugar consumption have to be controlled as important process parameters.

When the fermentation is completed after 40–60 h, the production strain may have accumulated more than 150 g $L^{-1}$ L-glutamic acid. After deactivation of the broth, product recovery is started by biomass separation using centrifuges or ultrafiltration units, concentration of the centrifugate or filtrate, followed by crystallization, filtration, and drying. As a novel aspect, it is proposed to use the mother liquor for feed supplementation (YOSHIMURA et al., 1994, 1995a, b). In the future, new approaches in strain development may continue. Moreover, there might be a potential

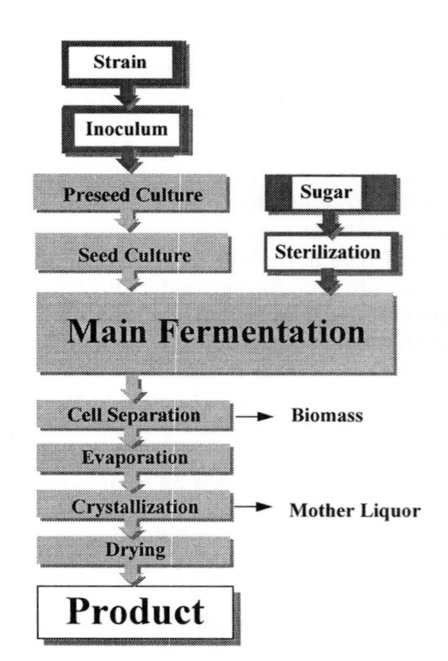


**Fig. 3.** Flow diagram of fermentation and downstream processing.

for improvements in L-glutamate fermentation by employing techniques such as computer supported control of the process (QIAN, 1989), fuzzy expert systems (KISHIMOTO et al., 1991; KITSUTA and KISHIMOTO, 1994), and the simulation of relevant effects in fermentation (NAGY et al., 1995).

### 3.2.2 L-Aspartic Acid

L-Aspartic acid is industrially manufactured by an enzymatic process in which aspartase (L-aspartate ammonia lyase, EC 4.3.1.1) catalyzes the addition reaction of ammonia to fumaric acid (CHIBATA et al., 1985; see Chapter 14b, Sect. 3.2.1). Advantages of the enzymatic production method are higher product concentration and productivity and the formation of fewer by-products. Thus L-aspartic acid can be easily separated from the reaction mixture by crystallization.

A process based on the continuous production of L-aspartic acid by means of carrier-fixed aspartase isolated from *Escherichia coli* was first commercialized in Japan (YOKOTE et al., 1978).

Competitive processes for L-aspartic acid are the use of resting or dried cells with high contents of aspartase. A 4.5% suspension of aspartase containing *B. flavum* cells could be recycled in seven repeated batches and achieved concentrations up to 166 g $L^{-1}$ L-aspartate (TERASAWA et al., 1985). In 1973, an immobilized cell system based on *E. coli* cells entrapped in polyacrylamide gel lattice was introduced for large-scale production (SATO et al., 1975). This example represents the first industrial application of immobilized microbial cells in a fixed-bed reactor. Further improvements, immobilization of the cells in $\kappa$-carrageenan, resulted in remarkably increased operational stability thus obtaining a biocatalyst half-life time of almost two years (CHIBATA et al., 1986). A column packed with the $\kappa$-carrageenan-immobilized system permits a theoretical productivity of 140 g $L^{-1}$ $h^{-1}$ L-aspartate.

The industry was prompted to develop improved processes by the fact that L-aspartic acid gained importance as intermediate for the manufacture of the dipeptide sweetener aspartame (methyl ester of L-aspartyl-L-phenylalanine).

For continuous production *E. coli* immobilized with polyurethane (FUSEE et al., 1981) or with polyethyleneimine and glass fiber support (SWANN and NOLF, 1986) and $\kappa$-carrageenan-immobilized *Pseudomonas putida* (MICHELET et al., 1984) have been reported and protected, respectively.

Less successful was the search for strains that overproduce L-aspartate by fermentation on sugar basis. A pyruvate kinase-deficient mutant of *B. flavum* accumulates up to 22.6 g $L^{-1}$ in a glucose-containing medium (MORI and SHIIO, 1984), not enough to be competitive with the enzymatic processes.

On the other hand, genetic recombination techniques helped to improve aspartase containing strains. An *asp*A gene bearing plasmid (pBR322::*asp*A-*par*) elevated aspartase formation in *E. coli* K-12 about 30-fold (NISHIMURA et al., 1987).

### 3.2.3 L-Phenylalanine

In previous large-scale production processes for L-phenylalanine enzymatic methods have been applied (see Chapter 14b, Sect. 3.2.2):

(1) Resolution of *N*-acetyl-D,L-phenylalanine by carrier-fixed microbial acylase: This process provided pharmaceutical-grade L-phenylalanine, but was hampered since the D-enantiomer had to be racemized and recycled.

(2) Stereoselective and enantioselective addition of ammonia to *trans*-cinnamic acid catalyzed by L-phenylalanine ammonia lyase (PAL, EC 4.3.1.5): PAL-containing *Rhodotorula rubra* has been used in an industrial process (MCGUIRE, 1986) to supply L-phenylalanine for the first production campaign of the sweetener aspartame. When continuously operated in an immobilized whole-cell reactor, the bioconversion reached concentrations up to 50 g $L^{-1}$ L-phenylalanine at a conversion rate of about 83% (EVANS et al., 1987a).

Other processes started from phenylpyruvate with L-aspartic acid as amine donor using immobilized cells of *E. coli* (CALTON et al.,

1986) or from α-acetamido cinnamic acid by means of immobilized cells of a *Corynebacterium equi* strain (EVANS et al., 1987b). In both cases L-phenylalanine concentrations of up to 30 g $L^{-1}$ and more have been achieved (molar yields as high as at least 98%). However, fermentation processes based on glucose consuming L-phenylalanine overproducing mutants of *E. coli* and coryneform strains turned out to be more economical.

The biosynthetic pathway for aromatic amino acids in bacteria is strictly regulated (PITTARD, 1987). L-Phenylalanine is formed in ten enzymatic steps starting from erythro-4-phosphate and phosphoenolpyruvate (Fig. 4).

The biosynthesis is governed by the first key enzyme 3-deoxy-D-arabinoheptulosonate-7-phosphate synthase (DAHPS) which is inhibited by both amino acids, L-phenylalanine and L-tyrosine, and repressed by L-tyrosine. The other important enzyme is prephenate dehydratase (PDT), also inhibited by L-phenylalanine, but stimulated by L-tyrosine. To overcome these regulatory mechanisms either auxotrophs of *C. glutamicum* have been constructed or L-phenylalanine analogs, e.g., *p*-aminophenylalanine and *p*-fluorophenylalanine, have been applied. The latter variant leads to resistant mutants of *B. flavum* or *B. lactofermentum* (SHIIO, 1986). These auxotrophic and regulatory mutants are able to produce more than 20 g $L^{-1}$ L-phenylalanine in a medium containing 13% glucose. Similar results can be obtained by tyrosine auxotrophic regulatory mutants of *E. coli* (HWANG et al., 1985).

With recombinant DNA techniques it was possible to improve overproducing strains of coryneform bacteria as well as of *E. coli.*

OZAKI et al. (1985) described amplification of the genes encoding chorismate mutase and PDT in *C. glutamicum.* Amplification of a deregulated DAHPS gene was achieved in a phenylalanine producer of *B. lactofermentum* (ITO et al., 1990).

SUGIMOTO et al. (1987) succeeded in isolating the genes encoding DAHPS and PDT in *E. coli* that are not effected by feedback inhibition (*aro*$F^{FBR}$ and *phe*$A^{FBR}$). A vector including these genes connected with a promotor and an operon of the bacteriophage λ was cloned into the tyrosine and thiamine deficient mutant *E. coli* AT2471. Recombinant *E. coli* strains with deregulated genes of DAHPS and CM-PDT were also investigated by FORBERG et al. (1988).

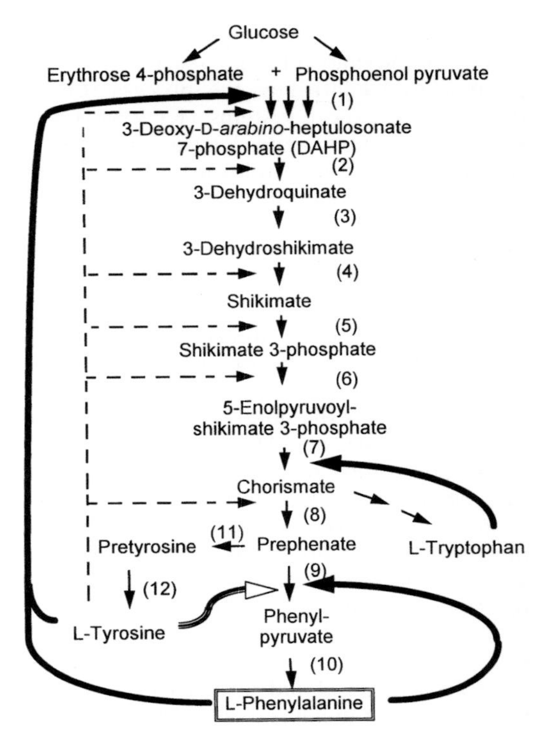


**Fig. 4.** Regulation of L-phenylalanine biosynthesis in *Corynebacterium glutamicum* (SHIIO, 1986). → feedback inhibition, ----→ repression, —▷ activation; enzyme (coding gene): 1 3-deoxy-D-arabinose-heptulosonate-7-phosphate synthase (*aro*F, *aro*G, *aro*H), 2 3-dehydroquinate synthase (*aro*B), 3 dehydroquinate dehydratase (*aro*D), 4 shikimate dehydrogenase (*aro*E), 5 shikimate kinase (*aro*L), 6 5-enolpyrovoylshikimate-3-phosphate synthase (*aro*A), 7 chorismate synthase (*thr*B), 8 chorismate mutase (*phe*A), 9 prephenate dehydratase (*phe*A), 10 phenylalanine aminotransferase (*phe*B), 11 pretyrosine dehydratase (*tyr*A), 12 tyrosine aminotransferase (*tyr*B).

For optimal production of L-phenylalanine in fed-batch cultivation it is necessary to pay attention to the critical specific glucose uptake rate. KONSTANTINOV et al. (1990) have concluded that the specific feed rate during fermentation should always be adjusted below a critical limit since, otherwise, the *E. coli* producer will be forced to excrete acetate. A suitable profile of the specific glucose feed rate prevents acetate formation and leads to improved L-phenylalanine production with final concentrations up to 46 g $L^{-1}$ and a corresponding yield of 18%. L-Phenylalanine is recovered from the fermentation broth either by two-step crystallization or by a resin process. The preferred cell separation technique is ultrafiltration; the filtrates may be treated with activated carbon for further purification. Instead of ion-exchange resins non-polar, highly porous synthetic adsorbents are recommended to get rid of impurities (OOTANI et al., 1986; VASCONCELLOS et al., 1989).

An alternative process in which a cell separator is integrated into the fermentation part thus allowing cell recycling has recently been suggested for L-phenylalanine production and may direct to prospective developments (HIROSE et al., 1994).

### 3.2.4 L-Cysteine

L-Cysteine is mainly produced via L-cystine from hydrolyzates of hair or other keratin. L-Cystine is easily recovered because of its weak solubility in aqueous solutions. The electrolytic reduction of L-cystine leads to L-cysteine. An enzymatic process for L-cysteine has been successfully developed using microorganisms capable to hydrolyze 2-amino-$\Delta^2$-thiazoline 4-carboxylic acid (ATC) which is readily available from methyl $\alpha$-chloroacrylate and thiourea (see Chapter 14b, Sect. 3.1.2). A mutant of *Pseudomonas thiazolinophilum* converts D,L-ATC to L-cysteine with a molar yield of 95% at product concentrations higher than 30 g $L^{-1}$ (SANO and MITSUGI, 1978).

## 3.3 Production of Amino Acids Mainly Used in the Feed Industry

The essential amino acids used as feed additives are D,L-methionine, L-lysine, L-threonine and, to a minor extent, L-tryptophan. L-Lysine revealed the most significant growth in the market during the last decade.

At present, L-lysine production is estimated to reach 250,000 t $a^{-1}$ (lysine*HCl) with a growth rate of 7%.

Since D,L-methionine is produced by chemical synthesis, L-lysine is the feed amino acid with the largest manufacturing capacities in fermentation.

### 3.3.1 D,L-Methionine

L-Methionine and the antipode D-methionine are of equal nutritive value, thus the racemate can directly be used as feed additive.

The most economic way for D,L-methionine production is the chemical process based on acrolein, methyl mercaptan, hydrogen cyanide, and ammonium carbonate. $\beta$-Methylthiopropionaldehyde, formed by addition of methyl mercaptan to acrolein, is the intermediate that reacts with hydrogen cyanide to give $\alpha$-hydroxy-$\gamma$-methylthio butyronitrile. Treatment with ammonium carbonate leads to 5-($\beta$-methylthioethyl)hydantoin that is saponified by potassium carbonate yielding D,L-methionine up to 95% calculated on acrolein (LÜSSLING et al., 1974).

### 3.3.2 L-Lysine

A strong competition between the major producers in Japan, Korea, and the USA has recently caused a remarkable decrease of the market price of L-lysine, thus stimulating application in the feed industry.

In order to be competitive it is necessary for a manufacturer to possess

- overproducing strains together with genetic, microbiological and biochemical know-how for further improvements,
- fermentation techniques and well equipped bioreactors to guarantee sterility, high conversion rates, high productivity, low side-product formation,
- efficient ideas for downstream processing facing ecological constraints and integration of environmental protection.

The most potent microorganisms to overproduce L-lysine are mutants derived from *C.*

*glutamicum,* a gram-positive bacterium first introduced as an L-glutamate producing microbe (Fig. 5); the wild strains themselves are not able to excrete L-lysine.

Mainly auxotrophic and regulatory mutants have been developed by classical breeding methods and mutagenesis (HILLIGER, 1991). These techniques have been applied to channel the metabolic pathway in the biosynthesis of amino acids (Tab. 3).

- Screening for auxotrophic mutants in order to release key enzymes, e.g., aspartate kinase from strict regulation by metabolites (feedback inhibition).
- Screening for regulatory mutants (TOSAKA and TAKINAMI, 1986) in which aspartate kinase is resistant to toxic analogs of L-lysine, such as *S*-(2-aminoethyl)-L-cysteine (AEC) or *O*-(2-aminoethyl)-L-serine ("oxalysine"). In such strains high lysine concentrations do not inhibit the enzymatic key step, the formation of 4-aspartylphosphate from L-aspartase catalyzed by aspartate kinase.
- Screening for mutants having amino acid auxotrophy combined with deregulation.
- Screening for regulatory mutants having additional enzyme defects (pyruvate kinase) and low level enzymes (citrate synthase).

As a modern technique, cell fusion by using the method of protoplast fusion (PEBERDY, 1980) has been successfully applied for breeding of industrial microorganisms (KARASAWA et al., 1986; NAKANISHI, 1986). This technique permits the combination of positive criteria of different strains, such as high selectivity and high productivity.

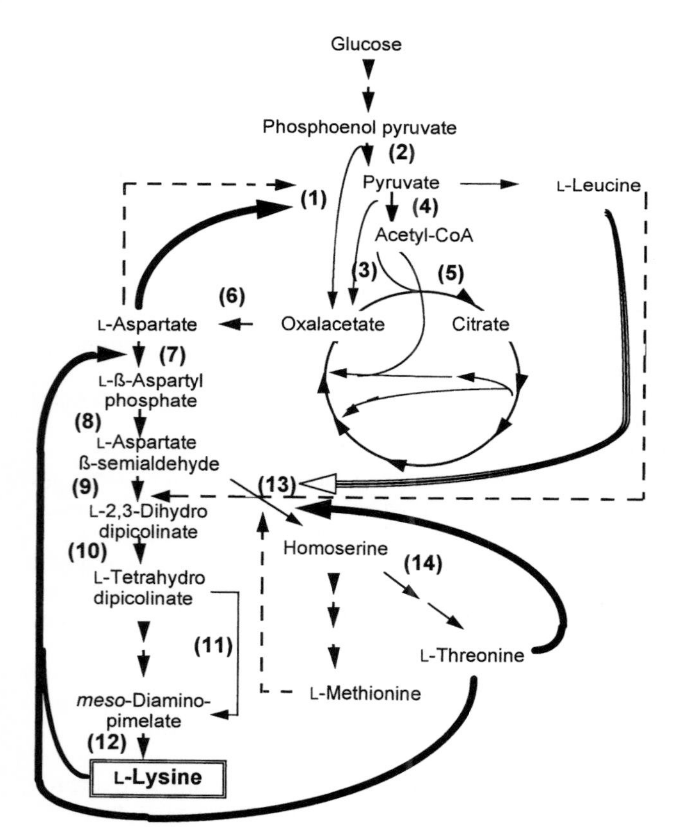


**Fig. 5.** Regulation of L-lysine biosynthesis in *C. glutamicum*
→ feedback inhibition,
---→ respression, —▷ activation;
enzyme (coding gene): 1 phosphoenolpyruvate carboxylase (*ppc*), 2 pyruvate kinase (*pyk*), 3 pyruvate carboxylase (*pyc*), 4 pyruvate dehydrogenase (*pdh*), 5 citrate synthase (*glt*A), 6 L-aspartate aminotransferase (*asp*B) or glutamate oxalacetate transaminase (*gota*), 7 aspartate kinase (*ask*) or (*lys*C), 8 aspartate-4-semialdehyde dehydrogenase (*asd*), 9 dihydrodipicolinate synthase (*dap*A), 10 dihydrodipicolinate reductase (*dap*B), 11 *m*-diaminopimelate dehydrogenase (*ddh*), 12 *m*-diaminopimelate decarboxyase (*lys*A), 13 homoserine dehydrogenase (*hom*), 14 homoserine kinase (*thr*B), 15 alanine transaminase, 16 $\alpha$-acetolactate synthase (*ilv*BN), 17 $\alpha$-isopropylmalate synthase (*leu*A).

**Tab. 3.** L-Lysine Producing Mutants of Coryneform Bacteria (Results in Batch Culture)

| Type of Mutant | Genetic Markers | L-Lysine Accumulation [g L$^{-1}$] | Yield (P/S) in [g 100 g$^{-1}$] | Reference |
|---|---|---|---|---|
| **Auxotrophic** | | | | |
| ATCC 13286 | Hse$^-$ | 14 | 28 | KINOSHITA et al. (1961) |
| T-36 | Thr$^-$ | 26 | n.d. | SANO and SHIIO (1967) |
| | Hse$^-$, Leu$^-$ | 34 | 34 | NAKAYAMA and ARAKI (1973) |
| | Thr$^-$, Met$^-$ | 34 | n.d. | SHIIO and SANO (1969) |
| **Regulatory** | | | | |
| | AEC$^r$ | 16 | n.d. | TOSAKA et al. (1978) |
| | LHX$^r$ | 36 | 20 | SMÉKAL et al. (1985) |
| AJ 3781 | AEC$^r$, Thiazolyl-Ala$^r$ | 35 | 35 | TOSAKA et al. (1976b) |
| | AEC$^r$, LHX$^r$ | 42 | 23 | SMÉKAL et al. (1985) |
| AJ-3856 | AEC$^r$, AHV$^r$ | 29 | 29 | TOSAKA et al. (1976a) |
| AJ 3610 | AEC$^r$, Met$^s$ | 45 | 45 | KUBOTA et al. (1974a) |
| ZIMET 11266 | AEC$^r$, CSS$^r$, Met$^s$ | 55 | 39 | MÜLLER et al. (1987) |
| AJ 12421 (FERM BP-2208) | AEC$^r$, Met$^r$, Ala-Glu$^r$ | 32 | 32 | TSUCHIDA et al. (1990) |
| P1–13 | AAH$^r$, PEN-G$^r$ | 50 | 37 | WANG et al. (1991) |
| AJ 12593 | AEC$^r$, TPL$^r$ | 25 | 31 | YOKOMORI et al. (1992) |
| FML 8611 | AEC$^r$, Rif$^r$ | 79 | 44 | FAN et al. (1988) |
| H-4412 (FERM BP-1069) | AEC$^r$, Rif$^r$, Str$^r$, AZA$^r$, $\beta$-Naphthoquinoline$^r$ | 52 | n.d. | YONEKURA et al. (1987) |
| H-4934 (FERM BP-1655) | AEC$^r$, Rif$^r$, Str$^r$, AZA$^r$, $\beta$-Naphthoquiniline$^r$, Iturin$^r$ | 54 | nd. | HIRAO et al. (1989a) |
| H-8241 (FERM BP-3954) | AEC$^r$, Rif$^r$, Str$^r$, AZA$^r$, $\beta$-Naphtoquinoline$^r$, Iturin$^r$, TIT$^r$ | 48 | n.d. | NAKANO et al. (1994a) |
| **Auxotrophic and Regulatory** | | | | |
| HA 42 | Hse$^-$, AEC$^r$ | 41 | n.d. | TANG et al. (1989) |
| TM-1101 (FERM BP-880) | Hse$^-$, LAC$^r$ | 49 | 45 | MATSUMOTO (1986) |
| MH20-22B | Leu$^-$, AEC$^r$ | 44 | 44 | SCHRUMPF et al. (1992) |
| | Ala$^-$, AEC$^r$ | 34 | 34 | TOSAKA et al. (1979) |
| AJ 3857 (FERM-P 2785) | Ala$^-$, AEC$^r$, AHV$^r$ | 45 | 45 | TOSAKA et al. (1976a) |
| AJ 3796 | Ala$^-$, AEC$^r$, CCL$^r$ | 47 | 47 | KUBOTA et al. (1976) |

| | | | | |
|---|---|---|---|---|
| AJ 3990 | Ala$^-$, AEC$^r$, CCL$^r$, ML$^r$ | 60 | 43 | Kubota et al. (1976) |
| AJ 11282 (FERM-P 4556) | Ala$^-$, AEC$^r$, CCL$^r$, Thienoyl-Met$^r$ | 46 | 46 | Ajinomoto (1982) |
| AJ 11204 | Ala$^-$, AEC$^r$, CCL$^r$, ML$^r$, FP$^s$ | 70 | 50 | Tosaka et al. (1985) |
| AJ 11657 | Ala$^-$, AEC$^r$, CCL$^r$, EG$^r$ | 60 | 40 | Shimazaki et al. (1983) |
| AJ 3429 (FERM-P 1857) | Ala$^-$, Leu$^-$, Nic$^-$, AEC$^r$ | 51 | 51 | Kubota et al. (1974b) |
| BL 25 ATCC 21526 | Hse$^-$, Leu$^-$, AEC$^r$ | 39.5 | 39.5 | Nakayama and Araki (1973) |
| | Hse$^-$, Leu$^-$, Pant$^-$, AEC$^r$ | 42 | 42 | Nakayama and Araki (1981) |
| CCM 4263 | Hse$^-$, Leu$^-$, Val$^-$, AEC$^r$, FP$^s$ | 55 | 43 | Plachy and Kratochvil (1992) |
| H 3106 (FERM BP 143) | Hse$^-$, Leu$^-$, AEC$^r$, Sulfamethazine$^r$ AZA$^r$ | 43 | 43 | Nakanashi et al. (1987) |
| AAH$^r$-492 | Hse$^-$, Leu$^-$, Ile$^-$, AAH$^r$ | 39 | 43 | Liu (1987) |
| MG from ATCC 21513 | Hse$^-$, Leu$^-$, AEC$^r$, $\varepsilon$-ML$^r$ | 76 | 44 | Hadj Sassi et al. (1990) |
| | Hse$^-$, CS*, PK*, AEC$^r$, Asp$^r$, Thr$^r$, Fp$^s$ | 51 | 51 | Ozaki and Shiio (1983) |
| BL-1 M76 | Ser$^-$, AEC$^r$, FP$^s$ | | 43 | Rogers et al. (1989), Satiawihardja et al. (1993) |
| CS-755 (FERM BP-2763) | Hse$^-$, Leu$^-$, AEC$^r$, AVH$^r$, ML$^r$, Thiazolyl-Ala$^r$, Canavanine$^r$, Cadaverine$^r$, Spermine$^r$, Spermidine$^r$, Putrescine$^r$, Hydroxyuridine$^r$, ArgHx$^r$, AZA$^r$, Fluoro-trp$^r$ | 34 | 45 | Oh et al. (1990) |
| **Regulatory, Having Enzyme Defects (CS)** | | | | |
| | AEC$^r$, CS$^L$ | 36 | 36 | Shiio et al. (1982a) |
| H-3-4 | HD$^-$, CS$^L$, PC$^R$ | 41 | 41 | Yokota and Shiio (1988) |
| 664-7 | AK$^R$, CS$^L$, PC$^R$ | 45 | 45 | Yokota and Shiio (1988) |
| | HD$^-$, CS$^L$, PC$^R$, PK$^-$ | 47.5 | 47.5 | Shiio et al. (1984) |
| AH-198 | AEC$^r$ (AK$^R$), CS$^L$, PC$^R$, PK$^-$ | 51 | 51 | Shiio et al. (1990a) |
| KB34 | (AK$^R$), $\alpha$-KB$^r$, AAH$^r$ | 42 | n.d. | Shiio et al. (1993) |
| AJ 12568 (FERM BP-3559) | AEC$^r$, Selys$^r$, PK$^-$ | 45 | 45 | Yokomoti et al. (1994) |
| ATCC 14067→ AK11842→AJ 12121 (FERM-P 7441) | PK$^-$, PDH$^-$ | 55 | n.d. | Shiio et al. (1985) |

n.d.: not determined.

**Tab. 3.** Continued

| Abbreviation | Genetic Marker | Abbreviation | Genetic Marker |
|---|---|---|---|
| $AAH^r$ | resistance to aspartic acid hydroxamate (=AspHx) | $Met^-$ | requirement of L-methionine for growth |
| $AEC^r$ | resistance to *S*-(2-aminoethyl)-L-cysteine | $Met^r$ | resistance to L-methionine |
| $AHV^r$ | resistance to $\alpha$-amino-$\beta$-hydroxyvaleric acid | $Met^s$ | sensitive to L-methionine |
| $AK^R$ | deregulated aspartate kinase | $ML^r$ | resistance to $\gamma$-methyllysine |
| | | $\varepsilon$-$ML^r$ | resistance to $\varepsilon$-methyllysine |
| $Ala^-$ | requirement of L-alanine for growth | Naphtho-$quinoline^r$ | resistance to naphthoquinoline |
| Ala-$Glu^r$ | resistance to dipeptide Ala-Glu | $Nic^-$ | requirement of nicotinic acid or nicotinamide for growth |
| $ArgHx^r$ | resistance to arginine hydroxamate | $Pant^-$ | requirement of pantothenic acid for growth |
| $Asp^r$ | resistance to L-aspartic acid | $PC^R$ | deregulated PEP carboxylase |
| $AZA^r$ | resistance to 6-azaurazil | $PDH^-$ | reduced pyruvate dehydrogenase activity |
| $Cadaverine^r$ | resistance to cadaverine | PEN-$G^r$ | resistance to penicillin-G |
| $Canavanine^r$ | resistance to canavanine | $PK^* = PK^-$ | reduced pyruvate kinase activity |
| $CCL^r$ | resistance to $\alpha$-chlorocaprolactam | $Putrescine^r$ | resistance to putrescine |
| $CS^* = CL^L$ | low activity of citrate synthase | $Selys^r$ | resistance to selenalysine |
| $CSS^r$ | resistance to cysteinesulfinic acid | | |
| $EG^r$ | resistance to ethylene glycol | $Ser^-$ | requirement of L-serine for growth |
| Fluoro-$trp^r$ | resistance to 6-fluorotryptophan | $Spermidine^r$ | resistance to spermidine |
| $FP^s$ | sensitive to fluoropyruvate | $Spermine^r$ | resistance to spermine |
| $HD^-$ | reduced homoserine dehydrogenase activity | $Str^r$ | resistance to streptomycin |
| $Hse^-$ | requirement of L-homoserine for growth | $Sulfamethazine^r$ | resistance to sulfamethazine |
| $Hydroxyuridine^r$ | resistance to 5-hydroxyuridine | Thiazolyl-$Ala^r$ | resistance to $\beta$-(2-thiazolyl)-D,L-alanine |
| $Ile^-$ | requirement of L-isoleucine for growth | Thienoyl-$Met^r$ | resistance to N-2-thienoylmethionine |
| $Iturin^r$ | resistance to iturin | $Thr^-$ | requirement of L-threonine for growth |
| $\alpha$-$KB^r$ | resistance to $\alpha$-ketobutyrate | $Thr^r$ | resistance to L-threonine |
| $LAC^r$ | resistance to L-$\alpha$-amino-$\varepsilon$-caprolactam | $TIT^r$ | resistance of 3,3′,5-triiodo-L-thyronine |
| $Leu^-$ | requirement of L-leucine for growth | $TPL^r$ | resistance to $N^\alpha$,$N^\alpha$,$N^\alpha$-trimethyl-$N^\varepsilon$-palmitoyl-D,L-lysine |
| $LHX^r$ | resistance to L-lysine hydroxamate | | |

In fermentations with media causing inhibitory osmotic stress the sugar consumption rate and the L-lysine production rate of some mutants can be stimulated by the addition of glycine (KAWAHARA et al., 1990).

Another attractive approach is the development of thermophilic strains. Recently, a mutant of *C. thermoaminogenes* was protected which is capable of growing at a temperature higher than 40 °C and of accumulating L-lysine in the culture medium (MURAKAMI et al., 1993).

New scientific findings in the metabolic characterization of lysine producing strains (KISS and STEPHANOPOULOS, 1991, 1992), especially analyses of carbon flux distributions (VALLINO and STEPHANOPOULOS, 1993, 1994) and studies of fluxes in the central metabolism by NMR spectroscopy (SONNTAG et al., 1995; MARX et al., 1996), will provide new impulses for a well-balanced optimization in strain development. Meanwhile, there is even detailed knowledge of the regulation of lysine excretion of overproducing strains (ERDMANN et al., 1994, 1995). Thus more attention should be paid to process development in the fields of molecular biology, biochemistry, and physiology and to find new approaches for improved overproducing mutants (KRÄMER, 1996).

The most specific and well-directed methods for strain development are offered by recombinant DNA techniques (DUNICAN and SHIVNAN, 1989). In principle, all genes encoding the relevant enzymes in L-lysine biosynthesis have been isolated, characterized, and amplified in coryneform bacteria to enhance L-lysine formation (JETTEN and SINSKEY, 1995) (Tab. 4).

As an example, amplification of the *dap*A gene that codes for the enzyme dihydrodipicolinate synthase in *C. glutamicum* resulted in a 35% higher overproduction of L-lysine compared to the parent strain (KATSUMATA et al., 1986). Another option for strain improvement is the transformation of the *dap*A gene together with a *lys*C gene coding for aspartate kinase with decreased feedback inhibition in *C. glutamicum* (CREMER et al., 1991).

In fed-batch culture the favorable mutants for lysine production are able to reach under appropriate conditions final concentrations of about 120 g $L^{-1}$ L-lysine, calculated as hydrochloride (OH et al., 1990).

It is a great challenge for microbiologists and engineers to realize the results found in laboratory experiments also on the industrial scale.

Fermentation processes performed in large tanks up to the 500 $m^3$ scale are strongly limited by mass transport phenomena like mixing and oxygen transfer. Tests with different pilot fermenters of a medium-sized scale may provide data for the production process. Less cost intensive and time consuming procedures are simulations of relevant limiting factors using tailor-made model reactors. Scale-up behavior may be described by mathematical models (NAGY et al., 1994), by a reactor cascade with well defined but different conditions in two reactors (OOSTERHUIS et al., 1983), or by special test reactors. In those reactors it is possible to simulate the fermentation process in a big tank and to characterize and minimize the limiting influences (PFEFFERLE et al., 1996).

An optimized feeding strategy practiced with a computer aided process control system may realize high conversion yield and productivity in large-scale fed-batch cultivation (LIU et al., 1993).

Alternative processes, such as the production of L-lysine in continuously operated fermenters (OH and SERNETZ, 1993) did not succeed so far on an industrial scale. However, examples are reported with encouraging results at the laboratory and the pilot scale (HIRAO et al., 1989b; LEE et al., 1995). On the other hand, new concepts for downstream processing will compete with conventional methods.

Apart from the specific fermentation know-how, inoculation, sterilization, and feeding strategy (VOSS et al., 1996), both the recovery process and the quality of the product can be decisive factors to guarantee competitiveness.

The conventional route of lysine downstream processing is characterized by

- removal of the bacterial cells from the fermentation broth by separation or ultrafiltration,
- absorbing and then collecting lysine in an ion exchange step (TOSA et al., 1987; JAFFARI et al., 1989),
- crystallizing or spray drying as L-lysine hydrochloride (FECHTER et al., 1995).

An alternative process consists of biomass separation, concentration of the fermentation solution, and filtration of precipitated salts.

**Tab. 4.** Genes of L-Lysine Biosynthesis Cloned in Mutants of Coryneform Bacteria

| Origin/Type of Mutant | Gene | Enzyme | Remark | Reference |
|---|---|---|---|---|
| *B. lactofermentum* ATCC 13869 | *ppc* | PC | plasmid pAJ 201 | SANO et al. (1985) |
| *C. glutamicum* ASO 19 | *ppc* | PC | 15-fold increase of enzyme activity | EIKMANNS et al. (1989) |
| *C. glutamicum* ASM1, *C. lactofermentum* 779SM1 | *ppc* | PC | inactivation | GUBLER et al. (1994a) |
| *C. lactofermentum* 21799 | *pyk* | PK | site-specific inactivation of *pyk* | GUBLER et al. (1994b) |
| *C. glutamicum M2* | *pyk* | PK | plasmid pMG 123 | JETTEN et al. (1994) |
| *C. glutamicum* ATCC 13032 | *aat, asd* | AAT, ASD | plasmid pAT1, plasmid pASD2 | KATSUMATA et al. (1987) |
| *C. glutamicum* ATCC 21543 | *lys*C | AK (feedback resistant) | AEC resistance, plasmid pAec5 | KATSUMATA et al. (1983) |
| *C. lactofermentum* JS231 | *lys*C | AK ($AEC^r$) | cloned in *E. coli*; shuttle vector pECCG117 | HAN et al. (1991) |
| *B. flavum* *CCRC* 18271 | *lys*C (*ask*) | AK | 4- to 11-fold higher specific activity | LU et al. (1994) |
| *C. glutamicum* ATCC 13032 mutant | *ask-asd* | AK/ASD ($AEC^r$) | AEC resistance | KALINOWSKI et al. (1989) |
| *C. lactofermentum* 21799 | *ask-asd* | AK/ASD | ASK-negative; 2 *ask* genes on the chromosome | JETTEN et al. (1995) |
| *C. glutamicum* ATCC 13032, *C. glutamicum* MH 20 | *lys*C, *dap*A | AK, DDPS ($AEC^r$) | plasmid pJC50 | CREMER et al. (1991) |
| *B. lactofermentum* ATCC 13869 | *ddh* | DDH | | ROH et al. (1994) |
| *C. glutamicum* RH6 | *dap*A, *dap*C | DDPS, THPS | plasmid pAC2 | KATSUMATA et al. (1986) |
| *C. glutamicum* ASO19 | *dap*A/*dap*B, *dap*D, *lys*A | DDPS, DDPR, THDS, DAPD | cosmid pSM71, pSM61, pSM531 | YEH et al. (1988) |
| *C. lactofermentum* JS231 | *dap*A | DDPS | cloned in *E. coli*; shuttle vector pECCG117 | OH et al. (1991) |
| *B. lactofermentum* 13869 | *lys*A | DAPD | plasmid pAJ 101 | SANO et al. (1989) |
| *C. glutamicum* DM368-3 | *hom* | HD | resistant to feedback inhibition | RENSCHEID et al. (1991) |

**Tab. 4.** Continued

| Abbreviation | Enzyme |
|---|---|
| AAT | aspartate aminotransferase |
| AEC$^r$ | resistance to *S*-(2-aminoethyl)-L-cysteine |
| AK | deregulated aspartate kinase |
| ASD | aspartate semialdehyde dehydrogenase |
| DAPD | *meso*-diaminopimelate decarboxylase |
| DDH | *meso*-diaminopimelate dehydrogenase |
| DDPR | L-2,3-dihydrodipicolinate reductase |
| DDPS | L-2,3-dihydrodipicolinate synthetase |
| HD | homoserine dehydrogenase |
| PC | deregulared phosphoenol pyruvate carboxylase |
| PK | pyruvate kinase |
| THDS | tetrahydrodipicolinate synthetase |
| THPS | tetrahydropicolinate succinylase |

The liquid product contains up to 50% L-lysine base that is stable enough to be marketed (LUCQ and DOMONT, 1993).

Recently, a new concept for lysine production was introduced, in which the lysine containing fermentation broth is immediately evaporated, spray dried, and granulated to yield a feed-grade product which contains lysine sulfate corresponding to at least 60% of L-lysine hydrochloride (ROUY, 1984; BINDER et al., 1993). Waste products usually present in the conventional L-lysine hydrochloride manufacture are avoided in this process.

The current price of L-lysine·HCl feed-grade is about US $ 3 per kg (in 1996). With the benefits provided by modern techniques for the useful design of metabolic pathways such as genetic engineering and with the potentials of the fermentation technology further improvements of the L-lysine process should be realized. Then, L-lysine would continue to be the most attractive feed additive produced by fermentation.

### 3.3.3 L-Threonine

Ten years ago, in 1986, L-threonine was mainly used for medical purposes, in amino acid infusion solutions, and nutrients. It was manufactured by extraction of protein hydrolyzates or by fermentation using mutants of coryneform bacteria in amounts of some hundred tons a year worldwide. The production strains were developed by classical breeding, auxotrophs and resistant to threonine analogs such as $\alpha$-amino-$\beta$-hydroxy-valerate (AHV), and reached product concentrations up to 20 g $L^{-1}$. These strains possess deregulated pathways with feedback inhibition-insensitive aspartate kinase and homoserine dehydrogenase (SHIMURA, 1995; NAKAMORI, 1986).

During the past decade, strain developments have been very successful in both cases using conventional methods and recombinant DNA techniques. Tab. 5 refers to potent mutants of the species *B. flavum, Providentia rettgeri, Serratia marcescens,* and *E. coli* which were selected by classical methods and are suggested for industrial production.

However, during the competition of the favorable candidates, strains of *E. coli* proved to be superior to other bacteria. Although the pathway of L-threonine biosynthesis in *E. coli* (Fig. 6) is much more regulated compared to that in *C. glutamicum* (see Fig. 2), new *E. coli* strains with excellent yields and productivity in threonine formation could be constructed by means of genetic engineering.

Currently, L-threonine has been successfully marketed as a feed additive with a worldwide demand of more than 10,000 t $a^{-1}$. Production strains are based on *E. coli* K12 constructs harboring plasmids containing the *thr* operon consisting of the genes *thr*A, *thr*B and *thr*C (DEBABOV et al. 1981).

**Tab. 5.** L-Threonine Producing Mutants Developed by Conventional Methods; Results in Batch (B) or Fed-Batch (FB) Culture

| Type of Mutant | Genetic Markers | Carbon Source/ Culture | L-Thr Accumulation [g L$^{-1}$] | Yield (P/S) [g 100 g$^{-1}$] | Productivity [g L$^{-1}$ h$^{-1}$] | Reference |
|---|---|---|---|---|---|---|
| *B. flavum* BD122 | AHV$^r$, DPS$^-$ | glucose/ B | 16.6 | n.d. | 0.23 | SHIIO et al. (1990b) |
| *B. flavum* AJ 12314 | AHV$^r$, Ile$^-$, MPA$^r$ | acetate/ FB | 25.0 | 17.2 | 0.35 | TSUCHIDA et al. (1993) |
| *P. rettgeri* NS-140 | Leu$^-$, Ile$^-$, AHV$^r$, Ethionine$^r$ | glucose/ FB | 26.0 | 17.3 | 0.34 | YAMADA et al. (1986) |
| *P. rettgeri* TP7-55 | Leu$^-$, Ile$^L$, AHV$^r$, Ethionine$^r$, Thiaisoleucine$^r$, AAH$^r$ | glucose/ B | 22.0 | 28.0 | 0.24 | YAMADA et al. (1987) |
| | | glucose/ FB | 72.0 | 38.0 | 1.23 | |
| *S. marcescens* P-200 | AEC$^r$, HN$^r$ (from transductant T-1026) | sucrose/ B | 41.0 | 27.0 | 0.47 | KOMATSUBARA et al. (1983) |
| *E. coli* H-4581 | DAP$^-$, Met$^L$, AHV$^r$, Thr-N$^-$, Rif$^r$, Lys$^r$, Met$^r$, Hse$^r$, Asp$^r$ | glucose/ B | 22.1 | n.d. | 0.31 | FURUKAWA and NAKANISHI (1989) |
| | | glucose/ FB | 70.5 | n.d. | 0.88 | |
| *E. coli* H-7256 | DAP$^-$, Met$^L$, AHV$^r$, Rif$^r$, CystinH$^r$ | glucose/ B | 22.3 | 31.9 | 0.31 | TAKANO et al. (1990) |
| | | glucose/ FB | 63.5 | n.d. | 0.79 | |
| *E. coli* H-7729 | Met$^L$, AHV$^r$, Thr-N$^-$, Rif$^r$, Lys$^r$, Met$^r$, Hse$^r$, Asp$^r$, Ser$^r$, Ethionine$^r$ | glucose/ FB | 38.4 | n.d. | 0.48 | KINO et al. (1991) |
| *E. coli* H-8311 | Met$^L$, AHV$^r$, Thr-N$^-$, Rif$^r$, Lys$^r$, Met$^r$, Hse$^r$, Asp$^r$, Ser$^r$, Ethionine$^r$, Phe$^r$ | glucose/ FB | 48.7 | n.d. | 0.61 | KINO et al. (1993a) |
| *E. coli* H-8460 | Met$^L$, AHV$^r$, Thr-N$^-$, Rif$^r$, Lys$^r$, Met$^r$, Hse$^r$, Asp$^r$, Ser$^r$, Ethionine$^r$, Phe$^r$, MAP$^r$ | glucose/ FB | 76.5 | n.d. | 1.09 | KINO et al. (1993b) |
| *E. coli* TF427 | AHV$^r$, Met$^-$, Ile$^L$ | glucose/ B | 18.3 | 26 | 0.38 | LEE et al. (1991) |
| | | glucose/ FB | 46.5 | n.d. | 1.06 | |

**Tab. 5.** Continued

| Abbreviation | Genetic Marker |
|---|---|
| AAH$^r$ | resistance to aspartic acid hydroxamate (=AspHx) |
| AEC$^r$ | resistance to *S*-(2-aminoethyl)-L-cysteine |
| AHV$^r$ | resistance to D,L-$\alpha$-amino-$\beta$-hydroxyvaleric acid |
| Asp$^r$ | resistance to aspartic acid |
| CystinH$^r$ | resistance to cystine hydroxymate |
| DAP$^-$, | requirement of diaminopimelic acid for growth |
| DPS$^-$ | dihydrodipicolinate synthase defect |
| Ethionine$^r$ | resistance to ethionine |
| HN$^r$ | resistance to $\beta$-hydroxynorvaline |
| Hse$^r$ | resistance to L-homoserine |
| Ile$^-$ | requirement of L-isoleucine for growth |
| Ile$^L$ | requirement of L-isoleucine for growth, leaky character |
| Leu$^-$ | requirement of L-leucine for growth |
| Lys$^r$ | resistance to L-lysine |
| MAP$^r$ | resistance to 6-methylaminopurine |
| Met$^-$ | requirement of L-methionine for growth |
| Met$^L$ | requirement of L-methionine for growth, leaky character |
| Met$^r$ | resistance to L-methionine |
| MPA$^r$ | resistance to mycophenolic acid |
| Phe$^r$ | resistance to L-phenylalanine |
| Rif$^r$ | resistance to rifampicin |
| Ser$^r$ | resistance to L-serine |
| Thiaisoleucine$^r$ | resistance to thiaisoleucine |
| Thr-N$^-$ | decreased ability to degrade L-threonine |

Further improvements resulted in strains capable to accumulate more than 80 g $L^{-1}$ during about 30 h with a conversion yield of more than 40% (DEBABOV et al., 1994). The strain stability could be further improved, e.g., by integrating the threonine operon into the chromosome (RICHAUD et al., 1989) (Tab. 6).

The recovery of feed-grade L-threonine is rather simple. After the fermentation is completed, cell mass is removed by centrifugation or by ultrafiltration, the filtrate is concentrated, depigmented, and L-threonine is isolated by crystallization (OOTANI et al., 1987).

### 3.3.4 L-Tryptophan

L-Tryptophan is one of the limiting essential amino acids required in the diet of pigs and poultry. A regular growing market for L-tryptophan as feed additive is still in development although many processes that have been proven on a production scale are available. However, high production costs so far prevent a tolerable price level that makes it easier to introduce L-tryptophan as a bulk product.

The most attractive processes are based on microorganisms used as enzyme sources or as overproducers:

- enzymatic production from various precursors,
- fermentative production from precursors,
- direct fermentative production from carbohydrates by auxotrophic and analog resistant regulatory mutants.

L-tryptophan has been synthesized from indole, pyruvate, and ammonia by the enzyme tryptophanase (YOKOTA and TAKAO, 1984) or from indole and L-serine/D,L-serine by tryptophan synthase (ISHIWATA et al., 1989; OGAWA et al., 1991) (see Chapter 14b, Sects. 3.4.3 and 3.4.5).

Although production in enzyme bioreac-

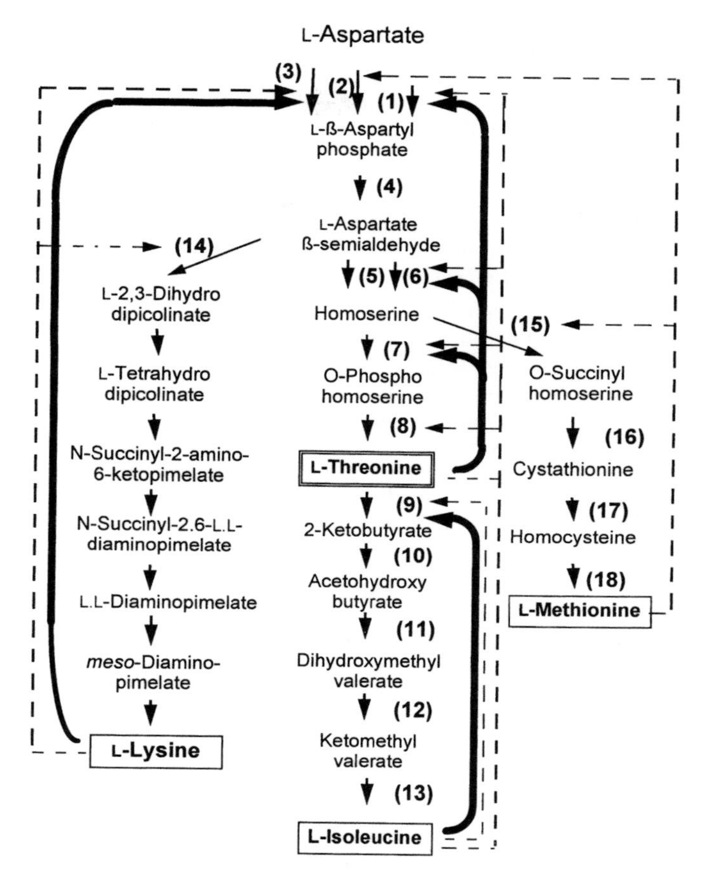


**Fig. 6.** Regulation of L-threonine and L-isoleucine biosynthesis in *E. coli*. —► feedback inhibition, ---► repression; enzyme (coding gene): 1 aspartate kinase I (*thr*A), 2 aspartate kinase II (*met*LM), 3 aspartate kinase III (*lys*C), 4 aspartate β-semialdehyde dehydrogenase (*asd*), 5 homoserine dehydrogenase I (*thr*A), 6 homoserine dehydrogenase II (*met*LM), 7 homoserine kinase (*thr*B), 8 threonine synthase (*thr*C), 9 threonine dehydratase (*ilv*A), 10 α-acetolactate synthase (*ilv*BN), 11 isomero reductase (*ilv*C), 12 dihydroxyacid dehydratase (*ilv*D), 13 transaminase (*ilv*E), 14 dihydrodipicolinate synthase (*dap*A), 15 homoserine succinyltransferase (*met*A), 16 cystathionine γ-synthase (*met*B), 17 cystathionine β-lyase (*met*C), 18 homocysteine methylase (*met*E, *met*H).

tors is quite efficient and concentrations of L-tryptophan up to 200 g $L^{-1}$ could be achieved by condensation of indole and L-serine (HAMILTON et al., 1985), these variants proved to be not economical due to the costs of starting materials.

The microbial conversion of biosynthetic intermediates such as indole or anthranilic acid to L-tryptophan has also been considered as an alternative for production. Whereas indole consuming mutants of *C. glutamicum* produced about 10 g $L^{-1}$ L-tryptophan (PLACHY and ULBERT, 1990), strains of *B. subtilis* and *B. amyloliquefaciens* reached final concentrations higher than 40 g $L^{-1}$ L-tryptophan with anthranilic acid as carbon source (Showa Denko, 1990; TAKINISHI et al., 1987). The process with anthranilic acid as precursor has been commercialized in Japan. However, the manufacturer using genetically modified strains derived from *B. amyloliquefaciens* IAM 1521 was forced to stop L-tryptophan production. L-Tryptophan produced by this process was not suitable because of the formation of by-products which caused a new serious disease termed eosinophilia–myalgia syndrome (EMS) (MAYENO and GLEICH, 1994). One of the problematic impurities, "Peak E", was identified as 1,1′-ethylidene-bis-(L-tryptophan), a product formed by condensation of one molecule of acetaldehyde with two molecules of tryptophan (SAKIMOTO and TORIGOE, 1994).

Other processes, i.e., direct fermentation using overproducing mutants and carbohydrates as carbon sources have not been restricted by formation of such impurities mentioned above (Tab. 7).

**Tab. 6.** L-Threonine Producing Strains Developed by Recombinant DNA Techniques; Results in Batch (B) or Fed-Batch (FB) Culture

| Type of Strain | Gene Amplified | Genetic Markers | Carbon Source/ Culture | L-Thr Accumulation [g $L^{-1}$] | Yield (P/S) [g 100 $g^{-1}$] | Productivity [g $L^{-1}$ $h^{-1}$] | Reference |
|---|---|---|---|---|---|---|---|
| *B. flavum* ATCC 21127 | *E. coli thr* operon (pEC711) | AHV$^r$ | sucrose/ B | 12.0 | 6.7 | 0.25 | PÁTEK et al. (1993) |
| *B. flavum* HT-16 | *E. coli thr* operon (pAJ514) | Ile$^-$, AHV$^r$ | glucose/ B | 27.0 | 27.0 | 0.38 | ISHIDA et al. (1989) |
| *B. flavum* HT-16 | *E. coli thr* operon (pAJ514) | Ile$^-$, AHV$^r$ | acetate/ FB | 64.6 | 19.9 | 0.7 | ISHIDA et al. (1993) |
| *B. lactofermentum* TBB-10 | *hom, Thr*B, *Thr*C (pDR345) | Ile$^-$, Leu$^-$, AEC$^r$, AHV$^r$, SMCS$^r$, | glucose/ FB | 57.7 | 25.9 | 0.58 | ISHIDA et al. (1994) |
| *S. marcescens* T-2000 | *thr* operon (pSK301) | mutant-type *thr* operon | sucrose/ FB | 100 | n.d. | 1.04 | MASUDA et al. (1993) |
| *E. coli* K-12 No. 29-4 | *thr* operon (*pBR322*) | Ile$^-$, Met$^-$, Pro$^-$, Thiamine$^-$, AHV$^r$ | glucose/ FB | 65.0 | 48.0 | 0.90 | SHIMIZU et al. (1995) |
| *E. coli* VNIIGenetika VL334 (pYN7) | *thr* operon (pYN7) | Pen$^r$, mutation *thr*C 1010, *ilv*A 442 | glucose/ FB | 20.0 | n.d. | 0.39 | DEBABOV et al. (1981) |
| *E. coli* BKIIM B-3996 | *thr* operon (pVIC40) | Sac$^+$, Thr$^r$, Hse$^r$ | sucrose/ FB | 85.0 | n.d. | 2.36 | DEBABOV et al. (1990) |
| *E. coli* BKIIM B-5318 | *thr* operon, λ-phage promoter, repressor C1 gene (pPRT614) | Ile$^+$, Sac$^+$, Thr$^r$, Hse$^r$ | sucrose/ FB | 82.0 | 41.0 | 2.9 | DEBABOV et al. (1994) |

During the past ten years striking progress has been made in the development of auxotrophic and deregulated mutants of *B. flavum, C. glutamicum,* and *B. subtilis.* The biosynthesis of L-tryptophan and its regulation have been reviewed in detail for the different species (NAKAYAMA, 1985; SHIIO, 1986; MAITI and CHATTERJE, 1991) (Fig. 7).

The precise knowledge of the structure of the *trp* operon in *E. coli* (SOMERVILLE, 1983,

**Tab. 6.** Continued

| Abbreviation | Genetic Marker |
|---|---|
| AEC$^r$ | resistance to *S*-(2-aminoethyl)-L-cysteine |
| AHV$^r$ | resistance to D,L-$\alpha$-amino-$\beta$-hydroxyvaleric acid |
| Hse$^r$ | resistance to L-homoserine |
| Ile$^+$ | requirement of L-isoleucine not necessary (prototroph) |
| Ile$^-$ | requirement of L-isoleucine for growth |
| Leu$^-$ | requirement of L-leucine for growth |
| Met$^-$ | requirement of L-methionine for growth |
| Pen$^r$ | resistance to penicillin |
| Pro$^-$ | requirement of L-proline for growth |
| Sac$^+$, | capable of utilizing saccharose |
| SMCS$^r$ | resistance to *S*-methylcysteine sulfoxide |
| Thiamine$^-$ | requirement of thiamine for growth |
| Thr$^r$ | resistance to L-threonine |

1988) comprising the *trp* promotor and the genes *trp*E, *trp*D, *trp*C, *trp*B, and *trp*A, which are coding for the enzymes anthranilate synthase (AS), phosphoribosyl anthranilate transferase (PRT), indole glycerol phosphate synthase (IGP), and tryptophan synthase (TS), respectively, was beneficial for further strain improvements.

Thus recombinant DNA techniques have been used to increase the capability of over-production, especially in strains of *C. glutamicum* and *E. coli* (Tab. 8).

One concept was realized successfully by amplification of the *trp* operon genes together with *ser*A, which codes for phosphoglycerate dehydrogenase. This key enzyme in L-serine biosynthesis should provide enough L-serine in the last step of L-tryptophan formation.

Currently, production strains are able to accumulate 30–50 g L$^{-1}$ L-tryptophan with yields higher than 20% based on carbohydrate. In general, crystals of tryptophan appear in the broth when concentrations are above the level of about 30 g L$^{-1}$ (IKEDA et al., 1994). Since L-tryptophan is sensitive to oxygen and heat (CUQ and GILOT, 1985), recovery from the fermentation broth without considerable product losses is still a challenge and somehow part of the sophisticated know-how of each specific process (KONO et al., 1991; SHINOHARA and OOTANI, 1994).

## 3.4 Production of Amino Acids Mainly Used in the Pharmaceutical and the Cosmetics Industry

Proteinogenic amino acids destined for pharmaceutical or cosmetic applications are expensive compared to the amino acids used in food and feed, and each of them is produced at a rate of hundreds of tons per year (Fig. 8, see Tab. 2).

They are available

- from protein hydrolyzates by extraction using ion exchange resins, except for L-methionine and L-tryptophan, which are sensitive to oxidation;
- from chemically produced precursors using enzymes (especially for L-alanine, L-aspartic acid, L-methionine, and L-valine);
- from carbohydrate-based fermentation technology.

### 3.4.1 L-Alanine

L-Alanine is produced industrially from L-aspartic acid by means of immobilized *Pseudomonas dacunhae* cells in a pressurized bioreactor (FURUI and YAMASHITA, 1983) (see Chapter 14b, Sect. 3.4.1).

In direct fermentation microorganisms usually accumulate D,L-alanine because of the

**Tab. 7.** L-Tryptophan Production by Antimetabolic Resistant Mutants; Results in Batch (B) or Fed-Batch (FB) Culture

| Type of Mutant | Genetic Markers | Carbon Source/ Culture | L-Trp Accumulation [g $L^{-1}$] | Yield (P/S) [g 100 $g^{-1}$] | Productivity [g $L^{-1}$ $h^{-1}$] | Reference |
|---|---|---|---|---|---|---|
| *B. flavum* AJ 11667 | Met$^-$, 5FT$^T$, AzaSer$^r$, | glucose/B | 9.1 | 7.0 | 0.13 | SHIIO et al. (1982b) |
| *B. flavum* S-225 | Tyr$^-$, 5FT$^r$, AzaSer$^r$, Sulfaguanidine$^r$, PFP$^r$ | glucose/B<br>sucrose/B | 19<br>17 | 14.6<br>17.1 | 0.26<br>0.24 | SHIIO et al. (1984) |
| *B. subtilis* AJ 11716 | 5FT$^r$, AzaSer$^r$, IM$^r$ | glucose/B | 7.1 | 8.9 | 0.07 | KURAHASHI et al. (1983) |
| *B. subtilis* K AJ 11982 | AzaSer$^r$, DON$^r$, CIN$^r$ | glucose/B | 21.5 | 10.8 | 0.26 | KURAHASHI et al. (1987) |
| *C. glutamicum* H-3656 | Phe$^-$, Tyr$^-$, PheHx$^r$, TyrH$^r$, PAP$^r$, PFP$^r$, Glyphosate-isopropylamine$^r$ | molasses/B | 18.2 | 12 | 0.25 | KINO et al. (1987) |
| *C. glutamicum* BPS-13 | Phe$^-$, Tyr$^-$, 5MT$^r$, 3BP$^r$ (PEPC$^L$) | glucose/B | 7.8 | 13 | 0.11 | KATSUMATA and KINO (1989) |
| *C. glutamicum* H-7853 | Phe$^-$, Tyr$^-$, 5MT$^-$, 3BP$^r$ (PEPC$^L$), Primaquine$^r$ | glucose/B | 8.7 | 14.5 | 0.12 | KINO et al. (1992) |
| *C. glutamicum* KY9229 | Phe$^-$, Tyr$^-$, Phe$^r$, Tyr$^r$ | cane molasses/ FB | 35.0 | n.d. | 0.44 | IKEDA and KATSUMATA (1995) |

| Abbreviation | Genetic Marker |
|---|---|
| 3BP$^r$ | resistance to 3-bromopyruvic acid |
| 5FT$^r$ | resistance to 5-fluoro-L-tryptophan |
| 5MT$^r$ | resistance to 5-methyl-L-tryptophan |
| AzaSer$^r$ | resistance to azaserine |
| CIN$^r$ | resistance to cinnamate |
| DON$^r$, | resistance to 6-diazo-5-oxo-L-norleucine |
| Glyphosate-isopropylamine$^r$ | resistance to glyphosate-isopropylamine |
| IM$^r$ | resistance to indolmycin |
| Met$^-$ | requirement of L-methionine for growth |
| PAP$^r$ | resistance to *p*-aminophenylalanine |
| PEPC$^L$ | low activity of phosphoenolpyruvate carboxylase |
| PFP$^r$ | resistance to *p*-fluorophenylalanine |
| Phe$^-$ | requirement of L-phenylalanine for growth |
| PheHx$^r$ | resistance to L-phenylalanine hydroxamate |
| Phe$^r$ | resistance to L-phenylalanine |
| Primaquine$^r$ | resistance to primaquine |
| Sulfaguanidine$^r$ | resistance to sulfaguanidine |
| Tyr$^-$ | requirement of L-tyrosine for growth |
| TyrHx$^r$ | resistance to L-tyrosine hydroxamate |
| Tyr$^r$ | resistance to L-tyrosine |

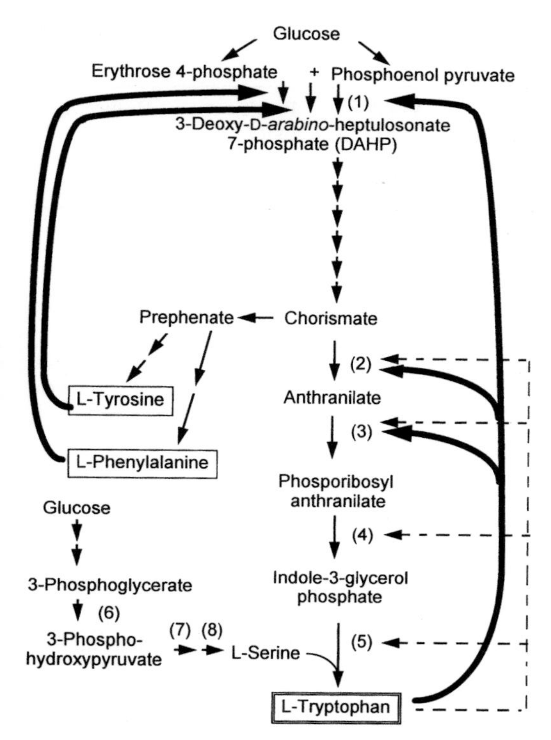


**Fig. 7.** Regulation of L-tryptophan biosynthesis in *E. coli.* ⟶ feedback inhibition, ---→ repression; enzyme (coding gene): 1 3-deoxy-D-arabinose-heptulosonate-7-phosphate synthase = DS (*aro*F, *aro*G, *aro*H), 2 anthranilate synthase = AS (*trp*ED), 3 phosphoribosyl anthranilate transferase = PRT ((*trp*D), 4 indole glycerol phosphate synthase = IGP (*trp*C), 5 tryptophan synthase = TS (*trp*AB), 6 phosphoglycerate dehydrogenase = PD (*ser*A), 7 phosphoserine aminotransferase (*ser*C), 8 phosphoserine phosphatase (*ser*B).

presence of alanine racemase. With a D-cycloserine-resistant mutant selected from *B. lactofermentum* it is possible to obtain 46 g $L^{-1}$ D-alanine with an enantiomeric excess (ee) of 95% (YAHATA et al., 1993). Recently, an alanine racemase-deficient mutant of *Arthrobacter oxydans* was reported producing 75 g $L^{-1}$ L-alanine from glucose with a yield of 52% and 95% ee (HASHIMOTO and KATSUMATA, 1994).

### 3.4.2 L-Arginine

L-Arginine is produced by extraction from protein hydrolyzates and by fermentation. Suitable strains are deregulated mutants derived from *C. glutamicum* and *B. subtilis* accumulating 25–35 g $L^{-1}$ L-arginine from glucose (JOSHIDA, 1986). Very potent strains of *S. marcescens* derived from mutants obtained by transduction that have feedback-insensitive and derepressive enzymes of arginine biosynthesis and show 6-azauracil resistance, are able to produce 60–100 g $L^{-1}$ L-arginine (CHIBATA et al., 1983).

### 3.4.3 L-Glutamine

L-Glutamine producers were selected from wild-type glutamate producing coryneform bacteria. A sulfaguanidine-resistant mutant *B. flavum* 1-60 accumulates 41 g $L^{-1}$ L-glutamine in 48 h from 10% glucose (TSUCHIDA et al., 1987). A yield of 44% was achieved by the mutant *B. flavum* AJ3409 (YOSHIHARA et al., 1992).

### 3.4.4 L-Histidine

Efficient L-histidine fermentation can be performed with strains of *C. glutamicum* and *S. marcescens.* A mutant of *C. glutamicum* resistant to 8-azaguanine, 1,2,4-triazole-3-alanine, 6-mercaptoguanine, 6-methylpurine, 5-methyltryptophan, and 2-thiourazil, produces 15 g $L^{-1}$. A strain of *S. marcescens* having both characteristics, feedback-insensitive and derepressed histidine enzymes combined with transductional techniques and 6-methylpurine resistance, accumulates 23 g $L^{-1}$ L-histidine (ARAKI, 1986). By amplification of the *his*G, *his*D, *his*B, and *his*C genes in *S. marcescens* the final concentration of L-histidine could be elevated from 28 g $L^{-1}$ to 40 g $L^{-1}$ (SUGIURA et al., 1987).

### 3.4.5 L-Isoleucine

An advantageous fermentation method was based on the use of chemically synthesized

**Tab. 8.** L-Tryptophan Producing Strains Developed by Recombinant DNA Techniques; Results in Batch (B) or Fed-Batch (FB) Culture

| Type of Strain | Gene Amplified | Genetic Markers | Carbon Source/ Culture | L-Thr Accumulation [g $L^{-1}$] | Yield (P/S) [g 100 $g^{-1}$] | Productivity [g $L^{-1}$ $h^{-1}$] | Reference |
|---|---|---|---|---|---|---|---|
| *C. glutamicum* BPS-13/ pCDtrp157 | genes coding for DS, AS, PRT, PRAI, InGPS, TS (pCDtrp157) | $Phe^-$, $Tyr^-$, $5MT^r$, $3BP^r$ ($PEPC^L$) | glucose/ B<br>sucrose/ FB | 11.0<br>32.6 | 18.0<br>n.d. | 0.15<br>n.d. | KATSUMATA and IKEDA (1989) |
| *C. glutamicum* BPS-13/ pDTS9901 | genes coding for DS, AS, PRT, PRAI, InGPS, TS, PGDH (pCDtrp157) | $Phe^-$, $Tyr^-$, $5MT^r$, $3BP^r$ ($PEPC^L$) | glucose/ B<br>sucrose/ FB | 11.7<br>35.2 | 19.5<br>n.d. | 0.16<br>n.d. | IKEDA and NAKANISHI (1990) |
| *C. glutamicum* KY9218/ pKW9901 | *ser*A (pKW9901) | $Phe^-$, $Tyr^-$, $Ser^-$ ($PGDH^-$) | sucrose/ FB | 50.0 | 20.0 | 0.63 | IKEDA et al. (1994) |
| *E. coli* JP4114 | *trp* operon on plasmid | $5MT^r$, $3FT^r$, mutant allele *trp*S | glucose/ FB | 23.5 | 17.2 | 0.71 | TEH et al. (1985) |
| *E. coli* EMS4-C25/ pTC576 integrated-16 | chromosome *trp* operon (copy no. 3) | $5FT^r$ | glucose/ B | 9.2 | 13.0 | n.d. | CHAN et al. (1993) |
| *E. coli* K-12 SV163/ pGH5/II | *ser*A allele *ser*A5 (pGH5/II) | mutation *trp*E6 ($AS^R$) feedback resistant | glucose/ FB | 24.1 | n.d. | 0.50 | BACKMAN et al. 1994) |

| Abbbreviation | Genetic Marker |
|---|---|
| $3BP^r$ | resistance to 3-bromopyruvic acid |
| $3FT^r$ | resistance to 3-fluoro-tyrosine |
| $5MT^r$ | resistance to 5-methyl-tryptophan |
| $AS^R$ | deregulated anthranilate synthase |
| $PEPC^L$ | low activity of phosphoenolpyrvate carboxylase |
| $PGDH^-$ | deficient of 3-phosphoglycerate dehydrogenase |
| $Phe^-$ | requirement of L-phenylalanine for growth |
| $Ser^-$ | requirement of L-serine for growth |
| $Tyr^-$ | requirement of L-tyrosine for growth |

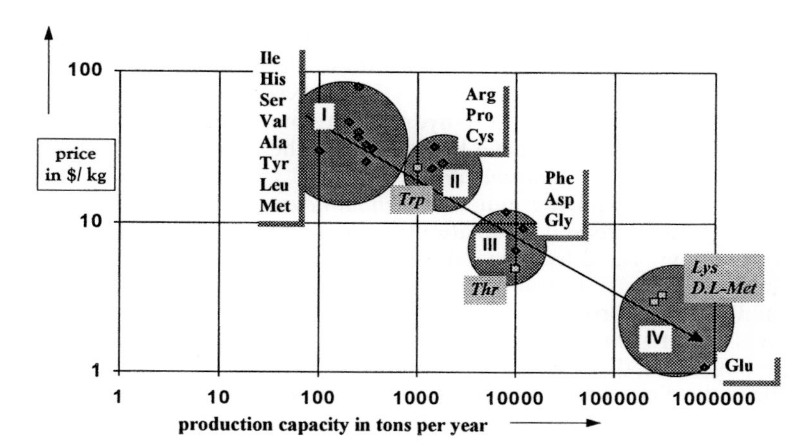


**Fig. 8.** Relationship of price and production capacity of amino acids.

**Tab. 9.** Pharmaceutical L-Amino Acids Produced by Direct Fermentation from Carbohydrates

| L-Amino Acid | Type of Mutant | Remarks | L-Amino Acid [g L$^{-1}$] | Yield (P/S) [g 100 g$^{-1}$] | Reference |
|---|---|---|---|---|---|
| L-Alanine (95% ee) | *A. oxydans* | glucose-nonrepressible alanine dehydrogenase; alanine racemase-deficient | 75.1 | 52 | HASHIMOTO and KATSUMATA (1994) |
| D-Alanine (95% ee) | *B. lactofermentum* | D-cycloserine-resistant from ATCC 13869 | 46.1 | n.d. | YAHATA et al. (1993) |
| L-Arginine | *S. marcescens* AU$^r$-1 | 6-azauracil-resistant | 100.2 | n.d. | CHIBATA et al. (1983) |
| L-Glutamine | *B. flavum* AJ 3409 | | 57.0 | 44 | YOSHIHARA et al. (1992) |
| L-Histidine | *S. marcescens* | amplification of genes *his*G, *his*D, *his*B, *his*C | 40.0 | n.d. | SUGIURA et al. (1987) |
| L-Isoleucine | *E. coli* H-8461 | resistance to thiaisoleucine, arginine hydroxamate, D,L-ethionine and 6-dimethylaminopurine | 30.2 | n.d. | KINO et al. (1993b) |
| L-Leucine | *B. flavum* AJ 3686 | | 19.5 | 15 | YOSHIHARA et al. (1992) |
| L-Methionine | *P. putida* VKPM V-4167 | resistant to D,L-ethionine, prototroph | 3.5 | n.d. | University of Odessa (1992) |
| L-Proline | *B. flavum* AP113 | isoleucine auxotroph, resistance to D,L-3,4-dehydroproline and osmotic pressure, incapable to degrade L-proline | 97.5 | n.d. | KOCARIAN et al. (1986) |
| L-Valine | *B. lactofermentum* AJ 12341 | | 39.0 | 28 | KATSURADA et al. (1993) |

substrates that only require a few steps to be converted to L-isoleucine. Among these the natural precursor 2-ketobutyrate (EGGELING et al., 1987) or D,L-2-hydroxybutyrate (SCHEER et al., 1988) have been used for the production with *C. glutamicum.*

A leucine-requiring mutant *C. glutamicum* DL-4, with increased D-lactate utilization and consuming D,L-2-hydroxybutyrate, accumulates 13.4 g $L^{-1}$ L-isoleucine. However, formation of by-products is hampering this process (WILHELM et al., 1989).

Sugar-based L-isoleucine processes have been developed with strains of *C. glutamicum* (YOSHIHARA et al., 1992), *S. marcescens* (KOMATSUBARA et al., 1980), and *E. coli* (LIVSHITS et al., 1992; NAKANO et al., 1994b; MÖCKEL et al., 1995). The mutant *E. coli* H-8225 resistant to thiaisoleucine, arginine hydroxamate, and D,L-ethionine accumulates 26 g $L^{-1}$ L-isoleucine in 45 h in a fed-batch process (KINO and KURATSU, 1994). Introduction of resistance to 6-dimethylaminopurine in strain H-8285 resulted in a mutant *E. coli* H-8461 that increased L-isoleucine accumulation to 30.2 g $L^{-1}$ (KINO et al., 1993b).

The biosynthesis of L-isoleucine has been investigated in detail at the level of involved genes (KEILHAUER et al., 1993), thus recombinant strains are being constructed with high productivity and selectivity. An appropriate balance of homoserine dehydrogenase and threonine dehydratase activities in the construct *C. glutamicum* DR17/pECM3::*ilv*A (V323A) with feedback-resistant aspartate kinase creates a specific productivity of 0.052 g L-isoleucine per g dry biomass per hour (MORBACH et al., 1995). The recombinant strain *E. coli* AJ13100 produces L-isoleucine from glucose with high selectivity with a yield of 30% (HASHIGUCHI et al., 1995).

### 3.4.6 L-Leucine and L-Valine

For L-leucine production, precursor fermentation was suggested: In a fed-batch culture of *C. glutamicum* ATCC 13032 it was possible to convert 32 g $L^{-1}$ 2-ketoisocaproate to 24 g $L^{-1}$ L-leucine by the transaminase B reaction (GROEGER and SAHM, 1987).

With the method of direct fermentation the branched chained amino acids L-leucine and L-valine can be produced by either $\alpha$-aminobutyric acid-resistant mutants of *S. marcescens* or by 2-thiazole-alanine-resistant coryneform strains (KOMATSUBARA and KISUMI, 1986).

*B. flavum* AJ3686 accumulates 19.5 g $L^{-1}$ L-leucine (15% yield) (YOSHIHARA et al., 1992), *B. lactofermentum* AJ12341 produces 39 g $L^{-1}$ L-valine (28% yield) (KATSURADA et al., 1993).

### 3.4.7 L-Methionine

The production method of choice for L-methionine is still the enzymatic resolution of racemic N-acetyl-methionine using acylase from *Aspergillus oryzae.* The production is realized in a continuously operated fixed-bed or enzyme membrane reactor (LEUCHTENBERGER and PLÖCKER, 1990).

Alternatively, L-methionine may be produced by microbial conversion of the corresponding 5-substituted hydantoin. With growing cells of *Pseudomonas* sp. strain NS671, D,L-5-(2-methylthioethyl)hydantoin was converted to L-methionine; a final concentration of 34 g $L^{-1}$ and a molar yield of 93% were obtained (ISHIKAWA et al., 1993).

L-Methionine biosynthesis and its regulation in bacteria is well known. Although some promising concepts, e.g., the choice of sulfate, sulfite, or thiosulfate as sulfur sources for microbes, have been suggested (LIEVENSE, 1993) it was not possible so far to develop strains able to excrete remarkable amounts of L-methionine into the culture medium.

### 3.4.8 L-Proline

L-Proline is produced by direct fermentation using analog-resistant mutants of coryneform bacteria and *S. marcescens* (YOSHINAGA, 1986).

An isoleucine auxotrophic mutant of *B. flavum* resistant to sulfaguanidine and D,L-3,4-dehydroproline (DP) accumulates 40 g $L^{-1}$ L-proline. *B. flavum* AP113 is claimed to produce 97.5 g L$^{-1}$ L-proline; this mutant is characterized by isoleucine auxotrophy, resistance to DP and osmotic pressure and is incapable of degrading L-proline (KOCARIAN et al., 1986).

A strain of *S. marcescens* lacking proline oxidase and showing resistance to DP, thiazoline-4-carboxylate, and azetidine-2-carboxylate overproduces 58.5 g L$^{-1}$ L-proline into the culture medium (SUGIURA et al., 1982). By amplification of the genes *pro*A and *pro*B in this type of regulatory mutant a construct was obtained, which yields 75 g L$^{-1}$ L-proline (IMAI et al., 1984).

### 3.4.9 L-Serine

L-Serine is available by extraction of protein hydrolyzates or by microbial/enzymatic conversion of glycine using immobilized resting cells or crude cell extracts.

*Hyphomicrobium* strains possess the serine pathway and are able to produce L-serine from methanol and glycine. Methanol is oxidized by methanol dehydrogenase to formaldehyde, which in turn is converted in an aldolytic reaction to L-serine catalyzed by serine hydroxymethyltransferase (SHMT) (IZUMI et al., 1993). *Hyphomicrobium* sp. NCIB10099 was found to produce 45 g L$^{-1}$ L-serine from 100 g L$^{-1}$ glycine and 88 g L$^{-1}$ methanol in 3 d (YOSHIDA et al., 1993). In an enzyme bioreactor with a feedback control system a crude extract from *Klebsiella aerogenes* containing SHMT was used to synthesize L-serine from glycine and formaldehyde in the presence of tetrahydrofolic acid and pyridoxal phosphate. In this bioreactor a serine concentration of 450 g L$^{-1}$ with an 88% molar conversion of glycine at a volumetric productivity of 8.9 g L$^{-1}$ h$^{-1}$ could be achieved under optimized conditions (HSIAO and WEI, 1986).

With whole cells of *E. coli* MT-10350 L-serine is produced by treatment of an oxygenated aqueous glycine solution (485 g L$^{-1}$) with aqueous formaldehyde for 35 h at 50 °C with a molar yield of 89% based on glycine (URA et al., 1995).

## 3.5 Future Prospects

Fermentation technology represents and will maintain the key position in the amino acid industry. Due to modern techniques such as genetic and enzyme engineering together with new ideas of processing, further improvement of established processes such as for L-lysine or L-threonine should be possible. Moreover, the amino acids that enter new markets with a growing demand will benefit from the existing experiences and the potential that the fermentation technology can offer today.

On the other hand, enzymatic processes will continue to play an important role for the amino acids that are manufactured from favorable raw materials by means of highly specific and well available biocatalysts in efficient bioreactor systems.

# 4 References

Ajinomoto (1982), Fermentative production of L-lysine, *Jpn. Patent* 57-94297.

AMIN, G. (1994), Continuous production of glutamic acid in a vertical rotating immobilized cell reactor of the bacterium *Corynebacterium glutamicum Bioresour. Technol.* **47**, 113–119.

ARAKI, K. (1986), Histidine, in: *Biotechnology of Amino Acid Production* (AIDA, K., CHIBATA, I., NAKAYAMA, K., TAKINAMI, K., YAMADA, H., Eds.), pp. 247–256. Tokyo: Kodansha, Amsterdam: Elsevier.

BACKMAN, K., WICH, G., LEINFELDER, W. (1994), Mikroorganismen für die Produktion von Tryptophan und Verfahren zu ihrer Herstellung. *Deutsche Offenlegungsschrift* 4 232 468.

BINDER, W., FRIEDRICH, H., LOTTER, H., TANNER, H., HOLLDORFF, H., LEUCHTENBERGER, W. (1993), Tierfuttersupplement auf Fermentationsbrühe-Aminosäurebasis, Verfahren zu dessen Herstellung und dessen Verwendung, *Eur. Patent* 533 039.

BÖRMANN, E. R., EIKMANNS, B. J., SAHM, H. (1992), Molecular analysis of the *Corynebacterium glutamicum gdh* gene encoding glutamate dehydrogenase, *Mol. Microbiol.* **6**, 317–326.

CALTON, G. J., WOOD, L. L., OPDIKE, M. H., LANTZ II, L., HAMMAN, J. P. (1986), The production of L-phenylalanine by polyazetidine immobilized microbes, *Biotechnology* **4**, 317–320.

CHAN, E.-C., TSAI, H.-L., CHEN, S.-L., MOU, D.-G. (1993), Amplification of the tryptophan operon gene in *Escherichia coli* chromosome to increase L-tryptophan biosynthesis, *Appl. Microbiol. Biotechnol.* **40**, 201–305.

CHIBATA, I., KISUMI, M., TAGAKI, T. (1983), *Jpn. Patent* (Kokai) 58-9692.

CHIBATA, I., TOSA, T., SATO, T. (1985), Aspartic acid, in: *Comprehensive Biotechnology,* Vol. 3 (BLANCH, H. W., DREW, S., WANG, D. I. C., Eds). Oxford: Pergamon Press Ltd.

CHIBATA, I., TETSUYA, T., SATO, T. (1986), Continuous production of L-aspartic acid, *Appl. Biochem. Biotechnol.* **13**, 231–240.

CREMER, J., EGGELING, L., SAHM, H. (1991), Verfahren zur fermentativen Herstellung von Aminosäuren, insbesondere L-Lysin, dafür geeignete Mikroorganismen und rekombinante DNA, *Eur. Patent Appl.* 435 132.

CUQ, J.L., GILOT, M. (1985), Influence of heat treatments on the stability of free or bound tryptophan, nutritional consequences, *Sci. Aliments* **5**, 687–697.

DEBABOV, V. G., KOZLOV, J. I., ZHDANOVA, N. I., KHURGES, E. M., YANKOVSKY, N. K., ROZINOV, M. N., SHAKULOV, R. S., REBENTISH, B. A., LIVSHITS, V. A., GUSYATINER, M. M., MASHKO, S. V., MOSHENTSEVA, V. N., KOZYREVA, L. F., ARSATIANTS, R. A. (1981), Method for preparing strains which produce amino acids, *U.S. Patent* 4 278 765.

DEBABOV, V. G., KOZLOV, J. I., KHURGES, E. M., LIVSHITS, V. A., ZHDANOVA, N. I., GUSYATINER, M. M., SOKOLOV, A. K., BACHINA, T. A., YANKOVSKY, N. K., TSYGANKOV, J. D., CHISTOSERDOV, A. J., PLOTNIKOVA, T. G., SHAKALIS, I. O., BELEREVA, A. V., ARSATIANTS, R. A., SHOLIN, A. F., POZDNYAKOVA, T. M. (1990), Strain of bacteria *Escherichia coli,* producer of L-threonine, *PCT Patent Appl.* WO 90/04636.

DEBABOV, V. G., KOZLOV, J. I., KHURGES, E. M., LIVSHITS, V. A., ZHDANOVA, N. I., GUSYATINER, M. M., SOKOLOV, A. K., BACHINA, T. A., CHISTOSERDOV, A. J., TSIGANKOV, J. D., YANKOVSKY, N. K., MASHKO, S. V., LAPIDUS, A. L., GAVRILOVA, O. F., RODIONOV, O. A. (1994), Novel L-threonine-producing microbacteria and a method for the production of L-threonine, *Eur. Patent Appl.* 593 792.

DUNICAN, L. K., SHIVNAN, E. (1989), High frequency transformation of whole cells of amino acid producing coryneform bacteria using high voltage electroporation, *Bio/Technology* **7**, 1067–70.

EGGELING, I., CORDES, C., EGGELING, L., SAHM, H. (1987), Regulation of acetohydroxy acid synthase in *Corynebacterium glutamicum* during fermentation of $\alpha$-ketoburyrate to L-isoleucine, *Appl. Microbiol. Biotechnol.* **25**, 346–351.

EIKMANNS, B. J., FOLLETTIE, M. T., GRIOT, M., SINSKEY, A. J. (1989), The phosphoenol pyruvate carboxylase gene of *Corynebacterium glutamicum*: molecular cloning, nucleotide sequence, and expression, *Mol. Gen. Genet.* **218**, 330–339.

EIKMANNS, B. J., KIRCHER, M., REINSCHEID, D. J. (1991), Discrimination of *Corynebacterium glutamicum, Brevibacterium flavum, Brevibacterium lactofermentum* by restriction pattern analysis of DNA adjacent to the *hom* gene, *FEMS Microbiol. Lett.* **82**, 203–208.

ERDMANN, A., WEIL, B., KRÄMER, R. (1994), Lysine secretion by *Corynebacterium glutamicum* wild type: Regulation of secretion carrier activity, *Appl. Mirobiol. Biotechnol.* **42**, 604–610.

ERDMANN, A., WEIL, B., KRÄMER, R. (1995), Regulation of lysine excretion in the lysine producer strain *Corynebacterium glutamicum* MH 20-22B, *Biotechnol. Lett.* **17**, 927–932.

EVANS, C. T., COMA, C., PETERSON, W., MISAWA, M. (1987a), Production of phenylalanine ammonia-lyase (PAL): isolation and evaluation of yeast strains suitable for commercial production of L-phenylalanine, *Biotechnol. Bioeng.* **30**, 1067–1072.

EVANS , C. T., BELLAMY, W., GLEESON, M., AOKI, H., HANNA, K., PETERSON, W., CONRAD, D., MISAWA, M. (1987b), A novel, efficient biotransformation for the production of L-phenylalanine, *Biotechnology* **5**, 818–923.

FAN, C., CHEN, L., ZHENG, S. (1988), Characteristics of L-lysine high yielding strain FML8611, *Weishengwuxue Zashi* **8**, 11–17.

FECHTER, W. L., DIENST, J. H., LE PATOUREL, J. F. (1995), Recovery of an amino acid, *PCT Patent Appl.* WO 95/14002.

FORBERG, C., ELIAESO, T., HAGGSTROM, L. (1988), Correlation of theoretical and experimental yields of phenylalanine from non-growing cells of a rec *Escherichia coli* strain, *J. Biotechnol.* **7**, 319–332.

FURUI, M., YAMASHITA, K. (1983), Pressurized reaction method for continuous production of L-alanine by immobilized *Pseudomonas dacunhae* cells, *J. Ferment. Technol.* **61**, 587–591.

FURUKAWA, S., NAKANISHI, T. (1989), Process for producing L-threonine by fermentation, *Eur. Patent Appl.* 301 572.

FUSEE, M. C., SWANN, W. E., CARLTON, G. J. (1981), Immobilization of *Escherichia coli* cells containing aspartase activity with polyurethane and its application for L-aspartic acid production, *Appl. Environ. Microbiol.* **42**, 672–676.

GROEGER, U., SAHM, H. (1987), Microbial production of L-leucine from α-ketoisocaproate by *Corynebacterium glutamicum, Appl. Microbiol. Biotechnol.* **25**, 352–236.

GUBLER, M., PARK, S. M., JETTEN, M., STEPHANOPOULOS, G., SINSKEY, A. J. (1994a), Effects of phosphoenol pyruvate carboxylase deficiency on metabolism and lysine-production in *Corynebacterium glutamicum, Appl. Microbiol. Biotechnol.* **40**, 857–863.

GUBLER, M., JETTEN, M., LEE, S. H., SINSKEY, A. J. (1994b), Cloning of the pyruvate kinase gene (*pyk*) of *Corynebacterium glutamicum* and site-specific inactivation of *pyk* in a lysine-producing *Corynebacterium lactofermentum* strain, *Appl. Environ. Microbiol.* **60**, 2494–2500.

HADI SASSI, A., COELLO, N., DESCHAMPS, A. M., LEBEAULT, J. M. (1990), Effect of medium composition on L-lysine production by a variant strain of *Corynebacterium glutamicum* ATCC 21513, *Biotechnol. Lett.* **12**, 295–298.

HAMILTON, B., K., HSIAO, H.-Y., SWANN, W. E., ANDERSON, D. M., DELENTE, J. J. (1985), Manufacture of L-amino acids with bioreactors, *Trends Biotechnol.* **3**, 64–68.

HAN, J.-K., OH, J.-W., LEE, H.-H., CHUNG, S., HYUN, H.-H., LEE, J.-H. (1991), Molecular cloning and expression of *S*-(2-aminoethyl)-L-cysteine resistant aspartokinase gene of *Corynebacterium glutamicum, Biotechnol. Lett.* **10**, 721–726.

HASHIGUCHI, K., KISHINO, H., TSUJIMOTO, N., MATSUI, H. (1995), L-leucine producing bacterium and method for preparing L-isoleucine through fermentation, *Eur. Patent Appl.* 685 555.

HASHIMOTO, S., KATSUMATRA, R. (1994), L-Alanine production by alanine racemase-deficient mutant of *Arthrobacter oxydans,* in: *Proc. Ann. Meet. Agric. Chem. Soc.* (Japan), 341.

HENKEL, H. J., JOHL, H. J., TROESCH, W., CHMIEL, H. (1990), Continuous production of glutamic acid in a three-phase fluidized bed with immobilized *Corynebacterium glutamicum, Food Biotechnol.* (N. Y.) **4**, 149–154.

HILLIGER, M. (1991), Biotechnologische Aminosäureproduktion, *BioTec* **2**, 40–44.

HIRAO, T., NAKANO, T., AZUMA, T., NAKANISHI, T. (1989a), Process for producing L-lysine, *Eur. Patent* 327 945.

HIRAO, T., NAKANO, T., AZUMA, T., SUGIMOTO, M., NAKANISHI, T. (1989b), L-Lysine production in continuous culture of an L-lysine hyperproducing mutant of *Corynebacterium glutamicum, Appl. Microbiol. Biotechnol.* **32**, 269–273.

HIROSE, Y., ENEI, H., SHIBAI, H. (1985), L-Glutamic acid fermentation, in: *Comprehensive Biotechnology,* Vol. 3 (BLANCH, H. W., DREW, S., WANG, D. I. C., Eds.). Oxford: Pergamon Press Ltd.

HIROSE, T., TSURUTA, M., TAMURA, K., UEHARA, Y., MIWA, H. (1994), Process for the production of an amino acid using a fermentation apparatus, *U.S. Patent* 5 362 635.

HSIAO, H.-Y., WEI, T. (1986), Enzymatic production of L-serine with a feedback control system for formaldehyde addition, *Biotechnol. Bioeng.* **28**, 1510–1518.

HWANG, S. O., GIL, G. H., CHO, Y. J., KANG, K. R., LEE, J. H., BAE, J. C. (1985), The fermentation process for L-phenylalanine production using an auxotrophic regulatory mutant of *Escherichia coli, Appl. Microbiol. Biotechnol.* **22**, 108–113.

IKEDA, K. (1908), A new flavor enhancer, *J. Tokyo Chem. Soc.* **30**, 820.

IKEDA, M., KATSUMATA, R. (1995), Tryptophan production by transport mutants of *Corynebacterium glutamicum, Biosci. Biotech. Biochem.* **59**, 1600–1602.

IKEDA, M., NAKANISHI, K. (1990), Process for producing L-tryptophan, *Eur. Patent Appl.* 401 735.

IKEDA, M., NAKANISHI, K., KINO, K., KATSUMATA, R. (1994), Fermentative production of tryptophan by a stable recombinant strain of *Corynebacterium glutamicum* with a modified serine-biosynthetic pathway, *Biosci. Biotech. Biochem.* **58**, 674–678.

IMAI, Y., TAKAGI, T., SUGIURA, M., KISUMI, M. (1984), *Abstr. Ann. Meet. Agr. Chem. Soc.* (Japan), 100.

ISHIDA, M., YOSHINO, E., MAKIHARA, R., SATO, K., ENEI, H., NAKAMORI, S. (1989), Improvement of an L-threonine producer derived from *Brevibacterium flavum* using threonine operon of *Escherichia coli* K-12, *Agric. Biol. Chem.* **53**, 2269–2271.

ISHIDA, M., SATO, K., HASHIGUCHI, K., ITO, H., ENEI, H., NAKAMORI, S. (1993), High fermentative production of L-threonine from acetate by a *Brevibacterium flavum* stabilized strain transformed with a recombinant plasmid carrying the *Escherichia coli thr* operon, *Biosci. Biotech. Biochem.* **57**, 1755–1756.

ISHIDA, M., KAWASHIMA, H., SATO, K., HASHIGUCHI, K., ITO, H., ENEI, H., NAKAMORI, S. (1994), Factors improving L-threonine production by a three L-threonine biosynthetic genes-amplified recombinant strain of *Bravibacterium lactofermentum, Biosci. Biotech. Biochem.* **58**, 768–770.

ISHIKAWA, T., WATABE, K., MUKOHURA, Y., KOBAYASHI, S., NAKAMURA, H. (1993), Microbial conversion of DL-5-substituted hydantoins to the corresponding L-amino acids by *Pseudomonas* sp. strain NS671, *Biosci. Biotech. Biochem.* **57**, 982–986.

ISHIWATA, K., YOSHINO, S., IWAMORI, S., SUZUKI, T., MAKIGUCHI, N. (1989), Cloned tryptophan synthase gene and recombinant plasmid containing the same, *Eur. Patent Appl.* 341 674.

ISHIZAKI, A., TAKASAKI, S., FURUTA, Y. (1993), Cell-recycled fermentation of glutamate using a novel cross-flow filtration system with constant air supply, *J. Ferment. Bioeng.* **76**, 316–320.

ITO, H., SATO, K., MATSUI, K., MIWA K., SANO, K., NAKAMORI, S., TANAKA, T., ENEI, H. (1990), Cloning and characterization of genes responsible for *m*-fluoro-D,L-phenylalanine resistance in *Brevibacterium lactofermentum, Agric. Biol. Chem.* **54**, 707–713.

IZUMI, Y., YOSHIDA, T., MIYAZAKI, S. S., MITSUNAGA, T., OHSHIRO, T., SHIMAO, M., MIYATA, A., TANABE, T. (1993), L-Serine production by a methylotroph and its related enzymes, *Appl. Microbiol. Biotechnol.* **39**, 427–432.

JAFFARI, M. D., MAHAR, J. T., BACHERT, R. L. (1989), Ion exchange recovery of L-lysine, *Eur. Patent Appl.* 337 440.

JETTEN, M. S. N., SINSKEY, A. J. (1995), Recent advances in the physiology and genetics of amino acid-producing bacteria, *Crit. Rev. Biol.* **15**, 73–103.

JETTEN, M. S. N., GUBLER, M. E., LEE, S. H., SINSKEY, A. J. (1994), Structural and functional analysis of pyruvate kinase from *Corynebacterium glutamicum, Appl. Environ. Microbiol.* **60**, 2501–2507.

JETTEN, M. S. N., FOLLETTIE, M. T., SINSKEY, A. J. (1995), Effect of different levels of aspartokinase on the lysine production by *Corynebacterium lactofermentum, Appl. Microbiol. Biotechnol.* **43**, 76–82.

KALINOWSKI, J., THIERBACH, G., BACHMANN, B., PÜHLER, A. (1989), Structural and functional analysis of lysine biosynthetic genes of *Corynebacterium gylutamicum, Forum Mikrobiol.* **1–2**, 40.

KARASAWA, M., TOSAKA, O., IKEDA, S., YOSHI, H. (1986), Application of protoplast fusion to the development of L-threonine and L-lysine producers, *Agric. Biol. Chem.* **50**, 339–346.

KATSUMATA, R., IKEDA, M. (1989), Process for producing L-tryptophan, *Eur. Patent Appl.* 333 474.

KATSUMATA, R., KINO, K. (1989), Process for producing amino acids, *Eur. Patent Appl.* 331 145.

KATSUMATA, R., OZAKI, A., MITSUKAMI, TO., KAGEYAMA, M., YAGISAWA, M., MITSUKAMI, TA., ITOH, S., OKA, T., FURUYA, A. (1983), Method for expressing a gene, *Eur. Patent Appl.* 88 166.

KATSUMATA, R., MITZUKAMI, T., OKA, T. (1986), Process for producing L-lysine, *Eur. Patent Appl.* 197 335.

KATSUMATA, R., MITSUKAMI, TO., OKA, T. (1987), Process for producing amino acids, *Eur. Patent Appl.* 219 027.

KATSURADA, N., UCHIBORI, H., TSUCHIDA, T. (1993), L-Valine producing microorganism and a process for producing L-valine by fermentation, *Eur. Patent* 287 123.

KAWAHARA, Y., YOSHIHARA, Y., IKEDA, S., YOSHII, H., HIROSE, Y. (1990), Stimulatory effect of glycine betaine on L-lysine fermentation, *Appl. Microbiol. Biotechnol.* **34**, 87–90.

KEILHAUER, C., EGGELING, L., SAHM, H. (1993), Isoleucine synthesis in *Corynebacterium glutamicum:* molecular analysis of the *ilv*B-*ilv*N-*ilv*C operon, *J. Bacteriol.* **175**, 5595–5603.

KIKUCHI, M., NAKAO, Y. (1986), Glutamic acid, in: *Biotechnology of Amino Acid Production* (AIDA, K., CHIBATA, I., NAKAYAMA, K., TAKINAMI, K., YAMADA, H., Eds.), pp. 101–116. Tokyo: Kodansha, Amsterdam: Elsevier.

KINO, K., KURATSU, Y. (1994), Process for producing L-isoleucine and ethionine by isoleucine analog resistant strains of *Escherichia coli. U.S. Patent* 5 362 637.

KINO, K., SUGIMOTO, M., NAKANISHI, T. (1987), A process for preparation of L-tryptophan, *Eur. Patent* 156 504.

KINO, K., TAKANO, J., KURATSU, Y. (1991), Process for producing L-threonine, *Eur. Patent Appl.* 445 830.

KINO, K., FURUKAWA, K., TOMIYOSHI, Y., KURATSU, Y. (1992), Process for producing L-tryptophan, *Eur. Patent Appl.* 473 094.

KINO, K., TAKANO, J., OKAMOTO, K., KURATSU, Y. (1993a), Process for producing L-threonine, *Eur. Patent Appl.* 530 803.

KINO, K., OKAMOTO, K., TAKEDA, Y., KURATSU, Y. (1993b), Process for the production of amino acids by fermentation, *Eur. Patent Appl.* 557 996.

KINOSHITA, S., TANAKA, K. (1972), Glutamic acid, in: YAMADA, K. et al. (Eds.). *The Microbial Production of Amino Acids*, pp. 263–324. New York: John Wiley & Sons.

KINOSHITA, S., UKADA, S., SHIMONO, M. (1957), Studies on the amino acid fermentation. I. Production of L-glutamic acid by various microorganisms, *J. Gen. Appl. Microbiol.* **3**, 139–205.

KINOSHITA, S., NAKAYAMA, K., KITADA, S. (1961), Method of producing L-lysine by fermentation, *U.S. Patent* 2 979 439.

KISHIMOTO, M., MOO-YOUNG, M., ALLSOP, P. (1991), A fuzzy expert system for the optimization of glutamic acid production, *Bioprocess Eng.* **6**, 163–172.

KISS, R. D., STEPHANOPOULOS, G. (1991), Metabolic activity control of the L-lysine fermentation by restrained growth fed-batch strategies, *Biotechnol. Prog.* **7**, 501–509.

KISS, R. D., STEPHANOPOULOS, G. (1992), Metabolic characterization of a L-lysine producing strain by continuous culture. *Biotechnol. Bioeng.* **39**, 565–574.

KITSUTA, Y., KISHIMOTO, M. (1994), Fuzzy supervisory control of glutamic acid production, *Biotechnol. Bioeng.* **44**, 87–94.

KOCARIAN, Š. M., KARAPETIAN, Z. V., KELEŠIAN, S. K., AZIZIAN, A. G., AKOPIAN, E. M., ARUŠANIAN, A. V., AMBRARCUMIAN, A. A. (1986), Verfahren zur Herstellung von L-Prolin, *Deutsche Offenlegungsschrift* 3 612 077.

KOMATSUBARA, S., KISUMI, M. (1986), Isoleucine, valine, and leucine, in: *Biotechnology of Amino Acid Production* (AIDA, K., CHIBATA, J., NAKAYAMA, K., TAKANAMI, K., YAMADA, H., Eds.), pp. 233–246. Tokyo: Kodansha, Amsterdam: Elsevier.

KOMATSUBARA, S., KISUMI, M., CHIBATA, I. (1980), Transductional construction of an isoleucine-producing strain of *Serratia marcescens, J. Gen. Microbiol.* **119**, 51–61.

KOMATSUBARA, S., KISUMI, M., CHIBATA, I. (1983), Transductional construction of a threonine-hyperproducing strain of *Serratia marcescens:* lack of feedback controls of three aspartokinases and two homoserine dehydrogenases, *Appl. Environ. Microbiol.* **45**, 1445–1452.

KONO, Y., ITOH, H., TANEDA, R., WATANABE, T. (1991), Process for purifying tryptophan, *Eur. Patent Appl.* 405 524.

KONSTANTINOV, B. K., NISHIO, N., YOSHIDA, T. (1990), Glucose feeding strategy accounting for the decreasing oxidative capacity of recombinant *Escherichia coli* in fed-batch cultivation for phenylalanine production, *J. Ferment. Bioeng.* **70**, 253–260.

KRÄMER, R. (1993), Mechanismen der Aminosäuresekretion bei *Corynebacterium glutamicum, BioEngineering* **9**, 51–61.

KRÄMER, R. (1994), Secretion of amino acids by bacteria: physiology and mechanisms, *FEMS Microbiol. Rev.* **13**, 75–93.

KRÄMER, R. (1996), Genetic and physiological approaches for the production of amino acids, *J. Biotechnol.* **45**, 1–21.

KUBOTA, K., YOSHIHARA, Y., KAWASAKI, K., HIROSE, Y. (1974a), Verfahren zur Herstellung von L-Lysin, *Deutsche Offenlegungsschrift* 2 350 647.

KUBOTA, K., YOSHIHARA, Y. K., HIRAKAWA, H., KAMIJO, H., NOSAKI, S., YOSHINAGE, F., OKUMURA, S., OKADA, H. (1974b), Method of producing L-lysine by fermentation, *U.S. Patent* 3 825 472.

KUBOTA, K., TOSAKA, O., YOSHIHARA, Y., HIROSE, Y. (1976), Fermentative production of L-lysine. *Jpn. Patent* 76 19 186.

KURAHASHI, O., KAMADA, M., ENEI, H. (1983), Process for producing L-tryptophan by fermentation, *Eur. Patent Appl.* 81 107.

KURAHASHI, O., NODA-WATANABE, M., TORIDE, Y., TAKENOUCHI, T., AKASHI, K., MORINAGA, Y., ENEI, H. (1987), Production of L-tryptophan by azaserine-, 6-diazo-5-oxo-L-norleucine- and cinnamate-resistant mutants of *Bacillus subtilis, Agric. Biol. Chem.* **51**, 1791–1797.

LEE, J.-H., OH, J.-W., LEE, H.-H., HYUN, H.-H. (1991), Production of L-threonine by auxotrophs and analogue resistant mutants of *Escherichia coli, Korean J. Appl. Microbiol. Biotechnol.* **19**, 583–587.

LEE, H.-W., PAN, J.-G., LEBEFAULT, J.-M. (1995), Characterization of kinetic parameters and metabolic transition of *Corynebacterium glutamicum* on L-lysine production in continuous culture, *Appl. Microbiol. Biotechnol.* **43**, 1019–1027.

LEUCHTENBERGER, W., PLÖCKER, U. (1990), Amino acids and hydroxycarboxylic acids, in: *Enzymes in Industry, Production and Applications* (GERHARTZ, W., Ed.). Weinheim: VCH.

LIEBL, W., EHRMANN, M., LUDWIG, W., SCHLEIFER, K. H. (1991), Transfer of *Brevibacterium divaricatum* DSM 2029T, *Brevibacterium flavum* DSM 20411, *Brevibacterium lactofermentum* DSM 20412 and DSM 1412, and *Corynebacterium lilium* DSM 20317 to *Corynebacterium glutamicum* and their distinction by rDNA gene restriction patterns, *Int. J. Syst. Bacteriol.* **41**, 225–235.

LIEVENSE, J. C. (1993), Biosynthesis of methionine using a reduced source of sulfur, *PCT Patent Appl.* WO 93/17112.

LIU, Y.-C. (1987), Studies on the fermentation production of L-lysine. 6. An improvement of L-lysine producing strain by mutation of regulatory gene, *Rep. Taiwan Sugar Res. Inst.* **116**, 39–54.

LIU, Y.-C., WU, W.-T., TSAO, J.-H. (1993), Fed-batch culture for L-lysine production via on-line state estimation and control, *Bioprocess Eng.* **9**, 135–139.

LIVSHITS, V. A., DEBABOV, V. G., GAVRILOVA, O. F., ZAKATAEVA, N. P., SHAKULOY, R. S., BACHINA, T. A., KHURGES, E. M. (1992), Bacterial strain having high productivity of an amino acid and method of constructing these strains, *Eur. Patent Appl.* 519 113.

LU, J.-H., CHEN, J.-L., LIAO, C.-C. (1994), Molecular breeding of a *Brevibacterium flavum* L-lysine producer using a cloned aspartokinase gene, *Biotechnol. Lett.* **16**, 449–454.

LUCQ, P., DOMONT, C. (1993), Procédé de séparation de la lysine sous la forme de solutions aqueuses et utilisation de ces solutions pour l'alimentation animale, *Eur. Patent Appl.* 534 865.

LÜSSLING, T., MÜLLER, K., SCHREYER, G., THEISSEN, F. (1974), Verfahren zur Gewinnung von Methionin und Kaliumhydrogencarbonat aus den im Kreislauf geführten Mutterlaugen des Kaliumcarbonat-Methioninverfahrens, *Ger. Patent* 2 421 167.

MAITI, T. K., CHATTERJE, S. P. (1991), Microbial production of L-tryptophan: a review, *Hind. Antibiot. Bull.* **33**, 26–61.

MARX, A., DE GRAAF, A. A., WIECHERT, W., EGGELING, L., SAHM, H. (1996), Determination of the fluxes in the central metabolism of *Corynebacterium glutamicum* by Nuclear Magnetic Resonance spectroscopy combined with metabolite balancing, *Biotechnol. Bioeng.* **49**, 111–129.

MATSUDA, M., TAKAMATSU, S., NISHIMURA, N., KOMATSUBARA, S., TOSA, T. (1993), Improvement of nitrogen supply for L-threonine production by a recombinant strain of *Serratia marcescens, Appl. Biochem. Biotech.* **37**, 255–265.

MATSUMOTO, S. (1986), A method for producing L-lysine, *Eur. Patent Appl.* 175 309.

MAYENO, A. N., GLEICH, G. J. (1994), Eosinophilia-myalgia syndrome and tryptophan production; a cautionary tale, *TIBTECH* **12**, 346–352.

MCGUIRE, J. C. (1986), Phenylalanine ammonia lyase-producing microbial cells, *U.S. Patent* 4 598 047.

MICHELET, J., DESCHAMPS, A., LEBEAULT, J. M. (1984), Production of L-aspartic acid from fumaric acid by bioconversion, *3rd Eur. Congr. Biotechnol.*, Vol. 2, pp. 133–138. Weinheim: Verlag Chemie.

MÖCKEL, B., EGGELING, L., SAHM, H. (1995), Herstellung von L-Isoleucin mittels rekombinanter Mikroorganismen mit deregulierter Threonindehydratase, *Ger. Patent* 4 400 926.

MOMOSE, H., TAGAKI, T. (1978), Glutamic acid production in biotin-rich media by temperature-sensitive mutants of *Brevibacterium lactofermentum,* a novel fermentation process, *Agric. Biol. Chem.* **42**, 1911–1917.

MORBACH, S., SAHM, H., EGGELING, L. (1995), Use of feedback-resistant threonine dehydratase of *Corynebacterium glutamicum* to increase carbon flux towards L-isoleucine, *Appl. Environ. Microbiol.* **61**, 4315–4320.

MORI, M., SHIIO, I. (1984), Production of aspartic acid and enzymatic alteration in pyruvate kinase mutants of *Brevibacterium flavum, Agric. Biol. Chem.* **48**, 1189–1197.

MÜLLER, M., HEYNE, B., WINKLER, B., LEBENTRAU, B., PETZOLD, M., KANSY, W., KRANZ, G., FUCHS, R. (1987), Verfahren zur mikrobiologischen Herstellung von L-Lysin, *GDR Patent* 269 167.

MURAKAMI, Y., MIWA, H., NAKAMORI, S. (1993), Method for the production of L-lysine employing thermophilic *Corynebacterium thermoaminogenes, U.S. Patent* 5 250 423.

NAGY, E., MAYR, B., MOSER, A. (1994), Bioprocess scale-up using a structured mixing model, *Comput. Chem. Eng.* **18** (Suppl.), 663–667.

NAGY, E., NEUBECK, M., MAYR, B., MOSER, A. (1995), Simulation of the effect of mixing, scale-up and pH value regulation during glutamic acid fermentation, *Bioprocess Eng.* **12**, 231–238.

NAKAMORI, S. (1986), Threonine and homoserine, in: *Progress in Industrial Microbiology* (AIDA, K., CHIBATA, I., NAKAYAMA, K., TAKINAMI, K., YAMADA, H., Eds.), pp. 173–182. Toyko: Kodansha, Amsterdam: Elsevier.

NAKANASHI, T., AZUMA, T., HIRAO, T., HATTORY, K., SAKURAI, M. (1986), Process for producing L-lysine by fermentation, *U.S. Patent* 4 623 623.

NAKANASHI, T., HIRAO, T., SAKURAI, M. (1987), Process for producing L-lysine by fermentation, *U.S. Patent* 4 657 860.

NAKANO, T., AZUMA, T., KURATSU, Y. (1994a), Process for producing L-lysine by iodothyronine resistant strains of *C. glutamicum, U.S. Patent* 5 302 521.

NAKANO, T., AZUMA, T., KURATSU, Y. (1994b), Process for producing L-isoleucine, *Eur. Patent Appl.* 595 163.

NAKAYAMA, K. (1985), Tryptophan, in: *Comprehensive Biotechnology*, Vol. 3 (MURRAY, M.-Y., Ed.), pp. 621–631. Oxford: Pergamon Press.

NAKAYAMA, K., ARAKI, K. (1981), *Jpn. Patent* 56-8692.

NAKAYAMA, K., ARAKI, K. (1993), Process for producing L-lysine, *U.S. Patent* 3 708 395.

NISHIMURA, N., KOMATSUBARA, S., KISUMI, M. (1987), Increased production of aspartase in *Escherichia coli* K-12 by use of stabilized *asp*A recombinant plasmid, *Appl. Environ. Microbiol.* **53**, 2800–2803.

OGAWA, S., IGUICHI, S., MORITA, S., KUWAMOTO, H. (1991), Process for producing L-tryptophan, *Eur. Patent Appl.* 438 591.

OH, N.-S., SERNETZ, M. (1993), Turnover characteristics in continuous L-lysine fermentation, *Appl. Microbiol. Biotechnol.* **39**, 691–695.

OH, J. W., KIM, S. J., CHO, Y. J., PARK, N. H., LEE, L. H. (1990), Nouveau microorganisme capable de produire de la L-lysine et procédé de fermentation l'utilisant pour produire de la L-lysine. *French Patent Appl.* 2 645 172.

OH, J.-W., LEE, J.-HO., NOH, K.-S., LEE, H.-H., LEE, J.-HE., HYUN, H.-H. (1991), Improved L-lysine production by the amplification of the *Corynebacterium glutamicum dap*A gene encoding dihydrodipicolinate synthetase in *E. coli, Biotech. Lett.* **10**, 727–732.

ONO, E., TOSAKO, O., TAKIMAMI, K. (1980), Fermentative production of L-glutamic acid, *Jpn. Patent* 5 521 762.

OOSTERHUIS, N. M. G., GROESBEEK, N. M., OLIVER, A. P. C., KOSSEN, N. W. F. (1983), Scale-down aspects of the gluconic acid fermentation, *Biotechnol. Lett.* **5**, 141–146.

OOTANI, M., SANO, C., KUSUMOTU, I. (1986), Refining phenylalanine, *U.S. Patent* 4 584 400.

OOTANI, M., KITAHARA, T., AKASHI, K. (1987), Procédé pour la purification de la L-thréonine. *French Patent Appl.* 2 588 016.

OZAKI, H., SHIIO, I. (1983), Production of lysine by pyruvate kinase mutants of *Brevibacterium flavum, Agric. Biol. Chem.* **47**, 1569.

OZAKI, A., KATSUMATA, R., OKA, T., FURUYA, A. (1985), Cloning of the genes concerned in phenylalanine biosynthesis in *Corynebacterium glutamicum* and its application to breeding of a phenylalanine producing strain, *Agric. Biol. Chem.* **49**, 2925–2930.

PÁTEK, M., HOCHMONNOVÁ, J., NESVERA, J. (1993), Production of threonine by *Brevibacterium flavum* containing threonine biosynthesis genes from *Escherichia coli, Folia Microbiol.* **38**, 355–359.

PEBERDY, J. F. (1980), Protoplast fusion – a tool for genetic manipulation and breeding in industrial microorganisms, *Enzyme Microb. Technol.* **2**, 23–29.

PFEFFERLE, W., BACHMANN, B., SCHILLING, B., DECKWER, F., LEUCHTENBERGER, W. (1996), Bioreaktor. *Ger. Patent Appl.* 19 520 485.

PITTARD, A. J. (1987), Biosynthesis of the aromatic amino acid, *Amer. Soc. Microbiol. Conf. Proc.*, pp. 368–394.

PLACHY, J., KRATOCHVIL, M. (1992), Kmen mikrooganismu *Brevibacterium flavum* CCM 4263 produkujici L-lysin, *Czech. Patent* 9 200 718.

PLACHY, J., ULBERT, S. (1990), Production of L-tryptophan, *Acta Biotechnol.* **10**, 517–522.

QIAN, Z. (1989), Computer model control in glutamic acid fermentation processes, *Huagong Yejin*, **10**, 52–58.

REINSCHEID, D. J., EIKMANNS, B. J., SAHM, H. (1991), Analysis of *Corynebacterium glutamicum hom* gene coding for a feedback-resistant homoserine dehydrogenase, *J. Bacteriol.* **173**, 3228–3230.

RICHAUD, F., JARRY, B., TAKINAMI, K., KURAHASHI, O., BEYOU, A. (1989), Procédé pour l'intégration d'un gene choisi sur le chromosome d'une bactérie obtenue par le dit procédé. *French Patent* 2 627 508.

ROGERS, P. L., SATIAWIHARDJA, W. H., CAIL, R. G. (1989), The potential for L-lysine production in Australia. A new assessment, *Austr. J. Biotechnol.* **3**, 126–142.

ROH, J. H., KIM, O. M., PARK, D. C., KIM, H. J., YUN, H. Y., KIM, S. D., LEE, I.-S., LEE, K. R. (1994), Cloning and functional expression of the *ddh* gene involved in the novel pathway of lysine biosynthesis from *Brevibacterium flavum, Mol. Cells* **4**, 295–299.

ROUY, N. (1984), Procédé de préparation de compositions pour alimentation animale à base de lysine, *Eur. Patent* 122 163.

SAKIMOTO, K., TORIGOE, Y. (1994), Study of mechanism of peak E substance formation in process for manufacture of L-tryptophan. *Curr. Prosp. Med. Drug Safety* 1994, 295–311.

SANO, K., MITSUGI, K. (1978), Enzymatic production of L-cysteine from D,L-2-amino-$\Delta^2$-thiazoline-4-carboxylic acid by *Pseudomonas thiazolinophilum.* Optimal conditions for enzyme formation and enzymatic reaction, *Agric. Biol. Chem.* **42**, 2315–2321.

SANO, K., SHIIO, I. (1967), Microbial production of L-lysine. I. Production by auxotrophs of *Brevibacterium flavum, J. Gen. Appl. Microbiol.* **13**, 349–358.

SANO, K., ITO K., MIWA, K., NAKAMORI, S. (1985), Recombinant DNA having a phosphoenol pyruvate carboxylase gene inserted therein, bacteria carrying said recombinant DNA and a process for producing amino acids using said bacteria, *Eur. Patent Appl.* 143 195.

SANO, K., ITO, K., MIWA, K., NAKAMORI, S. (1989), Coryneform bacteria carrying recombinant plasmids and their use in the fermentative production of L-lysine, *U.S. Patent* 4 861 722.

SATIAWIHARDJA, B., CAIL, R. G., ROGERS, P. L. (1993), Kinetic analysis of L-lysine production by a fluoropyruvate sensitive mutant of *B. lactofermentum, Biotechnol. Lett.* **15**, 577–582.

SATO, T., MORI, T., CHIBATA, I., FURUI, M., YAMASHITA, K., SUMI, A. (1975), Engineering analysis of continuous production of L-aspartic acid by immobilized *Escherichia coli* cells in fixed beds, *Biotechnol. Bioeng.* **17**, 1779–1804.

SCHEER, E., EGGELING, L., SAHM, H. (1988), Improved D,L-$\alpha$-hydroxybutyrate conversion to L-isoleucine with *Corynebacterium glutamicum* mutant with increased D-lactate utilization, *Appl. Microbiol. Biotechnol.* **28**, 474–477.

SCHRUMPF, B., EGGELING, L., SAHM, H. (1992), Isolation and prominent characteristics of an L-lysine hyperproducing strain of *Corynebacterium glutamicum, Appl. Microbiol. Biotechnol.* **37**, 566–571.

SHIIO, I. (1986), Tryptophan, phenylalanine, and tyrosine, in: *Biotechnology of Amino Acid Production* (AIDA, K., CHIBATA, I., NAKAYAMA, K., TAKINAMI, K., YAMADA, D., Eds.), pp. 188–206. Tokyo: Kodansha, Amsterdam: Elsevier.

SHIIO, I., OZAKI, H. (1986), Concerted inhibition of isocitrate dehydrogenase by glyoxylate plus oxalacetate, *J. Biochem.* **64**, 45–53.

SHIIO, I., SANO, K. (1969), *J. Gen. Appl. Microbiol.* **15**, 399.

SHIIO, I., UJIGAWA, K. (1978), Enzymes of the glutamate and aspartate synthetic pathways in glutamate producing bacterium *Brevibacterium flavum, J. Biochem.* **84**, 647–657.

SHIIO, I., UJIGAWA, K. (1980), Presence and regulation of α-ketoglutarate dehydrogenase complex in a glutamate-producing bacterium, *Brevibacterium flavum, Agric. Biol. Chem.* **42**, 1897–1904.

SHIIO, I., OZAKI, H., UJIGAWA-TAKEDA, K. (1982a), Production of aspartic acid and lysine by citrate synthethase mutants of *Brevibacterium flavum, Agric. Biol. Chem.* **46**, 101–107.

SHIIO, I., KAWAMURA, K., SUGIMOTO, S. (1982b), L-Tryptophan produced by fermentation, *UK Patent Appl.* 2 098 603.

SHIIO, I., SUGIMOTO, S., KAWAMURA, K. (1984), Production of L-tryptophan by sulfonamide-resistant mutants, *Agric. Biol. Chem.* **48**, 2073–2080.

SHIIO, I., SUGIMOTO, S., YASUHIKO, T. (1985), Fermentative production of L-lysine, *Jpn. Patent* 60 168 393.

SHIIO, I., YOSHINO, H., SUGIMOTO, S. (1990a), Isolation and properties of lysine-producing mutants with feedback-resistant aspartokinase derived from *Brevibacterium flavum* strain with citrate synthase- and pyruvate kinase-defects and feedback-resistant phosphoenolpyruvate carboxylase, *Agric. Biol. Chem.* **54**, 3275–3282.

SHIIO, I., SUGIMOTO, S., YOSHINO, H., KAWAMURA, K. (1990b), Isolation and properties of threonine-producing mutants with both dihydrodipicolinate synthase defect and feedback-resistant homoserine dehydrogenase from *Brevibacterium flavum, Agric. Biol. Chem.* **54**, 1505–1511.

SHIIO, I., SUGIMOTO, S., KAWAMURA, K. (1993), Isolation and properties of α-ketobutyrate-resistant lysine-producing mutants from *Brevibacterium flavum, Biosci. Biotech. Biochem.* **57**, 51–55.

SHIMAZAKI, K., NAKAMURA, Y., YAMADA, Y. (1983), Method for producing L-lysine by fermentation, *U.S. Patent* 4 411 997.

SHIMIZU, E., OOSUMI, T., HEIMA, H., TANAKA, T., KURASHIGE, J., ENEI, H., MIWA, K., NAKAMORI, S. (1995), Culture conditions for improvement of L-threonine production using a genetically self-cloned L-threonine hyperproducing strain of *Escherichia coli* K-12, *Biosci. Biotech. Biochem.* **59**, 1095–1098.

SHIMURA, K. (1995), Threonine, in: *Comprehensive Biotechnology,* Vol. 3 (BLANCH, H. W., DREW, S., WANG, D. I. C., Eds.). Oxford: Pergamon Press Ltd.

SHINOHARA, T., OOTANI, M. (1994), Method for recovering optically active tryptophan, *U.S. Patent* 5 329 014.

SHIRARSUCHI, M., KURONUMA, H., KAWAHARA, Y., YOSHIHARA, Y., MIWA, H., NAJAMORI, S. (1995), Simultaneous and high fermentative production of L-lysine and L-glutamic acid using a strain of *Brevibacterium lactofermentum, Biosci. Biotechnol. Biochem.* **59**, 83–86.

Showa Denko (1990), L-Tryptophan-producing microorganisms, and process for the production of L-tryptophan. *Jpn. Patent Appl.* 2 190 182.

SMÉKAL, F., ULBERT, S., BÁRTA, M. (1985), L-Lysine production with regular mutants of *Corynebacterium glutamicum, Kvasny Prum.* **31**, 282–283.

SOMERVILLE, R. L. (1983), Tryptophan: biosynthesis, regulation, and large-scale production, *Biotechnol. Ser.* **3**, 351–378.

SOMERVILLE, R. L. (1988), The *trp* promoter of *Escherichia coli* and its exploitation in the design of efficient protein production systems, *Biotechnol. Gen. Eng. Rev.* **6**, 1–41.

SONNTAG, K., SCHWINDE, J., DE GRAAF, A. A., MARX, A., EIKMANNS, B. J., WIECHERT, W., SAHM, H. (1995), $^{13}C$ NMR studies of the fluxes in the central metabolism of *Corynebacterium glutamicum* during growth and overproduction of amino acids in batch cultures, *Appl. Microbiol. Biotechnol.* **44**, 489–495.

SUGIMOTO, S., YABUTA, M., KATO, N., SEKI, T., YOSHIDA, T., TAGUCHI, H. (1987), Hyperproduction of phenylalanine by *Escherichia coli:* application of a temperature-controllable expression vector carrying the repressor–promoter system of bacteriophage lambda, *J. Biotechnol.* **5**, 237–253.

SUGIURA, M., TAKAGI, T., KISUMI, M. (1982), *Abstr. 31st Symp. Amino Acids Nucleic Acids* (Japan), 10.

SUGIURA, M., SUZAKI, S., KISUMI, M. (1987), Improvement of histidine producing strains of *Serratia marcescens* by cloning of mutant allele of the histidine operon on a mini-F plasmid vector, *Agric. Biol. Chem.* **51**, 371–377.

SWANN, W. E., NOLF, A. C. (1986), Immobilizing a biological material, *Eur. Patent Appl.* 197 784.

TAKANO, J., FURUKAWA, S., NAKANISHI, T. (1990), Process for producing L-threonine, *Eur. Patent Appl.* 368 284.

TAKINISHI, E., TAKAMATSU, H., SAKIMOTO, K., YAJIMA, Y. (1987), L-Tryptophan and its production with genetically engineered microorganisms, *Jpn. Patent* 62 186 786.

TANG, R. T., ZHU, X. Z., GONG, C. S. (1989), Effect of dimethyl sulfoxide on L-lysine production by a regulatory mutant of *Brevibacterium flavum, Can. J. Microbiol.* **35**, 668–670.

TEH, J., LEIGH, D., ALLEN, G., BURRILL, H., COWAN, P., CUMAKARIS, H., PITTARD, J. (1985), Direct production of tryptophan by *Escherichia coli* from simple sugars, *Biotech 85 Asia,* pp. 399–402, Online Publications, Pinner, UK.

TERASAWA, M., YUKAWA, H., TAKAYAMA, Y. (1985), Production of L-aspartic acid from *Brevibacterium* by the cell re-using process, *Process Biochem.* **20**, 124–128.

TOSA, T., KOGA, Y., MATSUISHI, T. (1987), Method for separating a basic amino acid, *U.S. Patent* 4 691 054.

TOSAKA, O., TAKINAMI, K. (1986), Lysine, in: *Biotechnology of Amino Acid Production* (AIDA, K., CHIBATA, I., NAKAYAMA, K., TAKINAMI, K., YAMADA, H., Eds.), pp. 152–172. Tokyo: Kodansha, Amsterdam: Elsevier.

TOSAKA, O., KUBOTA, K., HIROSE, Y. (1976a), Fermentative production of L-lysine, *Jpn. Patent* 51-61691.

TOSAKA, O., YOSHIHARA, Y., HIRAKAWA, H., KUBOTA, K., HIROSE, Y. (1976b), Fermentative production of L-lysine, *Jpn. Patent* 51-22884.

TOSAKA, O., TAKINAMI, K., HIROSE, Y. (1978), Production of L-lysine by leucine auxotrophs derived from AEC resistant mutant of *Brevibacterium lactofermentum, Agric. Biol. Chem.* **42**, 1181.

TOSAKA, O., HIRAKAWA, H., TAKINAMI, K. (1979), Effect of biotin levels on L-lysine formation in *Brevibacterium lactofermentum, Agric. Biol. Chem.* **43**, 491–495.

TOSAKA, O., MURAKAMI, Y., AKASHI, K., IKADA, S., YOSHII, H. (1981), L-Glutamic acid by fermentation with mutants, *Jpn. Patent* 56-92795.

TOSAKA, O., YOSHIHARA, Y., IKEDA, S., TAKINAMI, K. (1985), Production of L-lysine by fluoropyruvate-sensitive mutants of *Brevibacterium lactofermentum, Agric. Biol. Chem.* **49**, 1305–1312.

TSUCHIDA, T., MIWA, K., NAKAMORI, S., MIMOSE, H. (1981), L-Glutamic acid by fermentation with microorganisms obtained by genetic transformations, *Jpn. Patent* 56-148295.

TSUCHIDA, T., KUBOTA, K., YOSHIHARA, Y., KIKUCHI, K., YOSHINAGA, F. (1987), Fermentative production of L-glutamine by sulfaguanidine resistant mutants derived from L-glutamate producing-bacteria, *Agric. Biol. Chem.* **51**, 2089–2094.

TSUCHIDA, T., UCHIBORI, H., TAKEUCHI, H., SEKI, M. (1990), Process for producing L-amino acids by fermentation, *Eur. Patent Appl.* 379 903.

TSUCHIDA, T., KATSURADA, N., OHTSUKA, N., UCHIBORI, H., SUSUKI, T. (1993), Method for producing L-threonine by fermentation, *U.S. Patent* 5 188 949.

University of Odessa (1992), Novel strain of *Pseudomonas putita,* used as more efficient producer of L-methionine, *S.U. Patent* 1 730 152.

URA, D., HASHIMUKAI, T., MATSUMOTO, T., FUKUHARA, N. (1995), Process for the preparation of L-serine by an enzymatic method, *U.S. Patent* 5 382 517.

VALLINO, J. J., STEPHANOPOULOS, G. (1993), Metabolic flux distribution in *Corynebacterium glutamicum* during growth and lysine overproduction, *Biotechnol. Bioeng.* **41**, 633–646.

VALLINO, J. J., STEPHANOPOULOS, G. (1994), Carbon flux distributions at the glucose 6-phosphate branch point in *Corynebacterium glutamicum* during lysine overproduction, *Biotechnol. Prog.* **10**, 327–334.

VASCONCELLOS, A. M., NETO, A. L., GRASSIANO, D. M., DE OLIVEIRA, C. P. (1989), Adsorption chromatography of phenylalanine, *Biotechnol. Bioeng.* **33**, 1324–1329.

VOSS, H., WERNING, H., PFEFFERLE, W., LEUCHTENBERGER, W. (1996), Verfahren zur Herstellung von L-Aminosäuren durch Fermentation, *Ger. Patent Appl.* 19547361.

WALKER, T. E., HAN, C. H., KOLLMAN, V. H., LONDON, R. E., MATWIYOFF, N. A. (1982), $^{13}C$ Nuclear Magnetic Resonance studies of the biosynthesis by *Microbacterium ammoniaphilum* of L-glutamate selectively enriched with carbon-13*, *J. Biol. Chem.* **257**, 1189–1195.

WANG, J., KUO, Y., CHENG, W., LIU, Y. (1991), Optimization of culture conditions for L-lysine fermentation by *Brevibacterium* sp. P1–13, *Rep. Taiwan Sugar Res. Inst.* **134**, 37–48.

WILHELM, C., EGGELING, I., NASSENSTEIN, A., JEBSEN, C., EGGELING, L., SAHM, H. (1989), Limitations during hydroxybutyrate conversion to isoleucine with *Corynebacterium glutamicum,* as analysed by the formation of by-product, *Appl. Microbiol. Biotechnol.* **31**, 458–462.

WU, J. Y., WU, W. T. (1992), Glutamic acid production in an airlift reactor with net draft tube, *Bioprocess Eng.* **8**, 183–187.

YAHATA, S., TSUTSUI, H., YAMADA, K., KONEHARA, T. (1993), Fermentative production of D-alanine, in: *Proc. Ann. Meet. Agric. Chem. Soc.* (Japan), 92.

YAMADA, Y., SETO, A. (1988), Nouveaux microorganismes et procédé pour la production d'acide glutamique utilisant ces micro-organismes. *French Patent Appl.* 2 612 937.

YAMADA, K., TSUTSUI, H., YOTSUMOTO, K., SHIRAI, M. (1986), Process for producing L-threonine by fermentation, *Eur. Patent* 205 849.

YAMADA, K., TSUTSUI, H., YOTSUMOTO, K., TAKEUCHI, M., SHIRAI, M. (1987), Process for producing L-threonine by fermentation, *Eur. Patent* 213 536.

YEH, P., SICARD, A. M., SINSKEY, J. (1988), General organization of the genes specifically involved in the DAP-lysine biosynthetic pathway of *Corynebacterium glutamicum, Mol. Gen. Genet.* **212**, 105.

YOISHII, H., YOSHIMURA, M., TAKENAKA, Y., KAWAHARA, Y., NAKAMORI, S., HORIKOSHI, K., IKEDA, S. (1993a), Increased production of L-glutamic acid fermentation by control of trehalose formation, *Nippon Nogei Kagaku Kaishi* **67**, 949–954.

YOISHII, H., YOSHIMURA, M., NAKAMORI, S., INOUE, S. (1993b), L-Glutamic acid fermentation on a commercial scale by use of cane molasses with inverted sucrose, *Nippon Nogei Kagaku Kaishi* **67**, 955–960.

YOKOMORI, M., NIWA, T., TOTSUKA, K., KAWAHARA, Y., MIWA, H., OOSUMI, T. (1992), Procédé pour la production de L-lysine par fermentation, *French Patent Appl.* 2 673 645.

YOKOMORI, M., NIWA, T., TOTSUKA, K., KAWAHARA, Y., NAKAMORI, S., ESAKI, N., SODA, K. (1994), Process for producing L-lysine by fermentation with a bacteria having selenalysine resistance, *U.S. Patent* 5 362 636.

YOKOTA, A., SHIIO, I. (1988), Effects of reduced citrate synthase activity and feedback-resistant phosphoenolpyruvate carboxylase on lysine productivities of *Brevibacterium flavum, Agric. Biol. Chem.* **52**, 455–463.

YOKOTA, A., TAKAO, T. (1984), Conversion of pyruvic acid fermentation to tryptophan production by the combination of pyruvic acid-producing microorganisms and *Enterobacter aerogenes* having high tryptophanase activity, *Agric. Biol. Chem.* **48**, 2663–2668.

YOKOTE, Y., MAEDA, S., YABUSHITA, K., NOGUCHI, K., SAMEJIMA, H. (1978), Production of L-aspartic acid by *E. coli* aspartase immobilized by phenol-formaldehyd resin, *J. Solid-Phase Biochem.* **3**, 247–261.

YONEKURA, H., HIRAO, T., AZUMA, T., NAKANISHI, T. (1987), Procédé de préparation de L-lysine. *French Patent Appl.* 2 601 035.

YOSHIDA, H. (1986), Arginine, citrulline, and ornithine, in: *Biotechnology of Amino Acid Production* (AIDA, K., CHIBATA, I., NAKAYAMA, K., TAKINAMI, K., YAMADA, H., Eds.), pp. 131–143. Tokyo: Kodansha, Amsterdam: Elsevier.

YOSHIDA, T., MITSUNAGA, T., IZUMI, Y. (1993), L-Serine production using a resting cell system of *Hyphomicrobium* strains. *J. Ferment. Bioeng.* **75**, 405–408.

YOSHIHARA, Y., KAWAHARA, Y., YAMADA, Y. (1992), Process for producing L-amino acids by fermentation, *U.S. Patent* 5 164 307.

YOSHIMURA, M., KAWAKITA, T., YOSHIZUMI, T. (1994), Manufacture of feeds from glutamic acid fermented mother liquor. I. L-Glutamic acid mother liquor fermented from cane molasses as a feed additive, *Nippon Nogei Kagaku Kaishi* **68**, 1463–1473.

YOSHIMURA, M., WANCHAI, C., SOMMANEEWAN, C., VEARASILP, T. (1995a), Manufacture of feeds from glutamic acid fermented mother liquor. II. L-Glutamic acid mother liquor fermented from hydrolyzed cassava starch (glucose solution) as a feed additive, *Nippon Nogei Kagaku Kaishi* **69**, 337–345.

YOSHIMURA, M., KAWAKITA, T., WANCHAI, C., SOMMANEEWAN, C. (1995b), Manufacture of feeds from glutamic acid fermented mother liquor. III. Practical use of L-glutamic acid mother liquor as a feed additive, *Nippon Nogei Kagaku Kaishi* **69**, 347–356.

YOSHINAGA, F. (1986), Proline, in: *Biotechnology of Amino Acid Production* (AIDA, K., CHIBARA, I., NAKAYAMA, K., TAKINAMI, K., YAMADA, H., Eds.), pp. 117–120. Tokyo: Kodansha, Amsterdam: Elsevier.

# 14b Enzymology of Amino Acid Production

NOBUYOSHI ESAKI

Kyoto, Japan

SHIGERU NAKAMORI

Fukui, Japan

TATSUO KURIHARA

Kyoto, Japan

SETSUO FURUYOSHI

Kochi, Japan

KENJI SODA

Osaka, Japan

# 1 Introduction

An amino acid is generally defined as a compound that possesses one or more amino groups and one or more carboxyl groups. However, some of them, such as proline, are iminocarboxylic acids, and others contain a sulfonyl, sulfinyl, or phosphonyl group as an acidic group. Amino acids are the building blocks of proteins, and form a huge variety of complex copolymers. There are 20 naturally occurring $\alpha$-amino (or imino) acids, and 19 of them, except glycine, are optically active with the L-configuration. Humans require 8 essential amino acids for growth. In addition to these proteinous L-amino acids, several D-amino acids also occur in nature: D-alanine, D-glutamic acid, and *meso*-$\alpha,\varepsilon$-diaminopimelic acid are components of bacterial peptidoglycans. $\omega$-Amino acids also occur in nature in free and bound forms, and play important roles in metabolism.

The essential amino acids are used as formulations for parenteral feeding solutions. L-Glutamic acid is a taste component of Konbu (kelplike seaweed) and used as a flavor enhancer in foods at a world market size over one billion dollars. L-Lysine and D,L-methionine are also important nutritional additives to improve animal feedstuff. L-Phenylalanine and L-aspartic acid are produced in large quantities as the components of aspartame, a high-intensity sweetener. D-Phenylglycine and *p*-hydroxy-D-phenylglycine, unnatural D-amino acids, are used as side chains for $\beta$-lactam antibiotics. Various other amino acids are used as seasonings, flavorings, and starting materials for pharmaceuticals, cosmetics etc. Thus, markets for both naturally occurring and non-naturally occurring amino acids are being developed at significant growth rates.

Amino acids are produced by isolation from natural materials, by microbial or enzymatic procedures, or by chemical synthesis. The first two procedures give optically active (usually L-)amino acids, whereas the chemical methods in general produce the racemates, and an additional optical resolution step is necessary to obtain optically active amino acids. The enzymatic and chemical methods for the optical resolution of racemates have been developed simultaneously.

The microbial methods have been studied extensively since the pioneering work of KINOSHITA et al. (1957) who showed accumulation of L-glutamate in bacterial cultures. The amino acid producing microorganisms have been improved in order to increase their productivity of amino acids. Thus, L-lysine, L-glutamic acid and others are produced industrially based on microbial methods. Since an efficient method for the conversion of fumarate and ammonia into L-aspartate with bacterial aspartase was developed, various optically active amino acids have been prepared industrially by enzymatic procedures.

The microbial and enzymatic methods have their advantages and disadvantages. Starting materials for microbial methods are usually simple and cheap raw materials such as molasses, but the desired amino acids have to be produced not only by complicated and time-consuming fermentation processes but also by laborious isolation from various contaminating compounds. On the other hand, starting materials for enzymatic methods are in general more expensive than those for microbial methods, but the former procedures are less time-consuming and more efficient than the latter. In the following, examples of amino acid production by both microbial and enzymatic procedures are presented which have advanced in competition with each other and supplied us with various amino acids.

# 2 Microbial Production of Amino Acids

## 2.1 Basic Aspects

Since the great discovery by KINOSHITA et al. (1957) of a glutamate-producing coryneform bacterium, many excellent glutamate-producing bacteria have been found. Fermentation processes using these bacteria have been successfully commercialized, and today, the market of glutamate expanded to about

600,000 t a$^{-1}$ in the world. Thereafter, other amino acids have been obtained and produced industrially by deriving regulatory mutants, namely auxotrophs and mutants resistant to antagonists of amino acids.

Microbial overproduction of amino acids was essentially brought about by overcoming negative feedback controls on biosynthetic enzymes by end product amino acids. Glutamate production by a wild strain of coryneform bacteria, e.g., was explained by the release of glutamate regulation with efflux of glutamate through injured cell membranes by limitation of biotin and addition of penicillin or esters of fatty acids to the culture medium. Lysine production by homoserine auxotrophs was induced through the release from the concerted feedback inhibition by lysine plus threonine on aspartokinase, a key enzyme for lysine and threonine biosynthesis, by the limitation of threonine. Threonine production by threonine analog-resistant mutants was explained by the genetic alteration of homoserine dehydrogenase, a key enzyme for threonine biosynthesis, for insensitivity against the feedback inhibition by threonine (SHIIO and NAKAMORI, 1989).

Various trials have been made to enhance biosynthetic enzymes, to lower amino acid-degradative activities, to increase the supply of precursors, to stimulate excretion of amino acids to the medium etc. as a combination of these traits was reasonably effective for the improved production. In fact, many industrialized strains have been constructed by deriving corresponding amino acid analog-resistant mutants and auxotrophs.

Recombinant DNA techniques were most suitable for this purpose, and remarkable improvement has been reported by cloning and amplifying genes encoding enzymes of rate-limiting biosynthetic reactions by these techniques. Special efforts have been made to develop individual techniques such as host–vector systems, transformation and manipulation for each strain, except for *E. coli,* in which relevant techniques and information were "historically" abundant. These tests were reviewed by MARTIN (1989) and SHIIO and NAKAMORI (1989). Moreover, recent progress in manipulation techniques has occurred, especially in coryneform bacteria, however, practical production has not yet been carried out (JETTEN and SINSKEY, 1995). Although only few strains constructed with the techniques are industrialized now, these types of strains will increasingly be commercialized.

The following sections are reviewed comprehensively in this section: Technical aspects of the production of amino acids by microbes, mainly for large-scale industrial production, and recent progress in the use of wild type and genetically engineered strains.

## 2.2 Production of Individual Amino Acids

### 2.2.1 L-Glutamic Acid

Until the early 1970s, various coryneform bacteria, such as *Corynebacterium glutamicum, Brevibacterium flavum, Microbacterium ammoniaphilum,* and *Brevibacterium thiogenitalis* have been found (KIKUCHI and NAKAO, 1986). These bacteria commonly required biotin for their growth and produced more than 50 g L$^{-1}$ glutamate from 100 g of carbohydrates with limited biotin addition (less than 5 μg L$^{-1}$), but they did not produce with biotin additions of more than 10 μg L$^{-1}$. The name of the genus *Corynebacterium* was assigned to all of these bacteria because there were only minor differences in their characteristics (LIEBL et al., 1991). These bacteria also produced glutamate efficiently with the addition of $\beta$-lactam antibiotics such as penicillin and cephalosporin (SOMERSON and PHILLIPS, 1961) and with esters of fatty acids (TAKINAMI et al., 1964) in biotin-rich media such as those containing cane or beet molasses. Glutamate fermentation with these bacteria has been industrialized world-wide (KIKUCHI and NAKAO, 1986). Glutamate production from hydrocarbons had been studied in the 1960s, however, industrial production was prevented by the rise of the price of oil and by safety considerations.

The mechanisms of glutamate production were widely investigated. Glutamate was shown to inhibit the activity of glutamate dehydrogenase and to repress the synthesis of this enzyme, citrate synthase, and PEP carboxylase in *Brevibacterium flavum* (SHIIO and UJIKAWA, 1978). But this feedback control did not play an important role in practice, because accumulated glutamate was easily excreted into the medium. As to the mechanisms of glutamate excretion the three models "leakage", "inversion of uptake", and "glutamate-carrier" were proposed. The first explained excretion by the alteration or injury of the cell membrane by the limitation of biotin or by the addition of penicillin and surfactants. The second proposed inversion of general models for uptake of metabolites related to the chemical potential. The third proposed a specific carrier for glutamate (KRÄMER, 1994). Further investigations for isolation and characterization of the carrier substances would be necessary to elucidate the glutamate transport system.

Other raw materials for glutamate production such as acetate and ethanol could be used, but have not been industrialized. MATSUNAGA et al. (1988) reported the production by immobilized *Synecoccus* sp., a photosynthetic blue-green alga. The quantity of glutamate produced was far from a practical level. However, the study was expected as an approach to the use of cheap carbon sources. A trial for production by immobilized cells has been reported, however, its practical significance is unknown (CONSTANTINIDES et al., 1981). Accumulation of trehalose, which is a disaccharide composed of two glucose molecules with an $\alpha$-1,1 bond, was observed in the culture broth of a *Brevibacterium lactofermentum* mutant using a medium containing cane molasses, the main component of which is sucrose. The yield of glutamate was proportionally lower the higher the amount of trehalose. Hydrolysis of cane molasses with sulfuric acid and yeast invertase effectively reduced the amount of trehalose, increased the yield of glutamate, and the process has been successfully industrialized (YOSHII et al., 1993).

### 2.2.2 L-Lysine

Typical lysine production was obtained with a homoserine auxotroph of *C. glutamicum,* and with mutants of *B. flavum* and *B. lactofermentum* resistant to *S*-(2-aminoethylcysteine) (AEC), an analog of lysine. Fermentation processes using these strains are commercialized today worldwide. The industrialized strain of *B. lactofermentum* was constructed by deriving an alanine auxotroph and mutants resistant to lysine analogs other than AEC successively. They produced 60–70 g $L^{-1}$ of the monohydrochloride of lysine.

The mechanism of lysine production by a homoserine auxotroph was explained by the release of the concerted feedback inhibition by lysine plus threonine, by the limitation of threonine, and that of AEC-resistant mutants was explained by a genetic alteration of aspartokinase against the inhibition (TOSAKA and TAKINAMI, 1986). During this decade there have been quite a few remarkable advances regarding the construction of lysine producers, namely higher production as required for industrial production including genetically engineered strains. Meanwhile, production of about 20 g $L^{-1}$ was shown by a homoserine auxotrophic and lysine analog-resistant methylotrophic *Bacillus* sp. using methanol as a source of carbon (SCHENDEL et al., 1990). Amplification of the aspartokinase gene of the lysine producing *Pseudomonas acidovorans* strain in *E. coli* enhanced production of lysine and methionine (JACOBS et al., 1990). Similarly, introduction of the *dapA* gene from *C. glutamicum* into a threonine or a threonine and lysine producer of *E. coli* resulted in an increase of lysine (OH et al., 1991). In *C. glutamicum,* production was also increased by amplifying phosphoenolpyruvate (PEP) carboxylase, aspartokinase, *dapA* (CREMER et al., 1991), and lysine analog-resistant aspartokinase genes (HAN et al., 1991). Introduction of aspartate ammonia lyase genes of *E. coli* into *C. glutamicum* increased lysine production (MENKEL et al., 1989), indicating an effective supply of aspartate, a precursor of lysine. Attempts to improve the yield of lysine by varying culture conditions have been reported. The high osmotic pressure caused by increasing concentrations of lysine in the culture broth depressed bacterial growth and, therefore, the yield of lysine in *B. lactofermentum.* Addition of glycine betaine stimulated growth, sugar consumption, and as a result, lysine production (KAWAHARA et al., 1990). Simultaneous and high production of lysine and glutamate was found by cultivating a lysine producing *B. lactofermentum* using the glutamate production method, i.e., cultivation by adding esters of fatty acids or penicillin to the culture medium. The process is characterized by a 1.4-fold increase of productivity of total amino acids without ammonium sulfate in the medium. This suggests a cost reduction for wastewater treatment (SHIRATSUCHI et al., 1995).

## 2.2.3 L-Threonine

The first successful industrialization of an amino acid analog-resistant mutant as well as of a genetically engineered strain occurred in threonine production. All of the threonine producers so far obtained in *E. coli, B. flavum, B. lactofermentum, C. glutamicum,* and *Serratia marcescens* were derived as $\alpha$-amino-$\beta$-hydroxyvalerate(AHV)-resistant mutants. These mutants posessed commonly feedback inhibition-insensitive homoserine dehydrogenase (HD) and/or aspartokinase by threonine (NAKAMORI, 1986) and were used as hosts and DNA sources in investigations for the improved production by recombinant DNA techniques.

Biosynthetic pathways and enzymes responsible for aspartate-related amino acids are shown in Fig. 1.

DEBABOV (1982) reported an improved production in *E. coli* by cloning genes encoding threonine biosynthetic enzymes from an AHV-resistant threonine producer. The yield of threonine by the transformant was 55 g $L^{-1}$ in a jar fermentation. MIWA et al. (1983) also constructed a hyperproducer of *E. coli,* which harbored plasmids containing whole genes of the *thr* operon, *thrA, thrB, thrC* from another AHV-resistant threonine producer.

The yield was 65 g $L^{-1}$ on cultivation in a jar fermenter with glucose feeding and high oxygen supply (SHIMIZU et al., 1995). KURAHASHI et al. (1990) have succeeded in stabilized amplification and expression of the *thr* operon genes on the *E. coli* chromosome by a newly established defective Mu phage system. The constructed strain has been industrialized in France (HODGSON, 1994).

In *B. lactofermentum,* the feedback-insensitive HD gene from a threonine producer was cloned on a plasmid. In a transformant harboring the obtained plasmid, HD increased 2-fold and the yield of threonine was 1.5-fold higher (25 g $L^{-1}$) than that of the parent (17.5 g $L^{-1}$) (NAKAMORI et al., 1987). Production was further improved by introducing a compatible plasmid, on which the homoserine kinase (HK) gene had been cloned. The results indicate that a common precursor of lysine and threonine, aspartate-4-semialdehyde (ASA), was converted preferentially to threonine by a higher activity of HD compared to that of DDP synthase, which converted ASA to lysine. Furthermore, homoserine was also effectively converted to threonine by enhanced HK activity (MORINAGA et al., 1987). These results are shown in Tab. 1.

**Tab. 1.** Relation of Enzyme Activity and Amino Acid Production in Genetically Constructed *B. lactofermentum*

| Strain | Relative Activity of | | Production [g $L^{-1}$] of | | |
|---|---|---|---|---|---|
| 1 | HD | HK | L-Thr | Homoserine | L-Lys |
| Host | 1 | 1 | 17.5 | 0.5 | 12.1 |
| HD-recombinant | 3.1 | 1 | 25.0 | 6.0 | 4.2 |
| HK-recombinant | 0.6 | 3.15 | n.t. | n.t. | n.t. |
| HD, HK-recombinant | 4.5 | 14.2 | 33.0 | 1.1 | 2.5 |

HD homoserine dehydrogenase; HK homoserine kinase; L-Thr L-threonine; L-Lysine monohydrochloride: n.t. not tested.

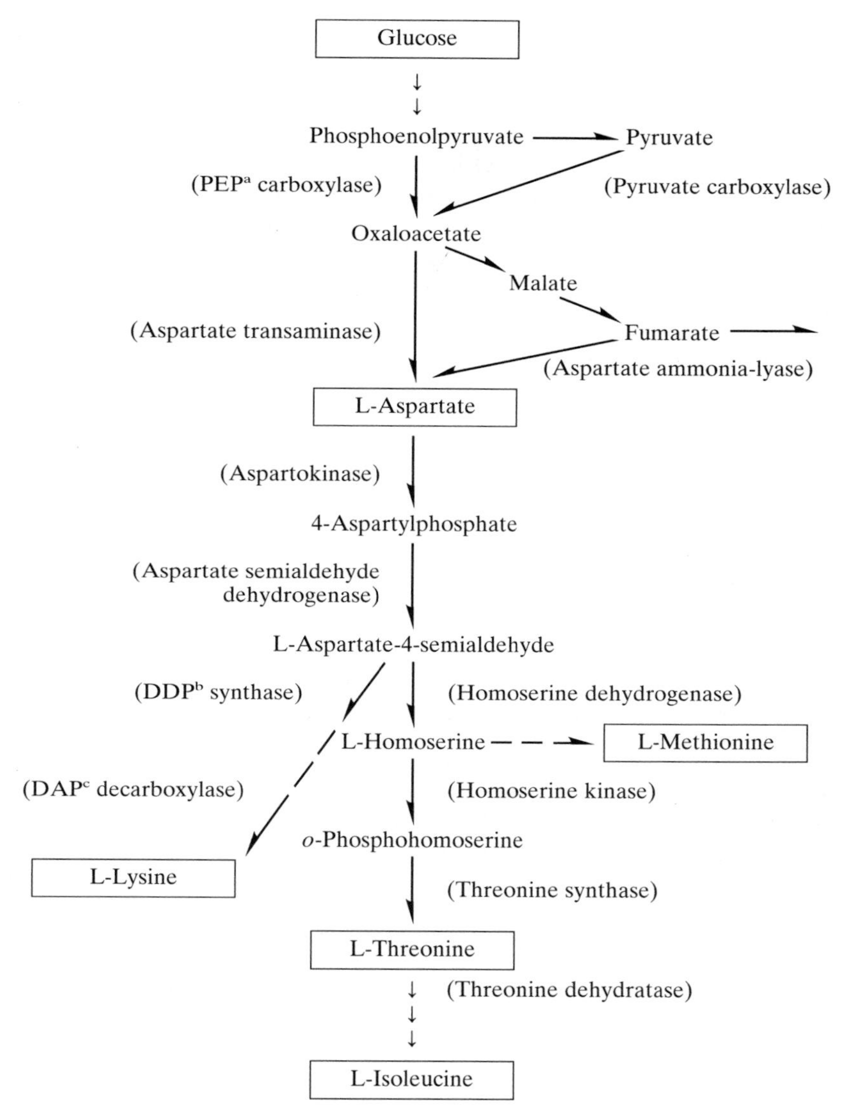


**Fig. 1.** Biosynthetic pathway of aspartate-related amino acids and corresponding enzymes.
[a] phophoenolpyruvate, [b] dihydrodipicolinate, [c] diaminopimelate.

The above *thr* operon gene of *E. coli* was introduced into a threonine producer of *B. flavum.* Production was 2.3-fold higher ($27 \text{ g L}^{-1}$) than that of the parent ($12 \text{ g L}^{-1}$) (ISHIDA et al., 1989). The wild type *thr* operon gene of *E. coli* was cloned and transformed into a threonine auxotroph of *C. glutamicum.* The deregulated HD gene was obtained by deriving AHV-resistant mutants from the transformant. The gene from *E. coli* was expressed in *C. glutamicum,* and the transformant produced $21 \text{ g L}^{-1}$, while the parent produced only $5.6 \text{ g L}^{-1}$ (KATSUMATA et al., 1984). Similarly, genes of HD, HK, and threonine synthase from the chromosomal DNA of *C. glutamicum* were introduced into a lysine producer of *C. glutamicum.* Lysine production (62 g $L^{-1}$) observed in the parent was converted in the transformant to that of threonine (51 g $L^{-1}$) with a jar fermentation using a medium containing 18% cane molasses (KATSUMATA et al., 1986). The explanation of this phenomenon was similar to that of *B. lactofermentum.* 60 g $L^{-1}$ of threonine were produced in a medium containing 15% sucrose in *S. marcescens* by the amplification of deregulated aspartokinase and HD (SUGITA et al., 1987). Amplification of PEP carboxylase, a key enzyme leading PEP to the TCA cycle, was effective for improved production in *B. lactofermentum* (SANO et al., 1987) and *S. marcescens* (SUGITA and KOMATSUBARA, 1989).

### 2.2.4 L-Isoleucine

Isoleucine producers were derived from *B. flavum* as AHV-resistant mutants. Potent producers were also constructed as *S. marcescens* mutants having feedback-insensitive and derepressed isoleucine synthetic enzymes as the result of transductional methods (KOMATSUBARA and KISUMI, 1986). Trials for improved production by recombinant DNA techniques have been carried out. By introducing *E. coli thr* operon genes into an isoleucine producer of *C. glutamicum,* production increased from 4.6 to 8.4 g $L^{-1}$. By amplifying the deregulated HD gene of *C. glutamicum* production also was enhanced from 4.6 to 11.2 g $L^{-1}$ (KATSUMATA et al., 1985).

### 2.2.5 L-Leucine and L-Valine

Leucine producers were obtained as isoleucine bradytrophic and $\alpha$-aminobutyrate (ABA)-resistant mutants of *S. marcescens,* and an isoleucine and methionine double auxotrophic and 2-thiazolealanine(TA)-resistant mutant of *B. lactofermentum.* Valine producers were also obtained as ABA-resistant *S. marcescens* mutants and TA-resistant *B. lactofermentum* mutants, respectively (KOMATSUBARA and KISUMI, 1986).

### 2.2.6 L-Phenylalanine

The aromatic amino acid biosynthetic pathway and its corresponding enzymes are shown in Fig. 2. 3-Deoxy-D-arabinoheptulosonate-7-phosphate (DAHP) synthase and chorismate mutase-prephenate dehydratase (CM-PDT) play an important role in phenylalanine production.

Phenylalanine producers were constructed as tyrosine auxotrophs of *C. glutamicum* and as mutants of *B. flavum* resistant to phenylalanine analogs, such as *p*-aminophenylalanine (PAP) and *p*-fluorophenylalanine (PFP). The highest yield was 25 g $L^{-1}$ with a *B. lactofermentum* mutant, which was resistant to PFP, 5-methyltryptophan and sensive to decoynin, an analog of purine, in a medium containing fumarate and acetate in addition to 13% glucose (SHIIO, 1986). OZAKI et al. (1985) reported the production of 19 g $L^{-1}$, which was 69% higher than that of the parent, by amplifying the CM and PDT genes in *C. glutamicum.* Production was also increased in a *C. glutamicum* mutant by amplifying the deregulated CM-PDT gene of *E. coli* (IKEDA et al., 1993). Amplification of the deregulated DAHP synthase gene of a phenylalanine producer of *B. lactofermentum* resulted in the production of 21.5 g $L^{-1}$ compared to 17.4 g $L^{-1}$ of the parent (ITO et al., 1990a). In *E. coli,* production (19.0 g $L^{-1}$) was obtained by constructing a strain having vectors, on which deregulated DAHP synthase and CM-PDT genes were cloned by connecting with promotors $P_R$ and $P_L$ from $\lambda$ phage, and by controlling the gene expression with a temperature-sensitive system (SUGIMOTO et al., 1990). A theoretical yield of phenylalanine was obtained by amplifying deregulated genes of DAHP synthase and CM-PDT in *E. coli* (FÖRBERG et al., 1988).

### 2.2.7 L-Tyrosine

A maximal yield of 17.6 g $L^{-1}$ by a *C. glutamicum* mutant has been obtained (SHIIO, 1986). The mutant was phenylalanine auxotrophic and resistant to 3-aminotyrosine, PFP, PAP, and tyrosine hydroxamate. Another tyrosine-producing mutant of *C. glutamicum*

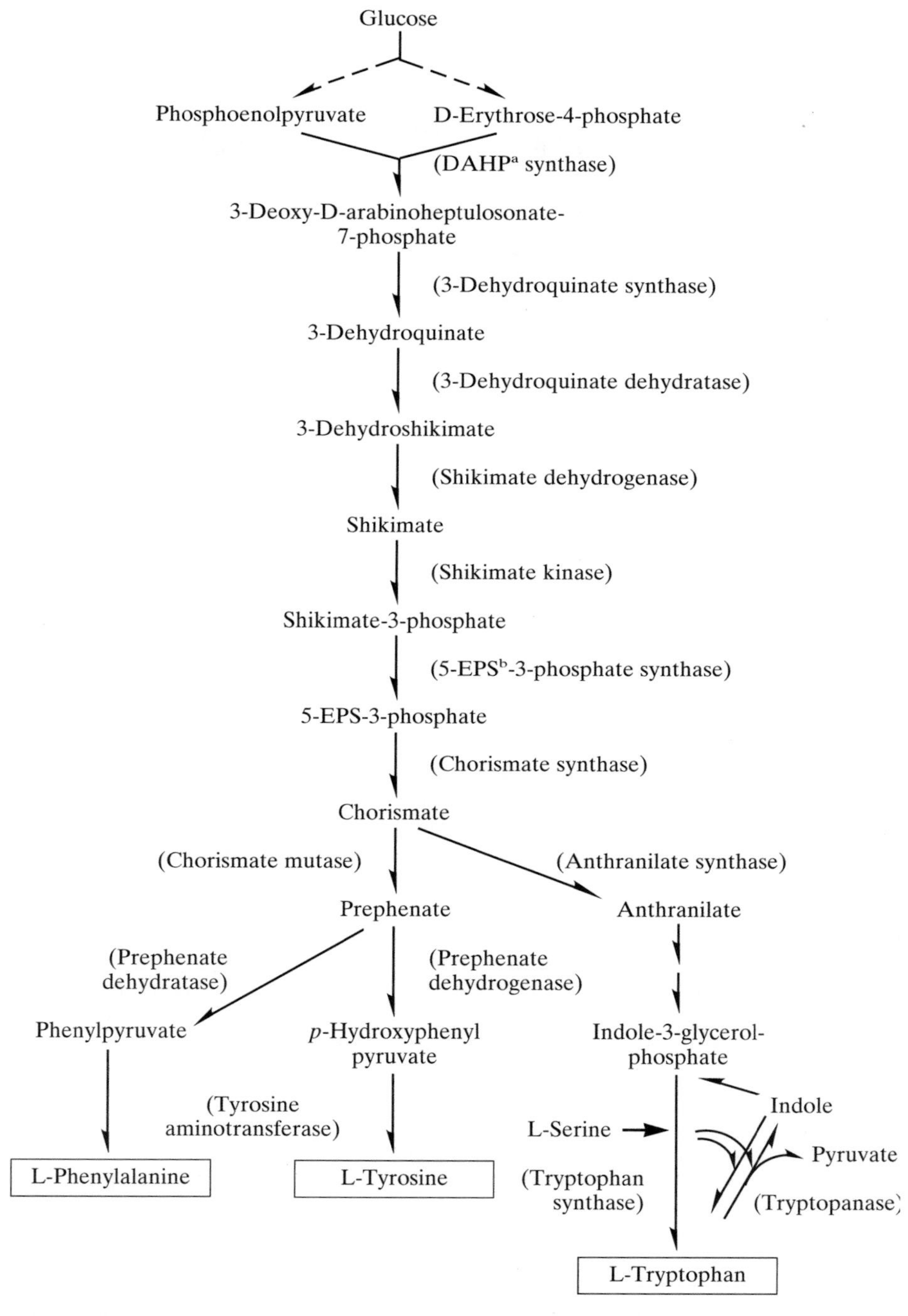


**Fig. 2.** Biosynthetic pathway of aromatic amino acids and corresponding enzymes. Enzymes are indicated in parentheses.
[a] 3-deoxy-D-arabinoheptulosonate-7-phosphate, [b] enolpyruvyl shikimate.

was transformed with a plasmid which contained a deregulated *aroF* gene from *E. coli.* One obtained transformant produced a yield of 9.0 g $L^{-1}$, 1.9 times higher than that of the parent, and it had a 30 times higher DAHP synthase activity (OZAKI et al., 1985). A tyrosine-producer of *B. lactofermentum* transformed with a plasmid containing the shikimate kinase gene of a phenylalanine producer of *B. lactofermentum* produced 21.6 g $L^{-1}$, while the parent produced 17.4 g $L^{-1}$ of tyrosine (ITO et al., 1990b).

## 2.2.8 L-Tryptophan

Tryptophan producers were derived from *B. flavum* as a tyrosine auxotrophic, 5-fluorotryptophan and azaserine resistant mutant, from *C. glutamicum* as a tyrosine and phenylalanine double auxotrophic and mutant resistant to several analogs (5-methyltryptophan, hydroxamates of tryptophan, tyrosine and phenylalanine, 6-fluorotryptophan, 4-methyltyrosine, PFP, and PAP), and from *B. subtilis* as a mutant resistant to 5-fluorotryptophan and indolmycin. These strains produced 10–15 g $L^{-1}$ of tryptophan, and the latter two were said to be used in practical fermentations (SHIIO, 1986). The first report which showed increased production with a genetically engineered strain was published by TRIBE and PITTARD (1979). They showed twice higher anthranilate synthase activity and tryptophan yield than the parent in an *E. coli* transformant, which harbored a Col V or F plasmid containing *trp* operon with a deregulated *trpE* gene. An *E. coli trpR* and *tnaA* mutant transformed with a plasmid containing the *trp* operon gene with deregulated *trpE* and *trpD* genes produced 6.2 g $L^{-1}$ of tryptophan by growing it on anthranilate, a precursor of tryptophan (AIBA et al., 1982). A transformant of *B. lactofermentum* which harbored plasmids containing genes of shikimate kinase, dehydroquinate synthase, and shikimate dehydrogenase produced 4.2 g $L^{-1}$, while the parent produced 1.6 g $L^{-1}$ (MATSUI et al., 1988). A tryptophan producer of *C. glutamicum* was transformed with a plasmid on which genes of deregulated DAHP synthase, other tryptophan synthetic enzymes, and wild-type 3-phosphoglycerate dehydrogenase – the first enzyme in the serine pathway – were cloned. Together with increased sugar consumption rate and plasmid stability, a 60% higher yield of tryptophan (50 g $L^{-1}$) than that of the parent was obtained from a medium containing 22–25% sucrose in a jar fermentation. In the transformant, indole which is a precursor of tryptophan and, at the same time, a growth inhibitor, was presumed to be converted effectively to tryptophan by an excess supply of serine (IKEDA et al., 1994).

Glutamate-5-semialdehyde, a precursor of proline, was produced by a proline auxotroph of *B. flavum.* The compound was converted chemically to tryptophan by direct addition of phenylhydrazine to the culture fluid (MORIOKA et al., 1989).

## 2.2.9 L-Glutamine

Glutamine producers were obtained from glutamate producing coryneform bacteria as mutants resistant to sulfaguanidine (TSUCHIDA et al., 1987).

## 2.2.10 L-Aspartic Acid

Aspartate is now manufactured from fumarate using bacterial aspartate-ammonia lyase. MORI and SHIIO (1984) obtained an aspartate producer of *B. flavum* from glucose. The strain was a pyruvate kinase-deficient mutant from a revertant of a citrate synthase-lacking mutant, and accumulated 22.6 g $L^{-1}$ in a medium containing 10% glucose supplemented with 3 μg $L^{-1}$ of biotin.

## 2.2.11 Alanine

All the microbially produced amino acids are of the L-form except alanine, which so far was obtained in the DL-form presumably converted intracellularly from L-alanine by alanine racemase. Practical demand for this amino acid is for the L-form (sometimes for the D-form). Thus, L-alanine is now manufactured from L-aspartate by using bacterial aspartate-β-decarboxylase. YAHATA et al.

(1993) reported production of 46.1 g L$^{-1}$ of D-alanine of 95% enantiomeric excess (ee) and 1 g L$^{-1}$ of L-alanine, respectively, by selecting D-cycloserine-resistant mutants from *B. lactofermentum* ATCC 13869. On the other hand, HASHIMOTO and KATSUMATA (1994) obtained an L-alanine producer by deriving an alanine racemase-deficient mutant of *Arthrobacter oxydans,* which had been selected as a strain possessing glucose-nonrepressible alanine dehydrogenase. Production of L-alanine was 75.1 g L$^{-1}$ with a 14.5% glucose medium, and the optical purity of the obtained L-alanine was 95% ee.

### 2.2.12 L-Histidine

Histidine producers were found in *C. glutamicum, B. flavum,* and *S. marcescens.* A *C. glutamicum* mutant, which was resistant to 1,2,4-triazole-3-alanine (TRA), 6-mercaptoguanine, 8-azaguanine, 2-thiouracil, 6-methylpurine, and 5-methyltryptophan, produced 15 g L$^{-1}$. A *B. flavum* mutant resistant to 2-thiazolealanine, sulfadiazine, AHV, ethionine, and 2-aminobenzothiazole, produced 10 g L$^{-1}$ of histidine in a 10% glucose medium. In *S. marcescens* two types of mutants, which had feedback-insensitive and derepressed histidine enzymes, were obtained as TRA-resistant mutants. A strain, which had both characters combined with transductional techniques and 6-methylpurine resistance, produced 23 g L$^{-1}$ in a 15% sucrose medium (ARAKI, 1986).

The *E. coli hisG* gene, which codes for ATP phosphoribosyl transferase, a key enzyme for histidine synthesis, was expressed in a histidine auxotroph of *C. glutamicum.* By deriving TRA-resistant mutants, which harbored the *hisG* gene, plasmids containing the deregulated *hisG* gene were obtained. Production by a transformant with the plasmid was 15.3 g L$^{-1}$ compared to 7.6 g L$^{-1}$ of the parent (MIZUKAMI et al., 1994). In *S. marcescens,* production rose from 28 g L$^{-1}$ to 40 g L$^{-1}$ by amplifying activities of *hisG, D, B, C* up to 2-fold higher than that of the parent (SUGIURA et al., 1987).

### 2.2.13 L-Proline

Proline producers were obtained from coryneform bacteria (NAKAMORI et al., 1982) and *S. marcescens* (SUGIURA et al., 1985a) as proline analog-resistant mutants. These strains are used in practical production. *ProBA* genes from a producer of *S. marcescens* were cloned on a mini-F plasmid. The transformant with the plasmid produced a 20% increase of proline (65 g L$^{-1}$ on a 22% sucrose medium) compared with that of the parent (SUGIURA et al., 1985b). Amplification of PEP carboxylase resulted in a 1.7-fold higher proline production than by the parent in *B. lactofermentum* (SANO et al., 1987).

### 2.2.14 L-Ornithine, L-Citrulline, and L-Arginine

The first successful application of auxotrophs for the production of an amino acid was the production of ornithine by growth of an arginine (citrulline) auxotroph of *C. glutamicum* under arginine limitation. Similarly, arginine auxotrophs of *B. subtilis* and *C. glutamicum* produced citrulline under limitation of arginine. An auxotroph derived from an arginine- and pyrimidine analog-resistant mutant showed genetic alterations of arginine biosynthetic enzymes. This mutant produced citrulline in the presence of arginine. Arginine producers were obtained from *C. glutamicum* as an isoleucine bradytrophic, D-serine-sensitive, and D-arginine-, arginine hydroxamate-, and 2-thiazolealanine-resistant mutant, from *B. flavum* as a 2-thiazolalanine- and sulfaguanidine-resistant mutant, and from *B. subtilis* as an arginine hydroxamate- and 6-azauracil-resistant mutant. These mutants produced 25–35 g L$^{-1}$ of arginine on an 8–10% glucose medium. A producer was also derived from an arginine-nondegradative *S. marcescens* mutant, which had feedback-insensitive and derepressed arginine synthetic enzymes and showed 6-azauracil resistance. The strain produced 60–100 g L$^{-1}$ of arginine (YOSHIDA, 1986).

### 2.2.15 L-Serine

Serine is manufactured by microbial conversion of glycine using a *C. glycinophilum* mutant with a molar yield of 33% (14 g $L^{-1}$) (Kubota, 1985). Cells of *Klebsiella aerogenes* were transformed with plasmids containing an *E. coli glyA* gene coding for serine hydroxymethyltransferase. The cells showed a high productivity of serine (5.2 g $L^{-1}$ $h^{-1}$) and produced 160 g $L^{-1}$ of L-serine after 30 h in a reaction mixture containing glycine, formaldehyde, tetrahydrofolate, and pyridoxal phosphate (Hsiao et al., 1986).

# 3 Enzymatic Synthesis of Amino Acids

Enzymes are useful catalysts for the enantioselective syntheses of amino acids, and a variety of enzymes have been used for the production of amino acids: hydrolytic enzymes, ammonia lyases, pyridoxal 5′-phosphate-dependent enzymes, $NAD^+$-dependent L-amino acid dehydrogenases, etc. Several amino acids are synthesized using a single enzyme and others using a combination of multiple enzymes. Most enzymes used for amino acid synthesis are obtained from microorganisms because of their easy, cheap, and constant availability. Many of the enzyme genes are cloned and overexpressed in *E. coli, Saccharomyces cerevisiae,* and other microbial strains. Enzymes are usually expensive, the limited stability, however, is the main disadvantage of enzymatic methods. In order to compensate for this disadvantage, various techniques including immobilization of enzymes have been developed. Thermostable enzymes derived from thermophilic microorganisms have been isolated and used for amino acid production. Recent advances in the enzymatic methods of amino acid synthesis are briefly described in this section.

## 3.1 Hydrolytic Enzymes

### 3.1.1 $\alpha$-Amino-$\varepsilon$-caprolactam Hydrolase

$\alpha$-Amino-$\varepsilon$-caprolactam (ACL) is a chiral heterocyclic compound synthesized from cyclohexene, which is a by-product in the industrial production of nylon. Fukumura (1976a, 1977a) established an enzymatic method to produce L-lysine from DL-ACL. The process is composed of two enzyme reactions: the selective hydrolysis of L-ACL to L-lysine, and the racemization of ACL (Fig. 3). The L-ACL-hydrolyzing enzyme ($\alpha$-amino-$\varepsilon$-caprolactam hydrolase, EC class 3.5.2) occurs in the cells of *Cryptococcus laurentii* and other yeasts, and is inducibly formed by DL-ACL. The enzyme purified to homogeneity from a cell extract of *C. laurentii* has a molecular mass of about 185,000 Da and is activated by $MnCl_2$ and $MgCl_2$ (Fukumura et al., 1978). L-ACL is the only substrate of the hydrolase. D-ACL and $\varepsilon$-caprolactam are not hydrolyzed.

ACL racemase was found in the cells of *Achromobacter obae* and other bacteria (Fukumura, 1977b) and is unique among racemases in acting exclusively on cyclic amides derived from $\alpha,\omega$-diamino acids. Ahmed et al. (1983b) purified the enzyme to homogeneity from the cell extract of *A. obae* and char-

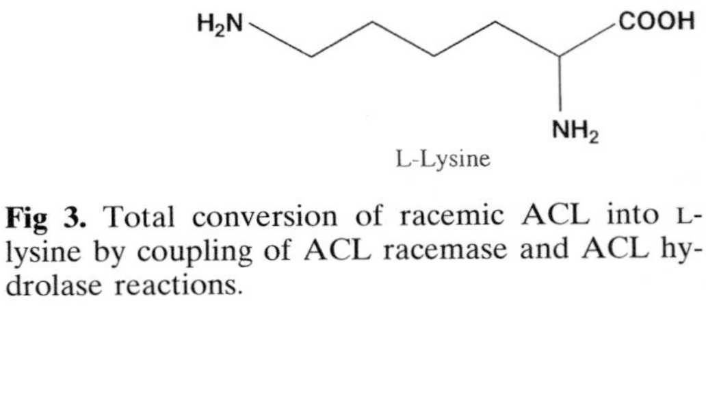

**Fig 3.** Total conversion of racemic ACL into L-lysine by coupling of ACL racemase and ACL hydrolase reactions.

acterized it. The ACL racemase gene was cloned from the chromosomal DNA of *A. obae,* and its complete nucleotide sequence revealed that the enzyme consists of 435 amino acids and has a molecular mass of 45,568 Da (NAOKO et al., 1987). The enzyme is composed of a single polypeptide chain and contains 1 mol PLP per mol of enzyme as a coenzyme. In addition to both isomers of ACL, D- and L-$\alpha$-amino-$\delta$-valerolactam also serve as effective substrates (AHMED et al., 1983a). The enzyme catalyzes the exchange of the $\alpha$-hydrogen of the substrate with deuterium or tritium during racemization in deuterium oxide or tritium oxide (AHMED et al., 1986). By tritium incorporation experiments, the enzyme was shown to catalyze both inversion and retention of the substrate configuration with a similar probability in each turnover. AHMED et al. (1986) have shown that the racemization catalyzed by ACL racemase proceeds through a single base mechanism.

### 3.1.2 2-Amino-$\Delta^2$-thiazoline-4-carboxylate Hydrolase

2-Amino-$\Delta^2$-thiazoline-4-carboxylate (ATC) is an intermediate in the chemical synthesis of DL-cysteine. SANO et al. (1977a) have found several bacterial strains that are capable of producing L-cysteine from DL-ATC: *Pseudomonas* sp., *E. coli, Bacillus brevis,* and *Micrococcus sodenensis* (SANO et al., 1977a). As shown in Fig. 4, L-ATC hydrolase, *S*-carbamoyl-L-cysteine hydrolase and ATC racemase participate in this pathway. *Pseudomonas thiazolinophilum* isolated from soil shows the highest activity of the enzymes participating in this pathway, which are inducibly formed by addition of DL-ATC to the growth medium.

L-Cysteine formed is decomposed by cysteine desulfhydrase also occurring in the cells. However, its activity was successfully inhibited by addition of hydroxylamine or semicarbazide to the incubation mixture. A mutant strain of *P. thiazolinophilum* lacking cysteine desulfhydrase was isolated and used to produce L-cysteine from DL-ATC with a molar yield of 95% and at a product concentration of 31.4 g $L^{-1}$ (SANO and MITSUGI, 1978). PAE et al. (1992) also isolated a mutant strain of *Pseudomonas* sp. CU6, another L-cysteine producer from DL-ATC, lacking in activity of L-cysteine desulfhydrase. The enzymatic formation of L-cysteine from DL-ATC suffered from product inhibition in both the original and the mutant *Pseudomonas* sp. CU6 strains. However, the inhibition was much weaker for the mutant strain (PAE et al., 1992). SANO (1987) isolated *P. desmolytica* AJ3872, another L-cysteine producer, and found that it lacks the ability to convert D-ATC to L-cysteine as it is an ATC racemase-deficient strain. However, little is known about the enzymological properties and the function of the racemase.

### 3.1.3 Hydantoinase

5-Substituted hydantoin derivatives have been used as precursors for D- and L-amino

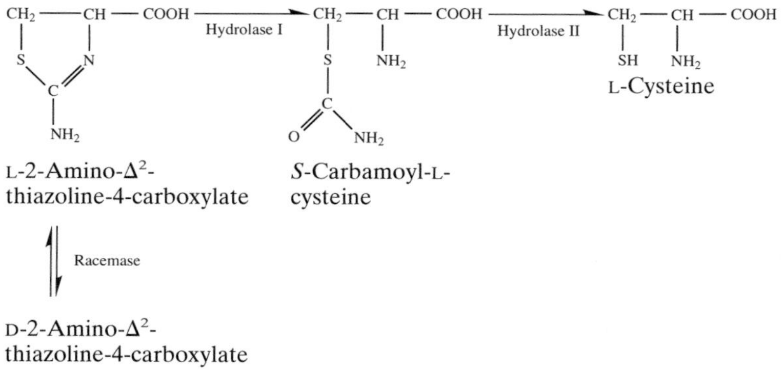

Hydrolase I
Hydrolase II
L-Cysteine
L-2-Amino-$\Delta^2$-thiazoline-4-carboxylate
*S*-Carbamoyl-L-cysteine
Racemase
D-2-Amino-$\Delta^2$-thiazoline-4-carboxylate

**Fig. 4.** Enzymatic synthesis of L-cysteine from DL-2-amino-$\Delta^2$-thiazoline-4-carboxylate.

acids in chemical synthesis. However, they are hydrolyzed enantioselectively by the enzymes named hydantoinases. Some of them act specifically on D-5-substituted hydantoins, and others on the L-isomers. *N*-Carbamoyl-amino acids formed are also hydrolyzed enantiospecifically by *N*-carbamoyl-amino acid amidohydrolases to produce D- or L-amino acids (Fig. 5). Kanegafuchi Chemical Industry, Japan, commercialized an enzymatic procedure for the production of D-*p*-hydroxyphenylglycine, which is a building block for the semisynthetic *β*-lactam antibiotic amoxycillin. Then, a variety of amino acids were produced by means of hydantoinases (SYLDATK et al., 1990, 1992a, b).

Since hydantoin was found to be hydrolyzed by extracts of mammalian livers and plant seeds, various microorganisms have been shown to utilize D- and L-5-substituted hydantoins as a sole carbon or nitrogen source by means of D- as well as L-specific hydantoinases inducibly formed (SYLDATK et al., 1990, 1992a, b).

Distribution of D-hydantoinase in microorganisms was shown by YAMADA et al. (1978). The enzyme is widely distributed in bacteria, in particular in *Klebsiella, Corynebacterium, Agrobacterium, Pseudomonas,* and *Bacillus,* and also in actinomycetes such as *Streptomyces* and *Actinoplanes.* Enzyme activity also occurs in eukaryotes: in yeasts, molds, plants, and mammals.

OLIVIERI et al. (1983) found that *Agrobacterium tumefaciens* cells grown on uracil as the sole source of nitrogen catalyze the complete conversion of racemic hydantoins to D-amino acids. Thus, the D-amino acid production was postulated to be due to the action of a series of enzymes involved in the pyrimidine degradation pathway (MOLLER et al., 1988). D-Hydantoinase was considered to be identical with dihydropyrimidinase (EC 3.5.2.2), and *N*-carbamoyl-D-amino acid amidohydrolase being identical with *β*-ureidopropionase (Fig. 6). Recently, however, RUNSER and MEYER (1993) showed occurrence of a D-hydantoinase without dihydropyrimidinase activity. OGAWA et al. (1993) also showed that *N*-carbamoyl-D-amino acid amidohydrolase of *Commamonas* sp. E222c acts exclusively on *N*-carbamoyl-D-amino acid but not on *β*-ureidopropionate. Accordingly, OGAWA et al. (1993) suggested that hydantoin and pyrimidine are not always degraded by the same series of enzymes (Fig. 6).

Other bacterial strains belonging to the genera of *Flavobacterium* (SANO et al., 1977b; NISHIDA et al., 1987), *Arthrobacter* (SYLDATK et al., 1987), *Pseudomonas* (YOKOZEKI et al., 1987; YOKOZEKI and KUBOTA, 1987; ISHIKAWA et al., 1993), and *Bacillus* (YAMASHIRO et al., 1988; ISHIKAWA et al., 1994) convert whole racemic 5-substituted hydantoins to the corresponding L-amino acids. In these bacteria, 5-substituted hydantoins are

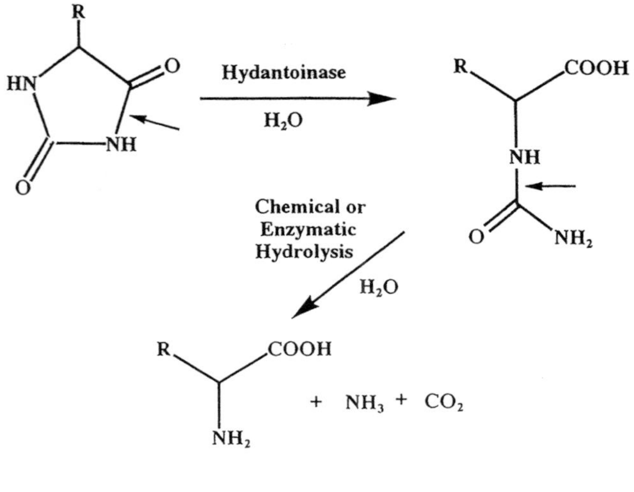

**Fig. 5.** Enzymatic synthesis of D- or L-amino acids from 5-substituted DL-hydantoins through *N*-carbamoyl-D- or L-amino acids. The arrows indicate the bonds to be enzymatically hydrolyzed.

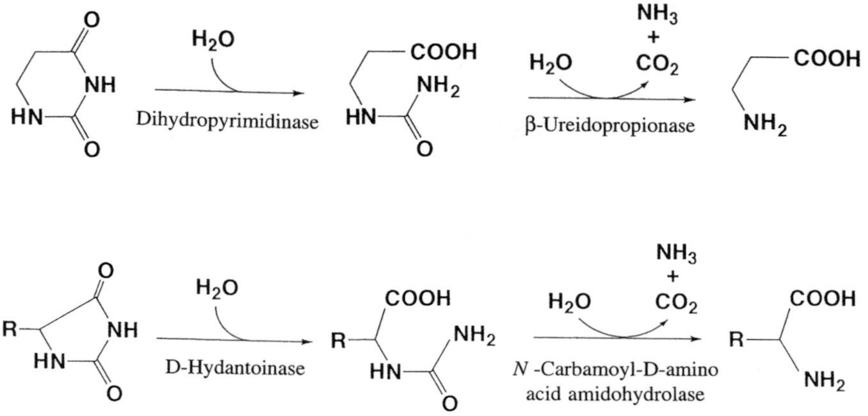

**Fig. 6.** Pyrimidine degradation pathway and D-amino acid production from D-5-substituted hydantoin.

hydrolyzed by L-hydantoinase to form *N*-carbamoyl-L-amino acids, which are hydrolyzed further to L-amino acids by *N*-carbamoyl-L-amino acid amidohydrolase in the same manner as D-hydantoinase and *N*-carbamoyl-D-amino acid amidohydrolase. MUKOHARA et al. (1993, 1994) cloned the genes of the thermostable L-hydantoinase and the thermostable *N*-carbamoyl-L-amino acid amidohydrolase from *Bacillus stearothermophilus*. *N*-Carbamoyl-L-aspartate amidohydrolase of *Clostridium oroticum* is identical with ureidosuccinase. Although $\beta$-ureidopropionase of rat liver hydrolyzes *N*-carbamoylglycine and *N*-carbamoyl-DL-alanine (TAMAKI et al., 1987), the enzymes from other sources act exclusively on either $\beta$-ureidopropionate or $\beta$-ureidoisobutyrate. The $\beta$-ureidopropionase of *P. putida* is another exception. It catalyzes hydrolysis of not only $\beta$-ureidopropionate but also of various *N*-carbamoyl-L-amino acids (OGAWA and SHIMIZU, 1994).

5-Substituted hydantoins are racemized spontaneously under weakly alkaline conditions. However, the chemical racemization proceeds only slowly (BATTILOTTI and BARBERINI, 1988), and hydantoin racemase participates in the total conversion (KNABE and WUMM 1980; BATTILOTTI and BARBERINI, 1988). WATABE et al. (1992a) isolated a plasmid which is responsible for the conversion of 5-substituted hydantoins to the corresponding L-amino acids from a soil bacterium, *Pseudomonas* sp. NS671. The genes involved in the conversion were cloned from the *Pseudomonas* plasmid, and functions of four genes named *hyuA, hyuB, hyuC,* and *hyuE* were identified. Both *hyuA* and *hyuB* are required for the conversion of D- and L-5-substituted hydantoins to the corresponding *N*-carbamoyl-D- and *N*-carbamoyl-L-amino acids, respectively, although the individual reactions catalyzed by the gene products have not yet been identified. *HyuC* codes for an *N*-carbamoyl-L-amino acid amidohydrolase, while *hyuE* is a hydantoin racemase gene (WATABE et al., 1992b). Significant nucleotide sequence similarity was found between *hyuA* and *hyuC* (43%), and also between *hyuB* and *hyuC* (46%). WATABE et al. (1992a) suggested that these genes have evolved from a common ancestor by gene duplication.

Wagner and coworkers purified the hydantoin racemase from *Arthrobacter* sp. DSM3747 and characterized it (SYLDATK et al., 1992b). WATABE et al. (1992b) also purified the enzyme from *E. coli* clone cells harboring a plasmid coding for the enzyme gene derived from *Pseudomonas* sp. NS671. The *Pseudomonas* enzyme is a hexamer composed of a subunit with a molecular mass of about 32,000 Da, which is consistent with the value deduced from the amino acid sequence. The D- and L-isomers of 5-(2-methylthioethyl)hydantoin and 5-isobutyrylhydantoin are racemized effectively. The hydantoin racemase of

*Arthrobacter* acts on aromatic and aliphatic hydantoin derivatives such as 5-indolylmethylhydantoin, 5-benzylhydantoin, 5-(*p*-hydroxybenzyl)hydantoin, 5-(2-methylthioethyl)-hydantoin, and 5-isobutylhydantoin (SYLDATK et al., 1992b). However, free amino acids, amino acid esters, and amides are inert.

WATABE et al. (1992c) found that the *Pseudomonas* enzyme is inactivated by a substrate, 5-isopropylhydantoin. Divalent sulfur-containing compounds such as methionine, cysteine, glutathione, and biotin effectively protected the enzyme from inactivation. *E. coli* cells expressing the racemase are capable of racemizing all of these hydantoin derivatives: the enzyme is protected from inactivation by divalent sulfur compounds occurring in the cells. Both *Pseudomonas* (WATABE et al., 1992c) and *Arthrobacter* (SYLDATK et al., 1992b) enzymes are inhibited strongly by $Cu^{2+}$. Various sulfhydryl reagents also inhibit the *Arthrobacter* enzyme. Therefore, the enzyme may contain essential cysteine residues, which are possibly modified by some activated intermediate derived from the particular substrates to lead to enzyme inactivation.

*E. coli* cells carrying a plasmid coding for *hyuA, hyuB, hyuC,* and *hyuE* convert D-5-(2-methylthioethyl)hydantoin only to L-methionine. On the other hand, *E. coli* cells harboring a plasmid coding for only *hyuA, hyuB,* and *hyuC* first convert L-hydantoin, and the D-isomer is hydrolyzed slowly when the L-isomer is depleted. D-5-(2-Methylthioethyl)hydantoin is probably only converted to L-methionine in the presence of the hydantoin racemase.

### 3.1.4 *S*-Adenosyl-L-homocysteine Hydrolase

*S*-Adenosyl-L-methionine is the important methyl donor in biological transmethylation to form *S*-adenosyl-L-homocysteine, which is hydrolyzed to adenosine and homocysteine by *S*-adenosyl-L-homocysteine hydrolase (EC 3.3.1.1) *in vivo.* However, equilibrium of the *S*-adenosyl-L-homocysteine hydrolase reaction favors the direction toward the synthesis of *S*-adenosyl-L-homocysteine. The enzyme acts on various adenosine analogs and the corresponding *S*-nucleotidyl-L-homocysteines (Fig. 7).

SHIMIZU and YAMADA (1984) developed a simple and efficient method for high-yield preparation of *S*-adenosyl-L-homocysteine and its analogs with *S*-adenosyl-L-homocysteine hydrolase of *Alcaligenes faecalis. S*-Adenosyl-L-homocysteine was produced at a concentration of about 80 g $L^{-1}$ with a yield of nearly 100%. The amino acid racemase with low substrate specificity acts on homocysteine, but not on *S*-adenosylhomocysteine

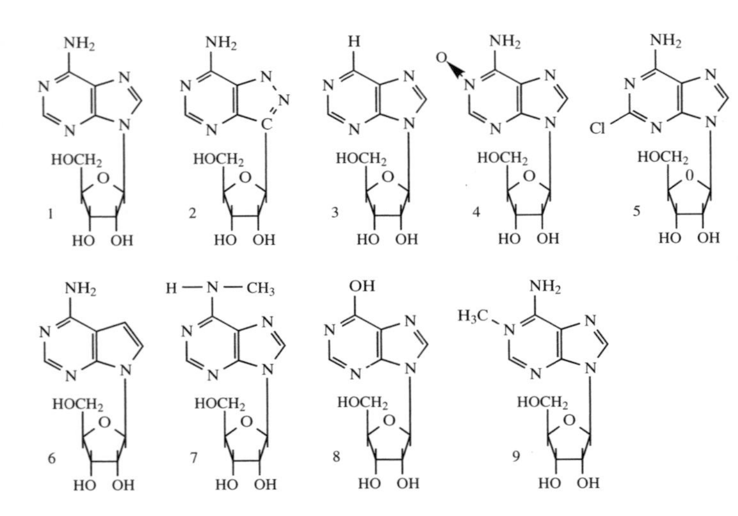


**Fig. 7.** Structures of adenosine and related nucleosides which serve as substrates for *S*-adenosyl-L-homocysteine hydrolase. (1) adenosine, (2) formycin A, (3) neburalin, (4) adenosine $N^1$-oxide, (5) 2-chloroadenosine, (6) tubercidine, (7) $N^6$-methyladenosine, (8) inosine, (9) 1-methyladenosine.

(SODA and OSUMI, 1971). Therefore, unreacted D-homocysteine was successfully converted to *S*-adenosyl-L-homocysteine with *Pseudomonas striata* (*P. putida*) cells. *A. faecalis* shows higher *S*-adenosyl-L-homocysteine hydrolase and lower adenosine deaminase activities than *P. striata*. Therefore, a mixture of both bacterial cells was used to produce 70 g $L^{-1}$ of *S*-adenosyl-L-homocysteine from DL-homocysteine and adenosine with a molar yield of nearly 100% (SHIMIZU and YAMADA, 1984). Various analogs of *S*-adenosyl-L-homocysteine were synthesized by total conversion of DL-homocysteine using both bacterial cells (SHIMIZU et al., 1984).

## 3.2 Ammonia Lyases

### 3.2.1 Aspartase

Aspartase (L-aspartate ammonia lyase, EC 4.3.1.1) catalyzes the reversible interconversion between L-aspartate and fumarate plus ammonia. The enzyme occurs in a variety of bacteria, plants, and animals, but its activity has not been found in mammals.

SUZUKI et al. (1973) purified aspartase from *E. coli* W to homogeneity. The enzyme was also purified from the overproducer cells carrying a plasmid coding for the enzyme gene by means of red A dye–ligand chromatography (KARSTEN et al., 1985). About 20% of the total soluble protein of the overproducer cells corresponds to the amount of aspartase in the cells. KOMATSUBARA et al. (1986) also cloned the aspartase gene of *E. coli* and constructed an overproducer of the enzyme. The level of aspartase was elevated 30-fold by the cloning, but the instability of the plasmid in the host cells prevented the clone cells from use for industrial production of L-aspartate (KOMATSUBARA et al., 1986).

Recently, SHI et al. (1993) crystallized the enzyme from *E. coli* by microdialysis, with polyethyleneglycol and sodium acetate as the precipitants. The crystals show the asymmetry of space group $P2_12_12$ with a = 156.5 Å, b = 147.6 Å, c = 102.5 Å, and diffract at the resolution of 2.8 Å.

Aspartase uses hydroxylamine and hydrazine as substrates in the reaction. However, D-aspartate, maleate, and mesaconate are inactive. The equilibrium constant of the deamination reaction catalyzed by the enzyme is 20 mmol $L^{-1}$ at 39 °C and 10 mmol $L^{-1}$ at 20 °C (pH 7.2). In contrast, the equilibrium of the phenylalanine ammonia lyase (EC 4.3.1.5) (see below) and the histidase (EC 4.3.1.3) reactions favor deamination.

The procedure for production of L-aspartate from fumarate with aspartase of *E. coli* K2 was established in 1960, as reviewed by CALTON (1992a). Crystalline ammonium fumarate was used as the substrate, which was solubilized as it was converted into L-aspartate. L-Aspartate was crystallized by acidification of the reaction mixture, and this process was named semi-transcrystallization. Thus, 56 g of L-aspartate per 100 mL was produced with a molar yield of 99% with 1 g of dried cells. Maleate, a stereoisomer of fumarate, is converted to fumarate by maleate *cis-trans*-isomerase (EC 5.2.1.1). TAKAMURA et al. (1966) produced L-aspartate from maleate and ammonium ions by coupling of the inducible maleate *cis-trans*-isomerase and aspartase reactions using resting cells of *A. faecalis*.

SATO et al. (1975) established a continuous procedure for L-aspartate synthesis with *E. coli* cells immobilized on polyacrylamide gel. They found that autolysis of the immobilized cells increases the amount of L-aspartate formed. This is probably due to an increase in permeability of the cell membrane to the substrates. The production of L-aspartate by means of immobilized cells has been industrialized in 1973 by Tanabe Seiyaku Co., Japan. They have developed κ-carrageenan as the efficient entrapping agent for immobilization of the cells, and significantly improved the productivity of L-aspartate with this material (UMEMURA et al., 1984).

### 3.2.2 Phenylalanine Ammonia Lyase

L-Phenylalanine ammonia lyase (EC 4.3.1.5) catalyzes the conversion of L-phenylalanine to *trans*-cinnamic acid and ammonia, and occurs in plants, fungi, and yeasts. The enzyme contains a dehydroalanine residue at the active site (HAVIR and HANSON, 1975). The genes of phenylalanine ammonia lyase have been cloned from *Rhodosporidium toruloides* (GILBERT et al., 1985), *Rhodotorula rubra* (FILPULA et al., 1988), and vaious plant sources such as tomato. The *R. toruloides* gene was expressed, and the active enzyme was produced in *E. coli* (ORUM and RASMUSSEN, 1992). FAULKNER et al. (1994) also expressed the enzyme gene from the same source in both *E. coli* and *S. cerevisiae.* Phenylalanine ammonia lyase accumulated to about 9% and 10% of the total soluble protein of *S. cerevisiae* and *E. coli,* respectively. The recombinant enzyme produced was shown to be fully active. Thus, the active-site dehydroalanine residue is probably formed by an autocatalytic mechanism, but not by chemical modification of the precursor protein. Two serine residues are conserved among all known phenylalanine ammonia lyases, and SCHUSTER and RETEY (1994) showed that one of them is essential for activity by site-directed mutagenesis. They suggested that serine-202 of parsely (*Petroselinum crispum*) is probably converted to the active-site dehydroalanine residue.

The equilibrium of the enzyme reaction favors the degradation of phenylalanine as described above (HANSON and HAVIR, 1972). Recently, DCUNHA et al. (1994) established a procedure for the direct, one-step conversion of *trans*-cinnamyl methyl ester to L-phenylalanine methyl ester by means of phenylalanine ammonia lyase from *Rhodotorula glutinis* in organic solvents. L-Phenylalanine methyl ester was obtained at a conversion rate of 70% in heptane.

## 3.3 Arginine Deiminase

L-Arginine iminohydrolase (arginine deiminase, EC 3.5.3.6) catalyzes the hydrolysis of the imino group of L-arginine to form L-citrulline and is applicable to the synthesis of L-citrulline. The enzyme genes were cloned from *Streptococcus sanguis* (BURNE et al., 1989), *Mycoplasma arginini* (KONDO et al., 1990; OHNO et al., 1990), and *P. putida* (WILSON et al., 1993), and sequenced. YAMAMOTO et al. (1974) selected *P. putida* as the best producer and immobilized the *P. putida* cells on polyacrylamide gel thus permitting the continuous production of L-citrulline (YAMAMOTO et al., 1974). The permeability of the cell membrane for L-arginine was increased by immobilization, and L-citrulline was produced with high efficiency.

## 3.4 Pyridoxal 5′-Phosphate Enzymes

Pyridoxal 5′-phosphate (PLP) functions as a coenzyme of various enzymes in the metabolism of amino acids. PLP enzymes are versatile and catalyze racemizations, transaminations, decarboxylations, eliminations, replacements, and other reactions. PLP at the active site reacts with the substrate amino acid by transaldimination to form a coenzyme–substrate Schiff base (I), in which a proton attached to the imine nitrogen holds the conjugated $\pi$-system plane (Fig. 8). The electron attraction by the protonated pyridine nitrogen which conjugates with the C=N bond permits electrons to flow from the substrate into the coenzyme. The apoenzyme determines which bonds to the $\alpha$-carbon atom of the activated substrate are cleaved to release (a) the $\alpha$-hydrogen, (b) the $\alpha$-carboxyl group, or (c) a side chain (Fig. 8). Several PLP enzymes show multicatalytic functions: tryptophan synthase (EC 4.2.1.20), tryptophanase (EC 4.1.99.1), $\beta$-tyrosinase (EC 4.1.99.2), cysteine desulfhydrase (EC 4.4.1.1), and methionine $\gamma$-lyase (EC 4.4.1.11). Optically active amino acids have been synthesized with various PLP enzymes.

### 3.4.1 Aspartate $\beta$-Decarboxylase

L-Aspartate 4-carboxy-lyase (aspartate $\beta$-decarboxylase, EC 4.1.1.12) catalyzes the removal of the $\beta$-carboxyl group of L-aspartate to form L-alanine. L-Aspartate is produced industrially from fumarate with aspartase as described above, and used in the industrial production of L-alanine with aspartate $\beta$-decarboxylase as reviewed by CALTON (1992b).

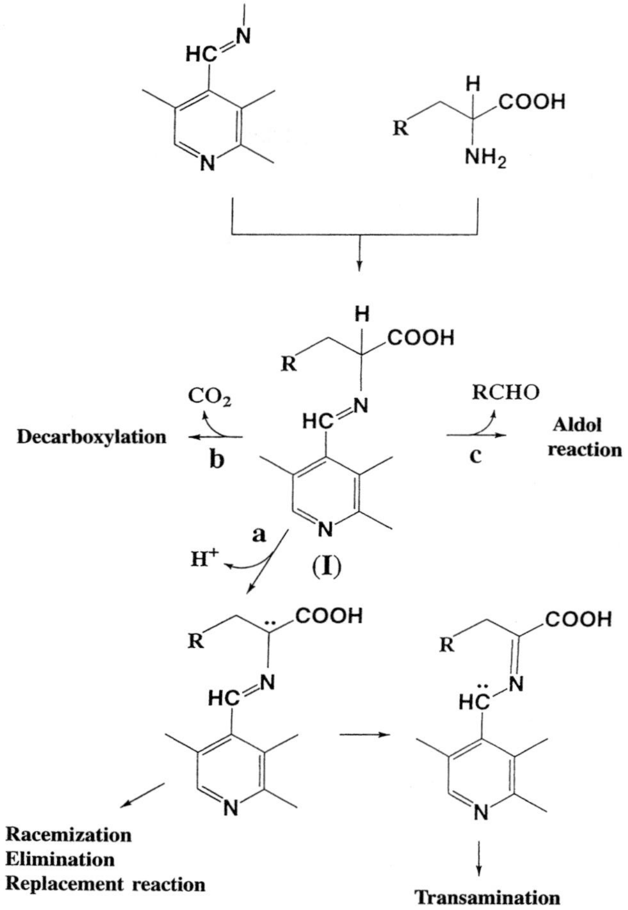

**Fig. 8.** Reactions catalyzed by PLP enzymes.

Tanabe Seiyaku Co. selected *Pseudomonas dacunhae* as the best producer of aspartate $\beta$-decarboxylase for industrial production of L-alanine. Crystalline L-asparate was added in an amount of 1 kg $L^{-1}$ of medium to the reaction mixture containing *P. dacunhae* cells and the surfactant Nikkol OP-10 (SHIBATANI et al., 1979). The crystals of the substrate were solubilized and converted to L-alanine as the reaction proceeded. Crystalline L-alanine was finally accumulated and obtained with a total yield of 93% (SHIBATANI et al., 1979).

CHIBATA and coworkers used *P. dacunhae* cells immobilized in the matrices of $\kappa$-carrageenan gels, which were cross-linked with glutaraldehyde (YAMAMOTO et al., 1980; TAKAMATSU et al., 1981). The main problem in this method is the evolution of a large amount of carbon dioxide, which prevents efficient operation of the immobilized $\kappa$-carrageenan gels due to substantial decrease in available surface area. Accordingly, FURUI and YAMASHITA (1983) developed a pressurized reactor system for the operation. Thus, an efficient industrial production of L-alanine was achieved with this system. L-Alanine can be synthesized in a single step from ammonium fumarate with a mixture of immobilized *E. coli* and *P. dacunhae* cells containing large amounts of aspartase and aspartate $\beta$-decarboxylase, respectively (SATO et al., 1982; TAKAMATSU et al., 1982).

### 3.4.2 β-Tyrosinase

L-Tyrosine phenol-lyase (deaminating, EC 4.1.99.2) catalyzes the α,β-elimination of L-tyrosine to produce pyruvate, phenol, and ammonia. The enzyme gene was cloned from *Citrobacter freundii* (IWAMORI et al., 1991), *Escherichia intermedia* (KURUSU et al., 1991), and *Erwinia herbicola* (IWAMORI et al., 1992; SUZUKI et al., 1993). BEPPU and coworkers carried out screening for β-tyrosinase-producing thermophiles and isolated a unique bacterial strain, *Symbiobacterium thermophilum*, which only grows by co-culture with a specific *Bacillus* strain (SUZUKI et al., 1988). *S. thermophilum* produces thermostable β-tyrosinase, 1992), the gene of which was cloned into *E. coli* (HIRAHARA et al., 1993). The cloned cells produced 375 times as much β-tyrosinase as the original *S. thermophilum* cells. The amino acid sequences deduced from the nucleotide sequences showed significant similarity to each other: 100% identity between *C. freundii* and *E. intermedia* enzymes; 90.6% identity between *E. intermedia* and *E. herbicola* enzymes, and 63.1% between *E. intermedia* and *S. thermophilum* enzymes.

β-Tyrosinase has been crystallized from *C. freundii* (DEMIDKINA et al., 1988). ANTSON et al. (1993) analyzed the three-dimensional structure of the apo enzyme from *C. freundii* at 2.3 Å resolution after refinement to an R-factor of 16.2%. The crystals belong to the space group $P2_12_12$ with a = 76.0 Å, b = 138.3 Å, and c = 93.5 Å. The β-tyrosinase molecule is a tetramer in which the subunits have 222 symmetry. There are two independent subunits in the asymmetric unit, and each subunit forms a dimer with another equivalent subunit. Thus, one pair of subunits is distinguishable from the other pair. Each subunit is composed of 14 α-helices and 16 β-strands forming a small and a large domain the architecture of which is similar to that found in aspartate aminotransferases (ANTSON et al., 1993). Most residues participating in binding with PLP in aspartate aminotransferases are conserved in the structure of β-tyrosinase.

β-Tyrosinase has multiple catalytic functions. In addition to L-tyrosine, D-tyrosine, L- and D-serine, *S*-methyl-L-cysteine, and β-chloro-L-alanine act as substrates in the α,β-elimination with the formation of pyruvate. The enzyme also catalyzes the β-replacement reaction between the substrates for α,β-elimination and phenol to yield L-tyrosine. When phenol is replaced by pyrocatechol, resorcinol, pyrogallol, and hydroxyhydroquinone, 3,4-dihydroxy-L-phenylalanine (L-DOPA), 2,4-dihydroxy-L-phenylalanine (2,4-DOPA), 2,3,4-trihydroxyphenyl-L-alanine (2,3,4-TOPA), and 2,4,5-trihydroxyphenyl-L-alanine were synthesized, respectively (NAGASAWA et al., 1981).

The synthesis of L-tyrosine from phenol, pyruvate, and ammonia (a reverse reaction of α,β-elimination) is also catalyzed by β-tyrosinase (NAGASAWA et al., 1981). The maximum reaction rate is 3.3 μmol min$^{-1}$ mg$^{-1}$ protein, which is about 1.5 times higher than that of the L-tyrosine α,β-elimination. When various derivatives of phenol are used as substituent donors, the corresponding tyrosine derivatives are synthesized (Tab. 2) (NAGASAWA et al., 1981).

The reaction mechanism of the multifunctional β-tyrosinase has been proposed by KIICK and PHILLIPS (1988a) (Fig. 9). Tyrosine forms a Schiff base via its α-amino group with the pyridoxal-P molecule, and thereby displaces the Schiff base between Lys257 and pyridoxal-P. The proton from the $C_\alpha$ is then abstracted by a base (p*K* ca. 7.6–7.8) to generate a stable quinoid intermediate. The second base (p*K* ca. 8.0–8.2) then abstracts the proton from the phenolic OH and a proton is donated to $C_\gamma$. The activated bond between $C_\beta$ and $C_\gamma$ is then cleaved, and phenol is released. The enzyme–α-aminoacrylate complex thus formed is a common key intermediate for all the α,β-eliminations, β-replacements, and reverse reactions.

Synthesis of L-DOPA by means of β-replacement between DL-serine and pyrocatechol was studied with *Erwinia herbicola* cells (ENEI et al., 1973a). Phenol derivatives, particularly at high concentrations, inhibit β-tyrosinase. L-DOPA was synthesized at a final concentration of 5.5 g per 100 mL of the reaction mixture by addition of limited concentrations of pyrocatechol throughout the reaction. β-Chloroalanine also served as substrate in the synthesis of L-DOPA with pyrocatechol

**Tab. 2.** Relative Velocities of Synthesis of L-Tyrosine-Related Amino Acids from Pyruvate, Ammonia, and Phenol Derivatives by $\beta$-Tyrosinase

| Phenol Derivatives | L-Amino Acid Synthesized[a] | Relative Rate | Phenol Derivatives | L-Amino Acid Synthesized[a] | Relative Rate |
|---|---|---|---|---|---|
| 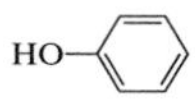 HO | 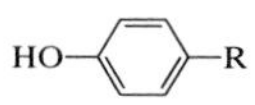 HO, R | 100 | 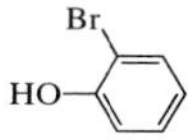 Br, HO | 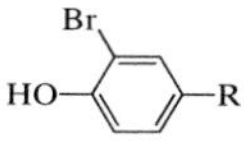 Br, HO, R | 2.0 |
| 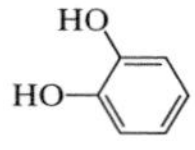 HO, HO | 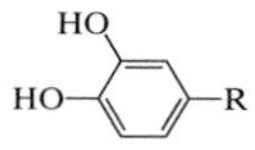 HO, HO, R | 60.0 | 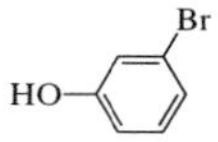 Br, HO | 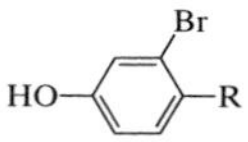 Br, HO, R | 4.2 |
| 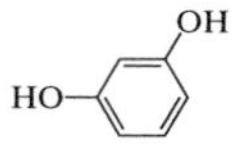 OH, HO | 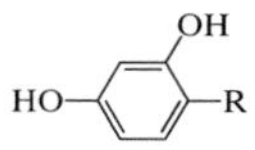 OH, HO, R | 58.2 | 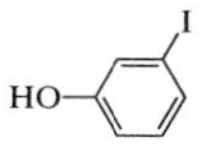 I, HO | 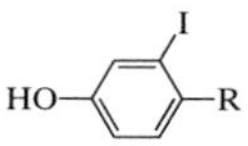 I, HO, R | 1.5 |
| 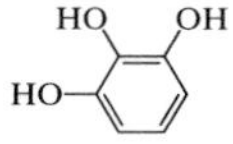 HO, OH, HO | 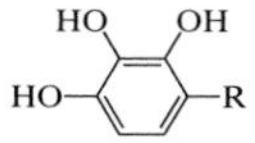 HO, OH, HO, R | 9.2 | 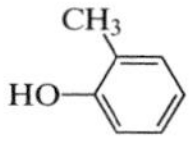 $CH_3$, HO | 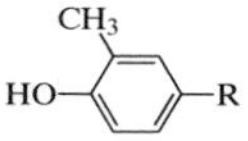 $CH_3$, HO, R | 2.0 |
| 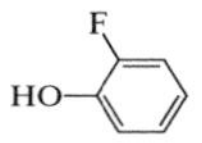 F, HO | 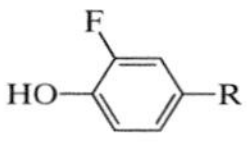 F, HO, R | 66.3 | 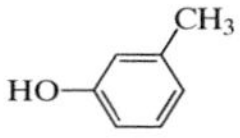 $CH_3$, HO | 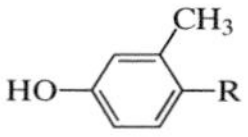 $CH_3$, HO, R | 9.8 |
| 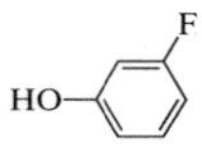 F, HO | 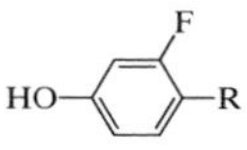 F, HO, R | 23.2 | 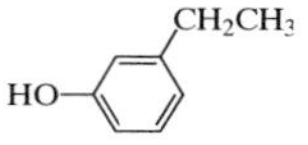 $CH_2CH_3$, HO | 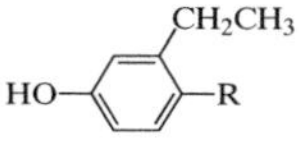 $CH_2CH_3$, HO, R | 2.0 |
| 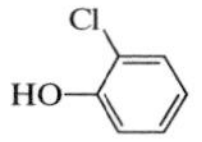 Cl, HO | 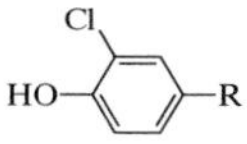 Cl, HO, R | 15.3 | 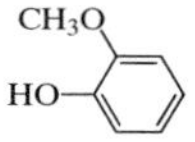 $CH_3O$, HO | 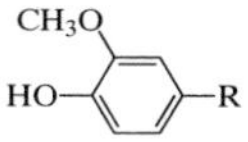 $CH_3O$, HO, R | 1.1 |
| 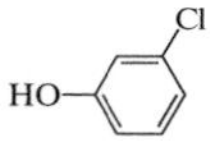 Cl, HO | 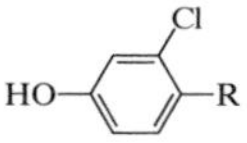 Cl, HO, R | 33.4 | 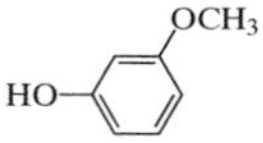 $OCH_3$, HO | 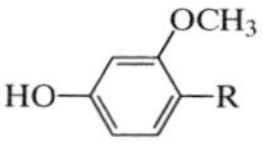 $OCH_3$, HO, R | 30.6 |

[a] R represents the L-alanyl moiety.

(NAGASAWA and YAMADA, 1986). More than 6.0 g of L-tyrosine per 100 mL of the reaction mixture was synthesized in the same manner from phenol, pyruvate, and ammonia (ENEI et al., 1973b).

IKEDA and FUKUI (1979) immobilized $\beta$-tyrosinase on Sepharose in order to synthesize L-tyrosine continuously. *Erwinia herbicola* cells, immobilized in collagen matrices, are more resistant to heat, contact with phenolic compounds, and changes in pH than the intact cells (YAMADA and KUMAGAI, 1978).

### 3.4.3 Tryptophanase

L-Tryptophan indole-lyase (deaminating) (tryptophanase, EC 4.1.99.1) is widely found in bacteria such as *E. coli, Bacillus albei, Aeromonas liquefaciens, Proteus rettgeri, Sphaerophorus funduliformis,* and *Vibrio* sp. Recently, BEPPU and coworkers found that an obligately symbiotic thermophile, *Symbiobacterium thermophilium,* produces thermostable tryptophanase (SUZUKI et al., 1991), and purified the thermostable enzyme to homogeneity.

The primary structure of tryptophanase from *E. coli* was determined by both protein sequencing (KAGAMIYAMA et al., 1972) and

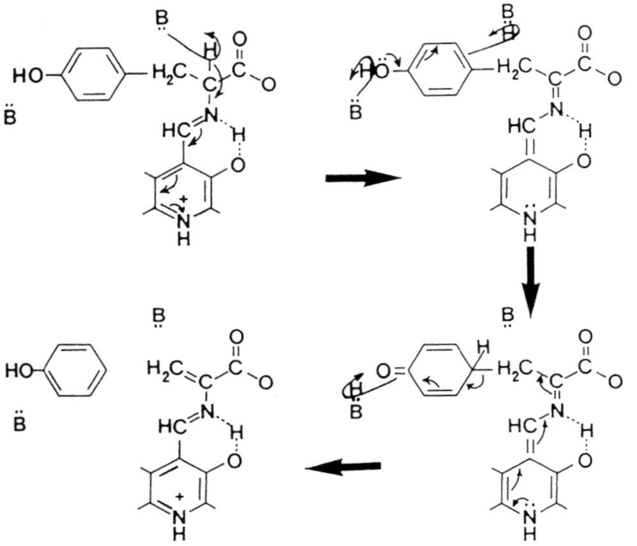

Fig. 9. Mechanism of tyrosine phenol-lyase reaction (reprinted from KIICK and PHILLIPS, 1988a).

nucleotide sequencing (DEELEY and YANOFSKY, 1981; TOKUSHIGE et al., 1989). Amino acid sequences of the *Proteus vulgaris* enzyme (KAMATH and YANOFSKY, 1992) and the *Enterobacter aerogenes* enzyme (KAWASAKI et al., 1993) were also deduced from the nucleotide sequence of the enzyme genes. The primary structures of the *E. coli* and *P. vulgaris* enzymes were 52% identical. Interestingly, they showed significant sequence similarity to the β-tyrosinase of *Citrobacter freundii* (KAMATH and YANOFSKY, 1992). The identity in the amino acid sequence of β-tyrosinase to the *E. coli* tryptophanase is 42%, and to the *P. vulgaris* enzyme is 50%. HIRAHARA et al. (1992) showed by gene cloning that two thermostable tryptophanase genes (*tna*-1 and *tna*-2) are located close to each other on the chromosome of *S. thermophilium*. Both tryptophanases shared 92% identical amino acids in a total of 453 amino acids. However, the two enzymes showed distinct differences in heat-stability and in activation energy in catalysis. *S. thermophilium* usually produces only Tna-2 enzyme; the *tna*-1 gene is silent. The thermostable β-tyrosinase produced by *S. thermophilium* also showed significant sequence similarity to the two tryptophanases over the entire amino acid sequence (HIRAHARA et al., 1993).

A large quantity of homogeneous preparations of tryptophanase are easily available by means of cloned cells. The *E. coli* B/lt7-A enzyme was overproduced at a quantity corresponding to more than 60% of the total soluble protein of the cloned cells; about 400 mg of purified enzyme was obtained from 1 L of medium (TANI et al., 1990). KAWATA et al. (1991) obtained crystals of the holoenzyme of *E. coli* B/lt7-A with the tetragonal space group $P4_12_12$ (a = b = 113.4 Å, c = 232.2 Å), which diffracted to 3 Å. The holoenzyme from *P. vulgaris* crystallized by DEMENTIEVA et al. (1994) was orthorrhombic ($P2_12_12_1$; a = 115.0 Å, b = 118.2 Å, c = 153.7 Å) and diffracted to 1.8 Å.

Tryptophanase catalyzes α,β-eliminations, β-replacements, and reverse reactions of α,β-eliminations, and all these reactions proceed through the enzyme-bound α-aminoacrylate intermediate. In addition to L-tryptophan, L-cysteine, *S*-methyl-L-cysteine, β-chloro-L-alanine, and L-serine serve as substrates for the α,β-elimination and the β-replacement reactions in the same manner as β-tyrosinase. L-Tryptophan is synthesized from indole, pyruvate, and ammonia by a reverse reaction of α,β-elimination; tryptophan derivatives such as 5-methyl-, 5-hydroxy-, and 5-amino-L-tryptophan are synthesized from the corresponding indole derivatives. Tryptophanase uses two catalytic bases in the same manner as $\beta$-tyrosinase: one with a p$K$ of 7.6 abstracts the proton at the 2-position of the substrate, and the other with a p$K$ of 6.0 abstracts the ring nitrogen proton from indole (KIICK and PHILLIPS, 1988b).

Synthesis of L-tryptophan based on the reverse reaction of $\alpha,\beta$-elimination was studied by YOSHIDA et al. (1974). Indole added to the reaction mixture was converted almost quantitatively to L-tryptophan with a yield of 100 g $L^{-1}$ (NAKAZAWA et al., 1972a). 5-Hydroxy-L-tryptophan (23.3 g $L^{-1}$) was synthesized from 5-hydroxyindole, pyruvate, and ammonia at a conversion rate of 57% based on 5-hydroxyindole (NAKAZAWA et al., 1972b).

### 3.4.4 L-Cysteine Desulfhydrase

L-Cysteine hydrogensulfide lyase (deaminating) (L-cysteine desulfhydrase, EC 4.4.1.1) catalyzes the $\alpha,\beta$-elimination of L-cysteine to produce hydrogen sulfide, pyruvate, and ammonia. The enzyme occurs in bacteria, yeasts, and plants, and was purified to homogeneity from *Salmonella typhimurium* (KREDICH et al., 1972) and *Aerobacter aerogenes* (KUMAGAI et al., 1974).

In addition to L-cysteine, L-serine, *S*-methyl-L-cysteine, and $\beta$-chloro-L-alanine serve as substrates for the $\alpha,\beta$-elimination (KUMAGAI et al., 1977). The enzyme also catalyzes the $\beta$-replacement reaction between $\beta$-chloro-L-alanine and various thiols to produce the corresponding *S*-substituted L-cysteines (KUMAGAI et al., 1977). The reverse reaction of $\alpha,\beta$-elimination is also catalyzed; L-cysteine is produced from pyruvate, ammonia, and sulfide (OHKISHI et al., 1981).

L-Cysteine was synthesized from $\beta$-chloro-L-alanine by a $\beta$-replacement reaction with *E. cloacae* cells (KUMAGAI et al., 1975). L-Cysteine thus synthesized forms a thiazolidine compound with acetone, which prevents L-cysteine from producing an adduct with the enzyme-bound $\alpha$-aminoacrylate intermediate. Accordingly, KUMAGAI et al. (1975) added acetone to the reaction mixture and succeeded to produce L-cysteine efficiently. Thus, $\beta$-chloro-L-alanine was converted to L-cysteine at a concentration of 50 g $L^{-1}$, giving a molar yield of more than 80%.

### 3.4.5 Tryptophan Synthase

Tryptophan synthase (EC 4.2.1.20) occurs widely in various bacteria, yeasts, molds, and plants, and catalyzes the last two reactions in the biosynthesis of L-tryptophan. The bacterial tryptophan synthase is composed of two kinds of proteins, $\alpha$ and $\beta$. PLP is bound to the $\beta$-subunit through a Schiff base. Two $\alpha$-subunits combine with one $\beta_2$-dimer to form an $\alpha_2\beta_2$-complex that catalyzes the total synthesis of tryptophan. Each subunit also catalyzes its own specific reaction (Tab. 3). The crystalline $\alpha_2\beta_2$-complex is obtained after a 6-fold purification from *E. coli trpR- trp*ED102/

**Tab. 3.** Reactions Catalyzed by Tryptophan Synthase

| Reaction | Catalyzed by |
|---|---|
| 1. Indole-3-glycerol phosphate + L-Serine → L-Tryptophan + D-Glyceraldehyde 3-phosphate + $H_2O$ | $\alpha_2\beta_2$ |
| 2. Indole-3-glycerol phosphate → Indole + D-Glyceraldehyde 3-phosphate | $\alpha$; $\alpha_2\beta_2$ |
| 3. Indole + L-Serine → L-Tryptophan + $H_2O$ | $\beta_2$; $\alpha_2\beta_2$ |
| 4. L-Serine → Pyruvate + Ammonia | $\beta_2$ |
| 5. $\beta$-Mercaptoethanol + L-Serine → *S*-($\beta$-Hydroxyethyl)-L-cysteine + $H_2O$ | $\beta_2$; $\alpha_2\beta_2$ |
| 6. $\beta$-Mercaptoethanol + L-Serine → *S*-Pyruvylmercaptoethanol + Pyridoxamine-P + $H_2O$ | $\beta_2$ |

F′ *trp*ED102. About 16% of the intracellular soluble protein of this mutant is the tryptophan synthase complex. ISHIWATA et al. (1989) cloned the gene of thermostable tryptophan synthase from *Bacillus stearothermophilus* into *E. coli,* and constructed an overproducer of the thermostable enzyme (ISHIWATA et al., 1990a).

The three-dimensional structure of the tryptophan synthase $\alpha_2\beta_2$ complex from *Salmonella typhimurium* was determined at 2.5 Å resolution (MILES et al., 1994). The four subunits are arranged in an extended order of $\alpha\beta\beta\alpha$ with an overall length of 150 Å (MILES et al., 1994). The active sites of neighboring $\alpha$- and $\beta$-subunits are connected by a "tunnel", which provides a pathway for internal diffusion of indole between the two active sites and prevents indole to escape from the active center. MILES and coworkers (1994) have studied in detail the mechanism of the enzyme by X-ray crystallography, single crystal microspectrophotometry, rapid scanning stopped flow spectrophotometry, steady-state kinetics, and site-directed mutagenesis.

ISHIWATA et al. (1990b) established a method to synthesize L-tryptophan from DL-serine and indole by means of the tryptophan synthase (EC 4.2.1.20) from *E. coli* and the amino acid racemase with low substrate specificity of *Pseudomonas striata* (*P. putida*). Both DL-serine and indole are available by chemical synthesis. Tryptophan synthase catalyzes the $\beta$-replacement reaction of L-serine with indole to produce L-tryptophan, and the amino acid racemase with low substrate specificity converts unreacted D-serine into L-serine. Because the racemase does not act on tryptophan, almost all DL-serine is converted to optically pure L-tryptophan. ISHIWATA et al. (1990b) successfully produced L-tryptophan at a concentration of about 110 g $L^{-1}$ of incubation mixture with a yield of 100% based on indole and 91% based on DL-serine.

ESAKI et al. (1983a) showed the enzymatic synthesis of various *S*-substituted-L-cysteines from L-serine and its derivatives (e.g., $\beta$-chloro-L-alanine and *O*-methyl-L-serine) with the $\alpha_2\beta_2$-complex (Tab. 4). Thiols such as $\alpha$-toluenethiol, 1-propanethiol, and 1-butanethiol are efficient *S*-substituent donors. When L-threonine and L-vinylglycine are used as *S*-substituent acceptors of thiols, the corresponding *S*-substituted $\beta$-methyl-L-cysteines are synthesized (Tab. 4). The enzyme also catalyzes the $\beta$-replacement reactions of L-serine with selenols to produce the corresponding *Se*-substituted L-selenocysteines (Tab. 4) (ESAKI et al., 1983b). *Se*-Benzyl-L-selenocysteine and *Se*-methyl-L-selenocysteine are synthesized from L-serine and $\alpha$-tolueneselenol or methaneselenol in a similar way with yields of 44% and 16%, respectively, based on L-serine. The relative activity of indole to methanethiol (100%) is approximately 150%. Those of methaneselenol and $\alpha$-tolueneselenol are 190% and 240%, respectively. The production of *Se*-methyl- and *Se*-benzyl-L-selenocysteines proceeds much more rapidly than tryptophan synthesis, the inherent reaction of the enzyme. L-Serine can be replaced with a variety of $\beta$-substituted L-alanines such as $\beta$-chloroalanine and *O*-acetylserine in the reaction system containing these selenols.

**Tab. 4.** Relative Reactivities of *S*-Substituent Acceptors in $\beta$-Replacement Reactions with $\alpha$-Toluenethiol by Tryptophan Synthase

| *S*-Substituent Acceptor | Relative Rate |
|---|---|
| L-Serine | 100 |
| *O*-Methyl-D-serine | 41 |
| $\beta$-Chloro-L-alanine | 35 |
| *O*-Acetyl-L-serine | 13 |
| *S*-Methyl-L-cysteine | 24 |
| *S*-Ethyl-L-cysteine | 12 |
| *Se*-Methyl-DL-selenocysteine | 84 |
| L-Cysteine | 0 |
| L-Threonine | 1 |
| L-*allo*-Threonine | 0 |
| L-Vinylglycine | 0.5 |

According to the general mechanism of the $\beta$-replacement reactions catalyzed by PLP enzymes, the nucleophilic addition of selenols occurs in an intermediate derived from the substrate. Although selenols are more nucleophilic than thiols, selenols are less reactive substituent donors than thiols in the enzymatic $\beta$-replacement reactions catalyzed by tryptophan synthase. This is compatible with the reactivities of $\alpha$-tolueneselenol and $\alpha$-toluenethiol, although methaneselenol is a slightly more effective substituent donor than methanethiol. Some physicochemical properties of methaneselenol and methanethiol, such as volatility and solubility, may affect their reactivity in the enzyme reaction.

### 3.4.6 *S*-Alkylcysteine α,β-Lyase

Various *S*-substituted cysteines and their sulfoxide derivatives occur in higher plants and bacteria. Alliin lyase (EC 4.4.1.4) catalyzes α,β-elimination of *S*-alkyl-L-cysteine sulfoxides to produce pyruvate, ammonia, and alkylsulfenate, and was purified to homogeneity from *Allium sativum* (MAZELIS and FOWDEN, 1973). *S*-Alkylcysteine α,β-lyase (EC 4.4.1.6) differs from alliin lyase in that it acts on both *S*-alkyl-L-cysteines and their sulfoxide derivatives. *S*-Alkylcysteine α,β-lyase was purified to homogeneity from etiolated seedlings of *Acacia farnesiana* (MAZELIS and CREVELINO, 1975) and from bacteria (KAMITANI et al., 1990a, b).

KAMITANI et al. (1990a) found that *P. putida* ICR 3640 produces *S*-alkylcysteine α,β-lyase abundantly by addition of *S*-metyl-L-cysteine to the medium. However, the addition of L-cysteine increases the enzyme activity only sligthly. In addition to *S*-methyl-L-cysteine, *O*-methyl-L-serine, and *Se*-methyl-DL-selenocysteine are also effective substrates. In particular, *Se*-methyl-L-selenocysteine is a better substrate than *S*-methyl-L-cysteine. Homologs of *S*-methyl-L-cysteine, such as *S*-ethyl-L-cysteine, *S*-benzyl-L-cysteine, and L-djenkolate, also serve as effective substrates. Conversion of the thioether of *S*-methyl-L-cysteine into sulfoxide does not affect the ability of the enzyme to utilize it as the substrate. *S*-Carboxymethyl-L-cysteine, *O*-acetyl-L-serine, L-cystathionine, L-cysteine, and L-cystine also serve as substrates, though slowly. The following amino acids are inert: *S*-methyl-D-cysteine, L-methionine, L-serine, L-alanine, L-tyrosine, L-tryptophan, L-phenylalanine, L-norvaline, and L-methionine sulfoxide.

*S*-Alkylcysteine α,β-lyase of *P. putida* ICR3640 catalyzes the β-replacement reaction between *S*-methyl-L-cysteine and various thiols to yield the corresponding *S*-substituted L-cysteines (KAMITANI et al., 1990a). In addition to *S*-methyl-L-cysteine, *S*-methyl-DL-selenocysteine, and *O*-methyl-L-serine also serve as substrates for the replacement reaction. The β-replacement reactivities between amino acids and various thiols are summarized in Tab. 5. Various sulfur amino acids can be prepared by the β-replacement reaction.

### 3.4.7 β-Chloro-D-alanine Hydrogenchloride Lyase

β-Chloro-D-alanine is a toxic compound for bacteria, because it inactivates an active transport system coupled to a membrane-bound D-alanine dehydrogenase, and also D-amino acid transaminase and alanine racemase, and thereby inhibits the biosynthesis of the peptidoglycan layer of the bacterial cell wall. However, NAGASAWA et al. (1982a) isolated a strain of *P. putida* CR1-1 that is able to grow in the presence of β-chloro-DL-ala-

**Tab. 5.** Substrate Specificity of *S*-Alkylcysteine α,β-Lyase

| Thiol (RSH) | Relative Activity on Amino Acid Substrates | | |
|---|---|---|---|
| | *Se*-Methyl-L-Selenocysteine | *S*-Methyl-L-Cysteine | *O*-Methyl-L-Serine |
| R= | | | |
| Ethyl | 100 | 121 | 91 |
| *n*-Propyl | 34 | 12 | 35 |
| Phenyl | 7 | 0 | 0 |
| Benzyl | 12 | 9 | 8 |

nine, and discovered a unique enzyme chloro-D-alanine hydrogenchloride lyase, which catalyzed α,β-elimination of β-chloro-D-alanine to form pyruvate, ammonia, and hydrogen chloride.

β-Chloro-D-alanine hydrogenchloride lyase is inducibly formed only by addition of β-chloro-D-alanine to the medium (NAGASAWA et al., 1982a). In addition to β-chloro-D-alanine, which is the preferred substrate, D-cysteine and D-cystine also serve as substrates. The enzyme catalyzes the β-replacement reaction between β-chloro-D-alanine and hydrogen sulfide to yield D-cysteine (YAMADA et al., 1981). NAGASAWA et al. (1982b) studied the conditions for the cultivation of *P. putida* CR 1-1 and for D-cysteine production by the β-replacement reaction with resting cells, and found that the preliminary treatment of *P. putida* CR 1-1 cells with phenylhydrazine effectively increases the optical purity of D-cysteine formed, because enzymes acting on the L-isomer of β-chloroalanine are inactivated by the treatment (NAGASAWA et al., 1983). Under the optimal conditions, 22 mg of D-cysteine per mL of reaction mixture was produced from β-chloro-D-alanine, with a molar yield of 88%.

### 3.4.8 D-Cysteine Desulfhydrase

NAGASAWA et al. (1985) found that a unique enzyme, D-cysteine desulfhydrase (EC 4.4.1.15), is produced by a mutant of *E. coli* (W3110 ΔtrpED 102/F′ΔtrpED 102). The wild-type strain also produces the enzyme, but the enzyme activity is about 10 times lower than the enzyme activity of the mutant strain.

The activity of D-cysteine desulfhydrase occurs not only in *E. coli* but also in *Klebsiella, Enterobacter, Citrobacter,* and other genera of enteric bacteria (NAGASAWA et al., 1985). The enzyme acts exclusively on D-enantiomers of cysteine, β-chloroalanine, and other amino acids in the same manner as β-chloro-D-alanine hydrogenchloride lyase. D-Cysteine desulfhydrase of *E. coli* is unique in its high thermostability, although both enzymes are immunologically distinct from each other. D-Cysteine desulfhydrase is produced constitutively. The enzyme also catalyzes the β-replacement reactions with various thiol compounds as substituent donors. Thiol compounds with a hydroxyl group, a carboxylic group, and other hydrophilic functional groups serve as good *S*-substituent donors. In particular, thioglycolic acid is the best substrate in the reaction to produce *S*-carboxymethyl-D-cysteine. However, straight-chain and branched-chain alkane thiols are inert in the β-replacement reaction.

### 3.4.9 D-Selenocystine α,β-Lyase

ESAKI et al. (1988) found a novel enzyme that catalyzes α,β-elimination of D-selenocysteine to produce pyruvate, ammonia, and elemental selenium in *Clostridium sticklandii* cells, and named it D-selenocystine α,β-lyase. The enzyme occurs also in *C. sporogenes,* but not in other *Clostridium* sp. The enzyme was purified to homogeneity from *C. sticklandii* which produced more enzyme than *C. sporogenes.*

In addition to D-selenocystine, D-cystine, D-lanthionine, *meso*-lanthionine, and D-cysteine serve as substrates, but D-selenocysteine is inert (ESAKI et al., 1988). The enzyme also catalyzes the β-replacement reaction between D-selenocystine and various thiols to yield the corresponding *S*-substituted D-cysteines (ESAKI et al., 1988). The rate of α,β-elimination in the presence of ethanethiol was determined to be only 12% of that of the β-replacement reaction; the β-replacement reaction proceeds preferentially in the presence of *S*-substituent donors. However, the rate of the β-replacement reaction was substantially the same as that of α,β-elimination determined in the absence of *S*-substituent donors. In addition, selenols also can serve as a substrate in the β-replacement reaction, and *Se*-substituted D-selenocysteines corresponding to the selenols used are produced (ESAKI et al., 1988).

### 3.4.10 L-Methionine γ-Lyase

Methionine γ-lyase (EC 4.4.1.11) catalyzes the conversion of L-methionine into α-keto-

**Tab. 6.** Reactions Catalyzed by Methionine $\gamma$-Lyase

| Reaction | Type |
|---|---|
| 1. $R-X-CH_2-CH_2-CH(NH_2)-COOH+H_2O \rightarrow R-XH+CH_3-CH_2-CO-COOH+NH_3$ | $\alpha,\gamma$-elimination |
| 2. $R-X-CH_2-CH_2-CH(NH_2)-COOH+R'-X'H \rightarrow R-XH+R'-X'-CH_2-CH_2-CH(NH_2)-COOH$ | $\gamma$-replacement reaction |
| 3. $R-X-CH_2-CH(NH_2)-COOH+H_2O \rightarrow R-XH+CH_3-CO-COOH+NH_3$ | $\alpha,\beta$-elimination |
| 4. $R-X-CH_2-CH(NH_2)-COOH+R'-X'H \rightarrow R-XH+R'-X'-CH_2-CH(NH_2)-COOH$ | $\beta$-replacement reaction |
| 5. $CH_2=CH-CH(NH_2)-COOH+H_2O \rightarrow CH_3-CH_2-CO-COOH+NH_3$ | deamination |
| 6. $CH_2=CH-CH(NH_2)-COOH+R-X'H \rightarrow R'-X'-CH_2-CH_2-CH(NH_2)-COOH$ | $\gamma$-addition reaction |

X=O, S, or Se; X′=S or Se.

butyrate, methanethiol, and ammonia (Tab. 6, reaction 1), and plays an important role in the bacterial metabolism of methionine. The enzyme is widely distributed in pseudomonads. It is inducibly formed by addition of L-methionine to the medium, and was purified to homogeneity from *P. putida* ICR3460, the best producer of the enzyme (NAKAYAMA et al., 1984a). NAKAYAMA et al. (1984b) isolated a strain of *Aeromonas* sp. which also produces methionine $\gamma$-lyase abundantly, and purified the enzyme to homogeneity. Recently, INOUE et al. (1995) cloned the enzyme gene from *P. putida* ICR3460 into *E. coli*, and determined its complete nucleotide sequence. The enzyme gene consists of 1194 nucleotides and encodes 398 amino acid residues corresponding to the subunit (*Mr* 43000) of the tetrameric enzyme. The amino acid sequence deduced from the nucleotide sequence showed homology with the following enzymes: cystathionine $\gamma$-lyases of *S. cerevisiae* and rat, cystathionine $\gamma$-synthase and cystathionine $\beta$-lyase of *E. coli*. The methionine $\gamma$-lyase gene was highly expressed in *E. coli* MV1184.

Methionine $\gamma$-lyase has multiple catalytic functions: it catalyzes $\alpha,\gamma$-elimination and $\gamma$-replacement reactions (Tab. 6, reactions 1 and 2) of L-methionine and its analogs, and $\alpha,\beta$-elimination and $\beta$-replacement reactions (reactions 3 and 4) of L-cysteine and its analogs (TANAKA et al., 1977). The enzyme also catalyzes the $\alpha,\gamma$-elimination of selenomethionine to yield $\alpha$-ketobutyrate, ammonia, and methaneselenol, and also $\gamma$-replacement reactions with various thiols to produce *S*-substituted homocysteines (TANAKA et al., 1977). Selenomethionine is a better substrate than methionine for $\alpha,\gamma$-elimination based on the $v_{max}$ and $k_m$ values, but is less effective for $\gamma$-replacement (TANAKA et al., 1977). In addition, L-methionine and its derivatives, which are substrates for the $\alpha,\gamma$-elimination, react with selenols to form the corresponding *Se*-substituted selenohomocysteines, although selenols are less efficient substituent donors than thiols. The enzymatic $\beta$-replacement reaction also occurs between *S*-substituted cysteines or *O*-substituted serines and selenols.

DAVIS and METZLER (1972) have proposed that a ketimine intermediate of PLP and vinylglycine (2-amino-3-butenoate) is the key intermediate of $\alpha,\gamma$-elimination and $\gamma$-replacement reactions catalyzed by PLP enzymes. Methionine $\gamma$-lyase catalyzes the deamination of vinylglycine (Tab. 7, reaction 5) to produce $\alpha$-ketobutyrate and ammonia, but is not inactivated by vinylglycine (ESAKI et al., 1977). The enzyme also catalyzes the $\gamma$-addition reaction of various thiols or selenols

**Tab. 7.** Kinetic Parameters of Reactions Catalyzed by *O*-Acetylhomoserine Sulfhydrylase

| Substituent Acceptor | Substituent Donor | Relative $v_{max}$ | $k_m$ (mM) of Substituent | |
|---|---|---|---|---|
| | | | Acceptor | Donor |
| *O*-Acetylhomoserine | NaHS | 100 | 4.1 | 0.52 |
| *O*-Acetylhomoserine | $Na_2Se_2$ | 17 | 5.3 | 8.9 |
| *O*-Acetylserine | NaHS | 14 | 2.5 | 0.70 |
| *O*-Acetylserine | $Na_2Se_2$ | 8.4 | 5.0 | n.d. |
| Serine *O*-Sulfate | NaHS | 5.1 | 4.0 | 0.70 |
| Serine *O*-Sulfate | $Na_2Se_2$ | 1.4 | 4.0 | 10 |
| Serine *O*-Sulfate | NaHSe | 1.3 | n.d. | 1.2 |

n.d.: not determined.

to yield the corresponding *S*- or *Se*-substituted homocysteines or selenohomocysteines (Tab. 7, reaction 6) (ESAKI et al., 1979).

### 3.4.11 *O*-Acetylhomoserine Sulfhydrylase

*O*-Acetylhomoserine sulfhydrylase (*O*-acetylhomoserine (thiol)-lyase, EC 4.2.99.10), a PLP enzyme, catalyzes the synthesis of cysteine and homocysteine from $H_2S$ with *O*-acetyl-L-serine and *O*-acetyl-L-homoserine, respectively. The enzyme of baker's yeast was purified and characterized. It is involved in the *in vivo* synthesis of cysteine (YAMAGATA and TAKESHIMA, 1976).

CHOCAT et al. (1985) have shown that *O*-acetylhomoserine sulfhydrylase catalyzes the $\beta$- and $\gamma$-replacement reactions between the *O*-acetyl groups of *O*-acetyl-L-homoserine and *O*-acetyl-L-serine and $Na_2Se_2$. Serine *O*-sulfate also serves as a substrate of the $\gamma$-replacement reaction, although its reactivity is lower than that of *O*-acetyl-L-serine (Tab. 7). The selenium amino acids produced were isolated and identified as L-selenocystine and L-selenohomocystine. The initial products of the enzyme reaction are probably Se-(selenohydryl)derivatives, i.e.
$-SeSeCH_2CH(NH_2)COOH$ and
$-SeSeCH_2CH_2CH(NH_2)COOH$
which are spontaneously oxidized to the corresponding diselenides.

### 3.4.12 Cystathionine $\gamma$-Lyase

Cystathionine $\gamma$-lyase is involved in the transsulfuration pathway with cystathionine $\beta$-synthase, and catalyzes the $\alpha,\gamma$-elimination of L-cystathionine to produce $\alpha$-ketobutyrate, cysteine, and ammonia. NAGASAWA et al. (1984) found that cystathionine $\gamma$-lyase is widely distributed in various bacterial strains. They purified the enzyme from the mycelial extracts of *Streptomyces phaeochromogenes* to homogeneity, and crystallized and characterized it. The enzyme catalyzes the $\alpha,\gamma$-elimination not only of L-cystathionine but also of L-homoserine. The $\gamma$-replacement reaction between L-homoserine and cysteine to produce L-cystathionine is also catalyzed. In addition to L-homoserine, L-vinylglycine and *O*-succinyl-L-homoserine also serve as substrates in the $\gamma$-replacement reaction (NAGASAWA et al., 1987). Recently, the cystathionine $\gamma$-lyase gene of the yeast *S. cerevisiae* was cloned and expressed in *E. coli* (YAMAGATA et al., 1993). The yield of cystathionine $\gamma$-lyase from the cloned cells was about 150 mg $L^{-1}$ of the culture. YAMAGATA et al. (1993) found that the yeast enzyme shows the activities of cystathionine $\beta$-lyase, cystathionine $\gamma$-synthase, and L-homoserine sulfhydrylase. ESAKI et al. (1981) also showed that cystathionine $\gamma$-lyase of rat liver catalyzes the $\alpha,\beta$-elimination of L-cystine in addition to the $\alpha,\gamma$-elimination of L-cystathionine. The $\alpha,\beta$-elimination of L-cystathionine proceeds much more slowly (3%) than the $\alpha,\gamma$-elimination. However, the $\alpha,\beta$-elimination of selenocystathionine proceeds at a comparable rate to the $\alpha,\gamma$-elimination of the same substrate. Cystathionine $\gamma$-lyase of rat liver can further eliminate both the amino acids formed from selenocystathionine and from cystathionine by elimination reactions, although slowly (ESAKI et al., 1981). All the selenium amino acids are decomposed 2.5–3 times more rapidly than the corresponding sulfur analogs.

YAMADA et al. (1984) studied the most preferred conditions for the synthesis of L-cystathionine and found that the substrate L-cysteine needs to be added to the reaction mixture at limited concentrations; high concentrations of L-cysteine inhibit the enzyme by an allosteric effect. The high-yield synthesis of L-cystathionine was achieved with *O*-succinyl-L-homoserine as the substrate: 11 g $L^{-1}$ of L-cystathionine was obtained at a 100% conversion rate (KANZAKI et al., 1986).

## 3.5 Serine Hydroxymethyltransferase

L-Serine hydroxymethyltransferase (EC 2.1.2.1) is a ubiquitous pyridoxal-P-dependent enzyme which acts on various $\alpha$-amino acids, and catalyzes retro-aldol cleavage/aldol condensation transamination, and decarboxylation reactions as shown in Tab. 8 (SCHIRCH, 1982). Reaction 1 is the physiological reaction for the generation of a one-carbon group, in the form of 5,10-methylenetetrahydrofolate, required in the biosynthesis of methionine, choline, thymidylate, and purines. In several microorganisms such as *Corynebacterium glycinophilum,* the enzyme is responsible for the production of L-serine from glycine (KUBOTA, 1985).

The enzyme has been demonstrated in a variety of tissues and organisms: mammalian liver, kidney and brain, yeasts, bacteria, plants, and insects (SCHIRCH, 1982). Eukaryotic cells contain both cytosolic and mitochondrial forms (SCHIRCH, 1982). The enzyme has been purified to homogeneity from various organisms and characterized in detail (SCHIRCH, 1982; SCHIRCH et al., 1985; MIYAZAKI et al., 1987). The primary structures of the enzymes from *E. coli* (PLAMANN et al., 1983), *Hyphomicrobium methylovorum* (MIYATA et al., 1993), *S. cerevisiae* (MCNEIL et al., 1994), *Pisum sativum* (garden pea) (TURNER et al., 1992), human cells (GARROW et al., 1993), etc. are predicted based on the nucleotide sequences of their genes. The primary structures of cytosolic and mitochondrial enzymes from rabbit liver were determined by amino acid sequencing (MARTINI et al., 1987, 1989).

The enzymes from different organisms show different substrate specificities. The enzyme from rat liver slowly catalyzes the interconversion of glycine and L-serine in the absence of tetrahydrofolate (SCHIRCH, 1982), but the enzymes from *E. coli* (SCHIRCH et al.,

**Tab. 8.** Reactions Catalyzed by Serine Hydroxymethyltransferase

| |
|---|
| 1. L-Serine + $H_4$-Folate ⇄ Glycine + 5,10-Methylene-$H_4$-folate |
| 2. L-$\alpha$-Methylserine + $H_4$-Folate ⇄ D-Alanine + 5,10-Methylene-$H_4$-folate |
| 3. L-*allo*-Threonine ⇄ Glycine + Acetaldehyde |
| 4. L-Threonine ⇄ Glycine + Acetaldehyde |
| 5. *erythro*-$\beta$-Phenylserine ⇄ Glycine + Benzaldehyde |
| 6. *threo*-$\beta$-Phenylserine ⇄ Glycine + Benzaldehyde |
| 7. $\varepsilon$-Trimethyl-3-hydroxylysine ⇄ Glycine + $\gamma$-Butyrobetaine Aldehyde |
| 8. D-Alanine + Pyridoxal-P ⇄ Pyruvate + Pyridoxamine-P |
| 9. Aminomalonate → Glycine + $CO_2$ |

$H_4$-folate represents tetrahydrofolate.

1985), *H. methylovorum* (MIYAZAKI et al., 1987), and *C. glycinophilum* (KUBOTA and YOKOZEKI, 1989) show an absolute tetrahydrofolate requirement for L-serine degradation to glycine and formaldehyde. The *E. coli* enzyme catalyzes both L-*allo*-threonine and L-threonine degradations in the absence of tetrahydrofolate (SCHIRCH and GROSS, 1968), but the enzyme from *C. glycinophilum* does not act on L-threonine (KUBOTA and YOKOZEKI, 1989).

Lys229 of the *E. coli* enzyme forms an internal aldimine with pyridoxal-P (SCHIRCH et al., 1985); Arg363 functions as the substrate carboxyl binding site (DELLEFRATTE et al., 1994), and His228 and Thr226 play critical roles in determining reaction and substrate specificity (STOVER et al., 1992; ANGELACCIO et al., 1992). The H228N mutant *E. coli* enzyme has a higher affinity for both D-alanine and L-alanine than the wild type, transaminates both alanine isomers and also catalyzes an alanine racemase reaction (SHOSTAK and SCHIRCH, 1988).

L-Serine can be produced enzymatically from glycine and formaldehyde, which are cheaply available starting materials. Also, a variety of $\beta$-hydroxy amino acids such as $\beta$-2-furylserine, $\beta$-hydroxyisohistidine, $\beta$-phenylserine, and $\beta$-hexylserine are prepared from the corresponding aldehyde substrates and glycine in reasonable yields by using the enzyme from pig liver (>20%) (SAEED and YOUNG, 1992). Stereospecificity at the $\alpha$-center is high, and L-amino acids are exclusively produced. However, stereospecificity is not high at the $\beta$-center.

## 3.6 L-Threonine Aldolase

L-Threonine acetaldehyde lyase ("L-threonine aldolase", EC 4.1.2.5) catalyzes the conversion of L-threonine into acetaldehyde and glycine (Eq. 1).
It occurs in various microorganisms, e.g., *Klebsiella, Escherichia, Arthrobacter, Bacterium, Xanthomonas, Proteus,* and *Candida* (YAMADA et al., 1971). These microorganisms grow in media containing L-threonine as the sole source of carbon, and the enzyme is induced by L-threonine. Strict anaerobes such

$$H_3C-CH(OH)-CH(NH_2)-COOH \rightleftharpoons H_3C-C(=O)-H + H_2C(NH_2)-COOH \quad (1)$$

as *Clostridium pasteurianum* and *Selenomonas ruminatium* also produce this enzyme, but it is synthesized constitutively and functions in the biodegradation of L-threonine into glycine in these microorganisms (DAINTY, 1970). Evidence has been obtained that microbial L-threonine aldolase and *allo*-threonine aldolase are identical, but distinct from L-serine hydroxymethyltransferase (Tab. 8, reaction 4) (BELL and TURNER, 1977). In mammals, however, L-threonine aldolase is identical to L-serine hydroxymethyltransferase (SCHIRCH, 1982). Tetrahydrofolate, a cofactor of L-serine hydroxymethyltransferase, is rather inhibitory to the microbial threonine aldolase (YAMADA et al., 1970).

YAMADA et al. (1970) have crystallized L-threonine aldolase from *Candida humicola,* which produces the enzyme abundantly, and have characterized it. The enzyme has a molecular mass of about 277,000 Da, and contains 6 mol of pyridoxal-P per mol of enzyme. In addition to L-threonine ($k_m$=0.55 mM), L-*allo*-threonine ($k_m$=0.39 mM), and L-serine ($k_m$=3.7 mM) serve as substrates producing glycine. These reactions proceed reversibly: L-threonine is synthesized from acetaldehyde and glycine, and L-serine from formaldehyde and glycine. Thus, with this enzyme, L-serine and L-threonine can be synthesized from cheaply available materials.

## 3.7 D-Amino Acid Aminotransferase

D-Amino acid aminotransferase catalyzes transaminations between various D-amino acids and $\alpha$-keto acids and is regarded as a key enzyme in bacterial D-amino acid metabolism. The enzyme has been demonstrated in bacteria of the genus *Bacillus, Rhodospiril-*

*lum rubrum, Rhizobium japonicum,* and also in germinating pea seedlings (SODA and ESAKI, 1985). It acts on various $\alpha$-keto acids as amino acceptor and is utilized to prepare various D-amino acids from the corresponding $\alpha$-keto acids.

$$\mathrm{R_1{-}CH(NH_2){-}COOH} + \mathrm{R_2{-}C(=O){-}COOH} \rightleftharpoons \mathrm{R_1{-}C(=O){-}COOH} + \mathrm{R_2{-}CH(NH_2){-}COOH} \quad (2)$$

A thermophile, *Bacillus* sp. YM-1, abundantly produces D-amino acid aminotransferase (TANIZAWA et al., 1989). The enzyme has a molecular mass of about 62,000 Da, and consists of two subunits with an identical molecular mass (30,000 Da). The enzyme is thermostable, and its activity was not lost by heating at 55 °C for more than 30 min. It is most active at 60 °C. Its gene has been cloned and expressed in *E. coli* C600 with the plasmid vector pBR322. The amount of enzyme produced is equivalent to about 9% of total protein in extracts. The enzyme can be purified efficiently by heat treatment of the extracts.

The enzyme from *Bacillus* sp. YM-1 has a low substrate specificity and can be used for the production of various D-amino acids from the corresponding $\alpha$-keto acids (SODA et al., 1988; NAKAJIMA et al., 1988b; ESAKI et al., 1989a). When D-alanine serves as an amino donor, the D-alanine regeneration system is used in combination with D-amino acid aminotransferase (SODA et al., 1988). This system contains D-amino acid aminotransferase, L-alanine dehydrogenase, alanine racemase, formate dehydrogenase, D-alanine, ammonia, $NAD^+$, formate, and an $\alpha$-keto acid corresponding to the desired D-amino acid (Fig. 10). The D-amino acid is produced from the corresponding $\alpha$-keto acid through transamination with D-alanine catalyzed by D-amino acid aminotransferase. Pyruvate produced from D-alanine is converted to L-alanine with L-alanine dehydrogenase, and L-alanine is racemized with alanine racemase to regenerate D-alanine. NADH consumed by the pyruvate reduction in the L-alanine dehydrogenase reaction is regenerated by coupling with the formate dehydrogenase reaction: The enzyme catalyzes an irreversible oxidation of formate to $CO_2$ with a concomitant reduction of $NAD^+$ to NADH. Therefore, D-alanine and NADH are regenerated continuously in the reaction system by consumption of ammonium formate. Although the reactions of D-amino acid aminotransferase, L-alanine dehydrogenase, and alanine racemase are reversible, the formate dehydrogenase reaction is irreversible. Consequently, the production of D-amino acids proceeds irreversibly.

According to the above procedure, D-selenomethionine has been synthesized from $\alpha$-keto-$\gamma$-methylselenobutyrate with D-alanine as an amino donor (ESAKI et al., 1989a). The yield was 80% based on $\alpha$-keto-$\gamma$-methylselenobutyrate. Other D-amino acids such as D-glutamate were also produced by this procedure (SODA et al., 1988). The initial rate of D-glutamate production was highest at 0.2 M $\alpha$-ketoglutarate, 1 M sodium formate, and 1 M ammonia. The most appropriate ratio of amounts of the four enzymes was formate de-

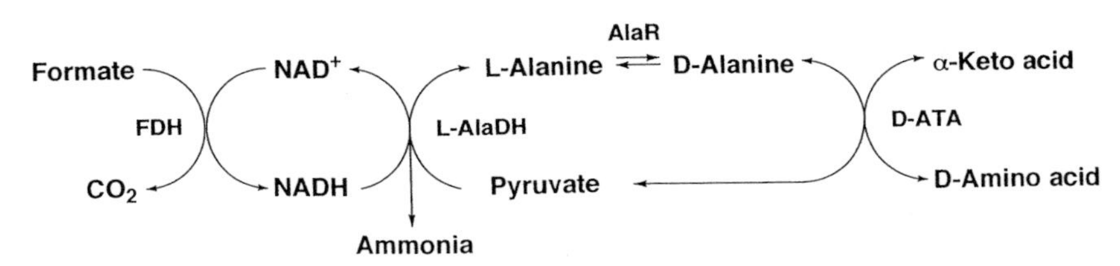


**Fig. 10.** Enzymatic synthesis of D-amino acids from $\alpha$-keto acids, formate, $NAD^+$, D-alanine, and ammonia with L-alanine dehydrogenase (L-AlaDH), formate dehydrogenase (FDH), alanine racemase (AlaR), and D-amino acid aminotransferase (D-ATA).

hydrogenase:alanine racemase:D-amino acid aminotransferase:L-alanine dehydrogenase =1:3:3:5 in units of each enzyme activity. Under these conditions, nearly 100% of the α-ketoglutarate initially added was converted to D-glutamate. In the same way, D-methionine, D-leucine, D-α-aminobutyrate, D-norvaline, and D-valine also were synthesized from their α-keto analogs with a high molar yield (more than 90%). Although α-keto analogs of histidine, arginine, tyrosine, and some other amino acids are poor amino acceptors, the yields of the corresponding D-amino acids could be raised to above 90% by addition of a large amount of D-amino acid aminotransferase.

Instead of D-alanine, D-glutamate also serves as an amino donor in the above system. In this case, L-glutamate dehydrogenase and glutamate racemase are used instead of L-alanine dehydrogenase and alanine racemase. The optimized conditions for the production of D-valine are as follows (NAKAJIMA et al., 1988b). The reaction mixture contains 0.1 M α-ketoisovalerate, 0.2 M ammonium formate, 10 mM L-glutamate, 1 mM $NAD^+$, and 50 μM pyridoxal-P. The most appropriate ratio for the four enzymes was found at glutamate racemase:D-amino acid aminotransferase:L-glutamate dehydrogenase:formate dehydrogenase=1:5:10:1 units of each enzyme activity. Under these conditions, D-valine was synthesized with a molar yield of about 100%. Various D-amino acids such as D-alanine, D-α-aminobutyrate, D-aspartate, D-leucine, and D-methionine were also synthesized by this procedure with a yield higher than 80% (NAKAJIMA et al., 1988b).

A procedure has been developed for the production of D-alanine from fumarate by means of D-amino acid aminotransferase, aspartase, and aspartate racemase (Fig. 11) (KUMAGAI, 1994). Aspartase catalyzes the conversion of fumarate into L-aspartate, which is racemized with aspartate racemase to form D-aspartate. D-Amino acid aminotransferase catalyzes transamination between D-aspartate and pyruvate to produce D-alanine and oxalacetate. Oxalacetate is decarboxylated spontaneously to form pyruvate in the presence of metals. Thus, the transamination proceeds exclusively toward the direction of D-alanine synthesis, and total conversion of fumarate to D-alanine is achieved.

The synthesis of D-α-aminobutyrate and α-keto-γ-methylthiobutyrate from cheaply available DL-methionine with D-amino acid aminotransferase from *Bacillus sphaericus* and methionine γ-lyase has also been reported (TANAKA et al., 1985). L-Methionine is converted to α-ketobutyrate with methionine γ-lyase. The transamination between α-ketobutyrate thus produced and D-methionine is catalyzed by D-amino acid amino transferase. Consequently, D-α-aminobutyrate and α-keto-γ-methylthiobutyrate are produced.

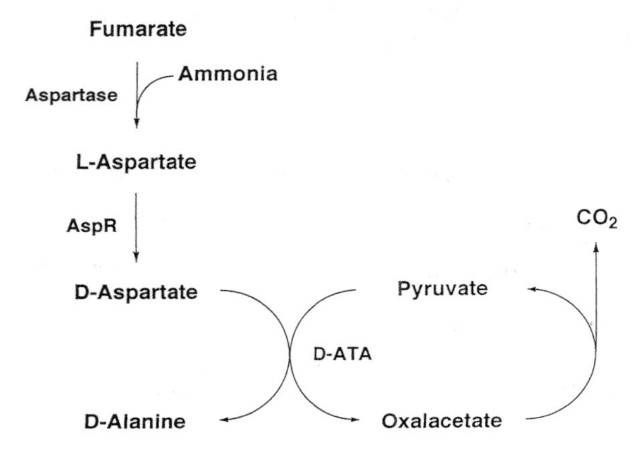


**Fig. 11.** Enzymatic synthesis of D-alanine from fumarate with aspartase, aspartate racemase (AspR), and D-amino acid aminotransferase (D-ATA).

## 3.8 Amino Acid Racemases

Racemases are useful for the total conversion of racemic starting materials into a particular stereoisomer of desired compounds. Half or more of starting materials and steps for separation of the products from the remaining starting materials can be saved and various processes using racemases have been developed. Racemic serine, e.g., was converted to L-tryptophan by means of an amino acid racemase with low substrate specificity and tryptophan synthase, as described below (MAKIGUCHI et al., 1987). Thus, racemases are very useful in the chemical industry when their reactions are coupled with some stereospecific reactions.

### 3.8.1 Alanine Racemase

Alanine racemase (EC 5.1.1.1) catalyzes the interconversion between L- and D-alanine. It requires pyridoxal-P as a coenzyme. The enzyme occurs ubiquitously in bacteria and provides D-alanine, which is the constituent of the peptidoglycan layer of the cell wall. Recently, the enzyme has also been found in eukaryotic cells: the enzyme acts as a key enzyme in cyclosporine biosynthesis (HOFFMANN et al., 1994). The enzymes from *Salmonella typhimurium* (ESAKI and WALSH, 1986; WASSERMAN et al., 1984), *Streptococcus faecalis* (BADET and WALSH, 1985), *B. subtilis* (YONAHA et al., 1975), and *B. stearothermophilus* (INAGAKI et al., 1986) have been purified and characterized. *S. typhimurium* produces two different types of alanine racemases: one (the *alr* gene product) is involved in the biosynthesis of D-alanine, and the other (the *dadB* gene product) participates in the catabolism of D-alanine.

The enzyme from *B. stearothermophilus* is thermostable and well-characterized (INAGAKI et al., 1986). The enzyme is a homodimer of a subunit with a molecular mass of about 39,000 Da. Each subunit contains one molecule of pyridoxal-P, and a Schiff base is formed from pyridoxal-P and the Lys residue of the enzyme. Enzymatic activity was not lost at all by heat treatment at 70 °C for 80 min in 10 mM potassium phosphate buffer (pH 7.2). The enzyme shows maximum reactivity at around 50 °C with a $v_{max}$ value of 1,800 units per mg (D- to L-alanine). At 37 °C in 100 mM 2-(*N*-cyclohexylamino)ethanesulfonic acid buffer (pH 9.0), the $v_{max}$ value (D- to L-alanine direction) is 1,400 units per mg, and the $k_m$ value for D-alanine is 2.67 mM. In the opposite direction, the $v_{max}$ value is 2,550 units per mg, and the $k_m$ value for L-alanine is 4.25 mM. Alanine is the exclusive substrate for the enzyme: L-serine, L-methionine, L-lysine, L-valine, L-homoserine, and L-$\alpha$-aminobutyrate (100 mM) are not racemized.

Alanine racemase is useful for the production of various D-amino acids when its reaction is coupled with that of D-amino acid aminotransferase, as described above (SODA et al., 1988) (see Fig. 10). This procedure is based on the low stereospecificity and the very high structural specificity of alanine racemase and on the strict stereoselectivity and low structural specificity of D-amino acid aminotransferase. D-Amino acids thus formed are not racemized by alanine racemase, and optically pure D-amino acids are obtained.

### 3.8.2 Amino Acid Racemase with Low Substrate Specificity

An amino acid racemase with low substrate specificity (EC 5.1.1.10) was discovered in *Pseudomonas striata* (*P. putida*) (SODA and OSUMI, 1971). The enzyme catalyzes racemization of various amino acids except for aromatic and acidic amino acids. The enzyme was also found in *Aeromonas punctata* (INAGAKI et al., 1987). In *Pseudomonas graveolens,* arginine racemase, which shows a broad substrate specificity, was found (YORIFUJI et al., 1971). These amino acid racemases do not act on threonine, valine, and their analogs, the $\beta$-methylene group of which is substituted. The enzyme recently found in *P. putida* ATCC 17642 catalyzes not only the racemization of various amino acids but also the epimerization of D- and L-threonine: stereoconversion occurs at the $\alpha$-position, and D- and L-threonine are converted into L- and D-*allo*-threonine, respectively (LIM et al., 1993). Pyridoxal-P is required as a coenzyme for these enzymes.

The enzyme from *P. putida* ATCC 17642 is composed of two subunits with identical molecular mass (about 41,000 Da). Lysine is the best substrate, and aliphatic straight-chain amino acids are also racemized efficiently (Tab. 9). In distinction to the enzymes from other bacterial cells, the enzyme from *P. putida* ATCC 17642 epimerizes D- and L-threonine.

A simple method for the synthesis of L-tryptophan from DL-serine and indole was developed by MAKIGUCHI et al. (1987). These compounds are cheaply available by chemical synthesis. The system contains the tryptophan synthase (EC 4.2.1.20) from *E. coli* and the amino acid racemase with low substrate specificity from *P. striata.* The tryptophan synthase catalyzes the $\beta$-replacement of L-serine with indole to produce L-tryptophan. The amino acid racemase with low substrate specificity converts unreacted D-serine into L-serine, and almost all DL-serine is converted to L-tryptophan. Since the racemase does not act on tryptophan, optically pure L-tryptophan is obtained. By this method, L-tryptophan was produced at a concentration of about $110\ g\ L^{-1}$ of incubation mixture with a yield of 100% based on indole and 91% based on DL-serine.

**Tab. 9.** Substrate Specificity of Amino Acid Racemase from *P. putida*

| Substrate | Relative Activity |
|---|---|
| L-Lysine | 100 |
| L-Ornithine | 80 |
| L-Ethionine | 83 |
| L-Arginine | 79 |
| L-Glutamine | 54 |
| L-Methionine | 66 |
| *S*-Methyl-L-cysteine | 3 |
| $\varepsilon$-*N*-Acetyl-L-lysine | 75 |
| L-Homocitrulline | 45 |
| L-Citrulline | 45 |
| L-Homoarginine | 24 |
| L-Norleucine | 45 |
| L-Leucine | 10 |
| L-Homoserine | 22 |
| L-Asparagine | 12 |
| L-Alanine | 6 |
| L-Serine | 10 |
| L-Histidine | 6 |
| L-Isoleucine | 0 |
| L-Cysteine | 0 |
| L-Threonine | 3 |
| L-Valine | 0 |
| L-Proline | 0 |
| L-Glutamic acid | 0 |
| L-Aspartic acid | 0 |
| L-Tyrosine | 0 |
| L-Tryptophan | 0 |
| L-Phenylalanine | 0 |

A simple and efficient method for the high-yield preparation of *S*-adenosyl-L-homocysteine from DL-homocysteine and adenosine was established by SHIMIZU and YAMADA (1984). The system uses *Alcaligenes faecalis* and *P. striata.* The former shows a high activity of *S*-adenosyl-L-homocysteine hydrolase, which catalyzes the condensation of L-homocysteine and adenosine to produce *S*-adenosyl-L-homocysteine. Since the amino acid racemase with low substrate specificity from the latter bacterial cells converts unreacted D-homocysteine into L-homocysteine, almost all DL-homocysteine is converted to *S*-adenosyl-L-homocysteine. A mixture of both bacterial cells was used to produce $70\ g\ L^{-1}$ of *S*-adenosyl-L-homocysteine with a molar yield of nearly 100%. Since various adenosine analogs also serve as substrates for *S*-adenosyl-L-homocysteine hydrolase, the corresponding *S*-nucleotidyl-L-homocysteines can be synthesized by this method (SHIMIZU et al., 1984).

### 3.8.3 Glutamate Racemase

Glutamate racemase (EC 5.1.1.3) catalyzes racemization of L- and D-glutamate. In distinction to alanine racemase and amino acid racemase with low substrate specificity, glutamate racemase does not require pyridoxal-P for its activity. In lactic acid bacteria, D-glutamate, a component of the peptidoglycan of the cell wall, is produced directly from L-glutamate through a glutamate racemase reaction (TANAKA et al., 1961; DIVEN, 1969). The enzyme activity has been found in *Pediococcus, Lactobacillus,* and *Leuconostoc* sp. (NAKAJIMA et al., 1988a).

The enzyme from *Pediococcus pentosaceus* IFO 3182 is composed of a single polypeptide chain with a molecular mass of about 29,000 Da. The enzyme acts specifically on glutamate, and the $k_m$ values for D- and L-glutamate are 14 mM and 10 mM, respectively. Other amino acids occurring in proteins and other glutamate analogs (homocysteate, $\alpha$-aminoadipate, glutamate $\gamma$-methyl ester, *N*-acetylglutamate, $\alpha$-hydroxyglutarate, and cysteine sulfinate) are not racemized. However, L-homocysteine sulfinate, a $\gamma$-sulfinate analog of glutamate, is racemized at a rate of about 10% of that of L-glutamate. The gene encoding this enzyme has been isolated, and its overexpression system has been developed.

The enzyme is useful for the production of various D-amino acids when its reaction is coupled with that of D-amino acid aminotransferase, as described above. When D-glutamate is used as an amino donor for the D-amino acid aminotransferase reaction, $\alpha$-ketoglutarate produced in this reaction is converted into L-glutamate by L-glutamate dehydrogenase, and D-glutamate is regenerated from L-glutamate by the action of glutamate racemase (NAKAJIMA et al., 1988b).

### 3.8.4 Aspartate Racemase

Aspartate racemase (EC 5.1.1.13) catalyzes the racemization of L- and D-aspartate. Aspartate racemase as well as glutamate racemase require no cofactors for their activity. D-Aspartate occurring in the peptidoglycan of the bacterial cell wall is formed from L-aspartate through the reaction of aspartate racemase (JOHNSTON and DIVEN, 1969). The enzyme has been found in various *Lactobacillus* and *Streptococcus* strains (OKADA et al., 1991). The enzyme from *Streptococcus thermophilus* IAM10064 is composed of two subunits with identical molecuar mass (about 28,000 Da). The $k_m$ values for L- and D-aspartate are 35 mM and 8.7 mM, respectively. Cysteate and cysteine sulfinate are racemized at a rate of 88% and 51%, respectively, of that of L-aspartate (YAMAUCHI et al., 1992). However, neither asparagine nor cysteine, serine, alanine, and glutamate are racemized. Aspartate racemase is used for the production of D-alanine from fumarate: aspartase and D-amino acid aminotransferase reactions are coupled with an aspartate racemase reaction, as described above (Fig. 11) (KUMAGAI, 1994).

## 3.9 Amino Acid Dehydrogenases

### 3.9.1 Leucine Dehydrogenase

Leucine dehydrogenase (EC 1.4.1.9) catalyzes the reversible deamination of L-leucine to $\alpha$-ketoisocaproate, and occurs mainly in *Bacillus* sp. (SODA et al., 1971; SCHUTTE et al., 1985; OHSHIMA et al., 1985a). The enzyme is also present in other species such as *Clostridium thermoaceticum* (SHIMOI et al., 1987) and *Thermoactinomyces intermedius* (OHSHIMA et al., 1994) (Eq. 3).

SODA et al. (1971) have crystallized the enzyme from *B. sphaericus* (IFO 3525), which produces the enzyme abundantly. The enzyme has a molecular mass of about 245,000 Da and consists of six identical subunits with a molecular mass of 41,000 Da each (OHSHIMA et al., 1978). In addition to branched-chain L-amino acids, which are the preferred substrates, straight-chain aliphatic L-amino acids also are effectively deaminated (Tab. 10) (OHSHIMA et al., 1978). All the $\alpha$-keto acids

$$(H_3C)_2CH-CH_2-CH(NH_2)-COOH + NAD^+ + H_2O \rightleftharpoons (H_3C)_2CH-CH_2-C(=O)-COOH + NH_3 + NADH + H^+ \qquad (3)$$

**Tab. 10.** Substrate Specificity for the Oxidative Deamination Catalyzed by Leucine Dehydrogenase from *B. sphaericus*

| Substrate | Relative Activity | $k_m$ [mM] |
|---|---|---|
| L-Leucine | 100 | 1.0 |
| L-Valine | 74 | 1.7 |
| L-Isoleucine | 58 | 1.8 |
| L-Norvaline | 41 | 3.5 |
| L-α-Aminobutyrate | 14 | 10.0 |
| L-Norleucine | 10 | 6.3 |
| γ-Methylallylglycine | 8.2 | |
| *tert*-DL-Leucine | 1.6 | |
| *S*-Methyl-L-cysteine | 1.4 | |
| 4-Azaleucine | 0.6 | |
| L-Methionine | 0.6 | |
| L-Cysteine | 0.3 | |
| L-Penicillamine | 0.2 | |
| *S*-Ethyl-L-cysteine | 0.1 | |

Inert: cycloleucine, L-cysteic acid, *S*-methyl-L-methionine, L-alanine, L-glutamate, L-threonine, L-serine, glycine, L-phenylalanine, L-ethionine, L-lysine, L-arginine, D-leucine, D-valine, ε-amino-*n*-caproate, 7-aminoheptanoate, β-alanine, β-aminovalerate.

**Tab. 11.** Substrate Specificity for the Reductive Amination Catalyzed by Leucine Dehydrogenase from *B. sphaericus*

| Substrate | Relative Activity | $k_m$ [mM] |
|---|---|---|
| α-Ketoisocaproate | 100 | 0.31 |
| α-Ketoisovalerate | 126 | 1.4 |
| α-Ketovalerate | 76 | 1.7 |
| α-Ketobutyrate | 57 | 7.7 |
| α-Ketocaproate | 46 | 7.0 |

Inert: pyruvate, α-ketoglutarate, phenylpyruvate, oxalacetate, glyoxylate.

resulting from the oxidative deamination reactions serve as good substrates for the amination reaction (Tab. 11). Ammonia is the exclusive substrate for the reductive amination of α-ketoisocaproate; hydroxyamine, methylamine, and ethylamine are inert. The enzyme requires $NAD^+$ as a natural coenzyme for the oxidative deamination of L-leucine, whereas $NADP^+$ is inert. A kinetic analysis was performed to elucidate the reaction mechanism for the leucine dehydrogenase reaction (OHSHIMA et al., 1978). The reductive amination proceeds through a sequentially ordered ternary-binary mechanism. NADH binds first to the enzyme, followed by α-ketoisocaproate and ammonia, and the products are released in the order of L-leucine and $NAD^+$. The *pro-S* hydrogen at C-4 of the dihydronicotinamide ring of NADH is exclusively transferred to the substrate by the enzyme without exchange of protons with the medium (OHSHIMA et al., 1978).

Leucine dehydrogenase has also been purified to homogeneity from a moderate thermophilic bacterium, *B. stearothermophilus* (OSHIMA et al., 1985a). The enzyme has a molecular mass of about 300,000 Da, and consists of six subunits with identical molecular mass of 49,000 Da. The enzyme of *B. stearothermophilus* is more stable than the *B. sphaericus* enzyme: it does not lose its activity by heat treatment at 70°C for 20 min and incubation in the pH range of 5.5–10.0 at 55 °C for 5 min. The *B. stearothermophilus* enzyme is stable in a wider range (pH 5.5–10.0) than that of *B. sphaericus* (pH 7–9). It is resistant to detergent and ethanol treatment. However, both enzymes have similar enzymological properties, except their stability. Thus, the *B. stearothermophilus* enzyme is preferred as a reagent.

The gene for leucine dehydrogenase from *B. stearothermophilus* has been cloned and sequenced (NAGATA et al., 1988). The open reading frame consists of 1,287 base pairs and encodes 429 amino acid residues (Mr 46,903). There is significant similarity between the amino acid sequence of leucine dehydrogenase and that of other pyridine nucleotide-dependent oxidoreductases, such as glutamate dehydrogenase, in the regions containing the coenzyme-binding domain and certain specific residues with catalytic importance (NAGATA et al., 1988). The ε-amino group of Lys80 is supposed to participate in catalysis as a general base, assisting the nucleophilic attack of a water molecule to the substrate α-carbon atom (SEKIMOTO et al., 1993). Lys68 is predicted to be located at the active site of the enzyme and involved in binding of the α-carboxyl group of the substrate through an ionic interaction (SEKIMOTO et al., 1994). The gene cloned into *E. coli* MV1184 with a vector plasmid, pUC119, was efficiently expressed, and the cells produced a large amount of the thermostable enzyme, which corresponds to about 60% of the total soluble protein (OKA et al., 1989). The enzyme was purified to more than 95% homogeneity by only one step, heat treatment of the cell extracts, with an average yield of 75 mg $g^{-1}$ of wet cells (obtained from 100 mL of the culture).

WANDREY and coworkers have established an enzymatic procedure for the continuous production of L-leucine from the corresponding keto analogs using a membrane reactor (WHICHMANN and WANDREY, 1981). The reactor consists of an ultrafiltration membrane, with leucine dehydrogenase, formate dehydrogenase of yeast (EC 1.2.1.43), and NADH covalently bound to polyethylene glycol. $NAD^+$ bound to polyethylene glycol is produced in the reductive amination of $\alpha$-ketoisocaproate, and is again reduced by the formate dehydrogenase reaction (Fig. 12). Thus $\alpha$-ketoisocaproate is effectively aminated to L-leucine with the simultaneous oxidation of formate to $CO_2$. The use of the thermostable enzyme from *B. stearothermophilus* is valuable in the membrane reactor because of its longer useful life (OSHIMA et al., 1985b). The average conversion rate was 87% with operation for 29 d, and more than 90% during the first 2 weeks. This value is higher than that obtained in the system containing the *B. sphaericus* enzyme (77%).

It has also been reported that L-leucine, L-valine, and L-isoleucine are produced from ammonia and the corresponding $\alpha$-keto acids with semipermeable nylon-polyethyleneimine artificial cells containing leucine dehydrogenase, alcohol dehydrogenase, and dextran-$NAD^+$ (GU and CHANG, 1990). In this system, the recycling of dextran-$NAD^+$ is achieved by alcohol dehydrogenase. When urease is encapsulated in addition to these enzymes, urea can be used instead of ammonia.

Another co-immobilization method, a "droplet gel-entrapping method", has been developed for the co-immobilization of leucine dehydrogenase and formate dehydrogenase (KAJIWARA and MAEDA, 1987). In this technique, these enzymes are freeze-dried with bovine serum albumin, dextrin, and stabilizers. The powder thus obtained is dispersed in methylcellosolve containing polyethyleneglycol(#4000)diacrylate, *N,N'*-methylenebisacrylamide, and 2-hydroxyethylacrylate, and the suspension is gelled with initiators. After cutting up the gel, the pieces are washed with buffer to remove the methylcellosolve and the dextrin inside. As a result, many droplets which retain leucine dehydrogenase, formate dehydrogenase, and bovine serum albumin are enclosed in the gel, and the diffused substrate is converted into the product in the droplets.

ESAKI et al. (1989b) found that leucine dehydrogenase from *B. stearothermophilus* acts on L-selenomethionine and its $\alpha$-keto analog, and established an efficient method for enantioselective synthesis of L-selenomethionine with the enzyme. L-Selenomethionine was produced in a 95% yield from $\alpha$-keto-$\gamma$-me-

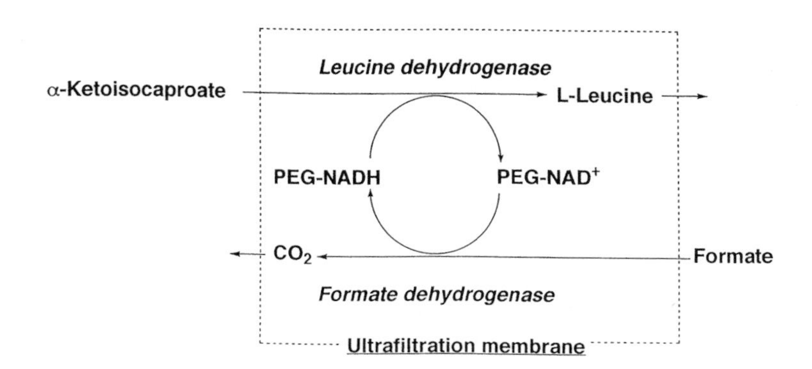


**Fig. 12.** Enzymatic synthesis of L-leucine form $\alpha$-ketoisocaproate and formate with leucine dehydrogenase, formate dehydrogenase, and $NAD^+$ covalently bound to polyethylene glycol.

thylselenobutyrate and ammonia by means of an NADH regeneration system with formate dehydrogenase.

HANSON et al. (1990) have shown that leucine dehydrogenase from *B. sphaericus* is useful for the production of L-$\beta$-hydroxyvaline, which is a key intermediate needed for the synthesis of tigemonam, an orally active monobactam antibiotic. $\alpha$-Keto-$\beta$-hydroxyisovalerate is converted into L-$\beta$-hydroxyvaline by reaction with NADH and ammonia catalyzed by leucine dehydrogenase. An apparent $k_m$ of 11.5 mM for $\alpha$-keto-$\beta$-hydroxyisovalerate is sufficiently low, and the apparent $v_{max}$ is 41% of the value for $\alpha$-ketoisovalerate, which is reported to be the best substrate for reductive amination.

**Tab. 12.** Substrate Specificity for the Oxidative Deamination Catalyzed by Alanine Dehydrogenase from *B. sphaericus*

| Substrate | Relative Activity | $k_m$ [mM] |
|---|---|---|
| L-Alanine | 100 | 18.9 |
| L-$\alpha$-Aminobutyrate | 2.2 | 330 |
| L-Serine | 1.4 | 39 |
| L-Norvaline | 0.14 | 14 |
| L-Valine | 0.07 | 20 |

Inert: L-threonine, $\beta$-alanine, L-cysteine, glycine, L-arginine, L-proline, L-methionine, L-lysine, L-histidine, L-isoleucine, L-phenylalanine, L-leucine, L-glutamate, L-aspartate, L-tryptophan, D-alanine, D-serine.

## 3.9.2 Alanine Dehydrogenase

Alanine dehydrogenase (EC 1.4.1.1) catalyzes the reversible deamination of L-alanine to pyruvate and has been found in vegetative cells, spores of various *Bacillus* sp. (YOSHIDA and FREESE, 1964; OHSHIMA and SODA 1979; PORUMB et al., 1987), and in some other bacteria (VALI et al., 1980; ITOH and MORIKAWA, 1983; BELLION and TAN, 1987).

$$H_3C-CH(NH_2)-COOH + NAD^+ + H_2O \rightleftharpoons H_3C-C(=O)-COOH + NH_3 + NADH + H^+ \quad (4)$$

It has been crystallized from *B. sphaericus* (IFO 3525) and characterized (OHSHIMA and SODA, 1979). The enzyme has a molecular mass of about 230,000 Da, and is composed of six identical subunits. It acts almost specifically on L-alanine, but shows comparatively low amino acceptor specificity (Tabs. 12 and 13). The reductive amination proceeds through a sequentially ordered ternary-binary mechanism. However, the order in which the substrates are bound to the enzyme is different from that of leucine dehydrogenase: NADH binds first to the enzyme followed by ammonia and pyruvate, and the products are released in the order of L-alanine and NAD$^+$. Alanine dehydrogenase shows the opposite stereospecificity to leucine dehydrogenase; the *pro-R* hydrogen at C-4 of the reduced nicotinamide ring of NADH is exclusively transferred to pyruvate (OHSHIMA and SODA, 1979).

**Tab. 13.** Substrate Specificity for the Reductive Amination Catalyzed by Alanine Dehydrogenase from *B. sphaericus*

| Substrate | Relative Activity | $k_m$ [mM] |
|---|---|---|
| Pyruvate | 100 | 1.7 |
| $\alpha$-Ketobutyrate | 103 | 11 |
| $\alpha$-Ketovalerate | 20.2 | 23 |
| $\beta$-Hydroxypyruvate | 3.8 | 80 |
| Glyoxylate | 3.4 | 12 |
| $\alpha$-Ketoisovalerate | 0.6 | 11 |

Inert: $\alpha$-ketoglutarate, $\beta$-phenylpyruvate, oxalacetate.

Alanine dehydrogenase from a moderate thermophile, *B. stearothermophilus,* has a molecular mass of about 240,000 Da and consists of six subunits of identical molecular mass (40,000 Da) (SAKAMOTO et al., 1990). The enzyme is not inactivated by heat treatment: at pH 7.2 and 75 °C for 30 min and at 55 °C and various pHs between 6.0 and 11.5 for 10 min. The enzymological properties are very similar to those of the mesophilic *B.*

*sphaericus* enzyme, except for thermostability. The *B. stearothermophilus* enzyme retained about 50% of its initial activity when heated at 85 °C for 5 min in 50 mM potassium phosphate buffer (pH 7.2), whereas the *B. sphaericus* enzyme lost the same activity when heated at only 65 °C for 5 min.

The genes encoding alanine dehydrogenases from *B. sphaericus* and *B. stearothermophilus* were cloned, and their complete DNA sequences were determined (KURODA et al., 1990). Each alanine dehydrogenase gene consists of a 1116-bp open reading frame and encodes 372 amino acid residues corresponding to the subunit (Mr 39,500–40,000) of the hexameric enzyme. The similarity of the amino acid sequence between the two alanine dehydrogenases is very high (>70%). There is significant similarity between the primary structures of both alanine dehydrogenases and those of other pyridine nucleotide-dependent oxidoreductases in the region containing the coenzyme-binding domain. Several catalytically important residues in lactate and malate dehydrogenases are conserved in the primary structure of alanine dehydrogenase at matched positions with similar mutual distances. The gene of *B. stearothermophilus* is efficiently expressed in *E. coli,* and a rapid and effective procedure for purification of the enzyme has been established based on its thermostability and overproduction by the clone cells (SAKAMOTO et al., 1990).

L-Alanine is produced continuously from pyruvate with a membrane reactor containing alanine dehydrogenase, formate dehydrogenase, and NADH covalently bound to polyethylene glycol (FIOLITAKIS and WANDREY, 1982). Pyruvate is effectively aminated to L-alanine with the simultaneous oxidation of formate to $CO_2$. The same method is applicable to the continuous production of 3-fluoro-L-alanine (OHSHIMA et al., 1989). 3-Fluoro-L-alanine is selectively and continuously produced from 3-fluoropyruvate and ammonium formate in the enzyme membrane reactor.

Amino acids labeled with $^{15}N$ are useful in investigating protein balance in human medicine because the patients are not exposed to the hazards associated with the administration of radioactive tracers. An enzymatic procedure for the preparation of L-[$^{15}N$]alanine has also been established by means of alanine dehydrogenase (MOCANU et al., 1982). The procedure is based on the coupling of alanine dehydrogenase and alcohol dehydrogenase reactions. Pyruvate and $^{15}NH_4^+$ are converted into L-[$^{15}N$]alanine by alanine dehydrogenase, and NADH consumed in this reaction is regenerated by alcohol dehydrogenase.

### 3.9.3 Phenylalanine Dehydrogenase

Phenylalanine dehydrogenase (EC 1.4.1.20) catalyzes the reversible oxidative deamination of L-phenylalanine using $NAD^+$ as a coenzyme. The enzyme has been found in a *Brevibacterium* sp. (HUMMEL et al., 1984), in *B. sphaericus* (OKAZAKI et al., 1988), *Rhodococcus* sp. (BRUNHUBER et al., 1994), *Thermoactinomyces intermedius* (TAKADA et al., 1991), and others.

$$C_6H_5{-}CH_2{-}CH(NH_2){-}COOH + NAD^+ + H_2O \rightleftharpoons C_6H_5{-}CH_2{-}C({=}O){-}COOH + NH_3 + NADH + H^+ \quad (5)$$

The enzyme from *T. intermedius* consists of six subunits of identical molecular mass (41,000 Da). The enzyme is highly thermostable: it is not inactivated by incubation at pH 7.2 and 70 °C for at least 60 min, and in the range of pH 5–10.8 at 50 °C for 10 min. The enzyme acts exclusively on L-phenylalanine and its $\alpha$-keto analog, phenylpyruvate, in the presence of $NAD^+$ and NADH, respectively. The $k_m$ values for L-phenylalanine, $NAD^+$, phenylpyruvate, NADH, and ammonia were 0.22, 0.078, 0.045, 0.025, and 106 mM, respectively. The enzyme shows the same stereospecificity to leucine dehydrogenase; the *pro-S* hydrogen at C-4 of the reduced nicotinamide ring of NADH is exclusively transferred to the substrate.

The gene encoding phenylalanine dehydrogenase of *T. intermedius* has been cloned in *E. coli* and sequenced. It consists of 1,098 nucleotides corresponding to 366 amino acid re-

**Tab. 14.** Synthesis of L-Amino Acids with Phenylalanine Dehydrogenase

| Substrate | Product | Yield [%] |
|---|---|---|
| Phenylpyruvate | L-Phenylalanine | 98 |
| 4-Hydroxyphenylpyruvate | L-Tyrosine | 99 |
| Indole-3-pyruvate | L-Tryptophan | 11 |
| $\alpha$-Keto-$\gamma$-methylthiobutyrate | L-Methionine | 87 |
| $\alpha$-Ketoisovalerate | L-Valine | 97 |
| $\alpha$-Ketoisocaproate | L-Leucine | 83 |
| DL-$\alpha$-Keto-$\beta$-methyl-$n$-valerate | L-Isoleucine | 48 |
| | *allo*-Isoleucine | 50 |

sidues. The predicted molecular mass is 40,488 Da. The deduced amino acid sequence was 56.0% and 42.1%, respectively, homologous to those of the enzymes of the mesophiles *B. sphaericus* and *Sporosarcina ureae*. There is a significant sequence similarity between phenylalanine dehydrogenase from *T. intermedius* and leucine dehydrogenase from *B. stearothermophilus:* the overall similarity is 47.0%.

Phenylalanine dehydrogenase is useful as a catalyst for the synthesis of L-phenylalanine and related L-amino acids from their $\alpha$-keto analogs (HUMMEL et al., 1986). The enzymatic synthesis of L-phenylalanine with transaminase has also been reported, but it has a drawback in that an $\alpha$-keto acid is formed as a by-product (CALTON et al., 1986). In contrast, the synthesis with phenylalanine dehydrogenase does not yield any $\alpha$-keto acid as a by-product. Furthermore, it does not require any other L-amino acid as an amino donor.

An enzyme membrane reactor system for the continuous production of L-phenylalanine from phenylpyruvate has been developed. A space time yield of 37.4 g $L^{-1}$ $d^{-1}$ L-phenylalanine can be achieved with L-phenylalanine dehydrogenase from *Brevibacterium* sp. (HUMMEL et al., 1986). ASANO and NAKAZAWA (1987) utilized phenylalanine dehydrogenase from *S. ureae* and formate dehydrogenase from *Candida boidinii,* and prepared several aromatic and hydrophobic aliphatic L-amino acids from their keto analogs (Tab. 14). Phenylalanine dehydrogenase from *B. sphaericus* is active toward 3-substituted pyruvate with bulky substituents. Optically pure L-phenylalanine and other amino acids, such as *S*-2-amino-4-phenylbutyrate and *S*-2-amino-5-phenylvalerate, were quantitatively synthesized from their keto analogs by using phenylalanine dehydrogenase from *B. sphaericus,* with a regeneration of NADH by formate dehydrogenase (ASANO et al., 1990).

## 3.10 Glutamine Synthetase

L-Glutamate:ammonia ligase ("glutamine synthetase", EC 6.3.1.2) catalyzes the following reaction:

$$\mathrm{HOOC{-}\underset{H_2}{C}{-}\underset{H_2}{C}{-}\underset{\underset{NH_2}{|}}{CH}{-}COOH + ATP + NH_3} \rightleftharpoons$$

$$\mathrm{NH_2{-}\underset{\underset{O}{\|}}{C}{-}\underset{H_2}{C}{-}\underset{H_2}{C}{-}\underset{\underset{NH_2}{|}}{CH}{-}COOH + ADP + Pi + H_2O} \quad (6)$$

It plays a key role not only in the biosynthesis of numerous compounds the nitrogen atoms of which are derived from glutamine, but also in the assimilation of ammonia by coupling with glutamate synthase (EC 1.4.1.13–14).

$$\text{2 Glutamate} + \text{NAD(P)}^+ \rightleftharpoons \text{Glutamine} + \alpha\text{-Ketoglutarate} + \text{NAD(P)H} + \text{H}^+ \quad (7)$$

Glutamine synthetase has been found in various organisms including mammals, plants, yeasts, and bacteria, and their primary structures have been analyzed (COLOMBO and VILLAFRANCA, 1986; TINGEY et al., 1987; MINEHART and MAGASANIK, 1992; KUO and DARNELL, 1989).

TACHIKI et al. (1981a, b) have developed an enzymatic procedure to produce L-glutamine by coupling the glutamine synthetase reaction to a sugar fermentation system of yeast. The synthesis of glutamine is driven by the energy released during fermentation. They have termed the procedure "Coupled Fermentation with Energy Transfer", and have applied it to the production of various nucleic acid-related compounds, such as nucleoside triphosphates, sugar nucleotides, cytidine coenzymes, and sugar phosphates. In this procedure, fructose 1,6-bisphosphate is accumulated as a result of ethanol fermentation in the first step. The degradation of fructose 1,6-bisphosphate is coupled to an endergonic reaction through an ATP–ADP system in the second step: ATP is produced during the degradation of fructose 1,6-bisphosphate and utilized in the endergonic reaction. Any endergonic reaction of baker's yeast or another microorganism can be used. TACHIKI et al. (1981a) have established the optimal conditions for L-glutamine production, which give 22 mmol $L^{-1}$ of L-glutamine in a reaction system composed of cell-free extracts of baker's yeast and glutamine synthetase from *Gluconobacter suboxydans,* with a molar yield of 92% based on glutamate. Glucose, glutamate, and ammonium ions are added to the reaction mixture as substrates. When resting cells of baker's yeast are substituted for the cell-free extract, L-glutamine is produced with an 80% yield based on glutamate (TACHIKI et al., 1981b).

# 4 Enzymatic Resolution of Racemic Amino Acids

## 4.1 Basic Aspects

The chemical synthesis of amino acids has made rapid and remarkable advances, and in fact several racemic amino acids have been chemically synthesized in industry. However, in spite of intensive investigations, only a few chiral amino acids were successfully synthesized from cheaper chirons. Therefore, the optical resolution of racemic amino acids is of industrial importance. Several methods have been developed for the preparation of amino acid isomers from the corresponding racemates; they may be classified into four categories: physicochemical methods, chemical methods, enzymatic methods, and microbial methods (CHIBATA, 1974).

The physicochemical methods include mechanical separation, selective crystallization, replacing crystallization, and chromatographic separation. By selective crystallization, e.g., the desired enantiomers are exclusively crystallized out of a supersaturated solution of the racemic amino acid by seeding with the crystalline enantiomer. Alternatively, they are crystallized out of a supersaturated solution containing the desired enantiomer in excess by preferential crystallization. Enantiomers of various amino acids are effectively separated by chromatography with chiral elutants (GIL-AV et al., 1980) and chiral ligands (DAVANKOV and SEMECHKIN, 1977), and also after their conversion to diastereomeric dipeptides (MANNING and MOORE, 1968). These chromatographic procedures are not suitable for industrial resolution, but for stereoselective analyses. Another example of the physicochemical method is ultrafiltration with a membrane prepared by introducing a chiron into a support membrane matrix (MASAWAKI et al., 1992), however, this method is not yet practical.

In the chemical resolution, racemic amino acids or their derivatives are treated with chiral acids or bases, and one of the diastereoisomeric salts formed is fractionally crystallized, as a result of the difference in solubility between the two diastereoisomeric salts. The less soluble salt is generally crystallized and purified. Alternatively, racemic amino acids are condensed with cheaper chirons, e.g., D-(+)-galactose, and one of the resultant diastereomers is fractionally crystallized. An optically pure amino acid is obtained from the diastereomer by cleavage of the chemical bond formed during condensation. The microbial method employs the observation, first made by PASTEUR, that certain microorganisms (yeasts, fungi, e.g., *Aspergillus niger,* and bacteria, e.g., *E. coli*), as a result of the specificity of their oxidases, decarboxylases and other stereospecific enzymes, metabolize only one enantiomer and leave the other, usually the D-enantiomer, unchanged. This method is not economical and requires a tedious purification of the isomer obtained from the impurities produced microbially.

## 4.2 Enzymatic Methods

At present the most useful and convenient methods for the resolution of amino acids are those employing stereospecific enzymes. The enzymatic resolution of racemic amino acids is outlined here. Additional methods have been extensively described by CHIBATA (1974) and NEWMAN (1981).

### 4.2.1 Resolution by Enzymatic Asymmetric Derivatization

The enzymatic methods of optical resolution rely on the structural and stereochemical substrate specificity of enzymes. There are two different methods for the resolution of amino acids. One is based on the asymmetric derivatization of amino acids with enzymes, usually papain, bromelain, and ficin, which catalyze the specific formation of amides of L-amino acids (DOHERTY and POPENOE, 1951; ALBERTSON, 1951). Incubation of *N*-acyl-DL-amino acids with aniline in the presence of papain, e.g., results in the formation of insoluble *N*-acyl-L-amino acid anilides, whereas the *N*-acyl-D-amino acids remain in solution. *N*-Acyl-L-amino acid anilides thus purified are hydrolyzed chemically to give the L-amino acids. The *N*-acyl-D-amino acids are separately hydrolyzed to D-amino acids (Fig. 13).

These methods are generally less useful for industrial application than the methods with amino acylase and amidase as described below.

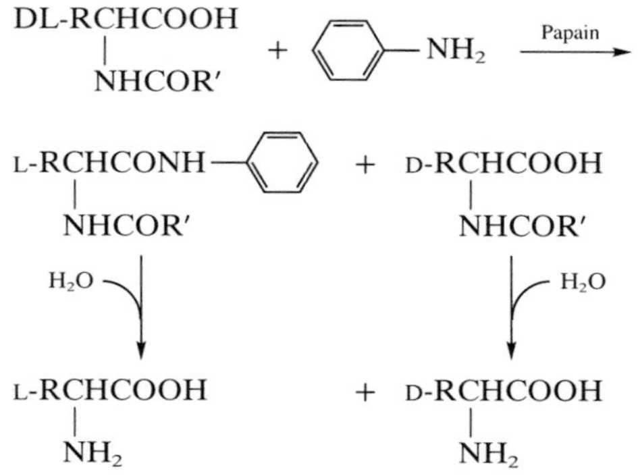

DL-RCH(NHCOR')COOH + $C_6H_5NH_2$ —Papain→ L-RCH(NHCOR')CONH–$C_6H_5$ + D-RCH(NHCOR')COOH

L-RCH(NHCOR')CONH–$C_6H_5$ —$H_2O$→ L-RCH($NH_2$)COOH

D-RCH(NHCOR')COOH —$H_2O$→ D-RCH($NH_2$)COOH

**Fig. 13.** Optical resolution of amino acids with papain.

### 4.2.2 Asymmetric Hydrolysis

The other enzymatic methods are based on the stereospecific hydrolysis of amino acid derivatives generating one enantiomer of the amino acid.

#### 4.2.2.1 Esterase Method

Various amino acid esters are hydrolyzed stereospecifically by chymotrypsin and other proteolytic enzymes which show esterase activity (Fig. 14).

The L-amino acids can be separated from the unreacted D-amino acid esters as they

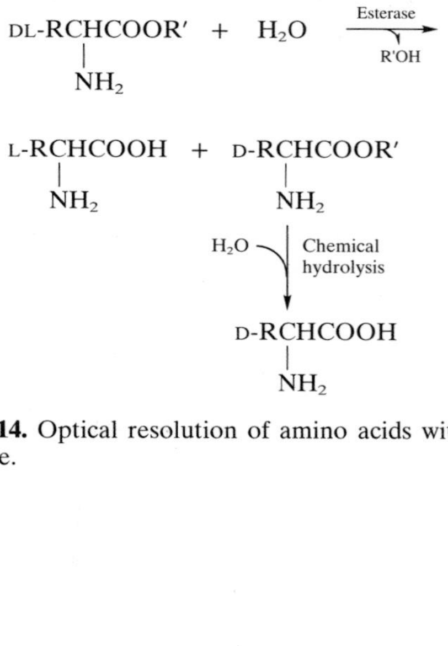

DL-RCH($NH_2$)COOR' + $H_2O$ —Esterase (− R'OH)→ L-RCH($NH_2$)COOH + D-RCH($NH_2$)COOR'

D-RCH($NH_2$)COOR' —$H_2O$, Chemical hydrolysis→ D-RCH($NH_2$)COOH

**Fig. 14.** Optical resolution of amino acids with esterase.

have different solubilities in solvents. Amino acid esters may be split by spontaneous hydrolysis which leads to a decrease in the optical purity of L-amino acids produced by this method.

#### 4.2.2.2 Amidase Method

L-Amino acid amides are hydrolyzed stereospecifically by amino acid amidases from mammalian kidney and pancreas, and from various microorganisms. L-Amino acids thus formed are separated from the unreacted D-amino acid amides by the differences in solubilities in various solvents (Fig. 15).

Since amino acid amides are less susceptible to non-enzymatic hydrolysis than the corresponding esters, amino acids of higher optical purity can be produced by this method.

#### 4.2.2.3 Aminoacylase Method

*N*-Acylamino acids are usually racemized much more readily than the corresponding free amino acids. Therefore, by combination of chemical racemization and enantioselective hydrolysis of *N*-acylamino acids, racemates of *N*-acylamino acids can be totally converted to the desired enantiomer of free amino acids according to the stereospecificity of the aminoacylases used (Fig. 16).

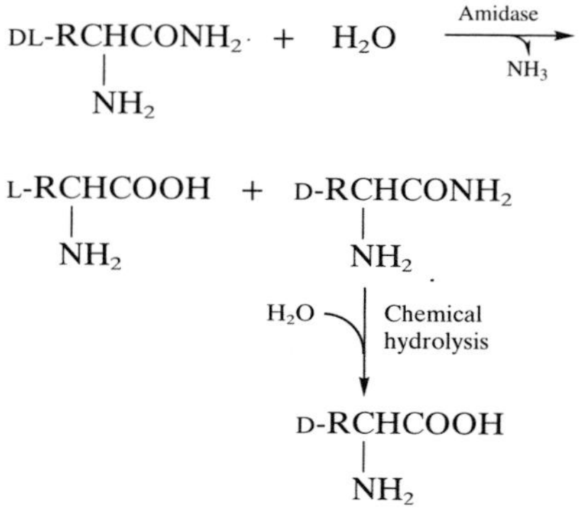

DL-RCH(NH$_2$)CONH$_2$ + $H_2O$ —(Amidase, −$NH_3$)→

L-RCH(NH$_2$)COOH + D-RCH(NH$_2$)CONH$_2$

D-RCH(NH$_2$)CONH$_2$ —($H_2O$, Chemical hydrolysis)→ D-RCH(NH$_2$)COOH

**Fig. 15.** Optical resolution of amino acids with amidase.

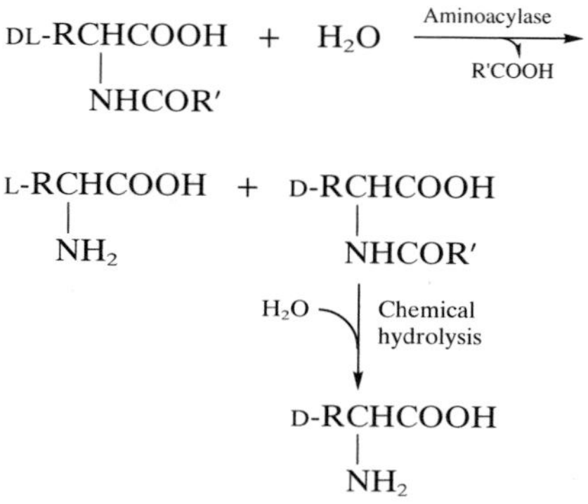

DL-RCH(NHCOR′)COOH + $H_2O$ —(Aminoacylase, −R′COOH)→

L-RCH(NH$_2$)COOH + D-RCH(NHCOR′)COOH

D-RCH(NHCOR′)COOH —($H_2O$, Chemical hydrolysis)→ D-RCH(NH$_2$)COOH

**Fig. 16.** Optical resolution of amino acids with aminoacylase.

L-Aminoacylases catalyze the hydrolysis of the amide bond of various *N*-acyl-L-amino acids, such as *N*-acetyl-, *N*-chloroacetyl-, and *N*-propionyl-L-amino acids. GREENSTEIN (1957) first studied the reactivity of a pig kidney enzyme and showed its application to optical resolution of racemic amino acids. CHIBATA et al. (1957) found that L-aminoacylase is produced abundantly by fungal species belonging to the genera *Aspergillus* and *Penicillium*. L-Aminoacylases were purified from pig kidney and *A. oryzae,* and their reaction mechanism and physiological function were studied (GENTZEN et al., 1980; KÖRDEL and SCHNEIDER, 1976; GALACV and SVEDAS, 1982; RÖHM and ETTEN, 1986). CHO et al. (1987) showed that various thermophilic *Bacillus* strains produce thermostable L-aminoacylase, and purified it to homogeneity from *B. thermoglucosidius* DSM 2542, which produces the enzyme most abundantly. L-Aminoacylases of pig kidney, *A. oryzae,* and *B. thermoglucosidius* share many features: they contain $Zn^{2+}$ as a prosthetic metal, they are strongly activated by $Co^{2+}$ and have a pH optimum in the range of 8.0–8.5. The substrate specificities of these enzymes are summarized in Tab. 16. The fungal enzyme effectively catalyzes the hydrolysis of the *N*-acyl derivatives of aromatic L-amino acids and has been used industrially for the resolution of tryptophan and phenylalanine racemates. When *N*-acetyl-DL-tryptophan is incubated with the fungal aminoacylase, *N*-acetyl-L-tryptophan is

**Tab. 15.** Substrate Specificity of Aminoacylase

| Substrate | Relative Rate of Aminoacylase from | | |
|---|---|---|---|
| | *A. oryzae* | Pig Kidney | *B. thermoglucosidius* |
| Chloroacetylglycine | – | – | 640 |
| Chloroacetylalanine | 100 | 100 | – |
| Chloroacetyl-DL-alanine | – | – | 620 |
| Chloroacetyl-DL-valine | – | – | 570 |
| Chloroacetyl-DL-serine | – | – | 440 |
| Chloroacetylmethionine | 400 | 480 | – |
| Chloroacetylnorleucine | 207 | 120 | – |
| Chloroacetyl-DL-norleucine | – | – | 640 |
| Chloroacetylleucine | 26 | 96 | 370 |
| Chloroacetylphenylalanine | 325 | 5 | – |
| Chloroacetyl-DL-phenylalanine | – | – | 52 |
| Chloroacetyltryptophan | 125 | 0 | – |
| Chloroacetyltyrosine | – | – | 230 |
| Acetylmethionine | – | – | 100 |
| Acetylglutamic acid | 0 | 21 | – |
| Acetylaspartic acid | 0 | 0 | – |
| Acetylglutamine | 13 | 4 | 27 |
| Acetylalanine | 14 | 7 | 78 |
| Acetyllysine | 3 | 0 | – |
| Acetly-DL-norleucine | – | – | 70 |
| Acetylvaline | – | – | 69 |
| Acetyl-DL-serine | – | – | 28 |
| Acetylglycine | – | – | 18 |
| Acetyltyrosine | – | – | 14 |
| Acetylleucine | – | – | 12 |
| Acetylphenylalanine | – | – | 3 |
| Dichloroacetylglycine | 0 | 1 | – |
| Dichloroacetylleucine | 0 | 3 | – |
| Dichloroacetylnorleucine | 4 | 69 | – |
| Dichloroacetylalanine | 0.7 | 2 | – |

selectively hydrolyzed to L-tryptophan, which is then crystallized from the solution. *N*-Acetyl-D-tryptophan in the mother liquor is racemized with acetic anhydride, and the racemate is again used as starting material. In principle, D- and L-amino acids can be produced from their corresponding *N*-acyl derivatives in the same manner.

$\varepsilon$-*N*-Acyl-L-lysine aminoacylase ($\varepsilon$-lysine acylase, $N^6$-acyl-L-lysine amidohydrolase, EC 3.5.1.17) is found in rat kidney (PAIK et al., 1957), in a *Pseudomonas* sp. (KAMEDA et al., 1958; WADA, 1959), and in *Achromobacter pestifer* (CHIBATA et al., 1970). It catalyzes the hydrolysis of $\varepsilon$-*N*-acyl-L-lysine exclusively; the other lysine *N*-acyl derivatives, $\delta$-*N*-acyl-L-ornithine and $\zeta$-*N*-acyl-L-homolysine, are not substrates. $\varepsilon$-*N*-Acyl-L-lysine aminoacylase is more useful industrially for the optical resolution of DL-lysine than the usual fungal, bacterial, and mammalian aminoacylases. KIKUCHI et al. (1983) reported another L-amino acid acylase, *N*-acyl-L-proline acylase. The enzyme was temperature-labile (rapid inactivation at 50°C) and thus not used for a practical purpose. Recently, a new *N*-acyl-L-proline acylase was isolated and characterized from *Comamonas testosteroni* DSM 5416 (GROEGER et al., 1990). The enzyme does not only exhibit good pH stability (4 weeks at room temperature in the pH range of 5–10) but also remarkable temperature stability (30 min at 70°C and pH 7.5). It is highly specific for *N*-acyl-L-proline and *N*-acyl-L-thiazolidine-4-carboxylic acid. Other *N*-acyl-L-amino acids are almost inert as substrate. GROEGER et al. (1992) demonstrated the stereoselective syntheses of L-proline and *N*-alkyl-L-amino acids with the *N*-acyl-L-proline acylase. *N*-Acylated secondary amines such as *N*-acylated proline or *N*-acylated *N*-alkylamino acids cannot be hydrolyzed by the other L-aminoacylases.

SUGIE and SUZUKI (1980) demonstrated the occurrence of D-aminoacylase, which hydrolyzes specifically the amide bond of *N*-acyl-D-amino acids in actinomycetes, and applied the enzyme to the production of D-phenylglycine. Recently, a new D-aminopeptidase was found in *Alcaligenes denitrificans* and which acts on various *N*-acyl-D-amino acids including *N*-acetyl-D-methionine (TSAI et al., 1988; MORIGUCHI and IDETA, 1988).

*N*-Acyl derivatives of desired amino acids are racemized chemically without a major loss by decomposition, however, the chemical racemization can be achieved only under extreme conditions for aminoacylases to be inactivated, and the enzymes must usually be saved for the subsequent cycles for reasons of economy. Therefore, the antipode of the substrate is separated from the enzyme and preferably from the product in order to avoid its possible racemization. TOSA et al. (1966) have developed a continuous resolution method now practiced in industry by means of the *Aspergillus* L-aminoacylase immobilized on DEAE-Sephadex. Recently, TOKUYAMA et al. (1994a) succeeded in finding a racemase that acts on *N*-acylamino acids but not on the corresponding free amino acids, and named it acylamino acid racemase. They have established a method of producing optically active $\alpha$-amino acids from the corresponding DL-*N*-acylamino acids by means of the acylamino acid racemase and aminoacylases.

Acylamino acid racemase occurs widely in various actinomycete strains belonging to the genera of *Streptomyces, Actinomadura, Actinomyces, Jensenia, Amycolatopsis,* and *Sebekia* (TOKUYAMA et al., 1994a). The enzyme was purified to homogeneity from *Streptomyces* sp. Y-53 (TOKUYAMA et al., 1994b), which shows the highest enzyme activity among the strains tested. The enzyme is composed of subunits identical in molecular mass (about 40,000 Da) and has a molecular mass of 200,000 Da in the native state. The enzyme is stable for at least 30 min at 40°C, but is inactivated at higher temperatures. However, $Co^{2+}$ stabilizes the enzyme against heat inactivation. Other divalent cations such as $Mg^{2+}$, $Mn^{2+}$, $Zn^{2+}$, and $Ni^{2+}$ are also effective. The enzyme catalyzes the reversible interconversion between L- and D-*N*-acylamino acids. The *N*-acetyl, formyl, chloroacetyl, and benzoyl derivatives of various amino acids serve as substrates. *N*-acyl derivatives of aromatic amino acids are good substrates of the enzyme.

TAKAHASHI and HATANO (1989) found that most of the acylamino acid racemase-producing strains produce not only acylamino acid racemase but also aminoacylases – one of either D- or L-aminoacylase or both of them. Moreover, acylamino acid racemase shows the optimum pH around 8.0, which is close to that of aminoacylases. Therefore, *N*-acylamino acid can be converted as a whole into L- and D-amino acids in one step by means of microbial cells of appropriate strains producing either L- or D-aminoacylase in addition to acylamino acid racemase.

# 5 References


AHMED, S. A., ESAKI, N., TANAKA, H., SODA, K. (1983a), Racemization of $\alpha$-amino-$\delta$-valerolactam catalyzed by $\alpha$-amino-$\varepsilon$-caprolactam racemase from *Achromobacter obae, Agric. Biol. Chem.* **47**, 1149–1150.

AHMED, S. A., ESAKI, N., TANAKA, H., SODA, K. (1983b), Properties of $\alpha$-amino-$\varepsilon$-caprolactam racemase from *Achromobacter obae, Agric. Biol. Chem.* **47**, 1887–1893.

AHMED, S. A., ESAKI, N., TANAKA, H., SODA, K. (1986), Mechanism of $\alpha$-amino-$\varepsilon$-caprolactam racemase reaction, *Biochemistry* **25**, 385–388.

AIBA, S., TSUNEKAWA, S., IMANAKA, T. (1982), New approach to tryptophan production by *Escherichia coli*: Genetic manipulation of composite plasmids *in vitro, Appl. Environ. Microbiol.* **43**, 289–297.

ALBERTSON, N. F. (1951), Asymmetric enzymatic synthesis of amino acid onilides, *J. Am. Chem. Soc.* **73**, 452–454.

ANGELACCIO, S., PASCARELLA, S., FATTORI, E., BOSSA, F., STRONG, W., SCHIRCH, V. (1992), Serine hydroxymethyltransferase: origin of substrate specificity, *Biochemistry* **31**, 155–162.

ANTSON, A. A., DEMIDKINA, T. V., GOLLNICK, P., DAUTER, Z., VON TERSCH, R. L., LONG, J., BEREZHNOY, S. N., PHILLIPS, R. S., HARUTYUNYAN, E. H., WILSON, K. S. (1993), Three-dimensional structure of tyrosine phenol-lyase, *Biochemistry* **32**, 4195–4206.

ARAKI, K. (1986), Histidine, in: *Biotechnology of Amino Acid Production* (AIDA, K., CHIBATA, I., NAKAYAMA, K., TAKINAMI, K., YAMADA, H., Eds.), pp. 247–256. Tokyo: Kodansha/Amsterdam: Elsevier.

ASANO, Y., NAKAZAWA, A. (1987), High yield synthesis of L-amino acids by phenylalanine dehydrogenase from *Sporosarcina ureae, Agric. Biol. Chem.* **51**, 2035–2036.

ASANO, Y., YAMADA, A., KATO, Y., YAMAGUCHI, K., HIBINO, Y., HIRAI, K., KONDO, K. (1990), Enantioselective synthesis of (*S*)-amino acids by phenylalanine dehydrogenase from *Bacillus sphaericus*: use of natural and recombinant enzymes, *J. Org. Chem.* **55**, 5567–5571.

BADET, B., WALSH, C. (1985), Purification of an alanine racemase from *Streptococcus faecalis* and analysis of its inactivation by aminoethyl phosphonic acid enantiomers, *Biochemistry* **24**, 1333–1341.

BATTILOTTI, M., BARBERINI, U. (1988), Preparation of D-valine from DL-5-isopropylhydantoin by stereoselective biocatalysis, *J. Mol. Catal.* **43**, 343–352.

BELL, S. C., TURNER, J. M. (1977), Bacterial catabolism of threonine: Threonine degradation initiated by L-threonine acetaldehyde-lyase (aldolase) in species of *Pseudomonas, Biochem. J.* **166**, 209–216.

BELLION, E., TAN, F. (1987), An $NAD^+$-dependent alanine dehydrogenase from a methylotrophic bacterium, *Biochem. J.* **244**, 565–570.

BRUNHUBER, N. M. W., BANERJEE, A., JACOBS, W. R. J., BLANCHARD, J. S. (1994), Cloning sequencing, and expression of *Rhodococcus* L-phenylalanine dehydrogenase. Sequence comparisons to amino-acid dehydrogenases, *J. Biol. Chem.* **269**, 16203–16211.

BURNE, R. A., PARSONS, D. T., MARQUIS, R. E. (1989), Cloning and expression in *Escherichia coli* of the arginine deiminase system of *Streptococcus sanguis* NCTC10904, *Infect. Immun.* **57**, 3540–3548.

CALTON, G. J. (1992a), The enzymatic production of L-aspartic acid, in: *Biocatalytic Production of Amino Acids and Derivatives* (ROZZELL, J. D., WAGNER, F., Eds.), pp. 3-21. München: Hanser.

CALTON, G. J. (1992b), The enzymatic preparation of L-alanine, in: *Biocatalytic Production of Amino Acids and Derivatives* (ROZZELL, J. D., WAGNER, F., Eds.), pp. 59–74. München: Hanser.

CALTON, G. J., WOOD, L. L., UPDIKE, M. H., LANTZ, L., HAMMEN, J. P. (1986), The production of L-phenylalanine by polyazetidine immobilized microbes, *Bio/Technology* **4**, 317–320.

CHIBATA, I. (1974), in: *Synthetic Production and Utilization of Amino Acids* (KANEKO, T., IZUMI, Y., CHIBATA, I., ITOH, T., Eds.), pp. 17–51.

CHIBATA, I., ISHIKAWA, T., YAMADA, S. (1957), Studies on amino acids. VII. Studies on the enzymatic resolution (VI). A survey of the acylase in molds., *Bull. Agr. Chem. Soc. Jpn.* **21**, 300–303.

CHIBATA, I., ISHIKAWA, T., TOSA, T. (1970), ε-Lysine acylase from *Achromobacter pestifer*, *Methods Enzymol.* **19**, 756–762.

CHO, H.-Y., TANIZAWA, K., TANAKA, H., SODA, K. (1987), Thermostable aminoacylase from *Bacillus thermoglucosidius*: purification and characterization, *Agric. Biol. Chem.* **51**, 2793–2800.

CHOCAT, P., ESAKI, N., TANAKA, H., SODA, K. (1985), Synthesis of selenocystine and selenohomocystine with *O*-acetylhomoserine sulfhydrylase, *Agric. Biol. Chem.* **49**, 1143–1150.

COLOMBO, G., VILLAFRANCA, J. J. (1986), Amino acid sequence of *Escherichia coli* glutamine synthetase deduced from the DNA nucleotide sequence, *J. Biol. Chem.* **261**, 10587–10591.

CONSTANTINIDES, A., BHATIA, D., VIETH, W. R. (1981), Immobilization of *Brevibacterium flavum* cells on collagen for the production of glutamic acid in a recycle reactor, *Biotechnol. Bioeng.* **23**, 899–916.

CREMER, J., EGGELING, L., SAHM, H. (1991), Control of the lysine biosynthesis sequence in *Corynebacterium glutamicum* as analyzed by overexpression of the individual corresponding genes, *Appl. Environ. Microbiol.* **57**, 1746–1752.

DAINTY, R. H. (1970), Purification and properties of threonine aldolase from *Clostridium pasteurianum, Biochem. J.* **117**, 585–592.

DAVANKOV, V. A., SEMECHKIN, A. V. (1977), Ligand-exchange chromatography, *J. Chromatogr.* **141**, 313–353.

DAVIS, L., METZLER, D. E. (1972), Pyridoxal-linked elimination and replacement reactions, *Enzymes* (3rd Edn.) **7**, 33–74.

DCUNHA, G. B., SATYANARAYAN, V., NAIR, P. M. (1994), Novel direct synthesis of L-phenylalanine methyl ester by using *Rhodotorula glutinis* phenylalanine ammonia-lyase in an organic aqueous biphasic systems, *Enzyme Microb. Technol.* **16**, 318–322.

DEBABOV, V. (1982), Construction of strains producing L-threonine, in: *Proc. 4th Int. Symp. GIM* (Kyoto), pp. 254–258. Tokyo: Kodansha.

DEELEY, M. C., YANOFSKY, C. (1981), Nucleotide sequence of structural gene for tryptophanase of *Escherichia coli* K-12, *J. Bacteriol.* **147**, 787–796.

DELLEFRATTE, S., IURESCIA, S., ANGELACCIO, S., BOSSA, F., SCHIRCH, V. (1994), The function of arginine 363 as the substrate carboxyl-binding site in *Escherichia coli* serine hydroxymethyltransferase, *Eur. J. Biochem.* **225**, 395–401.

DEMENTIEVA, I. S., ZAKOMIRDINA, L. N., SINITZINA, N. I., ANTSON, A. A., WILSON, K. S., ISUPOV, M. N., LEBEDEV, A. A., HARUTYUNYAN, E. H. (1994), Crystallization and preliminary X-ray investigation of holotryptophanases from *Escherichia coli* and *Proteus vulgaris, J. Mol. Biol.* **235**, 783–786.

DEMIDKINA, T. V., MYAGKIKH, I. V., ANTSON, A. A., HARUTYUNYAN, E. H. (1988), Crystallization and crystal data on tyrosine phenol-lyase, *FEBS Lett.* **232**, 381–382.

DIVEN, W. F. (1969), Studies on amino acid racemases: II. Purification and properties of the glutamate racemase from *Lactobacillus fermenti, Biochim. Biophys. Acta* **191**, 702–706.

DOHERTY, D. G., POPENOE JR., E. A. (1951), The resolution of amino acids by asymmetric enzymatic synthesis, *J. Biol. Chem.* **189**, 447–460.

ENEI, H., YAMASHITA, K., OKUMURA, S., YAMADA, H. (1973a), Culture conditions for preparation of cells containing high tyrosine phenol-lyase activity, *Agric. Biol. Chem.* **37**, 485–492.

ENEI, H., NAKAZAWA, H., OKUMURA, S., YAMADA, H. (1973b), Microbiological synthesis of L-tyrosine and 3,4-dihydroxyphenyl-L-alanine. 5. Synthesis of L-tyrosine or 3,4-dihydroxyphenyl-L-alanine from pyruvic acid, ammonia and phenol or pyrocathecol, *Agric. Biol. Chem.* **37**, 725–735.

ESAKI, N., WALSH, C. T. (1986), Biosynthetic alanine racemase of the *Salmonella typhimurium*: purification and characterization of the enzyme encoded by the *alr* gene, *Biochemistry* **25**, 3261–3267.

ESAKI, N., SUZUKI, T., TANAKA, H., SODA, K., RANDO, R. R. (1977), Deamination and $\gamma$-addition reactions of vinylglycine by L-methionine $\gamma$-lyase, *FEBS Lett.* **84**, 309–312.

ESAKI, N., TANAKA, H., UMEMURA, S., SUZUKI, T., SODA, K. (1979), Catalytic action of L-methionine $\gamma$-lyase on selenomethionine and selenols, *Biochemistry* **18**, 407–410.

ESAKI, N., NAKAMURA, T., TANAKA, H., SUZUKI, T., MORINO, Y., SODA, K. (1981), Enzymatic synthesis of selenocysteine in rat liver, *Biochemistry* **20**, 4492–4496.

ESAKI, N., TANAKA, H., MILES, E. W., SODA, K. (1983a), Enzymatic synthesis of *S*-substituted L-cysteines with tryptophan synthase of *Escherichia coli, Agric. Biol. Chem.* **47**, 2861–2864.

ESAKI, N., TANAKA, H., MILES, E. W., SODA, K. (1983b), Enzymatic synthesis of *Se*-substituted L-selenocysteines with tryptophan synthase, *FEBS Lett.* **161**, 207–209.

ESAKI, N., SERANEEPRAKARN, V., TANAKA, H., SODA, K. (1988), Purification and characterization of *Clostridium sticklandii* D-selenocystine $\alpha,\beta$-lyase, *J. Bacteriol.* **170**, 751–756.

ESAKI, N., SHIMOI, H., TANAKA, H., SODA, K. (1989a), Enantioselective synthesis of D-selenomethionine with D-amino acid aminotransferase, *Biotechnol. Bioeng.* **34**, 1231–1233.

ESAKI, N., SHIMOI, H., YANG, Y.-S., TANAKA, H., SODA, K. (1989b), Enantioselective synthesis of L-selenomethionine with leucine dehydrogenase, *Biotechnol. Appl. Biochem.* **11**, 312–317.

FAULKNER, J. D., ANSON, J. G., TUITE, M. F., MINTON, N. P. (1994), High-level expression of the phenylalanine ammonia lyase-encoding gene from *Rhodosporidium toruloides* in *Saccharomyces cerevisiae* and *Escherichia coli* using a bifunctional expression system, *Gene* **143**, 13–20.

FILPULA, D., VASLET, C. A., LEVY, A., SYKES, A., STRAUSBERG, R. L. (1988), Nucleotide sequence of gene for phenylalanine ammonia-lyase from *Rhodotorula rubra, Nucleic Acids Res.* **16**, 1381–1381.

FIOLITAKIS, E., WANDREY, C. (1982), Reaction technology of the enzymatically catalyzed production of L-alanine, p. 272–284, in: *Enzyme Technology, Proc. III. Rotenburg Ferment. Symp.* New York: Springer-Verlag.

FÖRBERG, C., ELIAESON, T., HÄGGSTRÖM, L. (1988), Correlation of theoretical and experimental yields of phenylalanine from non-growing cells of a *rec Escherichia coli* strain, *J. Biotechnol.* **7**, 319–332.

FUKUMURA, T. (1976a), Enzymatic conversion of DL-$\alpha$-amino-$\varepsilon$-caprolactam into L-lysine. 1. Screening, classification and distribution of L-$\alpha$-amino-$\varepsilon$-caprolactam-hydrolyzing yeasts, *Agric. Biol. Chem.* **40**, 1687–1693.

FUKUMURA, T. (1976b), Enzymatic conversion of DL-$\alpha$-amino-$\varepsilon$-caprolactam into L-lysine. 2. Hydrolysis of L-$\alpha$-amino-$\varepsilon$-caprolactam by yeasts, *Agric. Biol. Chem.* **40**, 1695–1698.

FUKUMURA, T. (1977a), Enzymatic conversion of DL-$\alpha$-amino-$\varepsilon$-caprolactam into L-lysine. 3. Bacterial racemization of $\alpha$-amino-$\varepsilon$-caprolactam, *Agric. Biol. Chem.* **41**, 1321–1325.

FUKUMURA, T. (1977b), Enzymatic conversion of DL-$\alpha$-amino-$\varepsilon$-caprotactam into L-lysine. 6. Conversion of D-$\alpha$-amino-$\varepsilon$-caprolactam into L-lysine using both yeast cells and bacterial cells, *Agric. Biol. Chem.* **41**, 1327–1330.

FUKUMURA, T., TALBOT, G., MISONO, H., TERAMURA, Y., KATO, K., SODA, K. (1978), Purification and properties of a novel enzyme, L-$\alpha$-amino-$\varepsilon$-caprolactamase from *Cryptococcus laurentii, FEBS Lett.* **89**, 298–300.

FURUI, M., YAMASHITA, K. (1983), Pressurized reaction method for continuous production of L-alanine by immobilized *Pseudomonas dacunhae* cells, *J. Ferment. Technol.* **61**, 587–591.

GALACV, I. V., SVEDAS, V. K. (1982), A kinetic study of hog kidney aminoacylase, *Biochim. Biophys. Acta* **701**, 389–394.

GARROW, T. A., BRENNER, A. A., WHITEHEAD, M. V., CHEN, X.-N., DUNCAN, R. G., KORENBERG, J. R., SHANE, B. (1993), Cloning of human cDNAs encoding mitochondrial and cytosolic serine hydroxymethyltransferases and chromosomal localization, *J. Biol. Chem.* **268**, 11910–11916.

GENTZEN, I., LOFFLER, H.-G., SCHNEIDER, F. (1980), Aminoacylase from *Aspergillus oryzae.* Comparison with the pig kidney enzyme, *Z. Naturforsch.* **35C**, 544–550.

GIL-AV, E., TISHBEE, A., HARE, P. E. (1980), Resolution of underivatized amino acids by reversed-phase chromatography, *J. Am. Chem. Soc.* **102**, 5115–5117.

GILBERT, H. J., CLARKE, I. N., GIBSON, R. K., STEPHENS, J. R., TULLY, M. (1985), Molecular cloning of the phenylalanine ammonia-lyase gene from *Rhodosporidium toruloides* in *Escherichia coli* K12, *J. Bacteriol.* **161**, 314–320.

GREENSTEIN, J. P. (1957), Resolution of DL mixtures of $\alpha$-amino acids, *Methods Enzymol.* **3**, 554–570.

GROEGER, U., DRAUZ, K., KLENK, H. (1990), Isolation of an L-stereospecific *N*-acyl-L-proline-acylase and its use as catalyst in organic synthesis, *Angew. Chem.* (Int. Edn. Engl.) **29**, 417–419.

GROEGER, U., DRAUZ, K., KLENK, H. (1992), Enzymatic preparation of enantiomerically pure *N*-alkyl amino acids, *Angew. Chem.* (Int. Edn. Engl.) **31**, 195–197.

GU, K. F., CHANG, T. M. S. (1990), Conversion of ammonia or urea into essential amino acids, L-leucine, L-valine, and L-isoleucine, using artificial cells containing an immobilized multienzyme system and dextran-NAD$^+$, *Biotechnol. Appl. Biochem.* **12**, 227–236.

HAN, J. K., OH, J. W., LEE, H. H., CHUNG, S., HYUN, H. H., LEE, J. H. (1991), Molecular cloning and expression of *S*-(2-aminoethyl)-L-cysteine resistant aspartokinase gene of *Corynebacterium glutamicum, Biotechnol. Lett.* **13**, 721–726.

HANSON, K. R., HAVIR, E. A. (1972), The enzymic elimination of ammonia, *Enzymes* (3rd Edn.) **7**, 75–166.

HANSON, R. L., SINGH, J., KISSICK, T. P., PATEL, R. N., SZARKA, L. J., MUELLER, R. H. (1990), Synthesis of L-$\beta$-hydroxyvaline from $\alpha$-keto-$\beta$-hydroxyisovalerate using leucine dehydrogenase from *Bacillus* species, *Bioorg. Chem.* **18**, 116–130.

HASHIMOTO, S., KATSUMATA, R. (1994), L-Alanine production by alanine racemase-deficient mutant of *Arthrobacter oxydans, Proc. Ann. Meet.* (Agric. Chem. Soc. Jpn.), p. 341. Tokyo: Japan Society for Bioscience, Biotechnology, and Agrochemistry.

HAVIR, E. A., HANSON, K. R. (1975), L-Phenylalanine ammonia-lyase: Maize, potato, and *Rhodotorula glutinis*: Studies of prosthetic group with nitromethane, *Biochemistry* **14**, 1620–1626.

HIRAHARA, T., SUZUKI, S., HORINOUCHI, S., BEPPU, T. (1992), Cloning, nucleotide-sequences, and overexpression in *Escherichia coli* of tandem copies of a tryptophanase gene in an obligately symbiotic thermophile, *Symbiobacterium thermophilum, Appl. Environ. Microbiol.* **58**, 2633–2642.

HIRAHARA, T., HORINOUCHI, S., BEPPU, T. (1993), Cloning, nucleotide-sequence, and overexpression in *Escherichia coli* of the $\beta$-tyrosinase gene from an obligately symbiotic thermophile, *Symbiobacterium thermophilum, Appl. Microbiol. Biotechnol.* **39**, 341–346.

HIROSE, Y., OKADA, H. (1979), Microbial production of amino acids, in: *Microbial Technology,* 2nd Edn., Vol. 1 (PEPPLER, H. J., PERLMAN, D., Eds.), pp. 211–244, New York: Academic Press.

HODGSON, J. (1994), Bulk amino-acid fermentation: Technology and commodity trading, *Bio/Technology* **12**, 152–155.

HOFFMANN, K., SCHNEIDER-SCHERZER, E., KLEINKAUF, H., ZOCHER, R. (1994), Purification and characterization of eukaryotic alanine racemase acting as key enzyme in cyclosporine biosynthesis, *J. Biol. Chem.* **269**, 2710–2714.

HSIAO, H. Y., WEI, T., CAMPBELL, K. (1986), Enzymatic production of L-serine, *Biotechnol. Bioeng.* **28**, 857–867.

HUMMEL, W., WEISS, N., KULA, M.-R. (1984), Isolation and characterization of a bacterium possessing L-phenylalanine dehydrogenase activity, *Arch. Microbiol.* **137**, 47–52.

HUMMEL, W., SCHMIDT, E., WANDREY, C., KULA, M.-R. (1986), L-Phenylalanine dehydrogenase from *Brevibacterium* sp. for production of L-phenylalanine by reductive amination of phenylpyruvate, *Appl. Microbiol. Biotechnol.* **25**, 175–185.

IKEDA, S., FUKUI, S. (1979), Immobilization of pyridoxal 5′-phosphate and pyridoxal 5′-phosphate-dependent enzymes on Sepharose, *Methods Enzymol.* **62**, 517–527.

IKEDA, M., OZAKI, A., KATSUMATA, R. (1993), Phenylalanine production by metabolically engineered *Corynebacterium glutamicum* with the *pheA* gene of *Escherichia coli, Appl. Microbiol. Biotechnol.* **39**, 318–323.

IKEDA, M., NAKANISHI, K., KINO, K., KATSUMATA, R. (1994), Fermentative production of tryptophan by a stable recombinant strain of *Corynebacterium glutamicum* with a modified serine-biosynthetic pathway, *Biosci. Biotech. Biochem.* **58**, 674–678.

INAGAKI, K., TANIZAWA, K., BADET, B., WALSH, C. T., TANAKA, H., SODA, K. (1986), Thermostable alanine racemase from *Bacillus stearothermophilus*: molecular cloning of the gene, enzyme purification, and characterization, *Biochemistry* **25**, 3268–3274.

INAGAKI, K., TANIZAWA, K., TANAKA, H., SODA, K. (1987), Purification and characterization of amino acid racemase with very broad substrate specificity from *Aeromonas caviae, Agric. Biol. Chem.* **51**, 173–180.

INOUE, H., INAGAKI, K., SUGIMOTO, M., ESAKI, N., SODA, K., TANAKA, H. (1995), Structural analysis of the L-methionine $\gamma$-lyase gene from *Pseudomonas putida, J. Biochem.* **117**, 1120–1125.

ISHIDA, M., YOSHINO, E. , MAKIHARA, R., SATO, K., ENEI, H., NAKAMORI, S. (1989), Improvement of an L-threonine producer derived from *Brevibacterium flavum* using threonine operon of *Escherichia coli* K-12, *Agric. Biol. Chem.* **53**, 2269–2271.

ISHIKAWA, T., WATABE, K., MUKOHARA, Y., KOBAYASHI, S., NAKAMURA, H. (1993), Microbial conversion of DL-5-substituted hydantoins to the corresponding L-amino acids by *Pseudomonas* sp. strain NS671, *Biosci. Biotech. Biochem.* **57**, 982–986.

ISHIKAWA, T., MUKOHARA, Y., WATABE, K., KOBAYASHI, S., NAKAMURA, H. (1994), Microbial conversion of DL-5-substituted hydantoins to the corresponding L-amino acids by *Bacillus stearothermophilus, Biosci. Biotech. Biochem.* **58**, 265–270.

ISHIWATA, K., YOSHINO, S., IWAMORI, S., SUZUKI, T., MAKIGUCHI, N. (1989), Cloning and sequencing of *Bacillus stearothermophilus* tryptophan synthase genes, *Agric. Biol. Chem.* **53**, 2941–2948.

ISHIWATA, K., SUZUKI, T., IWAMORI, S., YOSHINO, S., MAKIGUCHI, N. (1990a), Thermostable tryptophan synthase of *Bacillus stearothermophilus*: expression in *Escherichia coli, Biotechnol. Lett.* **12**, 185–190.

ISHIWATA K., FUKUHARA, N., SHIMADA, M., MAKIGUCHI, N., SODA, K. (1990b), Enzymatic production of L-tryptophan from DL-serine and indole by a coupled reaction of tryptophan synthase and amino acid racemase, *Biotechnol. Appl. Biochem.* **12**, 141–149.

ITO, H., SATO, K., MATSUI, K., MIWA, K., SANO, K., NAKAMORI, S., TANAKA, T., ENEI, H. (1990a), Cloning and characterization of genes responsible for *m*-fluoro-DL-phenyalanine resistance in *Brevibacterium lactofermentum, Agric. Biol. Chem.* **54**, 707–713.

ITO, H., SATO, K., ENEI, H., HIROSE, Y. (1990b), Improvement in microbial production of L-tyrosine by gene dosage effect of *AroL* gene encoding shikimate kinase, *Agric. Biol. Chem.* **54**, 823–824.

ITOH, N., MORIKAWA, R. (1983), Crystallization and properties of L-alanine dehydrogenase from *Streptomyces phaeochromogenes, Agric. Biol. Chem.* **47**, 2511–2519.

IWAMORI, S., YOSHINO, S. ISHIWATA, K., MAKIGUCHI, N. (1991), Structure of tyrosine phenol-lyase genes from *Citrobacter freundii* and structural comparison with tryptophanase from *Escherichia coli, J. Ferment. Bioeng.* **72**, 147–151.

IWAMORI, S., OIKAWA, T., ISHIWATA, K., MAKIGUCHI, N. (1992), Cloning and expression of the *Erwinia herbicola* tyrosine phenol-lyase gene in *Escherichia coli, Biotechnol. Appl. Biochem.* **16**, 77–85.

JACOBS, J. A., COURT, D., GUARNEROS (1990), Lysine and methionine overproduction by an *Escherichia coli* strain transformed with *Pseudomonas acidovorans* DNA, *Biotechnol. Lett.* **12**, 425–430.

JETTEN, M. S. M. SINSKEY, A. J. (1995), Recent advances in the physiology and genetics of amino acid-producing bacteria, *Grit. Rev. Biotechnol.* **15**, 73–103.

JOHNSTON, M. M., DIVEN, W. F. (1969), Studies on amino acid racemases: I. Partial purification and properties of the alanine racemases from *Lactobacillus fermenti, J. Biol. Chem.* **244**, 5414–5420.

KAGAMIYAMA, H., MATSUBARA, H., SNELL, E. E. (1972), The chemical structure of tryptophanase from *Escherichia coli.* III. Isolation and amino acid sequence of tryptic peptides, *J. Biol. Chem.* **247**, 1571–1575.

KAJIWARA, S., MAEDA, H. (1987), The improvement of a droplet gel-entrapping method: the co-immobilization of leucine dehydrogenase and formate dehydrogenase, *Agric. Biol. Chem.* **51**, 2873–2879.

KAMATH, A. V., YANOFSKY, C. (1992), Characterization of the tryptophanase operon of *Proteus vulgaris.* Cloning, nucleotide sequence, amino acid homology, and *in vitro* synthesis of the leader peptide and regulatory analysis, *J. Biol. Chem.* **267**, 19978–19985.

KAMEDA, Y., TOYOURA, E., KIMURA, Y., MATSUI, K. (1958), Studies on acylase activity and microorganisms. VIII. Enzymatic hydrolysis of 6-*N*-benzoyl-L-lysine., *Chem. Pharm. Bull.* (Tokyo) **6**, 394–495.

KAMITANI, H., ESAKI, N., TANAKA, H., IMAHARA, H., SODA, K. (1990a), Purification and characterization of *S*-alkylcysteine $\alpha,\beta$-lyase from *Pseudomonas putida, J. Nutr. Sci. Vitaminol.* **36**, 339–347.

KAMITANI, H., ESAKI, N., TANAKA, H., SODA, K. (1990b), Thermostable *S*-alkylcysteine $\alpha,\beta$-lyase from a thermophile: purification and properties, *Agric. Biol. Chem.* **54**, 2069–2076.

KANZAKI, H., NAGASAWA, T., YAMADA, H. (1986), Highly efficient production of L-cystathionine from *O*-succinyl-L-homoserine and L-cysteine by *Streptomyces* cystathionine $\gamma$-lyase, *Appl. Microbiol. Biotechnol.* **25**, 97–100.

KARSTEN, W. E., HUNSLEY, J. R., VIOLA, R. E. (1985), Purification of aspartase and aspartokinase-homoserine dehydrogenase I from *Escherichia coli* by dye-ligand chromatogaphy, *Anal. Biochem.* **147**, 336–341.

KATSUMATA, R., HARA, M., MIZUKAMI, T., OKA, T., FURUYA, A. (1984), Molecular breeding of threonine producers in *Corynebacterium glutamicum,* in: *Proc. Ann. Meet.* (Agric. Chem. Soc. Jpn.) p. 248. The Agricultural Chemical Society of Japan.

KATSUMATA, R., MIZUKAMI, T., HARA, M., OKA, T. (1985), Molecular breeding of isoleucine producers in *Corynebacterium glutamicum,* in: *Proc. Ann. Meet.* (Agric. Chem. Soc. Jpn.), p. 421. The Agricultural Chemical Society of Japan.

KATSUMATA, R. , MIZUKAMI, T., KIKUCHI, Y., KINO, K. (1986), Threonine production by the lysine producing strain of *Corynebacterium glutamicum* with amplified threonine biosynthetic operon, in: *Proc. 5th Int. Symp. GIM* (Split), pp. 217–226. Zagreb: B. Pliva.

KAWAHARA, Y., YOSHIHARA, Y., IKEDA, S., YOSHII, H., HIROSE, Y. (1990), Stimulatory effect of glycine betaine on L-lysine fermentation, *Appl. Microbiol. Biotechnol.* **34**, 87–90.

KAWASAKI, K., YOKOTA, A., OITA, S., KOBAYASHI, C., YOSHIKAWA, S., KAWAMOTO, S., TAKAO, S., TOMITA, F. (1993), Cloning and characterization of a tryptophanase gene from *Enterobacter aerogenes* SM-18, *J. Gen. Microbiol.* **139**, 3275–3281.

KAWATA, Y., TANI, S., SATO, M., KATSUBE, Y., TOKUSHIGE, M. (1991), Preliminary X-ray crystallographic analysis of tryptophanase from *Escherichia coli, FEBS Lett.* **284**, 270–272.

KIICK, D. M., PHILLIPS, R. S. (1988a), Mechanistic deductions from kinetic isotope effects and pH studies of pyridoxal phosphate dependent carbon–carbon lyases: *Erwinia herbicola* and *Citrobacter freundii* tyrosine phenol-lyase, *Biochemistry* **27**, 7333–7338.

KIICK, D. M., PHILLIPS, R. S. (1988b), Mechanistic deductions from multiple kinetic and solvent deuterium isotope effects and pH studies of pyridoxal phosphate dependent carbon–carbon lyases: *Escherichia coli* tryptophan indolelyase, *Biochemistry* **27**, 7339–7344.

KIKUCHI, M., NAKAO, Y. (1986), Glutamic acid, in: *Biotechnology of Amino Acid Production* (AIDA, K., CHIBATA, I., NAKAYAMA, K., TAKINAMI, K., YAMADA, H., Eds.), pp. 101–116. Tokyo: Kodansha/Amsterdam: Elsevier.

KIKUCHI, M., KOSHIYAMA, I., FUKUSHIMA, D. (1983), A new enzyme, proline acylase (*N*-acyl-L-proline amidohydrolase) from *Pseudomonas* species, *Biochim. Biophys. Acta* **744**, 180–188.

KINOSHITA, S., UDAKA, S., SHIMONO, M. (1957), Studies on the amino acid fermentation. I. Production of L-glutamic acid by various microorganisms, *J. Gen. Appl. Microbiol.* **3**, 193–205.

KNABE, J., WUMM, W. (1980), Racemic and optically-active hydantoins from disubstituted cyanoacetic acids, *Arch. Pharm. (Weinheim, Ger.)* **313**, 538–543.

KOMATSUBARA, S., KISUMI, M. (1986), Isoleucine, valine and leucine, in: *Biotechnology of Amino Acid Production* (AIDA, K., CHIBATA, I., NAKAYAMA, K., TAKINAMI, K., YAMADA, H., Eds.), pp. 233-246. Tokyo: Kodansha.

KOMATSUBARA, S., TANIGUCHI, T., KISUMI, M. (1986), Overproduction of aspartase of *Escherichia coli* K-12 by molecular cloning, *J. Biotechnol.* **3**, 281–291.

KONDO, K., SONE, H., YOSHIDA, H., TOIDA, T., KANATANI, K., HONG, Y. M., NISHINO, N., TANAKA, J. (1990), Cloning and sequence analysis of the arginine deiminase gene from *Mycoplasma arginini, Mol. Gen. Genet.* **221**, 81–86.

KÖRDEL, W., SCHNEIDER, F. (1976), Chemical investigation on pig kidney aminoacylase, *Biochim. Biophys. Acta* **445**, 446–457.

KRÄMER, R. (1994), Secretion of amino acids by bacteria: physiology and mechanism, *FEMS Microbiol. Rev.* **13**, 75–93.

KREDICH, N. M., KEENAN, B. S., FOOTE, L. J. (1972), Purification and subunit structure of cysteine desulfhydrase from *Salmonella typhimurium, J. Biol. Chem.* **247**, 7157-7162.

KUBOTA, K. (1985), Improvement production of L-serine by mutants of *Corynebacterium glycinophilum* with less serine dehydratase activity, *Agric. Biol. Chem.* **49**, 7–12.

KUBOTA, K., YOKOZEKI, K. (1989), Production of L-serine from glycine by *Corynebacterium glycinophilum* and properties of serine hydroxymethyltransferase, a key enzyme in L-serine production, *J. Ferment. Bioeng.* **67**, 387–390.

KUMAGAI, H. (1994), *Tanpakushitsu Kagaku,* Tokyo: Hirokawa Shoten.

KUMAGAI, H., SEJIMA, S., CHOI, Y., YAMADA, H. (1974), Crystallization and properties of cysteine desulfhydrase from *Aerobacter aerogenes, FEBS Lett.* **52**, 304–307.

KUMAGAI, H., CHOI, Y., SEJIMA, S., YAMADA, H. (1975), Formation of cysteine desulfhydrase by bacteria, *Agric. Biol. Chem.* **39**, 387–392.

KUMAGAI, H., TANAKA, H., SEJIMA, S., YAMADA, H. (1977), Elimination and replacement reactions of $\beta$-chloro-L-alanine by cysteine desulfhydrase from *Aerobacter aerogenes, Agric. Biol. Chem.* **41**, 2071–2075.

KUO, C. F., DARNELL, J. E. (1989), Mouse glutamine synthetase is encoded by a single gene that can be expressed in a localized fashion, *J. Mol. Biol.* **208**, 45–56.

KURAHASHI, O., BEYOU, A., TAKINAMI, K., JARRY, B., RICHAUD, F. (1990), Stabilized amplification of genetic information in Gram-negative bacteria with Mu phage and its application in L-threonine production, in: *Proc. 6th Int. Symp. GIM* (Strasbourg), p. 178. Paris: Société Française de Microbiologie.

KURODA, S., TANIZAWA, K., SAKAMOTO, Y., TANAKA, H., SODA, K. (1990), Alanine dehydrogenases from 2 *Bacillus* species with distinct thermostabilities: molecular cloning, DNA and protein sequence determination, and structural comparison with other $NAD(P)^+$-dependent dehydrogenases, *Biochemistry* **29**, 1009–1015.

KURUSU, Y., FUKUSHIMA, M., KOHAMA, K., KOBAYASHI, M., TERASAWA, M., KUMAGAI, H., YUKAWA, H. (1991), Cloning and nucleotide sequence of the tyrosine phenol-lyase gene from *Escherichia intermedia, Biotechnol. Lett.* **13**, 769–772.

LIEBL, W., EHRMANN, U., LUDWIG, W., SCHLEIFER, K. H. (1991), Transfer of *Brevibacterium divaricatum* DSM2029T, *Brevibacterium flavum* DSM20411, *Brevibacterium lactofermentum* DSM20412 and DSM1412, and *Corynebacterium lilium* DSM20317 to *Corynebacterium glutamicum* and their distinction by rDNA gene restriction patterns, *Int. J. Syst. Bacteriol.* **41**, 225–235.

LIM, Y. H., YOKOIGAWA, K., ESAKI, N., SODA, K. (1993), A new amino acid racemase with threonine alpha-epimerase activity from *Pseudomonas putida* – purification and characterization, *J. Bacteriol.* **175**, 4213–4217.

MAKIGUCHI, N., FUKUHARA, N., SHIMADA, M., ASAI, Y., NAKAMURA, T., SODA, K. (1987), in: *Biochemistry of Vitamin B6* (KORPELA, T., CHRISTEN, P., Eds.), 457 pp. Basel: Birkhäuser.

MANNING, J. M., MOORE, S. (1968), Determination of D- and L-amino acids by ion exchange chromatography as L-D and L-L dipeptides, *J. Biol. Chem.* **243**, 5591–5597.

MARTIN, J. F. (1989), Molecular genetics of amino acid-producing corynebacteria, in: *Microbial Products. A Practical Approach* (BAUMBERG, S., HUNTER, I., RHODES, M., Eds.), pp. 25–59, Society for General Microbiology Symposium 44. Cambridge: University Press.

MARTINI, F., ANGELACCIO, S., PASCARELLA, S., BARRA, D., BOSSA, F., SCHIRCH, V. (1987), The primary structure of rabbit liver cytosolic serine hydroxymethyltransferase, *J. Biol. Chem.* **262**, 5499–5509.

MARTINI, F., MARAS, B., TANCI, P., ANGELACCIO, S., PASCARELLA, S., BARRA, D., BOSSA, F., SCHIRCH, V. (1989), The primary structure of rabbit liver mitochondrial serine hydroxymethyltransferase, *J. Biol. Chem.* **264**, 8509–8519.

MASAWAKI, T., SASAI, M., TONE, S. (1992), Optical resolution of an amino acid by an enantioselective ultrafiltration membrane, *J. Chem. Eng. Jpn.* **25**, 33–38.

MATSUI, K., MIWA, K., SANO, K. (1988), Cloning of a gene cluster of *aroB, aroE* and *aroL* for aromatic amino acid biosynthesis in *Brevibacterium lactofermentum,* a glutamic acid-producing bacterium, *Agric. Biol. Chem.* **52**, 525–531.

MATSUNAGA, T., NAKAMURA, N., TSUZAKI, N., TAKEDA, H. (1988), Selective production of glutamate by an immobilized marine blue-green alga, *Synechococcus* sp., *Appl. Microbiol. Biotechnol.* **28**, 373–376.

MAZELIS, M., CREVELING, R. K. (1975), Purfication and properties of *S*-alkyl-L-cysteine lyase from seedlings of *Acacia farnesiana, Biochem. J.* **147**, 485–491.

MAZELIS, M., FOWDEN, L. (1973), Relationship of endogenous substrate to specificity of *S*-alkylcysteine lyases of different species, *Phytochemistry* **12**, 1287–1289.

MCNEIL, J. B., MCINTOSH, E. M., TAYLOR, B. V.,

ZHANG, F.-R., TANG, S., BOGNAR, A. L. (1994), Cloning and molecular characterization of 3 genes, including 2 genes encoding serine hydroxymethyltransferases, whose inactivation is required to render yeast auxotrophic for glycine, *J. Biol. Chem.* **269**, 9155–9165.

MENKEL, E., THIERBACH, G., EGGELING, L., SAHM, H. (1989), Influence of increased aspartate availability on lysine formation by a recombinant strain of *Corynebacterium glutamicum* and utilization of fumarate, *Appl. Environ. Microbiol.* **55**, 684–688.

MILES, E. W., AHMED, S. A., HYDE, C. C., KAYASTHA, A. M., YANG, X.-J., RUVINOV, S. B., LU, Z. (1994), Tryptophan synthase, in: *Molecular Aspects of Enzyme Catalysis* (FUKUI, T., SODA, K., Eds.), pp. 127–146. Tokyo: Kodansha.

MINEHART, P. L., MAGASANIK, B. (1992), Sequence of the *GLN1* gene of *Saccharomyces cerevisiae*: role of the upstream region in regulation of glutamine synthetase expression, *J. Bacteriol.* **174**, 1828–1836.

MIWA, K., TSUCHIDA, T., KURAHASHI, O., NAKAMORI, S., SANO, K., MOMOSE, H. (1983), Construction of L-threonine overproducing strains of *Escherichia coli* K-12 using recombinant DNA techniques, *Agric. Biol. Chem.* **47**, 2329–2334.

MIYATA, A., YOSHIDA, T., YAMAGUCHI, K., YOKOYAMA, C., TANABE, T., TOH, H., MITSUNAGA, T., IZUMI, Y. (1993), Molecular cloning and expression of the gene for serine hydroxymethyltransferase from an obligate methylotroph *Hyphomicrobium methylovorum* GM2, *Eur. J. Biochem.* **212**, 745–750.

MIYAZAKI, S., TOKI, S., IZUMI, Y., YAMADA, H. (1987), Purification and characterization of a serine hydroxymethyltransferase from an obligate methylotroph, *Hyphomicrobium methylovorum* GM2, *Eur. J. Biochem.* **162**, 533–540.

MIZUKAMI, T., HAMU A., IKEDA, M., OKA, T., KATSUMATA, R. (1994), Cloning of the ATP phosphoribosyl transferase gene of *Corynebacterium glutamicum* and application of the gene to L-histidine production, *Biosci. Biotech. Biochem.* **58**, 635–638.

MOCANU, A., NIAC, G., IVANOF, A., GORUN, V., PALIBRODA, N., VARGHA, E., BOLOGA, M., BARZU, O. (1982), Preparation of $^{15}$N-labeled L-alanine by coupling the alanine dehydrogenase and alcohol dehydrogenase reactions, *FEBS Lett.* **143**, 153–156.

MOLLER, A., SYLDATK, C., SCHULZE, M., WAGNER, F. (1988), Stereospecificity and substrate specificity of a D-hydantoinase and a D-*N*-carbamyl-amino acid amidohydrolase of *Arthrobacter crystallopoietes* AM-2, *Enzyme Microb. Technol.* **10**, 618–625.

MORI, M., SHIIO, I. (1984), Production of aspartic acid and enzymatic alteration in pyruvate kinase mutants of *Brevibacterium flavum, Agric. Biol. Chem.* **48**, 1189–1197.

MORIGUCHI, M., IDETA, K. (1988), Production of D-aminoacylase from *Alcaligenes denitrificans* ssp. *xylosoxydans* MI-4, *Appl. Environ. Microbiol.* **54**, 2767–2770.

MORINAGA, Y., TAKAGI, H., ISHIDA, M., MIWA, K., SATO, T., NAKAMORI, S., SANO, K. (1987), Threonine production by co-existence of cloned genes coding homoserine dehydrogenase and homoserine kinase in *Brevibacterium lactofermentum, Agric. Biol. Chem.* **51**, 93–100.

MORIOKA, H., MIWA, K., ETO, H., EGUCHI, C., NAKAMORI, S. (1989), Fermentative production of L-glutamic-γ-semialdehyde and its chemical conversion into L-tryptophan, *Agric. Biol. Chem.* **53**, 911–915.

MUKOHARA, Y., ISHIKAWA, T., WATABE, K., NAKAMURA, H. (1993), Molecular cloning and sequencing of the gene for a thermostable *N*-carbamyl-L-amino acid amidohydrolase from *Bacillus stearothermophilus* strain NS1122a, *Biosci. Biotech. Biochem.* **57**, 1935–1937.

MUKOHARA, Y., ISHIKAWA, T., WATABE, K., NAKAMURA, H. (1994), A thermostable hydantoinase of *Bacillus stearothermophilus* NS1122a: Cloning, sequencing, and high expression of the enzyme gene, and some properties of the expressed enzyme, *Biosci. Biotech. Biochem.* **58**, 1621–1626.

NAGASAWA, T., YAMADA, H. (1986), Enzymatic transformations of 3-chloroalanine into useful amino acids, *Appl. Biochem. Biotechnol.* **13**, 147–165.

NAGASAWA, T., UTAGAWA, T., GOTO, J., KIM, C., TANI, Y., KUMAGAI, H., YAMADA, H. (1981), Synthesis of L-tyrosine-related amino acids by tyrosine phenol-lyases of *Citrobacter intermedius, Eur. J. Biochem.* **117**, 33–40.

NAGASAWA, T., OHKISHI, H., KAWAKAMI, B., YAMANO, H., HOSONO, H., TANI, Y., YAMADA, H. (1982a), 3-Chloro-D-alanine chloride-lyase (deaminating) of *Pseudomonas putida* CR1-1, *J. Biol. Chem.* **257**, 13749–13756.

NAGASAWA, T., YAMANO, H., OHKISHI, H., HOSONO, H., TANI, Y., YAMADA, H. (1982b), Enzymatic synthesis of D-cysteine by 3-chloro-D-alanine resistant *Pseudomonas putida* CR1-1, *Agric. Biol. Chem.* **46**, 3003–3010.

NAGASAWA, T., HOSONO, H., YAMANO, H., OHKISHI, H., TANI, Y., YAMADA, H. (1983), Synthesis of D-cysteine from a racemate of 3-chloroalanine by phenylhydrazine-treated cells of *Pseudomonas putida* CR1-1, *Agric. Biol. Chem.* **47**, 861–868.

NAGASAWA, T., KANZAKI, H., YAMADA, H. (1984), Cystathionine $\gamma$-lyase of *Streptomyces phaeochromogenes*: The occurrence of cystathionine $\gamma$-lyase in filamentous bacteria and its purification and characterization, *J. Biol. Chem.* **259**, 10393–10403.

NAGASAWA, T., ISHII, T., KUMAGAI, H., YAMADA, H. (1985), D-Cysteine desulfhydrase of *Escherichia coli*: Purification and characterization, *Eur. J. Biochem.* **153**, 541–551.

NAGASAWA, T., OHKISHI, H., KAWAKAMI, B., YAMAMOTO, Y., HOSONO, H., TANI, Y., NAGASAWA, T., KANZAKI, H., YAMADA, H. (1987), Cystathionine $\gamma$-lyase from *Streptomyces phaeochromogenes, Methods Enzymol.* **143**, 486–492.

NAGATA, S., TANIZAWA, K., ESAKI, N., SAKAMOTO, Y., OHSHIMA, T., TANAKA, H., SODA, K. (1988), Gene cloning and sequence determination of leucine dehydrogenase from *Bacillus stearothermophilus* and structural comparison with other $NAD(P)^+$-dependent dehydrogenases, *Biochemistry* **27**, 9056–9062.

NAKAJIMA, N., TANIZAWA, K., TANAKA, H., SODA, K. (1988a), Distribution of glutamate racemase in lactic acid bacteria and further characterization of the enzyme from *Pediococcus pentosaceus, Agric. Biol. Chem.* **52**, 3099–3104.

NAKAJIMA, N., TANIZAWA, K., TANAKA, H., SODA, K. (1988b), Enantioselective synthesis of various D-amino acids by a multi-enzyme system, *J. Biotechnol.* **8**, 243–248.

NAKAMORI, S. (1986 ), Threonine and homoserine, in: *Biotechnology of Amino Acid Production* (AIDA, K., CHIBATA, I., NAKAYAMA, K., TAKINAMI, K., YAMADA, H., Eds.), pp. 173–182. Tokyo: Kodansha/Amsterdam: Elsevier.

NAKAMORI, S., MORIOKA, H., YOSHINAGA, F., YAMANAKA, S. (1982), Fermentative production of L-proline by DL-3, 4-dehydroproline resistant mutants of glutamate producing bacteria, *Agric. Biol. Chem.* **46**, 487–491.

NAKAMORI, S., ISHIDA, M., TAKAGI, H., ITO, K., MIWA, K., SANO, K. (1987), Improved L-threonine production by the amplification of the gene encoding homoserine dehydrogenase in *Brevibacterium lactofermentum, Agric. Biol. Chem.* **51**, 87–91.

NAKAYAMA, T., ESAKI, N., SUGIE, K., BEREZOV, T. T., TANAKA, H., SODA, K. (1984a), Purification of bacterial L-methionine $\gamma$-lyase, *Anal. Biochem.* **138**, 421–424.

NAKAYAMA, T., ESAKI, N., LEE, W.-J., TANAKA, I., TANAKA, H., SODA, K. (1984b), Purification and properties of L-methionine $\gamma$-lyase from *Aeromonas* sp., *Agric. Biol. Chem.* **48**, 2367–2369.

NAKAZAWA, H., ENEI, H., OKUMURA, S., YOSHIDA, H., YAMADA, H. (1972a), Bacterial synthesis of L-tryptophan and its analogs. 1. Synthesis of L-tryptophan from pyruvate, ammonia and indole, *Agric. Biol. Chem.* **36**, 2523–2528.

NAKAZAWA, H., ENEI, H., OKUMURA, S., YOSHIDA, H., YAMADA, H. (1972b), Enzymatic preparation of L-tryptophan and 5-hydroxy-L-tryptophan, *FEBS Lett.* **25**, 43–45.

NAOKO, N., OSHIHARA, W., YANAI, A. (1987), $\alpha$-Amino-$\varepsilon$-caprolactam racemase for L-lysine production, in: *Biochemistry of Vitamin B6* (KORPELA, T., CHRISTEN, P., Eds.) pp. 449–452. Basel: Birkhäuser.

NEWMAN, P. (1981), in: *Optical Resolution Procedures for Chemical Compounds*, Vol. 2, Part 2, pp. 1057–1074. Optical Resolution Information Center, Manhattan College, New York.

NISHIDA, Y., NAKAMACHI, K., NABE, K., TOSA, T. (1987), Enzymatic production of L-tryptophan from DL-5-indolylmethylhydantoin by *Flavobacterium* sp., *Enzyme Microb. Technol.* **9**, 721–725.

OGAWA, J., SHIMIZU, S. (1994), $\beta$-Ureidopropionase with *N*-carbamoyl-$\alpha$-L-amino acid amidohydrolase activity from an aerobic bacterium, *Pseudomonas putida* IFO12996, *Eur. J. Biochem.* **223**, 625–630.

OGAWA, J., SHIMIZU, S., YAMADA, H. (1993), *N*-Carbamoyl-D-amino acid amidohydrolase from *Comamonas* sp. E222C: Purification and characterization, *Eur. J. Biochem.* **212**, 685–691.

OH, J. W., LEE, J. H., NOH, K. S., LEE, H. H., LEE, J. H., HYUN, H. H. (1991), Improved L-lysine production by the amplification of the *Corynebacterium glutamicum dapA* gene encoding dihydrodipicolinate synthase in *E. coli, Biotechnol. Lett.* **13**, 727–732.

OHKISHI, H., NISHIKAWA, D., KUMAGAI, H., YAMADA, H. (1981), Microbiological synthesis of L-cysteine and its analogs. Part II. Synthesis of L-cysteine and its analogs by intact cells containing cysteine desulfhydrase, *Agric. Biol. Chem.* **45**, 259–263.

OHNO, T., ANDO, O., SUGIMURA, K., TANIAI, M., SUZUKI, M., FUKUDA, S., NAGASE, Y., YAMAMOTO, K., AZUMA, I. (1990), Cloning and nucleotide sequence of the gene encoding arginine deiminase of *Mycoplasma arginini, Infect. Immun.* **58**, 3788–3795.

OHSHIMA, T., SODA, K. (1979), Purification and properties of alanine dehydrogenase from *Bacillus sphaericus, Eur. J. Biochem.* **100**, 29–39.

OHSHIMA, T., MISONO, H., SODA, K. (1978), Properties of crystalline leucine dehydrogenase from *Bacillus sphaericus, J. Biol. Chem.* **253**, 5719–5725.

OHSHIMA, T., NAGATA, S., SODA, K. (1985a), Purification and characterization of thermostable leucine dehydrogenase from *Bacillus stearothermophilus, Arch. Microbiol.* **141**, 407–411.

OHSHIMA, T., WANDREY, C., KULA, M.-R., SODA, K. (1985b), Improvement for L-leucine production in a continuously operated enzyme membrane reactor, *Biotechnol. Bioeng.* **27**, 1616–1618.

OHSHIMA, T., WANDREY, C., CONRAD, D. (1989), Continuous production of 3-fluoro-L-alanine with alanine dehydrogenase, *Biotechnol. Bioeng.* **34**, 394–397.

OHSHIMA, T., NISHIDA, N., BAKTHAVATSALAM, S., KATAOKA, K., TAKADA, H., YOSHIMURA, T., ESAKI, N., SODA, K. (1994), The purification, characterization, cloning and sequencing of the gene for a halostable and thermostable leucine dehydrogenase from *Thermoactinomyces intermedius, Eur. J. Biochem.* **222**, 305–312.

OKA, M., YANG, Y.-S., NAGATA, S., ESAKI, N., TANAKA, H., SODA, K. (1989), Overproduction of thermostable leucine dehydrogenase of *Bacillus stearothermophilus* and its one-step purification from recombinant cells of *Escherichia coli, Biotechnol. Appl. Biochem.* **11**, 307–311.

OKADA, H., YOHDA, M., GIGAHAMA, Y., UENO, Y., OHDO, S., KUMAGAI, H. (1991), Distribution and purification of aspartate racemase in lactic acid bacteria, *Biochim. Biophys. Acta* **1078**, 377–382.

OKAZAKI, N., HIBINO, Y., ASANO, Y., OHMORI, M., NUMAO, N., KONDO, K. (1988), Cloning and nucleotide sequencing of phenylalanine dehydrogenase gene of *Bacillus sphaericus, Gene* **63**, 337–341.

OLIVIERI, R., FASCETTI, E., ANGELINI, L., DEGEN, L. (1983), Microbial transformation of racemic hydantoins to D-amino acids, *Biotechnol. Bioeng.* **23**, 2173–2183.

ORUM, H., RASMUSSEN, O. F. (1992), Expression in *Escherichia coli* of the gene encoding phenylalanine ammonia-lyase from *Rhodosporidium toruloides, Appl. Microbiol. Biotechnol.* **36**, 745–748.

OZAKI, A., KATSUMATA, R., OKA, T., FURUYA, A. (1985), Cloning of the genes concerned in phenylalanine biosynthesis in *Corynebacterium glutamicum* and its application to breeding of a phenylalanine producing strain, *Agric. Biol. Chem.* **49**, 2925–2930.

PAE, K. M., RYU, O. H., YOON, H. S., SHIN, C. S. (1992), Kinetic properties of a L-cysteine desulfhydrase-deficient mutant in the enzymatic formation of L-cysteine from DL-ATC, *Biotechnol. Lett.* **14**, 1143–1148.

PAIK, W. K., BLOCK-FRANKENTHAL, L., BIRNBAUM, S. M., WINITZ, M., GREENSTEIN, J. P. (1957), ε-lysine acylase, *Arch. Biochem. Biophys.* **69**, 56–66.

PLAMANN, M. D., STAUFFER, L. T., URBANOWSKI, M. L., STAUFFER, G. M. (1983), Complete nucleotide sequence of the *Escherichia coli* glyA gene, *Nucleic Acids Res.* **11**, 2065–2075.

PORUMB, H., VANCEA, D., MURESAN, L., PRESECAN, E., LASCU, I., PETRESCU, I., PORUMB, T., POP, R., BARZU, O. (1987), Structural and catalytic properties of L-alanine dehydrogenase from *Bacillus cereus, J. Biol. Chem.* **262**, 4610–4615.

RÖHM, K. H., ETTEN, R. L. V. (1986), Catalysis by hog-kidney aminoacylase does not involve a covalent intermediate, *Eur. J. Biochem.* **160**, 327–332.

RUNSER, S. M., MEYER, P. C. (1993), Purification and biochemical characterization of the hydantoin hydrolyzing enzyme from *Agrobacterium* species. A hydantoinase with no 5,6-dihydropyrimidine amidohydrolase activity, *Eur. J. Biochem.* **213**, 1315–1324.

SAEED, A., YOUNG, D. W. (1992), Synthesis of L-β-hydroxyaminoacids using serine hydroxymethyltransferase, *Tetrahedron* **48**, 2507–2514.

SAKAMOTO, Y., NAGATA, S., ESAKI, N., TANAKA, H., SODA, K. (1990), Gene cloning, purification and characterization of thermostable alanine dehydrogenase of *Bacillus stearothermophilus, J. Ferment. Bioeng.* **69**, 154–158.

SANO, K. (1987), Enzymatic production of L-cysteine from DL-2-aminothiazoline-4-carboxylic acid, in: *Biochemistry of Vitamin B6* (KORPELA, T., CHRISTEN, P., Eds.) pp. 453–456. Basel: Birkhäuser.

SANO, K., MITSUGI, K. (1978), Enzymatic production of L-cysteine from DL-2-amino-$\Delta^2$-thiazoline-4-carboxylic acid by *Pseudomonas thiazolinophilum.* Optimal conditions for enzyme formation and enzymatic reaction, *Agric. Biol. Chem.* **42**, 2315–2321.

SANO, K., YOKOZEKI, K., TAMURA, K., YASUDA, N., NODA, I., MITSUGI, K. (1977a), Microbial conversion of DL-2-amino-$\Delta^2$-thiazoline-4-carboxylic acid to L-cysteine and L-cystine. Screening of microorganisms and identification of products, *Appl. Environ. Microbiol.* **34**, 806–810.

SANO, K., YOKOZEKI, K., EGUCHI, C., KAGAWA, T., NODA, I., MITSUGI, K. (1977b), Enzymatic production of L-tryptophan from L-5-indolylmethylhydantoin and DL-5-indolylmethylhydantoin by newly isolated bacterium, *Agric. Biol. Chem.* **41**, 819–825.

SANO, K., ITOH, K., MIWA, K., NAKAMORI, S. (1987), Amplification of phosphoenolpyruvate carboxylase gene of *Brevibacterium lactofermentum* to improve amino acid production, *Agric. Biol. Chem.* **51**, 597–599.

SATO, T., MORI, T., TOSA, T., CHIBATA, I., FURUI, M., YAMASHITA, K., SUMI, A. (1975), Engineering analysis of continuous production of L-aspartic acid by immobilized *Escherichia coli* cells in fixed beds, *Biotechnol. Bioeng.* **17**, 1797–1804.

SATO, T., TAKAMATSU, S., YAMAMOTO, K., UMEMURA, I., TOSA, T., CHIBATA, I. (1982), Production of L-alanine from ammonium fumarate using 2 types of immobilized microbial cells, *Enzyme Eng.* **6**, 271–272.

SCHENDEL, F. J., BREMMON, C. E., FLICKINGER, M. C., GUETTLER, M., HANSON, R. S. (1990), L-Lysine production at 50 °C by mutants of a newly isolated and characterized methylotrophic *Bacillus* sp., *Appl. Environ. Microbiol.* **56**, 963–970.

SCHIRCH, V. (1982), Serine hydroxymethyltransferase, *Adv. Enzymol.* **53**, 83–112.

SCHIRCH, V., GROSS, T. (1968), Serine transhydroxymethylase, *J. Biol. Chem.* **243**, 5651–5655.

SCHIRCH, V., HOPKINS, S., VILLAR, E., ANGELACCIO, S. (1985), Serine hydroxymethyltransferase from *Escherichia coli*: purification and properties, *J. Bacteriol.* **163**, 1–7.

SCHUSTER, B., RETEY, J. (1994), Serine-202 is the putative precursor of the active site dehydroalanine of phenylalanine ammonia-lyase: site-directed mutagenesis studies on the enzyme from parsley *Petroselinum crispum, FEBS Lett.* **349**, 252–254.

SCHUTTE, H., HUMMEL, W., TSAI, H., KULA, M.-R. (1985), L-Leucine dehydrogenase from *Bacillus cereus, Appl. Microbiol. Biotechnol.* **22**, 306–317.

SEKIMOTO, T., MATSUYAMA, T., FUKUI, T., TANIZAWA, K. (1993), Evidence for lysine 80 as general base catalyst of leucine dehydrogenase, *J. Biol. Chem.* **268**, 27039–27045.

SEKIMOTO, T., FUKUI, T., TANIZAWA, K. (1994), Involvement of conserved lysine 68 of *Bacillus stearothermophilus* leucine dehydrogenase in substrate binding, *J. Biol. Chem.* **269**, 7262–7266.

SHI, W., KIDD, R., GIORGIANI, F., SCHINDLER, J. F., VIOLA, R. E., FARBER, G. K. (1993), Crystallization and preliminary X-ray studies of L-aspartase from *Escherichia coli, J. Mol. Biol.* **234**, 1248–1249.

SHIBATANI, T., KAKIMOTO, T., CHIBATA, I. (1979), Stimulation of L-aspartate $\beta$-decarboxylase formation by L-glutamate in *Pseudomonas dacunhae* and improved production of L-alanine, *Appl. Environ. Microbiol.* **38**, 359–364.

SHIIO, I. (1986), Tryptophan, phenylalanine, and tyrosine, in: *Biotechnology of Amino Acid Production* (AIDA, K., CHIBATA, I., NAKAYAMA, K., TAKINAMI, K., YAMADA, H., Eds.), pp. 188–206. Tokyo: Kodansha/Amsterdam: Elsevier.

SHIIO, I., NAKAMORI, S. (1989), Coryneform bacteria, in: *Fermentation Process Development of Industrial Organisms* (NEWAY, J. O., Ed.), pp. 133–168. New York, Basel: Marcel Dekker.

SHIIO, I., UJIGAWA, K. (1978), Enzymes of the glutamate and aspartate synthetic pathway in a glutamate-producing bacterium, *Brevibacterium flavum J. Biochem.* **84**, 647,–657.

SHIMIZU, S., YAMADA, H. (1984), Microbial and enzymatic processes for the production of pharmacologically important nucleosides, *Trends Biotechnol.* **2**, 137–141.

SHIMIZU, S., SHIOZAKI, S., OHSHIRO, T., YAMADA, H. (1984), High yield synthesis of *S*-adenosylhomocysteine and related nucleosides by bacterial *S*-adenosylhomocysteine hydrolase, *Agric. Biol. Chem.* **48**, 1383–1385.

SHIMIZU, E., OOSUMI, T., HEIMA, H., TANAKA, T., KURASHIGE, J., ENEI, H., MIWA, K., NAKAMORI, S. (1995), Culture conditions for the improvement of L-threonine production using a genetically self-cloned hyperproducing strain of *Escherichia coli, Biosci. Biotech. Biochem.* **59**, 1095–1098.

SHIMOI, H., NAGATA, S., ESAKI, N., TANAKA, H., SODA, K. (1987), Leucine dehydrogenase of a thermophilic anaerobe, *Clostridium thermoaceticum*: gene cloning, purification and characterization, *Agric. Biol. Chem.* **51**, 3375–3381.

SHIRATSUCHI, M., KURONUMA, H., KAWAHARA, Y., YOSHIHARA, Y., MIWA, H., NAKAMORI, S. (1995), Simultaneous and high fermentative production of L-lysine and L-glutamic acid using a strain of *Brevibacterium lactofermentum, Biosci. Biotech. Biochem.* **59**, 83–86.

SHOSTAK, K., SCHIRCH, V. (1988), Serine hydroxymethyltransferase: mechanism of the racemization and transamination of D- and L-alanine, *Biochemistry* **27**, 8007–8014.

SODA, K., ESAKI, N. (1985), Other microbial transaminases, in *Transaminases*, (CHRISTEN, P., METZLER, D. E., Eds.), pp. 463–467. New York: John Wiley & Sons.

SODA, K., OSUMI, T. (1971), Amino acid racemase (*Pseudomonas striata*), *Methods Enzymol.* **17B**, 629–636.

SODA, K., MISONO, H., MORI, K., SAKATO, H. (1971), Crystalline L-leucine dehydrogenase, *Biochem. Biophys. Res. Commun.* **44**, 931–935.

SODA, K., TANAKA, H., TANIZAWA, K. (1988), Thermostable alanine racemase and its application to D-amino acid synthesis, *Enz. Engineering* **542**, 375–382.

SOMERSON, N. L., PHILLIPS, T. (1961), Production of glutamic acid, *Belgian Patent* No. 593,807.

SOMERSON, N. L., PHILLIPS, T. (1962), *U.S. Patent* 3080297.

STOVER, P., ZAMORA, M., SHOSTAK, K., GAUTAM-BASAK, M., SCHIRCH, V. (1992), *Escherichia coli* serine hydroxymethyltransferase: The role of histidine 228 in determining reaction specificity, *J. Biol. Chem.* **267**, 17679–17687.

SUGIE, M., SUZUKI, H. (1980), Optical resolution of DL-amino acids with D-aminoacylase of *Streptomyces, Agric. Biol. Chem.* **44**, 1089–1095.

SUGIMOTO, S., YABUTA. M., TAKAYA, M., SEKI, T., YOSHIDA, T. (1990), Phenylalanine production by periodic induction of gene expression using a temperature-distributed dual fermentor system, *J. Ferment. Bioeng.* **70**, 376–380.

SUGITA, T., KOMATSUBARA, S. (1989), Construction of a threonine-hyper-producing strain of *Serratia marcescens* by amplifying phosphoenolpyruvate carboxylase gene, *Appl. Microbiol. Biotechnol.* **30**, 290–293.

SUGITA, T., KOMATSUBARA, S., KISUMI, M. (1987), Cloning and characterization of the mutated threonine operon (*thrA*$_1$ *5A*$_2$ *5BC*) of *Serratia marcescens, Gene* **57**, 151–158.

SUGIURA, M., TAKAGI, T., KISUMI, M. (1985a), Proline production by reguratory mutants of *Serratia marcescens, Appl. Microbiol. Biotechnol.* **21**, 213–219.

SUGIURA, M., IMAI, Y., TAKAGI, T., KISUMI, M. (1985b), Improvement of a proline-producing strain of *Serratia marcescens* by subcloning of a mutant allele of the proline gene, *J. Biotechnol.* **3**, 47–58.

SUGIURA, M., SUZAKI, S., KISUMI, M. (1987), Improvement of histidine producing strains of *Serratia marcescens* by cloning of mutant allele of the histidine operon on a mini-F plasmid vector, *Agric. Biol. Chem.* **51**, 371–377.

SUZUKI, S., YAMAGUCHI, J., TOKUSHIGE, M. (1973), Studies on aspartase. 1. Purification and molecular properties of aspartase from *Escherichia coli, Biochim. Biophys. Acta* **321**, 369–381.

SUZUKI, S., HORINOUCHI, S., BEPPU, T. (1988), Growth of a tryptophanase-producing thermophile, *Symbiobacterium thermophilum* gen. nov., sp. nov., is dependent on co-culture with a *Bacillus* sp., *J. Gen. Microbiol.* **134**, 2353–2362.

SUZUKI, S., HIRAHARA, T., HORINOUCHI, S., BEPPU, T. (1991), Purification and properties of thermostable tryptophanase from an obligately symbiotic thermophile, *Symbiobacterium thermophilum, Agric. Biol. Chem.* **55**, 3059–3066.

SUZUKI, H., NISHIHARA, K., USUI, N., MATSUI, H., KUMAGAI, H. (1993), Cloning and nucleotide sequence of *Erwinia herbicola* AJ2982 tyrosine phenol-lyase gene, *J. Ferment. Bioeng.* **75**, 145–148.

SYLDATK, C., COTORAS, D., DOMBACH, G., GROSS, C., KALLWASS, H., WAGNER, F. (1987), Substrate specificity and stereospecificity, induction and metallodependence of a microbial hydantoinase, *Biotechnol. Lett.* **9**, 25–30.

SYLDATK, C., LEUFER, A., MULLER, R., HOKE, H. (1990), Production of optically pure D- and L-$\alpha$-amino acids by bioconversion of DL-5-monosubstituted hydantoin derivatives, in: *Advances in Biochemical Engineering/Biotechnology,* Vol. 41 (FIECHTER, A., Ed.), pp. 29–75. Berlin: Springer-Verlag.

SYLDATK, C., MULLER, R., SIEMANN, M., KROHN, K., WAGNER, F. (1992a), Microbial and enzymatic production of D-amino acids from DL-5-monosubstituted hydantoins, in: *Biocatalytic Production of Amino Acids and Derivatives* (ROZZELL, J. D., WAGNER, F., Eds.), pp. 75–128. München: Hanser.

SYLDATK, C., MULLER, R., PIETZSCH, M., WAGNER, F. (1992b), Microbial and enzymatic production of L-amino acids from DL-5-monosubstituted hydantoins, in: *Biocatalytic Production of Amino Acids and Derivatives* (ROZZELL, J. D., WAGNER, F., Eds.), pp. 129–176. München: Hanser.

TACHIKI, T., MATSUMOTO, H., YANO, T., TOCHIKURA, T. (1981a), Glutamine production by coupling with energy transfer: employing yeast cell-free extracts and *Gluconobacter* glutamine synthetase, *Agric. Biol. Chem.* **45**, 705–710.

TACHIKI, T., MATSUMOTO, H., YANO, T., TOCHIKURA, T. (1981b), Glutamine production by coupling with energy transfer: employing yeast cells and *Gluconobacter* glutamine synthetase, *Agric. Biol. Chem.* **45**, 1237–1241.

TAKADA, H., YOSHIMURA, T., OHSHIMA, T., ESAKI, N., SODA, K. (1991), Thermostable phenylalanine dehydrogenase of *Thermoactinomyces intermedius*: Cloning, expression, and sequencing of its gene, *J. Biochem.* **109**, 371–376.

TAKAHASHI, K., HATANO, K. (1989), *Eur. Patent Appl.* 0304021A2.

TAKAMATSU, S., YAMAMOTO, K., TOSA, T., CHIBATA, I. (1981), Stabilization of L-aspartate $\beta$-decarboxylase activity of *Pseudomonas dacunhae* immobilized with carrageenan, *J. Ferment. Technol.* **59**, 489–493.

TAKAMATSU, S., UMEMURA, I., YAMAMOTO, K., SATO, T., TOSA, T., CHIBATA, I. (1982), Production of L-alanine from ammonium fumarate using 2 immobilized microorganisms: Elimination of side reactions, *Eur. J. Appl. Microbiol. Biotechnol.* **15**, 147–152.

TAKAMURA, Y., KITAMURA, I., IKURA, M., KONO, K., OZAKI, A. (1966), Studies on enzymatic production of L-aspartic acid from maleic acid. Part

II. Induction effect of malonic acid, *Agric. Biol. Chem.* **30**, 345–350.

TAKINAMI, K., OKADA, H., TSUNODA, T. (1964), Biochemical effects of fatty acid and its derivatives on L-glutamic acid fermentation. Part II. Effective chemical structure of fatty acid and derivatives on the accumulation of L-glutamic acid in biotin sufficient medium, *Agric. Biol. Chem.* **28**, 114–119.

TAMAKI, N., MIZUTANI, N., KIKUGAWA, M., FUJIMOTO, S., MIZOTA, C. (1987), Purification and properties of $\beta$-ureidopropionase from the rat liver, *Eur. J. Biochem.* **169**, 21–26.

TANAKA, M., KATO, Y., KINOSHITA, S. (1961), Glutamic acid racemase from *Lactobacillus fermenti*: purification and properties, *Biochem. Biophys. Res. Commun.* **4**, 114–117.

TANAKA, H., ESAKI, N., SODA, K. (1977), Properties of L-methionine $\gamma$-lyase from *Pseudomonas ovalis, Biochemistry* **16**, 100–106.

TANAKA, H., YAMADA, N., ESAKI, N., SODA, K. (1985), Synthesis of D-$\alpha$-aminobutyrate with methionine $\gamma$-lyase and D-amino acid aminotransferase, *Agric. Biol. Chem.* **49**, 2525–2527.

TANI, S., TSUJIMOTO, N., KAWATA, Y., TOKUSHIGE, M. (1990), Overproduction and crystallization of tryptophanase from recombinant cells of *Escherichia coli, Biotechnol. Appl. Biochem.* **12**, 28–33.

TANIZAWA, K., MASU, Y., ASANO, S., TANAKA, H., SODA, K. (1989), Thermostable D-amino acid aminotransferase from a thermophilic *Bacillus* species, *J. Biol. Chem.* **264**, 2445–2449.

TINGEY, S. V., WALKER, E. L., CORUZZI, G. M. (1987), Glutamine synthetase gene of pea encode distinct polypeptides which are differentially expressed in leaves, roots and nodules, *EMBO J.* **6**, 1–9.

TOKUSHIGE, M., TSUJIMOTO, N., ODA, T., HONDA, T., YUMOTO, N., ITO, S., YAMAMOTO, M., KIM, E. H., HIRAGI, Y. (1989), Role of cysteine residues in tryptophanase for monovalent cation induced activation, *Biochimie* **71**, 711–720.

TOKUYAMA, S., HATANO, K., TAKAHASHI, T. (1994a), Discovery of a novel enzyme, *N*-acylamino acid racemase in an actinomycete: Screening, isolation, and identification, *Biosci. Biotech. Biochem.* **58**, 24–27.

TOKUYAMA, S., MIYA, H., HATANO, K., TAKAHASHI, T. (1994b) , Purification and properties of a novel enzyme, *N*-acylamino acid racemase, from *Streptomyces atratus, Appl. Microbiol. Biotechnol.* **40**, 835–840.

TOSA, T., MORI, T., FUSE, N., CHIBATA, I. (1966), Studies on continuous enzyme reactions. I. Screening of carriers for preparation of water-insoluble aminoacylase, *Enzymologia Acta Biocatalytica* **31**, 214–224.

TOSAKA, O., TAKINAMI, K. (1986). Lysine, in: *Biotechnology of Amino Acid Production* (AIDA, K., CHIBATA, I., NAKAYAMA, K., TAKINAMI, K., YAMADA, H., Eds.), pp. 152–172. Tokyo: Kodansha/Amsterdam: Elsevier.

TRIBE, D. E., PITTARD, J. (1979), Hyperproduction of tryptophan by *Escherichia coli*: Genetic manipulation of the pathways leading to tryptophan formation, *Appl. Environ. Microbiol.* **35**, 181–190.

TSAI, Y.-C., TSENG, C.-P., HSIAO, K.-M. (1988), Production and purification of D-aminoacylase from *Alcaligenes denitrificans* and taxonomic study of the strain, *Appl. Environ. Microbiol.* **54**, 984–989.

TSUCHIDA, T., KUBOTA, K., YOSHIHARA, Y., KIKUCHI, K., YOSHINAGA, F. (1987), Fermentative production of L-glutamine by sulfaguanidine resistant mutants derived from L-glutamate producing bacteria, *Agric. Biol. Chem.* **51**, 2089–2094.

TURNER, S. R., IRELAND, R., MORGAN, C., RAWSTHORNE, S. (1992), Identification and localization of multiple forms of serine hydroxymethyltransferase in pea (*Pisum sativum*) and characterization of a cDNA encoding a mitochondrial isoform, *J. Biol. Chem.* **267**, 13528–13534.

UMEMURA, I., TAKAMATSU, S., SATO, T., TOSA, T., CHIBATA, I. (1984), Improvement of production of L-aspartic acid using immobilized microbial cells, *Appl. Microbiol. Biotechnol.* **20**, 291–295.

VALI, Z., KILAR, F., LAKATOS, S., VENYAMINOV, S. A., ZAVODSZKY, P. (1980), L-Alanine dehydrogenase from *Thermus thermophilus, Biochim. Biophys. Acta* **615**, 34–47.

WADA, S. (1959), Studies on the asymmetrical hydrolysis of $N^{\alpha}$, $N^{\varepsilon}$ diacyl-DL-lysine, *J. Biochem.* **46**, 445–452.

WASSERMAN, S. A., DAUB, E., GRISAFI, P., BOTSTEIN, D., WALSH, C. T. (1984), Catabolic alanine racemase from *Salmonella typhimurium*: DNA sequence, enzyme purification, and characterization, *Biochemistry* **23**, 5182–5187.

WATABE, K., ISHIKAWA T., MUKOHARA, Y., NAKAMURA, H. (1992a), Cloning and sequencing of the genes involved in the conversion of 5-substituted hydantoins to the corresponding L-amino acids from the native plasmid of *Pseudomonas* sp. strain NS671, *J. Bacteriol.* **174**, 962–969.

WATABE, K., ISHIKAWA, T., MUKOHARA, Y., NAKAMURA, H. (1992b), Identification and sequencing of a gene encoding a hydantoin racemase from the native plasmid of *Pseudomonas* sp. strain NS671, *J. Bacteriol.* **174**, 3461–3466.

WATABE, K., ISHIKAWA, T., MUKOHARA, Y., NAKAMURA, H. (1992c), Purification and characterization of the hydantoin racemase of *Pseudomonas* sp. strain NS671 expressed in *Escherichia coli, J. Bacteriol.* **174**, 7989–7995.

WICHMANN, R., WANDREY, C. (1981), Continuous enzymatic transformation in an enzyme membrane reactor with simultaneous NAD(H) regeneration, *Biotechnol. Bioeng.* **23**, 2789–2802.

WILSON, S. D., WANG, M. L., FILPULA, D. R., WHITLOW, M., SHORR, R. (1993), Cloning and sequencing of the gene for arginine deiminase from *Pseudomonas putida, FASEB J.* **7**, 176.

YAHATA, S., TSUTSUI, H. YAMADA, K., YONEHARA, T. (1993), Fermentative production of D-alanine, in: *Proc. Ann. Meet.* (Agric. Chem. Soc. Jpn.), p. 92. Japan Society for Bioscience, Biotechnology, and Agrochemistry.

YAMADA, H., KUMAGAI, H. (1978), Microbial and enzymatic processes for amino acid production, *Pure Appl. Chem.* **50**, 1117–1127.

YAMADA, H., KUMAGAI, H., NAGATE, T., YOSHIDA, H. (1970), Crystalline threonine aldolase from *Candida humicola, Biochem. Biophys. Res. Commun.* **39**, 53–58.

YAMADA, H., KUMAGAI, H., NAGATE, T., YOSHIDA, H. (1971), Formation of threonine aldolase by bacteria and yeasts, *Agric. Biol. Chem.* **35**, 1340–1345.

YAMADA, H., TAKAHASHI, S., KII, Y., KUMAGAI, H. (1978), Microbial transformation of hydantoins to amino acids. 1. Distribution of hydantoin hydrolyzing activity in microorganisms, *J. Ferment. Technol.* **56**, 484–491.

YAMADA, H., NAGASAWA, T., OHKISHI, H., KAWAKAMI, B., TANI, Y. (1981), Synthesis of D-cysteine from 3-chloro-D-alanine and hydrogen sulfide by 3-chloro-D-alanine hydrogen chloride-lyase (deaminating) of *Pseudomonas putida, Biochem. Biophys. Res. Commun.* **100**, 1104–1110.

YAMADA, H. KANZAKI, H., NAGASAWA, T. (1984), Synthesis of L-cystathionine by the $\gamma$-replacement reaction of cystathionine $\gamma$-lyase from *Streptomyces phaeochromogenes, J. Biotechnol.* **1**, 205–217.

YAMAGATA, S., TAKESHIMA, K. (1976), *O*-Acetylserine and *O*-acetylhomoserine sulfhydrylase of yeast: Further purification and characterization as a pyridoxal enzyme, *J. Biochem.* **80**, 777–785.

YAMAGATA, S., DANDREA, R. J., FUJISAKI, S., ISAJI, M., NAKAMURA, K. (1993), Cloning and bacterial expression of the *cys*3 gene encoding cystathionine $\gamma$-lyase of *Saccharomyces cerevisiae* and the physiological and enzymatic properties of the protein, *J. Bacteriol.* **175**, 4800–4804.

YAMAMOTO, K., SATO, T., TOSA, T., CHIBATA, I. (1974), Continuous production of L-citrulline by immobilized *Pseudomonas putida* cells, *Biotech. Bioeng.* **16**, 1589–1599.

YAMAMOTO, K., TOSA, T., CHIBATA, I. (1980), Continuous production of L-alanine using *Pseudomonas dacunhae* immobilized with carrageenan, *Biotechnol. Bioeng.* **22**, 2045–2054.

YAMASHIRO, A., YOKOZEKI, K., KANO, H., KUBOTA, K. (1988), Enzymatic production of L-amino-acids from the corresponding 5-substituted hydantoins by a newly isolated bacterium, *Bacillus brevis* AJ-12299, *Agric. Biol. Chem.* **52**, 2851–2856.

YAMAUCHI, T., CHOL, S. Y., OKADA, H., YOHDA, M., KUMAGAI, H., ESAKI, N., SODA, K. (1992), Properties of aspartate racemase, a pyridoxal 5′-phosphate-independent amino acid racemase, *J. Biol. Chem.* **267**, 18361–18364.

YOKOZEKI, K., KUBOTA, K. (1987), Enzymatic production of D-amino acids from 5-substituted hydantoins. 3. Mechanism of asymmetric production of D-amino acids from the corresponding hydantoins by *Pseudomonas* sp., *Agric. Biol. Chem.* **51**, 721–728.

YOKOZEKI, K., NAKAMORI, S., YAMANAKA, S., EGUCHI, C., MITSUGI, K., YOSHINAGA, F. (1987), Enzymatic production of D-amino acids from 5-substituted hydantoins. 2. Optimal conditions for the enzymatic production of D-amino acids from the corresponding 5-substituted hydantoins, *Agric. Biol. Chem.* **51**, 715–719.

YONAHA, K., YORIFUJI, T., YAMAMOTO, T., SODA, K. (1975), Alanine racemase of *Bacillus subtilis* var. *aterrimus, J. Ferment. Technol.* **53**, 579–587.

YORIFUJI, T., OGATA, K., SODA, K. (1971), Arginine racemase of *Pseudomonas graveolens, J. Biol. Chem.* **246**, 5085–5092.

YOSHIDA, H. (1986), Arginine, citrulline, and ornithine, in: *Biotechnology of Amino Acid Production* (AIDA, K., CHIBATA, I., NAKAYAMA, K., TAKINAMI, K., YAMADA, H., Eds.), pp. 131–143. Tokyo: Kodansha/Amsterdam: Elsevier.

YOSHIDA, A., FREESE, E. (1964), Purification and chemical characterization of alanine dehydrogenase of *Bacillus subtilis, Biochim. Biophys. Acta* **92**, 33–43.

YOSHIDA, H., KUMAGAI, H., YAMADA, H. (1974), Catalytic properties of tryptophanase from *Proteus rettgeri, Agric. Biol. Chem.* **38**, 463–464.

YOSHII, H., YOSHIMURA, M., NAKAMORI, S., INOUE, S. (1993), L-Glutamic acid fermentation on a commercial scale by use of cane molasses with inverted sucrose, *Nippon Nogeikagaku Kaishi* **67**, 955–960.

# 15 Nucleotides and Related Compounds

AKIRA KUNINAKA

Choshi, Chiba-ken 288

# 1 Introduction

Industrial production of nucleotides started in Japan in 1961, since some nucleotides were found to have flavor enhancing activity (KUNINAKA, 1960). Nucleotides and related compounds are now used not only as flavor enhancers but also as biochemicals and pharmaceuticals. Furthermore, it has been apparent from recent studies that the role of nucleotides in the diet is an extremely important issue.

For example, a roundtable symposium "*Nucleotides and Nutrition*" was held in New Orleans, March 1993. In the symposium, the role of nucleotides as dietary supplements, the biochemistry and physiology of nucleotides, sources of nucleotides in the diet of infants, nucleotide uptake and metabolism by intestinal epithelial cells, the action of nucleotides on humoral immune responses, effects of nucleotides on the cellular immune system, the intestinal system and the hepatic system, and the role of nucleotides in adult nutrition were discussed (WALKER, 1994). In addition, a workshop *"Nutritional and Biological Significance of Dietary Nucleotides and Nucleic Acids"* was also held in Granada, September 1993.

Usually, nucleotides as well as amino acids are produced by microorganisms today. Industrial microbial production of nucleotides has mostly been developed in Japan. The main purpose was to produce two flavor enhancers, 5′-inosinate (IMP) and 5′-guanylate (GMP). At present, in addition to flavor enhancers, a number of new pharmaceuticals are being produced from nucleotides.

In this chapter, nucleotides and related compounds are discussed as both flavor enhancers and pharmaceuticals, and recent research and developments in microbial production of nucleotides and related compounds are reviewed.

# 2 IMP and GMP as Flavor Enhancers

At present, three naturally occurring compounds are commercially available to enhance, potentiate, or improve the flavor of foods, i.e., monosodium L-glutamate (MSG), the disodium salt of IMP, and the disodium salt of GMP. In Europe and the U.S. they are generally called "flavor enhancers" or "flavor potentiators" and in Japan "umami compounds".

Umami or umami taste is a Japanese term that means deliciousness, savory taste, or palatable taste. As a matter of fact, the introduction of MSG, IMP, and GMP was based on studies on umami components of three Japanese traditional foods from which aqueous extracts have been prepared as soup stocks for seasoning Japanese meals for centuries. MSG, IMP, and GMP were isolated from dried sea tangle (kombu), dried bonito (katsuobushi), and dried Japanese mushroom (shiitake), respectively. The history of flavor enhancers was reviewed by KUNINAKA (1981).

The taste of the 5′-nucleotides (Fig. 1) is different from that of MSG. Furthermore, there is a marked synergistic action between a 5′-nucleotide and MSG (KUNINAKA, 1960, 1981). However, in common both tastes are called umami or umami taste because they are clearly different from four basic taste qualities sweet, sour, salty, and bitter. Thus,

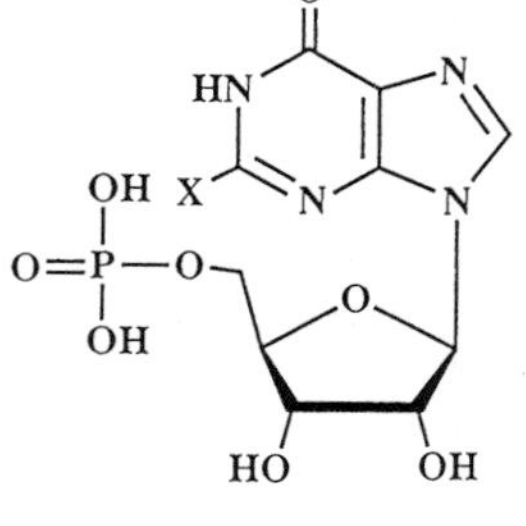

**Fig. 1.** Structure of flavor nucleotides (KUNINAKA, 1960). IMP: X=H, GMP: $X=NH_2$, and XMP: (xanthosine-5′-monophosphate, 5′-xanthylate) X=OH.

umami is ready to be internationally recognized as the fifth basic taste. The First *International Symposium on Umami* was held in Hawaii, October 1985 (KAWAMURA et al., 1987) and the second in Sisily, October 1990 (KAWAMURA et al., 1991). The *Umami International Symposium* was held in Tokyo, July 1993 (KAWAMURA and KURIHARA, 1993).

# 3 Nucleotide-Related Pharmaceuticals

## 3.1 Known Nucleotide-Related Pharmaceuticals

Pharmaceuticals derived from nucleotides are classified into two groups: biologically active substances (agonists or metabolic activators) and antimetabolites.

Typical examples of biologically active pharmaceuticals already approved are:

- CDP-choline (treatment of disturbance of consciousness, pancreatitis),
- ATP (cardiac insufficiency, cerebrovascular disorders, muscular atrophy, gastroptosis),
- adenosine (coronary disorder),
- FAD (vitamin $B_2$ deficiency).

Medical application of derivatives of cyclic adenosine 3′,5′-monophosphate (cAMP) as a cardiotonic agent has also been reported (ISHIYAMA, 1990).

Typical examples of pharmaceuticals already approved as antimetabolites are:

- cytarabine (ara-C) (treatment of acute leukemia),
- idoxuridine (corneal lesions),
- 5-fluorouracil (solid tumors),
- azidothymidine, dideoxyinosine, dideoxycytidine (AIDS).

Ribavirin is a nucleoside that appears to be the first synthetic, non-interferon-inducing, broad-spectrum antiviral agent. A number of nucleosides and nucleotides are now being developed as antitumor and antiviral agents (MARUMOTO, 1987; CHU and BAKER, 1993). Furthermore, there are more than 100 naturally occurring purine–pyrimidine-related antibiotics (SUZUKI, 1984).

## 3.2 Examples of New Nucleotide-Related Pharmaceuticals Developed in Japan

Since 1975, it has been attempted to create new drugs mainly from industrial degradation products of RNA, several examples are introduced here.

### 3.2.1 2-Octynyladenosine

Although adenosine has long been recognized as physiologically important, the therapeutic potential of adenosine is considered to be rather low. One of the reasons is its wide variety of effects.

There are two main subtypes of the cell surface-bound adenosine receptor: $A_1$ involved in the inhibition of adenylate cyclase, and $A_2$ involved in the stimulation of adenylate cyclase. The hypotensive action of adenosine occurs via two different mechanisms, $A_2$ receptor-mediated vasodilatation and $A_1$ receptor-mediated cardiac depression. As the cardiodepressive action is not acceptable for the treatment of hypertension, the development of vasoselective or $A_2$-selective adenosine derivatives was attempted.

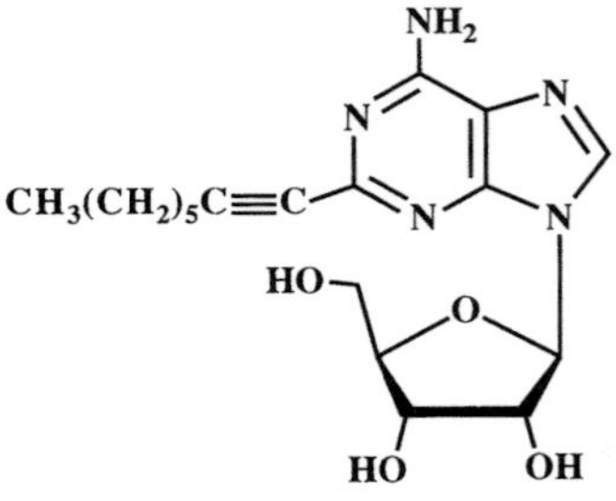

$CH_3(CH_2)_5C{\equiv}C$–, $NH_2$, N, HO, O, HO, OH

**Fig. 2.** 2-Octynyladenosine.

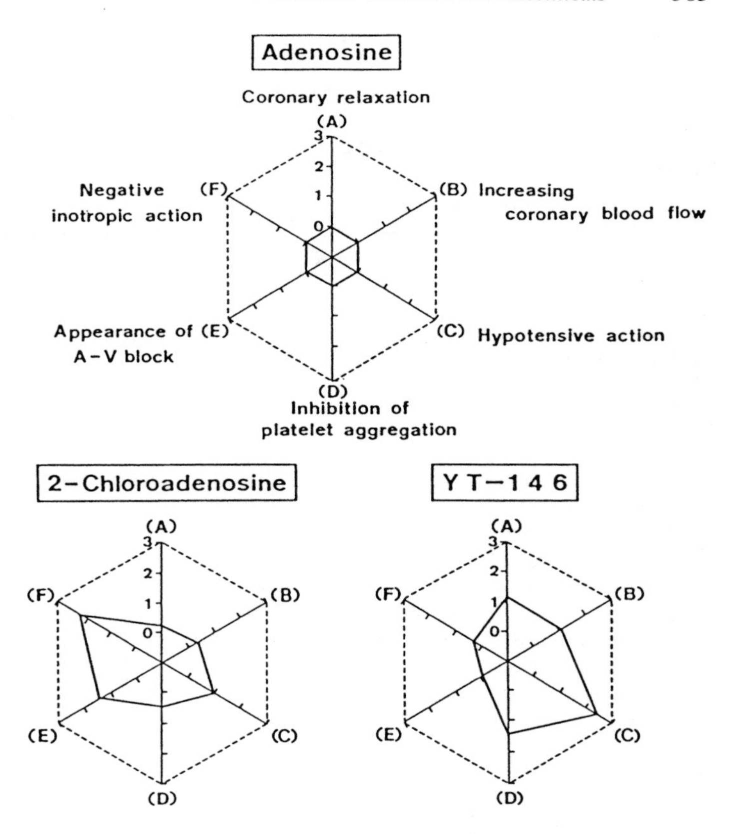


**Fig. 3.** Spectrum of selectivity for cardiovascular variables of 2-octynyladenosine (YT-146), adenosine, and 2-chloroadenosine. The relative potencies are given in terms of the logarithm (KOGI et al., 1991).

Among a large number of adenosine derivatives synthesized, 2-octynyladenosine (Fig. 2) showed a particularly selective receptor binding activity to the $A_2$ receptor. A potent and long-lasting antihypertensive effect was observed after oral administration.

Compared to adenosine and 2-chloroadenosine, a known nonselective adenosine agonist resistant to adenosine deaminase, 2-octynyladenosine was shown to be most potent in lowering blood pressure, causing coronary vasodilatation, increasing coronary blood flow, and inhibiting platelet aggregation (Fig. 3; KOGI et al., 1991).

Therefore, the coronary vasodilator 2-octynyladenosine is expected to be a potent, orally active, and long-acting hypotensive agent with a low cardiodepressive activity.

The compound is now under clinical study in Japan.

### 3.2.2 Cytarabine Ocfosfate – 1-β-D-Arabinofuranosylcytosine-5′-stearylphosphate (Starasid® cap.)

1-β-D-arabinofuranosylcytosine (ara-C) is an indispensable agent for the treatment of acute leukemia. In clinical use, however, the compound proved to be rapidly deaminated to an inert metabolite, 1-β-D-arabinofuranosyluracil (ara-U). To overcome this disadvantage, KODAMA et al. (1989) synthesized a number of ara-C derivatives and selected 1-β-D-arabinofuranosylcytosine-5′-stearylphosphate (C18PCA) (Fig. 4). The compound was officially named cytarabine ocfosfate and approved as a new drug (Starasid® cap.) for the treatment of acute non-lymphocytic leukemia and myelodysplastic syndrome in Japan in 1992.

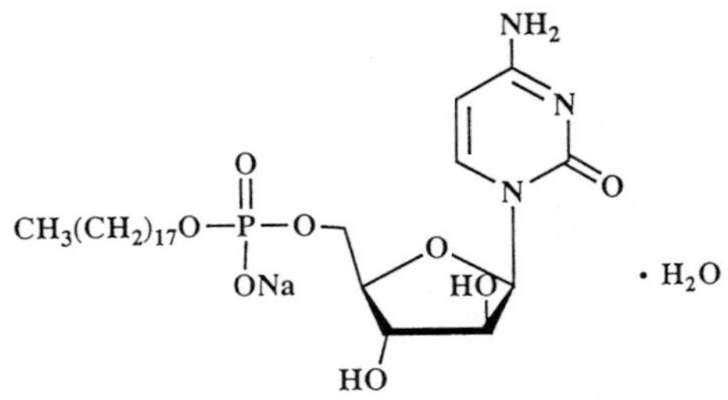

**Fig. 4.** Cytarabine ocfosfate (Starasid® cap.), approved in Japan for the treatment of acute nonlymphocytic leukemia and myelodysplastic syndrome in 1992.

After cytarabine ocfosfate was orally administered to rats, one of the metabolites, C-C3PCA, was observed in the liver at a constant level. C-C3PCA is gradually degraded to ara-CMP, and ara-CMP is rapidly degraded to ara-C (Fig. 5). Actually, the plasma concentration of ara-C has proved to be maintained satisfactorily in a patient with orally administered cytarabine ocfosfate (TAKAYAMA, 1993).

### 3.2.3 Sorivudine – 1-β-D-Arabinofuranosyl-E-5-(2-bromovinyl)uracil, BV-araU (Usevir® tab.)

Among the 5-substituted derivatives of 1-β-D-arabinofuranosyluracil, 1-β-D-arabinofuranosyl-*E*-5-(2-bromovinyl)uracil (BV-araU) (Fig. 6) most selectively inhibited replication of varicella–zoster virus (VZV).

BV-araU was more than 10 times as active as *E*-5-(2-bromovinyl)-2′-deoxyuridine (BV-DU), and inhibited almost completely plaque development of VZV at a concentration as low as 1 ng $mL^{-1}$. On the other hand, BV-araU did not affect the growth of human embryonic lung fibroblasts (HEL-F) even at a concentration higher than 800 µg $mL^{-1}$. The therapeutic index (50% inhibitory dose for human cells per 50% plaque reduction dose) was extraordinarily high (>3,100,000) (MACHIDA, 1986).

The marked inhibition of VZV replication by BV-araU is due to the strong inhibition of VZV DNA synthesis by BV-araU triphos-

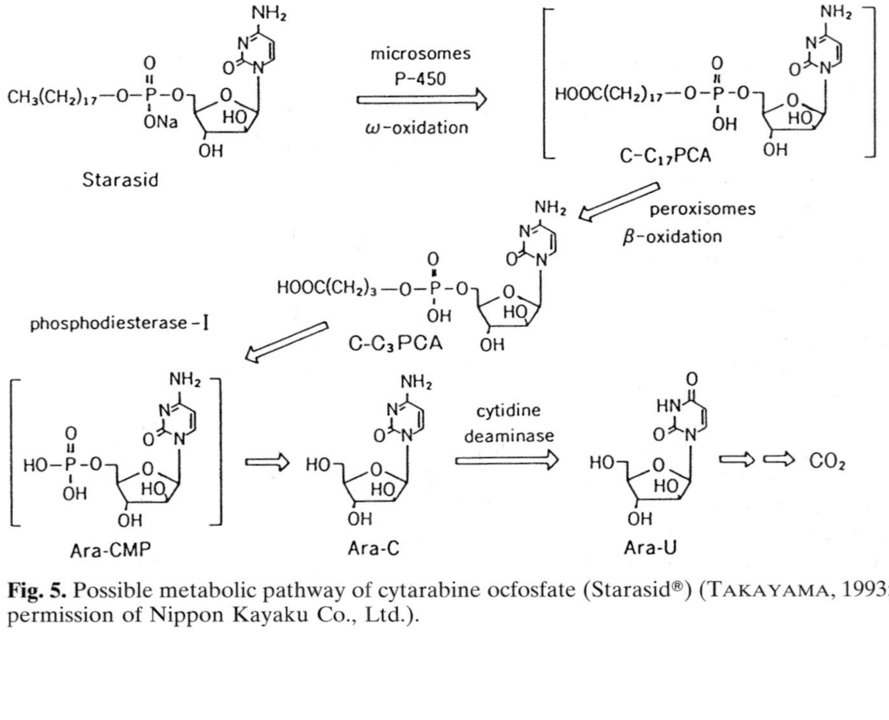

**Fig. 5.** Possible metabolic pathway of cytarabine ocfosfate (Starasid®) (TAKAYAMA, 1993; reproduced with permission of Nippon Kayaku Co., Ltd.).

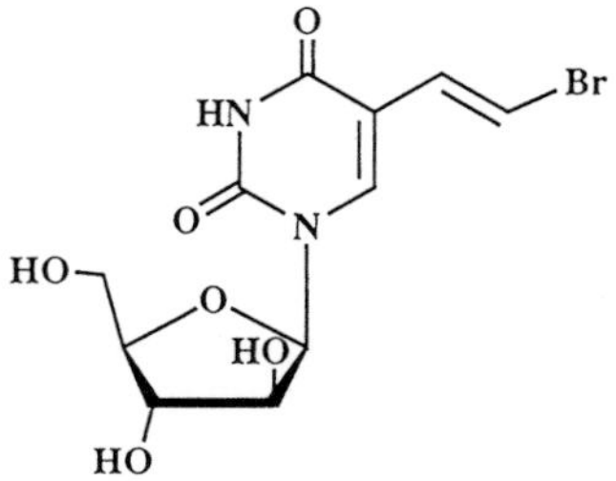

Fig. 6. Sorivudine (Usevir® tab.), approved in Japan for the treatment of herpes zoster in 1993.

phate after selective phosphorylation of BV-araU in thymidine kinase-positive VZV-infected cells without detectable incorporation into VZV DNA.

BV-araU was officially named sorivudine, and approved as a new drug (Usevir® tab.) for the treatment of herpes zoster in Japan in 1993.

### 3.2.4 DMDC – 2′-Deoxy-2′-methylidenecytidine, 1-(2-Deoxy-2-methylene-β-D-erythropentofuranosyl)cytosine

DMDC contains a double bond function in the 2′-position of 2′-deoxycytidine (Fig. 7). This compound, as well as cytarabine ocfosfate, was developed to overcome the disadvantages of ara-C.

DMDC is resistant to mouse kidney cytidine deaminase and has a unique *in vitro* antitumor spectrum different from that of the representative antitumor pyrimidine derivatives such as ara-C and 5-fluorouracil. It is noteworthy that DMDC is cytotoxic not only to leukemic cell lines but also to several solid tumor cell lines (YAMAGAMI et al., 1991).

DMDC is now under clinical study, and is expected to be a promising agent in the therapy of human cancer with a profile different from those of the known antitumor pyrimidine derivatives.

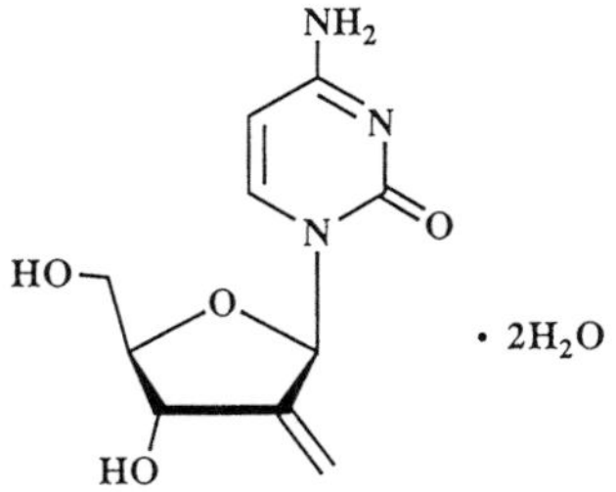

Fig. 7. 2′-deoxy-2′-methylidenecytidine (DMDC).

# 4 Production of 5′-Nucleotides

Since 1961, the flavor 5′-nucleotides IMP and GMP were industrially produced in Japan. The disodium salts of IMP and GMP and their combination with MSG have attained worldwide recognition as new, excellent flavor enhancers (KUNINAKA et al., 1964; KUNINAKA, 1981).

At present, the flavor 5′-nucleotides are microbiologically produced in Japan by four independent methods (Fig. 8):

- enzymic degradation of RNA,
- combination of fermentative production of inosine or guanosine and chemical phosphorylation thereof,
- direct fermentative production of IMP,
- direct fermentative production of XMP (see Fig. 1) and its conversion to GMP.

The RNA degradation method was the first to be established in 1961, and several years later the other three methods were developed industrially (KUNINAKA, 1988).

(1) RNA —Microbial enzyme→ 5′-Nucleotides

(2) Medium —Microorganism→ Inosine or Guanosine
↓ Chemical conversion
IMP or GMP

(3) Medium —Microorganism→ IMP

(4) Medium —Microorganism→ XMP
↓ Microbial enzyme
GMP

Fig. 8. Current processes for industrial production of flavor 5′-nucleotides.

Two of the four nucleotides obtained by RNA degradation are used as flavor enhancers: IMP (derived from AMP) and GMP; the two non-flavoring 5′-nucleotides, CMP and UMP, are used as starting materials for pharmaceutically valuable compounds.

## 4.1 Production of 5′-Nucleotides from Yeast RNA by Nuclease $P_1$

### 4.1.1 History

In 1955, it was found that among three isomers of inosinic acid only 5′-IMP had a flavor enhancing activity. Neither 2′-IMP nor 3′-IMP had this activity (Fig. 9). Therefore, screening of microorganisms capable of degrading RNA into 5′-nucleotides was carried out systematically and a strain of *Penicillium citrinum* was selected in 1957. Among the 5′-nucleotides purified from a *P. citrinum* nuclease digest of RNA, GMP and XMP as well as IMP were found to have flavor enhancing activity (see Fig. 1). The flavor enhancing activity of GMP was several times higher than that of IMP. Furthermore, there was a marked synergistic effect of glutamate and the flavor nucleotides. Thus, an economic basis for the production of IMP and GMP from RNA was established, and their isolation at an industrial scale by use of yeast RNA and the *Penicillium* enzyme began in 1961. At almost the same time, industrial degradation of RNA with a *Streptomyces* enzyme started; at present, however, this *Streptomyces* enzyme is no longer used.

The *Penicillium* enzyme was named nuclease $P_1$ and it is regarded as an enzyme similar to endonuclease $S_1$ (*Aspergillus*; EC 3.130.1). The crude preparation of nuclease $P_1$ is being used for the degradation of several tons of yeast RNA per day for the industrial production of 5′-nucleotides, while the commercially available purified preparation of the enzyme is used as an important biochemical tool worldwide. It may be said that the origin of the Japanese nucleotide industry was the discovery of nuclease $P_1$ (KUNINAKA, 1976).

### 4.1.2 Properties of Nuclease $P_1$

Nuclease $P_1$ cleaves substantially all 3′-5′-phosphodiester linkages of single-stranded polynucleotides and 3′-phosphomonoester linkages of mono- and oligonucleotides terminated by 3′-phosphate. The enzyme degrades both single-stranded DNA and RNA by endo- and exonucleolytic cleavage but does not actually attack double-stranded nucleic acids, especially in the presence of >400 mM sodium chloride. As shown in Fig. 10, nuclease $P_1$ does not attack *p*-nitrophenyl 5′-TMP, which is easily split by the snake venom phosphodiesterase, but splits *p*-nitrophenyl 3′-TMP into thymidine and *p*-nitrophenylphosphate (FUJIMOTO et al., 1974).

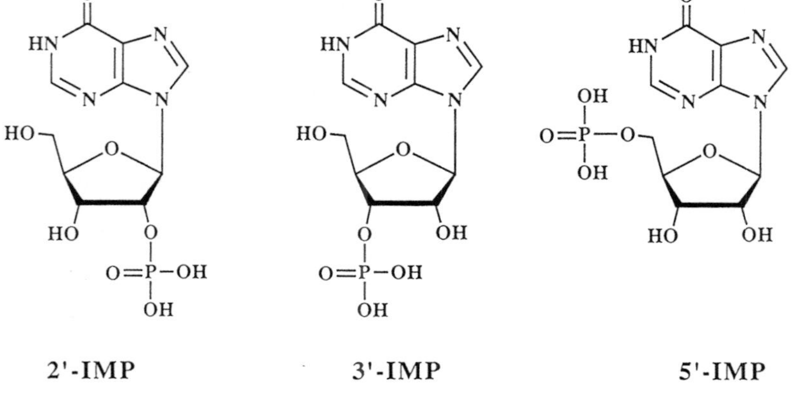

2′-IMP    3′-IMP    5′-IMP

**Fig. 9.** Three isomers of IMP. Only 5′-IMP has umami taste or flavor enhancing activity.

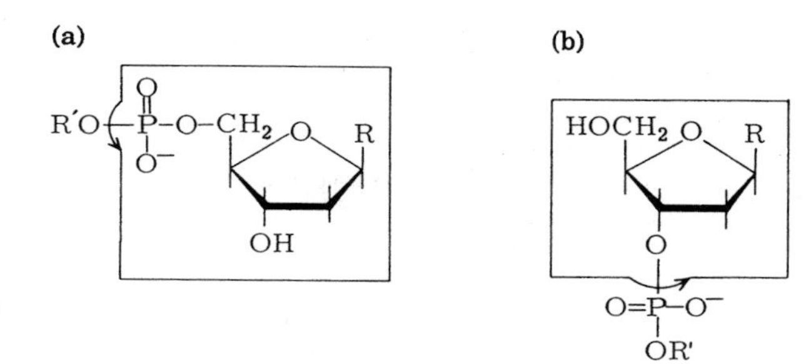


**Fig. 10.** Hydrolysis of synthetic substrates by snake venom phosphodiesterase (**a**) and nuclease $P_1$ (**b**); R: thymine; R′: *p*-nitrophenyl.

The optimum temperature is at about 70 °C. The optimum pH ranges from pH 4.0 to 8.5, depending on the kind of substrate: pH 4.0 for poly(U); pH 4.5–5 for deoxyribonucleoside 3′-monophosphates, poly(I), and poly(G); pH 5–6 for RNA and DNA; pH 6 for 3′-CMP, 3′-UMP, poly(A), and poly(C); pH 7–8.5 for 3′-AMP and 3′-GMP.

Nuclease $P_1$, is a typical zinc enzyme (COLEMAN, 1992). Its primary structure was elucidated by MAEKAWA et al. (1991) and its crystal structure by VOLBEDA et al. (1991).

### 4.1.3 Industrial Production of 5′-Nucleotides by Crude Nuclease $P_1$

A pigmentless, nuclease $P_1$-rich mutant of *P. citrinum* is grown on wheat bran. The aqueous extract of this culture contains not only thermostable nuclease $P_1$ but also thermolabile phosphomonoesterase capable of hydrolyzing 5′-nucleotides. The latter enzyme can completely be inactivated by heating without any loss of nuclease $P_1$.

An aqueous solution of yeast RNA is incubated with the heat-treated crude enzyme solution under optimal conditions. RNA is completely degraded into 5′-nucleotides. After incubation, the four different 5′-nucleotides AMP, GMP, CMP, and UMP are separated by anion exchange column chromatography and then purified. AMP is easily converted to IMP by *Aspergillus* adenyl deaminase.

If non-depolymerized, non-isomerized RNA in which the 5′-end is phosphorylated the 3′-end is not phosphorylated, and no 2′-5′-phosphodiester linkages are present, is used as substrate, 5′-nucleotides are quantitatively produced by the action of nuclease $P_1$. It should be noted that in technical-grade yeast RNA preparations each polynucleotide chain contains several percent of 2′-5′-phosphodiester linkages in addition to 3′-5′-phosphodiester linkages. The 2′-5′-linkages are formed as a result of isomerization of the internucleotide linkage (C3′-C5′→C2′-C5′) during extraction of RNA from yeast with hot water. As 2′-5′-phosphodiester linkages are rather resistant to nuclease $P_1$, an RNA preparation rich in such linkages is not suitable for industrial production of 5′-mononucleotides (KUNINAKA et al., 1980).

In addition, deoxyribonucleoside 5′-monophosphates are also quantitatively produced by incubating denatured DNA with nuclease $P_1$.

### 4.1.4 Application of Purified Nuclease $P_1$ as a Biochemical Tool

There are several applications of nuclease $P_1$ as a biochemical tool (KUNINAKA, 1993):

(1) Elucidation of the 5′cap structure of mRNA (FURUICHI and MIURA, 1975)
(2) Preparation of [$\alpha$-$^{32}$P] ribo- and deoxyribonucleoside triphosphates with a high specific radioactivity as follows (KIHARA et al., 1976; REEVE and HUANG, 1979):

$$\text{Ap} \xrightarrow[\text{Polynucleotide kinase}]{\text{p*ppA}} \text{p*Ap} \xrightarrow[\text{Nuclease P}_1]{} \text{p*A} \xrightarrow[\text{Kinase}]{} \text{ppp*A} \quad (1)$$

(3) Determination of the base composition of DNA (KUMAGAI et al., 1988; NOGUCHI et al., 1988)
(4) Elucidation of the structure–activity relationship of anti-HIV phosphorothioate oligonucleotides on the basis of the stereochemical course of nuclease $P_1$ reactions (STEIN et al., 1988)

In addition, nuclease $P_1$ was used for nucleotide sequence analysis (SILBERKLANG et al., 1977), elucidation of chromatin structure (FUJIMOTO et al., 1979), selective hydrolysis of single-stranded nucleic acids for shotgun cloning (SOEDA et al., 1984), nuclease $P_1$ instead of $S_1$ protection mapping of m-RNA (SAIGA et al., 1982), and identification of 2′-*O*-methylated oligonucleotides in ribosomal 18S and 28S RNA of a mouse hepatoma (HASHIMOTO et al., 1975).

Nuclease $P_1$ can also be used for preparing 2′-nucleotides from RNA:

$$\text{RNA} \xrightarrow{\text{Alkali}} 2'(3')\text{-Nucleotides} \xrightarrow{\text{Nuclease } P_1} 2'\text{-Nucleotides} + \text{Nucleosides} + P_1 \quad (2)$$

## 4.2 Fermentative Production of Individual 5′-Nucleotides and Related Compounds

### 4.2.1 IMP

Following the successful industrial production of amino acids by fermentation, extensive research was carried out in order to establish the fermentative production of the flavor nucleotides IMP and GMP. However, nucleotide fermentation differs in several respects from amino acid fermentation (TESHIBA and FURUYA, 1984a):

(1) Biosynthesis of purine nucleotides is comparatively complex: In addition to the *de novo* pathway, salvage pathways and mutual conversions are involved in purine nucleotide biosynthesis (Fig. 11).
(2) Enzymes capable of degrading nucleotides are widely distributed in microorganisms.
(3) Nucleotides do not readily permeate through cell membranes. Therefore, microorganisms can not easily secrete nucleotides into the culture medium.

The above complex problems have been overcome, and IMP is now produced in Japan at an industrial scale not only by enzymatic degradation of RNA but also by two types of fermentation:

- by direct fermentative production of IMP by *de novo* biosynthesis,
- by fermentative production of inosine and chemical conversion of inosine into IMP.

In this section, direct accumulation of IMP is described; fermentative production of nucleosides will be discussed in Sect. 5.

An adenine-requiring mutant of *Bacillus subtilis* is the first microorganism reported to accumulate IMP. However, the present industrial production of IMP by direct fermentation is mainly carried out with mutants of *Brevibacterium (Corynebacterium) ammoniagenes* (TESHIBA and FURUYA, 1984a, 1988). By the way, the present taxonomical name of *Brevibacterium ammoniagenes* is *Corynebacterium ammoniagenes.*

#### 4.2.1.1 Production by *Bacillus subtilis*

SAMP synthetase-deficient strains require adenine for their growth. Thus, UCHIDA et al. (1961) derived a number of adenine-auxotrophic mutants from various strains of wild-type microorganisms and found for the first time that an adenine-requiring mutant of *B. subtilis* IAM-1145 accumulated hypoxanthine derivatives. At the accumulation maximum of IMP, the concentrations of IMP, inosine, and hypoxanthine in the culture filtrate were 3.5 mM, 18.7 mM, and 3.6 mM, respectively.

AKIYA et al. (1972) derived strain 515, an adenine-auxotrophic strain with very weak 5′-nucleotidase activity from *B. subtilis* IAM-1145. This strain accumulated 9.82 mM IMP when grown at 40°C and the accumulation

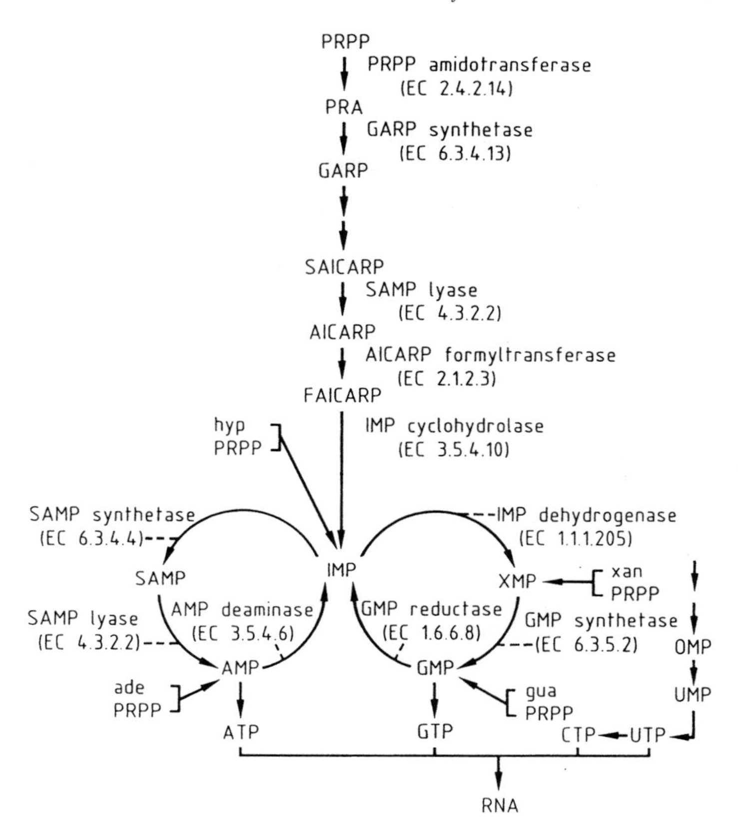


**Fig. 11.** Biosynthetic pathway of purine nucleotides.
PRPP: 5-phosphoribosyl-1-pyrophosphate (5-phospho-α-D-ribose-1-diphosphate); PRA: 5-phosphoribosyl-1-amine; GARP: glycineamide ribonucleotide; SAICARP: 5-aminoimidazole-4-succinocarboxamide ribonucleotide; AICARP: 5-aminoimidazole-4-carboxamide ribonucleotide; FAICARP: 5-formaminoimidazole-4-carboxamide ribonucleotide; SAMP: adenylosuccinate (succinoadenosine 5′-monophosphate).

was markedly increased by adding hypoxanthine. It accumulated 16.62 mM IMP when grown at 40 °C in the presence of 20 mM hypoxanthine.

### 4.2.1.2 Production by *Brevibacterium (Corynebacterium) ammoniagenes*

By using a mutant of *Brevibacterium (Corynebacterium) ammoniagenes* KY3454 (ATCC 6872), IMP could be produced industrially (TESHIBA and FURUYA, 1988).

*B. ammoniagenes* KY3454 (ATCC 6872) is regarded as "Aladdin's lamp" because this strain and its mutants produce various kinds of nucleotides and their derivatives according to *de novo* and salvage syntheses (Tabs. 1 and 2; FURUYA, 1978).

In order to establish fermentative production of IMP at the industrial scale, a series of mutants of *B. ammoniagenes* have been selected stepwise for increased IMP productivity after repeated mutation-inducing procedures such as UV irradiation and NTG (N-methyl-N′-nitro-N-nitrosoguanidine) treatment (Tab. 3; TESHIBA and FURUYA, 1982).

KY13102, an adenine-leaky auxotrophic mutant (ade[1]), produced considerable amounts of IMP (>12 g $L^{-1}$) when $Mn^{2+}$ was added to the culture broth at a concentration suboptimal for growth. Excessive amounts of $Mn^{2+}$ (>20 μg $L^{-1}$), however, stimulated growth and caused a drastic decrease in IMP accumulation and an increase in hypoxanthine accumulation. The $Mn^{2+}$ concentration range suitable for IMP accumulation was very narrow (about 10 μg $L^{-1}$). Furthermore, the cells were abnormally elongated and irregular in shape in media contain-

**Tab. 1.** *De novo* Synthesis of Nucleotides and Related Compounds by *Brevibacterium ammoniagenes* (adapted from FURUYA, 1978)

| Markers of Strain[a] | Product | Remark |
|---|---|---|
| $2FA^r$ | Adenine | |
| $ade^-$ | IMP, hypoxanthine | |
| $ade^-gua^-$ | IMP, hypoxanthine | |
| $ade^lMn^i$ | IMP, hypoxanthine | |
| $ade^l$perm | IMP, hypoxanthine | Permeable mutant |
| $Gua^-$ | XMP | |
| $gua^-ade^l$ | XMP | |
| Wild | XMP | Addition of psicofuranine or decoyinine |
| $ade^l6MG^r$ | Inosine | |
| $ade^lgua^-6MG^r6MTP^r$ | Inosine | |
| Wild | UMP | Addition of 6-azauracil |
| Wild | OMP[b] | Addition of 6-azauracil |
| $2FA^r$ | Cytosine | |
| Wild | Ribose-5-phosphate<br>PRPP | |

[a] $2FA^r$: 2-fluoroadenine resistant;
[b] OMP: orotidine-5′-monophosphate;
$gua^-$: guanine requiring;
$ade^l$: adenine leaky;
$Mn^i$: $Mn^{2+}$ insensitive;
$6MG^r$: 6-mercaptoguanine resistant;
$6MTP^r$: 6-methylthiopurine resistant

**Tab. 2.** Salvage Synthesis of Nucleotides and Related Compounds by *Brevibacterium ammoniagenes* (adapted from FURUYA, 1978)

| Precursor | Product |
|---|---|
| Hypoxanthine | IMP |
| Guanine | GMP, GDP, GTP |
| Adenine | AMP, ADP, ATP |
| Uracil, orotic acid | UMP |
| Orotic acid (+6-azauracil) | OMP |
| Purine analog | Purine analog nucleotide |
| Pyrimidine analog | Pyrimidine analog nucleotide |
| 6-Azauracil | 6-Azauridine |
| 1,2,4-Triazole-3-carboxamide + R-1-P | Ribavirin |
| Nicotinic acid + adenine | NAD |
| FMN + adenine | FAD |
| XMP (conversion by decoyinine$^r$ mutant) | GMP, GDP, GTP |
| XMP or GMP (conversion by decoyinine$^r$ mutant) | pppGpp, ppGpp, pGpp |

ing $Mn^{2+}$ at a concentration suitable for IMP accumulation. After addition of excessive amounts of $Mn^{2+}$, the cells changed to oval shapes or short rods. The fatty acid content of the $Mn^{2+}$-limited cells was higher than that of the $Mn^{2+}$-excessive cells, and the ratio of unsaturated to saturated fatty acids was higher in the former cells. This suggests that there is a close relationship between cellular morphology and IMP accumulation. In addition,

**Tab. 3.** Accumulation of IMP and Hypoxanthine (Hyp) by a Series of Mutants of *Brevibacterium ammoniagenes* KY3454 (ATCC 6872)[a] (adapted from TESHIBA and FURUYA, 1988)

| Strain (Marker) | IMP[b] [mg mL$^{-1}$] | Hyp[b] [mg mL$^{-1}$] | IMP+Hyp[b] [mg mL$^{-1}$] |
|---|---|---|---|
| KY3454 (wild) | – | – | – |
| ↓ UV | | | |
| KY13102 (ade$^{l}$) | 1– 2 | 8–10 | 9–12 |
| ↓ NTG | | | |
| KY13171 (ade$^{l}$Mn$^{i}$) | 7– 8 | 4– 6 | 11–13 |
| ↓ NTG | | | |
| KY13184 (ade$^{l}$Mn$^{i}$gua$^{-}$) | 8–10 | 7– 8 | 15–18 |
| ↓ NTG | | | |
| KY13198 (ade$^{l}$Mn$^{i}$Gu$^{-}$) | 9–12 | 5– 6 | 15–18 |
| ↓ NTG | | | |
| KY13361 (ade$^{l}$Mn$^{i}$gua$^{-}$) | 12–16 | 2– 3 | 15–19 |
| ↓ NTG | | | |
| KY13363 (ade$^{l}$Mn$^{i}$gua$^{-}$) | 18–20 | trace amounts | 18–20 |
| ↓ NTG | | | |
| KY13369 (ade$^{l}$Mn$^{i}$gua$^{-}$) | 20–27 | trace amounts | 20–27 |

[a] Cultivation was performed in the presence of 1 mg L$^{-1}$ $Mn^{2+}$.
[b] Expressed in terms of IMP $Na_2 \cdot 7.5H_2O$.

the accumulation of IMP was maximal in the presence of adenine at a suboptimal concentration for growth and was strongly depressed by excessive amounts of adenine. Adenine and $Mn^{2+}$ seem to affect IMP accumulation independently: adenine acts as a feedback regulator on purine nucleotide *de novo* synthesis and excessive amounts of $Mn^{2+}$ cause morphological changes and depress the membrane permeability for IMP.

Since strain KY13102 was not suitable for industrial production of IMP because of its high sensitivity to $Mn^{2+}$, the manganese-insensitive mutant KY13171 (ade$^{l}$ Mn$^{i}$) was derived. Cell growth and IMP accumulation of KY13171 increased as the concentration of $Mn^{2+}$ increased. Even in the presence of excessive $Mn^{2+}$, the strain did not show changes in cellular morphology and accumulated 7.5 g L$^{-1}$ IMP during 4 days of cultivation at 30°C. The optimum concentration range of adenine for IMP accumulation was rather broad, 0–100 mg L$^{-1}$. The intracellular nucleotide pool size was the same as that for KY13102. The permeability barrier for IMP excretion seems to be gradually changed by the mutation from KY13102 to KY 13171 cells.

From practical and economical points of view, IMP productivity of KY13171 still was not sufficient as purine nucleotide synthesis in *B. ammoniagenes* is regulated by adenine, guanine, and their derivatives (Tab. 4; NARA et al., 1969). Especially the key enzyme, PRPP amidotransferase, is strongly inhibited by GMP as well as ATP, ADP, and AMP. In order to improve IMP production in KY13171 an attempt was made to release feedback regulation in *de novo* purine biosynthesis by GMP. As a result, mutant KY13184, that is devoid of IMP dehydrogenase and that requires guanine (gua) for growth, was derived from KY13171. KY13184 produces IMP and hypoxanthine at higher levels than the parent strain, KY13171.

Other mutants were derived from KY13184 through repeated cycles of mutation and selection. Finally, an effective IMP producer, KY13369, was obtained. KY13369 accumulated 20–27 g L$^{-1}$ IMP without accumulation of hypoxanthine. Increased IMP production and decreased hypoxanthine production were

**Tab. 4.** Regulation of Purine Nucleotide Synthesis in *Brevibacterium ammoniagenes* (adapted from NARA et al., 1969)

| Main Enzymes Involved in Purine Nucleotide Synthesis | Repression by | Inhibition by |
|---|---|---|
| *De novo* synthesis | | |
| PRPP amidotransferase (PRPP→PRA) | Adenine | ATP, ADP, AMP, GMP |
| IMP dehydrogenase (IMP→XMP) | Guanine | GMP |
| Salvage synthesis | | |
| Phosphoribosyltransferase (Hyp→IMP) | – | GTP, ATP |
| Phosphoribosyltransferase (Gua→GMP) | – | GTP |

not caused by alterations in *de novo* purine nucleotide biosynthetic rate or by IMP degrading activity, but possibly by enhanced membrane permeability for IMP.

Mutants, that are altered in cell surface or cytoplasmic membrane characteristics, have been reported to exhibit variations in the sensitivity to a variety of drugs. The series of mutants of *B. ammoniagenes* also exhibited variations in sensitivity to antibiotics, detergents, dyes, and lysozyme. The increased IMP productivity was always accompanied by increased sensitivity to deoxycholate and lysozyme (TESHIBA and FURUYA, 1983a).

The series of mutants were further analyzed for the size of intracellular IMP and hypoxanthine pools. The pool size of the mutants corresponded fairly well to their IMP productivities. The direct excretion system of internal IMP *de novo* synthesized was stimulated by glucose. Thus, in the most effective producer, KY13369, all of the intracellular IMP was excreted without any degradation in the presence of glucose. Furthermore, the direct excretion system was confirmed to be stimulated by various energy sources and was completely inhibited by KCN, $NaN_3$, DNP, and carbonyl cyanide *m*-chlorophenyl hydrazone. Therefore, IMP excretion may depend on energy yielding reactions. The increased IMP productivity of KY13369 is due to the increased direct excretion system and to the change of the indirect excretion system by which IMP is excreted after being degraded to hypoxanthine. In addition, the indirect excretion system coupled with IMP-degrading activity was depressed by glucose. In KY13369, the indirect excretion system may be more susceptible to glucose inhibition than in other mutants (TESHIBA and FURUYA, 1983b, 1984b).

TOMITA et al. (1991) derived mutant strain KY10895 ($ade^l$, $Mn^i$, $gua^-$) via KY13371 and KY13374. This is the most effective producer of IMP derived from the original strain *Corynebacterium (Brevibacterium) ammoniagenes* KY13363. The productivity of strain KY10895 was about 1.3 times as high as that of strain 13363. The effect of amino acids on IMP production of strain KY10895 was examined and it was found that L-proline improved the productivity most efficiently. In a medium supplemented with 1–2% L-proline IMP productivity was about 170% compared to a medium without L-proline. The stimulatory effect of L-proline on IMP production decreased with an increase in fructose. The optimum glucose–fructose ratio for the stimulatory effect of L-proline on IMP production was 100/0 (TOMITA and KURATSU, 1992). According to TOMITA et al. (1992a) the specific IMP production (IMP produced–cell density) of KY10895 was promoted by increased osmotic strength in the medium, although growth was inhibited. As the intracellular concentration of L-proline highly increased with increased osmotic strength, the increase in intracellular L-proline concentration can be related to the increased IMP production. Isolation of an osmotolerant strain seems to be an effective way to obtain an IMP hyperproducer.

MUZZIO and ACEVEDO (1985) calculated theoretical yields for the biosynthesis of purine nucleotides, giving values of 1.105 g, 1.156 g, and 1.152 g nucleotide per g glucose for IMP, XMP, and GMP, respectively.

The following patent applications for IMP production have been filed recently:

(1) TSUCHIDA et al. (1984) isolated *B. subtilis* mutants for their improved IMP productivity by transformation with a recombinant plasmid pHE17 containing the PRPP amidotransferase gene. One of the transformed strains derived from AJ11914 produced 2.3 g $L^{-1}$ IMP vs. 1.2 g $L^{-1}$ for its parent strain (Tab. 5).
(2) IMP is produced at a good yield by *Corynebacterium equi, B. ammoniagenes,* or *B. subtilis* while adding both glucose and inosine and/or hypoxanthine during fermentation (ISHIBASHI et al. 1986). *C. equi* AJ11749 was cultivated in a medium containing 10% glucose, 1% yeast extract, 0.5% $(NH_4)_2SO_4$, 1.6% $KH_2PO_4$, 0.6% $MgSO_4$, 3% soybean protein hydrolyzate, and 2% glutamic acid at 34 °C for 5 d; 3% glucose and 0.5% hypoxanthine were added after 2 d. 26.9 g $L^{-1}$ disodium IMP hydrate were yielded vs. 16.3 g $L^{-1}$ without hypoxanthine.
(3) HAGIWARA et al. (1988) developed a method for selecting high-nucleotide producers. *B. ammoniagenes* P-3790 isolated from a log-phase culture, e.g., was treated for 16 h with 0.2 mg m $L^{-1}$ egg white lysozyme and 0.6 mg $mL^{-1}$ peptidase in a P3 hypertonic solution. The protoplasts were cloned in a hypertonic agar. The derived strain IP8 FERM BP-1258 produced 24.3 g $L^{-1}$ IMP vs. 20.8 g $L^{-1}$ for the parent strain after 7 days of fermentation.
(4) FUJIO et al. (1988a) cloned the *purF* gene encoding PRPP amidotransferase (EC 2.4.2.14) of *Escherichia coli* into plasmid pCE53 to create the expression plasmid pEF12. A transformed strain of *B. ammoniagenes* KY13184, KY13184/pEF12, showed 1.5-fold more PRPP amidotransferase activity than its parent strain, and produced 24.9 g $L^{-1}$ IMP vs. 20.1 g $L^{-1}$ for the parent strain (Tab. 6).
(5) FUJIO et al. (1988c) mixed inosine or inosine-containing culture broth, pretreated cells of an ATP-regenerating mi-

**Tab. 5.** Effect of Transformation with the Recombinant Plasmid pHE17 Containing the PRPP Amidotransferase Gene on Purine Biosynthesis by *Bacillus subtilis* (adapted from TSUCHIDA et al., 1984)

| pHE17 | Strain | | | |
|---|---|---|---|---|
| | AJ11913 | AJ11914 | AJ11915 | AJ11916 |
| | Product [g $L^{-1}$] | | | |
| | Inosine | IMP | Guanosine | GMP |
| – | 1.8 | 1.2 | 0.05 | 0.03 |
| + | 3.1 | 2.3 | 1.06 | 0.72 |

**Tab. 6.** Effect of Transformation with the Recombinant Plasmid pEF12 Containing the PRPP Amidotransferase Gene on Purine Biosynthesis by *Brevibacterium ammoniagenes* (adapted from FUJIO et al., 1988a)

| pEF12 | Strain | | |
|---|---|---|---|
| | ATCC21477 | KY13184 | ATCC21075 |
| | Product [g $L^{-1}$] | | |
| | Inosine | IMP | XMP |
| – | 6.2 | 20.1 | 14.8 |
| + | 9.4 | 24.9 | 19.5 |

croorganism (*Saccharomyces sake, Torulopsis psychrophilia, Candida zeylanoides*), pretreated cells of an inosine-phosphorylating microorganism (*Serratia marcescens*), a cheap phosphate donor and a cheap energy source for producing IMP effectively.

(6) FUJIO and IIDA (1990) cultured two microorganisms producing IMP from hypoxanthine: one is capable of synthesizing RPPP from a carbon source and the other synthesizes IMP from hypoxanthine and PRPP. *B. ammoniagenes,* a PRPP producer, and *E. coli* transformed with the plasmid pAI63, encoding the hypoxanthine phosphoribosyl transferase of *S. marcescens,* were suspended in a solution containing 100 mM hypoxanthine at pH 7.4 at a concentration of 200 mg m $L^{-1}$ and 50 mg m $L^{-1}$, respectively. The mixture was incubated at 32 °C for 24 h and yielded 18.7 mM IMP.

(7) TOMITA et al. (1990a) precultured *B. ammoniagenes* KY13184 in a medium containing glucose, peptone, meat extract, yeast extract, urea, L-cysteine, calcium pantothenate, biotin, thiamine-HCl, adenine, guanine, and salts at 30 °C for 30 h. The culture was supplemented with 1.0% L-proline and incubated for 80 h and yielded 8.7 g $L^{-1}$ IMP vs. 5.2 g $L^{-1}$ without L-proline.

(8) TOMITA and NAKANISHI (1990) cultured a *Corynebacterium* strain resistant to L-proline antagonists and capable of producing IMP. The strain accumulated 28.7 g $L^{-1}$ IMP vs. 26.3 g $L^{-1}$ of the parent strain KY13184.

(9) TOMITA et al. (1990b) derived a mutant of *C. ammoniagenes* KY13184 that was resistant to osmotic pressure of at least 2,000 mosmol $kg^{-1}$. The mutant strain produced 29.1 g $L^{-1}$ IMP vs. 26.5 g $L^{-1}$ for the parent strain.

(10) MORI et al. (1991) prepared plasmid pIK75 encoding *E. coli* inosine guanosine kinase and used it for transformation of *E. coli.* The transformants grown in a medium containing 50 g $L^{-1}$ inosine produced 100 g $L^{-1}$ IMP disodium salt (conversion rate >90%).

### 4.2.2 XMP

As reported by MISAWA et al. (1964), KY9978, a guanine requiring mutant of *Micrococcus glutamicus* (*Corynebacterium glutamicum*), accumulated 2.57 mg $mL^{-1}$ XMP. DEMAIN et al. (1965) also published similar results with auxotrophic mutants of a coryneform bacterium.

MISAWA et al. (1969) derived another guanine-requiring mutant, KY7450, from *Brevibacterium ammoniagenes* strain ATCC 6872. KY7450 accumulated higher amounts of XMP in the culture medium than *M. glutamicus* KY9978. An excessive addition of guanine derivatives suppressed XMP accumulation. The accumulation of XMP by KY7450 seems to be due to direct excretion of the *de novo* synthesized nucleotide, because XMP pyrophosphorylase activity (xanthine phosphoribosyl transferase) was very low or deficient, and exogenously supplied xanthine was not converted to XMP by growing cells.

In this fermentation, excess $Mn^{2+}$ in the medium markedly stimulated XMP accumulation. In the presence of 1,000 μg $L^{-1}$ $Mn^{2+}$, 6.46 mg $mL^{-1}$ XMP was produced, while in the presence of 10 μg $L^{-1}$ $Mn^{2+}$ only 3.18 mg $mL^{-1}$ was produced. This is in striking contrast with the results of IMP fermentation by adenine auxotrophs of the same species. Usually *B. ammoniagenes* cells do not excrete IMP directly, so isolation of permeability mutations was essential to derive IMP producing strains. Despite of that XMP producing mutants could easily be derived.

According to FUJIO and FURUYA (1988) *B. ammoniagenes* KY13215, an adenine-leaky, guanine-requiring mutant with weak nucleotidase activity, accumulated several ten grams of XMP per liter.

In addition, the accumulation of XMP together with IMP by adenine-requiring or adenine–guanine-requiring mutants of *B. subtilis* with weak 5′-nucleotidase activity was also reported by AKIYA et al. (1972).

The following patent applications for XMP production have been filed recently:

(1) HATTORI et al. (1985) cultured *B. ammoniagenes* T-2 being resistant to antibiotics inhibiting cell wall synthesis. This

strain was grown in a medium containing 100 g $L^{-1}$ glucose, 10 g $L^{-1}$ $K_2HPO_4$, 10 g $L^{-1}$ $KH_2PO_4$, and traces of salts, biotin, adenine, guanine, calcium pantothenate, and thiamin-HCl at pH 6.2 and 28 °C. Urea (3 g $L^{-1}$) was added at pH 6–7. After 4 d, the yield of XMP was 20 g $L^{-1}$.

(2) TOMITA et al. (1992b) cultured *Corynebacterium ammoniagenes* H-8077, a valine hydroxamate-resistant strain, in a medium containing glucose, biotin, adenine, guanine, calcium pantothenate, thiamin-HCl, and salts at 30 °C, with the addition of urea. After 4 d the yield of XMP was 19 g $L^{-1}$ vs. 15 g $L^{-1}$, for its parent strain *C. ammoniagenes* ATCC21075.

(3) See Tab. 6, FUJIO et al. (1988a).

### 4.2.3 GMP

Direct fermentative production of GMP has not practically been established. Instead, GMP can be produced industrially via XMP, by combining an XMP accumulating strain and an XMP converting strain (FUJIO and FURUYA, 1988).

Strain KY13510 derived from *Brevibacterium ammoniagenes* ATCC6872 by FUJIO et al. (1984a) is resistant to decoyinine, practically devoid of 5'-nucleotide degrading activity, and rich in GMP synthetase (XMP aminase, EC 6.3.5.2). Strain KY13510 ($DC^rNtase^w$) effectively converts XMP into GMP.

A surfactant, polyoxyethylene stearylamine (POESA), which enables XMP and GMP to permeate through the cell membrane is required for conversion. Significant amounts of GDP and GTP as well as GMP were formed from XMP when incubation was performed at 32 °C. On the other hand, at 42 °C GMP was formed exclusively (Tab. 7). Aeration was indispensable, the optimum pH was 7.4. The conversion reaction involves a complex enzymatic system mainly consisting of XMP aminase and ATP regenerating enzymes.

XMP aminase reaction:

$$XMP + NH_3 + ATP \xrightarrow{Mg^{2+}} GMP + AMP + PP_i \quad (3)$$

ATP regenerating enzyme system:

$$PP_i \rightarrow 2P_i \quad (4)$$

$$AMP + ATP \rightleftarrows 2ADP \quad (5)$$

$$ADP + P_i + Glucose \rightarrow ATP + H_2O + CO_2 + Acids \quad (6)$$

Probably, cellular ATP is consumed at the beginning of the conversion and almost all of the ATP consumed during the conversion reaction seems to be generated from AMP and glucose. If $Mg^{2+}$ and $PP_i$ form insoluble magnesium pyrophosphate ($Mg_2PP_i$), the above reactions cannot proceed smoothly. Fortunately phytic acid, a chelating agent, was able to prevent the formation of $Mg_2PP_i$.

GMP was produced from XMP as follows: XMP fermentation supernatant was concentrated and mixed with the converting cell culture. The mixture contained 50 mg $mL^{-1}$ (dry weight) of bacteria, 40 mg $mL^{-1}$ XMP·

**Tab. 7.** Conversion of XMP to GMP, GDP, and GTP by *Brevibacterium ammoniagenes* KY13510 at Different Temperatures for 8 h[a] (adapted from FUJIO et al., 1984a)

| Temperature [°C] | XMP Consumed [mg $mL^{-1}$] | GMP Formed [mg $mL^{-1}$] | GDP Formed [mg $mL^{-1}$] | GTP Formed [mg $mL^{-1}$] |
|---|---|---|---|---|
| 32 | 18.0 | 3.6 | 6.3 | 6.2 |
| 42 | 19.3 | 17.0 | 0 | 0 |

[a] The reacton mixture contained ($mL^{-1}$) 120 mg lyophilized cells; 40 mg XMP; 12 mg POESA; 50 mg glucose; 2 mg ammonium sulfate; and the basal mixture.

$Na_2 \cdot 7H_2O$, 50 mg mL$^{-1}$ glucose, 10 mg mL$^{-1}$ $Na_2HPO_4$, 12 mg mL$^{-1}$ POESA, and 10 mg mL$^{-1}$ phytic acid, with 5 mg mL$^{-1}$ $MgSO_4 \cdot 7H_2O$ added after 2 h and was incubated at 42°C for 12 h. During incubation 34.8 mg mL$^{-1}$ GMP·$Na_2 \cdot 7H_2O$ were produced from XMP, the conversion rate was 86.9%. Without addition of phytic acid and $MgSO_4 \cdot 7H_2O$, only 15.6 mg mL$^{-1}$ GMP·$Na_2 \cdot 7H_2O$, were produced. The conversion rate was then 39.0% (FUJIO and FURUYA, 1988).

MARUYAMA et al. (1986) attempted to increase the activity of the conversion enzyme, XMP aminase (GMP synthetase), in *E. coli*. They subcloned the *gua*A gene encoding XMP aminase on plasmid pLC34-10 into pBR322, and obtained plasmid pXAR33 by connecting the tryptophan (*trp*) promoter at a suitable position upstream of the *gua*A gene. XMP aminase activity of the resulting *E. coli* strain K294/pXAR33 was 83.5 times that of the host strain K294. The amount of recombinant enzyme in *E. coli* K294/pXAR33 was about 10% of the total cellular protein.

CHO and JUNG (1991) converted XMP to GMP by immobilized cells of *B. ammoniagenes* ATCC 19216, that were used as XMP aminase source.

CHO and CHU (1991) attempted intergeneric protoplast fusion between *B. ammoniagenes* (capable of producing XMP) and *Corynebacterium glutamicum* (rich in GMP synthetase) in order to develop a strain producing GMP directly.

The following patent applications for GMP production have been recently filed:

(1) SHIMIZU et al. (1983b) cultured strain AJ11870, a transformant of *Bacillus* containing purine analog resistance or decoyinine-resistance genes, at 34°C for 3 d. The culture filtrate contained 1.66 g L$^{-1}$ of GMP·$Na_2$.
(2) See Tab. 5, TSUCHIDA et al. (1984).
(3) FUJIO and FURUYA (1984) mixed 10 mL culture broth of *C. glutamicum* ATCC 21171 with a solution containing 40 g L$^{-1}$ XMP, an $NH_4$-donor, an energy source, a surfactant and/or an organic solvent, 10 g L$^{-1}$ phytic acid, and 5 g L$^{-1}$ $MgSO_4$. The mixture was incubated by shaking at 42°C for 12 h. Production of GMP·$Na_2 \cdot 7H_2O$ was 26. 5 g L$^{-1}$ vs. 13.2 g L$^{-1}$ in a culture without addition of phytic acid and $MgSO_4$.
(4) FUJIO et al. (1985) developed a new GMP producing microorganism by transformation of an *E. coli* strain that is capable of ATP regeneration from AMP in the presence of a non-phosphate energy donor, with a hybrid plasmid containing the IMP dehydrogenase gene, the GMP synthetase gene, and a DNA segment carrying the colicin $E_1$ gene derived from a donor *E. coli* strain. The transformed *E. coli* K294/pXA10 strain was suspended at 200 g L$^{-1}$ and incubated at 37°C for 6 h in a medium containing 40 g L$^{-1}$ XMP·$Na_2 \cdot 7H_2O$, 50 g L$^{-1}$ glucose, 2 g L$^{-1}$ Na phytate, 5 g L$^{-1}$ $Na_2HPO_4$, and 0.5 g/L $MgSO_4 \cdot 7H_2O$. After incubation the medium contained 23.8 g L$^{-1}$ GMP·$Na_2 \cdot 7H_2O$.
(5) FUJIO et al. (1988b) constructed a plasmid containing the *gua*A gene and transformed *E. coli* with this plasmid. Compared to the parent strain, the transformant produced about 400-fold more GMP synthetase. The transformed strain was added to a medium containing XMP, ATP, and $(NH_4)_2SO_4$, and incubated at 42°C for 3 h. 15.9 mg mL$^{-1}$ GMP were produced vs. only 4 mg mL$^{-1}$ for the parent strain.

### 4.2.4 AMP

Effective direct fermentative production of AMP as well as of GMP seems to be difficult.

POPOV et al. (1985) isolated a strain, *Erwinia herbicola* 47/3, from a natural source. The strain phosphorylated adenosine to AMP efficiently. The enzyme catalyzing adenosine phosphorylation was detected in the cell biomass and was completely absent in the culture supernatant. Centrifuged cell mass (50 mg mL$^{-1}$) was incubated with 12 mg mL$^{-1}$ adenosine and 96 mg mL$^{-1}$ *p*-nitrophenyl phosphate in 0.2 M acetate buffer, pH 4.5, at 50°C for 11–12 h. The yield of AMP was 75%.

### 4.2.5 UMP

NUDLER et al. (1991) reported that a mutant strain, *Brevibacterium (Corynebacterium) ammoniagenes* NG 513, produced 4 mg $mL^{-1}$ of UMP and 5 mg $mL^{-1}$ of uracil. The UMP-producing strain, NG 513, was derived from *B. ammoniagenes* AN 225-5. The parent strain was auxotrophic for adenine and biotin, permeable for nucleotides, and produced IMP and hypoxanthine. The mutant strain, NG 513, was bradytrophic for arginine, ornithine, or citrulline, oversensitive to adenine, and resistant to rifampicin and pyrimidine analogs such as 5-fluorouracil, 5-fluorouridine, azauracil, and thiouracil. As shown in Tab. 8, the activities of enzymes involved in UMP biosynthesis, i.e., orotate phosphoribosyltransferase (OPRTase, EC 2.4.2.19), OMP decarboxylase (OMP-DCase, EC 4.1.1.23), dihydroorotate oxidase (DHOase, EC 1.3.3.1), were 4-, 3.5- and 4.5-fold higher, respectively, in the mutant strain, NG 513, than in the parent strain, AN 225-5, when grown in a minimal medium. It should be noted that the synthesis of these enzymes in mutant cells was not repressed even in the presence of exogenous uracil.

NUDLER et al. (1991) suggest that the mutation that allows *B. ammoniagenes* NG 513 to produce UMP and uracil affects the regulation of pyrimidine biosynthesis *de novo* due to a defect in bacterial RNA polymerase.

MARUMO et al. (1992) derived pyrimidine analog-resistant and uracil-producing mutant strains, 214–17 and 214–27, from *B. ammoniagenes* ATCC 6872 by treatment with N-methyl-N′-nitro-N-nitrosoguanidine. While the parent strain ATCC 6872 was not able to grow in the presence of 100 μg $mL^{-1}$ or 300 μg $mL^{-1}$ 5-fluorouracil, 300 μg $mL^{-1}$ 6-azauracil, or 300 μg $mL^{-1}$ 2-thiouracil, both mutant strains, 214–17 and 214–27, were able to grow under the same conditions. The mutant strains were cultured in a medium containing glucose, corn steep liquor, yeast extract, urea, and $MgSO_4$. Strains 214–17 and 214–27 produced more than about 5 mg $mL^{-1}$ each of UMP and uracil (Tab. 9). The total concentration of the uracil derivatives, uracil, uridine, and UMP, was about 80 mM. On the other hand, the parent strain did not produce uracil, uridine, and UMP at a detectable level.

### 4.2.6 CMP

UMP is not converted to CMP directly. Amination proceeds at the triphosphate level by CTP synthetase (EC 6.3.4.2). Using suitable microorganisms, *de novo* synthesis of orotic acid (Sect. 5.9) and conversion of orotic acid to CDP-choline via UTP and CTP (Sect. 6.3) are now possible. Furthermore, mutants lacking cytidine deaminase (EC 3.5.4.5) and being resistant to pyrimidine analogs can produce cytidine at an industrial scale (Sect. 5.6). Therefore, CMP could be produced by selecting suitable microorganisms and conditions.

**Tab. 8.** Enzyme Activities Involved in UMP Biosynthesis of *Brevibacterium ammoniagenes* AN 225-5 (Parent) and NG 513 (Mutant) Grown in the Absence or Presence of Uracil (adapted from NUDLER et al., 1991)

| Strain[a] | Addition [0.1 mM] | Specific Activity of Enzyme [units per mg of protein] | | |
|---|---|---|---|---|
| | | OPRTase | OMP-DCase | DHOase |
| AN 225-5 | – | 15 | 18 | 11 |
| | Ura | 3 | 4 | 2 |
| NG 513 | – | 60 | 62 | 49 |
| | Ura | 57 | 60 | 46 |

[a] Strain NG 513 is a pyrimidine analog-resistant mutant derived from strain AN 225-5, and it produced 4 mg $mL^{-1}$ of UMP and 5 mg $mL^{-1}$ of uracil.

**Tab. 9.** Production of Uracil, Uridine, and UMP by *Brevibacterium ammoniagenes* ATCC 6872 and its Pyrimidine Analog-Resistant Mutants, 214-17 and 214-27 (adapted from MARUMO et al., 1992)

| Strain[a] | Products [mg mL$^{-1}$] | | |
|---|---|---|---|
| | Uracil | Uridine | UMP |
| ATCC6872 (5-FU$^s$ 6AU$^s$ 2TU$^s$) | 0 | 0 | 0 |
| 214-17 (5-FU$^r$ 6AU$^r$ 2TU$^r$) | 6.6 | 0.3 | 4.8 |
| 214-27 (5-FU$^r$ 6AU$^r$ 2TU$^r$) | 5.8 | 0.3 | 8.4 |

[a] 5-FU$^r$: 5-fluorouracil-resistant (300 μg mL$^{-1}$); 6-AU$^r$: 6-azauracil-resistant (300 μg mL$^{-1}$); 2-TU$^r$: 2-thiouracil-resistant (300 μg mL$^{-1}$).

### 4.2.7 Cyclic Adenosine 3′,5′-Monophosphate (cAMP)

Cyclic adenosine 3′,5′-monophosphate (cAMP) was first reported to be produced by microorganisms in 1963. Since then the fermentative production of cAMP has markedly been developed.

ISHIYAMA (1984) reviewed cAMP production by *E. coli*: Cloning of the *cya* gene coding for adenylate cyclase (EC 4.6.1.1), cAMP production by adenylate cyclase-enhanced *E. coli*, isolation, purification, and characterization of adenylate cyclase, structure analysis of adenylate cyclase and the *cya* gene, mechanism of cAMP production.

Until 1989, the following microorganisms were available for commercial production of cAMP (ISHIYAMA, 1990):

- *Microbacterium* sp. No. 205 (ATCC 21376), *Corynebacterium murisepticum* No. 7 (ATCC 21374) and *Arthrobacter* sp. No. 11 (ATCC 21375) producing 15–30 mM of cAMP via salvage biosynthesis from hypoxanthine or its derivatives,
- *Brevibacterium liquefaciens* ATCC 14929 and *Corynebacterium roseoparaffineus* ATCC 15584 producing about 5 mM of cAMP via *de novo* biosynthesis.

The wild-type strain *Microbacterium* sp. No. 205 (ATCC 21376) produces 15–30 mM of cAMP by salvage biosynthesis on supply of 30 mM hypoxanthine to the production medium. In order to improve cAMP productivity, ISHIYAMA (1990) derived mutants from the wild-type strain. The results are summarized in Tab. 10. The mutant produced a high level of cAMP even in the presence of excess amounts of both IMP and glucose, it was resistant to cAMP and had a low level of cAMP phosphodiesterase (CPDase, EC 3.1.4.17). It was also reported that there was a lower CPDase activity in some *E. coli* mutant strains resistant to catabolite repression.

The following patent applications for cAMP production have been filed recently:

(1) ISHIYAMA (1982) cultured a *Microbacterium* strain resistant to purine, pyrimidine, and methionine analogs in the presence of >0.02 mg L$^{-1}$ $MnCl_2 \cdot 4H_2O$, >10 mg L$^{-1}$ $FeCl_2 \cdot 7H_2O$, and/or >10 mg L$^{-1}$ $FeCl_3 \cdot 7H_2O$ that produced 26.5 mg mL$^{-1}$ cAMP.
(2) KIJIMA et al. (1985) cultured a strain capable of converting a cAMP precursor to cAMP in the presence of the precursor and tannin. Precultured *Microbacterium* No. 205 , e.g., was grown in a medium containing 12% glucose, 1% $MgSO_4 \cdot 7H_2O$, 100 mg L$^{-1}$ $ZnSO_4 \cdot 7H_2O$, 1% $(NH_4)_2SO_4$, 0.5% urea, 2% polypeptone, 1.5% inosine, and 120 mg L$^{-1}$ tannin at pH 7.5 for 90 h at 30 °C and produced 9.62 mg mL$^{-1}$ cAMP.
(3) OGUMA et al. (1989) cultured a *Microbacterium* strain capable of converting a cAMP precursor to cAMP in the presence of the precursor and an oligopeptide. As shown in Tab. 11, oligopeptides such as glycyl-DL-norleucine, L-phenylala-

**Tab. 10.** Production of cAMP in a Series of Mutants Derived from Wild-Type Strain No. 205 (ATCC 21376) on 100 mM IMP as a Precursor of cAMP (adapted from ISHIYAMA, 1990)

| Strain | CPDase Units[a] | Relevant Characteristics[b] | cAMP Production [mM] |
|---|---|---|---|
| 205 | 1,232 | | 0.3– 6 |
| 121-20 | | $IMP^{in}$ | 8 –15 |
| C-7 | | $IMP^{in}cAMP^{r}$ | 30 –40 |
| 72-5 | | $IMP^{in}cAMP^{r}Ana^{r}$ | 37 –43 |
| G-58 | 1,102 | $IMP^{in}cAMP^{r}Ana^{r}GLu^{in}$ | 45 –57 |
| 76-85 | 450 | $IMP^{in}cAMP^{r}Ana^{r}GLu^{in}CR$ | 56 –68 |
| 77-72 | 480 | $IMP^{in}cAMP^{r}Ana^{r}GLu^{in}Lac^{-}$ | 56 –75 |

[a] Formation of AMP (pmol $min^{-1}$ per mg protein of broken cells) from cAMP;
[b] $IMP^{in}$: insensitive to inhibition of cAMP production by an excess of IMP;
$cAMP^{r}$: resistant to cAMP;
$Ana^{r}$: resistant to purine and pyrimidine analogs;
$Glu^{in}$: insensitive to inhibition of cAMP production by glucose at higher concentrations;
CR: resistant to catabolite repression;
$Lac^{-}$: incapable of utilizing lactose.

**Tab. 11.** Effect of Oligopeptides on the Production of cAMP from 1.5% IMP at 30 °C for 96 h (adapted from OGUMA et al., 1989)

| Addition [% w/v] | IMP Remained [mg $mL^{-1}$] | cAMP Formed [mg $mL^{-1}$] |
|---|---|---|
| None | 9.97 | 1.73 |
| 0.5 % Betaine | 9.80 | 0.37 |
| 0.2 % Glycine | 8.43 | 2.31 |
| 0.35% DL-Norleucine | 9.85 | 2.76 |
| 0.20% Gly + 0.35% DL-nLeu | 8.40 | 3.00 |
| 0.5 % Gly-DL-nLeu | 0.00 | 8.31 |
| 0.5 % L-Phe-L-Leu | 0.03 | 7.08 |
| 0.5 % L-Leu-L-Leu-L-Leu | 0.03 | 6.11 |

nyl-L-leucine, and L-leucyl-L-leucyl-L-leucine were useful for the effective production of cAMP in the presence of IMP as precursor. Amino acids could not replace oligopeptides.

### 4.2.8 Cyclic Guanosine 3′,5′-Monophosphate (cGMP)

A recent patent application related to microbial production of cyclic guanosine 3′,5′-monophosphate (cGMP) is exemplified:

ISHIYAMA (1983) cultured an *Aspergillus* strain producing cGMP from 2-amino-cAMP that shows deaminase activity and lacks phosphodiesterase activity. Thus, 10 mL culture supernatant of *A. niger* IAM 2533 was added to 100 mL 0.2 M acetate buffer containing 2 g 2-amino-cAMP and incubated at 38 °C for 2 h. 1.9 g cGMP was produced.

# 5 Fermentative Production of Nucleosides and Related Compounds

## 5.1 Inosine

The biosynthetic pathway of purine nucleotides has been elucidated in general (Fig. 11). The following properties seem to be necessary for microorganisms to produce inosine (ENEI, 1988):

(1) Adenylosuccinate (succinoadenosine 5′-monophosphate, SAMP) synthetase and IMP dehydrogenase should be lacking or be present at a very low level.
(2) Nucleosidase and/or nucleoside phosphorylase should be lacking or be present at a very low level.
(3) Regulation of enzymes involved in IMP formation, such as PRPP amidotransferase, by AMP and GMP should be released.

Strains belonging to a number of genera, such as *Escherichia, Neurospora, Bacillus, Aerobacter, Brevibacterium, Corynebacterium, Streptococcus,* and *Saccharomyces,* are known to produce inosine. Among them *Bacillus subtilis* and *Brevibacterium ammoniagenes* have been studied in detail for the purpose of industrial production of inosine.

### 5.1.1 Production by *Bacillus subtilis*

As described in Sect. 4.2.1.1, UCHIDA et al. (1961) found that an adenine-requiring mutant of *B. subtilis* accumulated hypoxanthine derivatives, mainly inosine.

AOKI et al. (1963) derived amino acid-requiring mutants from *B. subtilis* No. 2, that required adenine and accumulated a small amount (0.21 g $L^{-1}$) of inosine. The mutant strain C-30 ($ade^-$ $his^-$ $tyr^-$) accumulated 6.3 g $L^{-1}$ inosine. The amino acid requirement may contribute to the elevation of inosine accumulation by lowering the degradation activity from inosine to hypoxanthine.

AOKI et al. (1968) were the first to scale-up the fermentative production of inosine by the mutant strain C-30 from a flask culture to a 50,000 L tank culture.

ISHII and SHIIO (1973) derived various purine auxotrophs, their revertants, and purine-sensitive mutants from *B. subtilis* K. The strains No. 231 and No. 102 were 8-azaguanine-resistant adenine auxotrophs ($ade^-$ $8AG^r$) in which enzymes in the common pathway of purine nucleotide synthesis, such as PRPP amidotransferase, are genetically derepressed. Adenine auxotrophs RDA-3 and RDA-16 ($ade^-$) and the genetically derepressed mutants No. 231 and No. 102 ($ade^-$ $8AG^r$) were grown in a medium containing 80 g glucose, 20 g $NH_4Cl$, 2 g of an amino acid mixture, 1 g $KH_2PO_4$, 0.4 g $MgSO_4 \cdot 7H_2O$, 2 g KCl, 2 ppm $Fe^{2+}$, 2 ppm $Mn^{2+}$, 300 mg L-tryptophan, 25 g $CaCO_3$, and 50–300 mg adenine per liter of distilled water adjusted to pH 7.0. Growth reached its maximal level at 200–300 μg $mL^{-1}$ adenine, and the maximal accumulation of inosine was at 100 μg $mL^{-1}$ adenine. The accumulation decreased sharply with increasing adenine concentration. The maximal accumulation was 9–11 g $L^{-1}$ in strains RDA-3 and RDA-16, and 17–19 g $L^{-1}$ in strains No. 231 and No. 102.

MATSUI et al. (1982) obtained a mutant of *B. subtilis* with high productivity for inosine. The parent strain *B. subtilis* AJ11100 required adenine and arginine for growth and was genetically deficient in the purine nucleoside degrading enzyme. From the parent strain, sulfaguanidine resistant mutants were derived. Sulfaguanidine resistance was frequently accompanied by xanthine requirement. From one of the xanthine-auxotrophic mutants, AJ11101, a revertant AJ11102 was obtained. AJ11100 (parent), AJ11101 ($xan^-$), and AJ11102 ($xan^+$) produced 16.0 g $L^{-1}$, 13.2 g $L^{-1}$, and 20.6 g $L^{-1}$ inosine, respectively, when grown for 72 h at 34 °C in a medium containing 80 g glucose, 15 g $NH_4Cl$, 0.5 g $KH_2PO_4$, 0.4 g $MgSO_4 \cdot 7H_2O$, 0.01 g $FeSO_4 \cdot 7H_2O$, 0.008 g $MnSO_4 \cdot 4H_2O$, 21 g KCl, 0.3 g DL-methionine, 32 mL Ajieki (soybean hydrolyzate), 0.8 g RNA, 20 g $CaCO_3$ in a total volume of 1 L distilled water, adjusted to pH 7.0. For growth of strain AJ11101 ($xan^-$), 200 mg $L^{-1}$ xanthine was supplemented. The level of the 5′-nucleotidase activity in AJ11102 was 2.5-fold higher than that in AJ11100, and the level of IMP dehydrogenase activity in AJ11102 was about 1/3 of that in AJ11100. PRPP amidotransferase activity in AJ11102 was at the same level as in AJ11100. The increase of specific activity of 5′-nucleotidase and the decrease of specific activity of IMP dehydrogenase by mutation seem to contribute to the increase of inosine production. The amount of inosine produced by AJ11102 (20.6 g $L^{-1}$) is equivalent to about 25% of the weight of consumed sugar (80 g $L^{-1}$).

In China, production of inosine by *B. subtilis* has also been studied: *B. subtilis* 301, a mutant derived from *B. subtilis* 717-9-1, produced 17.8 g $L^{-1}$ inosine (WANG et al., 1983). A mutant resistant to streptomycin, hydroxylamine, and sodium azide was derived from *B. subtilis.* The mutant produced 19.49 g $L^{-1}$ inosine, which was 38.4% more than by the parent strain (XU et al., 1992). An inosine-producing strain F22 was obtained by protoplast fusion of *B. subtilis* cells. The parent strain and F22 produced inosine at 9.95 g $L^{-1}$ and 24.58 g $L^{-1}$, respectively (WANG et al., 1993).

The following patent applications for inosine production by *B. subtilis* have been recently filed:

(1) SHIMIZU et al. (1983a) cultured an adenine-requiring *Bacillus* strain containing the cloned gene for resistance to purine analogs. The transformed strain AJ 11836 produced 3.1 g $L^{-1}$ inosine, compared to 0.7 g $L^{-1}$ for the parent strain AJ 11831.
(2) See Tab. 5, TSUCHIDA et al. (1984).
(3) NAKAZAWA and SONODA (1987) cultured *B. subtilis* AJ3772 (FERM-P2555) under aerobic conditions at 34 °C. The oxygen supply was controlled to maintain an acetoin concentration in the medium of 0.05–0.1 g $dL^{-1}$. Inosine formation reached 7.3 g $dL^{-1}$ compared to 4.6 g $dL^{-1}$ for the conventional method.
(4) YOSHIWARA et al. (1987) increased the yield of L-amino acids or nucleosides manufactured by microorganisms by adding N-methylglycine, N,N-dimethylglycine, N,N,N-trimethylglycine, and/or (2-hydroxyethyl)trimethylammonium to the culture medium. *B. subtilis* AJ11374 (FERM-P5026), e.g., was cultured in a medium containing 0.2 g $dL^{-1}$ of N,N,N-trimethylglycine at 34 °C for 16 h. The culture contained 18.5 g $L^{-1}$ inosine vs. 16.5 g $L^{-1}$ for the control without N,N,N-trimethylglycine.
(5) MIYAGAWA et al. (1988a) transformed *B. subtilis* NA-6128 with the plasmid pEX203 carrying the IMP dehydrogenase gene of *B. subtilis.* This gene is mutated by the insertion of the chloramphenicol acetyltransferase gene of *B. pumilus*. The resulting strain NA-6141 expresses an inactive IMP dehydrogenase. When cultivated in the presence of xanthine, NA-6141 produced 35 mg $mL^{-1}$ of inosine, while the parent strain NA-6128 produced only 19.8 mg $mL^{-1}$.
(6) YAMAZAKI et al. (1993) transformed an adenine-requiring *Bacillus* strain that is capable of producing guanosine, xanthosine, both inosine and guanosine, or both inosine and xanthosine, with a plasmid carrying the gene for the resistance to DNA gyrase inhibitors and a construct that permits the expression of the IMP dehydrogenase gene only at low levels. The transformed *B. subtilis* strain, NA6301, produced 27 mg $mL^{-1}$ inosine and 3.0 mg $mL^{-1}$ guanosine, while the parent strain, *B. subtilis* NA6128, produced 22 mg $mL^{-1}$ inosine and 8.0 mg $mL^{-1}$ guanosine.

### 5.1.2 Production by *Brevibacterium (Corynebacterium) ammoniagenes*

FURUYA et al. (1970) derived 6-mercaptoguanine (6MG)-resistant mutants from an adenine leaky, IMP producing strain KY13102 of *Brevibacterium ammoniagenes.* They found that 111 out of 206 strains that are resistant to 10–50 μg $mL^{-1}$ of 6MG produced a rather large amount of inosine while there was no inosine accumulating strain among 213 strains resistant to 150–500 μg $mL^{-1}$ of 6MG. Various conditions for accumulation of inosine by the inosine accumulating strain KY13714 were investigated. Accumulation was stimulated in this strain at low adenine concentrations (i.e., 25 mg $L^{-1}$), but depressed at high levels of adenine. Inosine accumulation was not inhibited by $Mn^{2+}$ as IMP accumulation in strain KY13102, but stimulated by addition of excessive amounts of $Mn^{2+}$. During 4 days of cultivation at 30 °C, strain KY13714 accumulated 9.3 mg $mL^{-1}$ inosine.

In Fig. 12, the inosine producing mutant KY13714 is compared to the IMP producing strain KY13102. They show similar IMP degrading activity (IMP→Ino) and IMP forming activity (Hyp→IMP), but are quite different in their inosine degrading activity (Ino →Hyp). The lack of purine nucleoside degrading activity in KY13714 seems to be involved in 6MG resistance. In wild-type cells (KY13102), 6MG is ribotylated to 6-methylguanine ribotide (6MG-RP), that is accumulated in the cells and strongly inhibits purine biosynthesis as a pseudo-feedback inhibitor. On the other hand, in cells lacking purine nucleoside degrading activity (KY13714), 6MG-PR formed from 6MG may be converted to 6-methylguanine riboside (6MG-R). 6MG-R, as well as inosine, is excreted by the cells. This is the reason why a 6MG-resistant strain lacks inosine degrading activity and accumulates inosine in the medium.

The accumulation of inosine by KY13714 was considerably depressed by 200–300 mg $L^{-1}$ adenine, probably because biosynthesis of purine nucleotides in the strain is regulated by adenine or its derivatives, such as AMP or ATP. Thus, KOTANI et al. (1978) derived various mutants from strain KY13714 that are resistant to the purine analogs 2-fluoroadenine, 6-mercaptopurine, and 6-methylthiopurine (6MTP). Compared to the parent strain KY13714, depression of inosine accumulation by adenine was weak in KY13735, a mutant resistant to 5–10 mg $mL^{-1}$ of 6MTP.

Furthermore, the mutant KY13738, that is devoid of IMP dehydrogenase and requires guanine for growth, was induced from KY13735, because a defect in the gene coding for IMP dehydrogenase was considered to be involved in the effective accumulation of IMP. The mutational process and the inosine productivity of the resulting mutants are summarized in Tab. 12. Introduction of the guanine-requiring character was most effective for accumulation of inosine.

Under optimal conditions, 31 mg $mL^{-1}$ inosine are accumulated during 42 h of cultivation of strain KY13761 ($ade^l$ $6MG^r$ $6MTP^r$ $gua^-$) in a 5L jar fermenter at 32 °C. Inosine

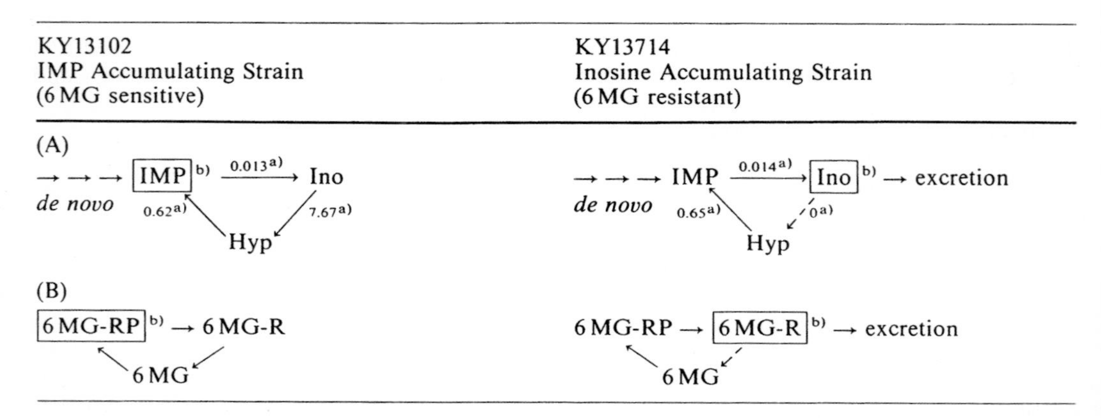


**Fig. 12.** Mechanisms for inosine accumulation (A) and 6-mercaptoguanine resistance (B) in *Brevibacterium ammoniagenes* KY13714 compared to YK13102 (adapted from FURUYA et al., 1970).
[a] Specific activity (μmol $mg^{-1}$ $h^{-1}$ protein).
[b] Compound to be accumulated.

**Tab. 12.** Inosine Productivity of Several Mutants Derived from *Brevibacterium ammoniagenes* KY 13102 (adapted from KOTANI et al., 1978)

| Strain (Marker) | Media Containing | | Inosine Produced [mg $mL^{-1}$] |
|---|---|---|---|
| | Adenine [mg $L^{-1}$] | Guanine [mg $L^{-1}$] | |
| KY13102 ($ade^l$) | | | |
| ↓ | | | |
| KY13714 ($ade^l 6MG^r$) | 100 | 0 | 5.3 |
| ↓ | | | |
| KY13735 ($ade^l 6MG^r 6MTP^r$) | 50 | 0 | 7.4 |
| ↓ | | | |
| KY13738 ($ade^l 6MG^r 6MTP^r gua^-$) | 200 | 100 | 14.9 |
| ↓ Monocolony isolation | | | |
| KY13745 ($ade^l 6MG^r 6MTP^r gua^-$) | 100 | 200 | 15.9 |
| ↓ Monocolony isolation | | | |
| KY13761 ($ade^l 6MG^r 6MTP^r gua^-$) | 100 | 200 | 16.3 |

accumulation, cell growth, and sugar consumption are linearly related to the incubation time. Thus a growth-associated type of accumulation was confirmed in inosine production by KY13761. The amount of inosine produced by KY13761 (31 mg $mL^{-1}$) from 17.5% of inverted molasses is equivalent to a yield of inosine >17% based on the consumed sugar.

The following patent applications for inosine production by *B. ammoniagenes* have been filed recently:

(1) See Tab. 6, FUJIO et al. (1988a).
(2) According to FUJIO and KITATSUJI (1990), *B. ammoniagenes* KT 813, a mutant resistant to 5-fluorouracil, accumulated 38.7 g $L^{-1}$ inosine vs. 31.7 g $L^{-1}$ of its parent strain KY4-7 in 24 h of culture at 30 °C.

## 5.2 Xanthosine

The following properties are necessary for microorganisms to produce xanthosine (ENEI, 1988):

(1) Both SAMP synthetase and GMP synthetase should be lacking or be present at a very low level.
(2) Xanthosine degrading enzymes (nucleosidase and nucleoside phosphorylase) should be present at a very low level.
(3) Feedback inhibition of IMP dehydrogenase by XMP should be weakened.
(4) Microorganisms should be grown in an adenine and guanine limited medium. (Enzymes involved in *de novo* biosynthesis of IMP, such as PRPP amidotransferase and SAMP lyase, are not regulated by XMP but AMP and GMP.)

FUJIMOTO et al. (1966) derived 15 guanine auxotrophs from *B. subtilis* IAM 1145. These strains accumulated 3.03–5.07 g $L^{-1}$ xanthosine and 0.41–0.78 g $L^{-1}$ xanthine. Adenine–guanine double auxotrophs were derived from one of the auxotroph strains, Gu-3 ($gua^-$), that accumulated 5.07 g $L^{-1}$ xanthosine and 0.68 g $L^{-1}$ xanthine. Their xanthosine productivity was higher than that of the parent strain, Gu-3. One of them, strain Gu-Ad-3-35 ($gua^-$ $ade^-$), accumulated 8.9 g $L^{-1}$ xanthosine.

ISHII and SHIIO (1973) also observed that an adenine–guanine double auxotroph strain of *B. subtilis,* AG-69 ($ade^-$ $gua^-$), accumulated 10 g $L^{-1}$ xanthosine together with inosine in an adenine and guanine limited medium. Xanthosine accumulation of AG-69 was about 2-fold higher than that of a guanine auxotroph and was depressed more strongly by guanosine than by adenine. Another adenine–guanine double auxotroph, No. 75-13 ($ade^-$ $8AG^{rr}$ $gua^-$), derived from the derepressed adenine auxotroph No. 102-75 ($ade^-$ $8AG^{rr}$) accumulated 17.5 g $L^{-1}$ of xanthosine in a medium containing 100 μg $mL^{-1}$ adenine and 150 μg $mL^{-1}$ guanosine.

## 5.3 Guanosine

The following properties are necessary for microorganisms to produce guanosine (ENEI, 1988):

1. Both SAMP synthetase and GMP reductase should be lacking or be present at a very low level.
2. Nucleosidase and/or nucleoside phosphorylase should be lacking or be present at a very low level.
3. Regulation of enzymes directly participating in GMP synthesis, IMP dehydrogenase and GMP synthetase, by GMP should be released.
4. Regulation of enzymes involved in *de novo* biosynthesis of IMP, such as PRPP amidotransferase, by AMP and GMP should be released.
5. Both IMP dehydrogenase and GMP synthetase should be more active than nucleotidase to avoid formation of inosine as byproduct.

KONISHI and SHIIO (1968) derived mutants resistant to several guanine analogs and purine analogs from *B. subtilis* AJ-1987, an inosine producing mutant requiring adenine for growth. As a result, strain AJ-1993 resistant to 8-azaguanine (8AG) was found to produce guanosine (4.3 mg $mL^{-1}$) and inosine (3.1 mg $mL^{-1}$) at the same time. Compared to the parent strain, strain AJ-1993 showed a high level of IMP dehydrogenase and lacked GMP reductase.

MOMOSE and SHIO (1969) derived strain 12-3 that is GMP reductase-negative, adenine-requiring, and 8-azaxanthine(8-AX)-resistant, from *B. subtilis* strain 38-3, an adenine-requiring inosine producer, by mutation and transduction as follows:

38-3($ade^-$) → TR-101($ade^-$ $try^-$)
→ TD-9($try^-$) → PU-17($try^-$ $pur^-$)
→ PD37($try^-$ $pur^-$ $red^-$)
→ TP-3($try^-$ $red^-$) → 30-12($try^-$ $red^-$ $ade^-$)
→ 12-3($try^-$ $red^-$ $ade^-$ $8AX^r$)

Tryptophan requirement was used as a genetic marker throughout the isolation process. Strain 12-3 accumulated up to 5 g $L^{-1}$ guanosine and about 10 g $L^{-1}$ inosine after 4 days of cultivation.

Independently NOGAMI et al. (1968) derived strain Nt 1071 ($8AG^r$ $his^-$ $red^-$ $ade^-$ $AAR^r$) from *Bacillus* sp. No. 102 by mutation and transformation. $AAR^r$ means adenine- and adenosine-resistant. Strain Nt 1071 produced 1.5 mg $mL^{-1}$ hypoxanthine, 3.3 mg $mL^{-1}$ inosine, and 5.4 mg $mL^{-1}$ guanosine during 4 days of cultivation at 37°C.

NOGAMI and YONEDA (1969) derived a streptomycin-resistant mutant, Nt 1011B ($8AG^r$ $his^-$ $red^-$ $ade^-$ $AAR^r$ $Sm^r$), from strain Nt 1071. This strain was superior to Nt 1071 in the productivity of hypoxanthine, inosine, and guanosine. They further derived nucleoside phosphorylase-lacking strains, T-180 and T-780 ($NP^-$), from Nt 1011B. T-180 and T-780 effectively produced inosine and guanosine without formation of hypoxanthine (Tab. 13).

KOMATSU et al. (1972) derived strain No. 20, an adenine, histidine, and threonine-requiring, adenase and GMP reductase-deficient, and 8-azaxanthine-resistant mutant, from *Bacillus* strain No. 1043-226 ($ade^-$). From strain No. 20 ($ade^-$ $his^-$ $thr^-$ $adenase^-$ $8AX^r$ $red^-$), a mutant partially deficient in purine nucleoside hydrolyzing activity, GnR176 ($ade^-$ $his^-$ $thr^-$ $adenase^-$ $8AX^r$ $red^-$ $NSase^{pd}$), was derived. GnR176 accumulated 10.6 mg $mL^{-1}$ guanosine, 1.86 mg $mL^{-1}$ inosine, and 2.81 mg $mL^{-1}$ xanthosine.

MATSUI (1984) succeeded in construction of an excellent guanosine producer accumulating more than 20 g $L^{-1}$ guanosine. From an inosine producing strain of *B. subtilis* 1411 ($ade^-$ $his^-$ $red^-$), methionine sulfoxide, psicofuranine, and decoyinine-resistant mutants were derived, as shown in Tab. 14.

A mutant resistant to 5 mg $mL^{-1}$ methionine sulfoxide, 14119 ($ade^-$ $his^-$ $red^-$ $MS^r$),

**Tab. 13.** Accumulation of Purine Nucleosides by Adenine–Adenosine-Resistant ($AAR^r$), Streptomycin-Resistant ($Sm^r$), Nucleoside Phosphorylase-Lacking ($NP^-$), and Adenine–Adenosine Deaminase-Lacking ($dea^-$) Mutants of *Bacillus* sp. No. 102 (adapted from NOGAMI and YONEDA, 1969)

| Strain[a] | | Products [mg mL$^{-1}$] | | |
|---|---|---|---|---|
| | | Hypo-xanthine | Inosine | Guanosine |
| Nt 729 | ($AAR^sSm^sNP^+$) | 2.8 | 0.8 | 1.1 |
| Nt 1071 | ($AAR^rSm^sNP^+$) | 1.5 | 3.3 | 5.4 |
| NT 1011B | ($AAR^rSm^rNP^+$) | 2.0 | 4.6 | 7.9 |
| T-180 | ($AAR^rSM^rNP^-dea^-$) | 0 | 12.6 | 5.4 |
| T-780 | ($AAR^rSm^rNP^-$) | 0 | 10.8 | 5.8 |

[a] The strains used had the following genetic markers in common: $8AG^r$ (8-azaguanine-resistant), $red^-$ (GMP reductase-lacking), and $ade^-$ (adenine-requiring).

**Tab. 14.** Production of Purine Nucleosides by Several Mutants Derived Successively from *B. subtilis* 1411 (adapted from MATSUI et al., 1979)

| Strain[a] | | Products [g L$^{-1}$] | | | |
|---|---|---|---|---|---|
| | | Inosine | Xantho-sine | Guano-sine | Total |
| 1411 | ($ade^-his^-red^-$) | 11.0 | 0 | 5.5 | 16.5 |
| 14119 | ($ade^-his^-red^-MS^r$) | 4.8 | 0 | 9.6 | 14.4 |
| AG169 | ($ade^-his^-red^-hMS^r$) | 0 | 6.0 | 8.0 | 14.0 |
| GP-1 | ($ade^-his^-red^-hMS^rPF^r$) | 0 | 3.4 | 10.6 | 14.0 |
| MG-1 | ($ade^-his^-red^-hMS^rPF^rDC^r$) | 0 | 0 | 16.0 | 16.0 |

[a] $MS^r$: resistant to methionine sulfoxide;
$hMS^r$: resistant to higher doses of methionine sulfoxide;
$PF^r$: resistant to psicofuranine;
$DC^r$: resistant to decoyinine.

accumulated more guanosine and less inosine than the parent strain 1411. A mutant resistant to 10 mg mL$^{-1}$ methionine sulfoxide, AG169 ($ade^-$ $his^-$ $red^-$ $hMS^r$), accumulated guanosine and xanthosine, but did not accumulate inosine. The methionine sulfoxide resistance mainly caused the decrease of specific activity of 5′-nucleotidase. The specific activity of 5′-nucleotidase for IMP in strain AG169 was about 1/8 of strain 1411. The methionine sulfoxide resistance also caused a slight elevation of the specific activity of IMP dehydrogenase and a partial loss of inhibition of the enzyme by GMP. Furthermore, both repression and inhibition of PRPP amidotransferase by guanine and GMP, respectively, and repression of SAMP lyase by guanine were lost (Tabs. 15 and 16). The above variation of enzyme activities caused by methionine sulfoxide resistance contributes to both the increase of guanosine and xanthosine productivity and the decrease of inosine productivity.

In AG169 the regulation of GMP synthetase by GMP is not released, because the strain accumulated a large amount of xanthosine together with guanosine. Therefore, mutants resistant to psicofuranine were derived. Growth of AG169 was almost completely inhibited by 0.5 mg mL$^{-1}$ psicofuranine or decoyinine and could not be recovered by xanthine, but were completely recovered by

**Tab. 15.** Represson of Four Enzymes of *B. subtilis* Mutants Grown in the Presence of Adenine by Guanine (adapted from MATSUI et al., 1979)

| Strain | Repression [%] by 500 mg $L^{-1}$ Guanine[a] | | | |
|---|---|---|---|---|
| | IMP Dehydrogenase | GMP Synthetase | PRPP Amidotransferase | SAMP Lyase |
| 1411 | 69.9 | 21.2 | 69.9 | 40.0 |
| 14119 | 55.0 | 10.0 | 66.0 | 27.6 |
| AG169 | 28.7 | 5.6 | 0 | 0 |
| GP-1 | 19.7 | 0 | 0 | 7.7 |
| MG-1 | 16.8 | 0 | 0 | 5.4 |

[a] Repression $[\%] = \frac{A-B}{A} \times 100\ [\%]$
A: Enzyme activity of the cells grown in the medium containing 50 mg $L^{-1}$ adenine, for SAMP lyase, 100 mg $L^{-1}$ adenine were used.
B: Enzyme activity of the cells grown in the medium containing 50 mg $L^{-1}$ adenine plus 500 mg $L^{-1}$ guanine, for SAMP lyase, 100 mg $L^{-1}$ adenine plus 500 mg $L^{-1}$ guanine were used.

**Tab. 16.** Inhibition of Four Enzymes of *B. subtilis* Mutants by GMP (adapted from MATSUI et al., 1979)

| Strain | Inhibition [%] by 4 mM GMP[a] | | | |
|---|---|---|---|---|
| | IMP Dehydrogenase | GMP Synthetase | PRPP Amidotransferase | SAMP Lyase |
| 1411 | 80.0 | 20.0 | 45.0 | 10.2 |
| 14119 | 55.0 | 25.0 | 33.9 | 4.3 |
| AG169 | 33.0 | 20.1 | 0 | 4.1 |
| GP-1 | 32.0 | 11.1 | 0 | 0 |
| MG-1 | 35.0 | 0 | 0 | 0 |

[a] In the case of SAMP lyase 0.08 mM GMP were used.

guanine. Therefore, both psicofuranine and decoyinine seem to specifically inhibit or repress GMP synthetase in AG169. When AG169 was mutated to be resistant to psicofuranine, with the resulting mutant, GP-1 ($ade^-$ $his^-$ $red^-$ $hMS^r$ $PF^r$), higher yields of guanosine and lower yields of xanthosine were achieved. In GP-1 the specific activity of GMP synthetase was about half of that in AG169, repression of the enzyme by guanine was completely lost, and the inhibition by GMP was partially weakened (Tabs. 15 and 16). Psicofuranine resistance mainly caused variations in GMP synthetase activity.

As growth of GP-1 was strongly inhibited by decoyinine, decoyinine-resistant mutants were derived from GP-1. Of these mutants, strain MG-1 ($ade^-$ $his^-$ $red^-$ $hMS^r$ $PF^r$ $DC^r$) exclusively accumulated guanosine with a high yield (16 g $L^{-1}$; weight yield: 20% of consumed sugar). Decoyinine resistance brought about complete loss of both repression and inhibition of GMP synthetase by guanine and GMP (Tabs. 15 and 16).

MATSUI concluded that the key to increased guanosine production is the loss of both repression and feedback inhibition of IMP dehydrogenase, GMP synthetase, SAMP lyase, and PRPP amidotransferase. It further depends on the degree of 5′-nueleotidase activity.

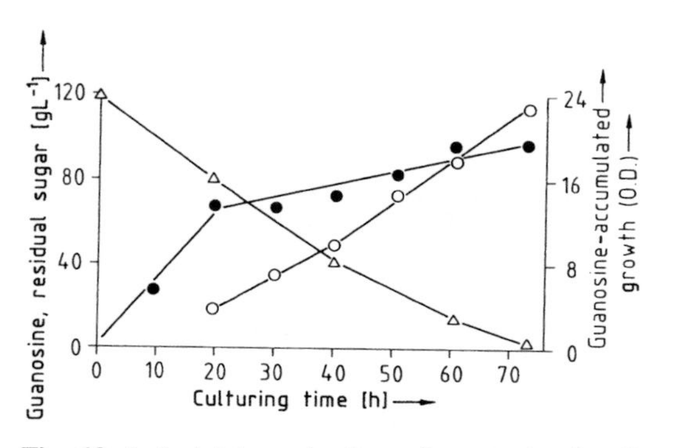


**Fig. 13.** Industrial production of guanosine by *B. subtilis* MG-1 (MATSUI, 1984).
○ guanosine, ● growth, △ residual sugar.

As shown in Fig. 13, strain MG-1 accumulated 23.2 kg per 1,000 L of guanosine during cultivation at 34°C for 72 h in 60,000 L of a medium containing 12% sugar (weight yield: 19.3% of consumed sugar).

MIYAGAWA et al. (1986) cloned the *B. subtilis* IMP dehydrogenase gene and applied it to increased production of guanosine. They used two strains of *B. subtilis*: Strain NA7821 predominantly produces guanosine, but the total amount of purine nucleosides produced is relatively small. The other strain, NA6128, produces large amounts of purine nucleosides, but they largely consist of inosine; this strain also retains a high level of GMP synthetase activity. The rate-limiting reaction for guanosine production in strain NA6128 is suggested to be the first step of the specific pathway for GMP biosynthesis from IMP, involving a reaction catalyzed by IMP dehydrogenase.

Thus, the gene encoding IMP dehydrogenase of *B. subtilis* NA7821 was cloned in pBR322 in *E. coli* and subcloned into the *B. subtilis* plasmid, pC194. The recombinant plasmid pBX121 consisting of the pC194 vector sequence and four *Hin*dIII insert fragments was introduced into the inosine–guanosine producing *B. subtilis* strain NA6128. Compared with the host strain NA6128, the transformed strain NA6128 (pBX121) showed high IMP dehydrogenase activity, high guanosine productivity, and low inosine productivity. In addition, the amount of guanosine produced by the transformant was 2.2-fold more than in NA7821, the donor DNA strain (Tab. 17).

Scale-up of a fermentation process requires a minimum number of runs in a pilot experiment in order to reproduce the shaking flask conditions in a stirred-tank fermentor. SUMITO et al. (1993) attempted to scale-up the nucleoside fermentation by a mutant strain of *B. subtilis* from a shaking flask to a stirred-tank fermentor. At the same temperature conditions the total amount of purine nucleosides produced in the stirred-tank reactor was almost the same as in the shaking flask, but the accumulation ratio of guanosine:total nucleosides was apt to be lower than that in the flask. When the pH of the stirred-tank culture was maintained at 6.9 by addition of ammonia water to keep the $NH_4^+$ level higher, the ratio was improved to the same level as that observed in the shaking flask culture. (Since urea could not efficiently be utilized in the

**Tab. 17.** Production of Inosine and Guanosine by a Recombinant Strain of *B. subtilis* (adapted from MIYAGAWA et al., 1986)

| Strain of *B. subtilis* | IMP Dehydrogenase Activity[a] | Productivity | | | |
|---|---|---|---|---|---|
| | | Inosine [mg mL] | Guanosine [mg/mL] | Total [mg/mL] | Guanosine/ Total [%] |
| NA6128(pBX121) | 15.8 | 5.0 | 20.0 | 25.0 | 80 |
| NA6128 | 1.5 | 19.0 | 7.0 | 26.0 | 27 |
| NA7821 | 9.8 | 1.0 | 9.0 | 10.0 | 90 |

[a] nmol of XMP formed per min and mg protein.

stirred-tank fermenter, the $NH_4^+$ concentration and the pH of the culture broth were lower during fermentation than in the shaking-flask culture.) It was concluded that a high $NH_4^+$ concentration gave a high value of guanosine/inosine by stimulating the key enzyme, IMP dehydrogenase.

TANG et al. (1992) improved the guanosine production of a *B. subtilis* mutant ($ade^-$ $his^-$) by using a fed-batch culture. After addition of 1.5% glucose, higher guanosine yields were obtained; the highest yield was 15.6 g $L^{-1}$.

The following patent applications for guanosine production have been recently filed:

(1) NAKAZAWA et al. (1982) cultivated a guanosine producing strain of *Bacillus* which is stored or precultured in a medium containing hydrazine or its derivatives. The yield of guanosine relative to the amount of consumed sugar was 27.5% when the inoculum was incubated in the presence, and 18.5% in the absence of hydrazine HCl.
(2) SUMINO et al. (1983) cultivated an inosine and guanosine-producing and adenine-requiring *Bacillus* strain in an $O_2$-enriched atmosphere; *Bacillus pumilus* IFO 12483, e.g., produced 17.7 g $L^{-1}$ guanosine. Under normal aerobic conditions the strain produced only 0.4 g $L^{-1}$ inosine and 0.6 g $L^{-1}$ guanosine, respectively.
(3) DOI et al. (1983) cultivated an inosine and/or guanosine-producing *B. pumilus* strain in a medium containing carbohydrate added continuously or intermittently to the medium in order to maintain its concentration at <1%. Under the above-mentioned conditions, the combined yield of inosine and guanosine was 34.4 g $L^{-1}$, a 21.5% yield based on the carbohydrate.
(4) See Tab. 5, TSUCHIDA et al. (1984).
(5) MIYAGAWA et al. (1985) improved a guanosine-producing strain of *B. subtilis* by transforming it with a vector containing the gene for IMP dehydrogenase. The transformed strain produced 18 mg $mL^{-1}$ of guanosine, while the host strain produced only 14 mg $mL^{-1}$.
(6) NAKAZAWA and SONODA (1987) cultured *B. subtilis* AJ3617 (FERM-P2313) at 34 °C under aerobic conditions in an $O_2$-enriched atmosphere. Guanosine formation reached 4.8 g $dL^{-1}$ compared to 2.9 g $dL^{-1}$ for the conventional method.
(7) YOSHIWARA et al. (1987) cultured *B. subtilis* AJ11315 (FERM-P4526) in a medium containing 0.2 g $dL^{-1}$ of N,N,N-trimethylglycine at 34 °C for 16 h. The culture contained guanosine 20.1 g $L^{-1}$ vs. 18. 3 g $L^{-1}$ for the control without N,N,N-trimethylglycine.
(8) MIYAGAWA et al. (1988b) enhanced the production of inosine and guanosine by a recombinant *B. subtilis* strain containing an insertion mutation in the adenylosuccinate (SAMP) synthetase gene that inactivates the enzyme to decrease feedback inhibition during inosine/guanosine synthesis. They constructed plasmid pPA350-1 by inserting the chloramphenicol acetyltransferase gene from *E. coli* into the *Hpa*I site of the SAMP synthetase gene of *B. subtilis* No.115 and transfected this construct to the SAMP synthetase-deficient *B. subtilis* BM 1032 strain. The transformant (BM 1051) produced 9.5 mg $mL^{-1}$ inosine and 5.0 mg $L^{-1}$ guanosine, while *B. subtilis* BM 1032 produced 6.3 mg $L^{-1}$ and 4.9 mg $mL^{-1}$, respectively.
(9) According to MIYAGAWA et al. (1990) inosine or guanosine production by *Bacillus* is greatly enhanced by the regulation of the expression of the IMP dehydrogenase gene with an exogenous promoter stronger/weaker than the endogenous promoter. With enhanced expression of the IMP dehydrogenase gene by a stronger promoter the production of guanosine is increased, while with reduced expression of the IMP dehydrogenase gene by a weaker promoter, the production of inosine is increased. *B. subtilis* was transformed with the expression plasmid pEX242 or pEX243 containing the SPO-1-26 or the P1 promoter in front of the IMP dehydrogenase gene. By standard fermentation for 84 h the production of guanosine by *B. subtilis*

harboring pEX242 was 3-fold higher than that by the wild-type *B. subtilis*; the production of inosine by *B. subtilis* harboring pEX243 was about 20% higher than that by the wild-type.

(10) MIYAGAWA and KANZAKI (1991) altered the promoter region of the *Bacillus* purine operon to enhance its expression in order to increase inosine and/or guanosine production. They cloned the purine operon promoter, digested it with *Bol*31 to remove a fragment of the promoter, and inserted an SP promoter at the restriction site. They transformed *B. subtilis* with the plasmid containing this chimeric promoter and selected recombinants containing this construct in the genomic DNA. One of the transformed clones produced 14 mg $mL^{-1}$ inosine and 2 mg $L^{-1}$ guanosine, while the parent strain produced 8.0 and 1.1 mg $mL^{-1}$, respectively.

(11) MIYAGAWA and KANZAKI (1992) modified the PRPP synthetase (PPS) gene of *Bacillus* at the ribosome binding site and/or the promoter to enhance the expression of the PPS gene. They cloned the PPS gene, and constructed plasmid pPSC32 containing the PPS gene fused to another promoter. *B. subtilis* containing this plasmid showed 2.8-fold more PPS activity than the untransformed parent. This transformant produced 14 mg $mL^{-1}$ inosine and 2.5 mg $mL^{-1}$ guanosine (compared to 8.0 mg $mL^{-1}$ and 1.1 mg $mL^{-1}$ for the parent strain).

(12) SUGIMOTO et al. (1993) enhanced the production of amino acids and/or nucleic acid-related compounds. In their method, a portion of the fermentation broth is continuously subjected to a treatment in a liquid cyclone in order to recover the crystallized product and reuse the treated fermentation broth. With this treatment, the production of crystalline guanosine by *B. subtilis* AJ11312 amounted to 0.43 kg $L^{-1}$ $h^{-1}$ vs. 0.28 kg $L^{-1}$ $h^{-1}$ without using a liquid cyclone.

## 5.4 Adenosine

The following properties are indispensable for microorganisms to produce adenosine (ENEI, 1988):

1. Both IMP dehydrogenase and AMP deaminase should be lacking or be present at a very low level.
2. Adenosine degrading enzymes (nucleosidase and nucleoside phosphorylase) should be present at a very low level.
3. Regulation of SAMP synthetase by AMP should be released.
4. Regulation of enzymes involved in *de novo* biosynthesis of IMP, such as PRPP amidotransferase and SAMP lyase, by AMP and GMP should be released.

Production of adenosine by microorganisms was reported by KONISHI et al. (1968) for the first time. They found that *B. subtilis* No. 717, an L-isoleucine-requiring mutant, accumulated 1.4 mg $mL^{-1}$ adenosine during cultivation in a chemically defined medium supplemented with 0.03% of L-isoleucine. The strain accumulated 2.5 mg $mL^{-1}$ adenosine in a glycerol–peptone medium.

HANEDA et al. (1971) attempted to derive adenosine-producing mutants from an inosine-producing *Bacillus* strain. They found that xanthine-requiring mutants lacking adenase produced a large amount of adenosine. Strain P53-18 ($his^-$, $thr^-$, $adenase^-$, $8AX^r$, $red^-$ ($GMP\text{-}reductase^-$), $xan^-$), e.g., produced 16.27 g $L^{-1}$ adenosine, 1.82 g $L^{-1}$ adenine, and 0.72 g $L^{-1}$ AICA riboside. Among the genetic characteristics of the adenosine-producing mutants, xanthine requirement was the most important factor, and adenosine productivity decreased significantly with an increasing number of revertants for this genetic marker. Thus HANEDA et al. (1972) derived mutant strains free from reversion of xanthine requirement.

For industrial production of adenosine, NISHIYAMA et al. (1993) derived an adenosine producer, *B. subtilis* A 3–5 ($ade^+$ (adenine revertant), $xan^-$, $8AG^r$, $sulfaguanidine^r$), from an inosine producer, *B. subtilis* AJ-11102 ($ade^-$, $xan^-$, $8AG^r$ $sulfaguanidine^r$).

The parent strain has been used for industrial production of inosine. Strain A 3–5 produced 9 g $L^{-1}$ adenosine from 80 g $L^{-1}$ glucose in 104 h.

Furthermore, NISHIYAMA et al. (1994) derived strain No. 43-35, a mutant resistant to adenine analogs, from *B. subtilis* A 3–5. The selection was performed by screening first for resistance to 2-thioadenine sulfate (0.5 mg $mL^{-1}$), then for resistance to 8-azaadenine (1 mg $mL^{-1}$), and finally for resistance to 6-methylaminopurine (1 mg $mL^{-1}$). Strain No. 43-35 produced 24 g $L^{-1}$ adenosine from 130 g $L^{-1}$ glucose in a glass jar fermenter with a working volume of 300 mL under appropriate conditions.

The following patent applications for adenosine production have been filed recently:

(1) NAKAMATSU et al. (1985a) cultivated a mutant of *Bacillus* producing 11.5 g $L^{-1}$ adenosine, requiring guanine, and showing resistance to 8-azaadenine and/or 8-methylaminopurine.
(2) NAKAMATSU et al. (1985b) cultivated an adenosine-producing strain *B. subtilis* AJ12050, *Corynebacterium glutamicum* AJ12130, or *Brevibacterium ammoniagenes* AJ12131 in a medium containing 0.02–0.2% $Mg^{2+}$. For example, *B. subtilis* AJ12050 produced 2.51 g $dL^{-1}$ adenosine when cultivated at 34 °C for 96 h in the presence of 0.1% $Mg^{2+}$ compared to 1.51 g $dL^{-1}$ adenosine in the presence of 0.004% $Mg^{2+}$.
(3) KARASAWA et al. (1992) cultivated an adenosine-producing and isoleucine-, histidine-, and arginine-dependent or pyrimidine analog-resistant *Bacillus* species. *B. subtilis* AJ12519, e.g., a mutant resistant to 4-aminopyrazolo[3,4-d]pyrimidine accumulated 14.6 g $L^{-1}$ adenosine, compared to 11.5 g $L^{-1}$ produced by the parent strain AJ12050.
(4) NISHIYAMA and NAKAMATSU (1993) derived strain AJ12708, a mutant strain requiring guanine and showing resistance to thioadenylsulfate, from AJ 12707, an adenosine-producing strain requiring guanine. Strain AJ12707 had been derived from *B. subtilis* ATCC 6643. Strain AJ 12707 and AJ12708 accumulated 4 g $L^{-1}$ and 10 g $L^{-1}$ adenosine, respectively, during cultivation at 34 °C for 72 h.

## 5.5 Uridine

Although a 2-fluoroadenine-resistant derivative of *Brevibacterium ammoniagenes* ATCC 6872 that accumulates a large amount of uracil was reported (3.9 mg $mL^{-1}$) by KOBAYASHI et al. (1988), mutants capable of producing high yields of uridine had not been obtained until DOI et al. (1988) prepared a mutant suitable for the industrial production of uridine.

Generally, growth of the wild-type strain of *B. subtilis* is inhibited by uracil analogs such as 2-thiouracil and 6-azauracil. Inhibition is reversed by adding uracil to the medium. DOI et al. (1988) derived mutants resistant to uracil analogs. A 2-thiouracil-resistant mutant, No. 258, was derived from a wild-type strain, *B. subtilis* No. 122. Then 6-azauracil-resistant mutants, No. 312, 417, and 508, were derived sequentially from strain No. 258 through exposure to N-methyl-N′-nitro-N-nitrosoguanidine. As shown in Tab. 18, strain No. 508 produced 46 mg $mL^{-1}$ uridine. A linear correlation was found between the level of resistance to 6-azauracil and the amount of uridine produced. The effective production of uridine by strain No. 508 seems to be due to the loss of pyrimidine repression of the six *pyr* genes encoding enzymes for pyrimidine nucleotide biosynthesis.

As the strain concomitantly accumulated 6 mg $mL^{-1}$ uracil it was subjected to further mutation to obtain mutants deficient in uridine phosphorylase activity. As a result, a useful mutant was obtained. Mutant No. 556 produced 55 mg $mL^{-1}$ uridine with no appreciable amount of uracil.

It is concluded that both relief of regulation of pyrimidine nucleotide biosynthesis and deficiency in uridine phosphorylase are indispensable for effective production of uridine.

The following patent applications for uridine production have been filed recently:

(1) YOKOZEKI et al. (1989a) microbiologically produced uridine from ribose-1-phosphoric acid or its salts and uracil in a hydrophilic solvent. *Microbacterium lacti-*

**Tab. 18.** Production of Uridine by Pyrimidine Analog-Resistant and Uridine Phosphorylase-Deficient Mutants Derived from *B. subtilis* No. 122 (adapted from DOI et al., 1988)

| Strain | Resistance to 6-Azauracil [mg mL $^{-1}$] | Relative Activity of Uridine Phosphorylase | Productivity [mg mL $^{-1}$] | |
|---|---|---|---|---|
| | | | Uridine | Uracil |
| No. 122 | – | | | |
| No. 258 | – [a] | | 10 | 7 |
| No. 312 | 1 | | 23 | 7 |
| No. 417 | 3 | | 38 | 6 |
| No. 508 | 5 | 100 | 46 | 6 |
| No. 556 | 5 | 3 | 55 | <1 |

[a] Resistant to 2-thiouracil (0.3 mg mL $^{-1}$).

*cum* ATCC-8180 cells, e.g., were suspended (5 g dL $^{-1}$) in 10 mL of 0.05 M Tris buffer, pH 7.2, containing 50 mM ribose-1-phosphoric acid and 50 mM uracil. The mixture was incubated for 16 h at 60 °C to obtain 628 mg dL $^{-1}$ uridine.

(2) ASAHI et al. (1989) cultivated uracil and uridine producing pyrimidine analog-resistant bacteria. They prepared plasmid pYRC100, e.g., which carried the pyrimidine analog-resistance gene of a mutant of *B. subtilis,* and transformed *B. subtilis* AA47 with pYRC100. The transformed strain, TFAA47, accumulated 3.2 mg mL $^{-1}$ uracil and 1.5 mg mL $^{-1}$ uridine in the culture medium.
(3) See Tab. 9, MARUMO et al. (1992).
(4) YAMAMOTO et al. (1992) cultivated a uridine producing, purine analog-resistant mutant strain derived from *B. subtilis* Y125 with UV irradiation. *B. subtilis* AG-34 (FERM-11758), e.g., accumulated 19.55 mg mL $^{-1}$ uridine in a shaking culture at 28 °C for 7 d. There was no accumulation of uridine in the parent strain.
(5) TOYODA et al. (1992) cultivated an uridine producing mutant strain of *B. subtilis* IFO 13719, that is resistant to orotic acid and thymidine analogs and incapable of decomposing uridine. During aerobic cultivation at 30 °C for 3 d the strain TL1 (FERM P-11951), e.g., produced 4 mg L $^{-1}$ uridine and 0.4 mg L $^{-1}$ uracil. The parent strain accumulated <0.01 mg mL $^{-1}$ uridine and uracil, respectively.

## 5.6 Cytidine

Industrial production of cytidine was first established by ASAHI et al. (1991). They successively derived mutant strains resistant to pyrimidine analogs from *B. subtilis* No. 122 as shown in Tab. 19.

The following properties seem to be indispensable for microorganisms to produce cytidine:

(1) Cytidine deaminase (cytidine→uridine) should be lacking or be present at a very low level.
(2) Regulation of enzymes involved in *de novo* biosynthesis of UMP by uracil derivatives should be released.
(3) Regulation of CTP synthetase (UTP→CTP) by cytidine derivatives should be released.

As shown in Tab. 19, strain CD-300 lacks cytidine deaminase (1). Strain No. 229 is resistant to 6-azauracil (2). Strain No. 344 is resistant to 5-fluorocytidine and strain No. 428 is resistant to 3-deazauracil (3). In addition, 3-deazaUMP is known to inhibit calf liver CTP synthetase.

At an industrial scale, strain No. 601 produced 27 mg mL $^{-1}$ cytidine during 90 h of cultivation in a medium containing 20% glucose.

The following patent applications for cytidine production have been filed recently:

**Tab. 19.** Production of Cytidine by Cytidine Deaminase-Deficient and Pyrimidine Analog-Resistant Mutants Derived from *B. subtilis* No. 122 (adapted from ASAHI et al., 1991)

| Strain | Productivity [mg mL$^{-1}$] | |
|---|---|---|
| | Cytidine | Uracil |
| No. 122 Wild-type | 0 | 0 |
| CD −300 Cytidine deaminase$^{-}$ | 0.1 | 0 |
| No. 229 6-Azauracil (0.2 mg mL$^{-1}$)$^{r}$ | 0.2 | 6 |
| No. 344 5-Fluorocytidine (0.5 mg mL$^{-1}$)$^{r}$ | 10 | 2 |
| No. 428 3-Deazauracil (2.0 mg mL$^{-1}$)$^{r}$ | 13 | <1 |
| No. 515 Transformant | 18 | <1 |
| No. 601 Homoserine$^{-}$ | 23 | <1 |

(1) FURUYA and OISHI (1991) cultivated a cytidine producing, 6-diazo-5-oxonorleucine-resistant mutant strain derived from *Bacillus natto* C-55. The mutant, *B. natto* D-18, was cultured at 39°C for 5 d and produced 9.6 mg mL$^{-1}$ cytidine vs. 4.7 mg mL$^{-1}$ for the parent strain.

(2) ASAHI and DOI (1991) cloned the CTP synthetase gene of a cytidine analog-resistant *Bacillus* strain. *B. subtilis* transformed with the expression plasmid pINV128 encoding the CTP synthetase gene produced 8.2 mg cytidine per mL of medium after 3 d at 37°C. The non-transformed parent strain produced only 0.1 mg mL$^{-1}$.

(3) ASAHI et al. (1993) prepared *B. subtilis* mutants lacking cytidine deaminase that are resistant to pyrimidine analogs and produce cytidine and deoxycytidine. *B. subtilis* FU-11 (FERM BP-908), e.g., was derived from *B. subtilis* ATCC 6051. The specific activity of cytidine deaminase in FU-11 was <0.01 unit per mg of protein, while that of the parent strain was 74.7 units per mg of protein. FU-11 was resistant to 100 µg mL$^{-1}$6-azauracil, 100 µg mL$^{-1}$ 2-thiouracil, 0.5 µg/mL$^{-1}$ 5-fluorouracil, and 0.5 µg mL$^{-1}$ 5-fluoroorotic acid, while the parent strain was sensitive to these analogs. During incubation at 37°C for 3 d, FU-11 produced 5.1 mg mL$^{-1}$ cytidine and 0.7 mg mL$^{-1}$ deoxycytidine, while the parent strain produced neither cytidine nor deoxycytidine at all.

## 5.7 Thymidine

YOKOZEKI et al. (1989b) microbiologically produced thymidine from 2′-deoxyribose-1-phosphoric acid or its salts and thymine in a hydrophilie solvent. *Sartina albida* FERM-P 7048 cells, e.g., were suspended (5 g dL$^{-1}$) in 10 mL 0.05 M Tris buffer, pH 7.2, containing 2′-deoxyribose-1-phosphoric acid and thymine (20 mM). The mixture was incubated at 60°C for 24 h to yield 329 mg dL$^{-1}$ thymidine.

## 5.8 Pseudouridine

FUJISHIMA and NAGATA (1982) isolated a new strain, *Pseudomonas* THD-9, from soil. This strain produced pseudouridine when grown at 28°C for 2 d in a bouillon medium supplemented with uracil and ribose. The molar conversion ratio of uracil to pseudouridine was 61.7%. Ribose could be replaced by arabinose, glucose, galactose, mannose, fructose, or sucrose for the production of pseudouridine. When arabinose was employed instead of ribose, 44.6% of uracil were converted to pseudouridine and 31.4% to 5′-O-acetylpseudouridine.

## 5.9 Orotidine and Orotic Acid

There have been many reports on the accumulation of orotic acid and orotidine by uracil-requiring mutant strains. MACHIDA and KUNINAKA (1969) and MACHIDA et al. (1970) reported that wild-type strains of *E. coli* K12 produced orotic acid but other *E. coli* strains failed to do so. WOMACK and O'DONOVAN (1978) concluded that this was caused by a mutation in the *pyr*F gene encoding OMP decarboxylase catalyzing the step 6 of UMP synthesis. Many of the observations considered to be a property of *E. coli* K12 wild-type strains are in fact the property of *E. coli* K12 strains with a defective *pyr*F gene.

As stated in Sect. 5.5, DOI et al. (1988) derived mutant No. 556 producing large amounts of uridine from *B. subtilis* No. 122. In this mutant strain all 6 enzymes of the *de novo* UMP biosynthesis were completely freed from regulation by uracil compounds. ASAHI and DOI (1993) further derived strains No. 122E and 122F from the parent strain No. 122, and No. 556E and 556F from strain No. 556. Strains No. 122E and 556E were deficient in orotic acid phosphoribosyltransferase (OPRTase), and No. 122F and 556F were deficient in OMP decarboxylase (OMP-DCase). As indicated in Tab. 20, strains No. 556E and 556F produced a large amount of orotic acid and/or orotidine compared to strains No. 122E and 122F. The high productivity may be due to the complete release from regulation of UMP synthetic enzymes. Tab. 20 also shows that accumulation of orotic acid alone may be due to the loss of OPRTase activity, while the simultaneous accumulation of orotic acid and orotidine is ascribed to a deletion of OMP-DCase.

As the mutant strains deficient in OPRTase or OMP-DCase derived from a uridine-producing mutant strain of *B. subtilis* accumulate large quantities of orotic acid and/or orotidine even in the presence of excess uracil, these strains may be used industrially.

The following patent application for production of orotic acid, e.g., has been filed recently: TAKAYAMA and MATSUNAGA (1989) derived mutant strains from *Corynebacterium glutamicum* ATCC 14275 resistant to pyrimidine analogs or resistant to both pyrimidine analogs and sulfa drugs. Several mutant strains obtained produced orotic acid more effectively than the parent strain. Strain T-30, e.g., is resistant to both 5-fluorouracil (100 $\mu g\ mL^{-1}$) and sulfaguanidine (500 $\mu g\ mL^{-1}$) and produced 56 $g\ L^{-1}$ orotic acid during cultivation in a medium containing 60 $mg\ L^{-1}$ uracil at 33 °C for 84 h. Under the same conditions, the parent strain ATCC 14275 produced 28.5 $g\ L^{-1}$ orotic acid.

## 5.10 Purine Arabinosides

YAMANAKA and UTAGAWA (1980) succeeded in enzymatically preparing adenine arabinoside (ara-A) from uracil arabinoside (ara-U) and adenine. Ara-U can be synthesized chemically from uridine that is commercially available. They screened microorganisms for their capability to catalyze the following transarabinosylation:

$$\text{ara-U} + \text{ade} \rightarrow \text{ara-A} + \text{ura} \qquad (7)$$

**Tab. 20.** Production of Orotic Acid and Orotidine by Mutants of *B. subtilis* (adapted from ASAHI and DOI, 1993)

| Strain | Enzyme Activity [units per mg of protein] | | Productivity [mM] | | |
|---|---|---|---|---|---|
| | OPRTase | OMP-DCase | Orotic Acid | Orotidine | Uridine |
| No. 122 | 8.0 | 0.3 | 0 | 0 | 0 |
| No. 122E | <0.01 | 0.2 | 87 | 0 | 0 |
| No. 122F | 7.2 | <0.01 | 55 | 27 | 0 |
| No. 556 | 98.3 | 6.2 | 0 | 0 | 226 |
| No. 556E | <0.01 | 5.5 | 229 | 0 | 0 |
| No. 556F | 88.9 | <0.01 | 117 | 114 | 0 |

It is quite interesting that in microorganisms the reaction could not be catalyzed under usual conditions, but in a variety of microorganisms this was possible at 60 °C.

Microorganisms capable of performing the above transarabinosylation are widely distributed. *Enterobacter aerogenes* AJ1125 was one of the most active strains. When a mixture of 30 mM ara-U and 10 mM adenine in a 30 mM potassium phosphate buffer, pH 7–7.5, was incubated with 5 g per 100 mL of wet cells at 60–65 °C for 5 h, 8–9 mM of ara-A were produced. At temperatures <40 °C there was no ara-A production because enzymes deaminating adenine and hydrolyzing ara-U were active up to about 40 °C. They were inactivated at temperatures >60 °C. Phosphate was essential for the reaction, as arabinose-1-phosphate is the intermediate:

$$\text{ara-U} + P_i \rightarrow \text{Arabinose-1-phosphate} + \text{ura} \quad (8)$$

$$\text{Arabinose-1-phosphate} + \text{ade} \rightarrow \text{ara-A} + P_i \quad (9)$$

The reaction shown in Eq. (8) is catalyzed by uridine phosphorylase (EC 2.4.2.3), and the reaction in Eq. (9) is catalyzed by purine nucleoside phosphorylase (EC 2.4.2.1). Phosphate is essential for the reaction in Eq. (8), and a high concentration of phosphate inhibits the reaction in Eq. (9). Therefore, a suitable concentration of phosphate (~20–30 mM) is necessary for an effective production of ara-A.

Uracil in ara-U can be replaced not only by adenine but also by various purine bases. The order of purine bases to be transarabinosylated is adenine (92%) > 2,6-diaminopurine (83%) > 2-methyladenine (72%) > 2-aminopurine (61%) > 2-methylhypoxanthine (55%) > hypoxanthine (50%) > adenine $N^1$-oxide (45%) > 2-chlorohypoxanthine (40%) > guanine (8%). The figures in parentheses are yields of purine arabinoside produced from 10 mM purine bases and 30 mM ara-U after 15 h at 60 °C.

It should be noted that the linkage between purine and arabinose in the above compounds is the biologically active 9-$\beta$-linkage.

As UTAGAWA and YAMANAKA (1981) further developed the application of nucleoside phosphorylases, the enzyme reactions shown in Fig. 14 are possible now.

## 5.11 Purine 3′-Deoxyribonucleosides

*Aspergillus nidulans* strain Y176-2 isolated by KODAMA et al. (1979) simultaneously produced 2′-deoxycoformycin and cordycepin (3′-deoxyadenosine) when cultured on wheat bran.

Microorganisms capable of forming 3′-deoxyguanosine from 3′-deoxyadenosine and GMP are widely distributed. Among them *Brevibacterium acetylicum* AT-6-7 effectively converted 3′-deoxyadenosine to 3′-deoxyguanosine, especially in the presence of 2′-deoxycoformycin that is known as an adenosine deaminase inhibitor. The molar conversion ratio was 92% after incubation at 55 °C, pH 7.0, for 18 h.

## 5.12 Ribavirin

Ribavirin is a an antiviral agent with a broad spectrum. UTAGAWA et al. (1986) synthesized ribavirin from inosine and 1,2,4-triazole-3-carboxamide (TC) by a two-step reaction catalyzed by the purified purine nucleoside phosphorylase of *Enterobacter aerogenes* AJ 11125:

$$\text{Inosine} + P_i \rightarrow \text{R-1-P} + \text{Hypoxanthine} \quad (10)$$

$$\text{R-1-P} + \text{TC} \rightarrow \text{Ribavirin} + P_i \quad (11)$$

TC (10 mM) was transformed to ribavirin with a 75% yield (molar basis) in the presence of 50 mM ribose-1-phosphate.

SHIRAE et al. (1988a) screened 400 strains of bacteria for their ability to produce ribose-1-phosphate and ribavirin, and selected *Erwinia carotovora* AJ 2992, the most active producer of ribose-1-phosphate from inosine (for the reaction see Eq. 10) and *Bacillus brevis* AJ 1282, an active producer of ribavirin from ribose-1-phosphate and TC (Eq. 11).

SHIRAE et al. (1988b) further screened microorganisms that produce ribavirin directly from pyrimidine nucleosides and TC, and se-

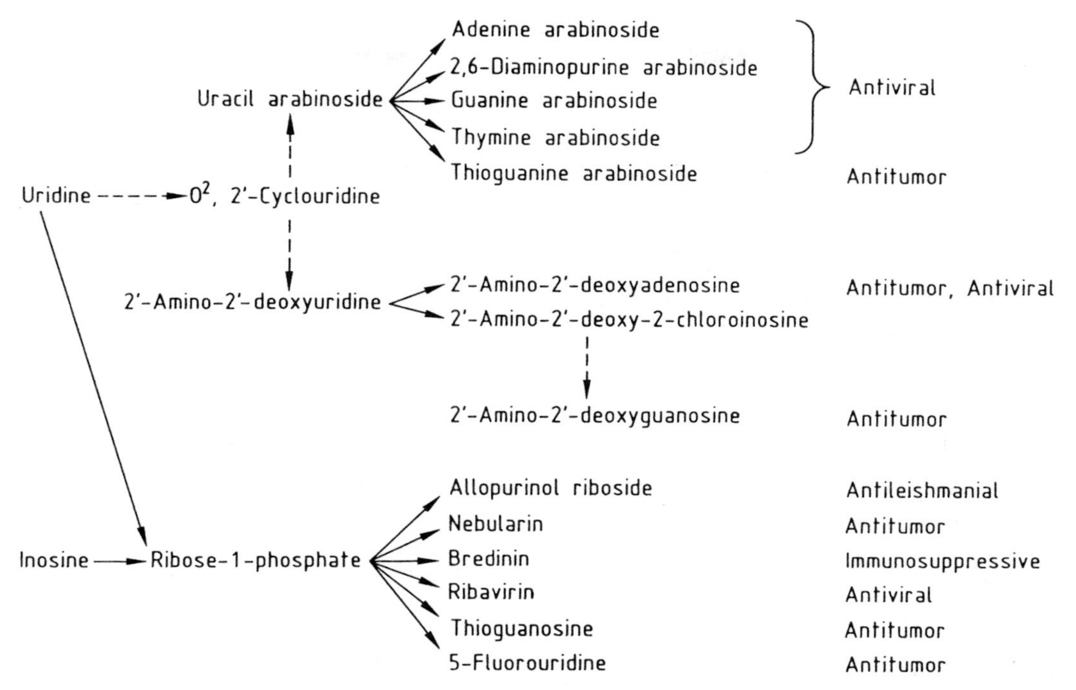


**Fig. 14.** Synthesis of biologically active nucleosides by nucleoside phosphorylase. The arrowed broken line indicates a chemical reaction (adapted from UTAGAWA and YAMANAKA, 1981).

lected *E. aerogenes* AJ 11125 as the most productive strain. When the reaction mixture consisting of 100 mM nucleoside, 100 mM TC, 300 mM potassium phospate buffer, pH 7.0, and intact cells (50 mg $mL^{-1}$, on a wet weight basis) in a total volume of 5 mL was incubated at 60 °C for 24 h. The following amounts of ribavirin were produced: 38.2 mM from uridine, 23.0 mM from cytidine, 12.6 mM from guanosine, 8.2 mM from 4-amino-5-imidazole-3-carboxylic acid riboside, 7.5 mM from xanthosine, 4.0 mM from adenosine, 1.6 mM from inosine, and none from orotidine.

SHIRAE et al. (1988c) further screened microorganisms that produce ribavirin directly from orotidine and TC, and selected *E. carotovora* AJ 2992 as the most effective strain. In the presence of intact cells (100 mg $mL^{-1}$ on a wet weight basis), 183 mM ribavirin were produced from 300 mM orotidine and 300 mM TC at 60 °C in 48 h.

SHIRAE et al. (1988d) further screened microorganisms that produce ribavirin directly from a purine nucleoside and TC, and selected *B. acetylicum* ATCC 954. Among the nucleosides tested, guanosine and xanthosine were the best substrates for ribavirin production. In the presence of intact cells (100 mg $mL^{-1}$ on a wet weight basis), 229 mM ribavirin were produced from 300 mM guanosine and 300 mM TC at 60 °C in 96 h.

DUDCHIK et al. (1990) synthesized ribavirin from guanosine and TC using whole cells of *E. coli* BM-11 as a biocatalyst. They transformed ribavirin into its 5′-monophosphate by the use of *p*-nitrophenylphosphate and whole cells of *Erwinia herbicola* 47/3.

KIM and WHITESIDES (1987) enzymatically synthesized ribavirin 5′-triphosphate from ribavirin 5′-monophosphate using phosphoenolpyruvate as phosphate donor and pyruvate kinase and adenylate kinase as catalysts.

The following patent applications for ribavirin production have been recently filed:

(1) YAMANAKA et al. (1982) claimed a process for microbiologically producing a ribofuranosyltriazole derivative from a 1,2,4-triazole derivative and ribonucleoside or ribose-1-phosphate. Microorganisms capable of producing ribofuranosyltriazole derivatives are widely distributed. *E. herbicola* ATCC 14536, e.g., produced 55 mg $dL^{-1}$ ribavirin from 2.0 g $dL^{-1}$ uridine and 0.2 g $dL^{-1}$ TC.
(2) FUJISHIMA and YAMAMOTO (1983) produced ribavirin from TC and a ribose donor by microbial action. A 250 mL cell suspension of *B. acetylicum* AT-6-7 (FERM-P 6350), e.g., was inoculated into 750 mL of a solution containing 66.7 mM TC, 66.7 mM inosine, and 100 mM $KH_2PO_4$. The mixture was incubated at 60°C for 24 h to produce ribavirin at a 74.9% yield.
(3) UENO et al. (1987) synthesized ribavirin using UV-irradiated, immobilized cells of *B. acetylicum* AT-6-7 (FERM-P 6350). The yield from TC, UMP, and $KH_2PO_4$ was 90%.
(4) YOKOZEKI et al. (1988) incubated *B. subtilis* IFO 3134 cells from a 5 mL culture mixed with 5 mL of 0.3 M phosphate buffer, pH 7.0, containing 50 mM orotidine and TC at 60°C for 5 h to obtain 22 mg $dL^{-1}$ ribavirin.
(5) In the production of ribavirin including the reaction of a ribose donor with TC in the presence of an enzyme preparation from *B. acetylicum,* POCHODYLO (1989) used a high concentration ($\geq$100 mM) of guanosine as ribose donor.
(6) In the production of ribavirin including the reaction of 20 mmol guanosine (a ribose donor) with 20 mmol of TC in the presence of *B. acetylicum* cell paste, LIEVENSE et al. (1989) used a high temperature ($\geq$65°C). The best yield of 17.2 g $L^{-1}$ ribavirin, i.e., 70% conversion, was obtained at pH 6.8 and 70°C.
(7) YAMAUCHI et al. (1990) synthesized various nucleosides with high efficiency and yield without microbial contamination by cultivating a newly isolated thermophilic strain of *Bacillus stearothermophilus* that is rich in nucleoside phosphorylase in the presence of a base, sugar, and phosphate donor. In synthesizing ribavirin from TC and a nucleoside, the thermophilic strain was much superior to prior ones and gave a yield of >94.1%.

## 5.13 Examples of Other Nucleosides

FURUYA et al. (1975) found that among the inosine producing mutants of *Brevibacterium (Corynebacterium) ammoniagenes,* strain KY 13761 produced 6-azauridine from 6-azauracil most effectively. As there was a competition between inosine and 6-azauridine accumulation due to scrambling for the ribose moiety, prototrophic revertants devoid of inosine productivity were induced from KY 13761. KY 13021, one of the revertants, produced 6-azauridine from 6-azauracil more effectively. The strain accumulated 12.4 mg $mL^{-1}$ 6-azauridine from 6 mg $mL^{-1}$ 6-azauracil.

Novel nucleosides are generated by exchange reactions between nucleosides and novel bases catalyzed by a number of phosphorylases or transferases. HUEY et al. (1990) incubated 1 mmol 7-methylguanosine hydroiodide and 0.25 mmol 3-deazaadenine in 10 mL 0.1 M phosphate buffer, pH 7, in the presence of 250 units of purine nucleoside phosphorylase at about 30°C for 2 d. During incubation 65% of the nucleoside was converted to 3-deazaadenosine.

HORI and UEHARA (1993) prepared 5-methyluridine from adenosine and thymine in the presence of potassium phosphate, adenosine deaminase, xanthine oxidase, and a crude extract containing purine nucleoside phosphorylase and pyrimidine nucleoside phosphorylase of *B. stearothermophilus.* Addition of adenosine deaminase and xanthine oxidase increased the yield of 5-methyluridine from 0.09 mM (the control) to 3.6 mM. The reason is as follows: Adenosine is first converted to inosine by adenosine deaminase. Inosine subsequently undergoes base exchange in the presence of phosphoric acid, purine nucleoside phosphorylase, and pyrimidine nucleoside phosphorylase to obtain 5-methyluridine and hypoxanthine. Hypoxanthine can be removed from the system by conversion to uric acid with xanthine oxidase.

# 6 Fermentative Production of Nucleoside Polyphosphates and Related Compounds

## 6.1 ATP

Recently, effective methods for microbial production of ATP have been developed. TANI (1984) summarized the data about microorganisms used as biological means and starting materials for ATP production until 1984 (Tab. 21). ATP production by methanol yeasts was reviewed by YONEHARA and TANI (1987).

FUJIO and FURUYA (1985) studied the effects of $Mg^{2+}$ and chelating agents on enzymic production of ATP from adenine. Enzymic formation of ATP from adenine by resting cells of *B. ammoniagenes* ceased within 6–8 h when 13 mg $mL^{-1}$ $ATP \cdot Na_2 \cdot 3H_2O$ were accumulated. Simultaneous addition of $Mg^{2+}$ and phitic acid, a chelator of divalent cations, allowed ATP formation to continue longer, and 24.2 mg $mL^{-1}$ $ATP \cdot Na_2 \cdot 3H_2O$ were accumulated in 10 h. However, ATP formation ceased thereafter. This second cessation was caused by the lack of $Mg^{2+}$ active as a cofactor ($Mg^{act}$). The $Mg^{act}$ is possibly the difference between soluble $Mg^{2+}$ ($Mg^{sol}$) and the ion chelated by an equimolar amount of ATP ($Mg^{ATP}$), namely:

$$Mg^{act} = Mg^{sol} - Mg^{ATP} \qquad (12)$$

In order to provide $Mg^{act}$, sufficient phytic acid was added at the beginning of the reaction and also $Mg^{2+}$ was added intermittently. Under these conditions ATP formation continued and the rate of ATP formation was increased; 37.0 mg $mL^{-1}$ $ATP \cdot Na_2 \cdot 3H_2O$ were accumulated in 13 h (Fig. 15). The conversion rate of adenine to ATP was approximately 82%.

**Tab. 21.** Representative Methods for ATP Production (adapted from TANI, 1984)

| Biological Means | Starting Materials | ATP Formed [g $L^{-1}$] | Molar Yield [%] |
|---|---|---|---|
| *Brevibacterium ammoniagenes* (resting cells) | Adenine, glucose | 10.9 | 83 |
| *Saccharomyces cerevisiae* (dried cells) | Adenosine, glucose | 102.5 | 50 |
| *Escherichia coli* (dried cells) | AMP, glucose | 14.0 | 92 |
| *Candida boidinii* (sorbitol-treated cells) | AMP, methanol | 30.0 | 60 |
| | Adenosine, methanol | 100.0 | 77 |
| | Adenine, methanol | 4.1 | 40 |
| *Rhodospirillum rubrum* (immobilized chromatophores) | ADP, light | 0.4 | 80 |
| *Mastigocladus laminosus* (intact cells) | ADP, light | 0.17 | 17 |
| *Candida tropicalis* mitochondria (immobilized adenylate kinase) | ADP | 0.24 | 60 |
| *Bacillus stearothermophilus* (acetate kinase, adenylate kinase) | ADP, acetylphosphate | 10.0 | 100 |

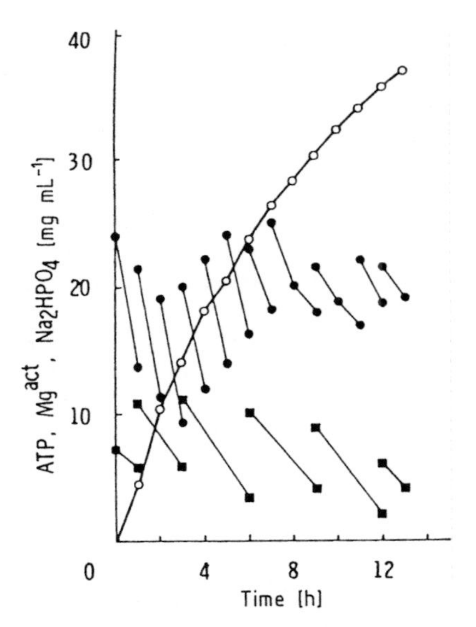


**Fig. 15.** Typical time course of ATP production by enzymatic conversion from adenine. The reaction was carried out at 32 °C for 13 h in a 5 L jar fermenter with the whole culture broth of *B. ammoniagenes* KY13510. The reaction mixture contained: 10 mg adenine, 80 mg glucose, 15 mg $Na_2HPO_4$, 10 mg $MgSO_4 \cdot 7H_2O$, 11 mg phytic acid per mL of broth. During incubation, 5 mg $MgSO_4 \cdot 7H_2O$ were added at 1, 3, 6, 9, and 12 h. The phosphate ion concentration was kept between 75–175 mM by intermittent addition of $Na_2HPO_4$ (FUJIO and FURUYA, 1985).
○ ATP ● $Na_2HPO_4$ ■ $Mg^{act}$

KIM and WHITESIDES (1987) enzymically synthesized ATP from AMP, using acetyl phosphate as phosphate donor and acetyl kinase from *B. stearothermophilus,* adenyl kinase, and inorganic pyrophosphatase as catalysts. Three reactions at a 150 mmol scale provided ATP as its barium salt in a yield of 82%.

MURATA et al. (1988) studied the distribution and properties of polyphosphate kinase, catalyzing the formation of polyphosphate from ATP. The *E. coli* B kinase also catalyzes the synthesis of ATP from ADP and metaphosphate.

WU et al. (1989) produced ATP from AMP using a system involving the oxidative phosphorylation by the methanol-utilizing yeast *Hansenula polymorpha* P5. The cells were pretreated with 1.5 mM KCl or sorbitol. After 16 h of incubation, 10.5 mg mL$^{-1}$ ATP and 1.8 mg mL$^{-1}$ ADP were produced under the following reaction conditions: pH 6.5, 25 °C, 60 mM AMP, 1,000 mM methanol, 10 mM $NAD^+$, 10 mM GSH, 600 mM sorbitol, and 200 mM potassium phosphate.

SODE et al. (1989) microencapsulated chromatophores – organelles for photophosphorylation in non-sulfur photosynthetic purple bacteria – and utilized them in continuous ATP production. ATP was produced at a rate of 14 μmol h$^{-1}$ L$^{-1}$ in 200 h. The yield of ATP from ADP was 35%. The total amount of ATP produced was 0.7 mM (μM Bchl)$^{-1}$.

KADOWAKI et al. (1989) designed a new process for ATP production from adenine using intact bacterial cells containing nucleoside phosphorylases and dried bakers' yeast cells containing an energy generating system (Fig. 16). It consists of (1) the formation of adenosine through trans-N-ribosylation between adenine and uridine, cytidine, or inosine, catalyzed by the nucleoside phosphorylases and (2) the phosphorylation of adenosine to ATP by alcoholic fermentation of the yeast cells. *Erwinia carotovora* was a better source of nucleoside phosphorylases than *E. coli* or *Aerobacter aerogenes.* Ionosine and uridine were effective as ribosyl donors. Cytidine was also usable in reactions performed with cell-free extracts. ATP production was greatly affected by the concentrations of inorganic phosphate and glucose in the reaction mixture. Immobilized phosphorylase was better in ATP production than the cell-free extract (Tab. 22). The nucleoside phosphorylases were stable at 60 °C but not the adenosine deaminating enzymes, and large amounts of adenine could be used due to the increase in solubility at that high temperature. Through a two-step reaction, i.e., transribosylation with intact cells of *E. carotovora* at 60 °C followed by fermentation with bakers' yeast at 28 °C, 98 mM ATP (about 50 g L$^{-1}$) was obtained from 150 mM adenine.

In addition, by coupling the energy generating system (alcoholic fermentation by bakers' yeast) and an energy utilizing system (deoxyribonucleoside monophosphate kinase of *E. coli* B), dATP, dGTP, dCTP, and dTTP

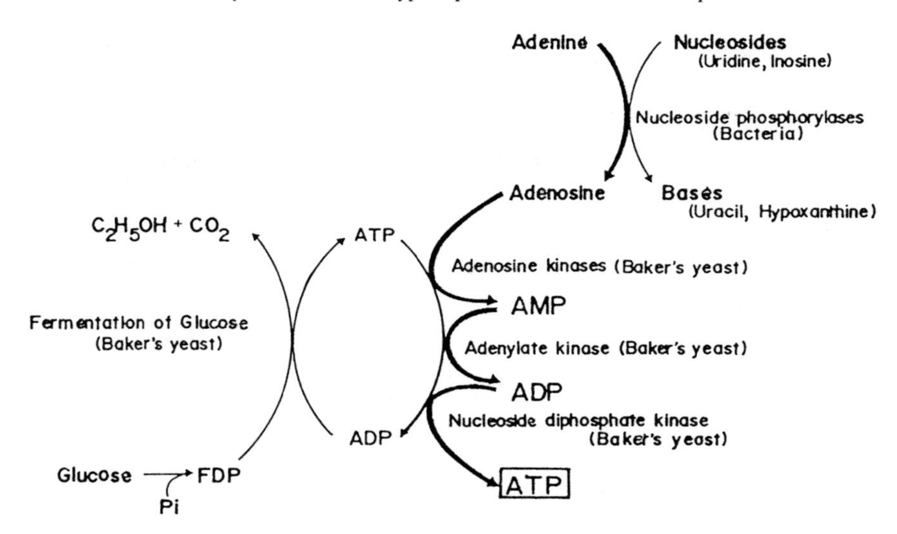


**Fig. 16.** ATP production by the coupled reaction of adenine transribosylation and sugar fermentation (KADOWAKI et al., 1989).

**Tab. 22.** Production of ATP with Cell-Free Phosphorlyase or Immobilized Phosphorylase from *E. carotovora* in the Presence of an Energy Generating System[a] (KADOWAKI et al., 1989)

| Enzyme Preparation | | Nucleotides Formed | | | |
|---|---|---|---|---|---|
| State | [units $mL^{-1}$] | AMP [mM] | ADP [mM] | ATP [mM] | Total [mM] |
| Cell free | 0.015 | 1.5 | 2.7 | 4.5 | 8.7 |
| | 0.05 | 3.1 | 8.9 | 18.7 | 30.7 |
| | 0.1 | 6.1 | 15.8 | 26.4 | 48.3 |
| Immobilized | 0.011 | 1.9 | 6.5 | 23.5 | 31.9 |
| | 0.022 | 2.6 | 7.6 | 35.6 | 45.9 |
| | 0.045 | 2.1 | 10.4 | 55.4 | 67.9 |

[a] The reaction mixture, pH 7.0, contained 75 mM adenine, 100 mM uridine, 600 mM glucose, 600 mM potassium phosphate, 10 mM $MgSO_4$, 25 mg $mL^{-1}$ bakers' yeast cells, and the indicated amounts of the enzyme preparation. Incubation was performed at 28°C for 24 h with the cell-free extract and for 48 h with the immobilized enzyme.

can be prepared from the corresponding deoxyribonucleoside 5′-monophosphates. TMP was not phosphorylated by the energy generating system alone but phosphorylated to dTDP and dTTP by the coupled fermentation, while dAMP was efficiently phosphorylated to dATP by the energy generating system alone (conversion ratio about 80%).

HAYNIE and WHITESIDES (1990) enzymatically synthesized a mixture of ATP, GTP, CTP, and UTP from RNA: RNA was hydrolyzed by nuclease $P_1$ to a mixture of nucleoside 5′-monophosphates that were converted into the nucleoside triphosphates by using a mixture of nucleoside monophosphate kinases (extracted from brewers' yeast) and acetate kinase, with acetyl phosphate as the ultimate phosphoryl donor. Conversion rates from nucleoside monophosphates to triphosphates in a mixture containing 0.34 mol total nucleoside phosphates were: 90% ATP, 90% GTP, 60% CTP, and 40% UTP.

FUJIO et al. (1992) reviewed the development of a microbial ATP regenerating system with *B. ammoniagenes* and its application in producing nucleotides. They discussed the practical procedure for microbial synthesis of ATP from adenine, of FAD from adenine and FMN, and of GMP from XMP.

The following patent applications for ATP production have been filed recently:

(1) FUJIO and FUJIOKA (1985) found that strains of *Escherichia* and *Staphylococcus* as well as *Brevibacterium, Micrococcus, Saccharomyces,* and *Candida* were able to synthesize ATP from adenine. For example, a mixture containing 200 g $L^{-1}$ (based on wet weight) freeze-dried bacterial cells, 50 g $L^{-1}$ glucose, 0.2 g $L^{-1}$ nicotinic acid, 14.4 g $L^{-1}$ $KH_2PO_4$, 5 g $L^{-1}$ $MgSO_4$, and 5 g $L^{-1}$ adenine was adjusted to pH 7.3 and stirred at 900 rpm and 32°C. The yields of ATP (as $Na_2ATP \cdot 3H_2O$) were 5. 77 g $L^{-1}$ for *E. coli* C600 ATCC 33525, 5.58 g $L^{-1}$ for *E. coli* B ATCC 11303, and 5.80 g $L^{-1}$ for *Staphylococcus aureus* ATCC 9510.

(2) IMAHORI et al. (1985) immobilized adenylate kinase and acetate kinase from *B. stearothermophilus* on CH-Sepharose 4B. The column was flowed with a buffer containing 10.1 mM AMP, 0.1 mM ATP, and 34 mM acetyl phosphate at pH 7.5, 190 mL $h^{-1}$ and 30°C. The flow-through contained 98.0–98.9% ATP and 1.1–2.0% ADP.

(3) YONEHARA and TANI (1986) cultivated a yeast that produced ATP from adenosine or adenine. *Hansenula capsulata* IFO 0974, e.g., was cultivated with methanol as carbon source and the cell suspension (80 g $L^{-1}$) was treated with 1.5 M sorbitol at 37°C. The suspension (0.2 mL) was shaken with a mixture containing 20 μmol adenosine or 5 μmol adenine, 100 μmol $K_2HPO_4$, 0.5 μmol NAD, 500 μmol methanol, 40 μmol $MgSO_4$, and 1.5 μmol ATP (in a total volume of 0.5 mL) at 25°C for 20 h to give 16 mmol $L^{-1}$ and 4 mmol $L^{-1}$ ATP from adenosine and adenine, respectively.

(4) HATANAKA and TAKEUCHI (1986) cultured microorganisms producing ATP and metabolizing methanol in a medium containing phosphate and methanol, but no adenine compounds. *Methylomonas probus,* e.g., produced 2–3 g $L^{-1}$ of ATP during cultivation at 30°C for 4 d.

(5) KIMURA et al. (1986) manufactured ATP by using genetically engineered yeasts. *Saccharomyces cerevisiae*, e.g., was genetically modified with recombinant plasmid pYH1 containing the enhanced hexokinase gene from a high-hexokinase yeast strain and the glucokinase gene from a high-glucokinase yeast strain. A medium containing 50 mg $mL^{-1}$ of dried cells of the resulting *S. cerevisiae* DKD-50-H, 1 M glucose, 1 M phosphate buffer, pH 8.0, 100 mM adenosine, and 30 mM $MgSO_4 \cdot 7H_2O$ was shaken at 28°C for 12 h. The medium contained 22.81 mg $mL^{-1}$ ATP vs. 4.06 mg $mL^{-1}$ of the control (parent strain).

(6) INATOMI (1987) used a continuous bioreactor containing a flat lipid bilayer consisting of ATPase, hydrogenase from thermophilic bacteria, and an electron transfer system from methane assimilating bacteria. In this system ATP and NAD(P)H were regenerated. Hydrogen was supplied by the electrolysis of water, energy was supplied by solar cells. ATP and NAD(P)H were produced at a yield of 80% in the bioreactor containing 10 lipid bilayers.

(7) FUJIO and MARUYAMA (1990) mixed ATP-producing bacteria with a solution containing adenine, phosphate donors, energy donors, and $NH_4^+$ with an initial concentration of less than 1 g $L^{-1}$. *Brevibacterium (Corynebacterium) ammoniagenes* ATCC 21170, e.g., was cultured in a medium with an initial $NH_3$ concentration of 0.05 g $L^{-1}$ at 32°C for 21 h and yielded 58.2 g ATP $L^{-1}$ vs. 10.1 g $L^{-1}$ at an initial $NH_3$ concentration of 2.03 g $L^{-1}$.

## 6.2 FAD

FAD is microbiologically prepared in a two-step process:

(1) Formation of ATP from an ATP precursor (adenine, adenosine, AMP, or ADP) in the presence of phosphate and energy donors,
(2) formation of FAD from ATP and FMN.

For example, FUJIO et al. (1984b) cultivated FAD-producing *Corynebacterium* or *Brevibacterium* cells: Wet cells (30 g) of *B. ammoniagenes* ATCC 21170 separated from 800 mL of the culture broth were suspended in 500 mL of a solution containing 50 g $L^{-1}$ glucose, 4 g $L^{-1}$ adenine, 15 g $L^{-1}$ $NaH_2PO_4$, 0.01 g $L^{-1}$ xylene, and 6 g $L^{-1}$ of a surfactant, and incubated at 32 °C for 10 h with stirring and aeration: the $H_3PO_4$ concentration was kept at 10 g $L^{-1}$ (as $NaH_2PO_4$) and the pH was adjusted to 7.4. A 20 mL aliquot of the reaction mixture, where ATP had accumulated, was supplemented with 3 g $L^{-1}$ Na-FMN. The mixture was incubated at 42 °C for 12 h to produce 2.34 g $L^{-1}$ FDA·$Na_2$.

## 6.3 CDP-Choline

As shown in Fig. 17, several reactions are involved in the biosynthesis of CDP-choline. A few years ago, CDP-choline had been produced from CMP and choline phosphate (phosphorylcholine) or choline chloride by the action of non-recombinant microbial enzymes involved in reactions 3, 4, 6, and 7. The following microorganisms, e.g., were used for this purpose: Acetone powder of *Brettanomyces petrophilum* ATCC 20224 (TAKEDA et al., 1981), brewers' yeast sludge (WANG et al., 1982), membrane-immobilized *Saccharomyces cerevisiae* (QUI et al., 1983), resting cells of *Candida* SP SVI 362 (ZHAN, 1984) or *Hansenula anomala* SVI 311 (ZHAN et al., 1984).

Recently, patent applications related to production of CDP-choline by recombinant microbial strains have been disclosed. It should be noted that the starting material in example (2) is orotic acid instead of CMP in example (1).

(1) YAMASHITA and KOMATSU (1988) transformed *Saccharomyces cerevisiae* (FERM P-9217) with plasmid pCC1 encoding the choline-phosphate-cytidylyltransferase (reaction 6). The transformed cells were cultured in a medium containing mo-

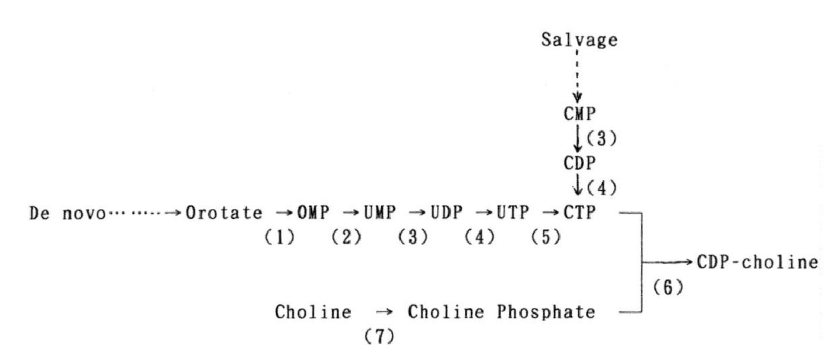


**Fig. 17.** Biosynthetic pathway of CDP-choline.
(1) Orotate phosphoribosyltransferase (EC 2.4.2.10),
(2) Orotidine-5′-phosphate decarboxylase (EC 4.1.1.23),
(3) Nucleoside-phosphate kinase (EC 2.7.4.4),
(4) Nucleoside-diphosphate kinase (EC 2.7.4.6),
(5) CTP synthetase (EC 6.3.4.2),
(6) Choline-phosphate cytidylyltransferase, (EC 2.7.7.15),
(7) Choline kinase (EC 2.7.1.32).
In addition to nucleoside-phosphate kinase (EC 2.7.4.4) cytidylate kinase (EC 2.7.4.14) is also known to catalyze the following reactions:
ATP + (d)CMP → ADP + (d)CDP
ATP + (d)UMP → ADP + (d)UDP

lasses, glucose, urea, and salts for 20 h. Water and toluene were added to the yeast cells (20 g dry weight). The suspension was treated with 2.4 g CMP, 15 g glucose, 1.6 g phosphorylcholine, and 0.38 g $MgCl_2$ in a 0.15 M phosphate buffer, pH 7.5, in a total volume of 240 mL at 30 °C for 5 h. During incubation 3.35 g of CDP-choline were produced: 92.3% of CMP on molar basis was converted to CDP-choline.

(2) MARUYAMA et al. (1993) cultivated two kinds of microorganisms, A and B, that produce CDP-choline from orotic acid and choline or phosphorylcholine:

(A) *E. coli*, containing the CTP synthetase (reaction 5) gene *pyr*G, the cholinephosphate cytidylyltransferase (reaction 6) gene, and the choline kinase (reaction 7) gene;

(B) *Corynebacterium ammoniagenes* capable of converting orotic acid to UTP (reactions 1–4) and that is rich in both PRPP supplying activity and ATP regenerating activity.

As for microorganism A, *E. coli* MM294 was transformed with plasmid pCKG55 containing the *E. coli* CTP synthetase (reaction 5) gene *pyr*G and the *S. cerevisiae* genes for choline-phosphate cytidylyltransferase (reaction 6) and choline kinase (reaction 7). The transformed *E. coli* strain MM294/pCKG55 (FERM BP-3717) was obtained. *C. ammoniagenes* strain ATCC21170 meets all the criteria for microorganism B.

*E. coli* MM294/pCKG55 and *C. ammoniagenes* ATCC21170 were precultured separately in 360 mL each. Both cultures were mixed in a 2 L tank. 100 g $L^{-1}$ glucose, 10 g $L^{-1}$ (47 mM final concentration) orotic acid, 8.4 g $L^{-1}$ (60 mM final concentration) choline chloride, 5 g $L^{-1}$ $MgSO_4 \cdot 7H_2O$, 20 mL $L^{-1}$ xylene, and water were added to a total volume of 800 mL. The mixture was incubated at 32 °C and pH 7.2, stirred at 800 rpm and aerated at 0.8 L $min^{-1}$. The phosphate concentration was kept at a level between 1–5 g $L^{-1}$ ($KH_2PO_4$). After 23 h 11.0 g $L^{-1}$ (21.5 mM) of CDP-choline had accumulated. *E. coli* MM294/pCKG55 alone produced only 0.7 g $L^{-1}$ (1.4 mM) of CDP-choline. With *C. ammoniagenes* ATCC21170 alone no CDP-choline was formed at all either.

# 7 Polyribonucleotides

GRUNBERG-MANAGO and OCHOA (1955) synthesized polynucleotides from ribonucleoside diphosphates by the action of *Azotobacter* polynucleotide phosphorylase (polyribonucleotide nucleotidyltransferase, PNPase, EC 2.7.7.8):

$$n\ \text{Ribonucleoside-P-P} \rightleftarrows (\text{Ribonucleoside-P})_n + n\ P_i \qquad (13)$$

The biological activities of polynucleotides, e.g., the interferon-inducing activity, have extensively been studied. However, their production at a large scale was quite difficult, because bacterial PNPases cannot effectively be extracted from cells.

KATOH et al. (1973) screened microorganisms suitable for industrial production of polynucleotides and found that several microorganisms belonging to the genera *Pseudomonas, Serratia, Xanthomonas, Proteus, Aerobacter, Bacillus,* and *Brevibacterium* were not only rich in PNPase easily extractable from cells, but also poor in both nuclease and nucleoside diphosphate-degrading enzymes.

From *Achromobacter* sp. KR 170-4 (ATCC21942) potent PNPase could be extracted most efficiently (ROKUGAWA et al., 1975). The enzyme from this strain required $Mn^{2+}$ instead of $Mg^{2+}$ to synthesize polynucleotides from nucleoside diphosphates. The optimum pH was 10.1, optimum temperatures were 46 °C for polymerization of ADP or IDP, and 43 °C for CDP or UDP. The enzyme was stable $<55$ °C at pH 9.2. The enzyme from *Achromobacter* sp. KR 170-4 has been used for producing poly(I) and poly(C) as components of an interferon inducer, poly(I)·poly(C).

*Achromobacter* PNPase was immobilized with glutaraldehyde and an aminopropyl spacer on porous glass (YAMAUCHI and MACHIDA, 1986). The specific activity of the immobilized enzyme was effectively increased by the addition of an appropriate ribonucleoside diphosphate on immobilization. A homopolynucleotide could be synthesized continuously by passing a nucleoside diphosphate solution through the column. The chain length of the product depended upon the temperature and the flow rate. Poly(I) was continuously synthesized with the immobilized enzyme for about one month without appreciable loss of activity.

MARUMO et al. (1993) established an efficient polyribonucleotide-producing system using a PNPase-overproducing *E. coli* strain. The region containing the *pnp* gene coding for PNPase was separated from chromosomal DNA of *E. coli* C600. After digesting the purified product with *Bam*HI, the resulting DNA fragment was inserted into the *Bam*HI site of plasmid pDR540 in order to overexpress the *pnp* gene under the control of the *tac* promoter. The resulting plasmid, pDR-PNP (Fig. 18), was introduced into *E. coli* JM105 cells. A selected clone was cultivated at 37 °C to a density of $5 \times 10^8$ cells $mL^{-1}$. Then isopropyl-β-D-thiogalactopyranoside (final concentration 1 mM) was added to induce expression of the recombinant *pnp* gene. After 18 h of induction the amount of PNPase was 10% of total cellular protein. The specific activity of PNPase was 3.17 units per mg protein in JM105[pDR-PNP], while in JM105[pDR540] it was 0.02 unit per mg of protein. By using this enzyme preparation, polynucleotides with a high molecular weight could be synthesized efficiently.

The following patent applications for polynucleotide production have been filed recently:

(1) MORAN (1989) immobilized PNPase on an epoxy-activated organic polymer. When incubated with IDP at 25 °C for 42 h, the immobilized PNPase catalyzed the poly(I) synthesis with a yield of 50–55%.

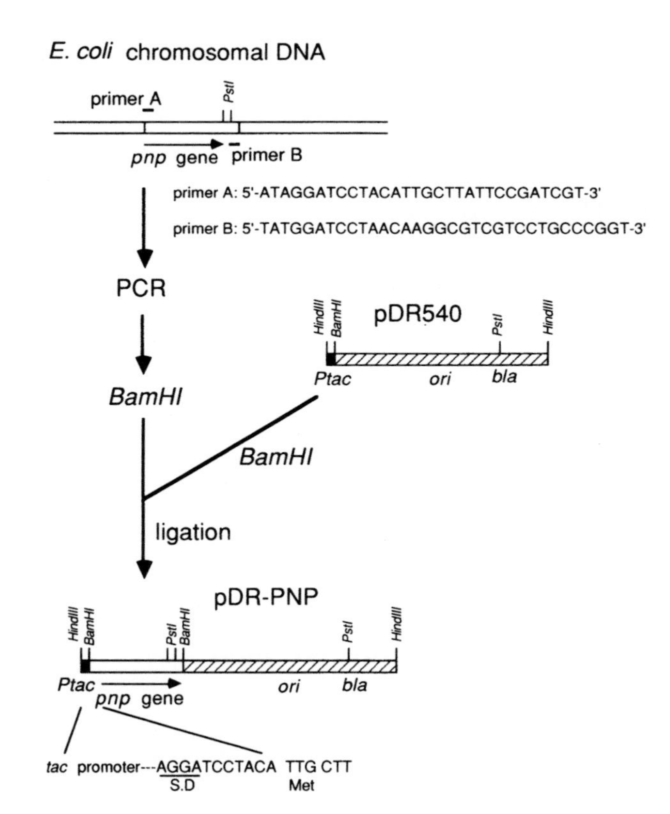


**Fig. 18.** Construction of plasmid pDR–PNP containing the *tac–pnp* gene; open box: *pnp* coding region, closed box: *tac* promoter, *bla:* β-lactamase gene, *ori:* replication origin of the plasmid (MARUMO et al., 1993).

**Tab. 23.** Production of Polynucleotides from Nucleoside Diphosphates with Recombinant Polynucleotide Phosphorylase[a] (adapted from MARUMO et al., 1993)

| Product | Conditions of Reaction | | | Yield [%] |
|---|---|---|---|---|
| | Substrate | Cation | Temperature | |
| Poly(A) | 24 mM ADP | 5 mM $Mg^{2+}$ | 37°C | 71.6 |
| Poly(C) | 24 mM CDP | 17 mM $Mg^{2+}$ | 37°C | 50.1 |
| Poly(G) | 24 mM GDP | 3 mM $Mn^{2+}$ | 65°C | 86.3 |
| Poly(U) | 24 mM UDP | 11 mM $Mg^{2+}$ | 45°C | 42.3 |

[a] Each nucleoside diphosphate was incubated with 0.33 U $mL^{-1}$ of the recombinant polynucleotide phosphorylase in the presence of a cation in 50 mM Tris·HCl, pH 9.0, for 3 h.

(2) NOGUCHI et al. (1993) obtained an enzyme preparation containing 900 units of PNPase from a 500 mL culture of *E. coli* JM105[pDR-PNP] after induction (MARUMO et al., 1993). They incubated ADP, CDP, GDP, or UDP with the enzyme preparation under the conditions shown in Tab. 23. More than 40% of nucleoside diphosphates were converted into the corresponding polynucleotides. Under almost the same conditions, 50% of IDP were converted to poly(I) at 32°C.

# 8 References


AKIYA, T., MIDORIKAWA, Y., KUNINAKA, A., YOSHINO, H., IKEDA, Y. (1972), Accumulation of 5'-inosinic acid and 5'-xanthylic acid by *Bacillus subtilis, Agric. Biol. Chem.* **36**, 227–233.

AOKI, R., MOMOSE, H., KONDO, Y., MURAMATSU, N., TSUCHIYA, Y. (1963), Studies on inosine fermentation – Production of inosine by mutants of *Bacillus subtilis.* I. Isolation and characterization of inosine producing mutants, *J. Gen. Appl. Microbiol.* **9**, 387–396.

AOKI, R., KONDO, Y., HIROSE, Y., OKADA, H. (1968), Inosine fermentation – Production of inosine by mutants of *Bacillus subtilis.* IV. Scale-up of inosine fermentation, *J. Gen. Appl. Microbiol.* **14**, 411–416.

ASAHI, S., DOI, M. (1991), *Eur. Patent Appl.* 458070.

ASAHI, S., DOI, M. (1993), Hyperproduction of orotic acid and orotidine by mutants of *Bacillus subtilis, Biosci. Biotech. Biochem.* **57**, 1014–1015.

ASAHI, S., TSUNEMI, Y., DOI, M. (1989), *Eur. Patent Appl.* 329062.

ASAHI, S., TSUNEMI, Y., DOI, M. (1991), Derivation of cytidine producing mutants, *Nippon Nogei Kagaku Kaishi* **65**, 475.

ASAHI, S., TSUNEMI, Y., DOI, M. (1993), *Jpn. Kokai Tokkyo Koho* 05292945.

CHO, J. I., CHU, M. (1991), Breeding of 5'-GMP producing microorganisms by intergeneric protoplast fusion between *Brevibacterium ammoniagenes* and *Corynebacterium glutamicum, Nonglim Nonjip* **31**, 25–31.

CHO, J. I., JUNG, S. W. (1991), Production of nucleotide by immobilized cell, *Han'guk Susan Hakhoechi* **24**, 111–116.

CHU, C. K., BAKER, D. C. (1993), *Nucleosides and Nucleotides as Antitumor and Antiviral Agents.* New York: Plenum Press.

COLEMAN, J. E. (1992), Zinc proteins: Enzymes, storage protein, transcription factors, and replication proteins, *Annu. Rev. Biochem.* **61**, 897–946.

DEMAIN, A. L., JACKSON, U., VITALI, R. A., HENDLIN, D., JACOB, T. A. (1965), Production of xanthosine-5'-monophosphate and inosine-5'-monophosphate by auxotrophic mutants of a coryneform bacterium, *Appl. Microbiol.* **13**, 757–761.

DOI, M. , SONOI, K., SUMINO, Y. (1983), *British Patent Appl.* 2100731.

DOI, U., TSUNEMI, Y., ASAHI, S., AKIYAMA, S., NAKAO, Y. (1988), *Bacillus subtilis* mutants producing uridine in high yields, *Agric. Biol. Chem.* **52**, 1479–1484.

DUDCHIK, N. V., ZINCHENKO, A. I., BARAI, V. N., MAURINS, J., SENTYUREVA, S. L., KVASYUK, E. I. (1990), Synthesis of 1-($\beta$-D-ribofuranosyl)1,2,4-triazole-3-carboxamide and its 5'-monophosphate using enzymes from microorganisms, *Vestsi Akad. Navuk BSSR, Ser. Khim. Navuk* **5**, 90–94.

ENEI, H. (1988), Purine nucleoside fermentation, in: *From Fermentation to New Biotechnology* (TOCHIKURA, T., Ed.), pp. 135–147. Tokyo: Japan Bioindustry Association.

FUJIMOTO, M., UCHIDA, K., SUZUKI, M., YOSHINO, H. (1966), Accumulation of xanthosine by auxotrophic mutants of *Bacillus subtilis, Agric. Biol. Chem.* **30**, 605–610.

FUJIMOTO, M., FUJIYAMA, K., KUNINAKA, A., YOSHINO, H. (1974), Mode of action of nuclease $P_1$ on nucleic acids and its specificity for synthetic phosphodiesters, *Agric. Biol. Chem.* **38**, 2141–2147.

FUJIMOTO, U., KALINSKI, A., PRITCHARD, A. E., KOWALSKI, D., LASKOWSKI, M., SR. (1979), Accessibility of some regions of DNA in chromatin (chicken erythrocytes) to single strand-specific nucleases, *J. Biol. Chem.* **254**, 7405–7410.

FUJIO, T., FUJIOKA, A. (1985), *Jpn. Kokai Tokkyo Koho* 60210995.

FUJIO, T., FURUYA, A. (1984), *Jpn. Kokai Tokkyo Koho* 59078697.

FUJIO, T., FURUYA, A. (1985), Effects of magnesium ion and chelating agents on enzymatic production of ATP from adenine, *Appl. Microbiol. Biotechnol.* **21**, 143–147.

FUJIO, T., FURUYA, A. (1988), Production of 5′-guanylic acid, in: *From Fermentation to New Biotechnology* (TOCHIKURA, T., Ed.), pp. 167–177. Tokyo: Japan Bioindustry Association.

FUJIO, T., IIDA, A. (1990), *PCT Int. Appl.* 9005784.

FUJIO, T., KITATSUJI, K. (1990), *Jpn. Kokai Tokkyo Koho* 02174689.

FUJIO, T., MARUYAMA, A. (1990), *Jpn. Kokai Tokkyo Koho* 02134394.

FUJIO, T., KOTANI, Y., FURUYA, A. (1984a), Production of 5′-guanylic acid by enzymatic conversion of 5′-xanthylic acid, *J. Ferment. Technol.* **62**, 131–137.

FUJIO, T., MARUYAMA, A., FUJIOKA, A., ARAKI, K. (1984b), *Jpn. Kokai Tokkyo Koho* 59132898.

FUJIO, T., MARUYAMA, A., NISHI, T., OZAKI, A., ITO, S., OZAKI, A. (1985), *PCT Int. Appl.* 8504187.

FUJIO, T., KATSUMATA, R., HAGIWARA, T. (1988a), *Jpn. Kokai Tokkyo Koho* 63248394.

FUJIO, T., NISHI, T., MARUYAMA, A., ITO, S. (1988b), *Eur. Patent Appl.* 263716.

FUJIO, T., TAKEICHI, Y., KITATSUJI, K., IIDA, A. (1988c), *Eur. Patent Appl.* 282989.

FUJIO, T., MARUYAMA, A., SUGIYAMA, K., FURUYA, A. (1992), Construction of practical ATP regenerating system and its application to production of nucleotides, *Nippon Nogei Kagaku Kaishi* **66**, 1457–1465.

FUJISHIMA, T., NAGATA, T. (1982), *Jpn. Kokai Tokkyo Koho* 57115193.

FUJISHIMA, T., YAMAMOTO, Y. (1983), *Eur. Patent Appl.* 93401.

FURUICHI, Y., MIURA, K. (1975), A blocked structure at the 5′ terminus of mRNA from cytoplasmic polyhedrosis virus, *Nature* **253**, 374–375.

FURUYA, A. (1978), Production of nucleic acid-related substances by *Brevibacterium ammoniagenes, Hakko to Kogyo* **36**, 1036–1048.

FURUYA, K., OISHI, N. (1991), *Jpn. Kokai Tokkyo Koho* 03262496.

FURUYA, A., ABE, S., KINOSHITA, S. (1970), Accumulation of inosine by a mutant of *Brevibacterium ammoniagenes, Appl. Microbiol.* **20**, 263–270.

FURUYA, A., KATO, F., NAKAYAMA, K. (1975), Accumulation of 6-azauridine by mutants of *Brevibacterium ammoniagenes, Agric. Biol. Chem.* **39**, 767–771.

GRUNBERG-MANAGO, M., OCHOA, S. (1955), Enzymatic synthesis and breakdown of polynucleotides; polynucleotide phosphorylase, *J. Am. Chem. Soc.* **77**, 3165–3166.

HAGIWARA, T., KONDO, R., KAWAHARA, S., FUJIO, T. (1988), *Jpn. Kokai Tokkyo Koho* 63185372.

HANEDA, K., HIRANO, A., KODAIRA, R., OHUCHI, S. (1971), Production of adenosine by mutants derived from *Bacillus* sp., *Agric. Biol. Chem.* **35**, 1906–1912.

HANEDA, K., KOMATSU, K., KODAIRA, R., OHSAWA, H. (1972), Stabilization of adenosine-producing mutants derived from *Bacillus* sp. No. 1043, *Agric. Biol. Chem.* **36**, 1453–1460.

HASHIMOTO, S., SAKAI, M., MURAMATSU, M. (1975). 2′-O-methylated oligonucleotides in ribosomal 18S and 28S RNA of a mouse hepatoma, MH 134, *Biochemistry* **14**, 1956–1964.

HATANAKA, M., TAKEUCHI, D. (1986), *U.S. Patent Appl.* 4617268.

HATTORI, K., KAWAHARA, S., HAGIWARA, T. (1985), *Jpn. Kokai Tokkyo Koho* 60156399.

HAYNIE, S. L., WHITESIDES, G. M. (1990), Preparation of a mixture of nucleoside triphosphates suitable for use in synthesis of nucleotide phosphate sugars from ribonucleic acid using nuclease $P_1$, a mixture of nucleoside monophosphokinases and acetate kinase, *Appl. Biochem. Biotechnol.* **23**, 205–220.

HORI, N., UEHARA, K. (1993), *Jpn. Kokai Tokkyo Koho* 05049493.

HUEY, W. C., HENNEN, W. J. (1990), *Eur. Patent Appl.* 391592.

IMAHORI, K., KONDO, H., NAKAJIMA, H., SHIOZAWA, S. (1985), *Jpn. Kokai Tokkyo Koho* 60203197.

INATOMI, K. (1987), *Jpn. Kokai Tokkyo Koho* 62138186.

ISHIBASHI, M., HIRAUMA, H., YAMADA, Y. (1986), *Jpn. Kokai Tokkyo Koho* 61132195.

ISHII, K., SHIIO, I. (1973), Regulation of purine nucleotide biosynthesis in *Bacillus subtilis, Agric. Biol. Chem.* **37**, 287–300.

ISHIYAMA, J. (1982), *Jpn. Kokai Tokkyo Koho* 57206396.

ISHIYAMA, J. (1983), *Jpn. Kokai Tokkyo Koho* 58013395.

ISHIYAMA, J. (1984), Production of cAMP by microorganisms, Current studies using *Escherichia coli, Hakko to Kogyo* **42**, 824–838.

ISHIYAMA, J. (1990), Isolation of mutants with improved production of cAMP from *Microbacterium* sp. no. 205 (ATCC21376), *Appl. Microbiol. Biotechnol.* **34**, 359–363.

KADOWAKI, S., YANO, T., TACHIKI, T., TOCHIKURA, T. (1989), Production of ATP from adenine by a combination of bacterial and bakers' yeast cells, *J. Ferment. Bioeng.* **68**, 417–22.

KARASAWA, M., ICHIUMI, T., NAKAMATSU, T. (1992), *Jpn. Kokai Tokkyo Koho* 04030797.

KATOH, Y., KUNINAKA, A., YOSHINO, H. (1973), Synthesis of polynucleotides by microorganisms, *Agric. Biol. Chem.* **37**, 1537–1541.

KAWAMURA, Y., KARE, M. R. (Eds.) (1987), *Umami: A Basic Taste.* New York: Marcel Dekker, Inc.

KAWAMURA, Y., KURIHARA, K. (Eds.) (1993), *Umami Int. Symp.* Tokyo: Society for Research on Umami Taste.

KAWAMURA, Y., KURIHARA, K., NICOLAÏDIS, S., OOMURA, Y., WAYNER, M. J. (Eds.) (1991), *Proc. 2nd Int. Symp. on Umami.* New York: Pergamon Press. (*Physiol. Behav.* **49**(5), special issue).

KIHARA, K., NOMIYAMA, H., YUKUHIRO, M., MUKAI, J. (1976), Enzymatic synthesis of [$\alpha^{32}$P]ATP of high specific activity, *Anal. Biochem.* **75**, 672–673.

KIJIMA, T., ISHIYAMA, J., MIZUSAWA, K., NASUNO, S. (1985), *Jpn. Kokai Tokkyo Koho* 60126094.

KIM, M. J., WHITESIDES, G. M. (1987), Enzyme-catalyzed synthesis of nucleoside triphosphates from nucleoside monophosphates. ATP from AMP and ribavirin 5′-triphosphate from ribavirin 5′-monophosphate, *Appl. Biochem. Biotechnol.* **16**, 95–108.

KIMURA, H., FUKUDA, Y., HASHIMOTO, H. (1986), *Jpn. Kokai Tokkyo Koho* 61124390.

KOBAYASHI, S., ARAKI, K., NAKAYAMA, K. (1988), Uracil-producing mutants of *Brevibacterium ammoniagenes, Nippon Nogei Kagaku Kaishi* **62**, 31–33.

KODAMA, K., KUSAKABE, H., MACHIDA, H., MIDORIKAWA, Y., SHIBUYA, S., KUNINAKA, A., YOSHINO, H. (1979), Isolation of 2′-deoxycoformycin and cordycepin from wheat bran culture of *Aspergillus nidulans* Y 176-2, *Agric. Biol. Chem.* **43**, 2375–2377.

KODAMA, K., MOROZUMI, M., SAITOH, K., KUNINAKA, A., YOSHINO, H., SANEYOSHI, M. (1989), Antitumor activity and pharmacology of 1-$\beta$-D-arabinofuranosylcytosine - 5′ - stearylphosphate: An orally active derivative of 1-$\beta$-D-arabinofuranosylcytosine, *Jpn. J. Cancer Res.* **80**, 679–685.

KOGI, K., UCHIBORI, T., AIHARA, K., YAMAGUCHI, T., ABIRU, T. (1991), Pharmacological profile of the 2-alkynyladenosine derivative 2-octynyladenosine (YT-146) in the cardiovascular system, *Jpn. J. Pharmacol.* **57**, 153–165.

KOMATSU, K., HANEDA, K., HIRANO, A., KODAIRA, R., OHSAWA, H. (1972), Derivation of guanosine-producing mutants from an adenine auxotroph of a *Bacillus* strain, *J. Gen. Appl. Microbiol.* **18**, 19–27.

KONISHI, S., SHIRO, T. (1968), Fermentative production of guanosine by 8-azaguanine resistant of *Bacillus subtilis, Agric. Biol. Chem.* **32**, 396–398.

KONISHI, S., KUBOTA, K., AOKI, R., SHIRO, T. (1968), Accumulation of adenosine by L-isoleucine-requiring mutant of *Bacillus subtilis, Amino Acid Nucleic Acid* **18**, 15–20.

KOTANI, Y., YAMAGUCHI, K., KATO, F., FURUYA, A. (1978), Inosine accumulation by mutants of *Brevibacterium ammoniagenes* strain improvement and culture conditions, *Agric. Biol. Chem.* **42**, 399–405.

KUMAGAI, M., FUJIMOTO, M., KUNINAKA, A. (1988), Determination of base composition of DNA by high performance liquid chromatography of its nuclease $P_1$ hydrolysate, *Nucleic Acids Res. Symp. Ser.* **19**, 65–68.

KUNINAKA, A. (1960), Studies on taste of ribonucleic acid derivatives, *Nippon Nogei Kagaku Kaishi* **34**, 489–492.

KUNINAKA, A. (1976), Hydrolysis of nucleic acids with *Penicillium nuclease,* in: *Microbial Production of Nucleic Acid-Related Substances* (OGATA, K., KINOSHITA, S., TSUNODA, T., AIDA, K., Eds.), pp. 75–86. Tokyo: Kodansha Ltd.

KUNINAKA, A. (1981), Taste and flavor enhancers, in: *Flavor Research* (TERANISHI, R., FLATH, R. A., SUGISAWA, H., Eds.), pp. 305–353. New York: Marcel Dekker, Inc.

KUNINAKA, A. (1988), Microbial production of nucleic acid-related substances – Discovery and progress, in: *From Fermentation to New Biotechnology* (TOCHIKURA, T., Ed.), pp. 20–46. Tokyo: Japan Bioindustry Association.

KUNINAKA, A. (1993), 35 years for nuclease $P_1$: From discovery of umami to development of antiviral agent, in: *From Molecular Biology to Biotechnology* (MIURA, K., Ed.), pp. 164–172. Tokyo: Kyoritsu Shuppan Co.

KUNINAKA, A., KIBI, M., SAKAGUCHI, K. (1964), History and development of flavor nucleotides, *Food Technol.* **18**, 287–293.

KUNINAKA, A., FUJIMOTO, M., UCHIDA, K., YOSHINO, H. (1980), Occurrence of 2′-5′-phosphodiester linkages in RNA preparations, *Agric. Biol. Chem.* **44**, 1821–1827.

LIEVENSE, J. C., SAWYER J. D., TERPOLILLI, A. J. (1989), *Eur. Patent Appl.* 307854.

MACHIDA, H. (1986), Comparison of susceptibilities of varicella–zoster virus and herpes simplex viruses to nucleoside analogs, *Antimicrob. Agents Chemother.* **29**, 524–526.

MACHIDA, H., KUNINAKA, A. (1969), Studies on the accumulation of orotic acid by *Escherichia coli* K12, *Agric. Biol. Chem.* **33**, 868–875.

MACHIDA, H., KUNINAKA, A., YOSHINO, H. (1970), Studies on the accumulation of orotic acid by *Escherichia coli* K12 II. Mechanism of the accumulation, *Agric. Biol. Chem.* **34**, 1129–1135.

MAEKAWA, K., TSUNASAWA, S., DIBO, G., SAKIYAMA, F. (1991), Primary structure of nuclease $P_1$ from *Penicillium citrinum, Eur. J. Biochem.* **200**, 651–661.

MARUMO, G., YAMAMOTO, Y., MIDORIKAWA, Y. (1992), *Jpn. Kokai Tokkyo Koho* 04158792.

MARUMO, G., NOGUCHI, T., MIDORIKAWA, Y. (1993), Efficient method for the preparation of *Escherichia coli* polynucleotide phosphorylase suitable for the synthesis of polynucleotides, *Biosci. Biotech. Biochem.* **57**, 513–514.

MARUMOTO, R. (1987), Nucleic acid-related chemotherapeutics, *Hakko to Kogyo* **45**, 974–984.

MARUYAMA, A., NISHI, T., OZAKI, A., ITO, S., FUJIO, T. (1986), Construction of a plasmid for high level expression of XMP aminase in *Escherichia coli, Agric. Biol. Chem.* **50**, 1879–1884.

MARUYAMA, A., FUJIO, T., TESHIBA, S. (1993), *Jpn. Kokai Tokkyo Koho* 05276974.

MATSUI, H. (1984), Studies on microbial production of guanosine by mutants of *Bacillus subtilis, Nippon Nogei Kagaku Kaishi,* **58**, 175–180.

MATSUI, H., SATO, K., ENEI, H., HIROSE, Y. (1979), Guanosine production and purine nucleotide biosynthetic enzymes in guanosine-producing mutants of *Bacillus subtilis, Agric. Biol. Chem.* **43**, 1317–1323.

MATSUI, H., SATO, K., ENEI, H., TAKINAMI, K. (1982), 5′-Nucleotidase activity in improved inosine-producing mutants of *Bacillus subtilis, Agric. Biol. Chem.* **46**, 2347–2352.

MISAWA, M., NARA, T., UDAGAWA, K., ABE, S., KINOSHITA, S. (1964), Accumulation of 5′-xanthylic acid by guanine-requiring mutants of *Micrococcus glutamicus, Agric. Biol. Chem.* **28**, 690–693.

MISAWA, M., NARA, T., UDAGAWA, K., ABE, S., KINOSHITA, S. (1969), Fermentative production of 5′-xanthylic acid by a guanine auxotroph of *Brevibacterium ammoniagenes, Agric. Biol. Chem.* **33**, 370–376.

MIYAGAWA, K., KANZAKI, N. (1991), *Eur. Patent Appl.* 412688.

MIYAGAWA, K., KANZAKI, N. (1992), *Eur. Patent Appl.* 465132.

MIYAGAWA, K., NAKAHAMA, K., KIKUCHI, M., DOI, M. (1985), *Eur. Patent Appl.* 151341.

MIYAGAWA, K., KIMURA, H., NAKAHAMA, K., KIKUCHI, M., DOI, M., AKIYAMA, S., NAKAO, Y. (1986), Cloning of the *Bacillus subtilis* IMP dehydrogenase gene and its application to increased production of guanosine, *Biotechnology* **4**, 225–228.

MIYAGAWA, K., KIMURA, H., SUMINO, Y. (1988a), *Eur. Patent Appl.* 273660.

MIYAGAWA, K., KANZAKI, N., SUMINO, Y. (1988b), *Eur. Patent Appl.* 286303.

MIYAGAWA, K., KANZAKI, N., HASEGAWA, T. (1990), *Eur. Patent Appl.* 393969.

MOMOSE, H., SHIIO, I. (1969), Effect of GMP reductase and purine analogue resistance on purine nucleoside accumulation pattern in adenine auxotrophs of *Bacillus subtilis, J. Gen. Appl. Microbiol.* **15**, 399–411.

MORAN, J. R. (1989), *Eur. Patent Appl.* 346865.

MORI, H., IIDA, A., TESHIBA, S., FUJIO, T. (1991), *PCT Int. Appl.* 9108286.

MURATA, K., UCHIDA, T., KATO, J., CHIBATA, I. (1988), Polyphosphate kinase: distribution, some properties and its application as an ATP regeneration system, *Agric. Biol. Chem.* **52**, 1471–1477.

MUZZIO, T., ACEVEDO, F. (1985), Theoretical yields in nucleotide production by fermentation, *Process Biochem.* **20**, 60–64.

NAKAMATSU, T., NISHIYAMA, T., KURASAWA, S., MAEDA, O., ICHIUMI, T., KURAHASHI, O. (1985a), *Jpn. Kokai Tokkyo Koho* 60078593.

NAKAMATSU, T., NISHIYAMA, T., ICHIUMI, T., SHIRATA, T. (1985b), *Jpn. Kokai Tokkyo Koho* 60176596.

NAKAZAWA, E., SONODA, H. (1987), *Jpn. Kokai Tokkyo Koho* 62014794.

NAKAZAWA, E., NISHIMOTO, Y., AKUTSU, E. (1982), *Jpn. Kokai Tokkyo Koho* 57110195.

NARA, T., KOMURO, T., MISAWA, M., KINOSHITA, S. (1969), Regulation of purine ribonucleotide synthesis by *Brevibacterium ammoniagenes, Agric. Biol. Chem.* **33**, 739–747.

NISHIYAMA, T., NAKAMATSU, T. (1993), *Jpn. Kokai Tokkyo Koho* 05317076.

NISHIYAMA, T., NAKAMATSU, T., MAEDA, O. (1993), Adenosine production by an adenine revertant derived from an inosine producer, *Nippon Nogei Kagaku Kaishi* **67**, 843–847.

NISHIYAMA, T., NAKAMATSU, T., SHIROTA, T. (1994), Adenosine production by a mutant of *Bacillus subtilis* resistant to adenine analogs, *Nippon Nogei Kagaku Kaishi* **68**, 809–814.

NOGAMI, I., YONEDA, M. (1969), Derivation of microorganisms producing guanosine and inosine, *Kagaku to Seibutsu* **7**, 371–377.

NOGAMI, I., KIDA, M., IIJIMA, T., YONEDA, M. (1968), Derivation of guanosine and inosine-producing mutants from a *Bacillus strain, Agric. Biol. Chem.* **32**, 144–152.

NOGUCHI, T., KUMAGAI, M., KUNINAKA, A. (1988), Analysis of base composition of sequenced DNA's by high performance liquid chromatography of their nuclease $P_1$ hydrolysate, *Agric. Biol. Chem.* **52**, 2355–2356.

NOGUCHI, T., MARUMO, G., OKUYAMA, K., MIDORIKAWA, Y. (1993), *Jpn. Kokai Tokkyo Koho* 05219978.

NUDLER, A. A., GARIBYAN, A. G., BOURD, G. I. (1991), The derepression of enzymes of *de novo* pyrimidine biosynthesis pathway in *Brevibacterium ammoniagenes* producing uridine-5-monophosphate and uracil, *FEMS Microbiol. Lett.* **82**, 263–266.

OGUMA, T., HONMA, S., KIJIMA, T., MIZUSAWA, K. (1989), *Jpn. Kokai Tokkyo Koho* 01086895.

POCHODYLO, J. M. (1989), *Eur. Patent Appl.* 307853.

POPOV, I. L., BARAI, V. N., ZINCHENKO, A. I., CHERNOV, S. P., KVASYUK, E. I., MIKHAILOPULO, I. A. (1985), Transformation of adenosine into adenosine 5′-monophosphate by intact cells of *Erwinia herbicola, Antibiot. Med. Biotekhnol.* **30**, 588–591.

QUI, W., WANG, W., ZHANG, X., YUMING, J. (1983), Study on the biosynthesis of cytidine diphosphocholine in the collagen membrane-immobilized *Saccharomyces cerevisiae* cells, *Yiyao Gongyr* **4**, 1–4.

REEVE, A. E., HUANG, R. C. C. (1979), A method for the enzymatic synthesis and purification of [$\alpha^{32}$-P] nucleoside triphosphates, *Nucleic Acids Res.* **6**, 81–90.

ROKUGAWA, K., KATOH, Y., KUNINAKA, A., YOSHINO, H. (1975), Polynucleotide phosphorylase from *Achromobacter* sp. KR 170-4, *Agric. Biol. Chem.* **39**, 1455–1460.

SAIGA, H., MIZUMOTO, K., MATSUI, T., HIGASHINAKAGAWA, T. (1982), Determination of the transcription initiation site of *Tetrahymena pyriformis* rDNA using *in vitro* capping of 35S prerRNA, *Nucleic Acids Res.* **10**, 4223–4236.

SHIMIZU, K., TSUCHIDA, T., KAWASHIMA, N., TANAKA T., ENEI, H. (1983a), *Jpn. Kokai Tokkyo Koho* 58158197.

SHIMIZU, K., TSUCHIDA, T., KAWASHIMA, N., TANAKA T., ENEI, H. (1983b), *Jpn. Kokai Tokkyo Koho* 58175492.

SHIRAE, H., YOKOZEKI, K., KUBOTA, K. (1988a), Enzymatic production of ribavirin, *Agric. Biol. Chem.* **52**, 295–296.

SHIRAE, H., YOKOZEKI, K., KUBOTA, K. (1988b), Enzymatic production of ribavirin from pyrimidine nucleosides by *Enterobacter aerogenes* AJ 11125, *Agric. Biol. Chem.* **52**, 1233–1237.

SHIRAE, H., YOKOZEKI, K., KUBOTA, K. (1988c), Enzymatic production of ribavirin from orotidine by *Erwinia carotovora* AJ 2992, *Agric. Biol. Chem.* **52**, 1499–1504.

SHIRAE, H., YOKOZEKI, K., UCHIYAMA, M., KUBOTA, K. (1988d), Enzymatic production of ribavirin from purine nucleosides by *Brevibacterium acetylicum* ATCC 954, *Agric. Biol. Chem.* **52**, 1777–1783.

SILBERKLANG, M., GILLUM, A. M., RAJBHANDARY, U. L. (1977), The use of nuclease $P_1$ in sequence analysis of end group labeled RNA, *Nucleic Acids Res.* **4**, 4091–4108.

SODE, K., SAITO, A., SUZUKI, M., KAJIWARA, K., KARUBE, I. (1989), Microencapsulated chromatophores for the production of ATP, *Biocatalysis* **2**, 309–316.

SOEDA, E., JIKUYA, H., NAKAYAMA, N., YASUDA, S. (1984), Shotgun DNA sequencing, *Kagaku to Seibutsu* **22**, 541–547.

STEIN, C. A., SUBASINGHE, C., SHINOZUKA, K., COHEN, J. S. (1988), Physicochemical properties of phosphorothioate oligodeoxynucleotides, *Nucleic Acids Res.* **16**, 3209–3221.

SUGIMOTO, N., SATO, K., YOKOYAMA, M., TAKENOUCHI, T., SEKI, M., IGARASHI, K., KISHINO, M. (1993), *Jpn. Kokai Tokkyo Koho* 05092944.

SUMINO, Y., SONOI, K., DOI, M. (1983), *Eur. Patent Appl.* 80864.

SUMINO, Y., SONOI, K., DOI, M. (1993), Scale-up of purine nucleoside fermentation from a shaking flask to a stirred-tank fermentor, *Appl. Microbiol. Biotechnol.* **38**, 581–585.

SUZUKI, S. (1984), Nucleic acid-related antibiotics, *Hakko to Kogyo* **42**, 847–858.

TAKAYAMA, H. (1993), Basic study on starasid, in: *The Science Lecture of Starasid* (KIMURA, K., Ed.), pp. 7–19. Tokyo: Nippon Kayaku Co., Ltd.

TAKAYAMA, K., MATSUNAGA, T. (1989), *Jpn. Kokai Tokkyo Koho* 01104189.

TAKEDA, I., WATANABE, S., SHIROTA, M. (1981), *Jpn. Tokkyo Kokohu* 56053359 (*Jpn. Patent* 1115735).

TANG, S., WANG, J., LIU, Z., ZHU, X., JU, N. (1992), Improvement of guanosine yield by fed-batch culture, *Gongye Weishengwu* **22**(2), 5–10.

TANI, Y. (1984), ATP production, *Hakko to Kogyo* **42**, 667–675.

TESHIBA, S., FURUYA, A. (1982), Genetical improvement of 5′-IMP productivity of a permeability mutant of *B. ammoniagenes, Agric. Biol. Chem.* **46**, 2257–2263.

TESHIBA, S., FURUYA, A. (1983a), Sensitivities of a series of mutants to various drugs (Mechanisms of 5′-inosinic acid accumulation by permeability mutants of *Brevibacterium ammoniagenes*), *Agric. Biol. Chem.* **47**, 1035–1041.

TESHIBA, S., FURUYA, A. (1983b), Intracellular 5′-IMP pool and excretion mechanisms of 5′-IMP, *Agric. Biol. Chem.* **47**, 2357–2363.

TESHIBA, S., FURUYA, A. (1984a), Production of 5′-inosinic acid, *Hakko to Kogyo* **42**, 488–498.

TESHIBA, S., FURUYA, A. (1984b), Excretion mechanisms of 5′-IMP, *Agric. Biol. Chem.* **48**, 1311–1317.

TESHIBA, S., FURUYA, A. (1988), Production of 5′-inosinic acid, in: *From Fermentation to New Biotechnology* (TOCHIKURA, T., Ed.), pp. 148–158. Tokyo: Japan Bioindustry Association.

TOMITA, K., T., KURATSU, Y. (1992), Influence of carbon sources on the stimulatory effect of L-proline on 5′-inosinic acid production, *J. Ferment. Bioeng.* **74**, 406–407.

TOMITA, K., NAKANISHI, T. (1990), *Jpn. Kokai Tokkyo Koho* 02312594.

TOMITA, K., NAKANISHI, T., TESHIBA, S., FURUYA, A. (1990a), *Jpn. Kokai Tokkyo Koho* 02234690.

TOMITA, K., KINO, K., NAKANISHI, T. (1990b), *Jpn. Kokai Tokkyo Koho* 02312595.

TOMITA, K., NAKANISHI, T., KURATSU, Y. (1991), Stimulation by L-proline of 5′-inosinic acid production by mutants of *Corynebacterium ammoniagenes, Agric. Biol. Chem.* **55**, 2221–2225.

TOMITA, K., NAKANISHI, T., KURATSU, Y. (1992a), Effect of osmotic strength on 5′-inosinic acid fermentation in mutants of *Corynebacterium ammoniagenes, Biosci. Biotech. Biochem.* **56**, 763–765.

TOMITA, K., OCHIAI, T., KURATSU, Y., NAKANISHI, T. (1992b), *Jpn. Kokai Tokkyo Koho* 04262790.

TOYODA, K., YOSHIOKA, H., MURAKAMI, K. (1992), *Jpn. Kokai Tokkyo Koho* 04252190.

TSUCHIDA, T., SHIMIZU, K., KAWASHIMA, N., ENEI, H. (1984), *Jpn. Kokai Tokkyo Koho* 59028470.

UCHIDA, K., KUNINAKA, A., YOSHINO, H., KIBI, M. (1961), Fermentative production of hypoxanthine derivatives, *Agric. Biol. Chem.* **25**, 804–805.

UENO, J., FUJISHIMA, T., IIDA, T., SAKAMOTO, M. (1987), *Eur. Patent Appl.* 233493.

UTAGAWA, T., YAMANAKA, S. (1981), Properties and application of nucleoside phosphorylase, *Hakko to Kogyo* **39**, 927–937.

UTAGAWA, T., NORISAWA, H., YAMANAKA S., YAMAZAKI, A., HIROSE, Y. (1986), Enzymatic synthesis of virazole by purine nucleoside phosphorylase of *Enterobacter aerogenes, Agric. Biol. Chem.* **50**, 121–126.

VOLBEDA, A., LAHM, A., SAKIYAMA, F., SUCK, D. (1991), Crystal structure of *Penicillium citrinum* $P_1$ nuclease at 2.8 Å resolution, *EMBO J.* **10**, 1607–1618.

WALKER, W. A. (1994), Nucleotides and Nutrition – Proceedings from a Roundtable Symposium held at the Hotel Inter-Continental New Orleans, LA, March 28, 1993, *J. Nutr.* **124**(1) Suppl., 120S–164S.

WANG, S., GAO, C., LIU, T., LU, Z. (1982), Synthesis of CDP-choline by the fermentation of choline chloride and CMP, *Shengwu Huaxue Yu Shengwu Wuli Jinzhan* **48**, 63–66.

WANG, J., WANG, G., LEI, Z., SHEN, G., CHEN, B., SHI, Y., WANG, W., WU, J. (1983), Studies on *Bacillus subtilis* No. 301, an inosine-producing strain, *Weishengwuxue Tongbao* **10**, 254–257.

WANG, Y., ZHANG, B., ZHOU, Y. (1993), Screening of inosine-producing strain by protoplast fusion, *Weishengwu Xuebao* **33**, 74–78.

WOMACK, J. B., O'DONOVAN, G. A. (1978), Orotic acid excretion in some wild-type strains of *Escherichia coli* K-12, *J. Bacteriol.* **136**, 825–828.

WU, L. M., LIAO, H. H., FANG, H. Y. (1989), ATP production from AMP by a methanol-utilizing yeast, *Hansenula polymorpha* P5, *Chung-kuo Nung Yeh Hua Hsueh Hui Chih* **27**, 12–21.

XU, J., CHEN, Y., FENG, P., SHI, S. (1992), Selection and cultivation of inosine-producing strain using resistant mutation technique, *Weishengwuxue Tongbao* **19**, 331–334.

YAMAGAMI, K., FUJII, A., ARITA, M., OKUMOTO, T., SAKATA, S., MATSUDA, A., UEDA, T., SASAKI, T. (1991), Antitumor activity of 2′-deoxy-2′-methylidenecytidine, a new 2′-deoxycytidine derivative, *Cancer Res.* **51**, 2319–2323.

YAMAMOTO, Y., MIDORIKAWA, Y., NOGUCHI, T. (1992), *Jpn. Kokai Tokkyo Koho* 04158793.

YAMANAKA, S., UTAGAWA, T. (1980), Purine arabinosides – Physiological function and enzymatic production, *Hakko to Kogyo* **38**, 920–929.

YAMANAKA, S., UTAGAWA, T., KOBAYASHI, T. (1982), *Jpn. Kokai Tokkyo Koho* 57146593.

YAMASHITA, S., KOMATSU, K. (1988), *Jpn. Kokai Tokkyo Koho* 63313594.

YAMAUCHI, H., MACHIDA, H. (1986), Continuous production of homopolynucleotides by immobilized polynucleotide phosphorylase, *J. Ferment. Technol.* **64**, 517–522.

YAMAUCHI, H., UTSUGI, H., MIDORIKAWA, Y. (1990), *PCT Int. Appl.* 9010080.

YAMAZAKI, E., KANZAKI, N., NAKATSUI, I. (1993), *Jpn. Kokai Tokkyo Koho* 05084067.

YOKOZEKI, K., SHIRAE, H., KUBOTA, K. (1988), *Jpn. Kokai Tokkyo* 63177797.

YOKOZEKI, K., SHIRAE, H., KOBAYASHI, K. (1989a), *Jpn. Kokai Tokkyo Koho* 01074998.

YOKOZEKI, K., SHIRAE, H., KUBOTA, K., SANO, C. (1989b), *Jpn. Kokai Tokkyo Koho* 01104190.

YONEHARA, T., TANI, Y. (1986), *Jpn. Kokai Tokkyo Koho* 61074592.

YONEHARA, T., TANI, Y. (1987), ATP production by methanol yeasts, *Nippon Nogei Kagaku Kaishi* **61**, 1333–1336.

YOSHIWARA, Y., KAWAHARA, Y., YAMADA, Y., IKEDA, S., YOSHII, H. (1987), *Jpn. Kokai Tokkyo Koho* 62061593.

ZHAN, G. (1984), Some results on the production of CDP-choline by *Candida* SP SVI 362, *Xibei Daxue Xuebao, Ziran Kexueban* **14**, 118–120.

ZHAN, G., ZHANG X., WANG. P., YE, M., TIAN, P. (1984), Condition of biosynthesis of CDP-choline by *Hansenula anomala, Zhenjun Xuebao* **3**, 224–231.

# 16 Extracellular Polysaccharides

IAN W. SUTHERLAND

Edinburgh, UK

# 1 Introduction

Industrial usage of polysaccharides (gums) still relies extensively on material obtained from plants or from marine algae. Such traditional commercial polysaccharides include starch, alginate, carrageenan, gum arabic, and the plant glycomannans – locust bean gum, gum guar, and konjac mannan. These substances are widely employed in the food and pharmaceutical industries as well as for other industrial purposes. Although they are valuable commercial products, they do have drawbacks including lack of an assured supply and variations in quality. Only starch is readily available from many alternative sources in virtually all parts of the world, but at least in theory, all are available from renewable resources. Problems arise through crop failure, drought, war, famine, and disease. Further, not all the available plant and algal polymers or their chemical derivatives necessarily possess the exact rheological properties required in a specific application at an acceptable cost.

The production of microbial polysaccharides provides a valid alternative, either through the development of products with properties almost identical to the gums in current use, which they can then replace, or as novel materials with unique or vastly improved rheological characteristics which can be applied for new uses. Microorganisms synthesize an enormous range of different polysaccharides even though the actual composition is restricted to a relatively small number of monosaccharides and other non-carbohydrate substituents such as acetate, pyruvate, succinate, and phosphate. Although the composition is thus restricted, the range of physical properties possessed by these polysaccharides is very large indeed. Some of the microbial polysaccharides yield very high-viscosity aqueous solutions, others form gels similar in their characteristics to agar or to carrageenan. Industrial production of the microbial polymers is not subject to crop failure, climatic conditions, or to marine pollution, although it does require inexpensive substrates, high technology equipment, well trained staff, and adequate power and water supplies. The products are less subject to variability and output can be carefully controlled and accurately costed. They are, however, expensive in relation to bulk products such as starch and can never be expected to displace it from all its many uses. Regulatory authorities must be entirely satisfied on all aspects of product safety and the customer must receive acceptable quality and good technical support. This all involves adequate research and development not only of the polysaccharides themselves but of the range of applications to which they may be put.

Microbial exopolysaccharides can be divided into homopolysaccharides and heteropolysaccharides. Most microbial homopolysaccharides are neutral glucans, while the majority of the heteropolysaccharides are polyanionic due to the presence of uronic acids. Further contributions to charge come from pyruvate ketals or succinyl half esters. Three types of homopolysaccharide structure have been demonstrated. Some are linear neutral polymers composed of a single linkage type. (Microorganisms do not synthesize the "mixed linkage" type of glucan found in cereal plants such as oats and barley.) The second group of homopolysaccharides as exemplified by scleroglucan, possess tetrasaccharide repeating units due to the 1,6-$\alpha$-D-glucosyl side-chains present on every third main chain residue. Finally, dextrans are branched homopolysaccharide structures. Almost all microbial heteropolysaccharides are composed of repeating units. Typically, these vary in size from disaccharides to octasaccharides and frequently contain a single mole of a uronic acid. This is commonly D-glucuronic acid but some heteropolysaccharides contain D-galacturonic acid; D-mannuronic acid is found in bacterial alginates and a few other polysaccharides. Very occasionally, two uronic acids are present in the repeat unit. The deduced uniformity of the repeat units is based on chemical studies and some irregularities may possibly be present, especially in the polymers which are composed of larger and more complex structures. The heteropolymers usually carry short side chains which may be from one to four sugars in length. In a few polymers of more complex structure, the side chains are also branched. Bacterial alginates differ from the other heteropolysaccharides as they are composed of D-mannuronic and L-guluronic acids in an irregular linear structure.

There are still relatively few commercially available microbial polysaccharides, but the number and their applications are gradually increasing. Some of the microbial polymers are already well established products of modern biotechnology with a sizeable market. Others have potentially useful chemical or physical properties. Xanthan, the polysaccharide from *Xanthomonas campestris* pv. *campestris*, has proved to be a very successful product with a wide variety of applications. It is currently the major microbial polysaccharide in commercial production. Its success has lead to the development of other microbial polymers and also to a much better understanding of the rheology of polysaccharides in solution. As a result, the general awareness of the potential for using microbial polysaccharides and of the subtle structural changes which can greatly affect physical properties has greatly increased studies in this area and will hopefully lead to the discovery of new polysaccharides with unique properties and new applications. No single polysaccharide is likely to fulfil all the potential industrial requirements. Current pressures for "green" technology and the use of water-based rather than solvent-based systems should also increase interest in and potential use of microbial exopolysaccharides. Only time can tell how many of these polymers will eventually be commercialized, but it should not be forgotten that microbial polysaccharides are already well-accepted as natural components of many fermented foods and are essential products of the microbial populations involved in, and essential for water and sewage purification!

Two approaches are possible when evaluating the potential of microbial polysaccharides. They can be examined empirically and, if the physicochemical properties appear useful, the polymers can then be subjected to development and testing. Alternatively, the requirements of a polysaccharide for specific or more generalized applications may be determined and known polymers can be evaluated to ascertain whether they might be suitable. Most of the biopolymers which have been developed for commercial applications have been found as a result of empirical experimentation and the full extent of their physicochemical properties has only been determined *after* the product has been made commercially available. In future, this empiricism may be replaced by a more pragmatic appraisal of the *properties required for specific applications*. From such a "blue-print", polysaccharides may then be sought or developed!

There are numerous recent reviews and symposia covering all aspects of interest relating to microbial exopolysaccharides (e.g., CRESCENZI et al., 1989; SUTHERLAND, 1990) and specifically to genetics (BOULNOIS and JANN, 1989; WHITFIELD and VALVANO, 1993), metabolic relationships (LINTON, 1990) and structure–function relationships (SUTHERLAND, 1994).

# 2 Isolation of Polysaccharide-Producing Microorganisms

Exopolysaccharide(EPS)-producing microorganisms occur very widely in nature in different types of habitat. Typically, they may be found in soil, in fresh water, in sea water, associated with plants or with food, and even as products of the Archae in high salt or high temperature or alkaline environments. Numerous human or animal pathogens are capable of synthesizing polysaccharides. As they present potential hazards to the industrial producer, they will not be considered here but it must not be forgotten that, in theory, the biosynthetic systems from such species could be transferred to non-pathogenic bacteria and used to express the products. Whether they could then be produced commercially is less certain. A small number of polysaccharides from pathogenic microorganisms are used as components of protective vaccines such as those for *Streptococcus pneumoniae* strains and *Haemophilus influenzae* type B.

As the actual amounts of polysaccharides are small, they will not be considered here.

Although polysaccharide-producing microorganisms can be found in so many different environments, the lack of knowledge about the physiology of the species and the difficulty in scaling-up production from bench to large-scale fermentation, has greatly reduced the number of potential new products, as indeed has the lack of identifiable new markets! It is not possible to generalize in respect of a single microbial species or an individual ecological niche – examples of *potential* as opposed to *actual* value may be found anywhere! Interestingly, a range of new polymers has been reported by the Kelco Division of Merck, who have had remarkable success in discovering bacterial cultures yielding polysaccharides with novel properties. Although considered initially to derive from various bacterial species, these isolates have all recently been designated as strains of *Sphingomonas paucimobilis* (POLLOCK, 1993).

The value of searching for new ecological niches can be seen in the recent isolation of several EPS-producing extreme halophiles. It has been suggested that this group of Archae might provide several useful products although one disadvantage is that each group of these bacteria probably requires specific media and cultural conditions. As yet, the only isolates reported to produce good yields of polysaccharides with relatively high viscosity and other potentially useful properties, have been *Haloferax* spp., which produce a sulphated polymer composed predominantly of D-mannose (ANTON et al., 1988).

## 2.1 Media and Culture Conditions

Although there are no selective media which can be used specifically for the isolation of polysaccharide-producing microbial species, use can be made of those physiological conditions which are known to favor polysaccharide synthesis in many of the bacterial, yeast and fungal species which have been examined. It is, of course also possible to use selective media for microbial groups such as *Azotobacter* or *Rhizobium* spp. which are known to be always exopolysaccharide producers. Most media used in the laboratory for polysaccharide production are based on high carbon substrate: limiting nutrient ratios, where nitrogen is usually the favored component to induce growth limitation and stimulate exopolysaccharide formation. These have now been shown to favor polymer production in the Enterobacteriaceae, in *Xanthomonas campestris, Pseudomonas* spp. and many other bacteria and also in many fungi. If the aim is to isolate species from natural environments such as fresh water, soil or plant material, we have found that a medium with the composition shown in Tab. 1, supports the growth of a wide range of polysaccharide-producing bacterial types. However, the lower organic nitrogen substrate and salt concentrations indicated in medium 2 may also be desirable for some microorganisms. Sea water-based media with high glucose or sucrose content and a similar concentration of yeast extract and of casein hydrolyzate or an ammonium salt may also prove useful. Recently, a complex medium containing organic substrates has been developed for polysaccharide-synthesizing extreme halophiles of the *Haloferax* group (Tab. 1). Glucose at concentrations of 2–5% (w/v) is usually the preferred carbon substrate as it is utilized by a very wide range of microbial species and is also widely available, the cost depending on the degree of purity employed. If production of levans or dextrans is sought sucrose must be used. For some microorganisms, glucose and sucrose are equally good carbon substrates either of which will provide good yields of exopolysaccharides, but other species may be more restrictive in their substrate requirements. Polysaccharide-producing bacteria capable of growth on other acceptable carbon substrates such as methanol have also been described, and the polymers from such species may have potentially interesting properties (CHOI et al., 1991). It has to be remembered that in the large-scale industrial process, the carbon substrate will almost certainly be a complex product, most probably obtained as a by-product from the processing of agricultural or other plant material (see, e.g., MILLER and CHURCHILL, 1986). Most polysaccharide-producing microorganisms are incubated at or near 30 °C, although incubation

**Tab. 1.** Media with High Carbon/Nitrogen Ratios [$gL^{-1}$]

| Medium | 1 | 2 | Czapeks | | Haloferax | |
|---|---|---|---|---|---|---|
| Casein hydrolyzate[a] | 1.0 | 0.1 | $NaNO_3$ | 2.0 | NaCl | 0.65 |
| $Na_2HPO_4$ | 10.0 | 1.0 | $K_2HPO_4$ | 0.35 | NaBr | 0.167 |
| $KH_2PO_4$ | 3.0 | 0.3 | KCl | 0.5 | KCl | 5.0 |
| $K_2SO_4$ | 1.0 | 0.1 | $FeSO_4$ | 0.01 | $CaCl_2$ | 0.723 |
| NaCl | 1.0 | 0.1 | Magnesium | | $MgSO_4 \cdot 7H_2O$ | 49.5 |
| $MgSO_4 \cdot 7H_2O$ | 0.2 | 0.02 | Glycerophosphate | 0.5 | $MgCl_2 \cdot 6H_2O$ | 34.6 |
| $CaCl_2$ | 0.02 | 0.001 | Sucrose | 30.0 | NaCl | 195.0 |
| $FeSO_4$ | 0.001 | 0.0001 | | | $KH_2PO_4$ | 0.15 |
| Yeast extract[b] | 1.0 | 0.1 | | | $NH_4Cl$ | 1.0 |
| Glucose[c] | 20.0 | 10.0 | | | | 10.0 |

[a] $NH_4Cl$ or $NH_4NO_3$ may be used in place of casein hydrolyzate.
[b] Yeast extract at 1.0 $gL^{-1}$ may be used to provide growth factors.
[c] Glucose may be replaced by sucrose or glycerol for some species.

at suboptimal temperatures frequently favors polysaccharide production. There would obviously be advantages in using thermophiles capable of growth at considerably higher temperatures and thus avoiding the necessity of expensive cooling systems for large-scale synthesis.

The effect of changes in the composition of the growth medium on the production of xanthan was clearly shown by SOUW and DEMAIN (1979). Sucrose was marginally better than glucose as the carbon source for polysaccharide production. However, at sucrose concentrations greater than 4% polysaccharide yields were reduced; this effect was not apparent when glucose was used. The presence of pyruvate, succinate or $\alpha$-ketoglutarate had a stimulatory effect on the yields from sucrose-based media, as did L-glutamic acid. A further advantage found following careful formulation of culture media has been the elimination of problems resulting from strain variability as the choice of growth medium can, under certain conditions, affect culture stability.

Polysaccharide yield and the rate of production are both affected by the oxidation state of the carbon source; both decrease when the substrate is at a greater or lesser oxidation level than hexose (LINTON et al., 1987a). More heat may also be generated and require increased cooling of the fermentation vessel. If hexose substrates are used, the metabolism of closely related compounds may differ as is seen for fructose and glucose in alginate production by *Pseudomonas* spp. (ANDERSON et al., 1987). Thus, fructose is incorporated into the polysaccharide directly as a 6-carbon unit, but glucose is converted into triose phosphates only one of which proceeds to gluconogenesis while the second is oxidized to $CO_2$.

Most media described for fungal culture also contain relatively high carbohydrate levels (BOOTH, 1971) but may not necessarily favor the polysaccharide-producing form of these microorganisms. Isolation may also require the addition of specialized media containing specific inhibitors of bacterial growth. For yeasts, Czapeks medium may be useful, but it does also support the growth of many bacterial types in addition to eukaryotes.

Under conditions employed for industrial production of microbial polysaccharides, the same principle of high carbon:nitrogen ratios is used, but the substrates utilized are the cheapest available form which will provide consistent yields and quality of product. Thus, corn steep liquor, distillers solubles, acid or enzymic hydrolyzates of starch, or other substrates have been used to form the bases for large-scale culture media. Some of these materials also provide sources of growth factors and amino acids required by the microorganisms. Also, whereas in the laboratory the water used for media preparation may be distilled or deionized, that used for industrial production of polysaccharides either comes from the public supply or from condensates. It thus has to be modified by the addition of salts or the adjustment of pH. Similarly, the industrialist normally uses acid or alkali to control pH during growth, whereas in the research laboratory heavily buffered media are frequently used.

Preliminary studies on several of the nutritional factors affecting polysaccharide production can be carried out in shake flasks, but the effects of culture aeration can only realistically be examined in stirred tank fermenters. Almost inevitably, yields are higher and more reproducible when cultures are subjected to well-defined culture conditions in a fermenter than in the flask cultures, which because of the steadily increasing viscosity during growth, are generally oxygen-limited. In the fermenter, pH control is also readily achieved and this is extremely important in cultures of many microorganisms including eubacteria, fungi or the extreme halophile *Haloferax* in which considerable amounts of acid are formed from the catabolism of carbohydrate substrates. In some fungal cultures the high shear rate achieved using Rushton turbine fermenters may be inhibitory to polysaccharide production compared to the lower shear conditions found in air lift vessels. It has to be remembered that cultures of polysaccharide-synthesizing microorganisms become increasingly viscous and non-Newtonian in their rheology. Fungal cultures also show viscosity due to the presence of mycelium but differ from bacterial cultures of similar viscosity in heat and mass transfer characteristics with heat transfer coefficients considerably higher than predicted values.

# 3 The Potential for Strain Development

There are various ways in which strain development could be useful in the production of microbial exopolysaccharides. Modern gene transfer methods could be applied and *in theory* it might be possible to develop new polysaccharide structures. In practise, as far as microbial exopolysaccharides of industrial potential are concerned, progress has only been made in simplifying the repeating unit structures of exopolysaccharides synthesized by *Xanthomonas campestris* and *Acetobacter xylinum* through reduction of the length of the polymer side chains. Although this has indeed yielded macromolecules with altered rheological properties, the yields have been much lower than those from the wild-type bacteria. Improved polymer yield from commercialized microbial strains has generally been achieved through altered fermenter design or changes in medium composition. Introduction of altered surface properties to the microbial cell may assist in the separation and recovery of the polysaccharide product. Deletion of genes for unwanted products such as enzymes and storage polymers may also be applicable in order to reduce processing costs and increase yields respectively.

It should be remembered that polysaccharide production is a complex process involving large numbers of enzymes, some of which are specific for exopolysaccharide synthesis while others are concerned in the metabolism or synthesis of compounds common to various types of cell component. The recent elucidation of the entire gene sequence necessary for succinoglycan biosynthesis by GLUCKSMANN et al. (1993a, b) revealed the number of enzymes needed for the production of a polymer with an octasaccharide repeating unit in which the only monosaccharides are D-glucose and D-galactose. Together with the information now available on the genes responsible for xanthan synthesis (BETLACH et al., 1987), such studies provide an insight into the complexity of the biosynthetic processes which yield microbial polysaccharides. Other recent work in *E. coli*, has shown the involvement of periplasmic proteins responsible for the transport of polysaccharides through the periplasm (SILVER et al., 1987; PAZZANI et al., 1993). It is not yet clear whether these or similar, relatively nonspecific proteins are needed by all exopolysaccharide-producing gram-negative bacteria, thus adding still further to the complexity of the biosynthetic process.

## 3.1 Removal of Unwanted Products

A number of microbial strains either produce more than one exopolysaccharide or additionally, form storage polymers such as glycogen, poly-$\beta$-hydroxybutyric acid or polyhydroxyalkanoates. Either multiple polysaccharide production or the synthesis of storage polymers effectively reduces the yield of exopolysaccharide and the substrate conversion rate. This may represent a considerable loss of carbon substrate and greatly increase production costs. Preferably, polymers will be produced from microbial strains limited to producing a single polysaccharide or from mutants in which synthetic capabilities for unwanted materials have been deleted.

The strains of *X. campestris* used for xanthan production generally also produce several extracellular enzymes including a cellulase. Although xanthan is effectively a cellulose molecule with trisaccharide substitutions (see Fig. 8), it is probably not susceptible to most cellulases unless it is in the disordered form (SUTHERLAND, 1984). However, the addition of xanthan contaminated with the bacterial cellulase to products containing carboxymethylcellulose or other cellulose derivatives could lead to undesirable effects. Selection of mutants incapable of cellulase synthesis may be desirable; alternatively, mild heat treatment of the polysaccharide during recovery can be used to destroy the enzyme activity. Similarly, the development of strains of *Streptococcus equi* and of *Streptococcus zooepidemicus* lacking hyaluronidase activity, is desirable for product improvement and increased yields of high molecular weight hyaluronic acid. The same type of strain development might be required if bacterial alginate production from either *Pseudomonas* spp. or *Azotobacter vinelandii* is to be developed commercially. The gradual release of the periplasmic enzymes degrading alginate results in reduced molecular mass in the final polysaccharide product (CONTI et al., 1994). A problem encountered at least in the *Pseudomonas* spp. is the presence of the alginase gene in close association with the genes responsible for exopolysaccharide synthesis. As yet, no strains devoid of alginate lyase activity and simultaneously capable of alginate production have been reported. Laboratory studies of the bacterial strains producing gellan and polysaccharides of related structure, have also indicated that a number of these microorganisms secrete a gellanase. The enzymes have now been identified in our laboratory as gellan lyases, showing greatest activity on the deacylated forms of the polymers gellan and rhamsan (SUTHERLAND, I. W. and KENNEDY, L., unpublished results). A selection procedure developed by LOBAS et al. (1992) used growth on nutrient media solidified with gellan to select mutants which lacked the enzyme and appeared to produce higher yields of the polysaccharide. A similar procedure was later used to obtain improved yields of another strain. In this case, the polysaccharide synthesized also appeared to differ from that of the original strain (LOBAS et al., 1994). Prolonged incubation of *Aureobasidium pullulans* cultures leads to recovery of pullulan with lower molecular mass, almost certainly due to the presence of degradative enzymes. Whether the genes for these enzymes can be deleted to yield cultures capable of increased polysaccharide production is not clear. Another unwanted product in several of the fungal cultures is melanin, which currently has to be removed in an additional step by charcoal treatment. In the production of bacterial cellulose by *A. xylinum*, mutant strains which synthesized increased quantities of cellulose, yielded much less gluconate (BYROM, 1991).

## 3.2 Derepression of Polysaccharide Synthesis

The wild-type exopolysaccharide-producing microorganism may not be the ideal strain for another reason. In a number of such systems, polysaccharide production is normally repressed and yields are low. GOVAN et al. (1981) demonstrated that strains of *Pseudomonas fluorescens*, *P. putida*, and *P. mendocina*, which initially yielded little polysaccharide, could be successfully converted to alginate production with an isolation frequency of 1 in $10^8$ cells through selection for carbenicillin resistance. However, not all the alginate-producing *Pseudomonas* species tested were amenable to this approach.

Intensive studies on the genetics of bacterial polysaccharide synthesis have revealed both common features and considerable differences in different species. Alginate production in *Pseudomonas aeruginosa* (and possibly in related species), is subject to complex regulatory systems (CHITNIS and OHMAN, 1993). The biosynthetic pathway is complex and it involves a considerable number of regulatory genes, some of which are responsive to environmental stimuli with the *alg*B gene belonging to the class of two component regulatory systems (GOLDBERG and DAHNKE, 1992). Among the structural genes is that for *alg*D, the gene product from which is the unusual bifunctional enzyme phosphomannose isomerase/GDP mannose pyrophosphorylase (SHINABARGER et al., 1991). The same is true for colanic acid production, which requires an operon of six genes *cps*A-F in *E. coli* (TRISLER and GOTTESMAN, 1984). The control of expression of the genes is complex, as they are regulated by three regulatory genes *rcs*A, *rcs*B, and *rcs*C together with the *lon* gene (BRILL et al., 1988; GOTTESMAN et al., 1985). The two positive regulators *rcs*A and *rcs*B are required for maximum polysaccharide production (GOTTESMAN and STOUT, 1991). The *lon* ATP-dependent protease normally rapidly degrades the *rcs*A gene product thus limiting the availability of this protein, *lon* acting indirectly as a negative regulator. The *rcs*B gene product is probably an effector in a two-component system. Homologs of several of these components have been identified in other exopolysaccharide-synthesizing gram-negative bacteria. In some genetic systems, such as that for succinoglycan synthesis, common features may be observed between different bacterial species capable of synthesizing the same exopolysaccharide (CANGELOSI et al., 1987).

## 3.3 Improvement of Polysaccharide Yield

The yields of the required polymer from wild type microorganisms may be low for reasons other than those considered above. If more than one polysaccharide is formed, elimination of the second is necessary, as it is usually impossible to separate the product mixtures. This was applied to strains of *Alcaligenes faecalis* var. *myxogenes* which originally produced both curdlan and succinoglycan. Spontaneous mutants incapable of succinoglycan synthesis were eventually obtained for curdlan production (AMEMURA et al., 1977). However, the synthetic processes for polysaccharides such as xanthan and succinoglycan are extraordinarily efficient, yielding high conversion rates from the commonly used carbon substrates (see, e.g., LINTON, 1990). Improvement of the yields of these polysaccharides may not be possible or may only increase polymer production by 2–3%. One possible avenue for study relates to the exact composition of polymers such as xanthan and the presence or absence of various acyl groups. RYE et al. (1988) showed that the energy for xanthan synthesis was derived from the oxidation of the carbon substrate to carbon dioxide or to products, some of which form integral components of the polysaccharide molecule. The amount of energy derived from the oxidation products depended on the polymer composition and on the ATP/oxygen quotient.

Few attempts have been made to determine the effect of nutrient uptake on polysaccharide production despite the observation by HERBERT and KORNBERG (1976) that in *E. coli* growth on various substrates was dependent on substrate uptake for compounds taken up by either active transport or group translocation mechanisms. The uptake of the carbon substrate is effectively the first major limitation on polysaccharide production. Increased efficiency of uptake might lead to improved polymer yields. Certainly, the investigation of nitrogen uptake in methanol-utilizing bacteria and replacement with a more efficient system from *E. coli* produced higher cell yields for the genetically engineered bacteria (WINDASS et al., 1980). A similar approach might increase cell and polymer yield in polysaccharide-producing bacteria. A study by CORNISH et al. (1988, 1989) on succinoglycan-synthesizing *Agrobacterium radiobacter* and *A. tumefaciens* bacteria revealed the presence of two distinct periplasmic binding proteins involved in glucose transport. Through the use of continuous culture under galactose or xylose limitation, mutants with enhanced uptake of either glucose + galactose or glucose + xylose were selected, in which hyperproduction of the binding proteins could be observed. Under the conditions used, succinoglycan production from one of the mutant strains was also enhanced. In a comparison of different carbon substrates for polysaccharide synthesis by these bacteria, LINTON et al. (1987b) concluded that although the rate of succinoglycan synthesis on glucose was greater than that on gluconate or xylose, the rate of ATP utilization for polymer synthesis was similar. It represented most of the respiratory activity in excess of that needed for biosynthesis of cell material. From such observations it was concluded that the rate of ATP turnover was of greater physiological importance than exopolysaccharide synthesis.

## 3.4 Enhancement of Strain and Polysaccharide Characteristics

One of the major problems encountered in the production of exopolysaccharides is the separation of microbial cells from a highly viscous or poorly soluble product. If the surface hydrophobicity is increased through mutation and cells are caused to autoagglutinate, it might be possible to modify the cell surface to render the microorganisms more easily separable from the polysaccharide. Conditional mutants capable of autolysis under permissive conditions following polysaccharide synthesis could be of some value but the polymer would have to be freed from nucleic acid liberated during cell lysis. Some polysaccharide-producing bacteria yield polymer in the form of capsules which are totally cell-associated. Such cultures can be converted to slime-forming derivatives in which the polymer is no longer cell-associated but is more readily recovered and retains its physical characteristics.

It may be desirable to alter the chemical or physical characteristics of a polysaccharide, but this can prove extremely difficult and is normally dependent on the strain and the polymer structure. Thus BETLACH et al. (1987) prepared polysaccharide from xanthan mutants in which the side chains had been abbreviated to remove the pyruvate and glucuronic residues (see Fig. 9). The resultant neutral polysaccharide with a single $\alpha$-D-mannosyl side chain yielded high viscosity in aqueous solutions. An alternative method of producing similar material was reported by TAIT and SUTHERLAND (1989), who applied enzymes to remove the terminal $\beta$-D-glucuronosyl residue from polysaccharide lacking the normal mannosyl side chain terminus. A similar product has been prepared from a mutant strain of *A. xylinum* which originally synthesized acetan (MACCORMICK et al., 1993). Xanthan with enhanced pyruvate content has been prepared from strains supplemented with a plasmid carrying part of the biosynthetic gene sequence (HARDING et al., 1987). Another modified xanthan totally lacks the pyruvate ketals attached to the mannose terminus (WERNAU, 1979, 1980); it has the further property that the bacterial strains from which it is produced, are non-pigmented.

In the case of alginates, whether of algal or bacterial origin, it is possible to alter the composition and hence the physical properties after polymerization. Consequently the application of the extracellular poly-D-mannuronate epimerase to alginates of low guluronate content can increase the amount of this uronic acid. Although the enzyme is inhibited by the protective effect of *O*-acetyl groups on mannuronosyl residues, these can be removed by alkali treatment prior to exposure of the polymers to the enzyme. The success of epimerase treatment in this way, using an enzyme prepared from *A. vinelandii* and various algal alginates as substrates, has been demonstrated by SKJÅK-BRÆK et al. (1985, 1986). The same concept may be applicable to other polymers including chondroitin and heparin which are subject to post-polymerization epimerization. An alternative approach may be available following studies on alginate synthesis and the identification of the specific genes and gene products in *Pseudomonas aeruginosa*. In these bacteria the *alg*G gene product is a poly-D-mannuronic acid epimerase (CHITNIS and OHMAN, 1990), while the *alg*F gene product is responsible for acetylation of the homopolymer (SHINABARGER et al., 1993). Thus deletion of these or similar genes could be used for the production of poly-D-mannuronic acid and non-acetylated alginate, respectively.

Polysaccharides with enhanced viscosity have been obtained from mutants from a range of polysaccharide-producing bacteria including *E. coli., Enterobacter aerogenes* and *Pseudomonas chlororaphis* (SUTHERLAND, 1979). The polymers isolated from the respective mutants were all identical in composition with the original material and were believed to differ only in their molecular mass. This procedure could, no doubt, be applied to other microbial species yielding polysaccharides with good physicochemical characteristics but low initial viscosity. The mutants can be recognized by their different colonial appearance and adherence to the surface of solid growth media. As attempts to obtain similar mutants from capsulate bacteria were unsuccessful, isolation of slime-forming mutants is probably an essential prerequisite. In lipopolysaccharide biosynthesis in enteric bacteria, a gene encoding chain length determination has been identified (BASTIN et al., 1993). It is possible that an analogous system was responsible for the high viscosity mutants described above.

# 4 Laboratory and Commercial Preparation of Microbial Polysaccharides

In the laboratory, exopolysaccharide producing microorganisms can be grown on solid media, in batch culture in liquid media, or in continuous culture. Dextrans, levans, and related polymers can even be prepared in cell-free systems. The methods applied in the laboratory provide much of the initial information both about the microorganisms and the polymers and are a prerequisite to large-scale preproduction evaluation. With the possible exception of material for vaccine preparation, solid cultures are only used for the initial screening and perhaps for the preparation of small amounts of polysaccharide.

## 4.1 Batch Culture

As batch culture is used industrially for polysaccharide production, it is also the method of choice for evaluating culture conditions, optimizing medium composition and pH. A large number of relatively small-scale cultures can be studied speedily, but shake flasks are difficult to oxygenate efficiently and the pH changes rapidly. The medium composition is constantly altering as growth proceeds and the different, changeable parameters can only be measured by removal of samples, introducing further changes in volume, etc. After the initial evaluation has been made, the system needs to be scaled up. The first stage is usually to study growth using a stirred tank fermenter with adequate instrumentation for measuring pH, *Eh*, stirrer speed and with provision of facilities for repeated sampling. Such systems offer the potential for control of the physical parameters important to the process and should be reasonably close in design to the equipment which will eventually be used for large-scale production. The effect of sheer and other physical parameters can be readily assessed. The pilot-scale operation will also give information on substrate conversion, yields, and quality as well as providing sufficient polysaccharide for a full study of the physical and chemical properties of the material and possible technical evaluation. Full-scale culture of the bacteria for polysaccharide production is probably in the range 10–200 $m^3$. The fermentation for bacterial species is normally continued for 48–72 h by which time the yield of a polymer such as xanthan or succinoglycan is $\sim$0.6–0.7g $g^{-1}$ glucose.

## 4.2 Continuous Culture

The use of continuous culture to study the production of microbial exopolysaccharides offers considerable advantages over other methods. Growth can be controlled through supply of a single nutrient, while other medium components are present in excess and do not vary as in batch culture. With adequate instrumentation, most growth parameters can be measured; air flow and impeller speed can be carefully controlled. Studies using these methods can be applied to determine the effects of different substrates and limiting nutrients, growth (dilution) rates, etc. DAVIDSON (1978) noted that synthesis of xanthan under magnesium or phosphorus limitation yielded a product with low pyruvate content (~1%), whereas the material from other limitations contained 5.5–8% ketal groups. These results contrasted with the marked differences in polymer composition and properties found at different stages and under different limitations in batch culture (TAIT et al., 1986).

One problem thought to be associated with continuous culture is the potential for selection of fast-growing mutants lacking the desired characteristics for polysaccharide production. While such variability has been observed by some workers in studies of xanthan production, other reports have indicated that sulphur limited cultures of the pathovar *juglandis* failed to show the presence of such mutants even after 83 days growth. Despite its apparent potential, continuous fermentation has not been adopted for commercial production processes because of the problems in maintaining sterility and the possible emergence of mutants.

## 4.3 Industrial Production of Polysaccharides

Little direct information on the equipment used for industrial production of microbial polysaccharides is available in the public domain despite the large number of papers discussing the theoretical problems associated with polysaccharide rheology in aqueous solution. In all polysaccharide-synthesizing fermentations, the hydrodynamic behavior of the broth changes greatly with time, as the polymer is produced. Adequate aeration and mixing have to be provided if polysaccharide synthesis is to be maintained and stagnant zones and hence oxygen limitation, avoided. Fermenter design, therefore, plays an important role in maintaining homogeneity and ensuring product quality. The process must ensure that there is adequate mixing, oxygen, mass and heat transfer. Because of the high final viscosity, air agitated vessels are not normally suitable for bacterial fermentations and most industrial processes are assumed to use conventional mechanically agitated fermentation tanks. As the turbulence and shear rate in vessels with rotating straight bladed turbines is higher near the impeller than in the bulk liquid, optimal design requires the provision of sufficient turbulence to give good gas dispersion and bulk mixing. The major requirements for microbial polysaccharide production on an industrial scale can be summarized as follows:

(1) For the fermentation process, water is needed both as a medium component and for cooling, while fuel is needed for steam generation and for sterilization. Electrical power is required for agitation and for pumping air and fluids.
(2) For product recovery, electricity is again needed for pumping, but also for filtration or centrifugation of the culture broth, and for milling the dried product. The fuel requirements at this stage are for pasteurization, solvent recovery, and drying.
(3) Finally it should not be forgotten that both water and fuel will be needed for cleaning the fermentation equipment and for effluent disposal.

The rheology of the microbial polysaccharides which are produced commercially is complex. Polymers such as pullulan and xanthan exhibit pseudoplastic flow behavior (shear thinning), but xanthan also displays a yield stress and viscoelasticity. Some alginate fermentations may be thixotrophic.

## 4.4 Product Recovery

The process of product recovery at the end of the fermentation represents a significant proportion of the total costs of polysaccharide production. The presence of bacterial cells is usually undesirable and their removal from highly viscous culture fluids presents considerable difficulties. When working with relatively small scale preparations in the laboratory, it is possible to centrifuge or ultracentrifuge culture supernatants prior to polysaccharide recovery. Thus, the preparations are essentially free of cells and associated material. It is also possible to remove all unbound low molecular weight material by dialysis or by passage through ion exchange columns. The exopolysaccharides are then precipitated from solution by the addition of polar solvents including ethanol, acetone, or isopropanol. Such procedures are not possible on a large industrial scale and alternative methods of purification and recovery must be applied. Processing of the fermentation products aims to concentrate the product into a form which is stable, easy to handle and transport and which can be readily redissolved for application (SMITH and PACE, 1982). During purification, non-polymeric solids should be removed so that the quality of the polysaccharide in terms of color, purity, etc. is enhanced, and undesirable enzymic activities should be destroyed. Typically, cells will be removed from the fermentation broth, polysaccharide will be precipitated, dewatered, dried, milled, and packaged.

Although high temperature pasteurization may cause degradation of certain polysaccharides, it can also in the case of xanthan, enhance the solution viscosity. Filtration to remove microbial cells and debris can be facilitated by dilution of the culture broth to reduce viscosity, or the solution can be heated above the transition temperature of the broth. Addition of salts may also reduce viscosity and make filtration easier. Cells can also be lysed by appropriate chemical or enzymic treatment; while alkali may be used it can, however, degrade the polymer and cause loss of ester-linked acyl groups. The cellulase present in xanthan preparations may be destroyed by treatment of the wet polymer mass with propylene oxide or propionolactone. Alkaline proteases may assist in cell lysis and remove extracellular proteins found in association with some exopolysaccharides. The enzymes also render particulate material more readily absorbable on silica, thus assisting its removal.

### 4.4.1 Broth Concentrates and Slurries

Much of the cost associated with the final product lies in the recovery and purification of the polysaccharide. Preparations which have been concentrated directly from the culture fluid may therefore have a price advantage. As such products are easily redissolved and the viscosity attained for a given solid matter content is usually greater than that from powders, they may also be especially useful for some applications. Thus, xanthan concentrates prepared by tangential flow filtration are commercially available in various industrial grades and can readily be diluted to the required working concentration. Typically concentrates contain between 8 and 12% dry polysaccharide to which appropriate antibacterial agents are added to prevent deterioration. Some other bacterial exopolysaccharides have also been marketed in this form. The concentrates have the added advantage that they can be pumped and accurately metered. Some deterioration on storage may, however, occur.

### 4.4.2 Recovery with Solvents

COTTRELL and KANG (1978) outlined a process for the recovery of xanthan from fermentation liquors in which the fermentation broth was first pasteurized then polysaccharide was recovered by precipitation with isopropanol. The precipitated xanthan was then dried and milled to yield a product which contained 11–12% moisture and 9–10% ash. The mesh size range was 40–200. Alcohol precipitation now appears to be the favored method and a range of products of mesh size 80 (177 μm) and 200 (74 μm) is marketed under the trade name Keltrol® (Kelco, 1992). Industrial grade materials (Kelzan®) are also produced to meet specific applications requirements. Similar isolation procedures have been reported for other polysaccharides with industrial potential. The solvent has the additional value of removing color and other impurities and can be recovered and recycled following redistillation. The addition of an electrolyte is frequently practised to assist precipitation of the polymers in the presence of miscible organic solvents which may be acetone, isopropanol, methanol, or ethanol (SANDFORD, 1979). The minimum alcohol concentration to precipitate the polysaccharide depends on ionic concentration and polymer concentration. A patent from Tate and Lyle Ltd. (1981) proposed the addition of isopropanol to the fermentation broth in amounts below that required for polymer precipitation. Solids were removed from the fluid by filtration at 100 °C, then more solvent was added in amounts which precipitated the polysaccharide. In all such methods, the organic solvent can be recovered by distillation in systems designed to minimize energy losses. A further benefit lies in the removal of any pigments which are soluble in the organic solvent. After precipitation, the polysaccharides are dried in either batch or continuous processes under vacuum or inert gases to a final moisture content of about 10%. The conditions used must not cause degradation or discoloration. Finally the dried polymer is milled to the desired mesh size which will facilitate dispersion and solution. In the powder form, xanthan may show lower viscosity than that seen in concentrates. It may also prove more difficult to redissolve.

The fungal polymer scleroglucan can be prepared by methods similar to those used for the polysaccharide products of bacterial growth (ROGERS, 1973). After heating to inactivate glucan-degrading enzymes and to kill the microorganisms, the culture is homogenized to separate polysaccharide from the fungal mycelium. The resultant viscous fluid is diluted and filtered at elevated temperature to remove the particulate material. The clear solution is then concentrated prior to polymer precipitation with isopropanol or methanol. The fibrous polysaccharide is drained, dried with hot air and ground to yield a fine powder containing 6% or less moisture and less than 5% ash. In preparation of pullulan, the presence of polymer-associated melanin pigments requires the additional step of decolorization with activated charcoal.

In the case of gellan gum, as with other polysaccharides, the fermentation broth is pasteurized to kill viable bacteria prior to product recovery. Thereafter, the gellan gum may be obtained in one of several ways (Kelco, 1992). If a highly acylated product is desired, the polysaccharide can be recovered directly from the fermentation broth. Alternatively the polymer can be deacylated by treatment with alkali then recovered to yield an industrial-grade product. A third alternative involves a clarification step subsequent to deacylation, and is the method favored for Kelcogel® and Gelrite®.

An alternative to precipitation with organic solvents is to remove water from the fermentation liquor by evaporation. This can be by freeze-drying (vacuum drying) at the laboratory scale or by drum or spray drying on a large scale. Care has to be taken to ensure that the conditions employed do not permit any degradation of the product or cause increased color or reduced solubility on rehydration. Assessments of the different costs attributed to the final product, and models for polysaccharide production processes have been made by MARGARITIS and PACE (1985).

# 5 Uses of Microbial Polysaccharides

Microbial polysaccharides find a very wide range of applications in the food industry, in non-food uses, in pharmaceuticals and cosmetics, and in the oil industry. In some of these applications, they can replace currently used polymers. In which case, they must show a cost advantage or have superior properties to the material presently employed. Alternatively, they may find completely novel applications. It has to be remembered that a microbial product such as xanthan costs £6–10 per kg compared to £4–6 for alginate or £1–2 for guar (1986 prices, from LINTON et al., 1991) (Tab. 2).

**Tab. 2.** Usage and Costs of Exopolysaccharides[a]

| Polymer | Estimated Consumption Worldwide [t] | Cost [$/kg] |
|---|---|---|
| Alginate | 23,000 | 5–15 |
| Cellulose (bacterial) | ? | 9–18 |
| Curdlan | | 940 |
| Dextran | 2,000 | 35–390 |
| Dextran derivatives | 600 | 400–2,800 |
| Gellan | – | 66–75 |
| Hyaluronic acid | 500 | 2,000–100,000 |
| Pullulan | | 11.5 |
| Rhamsan gum | – | 25 |
| Welan gum | – | 25 |
| Xanthan | 10–20,000 | 10–14 |
| Gum arabic | 25–40,000 | 2.8 |
| Gum guar | 10–15,000 | 0.9 |
| Gum tragacanth | | 25 |

[a] Production and price estimates derive from a number of sources.

**Tab. 3.** Polysaccharide Properties Used in Food

| Function | Application |
|---|---|
| Adhesive | Icings and glazes |
| Binding agent | Pet foods |
| Coating | Confectionary |
| Emulsifying agent | Salad dressings |
| Encapsulation | Powdered flavors |
| Film formation | Protective coatings, sausage casings |
| Fining (colloid precipitation) | Wine and beer |
| Foam stabilizer | Beer |
| Gelling agent | Confectionary, milk-based desserts, jellies, pie, and pastry fillings |
| Inhibitor (ice crystal formation) | Frozen foods, pastilles, and sugar syrups |
| Stabilizer | Ice cream, salad dressings |
| Swelling agent | Processed meat products |
| Syneresis inhibitor | Cheeses, frozen foods |
| Synergistic gel formation | Synthetic meat gels, etc. |
| Thickening agent | Jams, sauces, syrups, and pie fillings |

## 5.1 Food Usage

The use of polysaccharides in food probably accounts for just under 10% of the usage of *all* types of polysaccharides. Starch and modified starch are the most widely used polysaccharides in foodstuffs, but a range of other polysaccharides are also incorporated. The polysaccharides are added to foods to provide suitable rheological properties during processing and an appearance which will enhance consumer appeal during purchase and consumption. Their role may be as thickening or gelling agents; they are also used for their colloidal properties (Tab. 3). Starch is very widely used because of its low cost and its compatibility with many food ingredients, but is not suitable for all food applications. In foodstuffs which are frozen and thawed, polysaccharides are incorporated to control ice crystal formation. Regardless of the purpose of the polymer when it has been incorporated into foods, the polysaccharides must conform to the food additive regulations applying to that particular food product. There are, therefore, various factors which have to be taken into consideration before a decision can be taken to use a microbial polysaccharide in foods:

1. Type of application
2. Viscosity or gelation
3. Mouthfeel
4. Appearance (including clarity, etc.).
5. Emulsification capabilities
6. Compatibility with other food ingredients
7. Synergistic or other effects
8. Stability to physical and biological factors under the conditions of intended usage
9. Cost
10. Acceptability (legal, consumer and otherwise)
11. Quality – odorless, tasteless, consistent quality, etc.

In addition to their incorporation into foods for human consumption, polysaccharides are used in pet foods, especially in semi-moist preparations where they act as thickening agents and binders. When the polysaccharides are incorporated into human or animal foodstuffs, they must be compatible with the salts, pH and other components found naturally in the foods. In this respect, xanthan is highly satisfactory, having also good freeze and thaw stability (COTTRELL, 1980). An additional property of xanthan which is widely used in its food additive role, is the synergistic viscosity increase and gelation observed when it is mixed with gluco- and galactomannans derived from legumes. The maximal effect can be found in the viscosity of solutions containing approximately equal amounts of xanthan and galactomannan such as guar gum (from the legume *Cyamopsis tetragonoloba*). At higher concentrations of the two polymer types, use can be made of the gels formed; the total concentration of both the polysaccharides is generally less than the concentration of other gelling agents needed to form gels of similar strength.

The types of foodstuff into which polysaccharides such as xanthan can be incorporated range from salad dressings, sauces and syrups, to canned and frozen foods, dry mixes, and other types. At low concentrations, xanthan may be included in juice drinks and other beverages. In this type of product, the role of the microbial polysaccharide is either to suspend fruit pulp or to maintain foam on the head of the drink. In foods which are acid and also contain considerable quantities of multivalent cations, use of polysaccharides such as alginate may be limited because of precipitation or gel formation. Derivatives, such as propylene glycol alginate, do not gel until the pH falls below 3.0 and thus provide suitable alternatives. Alginates are widely used in foods such as dairy products and frozen desserts (COTTRELL et al., 1980). Their addition assists in the stabilization of such products as yoghurt, while in ice cream the polysaccharide controls ice crystal formation during freezing. Currently, the alginate used in food must be derived from the traditional marine algal sources, but technically it might be possible to replace such a product with material of bacterial origin.

## 5.2 Non-Food Uses of Polysaccharides

The range of uses for microbial polysaccharides is continually being extended, but some indications of the diversity of non-food applications are shown for xanthan alone in Tab. 4, while further examples can be found in various references (COTTRELL et al., 1980; KANG and PETTIT, 1993; LINTON et al., 1991). In many of the non-food applications, xanthan is used as a suspending agent, but it may also be employed for viscosity control, gelation, or flocculation.

Polysaccharide gels have numerous industrial applications. Agar is widely used in the formulation of microbiological culture media where its ability to form a brittle, thermoreversible gel of relatively high melting point (60–97 °C) and low setting point (32 °C) together with its resistance to enzymic attack make it particularly useful. Normally, the only microorganisms capable of producing agar-degrading enzymes (agarases) are marine bacteria. An alternative to agar, which has now found application in media for the isolation and culture of various microbial spe-

**Tab. 4.** Industrial Applications of Xanthan

| Physical Properties | Usage |
|---|---|
| Suspension | Laundry chemicals<br>Herbicides<br>Liquid feed supplements<br>Paper coatings<br>Flowable pesticides<br>Agrichemical sprays |
| Flocculant | Water clarification<br>Ore extraction |
| Shear thinning/ viscosity control | Oil drilling muds |
| Viscosity/cross linking | Hydraulic fracturing |
| Viscosity control | Abrasives<br>Jet printing |
| Stabilization | Thixotrophic paints |
| Foam stabilization | Beer, fire fighting fluids |
| Gelling agent | Explosives<br>Blockage of permeable zones in oil reservoirs |

cies, is gellan (gelrite) produced by a strain of *Sphingomonas paucimobilis*. The gels prepared from the deacylated commercial product, have the advantage that they are considerably clearer than corresponding agar gels (MOORHOUSE et al., 1981), thus being potentially useful for immune precipitation and similar tests involving optical assessment. Xanthan *per se*, does not form gels but in the form of complexes with borax or in the presence of $Cr^{3+}$, does so.

As suspending agents, polysaccharides find uses in preparations of agricultural chemicals – biocides, fungicides, and pesticides. In the ceramic industry, they perform a similar function for the ingredients of glazes; they can also be used for the suspension and stabilization of water-based and emulsion inks. An application proposed both for xanthan and other microbial polysaccharides, is as a replacement for hydroxyethyl cellulose (HEC) in water-based latex paints. While some of the polymers possess superior properties, they may still be at a cost disadvantage; others would need to be incorporated at relatively high concentration – 1 kg/166 L compared to 0.75 kg for HEC.

## 5.3 Oil Field Usage

Polysaccharides find various applications in the oil industry. Under some conditions, especially those in which high concentrations of salts are found, their properties and stability are considerably superior to synthetic polymers such as polyacrylamides.

Drilling muds are used to suspend the rock cuttings from oil wells and carry them to the surface. They also function as lubricants for the drill bit and as suspending agents for the dense chemicals used to maintain an overpressure and prevent blow outs. As the drilling muds are frequently formulated using brines or seawater, which are readily available on site, compatibility with high salt concentrations is needed for any polysaccharides included in their preparation. Xanthan has been used for drilling muds, while several other microbial polymers have been proposed. A range of properties essential for polysaccharides under consideration for oil field usage is seen as follows:

1. High viscosity in water and in concentrated salt solutions such as brine and connate water
2. High shear stability to enable pumping
3. Pseudoplasticity
4. Stability over a wide range of pH values
5. Stability for prolonged periods when exposed to high temperatures under anaerobic conditions
6. Freedom from particulate material; high injectivity
7. Low adsorption to reservoir rocks
8. Low cost

The stability of xanthan to the high temperatures and high salt environments encountered during drilling, coupled to its excellent recovery from shear have made it a realistic competitor to the cheaper guar and guar derivatives. Moreover, xanthan conforms in its properties, to many of the requirements desired for any polymer used in such applications; it has high viscosity in water and in concentrated salt solutions, high shear stability, it is pseudoplastic and is both stable over a wide pH range and is stable for prolonged periods when exposed to high temperatures. It is free from particulate material; has high injectivity and low adsorption to reservoir rocks.

Bacterial polysaccharides have also found applications in oil well development, as suspending agents for gravel packing materials used in well finishing and work over. The polysaccharide solution is claimed to give efficient gravel transport and placement. In gravel packing, the use of succinoglycan has also been proposed and it has been claimed to have the advantage of lower thermal stability and so is readily destroyed after it has been used to place the packing materials needed to stabilize the well.

Mobility control polymers are used in enhanced oil recovery (EOR) with the aim of increasing sweep efficiency and consequently, the yield of recoverable oil. The polymer solution is used in secondary or tertiary recovery processes to displace the oil which would otherwise remain *in situ*, being bypassed by the injection water flood. Polymers may also be used for micellar-polymer projects in which, as in polymer floods, the aim is to reduce the mobility of the injected water. Various authors have suggested the use of microbial polysaccharides for these purposes and have discussed their suitability relative to other types of available chemicals (e.g., UNSAL et al., 1979). The basic requirement is for a polymer of small enough molecular size to pass through the rock pores, yet yielding a high viscosity solution at low concentration and minimally affected by the high salt concentrations, high temperatures and high pressures encountered in the oil reservoir. The polymer must also withstand the high shear forces expected in pumping operations. Xanthan has intrinsic viscosity comparable with other polymers tested but has a higher molecular weight and greater molecular size. Although bacterial cells are present in commercial preparations of the polysaccharide, the development of cultural methods to reduce cell size and the use of heat and enzymes to destroy cell material, ensures that the xanthan will readily pass through rock pores of the size commonly found in oil reservoirs. Laboratory tests have indicated that xanthan can remain relatively undegraded for prolonged periods at the temperatures in the range (80–100 °C) frequently encountered in many deeper oil reservoirs, provided that precautions are taken to exclude oxygen (KIERULF and SUTHERLAND, 1988). Scleroglucan has also performed well in laboratory tests (DAVISON and MENTZER, 1980). Although other microbial polysaccharides including some of those structurally similar to gellan have shown higher thermostability than xanthan, they have lower injectability or are less compatible with high salt concentrations. The inclusion of biocides in any polysaccharide formulation is necessary to ensure that biological degradation of the xanthan does not occur either at the surface or underground after injection. In the latter case, the relatively shallow, low temperature reservoirs prevailing in the U.S. and other areas present this specific problem. In the deep, high temperature U.K. and Norwegian offshore oil reservoirs, the major difficulty is one of temperatures in excess of 80 °C in which chemical degradation of polysaccharides presents a greater problem, although care would still have to be taken to prevent degradation prior to or during injection as well as in areas of the reservoir near the injection site, where the temperature may be lower.

Selective pore blockage (profile modification) provides another application for xanthan in the oil industry. The process involves the injection of xanthan solutions into the reservoir rocks, where they are placed selectively in zones of high permeability and then cross-linked with $Cr^{3+}$. Once in place, the gels remain stable for prolonged periods. If there is a later need to reopen the structure and restore permeability, the xanthan gel can be destroyed in a controlled manner with oxidizing agents such as sodium hypochlorite or with acids. The use of thermostable xanthanases has also been proposed. The gelled xanthan is compatible with a wide range of salinities, over a pH range from 5–8 and at temperatures up to or even in excess of 65 °C. The temperature at which the gels remain stable will depend on the salinity of the environment in which they are placed. The gels will break and re-form when they are sheared. In this role as a gellant, the solubility of the polysaccharide in freshwater is exceedingly important, allowing the polymer to be selectively placed initially and the rate of gellation to be carefully controlled. Once the gel has been formed, connate water or seawater can be used. In the process of selective pore blockage, the aim is usually not to block the structure entirely, but to retain some residual permeability. It has been estimated that one well may require up to 1,360 tonnes of polymer. It was originally thought that the demand for xanthan from the oil industry would reach 60,000 tonnes per year by the mid 1990s, but the relatively low prevailing prices for oil have ensured that this forecast was overoptimistic.

# 6 Acceptability of Microbial Polysaccharides

Microbial polysaccharides have clearly to be acceptable to the potential user both in terms of their physical and chemical properties and their cost and quality relative to competing polymers. The producer and the user have to be certain that the polysaccharide conforms to the desired specifications and that it does not vary in its characteristics from batch to batch. A number of these microbial polymers have been very successfully produced and marketed and will no doubt continue to maintain their market share as well as finding new applications despite their relatively high cost when compared to synthetic polymers and to some of the traditional plant gums (see Tab. 2). The polysaccharides present no potential danger to the producer or user as none are produced from pathogenic species and steps are taken to ensure that live microorganisms are not likely to be present. However, some of the products may contain other microbial cell material and continued exposure of operatives could perhaps lead in a very few cases to sensitization. The onus of ensuring that the product is safe is primarily on the supplier and adequate testing is needed to ensure that the polysaccharide is non-irritant and lacks sensitizing activity under standard test conditions. The user must also ensure that the polymer is safe in the application proposed. In the case of potential food additives, it must also be shown to be free of adverse effects when fed to animals over several generations. An *acceptable daily intake* for xanthan has been issued by the Joint AF/WHO Expert Committee on Food Additives but that for gellan is *"not specified"*. Xanthan was tested to show the absence of irritant or sensitizing activity in standard tests and was fed to animals for prolonged periods over several generations (ROBBINS et al., 1964; WOODARD et al., 1973). Studies on the dietary effects of both gellan and xanthan in human volunteers have shown the absence of adverse effects and the lack of any change in enzymatic and other toxicological indicators (EASTWOOD et al., 1987; ANDERSON et al., 1988).

Government regulations normally apply restrictions to the use of non-indigenous plant pathogens. Fortunately, *X. campestris* pv. *campestris* from which xanthan is produced, is of very widespread occurrence and does not fall within such regulations. The use of the microbial polysaccharides as food additives brings them within the scope of food additive regulations. Two different approaches are employed (OVEREEM, 1979). Legislation may cover most food additives through the use of a single decree specifying additives and foodstuff, or separate regulations may cover different foodstuffs and additives. The Scandinavian countries and Belgium classify the food and the additives which are permitted, whereas Britain and the Netherlands use the second approach. There are distinct regulations for the different types of additives and also lists of each type of permitted additive. In the U.K., the schedule of "Emulsifiers and Stabilizers in Food Regulations" is promulgated by the Ministry of Agriculture, Fisheries and Food. This list includes alginic acid, plant gums such as guar, agar, and other emulsifiers. In the U.S., xanthan is permitted as a food ingredient under food additive regulations controlled by the U.S. Food and Drug Administration. The polysaccharide is on the list of compounds "Generally Regarded as Safe" (GRAS List), with approval for use as a stabilizer, emulsifier, foam enhancer, thickener, suspending and bodying agent. The EU has attempted to harmonize food additive regulations within its member countries through a "Directive of the Council of Ministers on Emulsifiers, Stabilizers, Thickeners and Gelling Agents for Food Use". This provides a list of the agents which are fully accepted for use in member states. These compounds are designated in Annex 1 of the EEC Directive 80/597/EEC with an appropriate serial number (E400 = alginic acid, E406 = agar, E415 = xanthan, etc.). Further details include the foodstuffs in which the use of the emulsifiers is permitted, together with criteria for purity and labeling. The attempted standardization of labeling procedures provides details of trade name, manufacture, designated number, etc. In practice many processed foods are clearly labeled with their contents and consumers can see for themselves if xanthan is one of these.

Most of the polysaccharides currently available are marketed as dry powders, following solvent precipitation, drying, and grinding. These may prove relatively difficult to redissolve and as a result formulations and specialized mechanical procedures have been developed to improve their water solubility and ease of resolution. If the polysaccharides are to be employed as food additives or to be included in pharmaceutical preparations, they must be of adequate purity and quality to conform to the appropriate government or other legislative regulations. Polymers which are to be incorporated into foodstuffs are subject to food additive regulations. Thus when employed as a food additive, xanthan must conform to the definition which is provided in the National Formulary of "... a high molecular weight polysaccharide gum produced by a pure culture fermentation of a carbohydrate with *X. campestris* ...". The food grade polysaccharide must also meet the specifications listed in the Food Chemicals Codex. These include freedom from *E. coli* and *Salmonella* spp., and limits on the content of arsenic (<3 ppm), heavy metals (<30 ppm) and isopropanol (<750 ppm).

Among the factors determining polysaccharide choice for food uses, the type of application will clearly be the most important one, i.e., is the polysaccharide required for its viscosity or gelation or for other purposes? However, other important qualities are mouthfeel, flavor release, appearance (including clarity, etc.), emulsification capabilities, and compatibility with other food ingredients. There may be synergistic or other effects as the polysaccharide will almost certainly interact with food ingredients and other food additives. Polysaccharides such as xanthan and gellan are usually incorporated into foodstuffs in mixtures with other plant or algal gums in order that a wide range of textural and other properties can be obtained. The microbial polysaccharide must be stable to physical and biological factors under the conditions of intended usage. If all these demands are satisfied, then final choice will depend on cost, acceptability (legal, consumer and otherwise) and quality – odorless, tasteless, consistency of quality, etc.

As an alternative to solid preparations, concentrates of xanthan and succinoglycan with 9–12% dry weight of polymer have been marketed for non-food use. These can relatively easily be diluted to yield the desired concentration, but require formulation including biocide to prevent degradation. In the case of xanthan, use of the liquid concentrate avoids or reduces the problem caused by *microgels*, small aggregates of poorly hydrated polysaccharide. These present problems in uses where the polysaccharide will be injected into oil reservoirs with low porosity of reservoir rocks.

As has been pointed out by AKSTINAT (1980), if a polymer is to be employed in *enhanced oil recovery* (EOR), the polymer solution must be readily filterable. Poor filterability would lead to partial or complete plugging of rock pores in the reservoir, with associated reduction in injection rate and fluid sweep. Thus, the concentrates mentioned above have been aimed specifically for oil industry usage, where the ease of formulation and filterability make up for the increased transport costs of the bulk fluids. Poor filterability may be due to the high molecular weight of the polysaccharide but is more commonly caused by microbial debris, or by interchain and intrachain associations leading to the formation of microgels. Strains or physiological conditions which will yield reduced bacterial cell size have been developed in order to diminish the first of these problems, but some of the problems are less easily solved. The use of bacteriolytic and other enzymes has also been proposed for the removal of cell debris.

Xanthan and other polysaccharides tend to form a gelatinous layer of partially hydrated polymer at the outside of the particle, when they are dispersed and rehydrated. This is unsatisfactory for many commercial applications and various formulations have been proposed to improve solubility and dispersability. The most widely used involves treatment with glyoxal (SANDFORD et al., 1981). Although this provides an answer at pH values of 7.0 or below, an alternative borate treatment was suggested for ensuring dispersion of xanthan gum in the alkaline pH range.

# 7 Commercialized Microbial Polysaccharides and Potential New Products

## 7.1 Bacterial Alginates

The alginates produced commercially are currently isolated from marine algae such as *Laminaria* and *Macrocystis* species. Bacterial alginates, linear polymers of irregular structure also composed of D-mannuronic acid and L-guluronic acid, have been proposed as alternatives to the algal products. Their synthesis differs from the biosynthetic mechanism found in the production of most bacterial polysaccharides. The primary product is thought to be an acetylated homopolymer, poly-D-mannuronic acid. This is modified in a post-polymerization reaction by the extracellular enzyme *polymannuronic acid epimerase* which converts some of the non-acetylated D-mannuronosyl residues to L-guluronic acid.

Structural investigations on alginates from the range of bacterial species now known to synthesize this type of polysaccharide have revealed a range of different polymer types – all with similar chemical composition. The bacterial alginates are all composed of the same two uronic acids as algal alginate, but in addition many of them are highly acetylated. The acetyl groups are present solely on D-mannuronosyl residues in the polymers. As can be seen from the details presented in Fig. 1 and Tab. 5, bacterial alginates are a family of exopolysaccharides which, despite their similar *composition*, vary considerably in their *structure*. The material from *Azotobacter vinelandii* bears the closest resemblance to algal alginate, although unlike the latter, it

*Azotobacter vinelandii*

→ [ 4 α L- Gul*p*A(1 → 4)β D- Man*p*A(1 → 4)β D- Man*p*A(1 → 4)β D- Man*p*A(1 → 4)α L- Gul*p*A(1 → 4)α L- Gul*p*A(1 → 4)α L- Gul*p*A 1 ] →
↑ (on the first D-Man*p*A)
$CH_3CO.O$

*Pseudomonas fluorescens* or *P. putida*

→ [ 4 β D- Man*p*A(1 → 4)β D- Man*p*A(1 → 4)β D- Man*p*A(1 → 4)β D- Man*p*A(1 → 4)β D- Man*p*A(1 → 4)β D- Man*p*A(1 → 4)α L- Gul*p*A 1 ] →
↑ ↑ ↑
$CH_3CO.O$ $CH_3CO.O$ $CH_3CO.O$

**Fig. 1.** The structures of typical bacterial alginates. Note that in bacterial alginates there is no regular structure and that *only* D-mannuronosyl residues carry *O*-acetyl groups; some are multiply acetylated. In *Pseudomonas* polymers there are only *single* L-guluronosyl residues, whereas *Azotobacter* has block structures of either type.

**Tab. 5.** Composition and Diad Frequencies of Some Bacterial Alginates Compared to Algal Material

| Source | $F_G$ [a] | $F_M$ [b] | $F_{GG}$ | $F_{MM}$ | $F_{GM,MG}$ | Acetyl |
|---|---|---|---|---|---|---|
| *L. hyperborea* | 0.665 | 0.335 | 0.558 | 0.228 | 0.107 | 0 |
| *M. pyrifera* | 0.41 | 0.59 | 0.24 | 0.42 | 0.17 | 0 |
| *A. vinelandii* 73 | 0.561 | 0.439 | 0.372 | 0.25 | 0.189 | 11 % |
| *A. vinelandii* 206 | 0.08 | 0.92 | 0.03 | 0.87 | 0.05 | 24 % |
| *P. aeruginosa* B | 0.16 | 0.84 | 0 | 0.68 | 0.16 | 37 % |
| *P. fluorescens* | 0.31 | 0.69 | 0 | 0.38 | – | 16.9% |
| *P. putida* | 0.22 | 0.78 | 0 | 0.56 | – | 21 % |

[a] G: L-guluronic acid
[b] M: D-mannuronic acid

→ [ 4 β D- ManpA(1 → 4)β D- ManpA(1 → 4)β D- ManpA(1 → 4)β D- ManpA(1 → 4)β D- ManpA(1 → 4)β D- ManpA 1 ] →

→ [4 α L- GulpA(1 → 4)α L- GulpA(1 → 4)α L- GulpA(1 → 4)α L- GulpA(1 → 4)α L- GulpA(1 → 4)α L- GulpA(1 → 4)α L- GulpA 1 ] →

→ [ 4 β D- ManpA(1 → 4)α L- GulpA(1 → 4)β D- ManpA(1 → 4)α L- GulpA(1 → 4)β D- ManpA(1 → 4)α L- GulpA(1 → 4)β D- ManpA 1 ] →

**Fig. 2** The "block" structures of typical algal alginates which define their physical properties.

does contain *O*-acetyl groups. Both *A. vinelandii* and algal alginates are composed of three types of structure: poly-(1→4)-*β*-D-mannuronic acid sequences, poly-(1→4)-*α*-L-guluronic acid sequences and mixed sequences (Fig. 2). These are sometimes termed "block structures". *Azotobacter chroococcum* synthesizes two polysaccharides, one of which is an acetylated alginate with high D-mannuronic acid content. Most of the alginates produced by *Pseudomonas* species also have a relatively high content of D-mannuronic acid but vary considerably in the degree of acetylation. The interesting feature of the *Pseudomonas* products is that, unlike the algal or *Azotobacter* alginates, there are no contiguous sequences of L-guluronic acid residues. Whereas the properties of *A. vinelandii* polysaccharide can therefore be expected to have much in common with the commercial algal products, the other bacterial alginates possess different and perhaps unique properties. Alginates from the plant pathogenic *Pseudomonas* species are generally of low molecular weight, relatively low acetyl content and high polydispersity and are unlikely to be exploited. In the highly acetylated polysaccharides from either *A. vinelandii* or *Pseudomonas aeruginosa*, some D-mannuronosyl residues may carry *O*-acetyl groups on the $C_2$ *and* $C_3$ positions. Between 3 and 11% of the mannuronosyl residues in such polysaccharides have been found to carry acetylation on both carbon atoms. As mentioned above alginate producing bacterial species all appear to possess alginate lyases. This can lead to considerable reduction in the molecular mass of the final product (CONTI et al., 1994) and possibly reduce its potential value. A possible solution to the problem was proposed by HACKING et al. (1983) who added proteases to the culture medium to destroy the alginase activity, with resultant high product viscosity.

The presence of acetyl groups in the bacterial alginates modifies the physical properties of the polysaccharides. The capacity for $Ca^{2+}$ binding and the selectivity for this ion are greatly reduced (GEDDIE and SUTHERLAND, 1994). In chemically acetylated alginates, the elastic modulus of the acetylated alginate in the gel form decreases as a function of the degree of acylation. Dry gel beads show greatly increased swelling after acetylation; the acyl substituents appear to reduce the number of junction zones and their strength, thus permitting the polymer network to swell greatly when it is rehydrated. The potential for bacterial alginates may reside in specialized applications for polymers of D-mannuronic acid, totally devoid of guluronic acid residues, or for polymers in which the percentage of guluronic acid has been greatly increased.

## 7.2 Cellulose

Cellulose is synthesized by a number of *Acetobacter* spp., selected strains of which have been used for industrial production of this material. In contrast to its role in the wall of plants, cellulose is produced as an exopolysaccharide by *Acetobacter xylinum* and possibly other, mainly gram-negative bacterial species. In cultures of these bacteria, it is excreted into the medium where it rapidly aggregates as microfibrils yielding a surface pellicle in shallow unshaken cultures. Alternatively, the bacteria can be grown in submerged culture although the design of the fermenter and the degree of aeration are important factors in optimizing cellulose yield. Low cost substrates such as molasses with a content of 55–60% sucrose, or corn steep liquor are preferable to glucose; the latter is converted to gluconic acid with consequent rapid fall in pH. Product yields of up to 28 g $L^{-1}$ dry polysaccharide have been reported.

Bacterial cellulose is essentially a speciality chemical with specific applications and usage. Some cellulose from bacteria is produced commercially as a source of highly pure polymer in the so-called cellulose I form, free from lignin and other related non-cellulosic material. Production may either be in the form of a pellicle from static cultures or from submerged, fed-batch culture in stirred tank reactors (BYROM, 1991). Low shear conditions are usually preferred. For such purposes, high yielding strains must be selected capable of at least 20% conversion of substrate to polysaccharide. The product has a number of potential uses including the manufacture of wound dressings for patients with burns, chronic skin ulcers, or other extensive loss of tissue. The cellulose acts as a temporary skin substitute and the high water capacity of the oxygen permeable film appears to stimulate regrowth of the skin tissue while at the same time assisting in preventing infection (JORIS and VANDAMME, 1993). Bacterial cellulose also forms the basis for high quality acoustic diaphragm membranes in which the distribution of the fibrils containing parallel orientation of the glucan chains results in fibers possessing high tensile strength. Other applications use bacterial cellulose as binders for ceramic powders and minerals and thickeners for adhesives (BYROM, 1991; CANNON and ANDERSON, 1991). Less pure material can be used for binding and coating purposes in the mining industry, but the cost remains relatively high in comparison with other, competing polymers. A further area for potential development lies in the applications for cellulose-synthetic copolymers.

For the preparation of material for use in wound dressings, the *Acetobacter* sp. was grown at 28 °C in an unshaken, complex medium supplemented with yeast extract. The product resembles hyaluronic acid (q.v.) in its very high water retention capacity (ca. 150 g $g^{-1}$). The disadvantage of this technique is the presence within the pellicle of the bacterial cell material. Growth in shaken, aerated cultures with glucose as carbon substrate, provides an alternative means of obtaining the polymer.

## 7.3 Curdlan

A number of bacterial strains, including *Agrobacterium* and *Rhizobium* species, each produce several exopolysaccharides. One of these is curdlan, while the other is succinoglycan (q.v.). Following examination of the relative yields of each polymer, strains effectively producing curdlan only were selected for the production of the neutral gel-forming 1,3-β-D-glucan. The strain was originally designated *Alcaligenes faecalis* var. *myxogenes* 10C3, but is probably an *Agrobacterium* sp. The polysaccharide produced by these bacteria is of relatively low molecular weight (~74,000) and is insoluble in cold water, but can be dissolved in hot water or in dimethyl sulphoxide (Fig. 3). Only low molecular weight material with DP below the range 30–45 is soluble in cold water. Media for curdlan production in high yield contain 4% glucose, together with citrate, succinate or fumarate, ammonium phosphate, and mineral salts (HARADA et al., 1966). Media with glucose concentrations up to 10% have also been used (HARADA et al., 1993). Production of curdlan differs from that of most other bacterial exopolysaccharides, both in terms of the physiological requirements and the physical characteristics of the culture broth. Whereas most bacterial exopolysaccharides are synthesized during growth, curdlan is formed in the stationary phase following depletion of nitrogen. Attempts to promote curdlan production by using limita-

Curdlan

→ [ 3 β D - Glc*p* (1 → 3)β D - Glc*p* (1 → 3)β D - Glc*p* 1 ] →

Scleroglucan

→ [ 3 β D - Glc*p* (1 → 3)β D - Glc*p* (1 → 3)β D - Glc*p* 1 ] →
1↑6
Glc*p*

**Fig. 3.** The structure of the homopolysaccharides curdlan and scleroglucan.

tion of other nutrients have not proved successful. As curdlan is insoluble in water, there is no great increase in culture viscosity as the polysaccharide is synthesized. It does however form a layer surrounding the bacteria and prevent free access of oxygen to the microbial cells. Culture, therefore, demands high oxygen transfer which can be achieved by the use of low shear, axial flow impellers rather than radial flow flat bed impellers (LAWFORD and ROUSSEAU, 1991, 1992). Another result of changing the fermenter design in this way, is that the polymer produced is of higher quality.

Curdlan forms a weak "low-set" gel on heating above 55 °C followed by cooling. Further heating to 80–100 °C increases the gel strength and produces a firm, resilient "high-set" gel, while autoclaving at 120 °C converts the molecular structure to a triple helix. The gels are intermediate in properties between the highly elastic gel of gelatine and the brittleness of agar gels. Those formed at higher temperature do not melt below 140 °C, but are very susceptible to shrinkage and syneresis but resistant to degradation by most $\beta$-D-glucanases. "Low-set" gels are also formed when alkaline solutions of curdlan are dialyzed against distilled water or are neutralized. Curdlan gels are stable over a wide range of pH values (3–10) and are formed even in the presence of sugars. The solubility of curdlan in alkali and its property of yielding gels on acidification has been proposed as a method for *in situ* gelation in oilfield development (VOSSOUGHL and BULLER, 1991).

In Japan, where use of curdlan in food is permitted following extensive safety testing, food grade curdlan is produced commercially for use in improving the texture of various foods such as bean curd (tofu), bean jelly, and fish pastes. It can also be used for water retention and for reconstitution and shaping of processed foods. In such processes, both "low-set" and "high-set" curdlan gels may be used. Possible non-food uses for curdlan include water purification, enzyme immobilization, and as a binder for reconstituted tobacco. Acetyl curdlan derivatives for the separation of monosaccharides and organic acids have been patented and enzyme digests of curdlan can be used as a source of the disaccharide laminaribiose. Current production of curdlan is estimated at 200 tonnes per year (HARADA et al., 1993).

## 7.4 Cyclodextrins

Strictly speaking, cyclodextrins (CDs) are oligosaccharides but they are valuable carbohydrate products obtained commercially using enzymes obtained from various *Bacillus* spp. and providing the raw material for a wide range of chemically modified products – the chemically modified cyclodextrins (CMCDs). CDs consist of 6–8 1,4-$\alpha$-linked glucopyranose residues, designated $\alpha$-, $\beta$- and $\gamma$-cyclodextrin, respectively. Each is capable of forming a stable ring structure in which the interior is apolar and hydrophobic, while the exterior of the molecule possesses sites available for hydrophilic interactions. Improved production of $\gamma$-cyclodextrins has recently been achieved using cloned enzymes from *Bacillus* spp. which excrete the enzymes poorly (SCHMID, 1991). Some of the bacteria which also synthesize curdlan and succinoglycan can be used to form cyclic $(1\rightarrow2)$-$\beta$-D-glucans but these do not yet appear to have been developed commercially.

The chemical conformation in which the central cavity of the structure has non-polar, electron-rich characteristics, determines the contrasting hydrophobic interior and hydrophilic exterior of the molecule. Chemical modifications at the 2, 3, and 6 hydroxyl groups of the glucopyranose residues greatly increase the aqueous solubility. Depending on the substituents introduced, the molecules can be made more hydrophilic or lipophilic. Most CMCD preparations are highly soluble in water (50% w/v or more) whereas the native molecules are much less soluble (1–15% w/v) (Tab. 6). Varying degrees of substitution can be achieved and the size and the distribution of the substituents contribute significantly to complex formation. Commercially available CMCDs include hydroxyethyl, hydroxypropyl, methyl, and sulphated derivatives.

The ability of CDs and CMCDs to interact with an extensive range of natural and synthetic chemicals provides a wide variety of potential uses. Commercial products contain-

**Tab. 6.** Properties of Natural Cyclodextrins

| | Alpha ($\alpha$) | Beta ($\beta$) | Gamma ($\gamma$) |
|---|---|---|---|
| Number of D-glucose | 6 | 7 | 8 |
| Molecular mass | 972 | 1135 | 1297 |
| Solubility (g 100 mL$^{-1}$) | 14.5 | 1.85 | 23.2 |
| Water mols in cavity | 6 | 11 | 17 |
| Cavity diameter (nm) | 0.47–0.53 | 0.6–0.65 | 0.75–0.83 |

ing CDs are acceptable in a number of countries for oral administration, as the CDs are regarded as enzymically modified starches. The major potential applications for the native and modified products are currently in drug delivery systems, including pharmaceutically important proteins and peptides, and diagnostic products (SZEJTLI, 1986). They also have potential applications in the development of cosmetics (AMANN and DRESSNANDT, 1993).

## 7.5 Dextrans

Almost all microbial exopolysaccharides are the products of intracellular catabolic conversion of substrates to intermediates which are then anabolized to form specific precursors such as sugar nucleotides and finally to polymer. Dextrans differ from these, the substrate does not enter the bacterial cell and it is converted extracellularly into the branched $\alpha$-D-glucan, dextran. Only sucrose is utilized as substrate for this reaction. Dextran can thus be produced either from whole cells in culture in a process similar to that used for other exopolysaccharides, or it can be derived from cell-free preparations of the enzyme complex *dextransucrase* ($\alpha$-1,6-glucan:D-fructose 2-glucosyltransferase, EC 2.4.1.5). The enzyme glycoprotein releases fructose from sucrose and transfers the glucose residue onto the reducing end of the growing dextran chain on an acceptor molecule, which is also bound to the enzyme (ROBYT and WALSETH, 1978, 1979):

$$(1,6\text{-}\alpha\text{-D-Glucosyl})_n + \text{Sucrose} \rightarrow (1,6\text{-}\alpha\text{-D-Glucosyl})_{n+1} + \text{Fructose}$$

The free energy of the glycosidic bond in the disaccharide is about 23 kJ, while that of the glucosidic bond in dextran is somewhat lower (12–17 kJ). Thus the reaction proceeds from left to right, accompanied by a drop in free energy. During polymerization, the growing dextran chain remains firmly bound to the enzyme, the degree of polymerization increasing until an acceptor molecule releases the polymer chain from the enzyme. The lowest members of the oligomeric saccharides are not very effective as acceptor molecules; the affinity to the enzyme increases with chain length. However, the diffusional limitations also grow with increasing molecular weight. Hence, under most experimental conditions, dextrans of very high molecular weight are formed, especially at low substrate concentrations. For the production of dextrans of medium molecular weight, the initial high molecular weight product must be hydrolyzed with acid or with enzymes and fractionated under carefully controlled conditions. Alternatively, polysaccharides of low and homogeneous molecular weight distribution may be taken as the starting acceptor material for a further reaction to produce dextrans of the desired degree of polymerization. High sucrose concentrations and possibly the use of maltose as starter, create conditions favorable to the formation of such low molecular weight, homogeneous primer dextrans. In the presence of 1 mg L$^{-1}$ low molecular weight dextran, yields of 33% clinical dextrans have been recorded but the presence of high molecular weight product as a contaminant makes purification difficult and few manufacturers appear to follow this type of procedure. Dextransucrase is stable over the pH range 5–6.5 and in whole cell cultures the enzyme activity is dependent on the presence of $Ca^{2+}$.

Most dextrans are exopolysaccharides composed predominantly of $(1\rightarrow6)\alpha$-linked D-glucosyl residues. In some dextran preparations there may be almost no other type of linkage. Alternatively, up to 50% of the glucose residues may be linked 1,2 or 1,3 or 1,4. Industrial dextran production outside Eastern Europe is mainly by batch fermentation from a strain of *Leuconostoc mesenteroides* designated NRRL B–512(F), which yields a polysaccharide with about 95% 1,6 linkages and 5% 1,3 linkages and has a molecular mass of about $4–5\times10^7$ kDa. For many purposes, the molecular weight is reduced by mild acid hydrolysis to yield polymer of the desired molecular size. In the resultant products, the degree of branching may fall to between 3 and 3.8%.

Dextran production demands good bacterial growth and optimal enzyme production. While cell growth requires complex media containing amino acids and growth factors and is optimal at 30°C, enzyme production is best at 23°C. If culture incubation is prolonged beyond 24 h in some strains including that normally used for commercial manufacture, the product may decrease in molecular mass. In technical production of dextrans, the maintenance of an adequate dissolved oxygen concentration poses potential problems, as the solution becomes highly viscous during fermentation. It is not an easy matter to calculate the relationship between the introduced mixing energy and the oxygen transfer rate in dextran solutions with non-Newtonian fluid behavior and in most cases the process engineer relies on experience (REUSS, 1980). The yield of dextran from sucrose depends on the substrate concentration and the exact conditions employed, but can be up to 42–46%.

The fermentation broth contains inorganic salts, corn steep liquor (2%), and up to 10% sucrose. The presence of the trace elements manganese and iron is extremely important for satisfactory production of dextran, while production of extracellular dextransucrase is optimal in the presence of 0.5% calcium chloride. The medium in solution is heat sterilized in a flow process, transferred to the fermentation vessel and after cooling, is inoculated from a seed culture. The fermentation size is gradually increased in several stages so that vessels up to 6,000 L capacity can be used with an inoculum of ~10%. The fermentation is carried out at 25°C, during which the pH value falls from 6.5–7.0 to 4.0, mainly due to the formation of lactic acid; the pH is adjusted by the use of alkali and further sucrose is added. The contents of the fermentation vessel are continuously stirred and aerated. As dextran is produced, the culture broth gradually increases in viscosity and the final product when fermentation is complete after 3–4 d is a highly gelatinous polymer. The levels of the enzyme dextransucrase and of polysaccharide can be increased through the use of fed-batch fermentation. Also present in the final fermentation broth are free fructose, lactic and acetic acids, ethanol and bacterial cells. The dextrans are transferred to a stirred tank and precipitated by the addition of either ethanol, acetone, or methanol. The supernatant fluid can be decanted and the alcohol or acetone recovered by distillation. Although LAWFORD et al. (1979) suggested the use of continuous culture for dextran production, such procedures have not been adopted. Dextran production differs from most other processes for exopolysaccharide production in that saleable by-products are obtained. These represent 75% of the original substrate and they include fructose, mannitol, and low molecular weight dextrans. The sugars can be used in the food industry, while the dextran is used for technical purposes.

High molecular weight dextran products are hydrolyzed with hydrochloric acid at 100–105°C; the progress of depolymerization can be followed through the reduction in viscosity. Cooling is applied to reduce the rate of hydrolysis and the reaction is terminated by the addition of sodium hydroxide. Finally, the solution is mixed with adsorbents and filtered over kieselguhr. The dextrans can then be fractionally precipitated with ethanol or methanol under careful temperature control and the recovered fractions redissolved, filtered, concentrated, and spray dried or lyophilized. Hydrolysis is difficult to control and fractionation often fails to yield polymer of the exact molecular mass required. Consequently, there have been a number of attempts to obtain direct synthesis of low molecular weight dextrans.

Dextrans of low molecular weight in the range 75,000 ± 25,000 are used for pharmaceutical purposes as blood expanders (plasma substitutes). Products with lower molecular weights are eliminated too rapidly from the circulation to be of therapeutic value, while products of too high molecular weight interfere with the coagulation of the blood. Solutions containing 6% of these dextrans (Dextran 70) have viscosity values and colloid-osmotic behavior which are very similar to those of human blood plasma. Since dextrans are not broken down by amylases in the human body, the possibility of liver damage occurring is lower than with other plasma substitutes. Dextrans are resorbed very efficiently by the body tissue and may thus be used in the application of many substances, e.g., iron, during treatment of anaemia in humans and in animals. For this purpose, a complex containing 5% iron and 20% dextran is prepared. Dextran with a molecular weight of 40,000 (Dextran 40) has also been employed as an antithrombitic agent in prophylactic treatments in high risk surgery (DE BELDER, 1990, 1993). Another feature of aqueous dextran solutions is the ability to form two-phase systems when combined with solutions of polymers such as polyethylene glycol (WALTER et al., 1985). These can be used in the purification of various biological macromolecules and also particulate material. The procedure can be rendered more selective through the chemical modification of the dextran to introduce affinity ligands (YALPANI and BROOKS, 1987) thus providing a further demand for dextrans.

A major technical use of dextran is in the form of the chemically cross-linked product Sephadex®. This product in the form of gels in a bead form, is widely used for the separation of proteins including the industrial fractionation of plasma proteins such as albumin, blood factors, and immunoglobulins. Subsequently, Sephadex-based anion- and cation-exchange materials were developed for the ion exchange separation of biological macromolecules. This was achieved by the attachment of functional groups by ether linkages to the glucose residues of dextran. In such products, dextran of high molecular weight is employed to ensure that high water regain is obtained. The reaction with dextran in alkaline solution uses epichlorhydrin to yield a gel, which is then ground and neutralized. Cooling has to be employed during the exothermic reaction. As the terminal reducing residues of the dextran molecules are highly reactive, and are subject to degradation in the polymerization reaction, they are reduced to sorbitol with sodium borohydride in the alkaline solution prior to addition of the epichlorhydrin. Anion and cation exchangers (DEAE – diethyl aminoethyl; CM – carboxymethyl, respectively) synthesized from cross-linked dextran, are available commercially. The ion-exchange products differ from cross-linked dextran *per se*, in their swelling behavior in aqueous solution. Whereas the initial material does not differ significantly in its swelling behavior in the presence or absence of ions, the ethers shrink at high ionic concentrations. Products with varying degrees of cross-linking provide adsorbents with differing porosities and fractionation range. The introduction of other chemical groups provides hydroxypropyl derivatives which have been developed for hydrophobic chromatography. These derivatives swell in organic solvents and are suitable for the separation of lipids.

Another derivative of dextran is dextran sulphate, prepared by the esterification of the neutral polysaccharide of molecular weight $5 \times 10^5$ with chlorosulphonic acid. The product is a polyelectrolyte with ca. 17% sulphur, i.e., an average of 2.3 sulphate groups on each glucose residue. It has various applications and has been the subject of a wide range of research studies. One application for which the dextran sulphate product can be used, is as a potential substitute for heparin in anticoagulant therapy. This and other dextran derivatives are relatively high value products, which require careful processing in their synthesis. The basic cost of such products partly reflects the cost of dextran production, but much of the additional cost represents the later, chemical processes needed for their production.

Although dextrans have been produced commercially for a much longer period than other microbial polysaccharides and their synthesis is a relatively simple procedure, they only comprise a small portion of the total polysaccharide market. The uses for dextrans, predominantly in the form of chemical derivatives, tend to be highly specialized and the market, currently estimated to be over 500 tonnes, appears unlikely to expand greatly.

## 7.6 Gellan and Related Polysaccharides

The discovery of *gellan*, a polysaccharide prepared from a bacterium initially identified as *Auromonas elodea*, was later followed by the observation that it was one of a series of 7 or 8 structurally related polysaccharides synthesized by a group of bacteria belonging to genera similar to *Pseudomonas*. A recent reassessment of the taxonomic status of all the strains involved in the synthesis of this group of polysaccharides, has shown that although they were initially allocated to various bacterial genera, they conform to a type named as *Sphingomonas paucimobilis* (POLLOCK, 1993). The polysaccharide is produced in aerobic submerged fermentation in medium with glucose as carbon source and ammonium nitrate and hydrolyzed soy protein as nitrogen sources. Potassium phosphate is used to buffer the medium, which also includes trace elements. Because of the viscosity which develops in the medium and the texture of the polysaccharide produced, there is a significant fall in the mass transfer coefficient during the fermentation.

In its native form, gellan carries both *O*-acetyl and glyceryl substituents on a linear polymer of 500 kDa composed of tetrasaccharide repeat units. Other polysaccharides (termed welan and rhamsan) with the same main-chain sequence were later obtained from other bacterial isolates. These carried differing side chains – a rhamnose-containing or a glucose-containing (gentibiosyl) disaccharide or, in one polysaccharide *either* L-rhamnose *or* L-mannose. This latter polysaccharide is thus highly unusual in containing L-mannose and a variable side chain. The ratio of L-rhamnose:L-mannose is approximately 2:1 and the L-mannosyl and L-rhamnosyl terminal groups are randomly distributed (JANSSON and WIDMALM, 1994). Two further polymers were even more unusual in that in both, one of the *main-chain* sugars varied. It could again be either L-mannose or L-rhamnose. One of the exopolysaccharides (S88) also contained 5% acetate. The structures proposed for the members of this series of polysaccharides are shown in Fig. 4.

Gellan is produced in batch fermentation under controlled aeration, pH, and temperature. Pasteurization is carried out prior to subsequent processing of the polysaccharide. The polymer recovery may be performed directly to yield a high acyl form or after deacylation and clarification. The deacylated gellan forms gels with properties common in some respects to those produced with agar, alginate, or carrageenan. In texture, they most resemble the gels formed with agar or κ-carrageenan. In common with agar gels, marked melting/setting hysteresis is observed. The native structure of the gellan polymer contains *O*-glycerate and ~6% acetate. As the glyceryl and acetyl groups in the native structure inhibit crystallization of localized regions of the gellan chains (CARROLL et al., 1983), weak, elastic thermoreversible gels are formed. When the polysaccharide is deacylated, extensive intermolecular association occurs and strong brittle gels are produced. A range of gel textures can be produced through control of the degree of acylation of the polymer. The difference between melting and setting temperatures is 45–60 °C (KANG and VEEDER, 1982). The deacylated polysaccharide interacts with various cations to form gels, the gelation being dependent on the identity of the cation and on the ionic strength. There is a lack of specificity among the alkaline earth cations and the temperatures at the midpoint of the sigmoidal transitions ($T_m$) increase only moderately with ionic strength (CRESCENZI et al., 1987). Gelling and melting temperatures increased when either the gellan or the salt concentration was increased within the concentration range 0.3–2.0%. It has been suggested that the *O*-acetyl groups on the native polysaccharide have only a weak effect on the aggregation of the gellan molecules, whereas the L-glyceryl residues are detrimental to crystal packing (CHANDRASEKARAN and THAILAMBAL, 1990). The potassium salt

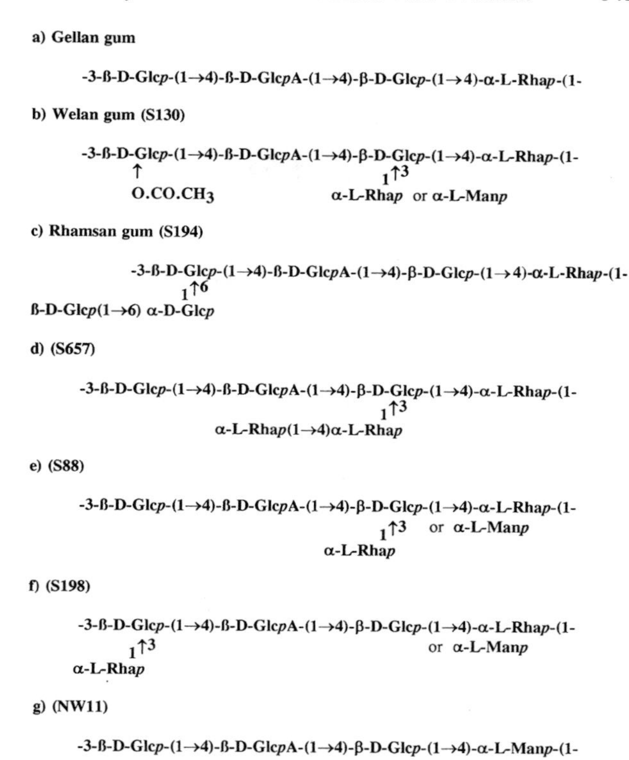


**Fig. 4.** The structures of gellan and related polysaccharides.

possesses a crystalline structure in which the L-glyceryl groups prevent the coordinated interactions of ions and carboxylate groups which are required for strong gelation (CHANDRASEKARAN et al., 1992). Once deacylated, the coordination of calcium between double helices is so strong that molecular aggregation occurs at very low cation concentrations.

As already mentioned, a series of microbial polysaccharides varying in the closeness of carbohydrate structure to gellan have also been discovered. Several of these carry carbohydrate side chains consisting of one or more sugar residues. However, none of these polysaccharides are capable of gelation, even when the acyl groups carried by several of them are removed. The polymer welan carrying $\alpha$-L-rhamnosyl or L-mannosyl side chains (S130) differs from gellan in its rheological properties. Solutions are highly viscous at low shear rates, with good thermal stability. A similar structure (rhamsan) in which disaccharide (gentibiosyl) side chains are present (S194), resembles S130 (Fig. 4) rather than gellan in its characteristics. TALASHEK and BRANT (1987) have indicated that the attractive interactions between side chains and backbone structures of this series of polysaccharides only slightly influence conformational characteristics of the backbone. The calculated unperturbed conformation of rhamsan in solution is slightly more extended than the other two polymers; welan is very similar to gellan. Consequently, the observed differences in physical properties are likely to derive from hydrogen bonding and interchain interactions. The failure of other polysaccharides in the series to gel can be ascribed to the presence of the various side chains which shield the carboxylate groups to a varying degree; shielding is greatest in welan gum and much lower for rhamsan. The interactions between side chains and main chains within the double helices preclude carboxylate-mediated aggregation of the double helices and thus inhibit gelation. In the case of welan gum, the double-helical conformation is adopted regardless of pH or ionic strength (CAMPANA et al., 1990). The remoteness of the gentibiosyl side chain of rhamsan gum from the carboxylic acid groups renders it more flexible and consequently the polysaccharide behavior to some extent resembles gellan (CAMPANA et al., 1992). In the presence of $Ca^{2+}$, gelation of deacetylated rhamsan has been demonstrated, but no such phenomenon was seen for welan or polymer S657. Glycosyl substituents exert a stabilizing effect on the ordered conformation and are more efficacious in welan and S657 in which intrachain hydrogen bonds are involved. For each of the polysaccharides in the gellan family, the intrinsic viscosity as a function of ionic strength depends on the structure and on the presence or absence of acetyl groups.

Gellan forms rigid, brittle, thermoreversible gels in the presence of almost all cations and, even at polysaccharide concentrations as low as 0.75%, provides high gel strength. As the affinity for various cations differs, the strength of the gel depends on the cation present. Affinity for divalent ions such as $Mg^{2+}$ and $Ca^{2+}$ is stronger than that for monovalent ions; consequently optimal gel strength is produced at lower concentrations of the divalent ions. The aqueous gels are very clear and have been used as the basis for the preparation of bacterial culture media; in many respects the gels resemble those prepared with agar, both polysaccharides being capable of forming media with a gel strength of 250–350 g $cm^{-2}$.

Aqueous solutions of welan are highly viscous, salt compatible and have very high thermostability. At lower temperatures, viscosity is little affected over the pH range 2–12. The other polysaccharides of the gellan "family" all appear to be dominated in their configuration by the backbone structure, while the side chains play a minor role (CAIRNS et al., 1991). A mutant strain recently obtained by LOBAS et al. (1994), although lacking uronic acid and being composed of a trisaccharide repeat unit, also formed gels after heating and cooling at alkaline pH. The gels were much weaker than those from gellan. Rhamsan and welan do not form gels even at concentrations of ~1%, but produce thermostable solutions with high viscosities which are retained at 105 and 140 °C, respectively.

The low acyl form of gellan marketed as Kelcogel® or Gelrite® can be used as a multifunctional gelling, stabilizing and suspending agent and as a structuring agent for a wide range of foods, either on its own or in combination with other food hydrocolloids. (A nonclarified product is also available for industrial usage). The deacylated polymer is marketed as a mixed salt form which, because of the low level of divalent cations present, is insoluble in cold water. For some purposes, sodium citrate must be added to chelate multivalent ions and thus permit hydration and solution. Gellan is considered to be a natural food additive in Japan and has been so used since 1988. It has now also been approved in the U.S. for food use as a stabilizer and thickener, while it is also included in a recent annex to the EC list of generally permitted additives. The brittle gellan gels give very good flavor release and are stable over the wide pH range found in food products. Particular care has to be taken in the formulation of gels with high contents of sugars, as gel strength is highly sensitive to the concentration of calcium in their presence. The brittleness of gellan gels can be reduced through the addition of mixtures of xanthan and locust bean gum. The high gel strength, good stability and high clarity of gellan gels provide the basis for a wide range of food applications for the bacterial polysaccharide and some of the potential uses can be seen in Tab. 7.

As a replacement for agar, gelrite can be incorporated into microbiological and cell culture media. In this role it has been suggested that it may even lead to some growth enhancement when compared to agar-based bacterial culture media. Tests on the growth of a wide range of bacterial species have indicated that media solidified with gellan compare favorably with agar-based preparations, while the high clarity of the gels may have distinct advantages. The high thermostability of gellan also makes it very useful for the culture of thermophilic microbial species. Similarly, gelrite-based plant cell culture media avoid the problems caused by impurities in agar. The bacterial polysaccharide also has the dis-

**Tab. 7.** Potential Food Uses for Gellan

| Type of Food | Products | Current Polysaccharides Used |
|---|---|---|
| Confectionary | Jellies, fillings, marshmallows | Pectin, starch, gelatine, agar, xanthan/lbg[a] |
| Dairy products | Ice cream, milk shakes, spreads | Carrageenan, alginate, guar, lbg |
| Icings, glazes | Bakery icings, frostings | Agar, pectin, xanthan/guar, starch |
| Gels | Deserts, aspics | Alginate, gelatine, agar, carrageenan |
| Jams, jellies | Low calorie jams, fillings | Alginate, pectin, carrageenan |
| Fabricated foods | Fabricated and textured products | Alginate, carrageenan/lbg |

[a] lbg: locust bean gum

tinct advantage that the concentrations required to provide a specific gel strength are much lower.

Other potential industrial applications for gellan are as gelling agents in dental and personal care toiletries. Deodorant gel products can also use much lower levels of gellan than of the polysaccharides (carrageenan/locust bean gum mixtures) currently used. Mixed with starch, native gellan can be used as a size in paper manufacture. Gellan has also been proposed as the starting material for chemical esterification to yield semisynthetic products with new physicochemical properties (CRESCENZI et al., 1992).

Rhamsan and welan provide high viscosity solutions with good temperature stability and pH compatibility. In the case of rhamsan, there is excellent suspending capacity, shear stability and compatibility with high concentrations of salts. The polysaccharide has potential applications in paints in which it can form blends with cellulose derivatives to give a range of rheological characteristics. Possible uses are found in ceramic glazes as well as solutions and suspensions of chemicals employed for crop protection. Welan may also be utilized to replace methylcellulose as a component of grouts and concretes. It provides greater pseudoplasticity and prevents separation of water during setting. It also has potential value in high temperature (deep) oil drilling as an additive for drilling fluids and as a component of workover fluids. The potential applications of this groups of polysaccharides has been discussed by BAIRD et al. (1983). The production of the welan and rhamsan polymers utilizes conditions similar to those used for gellan manufacture.

## 7.7 Hyaluronic Acid and Heparin

Although no microbial strains produce heparin, a strain of *Escherichia coli* serotype K5 does form a capsular polysaccharide in which the disaccharide repeat unit is essentially a form of *desulphatoheparin* (VANN et al., 1981). The polymer is composed of a repeating unit of 4-$\beta$-D-glucuronosyl-1,4-$\alpha$-N-acetyl-D-glucosamine (Fig. 5) and is thus similar to N-acetyl heparosan, a biosynthetic precursor of heparin. Desulphatoheparin from eukaryotic sources is normally composed of alternate disaccharides containing D-glucuronosyl amino sugar and L-iduronosyl-1,4-$\alpha$-N-acetyl-D-glucosamine. The bacterial product resembles type II glycosaminoglycuronan chains which are synthesized in eukaryotes in the Golgi complex and then polymerized onto core proteins. *After* polymerization the eukaryotic polymers may be modified through the action of uronosyl epimerase and sulphatotransferases. These enzymes introduce the L-iduronosyl and sulphate residues, respectively, both of which are absent from the bacterial product. Because of its structural similarity to heparin, this bacterial polysaccharide may prove useful in medical research and in determination of the specificity of heparinases and related enzymes. Typically, it is of relatively low molecular mass – ~50 kDa. Another interesting polymer of this general type from *E. coli* K4 possesses a chondroitin backbone to which $\beta$-D-fructofuranosyl residues are attached at the $C_3$ position of the D-glucuronic acid (Fig. 5). After the fructosyl residues have been removed by mild acid treatment, the polysaccharide is a substrate for both hyaluronidase and chondroitinase. Both polysaccha-

$$\rightarrow [\,3\ \beta\ \text{D-GalpNAc}\ (1 \rightarrow 4)\beta\ \text{D-GlcpA}\ 1\,] \rightarrow$$
$$\overset{\uparrow 3}{2}\ \ \beta\ \text{D-Fru}$$

*E. coli* K4

$$\rightarrow [\,4\ \alpha\ \text{D-GlcpNAc}\ (1 \rightarrow 4)\beta\ \text{D-GlcpA}\ 1\,] \rightarrow$$

*E. coli* K5

$$\rightarrow [\,4\ \beta\ \text{D-GlcpNAc}\ (1 \rightarrow 4)\beta\ \text{D-GlcpA}\ 1\,] \rightarrow$$

Hyaluronic acid from *Streptococcus equi* and other bacteria

**Fig. 5.** The structure of the exopolysaccharides from *Escherichia coli* strains K4 and K5 and of bacterial hyaluronic acid.

rides also provide sources of oligosaccharide fragments for use in the study of the biological activities of heparin and chondroitin. They may also prove valuable in studies on heparin biosynthesis (LIDHOLT et al., 1994).

Group A and Group C streptococci produce hyaluronic acid, apparently identical in chemical structure to that obtained from eukaryotic material. The polysaccharide is composed of repeating units of 1,4-$\beta$-linked disaccharides of D-glucuronosyl-1,3-$\beta$-*N*-acetyl-D-glucosamine (Fig. 5). The product from one Group C streptococcal strain forms a high molecular weight cell-bound product and a soluble exopolysaccharide of average molecular weight $2 \times 10^6$. As an alternative to the use of material from human or animal sources, bacterial production of hyaluronic acid has been commercialized using strains of *Streptococcus equi* and *Streptococcus zooepidemicus* to produce a high quality (and high cost) polysaccharide for use in surgery as well as in cosmetic preparations in both areas. Mutant strains lacking hyaluronidase have also been developed to ensure that degradation of the high molecular weight polymer does not occur. Yields of hyaluronic acid of the order of 6 g $L^{-1}$ from cultures with a cell density of 45 g $L^{-1}$ dry bacteria have been reported and the products provide advantages over the traditional sources in terms of cost, purity, and molecular mass. Continuous anaerobic culture studies of *S. zooepidemicus* in medium containing glucose and yeast extract, indicated that the highest polysaccharide concentrations were produced at low dilution rates. However, problems were encountered due to the emergence of bacterial variants possessing a higher $\mu_{max}$ than the native strain and showing loss of polymer production.

Hyaluronic acid molecules in aqueous solution display very limited configurational flexibility and thus show very marked stiffness. The hyaluronic acid finds two major applications. It is used as a replacement for the natural product in humans in surgery of the eye or of joints. A unique property of the polymer, its high water-binding and water-retention capacity, results in its inclusion in many cosmetic preparations. The water-binding capacity correlates with the molecular weight and products with very high molecular mass can retain up to 6 L water per gram polysaccharide. Additionally, hyaluronic acid is used in veterinary applications and as a biocompatible coating material for prostheses. Its non-pyrogenicity makes it ideal as a sheath for such implants. In therapeutic applications, the hyaluronic acid must be of high purity and high molecular mass if its benefit is to be long lasting. Further potential use of hyaluronic acid lies in the development of cross-linked polymer and its application as viscosurgical implants to control scarification and to prevent postoperative adhesions (Ciba, 1989). Ester derivatives of hyaluronic acid in the form of microspheres have also been proposed as vehicles for drug delivery (BENEDETTI et al., 1989).

## 7.8 Pullulan

The extracellular polysaccharide from *Aureobasidium pullulans* is an $\alpha$-D-glucan in which maltotriose and a small number of maltotetraose units (1,4 $\alpha$-linked) are coupled through 1,6 $\alpha$-bonds to form an essentially linear polymer for which the molecular mass is dependent on the physiological conditions; values between $10^3$ and $3 \times 10^6$ have been reported (TSUJISAKA and MITSUHASHI, 1993; WILEY et al., 1993). Although the polysaccharide is not degraded by most amylases, specific *pullulanase* enzymes can be readily isolated from a number of sources, including

*Enterobacter aerogenes,* and used to hydrolyze the polysaccharide to its component maltotriose (and maltotetraose) units. Similar products are formed by other fungal species including *Tremella mesenterica* and *Cyttaria harioti*.

The yields of pullulan from *A. pullulans* are affected by the nitrogen source and also by the design of the bioreactor, although it now seems that yields are independent of the cell morphology (GIBBS and SEVIOUR, 1992; SEVIOUR et al., 1992). An interrelationship between pH and nitrogen source appears to exist. At pH values close to 2.5, polysaccharide production is greatly decreased in the presence of $NH_4^+$, but not with glutamate (AUER and SEVIOUR, 1990). Molecular weight may decline if incubation is prolonged. Defined culture conditions for the production of pullulan with low ($<5\times10^5$), medium ($1-2\times10^6$) or high ($>2\times10^6$) $M_r$ have been developed (WILEY et al., 1993). Although it has been suggested that pullulan might be a secondary metabolite, this appears not to be the case as synthesis parallels increases in biomass. High cell densities are needed to ensure blastospore production as this morphological form of the microorganism is the major polymer producer. High carbon/nitrogen ratios, using sucrose as the carbon source provide the basis for economic production of pullulan. Alternatively, growth in fructose-based media yields high molecular mass product. Recovery of cell-free polysaccharide from the culture supernatant necessitates the dilution of the culture medium to reduce its viscosity and permit filtration, followed by treatment with activated charcoal to remove pigmentation. The actual molecular weight of the product is dependent on the physiological conditions employed, as well as on the culture strain used.

Pullulan is highly water soluble, forming stable, viscous solutions which are stable in the presence of most cations and do not form gels. Heating is not needed to achieve solution and polysaccharide solutions are stable for several hours at 100 °C in high NaCl concentrations. Unlike many of the other microbial polysaccharides it is very readily degraded by enzymes from many microbial species due to the widespread occurrence of a range of amylolytic and related $\alpha$-D-glucanases. Esterification can be used both to increase the range of physical properties and to reduce susceptibility to enzyme attack (TSUJISAKA and MITSUHASHI, 1993).

Pullulan can be used to form films which are oil resistant, water-soluble and have low oxygen permeability. This novel packaging material assists in flavor retention and maintains the fresh appearance of foods which can then be cooked directly. Solutions of the polymer can also be used to form odorless and tasteless coatings directly onto foodstuffs. While such applications have apparently been made in Japan, usage of the polymer in other countries appears to be much more limited. Pullulan can also be used for the preparation of fibers; it is also a good adhesive. In an application similar to one of those for dextrans, pullulan has also been proposed as a component along with polyethylene glycol, of aqueous two-phase systems (NGUYEN et al., 1988). The phase diagrams and the consequent separation of enzyme proteins depended greatly on the molecular weight of the polyethylene glycol (PEG) used. The phase diagrams for PEG 14000 and pullulan were similar to those for PEG 8000 and dextran, but as pullulan was cheaper to produce (see Tab. 2) it was considered to have a cost advantage. Two phase systems of this type also provided a model, semicontinuous system for cellulase production by *Trichoderma reesei.* As pullulan lacks side chains, molecular mass standards of high accuracy and low polydispersity have been prepared for calibration of HPLC columns used for the Size Exclusion Chromatography of water-soluble polymers. The properties of the native polysaccharide can be modified by preparation of chemical derivatives such as acetate esters.

## 7.9 Scleroglucan and Related Fungal (1→3)-$\beta$-D-Glucans

Scleroglucan is one of a group of closely related $\beta$-D-glucans which are produced by several fungal species including *Sclerotium rolfsii* and the wood rotting Basidiomycete *Schizophyllum commune*. Although they are structurally similar to the bacterial product curdlan the fungal products are soluble polysaccharides yielding highly viscous aqueous solutions in which the polymer molecules adopt a triple helical conformation. The main chain is composed of 1,3-$\beta$-D-linked glucose residues, to which are attached 1,6-$\beta$-D-glucosyl residues (Fig. 3). The side chains may be attached regularly or randomly with varying degrees of frequency. Those in the product from *Schizophyllum commune* are probably attached regularly on every third glucose in the main chain. The initial molecular weight can range from about $1.3 \times 10^5$ to $6 \times 10^6$. Some of the glucans, such as those from *Sclerotium glucanicum*, are of lower molecular weight (ca. 18,000) with glucosyl substituents on every fourth or every sixth main chain unit. The side chains may not always be regularly distributed on the main chain; distribution may possibly be random and wide variations can be found in the degree of branching of some of the polymers. This can also greatly affect their solubility.

The polysaccharide can be produced by batch aerobic submerged culture at 30°C of suitable strains of *S. rolfsii* in a medium containing corn steep liquor, glucose (3%) as carbon substrate, nitrate and salts. Polysaccharide production is coincident with mycelial growth and final yields can reach 10–20 g $L^{-1}$ after ~60 h. The mycelium can be removed from the highly viscous culture by filtration. A scheme for manufacture of clinical grade schizophyllan proposed by MISAKI et al. (1993) is indicated in Fig. 6.

Because of its high viscosity and good suspending properties together with compatibility with electrolytes, scleroglucan has been proposed for enhanced oil recovery in reservoirs with high salinity and temperature (DAVISON and MENTZER, 1980), but its relatively high cost has probably limited its adoption in practice. Solutions are pseudoplastic and the viscosity is not greatly affected over the temperature range 20–90°C. Various other applications in the food and cosmetic industries have been suggested, but no approval for food use appears to have been sought. As the polymer is neutral, the pseudoplastic behavior of scleroglucan is unaffected by the presence of various salts; this provides an advantage over polyanionic polysaccharides

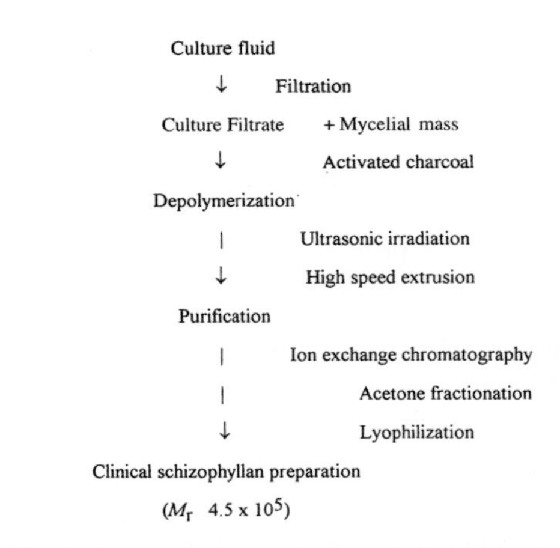


**Fig. 6.** Process for preparation of clinical grade schizophyllan (scleroglucan). Adapted from MISAKI et al. (1993).

such as xanthan or succinoglycan and its use in agricultural products has been suggested. It is, however, susceptible to degradation by a number of different enzyme preparations; it could consequently provide a source of the disaccharide gentibiose (BRIGAND, 1993).

The major potential value for several of the related fungal polysaccharides such as lentinan (CHIHARA et al., 1989), is as anti-tumor or immunomodulating agents. The mechanism of inhibition of tumor growth has not been fully established, but the most important features for this activity appear to be the triple-helical conformation adopted by the glucan, the molecular mass and the number and arrangement of the side chains (MISAKI et al., 1993). The chemical oxidation of side chains with periodate, followed by borohydride reduction, further enhanced activity – possibly due to increased solubility. The low activity of some of the fungal glucans could result from their poor solubility. For the immunomodulating role, the relatively high cost of preparation of the polysaccharides would probably be acceptable.

## 7.10 Succinoglycan

The name *succinoglycan* has been applied to a group of biopolymers from bacterial strains which also yield curdlan. As with other

→ [ 4 β D - Glc*p*(1 → 4)β D - Glc*p*(1 → 4) βD - Gal*p*(1 → 3)β D - Glc*p* 1 ] →
1↑6
β D - Glc*p*(1 → 3)β D- Glc*p*(1 → 3)βD - Glc*p*(1 → 6)β D - Glc*p* 1 ] →
4⇑6 ↑6
Pyr Succ

**Fig. 7.** The structure of the exopolysaccharide succinoglycan from *Rhizobium* and other bacterial species.

examples of microorganisms which can synthesize more than one polysaccharide, strain selection and choice of growth conditions can ensure high yields of one specific polymer. The strains most widely used for succinoglycan synthesis are *Agrobacterium radiobacter* utilized for the Shell product Enorflo-S and related species. Succinoglycans differ from curdlan in being water-soluble and are composed of octasaccharide repeat units, thus conforming to a pattern found in a number of other exopolysaccharides from *Rhizobium* species. There is close structural similarity to those other *Rhizobium* polysaccharides which possess highly conserved structures composed of octasaccharide repeat units. Succinoglycans contain D-glucose and D-galactose in the molar ratio 7:1. Attached to the sugars are three different acyl groups – acetate and succinate esters and pyruvate ketals (Fig. 7). Pyruvate is normally present in stoichiometric amounts, while the molar ratios of acetate and succinate are commonly of the order of 0.2 and 0.4–0.5, respectively. The monosaccharide components are of course neutral, but the pyruvate ketal situated on the side chain terminus (D-glucose) confers anionic properties to the polysaccharide as does the succinyl half-ester. Despite its 1,3, 1,4, and 1,6 linked β-D-glucosyl residues, succinoglycan is resistant to the action of the endo-β-glucanases which are available commercially but can be degraded using specific polysaccharide hydrolases.

→ [ 4 β D- Glc*p* (1 → 4)β D- Glc*p* 1 ] →
1↑3
α D-Man*p* - O.CO.$CH_3$
1↑2
β D-Glc*p*A
1↑4
β D-Man*p* = Pyr

**Fig. 8.** The structure of the exopolysaccharide from *Xanthomonas campestris* (Xanthan). Typically, the internal α-mannosyl residue is fully acetylated but only ca. 30% of the β-mannosyl termini are ketalated.

Studies by LINTON et al. (1987a) indicated that succinoglycan synthesis, like other bacterial exopolysaccharide formation, is a very efficient process with little energy loss during growth under carbon excess, and most of the ATP not required for the synthesis of cell material is applied to exopolysaccharide production.

The viscosity of succinoglycan solutions is higher than that for equivalent concentrations of xanthan and resistance to acid hydrolysis is also greater, despite the lower transition temperature. Succinoglycan does differ from a number of other microbial viscosifying agents in that above $T_m$, the viscosity is lost. It may be restored in part on cooling (LINTON et al., 1991). In xanthan solutions, the more flexible conformation above $T_m$ yields a lower but still significant viscosity. This property of succinoglycan, together with its good suspending capacity for clays and other solids, may be valuable for some oil field operations.

## 7.11 Xanthan

*Xanthan,* the exopolysaccharide from *Xanthomonas campestris* pv. *campestris* is one of the major commercial biopolymers produced. In tonnage terms total production from several different commercial sources probably exceeds >20,000 tonnes per year. The structure of xanthan is a pentasaccharide repeat unit, which is of particular interest as they represent trisaccharide side chains attached to a cellulosic backbone (Fig. 8). Alternate glucose residues in the backbone carry the side chains composed of D-mannose and D-glucuronic acid. Most commercial xanthan preparations are fully acetylated on the internal D-mannose residue, and carry pyruvate ketals on about 30% of the side chain terminal mannose residues. Recently, xanthan with a higher pyruvate content has become available commercially and strains totally lacking pyruvate have also been developed. Through the use of mutants, of different *X. campestris* pathovars and of different nutrient conditions, it is possible to obtain a range of different polysaccharides which all conform to the general structure of xanthan, but differ in the completeness of carbohydrate side chains and the extent of acylation. One group of pathovars yield xanthan in which very little pyruvate is present, but in which the internal mannosyl residue in the side chain carries *two* moles of acetate. Mutant polymers lacking the terminal D-mannose (and pyruvate) or lacking the terminal *disaccharide* are also available. The latter naturally differs from xanthan in being a neutral polymer, although it still possesses useful rheological properties. The various types of repeating unit to be found in different xanthan preparations are illustrated in Fig. 9. Typically the molecular mass of xanthan is in the range $1.2–1.8 \times 10^6$. Many of the rheological properties of xanthan appear to derive from the double helical ordered conformation adopted in solution. The trisaccharide side chains in this conformation are aligned with the cellulosic backbone, stabilizing the conformation by non-covalent interactions. The same conformation is also found in

→ [ 4 β D- Glc*p* (1 → 4)β D- Glc*p* 1 ] →
1↑3
α D-Man*p* - O.CO.$CH_3$
1↑2
β D-Glc*p*A

'Polytetramer'

→ [ 4 β D- Glc*p* (1 → 4)β D- Glc*p* 1 ] →
1↑3
α D-Man*p*
1↑2
β D-Glc*p*A

'Polytetramer' (Non-acetylated)

→ [ 4 β D- Glc*p* (1 → 4)β D- Glc*p* 1 ] →
1↑3
α D-Man*p* - O.CO.$CH_3$

'Polytrimer'

→ [ 4 β D- Glc*p* (1 → 4)β D- Glc*p* 1 ] →
1↑3
α D-Man*p*

'Polytrimer' (Non-acetylated)

**Fig. 9.** The structure of the exopolysaccharides from mutants of *Xanthomonas campestris.*

solutions of polymer lacking the terminal β-D-mannose residue. Although these are less viscous than the native polymer, further loss of the glucuronic acid residue leaving α-D-mannosyl residues only, yields a neutral polymer with viscosity comparable to xanthan.

As well as xanthan from wild type or mutant strains of *X. campestris* pv. *campestris* and related strains, several other polysaccharides have been shown to share part of the structure and also possess exploitable properties. The ability of mutant strains to produce "polytrimer" and "polytetramer" (i.e., xanthan molecules lacking the side chain disaccharide and monosaccharide termini, respectively) has been mentioned. In addition to these variants, several strains of *Acetobacter xylinum* yield apparently xanthan-like polysaccharides. As bacterial cellulose is a normal product of *A. xylinum*, the ability to modify cellulose by the addition of side chains is perhaps not unexpected. One product termed acetan has been characterized as having a cellulosic main chain together with a *pentasaccharide* side chain on alternate main chain sugars (JANSSON et al., 1993). This polymer closely resembled xanthan, the terminal β-D-mannosyl residue of the side chain being replaced by a L-rhamnosyl–gentibiosyl sequence. As is the case with xanthan, this polysaccharide is acetylated. Another strain of the same bacterial species is thought to synthesize a polysaccharide which differs only in the conformation of one of the side chain linkages. These xanthan-like polymers are produced in addition to cellulose in some strains or replacing it in others. They form highly viscous aqueous solutions but may not be as potentially valuable as xanthan and have not yet been produced commercially. Another polysaccharide of the *same composition*, but apparently different structure has also been found in a cellulose-negative *A. xylinum* strain. Bacterial mutants yielding polysaccharide with abbreviated side chains have also been reported.

Xanthan is produced commercially by aseptic culture using highly aerated stirred tank fermenters, which are seeded from a 10% inoculum at the intermediate stage then a 5% inoculum into the final fermenter. The medium composition is carefully controlled throughout the fermentation of approximately 80 h, to optimize polysaccharide production and quality. After the process is complete, the cultures are pasteurized to kill the bacteria prior to polymer recovery by alcohol or isopropanol precipitation.

Solutions of xanthan are highly pseudoplastic, rapidly regain viscosity on removal of shear stress and show very good suspending properties; they show high viscosity at low shear rates. No thixotrophy is observed under most conditions. The relatively rigid helical conformation of xanthan renders it stable for prolonged periods over a wide range of pH and ionic strength. The polysaccharide is also normally resistant to degradation by commercially available enzymes including cellulases, although it can be the substrate for specific enzymes – usually associations of endo-(1→4)-β-D-glucanases and xanthan lyases.

Many of the original industrial applications of xanthan were in the food industry. It received approval for food use from the U.S. Food and Drug Administration (FDA) in 1969 and it has been estimated that about 60% of the xanthan currently produced is probably food grade polymer. Polysaccharides are incorporated into foods in order to alter the rheological properties of the water present. The texture of the product can therefore be changed. Thus xanthan has found applications utilizing many of its physical properties (Tab. 8). In many foodstuffs xanthan possesses further useful attributes, including rapid flavor release, good mouthfeel and compatibility with other food ingredients including proteins, lipids, and polysaccharides. Most foodstuffs already contain polysaccharides such as starch or pectin in addition to proteins and lipids and it is important that any added polymer such as xanthan should be totally compatible with them. Some of the foodstuffs, including salad dressings, relishes and yoghurts, are of low pH. There is, therefore, an added requirement that any polysaccharide which is incorporated into foods of these types must be acid-stable. Xanthan, which is stable over a wide range of pH values, is very well suited to such applications. Indeed, the wide range of applications for which xanthan is used in the food industry, rely on its high viscosity, solubility, and stability over a wide range of pH values and temperatures, good suspending properties, and compatibility with many salts and other food ingredients. Xanthan may be used for the partial replacement of starch in foodstuffs to improve texture and flavor release. A full list of the applications of xanthan in foods and of its major attributes has been produced by Kelco (1992).

**Tab. 8.** Different Properties of Xanthan Used in Foods

| Function | Application |
|---|---|
| Adhesive | Icings and glazes |
| Binding agent | Pet foods |
| Coating | Confectionery |
| Emulsifying agent | Salad dressings |
| Encapsulation | Powdered flavors |
| Film formation | Protective coatings, sausage casings |
| Foam stabilizer | Beer |
| Stabilizer | Ice cream, salad dressings |
| Swelling agent | Processed meat products |
| Syneresis inhibitor | Cheeses, frozen foods |
| Thickening agent | Jams, sauces, syrups, and pie fillings |

As well as being compatible with other polysaccharides used as thickening agents, xanthan shows synergistic interaction with plant galactomannans and glucomannans such as guar gum and locust bean gum. The mixed polysaccharides reveal enhanced viscosity or gelation, properties which are widely employed in food processing. Neither of the components will gel on its own. The mixed solution must be heated above the transition temperature of the xanthan in order to denature the xanthan helix and then allowed to cool. Mixtures of xanthan and locust bean gum produce elastic, cohesive gels above total polysaccharide concentrations of ~0.03%. The interactions between the bacterial and plant polysaccharides are highly dependent on the acylation of the xanthan molecules and on the extent and distribution of D-galactosyl side chains present on the galactomannan. Acetyl groups on the xanthan molecules played an important role in inhibiting gelation of the polysaccharide mixtures and increase the polymer concentrations required

(SHATWELL et al., 1990a, b, 1991a, b, c). Removal of the *O*-acetyl groups on the bacterial polysaccharide enhanced gelation and reduced the total concentration of polysaccharides needed. Whereas the transition temperature of xanthan in solution is considerably affected by the ionic concentration, the melting temperature of xanthan and locust bean gum gels is apparently unaffected over a wide range of ionic strengths (ZHAN et al., 1993).

Many industrial applications of xanthan are unrelated to foods and beverages as can be seen in Tab. 5, but the same physical attributes which are used in foods are also of value in the pharmaceutical and cosmetic industries. An example is seen in the formulation of toothpastes, where the pseudoplastic behavior of xanthan allows the product to flow freely when a tube is squeezed, but regains its viscosity immediately the shear stress is removed.

## 7.12 Other Possible Polysaccharide Products

### 7.12.1 Emulsan and Related Emulsifying Polysaccharides (Biosurfactants)

The exopolysaccharides from *Acinetobacter calcoaceticus* are potentially useful emulsifying agents. Structural studies have shown considerable differences in the chemical structure of two of these polymers, despite their common functional properties. The polysaccharide of *A. calcoaceticus* BD4 is a heptasaccharide repeat unit composed of rhamnose, mannose, glucose, and glucuronic acid. The polysaccharide from strain RAG contains D-galactosamine, an aminouronic acid and an amino sugar as well as 15% fatty acyl *O*-esters. It has been marketed under the name *Emulsan*, proposed uses for which include both domestic and commercial applications. More detailed information on this group of microbial products can be found in Chapter 17.

The exopolymers synthesized by cyanobacteria such as *Phormidium* sp. differ from other *bacterial* exopolysaccharides in that they appear to be sulphated. The *Phormidium* exopolymer, also known as *emulcyan*, has attracted interest as it also possesses emulsifying properties. It has a molecular weight of $1.2 \times 10^6$ and contains rhamnose, mannose, galactose, an unidentified uronic acid (*not* galacturonic or glucuronic or mannuronic), protein, and sulphate. As yet, there is little information on its potential value as an emulsifying agent.

### 7.12.2 *Rhizobium* Heteroglycans

Some species of *Rhizobium* (*R. trifolii* and *R. leguminosarum*) form a neutral capsular heteropolysaccharide composed of D-glucose, D-galactose, and D-mannose in the molar ratio 1:3:2 forming a hexasaccharide repeat unit (Fig. 10) (GIDLEY, 1987). This polymer is produced in addition to soluble polyanionic material, the relative proportions of the two being dependent on the strain and on the growth conditions used. The main chain of the polysaccharide comprises a sequence of $\alpha$-D-glucosyl-1,3-$\alpha$-D-mannosyl-1,3-$\beta$-D-galactose. The glucose residue carries two side chains, one of a single D-galactosyl unit and the other a disaccharide composed of D-galactose. Like curdlan, this polysaccharide is insoluble in water at room temperature but can be dissolved on heating. On subsequent cooling, the polymer adopts an ordered conformation and forms thermoreversible gels (CESARO et al., 1987). The $T_m$ is ~49 °C on heating and 42 °C on cooling and despite the nonionic nature of the polymer, is influenced by the presence of salts (CHANDRASEKARAN et al., 1987).

```
            β D-Galp
              ↓1
             2
→ [ 4 α D- Glcp (1 → 3)α D- Manp(1 → 3)β D- Galp 1 ] →
             1↑6
            β D-Galp
             1↑4
            β D-Galp
```

**Fig. 10.** The structure of the neutral capsular exopolysaccharide from *Rhizobium meliloti*.

### 7.12.3 Sialic Acids

Strains of *Escherichia coli* and of *Neisseria meningitidis* secrete sialic acid-containing exopolysaccharides. The products from *E. coli* K1 and *N. meningitidis* group B are (2→8) α-linked. Group C meningococcal sialic acid is a homopolysaccharide in which the residues are (2→9) linked. Another sialic acid (from different strains of *E. coli*) is non-acetylated and is composed of *alternate* 2,8-α- and 2,9-α-linkages. In addition to these homopolymers, some strains of *N. meningitidis* synthesize heteropolysaccharides containing both sialic acid and D-glucose. Several of the bacterial sialic acids possess hydrophobic 1,2-diacyl glycerol groups at the polysaccharide chain termini, causing the polymers to aggregate in the form of micelles. Polymers such as these have potential value for vaccine development or as sources of sialyl derivatives, but are not likely to be developed on a large scale.

### 7.12.4 XM6

A polysaccharide with interesting gelation properties is produced by an *Enterobacter* strain (XM6); this polymer proved to resemble very closely two polysaccharides from *Klebsiella aerogenes* serotype 54 strains. Each polysaccharide was composed of the same tetrasaccharide repeat unit (Fig. 11). However, the two *Klebsiella* exopolysaccharides carried *O*-acetyl groups on every fucose or on *alternate* L-fucose residues, respectively, whereas the XM6 exopolysaccharide is devoid of acyl substituents. The *Enterobacter* XM6 polysaccharide forms gels with both monovalent and divalent ions, and has a very sharp melting point at ca. 30 °C. It can also form coacervates

-[ 4 Glc*p*(1 →4)- β D - Glc*p*A -(1 → 4)- α - Fuc*p* - 1]-
1↑3
β D - Glc*p*

**Fig. 11.** The structure of *Enterobacter* XM6 exopolysaccharide. In the *Enterobacter* polymer there are no *O*-acetyl substituents.

with gelatine (CHILVERS et al., 1988). The material from either of the type 54 strains did not form gels unless it had been chemically deacylated; following deacylation, the behavior of the polymers was identical with that of XM6.

Other possible new products could use microbial polysaccharides as starting materials yielding semisynthetic polymers. This is already the case for the chemically modified dextrans marketed as Sephadex® and could be applied to other neutral polysaccharides with appropriate properties; similar modifications can be made to polyanionic materials such as gellan or hyaluronic acid. The polymers could be esterified and fabricated to produce novel thermoplastics and hydrogels. Patents for processes modifying xanthan and other microbial products by carboxymethylation and other techniques have been filed, although none appears as yet to have been commercially developed. It has also been suggested that the heparin-like polysaccharide from *E. coli* K5 could be chemically modified by N-acetylation and N-sulphation to yield heparin-like compounds with antithrombin binding activity. Polysaccharides with very high contents of fucose or rhamnose could also be used as sources of the 6-deoxysugars for conversion into food flavor additives or in the case of pullulan as a substrate for enzymic conversion to maltotriose and other oligosaccharides.

# 8 The Future

The future development of microbial polymers will depend on a number of factors:

– Can they be produced economically? Currently xanthan sets the "bench mark" for cost. Substrate conversion is high, recovery costs are reasonable, and final treatment of the polymer is relatively simple. Although substrate conversion for gellan production is good, some is lost through synthesis of storage polymers. Gellan also requires more post-fermentation processing to remove acyl groups from the native polymer and to free the product from cellular material. In contrast to these two polysaccharides, conversion of substrate and resultant yields of hyaluronic acid are very low indeed.

– Is there a specific need? Some biopolymers have properties which are unique and so provide them with identifiable market share. Hyaluronic acid, although exceedingly expensive to produce, cannot be replaced by other materials and is in competition only with the polymer obtained from eukaryotic material.

– Can new markets be identified? It will be surprising if, in the current moves towards "greener" technology, the usage of biopolymers prepared from renewable resources does not increase. The physical properties of the polymers can be enhanced through chemical modification. As oil reserves diminish, use for enhanced oil recovery, drilling fluids and workover products could also greatly increase.

– Can microbiologists discover new biopolymers? Realism is required. Many papers purport to find polysaccharides with "useful physical properties" but fail to define their uniqueness. Xanthan is a very versatile and relatively low cost industrial chemical. Few of the supposed "new" polymers are superior for most applications. An alternative is to use the polysaccharides as base chemicals for modification by chemical procedures to introduce new physicochemical properties.

# 9 References

AKSTINAT, M. H. (1980), Polymers for enhanced oil recovery in reservoirs of extremely high salinities and high temperatures, *Soc. Petrol. Eng.* Paper **8979**.

AMANN, M., DRESSNANDT, G. (1993), Solving problems with cyclodextrins in cosmetics, *Cosmet. Toiletries* **108**, 90–95.

AMEMURA, A., HISAMATSU, M., HARADA, T. (1977), Spontaneous mutation of polysaccharide production in *Alacaligenes faecalis* var. *myxogenes, Appl. Env. Microbiol.* **34**, 617–620.

ANDERSON, A. J., HACKING, A. J., DAWES, E. A. (1987), Alternative pathway for the biosynthesis of alginate from fructose and glucose in *Pseudomonas mendocina* and *Azotobacter vinelandii. J. Gen. Microbiol.* **133**, 1045–1052.

ANDERSON, D. M. W., BRYDON, W. G., EASTWOOD, M. A. (1988), The dietary effects of gellan gum in humans, *Food Addit. Contam.* **5**, 237–250.

ANTON, J., MESEGUER, I., RODRIGUEZ-VALERA, F. (1988), Production of an extracellular polysaccharide by *Haloferax mediterranei, Appl. Env. Microbiol.* **54**, 2381–2386.

AUER, D. P. F., SEVIOUR, R. J. (1990), Influence of varying nitrogen sources on polysaccharide production by *Aureobasidium pullulans* in batch culture, *Appl. Microbiol. Biotechnol.* **32**, 637–644.

BAIRD, J. K., SANDFORD, P. A., COTTRELL, I. W. (1983), Industrial applications of some new microbial polysaccharides, *Bio/Technology* **1**, 778–783.

BASTIN, D. A., STEVENSON, G., BROWN, P. K., HAASE, A., REEVES, P. R. (1993), Repeat unit polysaccharides of bacteria: a model for polymerization resembling that of ribosomes and fatty acid synthetase, with a novel mechanism for determining chain length, *Mol. Microbiol.* **7**, 725–734.

BENEDETTI, L. M., TOPP, E. M., STELLA, V. J. (1989), A novel drug delivery system: microspheres of hyaluronic acid derivatives, in: *Biomedical and Biotechnological Advances in Industrial Polysaccharides* (CRESCENZI, V., DEA, I. C. M., PAOLETTI, S., STIVALA, S. S., SUTHERLAND, I. W, Eds.), pp. 27–33. New York: Gordon and Breach.

BETLACH, M. R., CAPAGE, M. A., DOHERTY, D. H., HASSLER, R. A., HENDERSON, N. M., VANDERSLICE, R. W., MARELLI, J. D., WARD, M. B. (1987), Genetically engineered polymers: manipulation of xanthan biosynthesis, in: *Industrial Polysaccharides* (YALPANI, M., Ed.), pp. 145–156, Amsterdam: Elsevier.

BOOTH, C. (1971), *Methods in Microbiology*, Vol. 4, New York: Academic Press.

BOULNOIS, G. J. JANN, K. (1989), Bacterial polysaccharide capsule synthesis, export and evolution of structural diversity, *Mol. Microbiol.* **3**, 1819–1823.

BRIGAND, G. (1993), Scleroglucan, in: *Industrial Gums* (WHISTLER, R. L., BEMILLER, J. N. Eds.), pp. 461–474. San Diego: Academic Press.

BRILL, J. A., QUINLAN-WALSHE, C., GOTTESMAN, S. (1988), Fine structure mapping and identification of two regulators of capsule synthesis in *Escherichia coli* K12, *J. Bacteriol.* **170**, 2599–2611.

BYROM, D. (1991), Microbial cellulose, in: *Biomaterials* (BYROM, D., Ed.), pp. 263–283. London: Macmillan.

CAIRNS, P., MILES, M. J., MORRIS, V. J. (1991). X-ray fibre diffraction studies of members of the gellan family of polysaccharides, *Carbohydr. Polym.* **14**, 367–372.

CAMPANA, S., ANDRADE, C., MILAS, M., RINAUDO, M. (1990), Polyelectrolyte and rheological studies on the polysaccharide welan, *Int. J. Biol. Macromol.* **12**, 379–383.

CAMPANA, S., GANTER, J., MILAS, M., RINAUDO, M. (1992), On the solution properties of bacterial polysaccharides of the gellan family, *Carbohydr. Res.* **231**, 31–38.

CANGELOSI, G. A., HUNG, L., PUVANESRAJAH, V., STACEY, G., OZGA, D. A., LEIGH, J. A., NESTER, E. W. (1987), Common loci for *Agrobacterium tumefaciens* and *Rhizobium meliloti* exopolysaccharide synthesis and their roles in plant interactions, *J. Bacteriol.* **169**, 2086–2091.

CANNON, R. E., ANDERSON, S. M. (1991), Cellulose, *CRC Crit. Rev. Biotechnol.* **17**, 435–447.

CARROLL, V. C., CHILVERS, G. R., FRANKLIN, D., MILES, M. J., MORRIS, V. J., RING, S. G. (1983), Rheology and microstructure of solutions of the microbial polysaccharide from *Pseudomonas elodea, Carbohydr. Res.* **114**, 181–191.

CESARO, A., PAOLETTI, S., DELBEN, F., CAVALLO, S., CRESCENZI, V., ZEVENHUIZEN, L. P. T. M. (1987), Thermoreversible gels of the capsular polysaccharide from *Rhizobium trifolii* strain TA1, in: *Industrial Polysaccharides* (STIVALA, S. S., CRESCENZI, V., DEA, I. C. M., Eds.), pp. 99–109. New York: Gordon and Breach.

CHANDRASEKARAN, R., THAILAMBAL, V. G. (1990), The influence of calcium ions, acetate and L-glycerate groups on the gellan double helix, *Carbohydr. Polym.* **12**, 431–442.

CHANDRASEKARAN, R., MILLANE, R. P., WALKER, J. K., ARNOTT, S., DEA, I. C. M. (1987), The molecular structure of the capsular polysaccharide from *Rhizobium trifolii,* in: *Industrial Polysaccharides* (STIVALA, S. S., CRESCENZI, V., DEA, I. C. M., Eds.), pp. 111–118. New York: Gordon and Breach.

CHANDRASEKARAN, R., RADHA, A., THILAMBAL, V. G. (1992), Roles of potassium ions, acetyl and L-glyceryl groups in native gellan double helix: an X-ray study, *Carbohydr. Res.* **224**, 1–17.

CHIHARA, G., MAEDA, Y. Y., SUGA, T., HAMURO, J. (1989), Lentinan as a host defence potentiator, *Int. J. Immunotherap.* **5**, 145–154.

CHILVERS, G. R., GUNNING, A. P., MORRIS, V. J. (1988), Coacervation of gelatin-XM6 mixtures and their use in microencapsulation, *Carbohydr. Polym.* **8**, 55–62.

CHITNIS, C. E., OHMAN, D. E. (1990), Cloning of *Pseudomonas aeruginosa algG*, which controls alginate structure, *J. Bacteriol.* **172**, 2894–2900.

CHITNIS, C. E., OHMAN, D. E. (1993), Genetic analysis of the alginate biosynthetic gene cluster of *Pseudomonas aeruginosa* shows evidence of an operonic structure, *Mol. Microbiol.* **8**, 583–590.

CHOI, J. H., OH, D. K., KIM, J. H., LEBAULT, J. M. (1991), Characteristics of a novel high viscosity polysaccharide, methylan, produced by *Methylobacterium organophilum, Biotechnol. Lett.* **13**, 417–420.

Ciba (1989), The biology of hyaluronan, *Ciba Foundation Symposium 143.* Chichester: Wiley.

CONTI, E., FLAIBANI, A., O'REGAN, M., SUTHERLAND, I. W. (1994), The production and properties of bacterial alginate by *Pseudomonas* sp., *Microbiology* **140**, 1128–1132.

CORNISH, A., GREENWOOD, J. A., JONES, C. W. (1988), Binding protein-dependent glucose transport by *Agrobacterium radiobacter* grown in glucose-limited continuous culture, *J. Gen. Microbiol.* **134**, 3099–3110.

CORNISH, A., GREENWOOD, J. A., JONES, C. W. (1989), Binding protein-dependent sugar transport by *Agrobacterium radiobacter* and *A. tumefaciens* grown in continuous culture, *J. Gen. Microbiol.* **135**, 3001–3013.

COTTRELL, I. W. (1980), in: *Fungal Polysaccharides* (SANDFORD, P. A., MATSUDA, K., Eds.), pp. 213–270. Washington: American Chemical Society.

COTTRELL, I. W., KANG, K. S. (1978), Xanthan gum, a unique bacterial polysaccharide for food applications. *Dev. Ind. Microbiol.* **10**, 117–131.

COTTRELL, I. W., KANG, K. S., KOVACS, P. (1980), in: *Handbook of Water-Soluble Gums and Resins* (DAVIDSON, R. L., Ed.), pp. 24.1–24.31. New York: McGraw Hill.

CRESCENZI, V., DENTINI, M., DEA, I. C. M. (1987), The influence of side-chains on the dilute solution properties of three structurally related bacterial anionic polysaccharides, *Carbohydr. Res.* **160**, 283–302.

CRESCENZI, V., DEA, I. C. M., PAOLETTI, S., STIVALA, S. S., SUTHERLAND, I. W. (1989), *Biomedical and Biotechnological Advances in Industrial Polysaccharides.* New York: Gordon and Breach.

CRESCENZI, V., DENTINI, M., CALLEGARO, L. (1992), Potential biomedical applications of partial esters of the polysaccharide gellan. *Proc. 9th. Int. Biotechnol. Symp.,* pp. 88–91. Washington: American Chemical Society.

DAVIDSON, I. W. (1978), Production of polysaccharide by *Xanthomonas campestris* in continuous culture, *FEBS Lett.* **3**, 347–349.

DAVISON, P., MENTZER, E. (1980), Polymer flooding in North Sea oil reservoirs, *Soc. Petr. Eng. J.* **22**, 353-362.

DE BELDER, A. N. (1990), *Dextran.* Uppsala: Pharmacia.

DE BELDER, A. N. (1993), Dextran, in: *Industrial Gums* (WHISTLER, R. L., BEMILLER, J. N., Eds.), pp. 399–425. San Diego: Academic Press.

EASTWOOD, M. A., BRYDON, W. G., ANDERSON, D. M. W. (1987), The dietary effects of xanthan gum in man, *Food Addit. Contam.* **4**, 17–26.

GEDDIE, J., SUTHERLAND, I. W. (1994), The effect of acetylation on cation binding by algal and bacterial alginates, *Biotechnol. Appl. Biochem.* **19**, **20**, 117–129.

GIBBS, P. A., SEVIOUR, R. J. (1992), Influence of bioreactor design on exopolysaccharide production by *Aureobasidium pullulans, Biotechnol. Lett.* **14**, 491–494.

GIDLEY, M. J. (1987), Structural studies of *Rhizobium trifolii* capsular polysaccharide by NMR spectroscopy, in: *Industrial Polysaccharides* (STIVALA, S. S., CRESCENZI, V., DEA, I. C. M., Eds.), pp. 129–137. New York: Gordon and Breach.

GLUCKSMAN, M. A., REUBER, T. L., WALKER, G. C. (1993a), Family of glycosyl transferases needed for the synthesis of succinoglycan by *Rhizobium meliloti, J. Bacteriol.* **175**, 7033–7044.

GLUCKSMAN, M. A., REUBER, T. L., WALKER, G. C. (1993b), Genes needed for the modification, polymerization, export and processing of succinoglycan by *Rhizobium meliloti*: a model for succinoglycan biosynthesis, *J. Bacteriol.* **175**, 7045–7055.

GOLDBERG, J. B., DAHNKE, T. (1992), *Pseudomonas aeruginosa* AlgB, which modulates the expression of alginate, is a member of the NtrC subclass of prokaryotic regulators, *Mol. Microbiol.* **6**, 59–66.

GOTTESMAN, S., STOUT, V. (1991), Regulation of capsular polysaccharide synthesis in *Escherichia coli* K12, *Mol. Microbiol.* **5**, 1599–1606.

GOTTESMAN, S., TRISLER, P., TORRES-CABASSA, A. (1985), Regulation of capsular polysaccharide synthesis in *Escherichia coli* K12: characterisation of three regulatory genes, *J. Bacteriol.* **162**, 1111–1119.

GOVAN, J. R. W., FYFE, J., JARMAN, T. R. (1981), Isolation of alginate-producing mutants of *Pseudomonas fluorescens, Pseudomonas putida* and *Pseudomonas mendocina, J. Gen. Microbiol.* **125**, 217–220.

HACKING, A. J., TAYLOR, I. W. F., JARMAN, T. R., GOVAN, J. R. W. (1983), Alginate biosynthesis by *Pseudomonas mendocina, J. Gen. Microbiol.* **127**, 3473–3480.

HARADA, T., FUJIMORI, K., HIROSE, S., MASADA, M. (1966), Growth and $\beta$-glucan production by a mutant 10C3K of *Alcaligenes faecalis* var. *myxogenes* in defined medium, *Agric. Biol. Chem.* **30**, 764–769.

HARADA, T., TERASAKII, M., HARADA, A. (1993), Curdlan, in: *Industrial Gums* (WHISTLER, R. L., BEMILLER, J. N., Eds.), pp. 427–446. San Diego: Academic Press.

HARDING, N. E., CLEARY, J. M., SMITH, D. W., MICHON, J. J., BRUSILOW, W. S. A., ZYSKIND, J. W. (1987), Genetic and physical analysis of genes essential for xanthan gum biosynthesis, *J. Bacteriol.* **169**, 2854–2861.

HERBERT, D., KORNBERG, H. L. (1976), Glucose transport as rate-limiting step in the growth of *Escherichia coli* on glucose, *Biochem. J.* **156**, 477–480.

JANSSON, P.-E., WIDMALM, G. (1994), Welan gum (S–130) contains repeating units with randomly distributed L-mannosyl and L-rhamnosyl terminal groups, *Carbohydr. Res.* **256**, 327–330.

JANSSON, P.-E., LINDBERG, J., WIMALSARI, K. M. S., DANKERT, M. A. (1993), Structural studies of acetan, an exopolysaccharide elaborated by *Acetobacter xylinum, Carbohydr. Res.* **245**, 303–310.

JORIS, K., VANDAMME, E. J. (1993), Novel production and application aspects of bacterial cellulose, *World Microbiol.* **1**, 27–29.

KANG, K. S., PETTIT, D. J. (1993), Xanthan, gellan, welan and rhamsan, in: *Industrial Gums* (WHISTLER, R. L., BEMILLER, J. N., Eds), pp. 341–397. San Diego: Academic Press.

KANG, K. S., VEEDER, G. T. (1982), *U.S. Patent* **4**, 326,053.

Kelco (1992), *Xanthan Gum: Natural Biogum for Scientific Water Control*, 4th Edn. San Diego: Kelco.

KIERULF, C., SUTHERLAND, I. W. (1988), Thermal stability of xanthan preparations, *Carbohydr. Polym.* **9**, 185–194.

LAWFORD, H. G., ROUSSEAU, J. D. (1991), Bioreactor design considerations in the production of high quality microbial exopolysaccharides, *Appl. Biochem. Biotechnol.* **28/29**, 667–684.

LAWFORD, H. G., ROUSSEAU, J. D. (1992), Production of $\beta$-1,3-glucan exopolysaccharide in low shear systems. The requirement for high oxygen tension, *Appl. Biochem. Biotechnol.* **34/35**, 1–17.

LAWFORD, G. R., KLIGERMAN, A., WILLIAMS, T. (1979), Dextran biosynthesis and dextransucrase production by continuous culture of *Leuconostoc mesenteroides, Biotechnol. Bioeng.* **21**, 1121–1131.

LIDHOLT, K., FJELSTAD, M., JANN, K., LINDAHL, U. (1994), Substrate specificities of glycosyltransferases involved in formation of heparin precursor and *E. coli* K5 capsular polysaccharides, *Carbohydr. Res.* **255**, 87–101.

LINTON, J. D. (1990), The relationship between metabolite production and the growth efficiency of the producing organism, *FEMS Microbiol. Rev.* **75**, 1–18.

LINTON, J. D., EVANS, M., JONES, D. S., GOULDNEY, D. N. (1987a), Exocellular succinoglucan production by *Agrobacterium radiobacter* NCIB 11883, *J. Gen. Microbiol.* **133**, 2961–2969.

LINTON, J. D., JONES, D. S., WOODARD, S. (1987b), Factors that control the rate of exopolysaccharide production by *Agrobacterium radiobacter* NCIB 11883, *J. Gen. Microbiol.* **133**, 2979–2987.

LINTON, J. D., ASH, S. G., HUYBRECHTS, L. (1991), Microbial polysaccharides, in: *Biomaterials* (BYROM, D., Ed.), pp. 216–261. London: Macmillan.

LOBAS, D., SCHUMPE, S., DECKWER, W.-D. (1992), The production of gellan exopolysaccharide with *Sphingomonas paucimobilis* E2 (DSM 6314), *Appl. Microbiol. Biotechnol.* **37**, 411–415.

LOBAS, D., NIMTZ, M., WRAY, V., SCHUMPE, S., PROPPE, C., DECKWER, W.-D. (1994), Structural and physical properties of the extracellular polysaccharide PS-P4 produced by *Sphingomonas paucimobilis* P4 (DSM 6418), *Carbohydr. Res.* **251**, 303–313.

MACCORMICK, C. A., HARRIS, J. E., GUNNING, A. P., MORRIS, V. J. (1993), Characterization of a variant of the polysaccharide acetan produced by a mutant of *Acetobacter xylinum* strain CR1/4, *J. Appl. Bacteriol.* **74**, 196–199.

MARGARITIS, A., PACE, G. W. (1985), Microbial polysaccharides, in: *Comprehensive Biotechnology,* Vol. 3 (MOO YOUNG, M., Ed.), pp. 1005–1044. OXFORD: Pergamon.

MILLER, T. L., CHURCHILL, B. W. (1986), Substrates for large-scale fermentation, in: *Manual of Industrial Microbiology/Biotechnology* (DEMAIN, A. L., SOLOMONS, N. A., Eds.), pp. 122–136. Washington, DC: American Society for Microbiology.

MISAKI, A., KISHIDA, E., KAKUTA, M., TABATA, K. (1993), Antitumour fungal (1→3)-β-D-glucans: Structural diversity and effects of chemical modification, in: *Carbohydrates and Carbohydrate Polymers. Analysis, Biotechnology, Modification, Antiviral, Biomedical and Other Applications* (YALPANI, M., Ed.), pp. 116–129. MT. PLEASANT: ATL Press.

MOORHOUSE, R., COLEGROVE, G. T., SANDFORD, P. A., BAIRD, J. S., KANG, K. S. (1981), in: *Solution Properties of Polysaccharides* (BRANT, D., Ed.), pp. 11–124. Washington, DC: American Chemical Society.

NGUYEN, A-L., GROTHE, S., LUONG, J. H. T. (1988), Applications of pullulan in aqueous two-phase systems for enzyme production, purification and utilization, *Appl. Microbiol. Biotechnol.* **27**, 341–346.

OVEREEM, A. (1979), in: *Polysaccharides in Food* (BLANSHARD, J. M. V., MITCHELL, J. R., Eds.), pp. 301–315. London, Boston, MA: Butterworths.

PAZZANI, C., ROSENOW, C., BOULNOIS, G. J., BRONNER, D., JANN, K., ROBERTS, I. S. (1993), Molecular analysis of region 1 of *Escherichia coli* antigen gene cluster: a region encoding proteins involved in the cell surface expression of capsular polysaccharide, *J. Bacteriol.* **175**, 5978–5983.

POLLOCK, T. J. (1993), Gellan-related polysaccharides and the genus *Sphingomonas, J. Gen. Microbiol.* **139**, 1939–1945.

REUSS, M. (1980), *10th Jubilee Conference of Chemical Engineering.* Lodz, Poland.

ROBBINS, D. J., MOULTON, J. E., BOOTH, A. N. (1964), Subacute toxicity study of a microbial polysaccharide fed to dogs, *Food Cosmet. Toxicol.* **2**, 545–552.

ROBYT, J. F., WALSETH, T. F. (1978), The mechanism of acceptor reactions of *Leuconostoc mesenteroides* B512F dextransucrase, *Carbohydr. Res.* **61**, 433–445.

ROBYT, J. F., WALSETH, T. F. (1979), Production, purification and properties of dextransucrase from *Leuconostoc mesenteroides. Carbohydr. Res.* **68**, 95–111.

ROGERS, N. E. (1973), in: *Industrial Gums* (WHISTLER, R. L., BEMILLER, J. N., Eds.), pp. 499–511. New York, London: Academic Press.

RYE, A. J., DROZD, J. W., JONES, C. W., LINTON, J. D. (1988), Growth efficiency of *Xanthomonas campestris* in continuous culture, *J. Gen. Microbiol.* **134**, 1055–1061.

SANDFORD, P. A. (1979), Exocellular microbial polysaccharides, *Adv. Carbohydr. Chem. Biochem.* **36**, 265–313.

SANDFORD, P. A., BAIRD, J., COTTRELL, I. W. (1981), in: *Solution Properties of Polysaccharides* (BRANT, D., Ed.), pp. 31–41. Washington, DC: American Chemical Society.

SCHMID, G. (1991), Preparation and application of γ-cyclodextrin, in: *New Trends in Cyclodextrins and Derivatives* (DUCHÊNE, D., Ed.), pp. 25–54. Paris: Editions de Santé.

SEVIOUR, R. J., STASINOPOULOS, S. J., AUER, D. P. F., GIBBS, P. A. (1992), Production of pullulan and other exopolysaccharides by filamentous fungi, *CRC Crit. Rev. Biotechnol.* **12**, 279–298.

SHATWELL, K. P., SUTHERLAND, I. W., ROSS-MURPHY, S. B. (1990a), Influence of acetyl and pyruvate substituents on the solution properties of xanthan polysaccharide, *Int. J. Biol. Macromol.* **12**, 71–78.

SHATWELL, K. P., SUTHERLAND, I. W., DEA, I. C. M., ROSS-MURPHY, S. B. (1990b), The influence of acetyl and pyruvate substituents on the solution properties of xanthan polysaccharide, *Carbohydr. Res.* **206**, 87–103.

SHATWELL, K. P., SUTHERLAND, I. W., ROSS-MURPHY, S. B., DEA, I. C. M. (1991a), Influence of the acetyl substituent on the interaction of xanthan with plant polysaccharides. I. Xanthan–locust bean gum systems, *Carbohydr. Polym.* **14**, 29–51.

SHATWELL, K. P., SUTHERLAND, I. W., ROSS-MURPHY, S. B., DEA, I. C. M. (1991b), Influence of the acetyl substituent on the interaction of xanthan with plant polysaccharides. II. Xanthan–guar gum systems, *Carbohydr. Polym.* **14**, 115–130.

SHATWELL, K. P., SUTHERLAND, I. W., ROSS-MURPHY, S. B., DEA, I. C. M. (1991c), Influence of the acetyl substituent on the interaction of xanthan with plant polysaccharides. III. Xanthan–Konjac mannan systems, *Carbohydr. Polym.* **14**, 131–147.

SHINABARGER, D., BERRY, A., MAY, T. B., ROTHMEL, R., FIALHO, A., CHARABARTY, A. M. (1991), Purification and characterization of phosphomannose isomerase–GDP mannose pyrophosphorylase, *J. Biol. Chem.* **266**, 2080–2088.

SHINABARGER, D., MAY, T. B., BOYD, A., GHOSH, M., CHARABARTY, A. M. (1993), Nucleotide sequence and expression of the *Pseudomonas aeruginosa algF* gene controlling acetylation of alginate, *Mol. Microbiol.* **9**, 1027–1035.

SILVER, R. P., AARONSON, W., VANN, W. F. (1987), Translocation of capsular polysaccharide in pathogenic strains of *Escherichia coli* requires a 60-kilodalton periplasmic protein, *J. Bacteriol.* **169**, 5489–5495.

SKJÅK-BRÆK, G., LARSEN, B. (1985), Biosynthesis of alginate: purification and characterisation of mannuron C-5 epimerase from *Azotobacter vinelandii, Carbohydr. Res.* **139**, 273–283.

SKJÅK-BRÆK, G., SMIDSROD, O., LARSEN, B. (1986), Tailoring of alginates by enzymatic modification *in vitro, Int. J. Biol. Macromol.* **8**, 330–336.

SMITH, I. H., PACE, G. (1982), Recovery of microbial polysaccharides, *J. Chem. Tech. Biotechnol.* **32**, 119–129.

SOUW, P. DEMAIN, A. L. (1979), Nutritional studies on xanthan production by *Xanthomonas campestris, Appl. Env. Microbiol.* **37**, 1186–1192.

SUTHERLAND, I. W. (1979), Enhancement of polysaccharide viscosity by mutagenesis, *J. Appl. Biochem.* **1**, 60–70.

SUTHERLAND, I. W. (1984), Hydrolysis of unordered xanthan in solution by fungal cellulases, *Carbohydr. Res.* **131**, 93–104.

SUTHERLAND, I. W. (1990), *Biotechnology of Microbial Exopolysaccharides.* Cambridge: Cambridge University Press.

SUTHERLAND, I. W. (1994), Structure– function relationships in microbial exopolysaccharides, *Biotechnol. Adv.* **12**, 293–348.

SZETJLI, J. (1986), Cyclodextrins in biotechnology, *Stärke* **38**, 388-393.

TAIT, M. I., SUTHERLAND, I. W. (1989), Synthesis and properties of a mutant type of xanthan, *J. Appl. Bacteriol.* **66**, 457–460.

TAIT, M. I., SUTHERLAND, I. W., CLARKE-STURMAN, A. J. (1986), Effect of growth conditions on the production, composition and viscosity of *Xanthomonas campestris* exopolysaccharide, *J. Gen. Microbiol.* **132**, 1483–1492.

TALASHEK, T. A., BRANT, D. A. (1987), The influence of side-chains on the calculated dimensions of three related bacterial polysaccharides, *Carbohydr. Res.* **160**, 303–316.

Tate and Lyle Ltd. (1981), *British Patent* 2065089A.

TRISLER, P., GOTTESMAN, S. (1984), *lon* transcriptional regulation of genes necessary for capsular polysaccharide synthesis in *Escherichia coli* K12, *J. Bacteriol.* **160**, 184–191.

TSUJISAKA, Y., MITSUHASHI, M. (1993), Pullulan, in: *Industrial Gums* (WHISTLER, R. L., BEMILLER, J. N., Eds), pp. 447–460. San Diego: Academic Press.

UNSAL, E., DUDA, J. L., KLAUS, E. (1979), Comparison of solution properties of mobility control polymers, in: *Chemistry of Oil Recovery* (JOHANSEN, R. T., BERG, R. L., Eds.), pp. 141–170. Washington, DC: American Chemical Society.

VANN, W. F., SCHMIDT, M. A., JANN, B., JANN, K. (1981), The structure of the capsular polysaccharide (K5 antigen) of urinary tract infective *Escherichia coli* O10:RK5:RH4. A polymer similar to desulfo-heparin, *Eur. J. Biochem.* **116**, 359–364.

VOSSOUGHL, S., BULLER, C. S. (1991), Permeability modification by *in situ* gelation with a newly discovered biopolymer, *SPEJ, Reservoir Engineering,* pp. 485–489.

WALTER, H., BROOKS, D. E., FISHER, D. (1985), *Partitioning in Aqueous Two-Phase Systems.* New York: Academic Press.

WERNAU, W. C. (1979), *British Patent* 2,008,600A.

WERNAU, W. C. (1980), *U.S. Patent* 4,282,321.

WHITFIELD, C., VALVANO, M. A. (1993), Biosynthesis and expression of cell surface polysaccharides in Gram negative bacteria, *Adv. Microb. Physiol.* **35**, 135–246.

WILEY, B. J., BALL, D. H., ARCDIACONO, S. M., SOUSA, S., MAYER, J. M., KAPLAN, D. L. (1993), Control of molecular weight distribution of the biopolymer pullulan produced by *Aureobasidium pullulans, J. Environ. Polym. Degrad.* **1**, 3–9.

WINDASS, J. D., WORSEY, M. J., PIOLOI, E. M., PIOLI, D., BARTH, P. T., ATHERTON, K. T., DART, E. C., BYROM, D., POWELL, K., SENIOR, P. (1980), Improved conversion of methanol to single-cell protein by *Methylophilus methylotrophus, Nature* **287**, 396–401.

WOODARD, G., WOODARD, M. W., MCNEELY, W. H., KOVACS, P., CRONIN, M. T. I. (1973), Xanthan gum: Safety evaluation by two year feeding studies in rats and dogs and three generation reproduction study in rats, *Toxicol. Appl. Pharmacol.* **24**, 30–36.

YALPANI, M., BROOKS, D. E. (1987), New dextran derivatives, in: *Industrial Polysaccharides* (STIVALA, S. S., CRESCENZI, V., DEA, I. C. M., Eds.), pp. 145–156. New York: Gordon and Breach.

ZHAN, D. F., RIDOUT, M. J., BROWNSEY, G. J., MORRIS, V. J. (1993), Xanthan-locust gum bean interactions and gelation. *Carbohydr. Polym.* **21**, 53–58.

# 17 Biosurfactants

NAIM KOSARIC

London, Ontario, Canada

# 1 Introduction

Microbial surface active agents (biosurfactants) are important products which find a wide application in many industries. Their properties of interest are in changing surface active phenomena such as a lowering of surface and interfacial tensions, wetting and penetrating actions, spreading, hydrophilicity and hydrophobicity, emulsification and de-emulsification, detergency, gelling, foaming, flocculating actions, microbial growth enhancement, metal sequestration and anti-microbial action.

Most of the applications today involve the use of chemically synthesized surfactants. Production of surfactants in the United States and worldwide is estimated at $3.4 \cdot 10^9$ kg and $7 \cdot 10^9$ kg in 1989, respectively. The U.S. surfactant industry shipments in 1989 were $3.65 billion (DESAI and DESAI, 1993). The applications are very wide in a variety of industries as shown in Tab. 1.

There are many advantages of biosurfactants as compared to their chemically synthesized counterparts. Some of those are:

- biodegradability
- generally low toxicity
- biocompatibility and digestibility, which allows their application in cosmetics, pharmaceuticals and as functional food additives
- availability of raw materials; biosurfactants can be produced from cheap raw materials which are available in large quantities; the carbon source may come from hydrocarbons, carbohydrates and/or lipids, which may be used separately or in combination with each other
- acceptable production economics; depending upon the application, biosurfactants can also be produced from industrial wastes

**Tab. 1.** Biosurfactant Uses and Effects

| Use | Effect of Surfactant |
|---|---|
| Metals | |
| Concentration of ores | Wetting and foaming, collectors and frothers |
| Cutting and forming | Wetting, emulsification, lubrication and corrosion inhibition in rolling oils, cutting oils, lubricants, etc. |
| Casting | Mold release additives |
| Rust and scale removal | In pickling and electrolytic cleaning |
| Plating | Wetting and foaming in electrolytic plating |
| Paper | |
| Pulp treatment | Deresinification, washing |
| Paper machine | Defoaming, color leveling and dispersing |
| Calender | Wetting and leveling, coating and coloring |
| Paint and protective coatings | |
| Pigment preparation | Dispersing and wetting of pigment during grinding |
| Latex paints | Emulsification, dispersion of pigment, stabilize latex, retard sedimentation and pigment separation, rheology |
| Waxes and polishes | Emulsify waxes, stabilize emulsions, antistat |
| Petroleum production and products | |
| Drilling fluids | Emulsify oil, disperse solids, modify rheological properties of drilling fluids for oil and gas wells |
| Workover of producing wells | Emulsify and disperse sludge and sediment in cleanout of wells |
| Producing wells | De-emulsify crude petroleum, inhibit corrosion of equipment |
| Secondary recovery | In flooding operations, preferential wetting |
| Refined products | Detergent sludge dispersant and corrosion inhibitor in fuel oils, crank-case oils and turbine oils |

**Tab. 1.** Continued

| Use | Effect of Surfactant |
|---|---|
| Textiles | |
| Preparation of fibers | Detergent and emulsifier in raw wool scoaring; dispersant in viscose rayon spin bath; lubricant and antistat in spinning of hydrophobic filaments |
| Dyeing and printing | Wetting, penetration, solubilization, emulsification, dye leveling, detergency and dispersion |
| Finishing of textiles | Wetting and emulsification in finishing formulations, softening, lubricating and antistatic additives to finishes |
| Agriculture | |
| Phosphate fertilizers | Prevent caking during storage |
| Spray application | Wetting, dispersing, suspending of powdered pesticides and emulsification of pesticide solutions; promote wetting, spreading and penetration of toxicant |
| Building and construction | |
| Paving | Improve bond of asphalt to gravel and sand |
| Concrete | Promote air entrainment |
| Elastomers and plastics | |
| Emulsion polymerization | Solubilization, emulsification of monomers |
| Foamed polymers | Introduction of air, control of cell size |
| Latex adhesive | Promote wetting, improve bond strength |
| Plastic articles | Antistatic agents |
| Plastic coating and laminating | Wetting agents |
| Food and beverages | |
| Food processing plants | For cleaning and sanitizing |
| Fruits and vegetables | Improve removal of pesticides, and in wax coating |
| Bakery and ice cream | Solubilize flavor oils, control consistency, retgard staling |
| Crystallization of sugar | Improve washing, reduce processing time |
| Cooking fat and oils | Prevent spattering due to super heat and water |
| Industrial cleaning | |
| Janitorial supplies | Detergents and sanitizers |
| Descaling | Wetting agents and corrosion inhibitors in acid cleaning of boiler tubes and heat exchangers |
| Soft goods | Detergents for laundry and dry cleaning |
| Leather | |
| Skins | Detergent and emulsifier in degreasing |
| Tanning | Promote wetting and penetration |
| Hides | Emulsifiers in fat liquoring |
| Dyeing | Promote wetting and penetration |

and by-products and this is of particular interest for bulk production (e.g., for use in petroleum-related technologies)

- use in environmental control; biosurfactants can be efficiently used in handling industrial emulsions, control of oil spills, biodegradation and detoxification of industrial effluents and in bioremediation of contaminated soil
- specificity; biosurfactants, being complex organic molecules with specific functional groups, are often specific in their action; this would be of particular interest in detoxification of specific pollutants, de-emulsification of industrial emulsions, specific cosmetic, pharmaceutical, and food applications
- effectiveness at extreme temperatures, pH and salinity

Most of the biosurfactants are high molecular-weight lipid complexes which are normally produced under highly aerobic conditions. This is achievable in their *ex situ* production in aerated bioreactors. When their large-scale application in petroleum and soil is encountered, their *in situ* production (and action) would be advantageous. Low oxygen availability under these conditions requires maintenance of anaerobic microorganisms and their anaerobic syntheses of biosurfactants, whereby other conditions for microbial growth are also most unfavorable (e.g., mixing, availability of substrate, mass transfer, availability of trace nutrients, etc.). Screening for anaerobic biosurfactant producers is of great importance for these conditions (MCINERNEY et al., 1990).

The biosurfactant sources, classes, and properties have been reviewed (FINNERTY and SINGER, 1983; CAIRNS et al., 1984; KOSARIC et al., 1983, 1987; HAFERBURG et al., 1986; ZAJIC and STEFFENS, 1984; KOSARIC, 1993; ISHIGAMI, 1993; LANG and WAGNER, 1987, 1993; GERSON, 1993). In general, biosurfactants can be classified as:

- glycolipids,
- hydroxylated and cross-linked fatty acids (mycolic acids),
- polysaccharide–lipid complexes,
- lipoprotein–lipopeptides,
- phospholipids,
- complete cell surface itself.

A list of various biosurfactants produced from different microbes is presented in Tab. 2.

# 2 Biosurfactant Classes

## 2.1 Bacterial Biosurfactants

Many bacteria have been identified as glycolipid producers. The glycolipids may contain various sugar moieties such as rhamnose, trehalose, sucrose, and glucose.

The rhamnolipids may contain one or two rhamnose units and, in general, two $\beta$-hydroxydecanoic acid residues. ITOH et al. (1971) proposed a rhamnolipid (RL2) con-

**Tab. 2.** Various Biosurfactants Produced by Microorganisms

| Microorganism | Type of Surfactant |
|---|---|
| *Torulopsis bombicola* | Glycolipid (sophorose lipid) |
| *Pseudomonas aeruginosa* | Glycolipid (rhamnose lipid) |
| *Bacillus licheniformis* | Lipoprotein (?) |
| *Bacillus subtilis* | Lipoprotein (surfactin) |
| *Pseudomonas* sp. DSM 2847 | Glycolipid (rhamnose lipid) |
| *Arthrobacter paraffineus* | Sucrose and fructose glycolipids |
| *Arthrobacter* | Glycolipid |
| *Pseudomonas fluorescens* | Rhamnose lipid |
| *Pseudomonas* sp. MUB | Rhamnose lipid |
| *Torulopsis petrophilum* | Glycolipid and/or protein |
| *Candida tropicalis* | Polysaccharide–fatty acid complex |
| *Corynebacterium lepus* | Corynomycolic acids |
| *Acinetobacter* sp. HO1-N | Fatty acids, mono- and diglycerides |
| *Acinetobacter calcoaceticus* RAG-1 | Lipoheteropolysaccharide (Emulsan) |
| *Acinetobacter calcoaceticus* 2CAC | Whole cell (lipopeptide) |
| *Candida lipolytica* | "Liposan" (mostly carbohydrate) |
| *Candida petrophilum* | Peptidolipid |
| *Nocardia erythropolis* | Neutral lipids |
| *Rhodococcus erythropolis* | Trehalose dimycolates |
| *Corynebacterium salvonicum* SFC | Neutral lipid |
| *Corynebacterium hydrocarboclastus* | Polysaccharide–protein complex |

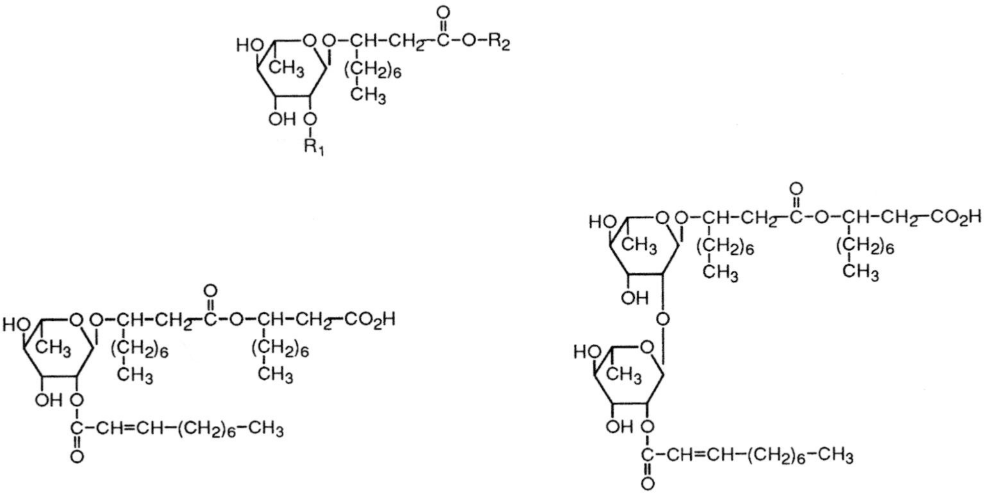

**Fig. 1.** Rhamnolipids from *Pseudomonas* sp.
RL 1: $R_1$ = L-$\alpha$-rhamnopyranosyl- $R_2$ = $\beta$-hydroxydecanoic acid
RL 2: $R_1$ = H
$R_2$ = $\beta$-hydroxydecanoic acid
RL 3: $R_1$ = L-$\alpha$-rhamnopyranosyl-
$R_2$ = H
RL 4: $R_1$ = H
$R_2$ = H

taining one molecule of rhamnose and two molecules of fatty acid while SYLDATK et al. (1985a) isolated the two rhamnolipids RL3 and RL4. The rhamnolipid RL1 was reported by BERGSTROM et al. (1947), JARVIS and JOHNSON (1949), EDWARDS and HAYASHI (1965), and HISATSUKA et al. (1971). Rhamnolipids containing one or two rhamnose units are presented in Fig. 1.

Trehalose lipids are, in most cases, cell-wall associated. The $\alpha$-branched $\beta$-hydroxy fatty acids are usually esterified with the 6 and 6′-hydroxyl groups of the trehalose unit. A trehalose lipid from *Rhodococcus erythropolis* (RAPP et al., 1979) is presented in Fig. 2.

Sucrose lipids were isolated from *Arthrobacter paraffineus* grown on sucrose (SUZUKI et al., 1974). Acyl glucoses were found when *Corynebacterium diphteriae* was grown on glucose as a carbon source (BRENNAN et al., 1970). This glycolipid is shown in Fig. 3.

Ornithin containing lipids have been reported to be produced by *Pseudomonas rubescens* and *Thiobacillus thiooxidans* (WIL-

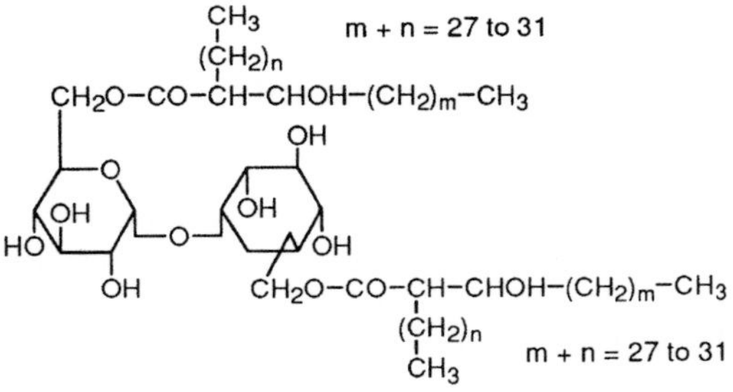

**Fig. 2.** Trehalose lipids from *Rhodococcus erythropolis.*

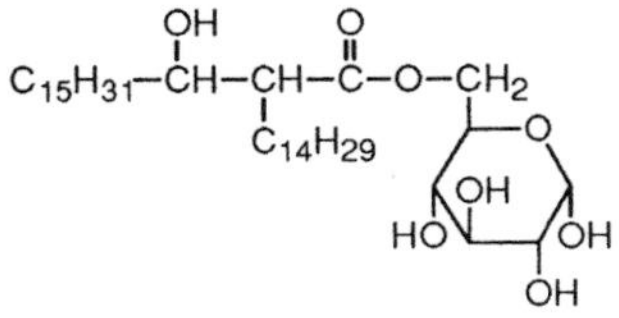

**Fig. 3.** Glucose-6-monocorynomycolate from *Corynebacterium diphtheriae.*

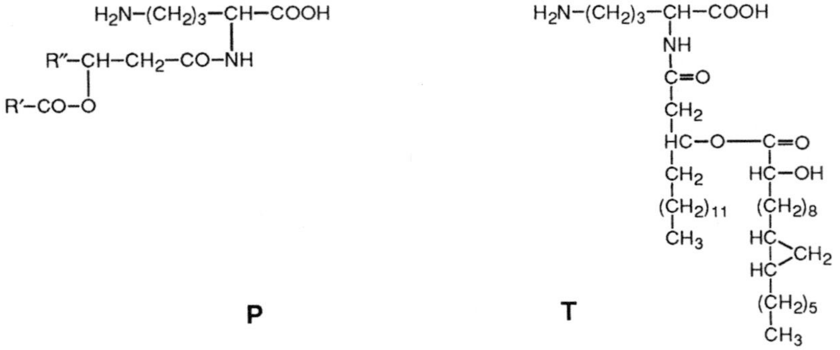

Fig. 4. Ornithin containing lipids from *Pseudomonas rubescens* (P) and from *Thiobacillus thiooxidans* (T).
R′, R″ = alkyl chains

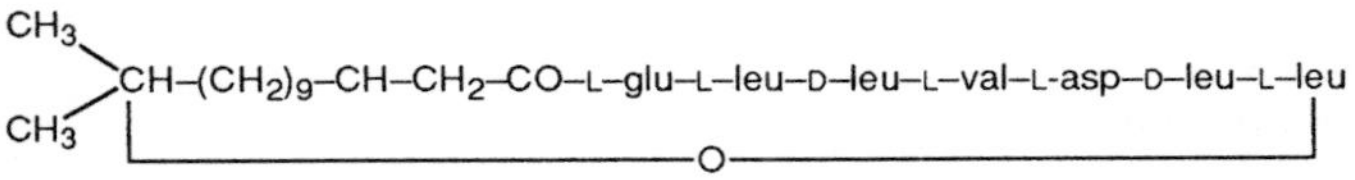

Fig. 5. Surfactin from *Bacillus subtilis.*

KINSON, 1972; KNOCHE and SHIVELY, 1972). Structures of these are shown in Fig. 4.

Another lipid of interest is surfactin produced by *Bacillus subtilis* (KAKINUMA et al., 1969), the structure of which is shown in Fig. 5.

## 2.2 Yeast Biosurfactants

Yeast biosurfactants are of particular interest for applications in food and cosmetics. *Torulopsis* species have been of most interest for the production of these lipids. As an example, *Torulopsis magnoliae (bombicola)*, when grown on glucose, yeast extract, and urea, produced predominantly two glycolipids containing sophorose in a lactonic and acidic configuration as shown in Fig. 6 (TULLOCH et al., 1968).

The main components of sophorose lipids consist of partly acetylated 2′-*o*-β-D-glucopyranosyl-D-glucopyranose units attached β-glycosidically to 17-L-hydroxyoctadecanoic and 17-L-hydroxy-9-octadecenoic acids.

## 2.3 Exopolysaccharide Bioemulsifiers

Microbial exocellular polysaccharides as high molecular-weight polymers show useful physical and mechanical properties such as

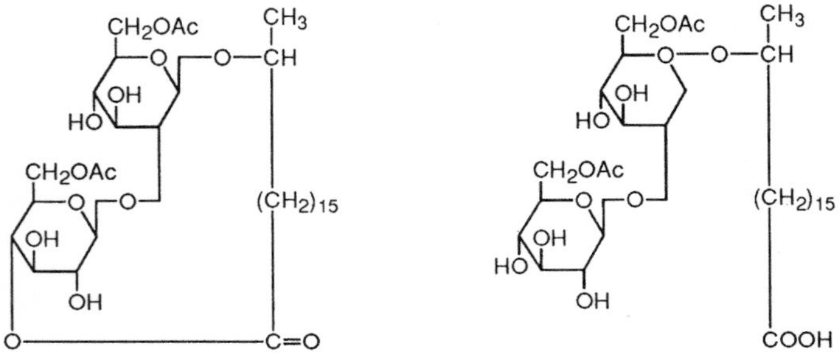

Fig. 6. Lactonic and acidic sophrose lipid from *Torulopsis magnoliae (bombicola).*

high viscosity, tensile strength, resistance to shear, etc. They are widely used in industry as gums, rheology modifiers, thickening agents, high viscosity stabilizers, gelling agents, etc. (BASTA, 1984; GABRIEL, 1979; HOYT, 1985; PACE, 1981). These materials are also believed to play an important role in the adaptation of microorganisms to specific ecological challenges such as in attachment and adsorption from surfaces, interference with cell surface receptors and in the utlization of specific carbon and energy sources.

The major characteristic of these bioemulsifiers appears to be their high affinity for oil–water interfaces, thus making them excellent emulsion stabilizers by forming a stable film around the oil droplet and preventing coalescence.

One of the best studied such bioemulsifier is Emulsan, produced by the oil degrading bacterium *Acinetobacter calcoaceticus* RAG-1 (KAPLAN and ROSENBERG, 1982).

Emulsan is a polyanionic heteropolysaccharide bioemulsifier (ZUCKERBERG et al., 1979) with a high molecular weight of $9.8 \times 10^5$ Da. The polymer is also characterized by a reduced viscosity of over 500 $cm^3\ g^{-1}$ which depends on the ionic strength, pH, and temperature. Emulsan does not appreciably reduce interfacial tension (10 $dyn\ cm^{-1}$) but binds tightly to the newly created interface and protects the oil droplets from coalescence. A comprehensive review of Emulsan properties and production has been written by GUTNICK and SHABTAI (1987).

# 3 Biosynthesis

Biochemical mechanisms for syntheses of biosurfactant lipids have been extensively elaborated by RATLEDGE (1988), HOMMEL and RATLEDGE (1993), RATLEDGE and BULL (1982) and reviewed by TURCOTTE and KOSARIC (1989). Fig. 7 shows the intermediary metabolism relating to biosurfactant syntheses from hydrocarbon substrates.

For lipogenesis, microbial culture conditions can be manipulated in order to achieve optimal yields of the biosurfactant. The microorganisms of interest must be oleaginous in nature, which means capable of accumulating lipids to an amount higher than 25% of their total dried biomass.

In batch culture, lipid accumulation normally occurs when the medium contains an excess of the principal carbon source over some other limiting nutrient. As shown in Fig. 8a, the lipid biosynthesis starts when nitrogen becomes limiting. During the initial phase of balanced growth, all nutrients are in ample supply and the biomass is accumulating according to the general growth curve. If, however, the limiting nutrient becomes exhausted, growth will slow down. The carbon, however, continues to be transported into the cell, where it is utilized for lipid biosynthesis. Consequently, a rise in lipid concentration is observed (BOULTON and RATLEDGE, 1987).

The limiting nutrient which can be easily manipulated is, in most cases, nitrogen though limitation of magnesium, iron, and phosphate may elicit a similar response (GILL et al., 1977).

Lipids may also be accumulated in continuous culture when a similar medium composition is applied. In addition, the dilution rate must be below a critical value at which microbial growth is just sufficient to fully assimilate the limiting nutrient at steady state. A typical pattern in continuous culture is shown in Fig. 8b (BOULTON and RATLEDGE, 1987).

# 4 Biosurfactant Production

Biosurfactants can be produced from both hydrocarbon and carbohydrate substrates. It is believed that the microorganisms capable of assimilating hydrocarbons excrete ionic surfactants which emulsify the hydrocarbon in the medium and thus facilitate its uptake by the cell. Characteristic of this mechanism are *Pseudomonas* species or the sophorolipids produced by different *Torulopsis* species (SYLDATK and WAGNER, 1987).

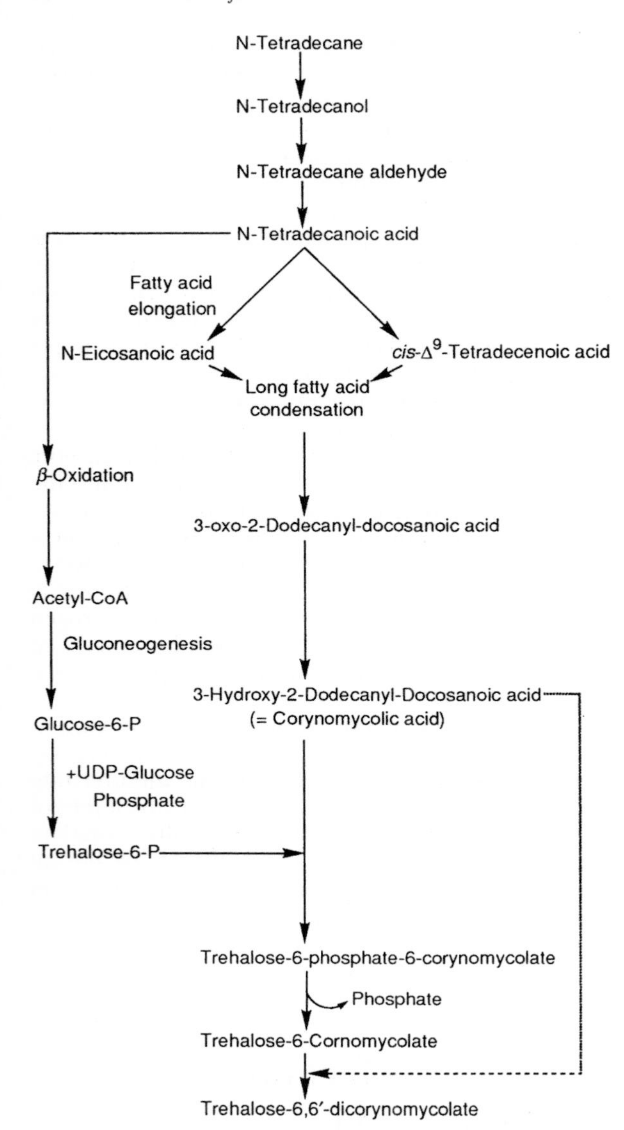


**Fig. 7.** Synthesis of trehalose biosurfactants form *n*-alkane (tetradecane).

On the other hand, some microorganisms change the structure of their cell wall by building their lipopolysaccharides or nonionic surfactants. To this group belong *Candida lipolytica* and *Candida tropicalis* which produce cell-wall bound lipopolysaccharides when growing on *n*-alkanes (OSUMI et al., 1975; FUKUI and TANAKA, 1981) and *Rhodococcus erythropolis* and several *Mycobacterium* and *Arthrobacter* species which synthesize nonionic trehalose corynomycolates (KRETSCHMER et al., 1981; RAPP et al., 1979; KILBURN and TAKAYAMA, 1981; SUZUKI et al., 1980).

Many biosurfactants are also synthesized by specific microorganisms when carbohydrates are present in the growth medium (BOULTON and RATLEDGE, 1987; SYLDATK

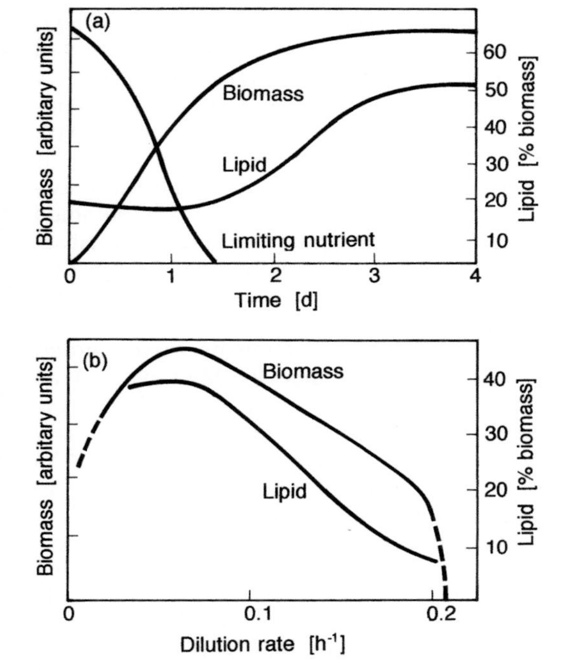


**Fig. 8.** Patterns of lipid accumulation in oleaginous microorganisms (**a**) in batch culture, (**b**) in continuous culture.

and WAGNER, 1987). This is of particular interest in the mass production of this material whereby the cost of substrate could be considerably reduced by use of cheap carbohydrates and even wastes (KOSARIC et al., 1984).

According to SYLDATK and WAGNER (1987), biosurfactants can be produced under growth associated and growth limiting conditions as well as by resting cells and by addition of precursors:

(1) Cell growth-associated production of biosurfactants:
   - induction of production by lipophilic substrates
   - increase of production by optimization of medium composition
   - increase of production by optimization of environmental influences as pH, temperature, aeration, agitation speed, etc.
   - increase of production by addition of reagents, which cause a change of cell wall permeability as penicillin, ethambutol, EDTA, etc.
   - increase of production by addition of reagents which cause a detachment of cell wall-bound biosurfactants into the medium as alkanes, kerosene, EDTA, etc.

(2) Biosurfactant production by growing cells under growth-limiting conditions:
   - production under N-limitation
   - production under limitation of multivalent cations
   - increase of production under growth-limiting conditions by a change of environmental conditions as pH or temperature

(3) Biosurfactant production by resting cells:
   - production by resting free cells
   - production by resting immobilized cells
   - production by resting immobilized cells with simultaneous product removal

(4) Biosurfactant production by growing, resting free, and resting immobilized cells with addition of precursors

The production of biosurfactants is often associated with foam formation, a lowering of the surface tension and interfacial tension in the medium, and by an emulsion of the lipophilic substrate in the culture broth. These characteristics can be utilized to show and evaluate quantitatively the biosynthesis of the biosurfactant.

The carbon source is quite important. As an example, *Pseudomonas* species show the best biosurfactant production when *n*-alkanes are present in the medium for rhamnolipid production (SYLDATK et al., 1985b; YAMAGUCHI et al., 1976b). However, glycerol, glucose, and ethanol gave lower biosurfactant yields. Sometimes the hydrocarbon chain length had also an effect. *Corynebacterium hydrocarboclastus* gave the best yields when linear alkanes of chain length $C_{12}$, $C_{13}$, and $C_{14}$ were present (RAPP et al., 1977). *Arthrobacter paraffineus* ATCC 19558 was grown on glucose and produced surface active agents by adding hexadecane to the medium in the stationary growth phase (DUVNJAK et al., 1982). An addition of suitable lipophilic compounds as long-chain acids, esters, hydrocarbons, or glycerides to growing cultures of *Torulopsis bombicola* increased glycolipid production 2- to 3-fold. Also, the same culture considerably increased glycolipid production when vegetable oils were added in the later exponential growth phase (COOPER and PADDOCK, 1984; ZHOU et al., 1992; ZHOU and KOSARIC, 1993).

As mentioned earlier, nitrogen limitation plays an important role. The nature of the N-source also affects biosurfactant production, as shown by *A. paraffineus* ATCC 19558 (DUVNJAK et al., 1983). The effects of different amino acids such as aspartic acid, asparagine, glycine, or glutamic acid in a mineral salts medium as well as yeast extract, peptone, bactotryptone, and nutrient broth were evaluated. The yeast extract yielded 6 times more surfactant than the control. Asparagine can serve as a sole nitrogen source. When inorganic salts were used as a nitrogen source, the microorganism preferred ammonium to nitrate forms of nitrogen. Urea could be utilized both as a sole source of nitrogen and in combination with an inorganic nitrogen salt, yielding relatively high surfactant concentrations in the broth.

Culture conditions such as pH, temperature, and ionic strength also influence biosurfactant production. The temperature was of great importance in the case of *A. paraffineus* ATCC 19558 (DUVNJAK et al., 1982), *R. erythropolis* (WAGNER et al., 1984), and *Pseudomonas* sp. (SYLDATK et al., 1985b).

SYLDATK and WAGNER (1987) showed a successful biosurfactant production by resting cells. These cells have been first grown in a suitable medium and were then centrifuged and washed from possibly disturbing by-products. The wet biomass was then used for production of biosurfactants (secondary metabolites) under specific conditions. The method was successfully applied to rhamnolipid biosynthesis from *Pseudomonas* sp. (SYLDATK et al., 1984), sophorolipid from *T. bombicola* (GOBBERT et al., 1984), cellobiose lipid from *Ustilago maydis* (FRAUTZ and WAGNER, 1984) and trehalose tetraester production from *Rhodococcus erythropolis* (WAGNER et al., 1984).

LIN et al. (1993) studied the production of lipopeptide biosurfactants by *Bacillus licheniformis.* This biosurfactant has been reported to possess excellent interfacial properties. A maximum concentration of 110 mg $L^{-1}$ lipopeptide was obtained in optimized media with 1.0% (w/v) glucose as the carbon source. It was also interesting to notice that the maximum amount of the surfactant was obtained in early stationary-phase cultures, but subsequently decreased rapidly and disappeared completely from the fermentation broth within 8 h.

Rhamnolipid production by *Pseudomonas aeroguiosa* was studied by MANRESA et al. (1991). The biosurfactant accumulation occurred in the exponential and stationary phases, and a maximum product yield (P/X) of 2.9 was detected when the carbon to nitrogen (C/N) ratio was 6.6.

*Bacillus licheniformis* has also been successfully used to produce biosurfactants under anaerobic conditions (JAVAHERI et al., 1985; MCINERNEY et al., 1990). Production of biosurfactants anaerobically is of particular interest for *in situ* enhanced oil recovery (EOR) applications. For the same applications, biosurfactants which are produced at high salt concentrations (halotolerant) have been investigated (JENNEMAN et al., 1983).

Trehalose lipid biosurfactants have also been shown to be of interest in enhanced oil recovery operations (SINGER and FINNERTY, 1990; SINGER et al., 1990; MARTIN et al., 1991; DATTA et al., 1991). *Rhodococcus* and *Corynebacterium* species have been investigated for this purpose.

Biosurfactants in EOR have also been studied by YULBARISOV (1990), and by FALATKO and NOVAK (1992).

Production of rhamnolipid biosurfactants from olive oil was studied by PARRA et al. (1990) using *Pseudomonas* strains.

Of interest are also biosurfactants produced from marine microorganisms (PASSERI et al., 1992). A novel glucose lipid synthesized by a marine bacterial strain MMI was reported and its structure was elucidated.

Spiculisporic acid biosurfactant produced by *Penicillium spiculisporicum* is also of interest (ISHIGAMI et al., 1988). A number of compounds can be derived from spiculisporic acid (4,5-dicarboxy-4-pentadecanolide) such as al-

**Tab. 3.** Production of Sophorose Lipids by *Torulopsis bombicola* in Batch Culture

| Medium (C and N Source) | Bioreactor Volume | Max. Yield [g $L^{-1}$] | Reference |
|---|---|---|---|
| 10% Glucose<br>14% palm oil<br>0.75% yeast N base<br>0.2% yeast extract | 500 mL | 120 | INOUE and ITO (1982) |
| 10% Glucose<br>9.5% sunflower oil<br>0.5% yeast extract | 7 L | 70 | COOPER and PADDOCK (1984) |
| 13.6% Oleic acid<br>1% yeast extract<br>0.1% urea | 30 L | 74 | ASHMER et al. (1988) |
| 10% Glucose<br>10.5% sunflower oil<br>0.25% yeast extract<br>0.2% urea | 20 L | 137 | ZHOU et al. (1992) |
| 10% Glucose<br>10% canola oil<br>0.4% yeast extract<br>0.1% urea | 1 L | 150–160 | ZHOU and KOSARIC (1993) |
| 10% Lactose<br>10% canola oil<br>0.4% yeast extract<br>0.2% urea | 1 L | 100–120 | ZHOU and KOSARIC (1993) |
| 10% Glucose<br>10.5% canola oil<br>0.25% yeast extract | 1 L | Fed batch<br>200 | ZHOU (1994) |

kylamine salts with multifunctional surface-active liquid crystal-, gel- and liposome-forming properties. New rhodamine-type dyes and a liposaccharide containing glucosamine residue were also produced from spiculisporic acid.

It is evident from the cited literature and supported by our own data, that for maximum yields of glycolipids from *T. bombicola,* two substrates (sugar and lipid or hydrocarbon) are required. Using inexpensive and commercially available substrates (molasses and soybean oil) it was calculated (COOPER and PADDOCK, 1984) that the substrates alone would contribute about C$1.00 to the production cost at a total production cost of about C$3.00 per kg. The price of commercial surfactants of the nonionic alcohol ethoxylate and alkylphenol ethoxylate types for use in EOR has been estimated at about C$1.40–$1.60 per kg. Sophorose lipids are now commercially available at a price of about $23 per kg (ISHIGAMI, 1993). According to our own estimate, the production cost, based on the high yields reported at 150 g sophorose lipids per liter of broth (ZHOU and KOSARIC, 1993) would amount up to C$1.0 per kg.

Data on the production of sophorose lipids in the author's laboratory are compared with available literature sources (Tab. 3). As shown, very high yields are obtainable which is of considerable commercial interest.

In order to reduce production costs, a multiorganism strategy for biosurfactant production was proposed (KOSARIC et al., 1984). The strategy is illustrated in Fig. 9. In this, an appropriate lipogenic bacterium or yeast *(Lipomyces)* and/or the appropriate lipogenic al-

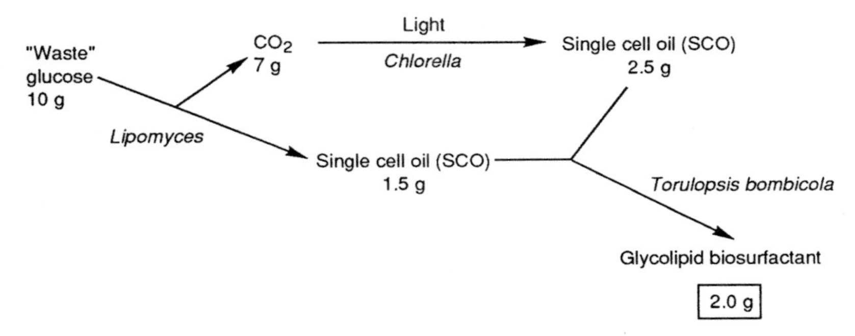


**Fig. 9.** Production of sophorolipids by use of *Lipomyces, Chlorella,* and *Torulopsis bombicola.*

**Tab. 4.** Lipid Accumulating Microbes which Could be Used to Produce Triglyceride Precursors to Biosurfactants

| Microorganism | Class | Substrate | Total Lipid as a % Biomass [% neutral lipid] | Lipid Yield [g lipid per 100 g sugar] |
|---|---|---|---|---|
| *Lipomyces lipofer* | yeast | glucose | 60–65 (80) | 22.0 |
| *Rhodotorula gracilis* | yeast | glucose | 55–65 (75) | 16.0 |
| *Chlorella vulgaris* | algae | $CO_2$ | 30 (65) | – |
| *Chlorella pyrenoidosa* | algae | $CO_2$ | 65–75 (70) | – |
| *Chlorella sorokinana* | algae | $CO_2$ | 45 (80) | – |
| *Arthrobacter* AK19 | bacteria | glucose | 80 (90) | 14.0 |

gae *(Chlorella)* are cultured to produce microbial single-cell oil in the form of triglycerides. The microbial triglycerides and the same sugar used to feed the lipogenic microbes are used by *T. bombicola* to produce the glycolipid. Some of the lipogenic microbes which could be used in this first phase are shown in Tab. 4. Wastes on which lipogenic microbes have been grown include rice hull hydrolyzate, starch waste liquors, whey, domestic waste, and potato processing wastes. Production processes, based on these wastes, would thus cut the production costs at least by 30% as the substrate would generally be available at minimal or no cost.

Another strategy has also been proposed (KOSARIC et al., 1984) which utilizes municipal sludge waste. This concept is presented in Fig. 10.

*Torulopsis* biomass from the biosurfactant production could be recycled to the front end of the entire process for treatment, or separated from the medium and sold as a yeast-rich feed supplement, the gaseous products ($CO_2 + CH_4$) can be used in the process. Five final products are obtained: treated water, high-grade biomass, low-grade biomass, biosurfactant and methane-rich gas. The other benefit is in utilization of the nuisance waste sludges which is environmentally beneficial.

# 5 Biosurfactant Properties of Interest for Cosmetics

The application of surfactants in cosmetics is very broad and the demands on surfactant properties differ substantially according to the application. In the following, a general overview of these applications is given for a variety of products.

- Surfactants function as:
  emulsifier, foaming agent, solubilizer, wetting agent, cleanser, antimicrobial agent, mediator of enzyme action.
- Forms of cosmetics:
  cream, lotion, liquid, paste, powder, stick, gel, film, spray.
- Cosmetic products using surfactants:
  insect repellent, antacid, bath products, acne pads, antidandruff products, contact lens solution, hair color and care products, deodorant, nail care, body massage accessories, lipstick, lipmaker, eye shadow, mascara, soap, tooth paste and polish, denture cleaner, adhesives, antiperspirant, lubricated condoms, baby products, food care, mousse, antiseptics, shampoo, conditioner, shave and depilatory products, moisturizer, health and beauty products.
- Surfactant groups:
  I. amphoteric surfactants: acyl-amino acids, N-acyl amino acids
  II. anionic surfactants: acyl-amino acids and derivatives, carboxylic acids, ester and ether carboxylic acids, phosphoric acid esters, sulfonic acids (acyl isothionates, alkylaryl sulfonates, alkyl sulfonates, sulfosuccinates) sulfuric acid esters (alkyl ether sulfates, alkyl sulfates)
  III. cationic surfactants: alkylamines, alkyl imidazolines, ethoxylated amines, quaternaries (alkylbenzyldimethylammonium salts, alkyl betaines, heterocyclic ammonium salts, tetraalkylammonium salts)
  IV. nonionic surfactants: alcohols, alkanolamides (alkanolamine-derived amides, ethoxylated amides), amine oxides, esters (ethoxylated carboxylic acid esters, ethoxylated glycerides, glycol esters, monoglycerides, polyglyceryl esters, polyhydric alcohol esters, sorbitol esters, triesters of phosphoric acid), ethers (ethoxylated alcohols, ethoxylated lanolin, ethoxylated polysiloxanes, propoxylated POE ethers)

A comprehensive review of cosmetic emulsions was represented by BREUER (1985).

Cosmetic products are generally designed to (1) deliver a functional benefit and (2) to enhance the aesthetic appeal of the product. Their function may be to impart some beneficial attributes by means of delivering active ingredients to the tissue. Cosmetic preparations may also provide a protective surface layer which either prevents the penetration of unwanted foreign matter, or moderates losses of water from the skin. Cosmetic emulsions may also deliver materials that screen out the damaging UV radiation of the sun (sunscreen creams). Unlike pharmaceutical preparations, where functionality is the main consideration, cosmetic products must also meet stringent aesthetic standards (e.g., texture, consistency, pleasing color, fragrance, etc.).

As cosmetic emulsions come into contact with human tissues for extended periods of time, they should be free of physiologically harmful effects and be completely free of allergens, sensitizers, and irritants. The choices of ingredients available to the cosmetic formulator are, therefore, much more limited than those for the preparation of other types of emulsions (e.g., paints, household cleaners, agricultural emulsions, etc.).

Most cosmetic emulsions contain about 20–40% oil, a few percent emulsifiers, some "active ingredients" dissolved either in the oil or water phases, dyes, perfumes, and deionized water. The oil phase is, in most cases, a mixture of several mineral or vegetable oils chosen in accordance with the particular requirements of the product. Generally, the oil phase of the product fulfils a "functional" role, e.g., it provides an occlusive layer covering the surface of the skin and acting as a barrier against water loss from the skin. The oil phase also serves as the carrier for perfumes and, often, for pigments, especially in liquid makeup preparations. In a similar manner, the aqueous phase also can contain active, functional materials: proteins, water-soluble vitamins, minerals, synthetic or natural water-soluble macromolecules, etc. To produce a stable emulsion from an aqueous solution and a mixture of oils, an emulsifier system is required.

Following are some common constituents of cosmetic emulsions:

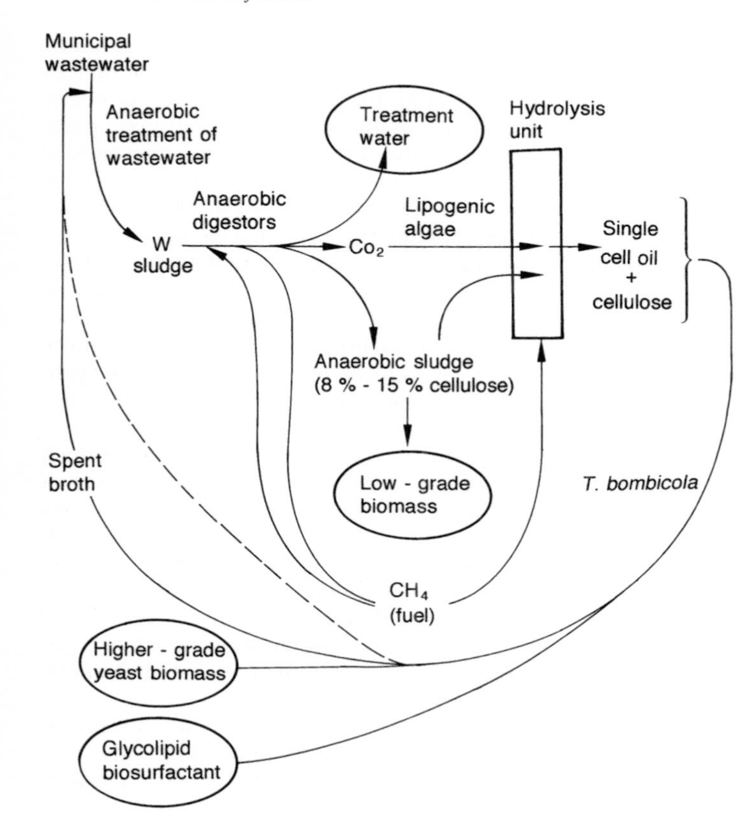


**Fig. 10.** Production of glycolipids from complex wastes incorporating anaerobic digestion of municipal wastes and use of *Torulopsis bombicola.*

## Lanolin and Its Derivatives

Lanolin or wool fat is extracted from raw wool. It shows some similarities in its properties to human sebum, and has been used since prehistoric times as an ingredient of cosmetic preparations. Unlike most lipids of animal origin, lanolin does not contain triglycerides: its main constituents are esters of fatty acids with sterols, triterpene alcohols, and long-chain fatty alcohols. It is also rich in $\alpha$- and $\omega$-hydroxy-fatty acids and in $\alpha,\beta$-alkane diols.

In addition to pure lanolin a great variety of lanolin fractions and chemically modified lanolin derivatives are also available for use in cosmetic emulsions. With the help of these "lanolides" a complete spectrum of W/O and O/W cosmetic emulsions can be prepared. Lanolin and lanolides are generally regarded as capable of imparting favorable attributes to skin. Consumers seem to accept them as materials of "natural" origin, and, as such, they are often used as ingredients in "natural" cosmetics. When formulating creams or lotions with lanolin and its derivatives, it is frequently necessary to use additional surfactants (anionic or nonionic types) in order to obtain emulsions of good quality and sufficient stability (PROSERIO et al., 1980).

## Nonionic Ethoxylates

In principle, any compound with a reactive hydroxyl group can give rise to ethoxylated surfactants by reacting it with ethylene oxide. Surfactants with very diverse HLB values can be prepared. The most frequently used starting materials are sugars, fatty acids, and fatty alcohols.

Ethoxylated surfactants are favored cosmetic emulsion ingredients. They are insensitive to pH and practically unaffected by salt concentrations, giving formulators a wide scope for constructing emulsion products. Nonionic surfactants also have a low skin sensitization potential. A precondition for a surfactant to become an irritant or a sensitizer is that its molecules should be capable of binding to proteins. Most of the binding sites on proteins are ionic groups (carboxyl or amine side chains). Consequently, nonionic surfactants bind only weakly, or not at all, to proteins. It was also suggested that the presence of nonionic ethoxylates in a cosmetic formulation reduces the irritation potentials of other ingredients, especially of ionic surfactants (GARRETT, 1965).

However, a side reaction of the ethoxylation process may lead to the formation of small quantities of dioxane, which is a known carcinogen. Therefore, ethoxylated surfactants for cosmetic use must be free of analytically detectable quantities of dioxane (BIRKEL et al., 1978). Another hazard of ethoxylated materials is their degradation due to oxidation and photooxidation processes (DAVIS, 1980) which is catalyzed by heavy metals. Therefore, emulsions containing ethoxylated surfactants should always contain antioxidants.

Ethoxylated surfactants can be prepared with different lengths of the polyethylene oxide chains on the surfactant molecule giving rise to a range of HLB values. This allows flexibility when they are used for formulating creams and lotions.

### Sucrose Esters

Sucrose esters, as nonionic materials, share many of the advantages of ethoxylated detergents as far as cosmetic applications are concerned (BROOKS, 1980). In addition, since their syntheses do not require ethylene oxide, there is no danger of dioxane contamination. Sucrose esters are obtained by esterification of the sugar hydroxyl groups with fatty acids, mainly lauric and stearic acids. Depending on the conditions used during the esterification reaction, mixtures containing predominantly mono, di-, or triesters of sucrose can be prepared. In this way, surfactants with HLB values ranging from 3 to 14 can be obtained.

The principal virtue of sucrose esters is their mildness toward skin. Since the molecules are nonionic, they interact only weakly or not at all with proteins of the skin and, on the whole, are neither irritants nor sensitizers.

### *Ortho*-Phosphoric Acid Esters

The main incentive for using *o*-phosphoric acid esters in cosmetic emulsions is their similarity to phospholipids, which are the natural building blocks of stratum corneum cell membranes. A large variety of *o*-phosphoric acid esters can be prepared, the properties of which will depend on the hydrophobic moiety, which is linked through the ester bond to the phosphoric acid. It has been claimed that emulsions prepared with *o*-phosphoric acid esters as emulsifiers have very narrow droplet size distributions and small mean droplet sizes (SKRYPZAK et al., 1980). Generally, better emulsion stability is achieved when *o*-phosphoric esters are used in combination with other anionic surfactants, especially when fatty acids and fatty alcohols are also added. It was found that they enhance the moisture retention ability of skin. A potential application of these surfactants might be in cosmetic products based on microemulsions. Phosphoric fatty acid esters in combination with fatty alcohols and glycol ethers appear to be good emulsifier systems for microemulsions (HOFFMAN, 1970).

### Glycerine Esters

Tryglycerides are important ingredients of sebum, the natural lubricant of skin. The reason for the popularity of triglycerides in cosmetic emulsions is twofold: (1) as naturally occurring surface-active agents, they are generally innocuous compounds and represent practically no medical hazards; (2) triglycerides can be prepared with a large variety of substituents and, consequently, their surface activity (HLB values) can be varied over large ranges.

### Protein Condensates

Some proteins are natural emulsifiers (e.g., casein in milk). Many more have the essential requirements for becoming emulsifiers due to their hydrophilic–hydrophobic character. In addition, as macromelecular substances, proteins absorb tenaciously at interfaces and, therefore, impart increased stability to colloidally dispersed systems (TADROS, 1982). The basic surface-active properties of proteins can be greatly enhanced and altered by chemical modifications. This can take two forms: hydrolysis to yield more manageable, lower molecular weight protein fragments (polypeptides), or chemical alteration of the reactive protein side chains (e.g., esterification or etherification of carboxyl and hydroxyl groups, acylation of the amine side chains, modification of the peptide bonds, etc.). Protein–sugar condensates are particularly suited for use in skin care emulsions. In addition to their emulsifying power, these compounds impart to the skin a lubricous feel and can be used as moisturizing agents.

### Block Copolymers

As mentioned above, macromelecular compounds possess considerable advantages for use as ingredients in cosmetic emulsions: They only penetrate skin in small amounts and, then, mainly into the nonliving outer layers of the stratum corneum. On the whole, polymers have much lower toxicity, sensitization, and irritation potentials than low molecular-weight compounds of similar chemical compositions. However, synthetic polymers are frequently contaminated with traces of the parent monomers. The latter compounds are chemically highly reactive materials and are often very toxic. Before synthetic polymers can be used as cosmetic ingredients, they must be freed of traces of monomer contaminants.

### Amine Oxides

So far, amine oxides have found limited application in cosmetic products as principal emulsifiers. However, they are used as auxiliary detergents in formulation of anionic, cationic, or glycerin ester surfactants. Amine oxides are mild to the skin and are compatible with cationic surfactants to even a larger extent than nonionic materials. Some reports suggest that amine oxides reinforce the action of some antibacterial preservatives, e.g., quaternary amines (LINKE et al., 1975).

### Acyl Lactylates

Acyl lactylates, i.e., substituted lactic acid derivatives, have been used as surfactants in the food industry, but to date have found very limited application in cosmetic emulsions. Acyl lactylates are obtained by the esterification of the hydroxyl group of lactic acid with a fatty acid (e.g., stearic acid, lauric acid, palmitic acid, etc.). Lactic acid has been shown to have a skin-softening effect and it is reasonable to expect, therefore, that its derivatives might have similar effects.

### Silicones

A large variety of silicone oils and compounds based on silicone chain backbones with a variety of functional chemical groups as side chains have been prepared and are becoming commercially available. They are also being used in cosmetic emulsions as highly volatile liquids to waxy solids. Volatile silicones have found particularly wide use in cosmetic products, owing to the pleasant, dry sensation they impart to the skin, evaporating without unpleasant cooling effects or without leaving a residue. Due to their low surface energy, silicones help spread the various active ingredients over the surface of hair and skin. The chemical structures of the silicone compounds used in cosmetic preparations varies; the silicone backbone can be either a straight-chain polymer or a ring structure (Fig. 11). The backbones can carry various attached "functional" groups (e.g., carboxyl, amine, sulfhydryl, etc.). While most silicone oils can be emulsified using conventional hydrocarbon surfactants, surface-active silicone materials which have been specially designed for emulsions containing silicone oils have also become available in commercial quantities (STARCH, 1984). Among these, the nonionics are the most common; however, silicone based detergents with anionic and cationic side chains are also being offered by various manufacturers. As with other detergents, the surface activity of these materials depends on the relative lengths of the hydrophobic silicone and the hydrophilic "functional" segments. Most silicones are innocuous materials and represent relatively small medical and environmental hazards (FRYE, 1983).

Phytosteroids and beeswax represent more "natural" surfactants. Phytosteroids are extracted from various plants and are used as emulsifiers for water-in-oil emulsions. Among the phytosteroids, the glucoside steroids, saponins constitute a distinct group. Owing to their high foam-producing capacites, they have been used in shampoos and bath emulsions (HUTTINGER, 1983).

Unlike other lipids, beeswax, a long-standing cosmetic ingredient, consists of esters of palmitic ($C_{16}$) and stearic ($C_{18}$) acids with long-chain wax-alcohols ($C_{26}$–$C_{28}$). It also contains about 15% free fatty acids, which act as emulsifiers. On its own, beeswax is a poor emulsifier and the emulsions prepared from beeswax break easily. This quality is utilized for the preparation of cold creams, where a fast emulsion breakdown and quick absorption of the ingredients into the skin are desired (HUTTINGER, 1983).

The above mentioned surfactants are obtained predominantly through chemical synthesis from materials which may have been derived from the petrochemical industry. However, many surfactants can be produced by use of microorganisms or by enzymatic transformations yielding biosurfactants which more or less possess properties required for a cosmetic product. Due to their advantages, as mentioned earlier, they can effectively penetrate the cosmetics market.

The cosmetics industry, in general, is protectionist and very little information is available in the literature as related to ingredients and processes which are in use. Concerning biosurfactants, which represent a novel very promising ingredient, this protectionism and secrecy are even more pronounced. Some information is available though and is presented further on. Biosurfactants in cosmetics have been reviewed by KLEKNER and KOSARIC (1993).

Of particular interest for cosmetics are biosurfactants derived from yeast, e.g., sophorose lipids (SLs) from *Torulopsis bombicola.*

Kao Co. (Japan) developed a process for fermentation and isolation of SLs (INOUE et al., 1979a), and the modification of SLs by esterification. Esterification was done in two steps. First a methyl ester was prepared (INOUE et al., 1979b) that could be easily transesterified to other esters in the presence of MeONa (INOUE et al., 1979c). Alkyl-SLs cover HLB values from approximately 7 (oleyl-SL) to 25+ (methyl-SL). Several applications of esters and hydroxylalkyl ethers of SLs in cosmetics were suggested, such as stick makeups (e.g., propolylated methyl ester; KANO et al., 1980) and especially moisturizers (e.g., octyl, lauryl, and oleyl esters or propolylated derivatives; ABE et al., 1980; TSUTSUMI et al., 1980) for skin and hair care products. Oleyl-SL and SLs modified by propylene oxide were reported to complement the natural skin moisturizing factor and to have found commercial application. Nevertheless, what is the extent of applicability of SLs or modified SLs in cosmetics remains unclear as no widespread discussion of these compounds or the formulation using them is available in the literature.

In addition, polymerized product of 1 mole of sophorolipid and 12 moles of propylene glycol have specific compatibility to the skin and have commercially been used as skin moisturizers (YAMANE, 1987). Antibiotic effects of biosurfactants have been reported by NEU et al. (1990) and LANG et al. (1989). It was also reported that biosurfactants have an inhibitory effect toward the growth of AIDS virus in white blood cells (MINTZ, 1990).

Useful surface-active properties and antitumor activity in mice (KOHYA et al., 1986) were shown by 6,6′-*O*-decanoyl-$\alpha,\alpha$ trehalose. Promising as excellent surfactants, dispersants, and stabilizers are also succinoyl trehalose lipids and their sodium salts (ISHIGAMI et al., 1988). An antiviral succinylated trehalose glycolipid efficient against *Herpes simplex* Type I virus was produced by *Rhodo-*

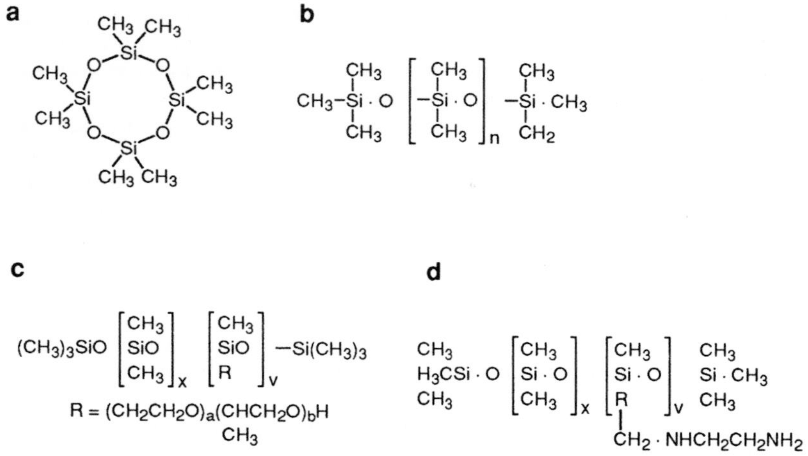

**a**

$$\text{cyclo-}[(CH_3)_2SiO]_4$$

**b**

$$CH_3\text{-}Si(CH_3)_2 \cdot O\left[\text{-}Si(CH_3)_2 \cdot O\right]_n\text{-}Si(CH_3)(CH_2) \cdot CH_3$$

**c**

$$(CH_3)_3SiO\left[\begin{matrix}CH_3\\SiO\\CH_3\end{matrix}\right]_x\left[\begin{matrix}CH_3\\SiO\\R\end{matrix}\right]_v\text{—}Si(CH_3)_3$$

$$R = (CH_2CH_2O)_a(CHCH_2O)_bH \quad (\text{with } CH_3)$$

**d**

$$\begin{matrix}CH_3\\H_3CSi \cdot O\\CH_3\end{matrix}\left[\begin{matrix}CH_3\\Si \cdot O\\CH_3\end{matrix}\right]_x\left[\begin{matrix}CH_3\\Si \cdot O\\R\end{matrix}\right]_v\begin{matrix}CH_3\\Si \cdot CH_3\\CH_3\end{matrix}$$

$$R: CH_2 \cdot NHCH_2CH_2NH_2$$

**Fig. 11.** Structural formulas of some typical silicone compounds used in cosmetic emulsions: (**a**) cyclic siloxane, (**b**) linear siloxane, (**c**) siloxane–polyethylene oxide copolymer, (**d**) siloxane–polyethylene amine copolymer.

*coccus erythropolis* on a glycerol-containing medium (KAWAI et al., 1988).

### Monoacylglycerides

Glycerol monostearate is still one of the most used surfactants in cosmetics. Recently, glycerolysis at low temperatures using lipase was accomplished (McNEILL and YAMANE, 1991). A yield of 70% MG (monoglyceride) was obtained at 42 °C when lipase from *Pseudomonas fluorescens* and glycerol/tallow (molar ratio 1.5:2.5) were used. With temperature programming, a yield of approximately 90% MG was obtained. In this case, the initial temperature was 42 °C for 8–16 h followed by incubation at 5 °C for up to 4 d. The procedure seems to be economically promising.

### N-Acylamino Acids

Ajimonoto Co. patented a preparation of N-long-chain acylamino acids, especially glutamates, by conversion of sodium salts using whole microbes (*Pseudomonas, Xanthomonas,* or *Gluconobacter*) (Ajinomoto Co., Inc., 1982). N-long-chain acyl glutamates can be used for cosmetic formulations of stable oil-in-water emulsions (HASEGAWA and MAENO, 1986).

Recently, commercially available immobilized lipase of *Mucor miehei* (Liposyme, Novo Industri) has been used for synthesis of N-acyl bonds (SERVAT et al., 1990; MONTET et al., 1989). N-ε-oleyl lysine, a new generation of surfactants, can be synthesized by transacetylation from triglycerides to lysine. A yield of up to 60% could be achieved without any solvent at 90 °C (MONET et al., 1990).

In this connection, it is interesting to note that N-acyl of basic amino acids as N-ε-cocoylornithine has been claimed to possess skin- and hair-moisturizing and protecting properties (SAGAWA, 1986).

### Phospholipids

Soy phospholipids, obtained as a by-product in the production of soybean oil, can be modified by enzymes yielding lysophospholipids. These compounds alone or in mixture with monoglycerides and fatty acids have shown significant surface activity (FUJITA and SUZUKI, 1990). The enzyme employed can be of pancreatic or microbial origin. When the pancreatic enzyme was used, mainly hydrolysis proceeded (EGI, 1987). Phospholipase D

(M or Y1) derived from actinomycetes functions rather like transphosphatidylase when a suitable acceptor is present. As an acceptor, it may serve primary, secondary, or phenolic hydroxyls, and so a broad spectrum of compounds can be synthesized. When glycerol is added, phosphatidyl glycerol can be formed (KUDO, 1988).

Apart from their direct use as surfactants, phospholipids can serve for construction of liposomes. Entrapment inside of liposomes of an active compound for delivery to the skin has been verified to be beneficial in cosmetics as well as in pharmaceuticals (LEUTENSCHLAGER, 1990a, b). The mildness of phospholipids and their compatibility with the skin make them primary sources for liposome creation. Liposomes can penetrate through skin layers and can be targeted to a specific layer by modifying their composition. As such, cosmetic applications may be considered dermatological, care must be taken to use pure compounds. This, together with demand for specific compounds that enable construction of specific liposomes, gives a new opportunity for cosmetic application of enzyme-modified phospholipids.

Sugar Esters

Sugars esterified with long-chain fatty acids can serve as surfactants, e.g., blends of sucrose esters and glycerol can cover entirely all HLB values (2 to 16+) demanded by the cosmetic industry (GUPTA et al., 1983). Moreover, they offer very low toxicity and irritancy.

Apart from that, sucrose cocoate has been used in wound-healing preparations (GOODE et al., 1988) that increased epidermal layer thickness and content of DNA, glucosamine glycan, and lipids. Sucrose esters have been used also for formulation of multiple emulsions (DELUCA et al., 1991). Also, esters of glucose have been studied as valuable cosmetic materials (DESAI, 1990).

Recently, pancreatic lipase was used for preparation of C-6 esters by transesterification from trichlorethyl carboxylates to unprotected monosaccharides and anhydrous pyridine (THERISOD and KLIBANOV, 1987). From these 6-*O*-acylmonosaccharides diesters can be prepared by next transesterification in tetrahydrofuran or methylchloride. Depending on the lipase used, C1 or C3 derivatives can be obtained. Sugar diesters can be selectively hydrolyzed in the C6 positon by *Candida cylindracea* lipase, and so pure C2 or C3 monoesters can be prepared (THERISOD and KLIBANOV, 1987).

# 6 Medical and Pharmaceutical Applications

A good review on medical and pharmaceutical applications of emulsions has been presented by DAVIS et al. (1985) while SCHURCH et al. (1993) presented properties and functions of alveolar and airway surfactants in particular. Summaries from these two reference sources are presented here.

## 6.1 Parenteral Emulsions

Emulsions are used as injectable (parenteral) systems in medical applications to provide vehicles for lipid-soluble materials for controlled drug release and for targeting of drugs to specific sites in the body.

Both oil-in-water (O/W) and water-in-oil (W/O) emulsions are used. More complex forms, including multiple emulsions (W/O/W or O/W/O) and emulsions containing dispersed solid microspheres, have also been considered. The type of emulsion employed is usually determined by the role that the emulsion vehicle plays and the intended route of administration. For example, W/O emulsion can be used as depot formulation to provide a sustained and controlled release of a drug following intramuscular administration. In contrast, O/W emulsions are normally given by the intravenous route as drug carriers or as medicinal agents. Multiple (or double) emulsions are used as depot systems and can be considered as a variation of the W/O type.

The range of oils and emulsifiers for intravenous administration in humans could involve purified paraffin oils for intramuscular use. Vegetable oils are being almost exclusively used intravenously with phosphatides (egg or soya) as the emulsifier.

The various oils of vegetable origin (soybean, sesame, safflower, and cottonseed) are regarded as being nontoxic if purified correctly and some are used in large quantities as parenteral systems for nutritional purposes. Similarly, a number of emulsified perfluorochemicals are thought to be acceptable as the disperse phases of artificial blood products. Oils derived from vegetable sources are biodegradable, whereas purified mineral oils such as liquid paraffin and squalene or squalane are not.

Particle size plays an important role in intravenous emulsions. Large oil droplets could give rise to blockages (emboli) in the body (particularly the fine capillaries of the lungs). In general, particles larger than 5 μm are excluded from O/W emulsions for intravenous use and in practice commercial systems usually have a mean size of the order of 200–500 nm with 90% or more particles below 1 μm. FUJITA et al., 1971) have shown that the larger the particle size and the broader the particle size distribution, the greater the toxicity of fluorocarbon emulsions. These authors proposed a limit of 6 μm. Factors such as viscosity of the external phase, phase volume ratios, droplet particle size, and the partition coefficient of the drug are all important. The release of drugs from W/O and O/W emulsion systems has been discussed by GOLDBERG et al. (1967).

Concerning the biological fate of these emulsions, the following general guidelines have been presented:

- Fine particle size emulsions are cleared more slowly than coarse particle size emulsions.
- Negatively charged and positively charged particles are cleared more quickly than neutral particles.
- Emulsions stabilized by low molecular-weight emulsifiers are cleared more rapidly than those stabilized by high molecular-weight emulsifiers.

Lipid emulsions for parenteral nutrition have been used for more than 100 years. The most normal method is to provide a hypertonic solution of carbohydrate (glucose), amino acids, electrolytes, and vitamins, administered through a large central vein rather than peripherally.

Some commercially available lipid emulsions are shown in Tab. 5.

Phospholipids (lecithins) have predominantly been used as natural emulsifiers. A wide range of nonionic materials have been investigated as potential emulsifying agents for intravenous fats: polyethylene glycol stereate TEM (diacetyltartrate ester of monoglyceride), Drewmulse (partially esterified polyglycerol), Myrj surfactants (polyoxyethylene sorbitan monoesters). However, all of these materials gave rise to toxic reactions on one form or another. Only one range of nonionics was found to be free from toxic effects: the poloxamers (polyoxyethylene–polyoxpropylene derivatives). These have been used widely for fat emulsions, either on their own or as coemulsifiers with lecithin (VELA et al., 1965).

The fat emulsion administered to the body is distributed rapidly throughout the circulation and then cleared. A number of processes are involved, such as uptake of cholesterol into the fat particle, exchange of phospholipids on the surface of the particle and endogenous phospholipids present in various lipoprotein fractions, and the adsorption of lipoproteins to the surface of the particle. Clinical use of fat emulsions in parenteral nutrition has been described in detail by JOHNSON (1983) and MENG and WILMORE (1976).

## 6.2 Perfluorochemical Emulsions as Blood Substitutes

Various alternatives to blood have been proposed including plasma and erythrocyte substitutes (GEYER et al., 1975). Nonpolar solvents such as perfluorochemicals and silicone fluid have the ability to dissolve large quantities of oxygen and carbon dioxide, and CLARK and GOLLAN (1966) found that mice could survive for extended periods of time when completely submerged in oxygenated perfluorochemicals, i.e., the mice undergoing "liquid breathing." Use of emulsified perfluorochemicals as a red blood cell substitute has been presented by GEYER (1973, 1974).

The advantages of such a system include (1) good shelf life, (2) good stability in surgical procedures, (3) no blood-group-incompatibility problems, (4) ready accessibility, and (5) no problem with hepatitis or AIDS infection. Thus, the emulsion could find use in organ perfusion and as a source of "blood" in times of disaster, wars, or national emergency. Thus produced emulsions would have to satisfy the following requirements: (1) low toxicity, (2) no adverse interaction with normal blood, (3) little effect on blood clotting, (4) safisfactory oxygen and carbon dioxide exchange, (5) satisfactory rheological properties, and (6) satisfactory clearance from the body. In addition, it should be able to take over from the natural material for a period long enough for new blood cells to be generated, i.e., it should have a lifetime of about 1–2 weeks in the circulation before being cleared.

The following oils have been examined for these emulsions: perfluorotributylamine, perfluorobutyltetrahydrofuran, perfluorodecalin, perfluoromethyldecalin, and perfluoromethylcyclohexane. The last three oils have been reported to have advantages as regards their lifetime in the body and are probably excreted through the skin and the lungs (YAMANOUCHI et al., 1975).

A formulation for an artificial blood preparation for total blood replacement in experimental animals is presented in Tab. 6 (GEYER, 1973). The emulsified fluorocarbon provides the necessary oxygen exchange and the hydroxyethyl starch acts as a plasma expander. The emulsifying agent is the Pluronic F68 (polyoxyethylene–polyoxypropylene copolymer).

A product from the Green Cross Corporation (Japan), Fluosol-DA, containing perfluorodecalin has been studied in humans. Administration of Fluosol-DA requires that the patient breathe high oxygen concentrations (HONDA et al., 1980; TREMPER et al., 1980).

The formulation of a suitable perfluorochemical emulsion has been difficult since those oils that provide stable emulsions are not cleared from the body in a satisfactory way and as with the case of fat emulsions for parenteral nutrition, the choice of surfactants is extremely limited (GEYER, 1976).

The size of perfluorochemical emulsion droplets can also have a pronounced effect on the biological results. The particle surface and surface charge could be of critical importance. The emulsion Fluosol-DA has a mean particle size in the region of 100–200 nm. The relation between toxicity and particle size for perfluorochemical emulsions has been presented by FUJITA et al. (1971). Particles larger than 6 μm in size gave rise to toxic manifestation, and GEYER (1976) has reported that particles of 1.0–1.5 μm in diameter were taken up by the reticuloendothelial system of rats (liver and spleen). The process of phagocytosis is known to be dependent on particle size, among other factors. Small par-

**Tab. 5.** Some Commercially Available Lipid Emulsions

| Trade Name | Oil Phase | Emulsifier | Other Components |
|---|---|---|---|
| Intralipid (Kabi-/Vitrum) | Soybean 10% or 20% | Egg lecithin 1.2% | Glycerol 2.5% |
| Lipofundin S (Braun) | Soybean 10% or 20% | Soybean lecithin 0.75% or 1.2% | Xylitol 5.0% |
| Lipofundin S (Braun) | Cottonseed 10% | Soybean lecithin 0.75% | Sorbitol 5.0% |
| Liposyn (Abbott) | Safflower 10% (and 20%) | Egg lecithin 1.2% | Glycerol 2.5% |
| Travemulsion (Travenol) | Soybean 10% or 20% | Egg lecithin 1.2% | Glycerol 2.5% |

**Tab. 6.** Typical Composition of a Perfluorochemical Emulsion

| Substance | Quantity |
|---|---|
| Perfluorocarbon oil | 20 mL |
| Pluronic F68 | 2.2 g |
| Hydroxyethyl starch | 3.0 g |
| Glucose | 100 mg |
| Potassium chloride | 32.0 mg |
| Magnesium chloride | 7.0 mg |
| Monobasic sodium phosphate | 9.6 mg |
| Sodium chloride | 54.0 mg |
| Calcium chloride | 18.0 mg |
| Sodium carbonate | to pH 7.44 |
| Water | to 100 mL |

ticles of 200 nm or less are cleared slowly by the reticuloendothelial system, especially when emulsified with very hydrophilic emulsifiers (ILLUM and DAVIS, 1982).

## 6.3 Lipid Emulsions for Drug Delivery

Problems can arise in the parenteral administration of drugs that have low water solubilities. The normal types of solution formulation comprise various cosolvents and/or surfactant mixtures (e.g., polyethylene glycol, propylene glycol, Cremophor, Pluronic).

The use of soybean oil emulsions as carriers for lipid-soluble drugs has been pioneered by JEPPSSON (1976) in Sweden. The drugs, dissolved in the oil phase, have included barbituric acids, cyclandelate, diazepam, and local anaesthetics and the routes used have been intravenous, intraarterial, subcutaneous, intramuscular and interperitoneal. The greatest interest has centered around the use of fat emulsions as vehicles for the intravenous administration of drugs (JEPPSSON, 1972).

Some parenteral lipid emulsions for drug administration are listed in Tab. 7 (WRETLIND et al., 1978).

Delivery of anticancer agents in lipid emulsions had been investigated by FORTNER et al. (1975) and LITTERST et al. (1974), and by EL-SAYED and REPTA (1983).

**Tab. 7.** Parenteral Lipid Emulsions for Drug Administration

| | |
|---|---|
| *Hexobarbital* | 3.75 g |
| Soybean oil | 10 g |
| Ethanol | 25 g |
| Egg phosphatides | 1 g |
| Myrj 52 | 0.5 g |
| Water | to 100 mL |
| *Phenylbutazone* | 2 g |
| Soybean oil | 10 g |
| Acetylated monoglycerides | 10 g |
| Glycerol | 2.5 g |
| Pluronic F68 | 0.5 g |
| Water | to 100 mL |
| *Diazepam* | 0.5 g |
| Soybean oil | 15 g |
| Acetylated monoglycerides | 5.0 g |
| Egg phosphatide | 1.2 g |
| Glycerol | 2.5 g |
| Water | to 100 mL |

MIZUSHIMA et al. (1982) have employed a lipid emulsion as a carrier for corticosteroids. Fat emulsions have also found commercial clinical use for the intravenous administration of diazepam (VON DARDEL et al., 1976).

## 6.4 Emulsions as Adjuvants for Vaccines

The antibody response by vaccination with bacterial or viral antigens can be increased by including a substance called an adjuvant. Adjuvants potentiate the immune responses nonspecifically: some preferentially stimulate the antibody response, whereas others can also stimulate cell-mediated hypersensitivity. The adjuvant methods of immunization have the advantages of achieving both the primary and secondary immune responses with only one injection of antigen and maintaining the antibody level over a long period. Adjuvants have been used in tumor immunotherapy (WEIR, 1977).

The mode of action of water-in-oil emulsions as adjuvants is not fully understood; the general opinion is that the emulsion acts as an inert depot which slowly releases antigen over a long period of time.

New possibilities have been opened up by developing multiple emulsions as adjuvants. These are less viscous than the W/O systems and gave a better antibody response (HERBERT, 1965). Water-in-oil-in-water (W/O/W) emulsions consist of water droplets inside oil droplets dispersed in an aqueous continuous phase, and an oil-in-water-in-oil (O/W/O) emulsion consists of oil droplets inside water droplets in an oily continuous phase. Such multiple-emulsion systems can be made by re-emulsifying the appropriate single-emulsion system. Multiple emulsions may also be produced when an emulsion undergoes phase inversion. A schematic of such emulsion is shown on Fig. 12.

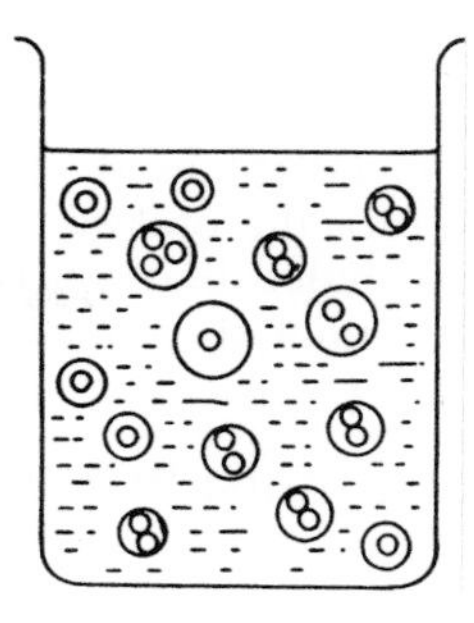

**Fig. 12.** Diagrammatic representation of a multiple emulsion.

Multiple emulsions are prepared by redispersing a water-in-oil emulsion, containing the antigen solution, in an aqueous phase which contains an emulsifier that promotes an oil-in-water emulsion.

Multiple emulsions were first investigated as vaccine adjuvants by HERBERT (1965). He reported that the multiple-emulsion vaccine had good *in vitro* stability, did not produce persistent subcutaneous depots, and was of much lower viscosity than simple water-in-oil emulsions, therefore being easier to inject.

TAYLOR et al. (1969) compared an aqueous influenza vaccine with an influenza vaccine comprising a simple mineral water-in-oil emulsion and another comprising a multiple-mineral-oil emulsion. They found that the multiple emulsion was less viscous than the simple emulsion and suggested that it would be less likely to produce local reactions, although there could be a weaker antigenic stimulus. They concluded that a multiple-emulsion vaccine would be at least as effective as standard water-in-oil emulsion preparations.

Oil-in-water emulsions have been used as adjuvant-type systems in cancer chemotherapy. Successful tumor regression has been achieved by intratumoral injections of viable Bacille-Calmette-Guerin (BCG) and a potential antitumor preparation has been constructed by combining a biosurfactant, trehalose dimycolate purified from BCG with aqueous soluble extracts of Re mutants of *Salmonella* sp. (Re glycolipid). An essential requirement of activity has been the formulation of an O/W emulsion. The O/W emulsions of mineral oil stabilized by Tween 80, containing aqueous soluble glycolipids with synthetic or naturally occurring fatty acid esters derived from mycobacteria, were found to be potent tumor regressive preparations.

## 6.5 Oral Emulsion Systems

Emulsions have been administered by mouth since the early 19th century to render oils such as cod-liver oil and liquid paraffin more palatable. Also, oil-in-water emulsions have been shown to facilitate the absorption of certain drugs from the gastrointestinal tract and may have application in the future development of drug delivery systems.

Orally administered emulsions are almost exclusively O/W systems, although the use of multiple emulsions of the O/W/O type has been reported.

The range of emulsifying agents is greater than that available for intravenous emulsions. Materials such as acacia, tragacanth gums, and methylcellulose, as well as various non-ionic surface-active agents have been employed in pharmaceutical formulations for oral use. The ionic surfactants are not normally used for internal preparations, but are reserved for topical and cosmetic applications.

Vegetable oils such as corn oil, peanut (arachis) oil, soybean oil, etc. are used as emulsion vehicles for drug delivery, while mineral oils (liquid paraffin) are emulsified to provide a laxative effect, but are not used for drug administration, unless the drug is intended for the same pharmacological effect.

ENGEL et al. (1968) used multiple emulsions (W/O/W) for the oral administration of insulin and concluded that this was a means of effecting intestinal absorption of the drug.

## 6.6 Topical Emulsion Systems

Emulsion systems can be administered to the external surfaces of the body and to body cavities as topical formulations in the form of O/W or W/O creams containing drugs. General review on the structure of the skin, the absorption of drugs through the skin, and pharmaceutical and cosmetic products for topical administration can be found in the relevant textbooks (VAN ABBE et al., 1969).

The pharmaceutical and cosmetic property of a topical product is related to its viscosity, which will affect its consistency, spreadability, and extrudability. Also, the viscosity will be of paramount importance in determining the rate at which the active drug can diffuse to the outer layers of the stratum corneum.

Surfactants present in emulsion systems may also influence the rate of release from the formulation as well as the rate of absorption, thus, the presence of soaps can enhance the penetration rates of poorly absorbed materials. The surfactants may also promote the diffusion from the base itself and hence influence therapeutic performance. It was also concluded that the surfactants had a direct action on the protein rather than the lipid components of the stratum corneum (DUGARD and SCHLENPLEIN, 1973).

The following types of emulsifying agent are used pharmaceutically:

- ionic, both anionic and cationic,
- nonionic,
- complex.

### Ionic Emulsifying Agents

Examples of anionic surfactants commonly used are the alkali salts of the higher fatty acids and sulfate esters of the higher fatty alcohols, e.g., sodium cetyl/stearyl sulfate. Cationic surfactants employed include the long-chain quaternary ammonium compounds, e.g., cetrimide. If ionic surfactants are included in a formulation, care must be taken that they do not interact with the active ingredient.

### Nonionic Emulsifying Agents

A large number of nonionic emulsifiers exists for stabilizing biphasic topical preparations. The higher alcohols (e.g., stearyl) exhibit mild emulsifying activity and are used to stabilize W/O emulsions. Cetyl polyoxyethylene glycol ether (Cetomacrogol) has a higher emulsifying capacity and is used in the preparation of O/W emulsions. The overall properties of nonionic surface-active agents depend on the relative proportions of the constituent hydrophilic and lipophilic moieties of the molecule (HLB value).

### Complex Emulsifying Agents

To obtain a stable emulsion, it is often useful to use mixed-surfactant systems. A combination provides a more rigid film at the lipid/aqueous interface and an increased reduction in the interfacial tension. An example of a complex emulsifier system is cetyl alcohol, sodium lauryl sulfate, and glyceryl monostearate.

The rheological properties of semi-solid emulsions made from mixed emulsifiers have been studied extensively by BARRY and ECCLESTON (1973). The system comprising oil/alcohol (usually a mixture of cetyl and stearyl alcohols) and surfactant (cetrimide, sodium lauryl sulfate, or Cetomacrogol) was found to produce a viscoelastic structure that gave a bodying effect to the formulation through the formation of a gel network of liquid crystals. The rheological properties were highly dependent on the quantity and nature of surfactant and the alcohol. Interestingly, the use of pure alcohols of different chain length gave rise to the formation of semisolid products with a poor stability.

There have also been reports on glycolipids possessing immunostimulating antigenic and antitumor activities. GORBACH et al. (1994) reported on new glycolipids (chitoligosaccharide derivatives) possessing immunostimulating and antitumor activities. These compounds with a low toxicity were synthesized from chitoligosaccharides of dp 2–4 containing both free and acylated amino groups. An induction of interleukin-1 and tumor necrosis factor by the immunocompetent cells and an augmentation by 140–180% of the mean life of mice with the Ehrlich carcinoma were observed.

The structure of an antigenic glycolipid (SL-IV) from *Mycobacterium tuberculosis* was studied by BAER (1993). The structure of the antigenic glycolipid was determined by mass spectrometry and nuclear magnetic resonance spectroscopy and was proposed to represent a sulfolipid consisting chiefly of 2,3-di-*O*-(hexadecanoyl/octadecanoyl)-$\alpha,\alpha$-trehalose 2 prime sulfate (designated as SL-IV).

### 6.7 Alveolar and Airway Surfactants

Pulmonary surfactants are produced by the body and play a very essential role for proper function of the lungs. The pulmonary air–liquid interfacial film reduces the surface tension in the parenchyma to less that 1 mN $m^{-1}$ on lung deflation. In addition to a low and stable surface tension, interdependence provided by the fibrous network enables the lung to maintain a large alveolar surface area (approximately 140 $m^2$ in the adult human lung) necessary for an efficient gas exchange.

Pulmonary surfactant of mammalian lungs consists of a variety of macromelecular complexes comprised of lipids and specific proteins. Surfactant is synthesized by the alveolar type II cell, in which it is stored in lamellar bodies. These lamellar bodies, when secreted into the alveoli, form tubular myelin, which appears to be the principal precursor of the surface film that lowers the surface tension. Lamellar bodies and tubular myelin both contain lipid and protein components of the surfactant. Surfactant obtained through bronchiolar lavage contains approximately 90% lipid, 10% protein, and small amounts of carbohydrate. Dipalmitoylphosphatidylcholine (DPPC), which accounts for approximately half the lipid in surfactant, is primarily responsible for the surface tension reducing property of the surfactant complex. The proteins associated with surfactant have been designated SP-A, SP-B, and SP-C by POSSMAYER (1988). SP-A is a variably glycosylated protein with a molecular mass of 18–35 kDa (reduced). SP-A is relatively water-soluble but SP-B and SP-C remain with the lipids extracted with organic solvents. SP-B has a molecular mass of 15 kDa (non-reduced) while SP-C has a molecular mass at 3–5 kDa in the non-reduced or reduced states.

Surfactant proteins directly affect the interfacial properties of surfactant lipids. Recent studies imply that surfactant-associated protein A and the recently identified SP-D have other important roles in the lung, including immunologic defense (WEAVER and WHITSETT, 1991). Rapid adsorption of surfactant phospholipids to the air–liquid interface is thought to be critical for maintaining the morphological integrity of the gas exchange region of the lung (GROSSMAN et al., 1986).

Preparations of surfactant lipids containing mixtures of SP-8 and SP-C were shown to increase lung compliance and preserve the morphological integrity in the distal airways in prematurely delivered ventilated fetal rabbits. Similar lipid extract surfactants have been widely tested in clinical trials and shown to improve oxygenation and decrease the need for respiratory support in infants suffering from respiratory distress syndrome (JOBE and IKEGAMI, 1987).

## 7 Biosurfactants in Foods

The preparation and processing of various foods, especially at an industrial scale, is often greatly facilitated by the inclusion of small amounts of selected additives. Surface-active agents play an important role, among them producing significant effects although present in minute amounts. Their importance will only grow with time, enabling the manufacture of new food products and the develop-

**Tab. 8.** Biotechnology Products Associated with Food Production and Preparation

| Product | Uses |
|---|---|
| Organic acids, their salts and derivatives | pH control agents, acidulants, preservatives, flavoring agents, flavor enhancers, adjuvants, color stabilizers, gelling enhancers, melt modifiers, turbidity reducers, etc. |
| Mono-oligosaccharides | Sweeteners for diet and health foods |
| Polysaccharides | Thickeners, water-binding agents, gellants, foaming agents, rheology modifiers, nutritive supplements |
| Amino acids, peptides | Constituents of protein hydrolyzates for flavoring, anti-microbial agents (nisin, bacterocins), monosodium glutamate as taste enhancer |
| Proteins | Single-cell proteins (SCP) as food and feed additives |
| Enzymes | Microbial rennets, meat tenderizers, flour modifying proteases, beer stabilizers/clarifiers, amylases, glycoamylases, and pullulanases for starch hydrolysis, glucose isomerase for fructose and high-fructose syrup production, pectin-degrading enzymes (fruit juice production), lipases as interesterification catalysts (e.g., in food surfactant production), glucose oxidase as oxygen scavenger, invertase for confectionery products |
| Lipids and derivatives | Specialty fats and oils, emulsifying and de-emulsifying agents, lubricants, die-releasing aids, wetting agents, fat-blooming preventers, etc. |
| Other substances of interest in food production | B-group vitamins, L-ascorbic acid, special flavors (vanilla, fruit, mushroom, mint, onion, etc.), coloring agents, taste enhancers (5′-nucleotides) |

ment of new technologies for the existing ones. Despite the highly efficient commercial synthetic food surfactants, a search for still newer and better products will continue.

Surfactants are used in food primarily to modify or improve functional properties of the product. A review related to use of surfactants (and biosurfactants) in food processing and in food products have been presented by KACHHOLZ and SCHLINGMAN (1987) and by VELIKONJA and KOSARIC (1993).

Properties of food surfactants:

- general surface-active properties: emulsifier, demulsifier, solubilizer, suspension agent, wetting agent, defoaming agent, thickener, lubricating agent, protecting agent
- food-specific properties: interaction with lipids, interaction with proteins, interaction with carbohydrates

Biotechnology products which are associated with food production and preparation are shown in Tab. 8. Functional properties of various food emulsifiers are listed in Tab. 9.

The food industry does not yet use biosurfactants as food additives at a large scale as many biosurfactant properties and regulations regarding the approval of new food ingredients have to be resolved. Issues that have to be addressed are related to nutritional, functional, sensory, biological, and toxicological properties of the new ingredient, production economics as compared to the synthetic surfactants for the same use, consumer acceptability of the new ingredient, legal regulations, and general eating habits and customs. However, it has to be pointed out that glycolipid biosurfactants are analogous to

**Tab. 9.** Functional Properties of Food Emulsifiers

| Functions | Product Examples |
|---|---|
| Emulsification (water-in-oil) | Margarine |
| Emulsification (oil-in-water) | Mayonnaise |
| Aeration | Whipped toppings |
| Improvement of whippability | Whipped toppings |
| Inhibition of fat crystallization | Candy |
| Softening | Candy |
| Antistaling | Bread |
| Dough conditioning | Bread dough |
| Improvement of loaf volume | Bread |
| Reduction of shortening requirements | Bread |
| Pan release agent | Yeast-leavened and other dough and batter products |
| Fat stabilizer | Food oils |
| Antispattering agent | Margarine and frying oils |
| Antisticking agent | Caramel candy |
| Protective coating | Fresh fruits and vegetables |
| Surfactant | Molasses |
| Viscosity control | Molten chocolate |
| Improvement of solubility | Instant drinks |
| Starch complexation | Instant potatoes |
| Humectant | Cake icings |
| Plasticizer | Cake icings |
| Defoaming agent | Sugar production |
| Stabilization of flavor oils | Flavor emulsification |
| Promotion of "dryness" | Ice cream |
| Freeze–thaw stability | Whipped toppings |
| Improved wetting ability | Instant soups |
| Inhibition of sugar crystallization | Panned coatings |

synthetic fatty acid esters of mono- and oligosaccharides, which are applied worldwide at an industrial scale. The successful application of sophorolipid biosurfactants and their acylated and alkoxylated derivatives in cosmetic preparations (INOUE, 1988) could make them competitive alternatives to conventional synthetic food surfactants, at least from the functional myogenic, and legal points of view.

A prerequisite of the use of new substances as additives in food is to make sure that no toxic intermediates or products of their metabolism appear in the organism during degradation or excretion. Similar products synthetically produced from sucrose and fatty acids have been known and used for a long time (OHATA and KOMATA, 1986). Besides, sophorose is already being used as a sweetening agent (KITAHARA et al., 1988).

In Japan, where legal restrictions concerning the use of novel naturally derived (or biotechnologically produced) components in the food industry are not as strict as elsewhere, sophorolipids are patented as flour additives for quality improvement and better shelf life of bakery goods (e.g., 0.2 parts of sophorolipids and/or their lactones to 70 parts of flour) (SHIGATA and YAMASHITA, 1986). Surface-active properties of hydrolyzed and freeze-dried yeast cell walls *(Saccharomyces uvarum)* have been patented as surfactants in margarine producton (OHATA and KOMATA, 1986).

Other glycolipids were also patented as additives for cosmetics and pharmaceuticals. Thus, rhamnolipids produced by *Pseudomonas* BOP 100 were used to obtain liposomes (ISHIGAMI et al., 1988). Emulsions suitable for cosmetic, food and pharmaceutical applications were prepared with a 2-*O*-(2-decenoyl)rhamnolipid. Rhamnolipids of *Pseudomonas aeruginosa* UI 29791 were produced in high yield (46 g $L^{-1}$ in batch culture, 6.4–10 $gL^{-1}$ $d^{-1}$ in semicontinuous culture on corn oil, (40 g $L^{-1}$) as the sole carbon source and recommended, among others, for use in foods (DANIELS et al., 1988; LINHARDT et al., 1989).

Some novel mannosylerythrytol lipids might be considered as potential surfactants. One example is 4-*O*-(di-*o*-acetyl-di-*O*-alkanoyl-$\beta$-D-mannopyranisyl) erythrytol and the monoacetyl derivative as the two major constituents ($\sim$80%) of the total lipids produced with a yield of approximately 40 g $L^{-1}$ by *Candida antarctica* T34 grown on soybean oil as the sole carbon source.

High molecular-weight emulsifiers from prokaryotes have also been proposed as food emulsifiers. For example, emulcyan, the extracellular complex of >700 kDa, consists of carbohydrates, fatty acids, and a protein moiety and is produced by the benthic cyanobacterium *Phormidium* J-1 (ATCC 39161) (SHILO and FATTOM, 1989).

The carbohydrate-emulsifying activity of lipopolyheterosaccharides of the genus *Acinetobacter* is well documented (GUTNICK and SHABTAI, 1987). In addition to the proposed use of emulsans and apoemulsans from *A. calcoaceticus* in personal care products, the microbial surfactant combined with synthetic emulsifiers (e.g., a polyoxyethylene oleyl ether) was patented also for use in food emulsions (MIYATA et al., 1987).

# 8 Biosurfactants for Environmental Control

## 8.1 Introduction

The accumulation and persistence of toxic materials in water and soil represents a major problem today. Various organics are generated either as by-products from various industries (e.g., petroleum and petrochemical, pulp and paper, chemical industries, etc.) which may be released into the environment, or are accidentally spilled. Of primary concern are aromatics and their chlorinated derivatives which are difficult to biodegrade and are toxic. Aromatics and their chlorinated derivatives are generated in chlorine bleaching of cellulose pulp (e.g., dioxins), pesticide and herbicides (e.g., chlorophenols), moth repellents and air deodorant (e.g., *p*-dichlorobenzene), petroleum and petrochemicals (e.g., naphthalene), transformer oils (e.g., polychlorinated biphenyls, PCB), chemical, plastics, iron and steel industries (e.g., phenols), wood preservation (e.g., pentachlorophenols, PCP), etc.

As the above chemicals are toxic and are proven carcinogens, their release to water and soil is prohibited. If, however, they do appear in industrial wastewaters, these must be treated and detoxified. Treatment of wastewaters is practiced worldwide utilizing a combination of methods (chemical, physical, and biological). Biological methods show many advantages, and many organics can be efficiently degraded by aerobic and anaerobic processes. However, for degradation of calcitrant pollutants, special and/or adapted microbial cultures are needed, which can (1) survive in the contaminated environment and (2) degrade the contaminant efficiently and completely.

While water treatment is relatively easy to perform, soil bioremediation is much more difficult and complex. The first problem arises due to difficulties in treating soil, especially when pollution is distributed over a large area. Thus, removal of soil from the contaminated site becomes a costly undertaking, even though such *ex situ* treatment might be well established. This could be accomplished in two ways:

(1) Addition of nutrients to the soil in form of nitrogen, phosphorous and, if necessary, carbon compounds, which would allow the native microbial population to develop and augment and thus provide more microorganisms for metabolism or cometabolism of the pollutant in question.
(2) Produce *ex situ* a microbial population which is adapted to the pollutant and is capable to metabolize it efficiently, and then add this population, along with necessary nutrients, to the polluted soil. The added biomass would, under proper conditions, be able to survive in the soil and to further degrade the objectionable organics.

Both methods are applied whereby method (1) seems to be more popular, but the strategy depends upon the type of pollutant, the environmental soil conditions, and the availability of the adapted culture.

## 8.2 Biodegradation Mechanisms

Many microorganisms are capable to metabolize and thus degrade aromatic hydrocarbons (ROCHKIND-DUBINSKI et al., 1987).

Biodegradation schemes for saturated and unsaturated aliphatic hydrocarbons are presented in Figs. 13 and 14. As an example, Pseudomonads oxidize aromatics according to the scheme presented in Fig. 15, producing catechol and its derivatives as one of the important intermediates.

Catechol can undergo further *ortho-* and *meta*-cleavage, as shown in Fig. 16. The final products in this cleavage are acetyl-CoA and succinate (for *ortho*) and acetaldehyde and pyruvate (for *meta*), which represent easy metabolizable substrates for common biochemical pathways.

For practical application, anaerobic biodegradation is of interest as anaerobic conditions would prevail in deeper layers of the soil. Microbial consortia (YOUNG, 1984) are capable to degrade anaerobically phenol and chatecol, as shown in Fig. 17, whereby the adaptation time for catechol degradation is generally longer.

When aromatics are chlorinated their degradation becomes more difficult. Depending on the starting material, oxidation and breakdown of the side chain may precede dechlorination which is the more difficult step. In general, chlorinated aromatic compounds are metabolized to chlorophenols. Fig. 18 pre-

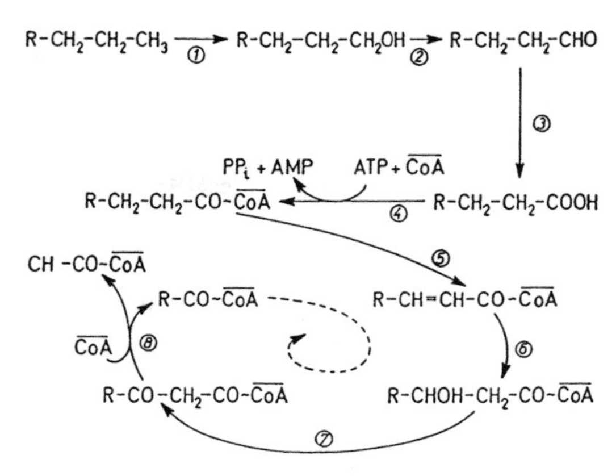


**Fig. 13.** Biodegradation of aliphatic hydrocarbons. Degradation of alkanes by terminal oxidation with mono-oxygenases (steps 1–3) and $\beta$-oxidation to acetyl-coenzyme A.
Participating enzymes: (1) mono-oxygenase (alkane-1-hydroxylase, (2) alcohol dehydrogenase, (3) aldehyde dehydrogenase, (4) acyl-CoA synthetase, (5) acyl-CoA dehydrogenase, (6) 3-hydroxyacyl-CoA hydrolase, (7) 3-hydoxyacyl-CoA dehydrogenase, (8) ketothiolase.

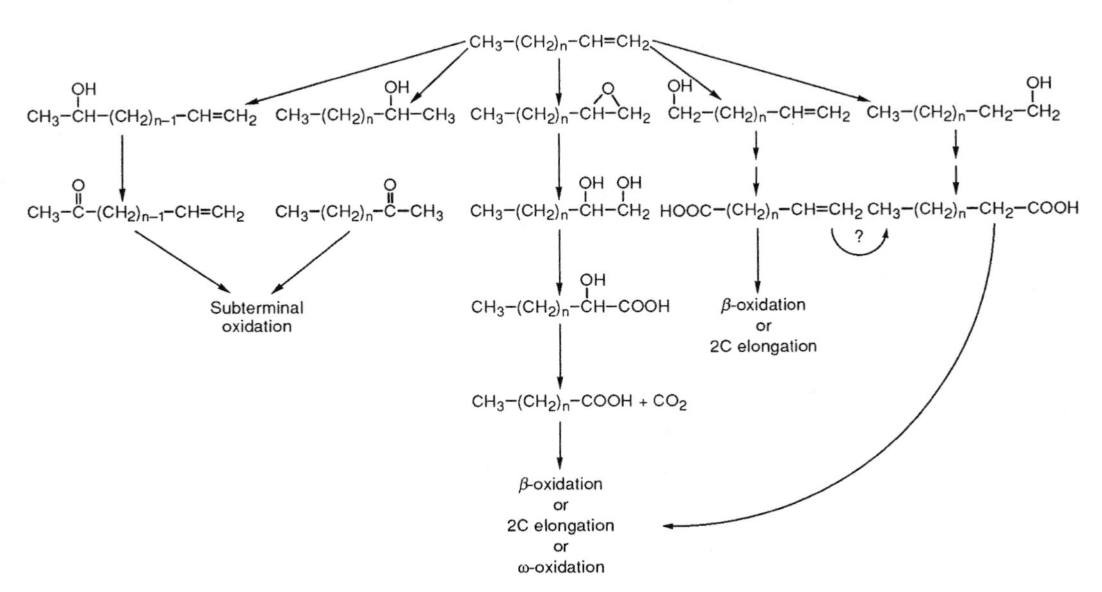


**Fig. 14.** Pathways for degradation of alkenes.

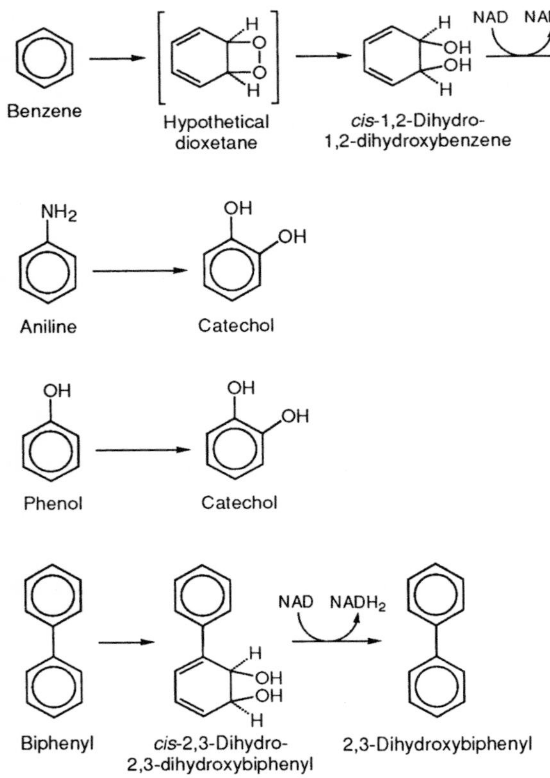

**Fig. 15.** Oxidation of aromatic molecules by bacteria.

sents possible pathways extrapolated from various studies.

Dechlorination coupled to oxidation of chlorinated aromatics has been demonstrated in both aerobic and anaerobic systems. As an example, *Brevibacterium* sp. can decarboxylate and dechlorinate trichlorobenzoates (HORVATH, 1971) according to the scheme in Fig. 19. Fig. 20 shows a reductive dechlorination of chlorobenzoates (HOROWITZ et al., 1983) by anaerobic microbial consortia.

The following summary can be presented for biodegradation of chlorinated aromatic compounds:

1) Biodegradability depends on:
   a) ionic state of the compound
   b) number, type and position of substituents
   c) general form of the molecule
   d) pH, temperature redox state, moisture, etc.
   e) availability of nutrients and growth factors
   f) concentration of toxic metabolites
2) Many microorganisms require a period of acclimation before biodegradation occurs.
3) Sometimes consortia of microorganisms are needed.
4) Fungi and bacteria metabolize most compounds by different biochemical pathways.
5) Predictions can be loosely made and experiments are needed.
6) Small-scale and scale-up performance data are required.
7) The breakdown products should be quantified and identified.

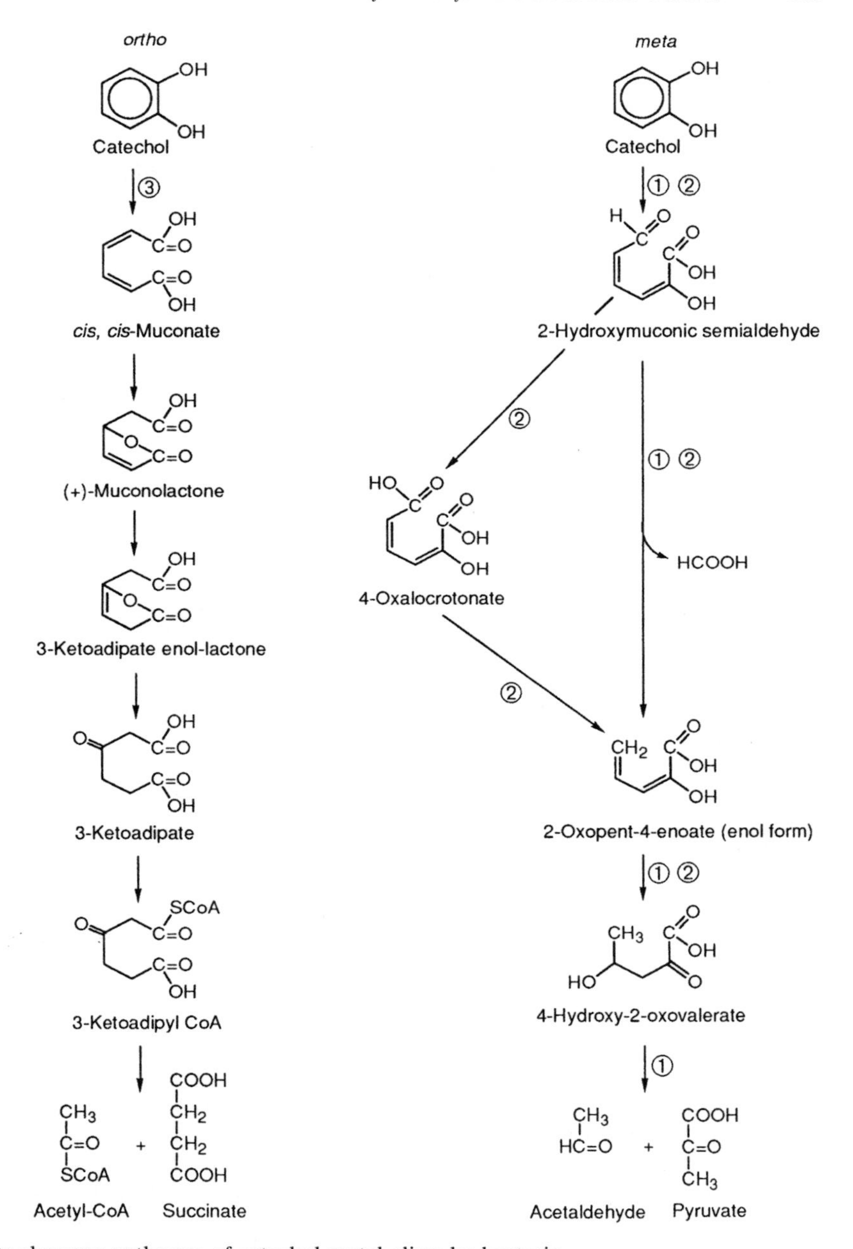


**Fig. 16.** *Ortho-* and *meta-*cleavage pathways of catechol metabolism by bacteria.

## 8.3 Soil Bioremediation

Bioremediation of soil contaminated with organic chemicals is a viable alternative method for clean-up and remedy of hazardous waste sites. The final objective in this approach is to convert the parent toxicant into a readily biodegradable product which is harmless to human health and/or the environment.

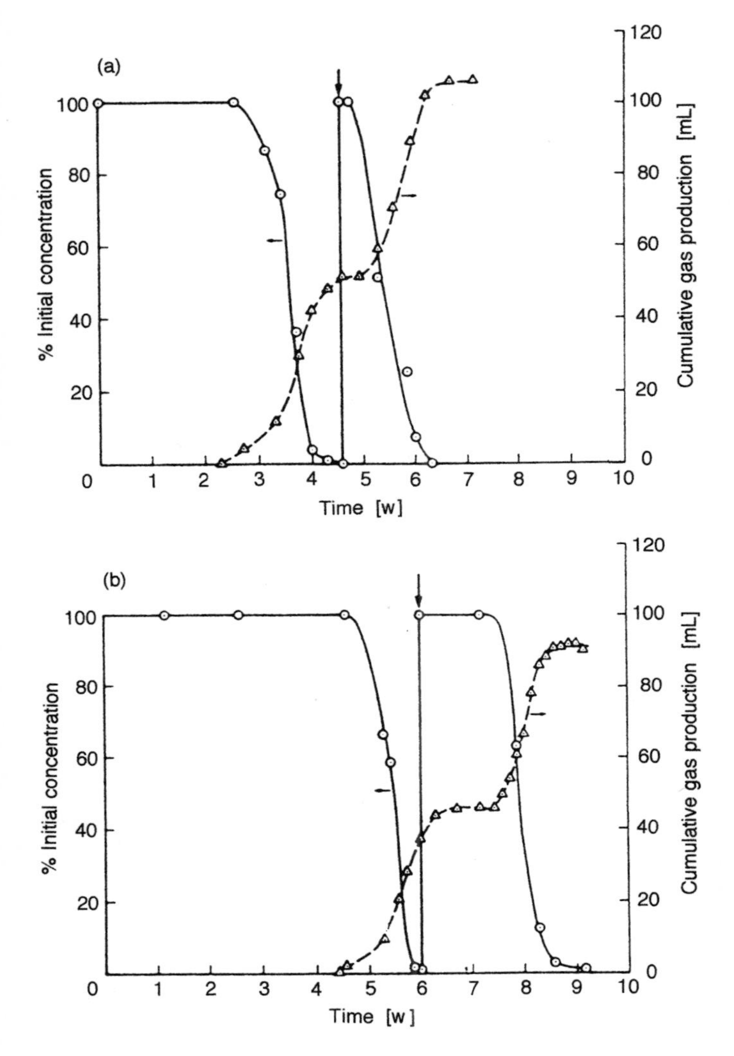


**Fig. 17.** Anaerobic degradation of phenol (top) and catechol (bottom) by a microbial consortium. At the point indicated by the arrow, fresh substrate was added to the culture vessels; ○–○ substrate; △–△ cumulative gas production.

It is essential to eliminate and make harmless such hazardous wastes as they may either enter into plants and thus contaminate this food source or be leached into the groundwater, the purification of which becomes even more complex and difficult.

The biological remediation processes can be performed (1) *in situ*, (2) in a prepared bed, and (3) in a slurry reactor system. *In situ* processes are usually accomplished by addition of microbial nutrients to the soil which allows considerable growth of soil microbial indigenous population. Thus increased microbial biomass in the soil results in faster biodegradation of contaminating organics. The soil can also be dug-out and treated off site in a similar way or it could be placed into a bioreactor to which water and nutrients are added and the biodegradation proceedes under continuous mixing which enhances the biodegradation process.

The other alternative is to selectively isolate and grow specific microbial cultures which are adapted to the toxicant and thus "trained" to degrade and utilize it as a substrate. Addition of surface-active agents, es-

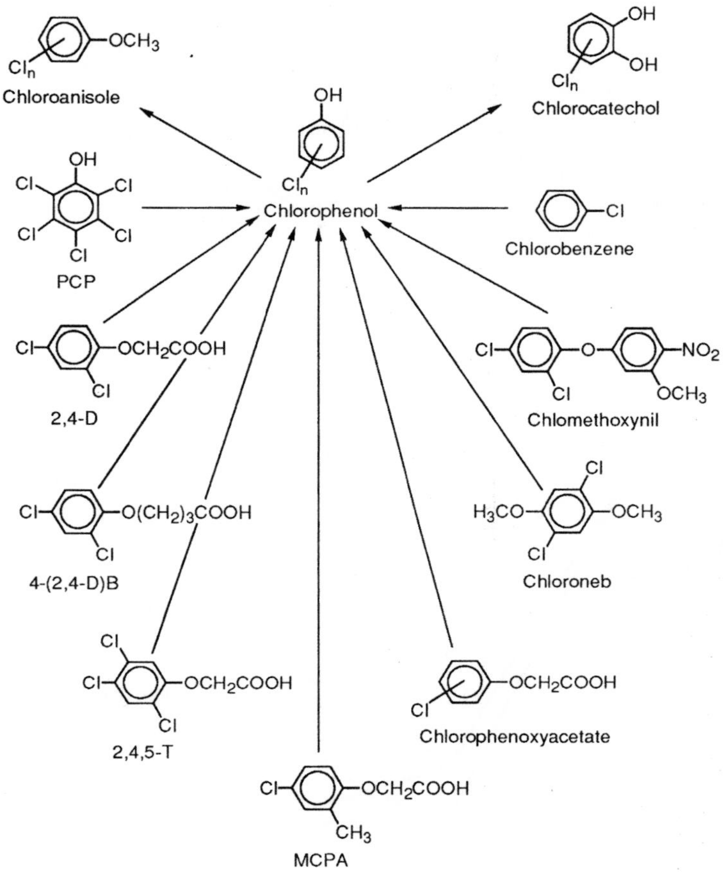

**Fig. 18.** Chlorinated aromatic compounds metabolized to chlorophenols. This figure presents possible pathways extrapolated from various studies. In actual environmental systems a given transformaton may be inhibited by a number of factors. Terminal compounds shown may be recalcitrant or insufficient research may exist on which to base a conclusion.

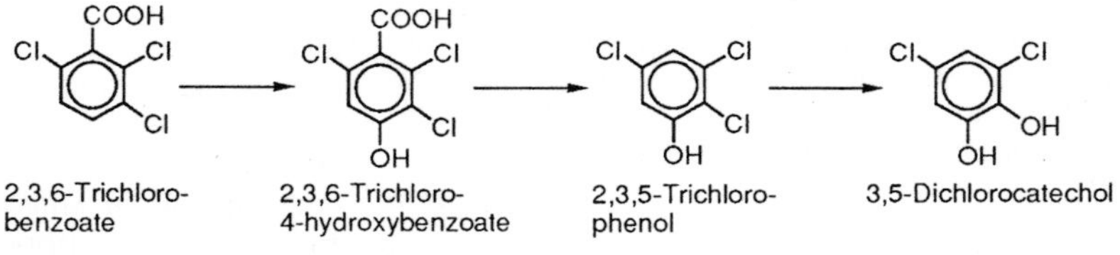

**Fig. 19.** Formation of 3,5-dichlorocatechol from 2,3,6-trichlorobenzoate by *Brevibacterium* sp.

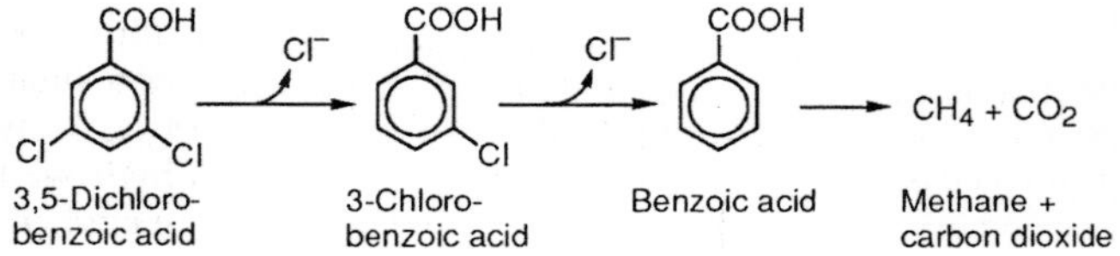

**Fig. 20.** Pathway for the reductive dechlorinaton of chlorobenzoates by anaerobic microbial consortia.

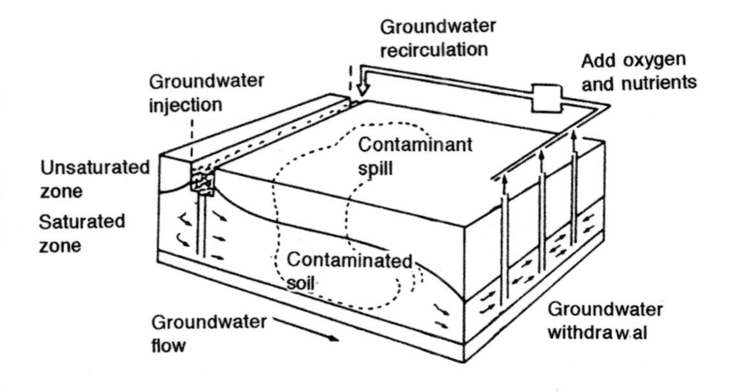


**Fig. 21.** *In-situ* treatment of contaminated saturated soil by recirculating amended groundwater through the contaminated region.

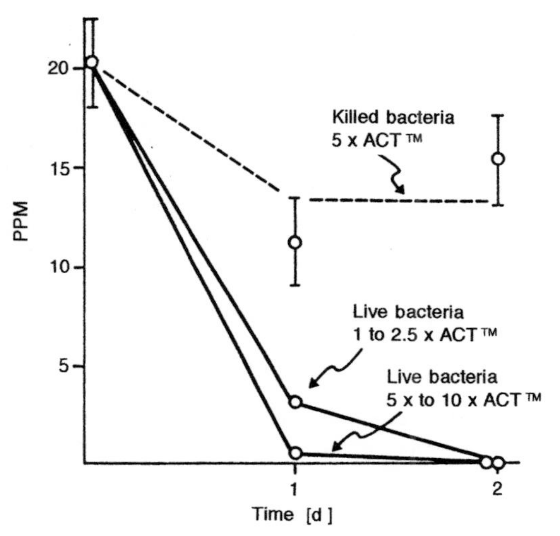


**Fig. 22.** Laboratory biodegradation documentation, groundwater and soil from the saturated zone of gasoline-contaminated site, in a sealed container with gasoline, ACT mineral nutrients, oxygen, and naturally present gasoline-degrading bacteria. Compounds analyzed were benzene, toluene, ethylbenzene, xylenes, and trimethylbenzene.

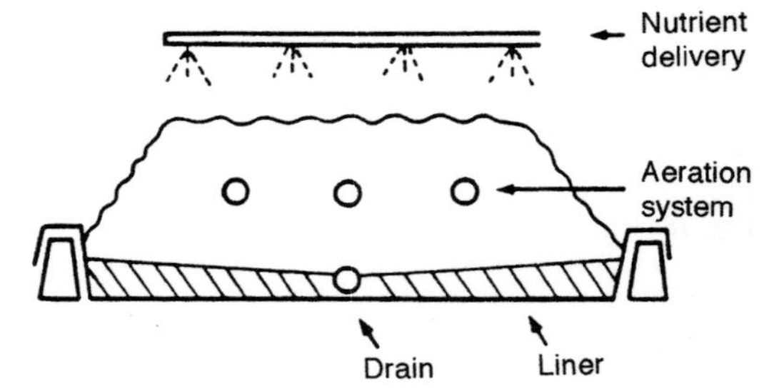


**Fig. 23.** On site treatment of excavated contaminated soil by CAA Bioremediaton System's Forced Aerabon Contamination Treatment FACT process.

**Tab. 10.** Cost of Soil Treatment

| Treatment process | Cost per Ton |
|---|---|
| Landfill disposal fees | \$140–200<br>+ taxes<br>+ transportation |
| Mobile incineration | \$150–140 |
| Stabilization/fixation | \$100–200 |
| Bioremediation | \$15–70 |

pecially when biodegradation of nonpolar compounds is encountered, helps in the uptake and metabolism of these compounds by the microbial population.

Comparing bioremediation with other available technologies for soil remediation, one can see a financial benefit when bioremediation is considered as shown in Tab. 10 (MOLNAR and GRUBBS, 1990).

A typical *in situ* approach is depicted in Fig. 21. In this approach, part of the groundwater can be collected at the underflow, pumped back onto the soil supplemented with nutrients and oxygen (FOGEL et al., 1990). It is well recognized that for biodegradation of petroleum 3 kg oxygen are required for every kg of petroleum hydrocarbon degraded. Sparging with oxygen can deliver only 40 ppm at the injection point while hydrogen peroxide can be dissolved and injected at concentrations >500 ppm and will gradually break down to oxygen during transport through the contaminated area.

Fig. 22 represents laboratory data when a proprietary microorganism mixture (ACT) was added to soil contaminated with gasoline and enriched with nutrients and oxygen. The effect of live bacteria on the degradation of the hydrocarbons is evident.

When the soil was excavated and then treated according to the scheme in Fig. 23, biodegradation data as shown in Fig. 24 were obtained. Stimulation of microbial growth by added nutrients results in almost complete biodegradation in a relatively short period of time.

Bioremediation of petroleum-contaminated soils using microbial consortia as inoculum is the so-called bioaugmentation in soil. The term biostimulation refers to enhanced biodegradation by indigenous soil bacteria due to increase of their population by addition of nutrients.

The nutrients include a nitrogen source, a phosphorous source, pH adjustment and a myriad of trace minerals.

In general, biodegradation of hydrocarbons at any given site will depend upon:

- indigenous soil microbial population,
- hydrocarbon variety and concentration,
- soil structure,
- soil reaction,
- nutrient availability,
- moisture content,
- oxygen availability.

Soil microorganisms reported to degrade hydrocarbons under favorable conditions include *Pseudomonas, Flavobacterium, Achromobacter, Arthrobacter, Micrococcus* and *Acinetobacter.* Hydrocarbons with less than 10 carbon atoms tend to be relatively easy to degrade as long as the concentration is not too high to be toxic to the organisms. Benzenes, xylene, and toluene are examples of gasoline components that are easily degraded. Complex molecular structures such as branched paraffins, olefins, or cyclic alkanes are much more resistant to biodegradation.

Soil structure, which is the form of assembly of the soil particles, determines the ability of that soil to transmit air, water, and nu-

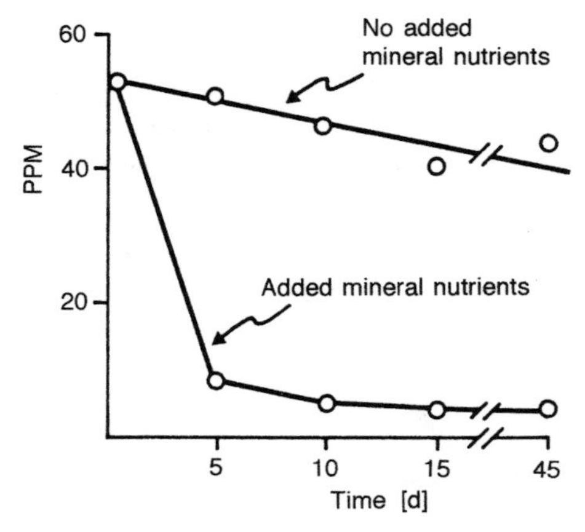


**Fig. 24.** Biodegradability test for diesel fuel in contaminated soil from the site of on-site soil pile treatment. Contaminated soil containing naturally occurring fuel-degrading bacteria was placed in sealed containers with excess mineral nutrients, oxygen, and water. Samples were analyzed after different amounts of time to show the progress of biodegradation. The data is given in terms of the sum of the 18 major constituents of diesel fuel, analyzed by gas chromatography with flame ionization detection.

trients to the zone of bioactivity. Another major controlling factor is the variety and balance of nutrients in the soil. Nitrogen and phosphorous are the most common additives and one could roughly estimate that to degrade 1 L gasoline, the microorganisms would need about 44 g nitrogen, 22 g phosphorous, and 760 g oxygen. Generally, optimum activity occurs when the soil moisture is 50–80% of saturation (moisture holding capacity). When the moisture content falls below 10% bioactivity becomes marginal.

Of particular interest, as mentioned earlier, is to establish to what extent biodegradation of hydrocarbons can proceed at low or no oxygen. At these low oxygen levels, denitrification will proceed if an alternate electron acceptor such as nitrate is available. When samples containing benzene, toluene, and xylene were incubated anoxically to which 500 mg $L^{-1}$ $NO_3^-$-N and 50 mg $L^{-1}$ $PO_4^{3-}$-P were added (GERSBERG et al., 1990), data as shown in Fig. 25 were obtained. The initial concentrations of benzene, toluene, and total xylene isomers were 13.3–13.7 mg $L^{-1}$, 33.7–

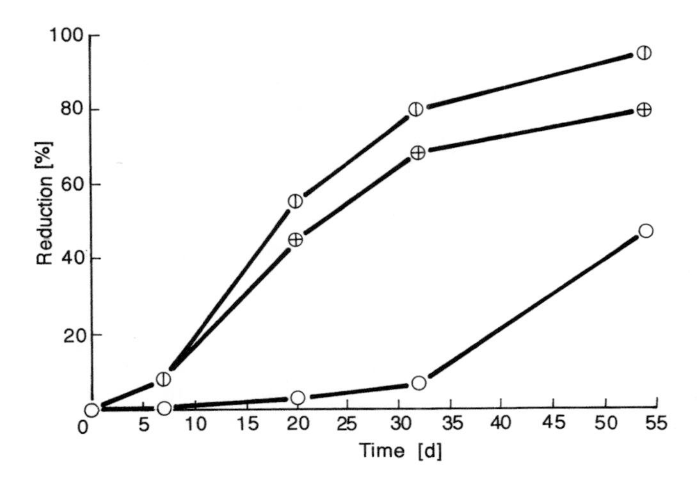


**Fig. 25.** Percent removal of benzene (⦶—⦶), toluene (⊕—⊕), and xylene isomers (○—○) in nutrient-enriched groundwaters incubated anoxically with 500 mg$^{-1}$ $NO_3^-$-N and 50 mg$^{-1}$ $PO_4^{3-}$-P. All removal values are calculated as compared to controls (no nutrients added).

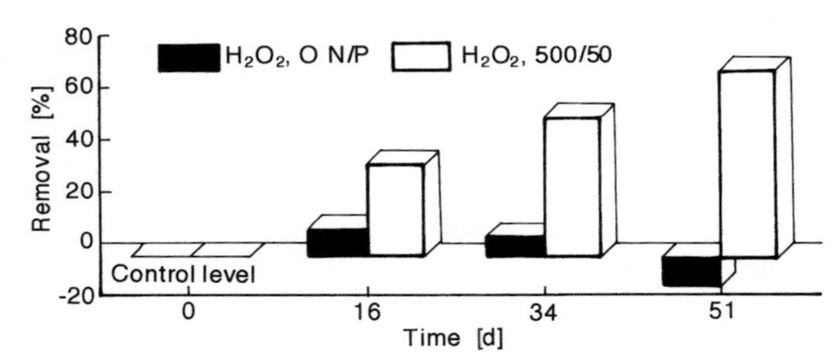


**Fig. 26.** Percent removal of total BTX (benzene, toluene, and xylene isomers) in groundwaters incubated aerobically with 50 mg L$^{-1}$ $H_2O_2$ (as an $O_2$ source) plus either nutrients (500 mg L$^{-1}$ $NO_3^-$-N and 50 mg L$^{-1}$ $PO_4^{-3}$-P) or no nutrients. All removal values are calculated as compared to controls (no $H_2O_2$ and no nutrients).

33.9 mg L$^{-1}$, and 15.4–23.2 mg L$^{-1}$, respectively. The effect of nutrients in the above system, as compared to a system containing 50 mg L$^{-1}$ $H_2O_2$ clearly show the difference in biodegradation (Fig. 26).

## 8.3.1 Effect of Biosurfactants in Soil Bioremediation

Biodegradation of hydrocarbons in soil can also efficiently be enhanced by addition or *in situ* production of biosurfactants (OBERBREMER, 1990; MEIER, 1990; MÜLLER-HURTIG et al., 1993). It was generally observed that the degradation time was shortened and particularly the adaptation time for the microbes.

Studies with chemical surfactants showed that the degradation of phenanthrene by a not identified isolate could be increased by a nonionic surfactant based on ethylene glycol (KÖHLER et al., 1990).

Adding rhamnolipids as biosurfactants, HISATSUKA et al. (1971) could only find a specific growth stimulation for 6 tested *Pseudomonas* strains producing also rhamnose lipids, but not for 11 strains of *Corynebacterium, Achromobacter, Micrococcus,* or *Bacillus* growing on *n*-hexadecane. The added concentration of 75 mg L$^{-1}$ of rhamnolipids may be toxic as the growth yield after 2 d was more than 30% lower for 3 of these 11 strains. Also, the stimulation of hydrocarbon degradation by added extracellular sophorose lipids – produced by *Torulopsis bombicola* growing on water-insoluble alkanes – was only found for *Torulopsis* yeasts but not for other typical alkane-utilizing yeasts. 43 synthetic nonionic surfactants with hydrophilic–lipophilic balance values ranging from 3.6–19.5 (from Kao Soap Co. Ltd., Tokyo) were unable to replace the glycolipid (ITO and INOUE, 1982). Furthermore, the biosurfactants of *Candida lipolytica* formed during the growth on hexadecane seemed to be specific for the degradation of hexadecane (GOMA et al., 1973); this means that specific biosurfactants should be produced by microorganisms growing on the contaminating crude oil.

The effect of various glycolipids on oil degradation was tested in soil model systems. In submerged culture with 10% soil content, the interfacial tension had to be lowered by the formation of glycolipids to enable the degradation of all added model oil components of low water solubility (tetradecane, pentadecane, hexadecane, pristane, phenyldecane, and trimethylcyclohexane). These hydrocarbons were degraded by an original soil population in a second degradation phase after a naphthalene degradation phase (OBERBREMER et al., 1990).

All tested glycolipid biosurfactants enhanced the degradation of the model oil when the initial interfacial tension was reduced from 21 mN to a range of 2–15 mN. Probably because the cells did not have to produce biosurfactants for a better uptake of these hydrocarbons of lower water solubility, all glycolipids shortened the time span of the second adaptation phase from 21–0–9 h. Furthermore, hydrocarbon elimination was increased from 81–93–99% (Tab. 11) and hydrocarbon mineralization by adding sophorose lipids from 17% $CO_2$ to 49% $CO_2$ (MÜLLER-HURTIG et al., 1990). Altogether, the degradation capacity was increased by the addition of biosurfactants from 16.3 to the range of 23.8–39.0 g hydrocarbon per kg soil (d.w.) per day. More effective were the more hydrophobic surfactants sophorose lipids, cellobiose lipids, and trehalose-6,6′-dicorynomycolates having HLB values from 4.05–8.0. Surprisingly trehalose-6,6′-dicorynomycolates and -2,3,4,2′-tetraesters were not most effective although approximately 50% of the selected soil population could produce similar trehalose lipids (OBERBREMER and MÜLLER-HURTIG, 1989).

Efficiently, all added biosurfactants except cellobiose lipids were degraded within the degradation phase of the hydrocarbons of lower solubility in water.

Under conditions without any oxygen limitation, the degradation capacity of the soil was 25.7 g hydrocarbon per kg soil d.w. (sdw) per day, that means 1–3 orders of magnitude higher than measured in soils with mineralization rates of 0.02–1.38 g hydrocarbon per kg soil (d.w.) per day (BOSSERT and BARTHA, 1984). The addition of biosurfactants could increase this rate up to 46.5 g hydrocarbon per kg soil (d.w.) per day, when sophorose lipids were added. The added surfactant concentration of 200 mg $L^{-1}$ was most effective. The stability of an emulsion of 1.35% model

**Tab. 11.** Influences of Glycolipids on the Degradation Efficiency

| Cultivation | Hydrocarbon Elimination | | HLB Value | Degradation Capacity [g hydrocarbon per kg soil (d.w.) per day] |
|---|---|---|---|---|
| | Durance [h] | Degree [%] | | |
| *Under oxygen limited conditions:* | | | | |
| Without surfactant | 114 | 82 | – | 16.3 |
| Trehalose-6,5′-corynomycolate | 71 | 93 | 4.05 | 37.2 |
| Sophorose lipids | 75 | 87 | 6.87 | 39 |
| Cellobiose lipids | 79 | 99 | 8 | 32.3 |
| Rhamnose lipids | 77 | 94 | 9.5 | 28.6 |
| Trehalose-2,3,4,1′-tetraester | | | 10.2 | 23.8 |
| *Without any oxygen limitation:* | | | | |
| Without surfactant | 79 | 89 | – | 25.7 |
| Sophorolipids | 57 | 95 | 6.87 | 46.5 |

oil in water was highest when the sophorose lipid concentration was above 150 mg $L^{-1}$ (OBERBREMER, 1990).

The biosurfactant emulsan, a polyanionic heteropolysaccharide from *Arthrobacter calcoaceticus* RAG-1 brings about a significant dispersion of crude oil (REISFELD et al., 1972). However, biodegradation of saturated and aromatic hydrocarbons from crude oil was reduced 50–90%, investigating a mixed population respectively of 8 isolated strains capable of degrading saturates, aromatics, or both. For some pure cultures, aromatic biodegradation was either unaffected or slightly stimulated by emulsification of the oil (FOGHT et al., 1989). Mutants of *Arthrobacter calcoaceticus* RAG-1 defective in emulsan production showed that the cell-bound form of emulsan is required for growth on crude oil because neither added emulsan nor emulsan-producing wild-type cells stimulated the growth of the mutant on these hydrocarbons (PINES and GUTNICK, 1986). The authors argued that emulsan-coated oil may prevent access of other biosurfactants that lower the interfacial tension for the formation of a macroemulsion, i.e., for hydrocarbon uptake.

WASKO and BRATT (1991) studied properties of a biosurfactant from *Ochrobacterium anthropii* isolated from contaminated fuel. The biosurfactant had a strong ability to emulsify hydrocarbon/water mixtures, which had a potential for the bioremediation of hydrocarbon-contaminated aquifers.

Genetically engineered microorganisms have also been studied for biosurfactant production (KHAN et al., 1994). A genetically engineered Tn 10 based transposon (Tn HH104) was successfully transferrred via conjugation from *E. coli* to *Pseudomonas aeruginosa* strain K3 using suicide delivery plasmid p SX2. Tetracycline resistant transconjugants were detected.

Biosurfactants produced by *Nocardia amarae* grown on hexadecane as a carbon source have been investigated for use in a surfactant/nonionic organic chemical sorption process. Removal and recovery of nonionic organics from aqueous solutions could be demonstrated.

Removal of hydrocarbon pollutants from contaminated soil has been a subject of investigation by numerous researchers (JAIN et al., 1992; University of Guelph, 1993; SCHEIBENBOGEN et al., 1994; SYLDATK and WAGNER, 1987; VAN BERNEM, 1984; KOSARIC, 1994, 1995, unpublished data).

The ability of rhamnolipid biosurfactants produced by *P. aeruginosa* UG2 to remove hydrocarbons from unsaturated soil was studied by SCHEIBENBOGEN et al. (1994). The results showed that UG2 biosurfactants have the potential for remediation of hydrophobic pollutants in unsaturated soil.

JAIN et al. (1992) demonstrated a biodegradation of tetradecane, hexadecane, and pristane (but not 2-methyl naphthalene) when either *P. aeruginosa* or UG2 or biosurfactant produced by this microorganism were added to the contaminated soil.

To reduce the cost of producing biosurfactants, the effect of glycolipid overproducing strains was investigated. The trehalosedicorynomycolates are the only tested biosurfactants that are overproduced under growing conditions (SYLDATK and WAGNER, 1987). However, the formed cell-bound trehalose lipid could not reduce the adaptation period of the soil population like the isolated form.

The elimination of the model oil was increased by biosurfactants added to the percolated medium with the same dependence on the hydrophilic–lipophylic balance values as in submerged culture. Added sophorose lipids caused 50% elimination within 6 d in comparison to 38% in the control experiment (Fig. 27). Furthermore, the extent of mineralization calculated from $H_2O_2$ use was 46% instead of 28.3% of the eliminated hydrocarbons.

This acceleration of biodegradation when adding sophorose lipids was confirmed in 3 or 4 different soil types (a light loamy sand, a loamy sand, and a silty loam with an annealing loss between 1.1% and 4.5). Only in a soil of high organic content (annealing loss 8.4%) there was no positive effect (MEIER et al., 1992).

In oil-contaminated mud flats, the elimination of polycyclic aromatics from the crude oil Arabian light was due to wave action or to microbial degradation. The chemical surfactant Finasol OSR-5 doubled the initial content of aromatics and decreased the amount

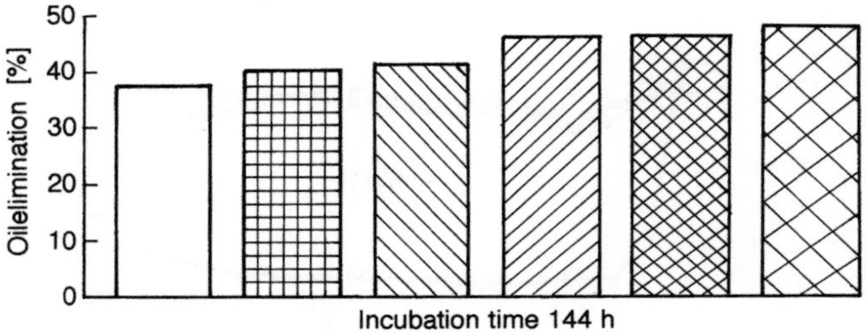

**Fig. 27.** Effect of biosurfactant addition to mineral salts medium on elimination of a model oil. □ without biosurfactant; ▦ trehalose tetraesters; ▧ trehalose dicorynomycolates; ▨ rhamnose lipids; ▩ cellobiose lipids; ⊠ sophorose lipids.

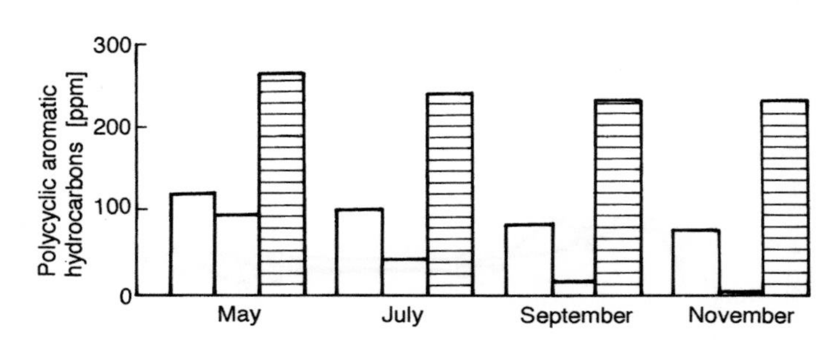


**Fig. 28.** Elimination of crude oil from tidal environments. 2 $m^2$ field contaminated 10× with 1 L Aramco crude oil; 10× with Aramco oil supplemented with 1 g $L^{-1}$ trehalose dicorynomycolates; 10× with 1 L Aramco oil and afterwards 10× with 100 mL Finaso OSR5 in 1 L seawater.

of aromatics removed after 6 months, whereas adding the biosurfactant Trehalose-5,5′-dicorynomycolates caused complete elimination within the period (Fig. 28; VAN BERMEN, 1984). The reason for that was not the difference in hydrophobicity, as is shown by the fact that biosurfactants of various hydrophobicites had a similar positive effect in contrast to the above chemical surfactant.

Recent data from the author's laboratory (KOSARIC et al., unpublished data) support other findings that sophorose lipids do enhance biodegradation of hydrocarbons and their chlorinated derivatives in contaminated soil. As an example, metalochlor [2-chloro-N-(2-ethyl-6-methyl phenyl)-N-(2-methoxy-1-methylethyl)-acetamide], a herbicide was significantly more degraded when sophorose lipids were added to a slurry bioreactor containing soil in suspension (Fig. 29). Another compound, 1-4-dichlorophenol was also considerably more degraded when sophorose lipids were added to the soil slurry (Fig. 30). Biodegradation of naphthalene in presence of sophorose lipids in a soil slurry bioreactor is shown in Fig. 31.

Interesting data were also obtained when soil in trays (about 20% moisture) contaminated with polycyclic aromatic hydrocarbons (PAH) was incubated (at room temperature) for 22 d and more. Nutrients, with sophorose lipids, were blended into the soil. Data are presented in Fig. 32. It is evident that many PAH were completely removed from soil while some were removed significantly.

# 9 Bio-De-Emulsifiers

## 9.1 Introduction

Microbes can act as de-emulsifying agents which have capabilities similar to the commercial solid phase de-emulsifiers (i.e., fibers, films). This de-emulsification process proceeds with agglomeration of the droplets of the emulsion discontinuous phase onto the bacterial cell surface, followed by spreading of the droplets over that surface and their coalescence as they contact each other during spreading. An overall mechanism for solid phase de-emulsification has been presented by KOSARIC et al. (1982). According to this mechanism, the droplets of the discontinuous phase preferentially wet the bacterial cell surface. In the case of an oil-in-water emulsion, the hydrophobic oil droplet has generally a

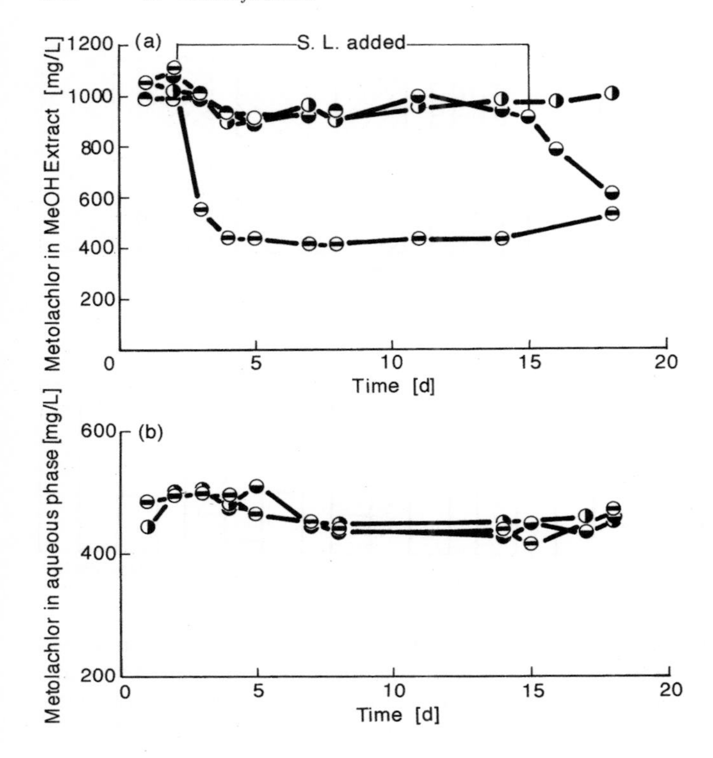


**Fig. 29.** Changes of metolachlor in MeOH extract of slurry (a) and aqueous phase of soil suspension (b). Soil: 5 g water: 150 mL Metolachlor 262 mg, sophorose lipids (S.L.): 38 mg. ◑—◑ sterilized soil −S.L.; ○—○ sterilized soil +S.L.; ⊖—⊖ non-sterilized soil −S.L.; ⊖—⊖ non-sterilized soil +s.L.

greater interaction with a hydrophobic cell surface and minimal interaction with a hydrophilic cell surface. On the other hand, for water-in-oil emulsions, a hydrophilic cell surface will interact better with the hydrophilic water droplet. Therefore, cell surface wettability is an important parameter in bacteria-induced de-emulsification.

To measure the properties of the bacterial cells in terms of their hydrophilic/hydrophobic character (and their wettability), contact angle measurements were performed. The contact angle between a filtered layer of cells and a water droplet placed on this layer was evaluated using a procedure modified from that of VAN OSS et al. (1975). According to this procedure, 5 mL of broth containing suspended bacterial cells were filtered onto a 0.8 micron cellulosic microsporous disc (Amicon) and then washed with 5 mL of distilled water. The filtered cells were then positioned in the optical system shown in Fig. 33 and the contact angle measured. Drops were applied to the filtered cells either by allowing 5 μL of aqueous phase to fall through the organic phase onto the cells (Mode II), by lightly touching a 5 μL drop to the cell surface (Mode III), or by extending upwards onto the cell surface a small organic phase drop from a syringe tip submerged in the aqueous phase (Mode I). In this latter method, the drop remained adherent to the syringe, and the contact angle was found to be independent of drop size over the range used. Fig. 34 shows the changes in contact angle of *Nocardia amarae* cells grown in various culture media (KOSARIC et al., 1983).

It is interesting to note that cells associated with the lag phase of growth on all media were hydrophilic (wettable) showing a low contact angle. As the culture aged, the cells became more hydrophobic and seemed to reach a certain constant level of hydrophobicity after a certain growth period which coincided with the stationary phase of growth. It is also obvious that the time course of increasing hydrophobicity depended upon the culture medium used. In further experiments, as shown below, it was demonstrated that younger cultures provided the hydrophilic

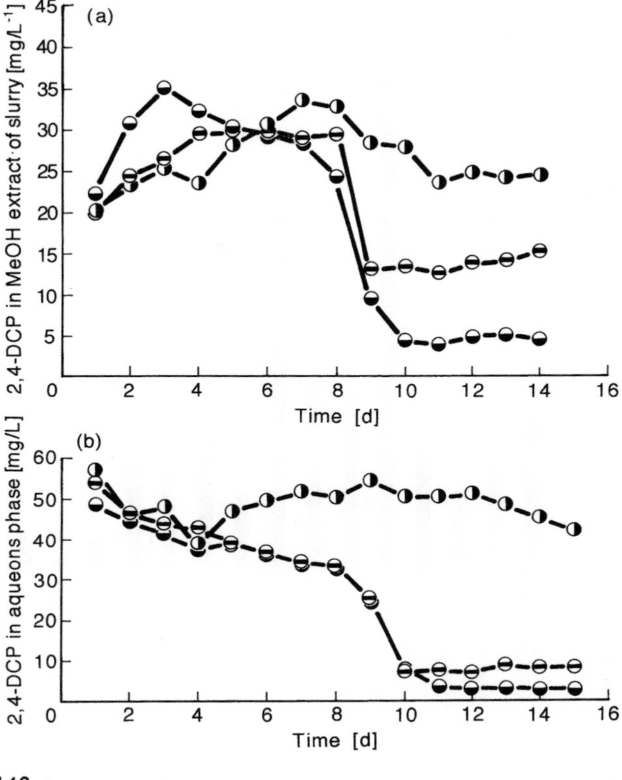


**Fig. 30.** Effect of sophorose lipids on 2,4-DCP in MeOH extract of slurry (a) and aqueous phase of soil suspensions (b).
Soil: 40 mg water: 60 mL 2,4-DCP: 6 mg sophorose lipids (S.L.): 38 mg. ◑—◑ sterilized soil −S.L.; ⊖—⊖ soil −S.L.; ◒—◒ soil +S.L.

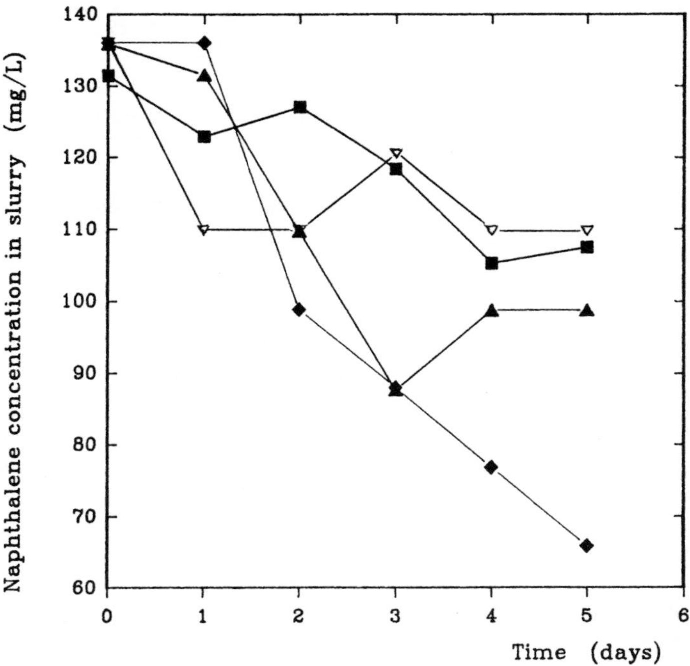


**Fig. 31.** Bioremediation of naphthalene-contaminated soil.
■ sterile soil; ▲ native soil;
▽ native soil +nutrients;
◆ native soil +sophorolipids.

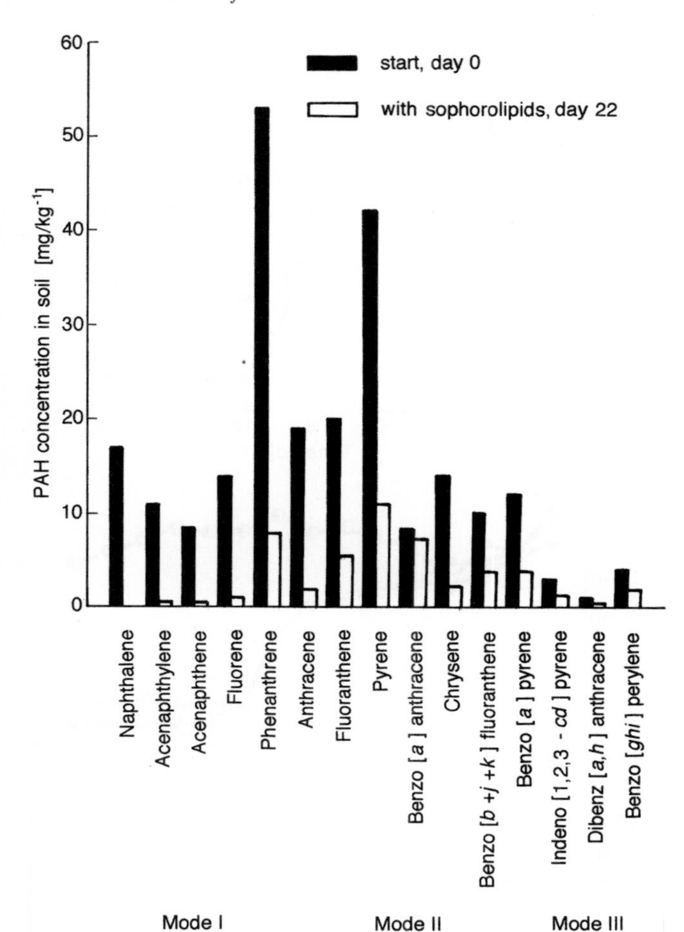


**Fig. 32.** Bioremediation of PAHs in soil; sophorolipids were added at time 0.

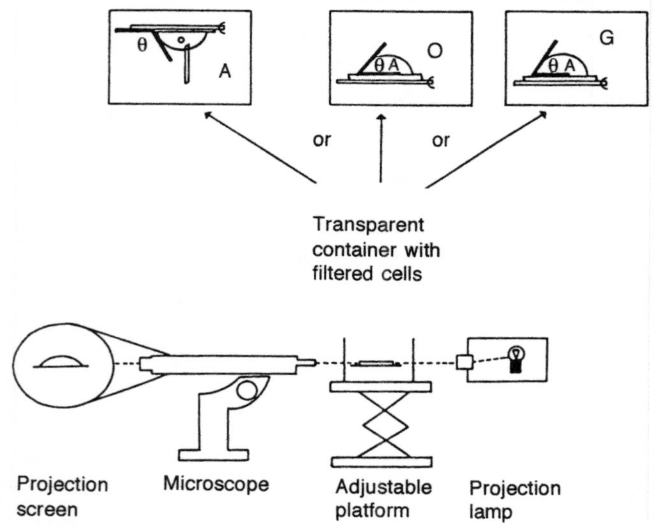


**Fig. 33.** Diagram of the optical system used to measure contact angle and of the three modes used to measure wettability of filtered bacterial cells (C).

Mode I: Organic drop (0) in bulk aqueous phase (A) (analogous to an O/W emulsion)

Mode II: Aqueous drop (A) in bulk organic phase (O) (analogous to a W/O emulsion)

Mode III: Aqueous drop (A) in air (G)

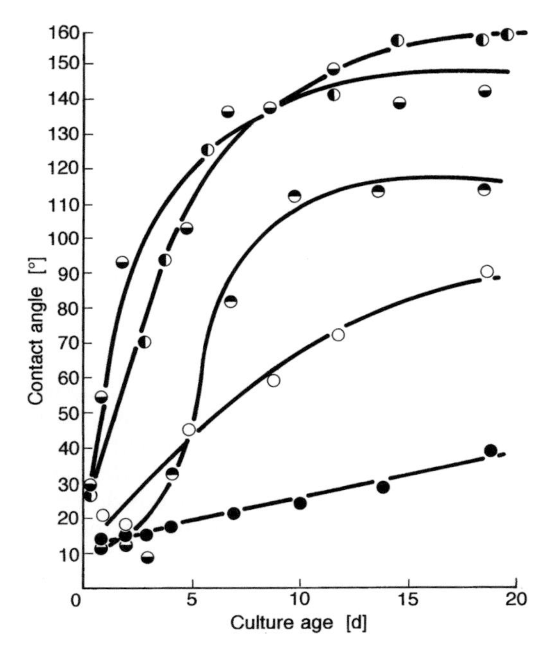


**Fig. 34.** Changes in the contact angle (O) of *Nocardia amarae* grown on various culture media. Symbols:
◐ 4% fructose and 0.8% yeast extract
◒ 4% glycerol and 0.8% yeast extract
◓ Artichoke juice stillage and 0.5% $Na_2$ $HPO_4$
○ 4% hexadecane and 0.8% yeast extract
● 2% glucose and 0.8% yeast extract

cells suitable for de-emulsifying W/O emulsions, and older cultures, especially those grown on hexadecane, fructose, or glycerol-containing media provided the more hydrophobic cells for breaking up O/W emulsions.

## 9.2 De-Emulsification of Model Emulsions

A number of bacteria, such as *Corynebacterium petrophilum, Nocardia amarae,* and *Rhodococcus aurantiacus* have been demonstrated to be potent de-emulsifiers of simple defined model O/W and W/O emulsions (COOPER et al., 1980). These model emulsions were kerosene/water type which were stabilized with commercial surfactants.

The standard W/O emulsions for testing were prepared using L92 pluronic surfactant from BASF (Wyandotte Corp). 4 mL of 0.068% aqueous solution of L92 was added to a test tube with 6 mL of kerosene and vortexed for about 1 min until a maximum emulsion was obtained.

The standard O/W emulsions were prepared in an analogous fashion. The aqueous phase in this system contained 0.072% Tween 60 and 0.028% Span 60. For each experiment, a plot was made of the logarithm of the percent of the volume which was an emulsion vs. the hour of measurement. This assumed that the emulsion breakdown could be approximated as a first order reaction.

$$\text{Rate} = -k(\%\ \text{Emulsion})$$

$$\log\left(\frac{\%\ \text{Emulsion}}{\text{Initial}\ \%\ \text{Emulsion}}\right) = \frac{-kt}{2.30}$$

The slopes of these plots could be used to calculate the half lives of the emulsions.

$$t_{1/2} = -\frac{0.301}{\text{Slope}}$$

Tab. 12 shows some data on de-emulsification with 3 different bacteria. Comparing to the control which represented the same system without bacteria, a considerable reduction of emulsion half life from >200 h down to less than 1 h was observed. This was a clear indication that bacterial cells do possess a surface-active property for de-emulsification.

## 9.3 Cell Surface as a De-Emulsifier

It was clearly noted that the activity was associated with the cells themselves and it was not due to an extracellular metabolite (KOSARIC et al., 1987b).

Another interesting feature was that the activity was very stable and could not be easily destroyed. The following procedures were applied:

**Tab. 12.** De-Emulsification of Kerosene–Water Emulsions by Whole 7-Day-Old Cultures of Bacteria Grown on Hydrocarbon-Containing Media

| | Neutral O/W | Anionic O/W | Cationic O/W | Neutral W/O |
|---|---|---|---|---|
| Control (culture medium) | >200 | >200 | >200 | >200 |
| *Nocardia* | < 1 | < 1 | < 1 | 2 |
| *Rhodococcus* | < 1 | 160 | 14 | < 1 |
| *Corynebacterium* sp. | < 1 | < 1 | < 1 | 14 |

HCl-Methanolysis

10 mg of dry chloroform–methanol (C-M) extracted *N. amarae* cells were suspended in 10 mL of 1% HCl–methanol at 100 °C for 20 h in a sealed tube. The non-extractable residue was extensively washed with methanol, the neutrality checked and then dried. 0.06 mg of the dried residue gave effectively instantaneous de-emulsification in the L92 W/O test emulsion, indicating no loss of activity compared with non-treated chloroform–methanol extracted cells which also gave instantaneous de-emulsification with 0.06 mg.

Trichloroacetic Acid (TCA) Extraction

10 mg of dry C-M extracted *N. amarae* cells were stirred in 10 mL of 10% w/v TCA for 24 h at 20 °C. The non-extractable residue was washed with distilled water and ethanol, neutrality checked and dried. 0.06 mg of dried residue had a $t_{1/2}$ of 1.2 h in the L92 W/O model emulsion; 0.3 mg had a $t_{1/2}$ of 0.5 h and 0.6 mg gave instantaneous de-emulsification. Some loss of activity was therefore achieved with TCA extraction.

Phenol Extraction

12 mg of dry C-M extracted *N. amarae* cells were stirred for 20 h at 20 °C in 10 mL of 0.9% w/v saline containing 2% w/v phenol. After 20 h, 35 mL was added of a solution containing 96% ethanol with 2% of a saturated solution of sodium acetate in water/alcohol (7/43, w/v). The supernatant was removed and dialyzed for 48 h with running water. The residue was washed extensively with acetone and dried. 0.06 mg of dried residue gave a $t_{1/2}$ of 5.4 h with the L92 model emulsion, and 0.6 mg gave instantaneous de-emulsification. No activity was found in the dialyzed extract. Phenol, like TCA, removes some activity, however, the majority of the activity still remains with the cell surface.

NaOH-Methanolysis

10 mg of dry C-M extracted *N. amarae* cells were suspended in 10 mL of 90% methanolic NaOH (0.3 N) for 20 h at 60 °C in a sealed tube. The residue was washed with water and methanol, neutrality checked and then dried. Both the residue and the extract (after neutralization with HCl) were tested for activity in the L92 W/O model emulsion. No de-emulsifying activity was found either in the residue (at concentrations up to 20 times that necessary for instantaneous de-emulsification using non-treated material) or in the extract (at concentrations up to 40 times that necessary for instantaneous de-emulsification).

According to these tests, alkaline methanolysis appears to be most effective in removing de-emulsifying ability from the cell surface. However, the activity is not recovered in methanol suggesting that the de-emulsifying activity is not likely due to any single compound that can be released by covalent bond cleavage, but to a number of compounds or chemical groupings present in the close proximity when the integrity of the cell surface is preserved.

**Tab. 13.** Effect of Autoclaving of Bacterial De-Emulsifiers (0.5 mL of Whole Broth Added)

| | $t_{1/2}$ of W/O | | $t_{1/2}$ of O/W | |
|---|---|---|---|---|
| | Unheated | Heated | Unheated | Heated |
| **1% yeast extract, 4% glucose:** | | | | |
| *N. amarae* | 6.8 | 22 | 8.4 | 18 |
| *R. aurantiacus* | 29 | 56 | 14 | 26 |
| *R. rubropertinctus* | 34 | 9.3 | 34 | 54 |
| **1% yeast extract, 4% hexadecane:** | | | | |
| *N. amarae* | 54 | 37 | 28 | 27 |
| *R. aurantiacus* | 36 | 34 | 26 | 37 |
| *R. rubropertinctus* | 124 | 154 | 120 | 150 |

Heat Stability

The 3 best de-emulsifying bacteria were grown 5 d on 1% yeast extract and 4% glucose and on 1% yeast extract and 4% hexadecane. Sterile samples were removed from each flask. The flasks were then autoclaved (120 °C) for 10 min and then cooled to room temperature. A comparison between the activity of the heat-killed and the unheated samples on the test emulsion is shown in Tab. 13. Very little change in the de-emulsifying ability of all 3 bacteria, after autoclaving, was observed. This is also an important observation which indicates that one does not have to use live cells for de-emulsification and that the de-emulsification is not associated with the active degradation of the oil phase and its metabolism by the live cells.

## 9.4 De-Emulsification of Complex W/O Field Emulsions

In order to assess the de-emulsification of complex field emulsions, a well head tar emulsion (ESSO 02-21-12) obtained in enhanced oil recovery field trials was used. *Nocardia amarae, Corynebacterium petrophilum* and *Rhodococcus aurantiacus* were tested. These bacteria have previously been grown in a 24 L bioreactor using 4% hexadecane and 0.8% yeast extract (*N. amarae*) or 2% glucose and 0.8% yeast extract (*C. petrophilum* and *R. aurantiacus*) in a mineral salts medium (CAIRNS et al., 1982, 1983; GRAY et al 1984; KOSARIC et al., 1981, 1987b).

A procedure for the evaluation of the de-emulsifying activity is presented in Fig. 35. The results are presented in Fig. 36 for the respective bacteria. It is clearly seen that the cells were much more active than the whole broth and that the activity was the largest with cells harvested at an early stage of growth (i.e., "young cells"). Of the 3 bacteria tested, *N. amarae* and *C. petrophilum* cells could accomplish nearly 100% de-emulsification of the diluted well head emulsion tested.

In all the above experiments, the contact time (the time during which the bacteria were in contact with the emulsion) was 24 h and the amount of bacteria added was 500 ppm. The de-emulsifying activity was directly proportional to the amount of cells used, as shown in Fig. 37.

This contact time was further investigated and it was shown that it can even be reduced to less than 12 h, as shown in Fig. 38.

In all previous studies, the well head emulsion was diluted 1 in 10 using toluene. However, this dilution later was found not to be totally necessary. Various dilutions (5 g emulsion diluted with toluene) of 1:1, 1:2, and 1:5 were set up and tested against the addition of 400 ppm *C. petrophilum* day 6 cells. After

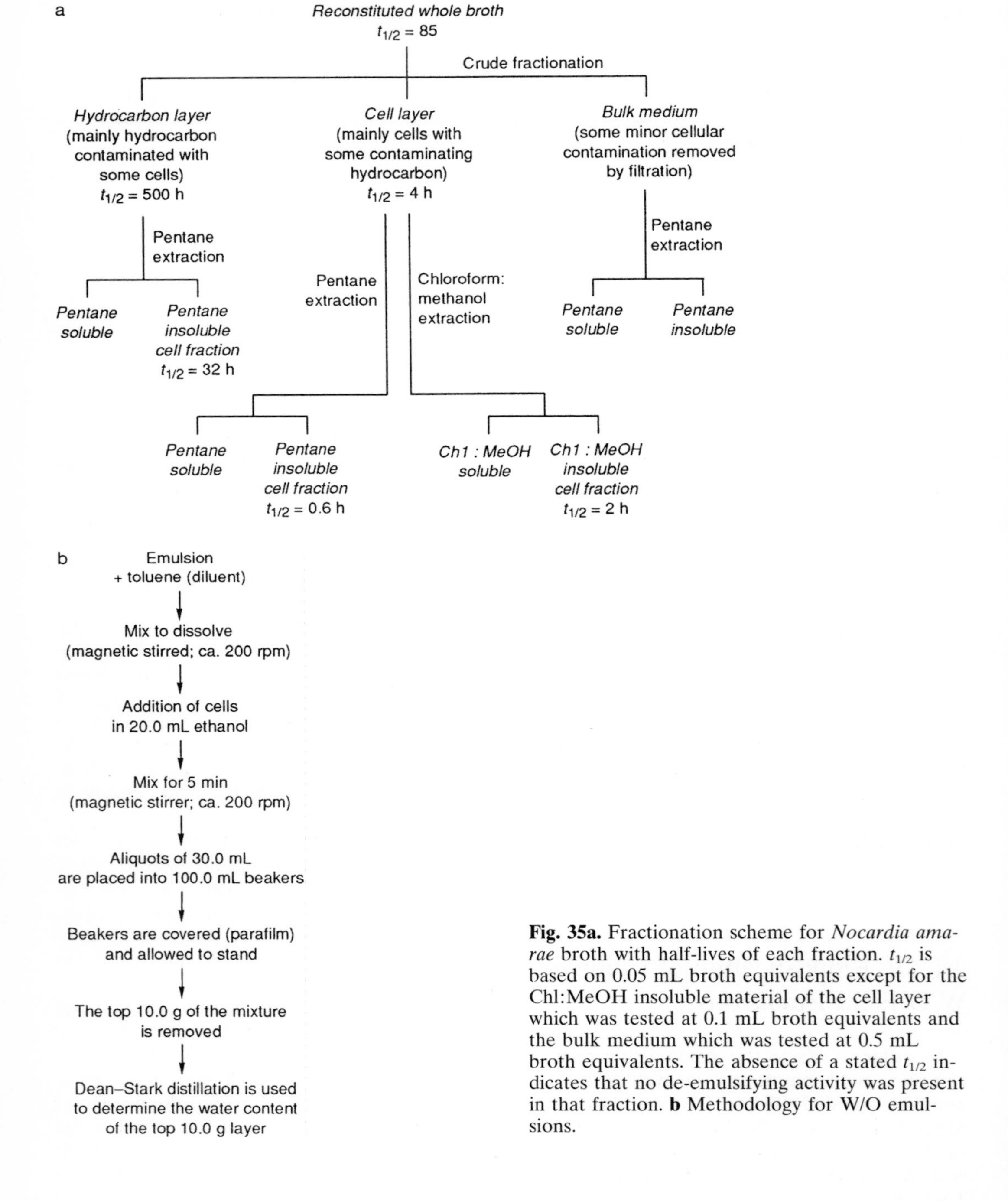


**Fig. 35a.** Fractionation scheme for *Nocardia amarae* broth with half-lives of each fraction. $t_{1/2}$ is based on 0.05 mL broth equivalents except for the Chl:MeOH insoluble material of the cell layer which was tested at 0.1 mL broth equivalents and the bulk medium which was tested at 0.5 mL broth equivalents. The absence of a stated $t_{1/2}$ indicates that no de-emulsifying activity was present in that fraction. **b** Methodology for W/O emulsions.

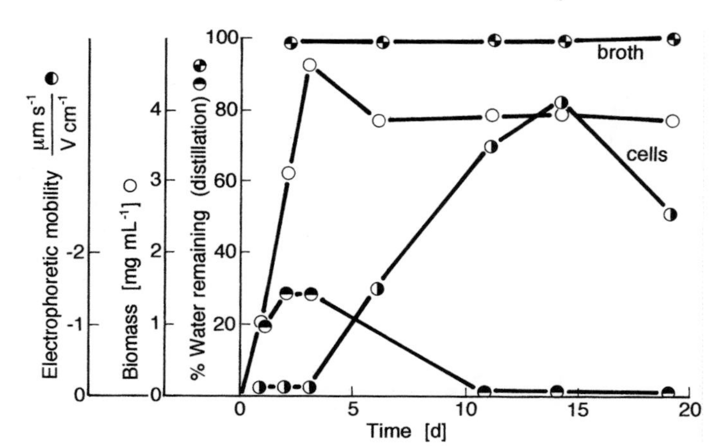


**Fig. 36.** The de-emulsifying activity of *C. petrophilum* cells and broth.

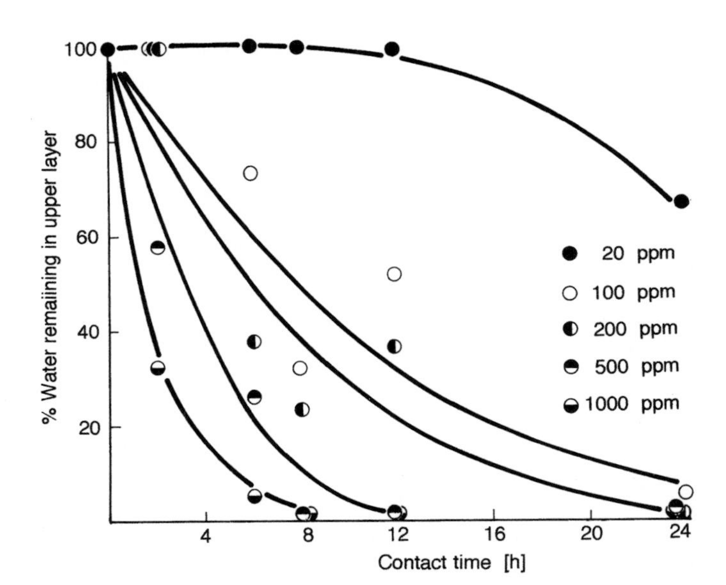


**Fig. 37.** The de-emulsifying ability of day 1 cells of *N. amarae* acting on emulsion #2 at various concentrations of bacteria.

24 h contact time (at room temperature) the amount of water in the top layer was measured by Dean-Stark distillation. As can be seen from Tab. 14, when *C. petrophilum* was not added, in all cases at least 90% of the water remained in the upper layer. With the addition of the bacteria, there was a significant drop in the water content of this upper layer. As the dilution factor became smaller, the level of de-emulsification increased. Based on these observations, the emulsions do not have to be diluted as much as shown in order to

**Tab. 14.** The De-Emulsifying Ability of *C. petrophilum* when the Water-in-Oil Emulsion is Diluted at Various Levels

| Dilution Factor | % Water Remaining in Top Layer (± standard error) | |
|---|---|---|
| | Control | + Bacteria |
| 1:1 | 90.0 ± 4.20 | 7.3 ± 1.60 |
| 1:2 | 91.0 ± 0.66 | 10.0 ± 0.87 |
| 1:5 | 92.0 ± 5.90 | 44.0 ± 2.87 |

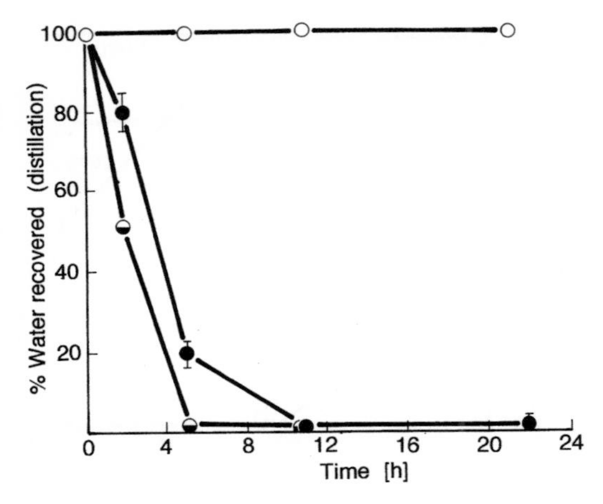


**Fig. 38.** A time course study of *C. petrophilum* acting on field emulsion #2 showing the contact time required for this bacterium to act as a de-emulsifier.
○ control; ● 500 ppm; ◒ 1000 ppm.

achieve the de-emulsifying ability and that, if they are, some of this ability may be lost.

Many other emulsions were also tested in the same system. These were complex field emulsions obtained in enhanced oil recovery operations and were supplied under codes mainly from ESSO and from the Alberta Oil Sands Research Authority (AOSTRA). Their characteristics, as determined in the author's laboratory, are shown in Tab. 15. Figs. 39 and 40 show the results when using emulsions from these two sources.

It was also interesting to note that the amount of water in these W/O emulsions influenced the bacterial de-emulsification process. A possible correlation is shown in Fig. 41.

A comparison of the de-emulsifying abilities of *C. petrophilum* with the commercial de-emulsifier Tretolite E-3453 on two differ-

**Tab. 15.** The General Characteristics of the Field Emulsions Tested

| Field Emulsion | Origin | Water[a,b] [%] | Clay[c] [%] | Density [g mL$^{-1}$] 22°C | Pour Temperature [°C] |
|---|---|---|---|---|---|
| P15 liquid<br>solid | Esso-well head<br>2.5:1 liquid/solid | 50.97 (6.1)<br>31.0 (0.12) | 0.20 (0.01)<br>0.20 (0.02) | 0.96 (0.015)<br>0.779 (0.09) | <20<br>38 |
| K9 liquid<br>solid | Esso-well head<br>2.5:1 liquid/solid | 74.3 (5.5)<br>41.3 (0.5) | 0.63 (0.06)<br>1.1 (0.07) | 1.0 (0)<br>1.16 (0.19) | <20<br>43 |
| L15 | Esso-well head | 31.3 (1.1) | 1.66 (0.3) | 1.12 (0.01) | <20 |
| R12 | Esso-well head | 10.3 (1.5) | 14. (0.2) | 1.22 (0.02) | <20 |
| FW15 | Treater sample Esso | 9.3 (1.2) | 3.1 (0.3) | 1.03 (0.9) | <20 |
| 1-JB-15-5 | Ath. Bitumen rec. with stream flood | 7.33 (0.8) | 1.2 (0.02) | 0.87 (0.2) | 55 |
| 1-JB-15-6 | Ath. Bitumen stream flood | 29.3 (2.3) | 0.9 (0.05) | 0.78 (0.12) | 68 |
| 1-JB-15-7 | Stream drive Lloydminster | 5.3 (0.9) | 2.03 (0.03) | 0.7 (0.3) | <20 |
| 1-JB-15-8 | Combustion Lloydminster | 29.3 (1.9) | 3.07 (0.06) | 0.96 (0.02) | < 2 |
| 1-JB-15-9 | Combustion Lloydminster | 7.9 (0.7) | 2.5 (0.05) | 0.95 (0.04) | 49 |

[a] Numbers represent the mean of 3 trials with the standard error in brackets.
[b] Dean-Stark distillation.
[c] Residue remaining after 16 h at 500 °C.
Ath.: Athabasca; rec.: recovered.

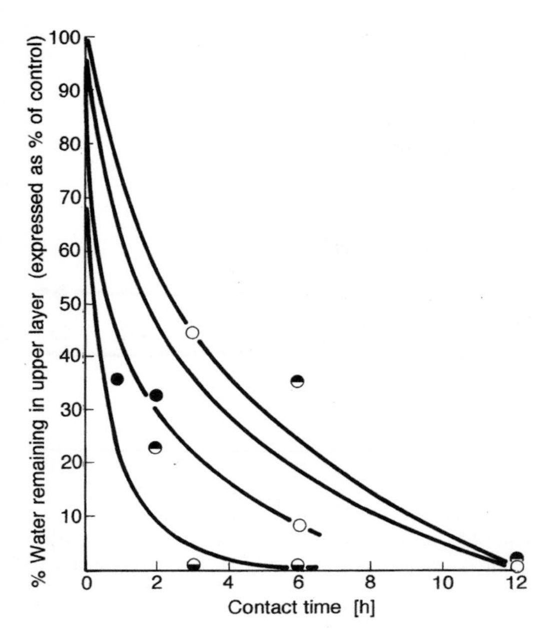


**Fig. 39.** The de-emulsifying ability of *C. petrophilum* (day 1) on the various W/O field emulsions supplied by ESSO.
○ R12; ◓ L15; ● P15; ◒ K9.

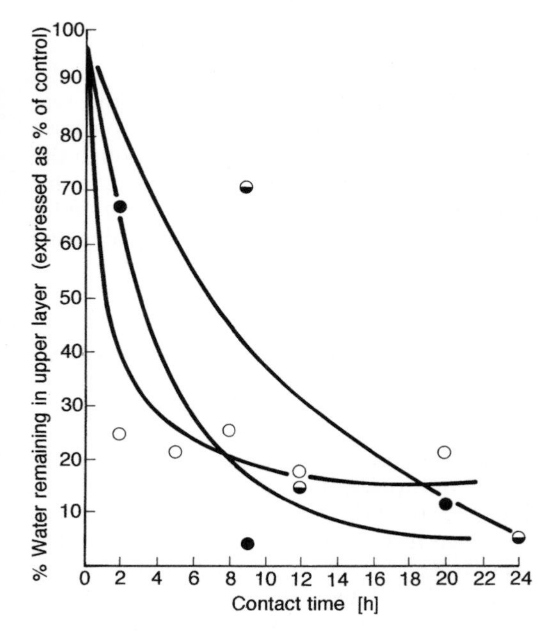


**Fig. 40.** The de-emulsifying ability of *C. petrophilum* (day 1) on the various W/O field emulsions supplied by AOSTRA.
◒ 1-JB-15-9; ○ 1-JB-15-8; ● 1-JB-15-5.

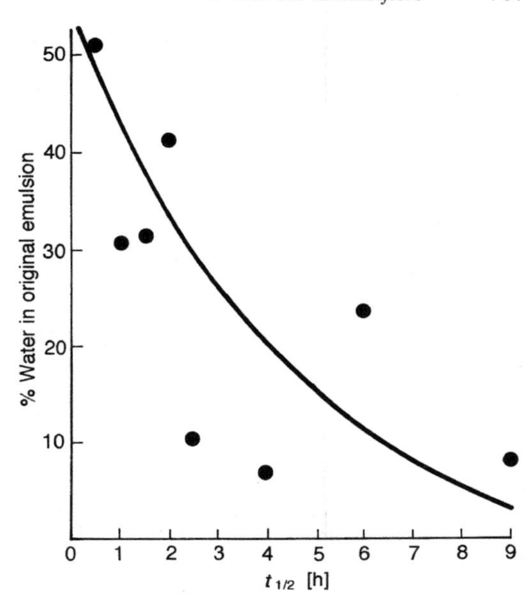


**Fig. 41.** Possible correlation between the amount of water in a W/O emulsion vs. bacterial de-emulsification (*C. petrophilum* day 1 cells).

ent emulsions (1-JB-15-8 and 1-JB-15-9) is shown in Figs. 42 and 43.

In the case of 1-JB-15-8 there was not much difference in the two rates of de-emulsification; neither could bring about more than 80% de-emulsification by 12 h. However, *C. petrophilum* (day 1, 500 ppm) was comparable to the commercial de-emulsifier (1000 ppm) even though only half as much was used.

A big difference was noted in the two rates of de-emulsification when the 1-JB-15-9 emulsion was tested. In this case, *C. petrophilum* accomplished almost complete de-emulsification within 4 h, while during the same time Tretolite F-3453 had only completed 60% de-emulsification.

Another W/O emulsion, L15 (not diluted) was also used for a comparison of the bacterial and commercial de-emulsifiers. It can be seen from Fig. 44 that neither de-emulsifier accomplished complete de-emulsification. The E-3453 seemed to have an overall greater effect, but 1000 ppm was required for this effect, compared to *C. petrophilum* (500 ppm).

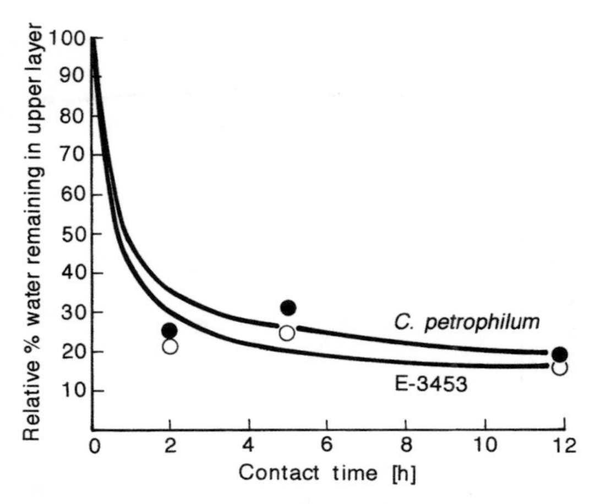


**Fig. 42.** A comparison of the de-emulsifying abilities of *C. petrophilum* (day 1; 500 ppm) and the commercial de-emulsifier Tetrolite E-3453 (1,000 ppm) acting on the field emulsion 1-JB-15-8.

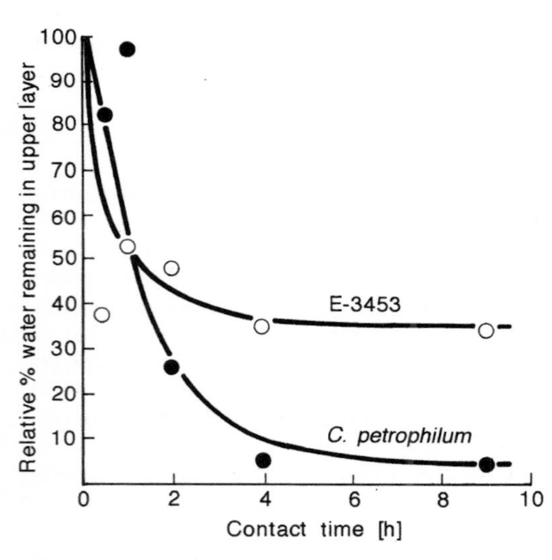


**Fig. 43.** A comparison of the de-emulsifying abilities of *C. petrophilum* (day 1; 500 ppm) and the commercial de-emulsifier Tetrolite E-3453 (1,000 ppm) acting on the field emulsion 1-JB-15-9.

The possibility of re-using the bacteria also exists. After a sample had been de-emulsified, the upper oil layer could be removed and a second amount of field emulsion added, ready to be de-emulsified. Preliminary results

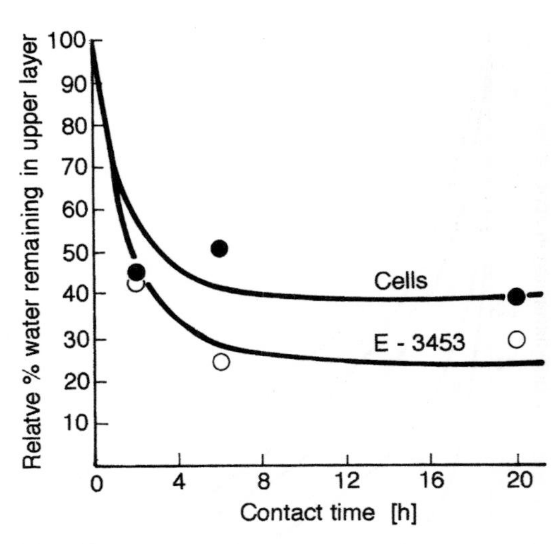


**Fig. 44.** A comparison of the de-emulsification abilities of *C. petrophilum* (day 1; 500 ppm) and the commercial de-emulsifier Tetrolite E-3453 (1,000 ppm) acting on the field emulsion L15 (not diluted with toluene).

(Fig. 45) indicate that the bacteria are re-usable without too much of a loss of activity. There is little variation in the rates of de-emulsification but some is indicated by the slight change in the slopes. One of the reasons may also be due to some loss of bacteria during recovery and recycle. Even though the bacteria have only been used 3 times in this particular experience, other studies are under way to determine the limit of recycling the bacteria.

## 9.5 De-Emulsification of Complex O/W Field Emulsions

The K9 and P15 emulsions (Tab. 15) are O/W emulsions obtained during tar sands bitumen extraction process. These emulsions were also tested with the same 3 bacteria (*N. amarae, C. petrophilum,* and *R. aurantiacus*). The methodology for quantifying the de-emulsification was somewhat different as (1) no dilution of the original emulsion was required, and (2) the amount of oil in the upper 1 mL layer was measured by weight and re-

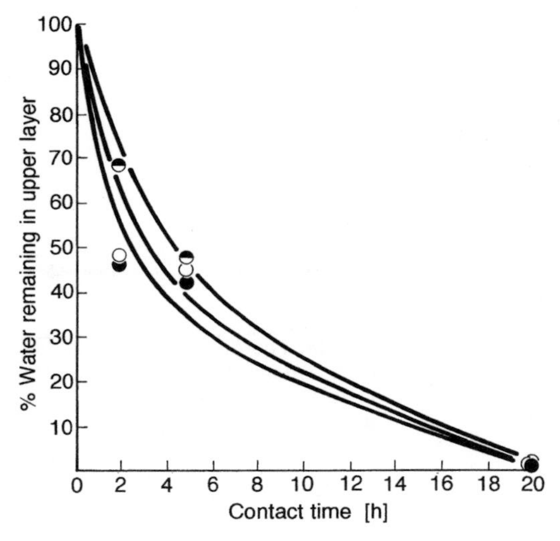


**Fig. 45.** The effect of re-using a bacterial de-emulsifier (500 ppm *C. petrophilum*; day 1 cells) on the JB-15-7 W/O emulsion.
● original de-emulsification; ○ 2nd de-emulsification; ◓ 3rd de-emulsification.

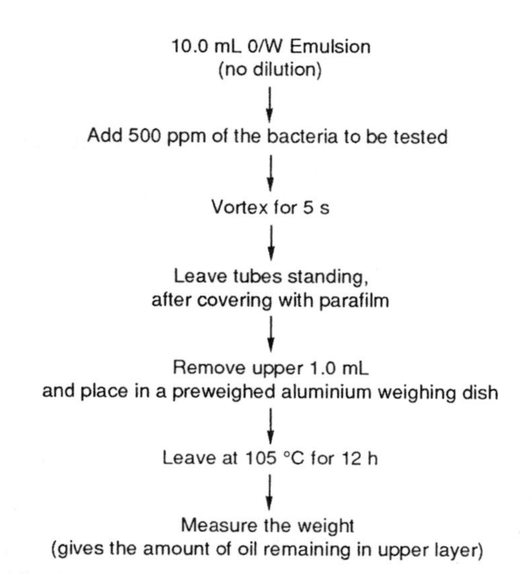


**Fig. 46.** Methodology for O/W emulsions.

ported as "oil remaining in the upper layer". The methodology is outlined in Fig. 46.

All 3 bacteria show some de-emulsifying activity (10–15%) when the "younger cells"

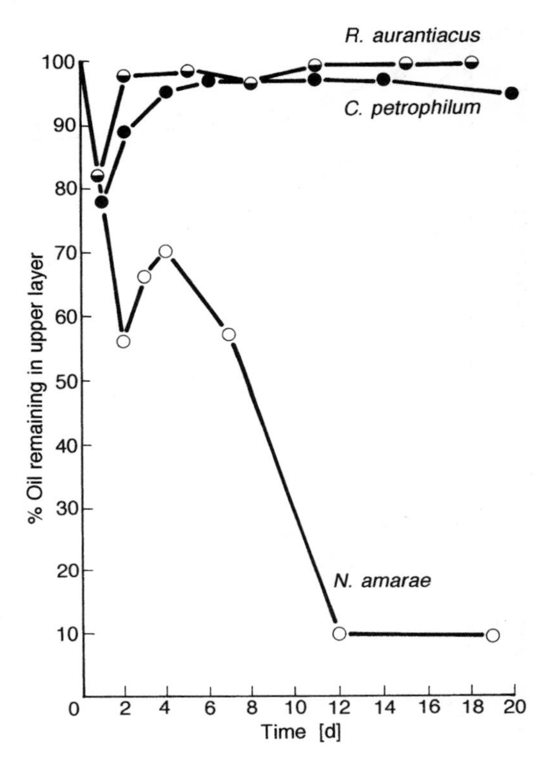


**Fig. 47.** The de-emulsifying ability of *R. aurantiacus, C. petrophilum,* and *N. amarae* on an O/W emulsion (K9).

were used against the K9 emulsion. However, the greatest de-emulsifying ability was exhibited by the "older" (day 12–19) *N. amarae* cells as shown in Fig. 47. The broths were also tested but showed only about 5% de-emulsification for all culture ages. The P15 emulsion was not so efficiently broken by these bacteria. The results are shown in Fig. 48.

A commercial flocculating agent, polynucleolite (Eusotech Inc., North Hollywood, CA), was compared to the day 19 *Nocardia* cells using the P15 emulsion. This flocculating agent worked better than the *Norcardia* cells (each being used at 500 ppm). After 24 h, the *Nocardia* cells were able to remove only ca. 35% of the oil from the upper layer while the flocculating agent removed ca. 55% of the oil from the upper layer (Fig. 49). However, when the bacteria and the flocculating agent

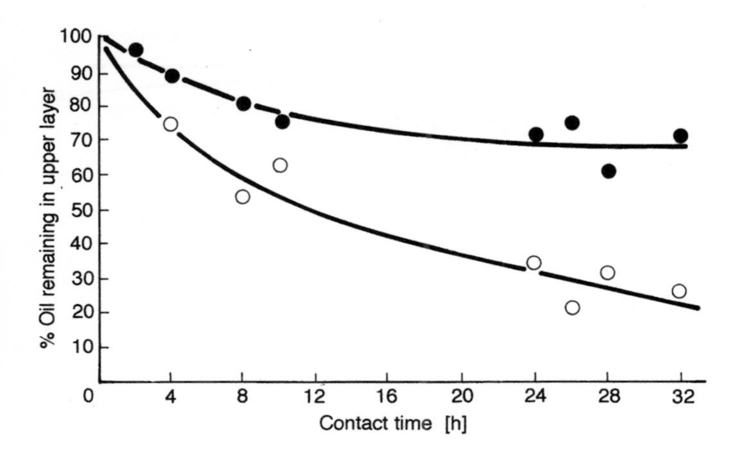


**Fig. 48.** The de-emulsifying ability of *N. amarae* (day 19) cells on the O/W field emulsion P15. ● 500 ppm; ○ 1000 ppm.

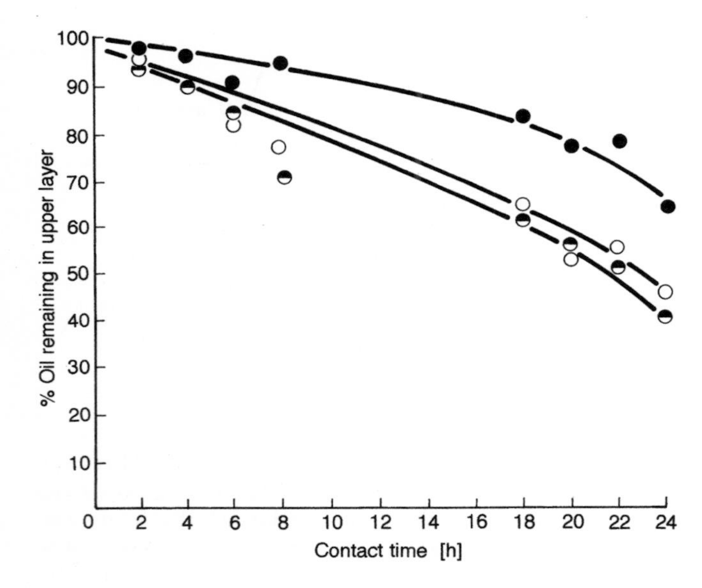


**Fig. 49.** The effect of the flocculating agent Polynucleolyte on the P15 O/W emulsion compared to the de-emulsifying abilities of *N. amare* (day 19). ● *N. amarae* (500 ppm); ○ Polynucleolyte (500 ppm); ◓ Polynucleolyte (250 ppm) + *N. amarae* (350 ppm).

were used together so that the total ppm was the same (500 ppm), ca. 60% of the oil was removed from the upper layer. Accordingly, bacterial agents could be efficiently used together with commercial flocculating agents. The same trend was also observed with K9 emulsion (Fig. 50).

The effect of a commercial de-emulsifying agent, Tretolite F-46 (Petrolite Corporation, Calgary, Alberta), was also compared. Day 19 cells of *N. amarae* (350 ppm) lowered the level of oil associated with the upper layer of both the K9 and P15 emulsions. On the other hand, F-46 (400 ppm) increased the level of oil in the upper layer (Fig. 51). Over the 32-h test period, the rate of increase (or decrease) of the oil in the upper layer was found to be higher for the F-46 de-emulsifier; however, if the test period was only 8 h, there was no significant difference between the commercial and the bacterial de-emulsifier.

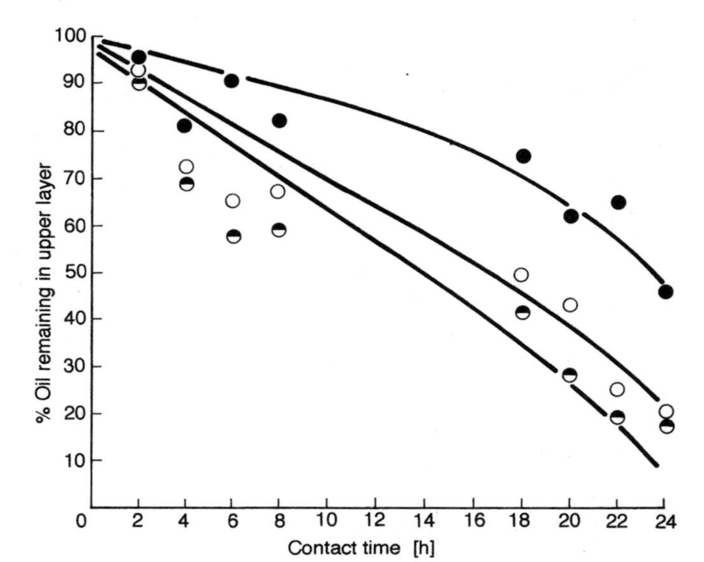


**Fig. 50.** The effect of the flocculating agent, Polynucleolyte on the K9 O/W emulsion compared to the de-emulsifying abilities of *N. amarae* (day 19).
● *N. amarae* (500 ppm); ○ Polynucleolyte (500 ppm); ◓ Polynucleolyte (250 ppm) + *N. amarae* (250 ppm).

## 9.6 De-Emulsification of Complex W/O Field Emulsions with Sludges

As selected, microbes have a demonstrated capability to de-emulsify simple and complex W/O and O/W emulsions, further investigation was performed to establish to what extent various sludges and biomass in general do have the same property (KOSARIC and DUVNJAK, 1987; KOSARIC et al., 1987b). If such a property exists, a large-scale de-emulsification process could be developed as an integral part of biological wastewater treatment which produces a large quantity of sludges, whose disposal is always a problem. Also, by coupling these processes, considerable cost savings could result as the de-emulsifier would be available at site, in large quantity, at minimal or no cost.

Using the same methodology as with the above mentioned 3 bacteria, various sludges were investigated (aerobic and anaerobic sludges from activated sludge and anaerobic digesters). The sludges were autoclaved at 120°C for 30 min prior to use, followed by centrifugation and drying, and were as such added to the emulsions tested. Fig. 52 shows the results obtained when ESSO, Husky, and Mobil Oil emulsions were tested with municipal aerobic sludge. A significant de-emulsification is accomplished with this sludge. Anaerobic flocculant and granular sludges showed also a similar effect.

## 9.7 Conclusions

It has been clearly demonstrated that pure bacterial cultures and sludges (heterogeneous mixed microbial populations) can efficiently de-emulsify both oil-in-water and water-in-oil complex industrial emulsions.

Bio-de-emulsification is comparable and in some cases better than the use of commercial synthetic de-emulsifiers for the same purpose.

Bio-de-emulsification appears to depend on bacterial species and on the emulsion characteristics. Cells harvested at an earlier stage of growth (e.g., "young cells") show better de-emulsification properties for W/O emulsions while cells harvested at the stationary broth growth stage ("older cells") seem to be better for breaking of O/W emulsions.

The de-emulsifying activity seems to be closely associated with the cell surface itself

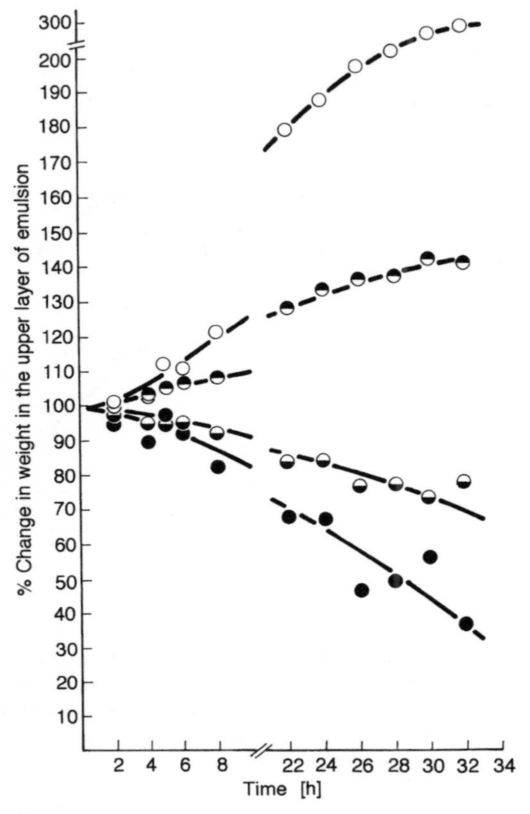


**Fig. 51.** The effect of a commercial de-emulsifier (Tretolite F-46) on both the K9 and P15 emulsions compared to a bacterial de-emulsifier (day 19 cells of *N. amarae*).
○ K9, F-46; ◓ P15, F-46; ◒ P15, *N. amarae*; ● K9, *N. amarae*.

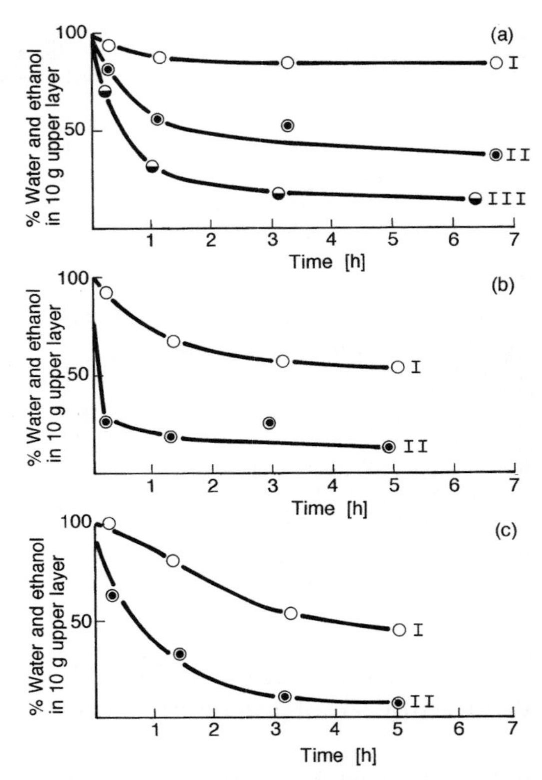


**Fig. 52.** De-emulsification of ESSO, Husky III and Mobil Oil emulsions with municipal sludge (aeration sludge)
(a) Test system contained (for ESSO emulsion) 150 g emulsion + 60 mL toluene and in:
I + 20 mL ethanol
II + 0.504 g (d.w.) sludge in 20 mL ethanol
III + 1000 g (d.w.) sludge in 20 mL ethanol
(b) Test system contained (for Husky III emulsion) 150 g emulsion 50 mL toluene and in:
I + 20 mL ethanol
II + 0.741 9 (d.w.) sludge in 10 mL ethanol
(c) Test system contained (for Mobil Oil emulsion) 150 g emulsion + 50 mL toluene and in:
I + 20 mL ethanol
II + 0.742 g (d.w.) sludge in 10 mL ethanol.

and is not appearing as a consequence of cell metabolism of the oil phase.

The activity is difficult to destroy even with drastic chemical treatments such as autoclaving (heat treatment), HCl methanolysis, trichloroacetic acid extraction and phenol extraction. Alkaline methanolysis, however, does destroy the de-emulsifying cell property. Consequently, the cells are active in the dry, dead form, which is a big advantage for safe handling during industrial applications.

Also, there is a strong indication that the cells can be re-used and recycled, which considerably reduces the overall cost of this de-emulsifier. This cost is also minimized when aerobic or anaerobic sludges are used as industrial de-emulsifying agents.

# 10 References


ABE, Y., INOUE, S., ISHIDA, A. (1980), *Ger. Patent* 2938383 to Kao Soap Co.

Ajinomoto Co. Inc. (1982), *Jpn. Patent* 57129696.

ASHMER, H. J., LANG, S., WAGNER, F., WRAY, V. J. (1988), *Am. Oil Chem. Soc.* **65,** 1460.

BAER, H. H. (1993), *Carbohydr. Res.* **240,** 1.

BARRY, B. W., ECCLESTONE, G. M. (1973), *J. Pharm. Pharmacol.* **25,** 244.

BASTA, N. (1984), *High Technol.* 67.

BERGSTROM, S., THEARELL, H., DAVIDE, H. (1947), *Ark. Kemi Mineral. Geol.* **23A**(13), 1.

BIRKEL, T. J., WARNER, C. R., FAZIOT, J. (1978), *Ass. Anal. Chem.* **62,** 931.

BOSSERT, I., BARTHA, R. (1984), in: *Petroleum Microbiology* (ATLAS, R., Ed.), 335. New York: Macmillan.

BOULTON, C. A., RATLEDGE, C. (1987), in: *Biosurfactants and Biotechnology* (KOSARIC, N., CHAIRNS, W. L., GRAY, N. C. C., Eds.), 47. New York: Marcel Dekker.

BRENNAN, P. J., LEHANE, D. P., THOMAS, D. W. (1970), *Eur. J. Biochem.* **13,** 117.

BREUER, M. M. (1985), in: *Encyclopedia of Emulsion Technology* (BECKER, P., Ed.), 385. New York: Marcel Dekker.

BROOKS, G. J. (1980), *Cosmet. Toiletries* **95,** 73.

CAIRNS, W. L., COOPER, D. G., ZAJIC, T. E., WOOD, J. M., KOSARIC, N. (1982), *Appl. Environ. Microbiol.* **43**(2), 362.

CAIRNS, W. L., COOPER, D. G., KOSARIC, N. (1983), in: *Microbial Enhanced Oil Recovery* (ZAJIC, T. E., COOPER, D. G., JACK, T., KOSARIC, N., Eds.). Tulsa, OK: Pennwell Publ. Co.

CAIRNS, W. L., WOODS, J. M., KOSARIC, N. (1984), in: *Biotechnology for the Oils and Fats Industry,* Vol. II (RATLEDGE, C., DAWSON, P., RATTRAY, T. B. M., Eds.), 269.

CLARK, L. C., GOLLAN, F. (1966), *Science* **152,** 1755.

COOPER, D. G., PADDOCK, D. A. (1984), *Appl. Environ. Microbiol.* **47,** 173.

COOPER, D. G., ZAJIC, J. E., CAIRNS, W. L., KOSARIC, N. (1980), *Biochemical Engineering Research Report,* Vol. VI, 1–178. University of Western Ontario, Canada.

DANIELS, L., LINHARDT, R. J., BRYAN, B. A., MAYERL, F., PICKENHAGEN, W. (1988), *Eur. Patent* 282942.

DATTA, A. K., TAKAYAMA, K., NASHED, M. A., ANDERSON, L. (1991), *Carbohydr. Res.* **218,** 95.

DAVIS, M. R. (1980), *Cosmet. Toiletries* **95,** 81.

DAVIS, S. S., HADGRAFT, J., PALIN, K. J. (1985), in: *Encyclopedia of Emulsion Technology* (BECHER, E., Ed.), 159. New York: Marcel Dekker.

DELUCA, M., ROCHA-FILHO, P., GROSSIORD, J. L., RABAROU, A., VANTION, C., SEILLER, M. (1991), *Int. J. Cosm. Sci.* **13,** 1.

DESAI, N. B. (1990), *Cosmet. Toiletries* **105,** 99.

DESAI, J. D., DESAI, A. J. (1993), in: *Biosurfactants* (KOSARIC, N., Ed.), 65. New York: Marcel Dekker.

DUGARD, P. H., SCHENPLEIN, R. J. (1973), *J. Invest. Dermatol.* **60,** 263.

DUVNJAK, Z., COOPER, D. G., KOSARIC, N. (1982), *Biotechnol. Bioeng.* **24,** 165.

DUVNJAK, Z., COOPER, D. G., KOSARIC, N. (1983), in: *Microbial Enhanced Oil Recovery* (ZAIJC, J. E., COOPER, D. G., JACK, T. R., KOSARIC, N., Eds.), 66. Tulsa, OK: Pennwell Publ. Co.

EDWARDS, J. R., HAYASHI, J. A. (1965), *Arch. Biochem. Biophys.* **III,** 415.

EGI, M. (1987), *Food Chem.* **1,** 51.

EL-SAYED, A. A. A., REPTA, A. J. (1983), *Int. J. Pharm.* **13,** 303.

ENGEL, R. H., RIGGI, S. S., FAHRENBACH, M. J. (1968), *Nature* **219,** 856.

FALATKO, D. M., NOVAK, J. T. (1992), *Res. J. Water Poll. Contr. Fed.* **64,** 163.

FINNERTY, W. R., SINGER, M. E. (1983), *Biotechnology* **1,** 47.

FOGEL, S., FINDLEY, M., MOORE, A. (1990), in: *Petroleum Contaminated Soils* (CALABRESE, E. J., KOSTECHY, P. T., Eds.), 102. Lewis Publishers.

FOGHT, J. M., GUTNICK, D. L., WESTLAND, D. W. S. (1989), *Appl. Environ. Microbiol.* **55,** 36.

FORTHNER, C. L., GROVE, W. R., BOWIE, D., WALKER, M. D. (1975), *Am. J. Hosp. Pharm.* **32,** 582.

FRAUTZ, B., WAGNER, F. (1984), in: Abstract Papers, *3rd Eur. Congr. Biotechnol.* Vol. 1. Weinheim: Verlag Chemie.

FRYE, C. L. (1983), *Soap Cosmet. Chem. Spec.* 118.

FUJITA, S., SUZUKI, K. (1990), *J. Am. Oil. Chem. Soc.* **67,** 1008.

FUJITA, T., SUMAYA, T., YOKOJAMA, K. (1971), *Eur. Surg. Res.* **3,** 436.

FUKUI, G., TANAKA, A. (1981), *Adv. Biochem. Eng.* **19,** 217.

GABRIEL, A. (1979), *Microbial Polysaccharides and Polysaccharases* (BERKELY, R. S., Ed.), 191. London: Academic Press.

GARRETT, H. E. (1965), *Trans. St. John's Hosp. Dermatol. Soc.* **51,** 160.

GERSBERG, R. M., DAWSEY, W. J., RIDGWAY, H. F. (1990), in: *Petroleum Contaminated Soils* (CALABRESE, E. J., KOSTECHY, P. T., Eds), 211. Lewis Publishers.

GERSON, D. F. (1993), in: *Biosurfactants* (KOSARIC, N., Ed.), 269. New York: Marcel Dekker.

GEYER, R. P. (1973), *N. Engl. J. Med.* **189,** 1077.

GEYER, R. P. (1974), *Bull. Parenter. Drug. Assoc.* **28,** 88.

GEYER, R. P. (1976), in: *Drug Design,* Vol. 7 (ARIENS, A. J., Ed.), 1. London: Academic Press.

GEYER, P. O., SHERER, P. B., WEILER, J. M. (1975), *Fed. Proc. Fed. Am. Soc. Exp. Biol.* **34,** 1428.

GILL, C. O., HALL, M. J., RATLEDGE, C. (1977), *Appl. Evnrion. Microbiol.* **33,** 231.

GOBBERT, U., LANG, S., WAGNER, F. (1984), *Biotechnol. Lett.* **6,** 225.

GOLDBERG, A. H., HIGUCKI, W. I., HO, N. F. H., ZOGRAFI, G. (1967), *J. Pharm. Sci.* **56,** 1432.

GOMA, G., PAREILLEUX, A., DURAND, G. (1973), *J. Ferment. Technol.* **51,** 616.

GOODE, S. T., LINTON, R. R., BAIOCCHI, F. (1988), *World Patent* 8806880 to R.I.T.A. Co.

GORBACH, V. J., KRASIKOVA, J. N., LYKYANOO, P. A., LOENKO, Y. N., SOLOVEVA, T. F., OVODOV, Y. S., DEEV, V. V., PIMENOV, A. A. (1994), *Carbohydr. Res.* **260**(1), 73.

GRAY, N. C. C., STEWART, A. L., CAIRNS, W. L., KOSARIC, N. (1984), in: *Biotechnology of Fats and Health* (RATLEDGE, C., DAWSON, P. S. S., RATTRAY, T. M. B., Eds.), 255–268. Champain, IL: Am. Oil Chem. Soc.

GROSSMANN, G., NILSSON, R., ROBERTSON, B. (1986), *Eur. J. Pediatr.* **145,** 361.

GUPTA, R. K., JANUS, K., SMITH, F. J. (1983), *J. Am. Oil. Chem. Soc.* **60,** 862.

GUTNICK, D. L., SHABTAI, Y. (1987), in: *Biosurfactants and Biotechnology* (KOSARIC, N., CAIRNS, W. L., GRAY, N. C. C., Eds.), 211. New York: Marcel Dekker.

HAFERBURG, D., HOMMEL, R., CLAUS, R., KLEBER, H. P. (1986), *Adv. Biochem. Eng. Biotechnol.* **33,** 53.

HASEGAWA, I., MAENO, K. (1986), *Jpn. Patent* 61271029 to Kanebo Ltd.

HERBERT, W. J. (1965), *Lancet* **ii,** 771.

HISATSUKA, K.-I., NAKAHARA, T. B., SANO, N., YAMADA, K. (1971), *Agr. Biol. Chem.* **35,** 686.

HOFFMANN, N. H. (1970), *Am. Perfum.* **85,** 37.

HOMMEL, R. K., RATLEDGE, C. (1993), in: *Biosurfactants* (KOSARIC, N., Ed.), 3. New York, Marcel Dekker.

HONDA, K., HOSHINO, S., SHAFI, M., USABA, A., MOTOKI, R., TSUBOI, M., INOUE, H., IWAYA, F. (1980), *N. Engl. J. Med.* **303,** 391.

HOROWITZ, A., SUFLITA, J. M., TIEDGE, J. M. (1983), *Appl. Environ. Microbiol.* **45,** 1459.

HORVATH, R. S. (1971), *J. Agric. Food Chem.* **19,** 291.

HOYT, J. W. (1985), *Trends Biotechnol.* **3,** 17.

HUTTINGER, R. (1983), *Seifen, Oele, Fette, Wachse* **109,** 475.

ILLUM, L., DAVIS, S. S. (1982), *J. Parenter. Sci. Technol.* **36,** 242.

INOUE, S. (1988), *Proc. World Conf. on Biotechnol. in Fats and Oils Ind.* (APPLEWHITE, T. H., Ed.), 206. Am. Oil Chem. Soc., Champaign, Il.

INOUE, S., ITOH, S. (1982), *Biotechnol. Lett.* **4,** 3.

INOUE, S., KIMURA, Y., KINDA, M. (1979a), *Ger. Patent* 1905118 to Kao Soap Co.

INOUE, S., KIMURA, Y., KINDA, M. (1979b), *Ger. Patent* 1905252 to Kao Soap Co.

INOUE, S., KIMURA, Y., KINDA, M. (1979c), *Ger. Patent* 1905295 to Kao Soap Co.

ISHIGAMI, Y. (1982), *INFORM* **4**(10), 1156.

ISHIGAMI, J., GAMA, Y., YAMAZAKI, S., SUZUKI, S. (1988), *Proc. Biotechnol. Fats Oils Ind.,* AOCS, 33.

ITO, S., INOUE, S. (1982), *Appl. Environ. Microbiol.* **43,** 1278.

ITOH, S., HONDA, H., TOMITO, F., SUZUKI, T. J. (1971), *Antibiotics* **24**(12), 855.

JAIN, D. K., LEE, H., TREVORS, J. T. (1992), *J. Ind. Mirobiol.* **10**(2), 87.

JARVIS, F. G., JOHNSON, M. J. (1949), *J. Am. Chem. Soc.* **71,** 4124.

JAVAHERI, M., JENNEMAN, G. E., MCINERNEY, J. J., KNAPP, R. M. (1985), *Appl. Environ. Microbiol.* **50**(3), 698.

JENNEMAN, G. E., MCINNERNEY, J. J., KNAPP, R. M., CLARK, J. B., FEERO, J. M., REVUS, D. F., MENZIE, D. E. (1983), *Soc. Ind. Microbiol.* **45,** 485.

JEPPSSON, R. (1972), *Acta Pharm. Suec.* **9,** 81.

JEPPSSON, R. (1976), *Acta Pharm. Suec.* **13,** 43.

JOBE, A., IKEGAMI, M. (1987), *Am. Rev. Respir. Dis.* **136,** 1256.

JOHNSON, I. D. A. (1983), *Advances in Clinical Nutrition,* MTP, Lancaster, England.

JULBARISOV, E. M. (1990), *Rev. Inst. Fr. Pet.* **45**(1), 115.

KACHHOLZ, T., SCHLINGMAN, M. (1987), in: *Biosurfactants and Biotechnology* (KOSARIC, N., GRAY, N. C. C., CAIRNS, W., Eds.), 183. New York, Marcel Dekker.

KAKINUMA, A., HORI, M., SUGINO, H. YASHIDA, I., ISOMO, M., TAMURA, G., ARIMA, K. (1969), *Agric. Biol. Chem.* **33**(10), 1523.

KANO, J., SUZUKI, T., INOUE, S., HAYASHI, S. (1980), *Jpn. Patent* 8043042 to Kao Soap Co.

KAPLAN, N., ROSENBERG, E. (1982), *Appl. Environ. Microbiol.* **44,** 1335.

KAWAI, A., KAYANO, M., FUNADA, T., HIRANO, J. (1988), *Jpn. Patent Kokai* 63-126, 493.

KHAN, Q. M., FAIZ, M., IQBAL, S., KHALID, Z. M., MALIK, K. A. (1994), *Proc. 2nd Int. Symp. Environ. Biotechnol.,* Brighton, U.K.

KITAHARA, S., OKADA, S., EDAKAWA, S. (1988), *Jpn. Patent Kokai* 63-166668.

KILBURN, J. D., TAKAYAMA, K. (1981), *Antimicrob. Agents Chemother.* **20,** 401.

KLEKNER, V., KOSARIC, N. (1993), in: *Biosurfactants* (KOSARIC, N., Ed.), 373. New York: Marcel Dekker.

KNOCHE, H. W., SHIVELY, J. M. (1972), *J. Biol. Chem.* **247**(1), 170.

KÖHLER, A., BRYNIOK, D., EICHLER, B., MACKENBOCK, F., FREIER-SCHRÖDER, KNACKMUSS, H. (1990), in: *Dechema Biotechnology Conferences* 4, 585. Weinheim: VCH.

KOHYA, H., ISHII, F., TAKANO, S., KATORI, T., EBINA, T., ISHIDA, N. (1986), *Jpn. J. Cancer Res.* **77** (6), 602.

KONO, J., SUZUKU, T., INOUE, S., HAYASHI, S. (1980), *Jpn. Patent* 8043042.

KOSARIC, N. (Ed.) (1993), *Biosurfactants-Production, Properties, Applications.* New York: Marcel Dekker.

KOSARIC, N., DUVNJAK, Z. (1987), *Water Poll. Res. J. Can.* **22**(3), 437.

KOSARIC, N., GRAY, N. C. C., STEWART, A. L., CAIRNS, W. L. (1981), *Biochem. Eng. Res. Report,* Vol. IX(1), 1–67. University Western Ontario, London, Canada.

KOSARIC, N., CAIRNS, W. L., COOPER, D. G., RUMBLE, R., WHITE, J., WOOD J. (1982), *J. Biochem. Eng. Res. Report* XIV, 1–119. University Western Ontario, London, Canada.

KOSARIC, N., GRAY, N. C. C., CAIRNS, W. L. (1983), Microbial Emulsifiers and De-Emulsifiers, in: *Biotechnology*, 1st Edn., Vol. 3 (REHM, H.-J., REED, G., Eds.), 575–592. Weinheim: Verlag Chemie.

KOSARIC, N., CAIRNS, W. L., GRAY, N. C. C., STACHEY, D., WOOD, J. (1984), *J. Am. Oil Chem. Soc.* **61**(11), 1735.

KOSARIC, N., GRAY, N. C. C., CAIRNS, W. L. (1987a), in: *Biosurfactants and Biotechnology* (KOSARIC, N., CAIRNS, W. L., GRAY, N. C., Eds.), 1. New York: Marcel Dekker.

KOSARIC, N., CAIRNS, W. L., GRAY, N. C. C (1987b), in: *Biosurfactants and Biotechnology* (KOSARIC, N., CAIRNS, W. L., GRAY, N. C. C., Eds.), 247. New York: Marcel Dekker.

KRETSCHMER, A., LANG, S., MARWEDE, G., RISTAN, E., WAGNER, F. (1981), in: *Advances in Biotechnology,* Vol. 3 (MOO-YOUNG, M., VEZINA, C., SING, K., Eds.), 475. Toronto: Pergamon Press.

KUDO, S. (1988), *Proc. World Conf. on Biotechnol. in Fats and Oils Ind.,* AOCS 195.

LANG, S., WAGNER, F. (1987), in: *Biosurfactants and Biotechnology* (KOSARIC, N., CAIRNS, W. L., GRAY, N. C. C., Eds.), 32. New York: Marcel Dekker.

LANG, S., WAGNER, F. (1993), in: *Biosurfactants* (KOSARIC, N., Ed.), 251. New York: Marcel Dekker.

LANG, S., KATSIWELA, E., WAGNER, F. (1989), *Fat. Sci. Technol.* **91,** 363.

LEUTENSCHLAGER, H. (1990a), *Cosmet. Toiletries* **105,** 89.

LEUTENSCHLAGER, H. (1990b), *Cosmet. Toiletries* **105,** 63.

LIN, S. C., SHARMA, M. M., GEORGION, G. (1993), *Biotechnol. Progr.* **9**(2), 138.

LINHARDT, R. J., BAKHIT, R., DANIELS, L., MAYERL, F., PICKENHAGEN, W. (1989), *Biotechnol. Bioeng.* **33,** 365.

LINKE, B., SORRENTINO, R., PETROCCI, A. A. (1975), *J. Soc. Cosmetic Chem.* **16,** 155.

LITTERST, C. L., MINMAUGH, E. G., COWLES, A. C., GRAM, T. E., GUARINO, A. M. (1974), *J. Pharm. Sci.* **63,** 1718.

MANRESA, M. A., BATISTA, J., MERCADE, M. E., ROBERT, M., DE ANDRES, C., ESPUNY, M. J., GUINEA, J. (1991), *J. Ind. Microbiol.* **8**(2), 133.

MARTIN, M., BOSCH, P., PARA, J. L., ESPUNY, M. J., VIRGILI, A. (1991), *Carbohydr. Res.* **220,** 93.

MCINERNEY, M. J., JAVAHERI, M., NAGLE, D. P. (1990), *Microgiology* **5,** 95.

MCNEILL, G. P., YAMANE, T. (1991), *J. Am. Oil Chem. Soc.* **68,** 6.

MEIER, R. (1990), *Doctoral Thesis.* Technical University, Braunschweig, Germany.

MEIER, R., MÜLLER-HURTIG, R., WAGNER, F. (1992), *Fachgespräch Umweltschutz,* Dechema, Vol. 9. Weinheim: VCH.

MENG, H. C., WILMORE (Eds.) (1976), *Fat Emulsions in Parenteral Nutrition.* Am. Med. Ass., Chicago.

MINTZ, G. (1990), *Catal. Rev. News Lett.* **2,** 8.

MIYATA, K., TSUSHIDA, T., TAWARA, K. (1987), *Jpn. Patent Kokai* 62-155931.

MIZUSHIMA, Y., HAMANO, T., YOKOYAMA, K. (1982), *J. Pharm. Pharmacol.* **34,** 49.

MOLNAR, B. A., GRUBBS, R. G. (1990), in: *Petroleum Contaminated Soils,* Vol. 2 (CALABRESE, E. J., KOSTECHY, P. T., Eds.), 219. Lewis Publishers.

MONTET, D., PINA, M., GRAILLE, J., RENARD, G., GRIMAUD, J. (1989), *Fett Wiss. Technol.* **91,** 14.

MONTET, D., SERVAT, F., PINA, M., GRAILLE, J., GALZY, R., ARNAUD, A., LEDON, H., MARCON, L. J. (1990), *Am. Oil Chem. Soc.* **67,** 771.

MÜLLER-HURTIG, R., OBERBREMER, A., MEIER, R., WAGNER, F. (1990), in: *Contaminated Soil '90* (ARENDT, F., HINSENVELD, M., VAN DEN BRINK, W. J., Eds.), 491. Dordrecht: Kluwer Academic Publishers.

MÜLLER-HURTIG, R., WAGNER, F., BLASZCZYK, R., KOSARIC, N. (1993), in: *Biosurfactants* (KOSARIC, N., Ed.). New York: Marcel Dekker.

NEU, T. R., HARTNER, T., PORALLA, K. (1990), *Appl. Microbiol. Biotechnol.* **32,** 518.

OBERBREMER, A. (1990), *Doctoral Thesis,* Technical. University, Braunschweig, Germany.

OBERBREMER, A., MÜLLER-HURTIG, R. (1989), *Appl. Microbiol. Biotechnol.* **31,** 682.

OBERBREMER, A., MÜLLER-HURTIG, R., WAGNER, F. (1990), *Appl. Microbiol. Biotechnol.* **32,** 485.

OHATA, K., KAMATA, K. (1986), *Jpn. Patent Kokai* 61-227, 827.

OSUMI, M., FAKUZUMI, F., YAMADA, N., NAGATANI, T., TERANISHO, Y., TANAKA, A., FUKUI, S. J. (1975), *Ferment. Technol.* **53,** 244.

PACE, G. W. (1981), *Advances Biotechnology,* Vol. 3 (MOO-YOUNG, M., VEZINA, C., SINGH, K., Eds.), 433. Toronto: Pergamon Press.

PARRA, J. L., PASTOR, J., COMELLES, F., MANRESA, M. A., BOSCH, M. P. (1990), *Tenside Surf. Deterg.* **27,** 302.

PASSERI, A., SCHMIDT, M., HAFFNER, T., WRAY, V., LANG, S., WAGNER, F. (1992), *Appl. Microbiol. Biotechnol.* **37,** 281.

PINES, O., GUTNICK, D. (1986), *Appl. Environ. Microbiol.* **51,** 661.

POSSMAYER, F. (1988), *Am. Rev. Respir. Dis.* **138,** 990.

PROSERIO, G., GATTI, S., GENESS, P. (1980), *Toiletries* **95,** 81.

RAPP, P., BOCK, H., URBAN, E., WAGNER, F. (1977), in: *Dechema Monographien,* Vol. 81 (REHM, H.-J., Ed.). Weinheim: Verlag Chemie.

RAPP, P., BOCK, H., WRAY, V., WAGNER, F. (1982), *J. Gen. Microbiol.* **115,** 491.

RATLEDGE, C. (1988), in: *Biotechnology for the Fats and Oils Industry* (APPLEWHITE, T. H., Ed.), 7.

RATLEDGE, C. (1982), in: *Progress in Industrial Microbiology,* Vol. 16 (BULL, M. J., Ed.), 119. Amsterdam: Elsevier Science Publ.

REISFELD, A., ROSENBERG, E., GUTNICK, D. (1972), *Appl. Microbiol.* **24,** 363.

ROCHKIND-DUBINSKY, M. L., SAYLER, G. S., BLACKBURN, J. W. (1987), New York: Marcel Dekker.

SAGAWA, K., YOKOTA, H., TAKEHARA, M. (1986), *Jpn. Patent* 61137808 to Ajinomoto Co., Inc.

SCHEIBENBOGEN, K., ZYTNER, R. G., LEE, H., TREVORS, J. T. (1994), *J. Chem. Technol. Biotechnol.* **50**(1), 53.

SCHURCH, S., GEISER, M., GEHR, P. (1993), in: *Biosurfactants Production, Properties, Applications* (KOSARIC, N., Ed.,), 187. New York: Marcel Dekker.

SERVAT, F., MONTET, D., PINA, M., GALZY, P., ARNAUD, A., LEDON, H., MARCON, L., GRAILLE, J. (1990), *J. Am. Oil Chem. Soc.* **67,** 646.

SHIGATA, A., YAMASHITA, A. (1986), *Jpn. Patent Kokai* 61-205449.

SHILO, M., FATTOM, A. (1989), *U.S. Patent* 4826624.

SKRYPZAK, W., RENG, A., QUACK, J. M. (1980), *Cosmet. Toiletries* **95,** 47.

STARCH, M. S. (1984), *Drug Cosmet. Ind.* **134,** 38.

SUZUKI, T., TANAKA, H., ITOH, S. (1974), *Agric. Biol. Chem.* **38**(3), 557.

SUZUKI, T., TANAKA, K., MATSUBARA, I., KINOSHITA, S. (1980), *Agric. Biol. Chem.* **33,** 1619.

SYLDATK, C., WAGNER, F. (1987), in: *Biosurfactants and Biotehnology* (KOSARIC, N., CAIRNS, W. L., GRAY, N. C. C., Eds.), 89. New York: Marcel Dekker.

SYLDATK, C., MATULOVIC, U., WAGNER, F. (1984), *Biotech. Forum* **1,** 53.

SYLDATK, C., LANG, S., WAGNER, F., WRAY, V., WITTE, L. (1985a), *Z. Naturforsch.* **40C,** 51.

SYLDATK, C., LANG, S., MATULOVIC, U., WAGNER, F. (1985b), *Z. Naturforsch.* **40C,** 61.

TADROS, T. F. (1982), *The Effect of Polymers on Dispersion Properties.* London: Academic Press.

TAYLOR, P. J., MILLER, C. L., POLLOCK, T. M., PERKINS, F. T., WESTWOOD, J. A. (1969), *J. Hyg.* **67,** 485.

THERISOD, M., KLIBANOV, A. M. (1987), *J. Am. Chem. Soc.* **109,** 3977.

TREMPER, K. K., LAPIN, R., LEVINE, E., FRIEDMAN, A., SHOEMAKER, W. C. (1980), *Crit. Care Med.* **8,** 738.

TSUTSUMI, H., KAWANO, J., INOUE, S., HAYASHI, S. (1980), *Ger. Patent* 2939519 to Kap Soap Co.

TULLOCH, A. P., HILL, A., SPENCER, J. F. T. (1968), *Can. J. Chem. 45,* 3337.

TURCOTTE, G., KOSARIC, N. (1989), in: *Advances in Biochemical Engineering/Biotechnology* (FIECHTER, A., Ed.), 73.

University of Guelph (1993), *J. Ind. Microbiol.* **11**(3), 163.

VAN ABBE, N. J., SPEARMAN, R. I. C., JARRETT, A. (1969), *Pharmaceutical Cosmetic Products for Topical Administration.* London: Heinemann.

VAN BERNEM, K.-H. (1984), *Senckenbergiana Marit.* **16,** 13.

VAN OSS, C. J., GILLMAN, C. F., NEUMANN, A. W. (1975), 160. New York: Marcel Dekker.

VELA, A. R., HARTWIG, O. L., ATIK, M., MARRERO, R. R., COHN, I. (1965), *Am. J. Clin Nutr.* **16,** 80.

VELIKONJA, J., KOSARIC, N. (1993), in: *Biosurfactants* (KOSARIC, N., Ed.), 419. New York: Marcel Dekker.

VOGT SINGER, M. E., FINNERTY, W. R. (1990), *Can. J. Microbiol. 36,* 741.

VOGT SINGER, M. E., FINNERTY, W. R., TUNECLID, A. (1990), *Can. J. Microbiol.* **36,** 746.

VON DARDEL, O., MEBIUS, C., MOSSBERG, T. (1976), *Acta Anaesthesol. Scand.* **20,** 221.

WAGNER, F., KIM, J. S., LI, Z. Y., MARVEDE, G., MATULOVIC, U., RISTAU, E., SYLDATK, C. (1984), in: Abstracts, *3rd Eur. Congr. Biotechnol.* Vol. 1. Weinheim: Verlag Chemie.

WASKO, M. P., BRATT, R. P. (1991), *Int. Biodeterior. Symp.* **27**(3), 265.

WEAVER, T. E., WHITSETT, J. A. (1991), *Biochem. J.* **273,** 249.

WEIR, D. M. (1977), *Immunology,* 4th Edn. London: Churchill.

WILKINSON, S. G. (1972), *Biochim. Biophys. Acta* **270,** 1.

WRETLIND, K. A., LJUNBERG, S., HAKANSSON, I., AJAXON, B. M. (1978), *U.S. Patent* 4073943.

YAMANE, T. J. (1987), *Am. Oil Chem. Soc.* **64,** 1657.

YAMAGUCHI, M., SATO, A., YUKUYAMA, A. (1976a), *Chem. Ind.* **4,** 741.

YAMAGUCHI, M., SATO, A., YUKUYAMA, A. (1976b), *Chem. Ind.* **17,** 741.

YAMANOUCHI, K., MURASHIMA, R., YOKOYAMA, K. (1975), *Chem. Pharm. Bull.* **23,** 1363.

YOUNG, L. Y. (1984), in: *Microbial Degradation of Organic Compounds* (GIBSON, D. T., Ed.), 487. New York: Marcel Dekker.

ZAJIC, J. E., STEFFENS, W. (1984), *CRC Crit. Rev. Biotechnol.* **1,** 87.

ZHOU, Q. H. (1994), *Ph. D. Thesis,* University of Western Ontario, London, Ontario, Canada.

ZHOU, Q. H., KOSARIC, N. (1993), *Biotechnol. Lett.* **15**(5), 477.

ZHOU, Q. G., KLEKNER, V., KOSARIC, N. (1992), *Am. Oil Chem. Soc. J.* **69**(1), 89.

ZUCKERBERG, A., DIVER, A., PEERI, Z., GUTNICK, D. L., ROSENBERG, E. (1979), *Appl. Environ. Microbiol.* **37,** 414.

# Index

**F**



**G**

**L**



**M**

**N**

**R**

**S**

**W**



**X**



**Y**



**Z**